Physical constants

Acceleration due to gravity (standard),

$g = 32.174$ ft/s^2 (9.80665 m/s^2)(value varies with latitude)

Avogadro's number, $N = 6.02283 \times 10^{23}$ molecules/g-mole

Boltzmann's constant, $k = 1.3805 \times 10^{-23}$ J/°K

British thermal unit (Btu at 60°F) = 778 ft-lb = 1054.54 J

Faraday's constant, $F = 96{,}487$ C(abs)/g-equiv.

Gas law constants (see Table C-1)

$R = 1543$ ft · lb/(lb-mol) · °R (universal)

$R = 53.3$ ft · lb/(lb-air) · °R (engineering gas constant for air)

$R = 0.729$ ft^3 · atm/(lb-mol) · °R

Latent heat of fusion of water (0°C and 1 atm) = 144 Btu/lb = 334 J/g

Latent heat of vaporization of water (100°C and 1 atm) = 971 Btu/lb = 2258 J/g

Molecular mass of dry air = 28.97 lb/lb-mol (g/g-mol)

One acre = 43,560 ft^2 (2.4698 hectare)

One angstrom, A = 10^{-10} m

One bar = 14.504 lb_f/in^2 (10^5 N/m^2)

One cubic foot = 7.48 gal (engineering value)

One horsepower = 550 ft · lb/s = 0.7457 kW

One pound per square inch = 2.31 ft of water

One torr = 1/760 standart atmosphere = 1 mm Hg

Specific weight of water (68°F) = 62.31 lb/ft^3

Standard atmosphere = 14.7 lb/in^2 [101.325 kPa (kN/m^2)]

= 33.899 ft (10.333 m) of water

= 29.92 in Hg = 760 mm Hg

Standard conditions

Specific gravity of solids and liquids is referred to water at 39.2°F (4°C)

Gases are referred to air free carbon dioxide and hydrogen at 32°F (0°C)

General scientific = 32°F and 14.7 lb/in^2

Compressors and blowers = 68°F, 14.7 lb/in^2, and 36% relative humidity

Natural gas industry = 60°F and 14.7 lb/in^2

Temperature (absolute)

Rankine, °R = 459.6 + °F

Kelvin, K = 273.0 + °C

Velocity of light, c = (2.99776 × 10^8 m/s)

Volume occupied by an ideal gas [32°F (0°C) and 1 atm] = 359 ft^3/lb-mol

= 22.4146 L/g-mol

HAZARDOUS WASTE MANAGEMENT

McGraw-Hill Series in Water Resources and Environmental Engineering

CONSULTING EDITOR

George Tchobanoglous, University of California, Davis

Bailey and Ollis: *Biochemical Engineering Fundamentals*
Bishop: *Pollution Prevention: Fundamentals and Practice*
Bouwer: *Groundwater Hydrology*
Canter: *Environmental Impact Assessment*
Chanlett: *Environmental Protection*
Chapra: *Surface Water-Quality Modeling*
Chow, Maidment, and Mays: *Applied Hydrology*
Crites and Tchobanoglous: *Small and Decentralized Wastewater Management Systems*
Davis and Cornwell: *Introduction to Environmental Engineering*
de Nevers: *Air Pollution Control Engineering*
Eckenfelder: *Industrial Water Pollution Control*
Eweis, Ergas, Chang, and Schroeder: *Bioremediation Principles*
LaGrega, Buckingham, and Evans: *Hazardous Waste Management*
Linsley, Franzini, Freyberg, and Tchobanoglous: *Water-Resources Engineering*
McGhee: *Water Supply and Sewage*
Metcalf & Eddy, Inc.: *Wastewater Engineering: Collection and Pumping of Wastewater*
Metcalf & Eddy, Inc.: *Wastewater Engineering: Treatment, Disposal, Reuse*
Peavy, Rowe, and Tchobanoglous: *Environmental Engineering*
Rittman and McCarty: *Environmental Biotechnology: Principles and Applications*
Rubin: *Introduction to Engineering and the Environment*
Sawyer, McCarty, and Parkin: *Chemistry for Environmental Engineering*
Tchobanoglous, Theisen, and Vigil: *Integrated Solid Waste Management: Engineering Principles and Management Issues*

HAZARDOUS WASTE MANAGEMENT

SECOND EDITION

Michael D. LaGrega

Phillip L. Buckingham
ERM

Jeffrey C. Evans
Bucknell University

Environmental Resources Management

Boston Burr Ridge, IL Dubuque, IA Madison, WI New York
San Francisco St. Louis Bangkok Bogotá Caracas Kuala Lumpur
Lisbon London Madrid Mexico City Milan Montreal New Delhi
Santiago Seoul Singapore Sydney Taipei Toronto

McGraw-Hill Higher Education

A Division of The McGraw-Hill Companies

HAZARDOUS WASTE MANAGEMENT
International Edition 2001

Exclusive rights by McGraw-Hill Book Co – Singapore, for manufacture and export. This book cannot be re-exported from the country to which it is sold by McGraw-Hill. The International Edition is not available in North America.

Published by McGraw-Hill, an imprint of The McGraw-Hill Companies, Inc., 1221 Avenue of the Americas, New York, NY 10020.

10 09 08 07 06 05 04 03 02
20 09 08 07 06 05 04 03 02 01
BJE BJE

Library of Congress Cataloging-in-Publication Data

LaGrega, Michael D.
Hazardous waste management / Michael D. LaGrega, Philip L. Buckingham, Jeffrey C. Evans.—2nd ed.
p. cm.
Includes bibliographical references and index.
ISBN 0-07-039365-6
1. Hazardous wastes—Management. I. Buckingham, Phillip L. II. Evans, Jeffrey C. III. Title.
TD1030. L.34 2001
363.72'876'0973—dc21 0-059453

www.mhhe.com

When ordering this title, use ISBN 0-07-118170-9

Printed in Singapore

To Our Children

DEDICATION

This second edition of *Hazardous Waste Management* is published in memory of the late Dr. Michael D. LaGrega (1944–1995). Dr. LaGrega, Professor of Civil and Environmental Engineering at Bucknell University and Principal of the firm Environmental Resources Management (ERM), was the senior author and motivating force for the first edition. Without his inspiration and leadership, the first edition simply would not have happened. We therefore dedicate this second edition to his memory.

Dr. LaGrega was raised on Long Island, New York, and earned his B.S. in Civil Engineering at Manhattan College in 1966. He subsequently earned his M.S. and Ph.D. in Civil Engineering at Syracuse University. He joined Bucknell University as assistant professor in the Department of Civil Engineering in 1974 and was promoted to full professor in 1985. He became affiliated with ERM in 1991.

The professional accomplishments of Professor LaGrega were many. He successfully combined professional practice (he was licensed professional engineer in six states) with academics (he was awarded the Lindback award for outstanding teaching in 1995). He found time for research (author of more than 75 publications) and public service (as Director of Hazardous Waste Planning for Pennsylvania's Department of Environmental Resources). His drive to educate is reflected in his many years of service on the organizing board for the Mid-Atlantic Industrial Waste conference (twice held at Bucknell University and in its 33rd year as of 2000).

Dr. LaGrega is survived by his wife Mary and two children, David and Diane.

ABOUT THE AUTHORS

Michael D. LaGrega (1944–1995) was a professor of Civil Engineering at Bucknell University. He received a Bachelor's degree in Civil Engineering from Manhattan College and a Master's degree and a Ph.D. in Environmental Engineering from Syracuse University. Dr. LaGrega was a licensed professional engineer in six states and was board certified in the specialty of hazardous waste management by the American Academy of Environmental Engineers. The author of more than seventy-five publications covering a wide variety of industrial waste topics, Dr. LaGrega also served as Director of Hazardous Waste Planning for the Pennsylvania Department of Environmental Resources in 1983–1984. Dr. LaGrega was a Principal for the firm of Environmental Resources Management, Inc. Dr. LaGrega was the senior author of the first edition.

Phillip L. Buckingham received a Bachelor's degree in Civil Engineering from North Carolina State University and a Master's degree in Sanitary Engineering from Cornell University. He is a licensed professional engineer in Florida, Pennsylvania, and Ohio. Mr. Buckingham is a founding Principal of Environmental Resources Management, Inc., and has directed numerous hazardous waste projects including state hazardous waste facilities plans and site remediation and risk assessment studies. He is the author of more than 20 technical publications dealing with a wide range of environmental engineering topics.

Jeffrey C. Evans is currently a Professor of Civil and Environmental Engineering at Bucknell University. He received a Bachelor's degree in Civil and Environmental Engineering from Clarkson University, a Master's degree in Civil Engineering from Purdue University, and a Ph.D. in Civil Engineering from Lehigh University. Dr. Evans is a licensed professional engineer in Michigan and, prior to joining the academic community, was affiliated with Woodward-Clyde Consultants for over 10 years. His research interests in the area of waste containment and environmental geotechnology are reflected in his publications, numbering over 50. In addition to his work as an academic, his consulting activities include a variety of hazardous waste remediation projects.

ABOUT ERM

For more than a quarter century, **Environmental Resources Management (ERM)** has been advising governments and businesses across the world on all aspects of environment and resource related issues. By providing insightful, cost-effective business solutions to environmental and resource-related issues, ERM has grown to become one of the most successful environmental health and safety management firms in the world.

ERM's management structure is based on one premise—people will perform best when they can control their own destinies and work for their own benefit. It has become almost a cliché to expound the virtues of a management structure that inspires creativity and ambition in its employees; however, since its inception, ERM has created a working environment where each individual can succeed with the support of clear accountabilities and performance-based rewards. The company seeks to attract and retain the best and brightest by providing growth opportunities and valuing creativity and personal service as highly as technical excellence.

Our 2,400 professionals in more than 120 locations worldwide form a global pool bringing together a blend of engineering, physical science, risk, social science, economics, planning and management expertise. In the United States, the company has 65 locations with at least one office in each EPA region. ERM also has locations in Argentina, Australia, Belgium, Brazil, Chile, Canada, China, France, Germany, Hong Kong, Hungary, India, Indonesia, Italy, Japan, Korea, Malaysia, Mexico, Netherlands, Poland, Portugal, Puerto Rico, Singapore, Spain, Sri Lanka, Sweden, Taiwan, Thailand, the United Kingdom and Vietnam. Employees can move from one region or country to another as their career plans and geographical preferences change.

ERM has significant consulting capabilities in a number of disciplines. Corestaff includes environmental, civil, structural, chemical, mechanical, geotechnical and construction engineers; acousticians; geologists; geophysicists;

political, computer and soil scientists; planners; business analysts; economists; geographers; hydrologists; hydrogeologists; biologists; toxicologists; chemists; industrial hygienists; regulatory analysts; and statisticians.

ERM understands today's business challenges and works to turn liabilities into assets. As a world provider of comprehensive environmental services, the company delivers innovative solutions that improve the corporate bottom line and facility operations.

In the public sphere, ERM applies its policy, management, and technology prowess in guiding governmental action to proactively improve the environment, public health and management of natural resources.

ERM's logo is based on the equiangular spiral—a universal form of nature found in a seashell, the curve of the inner ear, and the break of an ocean wave. The spiral backed by a grid represents the combination of science and nature.

ERM

CONTENTS

PART IV Site Remediation

FOREWORD

As a student and user of this textbook, you are undertaking the mastery of a very important subject. Hazardous waste is one of the most deleterious by-products of the Industrial Revolution.

While there is little doubt that humankind's increasing mastery of natural law has brought us increased life expectancy and a higher quality of life for more people than ever known before, it has also brought the risk of global calamity and impairment of human health. Many of the environmental problems we are dealing with today are an accumulation of 200-plus years of managing the residuals of our industrial output in ways that we now know to be inappropriate. Past practices have resulted in tens of thousands of contaminated sites in the United States and, in proportion to industrial output, elsewhere in the world. The global cost of restoring these sites to minimally acceptable levels will eventually run into the hundreds of billions of dollars. In addition, we are facing the cost of dealing with current and future production and use of products in such a way that they do not impair the environment or human health now or in the future.

Hazardous waste management is an international problem. It is a by-product of becoming a developed country. The more developed the country, the more legacy sites that require remediation. Developing countries do not need to repeat this cycle, but many will. Fortunately, most multinational companies who operate facilities in developing countries manage their current hazardous waste with the same standard of care worldwide, even when local laws and customs do not require it. However, remedial and current management options have to be aligned with local conditions.

Hazardous waste management is truly a multidisciplinary field. You will be called upon to utilize your previous training while at the same time gaining an expanded appreciation for other scientific and nontechnical disciplines. Hazardous waste management is also a multimedia issue because practitioners in this field must recognize and deal with water pollution, air pollution, solid waste, and groundwater aspects that affect the environment and human health.

Because the number of problem sites is high, the cost enormous, and the public's patience short, determining what gets cleaned up, to what extent, and who pays for it and how are not solely technical issues. They can and should invoke risk assessment as well as technical and economic feasibility. Clearly, education is crucial in providing technical tools for participants in this field—owners, operators,

regulators, consultants, contractors, and suppliers. However, in the final analysis, it is the political system, as much as science and technology, that will provide answers to these questions of what, who, and how.

Therefore, education is also crucial for the general public if technology and science are to be appropriately interwoven with the threads of public concern and the economic system in society.

You are encouraged to utilize technology and concepts gained in other fields. We need to continue to make progress in finding more cost-effective tools, both technical and nontechnical, to deal with the management of hazardous waste. This requires the adaptation of technology and concepts gained in other fields. This problem is, after all, a form of societal overhead. This is particularly true with respect to remediating the results of past practices. In our world, we have many other serious concerns related to education, housing, health care, and the building or rebuilding of our infrastructure, to name a few. Each of these areas is suffering from insufficient financial and human resources. Thus, the challenge is to find ways of producing more benefit for less cost.

The very real problem that hazardous waste has created and is creating is compounded by the commingling of technological obstacles; limited financial resources; a general public with an inadequate understanding of basic scientific concepts, including risk assessment; and an extremely complex time- and resource-consuming legal and regulatory process.

As the twenty-first century dawns we are seeing in most industrial sectors a dramatic reduction in the amount of hazardous waste produced per unit of product. This reduction is an expected result of the confluence of sky-rocketing remediation and disposal costs, regulatory enforcement, legal liability, exposure to public scrutiny, reputation, and image impacts. Source reduction, smarter production methods, material substitution, reuse, etc., have been cost-effective investments. As society, in general, and industry in particular, struggle to find a more sustainable extraction, production, use, and final disposition, we can all look forward to leaving a smaller residual footprint for future generations. But this outcome is dependent on continuing innovation in technology and in social and economic areas.

My hope is that you will find it worthwhile to study this subject. The future holds substantial challenge and reward for technical and managerial innovation. Finding better ways to do it will help realize competitive advantage in manufacturing and distribution by utilizing technology and management systems that result in higher net yield at a lower environmental cost. Inadequate knowledge and societal habits and attitudes dictated our past hazardous waste management practices. Through education and enlightened self-interest you can help change these attitudes and practices.

Paul H. Woodruff, P.E.
Chairman
The Environmental Resources Management Group, Inc.

PREFACE

OVERVIEW

The management of hazardous wastes has changed dramatically since the 1960s and continues to evolve as our knowledge of both the hazards and management methods grows. Since the first edition was published (1994), much of the emphasis has shifted: from site management to site remediation, from new facility siting to pollution prevention, and from a newly emerging market to a mature market.

This book provides a comprehensive introduction to a complex interdisciplinary field. The authors' objective is to provide a sufficient background to students so that they begin to think about how to approach hazardous waste problems rather than provide simple "cookbook" solutions. To achieve this objective, the book contains a number of unique elements:

- The Landmark Episodes section provides students with an historical understanding of the field of hazardous waste management (Section 1-3).
- An introduction to the important topic of environmental ethics is provided (Section 1-9).
- Our complex regulatory system is put in context and explanations of how the process works are provided instead of just tables of federal regulations.
- A comprehensive treatment of how toxic compounds move in the environment, including groundwater and air modeling, is provided (Chapter 4).
- A discussion of carcinogenesis details cancer-causing chemicals and the uncertainties involved (Section 5-5).
- An introduction to the methods that corporate management uses to minimize hazardous waste is offered.
- Coverage of the rapidly developing field of bioremediation is provided in depth (Chapter 10).
- Detailed case studies give students the necessary equations and design examples to evaluate the effectiveness of different treatment and containment technologies in addressing today's hazardous waste situation (throughout the text).

ORGANIZATION

This book is divided into four major parts:

- Part I provides background material for a complete understanding of hazardous wastes. Graduate students may have encountered much of the material in the first five chapters from other courses. Therefore, this section introduces basics, not comprehensive coverage, on water generation, law, how wastes move through the environment, and their effects on people and the environment.
- Part II examines the methods currently used by management in industry to understand the magnitude of hazardous waste problems and how to avoid many of the problems of the past.
- Part III contains a selection of treatment and disposal methods. Because literally hundreds of such methods exist, the authors chose some of those methods in contemporary practice and a few emerging technologies to permit an in-depth presentation for the processes selected.
- Part IV covers site remediation. This includes the characterization of a site, the assessment of risks that it poses, and the development and selection of remedies.

TO THE INSTRUCTOR

The information in this first edition is sufficient to support a two-semester or three-quarter course. The text is intended for engineering students; however, where possible, we have included background and basic equations to make the text usable by science students as well.

The level of the text is senior undergraduate or first-year graduate students. Draft copies of this text have been used at more than 100 universities, including Bucknell, Penn State (Harrisburg), and Villanova. Students and instructors have been very generous with their comments, and all chapters were rewritten to include requested information and/or topics.

As noted, this book cannot be adequately covered in a single semester so the authors have attempted to provide instructors with a great deal of flexibility in establishing an appropriate syllabus. Chapters have been written to be reasonably self-contained, and numerous references to other sections have been incorporated.

The following outline is for a suggested introductory one-semester course.

Topic	**Heading**
Fundamentals	
Introduction	Chapter 1
The Regulatory Process	Sections 2-1 through 2-4
Fate and Transport	Sections 4-1, 4-2
Toxicology	Sections 5-1, 5-2, and 5-5
Management Practices	
Environmental Audits	Chapter 6
Waste Minimization	Chapter 7

Treatment Methods	
Physicochemical Processes	Sections 9-1 through 9-3
Bioremediation	Sections 10-1 through 10-3, Section 10-6
Stabilization	Sections 11-1 through 11-3
Incineration	Sections 12-1 through 12-5
Land Disposal	Sections 13-1, 13-3, and 13-4
Site Remediation	
Risk Assessment	Sections 14-1 through 14-4
Remedial Investigations	Sections 15-1 through 15-5
Remedial Technologies	Chapter 16
Alternative Analysis	Chapter 17

This text has gone through numerous reviews and edits. The authors would greatly appreciate any criticisms or suggestions for improvement.

ACKNOWLEDGMENTS

First and foremost, the authors wish to thank Paul Woodruff and Environmental Resources Management for their support and counsel during the undertaking of both editions of this manuscript. Many individuals assisted in the review of numerous drafts of individual chapters, and each chapter concludes with a separate acknowledgments section.

The first edition would not have been completed without the exceptional assistance of Jeffrey Gorham, ERM's publications coordinator, and Karen L. Stein, ERM's technical editor. Marie Malmberg, Amaly MacIntyre, Paul Nordquist, Cathy Putz, Joan Plitt, and Sally Wilson provided word processing assistance at ERM. Bucknell University also provided extensive support to this effort and the authors would like to thank specifically Jai B. Kim, Chair, Civil Engineering Department; Thomas Rich, Dean of Engineering; and Larry Shinn, Vice-President for Academic Affairs; Lois Engle and Beverly Spatzer.

The second edition would not have been completed without the exceptional project management and coordination efforts of Gregory Lloyd, ERM's marketing communications manager for North America, who directed the entire review and editing process during the 2-year rewrite period.

Most of the figures are original art created by ERM's Graphics Department under the direction of Patricia Ackerman. The graphic artists were Angie Della Camera, Tom Hall, Erin McAllister, and Cindy Pomante.

The editors for McGraw-Hill for the first edition were B. J. Clark and John M. Morriss, who assisted in many ways throughout the project. The editors for the second edition were Maja Lorkovic and Eric Munson.

The authors also wish to acknowledge the numerous professors and students who used drafts of the text in classroom testing and provided a great number of suggestions for improvements in both editions.

In addition, we would like to thank the following reviewers who provided valuable critiques on the first-edition manuscript: Paul King, Northeastern

University; Ralph Kummler, Wayne State University; Raymond Letterman, Syracuse University; Howard Mettee, Youngstown State University; D. William Tedder; and Margrit von Braun, University of Idaho. We also wish to thank the following reviewers who critiqued the second edition: Frederic C. Blanc, Northeastern University; Sharon A. Jones, Rose-Hulman Institute of Technology; Ralph H. Ramsey III, Texas Tech University; Raymond W. Regan, The Pennsylvania State University; William E. Roper, George Washington University; Joseph N. Ryan, University of Colorado; and Margrit von Braun, University of Idaho.

Finally, we would like to give special acknowledgment to David L. Freedman, a professor of environmental engineering at Clemson University, who provided extensive comments regarding the book's content and accuracy.

Phillip L. Buckingham
Jeffrey C. Evans

ACRONYMS

2,4-D	Dichlorophenoxyacetic acid
2,4,5-T	2,4,5-Trichlorophenoxyacetic Acid
A/W	Air to Water Ratio (dimensionless)
AA	Atomic Adsorption
ABET	Accreditation Board for Engineering and Technology
ABS	Absorption into Bloodstream (decimal fraction)
acfm	Actual cubic feet per minute
ACS	American Chemical Society
ADI	Acceptable Daily Intakes
ADP	Adenosine diphosphate
AIC	Acceptable Intake Chronic exposures
AIChE	American Institute of Chemical Engineers
AIDS	Acquired Immune Deficiency Syndrome
ALA-D	Aminolevulinc Acid Dehydrase
ANS	American Nuclear Society
ANSI	American National Standards Institute
APC	Air Pollution Control
APCA	Air Pollution Control Association (Now AWMA)
API	American Petroleum Institute
AQCR	Air Quality Control Region
ARARs	Applicable or Relevant and Appropriate Requirements
ASCE	American Society of Civil Engineers
ASTM	American Society for Testing and Materials
AT	Averaging Time
atm	Atmospheres
ATP	Adenosine triphosphate
ATSDR	Agency for Toxic Substances and Disease Registry (U.S.)
AWFSO	Automatic Waste Feed Shutoff
AWMA	Air & Waste Management Association
AWQC	Ambient Water Quality Criteria
AWWA	American Water Works Association
BACT	Best Available Control Technology
BANANA	Build Almost Nothing Anywhere Near Anybody
BAT	Best Available Technology
BCF	Bioconcentration Factor
BDC	Business Development Corporation
BDST	Bed Depth Service Time
BEHP	Bis-2-ethyl hexyl plithalate
BIF	Boilers and Industrial Furnaces
BNA	Base, Neutral, and Acid compound
BOD	Biochemical Oxygen Demand
BOD_5	Five-day Biochemical Oxygen Demand
BTA	Benzyltriethyl Ammonium (clay)
BTU	British Thermal Unit(s)
CAA	Clean Air Act
CAAA	Clean Air Act Amendments
CAD	Computer Aided Design
CAG	Carcinogen Assessment Group (EPA)
Cal WET	California Wet Extraction Test
CASRN	Chemical Abstract Service Registry Number
CATNIP	Cheapest Available Technology Not Involving Prosecution
CDI	Chronic Daily Intake
CEC	Cation-exchange Capacity
CEM	Continuous Emission Monitors
CEO	Chief Executive Officer
CERCLA	Comprehensive Environmental Response Compensation and Liability Act (Superfund)
CERCLIS	CERCLA Information Service
CFC	Chlorofluorocarbons
CFM	Cubic feet per minute
CFR	*Code of Federal Regulations*
CFSTR	Continuous Flow Stirred-Tank Reactor
CGL	Comprehensive General Liability (insurance)
CH	High plasticity (clay)
CL	Low plasticity (clay)
CMA	Chemical Manufacturers Association
CMS	Corrective Measures Study
CMT	Carcinogenicity, Mutagenicity, and/or Teratogenicity
COD	Chemical Oxygen Demand
CPF	Carcinogen Potency Factor
CSF	Carcinogenic Slope Factor
CSPE	Chlorosulfonated Polyethylene
CWA	Clean Water Act
DCA	Di-chloroacetic acid
DCE	1,1 Dichloroethylene; 1,1 dichloroethane
DDD	1,1-*bis*(4-Chlorophenyl)-2,2 Dichloroethane
DDE	1,1-dichloro-2,2-*bis*(4-chloro-phenyl)ethene
DDT	Dichlorodiphenyltrichloroethane
DE	Destruction Efficiency
DEHPA	di-2-ethylhexyl phosphate
DEHP	(*bis*)diethyl hexyl phthalate
DI	Deionized

DMT	Dilatometer test
DNA	Deoxyribonucleic Acid
DNAPL	Dense Nonaqueous Phase Liquids
DOD	Department of Defense (U.S.)
DOE	Department of Energy (U.S.)
DOJ	Department of Justice (U.S.)
DOT	Department of Transportation (U.S.)
DRE	Destruction and Reduction Efficiency
DVB	Divinylbenzene
dscf	Dry standard cubic feet
dscfm	Dry standard cubic feet per minute
dscm	Dry standard cubic meters
dscmm	Dry standard cubic meters per minute
EA	Endangerment Assessment
ECAO	Environmental Criteria and Assessment Office
ED	Exposure Duration
EDB	Ethylene dibromide
EDC	Ethylene dichloride
EDTA	Ethylene diamine tetra acetic acid
EHS	Environmental Health and Safety
ENFLEX®	Regulatory Database
ENR	Engineering News Record
EP	Extraction Procedure (toxicity test)
EPA	Environmental Protection Agency (U.S.)
EPRI	Electric Power Research Institute
EPM	Equivalent Porous Media
EPTox	Extraction Procedure Toxicity test
ERM	The Environmental Resources Management Group
ESP	Electrostatic Precipitator
FAST®	Field Analytical Services Technology
FAG	Food and Agriculture Organization (U.N.)
FDA	Food and Drug Administration (U.S.)
FID	Flame Ionization Detector
FIFRA	Federal Insecticide, Fungicide, and Rodenticide Act
FML	Flexible-membrane Liner
FR	*Federal Register*
FS	Feasibility Study
FTIR	Fourier Transform Infra-red (Analysis)
GAC	Granular Activated Carbon
GAO	Government Accounting Office (U.S.)
GC	Gas Chromatograph
GC/MS	Gas Chromatograph Mass Spectrometry
GIS	Geographic Information Systems
GLP	Good Laboratory Practices
GPM	Gallons per minute
GPR	Ground penetrating RADAR
HAP	Hazardous Air Pollutant
HDPE	High-density Polyethylene
HEL	Higher Explosive Limit
HELP	Hydrologic Evaluation of Landfill Performance (Model)
HETP	Height Equivalent of a Theoretical Plate
HHV	Higher Heating Value
HI	Hazard Index
HMCRI	Hazardous Materials Control Research Institute
HP	Horsepower
HPLC	High-performance Liquid Chromatography
HRS	Hazard Ranking System
HSL	Hazardous Substances List
HSWA	Hazardous and Solid Waste Amendments
HTU	Height of a Transfer Unit in a packed column
HxCDDs	Hexachlorodibenzodioxins
IARC	International Agency for Research on Cancer (WHO)
ICAP	Inductively Coupled Argon-plasma (emission spectroscopy)
IPCS	International Programme on Chemical Safety
IRIS	Integrated Risk Information System (EPA database)
IS	Indicator Score
ISC	Industrial Source Complex (model)
ISV	*In Situ* Vitrification
ISWA	International Solid Waste Association
Kcal	Kilocalorie
kD	Kuderna-Danish (concentration gas chromatography)
kg	Kilograms
kg-cal	Kilogram-calories
kmol	Kilomoles
kW	Kilowatts
kWh	Kilowatt-hours
LAER	Lowest achievable emmision rate
LC	Liquid Chromatography
LCA	Life cycle analysis
LDPE	Low density polyethylene
LEL	Lower Explosion Limit
LEXIS®	Legal database
LF	Landfill
LFG	Landfill Gas
LHV	Lower Heating Value
ln	Natural logarithm
LNAPL	Light nonaqueous phase liquids
LOAEL	Lowest observed adverse effect level
LOEL	Lowest observed effect level
LPG	Liquid Propane Gas
MACT	Maximum achievable control technology
MCL	Maximum Contaminant Levels
MCLG	Maximum Contaminant Level Goals
MET	Maximum Exposed Individual
MEK	Methyl Ethyl Ketone
Me	Metal (cation)
MFO	Mixed function oxidases
MGD	Million Gallons per Day
MGLC	Maximum Ground Level Concentration
MIBK	Methyl Isobutyl Ketone
MIC	Methyl-isocyanate
MLSS	Mixed Liquor Suspended Solids
MLVSS	Mixed Liquor Volatile Suspended Solids
MMD	Methane monooxygenase
MSDS	Material Safety Data Sheets
MSW	Municipal Solid Waste
MT	Metric Tons
MTD	Maximum Tolerated Dose
MWC	Molecular Weight Cutoff
meq	Milliequivalents
mg	Milligrams
mgd	Million gallons per day
mm	Minutes
mL	Milliliters
mm	Millimeters
mo	Months
mol	Moles
ms	Milliseconds

NA	Not Applicable
NAAQS	National Ambient Air Quality Standards
NAPL	Nonaqueous phase liquids
NC	Noncarcinogen
NCP	National Contingency Plan
NFPA	National Fire Protection Agency
NIMBY	Not In My Back Yard
NIOSH	National Institute for Occupational Safety and Health
NMP	N-methyl-2-pyrrolidone
NOAA	National Oceanographic and Atmospheric Administration (U.S.)
NOAEL	No Observed Adverse Effect Level
NOEL	No Observed Effect Level
NOS	Not Otherwise Specified
NPDES	National Pollutant Discharge Elimination System (permit)
NPL	National Priorities List
NRTL	Nonrandom Two Liquid (Model)
NSPE	National Society of Professional Engineers
NSPS	New Source Performance Standard
NSR	New Source Review
NTIS	National Technical Information Service
NTU	Number of Transfer Units
NYCRR	New York Code Rules and Regulations
OAU	Organization of African Unity
OCPSF	Organic Chemicals, Plastics, and Synthetic Fibers
OECD	Organization for Economic Cooperation and Development
OERR	Office of Emergency and Remedial Response (EPA)
ORP	Oxidation Reduction Potential
OSC	On-scene Coordinator
OSHA	Occupational Safety and Health (Administration or Act)
OSWER	Office of Solid Waste and Emergency Response (EPA)
OTA	Office of Technology Assessment (U.S.)
OTS	Office of Toxic Substances (U.S.)
OVA	Organic Vapor Analyzer
PAC	Powdered Activation Carbon
PACT	Powdered Activated Carbon Treatment
PAH	Polynuclear Aromatic Hydrocarbons
Pa	Pascal = 1 Newton/meter^2
PBB	Polybrominated biphenyls
PC	Personal Computer
PCBs	Polychlorinated Biphenyls
PCC	Toxaphene
PCDD	Polychlorinated dibenzodioxins
PCDF	Polychlorinated dibenzofurans
PCE	Perchloroethylene
PCP	Pentachlorophenol
PFR	Plug Flow Reactor
PICs	Products of Incomplete Combustion
PID	Photoionization Detector
PL	Public Law
PM	Preventive Maintenance
PMN	Premanufacture Notification
PMT	Pressure Meter Test
PNAs	Polynuclear Aromatic (hydrocarbons)
POHCs	Principal Organic Hazardous Constituents
POTW	Publicly Owned Treatment Works
PP	Priority Pollutants
PPE	Personal Protective Equipment
PRPs	Potentially Responsible Parties
PSD	Prevention of Significant Deterioration
PVC	Polyvinyl Chloride
PVD	Pore Volume Displacement
PW	Present Worth
PWF	Present-Worth Factor
pH	Hydrogen-ion activity
ppb	Part(s) per billion
ppbv	Parts per billion by volume
ppm	Part(s) per million
ppmv	Parts per million by volume
psi	Pounds per square inch
psia	Pounds per square inch (absolute)
psig	Pounds per square inch (gage)
QA/QC	Quality Assurance/Quality Control
QSAR	Quantitative structural-activity relationship
R&D	Research and Development
RCRA	Resource Conservation and Recovery Act
RD/RA	Remedial Design and Remedial Action
RDF	Refuse-derived Fuel
RFA	RCRA Facility Assessment
RfC	Reference Concentration
RfD	Reference Dose
RFI	RCRA Facility Investigation
RI	Remedial Investigation
RI/FS	Remedial Investigation and Feasibility Study
RITZ	Groundwater model
RO	Reverse Osmosis
ROD	Record of Decision
RPM	Remedial Project Manager
RQD	Rock Quality Designation
s	seconds
SARA	Superfund Amendments and Reauthorization Act
SB	Soil Bentonite
SBR	Sequencing Batch Reactor
SCBA	Self-Contained Breathing Apparatus
SCC	Secondary Combustion Chamber
SCF	Supercritical Fluid
SCS	Soil Conservation Service (U.S.)
SCWO	Supercritical Water Oxidation
SDWA	Safe Drinking Water Act
SEM	Strategic Environmental Management
SF	Slope Factor
SI	International System of Units
SIC	Standard Industrial Code
SM	Percent of Soil Matrix
SNARL	Suggest-No-Adverse-Response Level
SPCC	Spill Prevention, Control and Countermeasures
SPHEM	Superfund Public Health Evaluation Manual (EPA)
SRT	Solids Retention Time
STP	Standard Temperature and Pressure
SVE	Soil Vapor Extraction
SVT	Solvent Vapor Transmissability
SWMU	Solid Waste Management Unit
TCDDs	Tetrachlorodibenzodioxins
TCE	Trichloroethylene
TCLP	Toxicity Characteristics Leaching Procedure
TDS	Total Dissolved Solids (mg/L)

TECE	Tetrachloroethylene
TEF	Toxic Equivalency Factors
THC	Total Hydrocarbons
THM	Trihalomethane
TLV	Threshold Limit Value
TMA	Tetramethyl ammonium (clay)
TNT	Trinitrotoluene
TOC	Total Organic Carbon
TOD	Total Oxygen Demand
TOSCA	Toxic Substances Control Act
TOX	Total Organic Halides
TPH	Total Petroleum Hydrocarbons
TRV	Thermal Relief Valve
TSCA	Toxic Substances Control Act
TSD	Treatment, Storage and Disposal
TSDF	Treatment, Storage, and Disposal Facility
TSDR	Treatment, Storage, Disposal, and Recycling
TSS	Total Suspended Solids
U.S. EPA	U.S. Environmental Protection Agency
U.S.C.	*United States Code*
UF	Ultrafiltration; urea-formaldehyde resin
U.S.G.S.	United States Geological Survey
UN	United Nations
UNEP	United Nations Environment Program
US	United States
USA	United States Army
USAF	United States Air Force
USC	United States Code
UST	Underground Storage Tank
UV	Ultraviolet
VES	Vapor Extraction System
VOC	Volatile Organic Compound
WAS	Waste Activated Sludge
WEF	Water Environment Federation
WES	Waterways Experiment Station (U.S.)
WESTLAW®	Legal database
WHO	World Health Organization
WMA	Waste Management Association
WPCF	Water Pollution Control Federation (Now WEF)
ZOI	Zone of Influence

SYMBOLS

∇	Del operator
∇	Water level
∂	Partial Derivative
[]	Molar concentration (moles/liter)

METRIC PREFIXES

E	exa	10^{18}
P	peta	10^{15}
T	tera	10^{12}
G	giga	10^{9}
M	mega	10^{6}
k	kilo	10^{3}
h	hecto	10^{2}
da	deka	10^{1}
d	deci	10^{-1}
c	centi	10^{-2}
m	milli	10^{-3}
μ	micro	10^{-6}
n	nano	10^{-9}
p	pico	10^{-12}
f	femto	10^{-15}
a	atto	10^{-18}

CHAPTER 1

Hazardous Waste

If we are going to live so intimately with these chemicals—eating and drinking them, taking them into the very marrow of our bones—we had better know something about their nature and their power.

Rachel Carson, *Silent Spring*

1-1 WORKING DEFINITION

What is **hazardous waste**? This chapter provides a working definition of hazardous waste and discusses its sources of generation and its classification. A review of the historical roots of hazardous waste issues explains why they became the dominant environmental program of the 1980s and continue as a major program into the twenty-first century. In this book, the two basic areas of hazardous waste management are presented: (1) management of currently generated waste and (2) remediation of sites contaminated by past practices. The chapter closes with a section on ethics, recognizing that technical decisions made in an arena of imperfect laws can affect the public. While the text's focus is on hazardous waste management, much of the science and engineering presented is applicable to nonhazardous waste as well.

The term **hazardous waste** gained acceptance starting about 1970 with the first national study of the issue, and it became vogue in the mid-1970s with the development of U.S. legislative initiatives to regulate it. Long before then the wastes that we now know as "hazardous" were referred to by such terms as special industrial waste or chemical waste. This is still often the case in Europe. The term hazardous waste by itself is ambiguous. A feature of any regulatory program is to provide a legal definition to determine what is and what is not a hazardous waste. Developing a legal definition can take considerable effort with much disagreement.

The U.S. Environmental Protection Agency (EPA) took nearly 4 years from the passage of the nation's first hazardous waste law in 1976 before promulgating regulations that defined hazardous waste. And even then the definition used rather broad terms, included a myriad of exceptions, and has shown the periodic need for refinement.

Other nations have had similar experiences. Each has developed its own administrative definition for identifying and classifying hazardous waste to the particular level of detail necessary to support its legal procedures. These definitions can lack scientific rigor. Each reflects the environmental, social, and political policies of a nation's government, and hence approaches, like governments, differ from each other. A practitioner must understand the current legal definitions that apply to a waste; these are introduced in Section 2-2. However, this chapter avoids dedicating a number of pages to explain the complexities of a legal definition because a legal definition would periodically change, and today varies from nation to nation. Instead, this chapter presents a practical, working definition.

Developing a working definition first requires defining the term **waste.** A terse definition of waste was prepared in a joint international study as follows[1]: "A waste is a moveable object which has no direct use and is discarded permanently." This definition implies solid waste. In fact U.S. federal hazardous waste laws tend to have stemmed from solid waste programs, and the EPA regulates hazardous waste as a subset of solid waste.

The use of the **solid waste** nomenclature would suggest that the definition covers only solids, sludges, tars, and the like. However, many liquids are considered hazardous waste, typically because of their high strength or because they are a mixture of a hazardous waste with water. Further, many liquid wastes are containerized upon generation and transported as such, thereby making them solid-like. Thus defined, hazardous waste can include solids, sludges, liquids, and containerized gases.

The form of a waste is not important in defining whether it is hazardous. Instead, the most critical part of any definition is to include those terms describing the characteristics that cause a waste to be considered "hazardous" (i.e., posing a substantial present or potential danger to human health or the environment). The potential for toxicity, particularly carcinogenesis, has resulted in the greatest public concern and heads any list of characteristics. However, a waste can be hazardous if it exhibits any of a variety of other characteristics such as being ignitable, flammable, reactive, explosive, corrosive, radioactive, infectious, irritating, sensitizing, or bioaccumulative. This textbook covers all such waste with the exception of radioactive and infectious waste because these two types require management and technical approaches typically different from those required for other types of hazardous waste.

Therefore, the following working definition of hazardous waste, virtually as prepared under the United Nations Environment Program auspices in December 1985, serves as a basis for this book:

> Hazardous wastes mean wastes [solids, sludges, liquids, and containerized gases] other than radioactive [and infectious] wastes which, by reason of their chemical activity or toxic, explosive, corrosive, or other characteristics, cause danger or likely will cause danger to health or the environment, whether alone or when coming into contact with other waste.[2]

1-2 HISTORICAL ROOTS

In the early 1980s, hazardous waste became the leading environmental issue of our society. Now, in the early twenty-first century, while scientific information shows some potentially devastating problems with global ecosystems, hazardous waste still continues to attract far more attention when measured by the amount of federal money spent for environmental programs. As another measure, hazardous waste accounted for about 50 percent of the \$8.2 billion environmental consulting market in 1991 (see Figure 1-1a). In comparison, remediation technology, including hazardous waste, in 1998 accounted for 36 percent of the \$11.1 billion environmental consulting and remediation market (see Figure 1-1b).[4]

Why has it commanded such importance?

As discussed earlier, a waste may be considered hazardous for any of a number of reasons. Of these reasons, the potential for some wastes to cause a toxic reaction in humans is preeminent among public concern. It is this concern (if not fear) compounded by misperception, neglect, mistrust, and politics that explains, in part, why hazardous waste dominates other environmental issues.

The problem did not emerge overnight. Environmental contamination by toxic substances from waste or other sources has a long history. It seems that many wealthy Romans suffered from lead poisoning two millennia ago; possibly the decline of the Roman Empire was due, at least in part, to lead-induced psychoses among the emperors. However, it is the rapid pace of our technological developments, beginning with the Industrial Revolution, that forms the real roots of the problem as we know it today.

The advent of the Industrial Revolution spurred progress on many fronts. Advances in medical science and public health reduced the death rate, thereby dra-

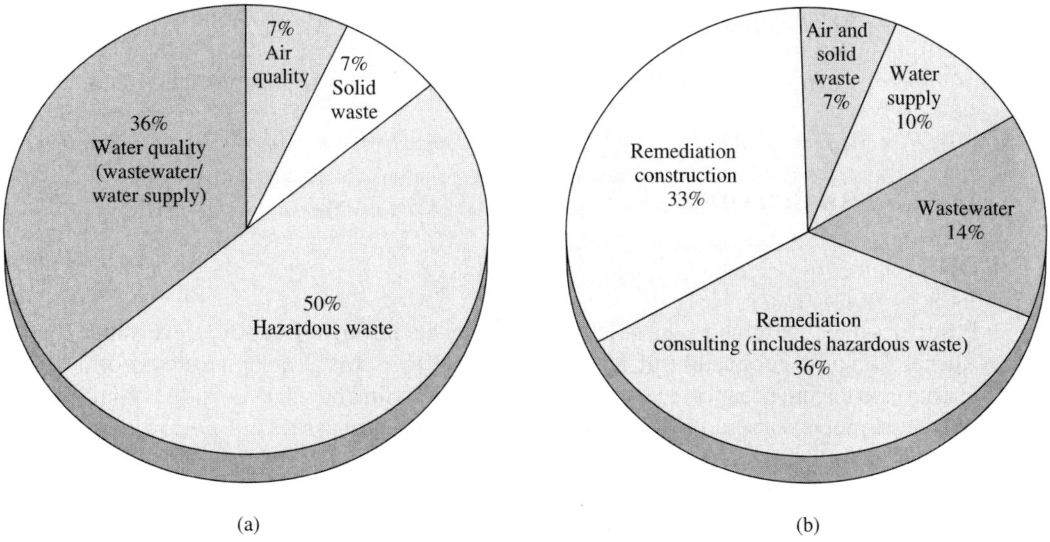

FIGURE 1-1
Hazardous waste share of the environmental (a) consulting and (b) remediation markets.[3,4]

matically increasing human population. Personal consumption also grew rapidly as expanding industrial production, resource extraction, and intensive agriculture supplied more goods. With these goods came toxic substances, sometimes as part of the goods themselves, that became residuals after use or were in the wastes generated in making the goods.

Until recently, government policies did not require appropriate precautions for waste, and few were taken. Simply disposing of waste was the standard. Relying on the seeming inexhaustibility of the environment was the method. Gradually, exposure to hazardous substances occurred outside the workplace via various environmental pathways. It started first with inorganic compounds such as lead and mercury, and expanded later, upon their introduction in the twentieth century, with synthetic organic compounds. Although technological progress has given us new products that raised our standard of living remarkably, it is the nature of technological development to outpace our knowledge of all that it brings and what that means. The failure in this instance is caused more by ignorance than negligence. The long-term toxic nature of many hazardous substances and the inability of the environment to completely assimilate them have done the rest.

The environmental and human health consequences of the residuals and wastes of our technological society were not understood or even recognized initially; it took years or decades for chronic effects to manifest themselves, and those cases were obscured by the fact that everyone is exposed to a wide number of chemicals. This changed when developments in the sciences of epidemiology, toxicology, and analytical chemistry enabled researchers to begin to recognize some noteworthy associations that had been overlooked earlier in our longer-term relationship with toxic chemicals. The landmark cases start not with a hazardous waste but with the effect of DDT residuals on bird populations. Soon afterward, with mercury poisoning of human populations in Japan and episodes involving PCBs, dioxin, and other organic substances, it became clear that humans were also at risk. Each episode received considerable press in a new and increasingly popular media form that came to be called environmental journalism. Each episode advanced public awareness, heightened public concern, and pushed ahead the environmental movement, eventually prompting the hazardous waste legislation that drives how waste is managed today.

1-3 LANDMARK EPISODES*

DDT

Pesticides have continued to be potentially sinister agents in the public eye ever since Rachel Carson's publication of *Silent Spring* in 1962.[5] Ms. Carson riveted world attention to the interconnected web of all life by recounting how DDT residues could be found in deep-sea squid, Antarctic penguins, and the fatty tissues of *Homo sapiens*. In aquatic birds, high levels of DDT were associated with lack of fertility. Later it would be found that DDT inhibits calcium deposition in avian ovaries, leading to egg shells that are too thin to withstand adult weight. In *Homo sapiens* it was not egg

*Section 1-3 of the first edition was prepared by Bruce Molholt, Ph.D., ERM.

shell thinning that was worrisome, but an insidious disease—in laboratory animals DDT exposure was associated with an increased frequency of cancer, one of the first instances associating cancer with exposure to a toxic chemical.

Mercury

Mercury has dramatically different toxicologic properties, depending upon its chemical state. As a liquid metal, it was once used to cure constipation, apparently with few adverse side effects. On the other hand, mercury salts, which were used to form felt in the Dutch hat industry, led to the neurologic disorder in workers renowned as being "mad as a hatter." Organic forms of mercury, such as methyl mercury, have proved to be even more pernicious, having caused hundreds of cases of paralysis and sensory loss along Minamata Bay[6] in Japan. Methyl mercury from a chemical plant was discharged and then bioaccumulated in shellfish. Because shellfish were the major protein source for much of the local population, this situation was an epidemic waiting to happen. Similar poisoning epidemics have occurred elsewhere (e.g., in Iraq and other countries where persons unknowingly ingested seed grain laced with an organomercury fungicide). However, it was the Minamata disaster and its graphic depiction through Katagiri's remarkable photographs in the late 1960s that heightened global awareness of industrial pollution.

PCB and PBB

Like DDT, polychlorinated biphenyls (PCBs) were produced at about 100 million pounds per year during the 1960s and 1970s. However, unlike DDT, which is useful only as an insecticide, PCBs had multiple uses such as transformer coolant, plasticizer, and in the manufacturing of carbonless paper. Two cases of accidental contamination of rice cooking oil in Japan and Taiwan in the late 1960s and mid-1970s exposed thousands of Asians to high concentrations of PCBs. Miscarriages and birth defects erupted within the exposed populations. Although later it would be shown that these prenatal health problems resulted not from PCBs themselves but from polychlorinated dibenzofurans (PCDFs) that formed in heating the contaminated rice cooking oil, the riveting environmental journalism made the world population vividly aware of the adverse human effects of PCB exposure.

About the same time in the United States, the contamination of Michigan cattle feed by polybrominated biphenyls (PBBs) not only caused widespread human exposures via milk and other dairy products, but also followed more complex pathways to reach humans.[7] PBBs were found in some Michigan mothers' breast milk. Afflicted cattle were rendered and then used to prepare chicken feed; thousands of human consumers were exposed to PBBs through eggs and egg products. One incident involved 24,000 crates of PBB-contaminated shortcakes which were confiscated in Alabama.

Love Canal

For many in the United States, Love Canal will always be the symbol of environmental contamination by hazardous waste. It was this pivotal event that eventually

resulted in the passage of the Superfund Act (Comprehensive Environmental Response, Compensation, and Liability Act) in 1980 by the U.S. Congress. Not useful as a canal, this channel was sealed off at the ends and used in the 1940s and 1950s by the then Hooker Chemical Co. and others as a hazardous waste disposal site. Later, Love Canal would be filled in and covered, and then sold to the Niagara Falls, New York, School District for \$1.00. The company informed the school district that the site had been used for the disposal of chemical wastes, warning against any excavation and underground construction. This information was added to the deed and copied into subsequent deeds when the school district sold parcels of the property.

The Niagara School District built an elementary school on the site that became quickly surrounded by hundreds of homes, many with young children. By the late 1970s a chemical odor was frequently discernible, especially in the basements of many of the homes, which, according to the local style, often contained the children's bedrooms. People began experiencing chemically induced health problems. Chemical analysis of samples showed various chemical substances at the disposal site, including dioxin.

A local reporter, Michael Brown of the *Niagara Gazette,* decided to follow up on the anecdotal reports of a few diseases among children that seemed to be linked to indoor fumes. Merely by walking around the neighborhood, he soon discovered more than 100 highly believable examples of chemically induced illness. Furthermore, in many basements he himself could smell the chemical fumes, sometimes to the point of nausea and tearing of his eyes. Brown's Pulitzer Prize–winning exposé catalyzed local politics and also reached the outside world.[8] Soon, the local community, led by Lois Gibbs, a housewife with two afflicted children, was demanding federal relief. In one pivotal meeting, the Love Canal activists, discontented with perceived inaction on the part of the EPA, locked up two officials overnight until they received assurances from Washington that action would be taken. Ultimately, New York State and the U.S. government (Federal Emergency Management Agency) purchased those homes in the vicinity of the canal. By this time drums were being discovered in many backyards and most homeowners were more than eager to move out (see Figure 1-2).

The EPA later declared Love Canal a Superfund site and eventually issued three Records of Decision (Chapter 2) describing the selected site cleanup methods. A cover system has been installed (Chapter 13) along with a groundwater pumping and treatment system (Chapter 16). By 1990, families began to move back into the area. Occidental Petroleum (the current corporation with the financial interests of

FIGURE 1-2
Boarded homes adjacent to Love Canal.

the original Hooker Chemical Co.) has agreed to pay \$98 million and \$129 million toward the cleanup costs incurred by the state of New York and the federal government, respectively. They have also assumed operation and maintenance of the chemical waste treatment plant operating at the site.

Times Beach

Another highly publicized episode in America's awakening concern with hazardous waste occurred in Times Beach, Missouri. As with Love Canal, dioxin would become the major toxic chemical of concern. In the late 1960s and early 1970s, wastes were taken from chemical plants near St. Louis and diluted into used crankcase oil (a legal practice at the time) and spread on dirt roads and horse farms for dust control (again, a legal practice at the time). In the early 1970s dioxins were unknown. However, when waste oil was sprayed at a horse farm in May 1971, many animals died. Even after 6 inches of topsoil was removed from contaminated areas, animals continued to die. Chemical tests showed dioxin contamination up to 100 ppm at the village of Times Beach. With the Seveso, Italy, chemical factory explosion in 1976, and the widespread toxic effects caused by the dioxin released by this explosion, dioxin (tetrachlorinated dibenzo(*p*)dioxin, or TCDD) arguably became recognized as the most toxic synthetic chemical yet discovered. The EPA eventually dealt with the Times Beach case by buying all properties in the community and permanently evacuating the residents.

Woburn

Municipal wells G and H operated by city of Woburn, Massachusetts, form the basis of another widely publicized environmental tragedy in the litany of hazardous waste management practices of the past that have come back to haunt us. While the wells were installed quite some time ago (1964 and 1967) and were shut down more than 20 years ago (1979), the events since that time offer several lessons for the student of hazardous waste management. This case provides insight to the nature of the impacts of waste management practices, society's ways of dealing with the harm, and perhaps even guidance to a safer way to make future decisions.

While residents complained of foul water soon after wells G and H were put into service, it wasn't until 1979 that the state confirmed the wells were contaminated with organic solvents including trichloroethylene (TCE) and tetrachloroethylene (PCE). Health problems including leukemia and cardiac arrythmia seemed unusually high in the area served by these wells. In 1982, a lawsuit was filed on behalf of six families against two corporations. Each of these families had a child who was undergoing treatment for or had died of leukemia. The lawsuit claimed the companies were liable for illnesses and death as a result of the contaminated well water. The case was later expanded to include two additional families and one additional corporation. In 1985, one corporation, Unifirst, settled out of court for just over \$1 million. In 1986, a jury cleared one of the corporations, Beatrice Co., but found W. R. Grace Co. liable. Rather than proceed with the second phase of the trial, where damages would be assessed, the parties settled out of court for \$8 million. With the out-of-court settlement essentially bringing the trial to a close, the judge threw out the jury verdict because of problems associated with the jury's following the judge's

instructions. (This case is discussed in an intriguing book by Jonathan Harr entitled *A Civil Action*.[9] Additional insight into the case is provided by Daniel D. Kennedy, a staff writer for the Woburn *Daily Chronicle* during the time of the trial.[10])

In parallel with the citizens' efforts to seek justice, the EPA was conducting its own studies. In 1990, a settlement was negotiated for additional studies and the cleanup of the municipal wells and contaminated properties. The amount of the settlement was $69.45 million and included W. R. Grace Company, Unifirst Corp., New England Plastics, Beatrice Co., Wildwood Conservation Corp., and John J. Riley, Jr.[11]

While there are many "facts" in this case study that are undisputed, there are also many on which interested parties do not agree. For example, it is agreed that there were some 20 cases of leukemia between 1964 and 1983. It was, in this case study (and in general), up to the citizens to "prove" that this was nothing more than chance. While a study at the Harvard School of Public Health demonstrated that there was a statistical link between the leukemia and the contaminated wells, this did not conclusively establish a cause and effect. In order to be successful in their lawsuit, the citizens also had to prove the source of the contaminants in the well water and that the companies involved were negligent in their handling and disposing of their wastes. As a result, substantial assistance was required from a large group of "experts" including engineers, geologists, medical experts, epidemiologists and the like. The burden of proof required of the citizens—and the resources available to large corporations to provide their own experts to refute those of the citizens—results in a process that deters such suits. This historical approach can be contrasted to a newly emerging principle for preventing harm to the environment and to human health termed the **precautionary principle.**[12] One component of the precautionary principle as it relates to this case study says that the burden of proof of harmlessness of a new technology, process, activity, or chemical lies with the proponents, not the general public. Another (relating to the city of Woburn's decision to deliver water from the wells to its citizens) notes that people in positions of responsibility have a duty to take anticipatory action to prevent harm.

1-4 REGULATORY INITIATIVES

As these and numerous other episodes evolved and received publicity, the public quickly became increasingly aware of the dangers posed to health and the environment by hazardous substances when not managed properly. In response to acute public perceptions, possibly spurred by environmental journalism, the U.S. Congress along with state and local governments, as well as other national governments, enacted sweeping legislation on two fronts:

1. Management of currently generated hazardous waste
2. Remediation of contaminated sites

These laws covered hazardous waste from a "cradle-to-grave" perspective. Prior to these legislative initiatives, economic considerations compelled most generators to dispose of their waste at a low, short-term cost with nominal regard given to the long-term impact. Although the actual risk may not have equaled the public's perceived risk, the hazardous waste management pendulum had swung dramatically with this legislation.

Currently Generated Waste

In 1970, the U.S. Congress first addressed hazardous waste by including with a solid waste law of minor significance a requirement that the Department of Health, Education, and Welfare (a predecessor to the EPA) conduct a comprehensive investigation of the storage and disposal of hazardous waste. No comprehensive program for regulating hazardous waste existed then. The programs that regulated wastewater effluents and air emissions addressed hazardous waste only in a peripheral manner; the existing solid waste program was small and only addressed municipal waste.

The EPA completed its report to Congress in 1973 and found that "the magnitude of the hazardous waste problem was larger than originally anticipated, and that current disposal practices are generally inadequate."[13] The report cited key examples where indiscriminate disposal of hazardous waste had caused adverse public health impacts, including the hospitalization of several people in 1972 in Minnesota after drinking well water that had been contaminated with arsenic wastes. With the impetus of the findings of this report and subsequent hearings, Congress included in the Resource Conservation and Recovery Act (RCRA) of 1976 Subtitle C empowering federal regulation of hazardous waste for the first time. The law focused on the recovery and recycling of solid waste; however, Subtitle C instituted a program to define which wastes are considered hazardous. Further, it required generators of hazardous waste to track the transportation of waste from the point of generation to its final disposition, to operate treatment and disposal facilities in accordance with established standards, and to otherwise manage hazardous waste properly. Thus, in the United States, hazardous waste regulatory initiatives started as just one small part of a small solid waste program. In a matter of a few years it overshadowed the EPA's air and water programs. (For more information on RCRA, see Chapter 2.)

The United States was not alone. Other developed nations took initiatives to address hazardous waste at about the same time. For example, the United Kingdom passed its Deposit of Poisonous Waste Act in 1972, in response to public outcry over drums of heat treatment cyanide wastes disposed on waste ground used by children for play. More comprehensive legislation (Control of Pollution Act) was introduced in 1974. Similar legislation was introduced in Denmark, Finland, France, Germany, and the Netherlands before 1980. A European Union directive on "toxic and dangerous waste" was issued in 1978 to harmonize control over hazardous waste across all its member states. Similar legislation was also introduced in Japan.

Contaminated Sites

In the late 1960s and early 1970s, analytical chemistry techniques advanced (and continue to advance) to levels of higher and higher resolution. When applied to the environmental field, these techniques allowed routine detection and measurement of contaminants at concentrations two or more orders of magnitude smaller than possible just a few years earlier. This new ability to detect compounds at the part per billion level or lower allowed the identification of the widespread presence of potentially toxic compounds in the environment.

Toxic contaminants, deriving from numerous human activities, were detected in the air, water, and land of practically any community. Toxic contamination was

no longer considered just limited to isolated cases where the adverse consequences of exposure had manifested themselves. Unfortunately, the scientific knowledge of the toxic effects of exposure to multiple compounds at low concentrations over a long period of time was (and is) not well understood and had not kept pace with the ability to detect and measure their concentration.

Although the toxic contamination of the environment has numerous sources, of which hazardous waste is just one, sites such as Love Canal where hazardous waste had been mismanaged prompted severe public reaction. Some of the sites contaminated by past practices were indiscriminate dumping grounds; others were bankrupt and abandoned manufacturing plants (see Figure 1-3). Many were former hazardous waste storage, treatment, or disposal facilities, such as Love Canal, that had been subsequently closed, if not abandoned, with minimal plans for further care. It was becoming clear in the late 1970s that potentially thousands of inactive hazardous waste sites existed throughout the nation. Whether these had caused an adverse health effect or had even resulted in exposure was not known; at best the public perceived them as "ticking time bombs."

Neither RCRA nor any other law provided the mechanisms to address inactive sites contaminated by past practices. The U.S. Congress changed that in 1980 with the passage of the Comprehensive Environmental Response, Compensation, and Liability Act (CERCLA), more commonly known as "Superfund." This program seeks to identify sites involving past releases of hazardous substances to the environment and to implement the remedial actions deemed necessary at each site to protect health and the environment. Additional detail on CERCLA is presented in Chapter 2.

Foundation of a Regulatory Structure

The two hazardous waste laws (RCRA and CERCLA) were based on the legal concept that generators are liable for the long-term impact of their waste management practices, including their past practices. This contrasted remarkably with the tra-

FIGURE 1-3
Site contaminated by past practices.

ditional policy that made short-term costs virtually the only economic factor. The enactment and implementation of laws with this foundation have greatly changed the impetus for waste management. They have also fostered a futuristic perspective: most generators endeavor to minimize waste, and many manufacturers even consider the life cycle of their products (from research through manufacturing to use by the consumer and eventually ultimate disposition). The regulatory programs born from RCRA and CERCLA have gained much experience in the years since their enactment, giving considerable insight into particular problem areas that the laws were intended to address. The programs have not slackened—even under considerable scrutiny that claims that they are overreactive, ponderous, costly, and time-consuming. Instead, the programs now carry even greater authority. The public's interest and concern with hazardous waste remain strong.

It should be noted that the RCRA and CERCLA regulatory programs evolved and operated separately. There are perhaps as many uncommon characteristics as common ones between the two programs. Indeed, there are scientific differences: Currently generated wastes (i.e., RCRA waste) are usually a residue of an industrial operation and tend to be more concentrated than when found in soils and groundwater contaminated by past practices (i.e., CERCLA waste). Even so, the substances in the waste in either case remain the same, as do the inherent hazard they pose to health and the environment. This book approaches the two areas (RCRA and CERCLA) together. The same basic scientific and engineering concepts apply whether the substance is in a waste generated today or was released to the environment years ago in a spill, though the regulatory requirements frequently differ.

In Europe, legislation has developed differently from the United States and even from country to country within Europe. Key issues in developing legislation include:[14]

- How to define soil pollution
- Acceptable risk levels and whether they are reasonably achievable
- Generic cleanup criteria or site-specific approaches
- Generic soil quality criteria or criteria related to land use
- Priorities and objectives to remediation

Along with this, European countries that have recognized soil and groundwater pollution from past practices as a significant problem are working to find the proper balance between generic criteria and site-specific criteria for site remediation.

Public Perception of Risk

While hazardous waste does pose a real risk to human health and the environment, there is a debate about the magnitude of the risk, especially the risk compared with other sources of environmental contamination. Some scientists assert that the public has exaggerated the real risks. In 1987, the EPA conducted a technical ranking of relative risks posed by 31 environmental threats and placed the health risk of hazardous waste sites below several other environmental problems.[15] The relative rankings were based on the comparative evaluations and judgments of a technical panel of 75 experts. In contrast, a concurrent survey conducted by the EPA showed

that the public perceived hazardous waste to represent a greater relative risk than most other environmental problems.[16] This is shown in Table 1-1, which compares the EPA experts' rankings with public perceptions of selected environmental problem areas.

Why do public perceptions of the risk posed by hazardous waste not correlate with expert assessments? The public's perceptions are influenced by a number of nontechnical factors such that a real risk is either perceived at a magnified level or simply is deemed unacceptable. Four of these factors are particularly applicable to hazardous waste generation or to a contaminated site, as follows:

- *Involuntarily encountered*—Because few people actually choose to live or work with hazardous waste, it is perceived as having been imposed upon a community or person. In contrast, smoking is known to be unhealthy, but it is a personal choice—it is voluntary—and thus those who smoke perceive the risk as acceptable.
- *Having little apparent benefit*—Few see any benefit in hazardous waste, ignoring that it results from some economic activity (e.g., manufacturing). In contrast, riding in an automobile poses greater risk but provides great direct benefit by getting one to where one wants or needs to go.
- *Uncontrollable or controlled by others*—The landmark contamination episodes such as Love Canal convinced many that hazardous waste is not controllable or at least those in control could not be trusted to manage it properly. In contrast, driving an automobile is controlled by the driver, and "Eighty-five percent of all drivers consider themselves better than average."[18]
- *Having unknown but substantial consequence*—It can be asserted that science has not developed the information to predict accurately the risk posed by hazardous waste. There is a virtual consensus, however, that birth defects and death are possibilities, however remote, from exposure to toxic substances in hazardous waste.

In aggregate, these four factors illustrate the great perceptual difference. The lack of agreement between public opinion and expert assessments does not suggest that either is wrong or even distorted. It is correctly pointed out that "... these

TABLE 1-1
Comparison of expert and public opinion regarding environmental problems[17]

	Relative risk	
Problem	**EPA expert panel**	**Public opinion**
Hazardous waste sites (active and inactive)	Low-medium	High
Pesticide residue on foods	High	Medium
Indoor air pollutants	High	Low
Consumer exposure to chemicals	High	Low

are not distortions of risk; they are part of what we mean by the term."[19] In the United States, regulatory programs are more closely aligned with public perception than with scientific assessments; that is, the political system bends to its constituency. Hence, the two federal hazardous waste regulatory programs (RCRA and CERCLA) continue to receive much of the nation's budget for environmental regulation, and together represent a massive and complex set of laws.

1-5
CLASSIFICATION

In the United States, the basis for determining if a waste is hazardous typically occurs in either of two ways. First, laboratory tests may indicate that it exhibits one or more of the characteristics deemed to make a waste hazardous. Second, it may be on a list of specific wastes compiled by the government because it is known or suspected of having the potential to exhibit hazardous characteristics. Further details of the U.S. system as well as international approaches to classification are discussed in Chapter 2.

Testing

In the United States, any of the following four characteristics will make a waste hazardous:

- Corrosivity (waste that is highly acidic or alkaline)
- Ignitability (waste easily ignited and thus posing a fire hazard during routine management)
- Reactivity (waste capable of potentially harmful, sudden reactions such as explosions)
- Toxicity (waste capable of releasing specified substances to water in significant concentrations)

The EPA has defined explicit laboratory procedures to analyze waste for these characteristics. The procedures include specific standards to determine whether the tested waste is hazardous (e.g., a corrosive waste is one with a pH ≤ 2 or ≥ 12.5).

Lists

The cornerstone of most regulatory programs is to itemize specific ha
into lists. Inclusion in such a list means that the waste is regulated
waste. Some programs have prepared "exclusive" lists describing th
are not hazardous, meaning anything not on the list is hazardous. This
the U.S. Food and Drug Administration's list of acceptable food add
approach to regulation was attempted in the United Kingdom but late

In addition to the designation of levels above which wastes are ch
hazardous, the EPA's Interim Final Regulations promulgated in 198

Section 3001 of RCRA provided for a separate designation as a *listed* waste for certain special situations. This additional set of criteria was designed for those substances that were considered to be a significant concern, even if test results for these wastes fell below the characteristically hazardous levels.

This additional system would seem at first glance to be contrary to the logic used to establish the characteristically hazardous levels. If there were wastes that pose a concern at levels below the thresholds, then perhaps the thresholds were merely set too high. In reality the intent of the EPA was to establish an analytical protocol to evaluate the majority of wastes, and then to use the listing process to handle extraordinary cases. Risk evaluation is at the heart of the listing evaluation process, and it may be helpful to view characteristically hazardous wastes as those determined to be an unacceptable risk by using default scenario assumptions. Alternatively, listed wastes would be those situations where default assumptions produced an acceptable result, but more site-specific assumptions resulted in the determination of an unacceptable risk consequence. This is the flip side of what is conventional practice in risk assessment—analyze first using default assumptions and then use site-specific assumptions if an unacceptable result is obtained from the default analysis.

In 40 CFR 261.11, the EPA establishes four criteria that it uses to decide if a waste should be subjected to a risk evaluation study to determine if it should be listed:

- It is a toxic waste because it contains a constituent that studies show has had "toxic, carcinogenic, mutagenic, or teratogenic effects on humans or other life forms," and the waste is capable of impacting receptors if mismanaged.
- It contains constituents that are acutely hazardous according to studies showing it can be fatal to humans or animals in low doses even if properly managed.
- It exhibits one of the hazardous characteristics (that is, it is listed and characteristic).
- It fits within the definition of hazardous waste developed by the U.S. Congress as part of RCRA. In some cases this has been interpreted to mean that the volume of waste generated is, in itself, a reason for concern.

Hazard codes are assigned to a waste on the basis of this analysis: It is toxic (T), it is acutely hazardous (H), or it exhibits one of the four hazardous waste characteristics (I for ignitability, C for corrosivity, R for reactivity, or E for toxicity characteristic). These hazard codes appear as part of the listing.

If a waste meets one of the EPA's listing criteria, a risk evaluation is performed to determine if the waste poses an unacceptable risk and therefore should be listed. This evaluation is based on typical waste management practices determined by performing a generator survey and, in selected cases, actual waste sampling. Results of the survey and other data collected are then used in a detailed risk evaluation that looks at the worst-case mismanagement scenarios. A large number of key variables that must be set in this evaluation include: the fate and transport setting, the proximity of receptors, the nature of the waste unit, other wastes placed in the unit (called codisposal), the waste unit area and volume, the toxicity of the waste itself, bioavailability and bioaccumulation, and background contamination. There has been a great deal of discussion over the years about how plausible a mismanagement scenario needs to be to be included in this evaluation. Interestingly, evaluations performed and accepted

since the process started in the 1980s have not been revisited, even though the science associated with toxicity, risk evaluation, sensitivity evaluation, and fate and transport has evolved dramatically over that period. It is unclear how determinations originally performed in 1980, both to list and not to list wastes, would conclude if they were to be performed today. Particularly with respect to the refining and chemical industries, these waste listing decisions have had significant economic consequences.

Four types of lists have been established for groups of related listed wastes.[21] These are:

- *F list*—Those wastes resulting generally from chemicals that have been used in a broad segment of chemical plants for their intended purposes and subsequently become wastes.
- *K list*—Those wastes from a group of 17 industries (wood preservation, inorganic pigment manufacturing, organic chemical manufacturing, inorganic chemical manufacturing, pesticides manufacturing, explosives manufacturing, petroleum refining, iron and steel production, primary copper production, primary lead production, primary zinc production, primary aluminum production, ferroalloys production, secondary lead processing, veterinary pharmaceuticals manufacturing, ink formulation, and coking) that are generated from specifically designated processes.
- *P list*—Discarded unused commercial chemicals that have been designated as acutely toxic.
- *U list*—Discarded unused commercial chemicals that have been designated as acutely toxic, but that also have one or more of the four hazardous waste characteristics.

To qualify for the F list, a solvent must be present in a mixture in at least 10 percent by volume (before use), and the solvent must have been used for its *solvent properties* (61 FR 42320). The EPA cites examples of solvent use that include degreasing, cleaning, fabric scouring, use as a diluent, use as an extractant, use as reaction and synthesis media, and use to dissolve or mobilize other constituents.

The K list is very specific to the designated manufacturing process. For example, sludges generated by cleaning out a petroleum refinery's American Petroleum Institute (API) separator in the wastewater treatment plant are designated as K051 because of toxicity, whereas similar sludges from an API separator located ahead of a wastewater treatment plant at an oil terminal do not meet this listing criteria.

To qualify for the P and U lists, a discarded unused chemical must be of technical grade (at or near 100 percent) and be the sole active ingredient.

Some nations have an inclusive listing of chemical substances, either specific or by classes, that indicate a hazard to human health or the environment.[22] Such an example would be PCBs (polychlorinated biphenyls). In this approach, regulations typically place a concentration limit on such constituents that, when exceeded, automatically classifies the waste as hazardous. The critical concentration limit may rely on leaching or volatility testing. Considering that there are a large number, possibly thousands, of chemical constituents that could render a waste hazardous, the implementation of such regulatory programs, including those in the United States, can be cumbersome.

Basel Convention

The Basel Convention grew out of a conference in Basel, Switzerland, in 1989 that resulted in the United Nations Environment Program (UNEP) document entitled "Basel Convention for the Control of Transboundary Movements of Hazardous Wastes and Their Disposal." (www.unep.ch/sbc/baselcon.html.) The convention was unanimously adopted by the 116 countries participating in the conference and the convention entered into force on May 5, 1992. The Basel Convention was the first global treaty addressing hazardous waste. The Basel Convention puts a limited ban on the transboundary movement of hazardous waste to countries that are producers of hazardous waste but do not have the capacity to dispose of the waste within their own borders. In keeping with the theme of this portion of Chapter 1 (Lists), the new Lists A and B were adopted at the Fourth Meeting of Contracting Parties in February, 1998. These are in Appendix A.

Waste Characterization Data

A waste generator may use process knowledge to characterize a waste, or may subject it to analytical testing. To decide how to proceed, it may be helpful to consult information from previous waste testing. There is a scarcity of waste characterization data available in the public sector; however, recent risk evaluations for potential waste listings have produced fairly comprehensive data on selected wastes. For example, the EPA's *Study of Selected Petroleum Refining Residuals* (August 1996) provides detailed process information, sampling details, potential mismanagement scenarios, and analytical test results for certain wastes that were the focus of this study (www.epa.gov/epaoswer/hazwaste/id/studies/study_p1.pdf).

In some cases the chemical nature of the process stream from which the waste was produced may be helpful. The EPA has collected process stream *speciation* data (a catalog of chemical contents by concentration) into a downloadable software package called *Speciate* (Version 1.5, October 1992—www.epa.gov/ttn/chief/software.html/#speciate). In addition, the EPA's *Locating and Estimating Air Emissions* (L&A) project contains source stream and some waste speciation data. For example, *Locating and Estimating Air Emissions from Sources of Benzene* contains speciation for typical refinery process streams (pp. 6–19, EPA-454/R-98-011, June 1998, www.epa.gov:80/ttnchie1/benzene).

By far the best source of chemical constituent data for process streams, intermediates, and products is the material safety data sheet (MSDS). A large number of these are available online through various manufacturers, manufacturers' cooperatives, and universities. A few of the websites that can be used to search for MSDSs are:

- The University of Maryland—
www.inform.umd.edu/CampusInfo/Departments/EnvirSafety/rtk/msds.html
- Vermont Safety Information Resources Institute (SIRI)—
http://siri.org/msds/index.html
- Cornell University, Planning, Design, and Construction (PDC)—
http://MSDS.PDC.CORNELL.EDU/issearch/msdssrch.htm

- Envirocare International Incorporated MSDS-Search—
  www.msdssearch.com/homeN.htm
- Interactive Learning Paradigms, Inc.—
  www.ilpi.com/msds/index.chtml

In December 1995 (60 FR 66344) the EPA proposed the Hazardous Waste Identification Rule (HWIR). Though subsequently withdrawn and slated for reproposal, this rule proposed an innovative way of implementing the language contained in Method 1311 [Toxicity Characteristic Leaching Procedure (TCLP)] regarding a *compositional analysis*. When a constituent of concern "is absent from the waste," or is present at "such a low concentration that the appropriate regulatory level could not be exceeded," a compositional analysis can be substituted for the TCLP. The heart of this analysis is the use of a dilution factor of 20 for the total waste analysis to produce a *pseudo TCLP* result that can then be compared to the regulatory threshold. For wastes containing liquid and solid phases, a theoretical equation was developed to normalize the pseudo TCLP result. Though this method is not technically in effect, it remains part of the tool kit that a generator has to help determine when analytical work is necessary.

The results of waste characterization, whether inferred or obtained from actual testing of the specific waste at the time of generation, are necessary for disposal site waste acceptance. Part of this process may also involve the delivery of appropriate safety information related to the waste. While wastes are technically exempted from the hazard communication requirements of 29 CFR 1910.1200 (Hazard Communication), from a practical standpoint a material safety data sheet (MSDS) is commonly developed for a waste so that waste handlers will be made aware of hazards. In some cases the best source of the safety information will be to assume that the waste has the characteristics of the process stream that generated it for the purposes of the MSDS. Where several streams mix together to form a waste, there is some potential for synergistic (additive or subtractive) effects that may need to be evaluated.

Other Classification Systems

Broad definitions and inclusive lists are very important in the regulation of hazardous waste, particularly to ensure that a hazardous waste does not escape the umbrella of regulatory coverage. However, a regulatory classification system has a number of limitations. It tends to ignore relative differences in hazards among waste. All wastes, once deemed hazardous, are accorded the same regulatory coverage. A regulatory classification system does not facilitate definition of hazard nor is it appropriate for evaluating treatment methods.

Consequently, some scientists and officials have proposed alternatives for defining hazardous waste based on a **degree of hazard** concept. This concept considers individual constituents, their concentration, and their mobility, together with a number of site-specific factors such as potential pathways for migration, exposure routes, dosages, and mitigation. Although such alternatives have been discussed, implementation is impractical.

Another limitation is that a regulatory classification system, because it is driven foremost by the hazard posed by a waste, is limited to identifying a waste's haz-

ardous characteristics and perhaps its constituents; the classification system does not describe adequately (in most cases) the concentration of the constituents nor all important physical and chemical characteristics. Without this information it is impossible to assess the treatability of the waste, its potential for recycling, or its likely fate if released to the environment. An example is the waste coded as D006 in the EPA system. (A waste determined to be hazardous based on testing is given a code, starting with the letter D, to indicate which specific characteristics caused it to be hazardous. In this case, D006 is a waste having a characteristic of toxicity due to the presence of hexavalent chromium.) Such wastes are often generated as sludges or liquids during electroplating operations; however, they are also generated during dust collection or deburring in the fabricated metals industry. These two examples of chromium waste vary considerably in their physical and chemical characteristics as do their risks if released to the environment.

EXAMPLE 1-1. WASTE CLASSIFICATION. A waste coded under the EPA classification system as D001 means the waste has been determined to have the characteristic of ignitability. In some states, this is the largest type of waste shipped off-site. What are some examples of wastes labeled as such?

Solution. A waste coded as D001 could be a spent but readily recoverable solvent, a combustible liquid suitable for energy recovery, a low-Btu paint sludge or paint filter suitable for incineration, or even a noncombustible sludge such as tank bottoms removed with the aid of gasoline. The potential treatability for each of these is significantly different.

Another alternative to the regulatory classification system is to classify a waste according to the following hierarchy:

1. Form or phase distribution (e.g., liquid or solid)
2. Organic or inorganic
3. Chemical class (e.g., solvents or heavy metals)
4. Hazardous constituent as it affects treatability (e.g., hexavalent chromium)

This system is effective for engineering purposes (e.g., grouping those wastes having similar physical and chemical characteristics and general treatment requirements). Table 1-2 shows the basic classes of an expandable system used in several statewide studies to define the need for new hazardous waste treatment and disposal facilities. The system is expandable.

EXAMPLE 1-2. WASTE CATEGORIES. How could the major category "Inorganic Aqueous Waste" listed in Table 1-2 be expanded?

Solution. It could be split up among the following subcategories:

- *Acid*—Inorganic acids; aqueous liquid with pH values of 2 or lower.
- *Acid with metals*—Acidic wastes with metal contamination (e.g., ferric chloride solution and spent pickle liquor).
- *Alkali*—Inorganic alkalis; aqueous liquids with pH of 12.5 or higher.
- *Alkali with metals*—Alkali wastes with metal contaminants.
- *Cyanide*—Spent cyanide solutions from plating, stripping, and cleaning.
- *Hexavalent chromium*—Aqueous liquids containing hexavalent chromium in excess of a specified limit.

TABLE 1-2
Engineering classification system for hazardous waste

Major category	Characteristics	Examples
Inorganic aqueous waste	Liquid waste composed primarily of water but containing acids/alkalis and/or concentrated solutions of inorganic hazardous substances (e.g., heavy metals, cyanide).	— Spent sulfuric acid from galvanizing — Spent caustic baths from metal finishing — Spent ammoniacal etchants from manufacturing electronic components — Rinse water from electroplating — Spent concentrates from hydrometallurgy
Organic aqueous waste	Liquid waste composed primarily of water but containing admixtures or dilute concentrations of organic hazardous substances (e.g., pesticides).	— Rinse water from pesticide containers — Washings of chemical reactors and formulation tanks
Organic liquids	Liquid waste containing admixtures or concentrated solutions of organic hazardous substances.	— Spent halogenated solvents from metal degreasing and dry cleaning — Distillation residues from production of chemical intermediates
Oils	Liquid wastes composed primarily of petroleum-derived oils.	— Used lubricating oils from internal combustion engines — Used hydraulic and turbine oils from heavy equipment operations — Used cutting oils from machinery manufacture — Contaminated fuel oils
Inorganic sludges/solids	Sludges, dusts, solids and other nonliquid waste containing inorganic hazardous substances.	— Wastewater treatment sludge from mercury cell process of chlorine production — Emission control dust from steel manufacture and smelters — Waste sand from coking operations — Lime sludge from coking operations — Dust from deburring of chromium parts in fabricated metal industry
Organic sludges/solids	Tars, sludges, solids and other nonliquid waste containing organic hazardous substances.	— Sludges from painting operations — Tar residues from production of dyestuff intermediates — Spent filter cake from production of pharmaceuticals — Distillation bottom tars from production of phenols — Soil contaminated with spilled solvents — Slop oil emulsion solids

1-6 GENERATION

Hazardous wastes can originate from a wide range of industrial, agricultural, commercial, and household activities. They are generated by manufacturers of many everyday products, by manufacturers of specialty articles, by both service and wholesale trade companies, and by universities, hospitals, government facilities, and households. After a waste is generated, the generator can either manage the waste on-site or transport it off-site for treatment, disposal, or recycling, typically to a commercial hazardous waste facility. Hazardous waste managed on the site where it is generated is termed **on-site waste**. Waste managed at a site other than where it is generated is termed **off-site waste** and requires, in the United States, the use of a document termed a **manifest** for tracking its transport on a "cradle-to-grave" basis (see Figure 2-5).

National Picture of Waste Generation

In the United States, generators are required to submit biennial reports of the generation, management, and final disposition of hazardous waste. The most recently analyzed data for the first edition of this book were those for 1987.[23] For this second edition, the most recently analyzed data are from 1995 (see Figure 1-4). The data provide an interesting overview of the hazardous waste generation and management dynamic as follows:

- For 1987, 17,677 **large-quantity generators** (those generating at least 1000 kilograms of waste per month) filed a report; for 1995 that number rose to 20,873.
- These establishments generated about 238 million tons of hazardous waste in 1987; for 1995, the waste generated decreased to 214 million tons. Changes in the toxicity characteristics rule in 1990 have resulted in an estimated 135 million tons of waste not regulated before 1990.
- For 1985, more than 90 percent of all hazardous waste was categorized as hazardous wastewater (it was aqueous liquid waste). For 1995, this grew to more than 99 percent. Most of the hazardous wastewater is regulated as a hazardous waste merely because it is corrosive (having a pH less than or equal to 2.0 or greater than or equal to 12.5). Typically, these wastes pose a low degree of hazard, and almost all of it is treated on-site, frequently by neutralization, and discharged to either a surface stream or a municipal sewer.
- A few of the generators accounted for most of the waste. More specifically:

	1987	1995
5 largest generators:	>57% of the waste	~ 39% of the waste
50 largest generators:	>90% of the waste	~ 83% of the waste

- In 1987, about 70 percent of the generators produced less than 100 tons each in 1987. Similarly, in 1995, 71 percent of the generators generated less than 113 tons each. This excludes the even greater number of generators, referred to as **small-quantity generators,** who are exempt from the biennial reporting

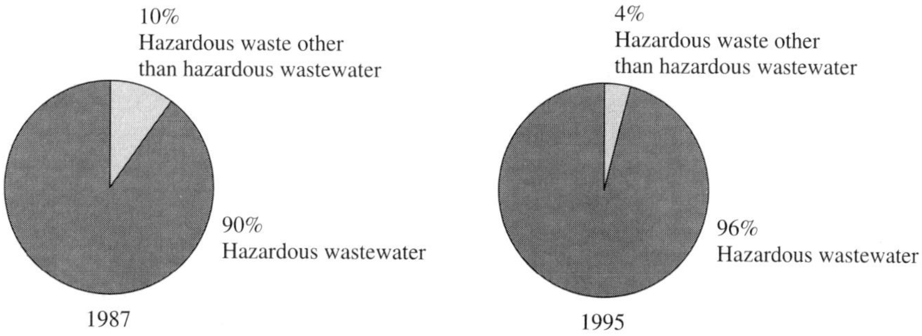

FIGURE 1-4
Generation of hazardous waste in the United States in 1987 and 1995 (by weight).

requirements because they generate less than 1000 kilograms per month (about 13 tons per year).

- In 1987, the nation's hazardous waste was managed at 3308 treatment, storage, and disposal facilities. About half of these facilities had capability only for storage. A vast majority of the facilities were on-site facilities. By 1995, the number of facilities had decreased to 1983.
- Of the total quantity of hazardous waste, 13 percent in 1987 and 11 percent in 1995 was disposed of in deep underground injection wells (mostly hazardous wastewater), less than 1 percent in landfills in both years, and 0.4 percent in 1987 and 2 percent in 1995 by incineration, with the remainder mostly discharged to sewers or water bodies after treatment.
- Only 3 percent in 1987 and 5 percent in 1995 of the generated hazardous waste was managed off-site, virtually all at commercial facilities.
- About 46 percent in 1987 and 50 percent in 1995 of the off-site quantities was transported to facilities in a state other than where the waste was generated.

The high volume of the hazardous wastewater category distorts the national picture. Most public concern is focused on the nonwastewater types of waste, particularly that portion shipped off-site to commercial facilities. These facilities typically accept waste from out-of-state generators. In response to the concerns of their citizens, some state governments have attempted to restrain the import of waste; however, the interstate commerce clause of the U.S. Constitution has prevented these actions. Also note that most other countries do not include hazardous wastewaters in their hazardous waste statistics. This difference in reporting distorts comparisons between the United States and most other countries where these wastes are regulated under water pollution regulations.

Overview of Sources and Types

As part of early United States hazardous waste management efforts to determine the need for new off-site facilities, many state governments analyzed in detail the waste generated within their borders and managed off-site. The data developed in these efforts provided a relatively accurate and detailed basis for examining waste generation patterns. Table 1-3 summarizes the specific industrial sources of waste

TABLE 1-3
Industrial sources of hazardous wastes generated and manifested in 1987 (percent of statewide total)

Standard industrial classification code	Industry	New Jersey[24]	Pennsylvania[25]
22	Textile mill products	0.1	0.1
24	Lumber and wood products	0.3	<0.1
25	Furniture and fixtures	<0.1	0.2
26	Paper and allied products	0.4	0.3
27	Printing and publishing	0.3	0.2
28	Chemicals and allied products	18.3	15.0
29	Petroleum and refining	7.3	1.5
30	Rubber and miscellaneous products	0.3	0.4
31	Leather and leather goods	0.1	<0.1
32	Stone, clay, glass, and concrete	0.7	1.5
33	Primary metals	5.0	39.5
34	Fabricated metals	6.1	5.8
35	Machinery, except electrical	2.0	1.0
36	Electrical and electronic machinery	1.7	14.7
37	Transportation equipment	3.0	0.8
38	Instrumentation	0.9	2.7
39	Miscellaneous manufacturing	0.6	0.2
40–47	Transportation services	5.6	–
48	Communications	0.2	<0.1
49	Electrical and gas services	1.8	0.2
50–59	Wholesale and retail trade	6.9	0.6
73	Business services	9.0	0.3
76	Miscellaneous repair services	12.8	7.9
–	Other/undetermined	16.5	6.9

manifested in two states with large amounts of hazardous waste generation. Table 1-4 examines the types of waste manifested in these same states in terms of the engineering classification system presented in Table 1-2, which proves to be a very suitable scheme for depicting the dynamic situation of hazardous waste generation.

While the data are dated, the two tables demonstrate the complex aspects of hazardous waste generation even within two states that border one another. Motor oils and other oils are generated in both states. However, they were regulated as hazardous waste in New Jersey but were exempt from regulation in Pennsylvania in 1987. Hence, the quantities of oil wastes manifested in Pennsylvania were nil.

Pennsylvania has a larger population and a larger industrial base than New Jersey and hence generates more nonoil hazardous waste. However, the types of waste generated and the sources of the waste vary considerably. New Jersey contains more chemical plants and refineries. Thus, the chemical products group and the petroleum and refining group are the two manufacturing industries with the largest percentage of the state's manifested waste (total of 25 percent). Pennsylvania, famous for its steel industry, had 40 percent of its manifested waste generated by the primary metals industry alone. The standard industrial classification (SIC) code 76 (miscellaneous repair services) includes commercial hazardous waste facilities, and

TABLE 1-4
Types of hazardous wastes generated and manifested in 1987 (quantities rounded to thousands of tons)

	New Jersey[26]	Pennsylvania[27]
Aqueous liquids		
Corrosive with metals	11	179
Contains cyanide	3	3
Contains hexavalent chromium	2	17
Corrosive, not otherwise specified (NOS)	16	82
Inorganics, NOS	2	< 1
Organics, NOS	46	< 1
Aqueous liquids, NOS	23	1
Total aqueous liquids	103	282
Organic liquids		
Nonhalogenated solvents	13	56
Halogenated solvents	6	14
Solvents, NOS	32	7
Combustible, NOS	52	36
Contains PCBs	3	< 1
Total organic liquids	106	113
Oils		
Automotive oils	54	0
Industrial oils	35	1
Fuel oils	21	0
Oils, NOS	12	0
Total oils	122	1
Organic sludges and solids		
With organics	50	8
Paint residues	< 1	0
Oily residues	43	10
Combustible, NOS	9	0
Sludges and solids, NOS	9	0
Total organic sludges and solids	111	18
Inorganic sludges and solids		
With heavy metals	66	192
Corrosive, NOS	3	0
Reactive, NOS	2	0
Toxic, NOS	< 1	–
Sludges and solids, NOS	6	3
Total inorganic sludges and solids	77	195
Others	20	19
Grand total	**539**	**628**

Table 1-3 indicates that this group is the second largest source in New Jersey and fourth in Pennsylvania. The wastes manifested by this group are either treatment sludges being sent to other facilities for final disposal, or they are wastes that have been consolidated and reprocessed at transfer stations prior to treatment or disposal. The industrial categories of retail trade and business services in New Jersey may include tank cleaning and remediation services and are accounted for differently in Pennsylvania.

Different categories of waste generators produce different types of waste. The large organic chemical industry based in New Jersey results in a larger portion of the waste in that state being organic, whether liquid, sludge, or solid. The large primary metals industry in Pennsylvania results in a different situation. In Pennsylvania, nearly half the wastes were aqueous liquids with most of the aqueous liquid being pickle liquor generated by the steel industry. Of the inorganic solid waste with metals, 44 percent in Pennsylvania came from electric arc furnace dust.

The foregoing analysis of the hazardous waste data for two adjacent states reveals the widely different wastes that are generated and their dependence on the nature of the industries. Similar differences would be expected from country to country. Further, the nature of the hazardous waste generated will undoubtedly change with time. For example, by 1995 the total hazardous waste generated in Pennsylvania was only 62 percent of that generated in New Jersey, down from 117 percent in the 1987 data. Among other factors, this change reflects the decline in the metals industry in Pennsylvania between 1987 and 1995.

Manufacturing Sources

What are the sources of hazardous waste? Most people associate the generation of hazardous waste with large chemical plants. While the chemical manufacturing industry represents a large source, Table 1-3 shows that other manufacturers (i.e., the chemical users) cumulatively produce far more. Primary manufacturers, those who prepare products from naturally occurring materials, may be thought of as the dominant source. Yet, waste generation extends beyond the primary and secondary manufacturing operations to include those companies that assemble and finish the final products demanded by society.

Virtually all manufacturing operations result in the generation of residuals because no production process can transform all input materials into products or services. Depending on economics and other factors, these "by-products" and "non-product output" can become wastes. Fashioning the steel, aluminum, plastic, and other components into automobiles and household items will generate abrasives and oils. Covering durable goods with attractive and protective finishes generates cyanides, solvents, concentrated acids, and paint sludges. Production of medicines yields organic solvents and other diverse residues, possibly containing toxic metals. Production of textiles generates heavy metal solutions, dyes, and solvents. The list of examples continues with electronic components, medical equipment, machinery, and publications.

Manufacturing waste may result from many different sources[28]

- *Spent material*—An input material that has been used and can no longer serve the purpose for which it was produced without reprocessing.
- *By-products*—Material that is generated in the specific process of making a product and has no use in its generated form without further processing.
- *Treatment*—Sludges from treating wastewater, controlling air emissions, or even from treating or recovering other hazardous waste.
- *Commercial chemical products*—An actual product that becomes a waste because of any of numerous reasons:

1. Cleanup of process equipment, sometimes with chemical cleaners, such as alkalis, that are hazardous by themselves.
2. Failure to meet manufacturer's specifications because of production startups and shutdowns, upsets, breakdowns, or other factors.
3. Accidental spills or leaks of raw material or product.
4. Residue from containers used for raw material or product.
5. Outdated shelf life.

The generation of waste usually correlates with production and technology. For example, in a comprehensive study of hazardous waste generation in Illinois, the following observations were made[29]:

- The manufacture of paint generated 4 to 6 percent of total production by weight as hazardous waste.
- The manufacture of steel generated 15 to 25 pounds of electric furnace dust per ton of steel produced.
- The manufacture of printing ink generated 1 percent of total production by weight as hazardous waste.

National data presented previously show that waste generation, including on-site waste, is skewed toward the largest generators. This holds true even for manifested wastes. In New York State, an analysis showed that nearly 50 percent of the total quantity of manifested waste came from just 35 out of about 1300 generators who manifested waste.[30] Similarly, in Missouri, only 28 generators in the entire state collectively accounted for 72 percent of the manifested waste.[31] In New Jersey, 93 percent of the manifested waste came from 7 percent of the generators in 1990.[32]

Small-Quantity Generators

A wide range of waste is generated by service industries such as dry cleaners, automobile maintenance shops, and photographic film processors. Analytical laboratories at research and educational institutions and even common households generate hazardous waste. The individual quantity of waste generated by such an operation is small. In fact, almost all generate less than 1000 kilograms/month (kg/mo), the level that exempts generators in the United States from many hazardous waste regulations. This exemption applies to both manufacturing sources and service industries, and such generators are referred to as **small-quantity generators**.

In the United States, there are actually two categories: (1) conditionally exempt small-quantity generators that generate less than 100 kg/mo and (2) small-quantity generators that generate between 100 and 1000 kg/mo. Conditionally exempt small-quantity generators are required to identify their hazardous waste, comply with storage quantity limits (<1000 kg at any one time), and ensure proper treatment and disposal of their waste (either by delivery to an off-site treatment or disposal facility or at a properly licensed on-site facility). Small quantity generators (between 100 and 1000 kg/mo) may be required by individual states to secure an EPA identification number and follow regulations for small-quantity generators. These include limits on accumulation of wastes and requirements for proper storage, handling, and disposal.

Waste from a small generator, if mismanaged, has the same potential for harm as does fully regulated waste from larger sources. It is often disposed of in municipal landfills, in sanitary sewers, or in other ways not intended for hazardous waste disposal. It has even resulted in the death of two boys who were playing in an industrial trash storage bin and inhaled toluene generated from cleaning rollers for printing presses.[33]

For many countries the vast majority of waste generators would be considered small-quantity generators by U.S. standards. As a result, the regulations do not exclude small-quantity generators from control in the same way as U.S. regulations.

Surveys of small generators in Missouri and New York provide the comparisons shown in Table 1-5 to indicate the magnitude of the small generator issue. Most of the small-quantity generator wastes are from automobile service establishments. In fact, 85 percent of the total waste in Missouri from small generators consisted of used motor oil and discarded lead-acid batteries, most of which were reclaimed. Other waste included spent solvents, acids/alkalis, and dry cleaning filtration residues. The cumulative quantity of hazardous waste from small generators can amount to more than 10 percent of the total quantity generated in a state, if not greater than 20 percent, as shown in Figure 1-5.

Household Hazardous Waste

The sources of hazardous waste extend even to the common household. A wide range of household products, when discarded, have the characteristics of hazardous waste. Pesticides, paint products, household cleaners, hobby chemicals, and auto-

TABLE 1-5
Number of generators of small quantities of hazardous waste

Quantity of waste generated	Number of generators of hazardous waste in 1985	
	New York State[34]	Missouri[35]
> 1000 kg/mo	1,300	500
100–1000 kg/mo	10,000	4000
< 100 kg/mo	27,000	9000

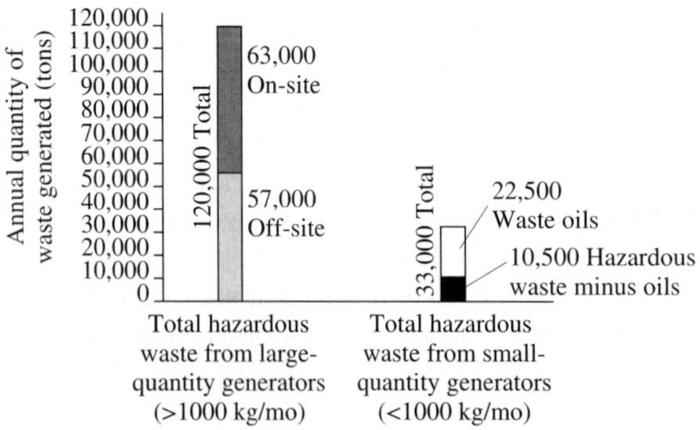

FIGURE 1-5
Comparison of sources of hazardous waste in Missouri.[36]

motive products frequently contain hazardous substances. Federal law specifically exempts household hazardous waste from regulation. Nevertheless, some local and state governments have implemented programs to educate the public about household hazardous waste and to operate waste collection programs. Data from various studies suggest that household hazardous waste collectively amounts to a significant quantity, perhaps as much as the cumulative amount from small generators.[37]

In Europe and in the United States, beginning in the 1980s, and gathering pace through the 1990s, efforts have been made to collect household hazardous waste separately so it could be treated with other hazardous waste. Extensive systems for separate collection are well established in most western countries.

Medical Wastes

Under RCRA as enacted in 1976 (see Chapter 2), hazardous waste was defined to include those wastes with infectious characteristics such as some of the waste generated by hospitals, doctors, dentists, biomedical researchers, and the like. By 1982, the EPA issued a draft guide to the management of these medical wastes, which was made final in 1986.[38] *Medical waste is defined as any waste or reusable material, other than a culture or stock of an infectious substance, that contains an infectious substance and is generated in the diagnosis, treatment, or immunization of human beings or animals*; in research pertaining thereto, or in the production or testing of biologicals, including but not limited to items such as soiled or blood-soaked *bandages,* discarded surgical gloves and surgical instruments such as *scalpels,* needles used to give shots or draw blood; removed body organs such as tonsils, appendices, etc.; and lancets, the little blades the doctor pricks your finger with to get a drop of blood.[39] All wastes must be "red-bagged" and identified with the symbol shown in Figure 1-6. Heightened concern and uncertainties associated with infectious diseases such as acquired immunodeficiency syndrome (AIDS) and the highly publicized cases of needles and syringes washing up on Atlantic Ocean beaches in 1988 ultimately resulted in regulations and guidance for medical waste storage, transportation tracking, and treatment. One of the more common treatment techniques is incineration (see Chapter 12) and management of the ash by land disposal (see Chapter 13).

Pollution Prevention

Much of this text is devoted to the management of hazardous waste, which is implicitly assumed to have been generated and thus requires management. It is important to note that there are many management steps that can be taken to reduce the generation

FIGURE 1-6
Biohazard symbol to be displayed on all medical waste.

of hazardous waste and thus reduce the long-term costs and liabilities associated with these wastes. Pollution prevention—activities undertaken to reduce the generation of hazardous wastes—includes such activities as waste minimization, recycling, reuse, waste reduction, and source reduction. Pollution prevention is the topic of Chapter 7 and is discussed in more detail there.

1-7 CONTAMINATED SITES

The preceding sections have dealt with that hazardous waste currently being generated. The management of this waste is now subject to extensive regulation; however, such was not always the case. In the past, much waste was dumped indiscriminately or disposed of in inadequate facilities. These problems went ignored, as did spills of product or leaks from tanks. These practices contaminated sites with hazardous substances that can pose a threat to human populations. A typical scenario is that contaminants released from a site migrated into the subsurface, where infiltrating rainfall, spilled solvents, or groundwater gradually transported the contaminants to points of groundwater use or surface discharge.

A great number of such sites exist in the United States and throughout the industrialized world. Many took decades to be created, and all have had years of neglect. Defining the problem, especially at sites with multiple contaminants or with a complex environmental setting, can take years and considerable effort before remediation can begin. Remediation can be even more demanding. In a survey of state governments, it was estimated that the average elapsed time per site for the investigation and construction of a remedy is 4.5 calendar years.[40] Implementation of some remedies can take decades.

The potential extent of the problem is illustrated with a few basic data. The Superfund law requires the EPA to identify contaminated sites and long-closed disposal sites, and to maintain a list (the National Priority List) of the sites deemed to represent the greatest potential threat. As of August, 1990, there were 1187 sites on this list, referred to as "Superfund" sites. The EPA culls its National Priority List on
continuing basis from a much larger list of suspected sites of contamination. This
er list is termed CERCLA Information Service (CERCLIS). The CERCLIS list
d with a total of more than 10,000 sites. Because newly discovered sites are
CERCLIS, it soon grew to 20,000 and was approaching 30,000 as of 1993.
numbers of newly discovered sites continued to grow since 1993, so has
of sites that have been evaluated and determined not to be appropriate
uperfund's National Priority List. As of April 1997, approximately
been removed from the CERCLIS list.[41]
st one program for addressing site contamination. Other U.S.

rivate parties to avoid government-regulated programs.
state Superfund programs—many states have their own
luded on the National Priority List.
d hazardous waste treatment, storage, or disposal
dous waste facility under RCRA—about 4600

such facilities exist (3000 active and 1600 closed) containing as many as 36,500 separate waste management units.[42]

- Sites remediated under federal programs other than the EPA's—both the Department of Defense and the Department of Energy have massive programs for remediating their properties.
- Property transfer–initiated remediation—the buyer of property may demand remediation of contamination; some states require this before the legal transfer can occur.
- Leaking underground storage tanks—there are perhaps as many as 1 to 2 million underground storage tanks in the United States and a significant percentage have leaked some of their product. Because these sites represent loss of product and not mismanagement of waste, most are remediated under separate federal and state programs and as such typically are not included on CERCLIS.

The estimated size of the total remediation effort is illustrated by the data in Table 1-6 developed by the New Jersey Department of Environmental Protection. In the intervening 9 years since these projections, the actual number of sites on the National Priority List is 123 (with two more proposed) as of October 1998. Importantly, during the intervening years remedial construction has been completed at 37 sites and 16 sites have been deleted from the list. Also, since the time of the estimates on Table 1-6, the number of underground storage tanks has been put at 31,804 at 11,367 facilities.[44]

In 1991 a university study estimated the cost of remediating contaminated sites in the United States. The estimate covered all government programs as well as privately funded efforts. The remediation of nuclear weapons complexes was included. The study placed the total cost over the next 30 years at $750 billion. This figure was based on the then current policies and technologies.[45]

The enormous cost of remediating sites raises an important point about the future extent of remediation programs. The question is not whether contaminated sites exist;

TABLE 1-6
Potential number of contaminated sites in New Jersey[43]

Program	Estimated total number of sites/facilities	Potential number of sites/facilities requiring remediation
Superfund's National Priority List	100	100
Current privately funded sites	107	107
Current state-funded sites	23	23
Planned additions to the above programs	71	71
CERCLIS	1,239	205
Hazardous waste treatment, storage, and disposal facilities	685	607
Property transfer–initiated remediation	–	3,000
Underground storage tanks	30,000	9,900

literally tens of thousands do exist. What is questionable is the level of risk posed by the sites and the degree of remediation necessary to reduce risks to acceptable levels. There is not a consensus that these sites pose, relative to other environmental problems, a level of risk that warrants the amount of societal investment necessary for their remediation. A policy decision to accept a higher level of risk would probably have two effects: at first, it would accelerate the pace at which sites are remediated, and second, it would eventually result in slackening the program. Nevertheless, site remediation would probably still represent a major environmental program even if it were slackened considerably.

As discussed above, the United States has put an enormous focus on the identification and remediation of sites contaminated by past industrial and waste management practices. This effort is out of proportion when compared to other developed (and developing) countries. The reasons for this difference are not clear. Some would argue that there are fewer problems elsewhere than in the United States. Alternatively, perhaps the United States has been more aggressive in looking for problems on sites with potentially contaminating uses. For example, in the early 1990s, the United Kingdom considered a registry of sites with potentially contaminating practices. The idea was abandoned over fears of potential loss in property value for any sites appearing on such a registry. Only time will tell whether the problems are less severe elsewhere or they have simply not yet become a national priority.[46]

1-8 FUTURE ENDEAVORS

John Quarles, the EPA's first general counsel, reviewed the nation's hazardous waste management strategy.[47] He observed that in 1970, when the nation's first Earth Day signaled the arrival of the modern environmental movement, hazardous waste management was an almost totally neglected subject. This has changed remarkably. He noted that the federal hazardous waste laws have "come to dominate environmental legal practice." He goes on to say that "At the current pace Superfund cleanups will require thirty to fifty years at a minimum," and that "Superfund sites are only one part of a much broader national waste contamination problem."

Environmental management, of which hazardous waste management is just one segment, is progressing through fundamental changes as shown in Figure 1-7. For example, the Environmental Health and Safety (EHS) function in most corporations was an isolated entity that primarily dealt with compliance with laws. Today this function typically reports directly to the CEO and is involved in all aspects of corporate activity.

Clearly, the management of hazardous waste represents a challenge and opportunity for the environmental engineering and science professions. While remediation of contaminated sites currently commands significant attention, the increasing costs of managing today's hazardous waste, in combination with the legal foundation that makes generators liable for contamination in the long term, even for past practices, has prompted a new way of managing hazardous waste. Great effort is being expended to reduce, at its source, the quantity of waste generated and its degree of hazard. Looking further ahead, attention should focus on **product life cycle,** a systematic approach for examining the environmental and health consequences

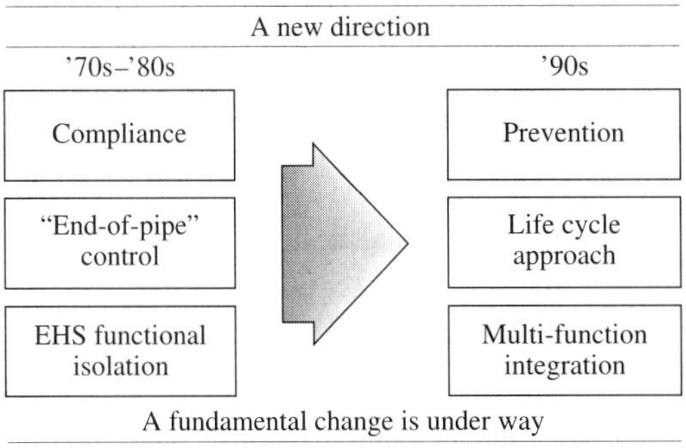

FIGURE 1-7
A new direction for corporations.

of a product at each stage of its life cycle and addressing such consequences in an integrated, cross-functional decision-making process. This results in products that are environmentally sound. Example considerations at each stage are shown in Figure 1-8. Indeed, hazardous waste management is creating great demand for specialized fields of study within and combining the areas of engineering, geology, chemistry, and biology.

While much of the focus of this text is on hazardous waste management in developed countries, and in the United States in particular, there is tremendous opportunity to develop environmentally sound hazardous waste management practices in developing areas. The opportunities in developing countries are not to export the U.S. system, or any other national system, but rather to help countries develop their own systems for control of hazardous wastes. This will remain a major professional challenge requiring expertise and assistance of experienced engineers and scientists for many decades to come.

	Life cycle stage					
	Product design	**Raw materials**	**Manufacture, sales, distribution**	**Packaging**	**Use**	**Final disposal**
Example considerations	• Chemical usage • Ease of recycling • Ease of reuse • Ease of disassembly (e.g., modular design) • Size	• Use of renewable resources • Reducing impact from transport	• Pollution prevention, waste minimization • Process hazards management	• Reduced packaging • Vegetable-based inks • Concentrating product	• Aerosol-free cleaning agents • Low-dosage pesticides	• Recycling • Degradability • Combustion products

FIGURE 1-8
Product life cycle considerations.

1-9 TOWARD AN ENVIRONMENTAL ETHIC

Why a section on ethics in a book on hazardous waste? The management of hazardous wastes is not simply a technical problem but is confounded by legal, social, political, and, yes, ethical considerations. Engineers and scientists formulate conclusions and make recommendations that can easily result in de facto decisions that affect society as a whole. Embracing ethical standards will ensure that the public trust is not betrayed. It is not possible, in this brief section, to provide the tools necessary for resolving the types of ethical problems that tomorrow's hazardous waste manager will confront. Rather our objectives here are more modest: to impress upon the reader that such considerations are of paramount importance and to show that help is available.

Ethical Theories

But where is this help to come from? Philosophers have been studying ethical questions for centuries. While they may not provide simple answers to the type of complex questions likely to be encountered, they do provide help in how to think about ethical questions.

The question of what is and is not ethical is not easily answered. In everyday life, one follows a broad set of personal rules that may include "do not cause pain" and "do not lie." Conflict among the rules may occur that raises an ethical test (e.g., what should an individual do when the only way to avoid causing pain to another is to lie?). How can an individual then do both—be honest and avoid causing pain?

In engineering, rules of conduct are presented as various codes of ethics covering professional practice. For example, the engineer is admonished to be truthful in all dealings with the public. A code will usually express that the engineer is to remain loyal to the client or employer. But what about the situation when the engineer is privy to confidential information about a client's past hazardous waste disposal practices and is asked by a regulatory agency to reveal what he or she knows. How does an individual remain loyal and retain integrity?

The answers to dilemmas such as these have been debated for more than two millennia, without definitive resolution. The search for the solutions to these questions has led to the development of a number of theories of ethics, most of them based on one of two general perspectives:

- Those focusing on the *act*
- Those focusing on the *consequences* of the act

In the former, one is expected to follow the strictures of moral action and never to deviate from them. The consequences are not of any importance. If lying is not accepted, then one should never lie, regardless of the circumstance. Under this type of theory, an individual is acting ethically if he or she does not perform an act that is considered unethical.

Much of the ethical theory based on the importance of the act is based on the work of a German philosopher, Immanuel Kant (1724–1804). Perhaps his most important contribution was his *Categorical Imperative*:

> There is . . . only a single categorical imperative and it is this: Act only on that maxim through which you can at the same time will that it should become a universal law.[48]

Under this theory, engineers would include in their set of ethical values only those rules that they are willing to observe and accept as inviolable (i.e., universal precepts that apply in all circumstances). If one can identify circumstances where a specific rule is inappropriate, the categorical imperative is not met and the rule is defective and lacking in ethical force.

The consequentialist ethical theory is credited in great part to the works of Jeremy Bentham (1748–1832) and John Stuart Mill (1806–1873). In this theory, sometimes termed **utilitarianism,** the consequences of the action are calculated in terms of how much pleasure and pain they produce for all people involved, and the ethical action is the one that has the greatest benefit for the greatest number of people. This is a very comfortable ethical system for engineers who are often placed in situations where it is necessary to calculate the benefits and costs of various projects, making it a "simple" matter to recommend the option having the highest benefit/cost ratio.

Unfortunately, the public does not always appreciate the beauty of utilitarian calculus. For an engineer, it may be reasonable to calculate the public health effect of a hazardous waste incinerator, yielding an estimate that it has the potential to cause one excess case of cancer in a population of 1 million over a 70-year period. But is this "cost" acceptable? The public may not care about these numbers but rather understands that there will be an additional case of cancer and thinks that this is not right. Who will be the one to bear the unfair burden of this facility? By the public's reasoning, the very *act* of causing one person to get cancer is not ethical, regardless of what the engineers think.

Such a divergence of ethical thinking is very difficult for many engineers and scientists to understand, and they will often attribute the reaction of the public to "technical illiteracy." It is not technical illiteracy that is the problem, but a divergence of value systems. The question is how to make decisions where values conflict. Perhaps some resolution can be found in Hans Jonas's *The Imperative of Responsibility*. He suggests that our first duty as citizens in a technological society is to visualize the long-range effects of technology, and this is followed by a second duty: "Summoning up a feeling appropriate to what has been visualized."[49] These are critical elements in what Jonas calls an "Ethic of the Future."

Environmental Ethics

The previous discussion of ethical considerations addresses our behavior toward other persons, both individually and collectively as a society. But what is environmental ethics? In 1949, Aldo Leopold provided an exposition of environmental ethics in an essay termed "The Land Ethic." In commenting on conservation attempts in Wisconsin, he bemoaned the fact that conservation decisions were being based solely on the economic value of the species in question:

> One basic weakness is a conservation system based wholly on economic value. Wildflowers and songbirds are examples. Of the 22,000 higher plants and animals native to Wisconsin, it is doubtful whether more than 5 percent can be sold, fed, eaten or otherwise put to economic use. Yet these creatures are members of the biotic community, and if (as I believe) its stability depends on its integrity, they are entitled to continuance.[50]

Is it sufficient to treat the impact on the environment as essentially value-free, unless that impact noticeably affects people? Gunn and Vesilind[51] have termed this approach **speciesism,** a term akin to racism or sexism. This anthropocentric view of the world is probably the most prevalent in today's society, but attitudes toward the intrinsic value of nature are changing. At the time of the first Earth Day in 1970, it would have been impossible to imagine that the construction of a multimillion dollar dam would be held up over concern for an endangered species, a minnow named the snail darter.

Although there is a greater acceptance of the intrinsic value of nature today than in Leopold's time, we probably are still making decisions that Leopold would consider foolish and shortsighted. One reason for this can be found in an essay entitled "The Tragedy of the Commons" by Garrett Hardin.[52] In it Hardin tells the story of a common pasture owned and used by farmers for grazing cattle. Because the commons is the property of all, there is nothing to prevent any farmer from adding an additional animal to the herd. All of the benefits (i.e., the value of an additional fatted calf) accrue to the farmer. The costs (the potential for overgrazing the commons) are not borne by the farmer alone but by all those using the commons. Each farmer being rational decides to maximize personal gain by maximizing the number of animals on the common. The eventual result of these individual decisions is a total destruction of the pasture through overgrazing. While Hardin directed his essay particularly at overpopulation and birth control, there is a clear message for the protection of all environmental resources. Protection is provided by laws; however, as shown earlier, law does not always suffice.

The Law

As will be seen in Chapter 2, much of the decision making in the hazardous waste field is driven by ever more restrictive laws and regulations. If in fact these laws do represent a set of standards that our society has agreed on, then why should we be concerned with ethical considerations? Isn't simply obeying the law enough?

The short answer is that the regulatory system presents a moving target. In addition, enforcement of all these laws is highly imperfect. For most corporations simply obeying the law is *not* enough. Informed corporations know that the laws will change and society will hold them to a higher standard. As discussed with site remediation, past practices done in compliance with laws in existence at the time can result in major future liabilities. Major environmental incidents, such as the Exxon Valdez oil spill in Alaska or the release of methyl isocyanate at the Union Carbide Plant in Bhopal, India, threatened the existence of these industrial giants. Smaller corporations know that such incidents could easily bankrupt them.

The current tendency is for corporations to move toward the use of corporate codes of conduct that require more than meeting the law. This is done not primarily out of altruism, but rather because of a heightened environmental awareness on the part of corporate officials and heightened environmental expectation of the public. In short it is simply good business. An example of this trend is illustrated in Section 7-3, Life Cycle Analysis. Using this method, corporations reevaluate all of their activities from product research and raw materials purchase through the manufacturing and packaging of the final product to determine the impact on the environment.

Engineers and scientists examine the health and environmental impacts of hazardous waste problems and generally have a good understanding of the degree to which mere compliance with the law will address the impacts. Sophisticated clients will appreciate learning whether the law will suffice. What if the client decides to just comply when more seems proper? Help to answer this and earlier ethical questions can be found expressly in formal codes adopted by a number of engineering societies for governing the conduct of their members.

The Code of Ethics of Engineers adopted by the Accreditation Board for Engineering and Technology (ABET) is illustrated in Figure 1-9. One characteristic of this and most other codes is the first canon: "To hold paramount the safety, health and welfare of the public."

The National Society of Professional Engineers (NSPE) expands on this fundamental canon by noting in Part II of its code (Rules of Practice):

a. Engineers shall at all times recognize that their primary obligation is to protect the safety, health, property and welfare of the public. If their professional judgment is

CODE OF ETHICS OF ENGINEERS
THE FUNDAMENTAL PRINCIPLES

Engineers uphold and advance the integrity, honor and dignity of the engineering profession by:

I. Using their knowledge and skills for the enhancement of human welfare
II. Being honest and impartial, and serving with fidelity the public, their employers and clients
III. Striving to increase the competence and prestige of the engineering profession
IV. Supporting the professional and technical societies of their disciplines

THE FUNDAMENTAL CANONS

1. Engineers shall hold paramount the safety, health, and welfare of the public in the performance of their professional duties.
2. Engineers shall perform services only in the areas of their competence.
3. Engineers shall issue public statements only in an objective and truthful manner.
4. Engineers shall act in professional matters for each employer or client as faithful agents or trustees, and shall avoid conflicts of interest.
5. Engineers shall build their professional reputation on the merit of their services and shall not compete unfairly with others.
6. Engineers shall act in such a manner as to uphold and enhance the honor, integrity, and dignity of the profession.
7. Engineers shall continue their professional development throughout their careers and shall provide opportunities for the professional development of those engineers under their supervision.

FIGURE 1-9
Engineering code of ethics.[54]

overruled under circumstances where the safety, health, property or welfare of the public are endangered, they shall notify their employer or client and such other authority as may be appropriate.[53]

The language in the ABET, NSPE, and other codes should make the engineers' obligations to public health unequivocal. Ethics are not something that can be forced on a person. However, these codes represent the feelings of individuals who have long ago experienced ethical tests and decided that their profession and society would be better served by such actions.

DISCUSSION TOPICS AND PROBLEMS

1-1. What are some of the problems encountered in finding a working definition for *hazardous waste*?

1-2. Is the form (e.g., solid, sludge, liquid, or gas) of a waste important in deciding if it is hazardous?

1-3. What is the purpose of grouping the different wastes according to characteristics (e.g., toxic) in classification lists? What are some of the problems with coding systems such as these?

1-4. Why is the generation of hazardous waste not only a result of large-scale industry? What are some other sources?

1-5. Why is it necessary to include a section on ethics in this book? What are some of the ethical considerations regarding decisions engineers have to make when working in the hazardous waste management field? Give examples.

1-6. In manufacturing a generic product, name five potential sources of hazardous waste generation.

1-7. Contrast Hardings's *Tragedy of the Commons* with Leopold's *Land Ethic*. Give an example of how they apply to an environmental problem in the 1990s.

1-8. Why does the definition of *hazardous waste* specifically exclude *radioactive waste*?

1-9. Contrast the incident at Minamata Bay with that at Love Canal, noting similarities and differences.

1-10. Select an environmental controversy (e.g., proposed landfill, Superfund site, suspected contamination) and comment on progress (or lack thereof) since the landmark cases such as Love Canal.

1-11. As a result of changing methods in compiling data, the pie charts in Figure 1-1 use slightly different categories. From your assessment of the data presented, comment on the changes in the market for each of the categories (air quality, solid waste, etc.). Cite one possible explanation for each.

1-12. Describe the impact advances in analytical chemistry have had on the hazardous waste management field.

1-13. In the United States, generators are responsible for hazardous waste they create from "cradle to grave." Explain what is meant by this phrase.

1-14. What factors might explain the dramatic difference between the public's perception of the relative risk of hazardous waste sites and the opinions of an EPA expert panel (Table 1-1)?

1-15. What are the four characteristics that make a waste hazardous in the United States? Can you suggest other characteristics that might be appropriate to such a listing?

1-16. Explain why "degree of hazard" is not a factor in determining whether or not a waste is hazardous in the United States.

1-17. Review the hazardous waste facilities plan for your state and prepare a one-page summary of the most serious problems.

1-18. What are *small-quantity generators*?

1-19. Review the summaries of generation in New Jersey and Pennsylvania (Tables 1-3 and 1-4). Explain the differences between these two contiguous states that cause the differences in waste generated.

1-20. Explain briefly the factors that might influence the amount of hazardous waste generated in a given year.

1-21. What is *household hazardous waste*?

1-22. What is the *National Priorities List* (NPL), and how does it differ from the CERCLIS list?

1-23. Explain what is meant by the *Product Life Cycle* approach to environmental management. Why have many corporations adopted this proactive approach?

1-24. Describe briefly the two general perspectives for dealing with ethical problems.

1-25. When it became known that a new process planned for a chemical plant was expected to produce a highly toxic waste, a plant environmental engineer wrote to the city newspaper expressing opposition to the action. Under what circumstances would the engineer's action be legal and/or ethical?

1-26. Discuss possible reasons for the difference between the United States and other selected countries with respect to the assessment and remediation of contaminated sites in the past.

REFERENCES

1. Batstone, R., J. E. Smith, Jr., and D. Wilson: "The Safe Disposal of Hazardous Waste," World Bank Technical Paper Number 983, Washington, D.C., 1989.
2. Ibid.
3. "Overview of the Environmental Consulting Industry and ERM, Inc.," presentation to Environmental Resources Management, Inc. shareholders, Alan L. Farkas, Farkas, Berkowitz and Company, Washington, D.C., May 15, 1992.
4. "The 11th Annual State-of-the-Industry Report," Alan L. Farkas, Farkas, Berkowitz & Company, Washington, D.C., September 1999.
5. Carson, Rachel: *Silent Spring*, Boston, Massachusetts, Houghton Mifflin, 1962.
6. Tsubaki, T., and K. Iru Kayama (eds.): *Disease Minimata*, Kodansha Ltd., Tokyo, Elsevier Scientific, Amsterdam, 1977.
7. Dunckel, A. E.: "An Updating on the Polybrominated Biphenyl Disaster in Michigan," *J. Am. Vet. Med. Assoc.*, vol. 167, pp. 838–43, 1975.
8. Brown, Michael: *Laying Waste,* Pantheon Books, New York, 1979.
9. Harr, Jonathan: *A Civil Action,* Random House, 1995, 492 pp.
10. Kennedy, Daniel D.: "Death and Justice: Environmental Tragedy and the Limits of Science," 1996, from www2.shore.net/~dkennedy/woburn_trial.html. Site from 1999.
11. U.S. Environmental Protection Agency, Wells G & H Settlement, url:http://www.epa.gov/region01/remed/sfsites/wellsgh/settle.html, revised Dec. 21, 1998.
12. Montague, Peter: "The Precautionary Principle," *Rachel's Environment and Health Weekly,* no. 586, Environmental Research Foundation, Annapolis, Md., February, 1998, 7 pp.
13. U.S. Environmental Protection Agency: *Report to Congress on Hazardous Waste Disposal*, June 30, 1973.
14. Visser, W. J. F. "Contaminated Land Policies in Europe," *Chemistry and Industry,* July 3, 1995, vol. 4, no. 13, p. 496.

15. U.S. Environmental Protection Agency: *Unfinished Business: A Comparative Assessment of Environmental Problems,* Office of Policy Analysis, Washington, D.C., February 1987.
16. Ibid.
17. Ibid.
18. U.S. Environmental Protection Agency: *Explaining Environmental Risk,* Office of Toxic Substances, Washington, D.C., November 1986.
19. Ibid.
20. Dowling, *Defining and Classifying Hazardous Waste,* op. cit., p. 36.
21. U.S. Code of Federal Regulations, 40 CFR261.1, Promulgated May 1980, and amended January 1985.
22. Dowling, *Defining and Classifying Hazardous Waste,* p. 36.
23. U.S. Environmental Protection Agency: "1987 National Biennial RCRA Hazardous Waste Report," Office of Solid Waste, Washington, D.C., July 8, 1991.
24. Environmental Resources Management: "New Jersey Hazardous Waste Facilities Plan Update," report prepared for New Jersey Hazardous Waste Facilities Siting Commission, Trenton, N.J., October 1989.
25. ERM: "Pennsylvania Hazardous Waste Facilities Plan," Pennsylvania Department of Environmental Resources, 1992.
26. ERM: "New Jersey Hazardous Waste Facilities Plan Update," op. cit.
27. ERM: "Pennsylvania Hazardous Waste Facilities Plan," op. cit.
28. U.S. Code of Federal Regulations, 40 CFR261.1, promulgated May 1980, and amended January 1985.
29. Environmental Resources Management: "Statewide Hazardous Waste Generation Study," prepared for Illinois Hazardous Waste Research and Information Center, Champaign, Ill., HWRIC RR 002, August 1985.
30. ERM: "New York State Hazardous Waste Facilities Needs Assessment," prepared for New York State Department of Environmental Conservation, Albany, New York, March 1985.
31. Environmental Resources Management: "Missouri Hazardous Waste Treatment and Resource Recovery Facility Feasibility Study," prepared for Missouri Environmental Improvement and Energy Resources Authority, Jefferson City, Missouri, January 1985.
32. Personal communication with Susan Boyle, executive director, New Jersey Hazardous Waste Facilities Siting Commission, May 27, 1992.
33. Rosen, Marty: "Intoxicating Solvent Cited in Bin Deaths," *St. Petersburg Times*, St. Petersburg, Florida, June 20, 1992.
34. ERM: "New York State Hazardous Waste Facilities Needs Assessment," op. cit.
35. Environmental Resources Management: "Analysis of Hazardous Waste Generation and Management by Small Generators and Households in the State of Missouri," prepared for Missouri Environmental Improvement and Energy Resources Authority, Jefferson City, Mo., August 1985.
36. Ibid.
37. Ibid.
38. U.S. Environmental Protection Agency, *EPA Guide for Infectious Waste Management* (EPA/530-SW-86-014), Washington, D.C., 1986.
39. U.S. Environmental Protection Agency, EPA Title 49 Part 173—Post-HM181, Washington, D.C., 1996.
40. Association of State and Territorial Solid Waste Management Officials: *State Programs for Hazardous Waste Site Assessments and Remedial Actions,* Washington, D.C., June 1987.
41. U.S. Environmental Protection Agency, *Archival of CERCLIS Sites*, EPA 50-F-97-089, Washington, D.C., April 1997.

42. U.S. General Accounting Office: *Hazardous Waste: Limited Progress in Closing and Cleaning Up Contaminated Facilities,* Report to the Chairman, Environment, Energy, and Natural Resources Subcommittee, Committee on Government Operations, House of Representatives, GAO-RCED-91-79.
43. ERM: "New Jersey Hazardous Waste Facilities Plan Update," op. cit.
44. New Jersey Department of Environmental Protection: *Site Remediation Program Annual Report*, February 1999.
45. "Researchers: Cleanups May Top $1 Trillion," *Superfund,* vol. 5, no. 25, Pasha Publications, Inc., December 13, 1991.
46. Wilson, D. C.: *Hazardous Waste Management in Developing Countries, ISWA 1993 Annual Conference,* Jon Koping, 1993.
47. Quarles, John: "In Search of a Waste Management Strategy," *Natural Resources and Environment*, vol. 5, Summer 1990.
48. Kant, Immanuel: *The Metaphysic of Morals [1797],* Bobbs-Merrill Publishers, Indianapolis, 1964.
49. Jonas, Hans: *The Imperative of Responsibility—In Search of an Ethics for the Technological Age,* University of Chicago Press, Chicago, p. 28, 1984.
50. Leopold, Aldo: *A Sand County Almanac, and Sketches Here and There*, Oxford University Press, Cary, North Carolina, 1949.
51. Gunn, A. S., and P. A. Vesilind: *Environmental Ethics for Engineers*, Lewis Publishers, Boca Raton, Florida, p. 22, 1986.
52. Hardin, Garrett: "The Tragedy of the Commons," *Science*, vol. 162, December 13, 1968.
53. Reprinted by Permission of the National Society of Professional Engineers, Code of Ethics for Engineers, NSPE Publication No. 1102, 1993.
54. Accreditation Board for Engineering and Technology, Code of Ethics for Engineers.

ADDITIONAL READING

Attfield, R.: *The Ethics of Environmental Concern*, Columbia University Press, New York, 1983.

Boraiko, A. A.: "Storing Up Trouble . . . Hazardous Waste," *National Geographic*, vol. 167, no. 3, 1985.

Carper, K. L.: "Engineering Code of Ethics: Beneficial Restraint on Consequential Morality," *J. Professional Issues in Engineering*, ASCE, vol. 117, no. 3, pp. 250–57, July 1991.

Dawson, G. W., and B. W. Mercer: *Hazardous Waste Management*, John Wiley & Sons, New York, 1986.

Epstein, S. S., L. O. Brown, and C. Pope: *Hazardous Waste in America*, Sierra Club Books, San Francisco, 1982.

Forester, W. S., and J. H. Skinner: *International Perspectives on Hazardous Waste Management,* Academic Press, London, 1987.

Freeman, H. M., ed.: *Standard Handbook of Hazardous Waste Treatment and Disposal*, McGraw-Hill, New York, 1989.

Hargrove, E.: *Foundations of Environmental Ethics*, Prentice-Hall, Englewood Cliffs, New Jersey, 1989.

Hoffman, W. M., R. Frederick, and E. S. Petry, Jr., eds.: *The Corporation, Ethics and The Environment*, Quorum Books, Westport, Connecticut, 1990.

Katz, E.: "Ethics and Philosophy of the Environment: A Brief Review of the Major Literature," *Environmental History Review*, vol. 15, pp. 79–86, Summer 1991.

Kiang, Y., and A. A. Metry: *Hazardous Waste Processing Technology*, Ann Arbor Science Publishers, Ann Arbor, Michigan, 1982.

Kolaczkowski, S. T., and B. D. Crittenden: *Management of Hazardous and Toxic Wastes in the Process Industries*, Elsevier Science Publishing, New York, 1987.

Lankford, P. W., and W. W. Eckenfelder, Jr.: *Toxicity Reduction in Industrial Effluents*, Van Nostrand Reinhold, New York, 1990.

Manahan, S. E.: *Hazardous Waste Chemistry, Toxicology and Treatment,* Lewis Publishers, Chelsea, Michigan, 1990.

Martin, E. J., and J. H. Johnson, Jr.: *Hazardous Waste Management Engineering*, Van Nostrand Reinhold, New York, 1987.

Martin, M. W.: *Ethics in Engineering*, McGraw-Hill, New York, 1983.

Mazmanian, D., and D. Morell, *Beyond Superfailure: America's Toxic Policy for the 1990's*, Westview Press, Boulder, Colorado, 1992.

Nash, R. F.: *The Rights of Nature: A History of Environmental Ethics*, University of Washington Press, 1989.

O'Brien & Gere Engineers, Inc.: *Hazardous Waste Site Remediation—The Engineer's Perspective*, Van Nostrand Reinhold, New York, 1988.

U.S. Bureau of the Census, Current Population Reports, Series P23-194, "Population Profile of the United States 1997," U.S. Government Printing Office, Washington, D.C., 1998.

Wentz, C. A.: *Hazardous Waste Management,* McGraw-Hill, New York, 1989.

White, L.: "The Historical Roots of Our Ecological Crisis," *Science*, vol. 155, pp. 1203–1207, March 10, 1967.

Windsor, D.: "Government Business Relationships in Environmental Protection," in *Business, Ethics and The Environment,* W. M. Hoffman, R. Frederick, and E. S. Petry, Jr., eds., Quorum Books, Westport, Connecticut, 1990.

ACKNOWLEDGMENTS

The authors wish to acknowledge the following individuals who assisted with the preparation of this chapter: Professor A. A. Friedman, *Syracuse University*; David Morell, Ph.D., *ERM*; Professor Rita Meyninger, *Brooklyn Polytechnic University*; Bruce Molholt, Ph.D., *ERM*; Susan Boyle, *New Jersey Hazardous Waste Facilities Siting Commission*; Dave Levy, *U.S. EPA Waste Management Division*; Professor Aarne Vesilind, formerly *Duke University* (now Bucknell University); Professor Richard Fleming, *Bucknell University*; and Kurt Becker, *Bucknell University*. For the first edition the principal reviewer for ERM was Neil L. Drobny, Ph.D., P.E.

The reviewer for the second edition was David C. Wilson, Ph.D., *ERM*. Section 1-5 was rewritten by John Beath, *ERM*.

CHAPTER 2

The Legal Framework

Ignorance of the law excuses no man.
John Selden

"If the law suppose that," said Mr. Bumble, "the law is a ass—a idiot."
Charles Dickens

Most actions taken today to manage hazardous waste are done in response to a legal requirement. In the United States, the legal requirements include the U.S. Constitution; federal, state, and local environmental statutes and regulations; and court decisions interpreting these laws.

The chapter begins with an introduction to the concept and language of environmental law in the United States and then highlights the two major federal statutes dealing with hazardous waste: the Resource Conservation and Recovery Act, which emphasizes wastes presently being generated, and the Superfund Law, which is primarily concerned with past practices including cleaning up abandoned hazardous waste sites. The regulations derived from these two laws are also discussed, followed by a brief description of other environmental statutes of the United States.

This body of law provides a legislative/regulatory control framework for the management of hazardous waste in the United States. The framework for one country can differ greatly from that of another. This chapter concludes with a lengthy section presenting an international perspective. It discusses the important elements of a legislative/regulatory control framework and explains on a comparative basis how other countries implement them.

While it is essential that engineers and scientists become familiar with environmental law, it is important to remember that the law will change with time and jurisdiction. Meeting current regulations provides a baseline or minimum standard for the design of solutions to hazardous waste problems. The overall objective should always be protection of human health and the environment through selection of the technology or management practice best suited to the problem at hand.

2-1 ENVIRONMENTAL LAW

To understand the laws governing hazardous waste management in the United States, engineers and scientists must be familiar with the concepts and language of the law. This section summarizes the fundamentals of environmental law and introduces some sources that can help determine how the law applies to a particular problem. For more information, the reader is referred to books on environmental law.[1–5]

The growth in public interest in environmental matters has spurred a concomitant expansion in environmental laws. Figure 2-1 illustrates the recent proliferation of major environmental laws, principally during the period 1968–1990. During this period, Congress enacted major laws governing all environmental media (e.g., air, surface water, groundwater) and sources of environmental contamination. After 1990, the growth of major new statutes slowed considerably. Nevertheless, the major laws have been repeatedly amended and the volume of regulatory law has expanded dramatically (Figures 2-2 and 2-3).

The System of Environmental Law

As in all countries, there is a hierarchy of law in the United States, principally as follows:

U.S. Constitution
Treaties
Federal statutes
Federal regulations
State statutes (State law is not necessarily subordinate to federal law.)

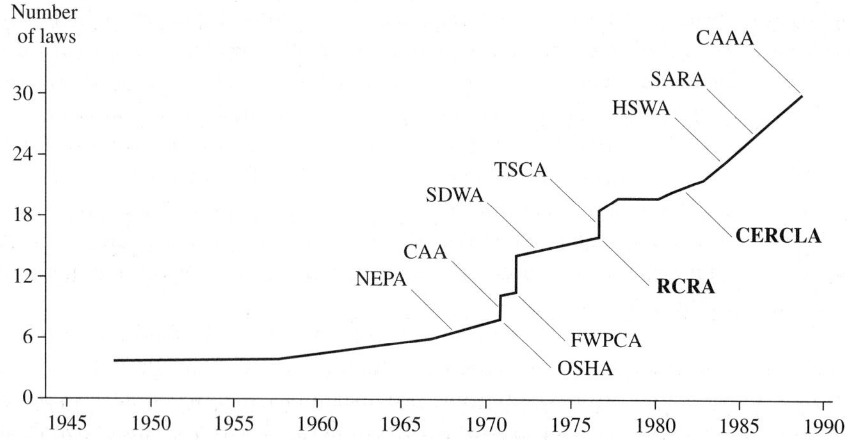

FIGURE 2-1
Growth of environmental laws in the United States.[6]

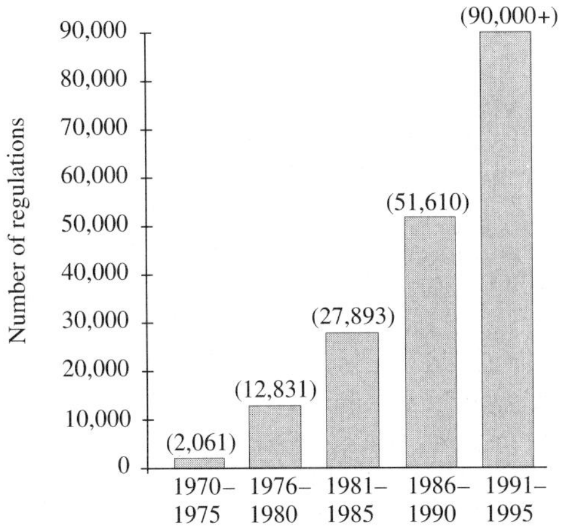

FIGURE 2-2
Changes to U.S. environmental regulations (federal and state).[6a]

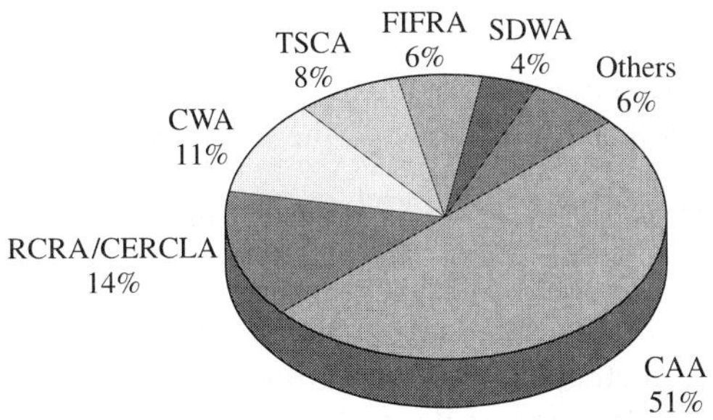

FIGURE 2-3
EPA's short-term agenda—458 regulations under development.[7]

State regulations
Local (city/county) laws, ordinances, and regulations
Case law (with its own hierarchy)

U.S. Constitution

The Constitution is the highest law of the land. The courts will eventually strike down any law that violates it. Environmental issues are rarely resolved on Constitutional grounds, but there are some notable exceptions. For example, laws regulating wetlands and those protecting endangered species are often challenged as violating the Constitution's prohibition on the taking of property without due compensation.

Treaties

Once ratified by the Congress, international treaties are second in priority only to the Constitution. Environmental issues, e.g., global warming and species and forest preservation, are increasingly viewed as international concerns. As a result, treaties are becoming a more important source of environmental law.

Federal statutes

Federal statutes are the foundation of all major environmental programs. The importance of federal law, compared to state and local law, has greatly increased over the past 50 years. Nevertheless, Congress's authority is not unlimited. The Constitution enumerates the powers given to the federal government; the rest are

reserved for the states. The significance of this distinction has lessened during this century but any federal law that directly usurps states' rights is subject to being ruled unconstitutional and therefore unenforceable. Federal authority to regulate the environment is most often grounded in Congress's constitutional authority to regulate interstate commerce. Federal laws are also subject to challenge if they are in conflict with other federal laws.

Statutes are referred to by legal designations assigned by the body that passed them. A federal law has several designations that can appear in legal documents or reports. First, when a federal bill becomes a law, it is assigned a Public Law number, which begins with the prefix PL attached to the number of the session of Congress in which the law was passed. A specific number then identifies the law. Thus, the Resource Conservation and Recovery Act (RCRA), passed in 1976 by the 94th Congress, was designated PL94-580. All federal statutory law is also codified (i.e., put together in a systematic listing) in the United States Code (USC). The USC is regularly updated with the latest amendments to statutes, and annotations are added describing case law interpretations of the statute. Thus, RCRA, as amended, is referenced as 42 USC 6901 et seq. (Title 42 of the United States Code, Section 6901 and the following sections). Finally, laws are frequently referenced by the law's common name, followed by the appropriate section, e.g., the Clean Water Act, Section 320 (or CWA §320).

Federal regulations

Over the past 50 years, there has been an explosion in the sheer volume of the law. Over 90 percent of the increase is attributable to the growth in regulatory law. Environmental law is no exception. Because of the technical complexity of environmental issues, Congress often drafts environmental laws in general terms, e.g., policies or goals. It is left to administrative agencies, such as the U.S. Environmental Protection Agency (EPA), to provide, through regulations, the technical details necessary to meet those goals and to implement the law. As a result, statutes consisting of only 40 pages may generate thousands of pages of implementing regulations. So long as the appropriate agency properly promulgates its regulations, they are in theory entitled to the same weight as the underlying law itself. There are, however, innumerable bases for challenging the legality of promulgated regulations. (A frequent challenge is the assertion that Congress passed a law so vague that it amounted to an unconstitutional delegation of its legislative function to the federal agencies.)

Regulations are published daily in the ***Federal Register.*** Under normal rule-making procedures, a federal agency will publish a Proposed Rule in the *Federal Register,* in which it explains its rationale for the rule and invites public comment. After the comment period expires, the agency will promulgate a Final Rule. Rules and regulations published in the *Federal Register* are codified in the **Code of Federal Regulations (CFR),** which is updated yearly. The CFR is divided into titles and parts. For example, the citation 40 CFR 261 refers to Title 40 (Protection of the Environment), Part 261 (Identification and Listing of Hazardous Waste).

State statutes and regulations

Most states follow procedures that parallel the federal system in promulgating, publishing, and codifying their laws and regulations. There are, however, numerous

exceptions, and a thorough knowledge of state-specific procedures is an absolute prerequisite to a sound understanding of the applicable law.

The role of a state in environmental law varies significantly from state to state and from one area of environmental law to another (for example, the regulation of nonhazardous solid waste is left principally to the states). In some areas, Congress has preempted the field; that is, federal law is the only law that applies. In other areas, Congress has passed a law without authorizing a further delegation to the states, but not prohibiting the states from passing a similar law. A good example is the Comprehensive Environmental Response Compensation and Liability Act (CERCLA, commonly known as Superfund). CERCLA authority is not delegable to the states, but the states are free to pass their own "Superfund laws" and many have. In the passage of most of the primary environmental laws (for example, the Clean Water Act, the Clean Air Act, RCRA), Congress has authorized a further delegation to the states. Typically, the EPA will publish rules governing such delegations. In general, the primary rule of delegation is that the state must first pass laws/regulations at least as stringent as the federal law. Once a state has received the delegation, the EPA will normally defer to the state as having **primacy** in the area. Delegation of authority may be in whole or in part, so it is important to identify the precise scope of the delegation a particular state received from the federal government.

While often of little significance in environmental law issues, there may be issues where the **state constitution** is controlling (property issues are one example).

Local (city/county) laws, regulations, and ordinances

The significance of local laws and regulations varies greatly from one jurisdiction to another. In some states, and in some environmental areas, the role of local government is critical.

Case law

In the United States, only legislative bodies are empowered to enact laws. Courts, on the other hand, are the final arbiters on the validity and interpretation of such laws. Their rulings create a body of law known as "case law." Laws are often subject to challenge in the courts on the basis that they are invalid, that is, that they violate the applicable constitution, that they conflict with other laws, or that they suffer from some other legally recognized infirmity. Far more prevalent, however, are court rulings regarding the interpretation of laws. All laws are subject to interpretation, that is, to determination of what the legislature meant in drafting the law. Perhaps the best known example is the seemingly straightforward phrase in NEPA (National Environmental Policy Act of 1969) describing when an environmental impact statement (EIS) is required—specifically, "major federal actions significantly affecting the quality of the environment." The meaning of every word in this phrase (except "of" and "the") has been the subject of scores of lawsuits, some terminating in the U.S. Supreme Court.

In their rulings, courts establish legal **precedents** that "lower" courts are bound to follow. In order to provide predictability and consistency in the law, courts routinely follow legal precedent even when they are not required to do so (for example, when the precedent comes from a court of equal authority). Occasionally, a court will overturn its own precedent, thereby establishing a new one.

Case law is also the source of U.S. **common law,** the term used to refer to the collection of well-established legal principles established by the courts over the

years. Common law reflects judicial concepts of fairness and equity that are applied in the absence of statutes and regulations. Modern environmental law is almost entirely codified and, therefore, resort to the common law is rare. There are notable exceptions. Laws governing environmental nuisance and trespass are often rooted in the common law. Most notable is the law of **negligence,** which is the primary body of law governing liability and is still significantly governed by common law principles. Many hazardous waste management practices are undertaken with an eye toward liability, a complex area of common law in which claims can be enormous and far reaching.

The U.S. Judicial System

U.S. Supreme Court

The U.S. Supreme Court is a court of very limited jurisdiction. This court is the final arbiter on the meaning of the Constitution and the interpretation of federal law. In most instances, it has the discretion to hear a case on appeal or not, and the vast majority of petitions to the Court are rejected. It normally hears cases in two situations: (1) it determines the case to be of constitutional significance or (2) it finds it necessarily to resolve disputes among the lower courts that are preventing stability and consistency in the law.

Federal courts of appeal (circuit courts)

These are the primary appellate courts in the federal system. They hear cases appealed from U.S. district courts. There are 10 circuits divided geographically among the United States and its territories. Their rulings establish precedent for federal courts in their circuits. Like the Supreme Court, their jurisdiction is limited and they generally are involved in ruling on U.S. Constitutional challenges and interpreting federal law.

Federal district courts

These are the initial or trial-level courts in the federal system. The vast majority of federal case law is established in the district courts. Although they are the lowest of the federal courts, the precedents of the district courts are often the only case law available to apply to a given environmental issue.

State courts

Rulings of state courts are of significance for a number of reasons. State courts also interpret federal law and in some instances, their precedents may be the only ones available. State courts, not federal courts, are the final arbiters of state law, which may be the controlling law on a given environmental issue (for example, a state Superfund law). Finally, state courts are the final arbiters on the interpretation of the state constitution.

Administrative "courts"

Many federal agencies have established administrative procedures, which must be complied with before a case may be filed in court. These procedures may involve hearings before administrative law judges, or similar forums. A failure to comply with these procedures may bar future legal action.

Researching the Law

Electronic resources have greatly simplified the process of finding the law. Access to the Internet over a simple modem is normally all that is required. Federal statutes and regulations are available on a number of pay and free sites. Two of the more common (and easy to use) are "FindLaw" at *www.findlaw.com* and LEXIS®-NEXIS® at *www.lexispublishing.com*. Virtually all federal agencies maintain a website with a library of publications available for downloading, covering everything from regulations to guidance documents. The *Federal Register* is available on line at *www.access.gpo.gov*.

Most state statutes and regulations are also available on line. Familiarity with the local method of citations (the abbreviations used in referencing the law) is usually necessary.

Finding case law, also available at a number of sites (including LEXIS®-NEXIS®), is a much more involved process. Finding the *applicable* law requires some level of legal interpretation and is therefore best left to legal professionals trained in the art of legal research.

Legal research is a methodical process of identifying the issues and finding the applicable law to apply to those issues. The following example is illustrative of the process. Assume an issue regarding whether a waste is "hazardous." The primary source would be the governing federal statute, RCRA. This will provide the statutory framework for defining hazardous waste. The next step is to determine whether the state in question has RCRA primacy, that is, whether it has received delegated authority from the EPA. If it has, state statutes, regulations, and case law must be researched because they are controlling. Assuming there has been no delegation, the next step would be to research the Code of Federal Regulations, wherein the EPA defines with specificity those categories of solid waste that are classified as hazardous waste. Next, the *Federal Register* must be researched to ensure no new rules have been promulgated since the last update of the CFR.

Next, case law must be researched to ascertain how the courts have interpreted the relevant statutes and regulations. In some instances, a particular statutory provision or regulation may have been ruled invalid and is therefore no longer binding. Case law itself must be interpreted, and the process is best left to legal professionals.

Finally, it should be remembered that the law should be reviewed as a whole. Statutes, regulations, and case law often create exceptions to their applicability or provide alternate standards of applicability. For example, statutes often provide special standards for application at federal facilities, and these standards may vary from one statute to another. Taking the law out of context will often result in a misapplication of the law.

2-2 RESOURCE CONSERVATION AND RECOVERY ACT

The Resource Conservation and Recovery Act (RCRA) of 1976 provided for a significant federal role in the management of hazardous waste for the first time.[8] Adopted as a series of amendments to the Solid Waste Act of 1965, RCRA set up a separate Office of Solid Waste within the EPA. This office was charged with

establishing a comprehensive regulatory program ranging from identifying which wastes are hazardous to establishing a manifest system for tracking wastes. A major outcome of RCRA is that generators are responsible from "cradle to grave" for the wastes they produce.

Under the **cradle-to-grave** concept, a generator of hazardous waste can no longer avoid liability by contracting with a third party to dispose of the waste. Even if it can be shown that the waste was mishandled through the actions of a third party, the original generator will remain liable for improper disposal. This compels generators to exercise care in the selection of the disposal companies they utilize.

To further assure compliance, Congress provided for **citizen suits,** wherein anyone could bring suit "against any person . . . alleged to be in violation of any permit, standard, regulation, condition, requirement, or order." Congress specifically permitted suits against the U.S. government and any other government agencies and permitted suits against the EPA for failure to perform any duty under the Act.

RCRA's title clearly announces Congress's intention that future management of hazardous waste should stress conservation and recovery of reusable materials (e.g., recycling) as opposed to disposal. RCRA is directed at the waste presently being generated, whereas the Superfund law is directed at correcting the bad practices of the past (see Section 2-4). This section discusses the provisions Congress included in the 1976 RCRA statute or in its major amendment, the Hazardous and Solid Waste Amendments of 1984. Regulations promulgated by the EPA under authority of RCRA will be discussed in Section 2-3.

Definitions

The first concern of Congress was to define the term **hazardous waste.** The following definitions are from Sec. 1004 of RCRA:

> The term "**hazardous waste**" means a solid waste or combination of solid wastes, which because of its quantity, concentration, or physical, chemical, or infectious characteristics may—
>
> **(A)** cause or significantly contribute to an increase in mortality or an increase in serious irreversible, or incapacitating reversible, illness; or
> **(B)** pose a substantial present or potential hazard to human health or the environment when improperly treated, stored, transported, or disposed of, or otherwise managed. . . .
>
> The term "**solid waste**" means any garbage, refuse, sludge from a waste treatment plant, water supply treatment plant, or air pollution control facility and other discarded material, including solid, liquid, semi-solid, or contained gaseous material resulting from industrial, commercial, mining, and agricultural operations, and from community activities, but does not include solid or dissolved materials in irrigation return flows or industrial discharges which are point sources subject to permits under Section 402 of the Federal Water Pollution Control Act, as amended (86 Stat. 880), or source, special nuclear, or byproduct material as defined by the Atomic Energy Act of 1954, as amended (68 Stat. 923).

RCRA defined hazardous waste but did not set precise standards for its disposal. Standards for disposal were left to regulatory development by the EPA. For example, RCRA required that the EPA establish criteria for sanitary landfills and gave the following general guidelines: "At a minimum, such criteria shall provide that a

facility may be classified as a sanitary landfill and not an open dump only if there is no reasonable probability of adverse effects on health or the environment from the disposal of solid waste at such a facility."

The Hazardous and Solid Waste Amendments

The 1976 Act was comprehensive and led to citizen expectations that hazardous waste problems would be solved in short order. Unfortunately, hazardous waste management problems are too complex to be amenable to simple answers, and the hopes of many were not met. The EPA was blamed by many for failing to carry out the clear mandate of Congress.

The Congressional response to a perceived lack of EPA action was to pass the **Hazardous and Solid Waste Amendments of 1984 (HSWA).**[9] This statute that amended RCRA was far more detailed than most other pieces of environmental legislation. For example, minimal technical requirements for land disposal facilities were included in HSWA, in contrast to the usual practice of leaving these technical details to the EPA. HSWA also specified standards for the issuance of permits, penalties for violating the law, and controls on underground storage tanks.

Congress, apparently not satisfied that the EPA would faithfully carry out its intent as reflected in RCRA, included many "**hammer provisions**" in HSWA. These provisions state that if the EPA did not meet specific deadlines for promulgating regulations, then extremely rigid regulations, contained within HSWA, would automatically become enforceable requirements.

A comprehensive summary of HSWA is beyond the scope of this text and the reader is referred to the articles in the references at the end of this chapter.[10, 11] The following subsections introduce several of the provisions of HSWA.

Land Disposal Prohibitions

HSWA prohibited the disposal of bulk (noncontainerized) liquid hazardous waste in landfills or surface impoundments. Additionally, disposal of containerized liquids is permitted only if the EPA determines on a case-by-case basis that:

- No reasonable alternative is available.
- Such disposal is environmentally acceptable.

Otherwise, the disposal of containerized liquids is banned. HSWA also prohibits land disposal of specific chemicals (e.g., halogenated organic compounds at concentrations greater than 1000 milligrams of chemical per kilogram of waste). Further, for other wastes not specifically banned, the EPA has issued regulations specifying treatment levels that do either of the following:

- Diminish the toxicity of the waste
- Reduce the likelihood of migration

Finally, HSWA required that the EPA must review each identified hazardous waste and determine if land disposal should be allowed. This provision allowed the EPA no discretion to consider other factors, such as EPA staffing available to perform such evaluations or the economic and social implications of such a comprehensive

ban. The clear intent was to make land disposal a last resort and to move hazardous waste management toward more "acceptable" technologies. The EPA has reviewed and made land ban determinations for many hazardous wastes.

In conjunction with the prohibition against land disposal for specific hazardous wastes, the EPA was required to establish treatment methods and/or treatment standards for the various hazardous constituents in these wastes. Wastes for which a treatment method is established must be treated by that method prior to disposal in a land-based unit. For many wastes, however, the EPA has specified treatment standards rather than treatment methods. The treatment standards are concentrations for individual chemical constituents in hazardous wastes. Treated hazardous wastes that meet the treatment standards given in 40 CFR 268.40, known as the Universal Treatment Standards (UTS), may be placed in land-based disposal units such as landfills, waste piles, land treatment units, surface impoundments, or injection wells. In a few select cases, the EPA has approved the disposal of otherwise land-banned wastes in land disposal units.

Minimum Technology Requirements

While Congress stopped short of banning land disposal outright, it determined that if land-based disposal facilities (e.g., landfills) were to exist, they would be state-of-the-art facilities. As an example, all new land disposal facilities for hazardous wastes must have:

- Double liners
- Leachate collection systems
- Groundwater monitoring
- Leak detection systems

Current practice is to design land disposal facilities with greater redundancy than the law requires (see Chapter 13).

New standards for incinerators in HSWA required destruction and removal efficiencies of:

- 99.99 percent for principal organic hazardous constituents
- 99.9999 percent for certain hazardous wastes from nonspecific sources (e.g., F-020 through F-023, to include a certain dioxin isomer)

Permitting

The 1984 amendments included many provisions relating to the permitting of treatment, storage, and disposal facilities (TSDFs). Since 1976, such facilities required a RCRA permit to remain in operation. For many manufacturing plants, closing the TSDF means closing the factory, because there is no method to make the products without generating hazardous waste. Therefore, these provisions provide a strong incentive for industry to remain in compliance. Permits issued after HSWA must:

- Incorporate all provisions of HSWA
- Provide for corrective action for releases

- Include a ban on the placement of liquids in landfills
- Include an exposure assessment (i.e., determine the risk involved if a release were to occur)
- Provide for a 5-year review of the permit requirements

RCRA Corrective Action Program

HSWA gave the EPA the authority to require corrective action (i.e., cleanup of contaminants in soil and groundwater) beyond the boundaries of a manufacturing facility. Similarly, the EPA can require corrective action for continuing releases of pollution, even if the releases do not leave the plant boundaries.

In September, 1988, the EPA proposed extensive requirements for corrective action at **solid waste management units (SWMUs)** for facilities permitted under RCRA.[12] These regulations are intended to provide a comprehensive program to correct deficiencies at existing landfills, lagoons, and surface impoundments at manufacturing plants throughout the United States. They require an analysis similar to that conducted at a Superfund site (see Section 2-4) whenever contamination was suspected. The basic elements of an RCRA Corrective Action Program are:

- RCRA Facility Assessment (RFA)
- RCRA Facility Investigation (RFI)
- Corrective Measures Study (CMS)
- Implementation

Inspection and Enforcement

While promulgation of laws, rules, and regulations form the foundation of a regulatory system, consistent compliance depends on adequate enforcement. Federal enforcement is the responsibility of the EPA, although when authorized by the EPA, a state can be responsible for RCRA programs. Additionally, state agencies will enforce state laws and regulations that may be more stringent than federal requirements. In some instances the EPA's task is primarily one of education—explaining complex regulations to the regulated community. In other instances, the agency must enforce the law. A Congressional perception that the EPA was not aggressive enough in enforcement of the 1976 law led to more stringent enforcement provisions in HSWA.

The 1984 Amendments make it a crime to knowingly:

- Transport hazardous waste without a manifest (see Figure 2-5)
- Omit relevant information on a manifest form
- Violate a regulation
- Fail to file a required report or application

The law provides for fines up to $50,000 and prison terms up to 2 years. HSWA defines **knowing endangerment** as violations that consciously place another person in imminent danger of death or serious bodily injury. Under circumstances constituting knowing endangerment, penalties include fines up to $250,000 and *imprisonment of up to 15 years.*

Enforcement actions include cleanup orders and compliance orders. The EPA can order a responsible party to undertake corrective action if a release of hazardous waste has occurred. Compliance orders may be issued for past violations of any requirement. Civil penalties for violations of these orders can range up to $25,000 per day.

Underground Storage Tanks

HSWA required the EPA to establish a comprehensive regulatory program for underground storage tanks. Underground storage tanks present a unique environmental problem in that their very presence is often unknown to people living in the vicinity until a problem occurs, and then it is too late for preventive measures. An example of the problems created by underground storage tanks is illustrated in Figure 2-4—the tank being removed has been corroded to the point that large holes were discovered when the tank was removed.

One EPA report indicated that state agencies were aware of more than 12,000 releases from underground tanks covering the period 1970–1984.[13] The majority of these incidents occurred at gasoline stations. Assuming that unreported incidents greatly exceeded those reported, Congress instituted a comprehensive program to assess and control leaking underground storage tanks. Within a phased-in 10-year period starting in 1988, existing underground tanks had to be evaluated and upgraded to meet the established standards. Although this deadline has passed, a large number of tanks have not been evaluated and/or upgraded.

To prevent future leaks, HSWA requires the owner of an existing underground storage tank to provide either a leak detection system or an inventory control with regular testing of tanks, and has imposed design standards to prevent corrosion. Owners are required to maintain detailed records of monitoring and tank testing, report releases, and take appropriate corrective action when leaks do occur. Owners are also responsible for maintaining liability insurance to cover the cost of any damage to third parties that might occur if a tank leaks.

FIGURE 2-4
Photograph of an underground storage tank removal.[14]

2-3
FEDERAL HAZARDOUS WASTE REGULATIONS UNDER RCRA

In RCRA, Congress required the EPA to institute one of the most far-reaching regulatory programs ever conceived. More than 500 pages of hazardous waste regulations were published in the *Federal Register* on May 19, 1980.[15] The U.S. Office of Management and Budget estimated that the paperwork burden on industry alone would consume 1.5 million hours annually.[16] These regulations defined which wastes were hazardous; established an EPA notification process for organizations producing waste; and set up detailed regulations covering the generation, transportation, treatment, storage, and disposal of hazardous waste. (Similar provisions are included in the regulations of many developed countries—see Section 2-6.) These regulations were modified by HSWA's requirements (as stated in Section 2-2) and continue to be updated regularly. This section discusses some of the more important provisions.

Definitions

What is a hazardous waste? There are three ways that a solid waste (i.e., solid, liquid, semisolid, or contained gas) can be considered "hazardous" under RCRA regulations:

1. The waste is specifically listed in any of four lists provided in EPA regulations. For example, spent halogenated solvents have been listed and given the EPA hazardous waste number F001.[17]
2. The waste is tested and meets one of four characteristics established by the EPA: **ignitable, corrosive, reactive, or toxic**.
3. The waste is declared hazardous by the generator (the entity producing the waste), on the basic of its knowledge of the waste.

The EPA is responsible for regularly updating the lists of hazardous wastes in 40 CFR 261. Among the criteria utilized in making this determination are the four characteristics mentioned in item 2 above and described in detail in the following paragraphs. Additional hazardous wastes include materials that have been shown to be fatal to humans in low doses. In the absence of human epidemiological evidence, animal tests may be used to infer human toxicity. This is discussed in greater detail in Chapter 5. The use of lists for many wastes precludes the necessity of expensive laboratory analyses.

The EPA selected the four characteristics of ignitability, corrosivity, reactivity, and toxicity on the basis of tests that met the intent of RCRA law, were reproducible, and were relatively easy to perform.

Ignitable wastes. Ignitable wastes are liquids with a flashpoint below 60°C or solids capable of causing fire under standard temperature and pressure. Ignitable wastes were assigned EPA Hazardous Waste No. D001.

Corrosive wastes. Corrosive wastes are aqueous wastes with a pH below 2 or above 12.5, or which corrode steel at a rate in excess of 0.25 inch per year. Corrosive wastes are classified D002.

Reactive wastes. Reactive wastes are normally unstable, react violently with air or water, or form potentially explosive mixtures with water. This category also includes wastes that emit toxic fumes when mixed with water and materials capable of detonation. Reactive wastes are classified D003.

Toxicity. The characteristic of toxicity is more difficult to define. The objective of this parameter is to determine whether toxic constituents in a solid waste sample will leach into groundwater if the waste is placed in a municipal solid waste landfill. If this is the case, then the waste will be declared hazardous.

The EPA has developed two tests for determining toxicity. The initial test, referred to as the EP Toxicity Test[18] (EP—Extraction Procedure), required grinding of solid samples, followed by rotary extraction in acetic acid for 24 hours. The material would then be filtered and the filtrate analyzed and compared to published standards (generally these were 100 times the concentration levels acceptable in drinking water for specified metals and organic compounds). If the concentration in the extract exceeded any one of the specified levels, the original solid waste was considered hazardous.

A number of shortcomings with the EP Toxicity Test led the EPA to develop a modified test referred to as the Toxicity Characteristic Leaching Procedure (TCLP).[19, 20] This test was more specific in many operational parameters, required a zero head-space extractor so that volatile organics would not be lost, and had more stringent quality control measures. The required chemical analyses and regulatory levels are provided in Table 2-1. If even one compound exceeds the level specified in this table, then the original waste is a hazardous waste. For some compounds, e.g., 2,4-dinitrotoluene and hexachlorobenzene, the quantitation limit (i.e., the minimum concentration that current laboratory methods can reproducibly measure) exceeds the calculated level based on toxicity. In these instances, the regulatory level is set at the quantitation limit. While TCLP replaced the EP Toxicity Test for most situations, the EP test was still used for several regulatory decisions, leading to a great deal of confusion.[21] A discussion of leaching tests is included in Chapter 11.

Generators and Transporters

Under the 1980 regulations, the EPA placed controls on all entities producing or transporting hazardous waste. Any generator producing waste must evaluate its waste and obtain an EPA identification number as a generator if the waste is hazardous. If the waste is stored at the generator's site for more than 90 days, the generator must also obtain a permit as a storage facility. Additional requirements include the use of appropriate containers and labels, maintenance of records of waste generated, training, preparation of contingency plans, and submission of annual reports.

Trucking companies that transport hazardous waste must obtain EPA identification numbers as transporters. Because trucking companies and railroads were already regulated by the U.S. Department of Transportation (DOT), the EPA chose to incorporate DOT regulations (49 CFR) into the EPA's regulatory scheme, and add additional requirements such as record keeping and cleanup of spills.

To assure that "cradle-to-grave" control of hazardous waste is achieved, Congress mandated that the EPA establish a **manifest system** that uses a form to track all waste

TABLE 2-1
Toxicity characteristic leaching procedure

Compound	ID no.	Regulatory level in TCLP extract (mg/L)
Arsenic	(D004)	5.0
Barium	(D005)	100.0
Benzene	(D018)	0.5
Cadmium	(D006)	1.0
Carbon tetrachloride	(D019)	0.5
Chlordane	(D020)	0.03
Chlorobenzene	(D021)	100.0
Chloroform	(D022)	6.0
Chromium	(D007)	5.0
o-Cresol*	(D023)	200.0
m-Cresol*	(D024)	200.0
p-Cresol*	(D025)	200.0
2,4-D	(D016)	10.0
1,4-Dichlorobenzene	(D027)	7.5
1,2-Dichloroethane	(D028)	0.5
1,1-Dichloroethylene	(D029)	0.7
2,4-Dinitrotoluene	(D030)	0.13†
Endrin	(D012)	0.02
Heptachlor (and its hydroxide)	(D031)	0.008
Hexachlorobenzene	(D032)	0.13†
Hexachloro-1,3-butadiene	(D033)	0.5
Hexachloroethane	(D034)	3.0
Lead	(D008)	5.0
Lindane	(D013)	0.4
Mercury	(D009)	0.2
Methoxychlor	(D014)	10.0
Methyl ethyl ketone	(D035)	200.0
Nitrobenzene	(D036)	2.0
Pentachlorophenol	(D037)	100.0
Pyridine	(D038)	5.0†
Selenium	(D010)	1.0
Silver	(D011)	5.0
Tetrachloroethylene	(D039)	0.7
Toxaphene	(D015)	0.5
Trichloroethylene	(D040)	0.5
2,4,5-Trichlorophenol	(D041)	400.0
2,4,6-Trichlorophenol	(D042)	2.0
2,4,5-TP (Silvex)	(D017)	1.0
Vinyl chloride	(D043)	0.2

*If o-, m-, and p-cresol concentrations cannot be differentiated, the total cresol concentration is used. The regulatory level for total cresol (D026) is 200 mg/L.
†Quantitation limit.
Source: 40 CFR 261.24, Table 1.

sent off-site from the generator through to ultimate disposal. A portion of the manifest form used by New York State is shown in Figure 2-5. The manifest has two sections, Part A—filled out by generator—and Part B—filled out by transporter or TSD facility. Six copies are distributed as follows:

48-14-1 (3/89)—7f

STATE OF NEW YORK
DEPARTMENT OF ENVIRONMENTAL CONSERVATION
DIVISION OF HAZARDOUS SUBSTANCES REGULATION

HAZARDOUS WASTE MANIFEST

Please print or type. Do not Staple — P.O. Box 12820, Albany, New York 12212 — Form Approved. OMB No. 2050-0039. Expires 9-30-91

In case of emergency or spill immediately call the National Response Center (800) 424-8802 and the N.Y. Dept. of Environmental Conservation (518) 457-7362

GENERATOR

UNIFORM HAZARDOUS WASTE MANIFEST

1. Generator's US EPA No. N Y D 1 2 3 4 5 6 7 8 9 — Manifest Document No. 3 8 7 5 2
2. Page 1 of 1 — Information in the shaded areas is not required by Federal Law.

3. Generator's Name and Mailing Address
XYZ Corp
1234 Fifth Ave. NY, NY 13201
4. Generator's Phone (212) 555-1234

A. State Manifest Document No. NY B 203875 2
B. Generator's ID

5. Transporter 1 (Company Name) ABC Trucking — 6. US EPA ID Number N J D 0 0 1 2 3 4 5 6 7
C. State Transporter's ID
D. Transporter's Phone (908)555-1234

7. Transporter 2 (Company Name) — 8. US EPA ID Number
E. State Transporter's ID
F. Transporter's Phone ( )

9. Designated Facility Name and Site Address
Chemical Disposal Services, Inc.
Syracuse, NY 13201
10. US EPA ID Number N Y D 0 0 0 0 1 2 3 4 5
G. State Facility's ID
H. Facility's Phone (315) 555-1234

11. US DOT Description (Including Proper Shipping Name, Hazard Class and ID Number)	12. Containers No.	12. Containers Type	13. Total Quantity	14. Unit Wt/Vol	I. Waste No.
a. Hazardous Solid Waste CEM-E NA 9/88	0 2 0	D M	2 1 2 0	K	EPA / STATE B007
b.					EPA / STATE
c.					EPA / STATE
d.					EPA / STATE

J. Additional Descriptions for Materials listed Above
a Soil/PCBs — a — b — b

K. Handling Codes for Wastes Listed Above
a — c — b — d

15. Special Handling Instructions and Additional Information

16. **GENERATOR'S CLASSIFICATION:** I hereby declare that the contents of this consignment are fully and accurately described above by proper shipping name and are classified, packed, marked and labeled, and are in all respects in proper condition for transport by highway according to applicable international and national government regulations and state laws and regulations.
If I am a large quantity operator, I certify that I have program in place to reduce the volume and toxicity of waste generated to the degree I have determined to be economically practicable and that I have selected the practicable method treatment, storage, or disposal currently avaliable to me which minimizes the present and future threat to human health and the environment; OR if I am a small generator, I have made a good faith effort to minimize my waste and select the best waste management method that is available to me and that I can afford.

Printed/Typed Name: John Smith, Agent for XYZ Corp. — Signature: John Smith — Mo. Day Year 10 6 3 0 9 2

TRANSPORTER

17. Transporter 1 (Acknowledgement of Receipt of Materials)
Printed/Typed Name — Signature — Mo. Day Year

18. Transporter 2 (Acknowledgement of Receipt of Materials)
Printed/Typed Name — Signature — Mo. Day Year

FACILITY

19. Discrepancy Indication Space

20. Facility Owner or Operator: Certification of receipt of hazardous materials covered by this manifest except as noted in Item 19.
Printed/Typed Name — Signature — Mo. Day Year

NY B203875 2

EPA Form 8700-22 (Rev. 9-88) Previous editions are obsolete.

COPY 1—Disposer State—Mailed by TSD Facility

FIGURE 2-5
Hazardous waste manifest, New York State Department of Environmental Conservation.

1. Sent to state agency where disposal occurs
2. Sent to state agency where generation occurs
3. Sent to generator by TSD facility, then retained by generator
4. Retained by TSD facility
5. Retained by transporter No. 1
6. Retained by transporter No. 2 (if required)

The generator is responsible for checking that it receives a copy of the manifest from the disposal facility for each shipment sent. By requiring quarterly and annual reports from both generators and TSD facilities, the system allows the states to track all hazardous waste shipments.

Because of the tremendous cost of paperwork involved in the manifest system, the EPA excluded small-quantity generators [i.e., those generating less than 1 metric ton (1000 kg or 2200 pounds) per month] from many of the 1980 regulations. In 1984 HSWA required the EPA to establish standards for this group. Small-quantity generators now must comply with most RCRA regulations, including using the manifest system and shipping only to permitted hazardous waste facilities. Generators producing less than 100 kg (approximately one-half of a 55-gallon drum) per month are exempt from most RCRA manifest regulations but are still required to meet minimum standards. This provision is unique to the United States because most developed countries do not provide a small-quantity exemption.

Treatment, Storage, and Disposal (TSD) Facilities

Federal regulations covering TSD facilities (40 CFR 264–265) represent the largest part of the RCRA regulatory package. Virtually any facility handling hazardous waste comes under the purview of these standards, including landfills, incinerators, surface impoundments, and storage areas used for more than 90 days.

Facility standards include general requirements for spill prevention procedures, contingency plans, and training programs. There are numerous reporting and paperwork requirements designed to assure compliance. For example, facilities must retain all manifests for 3 years and keep complete operating records until closure.

Some of the most important provisions of the regulations are those dealing with management of hazardous wastes in land-based units such as landfills (40 CFR 264.300–317 and 40 CFR 265.300–316), surface impoundments (40 CFR 264.220–232 and 40 CFR 265.220–231), waste piles (40 CFR 264.250–259 and 40 CFR 265.250–260) and land treatment units (40 CFR 264.270–283 and 40 CFR 265.270–282). Among other provisions, operators of landfills must meet minimum technology requirements (MTRs) for design and must construct covers or landfill caps at closure (the end of the useful life of the landfill) that minimize long-term migration of liquids through the closed fill. Additionally, a groundwater monitoring system must be established. Liquids in landfills were conditionally permitted in the 1980 regulations, but Congress banned the disposal of bulk liquids (i.e., those not in containers) in landfills in the 1984 HSWA.

Most requirements in the RCRA TSD regulations are directed toward ongoing management of hazardous wastes currently being generated. HSWA also required the EPA to develop a corrective action program for solid waste management units that are no longer used for the active management of hazardous wastes. TSD facilities must first determine whether these historical waste management sites have released hazardous constituents to soil or to groundwater. If so, the TSD facility must implement a cleanup program to remediate the release. However, strict application of RCRA regulations to hazardous wastes generated during cleanups has been perceived as discouraging remediation. To remove these impediments, the EPA has created a special set of regulations for corrective action in 40 CFR 264.552–554 that

apply only to management of remediation wastes. The major differences between the regulatory requirements for remediation wastes and currently generated wastes are:

- Land disposal restrictions do not apply to remediation wastes.
- MTRs do not apply to units used to manage remediation wastes.
- Permitting requirements are streamlined.

2-4 SUPERFUND

RCRA established an extensive regulatory program for newly generated hazardous wastes but did nothing to help correct the results of poor disposal practices and inadequate technology of the past. In some locations, owners and operators abandoned their plants and left drums and tanks of hazardous waste for others to deal with. (See Figure 2-6.) In 1978, a major episode occurred in the section of Niagara Falls, New York, known as the Love Canal, where toxic chemicals, buried years earlier, were found to be leaking into the basements of local homes (see Section 1-3). This and numerous similar events led Congress to pass the **Comprehensive Environmental Response, Compensation and Liability Act**[22] (CERCLA) of 1980. CERCLA was amended in 1986 by the **Superfund Amendments and Reauthorization Act** (SARA). The term **"Superfund"** is used to refer both to these laws and to the cleanup program they mandated. A few of the major provisions of Superfund are described in the following sections. For a more complete understanding of this law, refer to the references at the end of this chapter.[23–26]

The primary purpose of CERCLA is to clean up hazardous waste sites not regulated under RCRA. As the name indicates, Congress also provided the EPA with the ability to respond to spills and other environmental incidents, and set up a liability system to recover costs from those responsible for causing such incidents.

FIGURE 2-6
An abandoned plant becomes a Superfund site. *(Photo courtesy EPA.)*

CERCLA, 1980

CERCLA established a $1.6 billion fund (the Superfund), derived primarily from feedstock taxes on industry, to implement a massive environmental cleanup program over a 5-year period. Generators were required to report to the EPA any facility at which hazardous wastes are, or have been, treated, stored, or disposed. The intent was to identify and clean up hazardous waste sites first, and then to litigate to recover costs. This law gave the EPA strong powers to encourage potentially responsible parties (PRPs) to clean up sites. As in the case of RCRA, Superfund set unrealistically optimistic goals and deadlines and created public anticipation that sites would be rapidly cleaned up.

While the Congressional intent was clear, the incoming Reagan administration had campaigned on a platform that included significantly reducing governmental regulation of industry. The combination of unrealistic expectations and a deregulation attitude had the impact of slowing down an already difficult program. At the end of the 5-year period, environmental groups and legislators decried the fact that "only six sites had been cleaned up," and Congress debated whether to reauthorize Superfund at $5 billion or $10 billion.

SARA, 1986

Following nearly 2 years of debate, SARA[27] was passed in the final days of the 99th Congress in 1986. This statute was a complete rewrite of CERCLA and is four times as long. SARA created an $8.5 billion fund for cleaning up abandoned waste disposal sites and an additional $500 million for cleaning up leaking underground petroleum tanks. SARA increased the EPA's flexibility to perform removal actions, required attainment of **Applicable or Relevant and Appropriate Requirements** (ARARs) at the federal level or more stringent state standards, and established a preference for permanent remedies that reduce the volume, toxicity, or mobility of hazardous substances in selecting remedies for Superfund sites.

In response in part to the tragedy in Bhopal, India (see box, "The Bhopal Tragedy"), the **Community Right-to-Know** provisions of SARA require industries to plan for emergencies and inform the public of hazardous substances being used. It should be noted that the Bhopal tragedy was caused by hazardous materials, *not* hazardous waste. The American right-to-know policy should be contrasted to the European control of "major hazard industries."

SARA required that the Hazard Ranking System (HRS) that determines which sites are eligible for cleanup funding be totally revised. Pursuant to SARA, the EPA initiated Remedial Investigation/Feasibility Studies (see below) at 650 sites and remedial action at 375 sites within 5 years.

The Superfund Process

The procedures followed at a Superfund site are illustrated in Figure 2-7.

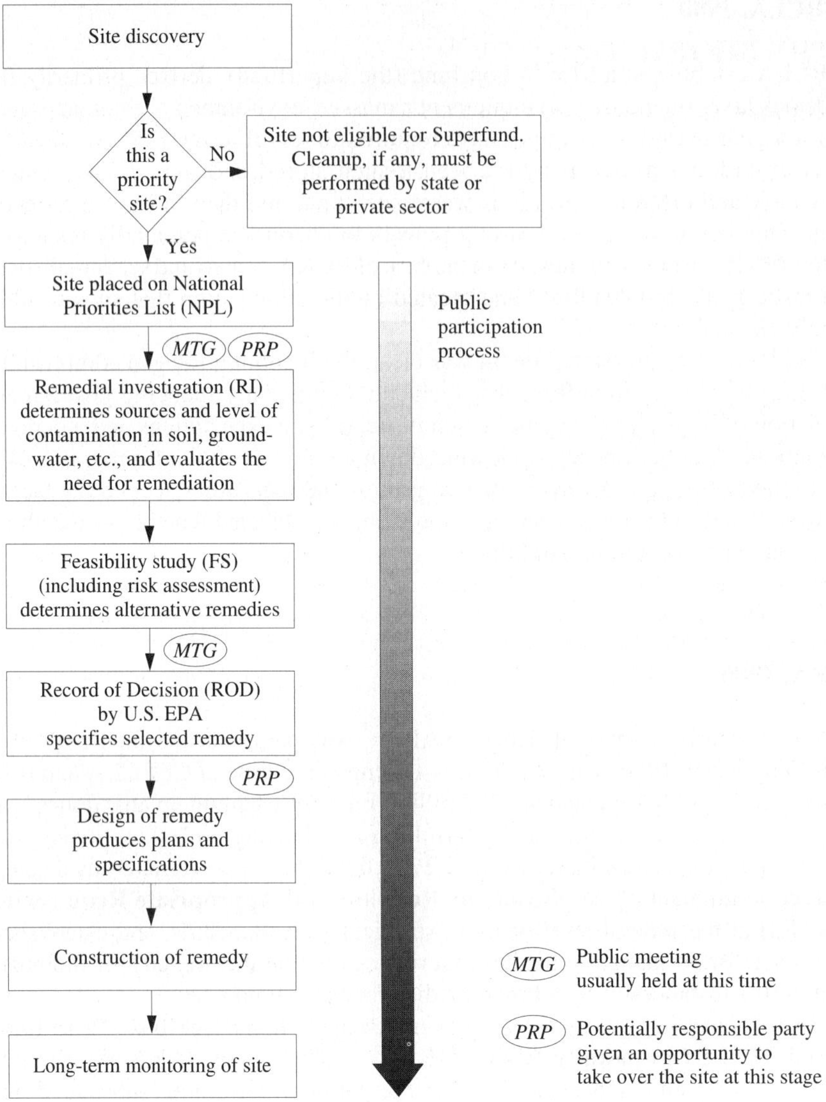

FIGURE 2-7
Activity schematic at a Superfund site.

The list of sites eligible for cleanup funding under Superfund is referred to as the **National Priorities List** (NPL). The EPA uses a method called the **Hazard Ranking System** (HRS) to estimate the degree of risk each suspect site poses to human health and the environment. Factors such as proximity to population, nature of the contaminants, and potential release and migration pathways are evaluated and then combined to develop a single number—the HRS score—that purports to represent a relative risk. If the HRS score exceeds a specified threshold value, the site is proposed for placement on the NPL. After a public comment period, the EPA determines whether

THE BHOPAL TRAGEDY[28]

At 1:00 A.M. on Sunday, December 3, 1984, approximately 45 tons of methyl isocyanate (CH_3NCO) (MIC) leaked from a storage tank at the Union Carbide India, Ltd., pesticide plant in Bhopal, India. Before what has been termed "the worst chemical accident in human history" was over, nearly 3000 people were dead and "about 200,000 injured."[29]

Methyl isocyanate is a precursor chemical used in the manufacture of pesticides. It is highly unstable and needs to be kept at low temperatures. It is extremely toxic, capable of causing severe bronchospasm and asthmatic breathing when inhaled. It is a severe eye, skin, and mucous membrane irritant and can be absorbed through the skin. Exposure to high concentrations can cause blindness, damage to lungs, emphysema, and ultimately death.[30]

The causes of the accident have been ascribed by the media to design flaws, operating deficiencies, maintenance failures, and inadequate training.[31] Union Carbide's internal investigation suggests that the accident was the direct result of employee sabotage.

The plant had a number of safety devices including a vent gas scrubber intended to neutralize any MIC that might escape and a flare tower capable of burning MIC to CO_2, H_2O, and NH_3. A detailed investigation by A. D. Little, Inc. suggests that direct water entry into the MIC tank was the likely cause of the accident.[32] The mixture of water and MIC produced high temperatures and pressures causing MIC gas to be generated at an extremely high rate.

In addition to technological failures, a number of social and political factors contributed to the severity of the incident. Bhopal had grown rapidly, had insufficient low-income housing, and had large numbers of migrants living in shantytowns despite the company's efforts to establish a greenbelt around the plant. Two of these areas were immediately across the street from the chemical plant, even though the area was not zoned for residential use. The state government was reluctant to impose strict environmental and safety standards on industry, for fear of losing badly needed jobs.

The U.S. Environmental Protection Agency was severely criticized because it had not listed MIC as a dangerous pollutant. Congressman Henry A. Waxman, Chairman of the House Health and Environment Subcommittee, noted: "It's a terrible thing to think we have to have these tragedies to get industry to come forward and say, 'Yes, maybe we should have thought these things through.'"[33] The fact that so tragic an accident could occur sobered many in Congress and led to the Right-to-Know provisions in SARA.

to place the site on the NPL. Sites placed on the NPL are eligible for funds for site remediation.

After a site has been listed, the EPA sends notices to the identified PRPs, giving them the opportunity to perform the required studies. If no agreement with the PRPs is reached, the EPA performs the work through the U.S. Army Corps of Engineers (COE) and outside consultants and contractors working for the COE. Once the work is completed, the EPA sues the PRPs to recover the costs spent from the fund. If the PRPs agree to perform the work, the EPA provides oversight and review of all

documents, usually through a consultant working for the EPA. These costs, paid from the fund, are also recovered in later claims against the PRPs.

The first activity after listing is the **remedial investigation** (RI). The RI work includes evaluating the site history, defining the contamination sources and the type and extent of the existing contamination, evaluating the fate and transport pathways of the contaminants, and finally performing a risk assessment to evaluate the risks posed by the site and whether it requires remediation. The techniques used to determine the sources, nature, and extent of the contamination are described in Chapter 15. A complete human health risk assessment is always conducted at Superfund sites (see Chapter 14). However, full ecological risk assessments are performed only if the ecological risk screening assessment indicates that the site is affecting the animal populations, plants, or a specific water body. No further studies or remediation are required if the RI work demonstrates that the site does not pose an unacceptable risk if no remediation were to be performed, and that federal or state ARARs are not exceeded.[34] Acceptable risks are defined to be a risk in the 10^{-6} to 10^{-4} range for carcinogens and a hazard index of 1 for noncarcinogens. Otherwise, a **feasibility study** (FS) is performed.

The FS is started concurrently with the RI to define potential remedial technologies and determine any specific field data or tests required to evaluate the effectiveness of remedial technologies. However, the FS is usually completed after the risk assessment (which helps define the areas that require remediation) is finished. During the FS, alternative solutions to the risk posed by the site are developed, screened, and evaluated (see Chapter 17).

Nine criteria are used in the feasibility study as a basis for selection of the appropriate remedy:

1. Overall protection of human health and the environment
2. Compliance with ARARs (cleanup standards)
3. Long-term effectiveness and permanence
4. Reduction of toxicity, mobility, or volume through treatment
5. Short-term effectiveness
6. Implementability
7. Cost (both capital and operation and maintenance)
8. State acceptance
9. Community acceptance

The first eight criteria are used to develop a selected remedy, which is then issued for comment by the public in the vicinity of the site. This is the second official opportunity for public participation, and provides the information required for evaluating the last criterion. Interested parties may (and have) become part of the entire decision process, from the RI to the remedy selection.

Because protection of human health and the environment was specifically spelled out as a primary objective in the original CERCLA statute, the first criterion is afforded special significance. Human health protection is typically determined by a quantitative risk assessment (see Chapter 14).

After the public comment period, the EPA issues a **Record of Decision** (ROD), specifying the selected remedy. At this time, the PRPs are again offered the opportunity of implementing the remedy. If no agreement is reached with the PRPs,

private consultants and contractors working under the COE implement the remedy with money from the fund. Details of the RI, risk assessment, and FS processes are presented in Chapters 14 to 17 of this book.

Cleanup Standards

One of the most difficult questions that arises in evaluating remedial measures at NPL sites is "How clean is clean?" For example, what standards should apply to contaminated soil? It is simply not possible to remove contaminants down to the last molecule. The 1986 amendments require remedial actions to meet **Applicable or Relevant and Appropriate Requirements**[35] (ARARs). ARARs are environmental regulations from programs other than Superfund that may be desirable to apply to activities at Superfund sites. For example, RCRA regulations require that when a landfill is closed, it must have a cap less permeable than the liner. Therefore, if waste is left in place at a Superfund site, the design of the cap must meet this RCRA requirement because the regulation is **applicable** to land disposal. Even if a regulation is not legally applicable, the EPA has the authority to decide that it is both **relevant and appropriate** to the situation and to use it in making remediation decisions. There are three types of ARARs:

- Chemical-specific (e.g., PCB level in soil < 50 mg/kg)
- Action-specific (e.g., if an on-site landfill is proposed, it should meet contemporary landfill standards.)
- Location-specific (e.g., prohibition of land disposal in a floodplain)

Where ARARs do not exist for important chemicals at the site, or where the EPA determines an environmental regulation is not appropriate to determine cleanup levels, such levels may be set through the use of quantitative risk assessment (see Chapter 14).

National Contingency Plan

The **National Contingency Plan** (NCP) is the EPA's blueprint for implementing Superfund. The 1982 revision of the NCP, issued in response to the 1980 CERCLA, provided a framework for the RI/FS process; specified criteria determining priorities among sites throughout the United States where releases have been reported or detected (the **Hazard Ranking System**); created a list of at least 400 of the nation's worst sites (the **National Priorities List**); and specified the roles of federal, state, and local governments and private parties in responding to releases of hazardous wastes.

The 1990 NCP,[36] issued in response to the 1986 SARA, established risk levels for the evaluation of whether sites required remediation (carcinogenic risk in the range of 10^{-6} to 10^{-4} and noncarcinogenic hazard index of 1); modified the FS process to provide for a preference in the selection of permanent remedies involving treatment to reduce the volume, toxicity, or mobility of hazardous substances; and established the **Superfund Innovative Technology Evaluation** (SITE) program, which allows treatment technology companies the opportunity to demonstrate the effectiveness of alternative or innovative treatment technologies for the remediation of hazardous waste releases.

Liability

The Superfund process includes liability provisions that are intended to replenish the fund and place the burden of cleaning up sites onto those entities responsible for the contamination. These provisions have, however, greatly complicated and delayed the cleanup process. Even though the fund comes from a tax on industry, Congress mandated that the EPA should recover costs for site cleanup from **potentially responsible parties** (PRPs). This is difficult because many companies do not agree that they are in fact responsible and are often unwilling to pay costs. Those companies that do acknowledge responsibility frequently question the extent of their responsibility. Both the EPA and the PRPs pursue lengthy litigation to advance their positions.

PRPs include:

- Present and past **owners** of the property in question
- **Operators** of the facility at the time of disposal
- **Generators** (i.e., persons whose wastes were disposed of at this site)
- **Transporters** who conveyed any hazardous substances to the site

Because of the "cradle-to-grave" nature of hazardous waste liability, generators are responsible for any waste they have produced, even if they contracted with a third party for proper disposal. CERCLA specifically applies this liability to waste produced prior to the enactment of CERCLA. Additionally, because it is rarely possible to ascertain which waste caused which part of the damage to the site, PRPs are jointly and severally liable for cleanup costs (see Section 2-1).

The EPA provides PRPs with an opportunity to conduct the Remedial Investigation and Feasibility Study and to undertake the cleanup. This practice releases fund dollars and EPA staff for work on other sites. Generally the EPA will not deal individually with entities listed as PRPs, but only with a PRP group. However, PRPs as a group are brought together only because the EPA intends to sue them for the cost of cleanup. It sometimes takes years for such a group of unwilling parties to organize themselves to speak with one voice. SARA, however, demands rapid action by the EPA and does not allow the agency to wait for a PRP response for more than 60 days. The complex question of corporate liability in the context of Superfund is discussed in the references at the end of the chapter.[37, 38]

2-5 OTHER FEDERAL STATUTES

While RCRA and Superfund are the major U.S. laws dealing with hazardous waste, other environmental laws have applications to hazardous waste management. Five of these laws are discussed in the following sections.

Toxic Substances Control Act

Congress passed the Toxic Substances Control Act (TSCA) in late 1976 in an attempt to greatly broaden the powers of the EPA to manage chemicals from production and

import through final disposal. This statute also gives the administrator of the EPA the power to take immediate action to remove a chemical from commerce that is determined to represent a hazard to human health or the environment. Congress also provided the EPA with comprehensive authority to regulate the manufacture, use, distribution in commerce, exposure, and disposal of chemical substances.

Through the premanufacture notification (PMN) provisions, manufacturers or importers of new chemical substances (i.e., chemicals not included on the TSCA Inventory of Chemicals in Commerce) can notify the EPA and provide information that is available on physical/chemical properties, health effects, environmental effects, quantities, uses, and exposures. The EPA then evaluates the PMNs and decides whether the chemical is safe to be placed into commerce or whether additional testing is required. If the chemical is placed in commerce with limitations, a Consent Order is prepared and negotiated.

The EPA also has the ability to require testing of existing chemicals under this statute to generate data needed to evaluate whether those materials represent potential exposure hazards. The EPA has greatly expanded its use of this authority in recent years and is currently in the early stages of implementing the High Production Volume (HPV) Testing Program. This effort will result in the generation of a basic set of data on over 2800 substances whose annual U.S. production/import volume exceeds 1 million pounds. Industry is being asked to undertake "voluntary" testing of these chemicals, but for those chemicals for which no industry volunteers, a mandatory test rule will be issued. This testing should be completed and the results made available by 2005.

TSCA also gives the EPA the power to regulate chemicals that are felt to represent a hazard but are not more appropriately regulated under another statute (e.g., Occupational Safety and Health Act or Clean Air Act). The EPA has regulated the following potential hazards: lead-based paint in residential structures, metalworking fluids, hexavalent chromium as a water treatment chemical, asbestos in schools, dibenzo-para-dioxins and dibenzofurans, and polychlorinated biphenyls (PCBs).

PCBs were in wide use in electrical equipment such as capacitors and transformers. However, it was determined that they were toxic and that exposure risks were high should the oil in the equipment leak or should a fire occur. The EPA therefore decided to ban the use of products containing greater than 50 ppm PCBs in equipment manufactured after 1979. They have also required labeling of PCB equipment containing greater than 50 ppm PCB, and if old equipment has not been tested, it must be assumed to be PCB-containing. In June 1998, the EPA published a major reworking of the PCB rules to bring them more in line with current practices. The rules on storage and disposal of PCB-containing equipment and PCB waste were made more explicit. They also rewrote the decontamination procedures. Among the modifications, the EPA added standards and procedures for disposing of PCB remediation wastes and for decontaminating spill areas, and a mechanism to streamline the process of obtaining approval for disposal of PCB remediation wastes. The PCB regulations issued under TSCA are considered ARARs (both chemical-specific and action-specific) at Superfund sites.

Finally, TSCA requires industry to keep records on the quantities of those chemicals it imports, manufactures, and uses and to report this information to the EPA every 4 years. Other mandatory information requirements include responding to EPA data call-ins when requests for industry to supply health and environmental data in

a company's possession on specific chemicals are published, usually twice per year. Companies are also required to report observed adverse effects of chemicals to the EPA immediately.

TSCA has lived up to its authors' expectations of broadening the EPA's information gathering and chemical hazard regulatory powers. The EPA has been expanding its use of this statute as the agency discovered its utility to close gaps in the EPA and other agencies' chemical regulations.

Occupational Safety and Health Act

The **Occupational Safety and Health Act** (OSHA) of 1970 addresses health and safety in the workplace. It created the **Occupational Safety and Health Administration** (also referred to as OSHA) at the same time that the EPA was created. The OSHA law requires the new agency to promulgate standards such that "no employee will suffer material impairment of health or functional capacity even if . . . [exposed] . . . for the period of his working life."[39] The OSHA law covers all industrial operations, including hazardous waste treatment, storage, and disposal (TSD) facilities and cleanup activities as Superfund sites.

OSHA instituted very stringent requirements for health and safety programs at hazardous waste sites (29 CFR 1910.120). Every hazardous waste disposal facility (as well as Superfund sites) must now establish a medical surveillance program, a site-specific health and safety plan, a minimum of 40 hours of health and safety training for all workers on site, and a decontamination program for workers as they leave the site (see Chapter 15).

Clean Water Act

The Clean Water Act (CWA), which underwent major amendments in 1972, 1977, and 1987, regulates discharges to surface water in the United States. The objective of the CWA is to restore and protect the integrity of the nation's waters. To accomplish the goal of fishable, swimmable waters throughout the nation, the 1972 CWA provided for a $5 billion per year federal grant program to finance construction of local sewage treatment plants [**publicly owned treatment works** (POTW)], and established a **National Pollutant Discharge Elimination System** (NPDES) program, which provides for the issuance of permits for discharges to surface waters. The NPDES program was to be based on **effluent limitations** that would be developed by the EPA in a staggered fashion, with the goal of achieving the elimination of the discharge of pollutants to the nation's waters by 1983. Recognizing that the 1972 CWA goal was too ambitious, the 1977 amendments to the CWA focused on effluent limitations for toxic pollutants, including those deemed by Congress to be "priority pollutants" (see Appendix A).

The goal of the 1987 amendments was to improve the quality of the receiving waters beyond that resulting from the application of the categorical effluent limitations. To this end, it required the development of regulations to establish facility-specific effluent discharge limits based on receiving water quality, as measured by **whole-effluent toxicity.** In addition, the 1987 amendments also expanded the NPDES pro-

gram to include the discharge of storm water and the use or disposal of wastewater treatment sludge that would result in any pollutant entering the navigable waters.

NPDES permits cover priority pollutants, **conventional pollutants** [biochemical oxygen demand (BOD), total suspended solids (TSS), fecal coliforms, oil and grease, and pH], and **nonconventional pollutants** [chemical oxygen demand (COD), total organic carbon, total solids, ammonia nitrogen, and nonpriority toxic pollutants]. Discharges may be **direct,** if the wastewater is conveyed to a surface water body directly, or **indirect,** if the wastewater is conveyed to a POTW for treatment. NPDES permits are issued for direct dischargers. Indirect dischargers are regulated by pretreatment standards, established as part of the effluent limitation process.

Provisions of the CWA which are considered ARARs for Superfund sites[40] include technology-based guidelines and standards, water quality criteria, water quality standards, ocean discharge regulations, and pretreatment standards.

Technology-based guidelines and standards

The EPA has issued effluent limitation guidelines for 51 specific categories of industries, based on the best conventional pollutant control technology (BCT) for conventional pollutants, and the best available economically achievable technology (BAT) for priority and nonconventional pollutants. These limits are based on the best-performing technology in the specific industry being regulated, taking into account the cost of the treatment system, and are usually given as mass of pollutant per unit of production. If effluent guidelines do not exist for a specific industry or discharge (such as a discharge from a site undergoing remediation), the permit writer determines BCT/BAT-based requirements on a case-by-case basis using best professional judgment. These best professional judgment requirements are given in terms of percent removal of individual pollutants by specific technologies, and are then used to develop numerical effluent discharge limits based on the maximum and average concentrations in the wastewater or discharge being regulated.

Water quality criteria

These are federal criteria for the protection of both human health and aquatic life. They are used by the states to set water quality standards for surface water. The EPA has issued over 82 federal water quality criteria documents. Water quality criteria are not enforceable.

Water quality standards

The states have developed water quality standards based on the federal water quality criteria to protect existing and attainable use or uses (public water supply, fishing, swimming, boating, etc.) of the receiving waters. NPDES permit limits may be more stringent than categorical effluent limitations if necessary to meet state water quality standards in the receiving waters.

Ocean discharge regulations

This set of regulations prohibits the discharge of pollutants into marine waters without an NPDES permit. The discharge limits must ensure that unreasonable degradation of the marine environment will not occur.

Pretreatment standards

These limits must be met by industries discharging to POTWs, and must ensure that the pollutants will not pass through or interfere with the POTW operations. A pollutant passes through a POTW if it is not susceptible to treatment by the POTW equipment. Interference refers to the inhibition of the operations, contamination of the sludge preventing its recycle and reclamation, or endangerment of the POTW workers' health. For those discharges that are not regulated by categorical pretreatment standards, the EPA has established general and specific limits, such as prohibition of pass-through or interference and prohibition of the discharge of pollutants that will create a fire or explosion hazard, cause corrosive structural damage, or obstruct flow in the sewer system.

Dredge and fill standards

These standards regulate the discharge of dredge and fill material into waters of the United States, to ensure that the proposed discharges do not impact the aquatic ecosystems. The related regulations apply to site remediation activities such as excavation of contaminated sediments.

Safe Drinking Water Act

The Safe Drinking Water Act (SDWA) was originally passed in 1974 and later amended in 1986 to protect human health from contaminants in drinking water. The 1974 SDWA required the establishment of drinking water standards and a program to control the contamination of underground sources of drinking water, including an **underground injection control** (UIC) permit program and a **sole source aquifer** program. The 1986 amendments required the establishment of drinking water standards for 83 specific contaminants, including toxic chemicals and viruses, criteria to determine when water utilities must filter surface water that is used for drinking water, regulations regarding disinfection of drinking water, and a **wellhead protection** program. All of these provisions are considered ARARs for Superfund sites.

The EPA has established primary (based on the protection of human health) and secondary (based on aesthetic quality) drinking water standards. Primary standards include **maximum contaminant levels** (MCLs) and **maximum contaminant level goals** (MCLGs). To date (1999), the EPA has established primary MCLs for over 60 contaminants (see Table 2-2). Whereas the MCLGs are purely health-based goals, the MCLs take into account the cost of removal of the pollutant from the drinking water source; however, MCLs are set as close as feasible to MCLGs. Secondary standards are not enforceable.

The UIC program established a requirement for obtaining permits to install and operate underground injection wells that requires the permit applicant to demonstrate that the underground injection will not endanger drinking water sources. The sole source aquifer designation was created to protect aquifers that were the sole or principal source of drinking water for an area, and provides that no federal financial assistance may be provided for projects that may contaminate a sole source aquifer. However, federal financial assistance may be provided to design a project to avoid contamination of a sole source aquifer. The Wellhead Protection program can be used by states to protect public drinking wells and springs from contaminants that

TABLE 2-2
Primary maximum contaminant levels for organic and inorganic parameters*

Parameter	Maximum contaminant level (MCL), μg/L	Parameter	Maximum contaminant level (MCL), μg/L
	Organics		
Alachlor	2	bis(2-Ethylhexyl)phthalate	6
Aldicarb	3	Glyphosate	700
Aldicarb sulfoxide	4	Heptachlor	0.4
Aldicarb sulfone	2	Heptachlor epoxide	0.2
Atrazine	3	Hexachlorobenzene	1
Benzene	5	Hexachlorocyclopentadiene	50
Benzo(a)pyrene	0.2	Lindane	0.2
Carbofuran	40	Methoxychlor	40
Carbon tetrachloride	5	Methylene chloride	5
Chlordane	2	Monochlorobenzene	100
2,4-D	70	Pentachlorophenol	1
Dalapon	200	Picloran	500
Dibromochloropropane	0.2	Polychlorinated biphenyls	0.5
o-Dichlorobenzene	600	Simazine	4
p-Dichlorobenzene	75	Styrene	100
1,2-Dichloroethane	5	2,3,7,8-TCDD (dioxin)	3×10^{-8}
1,1-Dichloroethene	7	Tetrachloroethene	5
cis-1,2-Dichloroethene	70	Toluene	1,000
trans-1,2-Dichloroethene	100	Total trihalomethanes†	100
1,2-Dichloropropane	5	Toxaphene	3
di(2-Ethylhexyl)adipate	400	2,4,5-TP	50
Dinoseb	7	1,2,4-Trichlorobenzene	70
Diquat	20	1,1,1-Trichloroethane	200
Endothall	100	1,1,2-Trichloroethane	5
Endrin	2	Trichloroethene	5
Ethyl benzene	700	Vinyl chloride	2
Ethylene dibromide	0.05	Xylenes (total)	10,000
	Inorganics		
Antimony	6	Copper	1,300‡
Arsenic	50	Cyanide	200
Barium	2,000	Lead	15‡
Beryllium	4	Mercury	2
Cadmium	5	Selenium	50
Chromium	100	Thallium	2

*From 40 CFR 141 and 143; excludes radioactive, bacteriological, and disinfection parameters.
†Includes bromodichloromethane, dibromochloromethane, bromoform, and chloroform.
‡Action level that triggers requirements for corrosion control treatment, source water treatment, lead service line replacement, and public education.

may produce adverse health effects on the public. Where states have designated sole source aquifers and wellhead protection areas, these designations are considered ARARs under Superfund.

Clean Air Act*

The Clean Air Act (CAA) was originally passed in 1970 and has been amended several times, including extensive amendments in 1977 and 1990. Generally, as amended through 1990, the Clean Air Act consists of major initiatives to attain and maintain the National Ambient Air Quality Standards to ensure that all new sources of potential atmospheric emissions are equipped with the "best available pollution control technology" and to ensure that emissions of hazardous air pollutants are controlled to the maximum extent possible on both existing and new emission sources. Additionally, the 1990 CAA Amendments mandated an extensive operating permit program applicable to all major emission sources in the country.

National standards

Since the early years of air quality management, air quality rules and regulations in the United States have been based on a set of air quality standards known as the **National Ambient Air Quality Standards,** or **NAAQS.** The NAAQS represent a maximum concentration or "threshold level" of a pollutant in the air, above which humans or the enivironment may experience some adverse effects. The actual threshold levels are based on years of epidemiological, health, and environmental effects research conducted by the EPA. There are two types of NAAQS: **primary standards,** which are set at levels that are designed to protect the public health, and secondary standards, which are designed to protect the public welfare (such as vegetation, livestock, building materials, and other elements of the environment).

The EPA has established NAAQS for six pollutants since the concept of NAAQS was established in the CAA Amendments of 1970. Each has been reviewed periodically, and several have been revised; however, no new pollutants have been added to the list. The six pollutants are referred to as "criteria pollutants" and their standards, which take the form of pollutant concentrations, are shown in Table 2-3 in micrograms per cubic meter ($\mu g/m^3$); sometimes the standards are expressed in parts per million (ppm).

The NAAQS specify the allowable concentration of these criteria pollutants in the ambient air. These standards are not directly enforced but are achieved through source emission limitations set within each state and on the degree of control predicted to be necessary to meet NAAQS.

Air quality control regions

States are divided into regions known as **air quality control regions** (AQCRs) to determine if the standards for criteria pollutants are being met. If the standards are being met, the area is considered to be in **"attainment"** for the specific NAAQS. If not, the AQCR is in **"nonattainment"** for affected pollutants. If an area is in nonattainment, the regulating authority (either a state or local air pollution control agency) must adopt regulations to limit facility emissions sufficient to ensure compliance with the NAAQS by a date specified by the CAA.

Because of the 1970 CAA, most areas of the United States are substantially in attainment with the NAAQS for all the criteria pollutants except ozone and

*This section prepared by Kenneth N. Weiss, P.E., air program director, ERM.

TABLE 2-3
National Ambient Air Quality Standards[41]

Criteria pollutant	Averaging period	Primary NAAQS, $\mu g/m^3$	Secondary NAAQS, $\mu g/m^3$
PM_{10} (particulate matter	Annual	50	50
<10 μm)	24 hours	150	150
$PM_{2.5}$ (particulate matter	Annual	15	15
<2.5 μm)	24 hours	65	65
SO_2 (sulfur dioxide)	Annual	80	–
	24 hours	365	–
	3 hours	–	1,300
NO_2 (nitrogen dioxide)	Annual	100	100
Ozone*	8 hours	157	157
	1 hour*	235	235
CO (carbon monoxide)	8 hours	10,000	10,000
	1 hour	40,000	40,000
Lead	Quarterly	1.5	1.5

*The EPA is phasing out and replacing the previous 1-hour primary ozone standard (health-based) with a new 8-hour standard to protect against longer exposure periods.

carbon monoxide. The 1990 Amendments include major programs designed to bring about ozone attainment through a reduction in the emissions of precursors to ozone: volatile organic compounds (VOCs) and oxides of nitrogen (NO_x). Carbon monoxide is addressed through tighter emission standards for motor vehicle tailpipe emissions.

Hazardous air pollutants

Section 112 of the Clean Air Act identifies a **hazardous air pollutant** (HAP) as "an air pollutant to which no ambient air quality standard is applicable and which ... causes or contributes to ... an increase in mortality or an increase in serious irreversible, or incapacitating reversible illness." Under the 1970 Act the regulation of HAPs was essentially a risk-based approach. After a HAP was listed for regulation by the EPA, the agency was mandated to adopt standards sufficient to protect the public health with an "ample margin of safety" defined by risk-based models. Defining an ample margin of safety for potential carcinogenic emissions proved to be very difficult for the EPA. As a result, the EPA was able to develop emission standards for only seven HAPs between 1970 and 1990: asbestos, beryllium, vinyl chloride, mercury, benzene, radionuclides, and arsenic.

The new HAP emissions control program is technology-based and requires the installation of state-of-the-art pollution control equipment. The 1990 Amendments list 188 chemicals to be regulated (see Appendix A). Stringent technology-based standards are being developed over a 10-year period for major sources of these HAP emissions. A major source is defined as any facility that potentially emits 10 tons per year of any single HAP or more than 25 tons per year of a combination of HAPs.

The EPA is in the process of issuing regulations for 174 categories of industrial sources that emit substantial quantities of HAPs. Emissions allowances reflecting **maximum achievable control technology** (MACT) are being established in each source category regulated. By definition, MACT can include process changes, material substitutions, or air pollution control equipment.

New source review

New source review (NSR) is the permitting process that governs the construction, modification, or reconstruction of major industrial air pollution emission sources throughout the United States. In areas where the ambient air quality standards are met, the program requires the **prevention of significant deterioration** (PSD) of air quality intended to ensure that air quality is not allowed to degrade in those areas of the country where the air quality is already better than required. Where standards are not met, **nonattainment** new source review provisions apply. The PSD provisions and nonattainment provisions apply separately to each regulated pollutant. Thus, facilities may be required to undergo both a PSD review and a nonattainment review if the proposed location is classified as attainment for one pollutant and nonattainment for a different pollutant.

Generally, the 1990 Amendments make the EPA's requirements to obtain a permit to construct a new major source or modify an existing major source more stringent in a nonattainment area. New sources or modifications in a nonattainment area are required to install "lowest achievable emission rate" (LAER) air pollution control technology and to offset remaining permitted emissions after LAER is installed. The concept of "offsetting" pollution is used to assure that the total emissions in a nonattainment area *after* construction of a new facility be *less* than before. This is accomplished by shutting down or adding more efficient air pollution control equipment to existing sources.

New Source Performance Standards

The 1970 Clean Air Act also required that the EPA promulgate New Source Performance Standards (NSPS) for major emission sources. NSPS represent nationwide emission standards for certain significant emission source categories (e.g., petroleum refining, VOC storage tanks).

Operating permits (Title V).

The 1990 Amendments established a comprehensive new operating permit program to be administered by state agencies in accordance with federal guidelines. The Amendments will require nearly all sources of even minor amounts of air emissions to apply for and obtain operating permits. The new operating permits are intended to lay out in one enforceable document all a facility needs to do to ensure compliance with all air quality regulations. The EPA and the states have been requiring Title V permits on a "rolling" basis (applying the requirement to a few industries at a time). Most facilities that are required to obtain the Title V permits have submitted their applications.

2-6 INTERNATIONAL PERSPECTIVES*

Philosophies of National Regulatory Systems

There are probably as many different approaches for a national system for regulation and control of hazardous waste as there are jurisdictions that have implemented such. Table 2-4 lists and summarizes the elements that may be incorporated in a national system. Despite some measure of agreement on the particular elements to be covered in a regulatory control system, the approaches are as diverse as the legal, political, social, and cultural differences of these countries. The differences correspond to how questions such as the following are answered:

- At what level of government is control exercised?
- Who retains long-term liability?
- Who selects treatment and disposal methods?
- Who provides facilities?

TABLE 2-4
Elements of a national regulatory control system for hazardous waste management

- *Definitions of terms,* particularly *wastes and hazardous* wastes. Related is a *classification system* to allow the unambiguous categorization of wastes.
- *Responsibilities placed on the waste generator,* which may include a requirement to *register* as a waste generator and a *duty of care* whereby the generator remains legally responsible for the waste till it reaches its proper destination. Some countries may require a generator to show that steps have been taken to *minimize waste generation.*
- *Registration or licensing* of those operators and owners involved in collection, transport, storage, treatment, or disposal of wastes.
- *Control over transport,* comprising a *manifest or tripticket* system for cradle-to-grave control; *standards* and *specifications* for vehicles, containers and packaging; labeling of vehicles and the waste containers; and *contingency planning for, and emergency response to, spills and accidents.*
- Controls over *import* and *export* of waste (between states or between countries). A particular issue is the application of such controls to materials destined for recycling—clearly recycling is to be encouraged, but this must be done in such a way as not to provide a loophole for those who wish to export wastes for dumping. The Basel Convention now prohibits the export of all hazardous wastes from developed to developing countries, whether for disposal or recycling.
- *Permitting* of facilities for the storage, recycling, treatment, or disposal of hazardous wastes, to ensure that facilities are designed and operated in an environmentally sound manner.
- A requirement on national, regional, and/or local authorities to prepare a *strategic plan* for the management of hazardous wastes, focusing in particular on (1) encouragement of pollution prevention and (2) assurance that adequate facilities are made available both now and in the future for the treatment and disposal of all hazardous wastes.
- *Prohibitions* of certain treatment and disposal options for particular wastes (or conversely, *mandatory* treatment requirements).
- Programs for dealing with *old or abandoned hazardous waste disposal sites.*

Source: ERM (Dr. David C. Wilson).

*Section 2-6 is based on the work of Dr. David C. Wilson of ERM in London and, in particular, on a 1999 report that he compiled on behalf of the International Solid Waste Association (ISWA) Working Group on Hazardous Waste.[42]

The issue of devolution of regulatory control from the national government to regional or local authority is very much a function of the cultural considerations and the type of legal system. In the United States, the federal government develops the hazardous waste regulatory system but must delegate implementation and enforcement to those states capable of operating the federal program. States may enact rules and regulations that are more stringent than federal rules, but not less stringent. In the United Kingdom, control was initially exercised through site licensing and permitting by nearly 200 local authorities with the central government having a largely advisory role. This changed in 1996 with the introduction of two nationwide environmental agencies that took over the waste regulation function. In Japan and Germany, primary responsibility is at the regional level, but the regions have more discretionary power than in the United States.

Table 2-5 shows the approaches used by 18 nations (mostly European) with regard to some of the regulatory control elements listed in Table 2-4. This section highlights some of their differences.

Definitions*

A necessary requirement of any hazardous waste regulatory control system is a legal definition of hazardous waste. Table 2-6 lists countries that have a legal definition of hazardous waste and compares some aspects of the definitions. Beyond that it provides no insight into the breadth of the underlying definitions, the number of substances considered, or the concentrations of various chemicals necessary to make a waste hazardous. All of these factors vary greatly from country to country.

It is extremely difficult to define a hazardous waste with any precision. Every national system differs both in the detailed method used for definition and in the breadth of waste included. The importance attached to a legal definition of waste depends both on its purpose and the use that will be made of it.

Most countries have developed a definition similar to the approach used by the United States: technical criteria and a series of lists. These are lists of particular types of hazardous wastes, lists of industrial processes from which the waste is defined as hazardous, and/or lists of substances, the presence of which is indicative of potential hazard.

In some cases, a listing on one or more of these criteria is judged to provide a sufficient definition. In other cases, a further reference is made to particular concentration levels for each of the substances, or to other indicative criteria. Many of the national definitions summarized in Table 2-6 have been amended over the past several years to fall in line with the requirements of the Basel Convention.

Classification systems

The problem with the Basel definition, and indeed with all regulatory definitions, is that they are in effect *descriptions of waste,* whereas for the purposes of running an effective, regulatory control system, one requires rather a definitive *list of wastes* or waste catalog. Much work over the last 5 years in many countries has been

*Adapted from *International Perspectives in Hazardous Waste Management: 1999 Edition,* to be published by ISWA.

TABLE 2-5

Elements in national regulatory control systems

	Austria	Belgium	Croatia	Czech Republic	Denmark	Finland	France	Germany	Hong Kong
Year of latest data update	1998	1998	1997	1997	1997	1998	1998	1998	1998
Date of main legislation	1983, 1990	1981, 1985	1995	1991	1972, 1996	1978, 1994	1975	1972, 1996	1980, 1992
Registration/licensing[1]									
Collectors/transporters	L	L, R	L	L	Note 2	L	R	L	L
Treatment/disposal contractors	L	L	L	L	L	L	R	L	L (waste producers R)
Control over transport									
Manifest system	Yes	Yes	Yes	Yes	Yes	Soon	New	Yes	Yes
Control over import	Yes	Yes	Yes	Yes	Yes	Yes	Yes	Yes	Yes
Control over export	Yes	Yes	Yes	Yes	Yes	Yes	Soon	Yes	Yes
Permitting of facilities									
Storage	Yes	Yes	Yes	Yes	Yes	Yes	Yes	Yes	Yes
Treatment	Yes	Yes	Yes	Yes	Yes	Yes	Yes	Yes	Yes
Disposal	Yes	Yes	Yes	Yes	Yes	Yes	Yes	Yes	Yes
Have all operating sites now been permitted?	Yes	Yes	Partly	Yes	Yes	Yes	Yes	Yes	Yes
Planning and establishment of facilities									
Is there a national strategy/plan?	Yes	Yes	Under preparation	Yes	Yes	Under preparation	No	No	Yes
Are authorities required to produce a plan?	Yes	Yes	Yes	No	Yes	Yes	No	Yes	Yes
Has this been done?	Yes	Yes	Not yet	No	Yes	Yes	No	Yes	Yes
Old or abandoned hazardous waste sites									
Is there a national inventory?	Yes	Yes	Under preparation	Yes	Yes	Yes	Yes	Yes	Yes
Is there a cleanup program?	Yes	Yes	No	No	Yes	Yes	Note 6	Note 6	Yes

[1] L = licensing scheme, implying investigation by the authorities; R = registration, implying simply being listed in a register.
[2] Mainly under the Trade Act, not under the Hazardous Waste Act.
[3] All permits are going to be revised in the near future.
[4] Officially yes, but doubtless some gaps.
[5] State responsibility.
[6] No formal nationwide cleanup program, but cleanup of individual sites is continuously proceeding.
[7] Some regions (local authorities) have prepared a cleanup program.

Reproduced with permission from the International Solid Waste Association report, "International Perspectives in Hazardous Waste Management: 1999 Edition."

TABLE 2-5
Continued

	Hungary	Italy	Japan	Netherlands	Norway	Spain	Sweden	United Kingdom	United States
Year of latest data update	1998	1997	1987	1998	1998	1998	1998	1998	1994
Date of main legislation	1981, 1996	1982, 1984	1970, 1979	1994	1994	1986, 1988	1975, 1985, 1995	1974, 1990	1976, 1984
Registration/Licensing[1]									
Collectors/transporters	L	R	L	L	Yes	Note 3	L	R	R
Treatment/disposal contractors	R	R	L	L	Yes	L	L	Sites, not contractors	R
Control over transport									
Manifest system	Yes	Yes	No	Yes	Yes	Yes	Yes	Yes	Yes
Control over import	Yes	Yes	Yes	Yes	Yes	Yes	Yes	Yes	Yes
Control over export	Yes	Yes	Yes	Yes	Yes	Yes	Yes	Yes	Yes
Permitting of facilities									
Storage	Yes	Yes	Yes	Yes	Yes	Yes	Yes	Yes	Yes
Treatment	Yes	Yes	Yes	Yes	Yes	Yes	Yes	Yes	Yes
Disposal	Yes	Yes	Yes	Yes	Yes	Yes	Yes	Yes	Yes
Have all operating sites now been permitted?	Yes[3]	Yes	Yes	Yes	Yes	No	Yes	Yes[4]	No
Planning and establishment of facilities									
Is there a national strategy/plan?	Under preparation	No	No	Yes	Yes	Yes	Yes	Yes	No
Are authorities required to produce a plan?	No, but from 1997 onward	Yes	Yes	Yes	Yes, locally	Yes	Yes, locally	Yes	Note 5
Has this been done?	1997 onward	No	Partial	Yes	Yes	Partial	Yes, locally	Not fully	Partial
Old or abandoned hazardous waste sites									
Is there a national inventory?	Partially	Yes	No	Yes	Yes	Yes	Yes	No	Yes
Is there a cleanup program?	Yes	No[7]	No	Yes	Yes	Yes	Yes	No	Yes

TABLE 2-6
Definitions of hazardous waste

	Austria	Belgium	Croatia	Czech Republic	Denmark	Finland	France	Germany	Hong Kong	Hungary	Italy	Japan	Netherlands	Norway	Spain	Sweden	United Kingdom	United States
Year of latest data update	1998	1998	1997	1997	1998	1998	1998	1998	1998	1998	1997	1987	1998	1998	1997	1998	1998	1998
Is there a legal definition?	Yes	Yes	Yes	Yes	Yes	Yes	Yes	Yes	Yes	Yes	Yes	Yes	Yes	Yes	Yes	Yes	Yes	Yes
Purpose of definition																		
Control over transport	Yes	Yes	Yes	Yes	Yes	Yes	Yes	Yes	Yes	No	Yes	No	Yes	Yes	Yes	Yes	Yes	Yes
Control over treatment/disposal	Yes	Yes	Yes	Yes	Yes	Yes	Yes	Yes	Yes	Yes	Yes	Yes	Yes	Yes	Yes	Yes	Yes	Yes
Type of definition																		
List of waste	Yes	Yes	Yes	Yes	Yes	Yes	Yes	Yes	Yes	Yes	Yes	Yes	Yes	Yes	Yes	Yes	Partly	Yes
List of substances	Yes	Yes	Yes	Yes	Yes	Yes	Yes	No	Yes	No	Yes	Yes	Yes	Yes	Yes	Yes	Partly	Yes
List of processes	No[2]	Yes	Yes	No	No	Yes	Yes	Yes	Yes	No	Yes	Yes	Yes	Yes	Yes	Yes	No	Yes
Concentrations	Yes[3]	Yes	No	No	Yes	No	No	No	Yes	No	Yes	No	Yes	Yes	No	Partly	Partly	Yes
Criteria																		
Toxicity of waste	Yes	Yes	Yes	Yes	Yes	Yes	Yes	Yes	Yes	Yes	Yes	No	No	Yes	Yes	Yes	Yes	Yes
Toxicity of extract	Yes[3]	Yes	No	Yes	Yes	Yes	Yes	No	No	Yes	Yes	Yes	Yes	No	Yes	No	No	Yes
Ignitability/flammability	Yes	No	Yes	Yes	Yes	Yes	Yes	Yes	Yes	Yes	Yes	No	Yes	Yes	Yes	Yes	Yes	Yes
Corrosiveness	Yes[3]	No	Yes	Yes	Yes	Yes	Yes	Yes	Yes	Yes	Yes	No	No	Yes	Yes	Yes	Yes	Yes
Reactivity	Yes[3]	Yes	Yes	Yes	Yes	Yes	Yes	Yes	Yes	Yes	Yes	No	No	Yes	Yes	Yes	Yes	Yes
Special rules																		
Mixing rule	Yes	Yes	No	Partial	Yes	Yes	Yes	Yes	Yes	Yes	Yes	No	Yes	Yes	No	Yes	Yes	Yes
Residue rule	Yes[4]	No	No	Partial	Yes	No	Yes	Yes	No	No	No	No	No	No	No	No	No	Yes
Exclusions																		
Small generators[5]	No	No	No	No	No	No	100 kg	No	No	No	Yes	Yes	No	No	Yes	No	No	100 kg
Wastewater	Yes	Yes	No	Yes	No	No	Yes	Yes	No	Yes	Yes	Yes	Yes	Yes	Yes	No	Yes	No[6]
Sewage sludge	Yes	No	No	No	No	No	No	No	No	Yes	No	Yes	No	Yes	Yes	Yes	No	No
Mining waste	Yes	No	No	Yes	No	No	Yes	Yes	No	Yes	Yes	Yes	No	Yes	Yes	Yes	Yes[1]	Yes[1]
Agricultural waste	Yes	No[7]	No	No	No	No	Yes	Yes	No	Yes, partial	Yes	Yes	No	No	Yes	Yes	Yes[8]	No

[1]Certain mining wastes are excluded from the definition of hazardous waste.
[2]Yes in some cases, within Austrian Standard on Hazardous Waste.
[3]In some federal regulations.
[4]Only for packaging.
[5]Quantity is that per month below which waste generator is exempt from the regulation.
[6]Some exclusions from the definition of hazardous waste are provided for certain specific wastewaters.
[7]Except animal slurry.
[8]Some specific agricultural wastes are excluded from the definition of hazardous waste.

Reproduced with permission from the International Solid Waste Association report "International Perspectives in Hazardous Waste Management: 1999 Edition."

devoted to the development of such waste catalogs, often in the broader context of developing a comprehensive classification system for all wastes, not just industrial wastes or hazardous wastes. One important aspect of a classification system is that it should be flexible, so that new types of waste do not evade control. This is a particular problem that must be addressed by any system relying mainly on listing specific types of waste or industrial processes. Denmark also cites it as the reason why that country does not prescribe concentration limits.

The European Union (EU) *Catalogue of Wastes,* published in 1994, was intended to fulfil this need; however, it appears to have raised as many problems as it has solved. Most EU member states have continued developing their own waste classification systems with the result that, while they share a number of common features, they also differ in the detail. The individual national systems are still in a state of flux.

Exclusions

In comparing definitions of hazardous waste, it is important to note those categories of waste specifically excluded from the control system. The reasons for such exclusions are generally concerned with problems of implementation, either because quantities are small or because the political or practical difficulties are too great.

- Hazardous wastes from households are outside the normal control system in all countries, although specific systems to encourage separation and delivery to special collection points have been or are being developed in many countries.
- Small-quantity generators are often placed outside the system. For example, in the United States, producers generating less than 1000 kg per month were excluded from control up to 1984, when a lower limit of 100 kg per month was implemented. It is estimated that this change increased the number of generators covered by the regulations by a factor of 10.
- Aqueous effluents discharged to sewers or treated on-site are controlled separately from hazardous wastes in most countries. For example, such wastewater is specifically excluded from the definition of "waste" in EU directives. However, in the United States, wastewater treatment in surface impoundments or lagoons is controlled within the hazardous waste regulations.
- Sewage sludge is specifically excluded in some countries but not others.
- Mining wastes are often excluded from control, even though the nature of the wastes may be very similar to other hazardous wastes.
- Similar exclusions may apply to agricultural waste.
- Certain "hazardous" wastes were already controlled in many countries under separate regulations and were therefore excluded from the new hazardous waste controls. Common examples are radioactive, pathogenic, and conventional explosive wastes.

Special rules

There can be special rules concerning mixtures of hazardous with nonhazardous waste and also residues from the treatment of hazardous waste. Where the definition of a hazardous waste depends on the concentration of specific substances, either in the waste itself or in an extract, it is in principle possible to evade control by mixing the hazardous with a nonhazardous waste and diluting the concentration. This is permitted in some countries but not in others. For example, in the United States, the so-called mixing rule states that if you mix a listed hazardous waste with anything

else, then the result is a hazardous waste. In France, special authorization is required before such mixing, while in the Netherlands, mixing is forbidden.

The treatment of hazardous waste generally results in a residue for final disposal. The question is whether these treatment residues can be regarded as a nonhazardous waste. Again, in the United States, the so-called residue rule states that any residue from treatment of a listed waste is also a hazardous waste unless specifically delisted, which is a lengthy and onerous process. In some countries, such as the Netherlands, the normal concentration limits or criteria are applied to residues in determining whether or not they are a hazardous waste. In others, such as Sweden, fly ash from municipal incinerators is classified as hazardous waste. In Germany, there is some confusion regarding the classification of fly ash from municipal incinerators, which some states define as a hazardous waste and others do not.

Although no acceptable international definition of hazardous waste exists, similarities exist among the definitions of many countries. Contrast the following three definitions with those presented in Sections 2-2 and 2-3.

Canada "Hazardous wastes are those wastes which due to their nature and quantity are potentially hazardous to human health and/or the environment and which require special disposal techniques to eliminate or reduce the hazard"

Philippines "Materials which are inherently dangerous to the human body or to animals, including, but not limited to, materials that are toxic or poisonous; corrosive; irritants; strong sensitizers; flammable; explosive"

United Nations Environment Program "Wastes other than radioactive wastes which, by reason of their chemical activity or toxic, explosive, corrosive or other characteristics causing danger or likely to cause danger to health or the environment."[43]

Responsibilities Placed on the Waste Generator

Registration

Most countries require waste generators to keep records documenting the generation and disposition (details of collection and destination) of the hazardous wastes. Also, generators in most countries must register with the authorities and make regular reports (a notable exception is the United Kingdom). The frequency of reporting varies, from annual, through three-monthly (Croatia), to monthly (Hong Kong).

Liability

One aspect which does vary widely between countries is what the waste generator is expected to do to fulfil its responsibility to ensure the proper management of the hazardous wastes. Critical to this is the point at which that responsibility is considered to be discharged. For example, this chapter explained earlier that liability is a major driving force of the regulatory control system in the United States, mainly because of the unique nature of the U.S. tort system. This is not the case in most other developed countries.

It is common that the generator be responsible for choosing a licensed transporter and/or having the waste transported to a facility that is licensed to handle the particular type of waste. An interesting question is whether the generator retains

some liability if either the transporter or the disposer should handle the waste in such a way as to cause environmental damage. Four basic approaches to this issue can be distinguished.

1. The generator has discharged its responsibility if the waste has been handed over to a licensed transporter, on the logic that it is then the authority's responsibility to ensure that the transporter delivers the waste to a properly licensed facility.
2. The generator has discharged its responsibility if the waste has been passed over to a properly permitted facility, the logic being that it is then the authority's responsibility to ensure that the facility operator complies with the permit conditions. In Denmark, this indemnification is available only if the waste is delivered to the national treatment facility.
3. The United Kingdom has pioneered the so-called *duty of care,* which is a statutory requirement placed on all waste generators, whether hazardous or not, to take all reasonable steps to ensure that their waste is handled in accordance with the law. This goes beyond passing the waste to an appropriately registered carrier (collector or transporter) or delivering it to a properly licensed facility. Generators can have certain obligations to investigate the companies that handle their wastes, such as ensuring that they are competent and that they handle or otherwise dispose of the waste properly. The extent to which a generator is obliged to follow through all stages of the chain depends on various factors such as the nature of the waste, the quantity, and the regularity with which it is generated. A generator must provide an adequate written description of the waste to enable the carrier and facility operator to assess properly what measures are required. Signed documentation created at each transfer provides proof that the handover took place under the declared conditions.
4. In some countries, a system of no-fault or contingent liability applies where the waste generator can be held liable for the environmental damage caused at a facility site to which it can be shown that the generator's waste was delivered, regardless if that site was properly licensed at the time for the receipt of that type of waste. This is the basis of the law in the United States. The high costs of cleanup of environmental damage at so-called uncontrolled hazardous waste sites, most dating from before current legislation, has caused many generators to focus on waste prevention and other alternatives to land disposal in order to reduce their environmental liability.

Minimize waste generation

In some countries, considerable recent attention has been paid to appropriate mechanisms to encourage or oblige waste generators to focus on waste avoidance, waste minimization, recovery, recycling, or reuse—that is, "moving waste management up the hierarchy." This is the subject of an entire chapter of this book (see Chapter 7). Not all countries have been moving in this direction, but among those countries that have taken initiatives, the exact nature of those initiatives has varied widely.

One general theme is to encourage or force companies to examine their waste management practices, in the hope that that knowledge, combined with high treatment and disposal costs, will encourage some form of pollution prevention. An example is making compulsory the formal environmental management systems under the new international standard ISO 14000. As another example, the Czech Republic requires a formal waste management program to be drawn up by a generator. A

generator that does not utilize the waste as a source of secondary raw materials or energy must offer waste to another party for utilization. Croatia requires generators to submit monthly reports to a *waste exchange scheme*.

A rather more formal approach is that pioneered by Germany under its *Waste Avoidance, Recycling and Disposal Act for Promoting Closed Substance Cycle Waste Management and Ensuring Environmentally Compatible Waste Disposal,* which took effect on October 7, 1996. This Act requires companies to take care that waste be avoided both during production and consumption of their products and, further, any wastes that are generated shall be recycled. The treatment or disposal of any waste is permitted only after submission of a detailed management certificate to the authorities, which must, inter alia, contain proof that the waste cannot be avoided or reused/recycled.

A completely different approach has been adopted in the United States. The incentive for pollution prevention is indirect, both via contingent liability for environmental damage from land disposal and also from the so-called **toxic release inventory** legislation which obliges U.S. companies to report publicly on the use and discharge of chemicals in annual reports. This so-called **community right-to-know** legislation has indirectly reduced the generation of hazardous wastes and of water and air emissions. The United States' approach to community rights to know is more formal and far reaching than in most countries. Nevertheless, the principle of public access to hazardous waste data is widespread.

In the spirit of the combined approach of "carrot" and "stick," it is also worth pointing to measures which the authorities can take to encourage and assist waste generators. The measures can help generators to comply with hazardous waste management legislation and more particularly to *move waste up the management hierarchy*. In general, such incentive schemes can fall under one of two headings, **information dissemination and use** or **economic and financial instruments.**

Examples under the heading of *information dissemination and use mechanisms* include information clearinghouse, provision of technical assistance, support for research and development, and support for demonstration projects. In many countries, such schemes are targeted at encouraging environmental technology in general, rather than specifically focused on hazardous wastes. Examples under the heading of *economic and financial instruments* (sometimes referred to as market-based instruments) include waste taxes/disposal levies (e.g., the landfill tax in the United Kingdom), charge/rebate schemes (e.g., a levy in France on water use is used to subsidize hazardous waste treatment), raw material taxes, product charges (e.g., Sweden taxes products whose manufacture generates hazardous waste), subsidies for waste prevention, accelerated tax depreciation, and emission trading mechanisms.

Registration or Licensing of Operators

A possible element in the regulatory control system is a registration or licensing system for waste collectors and transporters, and/or waste treatment and disposal contractors. This is separate from control over transport *operations* or over treatment and disposal *facilities*.

Registration is a relatively simple procedure for contractors. No endorsement of competence is required on the part of the authorities, although in some countries

any contravention of legal requirements may result in the company being removed from the register. **Licensing** is a more complex procedure, requiring the authorities to examine matters such as the professional, technical and financial competence of the contractor. Licensing of operators is more common than registration, although one or the other is now generally in place (a significant increase since 1987).

Control over Transport

Control over transport can take many forms. Critical in almost all countries is a system for **cradle-to-grave control,** through some kind of manifest or trip-ticket system. This system keeps track of shipments of wastes to ensure that they arrive at their designated destinations, thereby preventing abuses such as roadside dumping or disposal at unlicensed facilities. Despite the consensus on the need for a manifest system, the detailed working of such systems varies considerably between countries. For example, in only 60 percent of the countries does the regulatory authority receive a copy of every manifest. In the United Kingdom and the Czech Republic, the regulatory authority receives an advance copy, providing the agency the opportunity to witness the treatment. In the United States, primary responsibility for identifying lost shipments is placed on the generator, who must report exceptions to the authorities. In some other countries it is the government (local or regional) that does the checking of manifest forms to assure that shipments of waste have not been lost.

Other controls over transport include **standards and specifications** for design and construction of vehicles and containers, **labeling** of trucks and containers, and **contingency planning** for and emergency response to spills and accidents. These controls usually do not form part of specific hazardous waste management legislation. Rather, the general approach has been to adapt and extend legislation controlling hazardous materials (dangerous goods) in transit to cover hazardous wastes. For example, the international codes regulating the transport of dangerous goods drawn up under the auspices of the United Nations have been amended to explicitly include hazardous wastes.

Control over Import and Export

A further element of control over transport is control over import and export of hazardous wastes. Attention was originally drawn to this aspect in the early 1980s, both by a number of well publicized incidents where hazardous wastes have been exported from industrialized countries and dumped in low-income countries, and also by a specific incident in Europe, where 41 drums of dioxin waste, from the accident at Seveso in Italy, went missing, and were eventually discovered some time later in a warehouse in France. To address this problem, the *Basel Convention on the Control of Transboundary Movements of Hazardous Wastes and their Disposal* was agreed under the auspices of the United Nations Environment Program (UNEP) in 1989. More than 100 countries around the world have now signed and ratified the Convention, which means that they have national legislation in place to control export

and import (although not necessarily the infrastructure for proper implementation and enforcement).

In most countries, wastes move freely across internal boundaries. In the United States, states are prohibited from interfering with the transport of waste across their borders under the interstate commerce clause in the U.S. Constitution. Germany is an exception to the free movement of waste inside a country. A special permit, involving a time-consuming process to obtain, is required to cross state boundaries in Germany. The political and social climate in Germany is making it difficult to obtain such permits.

Permitting of Treatment, Storage, and Disposal Facilities

Essentially all developed countries require permitting or licensing of treatment, storage, and disposal (TSD) facilities; however, the requirements can vary. In the United States, industries may store hazardous waste for up to 90 days without obtaining a permit as a storage facility. In the United Kingdom, this is limited to 28 days and even this exemption is being eliminated in certain instances.

Many countries have adopted very detailed technical guidelines and standards to be met by different types of facilities, which are either legally binding or enforced through incorporation in the specific conditions of the permit for an individual facility. The U.S. TSD permit program is heavily driven by tight restrictions and large penalties (see Section 2-2). In many European countries, a more "cooperative" approach is utilized to keep facilities up-to-date and in compliance. For example, France provides subsidies to industries (funded by a levy on water use) to use the "best practicable" means of treatment and disposal. Belgium provides a zero tax on recycling and a maximum tax on landfills, taxing treatment facilities somewhere in between.

Another recent development has been to strengthen the requirements for training and qualification of personnel involved in the management, supervision, and operation of waste treatment and disposal facilities. For example, the 1990 legislation in the United Kingdom introduced the requirement for *certificates of professional competence,* to be held by all individuals managing or supervising waste treatment or disposal facilities.

Planning and Establishment of Facilities

Another important feature of regulatory control systems is the measures to ensure that adequate facilities are available both now and in the future for the treatment and disposal of all hazardous waste. It can be argued that, whatever method is used, it is important that coherent plans should be produced, taking into account the types and quantities of wastes expected to be generated over the next, say, 10 to 20 years and the facilities that will be available to deal with them. Preparation of such plans is required by legislation in a number of countries. However, even where this is the case, progress in actually producing the plans has often been poor.

Choice of Treatment or Disposal Options

Important parts of any strategic plan are the degree and methods of control over the choice of options available to a waste generator for the treatment or disposal of a specific waste. Table 2-7 shows how several countries regulate the choice of treatment and disposal options.

This can be done with prohibitions of certain options for particular wastes (particularly for land disposal), either broad or facility-specific, via permits. A more explicit measure is to direct how and/or where a waste is to be treated. In this case, central (or regional) government assumes powers to direct a specific type of waste to a particular treatment or disposal option or even to a particular site. In Denmark, for example, it is mandatory that local companies deliver their hazardous wastes to collection stations that are linked to the central state-owned treatment and disposal facility. There is a possible problem with this as it shifts the responsibility for deciding the most appropriate option for a particular waste from the generator to the state. In a larger number of countries, the treatment or disposal option to be used for certain waste types is specified in regulations. In Austria and Germany, recommendations are made in technical standards, but these are not mandatory.

Looking first at who provides the facilities, a number of cases may be distinguished. In Denmark, Finland, and some German states (and previously in Spain and Sweden), major treatment facilities are provided by a state utility, which is a cooperative venture between central government, local communities, and industry. In a number of other countries, the public and private sectors also collaborate in providing facilities, often with substantial investment from government (e.g., the Netherlands). In France, the facilities are entirely provided by the private sector, but the government provides substantial financing assistance on attractive terms. Of the countries with well-established systems, only the United Kingdom and United States rely entirely on the private sector for the provision of facilities, although in the United States many states have taken some responsibility for siting new facilities (see Section 8-4).

Of the 16 countries that have "high technology" facilities in place for hazardous waste treatment, all but two provided indirect subsidy of at least the initial capital investment. This eased the transition from cheap and uncontrolled disposal to the required use of relatively expensive treatment facilities. In many cases, the facilities have been operational now for more than 10, or even more than 20, years. Recurring capital investment in plant expansion or to meet new standards—for example upgrades in gas cleaning requirements for incineration—are now usually met on a purely commercial basis from the fees charged for waste treatment.

The two exceptions to the subsidy rule are the United States and the United Kingdom. In the United States, reliance is placed mainly on rigid regulations that may prohibit less costly options used for particular wastes, and also the liability instrument. It is then left to the private sector to provide the necessary facilities. In the United Kingdom, reliance has also been placed on market forces to provide treatment and disposal facilities. However, in this case regulatory control over the choice of options was initially weak. Some authorities took originally a more lenient view than others did as to what wastes might be allowed in landfills. This situation has now been largely corrected by the new, stricter, licensing permitting regime

TABLE 2-7

Regulating the choice of treatment of disposal options for a specific waste

	Austria	Belgium	Croatia	Czech Republic	Denmark	Finland	France	Germany	Hong Kong	Hungary	Italy	Japan	Netherlands	Norway	Spain	Sweden	United Kingdom	United States
Year of latest data update	1998	1998	1997	1997	1998	1998	1998	1998	1998	1998	1997	1987	1998	1998	1998	1998	1998	1998
Direction of waste																		
To particular sites	No	No	No[1]	Yes	Yes	No	No	Partial	Yes	Yes	No	No	Yes	Yes	No	No	Note 4	No
To particular options	Note 1	Yes	Yes	Yes	Yes	Yes	No	Note 1	Yes	Yes	No	Yes	Yes	No	No	Yes	Rarely	Yes
Powers exist, in reserve	Yes	Yes	Yes	–	–	–	Yes	Yes	–	No	No	Yes	–	–	Yes	–	Yes	No
Prohibition of certain options for particular wastes																		
National regulations	Yes	Yes	Yes[1]	Yes	Yes	Yes	Yes	Yes	Note 1	Yes	Yes	Yes	Yes	Yes	Yes	Yes	Reserve powers	Yes
Control via site permits																		
Strong national standards mean effective prohibition for certain waste	Yes	Note 2	System not yet in place	Yes	Yes	Yes	Yes	Note 2	Yes	Note 2	Yes[3]	Note 2	Yes	Yes	Yes	Yes	In some cases	Yes

[1]Recommendations are made in technical standards, but these are not mandatory.

[2]Strong controls exist in principle, but in practice there are wide local variations in what is/is not permitted at individual sites.

[3]Prohibition for asbestos and PCB.

[4]Powers exist but not usually done.

Reproduced with permission from the ISWA's report "International Perspectives in Hazardous Waste Management: 1999 Edition."

introduced under the 1990 legislation and the assumption in 1996 of the waste regulation function at the national level.

Site Cleanup

The failure of past policies to properly manage industrial wastes has left a legacy of abandoned hazardous waste sites that require some form of remedial action. As noted in Table 2-5, most developed countries have some program in place to address these sites. The size of the problem as reported by 10 countries is shown in Table 2-8.

The priority given to this element in the overall hazardous waste program depends on national circumstances and the extent of the perceived problems. Extensive remedial action programs, similar to those of the United States, are under way in Denmark and the Netherlands. France and Germany have cleaned up a number of sites on a case-by-case basis, without a national program aimed at all such sites. The costs of cleanup are paid though general revenues (Denmark, Japan, the Netherlands, and South Africa) or through a special tax or funding from the chemical industry (France, Germany, United States). Austria, Italy, and the United Kingdom have no special provisions to pay for the cleanup of these sites.

Developing Countries

Although environmental concerns in general, and hazardous waste management in particular, are not high on the political agendas of many developing countries, this is an active area of development, particularly in the newly industrializing countries of southeast Asia and Latin America and in the economies-in-transition of central and eastern Europe and the former Soviet Union.

Developing countries have a number of special characteristics that may make improper hazardous waste management a more serious problem than comparable practices in more developed countries:

- Lack of infrastructure that might ameliorate some of the risks of poor practices (e.g., proper treatment of drinking water);
- Lack of education in environmental risks;
- Unique subpopulations at special risk (e.g., scavengers who make their living sorting through refuse); and
- Inadequate health care systems.

All of these attributes argue for a more vigorous regulatory system in these countries. However, economic realities often dictate that environmental concerns not be permitted to stand in the way of development, and many problems are inadequately addressed, or not dealt with at all. Wilson and Balkau[44] identify several common constraints in developed countries, from which the developing countries might learn:

- Inadequate early identification and quantification of waste,
- Inadequate enforcement of existing pollution control laws so that the hazardous waste problem remains hidden,
- The inevitable long delay between recognizing the need for facilities and actually providing them,

TABLE 2-8
Uncontrolled hazardous waste sites

	Austria	Croatia	Czech Republic	Denmark	Germany	Italy	Netherlands	Norway	Sweden	United States
Year of latest data update	1997	1997	1997	1998	1997	1997	1997	1997	1997	1987[1]
Estimated total number of waste sites	4000	700	940	4048	No information	5000[2]	7500	3500	7000	–
Total number suspected of containing hazardous waste	2300	150	–	–	No information	150[2]	4350	2100	Approximately 1000	20,000[1]
Sites requiring immediate action	133	50	–	1385	Note 3	100[4]	1000	90	200	1,204[5]

[1]Early 1980s estimate.
[2]Estimated by the Ministry for the Interior and Civil Prevention.
[3]Waste sites which are suspected of being contaminated have been systematically listed by the Federal states. The particular investigations are still in progress.
[4]Estimated by regional authorities, according to cleanup programs.
[5]Final National Priorities List (NPL) sites, as of March 4, 1999.

Reproduced with permission from the International Solid Waste Association report "International Perspectives in Hazardous Waste Management: 1999 Edition."

- Loss of public confidence leading to opposition to the siting of essential new hazardous waste disposal facilities, and
- The complexity of the administrative response needed to control all parts of the waste cycle.

Another problem facing the less-developed nations is the import of wastes from developed countries. The lack of adequate facilities in most developed countries and the inability to site and build new ones (see Section 8-4), has led some generators to export wastes to developing countries. Often the importing countries have no capability to adequately dispose of these wastes but are unable or unwilling to stop the traffic. The response from some importers has been vehement:

> On May 25, 1987, for example, the Organization of African Unity (OAU) unanimously passed a resolution stating that illegal dumping of waste in Africa by other countries "is a crime against Africa and the African people."[45]

In large part because of such strong statements from developing countries, the United Nations Environment Program (UNEP) convened a conference in Basel, Switzerland, in 1989. Delegates from 116 nations approved a convention to control import and export of hazardous waste. Among its many provisions, the **Basel Convention**[46] permits export of waste to occur *only* if:

- The exporting country lacks adequate facilities for the proper disposal, or
- The wastes are to be used as raw material in recycling or recovery operations.

Other provisions dictate that permission of the importing country is required and that export is not permitted to countries that have banned imports. The convention makes clear that the exporting country is responsible to make certain that proper disposal occurs and that this responsibility cannot be transferred to the importing country.

DISCUSSION TOPICS AND PROBLEMS

2-1. Contrast the definition of hazardous waste included in RCRA law (Section 2-2) with that included in EPA regulations (Section 2-3). What are the differences, and why is there any difference at all?

2-2. Contact your congressional representative or U.S. senator and obtain a copy of one federal hazardous waste bill currently being considered. Prepare a brief synopsis of the bill, indicating your opinion as to whether it should be passed.

2-3. Contact your local state legislator and obtain a copy of one hazardous waste bill currently being considered in your state. Prepare a brief synopsis of the bill, indicating your opinion as to whether it should be passed.

2-4. Review recent copies of the *Federal Register* in the library and select a recently promulgated hazardous waste regulation for review. Prepare a brief synopsis of the regulation. Pay particular attention to the preamble and summarize the EPA's estimate of the impact of the regulation.

2-5. Repeat Problem 2-4 for a recent regulation promulgated in your state.

2-6. Provide a synopsis of a recent court decision involving hazardous waste, based on review of a case in **Environmental Reporter Cases**.

2-7. What is the National Contingency Plan? Where would you obtain a copy of the plan?

2-8. Explain what is meant by the term "applicable or relevant and appropriate requirements" (ARARs).

2-9. Explain the different types of applicable or relevant and appropriate requirements (ARARs). Provide an example for each.

2-10. Provide an example of how each of the six defenses to an action for negligence cited in Section 2-1 might arise.

2-11. One of the three ways for a solid waste to be classified as a hazardous waste is for a company to declare the waste hazardous. Because this will dramatically increase the cost of transportation and disposal, what is the incentive for industries to make such a declaration?

2-12. Using the *Federal Register*—Cumulative List of CFR Sections Affected in the library, determine the most recent changes in regulations for:

(*a*) Generators (40 CFR 262)
(*b*) Determining which wastes are hazardous (40 CFR 261)
(*c*) Landfills (40 CFR 265.300-316)

2-13. Determine what federal regulations would apply to:

(*a*) Insurance for hazardous waste landfills
(*b*) Disposal of batteries in landfills
(*c*) Design and operation of land disposal facilities
(*d*) Export of wastes containing PCBs
(*e*) Disposal of PCBs in landfills
(*f*) A list of extremely hazardous substances

2-14. Repeat Problem 2-13 for regulations in your state.

2-15. Define the following legal terms (use a law dictionary) and indicate their relevance to hazardous waste management:

(*a*) Res ipsa loquitur
(*b*) Strict liability
(*c*) Joint and several liability
(*d*) Sovereign immunity
(*e*) Negligence
(*f*) Contributory negligence
(*g*) Comparative negligence
(*h*) Ex post facto law
(*i*) Statute of limitations

2-16. Review the article "Summary and Analysis of the Hazardous and Solid Waste Amendments of 1984," by Beveridge and Diamond (Ref. 10) and prepare a one-page summary of the changes in the 1984 HSWA law that affect land disposal.

2-17. Suggest reasons why (or why not) each of the following might be considered hazardous under EPA regulations:

(*a*) A railroad tank car of fuming sulfuric acid
(*b*) Solvents, after use as cleaning agents
(*c*) Radioactive tracers after use in a university lab
(*d*) Mercury
(*e*) Soil that has had gasoline spilled on it

2-18. Review Section 3004(d) of RCRA (42 USC 6924), and prepare a brief synopsis summarizing your understanding of which wastes Congress intended to preclude from land disposal.

2-19. Review OSHA regulations [29 CFR 1910.120(b)-(p)], and determine what health and safety considerations are required at Superfund sites.

2-20. Write a one-page summary of Section 112 of the Clean Air Act.

2-21. "While it is ordinarily necessary to show negligence to prevail in court, activities involving hazardous waste fall under the concept of strict liability; i.e. proof of negligence is not required and merely being the generator of a waste is sufficient evidence to be liable for damages. This concept is more stringent than that used in common law in which courts usually apply the "reasonably prudent person" concept as the standard of care on which to judge an action alleged as negligent. Failing to do something that a reasonably prudent person would do to prevent harm or injury (or doing something such an individual would not do) is negligent conduct." Are there any instances in which a court might determine a person to be exempt from this standard and therefore not negligent, even if a supposedly negligent event occurred?

2-22. In a legal brief on negligence the following statement was made: "Where negligence can be established it is no defense that the negligent action was in full compliance with all government regulations and permit conditions." Explain why this defense may not be valid.

2-23. Suppose that you are an OSHA inspector sent to inspect an industry located in an extremely impoverished area. The industry also employs a large number of local people. In your inspection of the plant, you find a violation of OSHA regulations that could be potentially dangerous to the employees. You are told by the plant manager that correcting the flaw would require the plant to be shut down for several months. This would result in almost all of the company's employees being out of work. What factors should be considered in your response to this information?

2-24. Under the subsection entitled "Generators and Transporters," in Section 2-3, the regulatory requirement is discussed: "Generators producing less than 100 kilograms per month are exempt from most RCRA regulations but still required to meet minimum standards." Do you agree with this exemption? If so, why? If not, why not?

2-25. Explain why the statement "CERCLA specifically applies this liability to wastes produced prior to the enactment of CERCLA" does not violate the Constitutional prohibition against ex post facto laws?

2-26. Obtain a manifest form for hazardous waste from your own state regulatory agency. Fill it out for a waste from a local industry.

2-27. Explain how the principle of joint and several liability would apply to potentially responsible parties (PRPs) at a Superfund site.

2-28. What is the Federal Administrative Procedures Act?

2-29. Explain what is meant by "citizen suits."

2-30. Obtain an EPA Guidance Document related to the hazardous waste program from the government documents section of your library and provide a brief abstract of its provisions. Particularly note which laws and regulations provide authority to the guidance document. Finally explain why a guidance document was used rather than a new regulation.

2-31. Review the article by K. E. Aaron and A. J. Slap.[47] Explain how CERCLA may be in conflict with federal bankruptcy laws.

2-32. Review the article by J. Bartels, et al.[48] How does the European system differ from the manifest system in the United States?

2-33. Review the article by M. E. Doyle and C. Sweet.[49] Identify the various methods used in different states to establish cleanup goals.

2-34. Review the article by L. Ford, et al.[50] Explain what happens when the government attempts to regulate pollution at levels that cannot be consistently measured.

2-35. Review the article by J. Josephson.[51] Explain the various phases in the federal Superfund program.

2-36. Explain what is meant by "cradle-to-grave" control of hazardous waste.

2-37. Review the article by J. Nikmanesh, et al.[52] Explain what is meant by a de minimis PRP.

2-38. Review the article by D. C. Wilson and W. S. Forester,[53] noting the differences between European and American approaches to hazardous waste regulations.

2-39. Explain why the American definition of hazardous waste (and the chemical tests derived from that definition) would not be appropriate in Great Britain.

2-40. Contact your state environmental agency and find out what steps are required to obtain a permit for a hazardous waste land disposal facility.

2-41. Determine which departments generate hazardous waste at your university, and where the wastes are sent. Obtain a manifest form for a shipment of hazardous waste from the university.

2-42. Explain the Hazard Ranking System. What is the lowest score a site can have and be included on the NPL?

2-43. Find out from your state's regulatory agency where the nearest Superfund site is located and ascertain the status of cleanup.

2-44. The following is a list of potential hazardous waste generators. Explain the possible hazardous wastes generated from each and how each waste is classified as hazardous.

(*a*) Dry cleaners
(*b*) Automobile service stations
(*c*) "Do-it-yourself" garage mechanics
(*d*) Users of household pesticides (RAID, etc.)
(*e*) Metal finishers (e.g., ball bearing manufacturer)
(*f*) Hospitals
(*g*) Doctors' offices

REFERENCES

1. Bockrath, Joseph: *Environmental Law for Engineers, Scientists and Managers*, McGraw-Hill, New York, 1977.
2. Arbuckle, J. G., et al.: *Environmental Law Handbook*, 10th edition, Government Institutes, Inc., Rockville, Maryland, 1989.
3. Findley, R. W., and D. A. Farber: *Environmental Law—Cases and Materials*, 2nd edition, West Publishing, St. Paul, Minnesota, 1989.
4. Rogers, W. H., Jr.: *Environmental Law*, West Publishing, St. Paul, Minnesota, 1977.
5. Schoenbaum, T. J.: *Environmental Policy Law*, Foundation Press, Mineola, New York, 1985.
6. *Dupont Directions*, E. I. Dupont deNemours, Inc., vol. III, no. 2, n.d.

6a. U.S. Environmental Protection Agency: *EPA Regulatory Agenda*, Washington, D.C., 1989.

7. U.S. Environmental Protection Agency.
8. The Resource Conservation and Recovery Act of 1976, 42 USC §6901 et seq., 1976.
9. The Hazardous and Solid Waste Amendments of 1984, P.L. 98-616, November 9, 1984.
10. Beveridge & Diamond, P. C.: "Summary and Analysis of the Hazardous and Solid Waste Amendments of 1984," *Hazardous Waste & Hazardous Materials*, vol. 2, no. 1, pp. 113–130, 1985.
11. ERM: *The New RCRA—An Introduction to the Hazardous & Solid Waste Amendments Act of 1984*, Exton, Pennsylvania, 1985.
12. U.S. Environmental Protection Agency: *Corrective Action for Solid Waste Management Units (SWMUs) at Hazardous Waste Management Facilities—Proposed Rule*, September 12, 1988.
13. U.S. Environmental Protection Agency: *Summary of State Reports on Releases from Underground Storage Tanks*, EPA/600/M-86/020, August 1986.
14. Courtesy ERM/EnviroClean.

15. Federal Register, *Hazardous Waste and Consolidated Permit Regulations*, 45, no. 98, pp. 3066–3588, May 19, 1980.
16. Deland, M. R.: "EPA Regulates Hazardous Waste," *Environ. Sci. & Tech.*, vol. 14, no. 7, p. 777, July 1980.
17. 40 CFR 261.31.
18. U.S. Environmental Protection Agency: *Test Methods for Evaluating Solid Waste*, SW-846, 3rd edition, Sect. 7.4, November 1986.
19. *Federal Register*, March 29, 1990.
20. Kimmell, T. A.: "Development, Evaluation and Use of the Toxicity Characteristic Leaching Procedure (TCLP)," Chemical and Biological Characterization of Sludges, Sediments, Dredge Spoil, and Drilling Muds, ASTM STP 976, J. J. Lichtenberg, J. A. Winter, C. I. Weber, and L. Fradkin, eds., Amer. Soc. Test. Matls., Philadelphia, pp. 129–140, 1988.
21. "TCLP vs EP?—Confusion Reigns," *The Hazardous Waste Consultant,* p. 2.2, March/April 1991.
22. The Comprehensive Environmental Response, Compensation and Liability Act (Superfund), 42 USC §9601 et seq., 1980.
23. Hedeman, W. M., Jr., P. E. Shorb, III, and C. A. McLean: "The Superfund Amendments and Reauthorization Act of 1986: Statutory Provisions and EPA Implementation," *Hazardous Waste & Hazardous Materials*, vol. 4, no. 2, pp. 193–210, 1987.
24. ERM: *The New Superfund—CERCLA Amendments of 1986*, Exton, Pennsylvania, 1986.
25. U.S. House of Representatives, *Superfund Amendments and Reauthorization Act—Conference Report,* Report 99-962, October 3, 1986.
26. Stoll, R. G.: "The Comprehensive Environmental Response, Compensation and Liability Act (CERCLA or Superfund)" in Arbuckle, J. G., et al., *Environmental Law Handbook*, Government Institutes, Inc., Rockville, Maryland, 1987.
27. Superfund Amendments and Reauthorization Act, P.L. 99-499, October 17, 1986.
28. Shrivastava, Paul: *Bhopal—Anatomy of a Crisis*, Ballinger Publishing Co., Cambridge, Massachusetts, 1987.
29. Diamond, S., "The Bhopal Disaster: How It Happened," *The New York Times*, January 28, 1985.
30. Sax, N. I., and R. J. Lewis: *Dangerous Properties of Industrial Materials*, 7th edition, Van Nostrand Reinhold, New York, 1989.
31. Diamond, "The Bhopal Disaster."
32. Kalelkar, A. S.: "Investigation of Large Magnitude Incidents: Bhopal as a Case Study," presented at the Institute of Chemical Engineers Conference on Preventing Major Chemical Accidents, London, May 1988.
33. "Bhopal Disaster Spotlights Chemical Hazards Issues," *Chemical & Engineering News*, December 4, 1984.
34. Clay, D. R.: *Role of the Baseline Risk Assessment in Superfund Remedy Selection Decisions*, OSWER Directive 9355.0–30, 1991.
35. Stone, M. E., and A. B. Weissman: "Superfund Compliance with Applicable or Relevant and Appropriate Requirements," Superfund '88, *Proceedings of the 9th National Conference, Haz. Matls. Control Red. Inst.*, Washington D.C., pp. 8–11, 1988.
36. National Oil and Hazardous Substances Pollution Contingency Plan (NCP), Final Rule, 55 FR 8666, March 8, 1990.
37. LaGrega, M. D., A. Ertel, P. A. Marcus, and J. B. McIlvain: "Legal Responsibility of Industry for Superfund Sites," *Journal of Resource Management and Technology*, vol. 16, no. 1, pp. 14–20, April 1988.
38. Mott, Randy M.: *Hazardous Waste Litigation—1986*, Practicing Law Institute, New York, 1986.
39. Occupational Safety and Health Act of 1970, PL 91-596, 84 Stat. 1590, section 6(b)(5).

40. U.S. Environmental Protection Agency: *CERCLA Compliance with Other Laws Manual: Draft Guidance*, Washington, D.C., 1988.
41. Quarles, J., and W. H. Lewis, Jr.: *The New Clean Air Act—A Guide to the Clean Air Program as Amended in 1990*, Morgan, Lewis & Bockius, Washington, D.C., 1990.
42. Wilson, D.C.: "International Perspectives on Hazardous Waste Management: 1998 Edition," prepared by ERM on behalf of the ISWA Working Group on Hazardous Waste, May 1998.
43. Biswas, A. K.: "Environmental Aspects of Hazardous Waste Management for Developing Countries: Problems and Prospects," in *Hazardous Waste Management,* S. P. Maltezou et al., eds., United Nations Industrial Development Organization, Vienna, Austria, pp. 261–271, 1989.
44. Wilson, D. C., and F. Balkau: "Adapting Hazardous Waste Management to the Needs of Developing Countries—An Overview and Guide to Action," *Waste Management & Research*, vol. 8, pp. 87–98, March 1990.
45. Huntoon, B. D.: "Emerging Controls on Transfers of Hazardous Waste to Developing Countries," *Law & Policy in International Business*, vol. 21, pp. 247–271, 1989.
46. "Basel Convention on the Control of Transboundary Movements of Hazardous Wastes and Their Disposal," UNEP Document IG.80/3, March 22, 1989.
47. Aaron, K. E., and A. J. Slap: "The Chateaugay Decision: Collision of Bankruptcy and Environmental Law," *Toxics Law Reporter*, Bur. Natural Affairs, June 1990.
48. Bartels, J., et al.: "An Analysis of a European Hazardous Waste Transport Registration System (EARS)," *Waste Management & Research,* vol. 7, no. 229, 1989.
49. Doyle, M. E., and C. Sweet: "Soil Cleanup Goals for Total Petroleum Hydrocarbons" in *Superfund '90*, Haz. Materials Control Research Instit., Washington, D.C., pp. 21–24, 1990.
50. Ford, L., et al.: "Environmental Russian Roulette—Compliance At or Near the Detection Level," *Water Environment and Technology,* pp. 58–63, August 1990.
51. Josephson, J.: "Implementing Superfund," *Environ. Sci. Technol.*, vol. 20, no. 1, pp. 23–28, 1986.
52. Nikmanesh, J., et al.: "De Minimis Settlement—A Success Story" in *Superfund '89*, Haz. Materials Control Research Instit., Washington, D.C., pp. 190–193, 1989.
53. Op. cit. to reference 42.

ADDITIONAL READING

Anderson, D. R.: "Financing Liabilities Under Superfund," *Risk Anal.*, vol. 9, no. 329, p. 615, 1989.

Anon.: "Developments in the Law: Toxic Waste Litigation," *Harvard Law Review*, pp. 1459–1661, May 1986.

Anon.: "The Clean Air Act Amendment of 1990—A Detailed Analysis," *The Hazardous Waste Consultant*, pp. 4.1–4.30, January/February 1991.

Anon.: *Restatement of Torts*, 2nd ed., West Publishing, St. Paul, Minnesota, 1934.

Barr, L. M.: "CERCLA Made Simple: An Analysis of the Cases Under the Comprehensive Environmental Response, Compensation and Liability Act of 1980," *Business Lawyer*, vol. 45, pp. 923–1001, May 1990.

Berman, Eugene, and Jeffrey Howard, "How Should Federal Hazardous Waste Policy Address Potentially Responsible Local Governments," *Environment Reporter*, BNA, pp. 373–376, July 15, 1988.

Cooke, S. M., ed.: *The Law of Hazardous Waste: Management, Cleanup, Liabilities and Litigation*, Matthew Bender, Albany, New York, 1990.

Donovan, P. E.: "Serving Multiple Masters: Confronting the Conflicting Interests that Arise in Superfund Disputes," *Boston College Environmental Affairs Law Review*, vol. 17, pp. 371–403, 1990.

Ferrey, S.: "The Toxic Time Bomb: Municipal Liabilities for the Clean-Up of Hazardous Waste," *The George Washington Law Review*, vol. 57, pp. 197–278, December 1988.

Firestone, D. B., and F. C. Reed: *Environmental Law for Non-lawyers*, Ann Arbor Sci. Publ., 1983.

Forester, W. S., and J. H. Skinner, eds.: *International Perspectives on Hazardous Waste Management*, Academic Press, San Diego, California, 1988.

Frank, William Harris, and Timothy B. Atkeson: "Superfund Litigation and Cleanup," *Environment Reporter,* BNA, June 28, 1985.

Gelwan, L. A.: "PRP Access to Superfund Sites: A Primer," *Virginia Environmental Law Journal*, vol. 10, pp. 77–86, Fall 1990.

Gentry, Bradford S., "Regulation of Hazardous Waste in the United States and Great Britain," *Environment Reporter,* BNA, pp. 2321–2328, February 24, 1989.

Hammitt, J. K., and P. Renten: "Illegal Hazardous Waste Disposal and Enforcement in the United States: A Preliminary Assessment," *J. Hazard. Mater.*, vol. 22, no. 281, 1989.

Howard, J. H., and L. E. Benfield: "CERCLA Liability for Hazardous Waste Generators: How Far Does Liability Extend?" *Temple Environmental Law and Technology Journal*, vol. 9, pp. 33–49, Spring 1990.

Huntoon, B. D.: "Emerging Controls on Transfers of Hazardous Waste to Developing Countries," *Law and Policy in International Business*, vol. 21, pp. 247–271, 1989.

LaGrega, M. D., et al.: "Potentially Responsible Parties and Superfund—An Analysis of Issues," *Proceedings of the 41st Annual Industrial Waste Conference*, Purdue University, pp. 370–376, 1986.

Prosser, William: *Torts*, 3rd ed., West Publishing, St. Paul, Minnesota, 1964.

Quarles, J.: "In Search of a Waste Management Strategy," *Natural Resources and Environment*, pp. 3–5, Summer 1990.

Ruckelshaus, W. D.: "The Role of the Affected Community in Superfund Cleanup Activities," *Hazardous Waste*, vol. 1, no. 3, pp. 283–288, 1984.

Ryan, M. K.: "The Superfund Dilemma: Can You Ever Contract Your Liability Away?" *Massachusetts Law Review*, vol. 75, pp. 131–137, Summer 1990.

Travis, C. C., and C. B. Doty: "Superfund. A Program without Priorities," *Environ. Sci. Technol,* vol. 23, p. 1333, 1989.

U.S. Government Printing Office: *Code of Federal Regulations—Chapter 40—Protection of Environment*, 1988.

Wigmore, J.: *Evidence*, 3rd ed., Little, Brown & Co., 1961.

ACKNOWLEDGMENTS

The authors wish to acknowledge the following individuals who assisted with the preparation of this edition: Thomas F. McCaffery, III, Esq., *SmithKline Beecham*; J. Bradford McIlvain, Esq. and Joseph Dominguez, Esq., *Dilworth, Paxson, Kalish & Kauffman*; Holly Borcherdt, *Bucknell University*; Arnon E. Garonzik, Esq., David Morell, Ann Sirois, Esq., Kenneth Weiss, David C. Wilson, Ph.D., and Paul Woodruff, *ERM*. The principal reviewer for ERM was Peter Schneider.

Reviewers for the second edition included Mary Arndtsen, Jonathan Bonner, Annette Brewster, Judith Hushon, Ph.D., Elsie Millano, Ph.D., J. Edward Norris, Ph.D., Kenneth Weiss, David C. Wilson, Ph.D., and Deborah Woltjen, *ERM*. Section 2-1 was rewritten by Lawrence Ward, *ERM*. The revision of Section 2-6 drew largely from *International Perspectives in Hazardous Waste Management: 1999 Edition* by David C. Wilson, Ph.D., *ERM*.

CHAPTER 3

Process Fundamentals

Water, water, every where,
And all the boards did shrink;
Water, water, every where,
Nor any drop to drink.
Coleridge

This chapter is intended as an overview of fundamentals that many students using this book will have covered in previous science and engineering courses. For more background on the theory underlying the information presented here, refer to basic texts listed at the end of the chapter.

3-1 A LITTLE CHEMISTRY

Organic chemistry is the study of compounds containing carbon, which make up much of hazardous waste. A working knowledge of the properties and characteristics of organic compounds is essential to the management of hazardous waste. Additionally, a general understanding of chemical nomenclature is critical in the understanding of regulatory documents.

Nomenclature and Structure

In drawing chemical structures for organic molecules, a few rules should be kept in mind. Conceptually, electrons are arranged in concentric shells, or energy levels, surrounding the positively charged atomic nucleus. There is a maximum number of electrons that can be accommodated in each shell: two in the first (hydrogen), eight

in the second (carbon, oxygen, nitrogen). The greatest stability is reached when the outer shell is full, and covalent (and ionic) bonds arise from the tendency of atoms to attain this stable configuration. Because a vast majority of chemical bonds in organic molecules are covalent (electrons are shared between atoms), structures with bonds represented by dashes connecting atoms in a molecule can be drawn. This is known as a "dash structure." Since the most stable molecules will be formed when the outer (valence) shell of an atom is filled, in organic molecules hydrogen must have one dash (one pair of electrons) to fill the outer shell and carbon should have four shared pairs of electrons (four dashes) for the shell to be filled. A single bond is represented by a single dash, a double bond is represented by two parallel dashes, and a triple bond by three parallel dashes.

There are several other conventions for drawing organic molecules as seen in Example 3-1. Some are used to illustrate electron pairs; others are used as shorthand representation; still others serve to detail the three-dimensional aspects (stereochemistry) of the molecules.

EXAMPLE 3-1. VARIOUS WAYS OF REPRESENTING ORGANIC MOLECULES. Show different conventions for drawing the structure of methane.

Solution

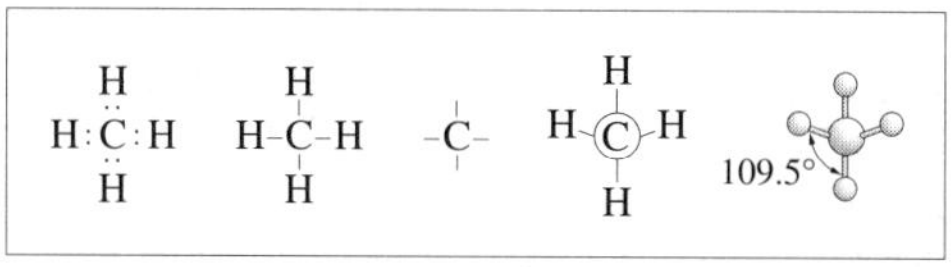

Organic compounds containing only hydrogen and carbon are collectively known as hydrocarbons. On the basis of electronic structure, hydrocarbons are divided into two main classes, aliphatic and aromatic. Aliphatic hydrocarbons are further divided into families: alkanes, alkenes, alkynes, and their cyclic analogs (cycloalkanes, etc.).

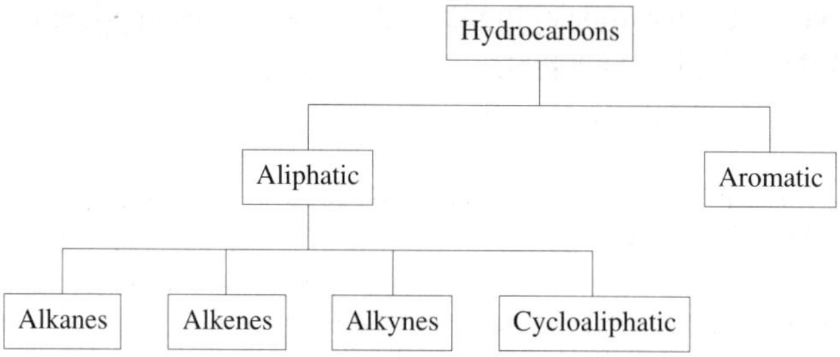

Alkanes are hydrocarbon compounds that contain only single bonds between carbon atoms. Alkanes, also referred to as saturated compounds, have the general formula C_nH_{2n+2} where n is any integer. Several examples of alkanes are listed in Table 3-1.

The names of many organic compounds are derived from the alkane series. Alkyl groups are derived from the parent alkane with a terminal hydrogen atom missing. For example, $CH_3–CH_2·$ is called ethyl, $CH_3–CH_2–CH_2·$ propyl, etc.

In general formulas, R is used to represent an unspecified alkyl group (e.g., methyl). An alkane may then be represented by R–H.

Normal alkanes are those that have a straight chain of carbon atoms and the name is proceeded by "*n*-." A branched alkane is composed of a normal alkane with one or

TABLE 3-1
Alkanes[1]

Name	Chemical formula	Structure
methane	CH_4	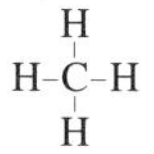
ethane	C_2H_6	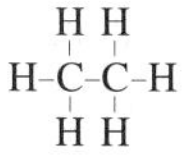
propane	C_3H_8	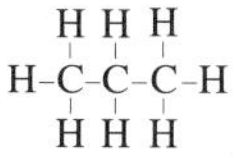
butane	C_4H_{10}	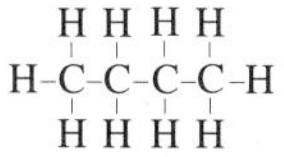
pentane	C_5H_{12}	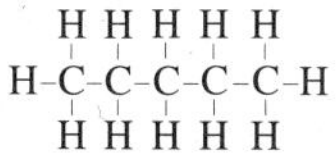

more alkyl groups attached at carbons other than the terminal carbons. The International Union of Pure and Applied Chemistry has developed a systematic method for naming branched hydrocarbons. To name a branched alkane, the procedure below must be followed:

1. Look for the longest straight carbon chain. That chain gives the base name.
2. Number the carbon atoms in the base chain. The carbons in this base chain must be numbered in a direction that has the lowest possible numbers assigned to carbons containing any branched alkyl groups.
3. Place the appropriate number before the alkyl group's name. Each group must have a number even if two groups are attached to the same carbon atom. When writing the name, numbers are always separated from letters by a hyphen. The name of the last group is joined to the base name as a continuous word and listed in alphabetical order regardless of numerical carbon assignment.

hexane:

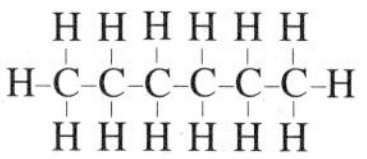

3-methyl-hexane:

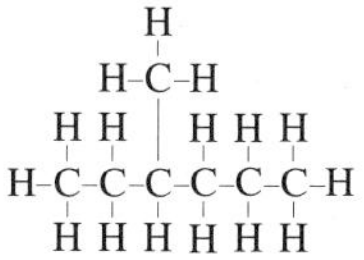

Many organic compounds have both a systematic and a common name. This is due in part to the way the science of organic chemistry developed. For example, the solvent used to extract base/neutral/acid extractables from an aqueous sample is

methylene chloride. The systematic name for this compound is 1,1-dichloromethane. Another example of the use of a more common name is xylene, which is actually the common name of three compounds (isomers) all having the same chemical formula (C_8H_{10}). Systematically, xylene is 1,2-dimethylbenzene, 1,3-dimethylbenzene, or 1,4-dimethylbenzene. Further complication arises when, because of the length of the systematic name for an organic compound, an abbreviation is used. For example, perchloroethylene (PCE), a common pollutant, is actually 1,1,2,2-tetrachloroethylene. Similarly, trichloroethylene (TCE) is systematically 1,1,2-trichloroethene.

Alkenes contain carbon-carbon double bonds. Alkenes have the general formula C_nH_{2n}. In naming alkenes, the following steps are taken:

1. Look for the longest chain that contains a double bond. Change the suffix of the corresponding alkane of the identical length from "*-ane*" to "*-ene*." This is the base name.
2. Number the chain beginning at the end nearer the double bond. Use the number of the first carbon of the double bond to designate the location of the double bond, and place this number before the base name using a hyphen.
3. Complete the compound name as outlined in Steps 2 and 3 for alkanes above.
4. If a double bond has two identical groups, the prefixes "*cis-*" and "*trans-*" are used. (*Cis-* denotes two identical groups on the same side of the molecule. *Trans-* is used if they are on opposite sides.)

cis-2,3-difluoro-2-butene: 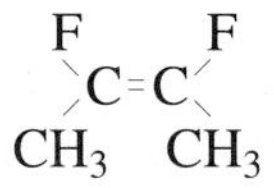

trans-2,3-difluoro-2-butene: 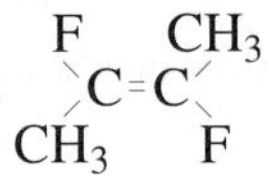

5. If more than one double bond is present, indicate the position of each and use diene, triene, etc.

1, 3-butadiene: $CH_2 = CH - CH = CH_2$

Alkynes contain carbon-carbon triple bonds. Alkynes have the general formula C_nH_{2n-2}. Alkynes are named in a manner similar to that used for alkenes except the longest chain must contain the triple bond and the -ene suffix is changed to -yne. There can be no *cis-* or *trans-* assignment to a carbon-carbon triple bond.

ethyne (common name acetylene): $HC \equiv CH$

The second general category of hydrocarbons is **aromatics.** The simplest aromatic ring is benzene:

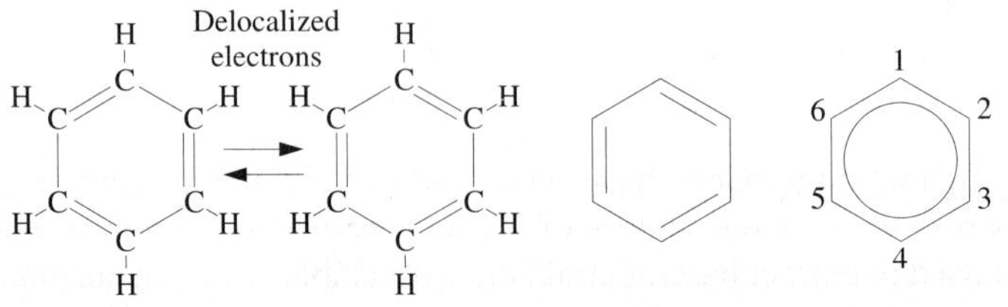

The term **aromatic** refers to the electronic structure of this group in which electrons are shared by more than one carbon-carbon bond. This gives rise to unusual stability. Note that the lower molecular weight aromatics such as benzene have high vapor pressures. Hence, they have an odor, or are aromatic.

Nomenclature of substituted benzenes is divided into two categories. For mono-substitute compounds, benzene is the base name and the group is designated as a prefix.

chlorobenzene: 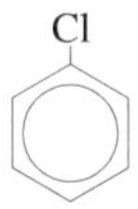

When two groups are present, the prefix "ortho-" (O), "meta-" (M), or "para-" (P), or a numerical designation, is used to designate position on the ring.

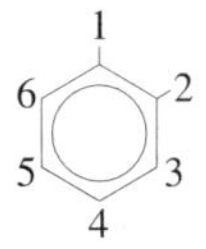

ortho- or 1,2-

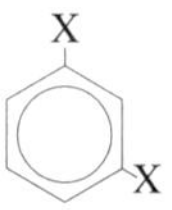

meta- or 1,3-

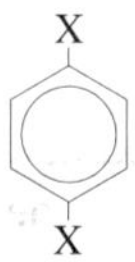

para- or 1,4-

When more than two groups are present, numbers must be used to indicate the positions. The ring is numbered so the groups have the lowest possible numbers.

1-chloro-3,4-dinitrobenzene: 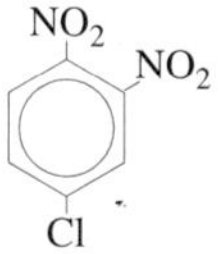

For some compounds, common names used most often are:

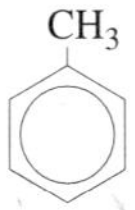

toluene

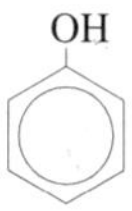

phenol

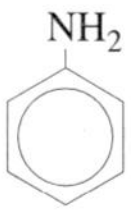

aniline

anisole

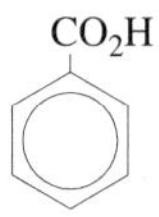

benzoic acid

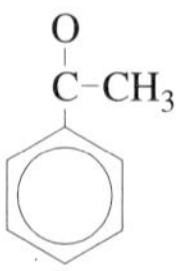

acetophenone

benzenesulfonic acid

Dimethylbenzenes are given the common name xylenes.

meta-xylene: 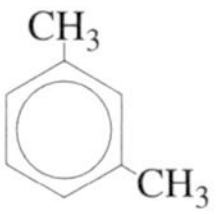

When a benzene ring is itself a substituent of a larger group, it is called phenyl.*

3,4′-dichlorobiphenyl: 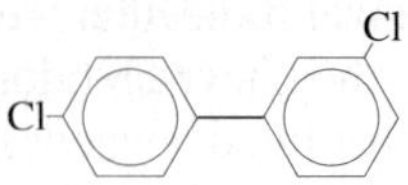

Polynuclear aromatic hydrocarbons (PAHs) are groups of aromatic rings containing only carbon and hydrogen. They may be considered as two or more benzene rings fused together with at least two common carbons.

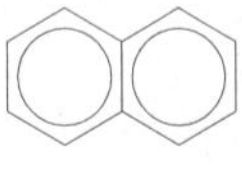

napthalene

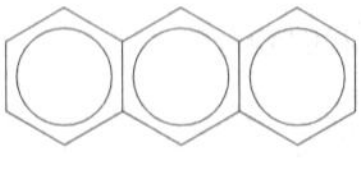

anthracene

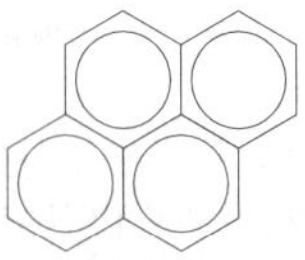

pyrene

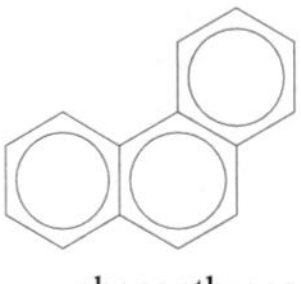

phenanthrene

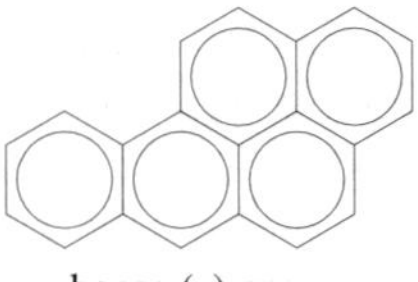

benzo (a) pyrene

Heteroatoms such as nitrogen, oxygen, and sulfur can be substituted into PAHs. Some of these compounds such as carbazole (see diagrams below), are characteristic of manufactured gas plant (MGP) wastes. Dibenzofuran and dibenzo-*p*-dioxin are unwanted combustion by-products.

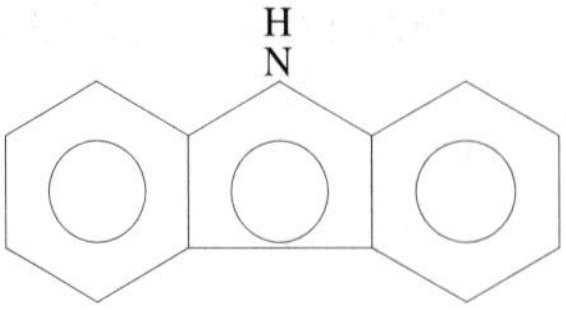

carbazole

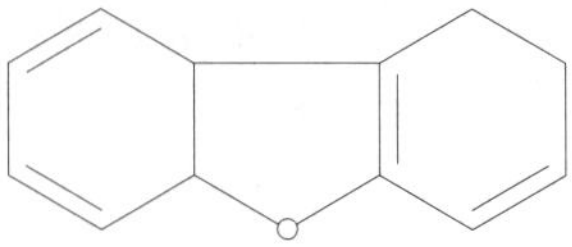

dibenzofuran

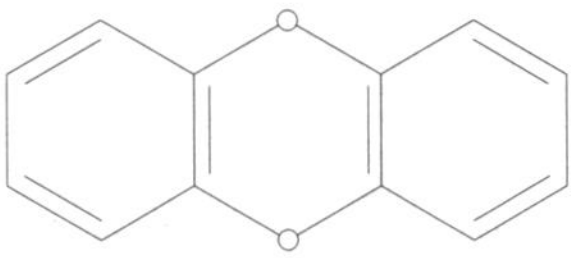

dibenzo-*p*-dioxin

Amines are considered organic derivatives of ammonia. They are classified as primary, secondary, or tertiary amines, by the number of attached alkyl groups.*

$R-NH_2$	$R-\underset{}{\overset{R'}{\overset{|}{N}}}-H$	$R-\overset{R'}{\overset{|}{N}}-R''$
Primary amine	Secondary amine	Tertiary amine

*The prime (′) indicates the presence of carbon atoms in side chains.

A nitrosamine has the nitroso group (–N=O), as one of the alkyl groups.

diphenylnitrosamine: 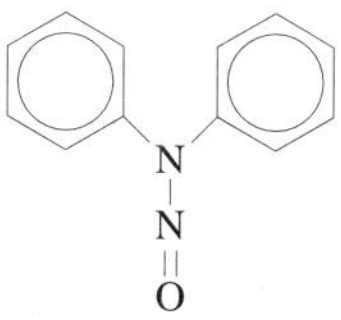

Ethers have the general formula R–O–R or R–O–R′, where R′ indicates a different alkyl group than R. In naming ethers, the two alkyl groups attached to the oxygen are named and the word ether is added.

bis-(2-chloroethyl) ether: Cl–CH_2–CH_2–O–CH_2–CH_2–Cl

(The term *bis*- is used to indicate a substituted substituent prefix or functional modification term.)

EXAMPLE 3-2. STRUCTURAL FORMULAS. (a) Draw the structure for ethyl propyl ether. (b) What name would you give to the structural formula $(CH_3)_2CHOCH_2CH_3$?

Solution

(a):

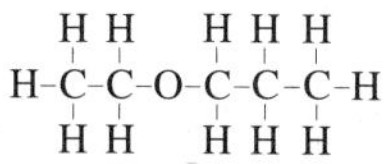

(b): Ethyl isopropyl ether. The prefix "iso" indicates a branched chain.

Esters have the general formula RCO_2R'. An ester is formed by reacting a carboxylic acid with an alcohol. To name an ester, the name of the alkyl radical (R′) from the alcohol comes first, followed by the name of the carboxylate ion or salt. The carboxylate ion has the general formula R–CO_2^-.

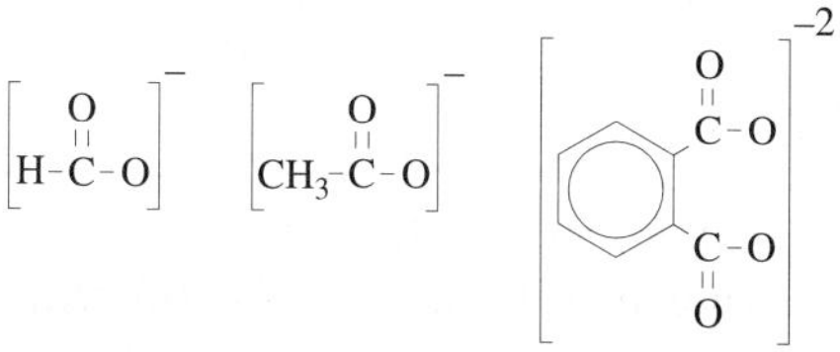

formate ion acetate ion phthalate ion

Phthalate esters contain the phthalate ion and any alkyl groups.

dimethyl phthalate: 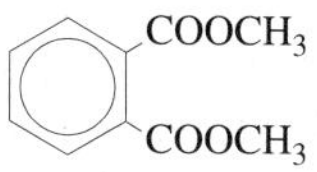

A somewhat different class of organic compounds are ketones. **Ketones** have the general formula

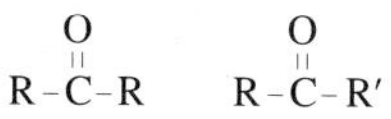

where R and R′ are different alkyl groups. The process of naming ketones involves naming each alkyl group and adding ketone to the end. The alkyl groups are given in alphabetical order.

From the perspective of environmental chemistry, organic chemicals can be divided into three basic categories related to volatility: volatile organics, semivolatile organics, and nonvolatile organics. As discussed briefly below, subcategories can be developed within each of the three main categories. However, no distinct lines can be drawn between what is considered to be volatile or semivolatile, or between semivolatile and nonvolatile. In environmental chemistry, compounds are typically assigned to these broad categories on the basis of which analytical method proves to be the best for the detection of that compound.

Volatile organics

Volatile organic compounds (VOCs) have a high vapor pressure that allows them to evaporate quickly. In general, they are relatively low molecular weight compounds. Their physical-chemical properties include high vapor pressure, high Henry's constant, and low organic carbon partition coefficient (K_{oc}). These properties are described in Section 3-2. Examples of volatile organics include:

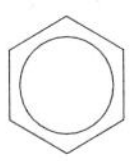
benzene

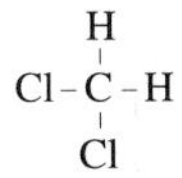
dichlormethane

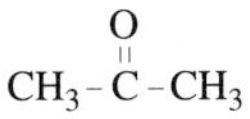
acetone

Analytically, detection of these compounds is performed by taking advantage of their high vapor pressure. By bubbling a pure gas (e.g., helium) through a water sample (or a soil sample mixed with water), volatile compounds are removed from the sample, are swept away with the evolving gas, and are finally trapped from the gas stream for instrumental analysis. This method of analysis is known as the **purge and trap technique**.

Semivolatile organics

Semivolatile organics include several broad classes of compounds with different physical and chemical properties. Because of these differences in properties, semivolatile organics are typically subcategorized as **base/neutral extractables** and **acid extractables**.

Unlike volatile organics, most of these compounds are not of sufficient volatility for use of the purge and trap technique. Therefore, the chemical analysis of these compounds must rely on the acid/base properties of these compounds to facilitate their isolation from a water sample into an organic solvent. When the aqueous sample is acidified, acid-extractable compounds such as phenol and cresol can be extracted with an organic solvent. When the sample is basic, base/neutral-extractable compounds are extracted by organic solvents. The insoluble solvent phases are separated from the water sample, and are then analyzed for semivolatile organics. This method of analysis is known as the **solvent extraction** technique.

Base neutrals include polyaromatic hydrocarbons (PAHs), nitrosamines, ethers, phthalate esters, and selected aromatics not containing hydroxyl (OH) or carboxyl (COOH) groups. Base/neutrals tend to have the following physical-chemical properties: low solubility, low Henry's constant, low vapor pressure, high K_{oc}, high K_{ow}, and high bioconcentration factor (BCF). These properties are described in Section 3-2.

In general, **acid extractables** are aromatic alcohols or phenols. Acid extractables have the following physical-chemical properties that depend on substituent and position: high to low water solubility, moderate vapor pressure, low Henry's constant, and variable K_{oc} and BCF. Examples of acid extractables include:

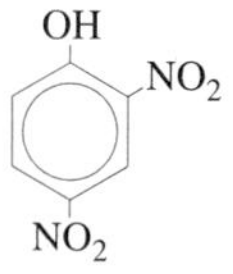

2,4-dinitrophenol

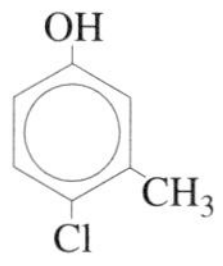

4-chloro-3-methyl phenol
or
4-chloro-3-cresol

Nonvolatile organics

Nonvolatile organics are typically characterized as organic compounds with little to no volatility even at elevated temperatures. Analytically, these compounds can be isolated from environmental samples by the same solvent extraction technique discussed for semivolatile organics analysis.

Pesticides and Herbicides

Pesticides and herbicides are chemical compounds used to control insects (insecticide), rodents (rodenticide), parasites (nematicide), fungi (fungicide), and undesirable plants (herbicide) (see Table 3-2). Both organic and inorganic compounds have been used as pesticides and herbicides. Organic pesticides can be generally grouped as chlorinated pesticides and organophosphate pesticides.

Chlorinated pesticides have been widely used as insecticides and fungicides. These pesticides have been found to have harmful side effects as they do not readily degrade in nature and tend to accumulate in the fatty tissues of most mammals.[3] Perhaps the most infamous of all chlorinated pesticides is **DDT** (see Table 3-2). Other chlorinated pesticides include lindane, dieldrin, chlordane, toxaphene, heptachlor, DDD, and DDE. Virtually all chlorinated pesticides have been scrutinized because of known or suspected hazards to human health.

Organophosphate pesticides are generally toxic to both humans and animals. One important organophosphate pesticide is parathion (see Table 3-2). Other organophosphate pesticides include malathion, systox, chlorthion, disyston, dicapthon, and metasystox.

Organic herbicides typically function by inhibiting photosynthesis, inhibiting plant chloroplast development, disrupting other phytochemical cycles, or by growth regulation. Photosynthetic disruptors include triazines such as atrazine and simazine, hydroxybenzonitriles such as bromoxynil, and carbamates such as propham and chlorpropham. Others include paraquat and diquat. Growth regulators include 2,4-D; 2,4,5-T; MCPA; and glyphosphate. 2,4,5-T is noteworthy because it contains significant amounts of dioxin as an impurity from its manufacture. Dioxin, also known as 2,4,7,8-tetrachlorodibenzo-*p*-dioxin (TCDD), has been shown to be teratogenic and can cause the skin disease chloracne in humans.[4]

TABLE 3-2
General household and field insecticides[2]

Common name	Remarks
lindane 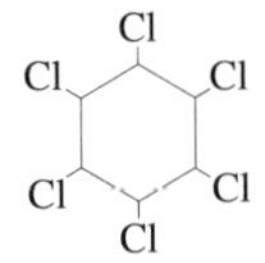	Particularly good against the boll weevil, the cotton aphid and the locust
m-delphene 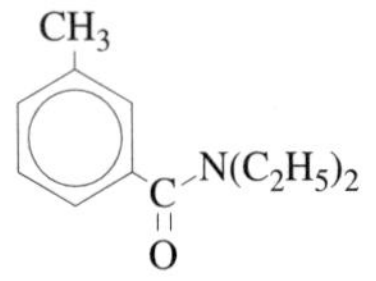	Good mosquito repellent
p-dichlorobenzene 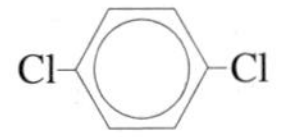	Common moth repellent
methoxychlor 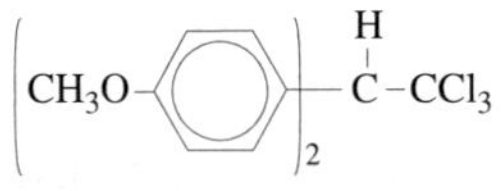	Methoxy analog of DDT; can be used in dairies without imparting toxicity or off-taste to milk
DDT 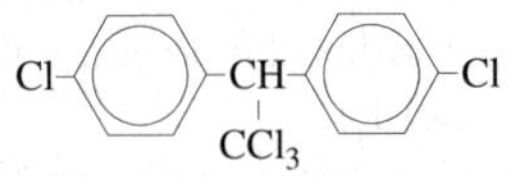	Fairly low toxicity to mammals
chlordane 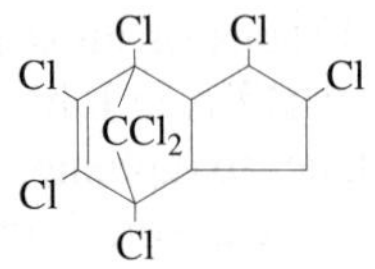	Moderate toxicity to mammals
diazinon 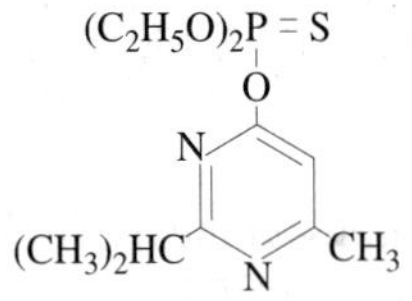	Moderate toxicity to mammals

Inorganic pesticides and herbicides include compounds of zinc, copper, arsenic, mercury, sulfuric acid, sodium arsenate, and sodium thiocyanate. They are highly toxic to a wide range of organisms and their toxic residues are persistent in

TABLE 3-2
General household and field insecticides (Continued)

Common name	Remarks
dieldrin 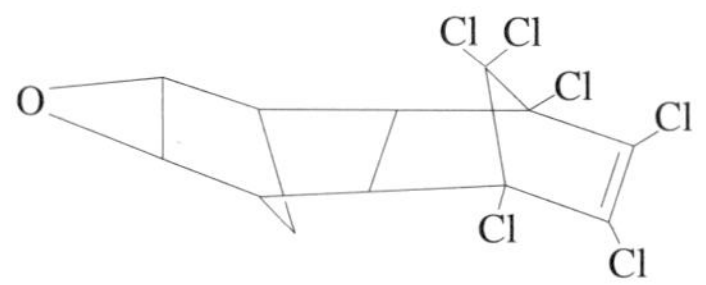	Fairly high toxicity to mammals
parathion 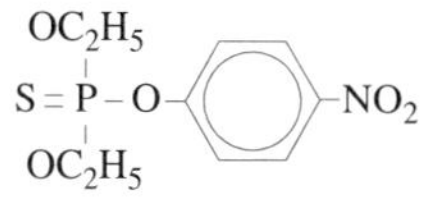	High toxicity to mammals

soil. The widespread use of inorganic pesticides and herbicides has been replaced by the more selective organic compounds. However, as organisms become resistant to the organic pesticides and herbicides, the use of inorganic compounds may again be necessary.[5]

Polychlorinated Biphenyls

Polychlorinated biphenyls (PCBs) contain a mixture of biphenyls with chlorine atoms attached at any of the carbons. There are 210 different PCB compounds, or isomers. Commercial mixtures of isomers are called arochlors, which are designated by their chlorine content, for example, Arochlor 1248 and Arochlor 1260; these mixtures typically have 40 to 60 percent chlorine by weight.[6]

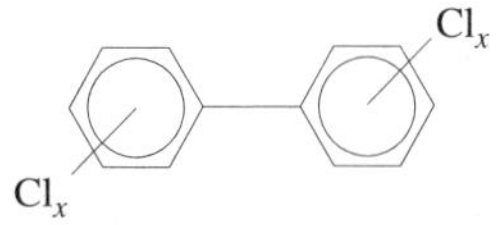

PCBs are a common pollutant because of their former widespread use in electrical transformers and capacitors. Since 1979, the EPA has banned their manufacture, processing, and distribution.

3-2
PHYSICAL-CHEMICAL PROPERTIES

Physical-chemical properties determine how a specific chemical will interact in the environment. These properties control how chemicals move in the environment, and how effectively they can be removed by various methods. Various physical-chemical properties for about 250 selected chemicals can be found in Appendix B.

TABLE 3-3
Constants for solubility equilibria[8]

Oxides and hydroxides	log K at 25°C
$H_2O(l) = H^+ + OH^-$	−14.00
$Cd^{+2} + H_2O = CdOH^+ + H^+$	−10.1
$Cd^{+2} + 2H_2O + Cd(OH)_2\ (aq) + 2H^+$	−20.4
$Cd^{+2} + 3H_2O = Cd(OH)_3^- + 3H^+$	< −33.3
$Cd^{+2} + 4H_2O = Cd(OH)_4^{-2} + 4H^+$	−47.4
$\beta\text{-}Cd(OH)_2\ (s) + 2H^+ = Cd^{+2} + 2H_2O$	13.65
$Hg^{+2} + H_2O = HgOH^+ + H^+$	−3.4
$Hg^{+2} + 2H_2O = Hg(OH)_2\ (aq) + 2H^+$	−6.2
$Hg^{+2} + 3H_2O = Hg(OH)_3^- + 3H^+$	−21.1
$HgO + 2H^+ = Hg^{+2} + H_2O$	2.56
$PbCO_3\ (s) = Pb^{+2} + CO_3^{-2}$	−13.1

Solubility

Solubility is the degree to which one substance, the solute, will dissolve into another, the solvent. The solubility of chemicals in water is a function of temperature and the particular chemical species. Solubility in water is usually expressed in units of milligrams per liter (mg/L) or parts per million (ppm). Because one liter of water weighs one million milligrams, in the dilute solutions typically encountered in environmental work, 1 mg/L = 1 ppm, and 1 μg/L = 1 part per billion (ppb).

All materials are soluble to some degree. Consider the precipitation of calcium hydroxide:

$$Ca^{+2} + 2OH^- \rightarrow Ca(OH)_2 \downarrow \tag{3-1}$$

The **solubility product**[7] for this reaction is:

$$K_{sp} = [Ca^{+2}][OH^-]^2 = 7.88 \times 10^{-6}$$

The square brackets indicate molar concentration (moles/liter).

Table 3-3 presents K_{sp} values for some heavy metals of interest in hazardous waste management.

EXAMPLE 3-3. SOLUBILITY PRODUCT. Nickel is removed from a plating waste stream by hydroxide precipitation at pH = 10. Determine the solubility of nickel at this pH in pure water at 25°C; $K_{sp} = 5.54 \times 10^{-16}$.

Solution

$$Ni^{+2} + 2OH^- \rightarrow Ni\,(OH)_2 \downarrow$$

$$-\log [OH^-] = pOH = 14 - pH = 4.0$$

$$[OH^-] = 1 \times 10^{-4}\ mol/L$$

$$[Ni^{+2}][OH^-]^2 = 5.54 \times 10^{-16}$$

$$[Ni^{+2}] = \frac{5.54 \times 10^{-16}}{(10^{-4})^2}$$

$$[Ni^{+2}] = 5.54 \times 10^{-8}\ mol/L$$

$$1 \text{ mol Ni} = 58.7 \text{ g} = 58{,}700 \text{ mg}$$

$$\text{Concentration} = 5.54 \times 10^{-8} \text{ mol/L} \times 58{,}700 \text{ mg/mol}$$

$$= \underline{\underline{3.25 \times 10^{-3} \text{ mg/L}}} = \underline{\underline{3.25 \mu\text{g/L}}}$$

While this is the theoretical solubility at pH = 10, it is not possible to achieve this low value in treating wastes by hydroxide precipitation.

Vapor Pressure

When a liquid is in contact with air, molecules will leave the liquid as a vapor, via evaporation. Equilibrium conditions exist when the rate of molecules leaving the liquid equals the rate of molecules being redissolved. The **vapor pressure** of a pure liquid is the pressure exerted by the vapor on the liquid at equilibrium. Vapor pressure is a function of the temperature of the liquid and is usually measured in atmospheres (atm) (1 atm = 760 mmHg = 760 torr = 1.0133×10^5 Pa). Vapor pressures vary greatly with temperature, and care must be exercised in using published values. Vapor pressures of various chemicals are tabulated in Appendix B.

An ideal solution of two liquids (liquid a and liquid b) at equilibrium can be described as:

$$p_a = p_{\text{vp},a} x_a \quad \text{and} \quad p_b = p_{\text{vp},b} x_b \tag{3-2}$$

where p_a = partial vapor pressure of component a over the solution
$p_{\text{vp},a}$ = vapor pressure of pure component a
x_a = mole fraction of component a in solution
x_a = mol a/(mol a + mol b)

$$p_{\text{total}} = p_a + p_b \tag{3-3}$$

These expressions of an ideal solution describe **Raoult's law,** which states: The vapor pressure of a liquid in a solution is directly proportional to the mole fraction of that liquid in solution.

In an ideal solution, neither component a nor component b has any significant molecular interactions. The ideal vapor-liquid equilibrium governed by Raoult's law may be approached by concentrated solutions and pure phase compounds.

EXAMPLE 3-4. ESTIMATE THE PARTIAL PRESSURE OF LNAPL COMPONENTS. A release of 500 lb of a mixture of 60% toluene and 40% ethyl benzene by weight at a temperature of 20°C has resulted in a pool of light nonaqueous-phase liquid (LNAPL) at the groundwater-vadose zone interface at a site. Estimate the partial pressure of toluene and ethyl benzene in the vadose zone soil gas immediately above the LNAPL pool.

Solution. From Appendix B:

	Toluene	**Ethyl benzene**
Molecular weight	92.15 g/mol	106.18 g/mol
	0.20 lb/mol	0.23 lb/mol
Vapor pressure	22 mmHg	7 mmHg

1. Calculate the mole fractions for each LNAPL component:

$$\text{mass}_{\text{toluene}} = 500 \text{ lb } \times 0.60 = 300 \text{ lb toluene}$$
$$\text{mass}_{\text{ethyl benzene}} = 500 \text{ lb } \times 0.40 = 200 \text{ lb ethyl benzene}$$
$$\text{moles}_{\text{toluene}} = 300 \text{ lb/(0.20 lb/mol)} = 1500 \text{ mol}$$
$$\text{moles}_{\text{ethyl benzene}} = 200 \text{ lb/(0.23 lb/mol)} = 869 \text{ moles}$$
$$x_a = \text{mol } a/(\text{mol } a + \text{mol } b)$$
$$x_{\text{toluene}} = 1500 \text{ mol}/(1500 \text{ mol} + 869 \text{ mol}) = 0.63$$
$$x_{\text{ethyl benzene}} = 869 \text{ mol}/(1500 \text{ mol} + 869 \text{ mol}) = 0.37$$

2. Using Raoult's law, estimate the partial pressure of each component immediately above the LNAPL pool:

$$p_a = p_{\text{vp},a} x_a$$
$$p_{\text{toluene}} = (22 \text{ mmHg})(0.63) = 13.9 \text{ mmHg}$$
$$p_{\text{ethyl benzene}} = (7 \text{ mmHg})(0.37) = 2.6 \text{ mmHg}$$

In reality, there are few miscible liquids that form solutions that come close to ideality. Nonideal liquid-liquid solutions deviate from Raoult's law because of strong molecular interactions, such as those found in mixtures of polar compounds (e.g., methanol-water) or dilute aqueous solutions.

Henry's Constant

In dilute aqueous solutions typical in environmental applications, the mole fraction of component a can be simplified to the form of:

$$x_a = \text{mol } a/\text{mol water}$$

When a volatile chemical is dissolved in water, a small amount of the chemical in gaseous form exists in the air immediately above the surface of the water. At equilibrium, as molecules of this gas pass into the water, an equivalent number of molecules of the chemical leave the water phase and become part of the vapor. In order to describe the equilibrium condition in dilute aqueous solutions, a special form of Raoult's law may be used, called Henry's law. Henry's law is used to describe the solubility of a gas in a liquid.

Henry's law states: Under equilibrium conditions, the partial pressure of a gas (volatile chemical) above a liquid is proportional to the concentration of the chemical in the liquid:

$$P_g = HC_L \tag{3-4}$$

where P_g = partial pressure of the gas
H = Henry's constant
C_L = concentration of chemical in the liquid

If P_g is expressed in atmospheres (atm) and C_L as a mole fraction, H has units of atm. If C_L is expressed as mol/m^3, H has units of (atm $\cdot$ m^3)/mol. This is the inverse of gas solubility at 1 atm. From the definition of partial pressure, Henry's

constant also represents the ratio of the concentration in the gas to the concentration in the liquid.

$$H = C_g/C_L \tag{3-5}$$

where C_g = concentration of a specified chemical in the gas phase (air)
C_L = concentration of a specified chemical in the liquid phase (water)

Henry's constant (also referred to as the Henry's law constant) may be expressed in a number of units. When Henry's law is developed for transferring from air into water, the constant typically has units of mg/L·atm, or mol/L·atm (the inverse of the Henry's constant used in Appendix B!). It is extremely important to check dimensions carefully before using tabulated values. A dimensionless form of the constant is obtained from

$$H' = H/RT \tag{3-6}$$

where H' = Henry's constant (dimensionless)
H = Henry's constant (atm · m^3/mol)
R = universal gas constant = 8.25×10^{-5} (atm · m^3)/(mol · K)
T = temperature (K)

Henry's constant is significantly affected by temperature and the chemical composition of the water. For example, Henry's constant for volatile hydrocarbons increases approximately threefold for a 10°C increase in temperature.[9] Regression equations for selected chemicals may be found in Appendix B. The tabulated values define Henry's constant as a function of temperature:

$$H = e^{(A-B/T)} \tag{3-7}$$

where H = Henry's constant [(atm · m^3)/mol]
A, B = regression coefficients
T = temperature (K)

In the absence of tabulated information, Henry's constant may be estimated as the vapor pressure divided by solubility, provided that both are measured at the same temperature.[10]

$$H = p_{\text{vp}}/S \tag{3-8}$$

It should be noted that tabulated H values are normally based on pure water, and the values for contaminated water can vary unpredictably.[11]

EXAMPLE 3-5. DIMENSIONLESS HENRY'S CONSTANT. Estimate Henry's constant for toluene in water at 20°C from vapor pressure and solubility data. Convert this to a dimensionless constant. Compare to the value in Appendix B.

Solution. From Appendix B: $p_{\text{vp}} = 22.0$ mmHg at 20°C
Solubility at 20°C = 5.15×10^2 mg/L
Molecular weight = 92.13 g/mol = 92.13×10^3 mg/mol

$$p_{\text{vp}} = (22.0 \text{ mmHg})/(760 \text{ mmHg/atm}) = 0.029 \text{ atm}$$

$$S = \frac{5.15 \times 10^2 \text{ mg/L}}{92.13 \times 10^3 \text{ mg/mol}} = 5.59 \times 10^{-3} \text{ mol/L} = 5.59 \text{ mol/m}^3$$

From Equation 3-8:

$$H = 0.029/5.59 = \underline{\underline{5.19 \times 10^{-3}\ (\text{atm} \cdot \text{m}^3)/\text{mol}}}$$

From Equation 3-6:

$$H' = \frac{5.19 \times 10^{-3}\ (\text{atm} \cdot \text{m}^3)/\text{mol}}{[8.205 \times 10^{-5}\ (\text{atm} \cdot \text{m}^3)/(\text{mol} \cdot \text{K})](20\ \text{K} + 273.2\ \text{K})}$$
$$= \underline{\underline{0.216}}\ (\text{dimensionless})$$

$H = e^{(A-B/T)}$

$A = 5.13$ (Appendix B)

$B = 3.02 \times 10^3$ (Appendix B)

$T = 20 + 273.2 = 293.2$ K

$$H = e^{[5.13-(3.02\times 10^3/293.2)]} = 5.68 \times 10^{-3}\ (\text{atm} \cdot \text{m}^3)/\text{mol}$$

Assuming this value to be "correct," the error in estimating H from vapor pressure and solubility data is

$$\frac{5.19 - 5.68}{5.68} = -0.086 \quad \text{or} \quad 8.6 \text{ percent}$$

Diffusion Coefficient

Contaminants in air or water tend to move from areas of high concentration to areas of lower concentration. **Diffusion** may be defined as the movement of a contaminant under the influence of a concentration gradient. The amount of a contaminant passing through a unit area in a unit time is controlled by **Fick's law:**

$$J = -D(dC/dx) \tag{3-9}$$

where $J =$ flux (mol/cm^2 · s)

$D =$ diffusion coefficient (cm^2/s)

$C =$ concentration (mol/cm^3)

$x =$ length in the direction of movement (cm)

Where tabulated values of the diffusion coefficient do not exist, they may be estimated by the methods illustrated below. A more extensive listing of estimation methods may be found in Tucker and Nelken.[12] Additionally, D may be estimated from data on similar chemical species from their molecular weights M:

$$D_1 = D_2\,(M_2/M_1)^{1/2} \tag{3-10}$$

EXAMPLE 3-6. ESTIMATE OF DIFFUSION COEFFICIENT. Estimate the diffusion coefficient of ethyl alcohol in water from tabulated values for methyl alcohol and n-butyl alcohol.

Compound	D (cm²/s)[13]	Mol. weight	Temperature
Methyl alcohol	1.75×10^{-5}	32.05	25°C
n-Butyl alcohol	0.56×10^{-5}	74.12	25°C
Ethyl alcohol	1.24×10^{-5}	46.07	25°C

Solution. Estimates for D for ethyl alcohol:
From methyl alcohol: $D = 1.75 \times 10^{-5}(32.05/46.07)^{1/2} = 1.46 \times 10^{-5}$
From n-butyl alcohol: $D = 0.56(74.12/46.07)^{1/2} = 0.71 \times 10^{-5}$

Average value = $\underline{\underline{1.09 \times 10^{-5}}}$ (12 percent lower than the tabulated value of 1.24×10^{-5})

The theoretical diffusion coefficient of a contaminant gas in air may be derived from the Boltzmann equation[14]:

$$D = 1.858 \times 10^{-3}\left(\frac{T^{1.5}M'}{P\sigma^2\Omega}\right) \tag{3-11}$$

where $D =$ diffusion coefficient (cm²/s)

$M' = \left(\dfrac{M_a + M_b}{M_a M_b}\right)^{1/2}$

$M_a =$ molecular weight of air (28.97 g/mol)
$M_b =$ molecular weight of contaminant (g/mol)
$P =$ pressure (atm)
$T =$ temperature (K)
$\sigma =$ collision diameter (Å)
$\Omega =$ collision integral (dimensionless function of temperature)

This coefficient is termed "theoretical" to distinguish it from the "apparent" coefficient measured in the environment. The calculation depends on obtaining estimates of σ and Ω, which are available in standard references.[15]

Values of diffusion coefficients in air for a number of compounds may be found in Appendix B. Diffusion coefficients in air are functions of temperature and pressure and tabulated values must be adjusted to ambient conditions:[16]

$$D_1 = D_2\,(P_2/P_1)\,(T_1/T_2)^m \tag{3-12}$$

where m theoretically is 1.5 and is measured as 1.75 to 2.0.

The diffusion coefficient in water may be estimated by the Wilke-Chang[17] method:

$$D = \frac{5.06 \times 10^{-7}T}{\mu V^{0.6}} \tag{3-13}$$

where $D =$ diffusion coefficient (cm²/s)
$T =$ temperature (K)
$\mu =$ viscosity of water (centipoises, cP)
$V =$ molal volume of contaminant (cm³/mol)

Values of the molal volume may be estimated from the method of LeBas,[18] using Table 3-4.

TABLE 3-4
Additive-volume increments* for the calculation of molal volumes V by method of LeBas†

	Increment, cm³/gmol
Carbon	14.8
Hydrogen	3.7
Oxygen (except as noted below)	7.4
In methyl esters and ethers	9.1
In ethyl esters and ethers	9.9
In higher esters and ethers	11.0
In acids	12.0
Joined to S, P, N	8.3
Nitrogen	
Doubly bonded	15.6
In primary amines	10.5
In secondary amines	12.0
Bromine	27.0
Chlorine	24.6
Fluorine	8.7
Iodine	37.0
Sulfur	25.6
Ring, three-membered	−6.0
Four-membered	−8.5
Five-membered	−11.5
Six-membered	−15.0
Naphthalene	−30.0
Anthracene	−47.5

*The additive-volume procedure should not be used for simple molecules. The following approximate values are employed in estimating diffusion coefficients: H_2, 14.3; O_2, 25.6; N_2, 31.2; air, 29.9; CO, 30.7; CO_2, 34.0; SO_2, 44.8; NO, 23.6; N_2O, 36.4; NH_3, 25.8; H_2O, 18.9; H_2S, 32.9; Cl_2, 48.4; Br_2, 53.2; I_2, 71.5.
†After Reid.[19]

EXAMPLE 3-7. ESTIMATION OF DIFFUSION COEFFICIENT. Determine the diffusion coefficient for methanol in water at 25°C.

Solution. Methanol: CH_3OH
$T = 25 + 273.2 = 298.2$ K
Viscosity of water $= \mu = 0.89$ cP

Molal volume [from method of LeBas (Table 3-4)] is determined by adding individual increments to each atom:

$$C = 1 \times 14.8 = 14.8$$
$$H = 4 \times 3.7 = 14.8$$
$$O = 1 \times 7.4 = \underline{7.4}$$
$$V = 37.0 \text{ cm}^3/\text{mol}$$

$$D = \frac{5.06 \times 10^{-7} \times T}{\mu V^{0.6}}$$
$$= \frac{5.06 \times 10^{-7}(298.2 \text{ K})}{(0.89)(37.0)^{0.6}}$$
$$= \underline{\underline{1.942 \times 10^{-5} \text{ cm}^2/\text{s}}}$$

Partition Coefficients

Partition coefficients are empirically derived constants that describe how a chemical distributes itself between two phases. They are expressed as a concentration ratio with the assumption that simple dissolution is the only interaction between the two phases. The phases can be a solid and a liquid, two solids, two immiscible liquids, a gas and a liquid, etc. Partition coefficients are useful in describing the environmental fate of organic compounds.

Four of the coefficients important in waste management are the octanol-water partition coefficient, the soil-water partition coefficient, the organic carbon partition coefficient, and the vapor-liquid partition coefficient. Section 4-3 discusses these coefficients regarding sorption of organics in the subsurface.

The **octanol-water partition coefficient** is a dimensionless constant defined as the ratio of a chemical's concentration in the octanol phase to its concentration in the aqueous phase of a two-phase octanol-water system. This ratio is expressed by:

$$K_{\text{ow}} = C_o / C_w \tag{3-14}$$

where C_o = concentration of chemical in octanol phase (mg/L or μg/L)
C_w = concentration of chemical in aqueous phase (mg/L or μg/L)

Interest in the K_{ow} parameter first developed with the study of structure-activity relationships, primarily with pharmaceuticals. Numerous studies showed that K_{ow} was useful for correlating structural changes of drug chemicals with the change observed in some biological, biochemical, or toxic effect.

The octanol-water partition coefficient has become an important parameter in studies of the environmental fate of organic chemicals. The octanol-water partition coefficient is significant because it provides an indication of how much of a chemical will be taken up by aquatic organisms. The coefficient is also useful in estimating fate and transport of chemicals and is related to soil adsorption coefficients, the bioconcentration factor (BCF), and the water solubility. K_{ow} is widely used because of the small data requirements needed to calculate it.

The octanol-water partition coefficient is not the same as the ratio of a chemical's solubility in octanol to its solubility in water. K_{ow} is measured at room temperature (20 or 25°C) at low chemical solute concentrations. Typically, the chemical being measured is added to a mixture of octanol and water. Very pure water and octanol are used. The mixture is gently shaken in a flask until equilibrium is achieved (15 minutes to more than 1 hour, depending on the chemical). Centrifugation, high-pressure liquid chromatography, and other analytical techniques are then used to determine the concentration of the chemical in the octanol and water phases.[20]

Values for K_{ow} have been measured as low as 10^{-3} and as high as 10^7, encompassing a range of 10 orders of magnitude. Thus values for K_{ow} are often reported as log K_{ow}, ranging from -3 to 7. Chemicals with high values of K_{ow} tend to be hydrophobic and partition to organic matter, lipids (fatty tissue), and soil. Chemicals with low values of K_{ow} tend to be hydrophilic, have low sediment/soil adsorption, remain in the aqueous phase, and are generally more mobile in the environment.

The **soil-water partition coefficient** alternatively K_{sw} or K_p, a dimensionless constant, describes the tendency of a chemical to be adsorbed by soil (or sediment)

in an aqueous suspension. K_{sw} is defined as:

$$K_{sw} = X/C \qquad (3\text{-}15)$$

where $X =$ concentration of chemical in soil (ppb or μg/kg)
$C =$ concentration of chemical in water (ppb or μg/kg)

K_{sw} is essentially equivalent to K_d, the Freundlich adsorption coefficient (see Chapter 9 and Section 4-3).

As shown in Figure 3-1, two linear isotherms are plotted to represent partitioning of two organic chemicals between soil and water. The two isotherms could also represent different affinities of a single organic chemical for two different soil types. The isotherm with the steeper slope indicates the chemical is more strongly attracted to the soil, and therefore has a high K_d. Conversely, the chemical with the lower slope has a lower K_d and will tend to stay in solution.

Sorption of organic chemicals to soils usually occurs on clay- or silt-sized particles. Almost all of the sorption of organic chemicals by a soil is due to the organic carbon component of the soil, even though the organic component is typically a minor amount of the sediment mass. The **organic carbon partition coefficient** K_{oc} is a key environmental fate and transport parameter. K_{oc} is defined as:

$$K_{oc} = C_{soil}/C_{water} \qquad (3\text{-}16)$$

where $C_{soil} =$ concentration of chemical in organic carbon component of soil (μg adsorbed/kg organic C, or ppb)
$C_{water} =$ concentration of chemical in water (ppb or μg/kg)

Values for K_{oc} are shown in Appendix B. Values for K_{oc} may be estimated from several other physical-chemical properties as shown in Table 3-5. K_{sw} may be estimated from K_{oc}:

$$K_{sw} = K_{oc} f_{oc} \qquad (3\text{-}17)$$

where $f_{oc} =$ fraction of organic carbon in the soil (dimensionless).

The relationship between the soil-water partition coefficient and f_{oc} is illustrated in Figure 3-2. The partition coefficients of various benzene species (y axis) are plotted

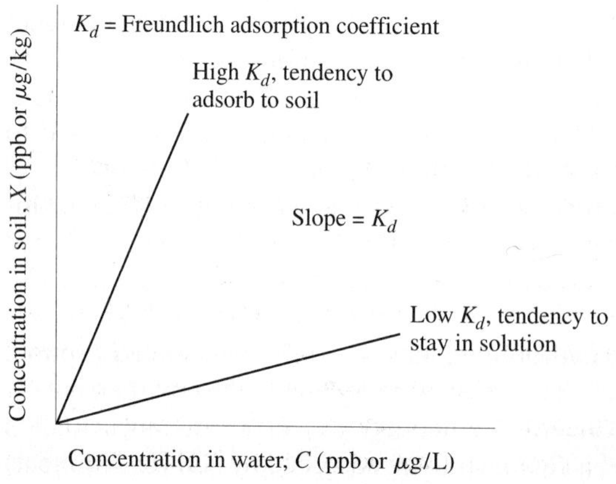

FIGURE 3-1
Distribution of two different organic chemicals between a soil phase and a water phase. (Modified from Deutsch, 1997.[21])

TABLE 3-5
Correlations between K_{oc} and chemical properties[23,24]

Class of chemicals	Number of chemicals	Equation	Remarks
Pesticides	45	$\log K_{oc} = 0.544 \log K_{ow} + 1.377$	
Aromatics	10	$\log K_{oc} = 1.00 \log K_{ow} - 0.21$	
Chlorinated hydrocarbons	15	$\log K_{oc} = -0.557 \log S + 4.277$	S in μmol/L
Aromatics	10	$\log K_{oc} = -0.54 \log S + 0.44$	S in mole fraction
Pesticides	106	$\log K_{oc} = -0.55 \log S + 3.64$	S in mg/L
Not stated	—	$\log K_{oc} = 0.681 \log \text{BCF} + 1.963$	BCF = bioconcentration factor

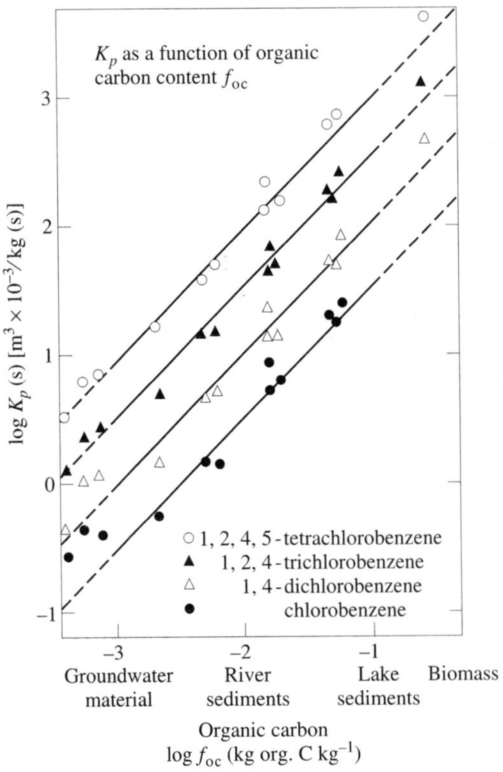

FIGURE 3-2
Relationship between organic carbon content of solid materials to soil-water partition coefficient for selected benzene compounds. *(After Stumms, 1992.*[22]*)*

as a function of their adsorption to solids with differing organic carbon content. The positive slope of the data on the graph indicates that the benzene species are more likely to adsorb to a soil with a greater organic carbon component.

K_{sw} and K_{oc} are commonly used in fate and transport modeling of specific chemicals in rivers, stormwater runoff, and soil/groundwater transport. Values for K_{sw} and K_{oc}, although relatively independent of soil type, may vary greatly because of influences by temperature, pH, particle size distribution and surface area, salinity, and the presence of dissolved organic matter or suspended particulate matter. Thus, care should be exercised in applying these parameters in fate and transport models.

The **vapor-liquid partition coefficient,** K_{vl}, is the ratio of the concentration of a compound in the vapor to the concentration in the liquid at equilibrium.

The formula for **vapor-liquid partition coefficient** may be defined as:

$$K_{\mathrm{vl}} = X_{\mathrm{eq,l}}/C_{\mathrm{eq,v}} \tag{3-18}$$

where $X_{\mathrm{eq,l}}$ = concentration in the liquid at equilibrium
$C_{\mathrm{eq,v}}$ = concentration in the vapor at equilibrium

This coefficient is a function of temperature, vapor pressure, atmospheric pressure, the composition of the liquid and vapor, and the specific compound. The lower the concentration of the compound, the greater the chances it will behave ideally, according to Raoult's law, or at least predictably, according to Henry's law.

Bioconcentration Factor

The **bioconcentration factor** (BCF) indicates the amount of a chemical that is likely to accumulate in aquatic organisms. It may be expressed mathematically as:

$$\mathrm{BCF} = C_{\mathrm{org}}/C \tag{3-19}$$

where C_{org} = equilibrium concentration in organism (mg/kg or ppm)
C = concentration in water (ppm)

The bioconcentration factor is an essential component in risk assessments. The lipid content of animal tissue is an important factor in determining the tendency of a chemical to bioconcentrate. The BCF varies from species to species. For example, fathead minnows tend to have higher values of BCF than trout.[25] Bioconcentration is also affected by metabolism and elimination systems in the test organism.

EXAMPLE 3-8. APPLICATION OF THE BIOCONCENTRATION FACTOR. The insecticide chlordane is found in lake water at a concentration approaching its solubility of 560 μg/L. Estimate the concentration in fish tissue.

Solution. From Appendix B: BCF (fish) = 14,000 L/kg.
From Equation 3-19:

$$\text{Concentration in fish} = 560\ (\mu\text{g/L}) \times 14{,}000\ (\text{L/kg}) = 7.84 \times 10^6\ \mu\text{g/kg}$$
$$= \underline{\underline{7840\ \text{mg/kg}}}$$

BCF is related to solubility, K_{ow} and K_{oc}, with K_{ow} being the most commonly utilized measure from which to estimate BCF. BCF should be distinguished from **biomagnification** which indicates an increase in the concentration of chemical through the food chain. BCF represents the tendency of a chemical to be absorbed by aquatic organisms, and therefore indicates the increase in concentration in the tissues of a single organism.

The BCF in fish is frequently correlated to K_{ow}:

$$\log \mathrm{BCF} = c_1 + c_2 \log K_{\mathrm{ow}} \tag{3-20}$$

where c_1, c_2 = empirical constants.

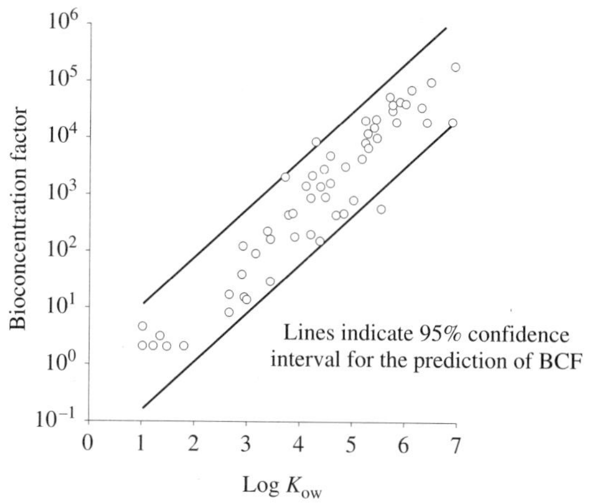

FIGURE 3-3
Relationship between the bioconcentration factor and K_{ow}.

Barron[25] indicates that the relationship between BCF and K_{ow} holds over a wide range of K_{ow}; it is not a very strong correlation over a narrow range. This is illustrated in the scatter of Figure 3-3. Banerjee and Baughman[26] indicate that this relationship does not hold for high K_{ow} compounds. Paustenbach[27] provides several examples of how BCF may be estimated from physical property data:

$$\log \text{BCF} = c_1 - c_2 \log S \tag{3-21}$$

where S = solubility.

$$\log \text{BCF} = c_1 \log K_{oc} - c_2 \tag{3-22}$$

It must be emphasized that relationships such as Equations 3-20 through 3-22 and Table 3-5 are valid only within the range of experimental conditions for which they were developed and cannot be used generally. Specifically, these relationships do not take into account many factors affecting BCF such as metabolism and excretion.

Sorption

Sorption is the process by which a component (the sorbate or contaminant) moves from one phase to another across some boundary.

Sorption is caused by interactions among three distinct molecules:

- Sorbent (e.g., activated carbon)
- Sorbate (e.g., the contaminant being removed)
- Solvent (e.g., water)

Driving forces behind sorption include the affinity of sorbate for sorbent:

- Electrical attraction
- Van der Waal forces
- Covalent bonds
- Hydrogen bonds

An additional driving force is the lyophobic behavior of sorbate (i.e., sorbates of lower solubility tend to sorb more easily).

In **absorption** the sorbate is taken into the sorbent (the sorbing phase). Weber et al.[28] suggest that the partitioning of a contaminant between the solvent (e.g., water) phase and the sorbent phase can be described in terms of the activity coefficient:

$$K = \frac{V_1 f_1}{V_2 f_2} \tag{3-23}$$

where K = phase partitioning coefficient for the specific solute
V_1, V_2 = molar volume for the solvent and sorbent phases respectively
f_1, f_2 = activity coefficient for the solvent and sorbent phases respectively

In **adsorption** the process takes place at a surface. A typical example of the adsorption process is activated carbon adsorption, discussed in Chapter 9.

Concentration

The concentration of a component is usually expressed in two terms: parts per million or milligrams per liter. There is a difference. The parts per million (ppm) of a component is the number of grams of that component in a million grams of a solution (or of a solid). Parts per million are used for trace amounts of chemicals. Weight percent, equivalent to any weight measure per 100 measures, can be used to express larger concentrations of a component. One weight percent of a component is equal to 10,000 ppm.

Chemists generally use molality rather than ppm as a measure of concentration because using ppm generally involves some approximation. Molality is defined as the number of moles of solute per 1000 grams of the solvent. It is the ratio of moles of solute to mass of solvent.

As an example, an aqueous solution that is 10^{-4} molal in calcium carbonate ($CaCO_3$) contains $0.0001 \times 100.0892 = 0.01001$ grams of $CaCO_3$ in $(1000 + 0.01001)$ grams of solution, or about 10^3 grams of solution.

Molality is expressed in parts per thousand (10^3 grams of solution), and the equivalent concentration of ppm (10^6 grams of solution), would be 10.01 grams.

In cases where data values in parts per million by volume (ppm_v) are desired, the conversion of vapor concentrations from μg/L to ppm_v can be achieved with the following equation:

$$C_{ppmv} = \frac{(C_{\mu g/l})(24.1)}{(mw)(P)} \tag{3-24}$$

where C_{ppmv} = vapor concentration in ppm_v
$C_{\mu g/L}$ = vapor concentration in μg/L
24.1 = molar volume at ambient room temperature (70°F) in (L · atm)/mol
mw = molecular weight in g/mol
P = pressure in atmospheres (typically assumed to be 1 atm)

For toluene, which has a molecular weight of 92.15, as an example: At ambient room temperature and 1 atm of pressure, 1 μg/L of toluene would be equivalent to 0.26 ppm_v.

3-3 ENERGY AND MASS BALANCES*

This section briefly presents thermodynamic properties and mass and energy balances that can be conducted to design treatment systems described elsewhere in this book. Additional information can be found in thermodynamics textbooks by Smith and Van Ness,[29] Sonntag and Van Wylen,[30] and Sandler.[31]

The notation for the thermodynamic properties presented in this section is as follows:

Capital letter alone, units of energy or volume (e.g., E_P = potential energy in J)
Lowercase letter, units of energy, mass, or volume per time (e.g., q = heat flow rate in J/s)
Capital letter with a prime symbol, units of energy per mass (e.g., E'_P = potential energy per unit mass in J/kg)
Capital letter with an asterisk, units of energy or volume per mol of substance (e.g., V^* = molar volume in m^3/gmol)

General Balance Equation

The **law of conservation of energy** states that, with the exception of nuclear reactions, energy can be neither created nor destroyed. Similarly, the **law of conservation of mass** states that, with the exception of nuclear reactions, mass can be neither created nor destroyed. These two principles are used in creating **energy and mass balances** to evaluate the energy requirements of processes and the conversion of reactants to products. In both laws, a **system** is defined as the region where the process takes place, while the **surroundings** are everything outside the system. An imaginary barrier around the system is referred to as the **system boundary.** An **open system** is one where mass is transferred from the system to its surroundings or vice versa. A **closed system** is one where mass does not enter or leave the system boundary, although energy can be transferred from or to the system. Both energy and mass balances are based on the following statement:

$$\text{Input} + \text{generation} - (\text{output} + \text{consumption}) = \text{accumulation} \qquad (3\text{-}25)$$

That is, the difference between the amount of a property that enters and is generated within the system and the amount that leaves and is consumed within the system is accumulated in the system.

Generalized Energy Balance

Energy can flow through a system boundary only as heat or work. **Heat** is energy that flows as a result of a temperature difference or a phase change. **Work** is energy that flows as a result of any other driving force, including a force, a torque, or a voltage. Heat and work are not contained within a system. The **first law of thermodynamics**

*Section 3-3 was prepared by Dr. Elsie Millano, ERM.

states that the energy lost by a system during a process must be transferred to the surroundings, as energy is neither destroyed nor created:

$$\text{Change in energy of system} - \text{change in energy of surroundings} = 0 \qquad (3\text{-}26)$$

Therefore, when energy disappears in one form, it appears in another. The total energy stored in a system is composed of internal, kinetic, and potential energies. The **internal energy** depends on the random motion of the molecules in a material in the system or on the chemical nature of the material. The **kinetic energy** is a result of the motion of the system. The **potential energy** is a result of the position of the system in a gravity field. The generalized expression for an energy balance, assuming that there is only one phase (i.e., gas, liquid, or solid) in the system, is:[32]

$$\frac{d}{dt}\left\{U + M\left(E'_{\text{kin}} + E'_{\text{pot}}\right)\right\} = \sum_{k=1}^{K} m_k \left(U' + E'_{\text{kin}} + E'_{\text{pot}}\right)_k + q - \left[w_s + P\frac{dV}{dt} - \sum_{k=1}^{K} m_k \left(PV'\right)_k\right] \qquad (3\text{-}27)$$

where
t = time (s)
U = internal energy (J)
M = system mass (kg)
E'_{kin} = kinetic energy per unit mass (J/kg)
E'_{pot} = potential energy per unit mass (J/kg)
K = number of entry ports or flow streams
m_k = mass flow rate of the kth flow stream (kg/s)
U' = internal energy per unit mass (J/kg)
q = rate of flow of heat into the system (J/s)
w_s = rate of flow of mechanical or electrical energy out of the system (J/s)
P = pressure exerted by the system at its boundaries, assumed to be uniform (Pa)
V = system volume (m^3)
V' = system volume per unit mass (m^3/kg)

The left term in Equation 3-27 is the change in the energy of the system. If there is more than one phase, the total energy is the sum of the energies of each phase. The first term on the right is the energy flow carried by each fluid element entering (positive) or leaving (negative) the system. The second term on the right, q, is the total rate of flow of heat into the system. The last three terms on the right represent the work done by the system: (1) the shaft work w_s, which is the rate of flow of mechanical or electrical energy out of the system that occurs without a deformation of the system boundaries; (2) the work resulting from the movement of the system boundaries, $P(dv/dt)$ (e.g., by expansion of the fluid); and (3) the work done by the system due to the movement of the fluid, represented by a change in pressure at constant volume. By convention, the flow of heat into the system is positive, while the work done on the system is negative.

This generalized equation can be simplified, depending on the system being evaluated and on whether the instantaneous rate of change or the total change are of importance. For example, for a closed system, $m_k = 0$ and the first and last terms on the right side of Equation 3-27 drop out. If the process is at steady state, the left-side term is equal to zero, and if no work is done by or on the system, the last three terms on the right side are zero.

EXAMPLE 3-9. SIMPLIFIED ENERGY BALANCE. Calculate the change in internal energy resulting when a liquid contained in a tank is stirred by a paddle wheel if the work input to the paddle wheel is W and the heat transferred from the tank is Q.

Solution. Consider the system to be the tank and the liquid. This system is closed because there is no flow of mass into or out of it. Also, there is no change in kinetic or potential energy, as the system is not moving and its elevation with respect to a reference point does not change.

For this system, $m_k = 0$, $dM/dt = 0$, $d/dt(E'_{\text{kin}} + E'_{\text{pot}}) = 0$, and the total change in internal energy, not the instantaneous rate change, is required. Because the work is done on the system (negative work) and the heat leaves the system (negative heat), Equation 3-27 reduces to:

$$\underline{\underline{\Delta U = +W - Q}}$$

Phase Equilibrium and Phase Rule

Thermodynamic properties are either intensive or extensive. **Intensive properties** are independent of the mass of material in the system. **Extensive properties** depend on the mass of material in the system. Extensive properties can be transformed into intensive properties by dividing by the total system mass. Mass and volume are extensive properties, while pressure, temperature, density, volume per unit mass (specific volume), and mass and mole fractions are intensive properties.

Intensive properties can be used to define systems at equilibrium. In thermodynamics, a system is in **equilibrium** if there is no change and no tendency of the properties of the system to change in a macroscopic scale. If a system in an **equilibrium state** returns to the equilibrium conditions after a small disturbance is removed, the system is defined as a **stable system. State properties** are defined as the intensive properties of a system in equilibrium. If a system deviates minimally from thermodynamic equilibrium during a process, such that all states through which the system passes can be considered equilibrium states, the process is called a quasi-equilibrium process, and the system is considered to have been in equilibrium throughout the process. For example, for the system shown in Figure 3-4, if the weights on the

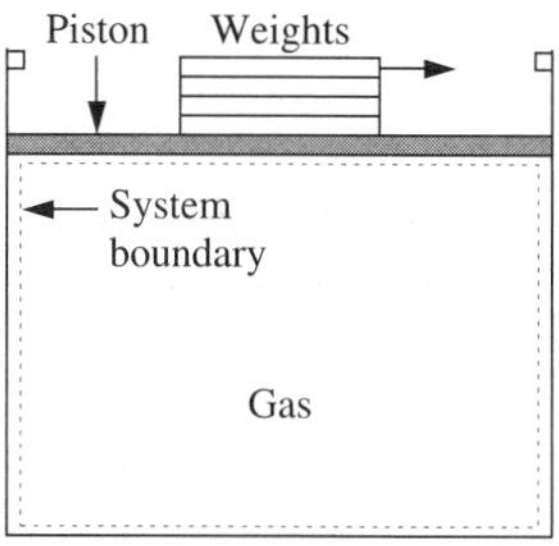

FIGURE 3-4
A system that may undergo a quasi-equilibrium process.[33]

piston are small and are taken off one by one, a quasi-equilibrium process will take place.

Intensive properties have been determined and summarized in tables for saturated and superheated pure substances.[34,35,36,37] The **saturation temperature** at a given pressure is the vaporization temperature. A **superheated** vapor has been heated beyond the saturation temperature. Example pages of a table for superheated steam in SI units are included in Appendix C.

To be able to duplicate an equilibrium condition, a certain number of intensive properties (i.e., temperature, pressure, or the composition of each phase) must be defined for a specific system, according to the **phase rule** for nonreacting systems[38]:

$$F = C + 2 - \phi \tag{3-28}$$

where F = degrees of freedom
C = number of components in the system
ϕ = number of phases in the system

EXAMPLE 3-10. USE OF THE PHASE RULE. Determine the degrees of freedom for each of the following systems in equilibrium: (1) pure liquid water; (2) liquid water and saturated vapor; and (3) a liquid solution of water and acetone and saturated vapor.

Solution

1. Pure liquid water: There is one component ($C = 1$) and one phase (liquid) ($\phi = 1$). Thus, $F = 1 + 2 - 1 = 2$. Therefore, for this system, two intensive properties must be given to define the equilibrium conditions.
2. Liquid water and saturated vapor: There is one component ($C = 1$) and two phases ($\phi = 2$). The value of F is 1. The tables with liquid-vapor equilibrium data require only one intensive property to define an equilibrium condition.
3. A liquid solution of water and acetone and saturated vapor: In this case, there are two components ($C = 2$) and two phases ($\phi = 2$). The number of degrees of freedom is two. To calculate the third intensive property, a relation between concentration and pressure must be used.

Ideal Gas Equation of State

There are equations relating each state variable to two others. One such equation is the **equation of state** for ideal gases. An **ideal gas** is defined as a gas of such low density that the interactions among its molecules are unimportant.[39] The equation of state for ideal gases relates the pressure, temperature, and molar specific volume of a system, as follows:

$$PV^* = RT \tag{3-29}$$

where P = pressure (Pa)
V^* = molar specific volume (m^3/gmol)
R = molar ideal gas law constant = 8.3144 $(\mathrm{Pa \cdot m^3})/(\mathrm{gmol \cdot K})$
T = absolute temperature (K)

P, V^*, and T are state properties. The ideal gas equation of state can also be expressed as $PV = NRT$ or $PV = MR'T$, where N and M are, respectively, the

number of moles and the mass of the substance in the system, and R' is the mass ideal gas law constant.

The **molar ideal gas law constant** R is also referred to as the **ideal gas law constant**, the **gas-law constant,** the **gas constant,** and the **universal gas constant.** Because it is given in units of moles, R is independent of the type of substance and can be used for any substance. Values of R with different units can be found in Appendix C. When the gas law constant is expressed in terms of mass units (designated R'), it is different for each compound. For example, the values of R' for ammonia and steam are 488 J/(kg · K) and 461.5 J/(kg · K), respectively.

EXAMPLE 3-11. USE OF THE IDEAL GAS EQUATION OF STATE. Calculate the temperature of 10 kg of air (molecular weight = 29) in a 10-m³ container, at a pressure of 100 kPa. Assume that air behaves as an ideal gas under these conditions.

Solution. The number of moles of air in the container is 10,000 g/29 = 345 gmol. Therefore, the molar specific volume of air in the container is 10 m³/345 gmol = 0.029 m³/gmol. Using the ideal gas equation of state:

$$T = \frac{PV^*}{R'} = \frac{100{,}000 \text{ Pa} \times (0.029 \text{ m}^3/\text{gmol})}{8.3144 \text{ (Pa} \cdot \text{m}^3)/(\text{gmol} \cdot \text{K})}$$

$$= \underline{\underline{348.8 \text{ K}}}$$

Enthalpy

Certain combinations of thermodynamic properties are frequently found when analyzing processes. One such combination is **enthalpy,** which is defined as:

$$H = U + PV \tag{3-30}$$

where H = enthalpy (J)
U = internal energy (J)

Specific enthalpy (i.e., enthalpy per unit mass) is tabulated in thermodynamics tables, as shown in Appendix C for superheated steam.

EXAMPLE 3-12. USE OF ENTHALPY DATA. Water at a flow rate of 1000 kg/h is evaporated in a boiler by addition of heat at a constant pressure of 100 kPa. The water temperature increases from the saturation temperature (100°C) to 300°C. The system is undergoing a quasi-equilibrium process at steady state (volume v remains constant with time). There are no heat losses to the surroundings, and the kinetic and potential energy changes between the entry and exit port are negligible. Calculate the heat transferred to the water.

Solution. The problem will be solved by using Equation 3-27. The system is the fluid inside the boiler. The term on the left side of Equation 3-27 is zero because the system is at steady state. Since the changes in kinetic and potential energies are negligible, the first term on the right reduces to $m(U'_{\text{in}} - U'_{\text{out}})$. Heat flows into the system and, therefore, q is positive. The term w_s is zero. The work resulting by expansion of the fluid at constant pressure is $P\,dV/dt = mP(V'_{\text{out}} - V'_{\text{in}})$, and is positive because it is done by the system (i.e., the fluid expands). Finally, the last term on the right-hand side of Equation 3-27, $\sum_{k=1}^{K} m_k(PV')_k$, is zero because there is no work being done by the system due to a change in pressure, as the pressure is constant. The terms remaining in Equation 3-27 are:

$$0 = m(U'_{\text{in}} - U'_{\text{out}}) + q - mP(V'_{\text{out}} - V'_{\text{in}})$$

Rearranging these terms,

$$\begin{aligned} q &= m\left[(U'_{\text{out}} - U'_{\text{in}}) + P(V'_{\text{out}} - V'_{\text{in}})\right] \\ &= m\left[(U'_{\text{out}} + PV'_{\text{out}}) - (U'_{\text{in}} + PV'_{\text{in}})\right] \\ &= m(H'_{\text{out}} - H'_{\text{in}}) \end{aligned}$$

From the superheated steam tables (Table C-2), the specific enthalpy (enthalpy per unit mass) for saturated water at 100 kPa and 100°C is 417.5 kJ/kg and for superheated steam at 100 kPa and 300°C, 3074.5 kJ/kg.

Therefore, the heat transferred to the water (positive heat) is:

$$q = (1000\ \text{kg/h})(3074.5\ \text{kJ/kg} - 417.5\ \text{kJ/kg}) = \underline{\underline{2.657 \times 10^6\ \text{kJ/h}}} = 2.52 \times 10^6\ \text{Btu/h}$$

Entropy

A process is **reversible** if it can be reversed such that no change is left in either the system or its surroundings. The process illustrated in Figure 3-5 is reversible if (1) the piston slides without friction; (2) heat is neither absorbed nor transmitted by the piston or the cylinder; and (3) the work done in taking off the weights is infinitesimal, for example, by having a platform for each weight level such that the transfer of the weight to the platform requires no work.

A **cycle** is a process in which a system undergoes changes of state or individual processes and at the end the system returns to the same initial state. For example, steam that is used to heat a substance, condensed, and then vaporized again to heat more substance undergoes a cycle.

Entropy is a state property defined as always having a positive rate of generation within a system, except when the system is at equilibrium, in which case the internal rate of generation is zero.[41] The entropy balance equation shown below (which can be used to evaluate processes) is a second, independent energy balance based on the **second law of thermodynamics,** which states that no device can operate in a cycle and produce only transfer of heat from a cool body to a hot body. The second law of thermodynamics is based on the fact that spontaneous processes occur only in the

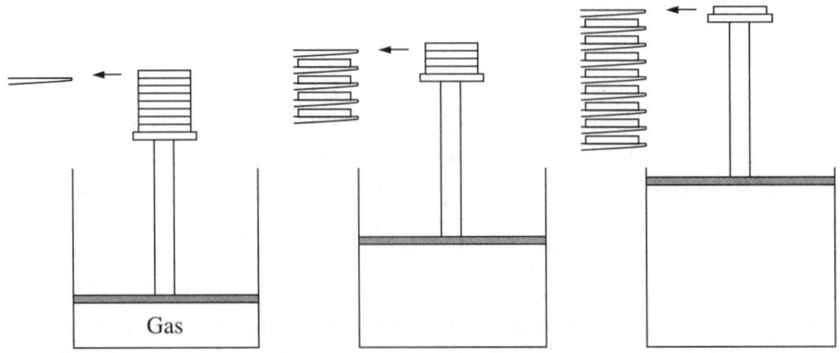

FIGURE 3-5
A system that may be considered to undergo a reversible process.[40]

direction that dissipates the gradients in a system, toward a state of equilibrium. The entropy balance equation is:

$$\frac{dS}{dt} = \sum_{k=1}^{K} m_k\ S'_k + \frac{q}{T} + s_{\text{gen}} \tag{3-31}$$

where $S =$ entropy of the system (J/K)
$m_k =$ mass flow rate of the kth flow stream (kg/s)
$S'_k =$ specific entropy of the kth flow stream [J/(kg · K)]
$s_{\text{gen}} =$ rate of generation of entropy within the system [J/(s · K)]
$q =$ rate of flow of heat into the system (J/s)

EXAMPLE 3-13. CALCULATION OF WORK BY USING ENTROPY VALUES. Steam enters a turbine at 3 MPa and 500°C and at a flow rate of 3 kg/s, expands in a reversible adiabatic (i.e., without gain or loss of heat) process, and exhausts at 300 kPa. Changes in kinetic and potential energy between the inlet and outlet of the turbine are negligible. Calculate the quantity of work produced in the process.

Solution. Assume the process is at steady state ($dS/dt = 0$) and that the system includes the turbine. In addition, from the problem definition, $q = 0$ (adiabatic process), and $s_{\text{gen}} = 0$ (reversible and adiabatic process).

From the entropy balance, Equation 3-31, $S'_{\text{in}} = S'_{\text{out}}$. The energy balance, Equation 3-27, becomes $w_s = m(H'_{\text{in}} - H'_{\text{out}})$, because $dM/dt = 0$ and $dU/dt = 0$ (steady-state process); $m[\Delta(E'_{\text{kin}} - E'_{\text{pot}})] = 0$; $q = 0$; $dV/dt = 0$ (constant turbine volume); and $H' = U' + PV'$ (there is a change in pressure).

According to the phase rule (Equation 3-28), for a system with one component and one phase, two state properties are needed to define the system. In the inlet, T and P are known, and the entropy and enthalpy can be found in the superheated steam tables. In the outlet, P and the entropy are known, since $S'_{\text{in}} = S'_{\text{out}}$. Thus, the enthalpy at the outlet conditions can also be found by using the superheated steam tables.

From Appendix C, for the steam entering the turbine, $H'_{\text{in}} = 3456.2$ kJ/kg and $S'_{\text{in}} = 7.2345$ kJ/(kg · K). With the value of $S'_{\text{out}} = 7.2345$ kJ/(kg · K), the value of enthalpy can be found in the superheated steam tables by interpolation.

For a pressure of 300 kPa, S'_{out} corresponds to a temperature between 175°C [$S' =$ 7.1990 kJ/(kg · K) and $H' = 2813.5$ kJ/kg] and 200°C [$S' = 7.3119$ kJ/(kg · K) and $H' = 2865.5$ kJ/kg]. The enthalpy at the outlet is, therefore:

$$\begin{aligned} H'_{\text{out}} &= 2865.5 - (2865.5 - 2813.5) \times [(7.3119 - 7.2345)/(7.3119 - 7.1990)] \\ &= 2829.9 \text{ kJ/kg} \end{aligned}$$

Finally, the work is calculated as:

$$w_s = (3 \text{ kg/s})(3456.2 \text{ kJ/kg} - 2829.9 \text{ kJ/kg}) = \underline{\underline{1878.9 \text{ kJ/s}}}$$

Because the work is positive, it is work done by the turbine.

Specific Heats

As mentioned before, several combinations of thermodynamic properties occur frequently in evaluating processes. Enthalpy has already been defined. Other state properties include the specific heat and the Gibbs free energy. The Gibbs free energy

will be defined later in this section. **Heat capacity** is defined as the quantity of heat a substance can absorb before a certain rise in temperature occurs:

$$C = dQ/dT \tag{3-32}$$

where C = heat capacity (J/K)
Q = flow of heat into the system (J)
T = absolute temperature (K)

Two forms of the heat capacity per mole, or **molar specific heat,** are state properties:

$$C_v^* = (\partial U^*/\partial T)_v \tag{3-33}$$

$$C_p^* = (\partial H^*/\partial T)_p \tag{3-34}$$

where C_v^* = specific heat at constant volume [J/(gmol · K)]
C_p^* = specific heat at constant pressure [J/(gmol · K)]

These definitions are based on an energy balance for a closed, steady-state system with no shaft work and occurring at constant volume ($q = N\ \Delta U^*$) or constant pressure ($q = N\ \Delta H^*$), respectively, where N = number of moles of the substance. However, because Equations 3-33 and 3-34 contain only state properties, the specific heats themselves are also state properties, and thus these equations can be used for any process. Partial derivatives are shown because the specific internal energy and specific enthalpy are functions of two variables, temperature and volume for U^* and temperature and pressure for H^*.

For the case of an ideal gas, where U^* and H^* are functions of temperature only, the specific heats are also functions of temperature only. For an ideal gas:

$$C_{v0}^*(T) = dU^*(T)/dT \tag{3-35}$$

$$C_{p0}^*(T) = dH^*(T)/dT \tag{3-36}$$

where $C_{v0}^*(T)$ = zero-pressure, constant-volume specific heat [J/(gmol · K)]
$C_{p0}^*(T)$ = zero-pressure, constant-pressure specific heat [J/(gmol · K)]

Equations 3-35 and 3-36 are based on the fact that gases approach ideal gas behavior as the pressure approaches zero. Formulas to calculate specific heats for several pure substances based on the temperature of the process are given in Table B-3.

EXAMPLE 3-14. USE OF SPECIFIC HEAT VALUE. Air (mol. wt.: 0.029 kg/gmol) enters a compressor at 300 kPa and 300 K at a flow rate of 500 kg/h and is discharged at an outlet pressure of 3000 kPa. The compression takes place adiabatically and at steady state. The work done on the system is 50 kJ/s. Assuming that air behaves as an ideal gas under these conditions, and that its constant-pressure specific heat, $C_{p0}^* =$ 29.3 J/(gmol · K), calculate the temperature of the gas at the outlet.

Solution. Assume that there is no change in the kinetic and potential energy of the system, which is the gas contained in the compressor. According to the system description, dU^*/dt and $dN/dt = 0$ (steady-state process), $q = 0$ (adiabatic system), and $w = w_s + P\ \Delta V$ (w_s is negative work, because it is done on the system). In addition,

$\Delta U^* + P\,\Delta V^* = \Delta H^* = C^*_{p0}(T_2 - T_1)$. Therefore, the energy balance, Equation 3-27, reduces to:

$$w_s = nC^*_{p0}(T_2 - T_1)$$

or

$$T_2 = T_1 + w_s/(nC^*_{p0})$$

where n = molar flow rate (mol/s).

$$T_2 = 300\text{ K} + \frac{(50\text{ kJ/s})(0.029\text{ kg/gmol})(3600\text{ s/h})}{[0.0293\text{ kJ/(gmol}\cdot\text{K)}](500\text{ kg/h})} = \underline{\underline{656\text{ K}}}$$

Latent Heats

Properties used to evaluate heat produced or absorbed during phase changes and reactions have been defined for a specific process (either a phase change or a reaction) carried out at constant pressure and temperature, in open, steady-state systems where no work is done, and where the kinetic and potential energies are negligible. From the energy balance, Equation 3-27, the heat produced or absorbed during such processes is equal to the change in enthalpy for each process. The **latent heat of vaporization** and **latent heat of fusion** are the phase change properties most commonly used:

$$\Delta H^*_{\text{vap}} = H^*_{\text{vap}} - H^*_{\text{liq}} \tag{3-37}$$

$$\Delta H^*_{\text{fus}} = H^*_{\text{liq}} - H^*_{\text{sol}} \tag{3-38}$$

where ΔH^*_{vap} = latent heat of vaporization (J/gmol)
H^*_{vap} = specific enthalpy of vapor at the temperature and pressure being evaluated (J/gmol)
H^*_{liq} = specific enthalpy of liquid at the temperature and pressure being evaluated (J/gmol)
ΔH^*_{fus} = latent heat of fusion (J/gmol)
H^*_{sol} = specific enthalpy of solid at the temperature and pressure being evaluated (J/gmol)

The latent heats for the reverse processes are defined as the negative of the latent heats listed above (i.e., $\Delta H^*_{\text{con}} = -\Delta H^*_{\text{vap}}$ where ΔH^*_{con} is the **latent heat of condensation**). The latent heat of vaporization and the latent heat of fusion for several substances at the normal boiling and melting point (i.e., at a pressure of 1 atm and the boiling or melting temperature of each substance) can be found in Table B-4. The latent heats at other temperatures can then be calculated by adding the changes in enthalpy required to (1) bring the substance from the process temperature to its normal temperature (boiling or melting point), (2) effect the phase change (i.e., the normal latent heat), and (3) bring the substance from its normal temperature back to the original process temperature.

EXAMPLE 3-15. CALCULATION OF HEAT OF VAPORIZATION. Calculate the heat to be added to an open, steady-state system where acetone enters as a liquid at a rate of 100 kg/h at 40°C and 101.315 kPa (1 atm) and exits as a vapor at 90°C. The pressure is held constant during the process and the kinetic and potential energy changes

are negligible. The following data are available from Appendix C: Molecular weight = 58.08, boiling point at 1 atm = 56°C, latent heat of vaporization at 56°C and 1 atm = 30.2 kJ/gmol, heat capacity of the liquid between −30°C and 60°C = $123.0 + 18.6 \times 10^{-2}T$ J/(gmol · °C) (T in °C), and heat capacity of the vapor between 0°C and 1200°C = $71.96 + (20.10 \times 10^{-2}T) - (12.78 \times 10^{-5}T^2) + (34.76 \times 10^{-9}T^3)$ J/(gmol · °C) (T in °C).

Solution. The energy balance, Equation 3-27, reduces to $q = \Delta H$, or

$$q = N(\Delta H^*_{40°-56°\text{C}} + \Delta H^*_{\text{vap}} + \Delta H^*_{56°-90°\text{C}})$$

where $\Delta H^*_{40°-56°\text{C}}$ = change in enthalpy when temperature is raised from 40° to 56°

N = number of moles of acetone entering the system = (100,000 g/h)/(58.08 g/gmol) = 1721.8 gmol/h

ΔH^*_{vap} = 30.2 kJ/gmol

From Equation 3-36, $H^* = \int C_p^*\, dT$

$$\Delta H^*_{40°-56°\text{C}} = \int_{40}^{56} (123.0 + 18.6 \times 10^{-2}T)\, dT = 123.0\,(56 - 40) + (18.6 \times 10^{-2})(56^2 - 40^2)/2 = 2100 \text{ J/gmol} = 2.1 \text{ kJ/gmol}$$

$$\Delta H^*_{56°-90°\text{C}} = \int_{56}^{90} (71.96 + 20.1 \times 10^{-2}T - 12.78 \times 10^{-5}T^2 + 34.76 \times 10^{-9}T^3)\, dT$$

$$= 71.96(90 - 56) + \frac{(20.1 \times 10^{-2})(90^2 - 56^2)}{2} - \frac{(12.78 \times 10^{-5})(90^3 - 56^3)}{3} + \frac{(34.76 \times 10^{-9})(90^4 - 56^4)}{4}$$

$$= 3000 \text{ J/gmol} = 3.0 \text{ kJ/gmol}$$

$$q = 1721.8 \text{ gmol/h} \times (30.2 + 2.1 + 3.0) \text{ kJ/gmol} = \underline{\underline{60{,}800 \text{ kJ/h}}}$$

Heats of Reaction, Formation, and Combustion

Additional state properties defined by using enthalpy values include the heats of reaction, formation, and combustion. These properties are used to evaluate the heat produced (negative value, exothermic reaction) or absorbed (positive value, endothermic reaction) in processes where chemical reactions take place under the same system conditions specified above for defining the latent heats.

The **heat of reaction** is, therefore, defined as the change of enthalpy resulting from a reaction occurring at constant pressure in an open, steady-state system, where no work is done, the kinetic and potential energies are negligible, and the reactants and products are at the same temperature. For a similar system, the **heat of formation** of a substance is the change of enthalpy resulting from the reaction that forms that substance from its constituent elements, and the **heat of combustion** of a substance is the change of enthalpy resulting from the combustion of that substance. **Combustion** is defined as the reaction of a substance with oxygen to produce the following set of

specified substances: gaseous carbon dioxide, liquid or gaseous water, and gaseous sulfur dioxide and nitrogen (N_2) if the original substance had any sulfur or nitrogen in it.

The heat of formation and the heat of combustion have been tabulated at the **standard states** (see Appendix C). Standard states have been defined for convenience, to allow definition of only a set of properties (specifically, enthalpy and Gibbs free energy of formation) at specific temperature and pressure that can then be used for calculation of the heat produced or absorbed during reactions at different temperatures and pressures.

To tabulate the heats of reaction, formation, and combustion, the standard state was defined as a pressure of 101.325 kPa (1 atm) and a temperature of 298 K (25°C). For gases, the standard state is the ideal-gas state. In the case of the heat of formation, a reference state, at which the value of enthalpy is zero, was defined to allow calculation of the change of enthalpy resulting from a formation reaction. For any atomic species, the reference state is the thermodynamically stable form at 1 atm and 25°C. For example, the thermodynamically stable form of oxygen is O_2, and not O (atomic oxygen) or O_3 (ozone). The following equation can be used to calculate the heat of reaction from the standard heats of formation[42]:

$$\Delta H_r^*(T, P_0) = \sum_p \nu_p \, \Delta H_{f,p}^{*\circ}(T_0, P_0) - \sum_r \nu_r \, \Delta H_{f,r}^{*\circ}(T_0, P_0) + \sum_i \nu_i \int_{T_0}^{T} C_{P_0,i}^{*\circ} \, dT \tag{3-39}$$

where
$\Delta H_r^*(T, P_0)$ = heat of reaction at the temperature T and 1 atm (kJ/gmol)
T = reaction temperature (°C)
P_0 = standard pressure, 101.315 kPa (1 atm)
ν_p, ν_r, and ν_i = stoichiometric coefficient of product p, reactant r, and substance i in the reaction, respectively (dimensionless)
$\Delta H_{f,p}^{*\circ}(T_0, P_0)$ = standard heat of formation of a product of the reaction at 25°C and 1 atm (kJ/gmol)
T_0 = standard temperature, 25°C
$\Delta H_{f,r}^{*\circ}(T_0, P_0)$ = standard heat of formation of a reactant of the reaction at 25°C and 1 atm (kJ/gmol)
$C_{P0,i}^{*\circ}$ = constant-pressure specific heat of substance i at 1 atm [kJ/(gmol · K)]

If only the standard heats of combustion are available, the heat of reaction can be calculated from the standard heats of combustion by using an equation similar to Equation 3-39, except that the heats of combustion of the reactants are positive and the heats of combustion of the products are negative.

EXAMPLE 3-16. CALCULATION OF STANDARD HEAT OF REACTION FROM STANDARD HEATS OF FORMATION. 50 kg/h of gaseous ammonia is oxidized in a process following the reaction: $4NH_3\ (g) + 5O_2\ (g) \rightarrow 4NO\ (g) + 6H_2O\ (g)$. Calculate the standard heat of

reaction for this process, given the following data: $\Delta H_f^{*\circ}[NH_3\ (g)] = -46.19$ kJ/gmol; $\Delta H_f^{*\circ}[O_2\ (g)] = 0$; $\Delta H_f^{*\circ}[NO\ (g)] = +90.37$ kJ/gmol; and $\Delta H_f^{*\circ}[H_2O\ (g)] = -241.83$ kJ/gmol.

Solution. The molar standard heat of reaction can be calculated from Equation 3-39, by setting the last term on the right-hand side to zero (i.e., $T = T_0$):

$$\begin{aligned} \Delta H_r^{*\circ} &= 4\{\Delta H_f^{*\circ}[NO\ (g)]\} + 6\{\Delta H_f^{*\circ}[H_2O\ (g)]\} \\ &\quad - 4\{\Lambda H_f^{*\circ}[NH_3(g)]\} - 5\{\Delta H_f^{*\circ}[O_2\ (g)]\} \\ &= 4(90.37) + 6(-241.83) - 4(-46.19) - 5(0) \\ &= -904.74 \text{ kJ/gmol} \end{aligned}$$

This is the molar standard heat of reaction for every 4 moles of ammonia that react. The number of moles of ammonia is $n = (50{,}000 \text{ g/h})/(17 \text{ g/gmol}) = 2941$ gmol/h, and the total standard heat of reaction for the process is:

$$\Delta H_r^\circ = n(\Delta H_r^{*\circ})/4 = 2941 \times (-907.74)/4 = \underline{\underline{-6.7 \times 10^5 \text{ kJ/h}}}$$

Because the change of enthalpy and, thus, the heat of reaction, are negative, the reaction is exothermic and heat is transferred from the system to its surroundings.

Gibbs Free Energy and Chemical Equilibrium

A frequently used state property is the **Gibbs free energy,** defined as:

$$G = H - TS \tag{3-40}$$

where $G =$ Gibbs free energy (J).

The Gibbs free energy is a property of importance in the evaluation of the thermodynamic equilibrium of systems where chemical reactions take place. For a closed, constant-pressure, constant-temperature system:

$$\Delta G^* = \Delta G_f^{*\circ} + RT \ln K \tag{3-41}$$

where $\Delta G_f^{*\circ} =$ Gibbs free energy of formation at the standard state (kJ/gmol)
$K =$ equilibrium constant (dimensionless)

A negative value of the Gibbs free energy indicates that there is a potential for the reaction to occur as written, while a positive value indicates that the opposite reaction is most likely to occur. The same reference state defined for the enthalpy of formation is used to define the zero Gibbs free energy of formation to allow calculation of the change in Gibbs free energy resulting from a formation reaction. Values of the Gibbs free energy of formation at the standard states can be found in Appendix B.

The Gibbs free energy for a reaction at any temperature can be calculated from the standard state value as:

$$\Delta G_T^{*\circ} = \sum_p \nu_p G_{T,p}^{*\circ} - \sum_r \nu_r G_{T,r}^{*\circ} \tag{3-42}$$

and, for each product p or reactant r,

$$G_T^{*o} = G_f^{*o} + \left(H_T^{*o} - H_{T0}^{*o}\right) - \left(T S_T^{*o} - T_0 S_{T0}^{*o}\right) \tag{3-43}$$

A combination of Equation 3-27 and 3-31 for this system at equilibrium gives $dG^*/dt = 0$, or

$$\Delta G^* = 0 \tag{3-44}$$

Therefore, at equilibrium, Equation 3-38 becomes:

$$\ln K = -\frac{\Delta G_f^{*\circ}}{RT} \tag{3-45}$$

For a gaseous reaction such as $\nu_a A + \nu_b B \rightleftharpoons \nu_c C + \nu_d D$, the equilibrium constant relates the molar ratios of the components by the equation:

$$K = \frac{(Y_c)^{\nu_c}\ (Y_d)^{\nu_d}}{(Y_a)^{\nu_a}\ (Y_b)^{\nu_b}} \left(\frac{P}{P_o}\right)^{\nu_c + \nu_d - \nu_a - \nu_b} \tag{3-46}$$

where $(Y_i)^{\nu_i}$ = mole fraction of component i in the gas (mol_i/total moles).

The equilibrium constants for several gaseous reactions have been tabulated. However, for chemical reactions in nonideal liquid solutions, the actual equilibrium concentrations are difficult to predict by using the thermodynamic equilibrium constant K. Therefore, in the case of nonideal liquid solutions, the approach is to determine experimentally the molar concentration equilibrium constant K_c (which uses the molar concentration of each constituent and thus is dependent on the molar concentrations of reactants and solvents) for a certain reaction at a specific temperature and then use it to determine the equilibrium concentrations under a different set of initial concentrations. An example of the calculation of equilibrium concentrations in a gaseous reaction by using the Gibbs free energy is presented below in the section Mass Balances on Reacting Systems.

Generalized Mass Balance Equation

Mass balances are based on the law of conservation of mass, defined above under General Balance Equation, which is represented by Equation 3-25:

$$\text{Input} + \text{generation} - (\text{output} + \text{consumption}) = \text{accumulation} \tag{3-25}$$

This equation can be used for any material that enters or leaves the system being evaluated, including compounds, molecules, and atoms. There are three different types of processes that can be evaluated: batch, semibatch, and continuous processes. A **batch process** is one where the constituents are added to the system at the beginning of the process, left in the system for a specified period of time, and then removed from the system. A **continuous process** is one where there is a continuous flow of constituents into and out of the system. A **semibatch process** is one where one or more constituents are continuously added or removed and one or more constituents are added at the beginning of the process and removed at the end of the process (i.e., a hybrid between a batch and a continuous process).

In addition, processes may be operated under steady-state or under transient conditions. The **steady-state** process variables (such as temperature, pressure, flow rates, concentrations, and extent of reaction) do not change appreciably with time, except for minor fluctuations about constant mean values. In a **transient or unsteady-state process,** one or more process variables change with time.

Finally, mass balances can be written as **integral balances** (i.e., determination of conditions over a finite period of time) or as **differential balances** (i.e., determination of instantaneous conditions). In integral balances, each term of the equation is given as a mass, whereas in differential balances, units are mass per unit time. In general, batch processes are evaluated with integral balances and continuous processes are evaluated with differential balances.

Procedure for Conducting a Mass Balance

In conducting a mass balance, the system and its components must be well defined (i.e., in terms of what constitutes the system, how many variables there are, and how many equations can be written to calculate the unknown variables). To avoid going through a lengthy calculation only to find that the system is undefined (i.e., the data provided are not enough to define the system), the problem must be organized to permit an evaluation of its degrees of freedom. The **degrees of freedom** of a system are the minimum number of variables that must be given to be able to define the system, and are calculated as[43]:

$$\mathrm{DF} = \mathrm{NV} - \mathrm{NE} \tag{3-47}$$

where DF = degrees of freedom of a system
NV = number of variables
NE = number of independent equations relating the variables

In general, to calculate all of the variables in a process, mass and energy balances must be combined. If $\mathrm{DF} > 0$, then DF variables must be defined before the rest of the variables can be calculated. If $\mathrm{DF} < 0$, the problem is undefined, and either all of the equations counted are not independent, all of the variables were not counted, or additional information is necessary. The procedure to perform a mass balance is as follows[44]:

1. Draw a flowchart of the process, with all of the variables shown. Include the values known from the definition of the problem, and assign names to each unknown variable.
2. Write equations relating the unknown variables, including:
   - Mass balances, which for a system with NV variables can be NV equations if there are no reactions and NV minus NE (the number of independent chemical reactions) if reactions occur.
   - Energy balances.
   - Problem specifications, such as relations between the variables (e.g., a definition that the outlet mass of a component is a fraction of its inlet mass).
   - Physical properties and laws, such as mass-volume relations (specific gravity), concentration-pressure relations (gas laws), or saturation or equilibrium conditions.

- Physical constraints, such as the relation between mole fractions (e.g., the sum of the mole fractions in a stream equals 1).

3. Check the degrees of freedom. If DF variables are specified (where DF > 0), proceed to solve the problem. If not, evaluate the flowchart and the problem definition again to determine what is wrong and then make the necessary corrections.
4. Convert all given information into the same unit basis, either mass or moles, and the same flow rate basis, if necessary.
5. Solve the equations.

Mass Balances on Nonreacting Systems

In the absence of reactions, the generation and consumption terms are zero and the general mass balance equation for nonreacting systems becomes:

$$\sum_{k_{ii}} \text{Input} - \sum_{k_{oi}} \text{output} = \text{accumulation} \tag{3-48}$$

where k_{ii}, k_{oi} = streams with component i entering or leaving the system.

For batch processes and steady-state continuous processes, accumulation = 0, and the general mass balance equation for nonreacting systems can be simplified as:

$$\sum_{k_{ii}} \text{Input} = \sum_{k_{io}} \text{output} \tag{3-49}$$

EXAMPLE 3-17. MASS BALANCE ON A NONREACTING SYSTEM. 50,000 L/day of groundwater contaminated with 5000 mg/L of benzene and 10,000 mg/L of trichloroethylene (TCE) are fed to a system where air is being injected at a rate of 100,000 L/day. The influent air has no organic contaminants and does not react inside the system (i.e., no air dissolves in the water and no water is picked up by the air). The process takes place at 25°C, the same temperature of the influent groundwater and air. The benzene concentration in the effluent groundwater is 1000 mg/L and the removal efficiency of TCE is 60 percent of the removal efficiency of benzene. Assume that the flow of groundwater and air does not change as a result of the process, and calculate the mass of benzene and TCE in the effluent air and groundwater streams.

A flowchart and mass balance is developed by noting that $M = QC$.

$$M = \text{mass (mg/day)} \; (\times 10^{-6} = \text{kg/day})$$

$$Q = \text{flow (L/day)}$$

$$C = \text{concentration (mg/L)}$$

The following subscripts are used:

w = water
b = benzene
t = TCE
a = air
1 = influent
2 = effluent to water
3 = effluent to air

The resulting flowchart is shown below.

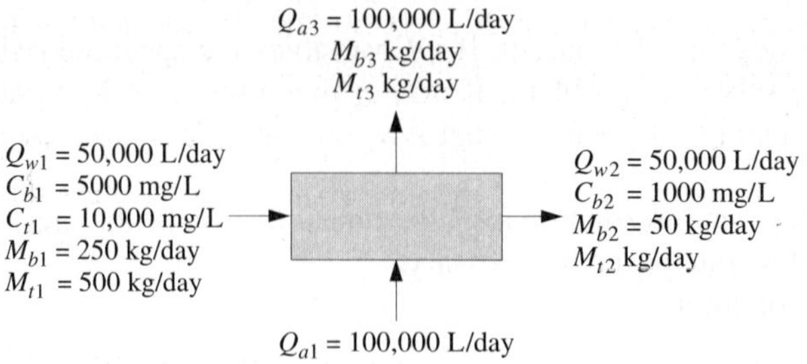

Solution. The problem has 13 variables: influent and effluent groundwater and airflow rates, influent and effluent groundwater benzene concentrations, influent TCE groundwater concentration (the problem does not require the definition of the effluent TCE concentration, but only the mass), influent and effluent benzene and TCE masses in the groundwater, and effluent benzene and TCE masses in the air (the influent air has no organics and, therefore, there is no need to define influent benzene and TCE mass variables in the influent air).

There are 8 relations among the variables: the effluent groundwater and airflow rates are equal to the influent groundwater and airflow rates, respectively; the groundwater influent and effluent benzene masses and the influent TCE mass are equal to the respective concentration multiplied by the groundwater flow rate; the masses of benzene and TCE in the effluent air are equal to the groundwater influent minus the groundwater effluent masses; and the efficiency of removal of TCE from the groundwater is 60 percent of the efficiency of removal of benzene from the groundwater.

Therefore, the problem has 5 degrees of freedom, and, in fact, 5 variables are given: the influent groundwater and airflow rates, the influent and effluent benzene concentrations, and the influent TCE concentration.

The previous flowchart already used some of the relations among variables given. The remaining relations are:

$$(M_{t1} - M_{t2})/M_{t1} = 0.6(M_{b1} - M_{b2})/M_{b1}$$
$$M_{b3} = M_{b1} - M_{b2}$$
$$M_{t3} = M_{t1} - M_{t2}$$

The subscripts b and t refer, respectively, to benzene and TCE. Solving these three equations gives the results:

$$\underline{\underline{M_{t2} = 260 \text{ kg/day}}} \qquad \underline{\underline{M_{b3} = 200 \text{ kg/day}}} \qquad \underline{\underline{M_{t3} = 240 \text{ kg/day}}}$$

Mass Balances on Reacting Systems

In reacting systems, the mass balance equation takes the form of Equation 3-25. The evaluation of transient or unsteady-state processes is presented in Section 3-4, Reactions and Reactors. For steady-state continuous-flow processes (accumulation = 0), the mass balance reduces to:

$$\sum_{k_{ii}} \text{Input} + \sum_{k_{ig}} \text{generation} = \sum_{k_{io}} \text{output} + \sum_{k_{ic}} \text{consumption} \tag{3-50}$$

where k_{ig}, k_{ic} = number of reactions where component i is generated or consumed.

Reactants are present in **stoichiometric proportions** if they are present in the same proportions shown in the balanced reaction equation. A **limiting reactant** is present at concentrations lower than the stoichiometric proportion with respect to the rest of the reactants, which are then **excess reactants.** For example, for the reaction $2A + B \rightarrow 3C + D$, if the influent has 2 moles of A per mole of B, the reactants are present in stoichiometric proportions. However, if the influent has 1 mole of A per mole of B, A becomes a limiting reactant and B an excess reactant. For any constituent:

$$n_{io} = n_{ii} + \beta \text{ER} \tag{3-51}$$

where n_{io}, n_{ii} = number of moles of constituent in the effluent and influent, respectively

$\beta = \nu$, if i is a product; $-\nu$, if i is a reactant; and 0, if i is inert

ν = stoichiometric coefficient of substance

ER = number of moles of the limiting reactant that react
$= f_{\text{lr}} \times n_{\text{lr}i}$ (where the subscript lr = limiting reactant)

f_{lr} = fractional conversion of the limiting reactant (moles reacted/moles fed)

The above method works only if the limiting reactant has $\upsilon = 1$. Otherwise, ER cannot be used in this manner.

EXAMPLE 3-18. MASS BALANCE ON A REACTING PROCESS. Ethylene is oxidized to produce ethylene oxide in accordance with the following reaction: $2C_2H_4 + O_2 \rightarrow 2C_2H_4O$. The feed to the reactor contains 100 kmol/h of ethylene and 200 kmol/h of molecular oxygen. The reaction proceeds to a fractional conversion of the limiting reactant of 40 percent. Determine the molar flow rates of all of the product gas constituents if the reaction proceeds under steady-state conditions.

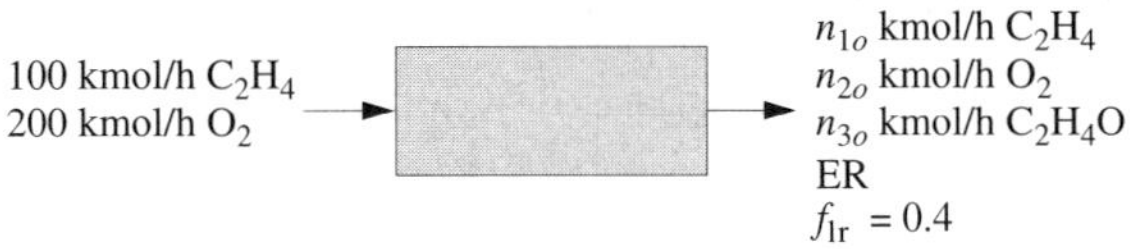

Solution. The problem has 7 variables and 4 relations [i.e., Equation 3-49 for each reactant and product, and the relation between the fractional conversion and ER (ER = moles fed × conversion)]. The degrees of freedom are 3, and 3 variables are given. Therefore, the problem is defined.

The equations are:

$$n_{1o} = 100 - \text{ER}$$
$$n_{2o} = 200 - \text{ER}/2$$
$$n_{3o} = 0 + \text{ER}$$
$$\text{ER} = f_{\text{lr}} \times n_{\text{lr}i}$$

For this problem, ethylene is the limiting reactant, because its ratio with respect to molecular oxygen in the feed is 0.5 and in the reaction is 2.0. Therefore, $n_{\text{lr}i} =$ 100 kmol/h. Solving the equations results in:

$\underline{\underline{\text{ER} = 40 \text{ kmol/h}}}$ $\underline{\underline{n_{1o} = 60 \text{ kmol/h}}}$ $\underline{\underline{n_{2o} = 180 \text{ kmol/h}}}$ $\underline{\underline{n_{3o} = 40 \text{ kmol/h}}}$

The next example illustrates the use of the Gibbs free energy (Equation 3.40) to calculate: (1) the equilibrium constant for a reaction occurring in the gaseous state and (2) the equilibrium concentrations based on the equilibrium constant.

EXAMPLE 3-19. USE OF THE GIBBS FREE ENERGY TO CALCULATE EQUILIBRIUM CONSTANT AND CONCENTRATIONS. Calculate the equilibrium constant K for the reaction $2CO_2 \rightleftharpoons 2CO + O_2$ at $T = 2000$ K. All constituents are gases. If 100 kmol/h of CO_2 enter a system where this reaction takes place essentially to equilibrium at $P = 1$ atm and under steady-state conditions, determine the equilibrium molar flow rate of all of the constituents in the effluent gas. The following data are given for the problem:

Component	$H^{*\circ}_{2000} - H^{*\circ}_{298}$	$S^{*\circ}_{2000}$	S°_{298}	$G^{*\circ}_{f}$
CO (g)	56,743	258.716	197.651	−137,163
O_2 (g)	59,176	268.748	205.148	0
CO_2 (g)	91,439	309.294	213.794	−394,389

The units of the values given in the table are: kJ/kmol for enthalpy, kJ/kmol · K for entropy, and kJ/kmol for Gibbs free energy of formation.

Solution. From Equations 3-42 and 3-43, the Gibbs free energy for the reaction can be calculated as:

$$\Delta G^{\circ}_{2000} = 2G^{*\circ}_{2000}(CO) + G^{*\circ}_{2000}(O_2) - 2G^{*\circ}_{2000}(CO_2)$$

$$G^{*\circ}_{2000} = G^{*\circ}_{f} + (H^{*\circ}_{2000} - H^{*\circ}_{298}) - (2000S^{*\circ}_{2000} - 298.15S^{*\circ}_{298})$$

Using the data given for the Gibbs free energy of formation, the difference in standard enthalpy, and the absolute entropy for each compound, the Gibbs free energy of each substance is calculated, and the Gibbs free energy for the reaction is:

$$\Delta G^{\circ}_{2000} = 2(-538{,}922) + (-417{,}155) - 2(-857{,}795) = 220{,}591 \text{ kJ/kmol}$$

Equation 3-45 is then used to calculate K:

$$(\ln K)_{2000} = \frac{-\Delta G^{*\circ}_{2000}}{RT} = \frac{-220{,}591}{(8.3144)\,(2000)} = -13.27$$

or $K = 1.73 \times 10^{-6}$

To determine the equilibrium molar flow rate of each constituent, a flowchart of the process is drawn as follows:

The molar ratios of the reactants and products are defined as:

$$Y_{1o} = n_{1o}/n_{to} \qquad Y_{2o} = n_{2o}/n_{to} \qquad Y_{3o} = n_{3o}/n_{to}$$

where $n_{to} = n_{1o} + n_{2o} + n_{3o}$. From Equation 3-47,

$$K = (Y_{2o})^2(Y_{3o})/(Y_{1o})^2$$

or

$$\ln K = 2\ln Y_{2o} + \ln Y_{3o} - 2\ln Y_{1o}$$

The values n_{io} can be defined in terms of the extent of reaction as:

$$n_{1o} = 100 - \text{ER} \qquad n_{2o} = 0 + \text{ER} \qquad n_{3o} = 0 + \text{ER}/2$$

Thus, $n_{to} = 100 + \text{ER}/2$ and the equilibrium constant equation becomes:

$$\ln K = 2 \ln [\text{ER}/(100 + \text{ER}/2)] + \ln [(\text{ER}/2)/(100 + \text{ER}/2)] - 2 \ln [(100 - \text{ER})/(100 + \text{ER}/2)]$$

By trial and error, ER = 1.502 kmol/h (i.e., at the indicated temperature, the reaction shifts to the left, and most of the carbon dioxide remains as such in the gas). Therefore, the molar concentrations of the components in the effluent gas are:

$$n_{1o} = 98.498 \text{ kmol/h} \qquad n_{2o} = 1.502 \text{ kmol/h} \qquad n_{3o} = 0.751 \text{ kmol/h}$$

3-4 REACTIONS AND REACTORS

Rate of Reaction

The equilibrium constant (Equation 3-46) provides no information on how fast a reaction will take place, or how much of the reactants will become products (on a finite amount of time). For a simple irreversible chemical reaction in a constant-volume system:

$$A + B \rightarrow AB \tag{3-52}$$

The **reaction rate constant** k may be defined for a constant-volume system as:

$$\begin{aligned} \text{Rate} &= -dC_A/dt \\ &= -dC_B/dt = dC_{AB}/dt = kC_AC_B \end{aligned} \tag{3-53}$$

where C_A, C_B = concentration of reactants
C_{AB} = concentration of product

For many reactions of interest in the environment, the reaction rate approximately doubles for every 10°C increase in temperature. Another relationship of importance is the **Arrhenius equation:**

$$k = Ae^{-E/RT} \tag{3-54}$$

where A = frequency factor
E = activation energy (J/mol)
R = universal gas constant [8.314 J/(mol · K)]
T = temperature (K)

The activation energy represents an energy barrier that must be overcome, if the reaction is to occur. The higher the activation energy, the lower the reaction rate.

Order of Reaction

The reaction rate also depends on the concentration of the chemical species involved. The **order of a reaction** refers to the powers to which concentrations are raised. Suppose A and B are possible reactants and the rate is defined:

$$\text{Rate} = k \times C_A^n \times C_B^m \tag{3-55}$$

The reaction is termed nth order with respect to A, and mth order with respect to B. Typical reaction orders are illustrated as follows:

$\text{Rate} = dC_A/dt = k$	Zero order
$dC_A/dt = -kC_A$	First order
$dC_A/dt = -kC_AC_B$	Second order
$dC_A/dt = -kC_A^2$	Second order

The rate constant for a first-order reaction may be determined by plotting $\ln C$ against time. For a second-order reaction, plot $1/C$ against time.

EXAMPLE 3-20. DETERMINING ORDER OF REACTION AND RATE CONSTANT. Analyze the following reaction data:

Time, min	0	12	24	36	48	60
C, mg/L	7.5	5.25	3.68	2.48	1.73	1.13

Solution. Calculate the logarithm and inverse of the concentration values:

$\ln C$	2.01	1.66	1.30	0.91	0.55	0.12
$1/C$	0.13	0.19	0.27	0.40	0.58	0.88

Plots of $\ln C$ and $1/C$ versus time are in Figure 3-6. The first plot is almost a straight line whereas the second is curved upwards, indicating that the data appear to fit a first-order model. The slope of this line provides an estimate of k:

$$k = -(0.12 - 2.01)/(60 - 0) = \underline{\underline{0.032\ \text{min}^{-1}}}$$

Regression analysis is recommended when significant scatter exists in the data.

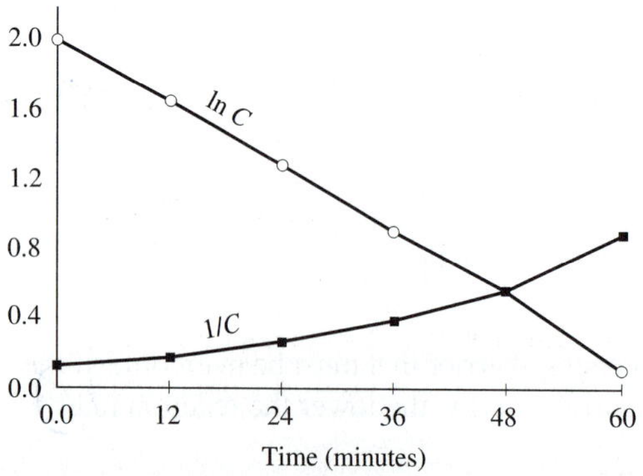

FIGURE 3-6
First- and second-order plots for Example 3-20.

Reactors

Chemical reactors are designated **continuous flow** or **batch** reactors depending on whether a fluid stream continuously flows through the reactor. For hazardous waste applications, the selection is generally based on the nature of the problem being solved. A batch reactor would be appropriate for a one-time cleanup or for intermittent operations, whereas a continuous flow reactor would be more cost-effective to treat a waste that is generated continuously.

The most commonly used reactor is the **continuous flow stirred-tank reactor** (CFSTR). This is a **completely mixed** reactor such that the concentration of materials in the reactor equals the concentration in the effluent. However, some of the particles entering the tank will exit immediately, while others will remain in the reactor for a very long time. This variation in particle detention (or residence) may result in effluent concentrations that do not meet environmental standards.

Alternatively, a **plug flow reactor** (PFR) is one in which mixing of the influent with the entire contents of the tank is kept to a minimum. This is often accomplished by using long narrow tanks. The actual detention time of all particles is equal to the theoretical detention time for a perfect plug with no diffusion/dispersion (volume divided by flow rate).

The impact the flow regime has on various inputs can be seen through dye tracer studies. The theoretical output (dye concentration) from a PFR and a CFSTR to step and impulse dye feeds is shown in Figure 3-7. The step input models the startup of a continuous operation, whereas the impulse can be conceived of as a batch dump of material into the reactor influent.

The effluent from a CFSTR may be analyzed by a mass balance as follows. Assuming no reaction takes place and a constant volume reactor:

$$V\,dC/dt = QC_{\text{in}} - QC_{\text{out}} \tag{3-56}$$

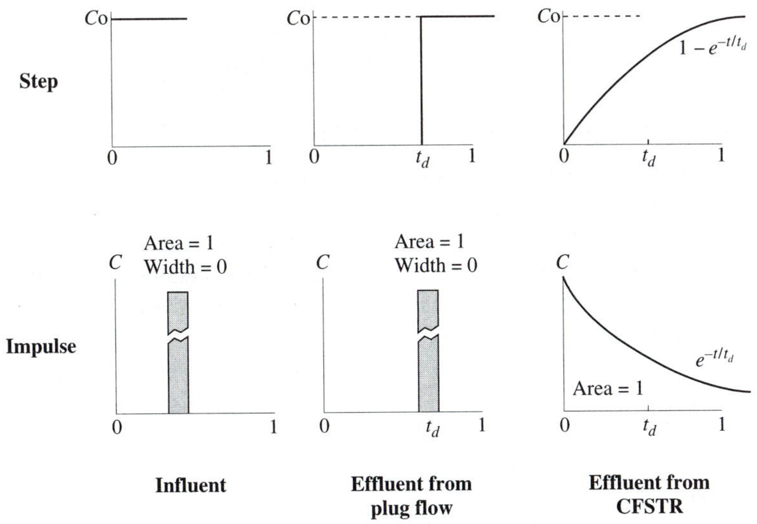

FIGURE 3-7
Response to step and impulse inputs.

where $V =$ volume of mixture in reactor
$dC/dt =$ rate of change of concentration in the reactor
$Q =$ flow rate into reactor
$C_{in} =$ concentration of contaminant entering (step input)
$C = C_{out} =$ concentration leaving = concentration in reactor

Equation 3-56 may be solved for the concentration as a function of time;

$$C = C_{in}(1 - e^{-t/t_d}) \qquad (3\text{-}57)$$

where $t_d =$ detention time $= V/Q$.

If we now assume an irreversible first-order chemical reaction takes place in the reactor:

$$VdC/dt = QC_{in} - QC - VkC_{out} \qquad (3\text{-}58)$$

where $k =$ first-order rate coefficient
$VkC_{out} =$ loss due to the chemical reaction

Equation 3-58 may be solved for concentration as a function of time. Equation 3-58 may be written:

$$\frac{dC}{dt} + \beta C = \left(\frac{Q}{V}\right) C_{in} \qquad (3\text{-}59)$$

where $\beta = k + Q/V$.

Equation 3-59 is a first-order linear differential equation. It may be solved by multiplying both sides by the integrating factor $e^{\beta t}$ and integrating. Noting that, at $t = 0, C = C_{in}$ yields:

$$C = \frac{QC_{in}}{V\beta}(1 - e^{-\beta t}) + C_{in}e^{-\beta t} \qquad (3\text{-}60)$$

Equation 3-60 is the non-steady-state solution to Equation 3-58.

For many environmental applications, the steady-state ($dc/dt = 0$) solution is of greater interest:

$$C = \frac{C_{in}}{1 + kt_d} \qquad (3\text{-}61)$$

Many treatment processes involve a recycle line, wherein part of the effluent stream is brought back to the beginning of the process. A schematic showing recycle is illustrated in Figure 3-8. One result of this modification is that the actual flow rate through the reactor is increased from Q to $Q + R$. If the concentration in the recycle stream is significantly lower than that in the feed stream, the concentration in the reactor is decreased by dilution.

This chapter has presented a review of some fundamental concepts that are assumed background for many chapters in this book. This review is necessarily

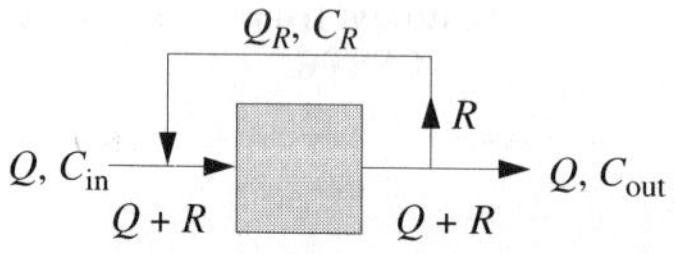

FIGURE 3-8
Recycle reactor.

brief, and students are referred to basic texts for more detailed information. Particularly recommended are the books by Smith and Van Ness,[45] Lyman et al.,[46] and Felder and Rousseau.[47]

3-5 GEOCHEMICAL MODELING AND ANALYSIS PROGRAMS

Several geochemical modeling and analysis programs are available from various organizations. Some of these programs are briefly described below.

NETPATH is an interactive Fortran 77 computer program used to interpret net geochemical mass balance reactions between an initial and final water along a hydrologic flow path.[48] Alternatively, NETPATH computes the mixing proportions of two to five initial waters and net geochemical reactions that can account for the observed composition of a final water. The program utilizes previously defined chemical and isotopic data for waters from a hydrochemical system. For a set of mineral and/or gas phases hypothesized to be the reactive phases in the system, NETPATH calculates the mass transfers in every possible combination of the selected phases that accounts for the observed changes in the selected chemical and/or isotopic compositions observed along the flow path. The calculations are of use in interpreting geochemical reactions, mixing proportions, evaporation, and/or dilution of waters, and mineral mass transfer in the chemical and isotopic evolution of natural and environmental waters. Distillation calculations are applied to each mass balance model that satisfies the constraints to predict carbon, sulfur, nitrogen, and strontium isotopic compositions at the end point, including radiocarbon dating. DB is an interactive Fortran 77 computer program used to enter analytical data into NETPATH, and calculate the distribution of species in aqueous solution.

NETPATH is available through the United States Geological Survey's (USGS's) Hydrologic Analysis Software Support Program.

PHREEQC is a computer program written in the C programming language that is designed to perform a wide variety of aqueous geochemical calculations.[49,50] PHREEQC is based on an ion-association aqueous model and has capabilities for (1) speciation and saturation-index calculations; (2) reaction-path and advective-transport calculations involving specified irreversible reactions, mixing of solutions, mineral and gas equilibria, surface complexation reactions, and ion-exchange reactions; and (3) inverse modeling, which finds sets of mineral and gas mole transfers that account for composition differences between waters, within specified compositional uncertainties.

A graphical user interface (GUI), PHREEQCI has been written to provide assistance in generating the program-input file.

PHREEQC and PHREEQCI are both available from USGS's Hydrologic Analysis Software Support Program.

TRIPLOT is a computer program that produces trilinear plots of the relative abundance of the major ions (Na^+, K^+, Ca^{+2}, Mg^{+2}, Cl^-, SO^4, CO^3, HCO_3) of groundwater analyses. This plotting technique is also referred to as the "Piper trilinear diagram."[51]

TRIPLOT is available from the Canada Centre for Inland Waters, located in Burlington, Ontario.

WATEQ4F is a program for the calculation of chemical equilibrium in natural waters. It models the thermodynamic speciation of major and important minor inorganic ions and complex species in solution for a given water analysis and *in situ* measurements of temperature, pH, and redox potential. From this model the states of reaction of the water with solid and gaseous phases are calculated. The examination of reaction states may suggest the origin of the dissolved constituents and assist in the prediction of the chemical effects of groundwater production, recharge, irrigation, and inorganic pollution. The original version (WATEQ) was written in 1973 by A. H. Truesdell and B. F. Jones in PL/1. WATEQ4F is the latest version of WATEQ and includes several revisions to the thermodynamic database including the addition of selenium and uranium species as well as the published values of Nordstrom and others (1990).[52]

The preprocessor WQ4FINPT is distributed with WATEQ4F, and allows the user to create and modify input files of analytical data. The revised documentation for WATEQ4F by Ball and Nordstrom (1991)[53] provides instructions for program operation, explanations of the input and output files, and all reactions and references for the thermodynamic database.[54]

WATEQ4F is distributed by the USGS's Hydrologic Analysis Software Support Program.

DISCUSSION TOPICS AND PROBLEMS

3-1. Draw the structure of the following compounds:

(*a*) benzene
(*b*) 1,1-dichloroethane
(*c*) methylene chloride (dichloromethane)
(*d*) bis (2-chloroethyl) ether
(*e*) 1,2-dichloro 2-butene

3-2. Name the following structures:

(*a*)
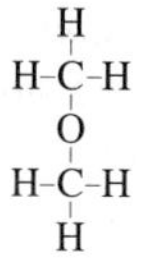

(*b*)
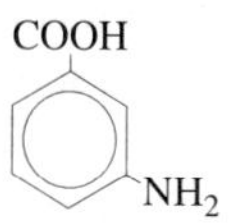

(*c*)
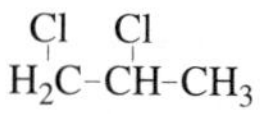

3-3. Determine the solubility of cadmium in equilibrium with cadmium hydroxide, $Cd(OH)_2$, as a function of pH. $K_{sp} = 5.33 \times 10^{-15}$.

3-4. Henry's constant is often tabulated in units of pressure (atm). Derive an expression to convert the value for trichloroethylene, 550 atm at 20°C to a dimensionless Henry's constant.

3-5. Convert Henry's constant for benzene, 5.5×10^{-3} atm · m^3 per mole at 20°C, to a dimensionless constant.

3-6. Estimate Henry's constant for benzene at 25°C from solubility and vapor pressure data in Appendix B.

3-7. Betterton and Hoffmann report the Henry's constant for formaldehyde in water at 25°C as 2.97×10^3 M atm^{-1} (molar/atmosphere). Convert this value to units of atm · m^3/mol.

3-8. Estimate the diffusion coefficient for ethylbenzene from tabulated values for phenol and toluene.

3-9. Estimate the diffusion coefficient for methyl phenyl sulfide in water at 25°C.

3-10. Explain the difference between bioconcentration and biomagnification.

3-11. Estimate the concentration of ethylbenzene in fish tissue in a lake with a concentration in water of 120 mg/L.

3-12. Estimate the organic carbon partition coefficient K_{oc} of benzene from (1) solubility and (2) log K_{ow} data in Appendix B. Compare your results to the measured value in Appendix B.

3-13. Derive an expression for the change in internal energy of a liquid recirculated by a pump in a tank. (Assume changes in potential energy are negligible and the system is at steady state.) Heat loss to the surroundings is negligible.

3-14. Determine the degrees of freedom for a system at equilibrium with water and benzene over an iron hydroxide layer. Saturated air is above the liquid.

3-15. A decanter for a steam stripper has two liquid layers, primarily toluene and water. How many degrees of freedom does this system have? (Hint: A saturated vapor is above the liquids.)

3-16. A 1-m^3 container holds 50 g of hydrogen. The temperature is 20°C. Determine the pressure.

3-17. Determine the molar volume of air at 20°C and atmospheric pressure. Compare this value to the molar volume of air at standard conditions (0°C and 1 atm). Assume ideal gas behavior applies.

3-18. A gas stream containing 30 wt % oxygen and 70 wt % nitrogen at 120 K and 1 atm enters a process. Determine the work required to raise the temperature to 235 K. Assume the process is adiabatic (i.e., $q = 0$) and the changes in potential and kinetic energy are negligible.

Component	Temperature, K	Enthalpy, J/gmol
Oxygen	120	3480
Oxygen	235	6795
Nitrogen	120	1625
Nitrogen	235	2460

3-19. 100,000 lbm/h of condensate (liquid water) at 310°F is fed to a wood-fired boiler. The superheated steam leaving the boiler is at 1200 psia and 900°F. How much heat must be transferred into the water in the boiler?

Temperature, °F	Enthalpy of water, Btu/(lb · mol)
310	280
900	1440.9

3-20. Steam enters a compressor at 101.325 kPA and 350°C and at a flow rate of 100 kg/min. The steam is compressed to 500 kPA. Changes in kinetic and potential energy between inlet and outlet of the compressor are negligible. Calculate the quantity of work produced in the process. (Assume the system is at steady state. Also assume the process is adiabatic and reversible.)

3-21. Determine the heat transferred to water in a boiler given the following data: Temperature is raised from 20°C to steam at 200°C; flow rate of water = 1 m^3/min. Assume atmospheric conditions.

3-22. A mixture of air and methane is to be heated from 20 to 600°C. The mixture is 50 moles air and 100 moles methane. How much energy is needed to achieve this temperature change? What temperature should be used in the temperature-dependent form for the heat capacity?

3-23. An open tank contains 50 kg of water at room temperature. The goal is to add steam at such a rate that all the steam condenses. If the steam supplied is at 2.8×10^5 Pa and

320°C, how much steam (in kg) should be added so that the temperature of the water reaches 90°C? (Assume no heat loss from the water.)

3-24. Calculate the energy of formation of aragonite ($CaCO_3$) at 298.15 K, given $H = -1207.0$ kJ/gmol and $S = 7.83$ kJ/(K · gmol).

3-25. A continuous and steady-state freezing process is used to purify seawater (3 wt % NaCl and 97 wt % water). The liquid brine concentration exiting the process is 10 wt % NaCl. Determine the mass rate of seawater required to produce 300 kg/h of pure water in the form of ice.

3-26. Water enters a 5-L tank at a rate of 4.00 kg/s and is withdrawn from the tank at a rate of 6.00 kg/s. The tank is initially four-fifths full.

(*a*) Write a mass balance for this process by assigning a variable function to the mass of water at time t. Describe the terms that are present in the balance; give reasons for deletion of any terms.
(*b*) Solve the balance to find how long it will take the tank to drain completely.

3-27. A distillation column is set up containing A and B. The feed to the column enters at 5000 kg/h and is in a 1:1 ratio. The top stream out of the distillation column is sent to a condenser at a rate of 3500 kg/h. At the condenser, A is taken out and separated into a product stream and a recycle stream. The product stream from the condenser contains 95 percent A. The bottoms stream contains 90 percent B. Find the ratio of recycle to product at top.

3-28. A reaction with the stoichiometric equation:

$$A + 3B \rightarrow 2C$$

is carried out with 30 percent conversion of A. The reactor feed stream contains 25 percent A and 75 percent B by weight and has a rate of 1000 kg/h. If the molecular weight of A is 32 and that of B is 2,

(*a*) Calculate the molecular weight of C.
(*b*) Calculate the composition of the exit stream on a weight basis.

3-29. (*a*) Write a balanced reaction for the reaction of 2 moles $CH_3CH_2CH_2NH_3$ with O_2 to produce CO_2, H_2O, and NO_2.
(*b*) Calculate the rate of production of CO_2, H_2O, and NO_2 if 5 mol/h of $CH_3CH_2CH_2NH_3$ is reacted with a stoichiometric amount of O_2.
(*c*) Calculate the rate of reaction.

3-30. The equilibrium constant K for the formation of nitrogen dioxide $N + O_2 \rightarrow NO_2$ (g) is 8.27×10^{-10} at 298.15 K. Calculate the standard free energy change for the reaction.

3-31. Cyclohexane is commercially produced from catalytic, high-pressure, and temperature hydrogenation of benzene. Given the Gibbs free energy values for cyclohexane, 31.91 kJ/gmol, and benzene, 129.72 kJ/gmol, at standard temperature and pressure (STP), determine whether this process is feasible at STP and to what extent.

3-32. Using the heat of combustion (ΔH_c°) value for DDT ($C_{14}H_9Cl_5$) of -1600 kcal/gmol, calculate the heat of formation (ΔH_f°), given the following heats of formation of the combustion products:

Compound	**CO_2**	**H_2O (l)**	**HCl (g)**
ΔH_f° (kcal/gmol)	−94.05	−68.32	−22.06

3-33. An environmental scientist wants to prepare some standards to calibrate a portable gas chromatograph to be used to conduct a screening of volatile organic compounds extracted from a soil vapor extraction system. Assuming ideal gas behavior (1 mole of gas occupies a 22.4-L volume at STP), what volume of neat (pure) liquid would be

needed to prepare a gas standard containing benzene (density = 0.877 g/mL) and tetrachloroethene (density = 1.62 g/mL) to produce standard concentrations of 100 μL/L (ppmv) and 500 μL/L (ppmv) respectively? The standard is prepared at STP and in a 1-L glass sampling bulb. Use molecular weight data provided in Appendix B.

3-34. An impoundment 20 × 12 feet is contaminated by tetrachloroethylene. The contaminated liquid comes in contact with the groundwater and a concentration gradient of 1.3×10^{-6} (mol/cm^3)/cm is established. Use Fick's law to determine the diffusive flux. Assuming diffusion is the only driving force, how many pounds of tetrachloroethylene will move toward the groundwater in 1 hour?

3-35. Derive Equation 3-57.

3-36. Derive Equation 3-60.

REFERENCES

1. Ferguson, Lloyd N.: *Organic Chemistry: A Science and Art*, Willard Grant Press, Boston, pp. 38–40, 1972.
2. Ibid.
3. Duffus, John H.: *Environmental Toxicology*, Edward Arnold Publishers, London, pp. 38–40, 1980.
4. Ibid, p. 67.
5. Zagula, S. J., and J. B. Risatti: "Cost-Effective Technologies for Remedial Action at Agrichemical Facilities," in *Proceedings of the 47th Annual Purdue Industrial Waste Conference*, Lewis Publishers, Ann Arbor, Michigan, pp. 93–104, 1992.
6. Solomons, T. W. Graham: *Organic Chemistry*, 4th edition, John Wiley and Sons, New York, p. 732, 1988.
7. Weast, R. C. (ed.): *CRC Handbook of Chemistry and Physics*, CRC Press, Boca Raton, Florida, pp. B-219–220, 1983.
8. Stumm, W., and J. J. Morgan: *Aquatic Chemistry: An Introduction Emphasizing Chemical Equilibria in Natural Waters*, John Wiley & Sons, New York, 1981.
9. Kavanaugh, Michael C., and R. Rhodes Trussell: "Design of Aeration Towers to Strip Volatile Contaminants from Drinking Water," *Journal AWWA*, December 1980.
10. Thomas, R. G.: "Volatilization from Water," in Warren J. Lyman, William F. Reehl, and David H. Rosenblatt, *Handbook of Chemical Property Estimation Methods*, Environmental Behavior of Organic Compounds, McGraw-Hill, New York, p. 15–11, 1982.
11. Yurteri, C., et al.: "The Effect of Chemical Composition of Water on Henry's Law Constant," *J. WPCF*, vol. 59, no. 11, pp. 950–956, 1987.
12. Tucker, W. A., and L. H. Nelken: "Diffusion Coefficients in Air and Water," Chapter 17 in Warren J. Lyman, William F. Reehl, and David H. Rosenblatt, *Handbook of Chemical Property Estimation Methods*, Environmental Behavior of Organic Compounds, McGraw-Hill, New York, 1982.
13. Reid, R. C., J. M. Prausnitz, and T. K. Sherwood: *Properties of Gases and Liquids*, 3rd edition, McGraw-Hill, New York, p. 570, 1977.
14. Reid, R. C., J. M. Prausnitz, and B. F. Poling: *The Properties of Liquids and Gases*, 4th edition, McGraw-Hill, New York, p. 581, 1987.
15. Perry, J. H.: *Chemical Engineers' Handbook*, 6th edition, McGraw-Hill, New York, pp. 3-285 to 3-287, 1984.
16. Thomas, R. G.: "Volatilization from Soil," in Warren J. Lyman, William F. Reehl, and David H. Rosenblatt, *Handbook of Chemical Property Estimation Methods*, Environmental Behavior of Organic Compounds, McGraw-Hill, New York, p. 10-16, 1982.
17. Reid, et al., *The Properties of Gases and Liquids*, p. 598.
18. Ibid., pp. 52ff.
19. Ibid., p. 53.

20. Lyman, W. J., W. F. Reehl, and D. H. Rosenblatt: *Handbook of Chemical Property Estimation Methods*, American Chemical Society, New York, pp. 1–2, 1990.
21. Deutsch, W. J.: *Groundwater Geochemistry: Fundamentals and Applications to Contamination*, Lewis Publishers, Boca Raton, Florida, 1997.
22. Stumms, W.: *Chemistry of Solid-Water Interface: Processes at the Mineral-Water and Particle-Water Interface in Natural Systems*, John Wiley & Sons, New York, 1992.
23. U.S. Environmental Protection Agency: "Transport and Fate of Contaminants in the Subsurface," EPA/625/4-89/019, Center of Environmental Research Information, Cincinnati, Ohio, September 1989.
24. Dragun, James: "The Fate of Hazardous Materials in Soil," *Hazardous Materials Control*, Hazardous Materials Control Research Institute, vol. 1, no. 5, September-October 1988.
25. Barron, M. G.: "Bioconcentration," *Environ. Sci. Technol.*, vol. 24, pp. 1612–1618, 1990.
26. Banerjee, S., and G. L. Baughman: "Bioconcentration Factors and Lipid Solubility," *Environ. Sci. Technol.*, vol. 25, pp. 536–539, 1991.
27. Paustenbach, D. J.: "A Survey of Health Risk Assessment," in *The Risk Assessment of Environmental and Human Health Hazards: A Textbook of Case Studies*, D. L. Paustenbach, ed., Wiley-Interscience, p. 56, 1989.
28. Weber, W. J., P. M. McKinley, and L. E. Katz: "Sorption Phenomena in Subsurface Systems: Concepts, Models and Effects on Contaminant Fate and Transport," *Water Research*, vol. 25, no. 5, pp. 499–528, 1991.
29. Smith, J. M., and H. C. Van Ness: *Introduction to Chemical Engineering Thermodynamics*, 4th edition, McGraw-Hill, New York, 1987.
30. Sonntag, R. E., and G. J. Van Wylen: *Introduction to Thermodynamics—Classic and Statistical*, 3rd edition, John Wiley & Sons, New York, 1991.
31. Sandler, S. I.: *Chemical and Engineering Thermodynamics*, John Wiley & Sons, New York, 1989.
32. Ibid., p. 33.
33. Sonntag and Van Wylen, *Introduction to Thermodynamics*, p. 20.
34. Smith and Van Ness, *Introduction to Chemical Engineering Thermodynamics*.
35. Sonntag and Van Wylen, *Introduction to Thermodynamics*.
36. Sandler, S. I., *Chemical and Engineering Thermodynamics*.
37. Perry, R. H., and D. W. Green: *Perry's Chemical Engineers' Handbook*, 6th edition, McGraw-Hill, New York, 1984.
38. Smith and Van Ness, *Introduction to Chemical Engineering Thermodynamics*, p. 38.
39. Sandler, S. I., *Chemical and Engineering Thermodynamics*, p. 43.
40. Sonntag and Van Wylen, *Introduction to Thermodynamics*, p. 167.
41. Sandler, S. I., *Chemical and Engineering Thermodynamics*, pp. 85–86.
42. Ibid., pp. 274–277.
43. Felder, Richard M., and Ronald W. Rousseau: *Elementary Principles of Chemical Processes*, 2nd edition, John Wiley & Sons, New York, pp. 482–483, 1986.
44. Ibid., pp. 103–106.
45. Smith and Van Ness, *Introduction to Chemical Engineering Thermodynamics*.
46. Lyman, Warren J., William F. Reehl, and David H. Rosenblatt: *Handbook of Chemical Property Estimation Methods*, Environmental Behavior of Organic Compounds, McGraw-Hill, New York, 1982.
47. Felder and Rousseau, *Elementary Principles of Chemical Processes*.
48. Plummer, L. N., Prestemon, E. C., and Parkhurst, D. L.: *An Interactive Code (NETPATH) for Modeling NET Geochemical Reactions along a Flow PATH—Version 2.0*, U.S. Geological Survey Water-Resources Investigations Report 94-4169, 130 pp., 1994.
49. Charlton, S. R., Macklin, C. L. and Parkhurst, D. L.: *PHREEQCI—A Graphical User Interface for the Geochemical Computer Program PHREEQC*, U.S. Geological Survey Water-Resources Investigations Report 97-4222, 9 pp. 1997.

50. Parkhurst, D. L.: *User's Guide to PHREEQC—A Computer Program for Speciation, Reaction-Path, Advective-Transport, and Inverse Geochemical Calculations*, U.S. Geological Survey Water-Resources Investigations Report 95-4227, 143 pp., 1995.
51. Piper, A. M.: "A Graphical Procedure in the Geochemical Interpretation of Water Analyses," *Trans. Amer. Geophys. Union*, vol. 25, pp. 914–923, 1944.
52. Nordstrom, D. K., L. N. Plummer, Donald Langmuir, Eurybiades Busenberg, H. M. May, B. F. Jones, and D. L. Parkhurst, *Revised Chemical Equilibrium Data for Major Water-Mineral Reactions and Their Limitations*, in Melchor, D. C., and R. L. Bassett, eds.: *Chemical Modeling in Aqueous Systems II*, American Chemical Society Symposium Series 416, Washington, D.C., pp. 398–413, 1990.
53. Ball, J. W., and D. K. Nordstrom: *User's Manual for WATEQ4F, with Revised Thermodynamic Data Base and Test Cases for Calculating Speciation of Major, Trace, and Redox Elements in Natural Waters*, U.S. Geological Survey Open-File Report 91-183, 189 pp. (revised and reprinted August 1992), 1991.
54. Nordstrom, Darrell Kirk, L. Niel Plummer, Donald Langmuir, Eurybiades Busenberg, Howard M. May, Blair F. Jones, and David L. Parkhurst: *Revised Chemical Equilibrium Data for Major Water-Mineral Reactions and Their Limitations*, in Melchior, Daniel C., and R. L. Bassett, eds., *Chemical Modeling of Aqueous Systems II*, ACS Symposium Series, 416, pp. 398–413, American Chemical Society, 196th national meeting, Los Angeles, Sept. 25–30, 1988.

ADDITIONAL READING

Alberty, R. A.: *Physical Chemistry*, 7th edition, John Wiley & Sons, New York, 1987.

Betterton, E. A., and M. R. Hoffmann: "Henry's Law Constants of Some Environmentally Important Aldehydes," *Environ. Sci. Technol.*, vol. 22, p. 1415, 1988.

Christian, G. D.: *Analytical Chemistry*, John Wiley & Sons, New York, 1986.

Dean, J. A., ed.: *Lange's Handbook of Chemistry*, 12th edition, McGraw-Hill, New York, 1979.

Ernst, W.: "Determination of the Bioconcentration Potential of Marine Organisms—A Steady-State Approach," *Chemosphere,* vol. 11, pp. 731–740, 1977.

Gossett, J. M.: Measurement of Henry's Law Constants for C1 and C2 Chlorinated Hydrocarbons, *Environ. Sci. Technol.*, vol. 21, pp. 202–208, 1987.

Howe, G. B., M. E. Mullins, and T. N. Rogers: "Evaluation and Prediction of Henry's Law Constants and Aqueous Solubilities for Solvents and Hydrocarbon Fuel Components—Vol I: Technical Discussion," Final Report, Engineering & Services Laboratory, Tyndall Air Force Base, Report No. ESL-86-66, 1987.

Jordan, D. G.: *Chemical Pilot Plant Practice*, Interscience Publishers, New York, 1955.

King, C. J.: *Separation Processes*, McGraw-Hill, New York, 1980.

Kishi, H., and Y. Hashimoto: "Evaluation of the Procedures for the Measurement of Water Solubility and N-Octanol/Water Partition Coefficient of Chemicals: Results of a Ring Test in Japan," *Chemosphere (G. B.)*, vol. 18, p. 1749, 1989.

Knoche, Herman W.: *Essentials of Organic Chemistry*, Addison-Wesley, Reading, Massachusetts, 1986.

Luyben, William L., and Leonard A. Wenzel: *Chemical Process Analysis: Mass and Energy Balances*, Prentice-Hall, Englewood Cliffs, New Jersey, E 1988.

Mackay, D., and W. Y. Shiu: "A Critical Review of Henry's Law Constant for Chemicals of Environmental Interest," *Journal Physical Chemical Reference Data*, vol. 10, p. 1175, 1981.

Mailhot, H., and R. H. Peters: "Empirical Relationships Between the l-Octanol/Water Partition Coefficient and Nine Physiochemical Properties." *Environ. Sci. Technol.*, vol. 22, p. 1479, 1988.

Manahan, S. E.: *Environmental Chemistry*, 4th edition, Lewis Publishers, Boca Raton, Florida, 1990.

Manahan, S. E.: *Hazardous Waste Chemistry, Toxicology and Treatment*, 4th edition, Lewis Publishers, 1990.

McCabe, Warren L., Julian C. Smith, and Peter Harriott: *Unit Operations of Chemical Engineering*, 4th edition, McGraw-Hill, New York, 1985.

Myers, Alan L., and Warren D. Seider: *Introduction to Chemical Engineering and Computer Calculations*, Prentice-Hall, Englewood Cliffs, New Jersey, 1976.

Nirmalakhandan, N. N., and R. E. Speece: "QSAR Model for Predicting Henry's Constant." *Environ. Sci. Technol.*, vol. 22, p. 1349, 1988.

Prausnitz, J. M.: *Molecular Thermodynamics of Fluid-Phase Equilibria*, Prentice-Hall, Englewood Cliffs, New Jersey, 1969.

Reklaitis, G. V.: *Introduction to Material and Energy Balances*, John Wiley & Sons, New York, 1983.

Sawhney, B. L., and K. Brown, eds.: *Reactions and Movement of Organic Chemicals in Soils*, Soil Science Society of America, 1989.

Sawyer, Clair N., and Perry L. McCarty: *Chemistry for Environmental Engineering*, 3rd edition, McGraw-Hill, New York, p. 159, 1978.

Smith, J. H., et al.: "Volatilization Rates of Intermediate and Low Volatility Chemicals from Water," *Chemosphere,* vol. 10, pp. 281–289, 1981.

Snoeyink, V. L., and D. Jenkins: *Water Chemistry*, John Wiley & Sons, New York, 1980.

ACKNOWLEDGMENTS

The authors wish to acknowledge the following individuals who assisted with the preparation of this chapter for the first edition: David Gallis, Ph.D.; Elsie Millano, Ph.D.; Cheri Pearson; Jill Sotsky; and Karen Winegardner; *Environmental Resources Management (ERM);* Stephen Galloway and Lori Hamilton, *Bucknell University;* and Professor William E. King, Ph.D. The principal reviewer for the first edition was Roy Ball, Ph.D., *ERM*.

Reviewers and contributors for the second edition included James Benko; Mark Bordelon, Ph.D.; Richard Brown, Ph.D.; Gregory Shkuda, Ph.D.; Gina Sperinde; Carl Stay; and Stan Zagula; *ERM*.

CHAPTER 4

Fate and Transport of Contaminants

I am the daughter of Earth and Water
and the nursling of the Sky;
I pass through the pores of the oceans and shores;
I change but I cannot die.

Percy Bysshe Shelley

The release of contaminants into the environment is inevitable. Contaminants are released through our manufacture and use of products and as a result of treatment and disposal of our wastes. Upon release to the environment, contaminants move and respond to a number of interrelated natural and synthetic factors. They may move quickly or ever so slowly to living receptors in their original or altered form. The pathway may be direct or tortuous. This movement is indicated in Figure 4-1 for the atmospheric and subsurface environment and in Figure 4-2 for the aquatic environment.

Understanding how contaminants are released and their transport and fate in the environment is necessary for successful hazardous waste management. In this context, this chapter defines key release mechanisms, presents the fundamental mechanics of contaminant transport, and describes the fate of contaminants in the environment. The reader should recognize that the understanding of contaminant transport builds on the knowledge of several diverse disciplines including earth science, geology, fluid mechanics, chemistry, physics, hydrology, and biology. The interrelationship of these sciences and the associated mathematical representations of transport phenomena make the understanding of contaminant behavior interdisciplinary in nature.

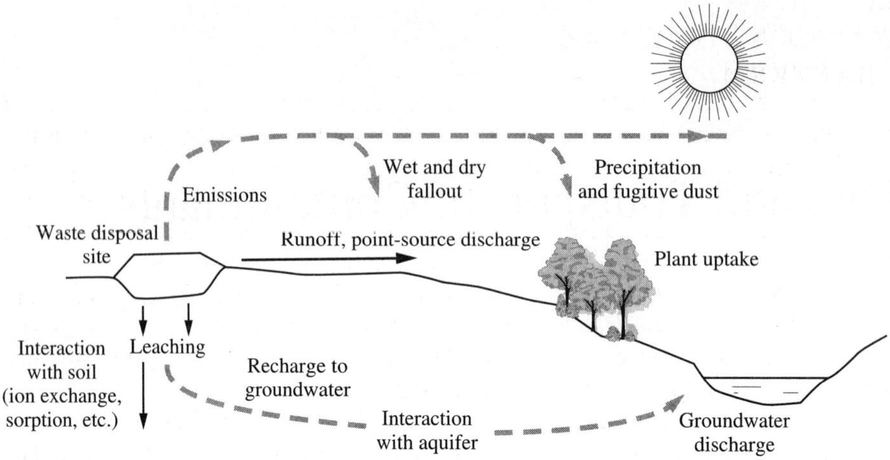

FIGURE 4-1
Some fate and transport processes in the subsurface and atmospheric environment.

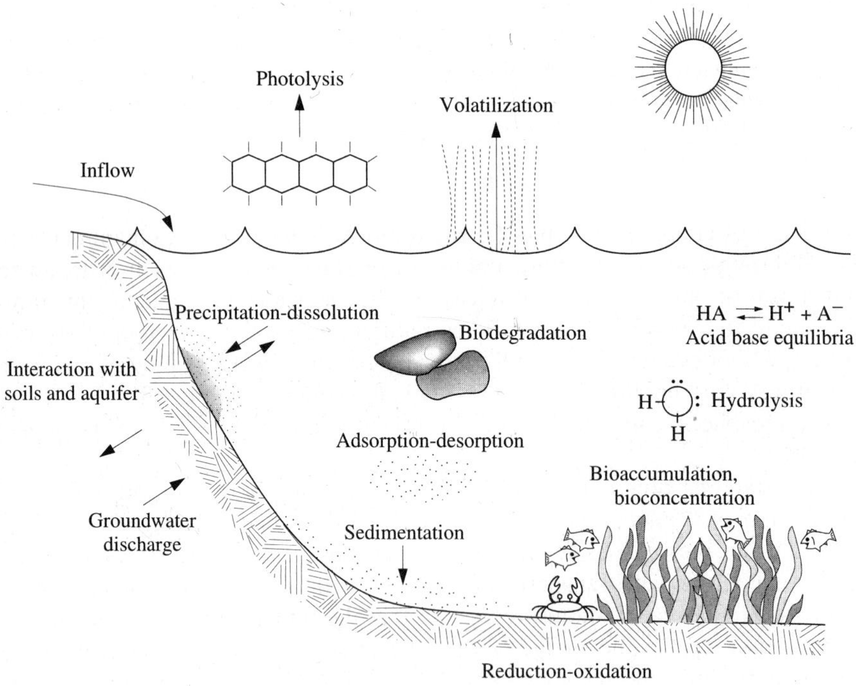

FIGURE 4-2
Some fate and transport processes in the aquatic environment.

4-1 CONTAMINANT RELEASE

The release of contaminants from hazardous waste sites occurs in one or more of three phases—liquid, solid, and gas. Liquid releases include such things as contaminated runoff, direct aqueous discharges to surface water, and leachate to ground water. Air emissions can include volatile emissions from lagoons and direct atmospheric discharges from stacks. Stack emissions can include gaseous emissions (e.g., products of combustion such as CO_2 and H_2SO_4 and products of incomplete combustion such as trace organic gases) and particulate emissions (e.g., fly ash potentially laden with toxic metals). Releases of contaminants in solid form occur in air (e.g., fugitive dust) and in water (e.g., suspended solids). These releases typically have contaminants adsorbed to solids or dissolved in fluids.

Contaminant releases are either controlled or uncontrolled. Controlled releases are an intrinsic part of our current hazardous waste management practices—part of virtually every manufacturing and waste treatment operation. The notion of controlled releases is that contaminant releases are managed (i.e., controlled) in such a way as to minimize harm to the environment. The predominant mechanisms supporting this approach are dilution and assimilation.

Uncontrolled releases are those that are not under the direct management of the unit operators. Some uncontrolled releases may be reasonably foreseeable, such as windblown fugitive dust or "puffs" from incinerators. Others may not, such as a break in a pipeline or a leak from an underground storage tank. Still others derive from closed, inactive sites not having adequate environmental control measures. In many scenarios, uncontrolled releases can result in major pollution incidents as exemplified by the large number of U.S. Superfund sites.

Air Emissions

Air emissions may occur in a variety of forms. The typical ways of characterizing air emissions sources are as point, line, area, volume, or puff (instantaneous) sources. A point source is a well-defined location typically emitting contaminants on a continuous basis. Line, area, and volume sources are considered fugitive in nature. Fugitive emissions are those emissions that could not reasonably pass through a stack, chimney, vent, or other functionally equivalent opening. A puff release is another type of fugitive emission that occurs instantaneously from such sources as spills or other accidental releases. The types of air emissions are illustrated by the following examples:

Point	Incinerator stack; landfill gas vent
Line	Dust from road; vehicle emissions from a roadway
Area	Volatile emissions from a lagoon
Volume	Volatile or particulate emissions from a building with open windows and doors
Puff	Volatile emissions from an accidental spill

Emissions from point sources are expressed in units of mass per time, line sources as mass per time per length of line, an area source as mass per time per two-dimensional area, and volume source as mass per time per three-dimensional volume. Instantaneous releases are commonly expressed as the mass of the total release.

Releases can be further classified as gas-phase emissions or particulate matter emissions. Gas-phase emissions consist primarily of organic compounds where the predominant release mechanism is volatilization. Gaseous emissions may also be produced by manufacturing or waste treatment processes. Particulate matter emissions derive principally from combustion, wind erosion, and mechanical processes, releasing fine particles that may have a variety of contaminants. These contaminants can include not only organics but metals and some normally immobile substances (e.g., PCBs and dioxins). Combustion processes can release gases (either as by-products of combustion or as a result of incomplete combustion) and particulate matter in the form of fly ash or uncombusted carbon.

Volatilization

Volatilization is the transfer of a chemical substance from a liquid phase to a gaseous phase. Volatilization is the predominant source of atmospheric emissions at most uncontrolled hazardous waste sites. Aboveground sources include storage tanks, pipe fittings, and the surfaces of lagoons. There are significant subsurface sources as well. It is noted later in this section how the migration of leachate from a landfill or lagoon releases contaminants to the groundwater. Organics may volatilize from the leachate and move to the ground surface. Emissions also can come from the volatilization of dissolved chemicals in contaminated groundwater, even from a plume that has migrated laterally a considerable distance from a site.

The volatilization rate depends in part on temperature, vapor pressure of the substance, and the difference in the concentration between the liquid and gas phases. Volatilized contaminants may go directly to the atmosphere, or they may traverse a tortuous path such as contaminants in the subsurface, as depicted in Figure 4-3. The depicted transfer depends on diffusion (see Section 3-2) through a porous medium, and variables such as soil porosity and soil moisture are important.

Contaminants that enter the atmosphere through volatilization can be measured in the field by various procedures such as through the use of organic vapor analyzers (see Section 15-6); however, these instruments may not be capable of differentiating among various organic compounds. Volatile emissions also can be predicted using mathematical models that have been developed for specific facility types such as

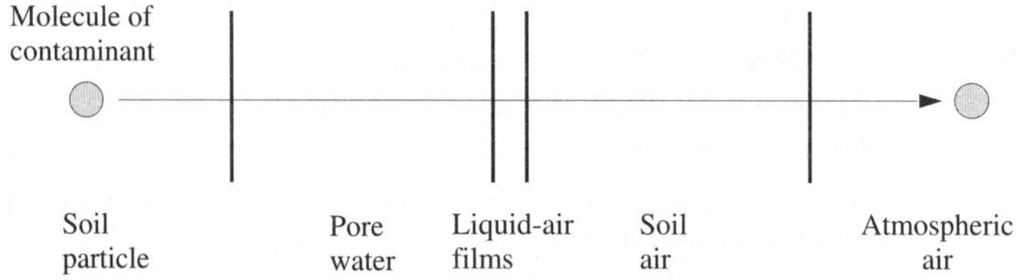

FIGURE 4-3
Contaminant transfer through soil-water-air interfaces.

TABLE 4-1
Volatile air emissions models[1]

Type of source	Number of models presented
Landfills without internal gas generation	5
Landfills with internal gas generation	3
Open landfills	3
Land treatment	2
Nonaerated lagoon	4
Aerated lagoon	2

landfills, lagoons, or land farms. The basis for most modeling includes Fick's law of diffusion (see Section 3-2).

A technical guidance manual, published by the U.S. Environmental Protection Agency for estimating baseline volatile air emissions at Superfund sites describes the number of models available for each of the different types of sources as listed in Table 4-1.

EXAMPLE 4-1. VOLATILE EMISSIONS. Describe the differences in the rate of volatile emissions from a landfill where codisposal of municipal solid waste and hazardous waste occurred versus a landfill used for the disposal of hazardous waste only.

Solution. Because of the large organic content of municipal solid waste, codisposal landfills generate considerable volume of decomposition gases, particularly methane. The convecting sweep of such gas creates a greatly accelerated flux and release of volatile emissions from the codisposed hazardous waste. Both the soil- and gas-phase diffusion become insignificant when compared with the convecting gas.[2]

At undisturbed, inactive sites, emissions of volatile organic compounds are typically considered insignificant even though the groundwater may show the presence of such compounds. Investigations at many Superfund sites have found such compounds in groundwater and even surface water but not in the air at detectable concentrations. The results of field measurements of baseline ambient volatile concentrations (prior to disturbances such as those from remediation) from various surface impoundments that have been reported in the literature are shown in Table 4-2.

The remediation of sites typically involves disturbances that may expose waste that was previously covered. Therefore, remediation activities have the potential to substantially increase the release of emissions from an inactive site. The source of the data for TSDF Site N shown in Table 4-2 reported that the flux of organic compounds, exclusive of methane, increased several thousandfold during remediation. Remediation of sites often requires treatment of contaminated groundwater and soils. Such treatment may feature pumping, air strippers, oil-water separators, soil venting, and other techniques. Removing volatile contaminants from soil and groundwater and allowing them to be released to the atmosphere without controls is simply transferring the contaminant from one medium to another. The ultimate fate of such contaminants needs to be considered. Generally, emissions from sources and the actions to remediate those sources can be estimated with models based on Henry's law constants, standard equilibrium chemistry, and local factors such as soil type, moisture content, wind speed, and lagoon surface area.

TABLE 4-2
Typical volatile organic emissions at undisturbed sites

Source of data	Identification of surface impoundment	Total emissions of nonmethane hydrocarbons prior to disturbance	
		μg/(m^2 · min)	g/(m^2 · s)
EPA[3]	NPL Site D	43	7×10^{-7}
	TSDF Site K	120	2×10^{-6}
	TSDF Site L	570	1×10^{-5}
	TSDF Site M	9–31	2×10^{-7} to 5×10^{-7}
	TSDF Site N	630	1×10^{-5}
de Percin[4]	Impoundment 1	690	1×10^{-5}
	Impoundment 2	3400	6×10^{-5}
	Impoundment 3	7700	1×10^{-4}
	Impoundment 4	56–190	9×10^{-7} to 3×10^{-6}

NPL: National Priorities List (Superfund).
TSDF: Treatment, storage, and disposal facility.

EXAMPLE 4-2. VOLATILE EMISSION SOURCES. Identify the sources of volatile emissions from a lagoon and a landfill.

Solution. The sources from a lagoon are:

- Lateral migration of volatiles through the sides of the lagoon
- Lateral migration of volatiles from leachate and contaminated soils in the vadose zone,
- Volatilization from the lagoon surface
- Volatilization of dissolved species in groundwater

A landfill would have the same sources. Release from its surface could occur via:

- Release through desiccation cracks in the cover
- Diffusion through the cover
- Release through the system used to vent methane generated by the biodegradation of organic waste
- Release through the system used to collect leachate generated within the landfill
- Volatilization of exposed waste prior to covering

Particulate matter emissions

Waste treatment operations generate particulate matter emissions. Incinerators, of course, directly discharge particulate matter. Fugitive particulate matter emissions generally originate from materials handling and surface areas:

- Roadways
- Open waste piles
- Staging areas
- Dry surface impoundments
- Landfills
- Land treatment operations
- Waste stabilization basins

Fugitive dust is usually a key concern at inactive hazardous waste sites. Two key parameters controlling fugitive dust are (1) the wind, which erodes and trans-

ports material from the site, and (2) disturbance of the soil by human activities (e.g., vehicular traffic). The dust has the same contaminants as those sorbed to the soil. Remediation of inactive sites generates dust from excavation, grading, and truck traffic. The amount of soil lost because of wind erosion at inactive sites is a function of wind velocity, vegetative cover, soil/waste properties such as texture and moisture, and the area of exposed soil surface. The Soil Conservation Service (SCS) of the U.S. Department of Agriculture has developed a graphical method for predicting annual average wind erosion and resulting soil loss.[5] The method uses annual average climatic data and cannot be used to analyze the impact of climatic extremes. According to the EPA, the model is not reliable for estimating short-term emissions.[6]

EXAMPLE 4-3. WIND EROSION SOIL LOSS. Qualitatively, explain what portion of the soil loss estimated by the SCS wind erosion procedure would represent a release of contaminants that could be transported to off-site receptors.

Solution. The SCS method estimates total soil loss, of which only a portion is **transportable** by the wind and **suspendible.** That transportable and suspendible amount is the portion of the total soil loss represented by particles $\leq$ 100 μm in diameter or smaller, depending on the location of the receptors (large particles could settle out of the air before moving off-site). The quantity transported to off-site receptors would have to be estimated from a grain size distribution of the parent material. If the analysis is concerned with just inhalable sizes, that would limit the analysis to the fraction of particulates 10 μm in diameter or smaller. Note, however, that contaminants are preferentially adsorbed to the smaller-size particles.

The generation of fugitive dust by vehicular traffic is a function of soil properties and vehicular characteristics such as vehicle speed, weight, and number of wheels. Dust generation from unpaved roads is particularly sensitive to soil moisture and soil particle size and can be estimated by the empirical formula:

$$E_{\mathrm{VT}} = 5.9\left(\frac{S_p}{12}\right)\left(\frac{V_v}{30}\right)\left(\frac{M_v}{3}\right)^{0.7}\left(\frac{W_v}{4}\right)^{0.5}\left(\frac{365 - D_p}{365}\right) \qquad (4\text{-}1)$$

where E_{VT} = emission factor (lb/vehicle-mile traveled)
S_p = silt content of road surface material (%)
V_v = mean vehicle velocity (mi/h)
M_v = mean vehicle mass (tons)
W_v = mean number of wheels
D_p = number of days per year with at least 0.01 inches of precipitation

Again, the percent of the generated dust that is wind transportable and respirable needs to be considered. The release of dust from unpaved roads can be as high as 1 kg/vehicle-kilometer (3.6 lb/vehicle-mile) if not more.[7]

By far the greatest source of fugitive dust is remediation involving soil handling. Contaminated soils frequently require remediation that may involve either excavation for treatment/disposal or containment using a cap. Excavation entails the following necessary operations, each of which generates large quantities of fugitive dust unless control measures are taken—transportation, dumping, storage in piles, conveyance to treatment unit, and treatment. Even if the soil does not require excavation, usually some form of surface grading is necessary and this generates fugitive dust.

To estimate particulate emissions from materials-handling operations such as soil loading and unloading, the following equation can be used.[8]

$$E = k(0.0032)\frac{\left(\frac{U}{5}\right)^{1.3}}{\left(\frac{M}{2}\right)^{1.4}} \tag{4-2}$$

where $E =$ emission factor [(lb of particulates released to air)/(ton of soil moved)]
$k =$ particle-size multiplier (dimensionless)
$U =$ mean wind speed (mi/h)
$M =$ material moisture content (%)

The emission factor E may be multiplied by the tons of material handled to achieve an estimate of total emissions. The particle-size multiplier k reflects the aerodynamic particle diameter, as follows:

Particle size	< 30 mm	< 15 mm	< 10 mm	< 5 mm	< 2.5 mm
k	0.74	0.48	0.35	0.20	0.11

The results of applying the preceding methods to estimating generation of fugitive dusts can be considered rough estimates at best. A site should be monitored downwind and upwind of its sources of release if there is any question as to whether the amount of release meets acceptable criteria for health and environmental risk.

Releases to Water

Water is an excellent medium for the transport of contaminants in the environment. How are contaminants released into the water? Controlled releases into surface water (e.g., streams, rivers, and lakes) are quite common. The varieties of aqueous waste streams with direct discharges to surface water are nearly endless. Examples include effluent from treatment works, cooling water scrubber blowdown from incinerators, treated leachate from landfills, and treated wash water from recycling/reuse facilities. Virtually every industrial and commercial facility generates a wastewater. Because our ability to clean up wastewater is and always will be less than 100 percent, the controlled discharge into surface waters will always be a source, however small, for some contaminant release into the environment.

Controlled releases of hazardous waste to groundwater are rare. In fact, most controlled contaminant releases involve materials other than hazardous waste and occur to the land above the groundwater. The application of pesticides directly to the land is one such example. Another is the use of on-site septic tanks and drain fields by rural homeowners. Controlled releases of hazardous waste can be rather controversial. One example is the use of deep well injection in which wastes are injected deep beneath the earth into rock formations where permanent containment is anticipated.

Uncontrolled releases occur to both surface and groundwater. Leachate and contaminated runoff from landfills are excellent examples of uncontrolled releases to ground and surface water. The releases are a result of natural mechanisms, that is, the impact of precipitation resulting in runoff and/or leachate generation. Uncontrolled releases as a result of human activity, such as spills, also occur.

Table 4-3 lists many principal releases to surface and groundwater, contrasting the volume of the release, the concentration of contaminants, and the types of factors affecting the nature of the release. Of these releases, the leakage of leachate through the base of a landfill to groundwater is one of the most studied. Characterization of leachate generation is concerned with the volume of liquid generated, its constituents, and the landfill construction.

Landfill Leachate

Generation

The release of leachate first requires the generation of leachate within the landfill. Figure 4-4 depicts the sources of liquid that may eventually become leachate. Most sources are water such as:

- Precipitation falling directly onto the landfill
- Surface flow that has run onto the landfill
- Groundwater inflow through a portion of the landfill lying below the groundwater table
- Putrescibles and other liquids in municipal wastes

In addition, there is the liquid fraction of the waste disposed in the landfill (past practices sometimes used landfills for disposal of liquid wastes, either in bulk form or in drums). With the exception of direct precipitation, these sources of leachate can be controlled with proper siting and engineering and the decomposition of organic matter. Infiltration can be minimized by constructing an engineered cover over the landfill as described in Chapter 13. Nevertheless, some infiltration inevitably occurs.

The calculation of infiltration through a landfill cover is done with a water budget model such as the widely used model HELP.[9] Such a model quantifies water movement according to the principle of conservation of mass. It balances the surface water input among the many possible routes and outputs as shown in Figure 4-5. Given climate, cover characteristics (e.g., slope, thickness, porosity, hydraulic conductivity), and other factors, the model is used to estimate the volumetric flux of water infiltrating into the waste on a time-varying basis. Contaminant transport through landfill barriers is presented in Chapter 13.

Contaminants

Water infiltrating into waste will first be absorbed by solid waste material if that material is below field capacity. **Field capacity** is the upper limit of a porous material's ability to absorb water. When the waste becomes saturated, water moves through the waste via gravity. This water will become contaminated with waste constituents in a partitioning process. The concentration of a constituent in the water depends on the amount of leachable material in the waste, the ease of mass transfer, and the ratio of the column depth (i.e., vertical thickness of waste in the landfill) to the infiltration rate. Other factors include contaminant solubility, surface area, contact time, and pH.[11]

For well-engineered landfills in which leachate generation is minimal, the concentration of a waste constituent in the leachate can approach the equilibrium solubility, given the availability of a sufficient mass of leachable material. This is because

TABLE 4-3
Releases of contaminants to surface and groundwater

Source	Volume of release	Contaminant concentration	Types of factors affecting release
Transport			
• Spills	Partial to entire contents of transport vessel	High (e.g., often pure product)	Traffic accidents; unloading mishaps
Storage			
• Spills	Partial to entire contents of storage vessels	High (e.g., often pure product)	Structural failure of tank; mishaps in handling
• Leaks	Minimal rate; yet could continue indefinitely, particularly if underground	High (e.g., often pure product)	Frequency of inspections and maintenance; age of storage facilities
Treatment			
• Effluent	Varies, often high	Low (required by regulatory permits)	Influent; design and operation of facility
Landfills			
• Runoff	Possibly large, depending on rainfall event	Low; typically contaminated sediment; nil if landfill is capped	Integrity of cap; slope; stormwater retention capacity
• Surface seeps	Minimal rate; yet could continue indefinitely	Medium to high	Characteristics of cap (slope and permeability); disposal of liquids; removal of leachate
• Leachate through base	Minimal to low rate for lined facilities; moderate to high rate for unlined facilities; continues indefinitely	Medium to high	Same as above plus permeability of base
Lagoons			
• Overflow, washout	Partial to entire contents	High (stored hazardous waste)	Structural failure; flooding
• Leakage	Minimal to low rate for lined facilities, moderate to high for unlined facilities; continues indefinitely	High (stored hazardous waste)	Permeability of base; liquid depth

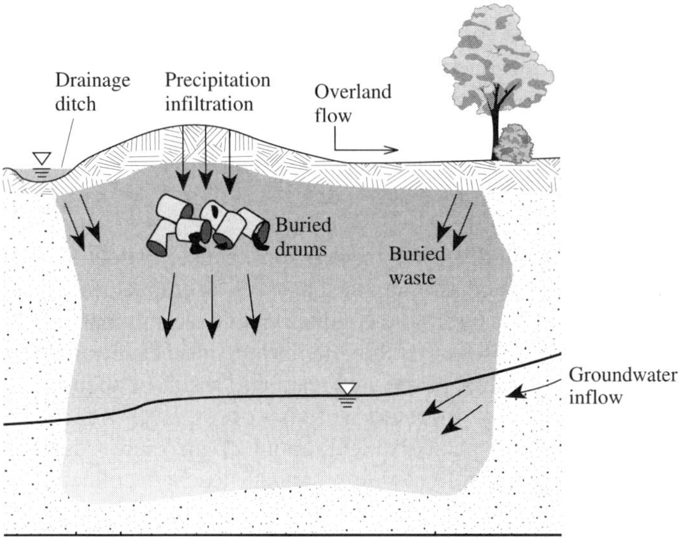

FIGURE 4-4
Sources of fluids for the generation of landfill leachate.

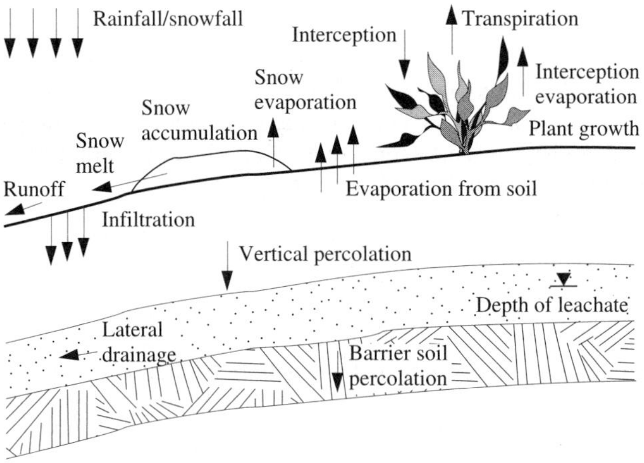

FIGURE 4-5
Water balance variables in the HELP model.[10]

the residence time in such landfills can be 20 years or more.[12] There are several complicating factors. Cosolvation (the dissolving of one contaminant in another) can result in higher concentrations of contaminants than possible for water alone, greatly exceeding equilibrium solubility. Further, retardation processes and chemical and biological transformations occur within the landfill that alter contaminant types and concentrations.

Because of these many factors, the presence of contaminants and their concentration in leachate varies widely, even within the same landfill. Analytical data

collected for leachate in a number of landfills that contained hazardous waste are shown in Tables 4-4 and 4-5 for inorganic and organic substances, respectively. The waste materials, landfill design, and climate varied widely among these landfills; hence the wide range of concentrations. Although there are analytical and computational procedures for predicting leachate concentrations, the only reliable method is to collect and analyze field samples from the actual site.

Release

A properly designed landfill will provide for leachate collection and removal. Even so, some leachate will leak through the base. For a compacted soil base, the rate of percolation is based on the hydraulic conductivity of the soil material. The case is different for a base that features a flexible membrane liner. Deterioration over time, pinholes, and improperly sealed seams and tears will result in openings in the liner and some leakage. Molecular diffusion will also occur. Even a modern, properly engineered liner system is expected to leak about 22 gal/(acre · day) through the primary liner. Much of this is collected in secondary leachate collection systems. A landfill without a liner will pass leachate as rapidly as it is generated.

EXAMPLE 4-4. LABORATORY LEACHING TEST. An engineer proposes that metal-plating sludges be disposed of in a municipal waste landfill. To demonstrate the safety of the proposal, the engineer predicted leachate concentrations based on a laboratory test. The test involved drying and pulverizing the sludge, adding distilled water (proportional to the ratio between the mass of sludge to be landfilled and the total pore volume of the landfill), and stirring the mixture for 90 days. Would this test be expected to yield conclusive predictions of worst-case leachate concentrations?

Solution. No, for at least two reasons. First, leachate in a municipal landfill will have a lower pH than distilled water because of the presence of acids from biological decomposition of the organic matter in municipal waste. Second, the 90-day test period is far less than the actual leachate residence time for a well-designed landfill. Chapter 11 provides additional detail on leaching tests.

In studies of hazardous waste management it is important to observe civilization's dependence on the need for potable water. In the United States in 1970 about

TABLE 4-4
Selected inorganic constituents detected in leachate from hazardous waste landfills

Constituent	1981 report[13] Reported concentration, μg/L	1981 report[13] Number of landfills reported	1977 report[14] Reported concentration, μg/L	1977 report[14] Number of landfills reported
As	11–<10,000,000	6	30–5,800	5
Ba	100–2,000,000	5	10–3,800	24
Cd	5–8,200	6	–	–
Cr	1–208,000	7	10–4,200	10
Cu	1–16,000	9	10–2,800	15
Hg	0.5–7	7	0.5–0.8	5
Ni	20–48,000	4	20–670	16
Pb	1–19,000	6	300–19,000	3
Se	3–590	4	10–590	21
CN^-	–	–	5–14,000	14

TABLE 4-5
Selected organic constituents in leachate from hazardous waste landfills[13]

Substance	Reported concentration, μg/L unless noted otherwise	Number of landfills reported
Acetone	0.1–62,000	3
Aldrin	< 2–< 10	2
Benzene	< 1.1–7,370	5
Chlorobenzene	4.6–4,620	5
Chloroform	0.02–4,550	4
Dichlorobenzene	< 10–517	2
1,1-dichloroethane	< 5–14,280	2
1,2-dichloroethane	2.1–4,500	5
Trans-1,2-dichloroethane	25–8,150	2
1,1-dichloroethylene	28–19,850	5
Dichloromethane	3.1–6,570	4
Ethyl benzene	3.0–10,115	4
Hexachlorobutadiene	< 20–109	2
Methylene chloride	< 0.3–184 mg/L	3
Methyl isobutyl ketone	2–10 mg/L	2
Perchloroethylene	ND–8,200	5
Phenol	< 3–17,000	4
Tetrachloroethene	< 1–89,155	3
Tetrachloromethane	< 1–25,000	3
TOC	10.9–8,700 mg/L	8
Toluene	< 5–100,000	7
1,1,1-trichloroethane	1.6 μg/L–590 mg/L	5
Trichloroethene	< 3–84,000	4
Trichloroethylene	< 3–260,000	4
Vinyl chloride	140–32,500	1

ND = Not detected

81 percent of our water supply was from surface water sources, and 19 percent is from groundwater sources.[15] More recently, the distribution is about 50/50. The distribution of water use varies significantly with region and population density. For example, many cities in the West, New Jersey, Florida, and other areas make more use of groundwater than surface water. Likewise, rural areas use groundwater more extensively than surface water. There is significant concern about the quality of groundwater and surface water and their suitability for potable water sources in terms of both quality and quantity.[16]

The fragile nature of our water supplies and the importance of releases to groundwater are best illustrated by the results of a calculation from a landfill designed and constructed in Canada.[17] In this study it was assumed that contaminant transport due to leachate leaking from the landfill could be reduced to zero and the only transport across the landfill liner system would be by diffusion (defined in detail in Section 4-2 and discussed in the context of landfill design in Section 13-6). Even assuming a very low concentration of benzene in the landfill leachate of 1 gram per liter (g/L), diffusion through the landfill liner would still amount to 19 kg/year. At this rate of 19 kg/year, combined with estimates of groundwater flow rates beneath the site, it was calculated that a total of 3.8 billion gallons of groundwater would annually be contaminated to the current drinking water standard of 5 parts per billion (5 μg/L)

for benzene. It is clear that very small quantities of toxic and hazardous chemicals have a significant impact on the quality of our water supplies.

How do we then protect our groundwater and surface water supplies for both our present and future use? The detrimental impact of contaminated groundwater as demonstrated above points to the philosophy that, where possible, pollutants should not be generated. If pollutants must be generated, they should be recycled in a safe manner. Where prevention or recycling is not possible, treatment in an environmentally safe manner should be used. Disposal or other release into the environment should be used only as a last resort, and then only in an environmentally safe manner.

4-2 TRANSPORT OF CONTAMINANTS IN THE SUBSURFACE

Hydrologic Cycle

In the subsurface environment, contaminant transport is strongly dependent on groundwater flow. More precisely, the spatial and temporal distribution of groundwater affects contaminant movement. Therefore, it is necessary to begin an understanding of contaminant processes with an understanding of the hydrologic cycle and its effect on groundwater. **Hydrology** is the science that deals with the properties and movement of water. The continuous circulation of water in the atmosphere, in the subsurface, and in surface water bodies is termed the **hydrologic cycle.** Figure 4-6 and the depicted conditions will be used to define a number of important concepts dealing with the water cycle, groundwater flow, and contaminant transport.

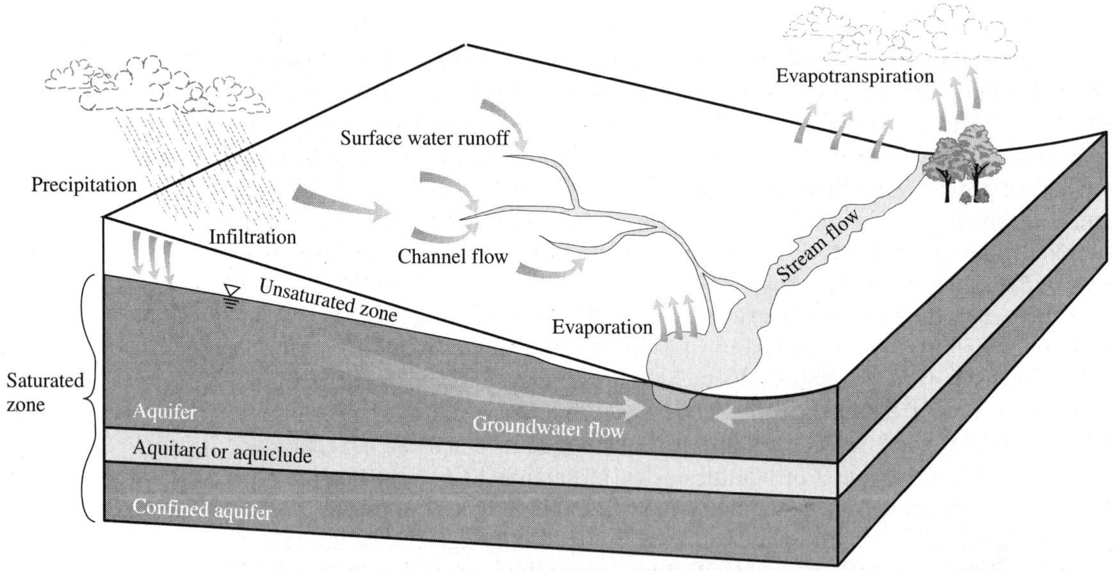

FIGURE 4-6
Hydrologic cycle.

The hydrologic cycle begins with **precipitation** that lands on the earth's surface and can be in the form of rain, snow, or sleet. From there, the water may spread along the ground surface as **surface water runoff** or **overland flow** or may seep into the ground and become groundwater. Overland flow may continue to concentrate as **channel flow**, and progressing as **stream flow**, continuing to concentrate in the form of streams and rivers until ultimately reaching the ocean.

Throughout this cycle the water may enter the subsurface environment through **infiltration** or may be returned to the atmosphere through **evaporation**. Water that does not evaporate or move on the earth's surface as overland flow enters the subsurface environment. Infiltrating water first encounters the **unsaturated zone**, in which the pores (voids) between soil particles contain soil, moisture, and air (a three-phase material). Water in the unsaturated zone may migrate further into the **saturated zone,** or may be held in the form of soil moisture to be returned to the atmosphere by plant and animal life via **evapotranspiration.**

Water reaching the saturated zone in the subsurface will flow from areas of high hydraulic head to areas of lower hydraulic head and may recharge other groundwater flow systems. The subsurface stratum, or layer, that readily permits this water movement is termed an **aquifer.** For example, a layer of sand and gravel may provide an excellent stratum to serve as an aquifer. If the upper surface of the saturated zone is free to move up or down (rise or decline) in the aquifer without being physically constrained by a stratum of lower permeability, the aquifer is termed an **unconfined aquifer.** Eventually groundwater may discharge to surface water bodies such as streams and lakes. In the migration in the subsurface the groundwater may encounter a zone that does not readily transmit water, termed an **aquitard** or a **confining layer.** A stratum of clay often serves as a confining layer. An aquifer located between confining layers is termed a **confined aquifer.** It is in the context of this hydrologic cycle that we will study the force systems that cause the migration of water and associated transport of contaminants.

Before going further it is necessary to define additional aquifer characteristics. In this context, **soil** will represent all of the unconsolidated sediment or **regolith** above the competent rock. This usage is consistent with that in the soil mechanics field of civil engineering (geotechnical engineering) and encompasses much more material than the term soil denotes in soil science (i.e., that shallow material subjected to biological and physical weathering). A **homogeneous stratum** is one in which the properties of the material (soil or rock) do not vary with location in the layer. Thus, a porous material is homogeneous with respect to permeability if the permeability is the same at all locations within the formation. A stratum whose material properties are different at different locations within the layer is said to be **heterogeneous.** An **isotropic** stratum is one in which the soil properties do not vary with orientation (direction). For example, the materials of a sand and gravel stratum may have a hydraulic conductivity higher in the horizontal direction than in the vertical direction; that is, water may flow through the soils more easily in the horizontal direction. Such a stratum is anisotropic. As an example of the possible combinations, a clay deposit may be homogeneous and isotropic with respect to permeability—the permeability is the same in the vertical and horizontal directions and is the same everywhere in the formation. A clay of lacustrine (lake) origin may be homogeneous and anisotropic with respect to permeability. In this case, the permeability is different (likely lower) in the vertical direction than in the horizontal direction. However, because the example

is considered homogeneous and anisotropic, the horizontal hydraulic conductivity is the same everywhere in the formation, and, similarly, the vertical hydraulic conductivity is the same everywhere in the formation (although they are not equal to each other).

In the real world of subsurface conditions, materials are almost always anisotropic and heterogeneous. To strictly account for this heterogeneity and anisotropy theoretically requires an infinite amount of data. Therefore, for the purposes of analysis of groundwater flow it is usually necessary to assume that a formation, or a portion of a formation, is homogeneous.

The information provided in this section of the chapter allows the student to understand the nature of contaminant fate and transport. It is important to recognize that this chapter on contaminant fate and transport is only an introduction to a complex aspect of hazardous waste management. Readers are referred to several useful texts on groundwater flow and contaminant transport including Freeze and Cherry, 1979;[18] Todd, 1984;[19] and Fetter, 1988.[20]

Groundwater Flow

Flow of groundwater in the subsurface is complex and necessitates some simplifying assumptions. The following subsections of this chapter deal with "porous media" flow, which is taken to be typical of many formations. For subsurface conditions that cannot be idealized as porous media flow, it is necessary to consider the added complexities of fractured media flow. Nonidealized flow conditions such as fracture flow are discussed later in the chapter.

To assist in our understanding of groundwater flow, let us first characterize the subsurface as a porous medium (the term **porous media model** is often used). As such, **porosity** is a dimensionless ratio defined (in percent) as the ratio between the volume of the void to the total volume. For example, if a given cubic meter of soil is half soil and half void, the porosity would be 50 percent. Another useful concept is that of **bulk density,** defined as the mass per unit volume. Hence, terms like porosity and density help us to quantitatively describe the nature of the subsurface. At this point the concept of head needs to be introduced, although it is discussed in more detail (and more quantitatively) later in this chapter. **Hydraulic head** is the term used to represent the energy of water and can be in the form of elevation head, pressure head, or velocity head. For most groundwater problems, velocity head is negligible. The total hydraulic head is thus the sum of the pressure head and the elevation head. It is the difference in total hydraulic head from point to point that is the driving force causing water to flow. With this brief introduction, we can begin our study of groundwater flow and contaminant transport. The reader is cautioned that the notation for some of the variables is different among the different academic disciplines working in this area (for example, the use of K and k for hydraulic conductivity). This text, written primarily by three civil and environmental engineers, will use notation common to the civil and environmental engineering discipline (and in particular the geotechnical engineering subdiscipline).

Darcy's law

Darcy's law forms the basis for much of our understanding of groundwater flow in the subsurface environment. Darcy developed the law in the mid-nineteenth

century in France on the basis of a series of experiments to understand water filtration. Figure 4-7 is a schematic of Darcy's experiment to determine the laws of the flow of water through clean sand.

In this experiment Darcy measured the flow of water through a column of sand and found that the flow rate was proportional to the change in hydraulic head and the cross-sectional area of flow, and inversely proportional to the length of the sand column. He found that to relate the flow rate to the hydraulic head and area of flow required a constant of proportionality that he termed k. In this text we will refer to k as the hydraulic conductivity. It has units of velocity. Note that the value is a function of both the porous medium and the fluid. It is also commonly known as the permeability. His findings describing flow through porous media are presented in algebraic form as Equation 4-3:

$$q = kiA \tag{4-3}$$

where q = flow rate (cm^3/s)
k = hydraulic conductivity (cm/s)
i = hydraulic gradient (cm/cm)
A = cross-sectional area of flow measured perpendicular to the flow direction (cm^2)

The **hydraulic gradient** i describes the rate of change in which the head (or energy) is lost as water flows through the porous materials and is defined in algebraic form as:

$$i = (h_1 - h_2)/l \tag{4-4}$$

where h_1 = head at location 1 (cm)
h_2 = head at location 2 (cm)
l = length of sand column (cm)

Note the units of $h_1 - h_2$ (change in head) and l (flow distance) are both in units of length, resulting in a gradient that is dimensionless.

Darcy's law is empirical—developed as a result of experiments. It has been verified by subsequent experiments and is considered valid over a wide range of subsurface flow conditions. For materials of high hydraulic conductivity and/or with high

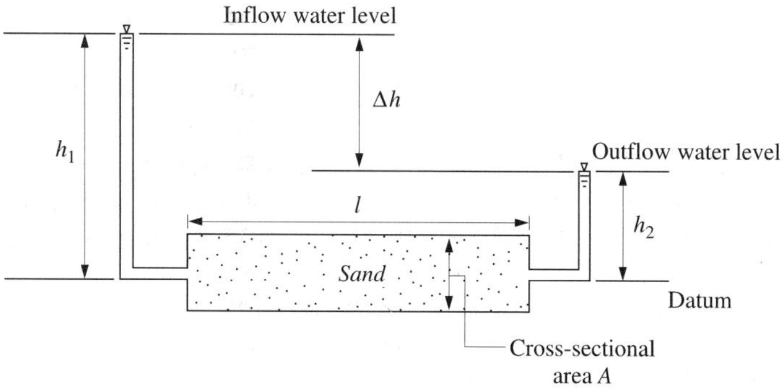

FIGURE 4-7
Darcy's experiment.

hydraulic gradients, Darcy's law is valid as long as the flow is laminar. Groundwater flow in porous media is almost always laminar. For fracture flow, discussed earlier, flow is often turbulent. For materials with low hydraulic conductivity, a threshold gradient may be required before flow can take place. For most of the conditions encountered in the subsurface, Darcy's law has been demonstrated through experience to be valid. Darcy's law is, however, macroscopic; therefore, the parameters and variables are interpreted as averages over the representative elementary volume.

By rearranging terms in Equation 4-3, Darcy's law can be expressed differently by dividing both sides of the expression by the cross-sectional area of flow to yield Q/A with units of velocity as:

$$Q/A = ki \tag{4-5}$$

where Q/A = specific discharge, Darcy flux, or Darcy velocity, v. The equation can be rewritten as:

$$v - ki \tag{4-6}$$

where v = specific discharge, Darcy flux, or Darcy velocity (cm/s). Because gradient is dimensionless, the hydraulic conductivity has units of velocity (cm/s).

Flow is in the direction of declining hydraulic head and, according to the sign convention for groundwater flow calculations, a negative sign is required. In differential form, Darcy's law can be written as:

$$v = -kdh/dl \tag{4-7}$$

where dh/dl = hydraulic gradient.

EXAMPLE 4-5. DARCY'S LAW. Using Darcy's law from Equation 4-6, can we now calculate the actual velocity of a "particle" of water, that is, the seepage velocity? Referring to Figure 4-8 and using the dimensions of $(h_1 - h_2) = 0.1$ m and $l = 0.4$ m and the hydraulic conductivity is 1×10^{-2} cm/s or 1×10^{-4} m/s (a representative value of hydraulic conductivity for medium to fine sand), what is the Darcy velocity? Is this the seepage velocity?

Solution

$$\begin{aligned} v &= ki \\ &= 1 \times 10^{-2}(0.1/0.4) = 2.5 \times 10^{-5} \text{ m/s} \end{aligned}$$

Can we assume that the time, t, required for a particle of water to migrate from one end to the other under these hydraulic head conditions is equal to the sample length divided by the velocity?

$$t = l/v = (0.4 \text{ m})/(2.5 \times 10^{-5} \text{ m/s}) = 4.4 \text{ hours?} \quad \textbf{NO!}$$

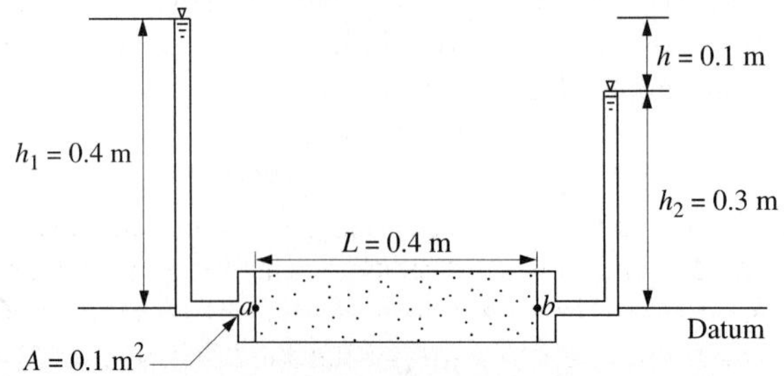

FIGURE 4-8
Schematic of saturated flow in laboratory experiment.

The specific discharge or Darcy velocity is *not* the **seepage velocity.** The seepage velocity is always faster than the Darcy velocity. The travel time cannot be directly computed by using the specific discharge.

Figure 4-8 is a schematic of the saturated water flow in Darcy's experiment, where some measured discharge Q flows through some gross cross-sectional area A at some measured hydraulic gradient i. The cross-sectional area of flow A includes the area of both solids and voids. The water migrates through only the voids and must follow flow paths around the solids. Porosity n is defined as the volume of voids V_v divided by the total volume v_t, or:

$$n = (v_v/v_t) \times 100(\%)$$

Thus, we can set Q equal to the velocity of seepage V_s times the area of voids, and invoke the equation of continuity and our definition of porosity to derive:

$$v_s = v/n \tag{4-8}$$

The travel time of a particle of water can be estimated as the linear length of the flow path, l, divided by the seepage velocity, as Example 4-6 shows.

EXAMPLE 4-6. SEEPAGE VELOCITY. Using a porosity of 0.25 and the information from Example 4-5, calculate the travel time t.

Solution

$$t = \frac{l}{v/n} = \frac{0.4 \text{ m}}{(2.5 \times 10^{-5} \text{ m/s})/0.25} \times \frac{\text{min}}{60 \text{ s}} \frac{\text{h}}{60 \text{ min}} = 1.1 \text{ h}$$

Hence, the actual travel time is less than the time predicted by using the Darcy velocity. The water particles migrate through the voids and not through the cross-sectional area represented by the Darcy velocity. The estimation of contaminant travel time through porous media uses the seepage velocity as described above. As Equation 4-8 shows, the seepage velocity depends on the porosity. More precisely, the seepage velocity depends on the **effective porosity.** In some materials, not all of the void space is available for flow and thus the effective porosity is less than the total porosity. For example, in clays some of the pores may be dead ends. Also in clays, the "bound water" in the diffuse ion layer may result in only a portion of the pore space being available for flow. The net result is that seepage velocity is underestimated for conditions where the effective porosity is less than the total porosity.

In summary, Darcy's law is empirical, derived from a series of experiments. Further, it is a porous-media model on a macro scale. Darcy's law describes flow in the subsurface in a macroscopic way and does not address the tortuous path in which an individual particle of water moves from void to void.

Hydraulic head

In Darcy's experiment, flow is induced by a change in total hydraulic head. Hydraulic head is an **energy potential**. Examples of other potentials include voltage and temperature. For example, heat moves from a region of high temperature to a region of a lower temperature. The rate at which heat is transmitted depends on the

temperature difference, in other words, the temperature potential. A chemical potential, described in more detail later in this section, drives the migration of chemicals from areas of high concentration to areas of low concentration. The hydraulic head is a fluid potential, which drives the flow of fluid from areas of higher hydraulic head to areas of lower hydraulic head. This fluid potential is described in terms of the Bernoulli energy equation, as shown in Figure 4-9:

$$z_1 + p_1/\rho_w g + v_1^2/2g = z_2 + p_2/\rho_w g + v_2^2/2g = \text{constant} \tag{4-9}$$

where $z_1, z_2 =$ distance above datum of measuring locations 1 and 2 (m)
$v_1, v_2 =$ fluid velocity at points 1 and 2 (m/s)
$p_1, p_2 =$ fluid pressures exerted by the water column at points 1 and 2 (Pa)
$g =$ acceleration of gravity (m/s^2)
$\rho_w =$ mass density of water (kg/m^3)

The Bernoulli equation presented previously defines the energy per unit weight of fluid for incompressible fluid and steady flow (units: ft · lb/lb or ft; N · m/N or m). Observe that the terms of the total energy in Equation 4-9 include contributions from velocity, pressure, and elevation. Further, the Bernoulli equation written in terms of energy per unit weight of fluid, as in Equation 4-9, results in energy expressed in units of length. The three terms, when expressed as energy per unit weight of fluid, are termed the **elevation head, pressure head,** and **velocity head.** Bernoulli's equation (Equation 4-9) can be rewritten for any point in the flow system as:

$$h_e + h_p + h_v = h_t \tag{4-10}$$

where $h_e =$ elevation head (m)
$h_p =$ pressure head (m)
$h_v =$ velocity head (m)
$h_t =$ total hydraulic head (m)

It has been found that for the range of velocity normally encountered in problems dealing with groundwater flow, the velocity head is negligible compared to the pressure and elevation heads. Hence, for groundwater flow the problem can be studied in terms of pressure and elevation head components of the total hydraulic head as follows:

$$h_p + h_e = h_t \tag{4-11}$$

The elevation head is the distance above (or below) an arbitrary datum. The pressure head is the pressure exerted by the fluid per unit weight. The **gauge pressure** p is

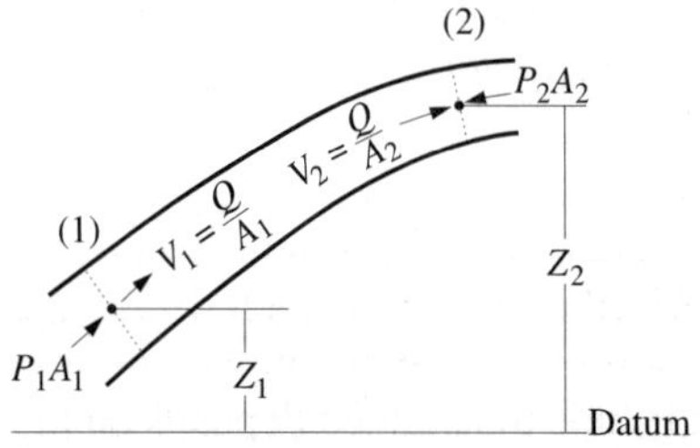

FIGURE 4-9
Bernoulli's equation for flow through a pipe.

the pressure head multiplied by the unit weight of the fluid:

$$p = \rho_w g h_p \tag{4-12}$$

where $p =$ gauge pressure (N/m^2)
$\rho_w =$ fluid density (kg/m^3)
$g =$ acceleration due to gravity (m/s^2)
$h_p =$ pressure head (m)
$\rho_w g =$ unit weight of the fluid (N/m^3)

Looking again at Darcy's experiment, we can solve for the elevation, pressure, and total head using the dimensions shown in Figure 4-10. As shown in the figure, locations or measuring points placed throughout the flow system in Darcy's experiment have been identified by the lowercase letters *a* through *g*. At these locations we can examine the head conditions in the experiment. First, an arbitrary **datum** (an elevation from which all vertical distances are measured) is assumed through the centerline of the sample in the experiment. This is assigned the notation of datum 1 to permit later examination of the effects of assuming the datum elsewhere in the system.

A table identifying selected locations in the flow field (measuring points *a* through *g*) is created in Table 4-6 as described in the following paragraphs.

In reference to Figure 4-10, measuring point *a* is 0.40 m above datum 1, yielding an elevation head of 0.40 m. Point *a* is open to the atmosphere so there is no fluid pressure; thus, the pressure head is zero (see Table 4-6). By adding the elevation head and the pressure head, the total hydraulic head at point *a* is determined to be 0.40 m.

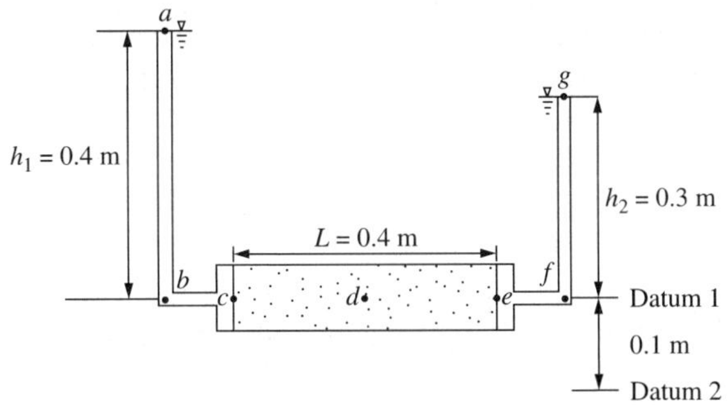

FIGURE 4-10
Hydraulic heads in Darcy's experiment.

TABLE 4-6
Values of hydraulic head from Darcy's experiment—datum 1

Location	h_e	+	h_p	=	h_t
a	0.40		0.00		0.40
b	0.00		0.40		0.40
c	0.00		0.40		0.40
d	0.00		0.35		0.35
e	0.00		0.30		0.30
f	0.00		0.30		0.30
g	0.30		0.00		0.30

At measuring point b, located on datum 1, the elevation head is zero. However, there is a column of water 0.40 m high above measuring point b exerting a fluid pressure; thus the pressure head is 0.40 m. By adding the elevation and pressure heads, the total hydraulic head is again found to be 0.40 m. Similarly, at measuring point c, on datum 1, the elevation head is zero. The pressure head (as a result of a column of water above measuring point c of 0.40 m) is 0.40 m. By addition, the total hydraulic head at measuring point c is 0.40 m. It is important to note that the total hydraulic head across the column from measuring points a to c is 0.40 m; there is no change in total hydraulic head. No head is lost in the portion of the column that does not contain soil (the open reservoir portion in the experiment shown in Figure 4-10). In this experiment, and in groundwater flow systems, total hydraulic head is lost as a result of friction on the water flowing through the porous media.

Continuing with the analysis of the downstream head conditions, point g in Figure 4-10 is examined as representative of the head in the downstream (outlet) portion of the experiment. The elevation of measuring point g is 0.30 m above the datum, that is, the elevation head is 0.30 m. Measuring point g is open to the atmosphere so there is no fluid pressure and the pressure head is zero. Summing, the total hydraulic head is equal to 0.30 m. Similarly, points f and e have a total hydraulic head equal to 0.30 m. On the outlet, or downstream, side of the experiment, the total hydraulic head is found to be constant and equal to 0.30 m.

Now let us determine hydraulic head conditions within the sand of the experiment. To calculate the head within the soil column, such as at measuring point d, it is first necessary to calculate the hydraulic gradient. In this case the change in hydraulic head across the sample column is the head at measuring point c (h_c) minus the head at measuring point e (h_e) or $h_c - h_e$; $0.40 - 0.30$ equals 0.10 m. The linear length of flow, L, through which the head changes is equal to 0.40 m. The gradient, $(h_c - h_e)/L$, is thus equal to 0.25. The hydraulic head at measuring point d is thus equal to the hydraulic head at point c minus the head loss as water flows a distance of 0.20 m (at a head loss rate of 0.25 m/m) from measuring points c to d. Therefore the total hydraulic head at point d is $(0.4) - (0.25 \times 0.2) = 0.35$ m. Having found the total hydraulic head at measuring point d to equal 0.35 m and knowing the elevation head h_e to equal 0 m, it is possible to solve for the pressure head to equal 0.35 m, as shown in Table 4-6.

These calculations show that the energy lost because of flow is energy measured in terms of total hydraulic head. The hydraulic gradient must be calculated in terms of total hydraulic head.

The hydraulic head values were also determined in Darcy's experiment by using datum 2 in Figure 4-10. The results are presented on Table 4-7.

What can we learn about the nature of hydraulic head by these calculations using datum 1 and datum 2? First, note that the values of total hydraulic head for each measurement in Table 4-7 are different from those in Table 4-6. The total hydraulic head depends on the selection of the datum. However, the change in total hydraulic head from points c to d to e is the same (independent of the datum), as is the total hydraulic head difference causing flow (the head difference between measuring points). Thus, although the values of total hydraulic head depend on the datum, the head difference is independent of datum, and as a result, the gradient is independent of datum. The elevation head depends on the datum. Importantly, the pressure head is independent of the datum.

TABLE 4-7
Hydraulic heads from Darcy's experiment—datum 2

Location	h_e	+	h_p	=	h_t
a	0.50		0.00		0.50
b	0.10		0.40		0.50
c	0.10		0.40		0.50
d	0.10		0.35		0.45
e	0.10		0.30		0.40
f	0.10		0.30		0.40
g	0.40		0.00		0.40

Two-dimensional flow

To illustrate the principles and terms associated with two-dimensional flow, Figure 4-10 is redrawn with the dimensions shown in Figure 4-11, and measuring points *r* through *z* have been labeled. By using the Bernoulli equation and the concepts of one-dimensional flow previously described, the elevation head, pressure head, and total hydraulic head have been determined and are shown on Table 4-8.

Measuring points selected along cross sections are labeled *r–s–t*, *u–v–w*, *x–y–z*. On examination of the data generated in Table 4-8, it is found that the measuring points on each of the cross sections, *r–s–t*, *u–v–w*, and *x–y–z*, have a constant total

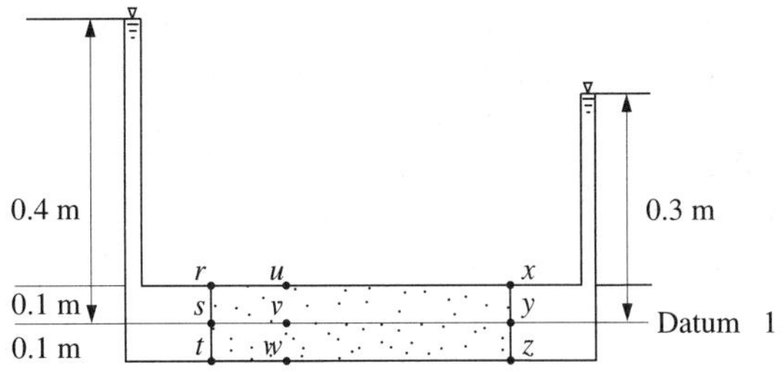

FIGURE 4-11
Darcy's experiment revisited.

TABLE 4-8
Hydraulic heads from Darcy's experiment—datum 1

Location	h_e	+	h_p	=	h_t
r	0.10		0.30		0.40
s	0.00		0.40		0.40
t	−0.10		0.50		0.40
u	0.10		0.275		0.375
v	0.00		0.375		0.375
w	−0.10		0.475		0.375
x	0.10		0.20		0.30
y	0.00		0.30		0.30
z	−0.10		0.40		0.30

hydraulic head (e.g., the total hydraulic head is equal to 0.30 at x–y–z, 0.375 at u–v–w). Contouring the values of equal total hydraulic head in Figure 4-11 results in Figure 4-12, which shows the contours of equal total hydraulic head as dotted lines.

These lines of equal total hydraulic head are known as **equipotentials;** that is, the potential energy at any point on one of these lines is the same. **Flow lines,** by definition, are perpendicular to the equipotentials, as shown in Figure 4-12. The Laplace equation will be derived later in this chapter as the solution for flow in isotropic, homogeneous porous media.

Although we can draw flow lines and equipotentials for the Darcy experiment, it should be noted that all flow is along the axis of the tube of porous media. In this case, all flow is in the x direction with the flow in the y direction equal to zero. The more general case of two-dimensional flow is illustrated by an example of flow through a dike surrounding a liquid impoundment shown in Figure 4-13.

Hydraulic Conductivity of Geologic Materials

The hydraulic conductivity k in Equation 4-3 is a measure of the capability of a material to transmit water. It is an "order-of-magnitude" parameter. The most imper-

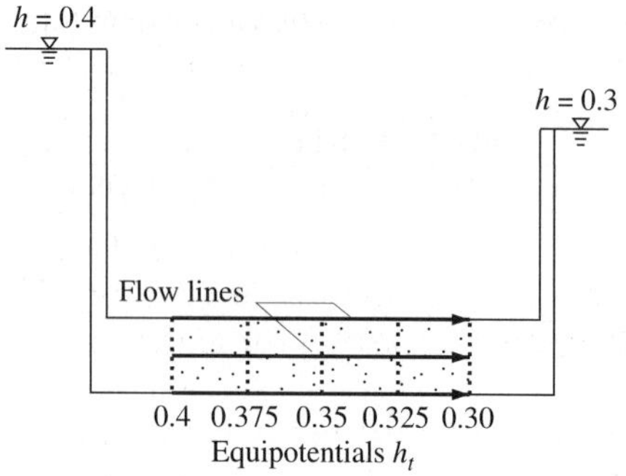

FIGURE 4-12
Flow lines and equipotentials.

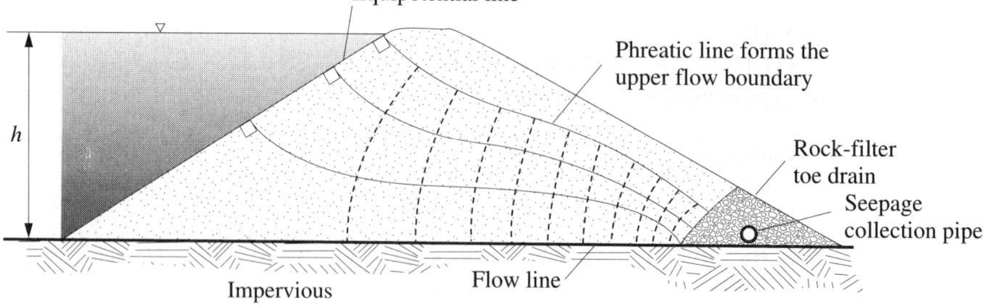

FIGURE 4-13
Flow net for steady-state flow through a homogeneous embankment. *(After Cernica, 1982.)*[21]

meable unconsolidated material (unconsolidated material is synonymous with soil in civil engineering terminology) may have a hydraulic conductivity of 1×10^{-9} cm/s. At the other extreme, open-worked clean gravel may have a hydraulic conductivity of 1×10^{5} cm/s. Stated another way, the gravel is 100,000,000,000,000 times more permeable than clay. Presented on Table 4-9 is the range of expected values for the hydraulic conductivity of various geologic materials. The values provide a starting point for the assessment of groundwater flow. There are two facts to remember about the values in this table.

First, under the best investigatory conditions, hydraulic conductivity can be estimated only within order-of-magnitude accuracy. The ramifications of being an order of magnitude off in a groundwater flow calculation must be considered. For example, if the flow to a groundwater extraction well used in site remediation (see Chapter 16) is calculated to be 10 gal/min in a clean sand estimated to have a permeability of 1×10^{-2} cm/s, the flow could be 100 gal/min should the actual permeability of the formation be 1×10^{-1} cm/s. Could the remedial groundwater pumping and treatment system accommodate this variation? If not, extreme care and adequate testing must be employed to reduce the risk of overestimating or underestimating the hydraulic conductivity (see Chapter 15).

The second aspect of the values in Table 4-9 to bear in mind is that the value of hydraulic conductivity for a given material type may be quite different than that for the entire formation. For example, consider that a sample of glacial clay may have a hydraulic conductivity of 1×10^{-7} cm/s. However, should the clay contain interconnected seams of sand, the formation hydraulic conductivity may be several orders of magnitude higher, even if the sand seams constitute only a minute percentage of the overall deposit. Similarly, should the clay be highly fractured, as glacial tills often are, the secondary structure (i.e., the fractures) would allow for considerably greater flow than the portion of the clay that is not fractured. Given the heterogeneity and anisotropy that exist in the natural subsurface environment, it is little wonder that the accurate prediction of groundwater flow and contaminant transport in the subsurface environment requires considerable investigation, analysis, skill, experience, and luck.

Darcy's law (Equation 4-3) shows that the flow is proportional to both the hydraulic conductivity of the medium as well as the cross-sectional area of flow. It is convenient (for both mathematical modeling and analysis of field data) to define the **transmissivity** of the formation as the hydraulic conductivity times the thickness of

TABLE 4-9
Typical ranges of hydraulic conductivity for various soils

Soil type	Hydraulic conductivity, cm/s
Clean gravel	1×10^{5} to 1.0
Clean sand or sand + gravel mixtures	1.0 to 1×10^{-3}
Fine sands and silts	1×10^{-2} to 1×10^{-6}
Silty clay and clay	1×10^{-5} to 1×10^{-9}

the formation. In equation form, we have

$$T = kt \tag{4-13}$$

where $T =$ transmissivity (cm^2/s)
$k =$ hydraulic conductivity (cm/s)
$t =$ aquifer thickness (cm)

Thus, transmissivity is defined as the rate of water flow through a vertical section of an aquifer (over its full saturated thickness, one unit wide), under a unit hydraulic gradient.

To better understand the interaction between hydraulic head and flow from an aquifer it is necessary to describe the storage capacity of the aquifer. As water is pumped from an unconfined aquifer and the water level drops, the saturated portion of the aquifer becomes unsaturated. Unsaturated does not imply dry. An unsaturated material is one in which the void space is not completely filled with water. The volume of water that drains from the saturated soil pores as the water table drops is termed the **specific yield.** For unconfined conditions as just described, the specific yield is synonymous with the **storativity.**

The storativity for a confined aquifer situation is more complex than for the unconfined case described above. As water is pumped from a confined aquifer, the aquifer remains saturated. However, as the hydraulic head decreases, the **effective stress** increases, causing a small but real compression in the aquifer. In addition, at the lower water pressure level there is a small, but real, expansion in the water (although we often assume water is incompressible, a given mass of water will expand or contract with changes in pressure). The combination of these two phenomena is responsible for the storativity of confined aquifers. Thus, for confined aquifers, storativity is not synonymous with specific yield.

Coupled flow phenomena

In preceding sections, flow induced by hydraulic gradients has been described. In some cases, groundwater may flow in response to gradients other than, or in addition to, hydraulic gradients. These other gradients include voltage, chemical concentration, and temperature gradients. Flow induced by more than one gradient is termed coupled flow. A general flow law would include a specific discharge that includes flow in response to all applicable gradients.[22] The importance of coupled flow processes with respect to hazardous waste management is illustrated by the proposed use of electrokinetics for the cleanup of contaminated groundwater.[23] The derivation and a complete discussion of coupled flow phenomena are beyond the scope of this text, although the description and use of electrokinetics for groundwater remediation are subsequently detailed in Chapter 16.

The Unsaturated Zone

The foregoing discussions describe flow in the zone below the water table. How do the processes differ in the unsaturated zone? First, a few terms need to be defined:

Water table A surface on which the fluid pressure (pressure head) is exactly atmospheric; this is sometimes called the phreatic surface and applies to unconfined aquifers.

Phreatic zone The zone located below the water table where the fluid pressure is positive (in compression).

Vadose zone The zone located above the water table where the fluid pressure is negative (in tension). The zone is divided into a saturated portion, due to capillary rise, and an unsaturated or partially saturated zone.

Tension saturated zone The zone located above the water table where the pressure head has a negative value and the void spaces are filled with water because of capillary suction (also known as capillary fringe).

Piezometric (potentiometric) surface A surface that results from plotting and contouring hydraulic head in a confined aquifer (hydraulic heads in a confined aquifer can be obtained from piezometers as discussed in Chapter 15).

In the vadose zone, the pressure head is termed the **suction head,** ψ, and is negative. The total hydraulic head remains the sum of the elevation head and the pressure head. As noted above, there is some water in the unsaturated zone, and the pressure head depends on this soil moisture content, θ. Further, the relationship between θ and ψ is hysteretic; that is, it is different depending on whether the soil is being wetted or dried. The hydraulic conductivity in the unsaturated zone is similarly affected. A typical relationship is shown in Figure 4-14. As a result of the changing hydraulic conductivity and pressure head with changing water content, the flow models are considerably more complex for the unsaturated zone.

The unsaturated zone, also known as the **vadose zone,** extends from the ground surface down to the water table. Since this is a three-phase system (air, water, and soil), the flow of water and the transport of contaminants is often more complex than for the two-phase (water and soil) saturated zone. For example, as we have just seen, the hydraulic conductivity depends on the suction head. The importance of the unsaturated zone in the field of hazardous waste management is further highlighted by the recognition that much of the current subsurface contamination originates at or near the ground surface, frequently in the unsaturated zone. Many of the concepts discussed in this chapter relating to contaminant transport are relevant to the unsaturated zone. But for a detailed look at flow and transport in the unsaturated zone, the reader is directed elsewhere.[25] In Chapters 10 and 16, this text will address some of the means by which the unsaturated zone can be remediated.

Contaminant Transport Mechanisms

Contaminants are found in the subsurface environment in various forms (or phases). Contaminants that are dissolved in water are the **solutes,** and water is the **solvent.** Water having dissolved contaminants (solutes) flowing in a porous medium is schematically shown in Figure 4-15. As the water flows, the dissolved contaminants are transported with the water. This method of contaminant transport is termed **advection.** Advection is the transport of solutes along stream lines at the average linear seepage flow velocity. On a macroscopic scale, the porous medium model governs the flow rate and direction. However, on a microscopic scale, the porous medium is composed of discrete solid particles and pore spaces or voids. Water flows, not through the solid particles but around them through the interconnected pore space. These concepts are shown schematically in Figure 4-15.

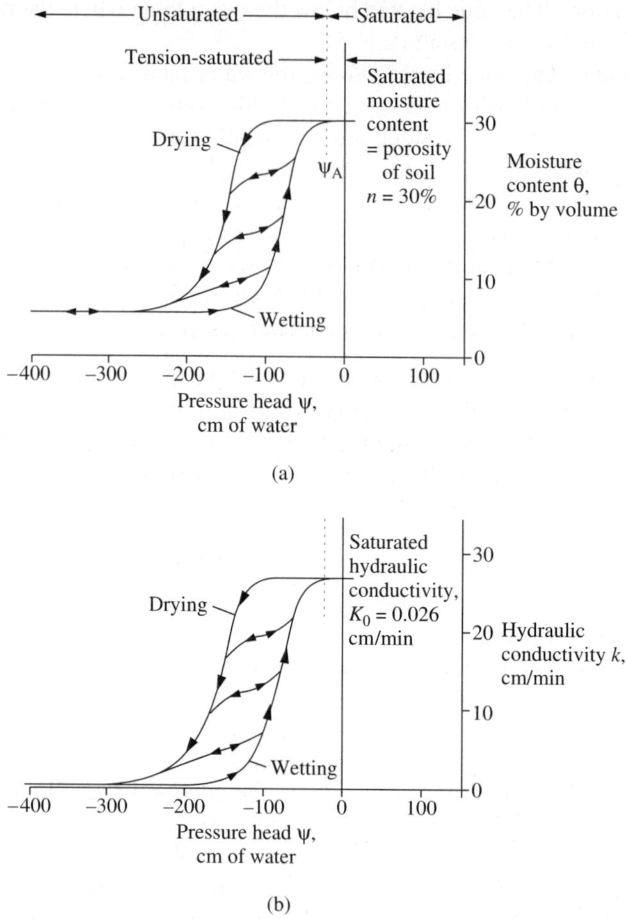

FIGURE 4-14
Characteristic curves relating hydraulic conductivity and moisture content to pressure head for a naturally occurring sand soil. *(After Liakopoulos.)*[24]

Flowing water, on encountering the solid particles of the porous media, must alter its course by flowing around the solids, either to the right or left of the solid or by spreading and migrating to both sides. This process is repeated millions of times by millions of water "particles." The result is a mixing of the flowing water by mechanical means termed **mechanical dispersion** (also known as **hydrodynamic dispersion** or just **dispersion**). The most important effect of dispersion is to spread the contaminant mass beyond the region it would occupy without dispersion.

If the contaminants can be assumed to enter the porous medium "instantaneously" at a discrete location with known concentration of contaminants, then this mass of water and dissolved contaminants is termed a **slug,** and the discrete location is termed a **point source.** The contaminant concentration will decrease in

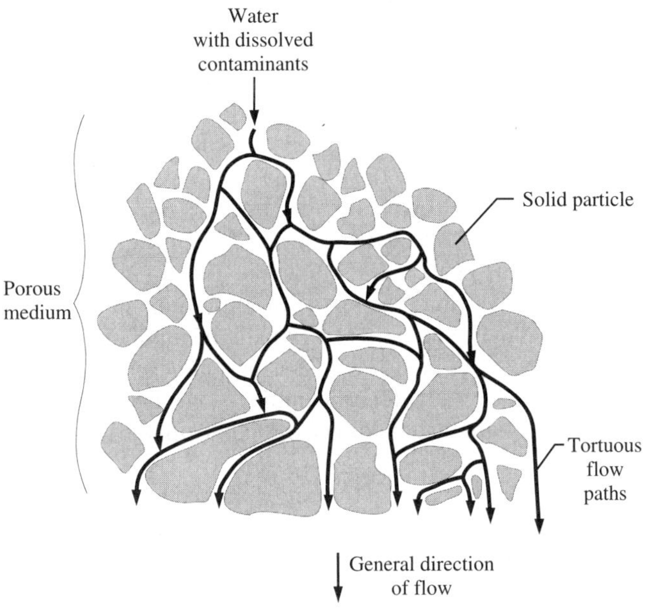

FIGURE 4-15
Schematic of mechanical dispersion.

the slug as it migrates through the formation as a result of the mechanical dispersion spreading the mass of contaminants over a larger and larger volume and mixing with water without the contaminants. The result is a **dilution** or reduction in contaminant concentration. Thus, contaminants are transported primarily by advection, and their concentration changes as a result of dispersion.

To investigate further the transport of contaminants, a laboratory experiment similar to the Darcy experiment could be set up as shown in Figure 4-16. In this configuration the contaminant is introduced not as a one-time slug described above, but continuously and at the same concentration. This is termed a **continuous source**

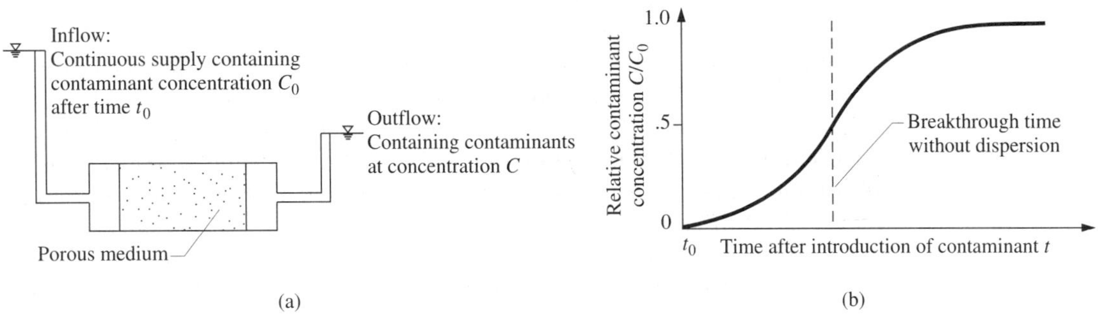

FIGURE 4-16
Effect of dispersion on contaminant transport.

and is representative of releases such as that from an underground storage tank with a small leak that goes undetected. The concentration of the contaminant can be measured as a function of time at the down-gradient end of the column experiment. Without mechanical mixing (dispersion), the water would be free of contaminants until the contaminant **front** reaches the outflow end of the sample. The time required would be determined by the average linear seepage velocity previously defined. However, mechanical dispersion does occur, and as a result, contaminant arrival at the outflow (termed **breakthrough**) occurs more rapidly than the seepage velocity would indicate. However, at the time the contaminant is first detected, the concentration is much lower than the input concentration. The concentration at the outflow, C, is presented in Figure 4-16 as a ratio to the input concentration as C/C_0. The contaminants arrive at the outflow faster than indicated by the average linear seepage velocity. The time required for the outflow concentration to equal the inflow concentration is much longer than predicted directly from the average linear seepage velocity.

The concept of dispersion can be further explained by examining the nature of the contaminant distribution with time in a homogeneous and isotropic formation. The distribution and extent of contaminants migrating in the subsurface is termed the **plume.** Shown in Figure 4-17 is the plan view of a plume at various times. This figure then depicts the effects of dispersion on contaminant transport for two separate cases. In Figure 4-17a the source is a continuous point source loading such as a leaking underground storage tank described earlier. At time t_1, the contaminants have migrated down-gradient and have spread laterally by mechanical dispersion. With increasing time, the plume has migrated farther down-gradient and continues to spread laterally. Alternatively, a plume resulting from a one-time event (a slug) such as a spill is shown in Figure 4-17b. In this case the slug migrates down-gradient, spreading laterally to encompass a larger and larger area, albeit at lower and lower concentrations. Observe that dispersion spreads the plume both laterally (width) and longitudinally (length), although the longitudinal dispersion is masked by the advective transport. The longitudinal dispersion is also illustrated by Figure 4-16.

Contaminants (particularly ionic and molecular constituents) will also move in response to their chemical kinetic activity. The movement is from areas of higher concentration to areas of lower concentration and is termed diffusion. The terms **molecular diffusion, ionic diffusion,** and **self-diffusion** are often used synonymously with diffusion. The difference in contaminant concentration taken with respect to the distance over which the concentration difference occurs is referred to as the **concentration gradient** (see Equation 3-9).

Shown in Figure 4-18 is the Darcy laboratory experiment in which there is no hydraulic head difference from one end of the sample to the other. At time t_0, there is a concentration C_0 at the left-hand end of the porous media column and a concentration equal to zero at the right-hand end of the column, thus creating a concentration difference from one end of the sample to the other. Because there is no hydraulic head difference, the hydraulic gradient, Darcy flux, and advective transport are all equal to zero. Yet, as shown in Figure 4-18, contaminants migrate through the sample in response to the chemical concentration gradient. The final concentration ratio at the right-hand side of the column is less than 1.0. The concentration ratio depends on the initial concentration and the total volume of the sand and water

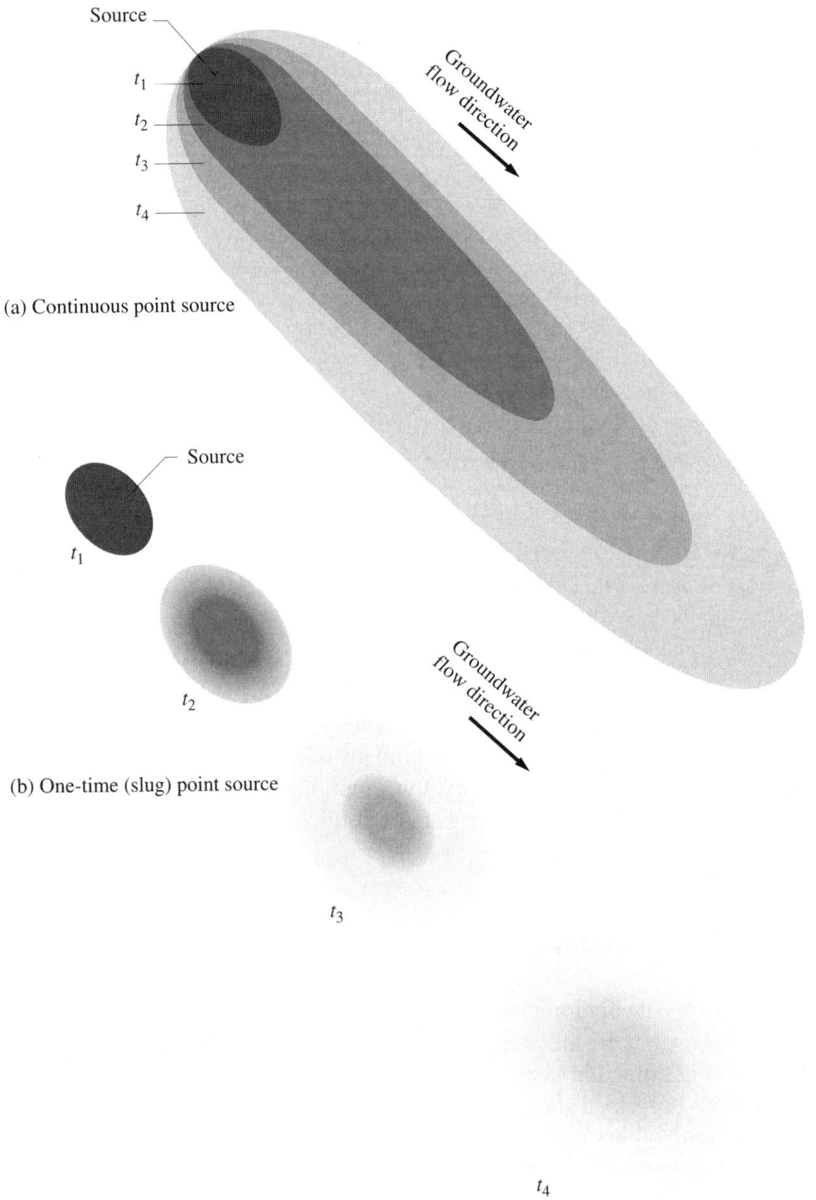

FIGURE 4-17
Plume migration affected by dispersion and source type. The darker the area, the higher the contaminant concentration.

in the laboratory experiment. If contaminants were added to the left-hand side of the sample during the diffusion to maintain the initial concentration C_0, then the equilibrium concentration ratio would be 1.0 on the right.

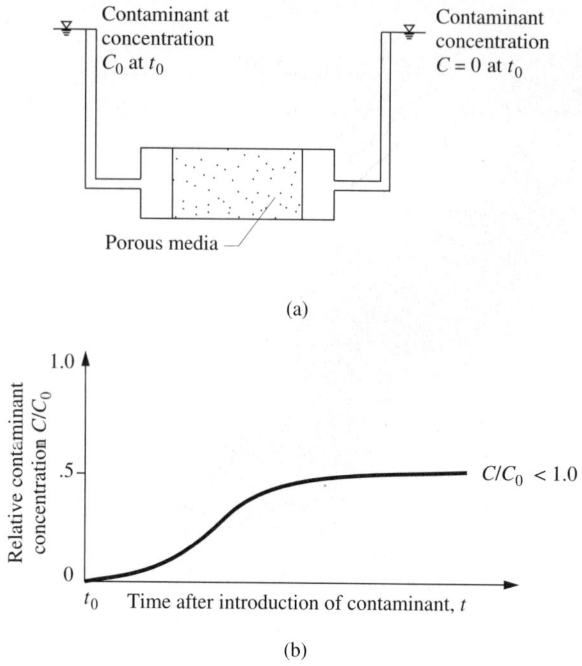

FIGURE 4-18
Effect of diffusion on contaminant transport with no advective transport.

It can be generally stated that advective transport and the associated mechanical dispersion processes dominate the contaminant transport system in formations of medium-to-high hydraulic conductivity. In formations of low hydraulic conductivity, including clay liners in waste management facilities, diffusive transport is frequently the controlling mechanism.

Nonidealized (Real-World) Contaminant Transport

The basic concepts of contaminant transport in the subsurface environment have been described. However, there are many mechanical, chemical, and biological processes by which the contaminant transport differs from that predicted by using the concepts and assumptions described above. Some processes will retard the rate of contaminant transport whereas others can speed up the rate. Some natural processes affecting contaminant transport are summarized in Table 4-10. The chemical and biological processes are discussed in more detail in Section 4-3, Fate of Contaminants in the Subsurface.

There are also some mechanical processes and special conditions that depart from idealized flow including fractured media, heterogeneity, and nonaqueous-phase liquid flow.

Fractured media flow

Up to this point, the porous media model has been used as the basis for our understanding of groundwater flow and contaminant transport. Indeed, many sites can be adequately characterized by a porous media model. However, there are also many sites where the flow is better characterized as fractured flow. For example,

TABLE 4-10
Some natural processes affecting contaminant transport

Process type	Process
Mechanical (physical) processes	Advection
	Dispersion
	Diffusion
	Density stratification
	Nonaqueous-phase fluid flow
	Fractured media flow
Chemical processes	Oxidation-reduction reactions
	Ion exchange
	Complexation
	Precipitation/dissolution
	Immiscible phase partitioning
	Sorption
	Hydrolysis
Biological processes	Aerobic degradation
	Anaerobic degradation
	Cometabolism
	Biological uptake

Figure 4-19 illustrates the subsurface stratigraphy at a site showing the bedding planes and vertical fractures in the underlying limestone bedrock.

Flow and associated contaminant transport will occur predominantly along the joints and fractures and not in the bulk rock. In essence the flow is more like pipe flow

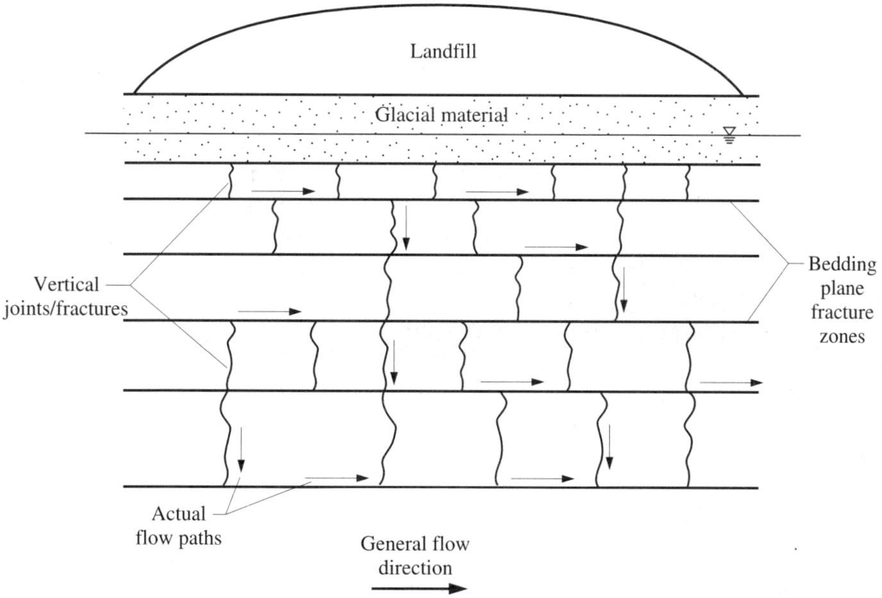

FIGURE 4-19
Schematic of fractured flow.

than porous media flow. Efforts are currently under way to develop readily usable techniques to enable predictions of groundwater flow and contaminant transport under fractured flow conditions, although they have not as yet been widely employed in the hazardous waste management field. Present approaches to describe flow and transport in fractured rock systems are:

1. Fractured rock system as an equivalent porous medium (requires adequate density of fractures)
2. Discrete fracture approach where channels are represented as channels with parallel sides and individual fractures are combined into fracture networks

Heterogeneity

Heterogeneity in porous media also alters the idealized transport processes described above. As defined previously, heterogeneity is a spatial variation in hydraulic and transport properties originating from the nature of the geologic properties. Figure 4-20 illustrates the effect of a lens of higher hydraulic conductivity within an otherwise homogeneous and isotropic formation. The Darcy flux is greater in the zone of higher hydraulic conductivity than in the adjacent zones under the same gradient. Correspondingly, the linear seepage velocity and the rate of advective transport are greater in the more permeable lens.

Nonaqueous-phase liquid (NAPL)

The problems of contaminant transport predictions are further complicated by the presence of **nonaqueous-phase liquid** (NAPL). Many organic liquid compounds are characterized as **immiscible,** that is, they have a very low **solubility** in water. As a result these nearly insoluble organic compounds frequently exist as a separate

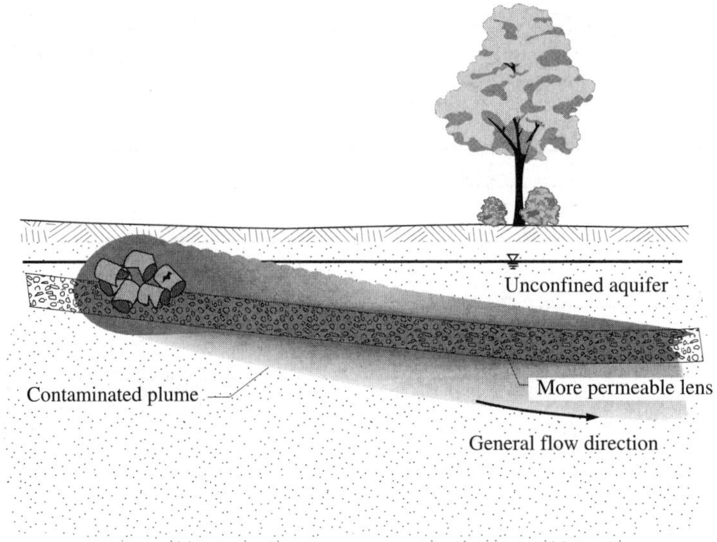

FIGURE 4-20
Effect of high-permeability zone on contaminant transport.

liquid phase in the subsurface. This separate liquid phase typically has a different density and viscosity than water. A NAPL with a density greater than water is termed a **sinker** [dense nonaqueous-phase liquid (DNAPL); e.g., chlorinated solvents], and a NAPL with a density lighter than water is termed a **floater** [light nonaqueous-phase liquid (LNAPL); e.g., petroleum products]. The transport of NAPL is influenced by gravitational forces as well as differences in hydraulic head. As with fractured flow, we are only beginning to understand transport of NAPL,[26] and analytical techniques have not been widely employed. The migration of LNAPL is largely dependent on the slope of the water table. The migration of DNAPL is strongly dependent on the subsurface stratigraphy, particularly the distribution of zones of low permeability that act as barriers. Typically, the volume of a groundwater plume contaminated with dissolved NAPL constituents is many times greater than a NAPL plume, although the greater mass of contaminants may be in the NAPL plume. Figure 4-21a is a schematic of the distribution of DNAPL from a leaking underground storage tank near the surface. Although the hydraulic gradient is from left to right, the DNAPL has predominantly migrated downward and down-slope (even if hydraulically upgradient).

Figure 4-21b is a schematic of the distribution of LNAPL from a leaking underground storage tank near the surface. As shown, LNAPL floats on the groundwater and spreads laterally as well as being transported down-gradient.

Selected Flow Equations

In this section of the chapter, we will look at the equations that govern groundwater flow and thus impact contaminant transport. Derivations, where they enhance our understanding of the processes, will also be presented by making simplifying assumptions, and the solutions are extended to more complex boundary conditions.

Figure 4-22 illustrates a two-dimensional control volume. The Darcy flux into the control volume in the x and y directions is v_x and v_y, respectively. The dimensions of

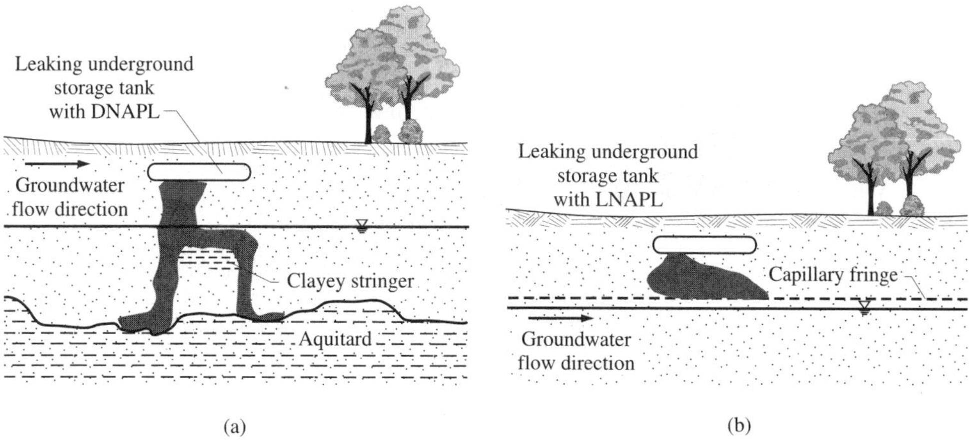

FIGURE 4-21
Movement of NAPL.

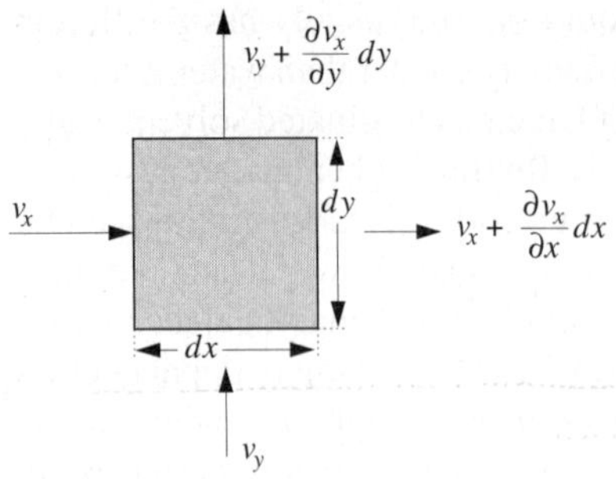

FIGURE 4-22
Two-dimensional control volume.

the element are dx and dy. Assuming no change in the control volume, the **equation of continuity** (what goes in, must come out) tells us:

$$Q_{\text{in}} = Q_{\text{out}} \tag{4-14}$$

where Q_{in} = flow rate into control volume
Q_{out} = flow rate out of control volume

The volumetric flow rate is equal to the Darcy flux times the area of flow:

$$Q_{\text{in}} = V_{\text{in}} A \tag{4-15}$$

where Q_{in} = volumetric flow rate into control volume (cm^3/s)
V_{in} = Darcy velocity (cm/s) into control volume
A = area of flow (cm^2)

Thus, the flow into the control volume is:

$$Q_{\text{in}} = v_x\, dy + v_y\, dx \tag{4-16}$$

and the flow out of the control volume is:

$$Q_{\text{out}} = \left(v_x + \frac{\partial v_x}{\partial x}\, dx\right) dy + \left(v_y + \frac{\partial v_y}{\partial y}\, dy\right) dx \tag{4-17}$$

where $\partial v_x/\partial x$ = the rate of change in the Darcy flux in the x direction
$\partial v_y/\partial y$ = the rate of change in the Darcy flux in the y direction

With no net change in our control volume, the inflow and outflow volumes are equal, resulting in:

$$0 = (\partial v_x/\partial x)\, dx\, dy + (\partial v_y/\partial y)\, dy\, dx \tag{4-18}$$

For a square element with unit dimensions ($dx = dy = 1$), Equation 4-17 reduces to:

$$0 = \partial v_x/\partial x + \partial v_y/\partial y \tag{4-19}$$

Darcy's law can be written for the flux in the x and y directions as:

$$v_x = k_x\, \partial h/\partial x \tag{4-20}$$

where k_x = hydraulic conductivity in the x direction, and k_y = hydraulic conductivity in the y direction,

$$v_y = k_y \, \partial h/\partial y \tag{4-21}$$

Substituting into Equation 4-19 yields:

$$0 = k_x \, \partial^2 h/\partial x^2 + k_y \, \partial^2 h/\partial y^2 \tag{4-22}$$

Finally, for isotropic flow conditions $k_x = k_y$, Equation 4-28, the **Laplace equation,** for a two-dimensional flow field can be written as:

$$\partial^2 h/\partial x^2 + \partial^2 h/\partial y^2 = 0 \tag{4-23}$$

The Laplace equation describes the value of hydraulic head at any point in the two-dimensional flow field. For a three-dimensional situation the Laplace equation becomes:

$$\partial^2 h/\partial x^2 + \partial^2 h/\partial y^2 + \partial^2 h/\partial z^2 = 0 \tag{4-24}$$

Note that the value of hydraulic head is a function only of position in the flow field. The value of hydraulic head is not a function of the hydraulic conductivity under isotropic flow conditions. Implicit in the derivation of the Laplace equation is the assumption that the flow is **steady** (i.e., not changing with time). The equations of groundwater flow are considerably more complex for **unsteady** or **transient** flow conditions. The Laplace equation can be rewritten in three dimensions for unsteady flow in a saturated confined aquifer as follows:

$$\partial^2 h/\partial x^2 + \partial^2 h/\partial y^2 + \partial^2 h/\partial z^2 = [\rho g(\alpha + n\beta)/k](\partial h/\partial t) \tag{4-25}$$

where x, y, z = positions in the flow system
h = hydraulic head (cm)
ρ = density of water (g/cm^3)
g = acceleration of gravity (cm/s^2)
α = aquifer compressibility (cm^2/N)
n = aquifer porosity (dimensionless)
β = compressibility of water (cm^2/N)
k = hydraulic conductivity (cm/s)
t = time (s)

An examination of the transient form of the Laplace equation reveals that the head is a function of four variables (x, y, z, and t) as well as six constants (ρ, g, n, k, a, β).

The Laplace equation for horizontal unsteady flow in a saturated unconfined aquifer is:

$$\partial^2 h/\partial x^2 + \partial^2 h/\partial y^2 + \partial^2 h/\partial z^2 = (S/T)(\partial h/\partial t) \tag{4-26}$$

where x, y, z = spatial positions in the flow system
h = hydraulic head (cm)
S = aquifer storativity (g/s)
T = aquifer transmissivity (cm^2/s)
t = time (s)

Solutions to the Laplace equation

The Laplace equation describes the steady-state hydraulic head conditions throughout the groundwater flow regime. The solution of this partial differential equation can be by analytical, graphical, electrical analog, or numerical methods. Analytical methods are possible for only relatively simple boundary conditions. Graphical methods (flow nets) provide good, relatively rapid approximations to the flow conditions and are useful for first approximations or "back of the envelope" checks of the results from more sophisticated numerical methods. Electrical analog methods are time-consuming and expensive.

Most complex site and subsurface conditions must depend on the use of a numerical method that employs a system of equations solved by an iteration method. The method yields, by successive approximation, an approximate solution to the system of mathematical equations. The two commonly employed numerical approximation methods are finite element and finite difference methods. In the finite element approach, the partial differential equations are approximated with integral representations. The finite difference method replaces the continuous flow field with discrete units and describes the governing partial differential equations by methods used in differential calculus.

The graphical method (flow nets) and the finite difference method are described below. For additional information regarding these and other numerical methods as applied to groundwater modeling, the reader is referred to the References section.[27–30]

Flow nets

Flow nets are two-dimensional graphical representations of the hydraulic head conditions in the subsurface. A flow net is composed of equipotentials (lines of equal hydraulic head) and flow lines (lines of flow normal to equipotentials representing the average path a particle of water takes as it flows in the subsurface). The flow net is a solution to the two-dimensional steady-state flow conditions under the specified boundary conditions. Why draw a flow net in this age of computer solutions? For the inexperienced, construction of a flow net develops a sense of appreciation of the nature of groundwater flow and, for the experienced, a flow net provides an independent validation of results obtained from more sophisticated methods.

Perhaps the easiest example to visualize is the flow through a homogeneous, isotropic embankment on an impervious foundation surrounding a liquid impoundment, as shown earlier in Figure 4-13. The figure illustrates several principles regarding flow net construction. First, the flow lines are normal to the equipotentials resulting in "squares" bounded by equipotentials and flow lines. Also, the impervious boundary is a flow line and the free water boundary is an equipotential. A flow channel is that flow path bounded by two adjacent flow lines. The fundamental properties of flow nets can be summarized as follows:[31]

- The head difference between any pair of adjacent equipotentials is the same as any other pair.
- Flow lines intersect equipotentials at right angles.
- Figures enclosed by adjacent pairs of equipotentials and flow lines are essentially squares.

- The spacing of equipotentials is inversely proportional to the hydraulic gradient (and thus the Darcy velocity).
- Every flow channel transmits the same quantity of seepage.

It is possible to determine the hydraulic gradient and the quantity of seepage from a well-constructed flow net. The seepage quantity under homogeneous, isotropic flow conditions can be shown to be:[32]

$$q = khn_f/n_d \tag{4-27}$$

where q = flow rate per unit length [cm^3/(s · cm)]
k = hydraulic conductivity (cm/s)
h = hydraulic head (cm)
n_f = number of flow channels
n_d = number of equipotential drops

The hydraulic gradient between two equipotentials can be determined as:

$$i = (h/n_d)/l \tag{4-28}$$

where i = hydraulic gradient
h = hydraulic head (m)
n_d = number of equipotential drops
l = distance along flow path between the two equipotentials under consideration (m)

In summary, flow nets can provide a relatively rapid means to examine accurately the hydraulic head and flow conditions in the subsurface. Flow nets can be constructed for two-dimensional flow through a cross section or in plan view. The following example illustrates a flow net constructed in plan and demonstrates the calculations for seepage quantity and gradient.

EXAMPLE 4-7. FLOW NET FOR A WATER TABLE AQUIFER. The following figure is a flow net constructed in plan view for a water table aquifer discharging to a stream (looking down on the stream valley from above).[33] Calculate the subsurface flow per meter of stream using the entire flow net shown in the figure and assuming $k = 1 \times 10^{-2}$ cm/s.

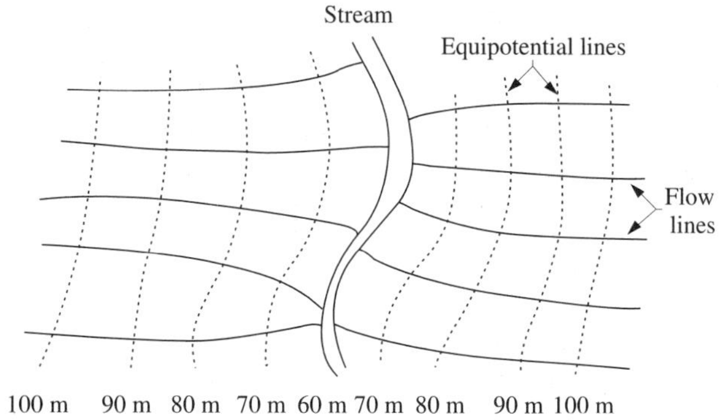

Solution. $Q = khn_f/n_d$
By counting from the figure we find $n_f = n_d = 4$; substituting and converting cm to m we have:

$$k = 1 \times 10^{-2} \text{ cm/s}$$
$$h = 100 \text{ m} - 60 \text{ m} = 40 \text{ m}$$
$$n_f = 4$$
$$n_d = 4 \text{ (100 m to 90 m = 1 drop; 100 m to 60 m = 4 drops)}$$
$$Q = 1 \times 10^{-2} \text{ cm/s } \times 40 \text{ m} \times 4/4 \times 1 \text{ m}/100 \text{ cm} = 4 \times 10^{-3} \text{ m}^3/(\text{s} \cdot \text{m})$$

The units may appear confusing, but in fact represent the volumetric flow rate per meter of stream length.

Finite difference method

Figure 4-23 illustrates a numerical grid representing a two-dimensional flow regime. The abscissa and ordinate axes of the finite difference grid use i, j notation. Each nodal point is modeled to represent the average head for the area enclosed by the dotted square.

The finite difference method further assumes that the head at a particular node, $h_{i,j}$, is the average of the head at the four surrounding nodes. For an interior node, the head at any point is:

$$h_{i,j} = (h_{i,j-1} + h_{i+1,j} + h_{i,j+1} + h_{i-1,j})/4 \tag{4-29}$$

The finite difference numerical technique first requires establishing the boundary conditions, either as **constant-head** boundaries or as **no-flow** boundaries. Numerical values for head are then assigned. Next, the heads are recalculated systematically by using the finite difference equation for each node. This trial and adjustment method is the **relaxation technique**. The head at each node is then recalculated throughout the system; each pass through the system is known as an **iteration.** The iterations are repeated until the change in head from one iteration to another is less than the **tolerance** that is selected by the groundwater modeler, which may be on the order

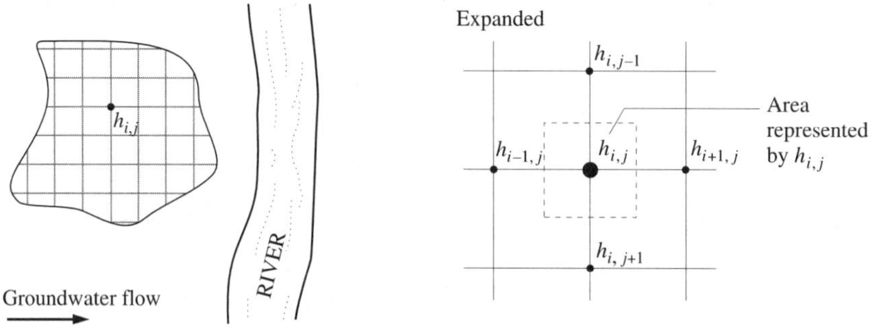

FIGURE 4-23
Finite difference.

of 2 cm for typical hazardous waste sites. The technique is illustrated for the Darcy experiment in Example 4-8.

EXAMPLE 4-8. FINITE DIFFERENCE SOLUTION FOR DARCY'S EXPERIMENT. For the boundary conditions in Figure 4-10 use the finite difference technique to determine the hydraulic heads in the flow regime.

Solution

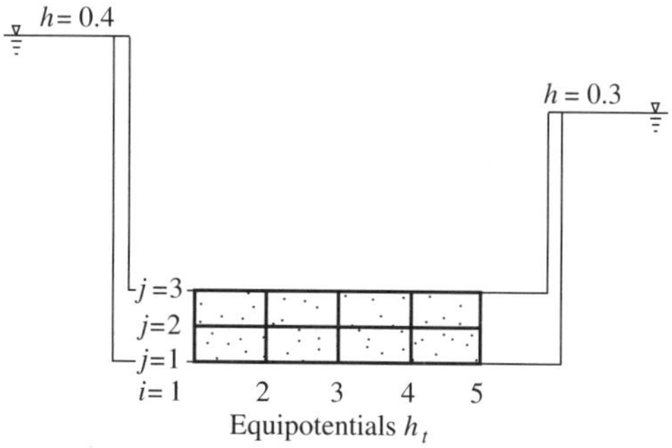

The initial head matrix in terms of variable name $h_{i,j}$ is:

$$
\begin{matrix}
h_{1,3} & h_{2,3} & h_{3,3} & h_{4,3} & h_{5,3} \\
h_{1,2} & h_{2,2} & h_{3,2} & h_{4,2} & h_{5,2} \\
h_{1,1} & h_{2,1} & h_{3,1} & h_{4,1} & h_{5,1}
\end{matrix}
$$

To begin the process, enter the values of head into the initial head matrix under the known boundary conditions and estimate the head at intermediate points. In our example we know that $h = 0.4$ at the upstream boundary and $h = 0.3$ at the downstream boundary. Thus, we know that the head is < 0.4 and > 0.3 in the remainder of the flow regime. Therefore, we may make the initial estimate of the head matrix as:

$j = 3$	0.4	0.35	0.35	0.35	0.3
$j = 2$	0.4	0.35	0.35	0.35	0.3
$j = 1$	0.4	0.35	0.35	0.35	0.3
	$i = 1$	$i = 2$	$i = 3$	$i = 4$	$i = 5$

The next step is to iterate the matrix by systematically recalculating the heads using Equation 4-29: $h_{i,j} = (h_{i,j-1} + h_{i+1,j} + h_{i,j+1} + h_{i-1,j})/4$

Beginning at $h_{2,1}$ we have

$$h_{2,1} = (h_{1,1} + h_{3,1} + 2h_{2,2})/4$$
$$= [0.5 + 0.35 + 2(0.35)]/4 = 0.3875$$

and

$$h_{2,2} = (h_{1,2} + h_{3,2} + h_{2,3} + h_{2,1})/4$$
$$= (0.4 + 0.35 + 0.35 + 0.3875)/4 = 0.3719$$

and

$$h_{2,3} = (h_{1,3} + h_{3,3} + 2h_{2,2})/4$$
$$= [0.4 + 0.35 + 2(0.3719)]/4 = 0.3735$$

Continuing through the matrix in this fashion, the values for head at the end of the first iteration are:

$j = 3$	0.4	0.3735	0.3598	0.3334	0.3
$j = 2$	0.4	0.3719	0.3578	0.3369	0.3
$j = 1$	0.4	0.3875	0.3594	0.3399	0.3
	$i = 1$	$i = 2$	$i = 3$	$i = 4$	$i = 5$

The iterations continue (normally using a computer spreadsheet works very well) until the change in head between successive iterations is less than a specified value; in this case a value of 0.001 m would be reasonable. Thus, the final values of head are:

$j = 3$	0.4	0.375	0.35	0.325	0.3
$j = 2$	0.4	0.375	0.35	0.325	0.3
$j = 1$	0.4	0.375	0.35	0.325	0.3
	$i = 1$	$i = 2$	$i = 3$	$i = 4$	$i = 5$

Groundwater modeling

The attractiveness of groundwater modeling is in its use in evaluating alternatives. As described in Chapter 16, the remediation of contaminated hazardous waste sites frequently means the remediation of contaminated groundwater, and there are numerous alternatives to groundwater remediation including pump and treat, *in situ* bioremediation, and containment. In selecting remedial measures, questions arise regarding the performance of potentially suitable alternatives. How is the groundwater flow and contaminant transport altered by the selected remedial alternative? Which alternative provides the best control of groundwater flow and contaminant transport at a particular site? If a pumping system is proposed, is it more effective to use 10 wells pumping at 50 gal/min or 20 wells pumping at 25 gal/min? What should the well spacing be? These and many other questions need to be answered in the evaluation of remedial alternatives. Groundwater modeling can help the decision maker by providing insight into the relative impacts of various alternatives on groundwater.

What is groundwater modeling? A model is a replica of some real-world object or system. A groundwater model must therefore simulate the actual groundwater conditions. The first step is to develop a conceptual model of the conditions. This conceptual model is generally macro scale over a large area and is often a graphical representation, allowing a visualization of the actual conditions. Subsequently, the simulation may also be a complex digital three-dimensional groundwater model solvable only through the application of numerical methods. Most applications of groundwater modeling use a mathematical model. A mathematical model tries to simulate the actual behavior of a system through the solution of mathematical equations. Specifically, a numerical groundwater model solves for the hydraulic head conditions throughout the flow regime (governed by the Laplace equation) by using numerical techniques.

How is a numerical groundwater model employed for hazardous waste management applications? A complete treatise is beyond the scope of this chapter but general guidance is provided. On selection of a numerical groundwater model, the required data regarding subsurface conditions and physical properties are **digitized.**

In this step physical parameters, such as hydraulic conductivity, porosity, aquifer thickness, and extent, are converted into numerical inputs. For example, the location and elevation of the groundwater are input as x, y, and z coordinates. The next phase of groundwater modeling is termed **calibration.** Calibration of the numerical model is the process of adjusting model inputs until the resultant predictions give a reasonably close fit to the observed data. For example, water level elevations may be known at selected locations in the subsurface flow regime. The known observed water levels are compared with the simulated water levels predicted by the numerical model. If the observed and predicted conditions compare reasonably well, the model is said to be calibrated. If not, the modeler must examine the material properties and boundary conditions and make adjustments accordingly. A simplified flow diagram representing the general steps in developing a groundwater and contaminant transport model is shown in Figure 4-24.

Once the model is calibrated, it can be used in hazardous waste management as follows:

- As a tool in understanding groundwater flow and contaminant transport processes at a site
- As a tool for evaluating the adequacy of the available field and laboratory data and for identifying deficiencies in the database
- As a tool for estimating flow and transport parameters
- In the prediction of future patterns of groundwater flow and contamination migration
- For assessment of groundwater flow and contaminant migration patterns in response to proposed remedial actions.

Groundwater modeling for remedial investigations and feasibility studies could be used to:

- Guide the placement of monitoring wells
- Predict contaminant concentrations in groundwater for an assessment of present and future risks at the site
- Assess the feasibility and efficacy of remedial alternatives
- Predict the residual contaminant concentration after implementation of preferred remedial action

Considering the complexity of the mathematical issues associated with groundwater flow and contaminant transport, it should come as no surprise that computer modeling is used to solve all but the simplest problems. Having said that, understanding the concepts developed in this chapter often permit "back of the envelope" calculations that are used as part of the reality check of computer output and/or to begin to get a handle on a given problem. While anything written in a textbook regarding computer models is likely to be outdated within a short time, it is useful to mention two widely used programs the reader is likely to encounter. POLLUTE[34] is a computer program that solves the one-dimensional advection-dispersion equation and is particulary well suited for the study of landfill liner systems (see Chapter 13). The program handles layered systems and can calculate the concentration of contaminants at any time or depth as desired by the user. The program has many features including the ability to consider a constant or time-varying source, some or all layers as fractured rather than porous media, geochemical reactions including both linear

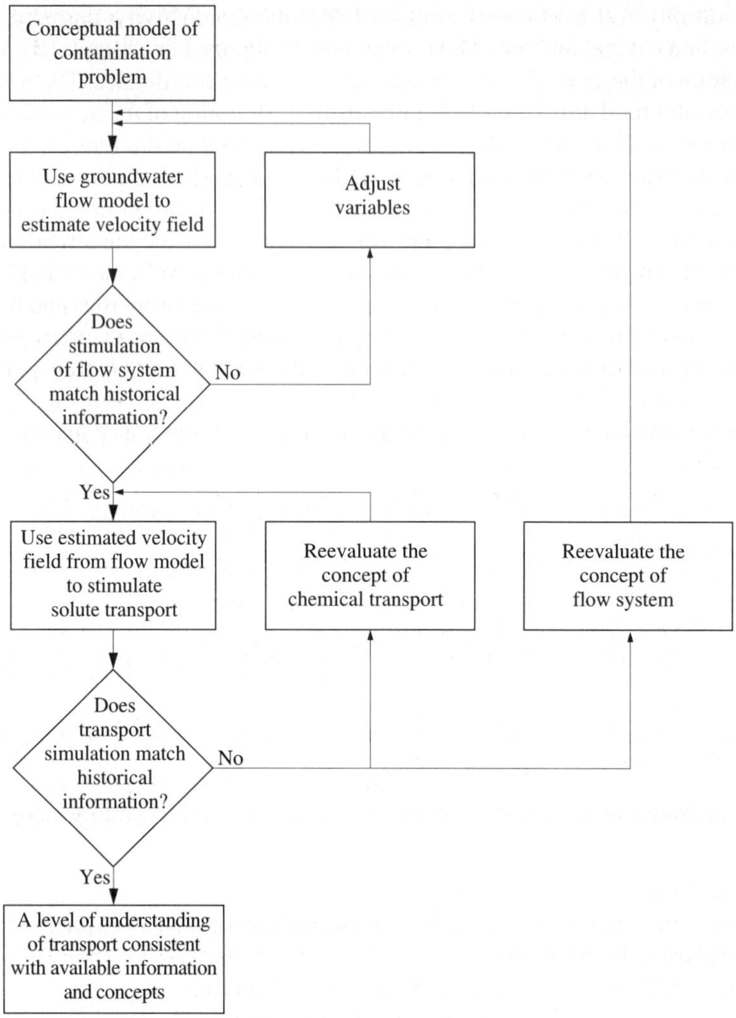

FIGURE 4-24
Simplified flow diagram representing general process of developing a transport model.

and nonlinear sorption, and passive sinks such as a secondary leachate collection system. MODFLOW[35–37] is a three-dimensional groundwater flow model that has been widely used for over 15 years and has seen numerous upgrades and revisions since the original. It has a modular structure that allows programmers to modify the code for particular applications. The program can simulate both steady and transient flow in aquifers of significant complexity including shape, confining layers, wells, evapotranspiration, and recharge. As a result this program is well suited for studies of regional aquifers or large sites with complex subsurface conditions. In addition to these two computer programs, there are countless others.

Groundwater modeling represents a formidable challenge for data collection, parameter estimation, model formulation, numerical methods, and computational

speed and power. As a result, modeling is an art that requires the combined skills of a computer scientist, mathematician, geologist, groundwater hydrologist, geochemist, and engineer.

4-3 FATE OF CONTAMINANTS IN THE SUBSURFACE

Section 4-2 has detailed how contaminants move in the subsurface in response to advection, hydrodynamic dispersion, and molecular diffusion. In addition to these three transport processes, a myriad of complex chemical and microbiological processes affect the **fate** of contaminants in the subsurface. They **retard** the movement of contaminants, if not **attenuate** their concentration. Table 4-11 summarizes the various effects of subsurface processes on the fate of hazardous constituents.

Retardation refers to processes that impede the transport of contaminants by removing or immobilizing them from a free state (i.e., aqueous solution or vapor). Examples of chemical reactions that result in retardation include sorption and precipitation. It is important to note that in retardation processes the immobilized contaminants are not transformed and the processes are reversible. Reversal tends to occur when the concentration of a contaminant in the groundwater declines and the quantity of retarded contaminants is high. Reversal would return retarded contaminants to solution over an extended period, giving the plume of contamination a long "tail" over time.

Attenuation refers to two types of processes: (1) irreversible removal and (2) transformation. Removal by an attenuation process differs from retardation in that it reduces the mass of a substance (i.e., the process functions as a sink). A common example is a process that transfers the contaminant to another medium (e.g.,

TABLE 4-11
Summary of natural processes affecting the fate of hazardous constituents in the subsurface

Process	Class of chemical	Effect
Sorption	Organic/inorganic	Retardation
Precipitation	Inorganic	Retardation
Ion exchange	Inorganic	Retardation
Filtration	Organic/inorganic	Retardation
Chemical oxidation-reduction	Organic/inorganic	Transformation/retardation
Biological uptake	Organic/inorganic	Retardation
Biodegradation	Organic	Transformation
Hydrolysis	Organic	Transformation
Volatilization	Organic	Elimination by intermedium transfer
Dissolution	Organic/inorganic	Mobility enhancement
Cosolvation	Organic	Mobility enhancement
Ionization	Organic	Mobility enhancement
Complexation	Inorganic	Retardation mobility enhancement
Immiscible phase	Organic	Various partitioning

Note: If humic substances are in the subsurface in significant amounts, significant compression and ion exchange can occur with organic chemicals.

volatilization). The more common type of attenuation process is one that transforms the molecular structure of the substance, an example being oxidation-reduction reactions.

Some processes actually increase the mobility of chemical substances in the subsurface. Examples include dissolution of organic substances and complexation of metallic ions. Such processes are categorized as **mobility enhancement** phenomena.

Subsurface Environment

Understanding the fate of contaminants in the subsurface requires a more exact depiction of the subsurface environment than that used to describe the movement of groundwater in response to advection and hydrodynamic dispersion. The model presented in Section 4-2 depicts the subsurface as a porous medium through which groundwater flows. Describing the subsurface simply as a porous medium is a "macroscale" approach when "microscale" is needed.

The porous medium consists of both consolidated (rock) and unconsolidated geologic formations. Civil engineers commonly call all unconsolidated sediment "soil" and the study of its properties "soil mechanics." To others, only the uppermost covering mantle of material is referred to as **soil**. The formation of soil begins with weathering of parent rock or unconsolidated sediment followed by the transport, deposition, and accumulation of this material. Frost action, geochemical processes, water movement, and biological activity cause further change. The physical and chemical characteristics of soil vary widely with location, depth, and time, dependent primarily on the parent material, climate (temperature, wind, and water), and topography (relief and slope).

Soil is a mixture of different inorganic and organic materials. The inorganic fraction consists mostly of fine mineral grains, subdivided according to size as shown in Table 4-12. The percentages by weight of gravel, sand, silt, and clay provide a basis for classifying soil by texture. The mineralogy of clays varies, but most are formed from silicates. The most common clays are termed kaolinite, illite, and smectite (montmorillonite). Individual, plate-shaped, colloidal clay particles typically agglomerate to form clay aggregates.

The predominant inorganic elements in soil are silicon, aluminum, and iron, with a great number of micro and trace elements. It is interesting to note that some

TABLE 4-12
Classification of mineral grains in soil

Classification	Description	Size
Clay	Microscopic mineral particles of a colloidal nature, laminated as layers of plates	<0.002-mm diameter (effective diameter)
Silt	Fine particles composed of minerals from the parent formation	>0.002-mm diameter <0.075-mm diameter
Sand	Granular particles composed of minerals from the parent formation	>0.075-mm diameter <2-mm diameter
Gravel	Granular particles composed of minerals from the parent formation	>2-mm diameter <75-mm diameter

of the naturally occurring and commonly found trace elements in soils are hazardous substances. Average concentrations in soils for such examples are as follows:[38]

Element	**Average concentration in soil, μg/kg**
Arsenic	6
Cadmium	10
Nickel	40
Lead	10
Selenium	0.2

These naturally occurring elements differ from the same elements introduced with hazardous waste in that natural elements are essentially in an insoluble form and release very slowly with weathering. Waste constituents frequently are in a soluble form. Furthermore, average values may be misleading and are presented only to show the reader that hazardous substances may indeed be found in naturally occurring environments.

Soils typically contain appreciable amounts of organic material, basically consisting of decomposed plant matter or humus. Consequently, organic content normally decreases with depth, varying more in the vertical than horizontal dimension. The organic content of most soils typically would fall in the range of 0.2 to about 3 percent[39] and is composed of a wide mixture of organic chemicals (e.g., humic acid). The organic fraction can have a colloidal nature and has been described as organic polymer matrixes.[40]

The organic matter in soil usually acts as a stabilizer that binds inorganic particles together as **aggregates.** Thus, the soil mantle consists of aggregates, each containing sand, silt, clay, and organic matter. It should be noted that the size and structure of aggregates can vary considerably (see Figure 4-25).

With depth, a transition occurs from soil to the underlying geological formations, whether unconsolidated deposits or rock. The soil mantle, the underlying permeable

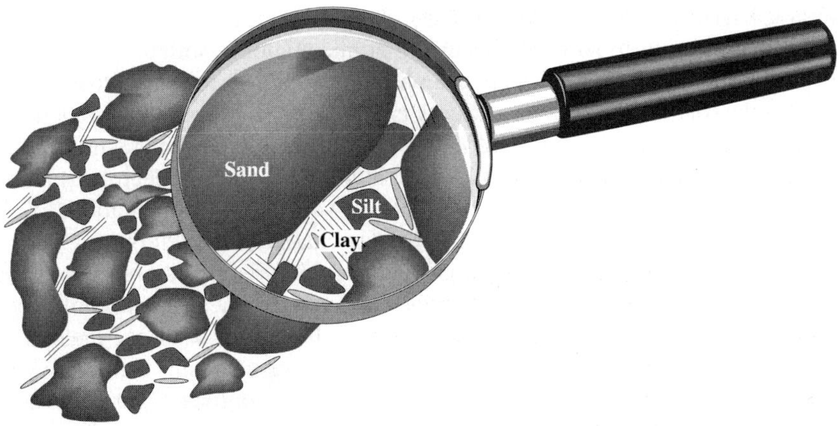

FIGURE 4-25
Soil aggregates in subsurface domain.

formations, and transitional material between the two constitute the porous medium. The three are sometimes referred to collectively as the subsurface matrix. Movement of groundwater through porous media, whether saturated or unsaturated, thus is a combination of transport both through the interstitial openings between aggregates and through the pores *within* an aggregate.

Within aggregates are found much of the colloidal inorganic and organic surfaces, the most active portions of the subsurface matrix.[41] It is within this microscale domain where much of the retardation and attenuation processes occur. These processes depend on intra-aggregate transport of groundwater (i.e., through pores within the aggregate to reach internal surfaces). The effective diameter of pores within the crystalline mineralogical structure of clay can be 100 to 1000 nm (10^{-6} to 10^{-7} inches), thus severely restricting groundwater advection and limiting intra-aggregate contaminant transport mostly to molecular diffusion.[42] Nevertheless, the structural arrangement of clay provides for an extremely large internal surface area, much greater than its external area,[43] and certainly great in relation to its mass.

Retardation Processes

Retardation processes consist of sorption of organic substances and ion exchange and precipitation of inorganic substances. These reactions result in a reduced rate of movement of contaminants compared to the average rate of groundwater flow. The processes will have an important impact on "pump and treat" remediation, discussed in Chapter 16.

Sorption

Sorption is a process that includes the partitioning of **sorbate** (chemical constituents) among phases [i.e., movement of **solutes** from a **solvent** to a **sorbent** (or sorbing material)], as explained in Section 3-2. For the specific case of the fate of hazardous waste contaminants in the subsurface, sorption is primarily the accumulation of chemicals on soil surfaces.[44] More specifically, it is the partitioning of dissolved organic contaminants (the solutes) from ground or surface water (the solvent) to natural subsurface materials (the sorbent). An important example is the adherence of organic molecules to naturally occurring humic matter in soil.

The partitioning of solutes between phases depends on the relative affinity of a solute for the solvent and sorbent. Affinity is primarily a molecular phenomenon and is a function of multiple chemical, physical, and electrostatic mechanisms. Important examples of attractive mechanisms include hydrogen bonding (chemical), van der Waals forces (physical), and coulombic attractive forces (electrostatic).[45,46] Two other examples are charge transfer and ligand exchange with metal ions.[47]

Any of these mechanisms may operate, depending on the nature of the organic chemical and the properties of the subsurface materials. In fact, several could operate simultaneously. The sum of all mechanisms and all active subsurface materials accounts for the effect of sorption. Cumulatively, these mechanisms can easily yield a nonlinear relationship between solute and sorbed concentrations. Figure 4-26 depicts both a linear and a nonlinear relationship.

Depending on the mechanism and the organic chemical, the rate of sorption can be rather rapid, with much of the sorption occurring within 48 hours. A two-stage

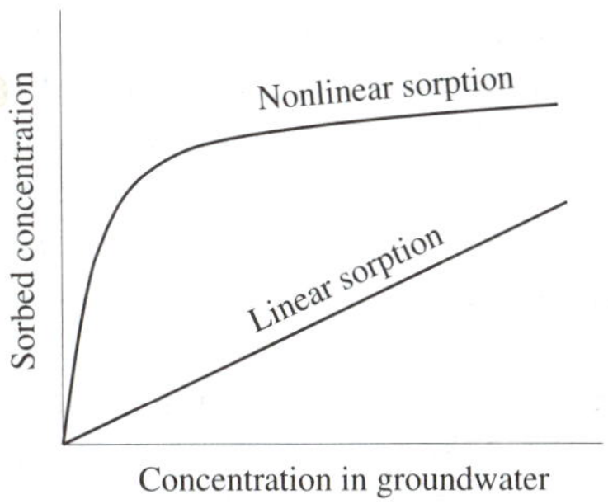

FIGURE 4-26
Partitioning of sorbate between solvent and sorbent.

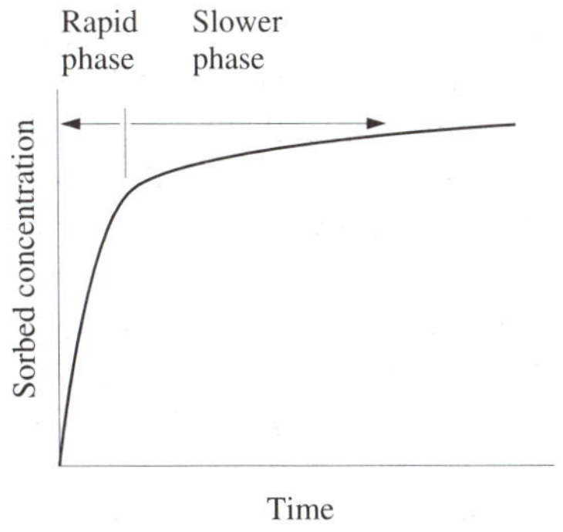

FIGURE 4-27
Two-stage sorption model.

model is usually considered for this nonlinear relationship, consisting of a rapid, short-lived stage, followed by a slower, longer stage approaching equilibrium (see Figure 4-27).

The rate of sorption can be slow, given the tortuous path of diffusion from the bulk aqueous solution to pore water and then through the pores of the aggregate across the fixed water film to sorption sites in organic or mineral matter. In fact, many contaminants of environmental concern have sorption equilibrium times of days, weeks, or even years, as for the case of polychlorinated biphenyls (PCBs).[48]

It is important to note that sorption is reversible (a process termed **desorption**). This can occur when the concentration of solute has decreased and the quantity of chemicals sorbed to the sorbent is high. A hysteresis effect, as shown in Figure 4-28, can occur. Desorption can take an extremely long time, thus giving the plume shape a "long tail of decreasing contamination."[49] This phenomenon limits the use of "pump and treat" cleanup technology discussed in Section 16-7. On cessation of pumping, concentrations in groundwater will return to equilibrium levels.

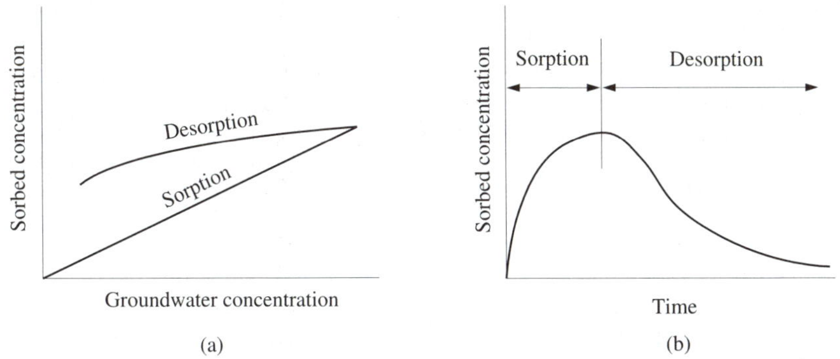

FIGURE 4-28
Desorption of sorbate.

As explained earlier in this section, a microscopic examination of the porous medium composing the subsurface shows that it has both organic and mineral surfaces. Many organic chemicals of environmental concern are hydrophobic. At most hazardous waste sites, nonpolar species such as trichloroethene, benzene, and chlorinated hydrocarbon pesticides account for much of the risk to health. For hydrophobic organic compounds, several researchers have shown that the dominant sorption mechanism is the affinity of such solutes for natural organic material in the aquifer.[50] In fact, researchers have correlated the sorption capability of various soils with their organic matter content.[51] The importance of the organic matter content of the surrounding geologic material is shown in Figure 4-29. This figure shows the amount of lindane sorbed by a soil under two conditions: natural soil and the same soil stripped beforehand of its organic matter.

It should be noted that hydrophobic organics can adsorb to mineral surfaces, especially to clay-sized particles. Clay particles have a net negative charge due to isomorphous substitutions on the internal surfaces between the plates of silicate clay crystals as well as the external surfaces.[52] Examples of organic chemicals adsorbing to these sites include those with polar organic functional groups (via ion-dipole forces) and large organic molecules (via hydrogen bonding).[53] Cation exchange is also possible.[54] Even hydrophobic organics can accumulate on clay surfaces.[55] This can be significant, particularly for soils with organic matter content below 1 percent.[56]

It is clear that a variety of molecular mechanisms can act to sorb organic chemicals to natural materials in the subsurface. The magnitude of the mechanisms' effect varies, depending on the nature of the solute, solvent, and sorbent. Their combined net effect determines the distribution between phases. Depending on conditions, the degree of sorption can be high, with the sorbed phase exceeding the concentration in groundwater by a ratio as high as 5 to 10:1, if not higher. The ratio can also be less than 1.0.

The potential retention capability, which includes sorption, of soil can be estimated by saturating undisturbed soil samples with a liquid contaminant and then allowing the sample to drain. The retained contamination is termed residual saturation or retention capacity and is shown in Table 4-13 for fuels. Obviously, retention capacity for fuels represents partitioning from the bulk phase as occluded droplets rather than as solutes to the soil matrix. It is not true molecular sorption as discussed here; however, the values indicate the degree to which bulk contaminants can partition to soil.

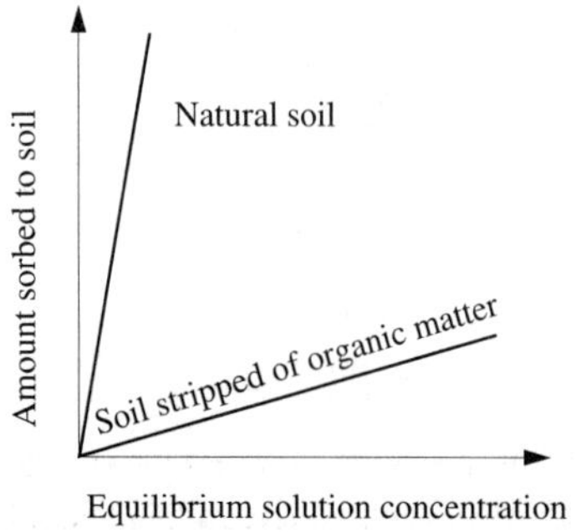

FIGURE 4-29
Sorption of lindane by unstripped and stripped soil.[57]

TABLE 4-13
Retention capacity of various soil types for common fuels (ppm of fuel in soil)[58]

Fuel	Gravel	Coarse sand	Medium sand	Fine sand
Gasoline	1000	1600	3,200	5,000
Diesel fuel	2000	3200	6,400	10,000
No. 6 fuel oil	4000	6400	12,800	20,000

EXAMPLE 4-9. RETENTION CAPACITY. On the basis of the data provided in Table 4-13, what are two physical characteristics that control retention capacity?

Solution. For a given soil type, retention of a fluid increases as the viscosity of the fluid increases. For a given contaminant, retention by soil increases as the texture of the soil becomes finer.

The properties of the sorbent are also very important. Key factors influencing sorption include the following:

- Molecular size (for organics, generally the larger the molecule, the more readily it is adsorbed)
- Hydrophobicity (the adsorption of nonionic organic chemicals onto soil particles correlates well as an inverse function with water solubility and octanol-water partition coefficients, two surrogates for hydrophobicity)
- Molecular charge (for some organics, it can be an uncharged polar molecule at a certain pH and an anion at a higher pH)[59]
- Structure (one isomer may be more readily adsorbed than another)[60]

Suffice it to say, the mass transfer and reaction rate interrelationships of the various forces are very difficult to formulate. Furthermore, in the field, subsurface materials exhibit a high degree of complexity and heterogeneity. This necessitates that sorption be determined from tests of the contaminant and subsurface materials and the net effect be expressed using simplified, empirical models.

Linear sorption model

For saturated conditions and nonpolar organic constituents, sorption from the aqueous phase to the porous media of the subsurface can be "... treated as an equilibrium-partitioning process.... When solute concentrations are low (e.g., either $\leq 10^{-5}$ molar, or less than half the solubility, whichever is lower) partitioning often can be described using a linear isotherm"[61] as discussed in Sections 3-2 and 9-2. The linear isotherm relates the concentration of the organic substance in groundwater and the sorbent at constant temperature as follows:

$$S = K_d C \tag{4-30}$$

where S = mass sorbed per mass of sorbent (mg/kg)
K_d = partition or distribution coefficient
C = concentration in groundwater at equilibrium (mg/L)

Such a linear relationship was shown earlier in Figure 4-26. The slope in this figure represents K_d. (*Note:* K_d is used in this section to keep notation consistent

with that normally associated with groundwater literature. The term K_p was used alternatively with K_d in Section 3-2.) It represents the ratio of the mass concentration of a contaminant sorbed to soil to its concentration in the surrounding groundwater. It would vary with the specific chemical constituent and the subsurface material. The slope can be determined from batch equilibrium tests conducted with samples of actual subsurface material. As discussed earlier, such tests frequently show that the greater the organic content of the aquifer, the greater the sorption potential. Thus, sorption can be characterized in many cases, particularly for hydrophobic organics, on the basis of the organic content of the porous medium of the aquifer by the following linear expression (Equation 3-17):

$$K_d = K_{oc} f_{oc} \tag{4-31}$$

where f_{oc} = organic carbon fraction of the porous medium
K_{oc} = organic carbon partition coefficient of the chemical constituent

Experience has shown that this method produces a reasonable estimate of sorption only for subsurface materials having the following organic content:

Minimum:	"Greater than 0.1 percent"[62]
	"Few tenths of a percent"[63]
	1.0 percent[64]
Maximum:	20 percent[65]

As stated earlier, the organic content of most soils falls in the range of 0.2 to about 3 percent. Thus, this approach may not be applicable for many situations. Where applicable, the variable f_{oc} can be determined from field sampling and analyses, and the variable K_{oc} can be found in Appendix A for many chemicals.

When unknown, K_{oc} sometimes can be derived by correlation with a property indicating the degree of the chemical's hydrophobic nature (e.g., octanol-water partition coefficient).[66,67] A commonly used correlation is as follows:[68]

$$K_{oc} = 0.63 K_{ow} \tag{4-32}$$

where K_{ow} = octanol-water partition coefficient for the chemical.

However, the actual value can vary considerably from this approximation as shown by various studies that have developed similar regression equations for different classes of compounds (see Table 3-5). These studies note that some large errors are possible, particularly for a class covering a broad range of chemicals.[69] In using these correlations, it is important to consider factors such as chemical structure similarity, molecular weight, organic carbon content of soil, contribution of clay mineral surfaces, solute concentration, and others.[70]

When applicable, an estimate of linear sorption can be developed simply on the basis of Equation 4-30, the correlation K_{oc}, and the determination of the organic fraction of the porous medium. This linear sorption equilibrium model thus provides a rather simple method for estimating sorption. However, it is not adequate for many conditions, and probably best applies when the concentration of the solute is "very low,"[71] at least not higher than 10 percent of the maximum solubility of the contaminant in water.[72] A high concentration may saturate if not exceed the sorption capacity of the subsurface. Thus, application of the linear model to any situation requires care.

EXAMPLE 4-10. SORPTION. Groundwater near a leaking underground tank contains 0.5 mg/L of benzene. What is the expected concentration of benzene sorbed to a silty soil containing 2 percent organic carbon? Assume sorption follows a linear model.

Solution. From Appendix B, K_{oc} for benzene is 83 mL/g.
From Equation 4-31,

$$K_d = K_{oc} f_{oc}$$
$$= (83)(0.02)\ \frac{\text{mL}}{\text{g}}$$
$$= 1.66\ \frac{\text{mL}}{\text{g}}$$

From Equation 4-30, the sorbed concentration is estimated as follows:

$$S = K_d C$$
$$= \frac{166\ \text{mL}}{\text{g}} \times \frac{0.5\ \text{mg}}{10^3\ \text{mL}} \times \frac{10^3\ \text{g}}{\text{kg}}$$
$$= \underline{\underline{0.83\ \text{mg/kg}}}$$

Nonideal sorption conditions

There are a number of conditions for which sorption does not follow the ideal pattern as expressed by the linear equilibrium approach. Some example conditions are as follows:

- The rate of groundwater movement may be so rapid that the relatively slower rate of sorption becomes limiting, thereby negating the assumption of equilibrium.
- Sorption may follow second-order kinetics, necessitating a nonlinear model, particularly at high solute concentrations.
- Organic chemicals may adsorb to suspended microparticles of organic matter in the groundwater; as such, adsorbed organic chemicals would move with groundwater and yield less apparent sorption than predicted.
- In groundwater with multiple organic chemicals, competition for limited sorption sites will occur, thereby yielding less than the additive amount of sorption determined for an individual species.

The linear partitioning model assumes equilibrium between sorbed constituents and solute. This is deemed adequate for many hydrophobic organic chemicals if the solvent-sorbent contact time is on the order of "days to months."[73] Nevertheless, for some compounds, such times may not suffice for reaching equilibrium.

Nonlinear sorption is frequently observed. For example, in one case involving the sorption of trichloroethene onto a glacial till, the K_d decreased from 110 at 1 ppm TCE to less than 10 at 4 ppm (see Figure 4-30). There are two models frequently used to describe nonlinear adsorption, the Langmuir model (see Equation 9-33) and the largely empirical Freundlich isotherm (see Equation 9-34). The latter is used more often. It is very similar to the linear sorption model with the addition of the empirical constant N as follows:

$$S = K_d C^N \tag{4-33}$$

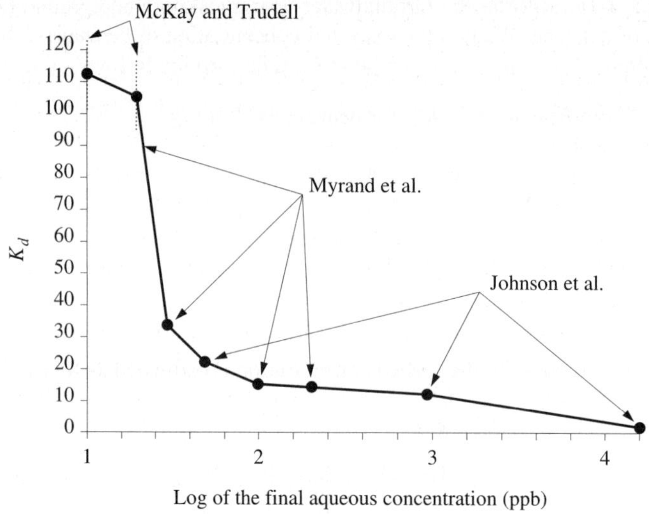

FIGURE 4-30
Variable sorption of trichloroethene on glacial till.[74]

The use of such models requires that they be carefully calibrated using data from batch, if not flow-through tests, of the specific contaminants with samples of the subsurface materials. Even then, it is important to have verified the selected model for the conditions measured in the field at the site under investigation.

Ion exchange

Ion exchange involves the sorption of ions in solution onto oppositely charged, discrete sites on the surface of a soil particle. It is driven by the attractive force of maintaining electrostatic neutrality: the electric charges of the surface of the soil are balanced by equivalent free ions of opposite charge.[75] As such, a previously held ion of weaker affinity is exchanged by the soil for the ion in solution. For example, calcium ions in groundwater may be exchanged onto the clay surface, replacing existing sodium ions.

Ion exchange can be considered as a subcategory of sorption involving electrostatic interactions. However, a separate distinction can be made in that ion exchange is generally thought to apply to metals while sorption applies to organics. Most metals in aqueous solution occur as charged cations. Thus, metal species adsorb primarily in response to electrostatic attraction, but such is only one type of interaction resulting in the sorption of organics. However, many use the term sorption as an umbrella term to include all forms of reversible interactions between the solute and solid.

In contrast to sorption (as narrowly defined), which occurs mostly on organic surfaces, both organic and inorganic surfaces are important with ion exchange. The capability of a soil to retain and exchange cations is quantified as the **cation exchange capacity**. For example, the cation exchange capacities of humus and the two clay minerals, vermiculite and montmorillonite, are in the order of 200, 150, and 100 meq per 100 g.[76] Although the organic and inorganic fractions together account

for the cation exchange capacity of a soil, the clay fraction is usually more important because soils frequently have a far greater clay than organic content.

Clay has a much higher cation exchange capacity than other inorganic particles because of its extremely large surface area, abounding with negative charge sites. The exposed surface area of a unit mass of colloidal clay is at least 1000 times that of coarse sand.[77] It should be noted that the cation exchange capacity of clays varies considerably, depending on the clay mineralogy. Table 4-14 presents the cation exchange capacity of commonly encountered clay minerals.

In the process of ion exchange, ions compete for the exchange sites and displace a previously held cation, termed cation exchange selectivity. Typically displaced cations include sodium and calcium. Exchange affinity depends on electric charge, hydrated radius, and molecular configuration.[78] Generally, divalent cations are more strongly adsorbed than monovalent ions, and smaller cations tend to replace larger cations. Cations are typically replaced in the following order:[79]

$$Na^+ < Li^+ < K^+ < Rb^+ < Cs^+ < Mg^{+2} < Ca^{+2} < Ba^{+2} < Cu^{+2} < Al^{+3} < Fe^{+3} < Th^{+4} \quad (4\text{-}34)$$

These guidelines are not absolute. For example, at high concentrations a cation with low replacing power, such as Na^+, can replace a cation higher in the series.

Ion exchange capacity is very dependent on pH. As the pH is lowered below neutrality, hydrogen ions readily replace metal ions. As the pH increases, the hydrogen ions are less resistant to replacement and the net result is an increase in the exchange capacity.

A key point is that ion exchange of metallic ions with soil may be partially reversible; saturated exchange sites can release cations as concentrations of contaminants decline in groundwater[80] and in response to pH changes. Thus, ion exchange represents a retardation rather than an attenuation process. Further, the exchange capacity of a subsurface material can be saturated such that eventually the transport of contaminants is unaffected by ion exchange and goes unretarded.

Precipitation

Precipitation is the converse of dissolution: The concentration of a solute exceeds the solubility of that particular compound, and any excess solute changes to a solid and thus precipitates out of the solution. That part which separates out of solution is termed **precipitate**. The process is reversible. If the concentration of a solute drops below its solubility, dissolution of a previous precipitate could occur. Precipitation

TABLE 4-14
Cation exchange capacity of clay minerals

Clay mineral	Cation exchange capacity, meq/100 g
Kaolinite	3–15
Illite	10–40
Smectite (montmorillonite)	80–150
Palygorskite (commonly called attapulgite)	20–30

is particularly applicable to heavy metals such as nickel, mercury, chromium, and lead. Not all thermodynamically favored precipitation/dissolution reactions occur. Since reactions proceed at a finite rate, the time available for the reaction will affect the extent of the reaction.

Precipitation can occur when a chemical reaction transforms a solute to a much less soluble form, typically by mixing a precipitant with the solution (e.g., mixing of a lead solution with high-carbonate waters can produce the relatively insoluble form $PbCO_3$). Also, changes in oxidation state can promote precipitation (e.g., oxidation of ferrous solutes will yield the much less soluble ferric forms).

Precipitation depends greatly on pH. Most metals precipitate at high pH levels as hydroxides. However, continued elevation of pH will increase the solubility of amphoteric metals such as nickel. This is shown in Figure 4-31. In addition, redox potential is important (see next page).

Filtration

Filtration is a physical form of retardation, resulting from the clogging of pore spaces. This occurs from solid particles becoming trapped in the pores. It also occurs from precipitation and accumulation of dissolved matter.

Attenuation Processes

The significant **attenuation** processes that transform substances are the oxidation-reduction reactions, whether initiated via chemical or biological pathways; hydrolysis which can also transform substances; and volatilization which results in transfer of chemical substances to the atmosphere.

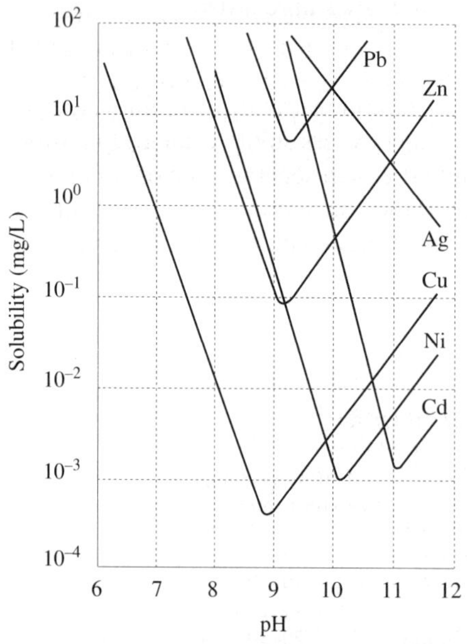

FIGURE 4-31
Solubilities of metal hydroxides as a function of pH.

Chemical oxidation-reduction

Chemical reduction-oxidation (redox) reactions can occur with organic and inorganic chemicals and classically involves the gain or loss of electrons. A reaction that removes electrons from a compound is oxidation [e.g., $FeSO_4$ oxidizes to $Fe_2(SO_4)_3$]. Likewise, addition of electrons is termed reduction. Oxidation-reduction reactions do not necessarily involve oxygen; oxygen is an important electron acceptor but not the only one.

Oxidation is defined as a reaction that results in the loss of electrons by a chemical (the chemical donating the electron is oxidized). The converse, gaining an electron, is reduction. For inorganic solutes, oxidation again involves the loss and acceptance of electrons, but is more accurately defined as a reaction that increases the oxidation state of an atom (i.e., the charge on the atom becomes more positive). For the example mentioned in the preceding paragraph, Fe^{+2} is oxidized (i.e., donates electrons) to become Fe^{+3}. The converse is reduction as in the following example in which Cr^{+6} is reduced to Cr^{+3} by the acceptance of electrons:

$$2H_2CrO_4 + 3SO_2 \rightarrow Cr_2(SO_4)_3 + 2H_2O \tag{4-35}$$

Ions having multiple oxidation states (i.e., more than one unit of charge, as with the example of Cr^{+3} and Cr^{+6} above) are termed **redox couples.** Depending on the oxidation state, the implications for fate in the subsurface can vary widely, as indicated by the examples in Table 4-15. For natural waters, oxidation and reduction of inorganic solutes of hazardous substances may occur over geologic time but some are potentially quite rapid.

Oxidation and reduction of organic chemicals is principally a result of biological activity (as discussed in Chapter 10). Some chemical oxidation and reduction reactions occur between organic chemicals and soil materials. The role of soil materials in the oxidation of organic chemicals is complex and incompletely known. Researchers have reported the free radical oxidation of substituted aromatic chemicals at soil and clay surfaces, catalyzed by adsorbed oxygen and trace metals, particularly in aerated, unsaturated soil.[82] However, the same research suggests that many chlorinated organics likely will not undergo such oxidation, and for organics that do, "not all soils and clays will catalyze these reactions." Clay can also initiate some reduction reactions, particularly in anaerobic conditions. Generally, research shows that redox

TABLE 4-15
Multiple oxidation states[81]

Element	Ion	Comment
Iron	Fe^{+3}	Low solubility, and hydroxide precipitate can adsorb other metals
	Fe^{+2}	"Very soluble"
Chromium	Cr^{+6}	Toxic; "relatively mobile"
	Cr^{+3}	Precipitated as relatively insoluble $Cr(OH)_3$ or other chemical; "strongly adsorbs to surfaces"
Selenium	Se^{+6}	More "mobile, but less toxic"
	Se^{+4}	"More toxic, but less mobile"

conditions can vary widely over short distances, with saturated conditions being more homogeneous than unsaturated conditions.

Biological oxidation-reduction

As discussed in Chapter 10, most organic substances in the subsurface will undergo transformation to smaller molecules via oxidation and reduction mechanisms induced by the metabolic activity of native microorganisms. Such transformation is termed **biodegradation**. Anaerobic biodegradation generally occurs at a very slow rate in the subsurface; however, it favors dehalogenation of chlorinated compounds that typically resist aerobic degradation.

A classic example is the dechlorination pathway shown in Figure 4-32. Under anaerobic conditions, PCE has been reductively dechlorinated to trichloroethene (TCE), dichloroethene (primarily the *cis*-1-2 isomer, cDCE), vinyl chloride (VC), ethene (ETH),[83] and ethane (ETHA).[84] More recent work involving mixed cultures obtained from anaerobic digesters of wastewater treatment facilities indicates that PCE dechlorination to ETH has become a more common observation.[85–88] Furthermore, numerous field studies have shown that biological dechlorination of PCE and TCE can occur under natural conditions. Because of this, natural bioattenuation has received much interest as a potential treatment technology for the remediation of groundwater contaminated with chlorinated ethenes. However, field data sometimes show that dechlorination is incomplete, and either DCE or VC persists.[89–96] Conversion of PCE to VC is problematic because toxicological data indicate that VC poses a greater health risk than the other chlorinated ethenes. The persistence of VC at many sites indicates a need for greater understanding of the factors that preclude complete dechlorination in the field.

A recent study has suggested that complete dechlorination of PCE may be difficult at some sites because of differential retardation of chlorinated ethenes.[97] These results may provide an important consideration for natural or stimulated bioattenuation of PCE-contaminated groundwater. In aquifers with significant organic content, transport of chlorinated ethenes will be retarded relative to groundwater. Further-

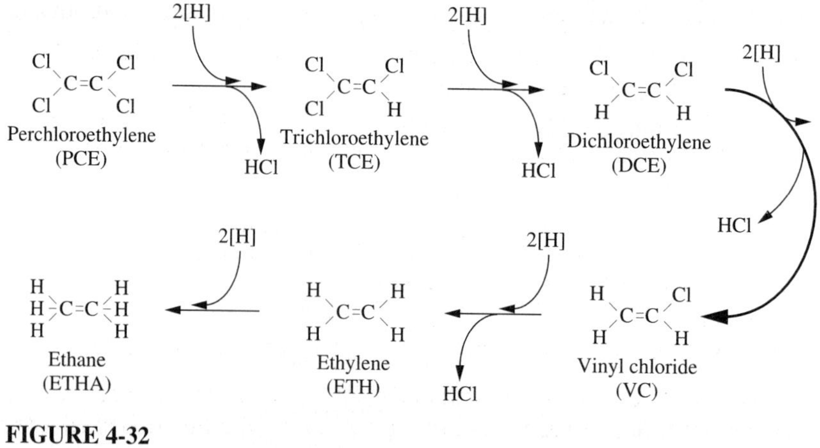

FIGURE 4-32
Biological transformation of PCE under anaerobic conditions.[83,84]

more, retardation becomes less pronounced for each successive daughter product of PCE dechlorination. This is indicated by comparison of octanol/water coefficients (K_{ow}) for the chlorinated ethenes. For each chlorinated ethene, log K_{ow} values are as follows: PCE (2.88), TCE (2.42),[98] *cis*-DCE (1.86), VC (0.60).[99] Therefore, less-chlorinated ethenes will move more quickly in an aquifer than more highly chlorinated predecessors. It is therefore possible that, in an aquifer with significant organic content, partial dechlorination of PCE or TCE to VC would likely result in separation of daughter products—i.e., VC will be at the leading edge of the plume, followed by DCE, etc. This phenomenon has been observed under field conditions.[95] For sites that achieve only partial dechlorination, aerobic biotransformation of DCE or VC may be possible.[100–102]

Hydrolysis

Chemical substances may react with water molecules, a chemical process termed hydrolysis. The exothermic reaction of some types of reactive waste is an example. Hydrolysis is frequently depicted as an exchange between OH^- and an anionic group X of the chemical compound, resulting in decomposition of the compound as follows:

$$RX + HOH \rightarrow ROH + HX \tag{4-36}$$

For most chemicals, hydrolysis has a relatively insignificant effect compared to other processes for transforming organic chemicals (e.g., biodegradation). However, for chlorinated organics, which typically are not readily transformed by biodegradation, hydrolysis may be significantly dependent on other factors.[103] Hydrolysis of chlorinated organics involves exchange of the hydroxyl group with an anionic X on a carbon atom.[104] The reaction typically forms an alcohol or alkene as follows:[105]

$$\overset{\displaystyle H}{\overset{|}{C}}-\overset{\displaystyle X}{\overset{|}{C}} \xrightarrow[OH^-]{H^+ \text{ or}} C=C + HX \tag{4-37}$$

Actually, the hydroxyl reactant may come from a water molecule or another source of a hydroxyl ion. The reaction can be catalyzed by an acidic or basic intermediate.[106] Therefore, the rate of hydrolysis depends on the pH and temperature of the water as well as the chemical compound. Figure 4-33 shows the effect of pH on the hydrolysis of two alkyl chlorides. This figure also illustrates the strong dependence of the rate of hydrolysis on the compound structure.

A simple but not always realistic approach is to depict hydrolysis as a first-order reaction under constant pH and temperature, using a form of Equation 3-55 as follows:

$$\frac{dC}{dt} = -KC \tag{4-38}$$

where C = concentration (mass/volume)
t = time
K = first-order reaction rate at constant temperature and pH (time)

The value of K can be determined in the laboratory. Nevertheless, it is difficult to develop reliable data because competing reactions mask the individual effect of hydrolysis.

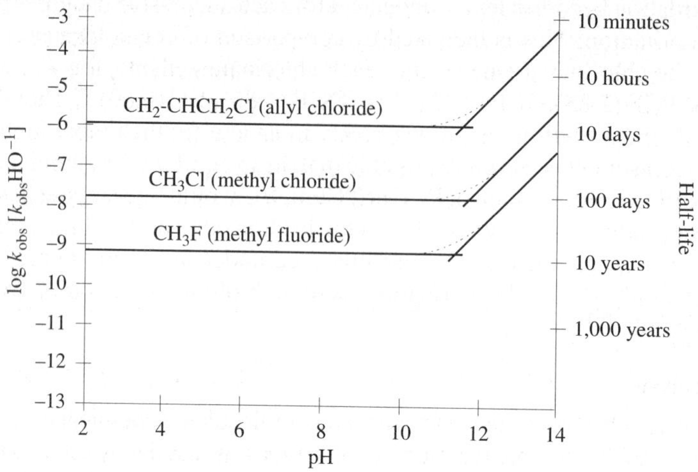

FIGURE 4-33
Hydrolysis of chlorinated alkyl compounds.[107]

Volatilization

Volatilization is the conversion of volatile chemical constituents in groundwater to vapor, which ultimately results in the transfer to the atmosphere. This process was covered earlier in Section 4-1.

Volatilization can occur from three sources in the subsurface—free product, vadose water, and groundwater. The key variables are vapor pressures and the area of contact between soil air and the three sources. The porosity of the subsurface materials less moisture (i.e., the volume of soil air) provides a reasonable indication of the degree of contact. Thus, vadose water and residual free product trapped in the unsaturated zone both have a much greater area of contact with soil air than groundwater, limited to the boundary at the groundwater table. Low-density immiscible fluids that move to the groundwater table and float there in pools because of lower specific gravity would also result in a large relative area of contact with soil air. However, the area is perhaps not as significant as the residual fluid trapped in and on soil particles in the unsaturated zone.

Mobility Enhancement

Retardation and attenuation processes alter the transport of contaminants to yield a lower concentration than would be predicted for advection and dispersion alone. Some processes actually work counter to these processes to increase the mobility of a substance. These processes principally consist of cosolvation, ionization, dissolution, and complexation.

Cosolvation

Cosolvation is the result of the introduction of bulk quantities of an organic solvent in the subsurface. The properties of an organic solvent differ from water

such that a mixture of the two behaves differently than water alone. The presence of bulk solvents promotes greater interaction between a solute and the solvent than would occur with water. The resultant mixture can dramatically increase the mobility of constituents in comparison with the case in which water is the sole solvent. In particular, the solubility of an organic constituent can increase and the sorption capacity of the subsurface materials can decrease.

Organic solvents affect the sorption capacity of subsurface materials in various ways[108]:

- Competition for sorption sites
- Removal of organic matter from the aggregate, which is the primary site for sorption to occur
- Physical alteration of soil, thereby changing intra-aggregate flow patterns to bypass sorption sites

Figure 4-34 shows how adding the cosolvent methanol to three different soils reduces sorption capacity. Besides reducing sorption, solvents may alter the properties of soil such as to increase the hydraulic conductivity of the subsurface.[109]

The effect of cosolvation on solubility can be just as dramatic. The introduction of solvents in water "in the range of 20 percent by volume or greater" can increase solubility of hydrophobic compounds by more than one order of magnitude.[110] The literature is replete with examples where the presence of solvents has prompted the migration of compounds that otherwise are deemed relatively insoluble.

Ionization

Organic acids (e.g., phenols and aliphatic acids) have the ability to donate protons when in aqueous solutions. In a process termed ionization, acids losing a proton

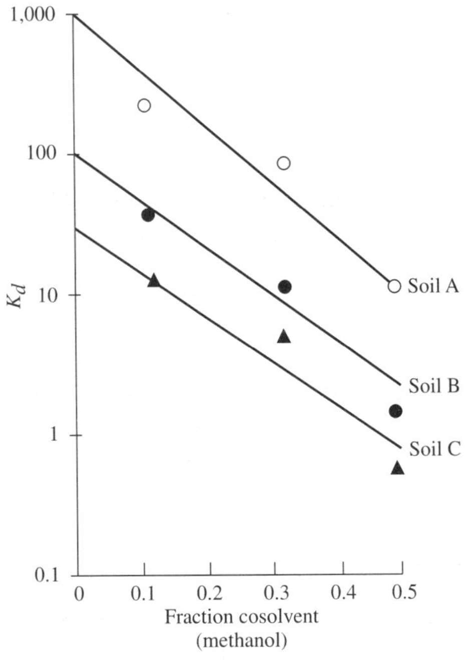

FIGURE 4-34
Reduction of sorption by cosolvation.[112]

would become anions and their solubility in water would increase significantly. An example is 2,4,5-trichlorophenol for which the K_{oc} would "decrease from 2330 for the phenol, to near zero for the phenolate."[111]

Dissolution

Dissolution is the dissolving of chemical substances from free product or solid forms of hazardous waste to solutes in the groundwater. The solutes can be inorganic cations and anions, polar organics, and nonpolar organics. Leachate, as discussed in Section 4-1, is an example of a liquid that dissolves constituents as it percolates through a landfill.

Complexation

Complexation, or chelation, is the formation of a coordinate bond between a metal ion and an anion known as a ligand. The ion-ligand arrangement is termed a complex, in which the ligands surround the ion. Complexation increases the potential mobility of a metal because (1) the complexed metal is effectively more soluble and (2) the **complex** ties up what would otherwise be free metallic ions, thus decreasing the opportunity for adsorption or precipitation of such ions. An example is mercury and chloride salts; increasing chloride concentrations may drive mercury from Hg^{+2} to a neutral complex.[113]

A ligand is often composed of multiple elements functioning together. Ligands can be inorganic (e.g., many of the common anions such as CO_3^{-2} and OH^{-1}) or organic.[114] Some synthetic organic ligands may be present in groundwater from waste sources. Natural organic ligands may also be present, having derived from humic materials. It is noted that in shallow aquifers, organic ligands from humic materials can be present in significant concentrations and dominate metal chemistry.[115] Complexed metals formed with organic ligands can be fairly stable, requiring a low pH (or for some metals a high pH) to break the bond. Because of the variety of ligands that can be found in groundwater, a metal can be complexed in different ways; it is difficult to predict the outcomes.

Immiscible Phase Partitioning

Immiscible fluids, such as gasoline, that have migrated into the subsurface may exist in three phases: (1) free product, (2) sorbed to soil particles and interstitial openings in the subsurface matrix, and (3) dissolved in ground and soil water. The amount in each phase depends on the properties of the subsurface and the immiscible substance. The partitioning to the sorbed phase follows the same principles as explained previously for sorption. In most cases, the mass in the solute phase is small in comparison with the mass in either the sorbed or free product phases. As such, the latter two represent an extended and continuous source of release of contaminants to the groundwater. The free product phase is termed **nonaqueous-phase liquid** (NAPL).

Inclusion of Fate in Transport Equations

Some retardation processes, particularly sorption and ion exchange, can have a significant effect on transport rates. The approach to these processes is to represent

them in groundwater transport equations by the addition of a retardation factor on an empirical basis.

Recall in Section 4-2 that the transport of contaminants in the subsurface was in response to hydraulic gradients (advection) and chemical gradients (diffusion). The one-dimensional advection-dispersion equation for nonreactive dissolved constituents in a homogeneous, isotropic aquifer under steady-state, saturated flow conditions is:

$$D_l \frac{\partial^2 c}{\partial l^2} - \bar{v}_l \frac{\partial c}{\partial l} = \frac{\partial c}{\partial t} \tag{4-39}$$

where D_l = longitudinal coefficient of hydrodynamic dispersion
c = solute concentration
$\bar{v}_l$ = average linear groundwater velocity
l = distance along the flowline
t = time

While two- and three-dimensional cases are beyond the scope of this chapter, their existence should be acknowledged. These cases include additive terms that are directionally dependent.

The coefficient of hydrodynamic dispersion D_l is a result of two mechanisms, mechanical dispersion and molecular diffusion, and can be expressed as

$$D_l = \alpha_l \bar{v}_l + D^* \tag{4-40}$$

where α_l = longitudinal dispersivity
D^* = coefficient of molecular dispersion

At this juncture, the transport equation has not considered the attenuation, retardation, and mobility enhancement processes discussed earlier in this section. It is clear that these phenomena collectively can reduce the concentration of inorganic and organic constituents in groundwater to levels less than that estimated on the basis of merely mechanical forces. The converse is also true. Each phenomenon is individually complex, much more so if considered in a combined and interdependent manner. Because of this mosaic of great complexity, the processes are typically accounted for in a varied manner. The general, simple approach is to quantify the influence of these processes in a collective, gross manner as a retardation factor R, expressed as:

$$R = 1 + \frac{\rho_b}{n} K_d \tag{4-41}$$

where ρ_b = bulk mass density of the porous medium
n = porosity
K_d = distribution coefficient

Equation 4-39 can be rewritten to accommodate retardation as follows:

$$\frac{1}{R}\left(D_l \frac{\partial^2 c}{\partial l^2} - \bar{v}_l \frac{\partial c}{\partial l}\right) = \frac{\partial c}{\partial t} \tag{4-42}$$

How then does retardation alter the contaminant plume? Shown in Figure 4-35 is a section through a shallow groundwater system with a continuous source of contaminants and steady-state flow. In Figure 4-35a and b there is no retardation

whereas the retardation increases (as K_d goes from 0 to 1 to 10) in Figure 4-35b, c, and d. The effect is dramatic. After a transport time of 60 years there is little contaminant migration beyond the source for the retardation case ($K_d = 10$) compared to the case having no retardation ($K_d = 0$).

By using only the variable K_d, this approach is limited to sorption and ignores other fate processes. Nevertheless, this approach has been shown to estimate the velocity of the transport of organic contaminants within a factor of 2 to 3 of the actual velocity observed in the field.[116] Thus, this approach can provide reasonable values. Part of the success of this approach derives from the fact that sorption can be the dominant process in many cases. There are other mathematical approaches, including nomographs, to predicting the mobility of organic chemicals.[117]

The **advection-dispersion-reaction equation** (Equation 4-42) is widely used as a basis for contaminant transport calculations. One common modification to this equation is the addition of a sink/source term. For example, NAPLs already discussed can provide a steady **source** of contaminants for transport within the system. Conversely, biodegradation can remove contaminants (a **sink**) from the system, resulting in lower contaminant concentration than would otherwise be predicted. When appropriate, these source-sink terms can be added to the advection and dispersion terms in Equation 4-42. Of course, the rate at which contaminants are added or removed from the system must be considered in any source or sink term. To complicate matters further, these rates may vary with time.

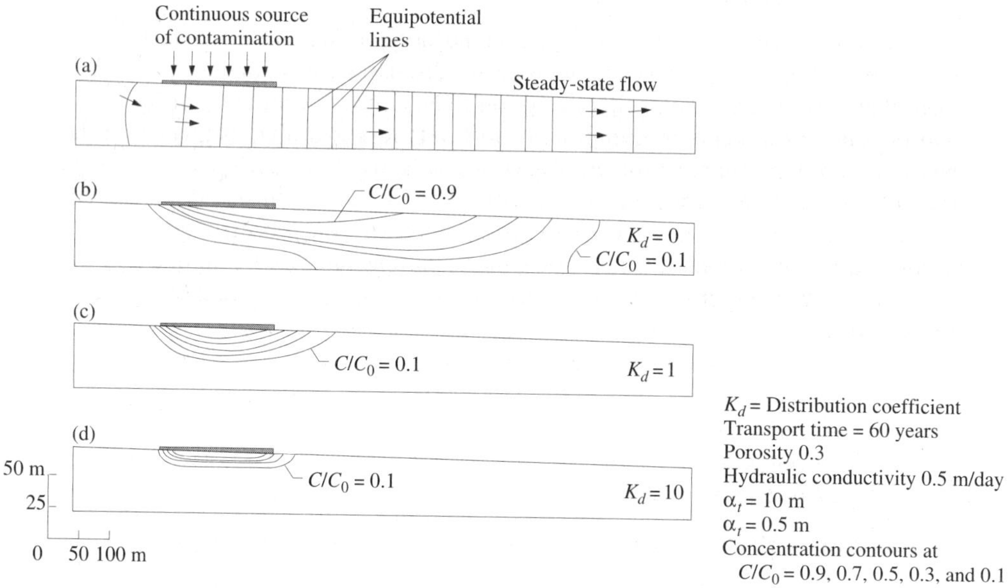

FIGURE 4-35
Effect of the distribution coefficient on contaminant retardation during transport in a shallow groundwater flow system. *(Adapted from Freeze and Cherry.[118])*

4-4 ATMOSPHERIC TRANSPORT AND DISPERSION OF CONTAMINANTS

Contaminants released to the outdoor air are transported by the wind and dispersed as a function of many variables, including characteristics of the atmosphere, of the surrounding terrain, of the contaminant (which can be in gaseous or particulate form), and of the source of the release. Concerns regarding air releases from hazardous waste sites focus not only on the quantity of material that becomes airborne, but more importantly on the concentration of the contaminant when it reaches downwind receptors. Health-related, concentration-based air quality standards have been established for many contaminants, and it is frequently of interest whether contaminants released from hazardous waste sites will exceed applicable standards at locations where people live or work.

The scientific basis for making estimates of pollutant concentrations at downwind locations has its roots in theoretical expressions of dispersion tracing back to the 1920s,[119] in field experiments conducted in the 1950s and early 1960s,[120] and theoretical work conducted in the 1960s.[121] Although much work has been done in recent years to improve the scientific basis of models that are used for estimating the impact of air pollution sources, air quality models that are currently in widespread use are nonetheless based on the fundamental principles developed in the 1960s.

Through the careful application of modern air quality models, estimates of the impact of releases of air contaminants from hazardous waste sites can provide useful information when questions are raised and decisions are required relative to those impacts. Air quality models are widely regarded as capable of providing conservative estimates (i.e., estimates of impacts that are likely to be higher than actual impacts). As a general rule, modeled impacts are more accurate for the following: longer averaging periods (e.g., annual averages as opposed to maximum 1-hour concentrations), sources with well-defined emissions and release characteristics, and receptor areas that do not experience steep concentration gradients (i.e., concentrations that change rapidly within a short distance).

Air quality models are also not any better than the data that are used as input. Source parameters must be carefully evaluated and specified as accurately as possible. Contaminant emission rates that are grossly underestimated will produce modeled impacts that are also grossly underestimated. Since the estimation of emissions is sometimes many times more uncertain than the models themselves, these uncertainties should also be taken into account when reviewing and taking action based on model predictions.

In some circumstances, air quality monitoring can be used to verify model predictions. Monitored air quality data have the advantage of being more inherently credible than model predictions; however, monitoring cannot provide satisfactory information in every instance. The effect on air quality of proposed changes at a site, including a comparison of different remediation alternatives, is virtually impossible to determine without the use of predictive models. Air quality impact issues can frequently be established by a combined approach of using models to predict and installing carefully sited monitors to verify.

This section provides a discussion of the basics of transport and dispersion, describes the use of mathematical models to characterize air quality impacts, and concludes with a look at some special topics, including a brief description of the basis of more advanced modeling techniques.

Transport and Dispersion: Basics

The discussion that follows provides a simplified description of atmospheric processes, emphasizing a practical understanding of underlying principles. Several excellent textbooks exist that are devoted entirely to the discussion of atmospheric processes that affect air pollution, most notably Pasquill and Smith (1983)[122] and Panofsky and Dutton (1985).[123]

A simplified, expanding box model

As an air contaminant moves with the wind, its concentration decreases as a function of the action of atmospheric turbulence. A useful way of visualizing this process is to think of the release and subsequent dispersion occurring in an expanding box as shown in Figure 4-36. This figure depicts a highly simplified description of dispersion, with the contaminant from an elevated source mixing uniformly in both the horizontal and the vertical direction. If the contaminant source depicted in this figure is a continuous source, the concentration at any point on the x axis can be calculated as:

$$C(x) = \frac{Q}{yzu} \tag{4-43}$$

where $C(x)$ = concentration (mass/volume) at distance x
Q = contaminant emission rate (mass/time)
y, z = horizontal and vertical dimensions of the box (distance)
u = wind speed (distance/time)

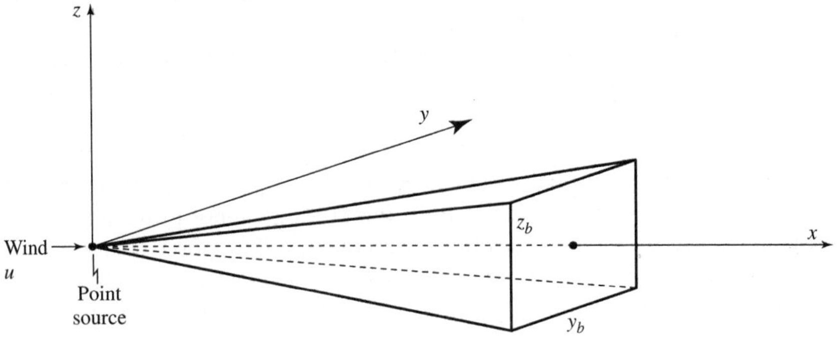

FIGURE 4-36
A simplified, expanding box model.

The principal characteristics of this simplified representation are that lower concentrations result both from higher wind speeds and from a larger box (i.e., more dispersion), and that concentrations within the expanding box are uniform across an entire cross section at any point on the x axis. Although this model is not a particularly accurate representation of the action of atmospheric turbulence, it provides a readily understandable model that can be modified to introduce concepts that more accurately describe pollutant transport and dispersion.

Atmospheric turbulence

The dispersion of pollutants in horizontal and vertical directions is caused by turbulence. There are two sources of turbulence. **Mechanical turbulence** is caused by the wind (due to random fluctuations and to passing over surface irregularities) and by vertical wind shear (the increase of wind speed with height). **Convective turbulence** is generated by heating of the ground surface during the day and the generation of "thermals" (pockets of air that rise because of the lower buoyancy of warm air) and the descent of cooler air to replace the air that is rising. Of these sources, wind fluctuations give rise to turbulent eddies that do most of the work of dispersing a pollutant cloud. Figure 4-37 depicts the action of turbulent eddies

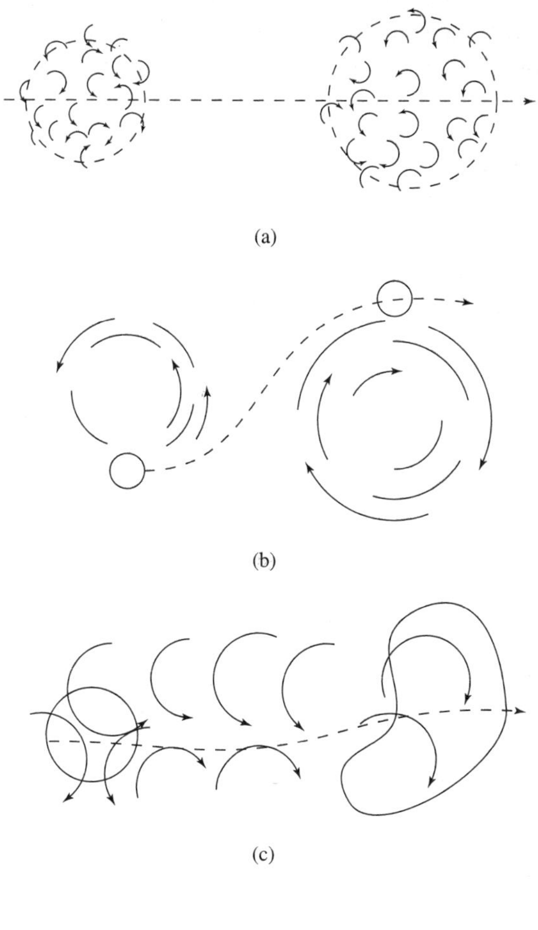

FIGURE 4-37
The effect of turbulent eddies on plumes (after Pendergast, 1984).[124] (a) A large cloud in a uniform field of small eddies, (b) a small cloud in a uniform field of large eddies, (c) a cloud in a field of eddies of the same size as the cloud.

of different sizes on a pollutant cloud. The growth of the cloud by turbulence is analogous to the expanding box depicted earlier.

The atmospheric boundary layer is the layer closest to the earth's surface that is affected by turbulence. The boundary layer has a distinct diurnal, or daily, pattern of evolution and decay. During the daytime, the boundary layer can grow—fueled by energy from the sun—to reach up to several kilometers above the surface. At night, the turbulent boundary layer typically becomes virtually nonexistent as winds diminish and the atmosphere becomes stable.

Atmospheric stability

Atmospheric stability is a surrogate term used by modelers for describing the capacity of the boundary layer to disperse pollutants. Atmospheric stability is described by examining what would happen to a parcel of air if it were elevated and allowed to expand as a result of the decreased atmospheric pressure. If this occurred adiabatically (with no exchange of heat between the parcel and surrounding air), the temperature of the parcel would cool. If the resultant temperature were less than that of the surrounding air at the higher altitude, the parcel would sink. This is defined as stable air (i.e., the atmospheric thermal gradient is such as to retard upward motion). Unstable conditions exist when the converse occurs (i.e., the adiabatic temperature of the parcel is higher than the atmospheric temperature, thus tending to induce vertical motion). Neutral conditions occur when the adiabatic temperature of lifted air equals the temperature of the surrounding air.[125]

An atmospheric stability classification scheme, developed in the 1960s and still in widespread use, assigns letters from A (very unstable) to D (neutral) and to F (very stable), indicating the capacity of the atmosphere to disperse pollutants. In simplistic terms, unstable conditions cause a plume to expand rapidly and stable conditions cause a plume to expand slowly. This scheme is called the Pasquill-Gifford-Turner (PGT)[126] classification scheme, and can be used to assign stability categories on an hourly basis by relying on readily available observations including wind speed, cloud cover, and ceiling height. These parameters are routinely recorded at the U.S. National Weather Service (NWS) stations located at airports across the country, and have served as the basis for archiving databases that include hourly observations spanning many years.

The PGT stability classification scheme is shown in Table 4-16. Several features of this table are worth noting. The classification scheme indicates that during the day, the relative rate of solar heating influences stability—with increased heating, the atmosphere becomes increasingly unstable. The rate of incoming solar radiation in turn depends on the solar angle (which is a function of the time of year and time of day) and the amount of cloud cover. At night, the amount of ground cooling is influenced by the amount of cloud cover—clouds tend to reduce the rate of heat radiated from the ground. If the sky is completely overcast, the amount of heating or cooling during the day or night is assumed to be minimal and a neutral category is assigned. Day or night, the surface wind speed determines the degree of mechanical and wind shear–based turbulence—with higher wind speeds, the effects of cooling and heating are damped and the atmosphere tends toward neutral stability.

The vertical temperature structure of the atmosphere strongly influences stability and is described in terms of a lapse rate, which refers to the change in temperature

TABLE 4-16
Pasquill chart for classifying atmospheric stability (after Turner)[126]

Surface wind speed at 10 m, m/s	Day: Incoming solar radiation: Strong	Day: Incoming solar radiation: Moderate	Day: Incoming solar radiation: Slight	Night: Mostly overcast, ≥ 4/8 cloud cover	Night: Mostly clear, ≤ 3/8 cloud cover	Overcast (day or night)
< 2	A	A–B	B			D
2–3	A–B	B	C	E	F	D
3–5	B	B–C	C	D	E	D
5–6	C	C–D	D	D	D	D
> 6	C	D	D	D	D	D

Key: A Extremely unstable
B Unstable
C Slightly unstable
D Neutral
E Slightly stable
F Stable to extremely stable

with height above the ground. The so-called adiabatic lapse rate of approximately -1^0C/100 m (-5.4^0F/1000 ft) represents the undisturbed condition of the atmosphere, where the temperature decrease with height balances the lower pressure at levels above the ground and no external heating or cooling is affecting the temperature structure. A typical daily cycle in the boundary layer starts with heating of the surface of the ground on clear days. The heating contributes to the formation of a "superadiabatic," or unstable, lapse rate where vertical motions tend to be enhanced—conditions associated with an "unstable" classification. On clear nights when the ground surface cools, a "subadiabatic" lapse rate, or temperature inversion, frequently forms that suppresses vertical motions—conditions associated with a "stable" classification. Overcast skies, during either day or night, tend to reduce the rate of heating or cooling and promote the maintenance of a lapse rate that is close to adiabatic—conditions associated with a "neutral" classification. Figure 4-38 illustrates stable versus unstable lapse rates.

The Gaussian Equation

The spatial and temporal variations in turbulence create nonuniform concentrations of contaminants in a dispersing plume, instead of the uniform distribution depicted by the expanding box model. A concept that is at the heart of most air quality models is that concentrations within a dispersing plume follow a gaussian distribution. Figure 4-39 displays a gaussian distribution (also familiar as the "bell-shaped" curve). Only two parameters—the value at the vertex of the curve and the standard deviation [depicted by the Greek lowercase sigma (σ)]—are needed to describe the entire distribution. A pollutant plume in the shape of a "bivariate" gaussian distribution (i.e., a concentration distribution that follows a gaussian shape in both horizontal and vertical dimensions) is shown in Figure 4-40. Maximum concentration occurs

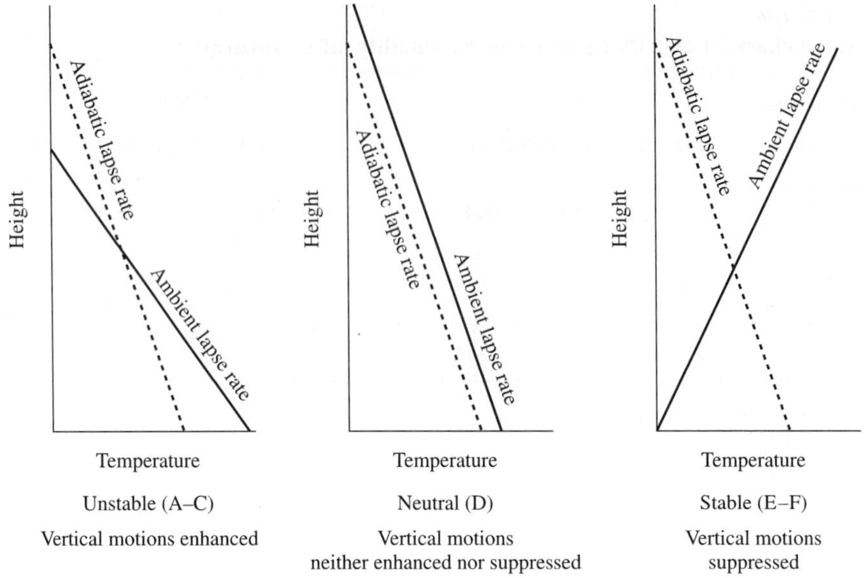

FIGURE 4-38
Stable and unstable lapse rates.

at the center of the plume, which can be calculated using the following formula:

$$C(x) = \frac{Q}{2\pi \sigma_y \sigma_z u} \tag{4-44}$$

where $C(x) =$ plume centerline concentration (mass/volume) at distance x
$Q =$ contaminant emission rate (mass/time)
$\sigma_y =$ standard deviation of horizontal concentration distribution (distance)
$\sigma_z =$ standard deviation of vertical concentration distribution (distance)
$u =$ wind speed (distance/time)

In this equation, the parameters σ_y and σ_z represent the standard deviation of the gaussian distribution in the horizontal and the vertical, respectively, and replace the horizontal and vertical dimensions of the expanding box. The two parameters are sometimes referred to as the horizontal and vertical dispersion coefficients. The horizontal coefficient can be considered the crosswind coefficient. The concentration at any distance perpendicular to the centerline can be determined (for any gaussian distribution) by using the following equation:

$$C(d) = C(o)\exp\left[-\frac{1}{2}\left(\frac{d}{\sigma}\right)^2\right] \tag{4-45}$$

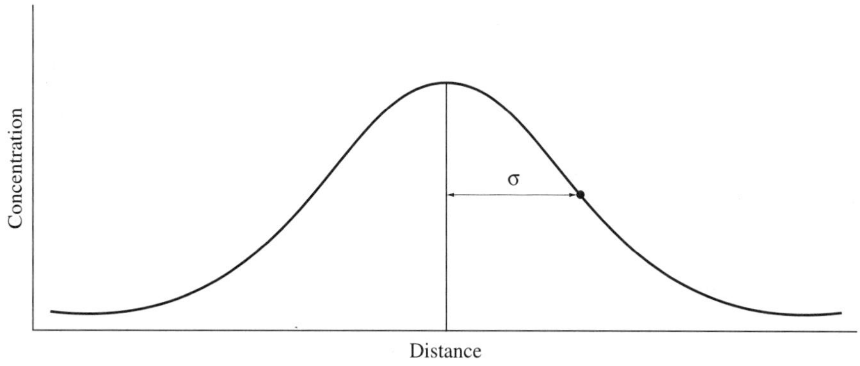

FIGURE 4-39
A gaussian distribution.

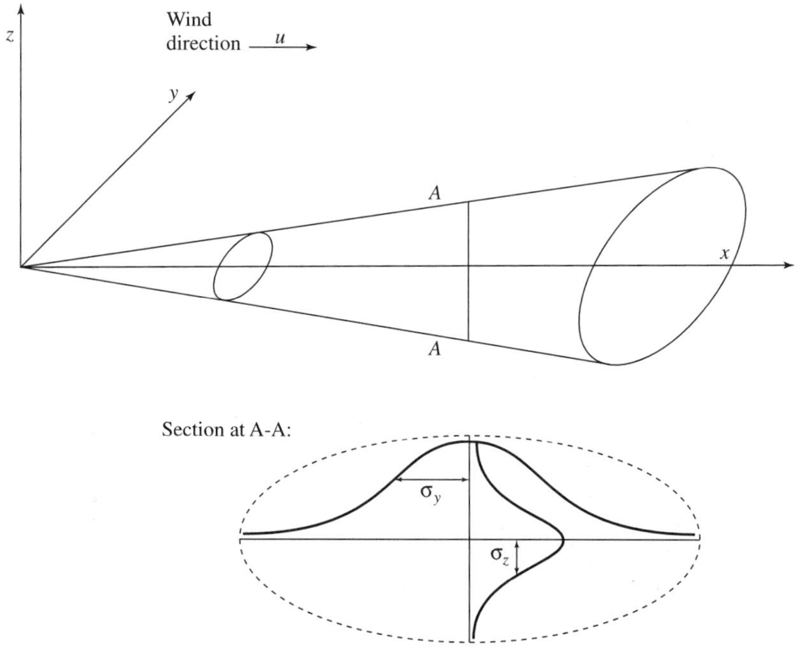

FIGURE 4-40
Bivariate plume and plume cross section.

where $C(d) =$ concentration at distance d from centerline
$C(o) =$ concentration at centerline
$d =$ distance from centerline
$\sigma =$ standard deviation of distribution

EXAMPLE 4-11. GAUSSIAN DISTRIBUTION CHARACTERISTICS. How much does a parameter concentration decrease from the centerline concentration at a distance equal

to one standard deviation from the centerline? Two standard deviations? How many standard deviations from the center of the distribution result in a parameter value of 10 percent of the centerline value?

Solution. Using Equation 4-46, at one standard deviation,

$$C(d)/C(o) = \exp\left[-\frac{1}{2}(1/1)^2\right] = 0.606 = 60.6 \text{ percent}$$

the parameter value is 60.6 percent of its centerline value; at two standard deviations, 13.5 percent. It can be determined that at 2.15 standard deviations, the parameter value decreases to 10 percent of its centerline value.

Of primary interest in air quality assessments is the concentration of a contaminant at the ground surface where exposure is most likely. The following equation describes the full three-dimensional concentration distribution for a bivariate gaussian plume with its centerline elevated at height H above the ground surface:

$$C(x, y, z) = \frac{Q}{2\pi\sigma_y\sigma_z u}\exp\left[-\frac{1}{2}\left(\frac{y}{\sigma_y}\right)^2\right]\left\{\exp\left[-\frac{1}{2}\left(\frac{z-H}{\sigma_z}\right)^2\right] + \exp\left[-\frac{1}{2}\left(\frac{z+H}{\sigma_z}\right)^2\right]\right\} \quad (4\text{-}46)$$

where C = concentration of pollutants at coordinates x, y, z (μg/m^3)
Q = emission rate of pollutants (μg/s)
σ_y, σ_z = horizontal (crosswind) and vertical standard deviations of pollutant concentration along the centerline of the plume (i.e., dispersion coefficient) (m) (See Figure 4-41.)
u = mean wind velocity (m/s)
x = downwind distance along the centerline of the plume (m)
y = horizontal distance from the centerline of the plume (m)
z = vertical distance from ground level (m)
H = plume height (m)

This equation relies on several assumptions:

- Wind velocity and direction are constant over height and over the averaging period.
- The emission rate is constant.
- The plume reflects completely at the ground (i.e., no deposition).
- No diffusion occurs in the direction of the plume travel.

In this equation, the x axis is aligned with the direction in which the plume is traveling, and the vertical term allows for concentration calculations at locations above the ground surface at a height z. The two terms involving $z - H$ and $z + H$ account for plume reflection from the ground surface. The more complete formulation found in air quality models includes additional terms to account for plume reflection from the top of the boundary layer in addition to reflection from the ground surface. These additional terms generally do not add significantly to the total predicted concentration, especially for low-level releases (50 ft or less).

EXAMPLE 4-12. GAUSSIAN DISPERSION EQUATION. Simplify Equation 4-44 to estimate concentration along the plume centerline at ground level.

Solution. In this case, y and z both equal 0. Equation 4-46 thus reduces to the following form:

$$C_{x,0,0} = \frac{Q}{\pi \sigma_y \, \sigma_z u} \exp\left[-\frac{1}{2}\left(\frac{H}{\sigma_z}\right)^2\right] \tag{4-47}$$

EXAMPLE 4-13. GAUSSIAN DISPERSION EQUATION. Further simplify Equation 4-47 to describe a ground level source with no thermal or momentum flux, the typical release that occurs at a hazardous waste site. In this situation the effective plume height H is essentially 0.

Solution. With $H = 0$, Equation 4-47 thus reduces to the following form:

$$C_{x,0,0} = \frac{Q}{\pi \sigma_y \sigma_z u} \tag{4-48}$$

Note the similarity between Equations 4-44 and 4-48. The primary difference is that Equation 4-48 accounts for reflection from the ground surface and is therefore a factor of 2 greater than the concentration calculated in Equation 4-44.

The link between stability classification and the calculation of concentration in a pollutant plume is made by the use of curves that provide values of σ_y and σ_z as a function of downwind distance and stability category. Curves that represent the values used in most of the widely used models are presented in Figures 4-41 and 4-42. The values presented in these curves were based on measurements taken during the Prairie Grass experiment conducted in 1957. The measurements were limited to a few hundred meters from the source, and the source itself was a ground level release. The curves therefore represent values extrapolated out to the distances shown, and also most accurately represent values typical of a ground release.

It should be noted that in reality an instantaneous measurement of concentrations within a pollutant plume will not display a gaussian distribution. Atmospheric turbulence is a process that is stochastic in nature; i.e., when viewed from an instantaneous perspective it appears to be a random process that produces random variations in pollutant concentration. The gaussian distribution is intended to represent an average characteristic over a suitable averaging interval. The basis of the PGT curves is nominally 10 minutes, and the dispersion parameters derived from the curves are used in models to represent hourly averages.

EXAMPLE 4-14. CONCENTRATION ESTIMATES/GROUND LEVEL VERSUS ELEVATED RELEASE. Use Equations 4-47 and 4-48 to estimate ground level concentrations due to a point source emitting 1 g/s of a contaminant (approximately 8 lb/h), using the following information: stability class C, wind speed 2.0 m/s. Estimate concentrations at 100 m downwind of the source, first assuming a ground level release and then assuming a release (plume) height of 10 m.

Solution. From Figures 4-41 and 4-42, σ_y and σ_z are approximately 15 and 8 m, respectively. Equation 4-47 can be used to calculate concentration for the ground level source of 1326 μg/m^3 and Equation 4-48 is used to calculate concentration from the elevated source of 607 μg/m^3. These concentration values represent 1-h averages.

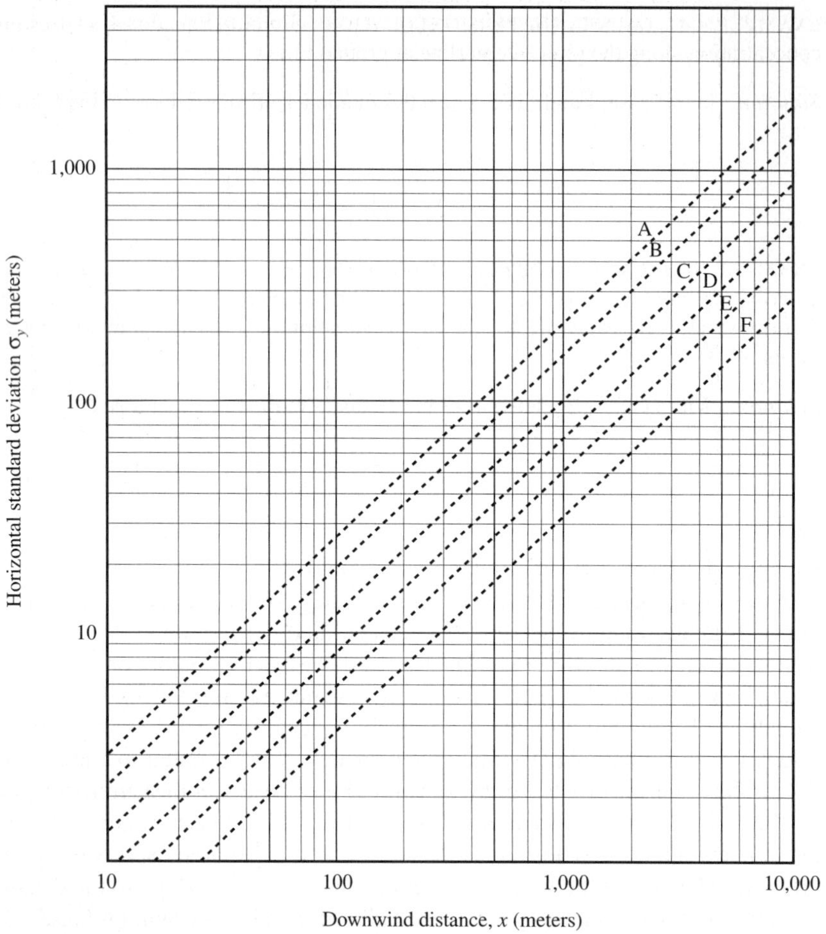

FIGURE 4-41
PGT horizontal (crosswind) dispersion coefficient as a function of stability category and downwind distance.[127]

Source Types

The release of air contaminants at a hazardous waste site can result from different types of sources. Air strippers and incinerators are two examples of sources that may be present at a hazardous waste site that are classified as a **stack source** (an elevated point source). In addition, sources of air contaminants at hazardous waste sites are frequently fugitive in nature—i.e., they are released at ground level and not exhausted from a stack. A fugitive source can be a **volume source** (e.g., release of fugitive dust when soil is excavated) or an **area source** (e.g., volatilization from a lagoon). Each type of source can be evaluated by using the preceding equations.

Elevated sources

The height to which a plume rises from a stack before bending horizontally and dispersing is important because the higher the rise, the lower the contaminant

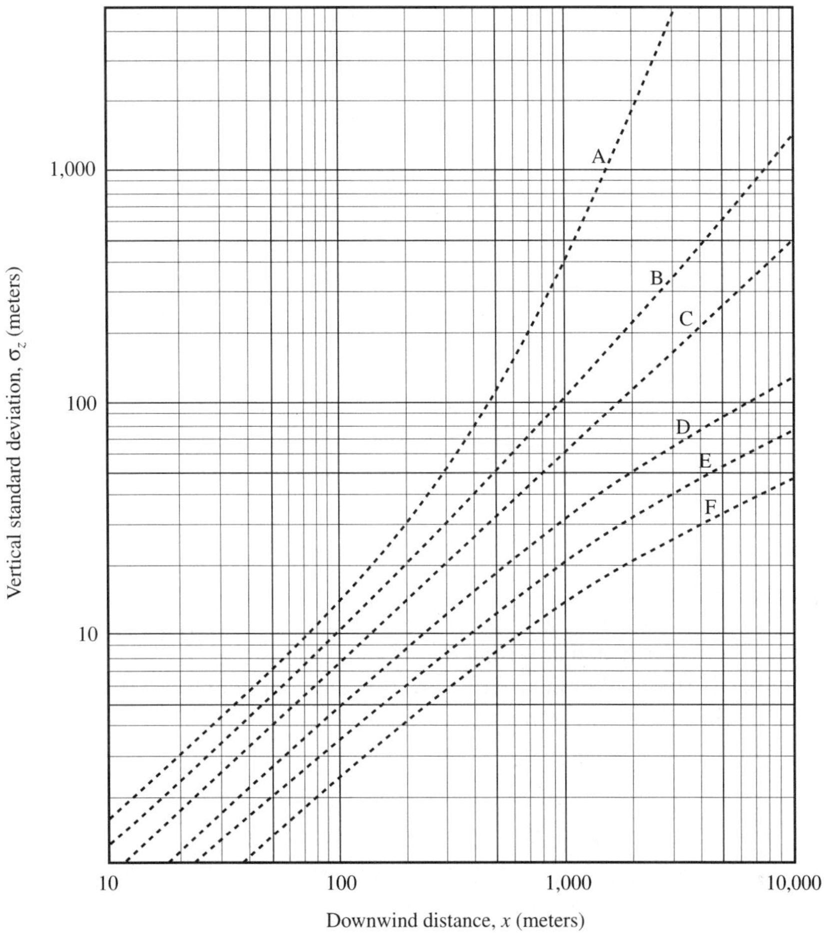

FIGURE 4-42
PGT vertical dispersion coefficient as a function of stability category and downwind distance.[128]

concentration at downwind, ground level receptor points. The degree to which the plume from a stack rises as it is being transported downwind is addressed in air quality models by calculating the source's **plume rise,** and then adding the plume rise to the physical stack height prior to calculating dispersion and concentration. Figure 4-43 depicts the determination of final plume height by first calculating plume rise and then adding the resultant value to the stack height.

A plume rises in response to its momentum and buoyancy. The exit velocity from a stack imparts **momentum** to the plume. Heated stack gases, especially those from combustion processes, are considerably warmer than ambient air, giving the plume a lower density and thereby making it **buoyant** in ambient air. Momentum and buoyancy propel the plume upward. The force of any wind counteracts this upward motion by bending the plume to the horizontal. Ambient air also mixes with the plume in a process called **entrainment** that involves phenomena such as turbulence and velocity shear. Entrainment speeds the cooling of the plume, thus dissipating its

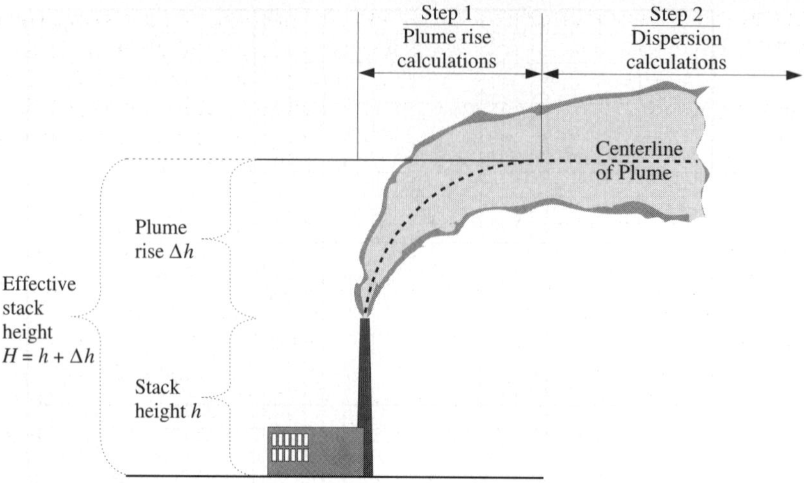

FIGURE 4-43
Two steps of estimating dispersion.

buoyancy and impeding its rise. As a plume becomes horizontal, the momentum and buoyancy flux dissipate and atmospheric turbulence becomes the dominating factor in the plume dispersion. For heated releases (e.g., incinerator stack), momentum can be considered negligible in comparison with buoyancy.

Buoyancy flux is calculated as follows:

$$F_b = g V_s r_s^2 (1.0 - T_A/T_s) \tag{4-49}$$

where F_b = buoyancy flux (m^4/s^3)
g = acceleration of gravity (9.80616 m/s^2)
V_s = stack exit velocity (m/s)
r_s = stack radius (m)
T_s = stack temperature (kelvins)
T_A = ambient temperature (kelvins)

and final plume rise (to the point that the plume levels off) can be calculated using one of the following two equations:

$$\Delta h_b = \frac{21.425 F_b^{3/4}}{u_s} \tag{4-50}$$

$$\Delta h_m = \frac{3.0 d_s V_s}{u_s} \tag{4-51}$$

where Δh_b = plume rise due to buoyancy (m)
Δh_m = plume rise due to momentum (m)
u_s = wind speed at stack top (m/s)
d_s = stack diameter (m)
V_s = stack exit velocity (m/s)

These equations are valid for unstable and neutral atmospheric conditions (stability categories A–D), and when wind speeds are greater than about 1.0 m/s.

EXAMPLE 4-15. CONCENTRATION ESTIMATES FOR BUOYANT VERSUS NONBUOYANT STACKS. Using the emission and meteorological parameters identified in Example 4-14, calculate the ground level concentration from two different stacks. The first is an air stripper with an exhaust temperature equal to ambient (293 K); the second is an incinerator with an exhaust temperature of 400 K. Exit parameters are identical, stack exit velocity $V_s = 5.0$ m/s, stack diameter = 0.5 m, stack height is 10 m.

Solution. For the incinerator, buoyancy is the dominant factor. Buoyancy flux from the incinerator stack is calculated from Equation 4-49; $F_b = 0.82\ \text{m}^4/\text{s}^3$. Plume rise is calculated from Equation 4-50 and is 9.2 m for a total plume height of 19.2 m. Equation 4-47 is used to calculate a concentration of 73.8 μg/m^3 from the incinerator stack. Since the air stripper stack is nonbuoyant, Equation 4-51 is used to calculate a plume rise of 3.8 m, which yields a plume height of 13.8 m. Equation 4-47 is then used to calculate a concentration of 300 μg/m^3 from the air stripper stack. This difference in concentration estimates is due entirely to the buoyancy from the incinerator stack, yielding a higher plume rise.

Volume and area sources

The basic concepts applicable to point sources can be applied to other types of sources, with some modifications. Plume rise is usually negligible because the releases have minimal momentum and a nominal difference in temperature. Nevertheless, the dispersion parameters σ_y and σ_z must be modified in a way that reflects the characteristics of the release. For example, the release of fugitive dust as a consequence of excavating contaminated soil and loading it into a truck is influenced by the additional turbulence caused by the action of the equipment and the action of the wind as it passes over the equipment. This initial dispersion is not accounted for in the dispersion curves discussed earlier (see Figures 4-42 and 4-43). This type of source can be modeled as a "volume" source, where the initial vertical and lateral dispersion is characterized by scaling the initial plume to the size and action of the equipment causing the enhanced turbulence. This initial plume size can then be used to adjust the atmospheric dispersion indicated by the dispersion curves. This adjustment can be accomplished by simply moving to a point on the curve corresponding to the estimated initial plume size, and proceeding from that point to determine effective dispersion parameters as a function of distance.

Sources that are characterized by a large surface area can be modeled by increasing the initial horizontal dispersion prior to calculating concentrations downwind. Different models accomplish this in different ways, but the net effect is the same—emissions over a wide area start out with an initial dispersion commensurate with the size and orientation of the area source.

EXAMPLE 4-16. CONCENTRATION ESTIMATES FOR VOLUME VERSUS POINT SOURCE. Use the data from Example 4-14 to estimate the effects of assuming a volume source configuration instead of the elevated point source. Initial plume size is estimated to be 10 m horizontally and 5 m vertically.

Solution. From Figures 4-41 and 4-42, the "enhanced" σ_y and σ_z are approximately 22 and 12 m, respectively. Equation 4-48 can be used to calculate a concentration for the volume source of 425 μg/m^3. (This is compared to the concentration for the point source of 607 μg/m^3.)

Equation 4-46 can be used to show relative ground level concentration with respect to distance as in Figure 4-44 (actual concentration is estimated by multiplying ordinate values by Q/u). The effect of different stability classes is shown in Figure 4-45 where apparently the less stable the atmosphere, the higher the ground level concentration within close proximity to a source. The influence of various effective stack heights is shown in Figure 4-46. It is apparent from the figure that effective stack height has a great influence on maximum ground level concentration.

Maximum ground level concentrations

As stated previously, most concern focuses on ground level concentrations, particularly the maximum concentration and its location and duration. The maximum ground level concentration is referred to as MGLC. For ground level releases, stable atmospheric conditions obviously create the greatest concentrations, all else being equal. The situation is more complicated for elevated releases. For an elevated re-

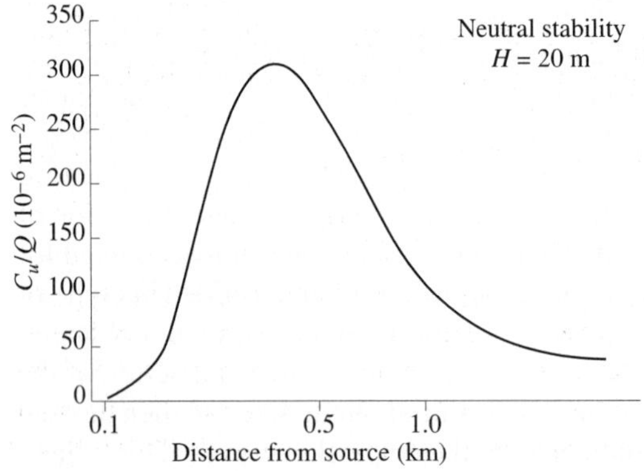

FIGURE 4-44
Relative ground level concentration versus distance.

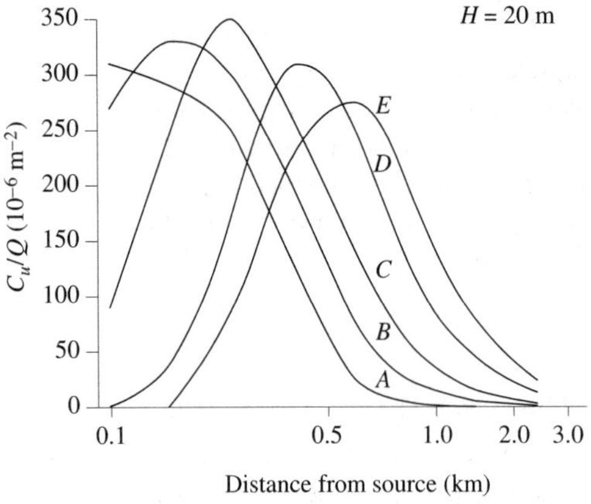

FIGURE 4-45
Relative ground level concentration versus distance for various stability classes.

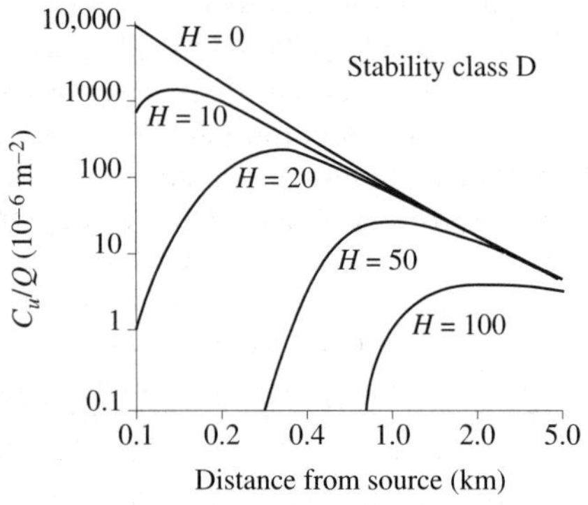

FIGURE 4-46
Relative ground level concentration versus distance for various effective stack heights.[129]

lease of given rate and duration, the absolute value of the maximum ground level concentration depends on effective stack height, atmospheric stability, and wind velocity.

EXAMPLE 4-17. VARIABLES AFFECTING MAXIMUM CONCENTRATION. Using Figures 4-45 and 4-46, determine the relative influences on MGLC of the variables—wind velocity, stability class, and stack height.

Solution. Doubling the effective stack height from 10 m to 20 m and from 50 m to 100 m results in a 75 percent and 84 percent reduction in the MGLC, respectively. In comparison, doubling wind velocity for a constant effective stack height would reduce MGLC by 50 percent. Thus, *effective* stack height has a greater effect than wind velocity. However, effective stack height equals stack height plus plume rise, and the latter is inversely proportional to wind velocity. A doubling of wind velocity for a constant stack height would thus result in *more* than a 50 percent reduction in MGLC. An increase in wind velocity also tends to stabilize the atmosphere. Thus, it is unclear from these figures which has a greater effect—wind velocity or stack height. It is clear that both have a much greater effect than stability class.

Given values for these three variables (effective stack height, wind speed, and stability class) and an emission rate, the MGLC and its distance from a source can be calculated by using the previously presented equations. Even though wind dilutes continuous releases, maximum ground level concentrations for releases with thermal or momentum flux tend not to occur at low wind velocities because such conditions promote a higher plume rise. Turner states that maximum ground level concentrations "occur at intermediate wind speed" which represents a balance between the countervailing effects of wind on dilution and plume rise.[130]

EXAMPLE 4-18. DUAL STACKS. The gases from two proposed side-by-side furnaces are each to be treated by a specific technology that is considered the "best available." The treated gases are then to be released from dual 80-ft stacks, one for each furnace. The resultant maximum ground level concentrations (MGLCs) are estimated to exceed allowable standards. The height of the stacks cannot be increased because of land use restrictions. What can be done other than additional treatment (which perhaps cannot be accomplished except at prohibitive cost) to reduce MGLCs?

Solution. Combining the two stacks as one will increase the effective stack height. Even if the cross-sectional area of the new stack equals the sum of the two proposed stacks, the circumference of the plume on release will be less. A smaller circumference will result in less horizontal dispersion and less entrainment of ambient air, thereby slowing the cooling of the plume. The plume will retain its buoyancy longer and thereby rise higher. Alternative ways to achieve a similar effect include increasing either the exit velocity from the stack or the temperature of the stack gas.

Computer modeling

As a practical matter, the mathematical computations required to estimate downwind concentrations are sufficiently complex that it is best in most instances to utilize a computer model. Models for many different source types are widely available, and a body of literature and guidance has been developed to guide their use.[131,132] The SCREEN3[133] and ISCST3[134] models were developed by the EPA and are in widespread common use. Both of these models are applicable to the assessment of many types of contaminant releases.

Depending on the situation, screening or refined modeling can be used, if not both. Generally, a **screening model** is easier to use and produces conservative estimates (i.e., concentration estimates that are higher than would result from more refined techniques). Screening models utilized "synthesized" meteorological data, i.e., data that represent a wide range of possible conditions but not actual measurements. A screening model calculates concentrations at selected downwind distances from a source, and estimates the maximum concentration at each receptor location over all meteorological conditions examined. The concentrations are expressed as hourly averages. However, regulatory standards and thresholds are frequently expressed as longer-term averages, e.g., 24-h up to annual averaging periods. Screening model outputs can be compared to longer averaging times through the use of scaling factors intended to represent the effect on concentration averages of atmospheric variability over time. Typical values of scaling factors are 0.4 for 24-h averages and 0.08 for annual averages multiplied times hourly averages.

EXAMPLE 4-19. CONCENTRATION ESTIMATES FOR HOURLY VERSUS LONG-TERM AVERAGES. SCREEN3 was used to estimate the maximum off-site hourly concentration due to an area source of fugitive particulate matter. The value determined by SCREEN3 was 45 $\mu g/m^3$. Determine 24-h and annual average concentrations.

Solution. For scaling factors of 0.4 and 0.08, off-site concentrations are calculated to be 18 and 3.6 $\mu g/m^3$ for 24-h and annual averages, respectively.

A **refined model** is distinguished from a screening model primarily through its use of actual meteorological data on an hourly basis, typically for a period of at least 1 year. The model can use actual meteorological data measured on-site or that derived from other sources (such as the NWS archive) if representative of the site. The use of actual meteorology allows a refined model to produce a more complete set of concentration estimates spatially (i.e., with a more extensive receptor grid) and a more accurate estimate of long-term averages.

Emerging approaches

Gaussian-based models have a number of limitations that have been recognized for some time. Some of these limitations include: the use of discrete stability categories when meteorological parameters (including turbulence levels and wind speed) are actually continuous variables; the use of dispersion rates (the PGT curves), based on measurements of surface releases, for modeling elevated (i.e., stack) sources; the assumption of a homogeneous (uniform with height) boundary layer in determining downwind dispersion; and the use of a gaussian plume distribution for unstable conditions (associated with clear, sunny days), when measurements suggest a skewed distribution dependent on the relative contribution of updrafts and downdrafts in the unstable boundary layer.

In the early 1990s the American Meteorological Society (AMS) and the U.S. EPA embarked upon the development of a new dispersion model that would address these deficiencies. The new model, the *AMS/E*PA *R*egulatory *Mod*el (AERMOD), has been developed to eventually replace gaussian-based models currently in widespread use. Cimorelli[135] provides a detailed description of the basis for AERMOD. It is anticipated that current models (including ISCST3 and SCREEN3) will eventually be replaced by AERMOD and a screening version of AERMOD.

There are a number of situations in which the mathematical expressions presented in this section may not apply, even with modification. These include accidental releases, chemical transformations, deposition, building downwash, and complex terrain.

Accidental releases and accidental release modeling

The foregoing discussion has focused primarily on releases from contaminant sources at hazardous waste sites that can be reasonably anticipated and for the most part are a consequence of routine activities at the site. Determining the consequences of contaminant releases associated with unanticipated events, such as spilling a drum or the rupture of a containment vessel, requires specialized modeling approaches that are generally beyond the capabilities of models used to estimate the impacts of routine releases. Accidental release types that affect air concentrations fall into three broad categories: toxic liquid spills followed by evaporation of the liquid, releases of gases liquefied under pressure and stored in cylinders or other types of vessels, and pure gaseous releases. The American Institute of Chemical Engineers (AIChE)[136] has published a guideline that provides comprehensive coverage of these types of releases and the models that are suitable for use in determining downwind consequences.

Chemical transformations

The significance of chemical and biological processes in transforming hazardous substances in the subsurface were discussed earlier in this chapter (see Section 4-3). Chemical transformations likewise occur in the atmosphere. The conversion of emissions of nitrogen oxides (NO_x) and volatile organic compounds (VOCs) in the presence of sunlight into the ground level pollutant ozone (O_3) is one example. The contribution of sulfates and nitrates, formed from emissions from cars and fossil fuel–fired power plants, to ambient concentrations of fine particulate matter, is another example. Transformations occur, however, over time and distance scales that are much longer and much farther than are typically encountered at hazardous waste sites. Therefore, unless a specific contaminant is released that transforms quickly into another that has different health consequences, pollutant transformations are not routinely considered at hazardous waste sites.

Deposition

Contaminants released from a source can not only affect air concentrations downwind of the source, but can also be deposited onto the ground surface. For those contaminants that bioaccumulate and can affect receptors through this pathway, an estimate of deposition should be made to estimate exposure. Deposition can occur via wet and dry processes. Wet deposition is a result of precipitation reacting with gases, or physically impacting particles. Obviously, it is significant only to the degree that emissions occur when it is raining. Dry deposition is a function both of the size distribution of particles being emitted (most particles greater than 30 μm in diameter settle out by gravitation within a few hundred meters of a source), and of the degree of interaction of the particles or of the gaseous contaminant with the underlying ground cover. The ISCST3 model contains algorithms to estimate the wet and dry deposition rates associated with both particulate and gaseous pollutants.

Building downwash

Buildings or other features such as a grove of trees can affect turbulence levels and suppress plume rise from stacks that are located in close proximity to the features. A "good engineering practice" (GEP) stack height is defined as follows:

$$h_{\mathrm{GEP}} = h_b + 1.5L \tag{4-52}$$

where $h_{\mathrm{GEP}} =$ good engineering practice stack height
$h_b =$ building height
$L =$ lesser of building height or width

Stacks that are less than h_{GEP} are potentially subject to downwash. Models that address the impact of downwash utilize the rule of thumb that stacks can be influenced by structures that are as close as 5 times the height (or width, whichever is less) of the structure itself. The net effect of downwash is to increase ground level concentrations at nearby locations. For this reason, stack sources need to be examined in terms of the potential for downwash from structures. The SCREEN3 and ISCST3 models contain algorithms to estimate impacts in the presence of downwash.

EXAMPLE 4-20. BUILDING DOWNWASH. A 20-ft stack is located 50 ft from a 15-ft-tall building, and 120 ft from a 20-ft-tall building. Is the stack potentially subject to downwash? Both buildings are wider than they are tall.

Solution. The 20-ft building has a radius of influence of 100 ft (5 times its height); the 15-ft building has a radius of influence of 75 ft. The stack is influenced by the 15-ft building but not the 20-ft building. The GEP stack height based on the smaller structure is $15 + 1.5 \times 15$ or 37.5 ft. Since the stack height is less than the GEP stack height, it will be subject to downwash.

Complex Terrain

Topographical features on the earth's surface influence turbulence and in turn affect dispersion. Ridges, hills, and mountains divert the wind, thereby altering wind direction and increasing turbulence. Plumes from stack sources can interact with elevated terrain receptors, causing concentration predictions to increase over predictions in the absence of terrain. The potential for plume impaction and significant effects on predicted concentrations is much greater for tall, buoyant stacks associated with large industrial sources and is not often a factor for typical sources at a hazardous waste site. The presence of significant terrain in the vicinity of the site, however, should be incorporated into any consequence modeling through the use of terrain elevations as part of model input preparation. Complex terrain can affect the wind field and make it much less likely that meteorological data collected anywhere but on-site will be suitable for refined modeling. Sites with complex terrain are thus candidates for screening level modeling that relies on synthesized meteorological conditions.

DISCUSSION TOPICS AND PROBLEMS

4-1. Rewrite Equations 4-16, 4-17, and 4-22 for three dimensions.

4-2. Determine the elevation, pressure, and total hydraulic heads at all points (a–g) labeled in Figure P-1 using datum 1.

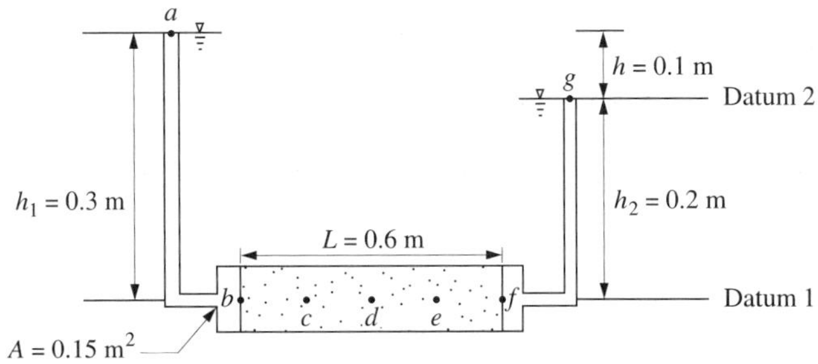

4-3. The "contaminant" concentration has been found to be 200 mg/L for the area of a site represented in plan view in Figure P-2. The groundwater contours are spaced at intervals of about 50 m. From test borings, the average thickness of the aquifer is 5 m and the hydraulic conductivity is estimated to be 1×10^{-2} cm/s. For each meter of length as measured parallel to the river, what is the contaminant loading rate in kg/year?

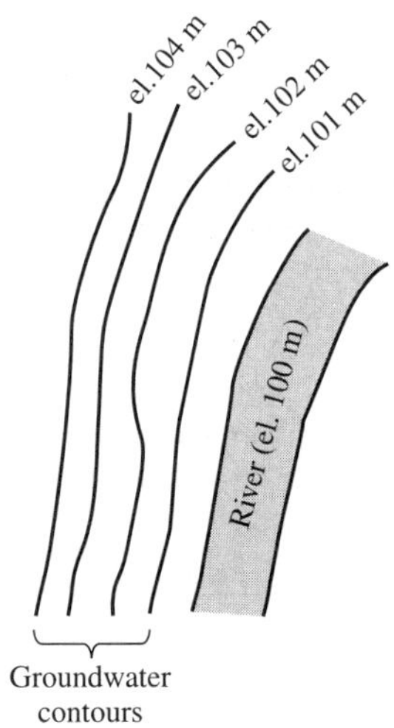

4-4. Determine the hydraulic conductivity for the experiment in Figure P-1 using datum 1 if the measured flow rate is 100 mL in 15 minutes.

4-5. Redo Problems 4-2 and 4-3 using datum 2.

4-6. Draw a flow net for the sample and head conditions in Problem 4-2.

4-7. Calculate the rate of flow through the sample shown in Figure 4-8, assuming the hydraulic conductivity is 5×10^{-3} cm/s.

4-8. Perform a triaxial permeability test on a cylindrical compacted clay sample that is 2.8 in. in diameter by 6 in. high. It is planned to use a backpressure of 50 psi to aid in saturation and average effective stress of 10 psi. The inflow and outflow burettes have an inside diameter of 0.5 cm.

(*a*) What is the cell pressure you will use in this test?

(*b*) If the hydraulic head causing flow is to be applied by elevating the water level in the inflow burette a distance of 20 cm above the water level in the outflow burette, what is the gradient?

(*c*) If the hydraulic head causing flow is to be applied by elevating the water level in the inflow burette a distance of 10 cm above the water level in the outflow burette after 8 hours, what is the hydraulic conductivity of the sample?

4-9. A laboratory experiment similar to the one conducted by Darcy is shown in Figure P-3.

(*a*) What is the pressure head at point *A* in sand of the laboratory experiment shown below?
(*b*) Assuming a porosity of 0.5 and $k = 1 \times 10^{-2}$ cm/s, how long will it take a molecule of water to pass through from one end of the sand to the other?

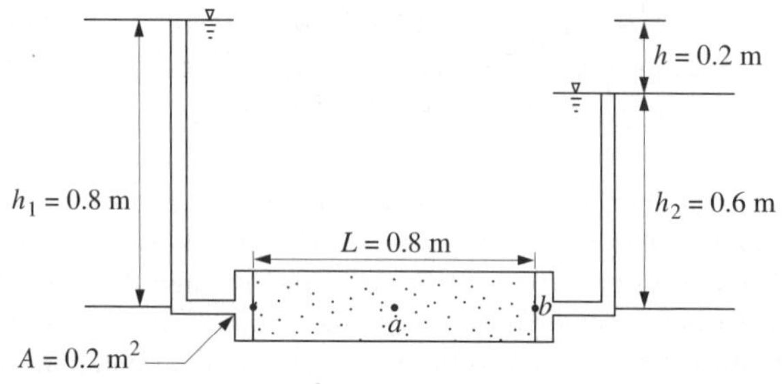

4-10. The boundary conditions and material properties for a flow regime are shown in Figure P-4.

(*a*) For the conditions shown in Figure P-4, sketch the flow net and calculate the flow rate through the system.
(*b*) Solve for the heads on 0.5-m centers within the flow regime of Figure P-4 using the finite difference mathematical approximation.

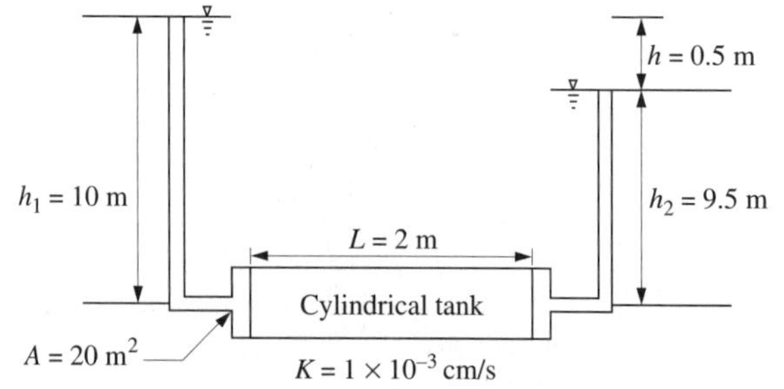

4-11. Derive the relationship shown in Equation 4-8 (*Hint:* Consider the flow in the experiment shown on Figure 4-8 and make use of the equation of continuity.)
4-12. Differentiate between hydrogeology and groundwater hydrology.
4-13. Discuss your opinion about the use of the terms aquitard and aquiclude in the context of hazardous waste management from the perspective of:

(*a*) A landfill operator
(*b*) A state environmental regulator
(*c*) An environmental organization spokesperson
(*d*) An "average" citizen

4-14. Differentiate between water table and piezometric surface.
4-15. The derivation of the Laplace equation assumed steady flow and homogeneous, isotropic conditions and showed that hydraulic head depended only on position within the aquifer. What would the dependent variables be for:

(*a*) Anisotropic conditions
(*b*) Heterogeneous conditions
(*c*) Transient conditions

4-16. From a contaminant transport viewpoint, describe the advantages and disadvantages of selecting *in situ* bioremediation to clean a groundwater supply aquifer contaminated with petroleum hydrocarbons from a leaking underground storage tank. Benzene is the compound of concern while xylene and ethylbenzene are present at lower concentrations.

4-17. Critically comment on the possible negative aspects of the use of computer models.

4-18. Sample A has a porosity of 40 percent. Sample B has a porosity of 8 percent. Show the relationship between seepage velocities for samples A and B under the same gradient.

4-19. Using Darcy's law and porosity, show that the seepage velocity is not accurate at extreme values.

4-20. Determine the pressure, elevation, and total heads at points a, b, and c in Figure P-5.

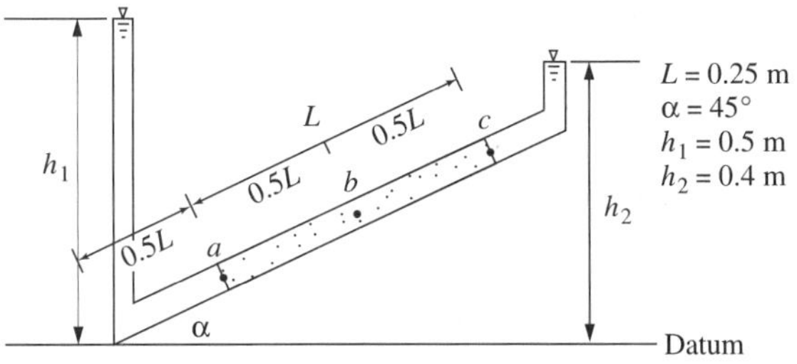

4-21. Determine the pressure, elevation, and total heads at points A, B, and C in Figure P-6.

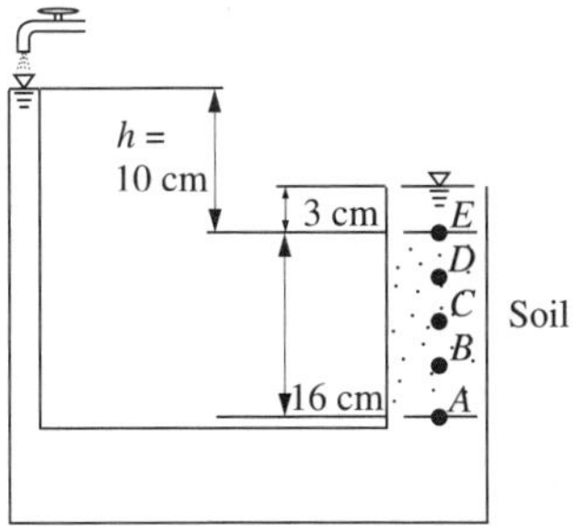

4-22. Using Darcy's law, calculate the quantity of groundwater (L/day), the seepage velocity (m/s), and the contaminant loading (kg/day) to the river (assuming no retardation) for the following scenario. Be sure to include both a plane view and cross-sectional sketches of the conditions analyzed.

Groundwater is flowing through an unconfined, uniform, 3-m-thick sand aquifer and discharging to a river. Two wells located 50 m apart and perpendicular to the flow direction are representative of the formation flow and show a difference in water level of 1 m. The hydraulic conductivity of the aquifer is 1×10^{-3} cm/s and the average contaminant concentration throughout the aquifer in the subject area is 200 mg/L. The zone of flow and contamination is 100 m long as measured parallel to the river.

4-23. Assuming the flow net in Figure 4-13 represents that for a dike around a liquid waste impoundment having a perimeter of 1000 m and a height of liquid of 3 m, calculate the leakage rate (m^3/yr).

4-24. Given a region found to have a glacial till overlying a thin lacustrine clay layer overlying a competent sandstone, how would you model groundwater flow through the system? In other words, for each layer what range of values for hydraulic conductivity would

you expect and how would you diameterize the layer (as unconfined aquifer, aquitard, etc.)?

4-25. What provides soil the capacity to buffer acidic water?

4-26. For what conditions would ideal sorption not apply?

4-27. What are two properties that can be measured in a laboratory that indicate the "hydrophobicity" of an organic chemical?

4-28. What are five important characteristics of soil affecting its molecular sorption capacity?

4-29. Which, if any, of the five characteristics named above apply more to inorganics than organics?

4-30. What is the difference between retention capacity of a soil and its molecular sorption capacity?

4-31. What indirect effect would depth to groundwater likely have on the sorption of organic contaminants?

4-32. How do the releases from an explosion differ from a continuous release? How would this be handled mathematically?

4-33. Frequent shifts of wind direction would cause a plume of air contaminants to do what?

4-34. A stable layer of air above an unstable layer would have what effect on the dispersion of a ground release of air emissions? On the dispersion of emissions from a stack sufficiently tall to penetrate into the stable layer?

4-35. Equation 4-46 ignores diffusion in the direction of transport (x axis). What assumption regarding emission duration must this be based upon?

4-36. What differences would occur in effective stack height for two releases, both at the same temperature but one containing water droplets from having passed through a cooling tower?

4-37. Determine the concentration of TCE 5.0 km downwind of a point source at ground level that releases 3000 g/s TCE on a continuous basis. The following atmospheric conditions are given: wind velocity = 4.0 m/s; stability class B.

4-38. Figure 4-35 depicts the effect of K_d on contaminant transport for 2-D flow. For 1-D flow as represented in Figure 4-13, sketch the relative concentration (C/C_0) as a function of time for a point at the midpoint along the length of the column for a continuous supply of contaminant under the following conditions:

(*a*) No dispersion or retardation
(*b*) With dispersion but no retardation
(*c*) With dispersion and retardation
(*d*) With dispersion and retardation greater than that of (*c*) above.

REFERENCES

1. U.S. Environmental Protection Agency: *Procedures for Conducting Air Pathway Analyses for Superfund Applications*, Volume II, Office of Air Quality Planning and Standards, Research Triangle Park, N.C., January 1989.
2. U.S. Environmental Protection Agency: *Superfund Exposure Assessment Manual*, EPA/540/1-88/001, Office of Emergency and Remedial Response, Washington, D.C., April 1988.
3. U.S. Environmental Protection Agency, *Procedures for Conducting Air Pathway Analyses for Superfund Applications,* Volume II.
4. de Percin, P. R.: "Air Pollution from Land Disposal Facilities," *Standard Handbook of Hazardous Waste Treatment and Disposal,* H. M. Freeman, ed., McGraw-Hill, New York, pp. 10.53–10.102, 1988.
5. Skidmore, E., and N. Woodruff: *Wind Erosion Forces in the United States and Their Use in Predicting Soil Loss*, Agriculture Handbook No. 346, U.S. Department of Agriculture, Washington, D.C., 1968.

6. U.S. Environmental Protection Agency, *Procedures for Conducting Air Pathway Analyses for Superfund Applications,* Volume II.
7. U.S. Environmental Protection Agency: *Procedure for Conducting Air Pathway Analyses for Superfund Applications*, Volume III, Office of Air Quality Planning and Standards, Research Triangle Park, N.C., January 1989.
8. U.S. Environmental Protection Agency: "Compilation of Air Pollutant Emission Factors," Manual AP-42.
9. Schroeder, P. R., A. C. Gibson, and M. D. Smolen: *The Hydrologic Evaluation of Landfill Perfomance (HELP) Model*, EPA/830-SW-84-010, 1984.
10. Ibid.
11. "Management of Hazardous Waste Leachate," report prepared for U.S. EPA by Touhill, Shuckrow and Associates, Pittsburgh, September 1980.
12. U.S. Environmental Protection Agency: *Superfund Exposure Assessment Manual*, U.S. EPA, EPA/540/1-88/001, Washington, D.C., April 1988.
13. "Management of Hazardous Waste Leachate," report SW-871, prepared for U.S. EPA by Shuckrow, Touhill and Associates, Pittsburgh, 1981.
14. "Prevalence of Subsurface Migration of Hazardous Chemical Substances at Selected Industrial Waste Land Disposal Sites," report EPA/530/SW-634, prepared for U.S. EPA by Geraghty and Miller, 1977.
15. Murray, C. L.: "Water Use, Consumption, and Outlook in the U.S. in 1970," *J. American Water Works Association,* vol. 65, pp. 302–308, 1973.
16. Pye, V. I., R. Patrick, and J. Quarles: *Groundwater Contamination in the United States,* University of Pennsylvania Press, Philadelphia, 1983.
17. Johnson, R. L., et al.: "Diffusive Contaminants Transport in Natural Clay: A Field Example and Implications for Clay-lined Waste Disposal Sites," *Environ. Sci. Technol.,* vol. 23, p. 340, 1989.
18. Freeze, R. A., and Cherry, A. A.: *Groundwater*, Prentice-Hall, Englewood Cliffs, N.J., 1979.
19. Todd, D. K.: *Groundwater Hydrology*, 2nd ed., John Wiley and Sons, New York, 1980.
20. Fetter, C. W.: *Applied Hydrogeology*, 2nd ed., Merrill Publishing, Columbus, Ohio, 1988.
21. Cernica, J. N.: *Geotechnical Engineering*, CBS College Publishing, 1982.
22. Domenico P. A., and F. W. Schwartz: *Physical and Chemical Hydrogeology,* John Wiley and Sons, New York, 1990.
23. Acar, Y. B.: "Electrokinetic Soil Processing (a Review of the State of the Art)." Geotech. Special Publication No. 30, ASCE, New York, pp. 1420–1432, 1992.
24. Liakopoulos, A. C.: "Variation of the Permeability Tensor Ellipsoid in Homogenous Anisotropic Flow," *Water Resources Res.,* vol. 1, pp. 135–141, 1965.
25. Tindall, James A., and James R. Kunkel: *Unsaturated Zone Hydrology for Scientists and Engineers*, Prentice-Hall, Upper Saddle River, N.J., 1999.
26. Palmer, C. D., and R. L. Johnson: "Physical Processes Controlling the Transport of Non-Aqueous Phase Liquids in the Subsurface," *Transport and Fate of Contaminants in the Subsurface*, EPA/625/4-89/019.
27. Bear, Jacob, and Arnold Verruijt: *Modeling Groundwater Flow and Pollution,* D. Reidel Publishing, Dordrecht, Holland, p. 414, 1987.
28. Huyakorn, Peter S., and George F. Pinder: *Computational Methods in Subsurface Flow*, Academic Press, Orlando, Fla., p. 472, 1983.
29. Walton, William C.: *Analytical Groundwater Modeling*, Lewis Publishers, Chelsea, Mich., p. 173, 1989.
30. Walton, William C.: *Numerical Groundwater Methodology*, Lewis Publishers, Chelsea, Mich., p. 272, 1989.
31. Spangler, M. G., and R. L. Hardy: *Soil Engineering*, 4th ed., Harper & Row, New York, 1982.

32. Cedergren, H. R.: *Seepage, Drainage and Flow Nets*, 3rd ed., John Wiley & Sons, New York, 1989.
33. Driscoll, F. G.: *Groundwater and Wells*, 2nd ed., Johnson Division, St. Paul, Minn., 1986.
34. Rowe, R. K., J. R. Booker, and M. J. Fraser, *POLLUTEv6 and POLLUTE-GUI User's Guide*, GAEA Environmental Engineering Ltd., Ontario, Canada.
35. Harbaugh, A. W., and M. G. McDonald: "User's documentation for MODFLOW-96, an update to the U.S. Geological Survey modular finite-difference ground-water flow model," U.S. Geological Survey Open-File Report 96-485, 1996.
36. Harbaugh, A. W., and M. G. McDonald: "Programmer's documentation for MODFLOW-96, an update to the U.S. Geological Survey modular finite-difference ground-water flow model, "U.S. Geological Survey Open-File Report 96-486, 1996.
37. McDonald, M. G., and A. W. Harbaugh: "A modular three-dimensional, finite-difference ground-water flow model," U.S. Geological Survey Techniques of Water-Resources Investigations, Book 6, Chap. A1, 1988.
38. *McGraw-Hill Encyclopedia of Science and Technology*, McGraw-Hill, New York, 1982.
39. Brubaker, G. R.: "Lecture No. 1—Introduction to In Situ Bioremediation and Contaminant Transport," *HMCRI Superfund Conference*, November 28, 1990.
40. Weber, W. W.: "Sorption Phenomena in Subsurface Systems: Concepts, Models, and Effects on Contaminant Fate and Transport," *Water Research*, Pergammon Press, vol. 25, no. 5, 1991.
41. Brady, N. C.: *The Nature and Properties of Soils,* 8th ed., McMillan, New York, 1974.
42. Pignatello, J. J.: "Sorption Dynamics of Organic Compounds in Soils and Sediments," *Reactions and Movement of Organic Chemicals in Soils,* B. L. Sawhney and K. Brown, eds., Soil Science Society of America, Inc., Madison, Wis., 1989.
43. Brady, N. C.: *The Nature and Properties of Soils.*
44. Dragun, James: "The Soil Chemistry of Hazardous Materials," Hazardous Materials Control Research Institute, Silver Springs, Md., 1988.
45. Voice, T. C.: "Activated-Carbon Adsorption," *Standard Handbook of Hazardous Waste Treatment and Disposal*, Harry M. Freeman, ed., McGraw-Hill, New York,1989.
46. Dragun, "The Soil Chemistry of Hazardous Materials."
47. Pignatello, "Sorption Dynamics of Organic Compounds in Soils and Sediments."
48. Ibid.
49. U.S. Environmental Protection Agency: "Superfund Exposure Assessment Manual."
50. U.S. Environmental Protection Agency: "Transport and Fate of Contaminants in the Subsurface," EPA/625/4-89/019, Center for Environmental Research Information, Cincinnati, September 1989.
51. Dragun, "The Soil Chemistry of Hazardous Materials."
52. Brady, N. C.: *The Nature and Properties of Soils.*
53. *McGraw-Hill Encyclopedia of Science and Technology.*
54. Pignatello, "Sorption Dynamics of Organic Compounds in Soils and Sediments."
55. Dragun, "The Soil Chemistry of Hazardous Materials."
56. Piwoni, M. D., and J. W. Keeley: "Basic Concepts of Contaminant Sorption at Hazardous Waste Sites," EPA/540/4-30/053, U.S. EPA, Robert S. Kerr Environmental Research Laboratory, Ada, Okla., October 1990.
57. U.S. Environmental Protection Agency: "Transport and Fate of Contaminants in the Subsurface,"
58. Brubaker, G. R.: "Lecture No. 1—Introduction to In Situ Bioremediation and Contaminant Transport," *HMCRS Superfund Conference*, November 28, 1990.
59. See note 56 above.
60. Dragun, "The Soil Chemistry of Hazardous Materials."
61. U.S. Environmental Protection Agency, "Transport and Fate of Contaminants in the Subsurface."

62. Ibid.
63. Pignatello, "Sorption Dynamics of Organic Compounds in Soils and Sediments."
64. Dragun, "The Fate of Hazardous Material in Soil."
65. U.S. Environmental Protection Agency, "Superfund Exposure Assessment Manual."
66. U.S. Environmental Protection Agency, "Transport and Fate of Contaminants in the Subsurface."
67. Dragun, "The Fate of Hazardous Materials in Soil."
68. Brubaker, "Lecture No. 1—Introduction to In Situ Bioremediation and Contaminant Transport."
69. U.S. Environmental Protection Agency, "Transport and Fate of Contaminants in the Subsurface." op. cit.
70. Dragun, "The Soil Chemistry of Hazardous Materials."
71. Weber, "Sorption Phenomena in Subsurface Systems: Concepts, Models, and Effects on Contaminant Fate and Transport," op. cit.
72. Brubaker, "Lecture No. 1—Introduction to In Situ Bioremediation and Contaminant Transport."
73. Wu, S. C., and P. M. Gschwend: "Sorption Kinetics of Hydrophobic Compounds to Natural Sediments," *Environmental Science and Technology*, American Chemical Society, vol. 20, no. 7, 1986.
74. U.S. Environmental Protection Agency, "Transport and Fate of Contaminants in the Subsurface," op. cit.
75. Dragun, "The Fate of Hazardous Materials in Soil," p. 154.
76. Brady, *The Nature and Properties of Soils*.
77. Ibid.
78. Weber, "Sorption Phenomena in Subsurface Systems," op. cit.
79. Grim, R. E.: *Clay Mineralogy*, McGraw-Hill, New York, 1960.
80. U.S. Environmental Protection Agency, "Transport and Fate of Contaminants in the Subsurface."
81. Ibid.
82. Dragun, "The Fate of Hazardous Materials in Soil," op. cit., p. 307.
83. Freedman, D. L. and J. M. Gossett: "Biological reductive dechlorination of tetrachloroethylene and trichloroethylene to ethylene under methanogenic conditions," *Appl. Environ. Microbiol.*, vol. 55, no. 9, pp. 2144–2151, 1989.
84. De Bruin, W. P., et al.: "Complete Biological Reductive Transformation of Tetrachloroethene to Ethane," *Appl. Environ. Microbiol.*, vol. 58, no. 6, pp. 1996–2000, 1992.
85. Carter, S. R., and W. J. Jewell: "Biotransformation of Tetrachloroethylene by Anaerobic Attached-films at Low Temperatures," *J. Environ. Eng.*, vol. 27, no. 4, pp. 607–615, 1993.
86. Dudiak, K. A. and T. D. DiStefano: "Reductive Dechlorination of PCE by a Methanogenic Culture Taken from an Aerobic Fluidized-Bed Reactor," *The Twenty-Ninth Mid-Atlantic Industrial and Hazardous Waste Conference*, Blacksburg, Va., Technomic, 1997.
87. Long, J. L., et al.: "Anaerobic and Aerobic Treatment of Chlorinated Aliphatic Compounds," *J. Environ. Eng.*, vol. 119, no. 2, pp. 300–320, 1993.
88. Speece, R. E., et al.: "The Role of Process Configuration in the Performance of Anaerobic Systems," *Wat. Sci. Tech.* (GB), vol. 36, nos. 6–7, pp. 539–547, 1997.
89. Cox, E., et al.: "Intrinsic Biodegradation of Trichloroethene and Trichloroethane in a Sequential Anaerobic-Aerobic Aquifer," *Third International In Situ and On-Site Bioreclamation Symposium*, San Diego, Calif., Battelle Press, 1995.
90. Graves, R. W., et al.: "Natural Attenuation of Chlorinated Compounds in Six Plumes at Hill AFB," *Fourth International In Situ and On-Site Bioremediation Symposium*, New Orleans, La., Battelle Press, 1997.

91. Johnson, J. A., et al.: "Heterogeneous Degradation within a Dissolved Chlorinated Solvent Plume," *Fourth International In Situ and On-Site Bioremediation Symposium*, New Orleans, La., Battelle Press, 1977.
92. Lee, M. D., et al.: "Intrinsic In Situ Anaerobic Biodegradation of Chlorinated Solvents at an Industrial Landfill," *Third International In Situ and On-Site Bioreclamation Symposium*, San Diego, Calif., Battelle Press, 1995.
93. Lorah, M. M., L. D. Olsen, and B. L. Smith: "Natural Attenuation of Chlorinated Hydrocarbons in a Freshwater Wetland," *Fourth International In Situ and On-Site Bioremediation Symposium*, New Orleans, La., Battelle Press, 1997.
94. Major, D., et al.: "Intrinsic Dechlorination of Trichloroethene to Ethene in a Bedrock Aquifer," *Third International In Situ and On-Site Bioreclamation Symposium*, San Diego, Calif., Battelle Press, 1995.
95. Major, D. W., W. W. Hodgins, and B. J. Butler: "Field and Laboratory Evidence of In Situ Biotransformation of Tetrachloroethylene to Ethene and Ethane at a Chemical Transfer Facility in North Toronto," *On-Site Bioreclamation*, San Diego, Calif., Butterworth-Heinemann Stoneham, 1991.
96. McCarty, P. L. and J. T. Wilson: "Natural Anaerobic Treatment of a TCE Plume, St. Joseph, Michigan, NPL site," in *Bioremediation of Hazardous Wastes*, U.S. Environmental Proection Agency, EPA/600/R-92/126, Office of Research and Development, Washington, D.C., 1992.
97. DiStefano, T. D.: "The Effect of Tetrachloroethene on Biological Dechlorination of Vinyl Chloride: Potential Implication for Natural Bioattenuation," *Wat. Res.*, vol. 33, no. 7, pp. 1688–1694, 1999.
98. Veith, G. D., D. J. Call, and L. T. Brooke, "Structure-Toxicity Relationships for the Fathead Minnow, *Pimephales Promelas:* Narcotic Industrial Chemicals." *Can. J. Fish Aquat. Sci.*, vol. 40, pp. 743–748, 1983.
99. Callahan, M. A., et al.: "Water-Related Environmental Fate of 129 Priority Pollutants," Vol. II, *Halogenated aliphatic hydrocarbons, halogenated ethers, monocyclic aromatics, phthalate esters, polycyclic aromatic hydrocarbons, nitrosamines and miscellaneous compounds*, EPA-440/4-79-029b, 1979.
100. Freedman, D. L., and S. D. Herz: "Use of Ethylene and Ethane as Primary Substrates for Aerobic Cometabolism of Vinyl Chloride." *Water Environ. Res.*, vol. 68, no. 3, pp. 320–328, 1996.
101. Hartmans, S., et al.: "Bacterial Degradation of Vinyl Chloride," *Biotechnol. Lett.*, vol. 7, no. 6, pp. 383–388, 1985.
102. Hartmans, S., and J. A. M. de Bont: "Aerobic Vinyl Chloride Metabolism in *Mycobacterium aurum* L1," *Appl. Environ. Microbiol.*, vol. 58, no. 4, p. 1220, 1992.
103. U.S. Environmental Protection Agency, "Superfund Exposure Assessment Manual."
104. U.S. Environmental Protection Agency, "Transport and Fate of Contaminants in the Subsurface."
105. Dragun, "The Fate of Hazardous Materials in Soil."
106. Ibid.
107. U.S. Environmental Protection Agency, "Transport and Fate of Contaminants in the Subsurface."
108. Dragun, "The Fate of Hazardous Materials in Soil."
109. Brown, K. W., and D. C. Anderson: "Effects of Organic Solvents on the Permeability of Clay Soils," EPA 600/2-83-016, U.S. Environmental Protection Agency, Cincinnati, Ohio, p. 45, 1983.
110. U.S. Environmental Protection Agency, "Transport and Fate of Contaminants in the Subsurface," p. 45.
111. Ibid.
112. Ibid.
113. Piwoni, M. D., and J. W. Keeley, "Basic Concepts of Contaminant Sorption at Hazardous Waste Sites."

114. Ibid.
115. Freeze, R. A., and J. A. Cherry: *Groundwater*, Prentice-Hall, Englewood Cliffs, N.J., 1979.
116. Dragun, "The Soil Chemistry of Hazardous Materials."
117. Ibid.
118. Freeze and Cherry, *Groundwater,* op cit. p. 406.
119. Taylor, G. I.: "Diffusion by Continuous Movements," *Proc. London Math. Soc.*, ser. 2, no. 20, pp. 196–211, 1921.
120. Haugen, D. A. (ed.): "Project Prairie Grass, A Field Program in Diffusion," Geophysical Research Paper, no. 59, vol. III, Report AFCRC-TR-58-235, Air Force Cambridge Research Center, 1959.
121. Pasquill, F.: "The Estimation of the Dispersion of Windborne Material," *Meteorology Magazine*, vol. 90, 1961.
122. Pasquill, F., and F. B. Smith: *Atmospheric Diffusion*, Ellis Horwood, Chichester, U.K., 1983.
123. Panofsky, H. A., and J. A. Dutton: *Atmospheric Turbulence*, John Wiley & Sons, New York, 1985.
124. Pendergast, M. M.: "Meteorological Fundamentals," *Atmospheric Science and Power Production*, Randerson, D. (ed.), Office of Scientific and Technical Information, U. S. Department of Energy, Washington, D.C. 1984.
125. Shrock, J. E.: "Overview of Dispersion Models and Applications," *Air Pollution: Environmental Issues and Health Effects*, Pennsylvania Academy of Science, Easton, Pa., 1991.
126. Turner, D. Bruce, *Workshop of Atmospheric Dispersion Estimates*, U.S. Environmental Protection Agency, Office of Air Programs, Publication No. AP-26, Research Triangle Park, NC, 1970.
127. Ibid.
128. Ibid.
129. Ibid.
130. Ibid.
131. U.S. Environmental Protection Agency, *Guideline on Air Quality Models*, Appendix W to 40 CFR Part 51. U.S. Government Printing Office, Washington, D.C., 1997.
132. U.S. Environmental Protection Agency, *Procedures for Conducting Air Pathway Analyses for Superfund Applications*, Volumes I–IV, Office of Air Quality Planning and Standards, Research Triangle Park, N.C., July 1989.
133. U.S. Environmental Protection Agency, *SCREEN3 Model User's Guide*, Office of Air Quality Planning and Standards, Research Triangle Park, N.C., publication no. EPA-454/B-95-004, September 1995.
134. U.S. Environmental Protection Agency, *User's Guide for the Industrial Source Complex (ISC3) Dispersion Models*, Office of Air Quality Planning and Standards, Research Triangle Park, N.C., publication no. EPA-454/B-95-003a, September 1995.
135. Cimorelli, A. J., et al.: "Current Progress in the AERMIC Model Development Program," in *Proceedings 89th Annual Meeting, Air and Waste Management Association*, 96-TP24B.04, Air and Waste Management Association, Pittsburgh, 1996.
136. American Institute of Chemical Engineers, *Guidelines for Use of Vapor Cloud Dispersion Models*, Center for Chemical Process Safety, New York, 1996.

ADDITIONAL READING

Bahr, L. M.: "Analysis of Nonequilibrium Desorption of Volatile Organics during Field Test of Aquifer Decontamination," *J. Contaminant Hydrology*, vol. 4, no. 3, p. 205, 1989.

Boucher, P.: "Ecological Assessment and Modeling of a Contaminated Wetland," *Superfund '90*, Haz. Materials Control Research Instit., Washington, D.C., pp. 148–152, 1990.

Charbeneau, R. J., et al.: "Kinematic Modeling of Multiphase Solute Transport in the Vadose Zone," NTIS PB89-207948, Robert S. Kerr Environ. Res. Lab., Ada, Okla., 1989.

Cooper, C. D., and Alley, F. C.: *Air Pollution Control*, PWS Publishers, Boston, Mass., 1986.

Devinny, J., et al.: *Subsurface Migration of Hazardous Wastes*, Van Nostrand Reinhold, New York, 1989.

Dupont, R. R., et al.: "Evaluation and Modeling of Volatile Organic Vapor Transport in the Unsaturated Zone for Groundwater Quality Protection," *Water Resources Division*, NTIS PB90119736, U.S. Geologic Survey, Reston, Va., 1989.

Hauptmann, M. G., et al.: "Use of Groundwater Modeling During Superfund Cleanup," *Superfund '90*, Haz. Materials Control Research Instit., Washington, D.C., pp. 110–116, 1990.

Hosseinipour, Z., and K. Brown: "Application of Hydrodynamic and Water Quality Models at a Superfund Site in Montana," *Proc. 1989 Natl. Conf. Hydraul. Eng.*, ASCE, New York, N.Y., p. 466, 1989.

Johnson, R. L., et al.: "Diffusive Contaminants Transport in Natural Clay: A Field Example and Implications for Clay-lined Waste Disposal Sites," *Environ. Sci. Technol.*, vol. 23, p. 340, 1989.

Kapila, S., et al.: "Field and Laboratory Studies on the Movement and Fate of Tetrachlorodibenzo-p-dioxin in Soils," *Chemosphere*, vol. 18, p. 1297, 1989.

Kincare, K. A., and S. M. Aulenbach: "Geostatistical Decision Making Process for Plume Modeling in Cadillac, MI," *Superfund '89*, Haz. Materials Control Research Instit., Washington, D.C., pp. 146–151, 1989.

Koch, D., et al.: "Using a Three-Dimensional Solute Transport Model to Evaluate Remedial Actions for Groundwater Contamination at the Picatinny Arsenal, NJ," *Superfund '89*, Haz. Materials Control Research Instit., Washington, D.C., pp. 152–156, 1989.

Li, W. W., et al.: "Modeling of On-Site Air Concentrations at Superfund Sites," *Superfund '90*, Haz. Materials Control Research Instit., Washington, D.C., pp. 117–122, 1990.

McCarty, J. F., and J. M. Zachara: "Subsurface Transport of Contaminants," *Environ. Sci. Technol.*, vol. 23, p. 496, 1989.

Militana, L. M., and S. C. Mauch: "Statistical Modeling of Ambient Air Toxics Impacts During Remedial Investigations at a Landfill Site," *Superfund '89*, Haz. Materials Control Research Instit., Washington, D.C., pp. 157–162, 1989.

Olsen, R. L., and A. Davis: "Predicting the Fate and Transport of Organic Compounds in Groundwater," *Superfund '89*, Haz. Materials Control Research Instit., Washington, D.C., p. 145, 1989.

Ozbilgin, M. M., et al.: "Evaluation of Groundwater Extraction and Treatment Scenarios Using Quasi-Three-Dimensional Transport Model," *Superfund '90*, Haz. Materials Control Research Instit., Washington, D.C., pp. 386–397, 1990.

Reinhart, D. R.: "Fate of Selected Organic Pollutants During Landfill Codisposal with Municipal Refuse," *Diss. Abstr. Int. B*, vol. 50, p. 1830, 1989.

Ridgway, H. F., et al.: "Investigation of the Transport and Fate of Gasoline Hydrocarbon Pollutants in Groundwater," NTIS PB89151450, U.S. Geologic Survey, Reston, Va., 1989.

Roffman, A., and R. Stoner: "Air/Superfund National Technical Guidance Study Series. Volume 4. Procedures for Dispersion Modeling and Air Monitoring for Superfund Air Pathway Analysis." Office of Air Qual. Plann. and Stand., NTIS PB90113382, U.S. EPA, Research Triangle Park, N.C., 1989.

Rudy, R., et al.: "RANDOM-WALK Modeling or Organic Contaminant Migration from the Theresienfeld Landfill Located in the Vienna Basin Aquifer of Austria," *Superfund '89*, Haz. Materials Control Research Instit., Washington, D.C., pp. 163–180, 1989.

Stein, C. L., and D. F. McTigue: "Chromium Distribution and Transport Beneath a Contaminated Site," NTIS DE89009696, U.S. Department of Energy, Washington, D.C., 1989.

Stevens, D. K., et al.: "Simulation of Oxygen Dynamics in the Vadose Zone," *Environ. Eng. Proc. 1989 Specialty Conf.,* ASCE, New York, p. 694, 1989.

Stevens, D. K., et al.: "Sensitive Parameter Evaluation for a Vadose Zone Fate and Transport Model," NTIS PB89-213987, Robert S. Kerr Environmental Research Laboratory, Ada, Okla., 1989.

Stoner, R.: "Air/Superfund National Technical Guidance Study Series. Volume 1. Application of Air Pathway Analyses for Superfund Activities." Office of Air Qual. Plann. and Stand., NTIS PB90-1 13374, U.S. EPA, Research Triangle Park, N.C., 1989.

Study of Dredging and Dredged Material Disposal Alternatives, Report 5, "Evaluation of Leachate Quality," Army Eng. Waterways Exp. Sta. Rep. No. WES/TR/EL-88-15-5, 1989.

Thomann, R. V.: "Bioaccumulation Model of Organic Chemical Distribution in Aquatic Food Chains," *Environ. Sci. Technol.*, vol. 23, p. 699, 1989.

Thorne, D. J., and A. S. Rood: "Surface Water for Site Closure of a Remediated Dioxin (TCDD) Site," *Superfund '89*, Haz. Materials Control Research Instit., Washington, D.C., pp. 117–121, 1989.

Tindall, James A., and Kunkel, James R., *Unsaturated Zone Hydrology for Scientists and Engineers*, Prentice-Hall, Upper Saddle River, N.J., 1999, p. 624.

Ulirsch, G. V., et al.: "Role of Environmental Fate and Transport Data in Health Assessments: Four Case Studies," *Superfund '90*, Haz. Materials Control Research Instit., Washington, D.C., pp. 128–132, 1990.

Walton, J. C., et al.: "Model for Estimation of Chlorinated Solvent Release from Waste Disposal Sites," *J. Hazardous Mater.,* vol. 21, no. 1, p. 15, 1989.

Wark, K., and Warner, C. F.: *Air Pollution: Its Origin and Control*, 2nd ed., Harper & Row, New York, 1981.

Whetzel, N.: "SLAPMAN Model for Groundwater Pathway," Office of Water Regulations and Standards, NTIS PB89-138747, U.S. EPA, Washington, D.C., 1989.

ACKNOWLEDGMENTS

The authors wish to acknowledge the following individuals who assisted with the preparation of this chapter for the first edition: Mike De Busschere, Nick DeSalvo, Paul Gruber, John Harsh, Kathryn Z. Klaber, Ken Weiss, ERM; Tim Mead, William Reichardt, Charles Johnson, Michael Costa, and Kevin Grega, Bucknell University; and David Garg, Pennsylvania Department of Environmental Resources. The principal reviewer for ERM for the first edition was Roy O. Ball, Ph.D., P.E.

Reviewers for the second edition were Mark Garrison and Sam Upchurch, P.G., ERM.

CHAPTER 5

Toxicology

Was ist das nit gifft ist?
Alle ding sind gifft und nichts ohn gifft
Allein die dosis macht das ein ding fein gifft ist.

What is there that is not poison?
All things are poison and nothing without poison.
Only the dose determines that an agent is not a poison.
Paracelsus, c. 1538

A waste is considered hazardous if it exhibits any of four characteristics—reactivity, corrosivity, ignitability, or toxicity. Of these, the potential for toxicity, both to humans and to other biota (ecotoxicity), has caused the greatest public concern and has prompted massive regulatory initiatives in hazardous waste management. It follows that the management of hazardous waste has as its fundamental objective the protection of human and ecological health by reducing the risk, if not the toxicity, of the substances in hazardous wastes to acceptable levels. An understanding of toxicology is therefore needed to determine whether the management of hazardous waste or the remediation of contaminated sites meets this objective.

Chapter 4 covered the fate and transport of substances from their release into the environment up to their contact with receptors. This chapter addresses the toxic effects caused by hazardous substances on gaining access to the human body and other living organisms. Toxic effects can range from mild allergic reactions to death. Indeed, the constituents of some hazardous waste can kill humans outright as evidenced by graphic accounts. Some constituents can cause death in insidious ways such as cancer.

What is toxicology?

Toxicology deals with the adverse effects caused by exposure of living organisms to chemical substances. Although poisonous plants and animal venoms have been

known since early human history, the emergence and recognition of toxicology as an independent scientific discipline did not occur until well into the twentieth century. It has evolved as a discipline that incorporates areas of several other basic sciences such as physiology, pharmacology, biochemistry, molecular biology, and epidemiology. Being basically a conglomerate of biological disciplines, toxicology is, thus, not an exact science in the same sense as physics or chemistry. Much remains unknown, particularly the basic mechanisms by which chemicals produce their deleterious effects.

What is the single most important thing for an engineer to know about toxicology?

The engineer cannot and should not practice toxicology. An engineer's education and experience falls far short of what is needed to make toxicological evaluations. However, a general understanding of how a toxicologist makes decisions will open meaningful communication between the two disciplines that share the responsibility to decide whether an engineering solution offers adequate protection.

What is the key limitation of applying toxicology?

Because the fundamental mechanisms that cause toxic responses are not fully understood, toxicological findings are largely based on observations, only partially derived directly from humans. Toxicology is to a large extent supported by information obtained from experiments on animals. When extrapolated to humans, such test data carry a considerable degree of uncertainty. This lack of precision is not only evident in using information obtained in one animal species (e.g., the mouse) to predict toxic action in another (e.g., humans), but also in extrapolating from the high exposures used in an animal experiment to the low-dose situation usually encountered with respect to exposure of the general population. Depending on the mathematical mode used for such extrapolations, the risk estimates obtained may differ by orders of magnitude. Thus, quantification of the toxicity of a chemical is associated with a large degree of uncertainty, typically dwarfing the uncertainty inherent in a scientific analysis of the fate and transport in the environment of the same chemical.

Having said this about the uncertainty associated with toxicological data, it should also be stressed that, for many chemicals, the toxicological database is sufficient to predict the risks associated with their presence in the environment. For chemicals without a sufficient database, generally accepted approaches exist to make a reasonable approximation to provide the basis for selecting appropriate strategies protective of human and ecological health. Chapter 14 of this textbook presents a systematic approach for using toxicological and site data to assess in a quantitative fashion the risk posed by hazardous waste at a site.

5-1 EXPOSURE

Chapter 4 covered the release of hazardous waste constituents and their fate in the environment. The series of steps starting with a release of waste constituents and concluding with the interface with a human body is referred to as an **exposure pathway.** Gaining entry into the body is referred to by toxicologists as the **exposure**

route. There are three exposure routes by which environmental contaminants may enter the body: **inhalation** (respiratory tract), **ingestion** (gastrointestinal tract), and **dermal contact** (skin) (see Figure 5-1).

Overview of Fate of Substances

Except for corrosive agents, most toxic substances do not cause harmful effects at the point of entry. Instead, exposure marks the beginning of a rather tortuous path as the physiological (metabolic) processes of the human body interact to absorb, distribute, store, transform, and eliminate a substance. To produce a toxic effect, the chemical agent or its biotransformation products must reach the critical site of action in a **target organ** at a sufficiently high concentration and for a sufficient length of time. A target organ is the "preferential anatomical site" for the expression of the toxic effects caused by a substance.[2] In Figure 5-2 a schematic representation is given of the fate of a toxicant on uptake.

Distribution data obtained from humans exposed to benzene vapors[3] may serve as an illustration of the pathways depicted in Figure 5-2. The amount of benzene taken up through the lungs into the blood and retained is approximately 30 percent of the inhaled dose. The rest is exhaled, 50 percent immediately and another 20 percent excreted unchanged via the lungs. Studies in rodents indicate that benzene is also readily (>90 percent) taken up by ingestion. Finally, although much less effective, dermal exposure may also give rise to some absorption (about 0.2 percent of the applied dose). Once taken up, over half of the benzene absorbed is rapidly translocated to organs with a rich blood supply, such as the liver and the kidneys, as well as to tissues rich in fat such as adipose tissue, brain, and bone marrow. Benzene may also be found in the placenta and fetus following exposure of pregnant women to benzene. Of importance is the fact that benzene is only slowly released from fatty tissues. Coincidentally, complex transformation processes convert the nonpolar, fat-soluble benzene into a host of polar water-soluble metabolites that are excreted in the urine. For several organic compounds there is a substantial database relative to metabolism, but for most substances our knowledge is incomplete.

Absorption

Epithelial layers of cells (membranous tissue) cover the mammalian body at any port of entry regardless of whether the exposure route is inhalation, ingestion, or dermal contact. Having gained entry to the human body, the toxic agent must pass through this interface to enter the blood stream. If ingested, toxic agents must pass through the mucosa (mucous membrane) which lines the gastrointestinal tract. If inhaled, they must pass through the thin lining of cells covering the inside of the alveoli in the lungs, representing the barrier between blood and the air which may carry gaseous toxicants. In the case of dermal exposure, the agents must pass through the stratified layers of the skin. The transport across any body barrier (i.e., skin, lung cells, the lining of the gastrointestinal tract) is referred to as **absorption.**

Whether in the lungs, skin, or elsewhere, these barriers consist basically of cellular membranes having a similar structure of a two-molecule-thick inner layer of various fatty substances (lipids) blanketed between two outer layers of proteins.

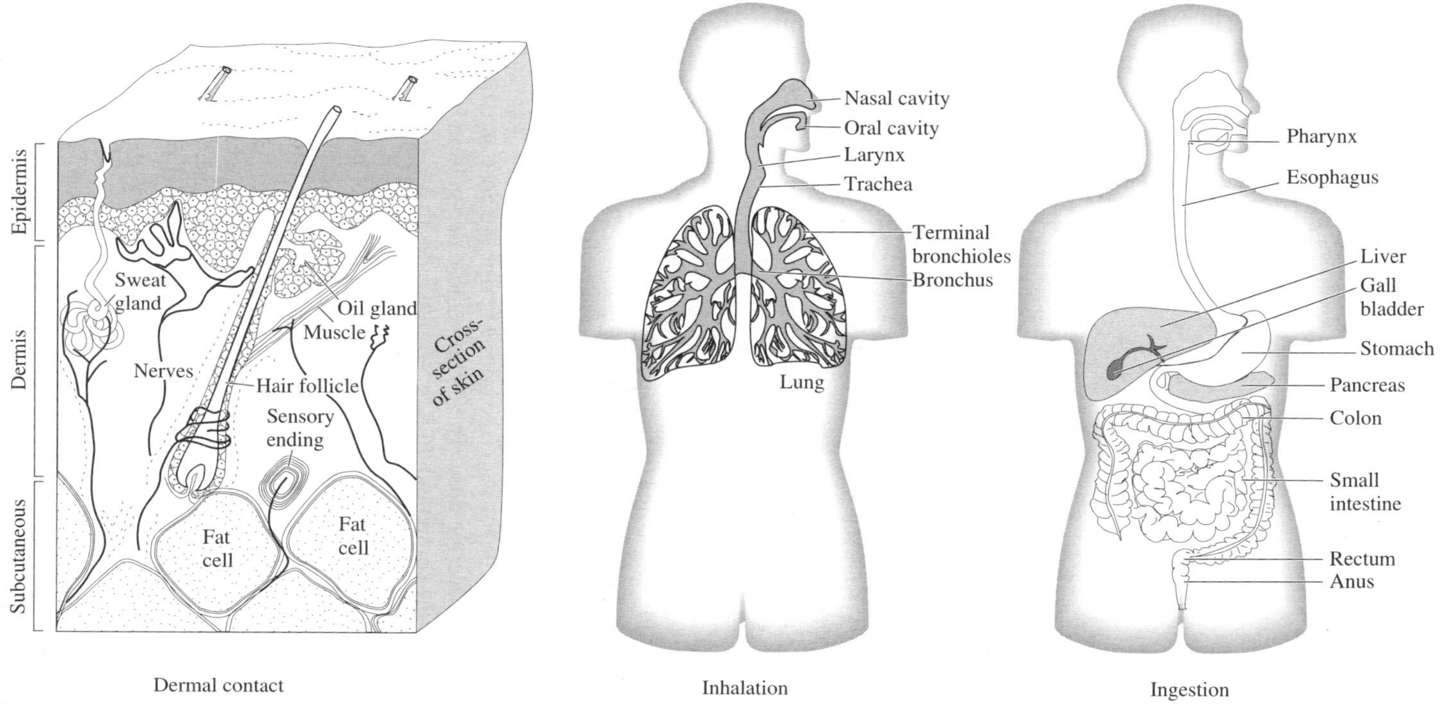

FIGURE 5-1
Exposure routes for chemical agents in hazardous waste.[1]

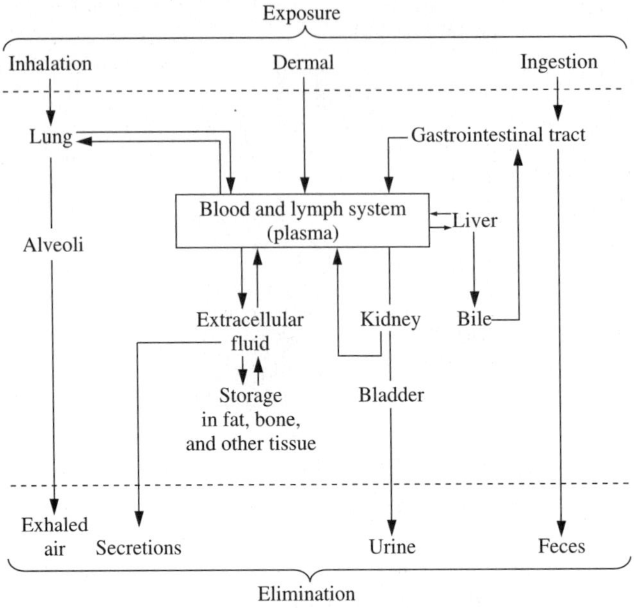

FIGURE 5-2
Overview of absorption, distribution, storage, transformation, and elimination of a toxic substance in the human body.[4]

However, the chemical composition and physical properties of epithelial linings can vary depending on the function of the organ they cover. They can be almost fluid-like at some sites, possessing pores of 4 to 70 angstroms in diameter.[5]

The amount of a particular toxic agent to which a body is exposed is termed the **administered dose.** Although the administered dose is important, the amount of the toxic chemical absorbed by the body and the amount reaching the target organ are of greater concern. The former is the **intake** or **uptake dose** and the latter can be termed the **target** or **effective dose.** The effective dose depends on how the body's systems interact with the compound. Figure 5-3 illustrates these definitions for contaminated soil.

The same processes that govern the absorption of nutrients, oxygen, and other elements and chemicals vital to life also govern the absorption of toxic agents. The key variables are the properties of the compound, the exposure route, and the susceptibility (or physiological state) of the individual. Absorption through the skin, for example, depends very much on the medium in which the toxic substance is contained. Soil contaminated with organics would result in less of an intake dose than the same organics in a solvent base. Other factors include the humidity of the skin, the location on the body (penetration by chemicals is especially rapid on the scrotum), and whether the skin is damaged.

In many cases, absorption by more than one exposure route must be considered. For example, about 30 to 50 percent of airborne particulate lead inhaled by an adult can be deposited in the respiratory tract, and virtually all deposited lead is absorbed over time. For ingested lead, absorption is about 50 percent in children and 8 to 15

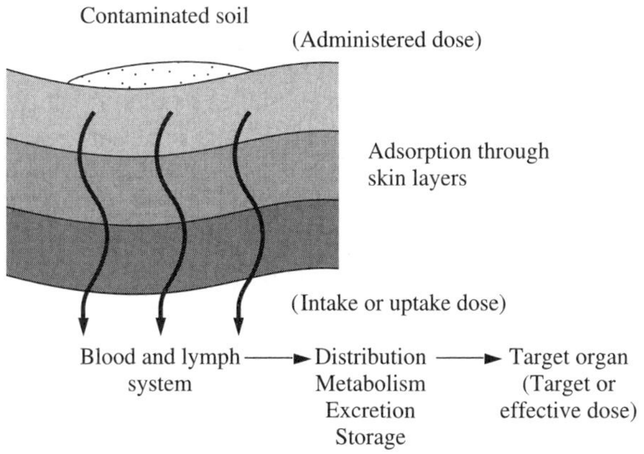

FIGURE 5-3
Types of doses for dermal contact with contaminated soil.

percent in adults, with solubility of the particular lead salt in gastric acid being an important variable. Absorption of lead via dermal contact is insignificant except at extremely high concentrations.[6] However, some other substances readily penetrate the skin (e.g., 15 to 60 percent of the total applied dose of lindane is taken up by the skin[7]). This route thus would become important.

Absorption mechanisms

Chemical substances may pass in different ways through the cellular linings (epithelium) that make up the protective external and internal barriers of the organism: (a) passage through cellular membranes by passive transport (e.g., simple diffusion) or by active transport, or (b) passage through pores or channels present between the epithelial cells. Most toxic agents pass through cell membranes by simple diffusion.

Diffusion

The rate of diffusion depends on the chemical and physical properties of a toxic agent, particularly its degree of ionization, lipid solubility, protein binding, and water solubility. For passing through cell membranes, lipid solubility is the most important property. Lipid solubility is a measure of the affinity of a chemical substance for fat-like solvents (e.g., lipids) versus water-like solvents (e.g., blood, urine). The property is related to the polarity of a chemical substance. Polar substances readily dissolve or ionize in water and thus can be thought of as **hydrophilic.** Similarly, nonpolar substances can be termed hydrophobic, or **lipophilic.** The property of lipid solubility is indicated by an octanol/water partition coefficient (see Section 3-2) above 1.0 (the higher the coefficient, the greater the affinity for lipids). Many organic chemicals (e.g., benzene) are nonpolar and highly soluble in lipids. Thus, the lipid-rich composition of a membranous lining facilitates rapid penetration and transport of nonpolar compounds; the higher the octanol/water partition coefficient, the more rapid the movement through membranes to enter the blood stream. However, increasing molecular weight reduces the movement through membranes.[8]

It follows that the lipid structure of cellular membranes may inhibit the movement of polar or charged particles. Even for cellular membranes, small molecules of polar compounds may still move as ions in channels with a diameter on the order of 4 angstroms, allowing for passage of only molecules with molecular weights below 200.[9]

The description of the properties of cell membranes given above will to a large extent also hold for the so-called blood-brain barrier, a complex structure involving several cell types that separate the central nervous system from the circulating blood and lymph. Although most analgesics, anesthetics, and tranquilizers penetrate readily, the blood-brain barrier is much less permeable to toxic substances than other areas of the body. However, it is important to point out that the nature of biological membranes from one part of the body to another may vary as does the composition of the cellular layers that make up the epithelial linings of the organism. The permeability properties of the different parts of the gastrointestinal tract, for example, present a complex picture, allowing for the uptake also of larger polar molecules (e.g., of alkaloids like codeine sulfate with molecular weights in the range 300 to 400). In the kidneys the membranous lining is very porous in some places allowing for passage of molecules with molecular weights up to about 60,000. Here, passage occurs through interstitial pores between cells of the membrane.

Specialized absorption

In addition to the primary mechanism of diffusion, some absorption may take place by specialized and complex transport systems. A variety of such mechanisms exists and works selectively, for example, to absorb sugars and other hydrophilic, nutritional compounds. The same systems can also selectively transport some toxicants. Since these mechanisms work in both directions, some of the specialized systems are important in the elimination of toxic agents from the body.

Examples include processes such as phagocytosis in which the cell membrane flows around and engulfs particles. This type of absorption is important for the removal of particles from alveoli by specialized white blood cells, so-called alveolar macrophages, but may also occur elsewhere to a limited extent (e.g., in the duodenum)[10] accounting for the uptake of high molecular weight toxins like botulinum toxin (the toxic agent causing botulism).

Absorption via the ingestion route

Absorption of toxic substances may occur along the whole length of the gastrointestinal tract, but tends to differ in the stomach compared to the intestine due to the lower pH in the former. Both passive diffusion as well as absorption by specialized transport systems occur. Many factors may alter gastrointestinal absorption. As examples, the presence of chelating agents may increase absorption of heavy metals, the presence of zinc may decrease the uptake of cadmium, and decreased gastric mobility has a general promotive effect on absorption.

Absorption via the inhalation route

In the gaseous state, the inhaled compound is transferred into the liquid layer (mucous) lining the airway wall by diffusion. Transfer through mucous depends on the gas-diffusion coefficient, layer thickness, and the gas concentrations at the boundaries of this layer. Capillary blood flow removes the dissolved gas on the other side of the mucous and the separating tissue layers. Alveolar uptake of gases is governed by air-to-blood partitioning, ventilation/perfusion, and air and blood concentrations. For readily soluble gases, uptake is linearly related to solubility. In general, the tissue-gas partition coefficient linearly correlates with its fat-gas and blood-gas partition coefficients, providing an approach for estimating these param-

eters. However, the relation between absorption in the lung and air concentrations in itself may be nonlinear as is the case for a poorly soluble gas like butadiene.[11]

The site of deposition of inhaled particles varies and depends largely on particle size. Particles with a diameter of 2 μm and above are deposited in the upper respiratory tract, removed by the mucous present in the airways, coughed up, and swallowed. Particles of smaller diameter **(respirable particles)** penetrate to the alveoli of the lung where they may be deposited.[12]

Different animal species when exposed to the same levels of particles or gas in the inhaled air will not receive the same doses in comparable regions of the respiratory tract. Thus, the derivation of human equivalent doses on the basis of animal inhalation experiments is particularly difficult, especially for aerosols. The regional deposition pattern is determined by anatomical-physiological features specific to various mammalian species, as well as the physicochemical characteristics of the inhaled agent. For particles, deposition is controlled by impaction, sedimentation, diffusion, and electrostatic precipitation; for gases the important parameters are convection, diffusion, chemical reaction, and solubility. Retention is governed by the relative rates of deposition and clearance. Whereas deposition data are usually obtainable, the determination of clearance rates is technically difficult to perform and often influenced by the methodology used. Clearly, species' differences in relative length of the nasal passage, alveolar ventilation, and cardiac output are critical determinants of the absorbed dose.[13]

Absorption via the dermal contact route

Skin is an excellent barrier to many toxic compounds. For most substances, the body has the potential to tolerate 100 to 1000 times the exposure amount by dermal contact versus ingestion or inhalation. Many exceptions exist, however, as exemplified by corrosive agents like phenol and a number of lipophilic compounds like carbon tetrachloride and lindane. Also, certain relatively hydrophilic organophosphorus compounds like sarin are highly toxic upon contact with the skin.

Effect of exposure route on toxic response

Clearly the exposure route has a significant effect on uptake. This and subsequent factors manifest themselves such that the magnitude and type of toxic effect will show great variation, depending on route of exposure. For example, silica can cause a pulmonary disease known as "silicosis" after lung deposition of dust particles over a protracted period of time, but is virtually harmless on ingestion. Other examples are asbestos, hexavalent chromium, and nickel, which act as carcinogens if inhaled but do not seem to induce tumors in humans by the oral route. Many substances are much more toxic by the inhalation route than by oral or dermal administration. This is due in part to the rapid uptake in the lung that is directly transferred into the blood stream. In addition, ingested toxins may be detoxified by metabolic enzymes of the liver prior to reaching target organs. Calculated per amount absorbed, the toxicity by the inhalation is often comparable to that seen in intravenous administration. If a substance is found to have a low toxicity by oral administration, but is significantly more toxic by inhalation in the same animal, one might suspect that the low oral toxicity may be due to metabolic inactivation or poor uptake in the gastrointestinal tract.

Biotransformation and Elimination

On reaching an organ, a toxicant can cause three events other than storage, as follows:

1. Biotransformation—enzyme-rich organs, such as the liver, metabolize the toxicant to other molecular species, termed metabolites, which may not necessarily be less toxic than the parent compound.
2. Elimination—toxicants are not stored and, along with most metabolites, are eliminated from the body.
3. Formation of a chemical-receptor complex—many toxic agents attack only target organs that have specific receptor proteins.

The body has a remarkable capacity for elimination of a vast array of chemical agents absorbed through dietary and environmental exposures. A toxic agent, particularly polar compounds, may pass quickly from the body in urine, bile, feces, or secretions. Elimination of nonpolar, nonvolatile compounds is more difficult and usually occurs only after their metabolic transformation (biotransformation) by the organism to more polar, and thus more water-soluble, metabolites that can be excreted by urine.

Biotransformation of nonpolar, nonvolatile toxicant molecules can be described as a two-phase biochemical reaction.[17] In the first phase, the metabolic enzymes introduce a polar functional group to the toxicant. Examples of first-phase reactions include oxidation, reduction, hydrolysis, and dehalogenation. The second phase typically involves the linkage of polar molecules (conjugation) to yield products which are more water-soluble, thus making them more easily eliminated. Conjugating agents are endogenous to the body, and key examples are glucuronic acid and sulfate.

The two most important organs for biotransformation and elimination of toxicants are the liver and the kidneys. Blood from the stomach and intestines flows to the liver, carrying toxicants which have been absorbed through the gastrointestinal tract. The liver is rich in biotransformation enzymes and can quickly transform substances it absorbs. Some substances are rendered water-soluble and transported by blood flow to the kidneys, which filter or otherwise remove polar toxicants and metabolites from blood and transport them as concentrated solutes to the bladder for elimination in urine. Some toxic agents like lead, mercury, and some other heavy metals, as well as many organic substances, are excreted in the bile, often at concentrations much higher than in the blood, to the intestinal tract for elimination in feces. Still, some hydrophobic toxicants, or their biotransformation products, remain concentrated in the liver.[18]

Considering benzene as an example,[19] it is metabolized primarily in the liver where much of it is oxidized to benzene oxide. Benzene oxide is unstable and converts to phenolic compounds, which are excreted in urine together with other metabolites. Analyses of specific metabolites in the urine are frequently used as quantitative indicators of exposure to substances. This is termed **biological monitoring** and the metabolite a **biomarker.** Phenol is thus a biomarker for benzene exposure. Figure 5-4 shows the direct increase in urinary phenol as the concentration of benzene in inhaled air increases.

Biotransformation does not necessarily reduce the toxicity of a substance. On the contrary, some metabolites are many times more toxic than the parent compound.

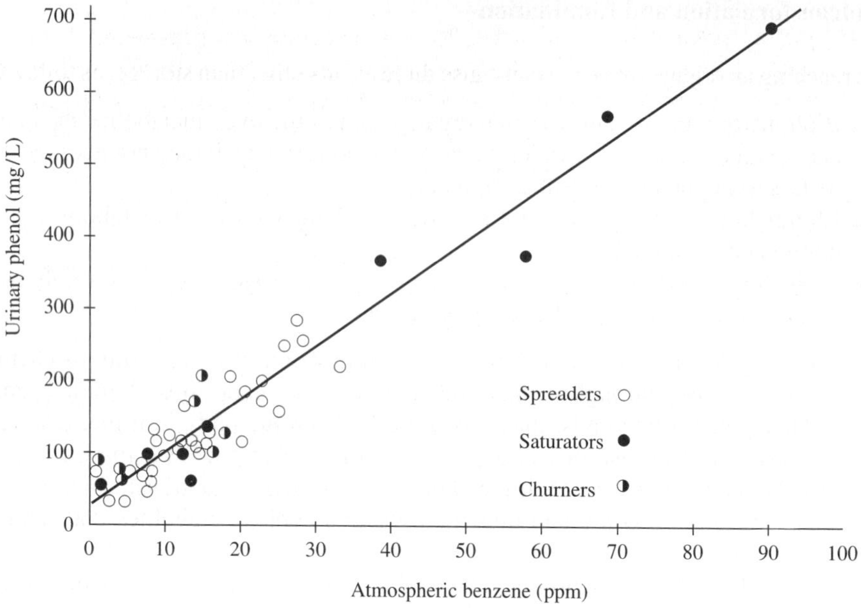

FIGURE 5-4
Correlation of atmospheric benzene concentrations with urinary phenol levels in workers in a rubber coating plant.[20]

For example, benzene must undergo metabolic transformation to produce its toxic effects in bone marrow. As another example, benzo(a)pyrene, a substance in tobacco smoke, is noncarcinogenic; however, it is biotransformed to its diol epoxide which is a potent carcinogen. Other examples of such a toxic transformation are the conversion of parathion to paraoxon and of the pesticide DDT to DDE (responsible for eggshell thinning).

Pharmacokinetics

The rates at which toxicants are absorbed, metabolized, and eliminated are very critical. If the body eliminates a particular toxicant rapidly, it may tolerate an otherwise toxic dose when received in fractionated administrations. If the body eliminates a toxicant fairly slowly, a low dose over a long period could result in the accumulation of the toxicant to a critical concentration. For these reasons, the duration, concentration, and frequency of exposure are important.

Pharmacokinetics is the quantitative description of the time course for uptake, transformation, storage, and transport of compounds (e.g., a hazardous substance).[21] Sometimes referred to as **toxicokinetics,** it is the quantitation in time of all the processes that take place subsequent to uptake in the organism. The various steps and processes involved between exposure and reaching the target organ can often be predicted in a general manner by **pharmacokinetic modeling.** This approach offers a simpler (and less accurate) alternative to the cumbersome analytical and biochemical

techniques involved in the elucidation of the pharmacokinetics of a toxicant. Many of the pharmacokinetic models of such actions rely on zero-order, first-order, or second-order reaction kinetics combined with other mathematical approaches familiar to engineers. Such calculations yield values for parameters, such as adsorption rate, half-life of substance, and metabolic saturation concentration. By pharmacokinetic evaluation the intake (uptake) can be related to the target or effective dose. For the example of air pollutants, it would consist of defining the relation between concentrations in inhaled air to the amounts actually taken up and transformed by the organism and, ultimately, its relation to the dose delivered to the sensitive target. Simple models are illustrated in Examples 5-1 and 5-2.

EXAMPLE 5-1. ELIMINATION OF ETHANOL. Metabolism of ethanol is a zero-order reaction (i.e., elimination is constant regardless of concentration). If an average human is able to metabolize 10 mL of ethanol per hour,[22] how long a time is required to eliminate a six-pack of beer containing 5 percent alcohol?

Solution

$$t = \frac{\text{total dose}}{\text{elimination rate}}$$

$$= \frac{(6 \text{ beers})(436 \text{ mL/beer})(5\% \text{ alcohol})}{10 \text{ mL/h}}$$

$$= \underline{\underline{13.1 \text{ hours}}}$$

EXAMPLE 5-2. CALCULATION OF THE HALF-LIFE FOR ELIMINATION OF METHYL MERCURY FROM THE HUMAN BODY. Many metabolic processes follow first-order kinetics (i.e., the formation or elimination of a substance depends on the absorbed concentration) and the level of the substance changes by some fraction per unit time, the first-order rate constant k. The time required for the concentration to decrease by one-half is known as the half-life ($t_{1/2}$). The half-life of a chemical obeying first-order kinetics is constant and independent of administered dose. From determination of the levels of methyl mercury in an individual surviving poisoning by a fungicide in Iraq,[23] the average excretion rate was found to be 1.75 percent of the total body burden per day. What value for the half-life can be determined on the basis of these observations?

Solution. From Section 3-4, a first-order reaction may be written as:

$$C = C_0 e^{-kt} \tag{5-1}$$

where C = concentration at time t (mg/L)

t = time (hours)

C_0 = initial concentration (mg/L)

$$t = -\frac{\ln C/C_0}{k}$$

$$t_{1/2} = -\frac{\ln(0.5)}{0.0175/\text{day}} = \underline{\underline{39.6 \text{ days}}}$$

Pharmacokinetic modeling often presents a formidable challenge because most metabolic reaction sequences are exceedingly complex. Many critical reaction variables exist that can only be roughly approximated because of the complexity of the human body and how its components function as a living system in a dynamic

environment. Pharmacokinetic modeling is also complicated by the fact that the body cannot be considered as one single compartment with respect to the absorbed compound. This is the case for many lipophilic solvents (e.g., methylene chloride, chloroform, and trichloroethylene), which are characterized by a complex distribution pattern and metabolism. The fate of such compounds can be predicted by using physiologically based pharmacokinetic (PB-PK) models, where attempts are made to identify the various distribution compartments and the different kinetic components in biologically relevant terms.[24]

Exposure Period

The period over which a dose is administered is obviously important. In humans the following criteria have been used for characterizing exposure:

Acute	One day
Subacute	Ten days
Subchronic	Two weeks–seven years
Chronic	Seven years–lifetime

When an acute toxic dose is fractionated into smaller portions and administered over a longer period of time, the toxic effect usually decreases. This will be true if there is an appreciable metabolic detoxification and/or excretion of the agent between exposures. If the deactivation/elimination is sufficiently rapid in comparison with the interval between doses, no toxic effects will be noted.

Figure 5-5 illustrates the case of repeated exposure to a compound at two different administered doses, neither dose being sufficiently high to cause a toxic effect after a single exposure. The interval between exposures is the same for both doses. If the elimination rate allows the toxic chemical to decline to preexposure levels, as in the case of the lower doses, it prevents the total body burden from reaching the toxicity threshold. However, if the elimination rate cannot keep pace with uptake, as in the case of the higher exposure, this toxicity threshold is eventually reached.

Occupational exposure to TCE constitutes an illustration of the situation depicted in Figure 5-5. The slow elimination from fat explains why repeated fluctu-

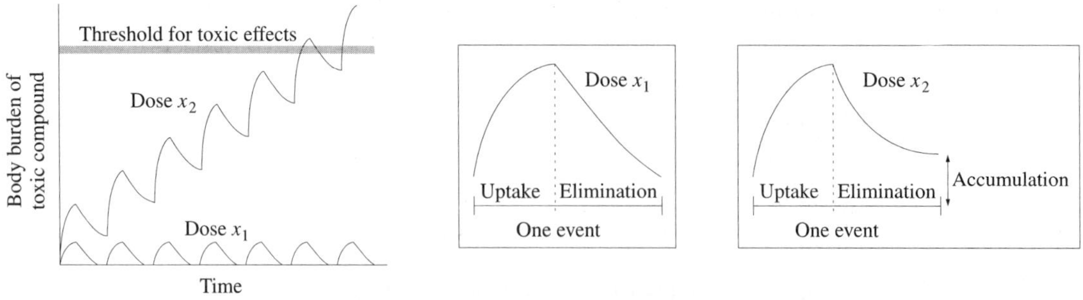

FIGURE 5-5
Effect of dose fractionalization on accumulation of a toxic compound.

ating concentrations at moderate exposure levels (100 ppm) during a workday, 5 days/week, results in an accumulated concentration as TCE from each day's new exposures adds to the residual concentration from previous exposures until a steady state is reached. This equilibrium level attained toward the end of the week represents a relatively high total body burden of TCE of toxicological significance and has led to a revision of previously adopted standards in occupational hygiene. Assuming first-order elimination kinetics, the "average" body burden at steady state (X_ϕ) after repeated exposure may be approximated by the following equation[25]:

$$X_\phi = 1.44 \times t_{1/2} \times f \times D/\tau \qquad (5\text{-}2)$$

where $t_{1/2}$ = half-life of the substance (days)
f = fraction absorbed
D = dose taken up during each exposure interval (mg/kg of body weight)
τ = constant time interval between exposures (days)

Regardless of the elimination rate, toxic effects will eventually occur with repeated small doses if either: (a) there remains a "residual injury" after each exposure, or (b) there is a gradual accumulation of the toxicant. As an example of the first, repeated exposure to tri-o-cresyl phosphate at doses below the acutely toxic level will result in accumulation of irreversible nerve damage which later will show up clinically.[26]

5-2 TOXIC EFFECTS

The events leading to a toxic effect can be grouped into three phases: **exposure, pharmacokinetic,** and **manifestation.** The previous pages have covered the first two, which conclude with a toxic substance or its metabolite reaching a target organ. This marks the start of the third phase, manifestation, which eventually results in the clinically overt symptoms of toxic action. The general sequence of this phase is depicted in Figure 5-6.

The mechanisms and specific sequence of actions that result in a toxic effect are often not well understood. Generally, it is believed that after the movement of a toxic chemical from the point of absorption to the target organ, the toxic chemical joins, if not binds, with a molecule or group of molecules from a cell of the target

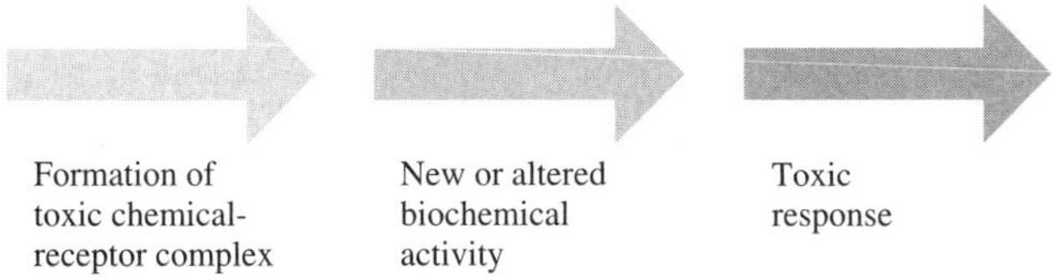

FIGURE 5-6
Sequence of events leading up to a toxic response.

organ.[27,28] The affected molecule or group of molecules from the target organ is termed the **receptor.** The typical receptor is an enzyme. However, it can also be a protein of the cell membrane or some other molecule from the cell, such as DNA, as depicted in Figure 5-7.

The target molecule(s) may be a specific receptor for endogenous or "natural" substrates located on an enzyme catalyzing a specific biochemical reaction; it (they) may be a receptor on a cell membrane designed to combine with a "signal substance" like a hormone having a central regulating function for various metabolic processes. If the target molecule is DNA, this may result in hereditary alterations (mutation) if this occurs in sex cells (germinal cells) or initiation of cancer in any other cell (somatic cell).

Reactions at the target may also be unspecific in nature. For example, toxic agents like arsenic and some heavy metals bind to reactive groups, such as sulfhydryl (SH) groups, in a wide variety of biological constituents. Since the function of many enzymes depends on an intact SH group, these agents will act as general enzyme inhibitors. Except for a few biochemical functions it is virtually impossible to identify the affected target molecules that are involved relative to the wide range of toxicological effects eventually seen clinically. A common sequel to unspecific toxic reactions is cell death.

Concept of Structural Affinity

A primary factor governing toxic action is the affinity of a substance for a specific receptor in the cells of an organism, or sometimes lack thereof. This affinity derives from the structure of the compound (i.e., its three-dimensional spatial arrangement

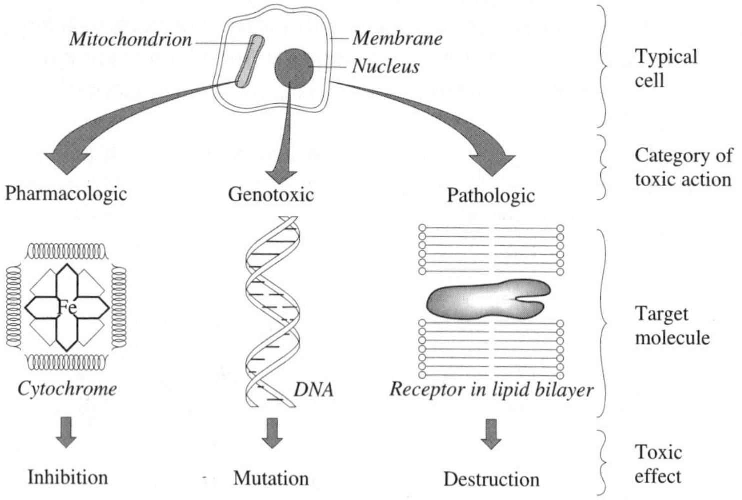

FIGURE 5-7
Representative cellular targets for toxic action.

of atoms or its stereochemical molecular configuration). An affinity exists if the structure of the toxicant is complementary with that of a cellular receptor. Basically, an affinity shared between the structure of a cellular component and that of a xenobiotic molecule causes the two to join, if not bind, together on contact and recognition of this affinity (see Figure 5-8). An analogy is the "key within a lock."

A mechanism of many toxic chemicals for toxic action is to bind to stereospecific clefts in receptor macromolecules such that the receptors no longer can perform their normal cellular function. This binding can result in interference with enzymatic function, or with some other process. Binding does not necessarily result in toxic action; it can, for example, prompt passage through a membrane, storage in particular organ tissue, or induction of enzymes that transform the toxicant. In some cases, such as the binding of carbon monoxide to hemoglobin with 200 times the affinity of oxygen, the "key" fits the same "keyhole" used by the normal substrate, oxygen. Thus, the toxic chemical competes with, if not displaces, beneficial substances. In other cases, insertion of the key takes place elsewhere, distorting the receptor molecule sufficiently to make its keyhole inoperative. Hypothetically, the more exact the fit, the more potent the toxic response.[30] Xenobiotics not sharing an affinity with cellular molecules are likely nontoxic and mostly excreted; however, exceptions exist for nonspecific agents.

The existence of structural affinity is indicated by examples of numerous isomers (isomers are compounds having the same number and kind of atoms but different structure) exhibiting much different toxicity. An example is represented in Figure 5-9, which shows two isomers of hexachlorinated biphenyl, a class of polychlorinated biphenyls (PCBs) with 6 chlorines. The two phenyl rings of PCBs are free to rotate about the carbon-carbon bond that joins them when there are no adjacent chlorine atoms. However, chlorination of the internal biphenyl carbons (0 position), as shown in the upper molecule, inhibits rotation such that the two phenyl rings cannot achieve coplanarity. Because coplanarity is required for binding to certain cellular receptors and toxicity, the upper PCB isomer is much less toxic than the lower by a ratio of about 1:20.[31]

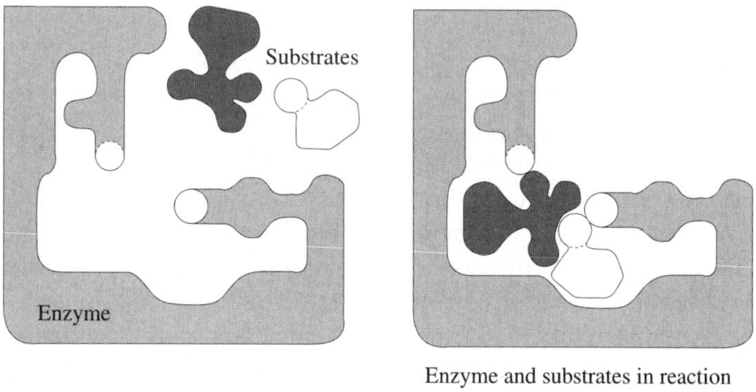

FIGURE 5-8
Lock-and-key model for toxic action.[29]

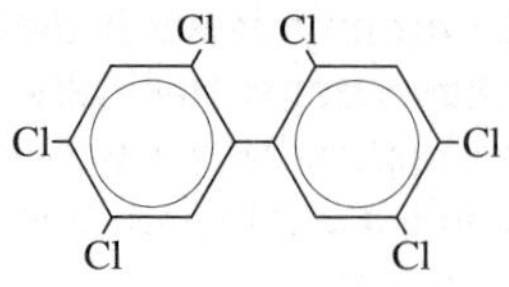

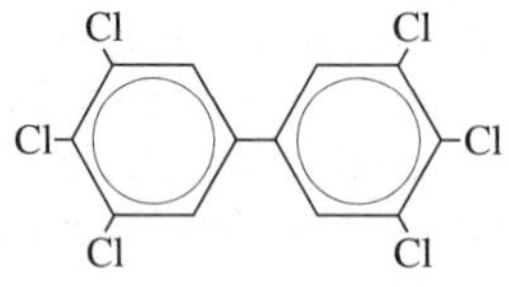

FIGURE 5-9
Example of compounds with identical molecular formula but different toxicity.[33]

The concept of a toxic chemical-receptor molecule complex based on molecular structure has been the focus of considerable research in the past. When better understood, by means of so-called **quantitative structural-activity relationships (QSARs),**[32] it should eventually be possible to predict the toxicological properties of new substances before submitting them to full laboratory tests.

Mechanism of Toxic Action and the Manifestations of Toxic Effects

As discussed above, the binding of toxic substances to enzymes, or to other types of receptors, initiates a chain of events that are not fully understood. Nevertheless, they interfere with the normal beneficial biochemical reactions of the human body or initiate abnormal metabolic reactions, resulting in toxic response. The response may consist of only short-term effects such as headaches or nausea. They could also be fatal as is the case with cyanide, which decreases oxygen exchange at the cellular level to the point of disrupting the nervous system function, leading to respiratory failure.

The mechanisms of toxic action and the manifestation of toxic effects can be illustrated by examples. The toxic effects of organophosphorus pesticides can be adequately explained in terms of their primary mechanism of action as acetylcholinesterase inhibitors. At nerve endings, the release of acetylcholine (ACh) triggers constriction of muscle fibers, but also activates certain parts of the autonomous nervous system.[34] To prevent prolonged and excessive action, the enzyme acetylcholinesterase present at the nerve endings cleaves acetylcholine into inactive acetic acid and choline. Exposure to an organophosphorus compound, like paraoxon (a metabolite of parathion), can inhibit the action of acetylcholinesterase (AChE), thus allowing the accumulation of acetylcholine at nerve endings. This can result in bronchoconstriction, increased salivation, contraction of smooth muscle (i.e., those performing involuntary functions), and twitching and cramps of skeletal muscles.[35] This is illustrated in Figure 5-10, showing normal hydrolysis of ACh and inhibition by the pesticide paraoxon.

Allergic reactions, although only occasionally lethal (in association with anaphylactic shock), nevertheless represent the most widespread manifestation of toxicity in humans that is caused by chemical agents. It is basically a dysfunction of

the immune system, and the target organs usually include the respiratory tract—for so-called immediate type of reactions—or the skin for contact allergies (delayed). Whereas high molecular weight agents like polysaccharides (e.g., pollen) are most prone to induce immediate allergies, a number of low molecular weight substances, including some metal ions (nickel, chromium, beryllium, mercury, platinum), can cause contact allergy. Similarly, low molecular weight organic compounds such as formaldehyde, acrylic acid, and toluene diisocyanate may act as contact allergens. To produce allergic sensitization, these low molecular weight substances **(haptens)** must first bind to an endogenous protein to form an antigen that can activate the memory cells of the immune system. Subsequent exposure to the chemical will then induce an antigen-antibody reaction that elicits the typical symptoms of an allergic reaction.

Even when the primary molecular receptor of a toxic substance is known, the sequence of events that results in the clinical manifestation of toxicity is rarely well understood. Although arsenic acts as an inhibitor of enzymes that depend on free SH groups for their function, the plethora of divergent toxic effects elicited by this element is difficult to explain on the basis of just this basic inhibitory mechanism. Arsenic's toxic effects include nausea and vomiting, fever, insomnia, progressive general weakness, liver swelling, disturbed heart function, peripheral nervous disturbances, and hyperkeratosis of palms and soles, as well as melanosis, cancer, anemia, and leukopenia.[36]

The biochemical role of riboflavin (vitamin B_2) in intermediate metabolism is well characterized. However, the symptoms of riboflavin deficiency induced by

FIGURE 5-10
Inhibition of enzyme acetylcholinesterase (AChE) by organophosphorus.[37]

vitamin B_2 antagonists—inflammation of the tongue, lesions of the mucocutaneous juncture of the mouth, corneal vascularization, photophobia, etc.—likewise cannot be readily related to the primary receptor.[38]

This lack of understanding of basic mechanisms underlying toxic responses seems less surprising when one considers that the mechanism of action of many commonly used beneficial drugs for which a long history exists (e.g., sodium acetyl-salicylate, aspirin) are not well understood either. Suffice it to state that toxicologists cannot delineate the whole sequence of events but can identify the organs that are affected by toxic chemicals, the symptoms expressed, and the range of the severity of the symptoms in response to the different doses.

Classification of Toxic Actions and Effects

Toxic responses are manifested in behavioral and physiological terms and can range from headaches and nausea to convulsions and death. Obviously, depending on the context, there are a number of ways in which a toxic chemical can be classified.

Classification by end point

A simple but apt and commonly used way of classifying the toxic action of a chemical substance is according to toxic end point. Because of the inherently different dose-response relationships in risk management, toxic effects are often subdivided into the two categories: **carcinogenic** effects with tumor induction as the end point and **noncarcinogenic** effects comprising all other effects. This classification creates some confusion because it means that noncarcinogenic effects would encompass mutagenic effects, yet mutation is more related to tumor induction. Nevertheless, this classification system is widely used, and it provides a way to illustrate toxic effects for lead and benzene (see Table 5-1) and other hazardous substances (see Table 5-2). The term "**genotoxic**" more precisely groups carcinogenic initiator and mutagenic effects. This term would include all chemical substances that can alter DNA or chromosomes, possibly initiating cancer in somatic cells (all cells other than reproductive cells) or leading to heredity changes due to action on germ cells (reproductive cells).

The toxic effects occurring from fertilization until sexual maturation to an adult can be grouped as **developmental effects** with special subcategories. Adverse effects associated with the process of reproduction starting from the production of

TABLE 5-1
Toxic effects of lead[39] and benzene[40]

Toxic substance	Carcinogenic effects	Noncarcinogenic effects
Lead	Kidney tumors (in test animals)	Reduced birth weight, anemia, increased blood pressure, brain and kidney damage, IQ impaired, decreased learning
Benzene	Leukemia (in humans)	Drowsiness, dizziness, headaches, anemia, suppressed immunity, fetotoxicity

TABLE 5-2
Example of adverse effects caused by selected toxic substances

Carcinogenic chemical	Potential carcinogenic effects	Potential noncarcinogenic effects
Metals (by inhalation)		
Arsenic	Lung cancer	Liver damage, lung fibrosis, neurological damage
Cadmium	Lung cancer (in test animals)	Kidney damage, osteoporosis, anemia
Chromium^{+6}	Lung cancer	Bronchitis, liver and kidney damage
Organophosphorus pesticides		Neurotoxicity, systemic poisoning
Chlorinated organic compounds	Liver cancer (in test animals)	Liver damage, neurological effects
Polycyclic aromatic hydrocarbons	Lung cancer, stomach cancer (ingested), skin cancer (dermal contact)	Liver damage, dermatitis

germ cells via fertilization, implantation, and development of the embryo to fetal growth and ending with birth are usually grouped under **reproductive** effects. **Teratogenicity** (from the Greek work "teras" connoting monster) refers specifically to the subcategory of pathological effects during the sensitive development phase after implantation of the embryo up to the first 3 months of pregnancy in humans when major tissues and organs differentiate and develop. **Fetotoxicity** represents various disturbances, such as growth depression, that may occur to a developing fetus during the later stages of pregnancy. No major skeletal malformations or malformations of organs can occur during this period.

Postnatal development is another period of concern because the rapid growth of a child makes it more susceptible to some toxic agents than the adult body. Lead is an example, with its capacity to retard precognitive development in young children. Exposures of moderate duration, or even a single event, during gestation or prior to sexual maturation could result in adverse effects that would rarely occur in an adult.

In recent years a category of environmental contaminant known as **endocrine disruptors** has captured the attention of regulatory agencies regarding both human and ecological health. These substances, also known as "hormonally active agents" (HAAs), include chemicals such as the polychlorinated biphenyls (PCBs), 1,1-dichloro-2,2-*bis* (*p*-chlorophenyl)ethene (DDE), methoxychlor, dibutyl phthalate, and dioxin. These compounds are suspected to mimic or modulate the action of naturally occurring reproductive and thyroid hormones. Because of their hormonal nature, these substances have demonstrated a wide range of effects in humans, laboratory animals, and wildlife, including reproductive, immunological, neurological, and developmental effects. One proposed mechanism of action is that the HAAs bind to the natural hormone receptor in the cell.

The wide distribution and environmental persistence of these substances, as well as their potential to induce profound systemic effects in humans and wildlife, has prompted the U.S. Environmental Protection Agency to establish the Endocrine

Disruptor Screening and Testing Advisory Committee (EDSTAC). This group has issued a report that proposes a tiered approach for identifying and characterizing endocrine disruptors. This approach involves a screening step based on an evaluation of a substance's chemical and physical properties that suggest potential hormone-like action, followed by a series of *in vitro* tests (e.g., receptor-binding and cell proliferation assays) and, if necessary, *in vivo* assays such as the two-generation mammalian reproductive toxicity study. It is believed that approximately 86,000 chemicals are candidates for screening and evaluation. The ability of the HAAs to exert toxic effects at low environmental concentrations is a matter of considerable debate. Much more research is needed to adequately characterize the health risk of environmental HAAs to humans and wildlife. For a comprehensive discussion of the HAAs refer to the National Research Council's report *Hormonally Active Agents in the Environment* (July 1999) and to the EDSTAC *Final Report* (August 1998).

Classification by target organ

In some cases, a specific target organ, or tissue, may be distinguished and used as a basis for classification (e.g., the kidney in case of cadmium toxicity, the bone marrow for benzene, the brain for methyl mercury, the liver for carbon tetrachloride, the lung for the pesticide paraquat, the eye for the antimalarial drug chloroquine).

It should be stressed, however, that it may be difficult to single out a characteristic target for toxic action in some cases. Several substances induce generalized symptoms of intoxication, where nausea with abdominal distress, loss of appetite, headache, drowsiness, etc., may be the only obvious symptoms. Oral intake of toxic, but nonlethal, concentrations of inorganic arsenic compounds provides an illustrative example.

Toxicological effects may also be characterized in the following manner:

- Immediate versus delayed effects
- Irreversible versus reversible effects
- Local versus systemic effects

Classification by immediate versus delayed effects

The ingestion of a lethal dose of a soluble cyanide results in death within a few minutes, providing illustration of an immediate toxicological action. In contrast, carcinogenic agents induce tumors in humans only after a latency period possibly extending 10 to 30 years.

A way of characterizing toxic action according to the latency periods is to use the terms acute and chronic. **Acute or immediate toxicity** results shortly after a single exposure; the typical concern is with a single exposure of a magnitude to overcome existing protective mechanisms. **Delayed toxic reactions** feature long latency periods, perhaps measured in years. It should be noted that an acute exposure may result in a delayed toxicological effect or chronic illness.

Classification by irreversible versus reversible effects

The beneficial pharmacological action of drugs is usually reversible. This may not be the case for nontherapeutic toxic effects, depending on the nature of the action and capacity for the target tissue to regenerate. Skin and liver have a high repair capacity; consequently, moderate damage induced at these sites is often reversible. Injuries of the central nervous system caused by chemicals are, on the other hand,

mostly irreversible because of the slow or nonexisting regenerative power of these tissues.

Classification by local versus systemic effects

In toxicology a distinction is made between local toxic effects and systemic action. Reactive chemicals, such as acid anhydrides, epoxides, strong acids and bases, acyl chlorides, and corrosive agents, cause toxic effects, such as cell necrosis, immediately on coming into contact with tissue (e.g., skin and eyes, mucous membranes of the respiratory tract). However, such agents are usually not taken up appreciably in the organism. In contrast, systemic poisons may exert their toxic action at a point distant from the site of absorption (e.g., tri-o-cresyl phosphate, which induces damage to the peripheral nervous system on ingestion of small amounts over extended periods of time).

5-3
DOSE-RESPONSE RELATIONSHIPS

A number of factors influence the degree of toxic action. It is readily apparent that the dose represents the major determinant of the extent of toxic action. Toxic effects, as detailed in Section 5-1, may show great variation, depending on the route of exposure, with toxic response that may vary between different animal species. A relationship exists that the greater the dose, the more severe the response (i.e., the greater the number of chemical-receptor complexes, the more severe the response).

Nature of Dose-Response Relationships

Whether a toxic effect occurs, and how severe it is, depends not only on dose but also on the susceptibility of the exposed individual. The susceptibility of an individual is primarily a function of the individual's age, sex, diet, health status, genetics, prior exposures to the agent, and exposures to other agents. Even when variations of these key variables are reduced, as in the case of carefully inbred laboratory animals, other differences will yield variable responses among individuals.

When the incidence (or frequency) of a specified toxic end point among individuals (e.g., proportion of animals with tumors, proportion of dead animals) of a population is recorded as a function of dose, this correlation is called a **dose-response relationship.** The classic example of a dose-response relationship derives from exposing a population of living organisms to increasing doses of a toxic compound. The exposed population will typically experience no deaths at low doses, a few deaths as the dose increases, and more deaths with higher doses until all the organisms die. Most plots of log dose versus cumulative mortality will show the classic nonlinear S-shaped curve in Figure 5-11a. This concept was first presented in the literature in 1927[41] to introduce the term **median lethal dose** (LD_{50}). This is the dose, usually expressed as milligrams per kilogram body weight, at which only 50 percent of the organisms remained alive. It is sometimes referred to as a quantal dose-response relationship because the elicited response is an all-or-none effect (i.e., dead or alive). The equivalent for inhaled substances is the **median lethal concen-**

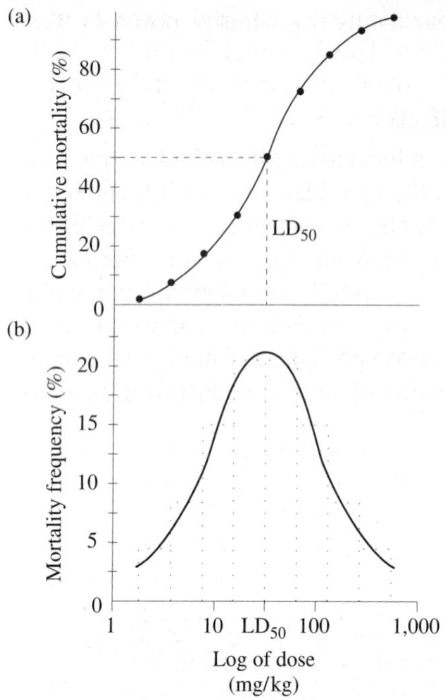

FIGURE 5-11
Dose-response relationship (dose versus mortality).[42]

tration (LC_{50}), where the dose is expressed as the concentration of the substance present in a volume of inhaled air.

To make comparisons of the acute toxicity of chemicals on a relative scale, the acute oral LD_{50} in rodents is commonly used. In Table 5-3 the oral LD_{50} for the rat is shown for selected chemicals.

Because of the wide variation of susceptibility among individuals, dose-response relationships are frequently plotted with dose on a log scale and response as mortality

TABLE 5-3
Acute oral LD_{50} in the rat for selected compounds

Substance	LD_{50} (mg/kg)
Ethanol[43]	13,000
Table salt[44]	3800
Malathion (insecticide)[45]	2800
Aspirin[46]	1500
Lindane (insecticide)[47]	88–270
Sodium fluoride[48]	180
DDT[49]	113–118
Nicotine[50]	50–60
Sodium arsenite[51]	40
Parathion[52]	2
Dioxin (TCDD)[53,54]	0.02–0.05

frequency on an arithmetic scale, as depicted in Figure 5-11. The bell-shaped form in Figure 5-11b is that of a log-normal distribution. This is typical for the variation in susceptibility of a population with respect to a majority of toxicological effects.

Developing a dose-response relationship first requires selecting a toxicological end point that is caused by exposure to the selected chemical. Preferably, the end point should be clear, definitive, and directly measurable, such as death or organ dysfunction. Having to use an equivocal end point, such as retardation of mental development, makes the interpretation of the results more susceptible to question.

The ordinate (i.e., response) of a dose-response relationship is usually measured as incidence of the selected end point (e.g., percent mortality as in Figure 5-11). Similarly, it could be reported as the probability that the end point will occur. This latter approach is used more for carcinogens, which are typically reported in terms of probability of cancer over a lifetime of exposure.

Clinical chemistry data, when available and well understood, can serve as indicators of toxic action. The selected parameter must be specific, quantifiable, and functionally related to the toxicological action of the chemical. In humans, lead causes anemia by inhibiting the synthesis of hemoglobin from various precursors. This results in the accumulation of the precursor protoporphyrin in the erythrocytes (red blood cells). Thus, erythrocyte protoporphyrin in blood can be used as a biomarker for measurement of the "toxic" action of lead and has been found to constitute a sensitive and reliable indicator of low-level exposure to lead. In Figure 5-12

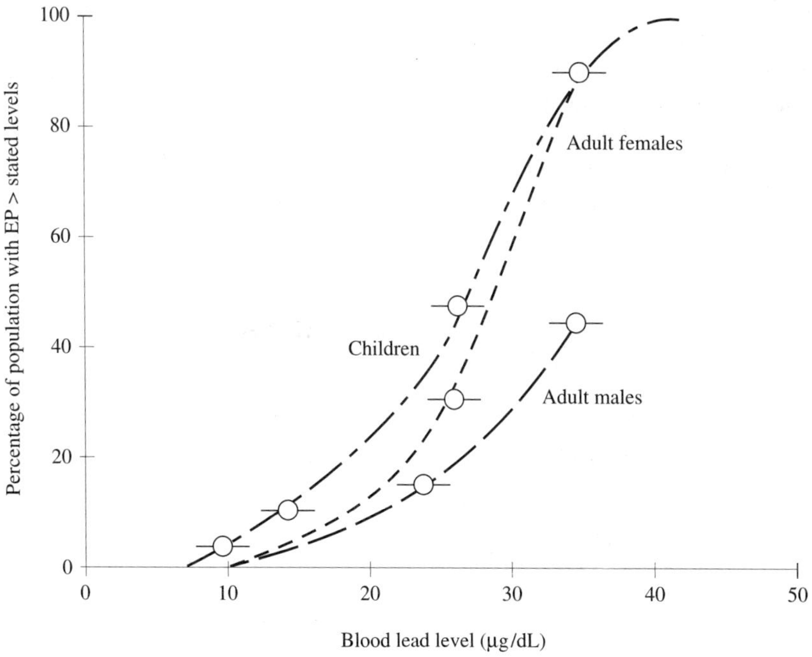

FIGURE 5-12
Dose-response relationship for erythrocyte protoporphyrin (EP) in blood as a function of the blood lead level in subpopulations.[56]

the dose-response relationship for the levels of as a function of the concentration of lead in blood derived from inhalation exposure is given for different segments of a human population.[55]

The abscissa (i.e., dose) of a dose-response relationship is straightforward. Doses are most frequently expressed on a body weight basis (e.g., milligrams or moles of chemical per kilogram of body weight). Reporting on a body size basis allows the extrapolation from small test animals to humans. The abscissa may not be expressed as a dose in some cases but as an ambient concentration; this is particularly valid in evaluating exposure by inhalation or in testing the response of fish to aquatic contaminants.

Dose-Effect Relationship

A toxic chemical will cause adverse health effects at doses significantly below the lethal level. When instead of incidence among a population of individuals, the intensity (or degree) of the toxic effect in a single individual (e.g., increased blood pressure, retardation of growth, increase in plasma levels of indicators for liver damage) is plotted as a function of dose, the correlation is customarily termed a **dose-effect relationship.** Glycol ethers (an industrial solvent) produce testicular damage in rodents. Figure 5-13 shows the dose-effect relationship between orally administered ethylene glycol monomethyl ether (EGM) and the relative organ weights for the testes and prostate in rats after 11 days of treatment.[57] The dose-effect graph of Figure 5-13 can be converted into a dose-response curve by first determining a cutoff value for defining a significant effect, and then plotting the percent of the population that meets this value for each dose tested.

Figures 5-14 and 5-15 provide an alternative manner for demonstrating in a condensed fashion the dose ranges for induction of different toxicological responses by lead and benzene in humans.

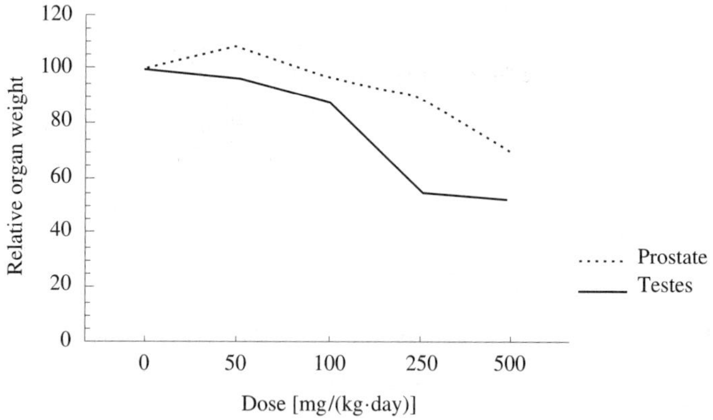

FIGURE 5-13
Effect of EGM on the relative organ weights of testes and prostate.[58]

Toxicological data are based largely on observations of (a) human populations which have been exposed to a particular substance, typically as part of their workplace, and (b) laboratory animals which have been tested under controlled conditions. Of these, epidemiologic studies of human exposure to a toxic substance represent very convincing evidence for establishing a positive association between a toxic substance and adverse human health effects. For ethical reasons, the cases embraced by these studies have been limited to occupational exposures or to single catastrophic exposure incidents involving the general population.

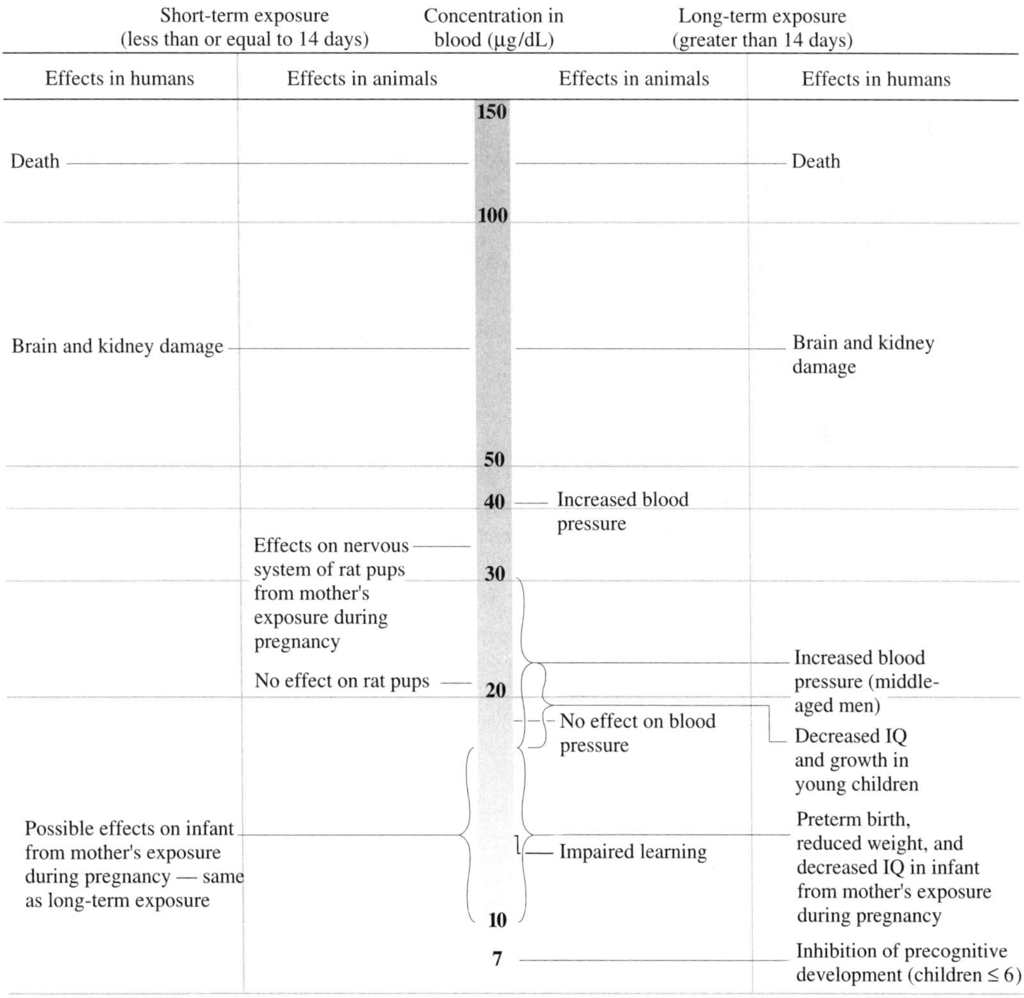

FIGURE 5-14
Health effects from breathing and/or ingesting lead.[59]

Types of epidemiological studies

Out of several types of epidemiological study designs, three will be mentioned here: **cohort** studies, **case-control** studies, and **cross-sectional** studies. There are two types of cohort studies, prospective and retrospective. In **prospective cohort studies,** a defined population, or cohort, is followed over a period of time to measure exposure and a particular toxic end point. This type of study design has the advantage that definite dose-response relationships can be established. On the other hand such studies require a long period of time to complete, and they are characterized by a low power of resolution; i.e., unless the exposures are high, a large number of individuals will have to be studied in order to detect a statistically significant effect. When such cohorts are investigated in retrospect **(retrospective cohort study),** the estimation of dose-response relationships requires favorable circumstances. Frequently, however, bias is introduced by underreporting, difficulties in reconstructing

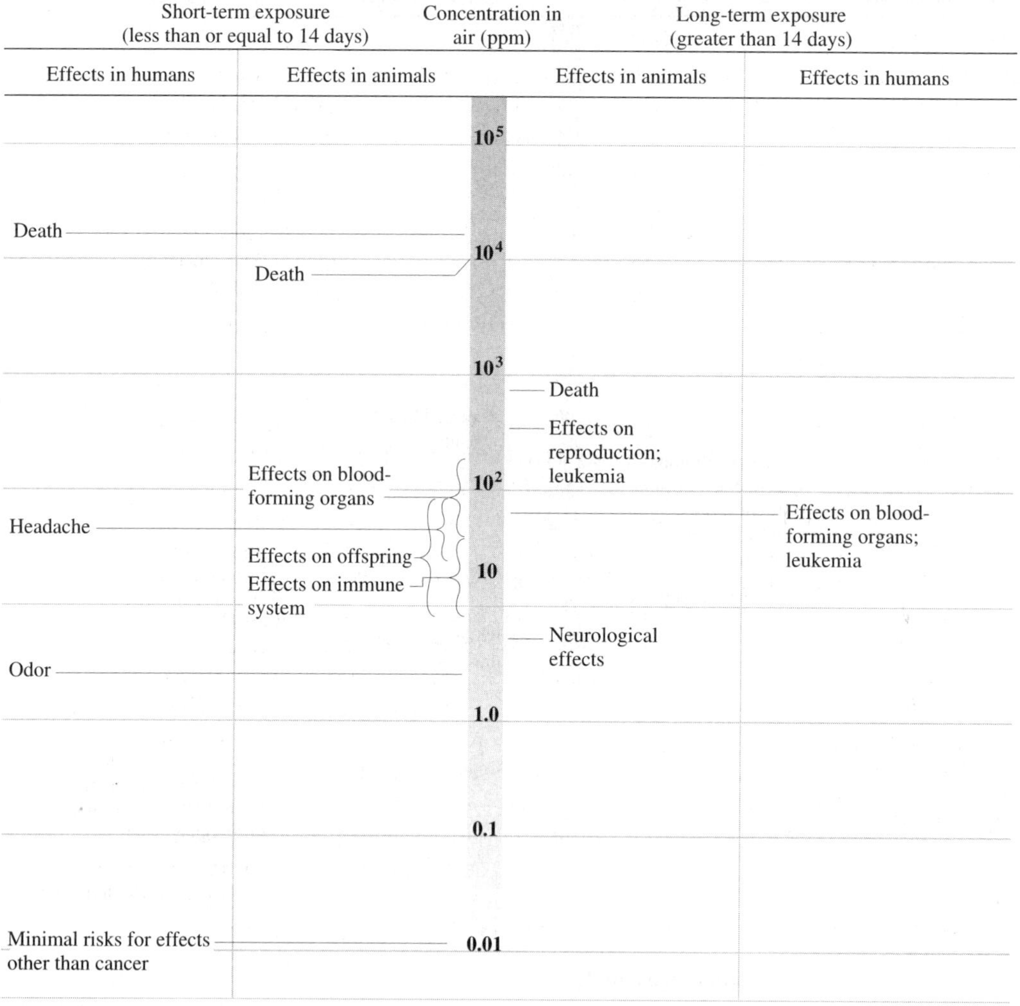

FIGURE 5-15
Health effects from breathing benzene.[60]

exposure situations, not having complete information for mixed exposure situations, and other factors.

In **case-control studies,** a different approach is used. In retrospect, individuals with a specific disease are selected from a larger population. Each person with the specific disease represents a case, and the exposure to a suspected causative agent is evaluated for these cases. The incidence of exposure to the selected agent for cases is then compared with the exposure in controls who have not experienced the studied disease. The finding of a more frequent occurrence of exposure to a certain agent among cases than among controls could thus indicate a causal association between disease and exposure. This method requires careful matching with respect to other factors, such as age, sex, race, and smoking. Because the cases and controls may be drawn from a very large population, the methodology is much more sensitive than cohort studies, but suffers from the disadvantage that bias is difficult to avoid. Case-control studies are in general only feasible when the studied toxicological end point is relatively rare. When properly conducted, such studies may provide important clues as to the causes of human disease (i.e., the etiology), but rarely afford reliable estimates of dose-response relationships.

In **cross-sectional studies** (prevalence studies) information on the prevalence of a certain disease or toxic effect at a given time is collected and related to an exposure situation. For instance, the following question could be raised: Is anemia more common among workers in copper smelters using sulfide ores with a high content of arsenic than among healthy males in the general population? The current arsenic exposure among all smelter workers could be estimated by measuring arsenic in urine, and exposures over a longer time by determining arsenic in hair and/or nails. The same measurements would be made of controls carefully selected so as to be comparable with respect to age, smoking habits, dietary factors, health status, etc. The collected data will answer the posed question.

As an example of the use of epidemiology in evaluating hazardous waste sites, a case-control study was conducted in New York to determine if communities adjacent to such sites demonstrated increased rates of central nervous system or musculoskeletal birth defects. Approximately 3800 birth defects were identified from medical records over an 18-county area. Controls consisted of 12,436 randomly selected normal births. This 18-county study area contained 643 hazardous waste sites. The investigators then assigned a low, medium, or high probability of exposure by air and groundwater to solvents, metals, and pesticides for each birth address, depending on proximity to a site. Odds ratios (ORs) were then calculated for each birth defect and exposure to one of the three contaminant groups. Almost all categories showed an OR of less than 1.0, indicating no statistically significant association between the health effect and the chemical exposure. Two categories showed an OR of greater than 1.0: central nervous system defects and air exposure to solvents (OR = 1.3) and to metals (OR = 1.4), suggesting increased risk. However, the conclusions of this study were hindered by lack of adequate exposure information and by a low proportion of subjects categorized as medium or high probability of exposure. The investigators acknowledge that possible associations between central nervous system defects and industrial emissions of solvents and metals merit further research.*

*"Maternal residential exposure to hazardous wastes and risk of central nervous system and musculoskeletal birth defects," *Arch. Environ. Health*. vol. 52, no. 6, pp. 416–425, 1997.

EXAMPLE 5-3. EPIDEMIOLOGICAL STUDIES AT A HAZARDOUS WASTE SITE. How would an epidemiologist approach the study of a site historically contaminated by illegal dumping of hazardous waste? What types of problems would such a study face?

Solution. The epidemiologist would use the retrospective or case-control method. He/she would first try to obtain data on the extent of environmental contamination to assess potential human exposure to specific chemical contaminants at representative sampling sites, review the literature to determine the types of health effects expected for exposure to such substances at the levels found, and survey the exposed population for such effects. In another approach, the epidemiologist could identify any unusually high incidence of a disease in the surrounding community and then seek to establish a relationship between exposure and the disease by means of a case-control or a cross-sectional study.

It should be realized, however, that an epidemiological investigation of human populations living close to a hazardous waste site represents a formidable task. One of the most difficult problems will be to secure a satisfactory assessment of exposure, which, most certainly, will not be confined to one substance. Also, it will not be sufficient to obtain representative analytical data at the time of the investigation; an assessment will also have to be made with respect to previous exposures. This requires examining historical records on morbidity and mortality of residents, as well as tracing residents who have moved to and from the area, both contributing further to the complexity of the problem. Finally, the population size may be too small or toxicities too weak to produce statistically valid results.

Limitations of epidemiological studies

It is important to keep in mind that statistically based epidemiological studies by themselves can never prove the existence of a cause-and-effect relationship. However, such observations may be used to generate or to test a hypothesis. Many possibilities exist for introducing bias in this type of investigation, and statistical epidemiological correlations are fortuitous. It is, therefore, of utmost importance that any such correlation is supported by a biological plausibility based on existing toxicological data and other information. Here, using common sense is a logical first step. Even for the case of a well-established cause-and-effect relation, confounding factors often impede the establishment of a reliable dose-response (effect) relationship.[61]

Epidemiological investigations of the general population are especially difficult to conduct. This is particularly true when an investigation of incidence of disease (e.g., cancer) is based on simple comparisons of morbidities for exposed populations of a certain geographical area against "normal" incidences for the general population. The main objection typically raised with such studies is the bias introduced by inappropriate matching between exposed individuals and controls. Important confounding factors are smoking and other "lifestyle"-associated factors that influence exposure to a melange of chemicals as well as the quality of the diagnostic routines employed in various hospitals. For these reasons, only a limited number of toxic substances have sufficient human data to support the development of a quantitative relationship between dose and incidence of adverse health effects. This situation is illustrated in Figure 5-16 for benzene, a ubiquitous substance that has been the focus of numerous and extensive epidemiological studies.[62]

Toxicological Data—Experimental Animals

Because of the lack of sufficient human data, most determinations of the risk of toxic chemicals must depend on testing of experimental animals, particularly small

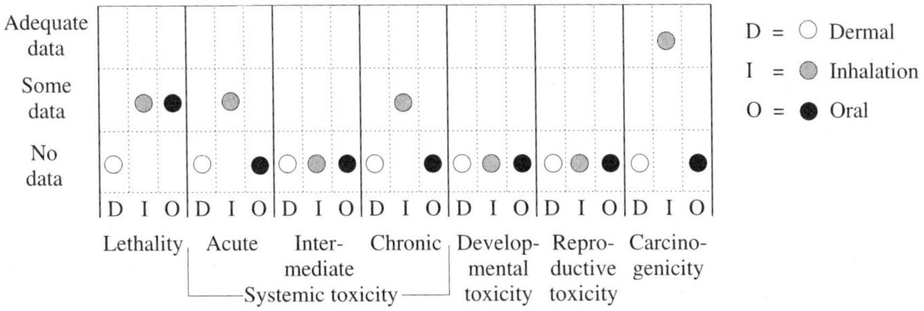

FIGURE 5-16
Adequacy of human exposure data on health effects of benzene.[63]

mammals. In many cases, experimental animals constitute adequate models for the prediction of adverse reactions induced by chemicals in humans, making them indispensable in toxicological and pharmacological research. The interpretation of data from animal studies requires high-level expertise mainly in the fields of biochemistry, toxicology, and veterinary pathology, and this can encounter substantial differences of opinion among similarly trained professionals. Hence, peer review of such information by a qualified panel of experts is critical. For much of the published information, this has, unfortunately, not always been the case.

Problems associated with animal test data interpretation arise especially in the following areas:

- Interpretation of the significance of certain types of pathological findings for humans
- Low degree of statistical resolution because of the limited number of animals used per dose group
- Design of carcinogenicity studies, especially use of maximum tolerated doses
- Extrapolation of dose-response relationships for certain types of effects, in particular carcinogenic effects and induction of allergic reactions

Interpretation of findings

As to the first item, a recent international evaluation concluded that (1) the introduction of international guidelines for the conduction of toxicological testing [e.g., the Organization for Economic Cooperation and Development (OECD) Guidelines[64]] has streamlined the production of experimental data, and (2) interpretations of most findings are generally consistent among qualified toxicologists. Disagreement does occur. Notable examples involve pathological changes that are notoriously difficult to classify (e.g., adrenal medullary tumors) and certain minor pathological findings for which it is not clear whether they should be interpreted as signs of toxicity, or only as signs of exposure (e.g., induction of certain liver enzymes).[65]

Limited number of animals

The OECD Guidelines for testing of chemicals, which are legally binding for all of the OECD countries (including the United States), require a minimum of 50 animals per sex and dose group for long-term toxicity testing in rodents. The statistical limitations with respect to the possibility of detecting a significant increase

of a certain toxic effect in a test group of 50 animals, using control groups of varying size, are demonstrated in Table 5-4. As is generally the case, the controls also exhibit a varying incidence of the toxicological end point being studied.

The severe consequences of missing an incidence of a severe toxic reaction (like cancer) induced by chemical exposure of the general population, and which could occur at a possible incidence of 5 percent in the general population, can readily be visualized. To cope with this problem, the doses used in such experiments as described above are, therefore, increased to the so-called **maximum tolerated dose (MTD)**, a dose that is high enough to induce signs of toxicity in the animals, but not so high as to increase mortality.[67]

Selection of surrogate species

Toxic effects can vary considerably between species even after adjustment for body weight. This derives from differences in metabolism and pharmokinetics. For example, absorption after ingestion may vary between species, thus necessitating caution in extrapolating an oral LD_{50} to humans. In the absence of information from humans or other primates, it is difficult to know which of the common experimental animals will resemble humans the most: the mouse, rat, rabbit, or guinea pig. The toxicity of dioxin (2,3,7,8-tetrachlorodibenzo-*p*-dioxin, TCDD) is shown in Table 5-5 for different animal species to demonstrate the unusual differences in sen-

TABLE 5-4
Smallest percentage of treated animals exhibiting a toxic effect (F_t) that is considered statistically significant when compared to control incidences (F_c); based on Fisher's exact test at $p \leq 0.05$ (one-sided)[66]

Incidence, %, of effect in controls (F_c)	Number of animals in controls	Number of animals in treated group	Smallest statistically significant incidence, %, of effect in treated animals (F_t)
0	50	50	10
	500	50	4
10	50	50	26
	500	50	20
30	50	50	50
	500	50	44
50	50	50	70
	500	50	64

TABLE 5-5
LD_{50} for dioxin (TCDD)

Animal	LD_{50}, mg/kg
Guinea pig[68]	0.001
Rat (Sprague Dawley or Long-Evans)[69,70]	0.02–0.05
Rat (Han/Wistar)[71]	>3
Hamster[72]	5

sitivity found for this substance. Further, the toxicity of TCDD to strains of rat shows that similar variations also occur between strains of the same species.

Other tests

Today's legal requirements stipulate for chemicals such as pesticides and pharmaceuticals that, in addition to investigations of toxic reactions in whole animals, studies on metabolism and pharmacokinetics be performed. Such studies are preferably carried out in whole animals, but tissue culture—the maintenance and growth of an explanted tissue in a culture away from the source organism—of perfused isolated organs (e.g., liver) are sometimes utilized. Such data are indispensable for assessing parameters such as uptake, accumulation, and other factors. Although rarely performed, comparative studies of human metabolism and the metabolism of an animal which has shown susceptibility to a toxic agent will further facilitate the interpretation of available animal toxicity data. In some instances, experiments in whole animals may be difficult to perform, and microorganisms or mammalian cell cultures may provide important additional information, especially in the field of mutagenicity testing.

Adequacy of Data

In most cases, the scientific database is not adequate to describe, with a high degree of confidence, the quantitative risk of all health effects from exposure to toxic substances. Figure 5-16, presented earlier, showed that even for benzene, one of the more thoroughly studied toxic substances, human data are inadequate for most exposure situations. Many inadequacies still exist even after including animal test data, as depicted in Figure 5-17 for benzene. Toxicologists, when assessing toxicity, should bracket their estimates to reflect the inherent uncertainty of such endeavors. On the other hand, it is important to realize that, depending on the situation, not all toxicological effects are equally essential to making decisions in the hazardous waste management field. For episodes involving spills, acute hazards are usually the most important. In the remediation of a hazardous waste site, where continuous low-level exposure is generally of highest concern, efforts should be directed toward identifying and defining dose-response relationships for those serious chronic effects that appear at the lowest doses.

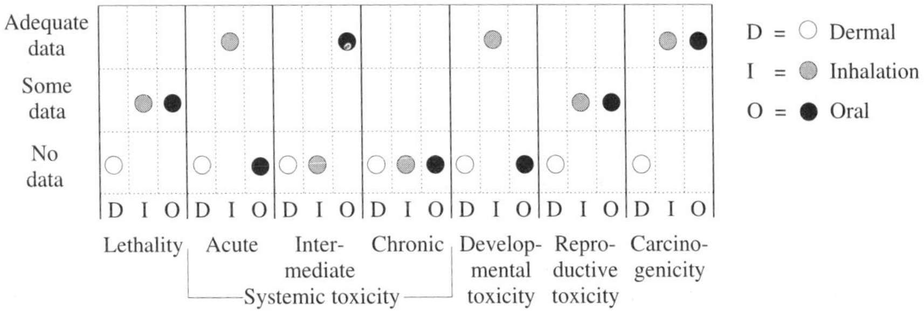

FIGURE 5-17
Adequacy of animal test data on health effects of benzene.[73]

When reviewing toxicological information, it is very important to make a clear distinction between toxicological data appropriate for risk assessment and administrative "standards" that are associated with risk management. The former is a scientific procedure. Risk management, on the other hand, comprises a range of actions taken to minimize, reduce, or otherwise control specific risks posed by a certain situation; it contains elements of policy. An LD_{50} derived from an animal experiment is an example of basic information that can be used in risk assessment. In contrast, a threshold limit value (TLV) for occupational exposure, although partly based on scientific data, represents an administrative decision made by regulatory agencies. Their decisions have taken into account, to a greater or lesser extent, a number of nonscientific factors, such as technical feasibility, economic impact, and other factors. Such decisions thereby involve policy judgments.

5-4 NONCARCINOGENS

Noncarcinogenic effects literally include all toxicological responses other than the induction of tumors. The toxic end points of noncarcinogenic agents vary widely, as do mechanisms by which they cause their toxic effects. Chemical agents may combine with cell membranes, thereby reducing the permeability of the membranes. Some can chemically displace elements important to a cell performing its normal activity. Nevertheless, the most prevalent and typically the most toxic noncarcinogenic responses are those in which an agent affects enzymes. All enzymes are predisposed to perform specific physiologic functions. However, binding with a toxic substance may interfere with, or block, these normal functions, prompting an altered series of biochemical activities that eventually expresses itself as a toxic response.

Concept of Threshold

The assessment of the toxicity of noncarcinogenic chemical substances is based on the concept of threshold. That is, for toxicological end points other than mutations and cancer initiation, the dose-effect or dose-response relationship is generally characterized by a threshold below which no effects can be observed on the cellular, subcellular, or molecular level. Most structures that have critical functions in the organism exist in a large number of identical copies. The destruction of a few units (e.g., enzyme molecules) will have little effect on overall performance, especially in the presence of efficient repair mechanisms. Only when a significant fraction of the targets have been eliminated by toxic action, that is, above a certain threshold for the target dose, will a toxic effect occur.

The toxic effects of lead provide an example. At a blood level in the range of 7 to 30 μg/100 mL, lead can cause inhibition of precognitive development in preschool children. Until this threshold concentration of lead is reached, the developing brain will function satisfactorily. As another example, carbon tetrachloride will result in pathologic damage to liver tissue; however, the liver will continue to function satisfactorily and its remarkable ability to grow new cells will eventually replace the lost cells. At some threshold, the liver will become dysfunctional, and the damage

may not be reversible. Therefore, the body can tolerate a wide range of exposures between negligible intake and the threshold value for a discernible effect.

No Observed Adverse Effect Level (NOAEL)

The threshold value for a toxic substance cannot be identified precisely. Instead, it can only be bracketed on the basis of analysis of epidemiological data and animal tests. In practice, hazard evaluations are often based on controlled animal experiments, where the highest dose that will not cause an effect, the "no observed effect level" (NOEL) is sought. Because we are usually only interested in toxicologically significant effects, a variant of NOEL termed the "no observed adverse effect level" (NOAEL) is often utilized. Two other terms used in this context are:

- *LOEL (lowest observed effect level)*—The lowest dose tested for which effects were expressed; typically used when an effect is expressed at all doses
- *LOAEL (lowest observed adverse effect level)*—A stricter version of LOEL to address only adverse effects

Figure 5-18 shows NOAEL and LOAEL doses bracketing the threshold for a typical noncarcinogenic dose-response curve. The closer that the NOAEL and LOAEL doses are in experimental or epidemiologic studies, the more precisely the threshold dose is defined.

In using data like LOELs (LOAELs) or NOELs (NOAELs), it is important to be aware of the limitations of such values.[74] A NOEL, for instance, is determined simply by selecting the lowest dose in a study that does not seem to cause any effect. Thus, in deriving such a parameter, statistical uncertainty due to the limited number of animals used in the critical study is not accounted for. As mentioned before, the variation in sensitivity in a population for a majority of toxicological effects follows a log-normal distribution.

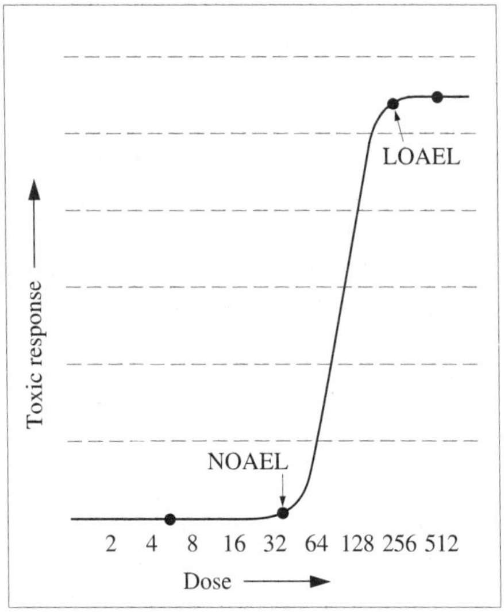

FIGURE 5-18
Location of NOAEL and LOAEL doses with respect to the threshold on a typical dose-response curve.

A NOEL (as well as an LD_{50}) gives no indication of such individual variations in susceptibility. However, without performing any kind of sophisticated analysis, a qualitative judgment can sometimes be made simply by determining the slope of the dose-response curve; a steep curve usually indicates a uniform response curve (e.g., cyanide) with little variation between individuals, and the NOEL may be determined with good precision. A flattened response curve, as for ethylene glycol, tends to reveal large differences in susceptibility among individuals, thus decreasing the confidence in the NOEL.

Acceptable Daily Intakes and Reference Doses

Instead of establishing a threshold value, toxicologists use the term **acceptable daily intake (ADI)** to represent the level of daily intake of a particular substance which should not produce an adverse health effect. ADIs are based on NOAELs and should not be construed as a strict physiological threshold that when exceeded will result in adverse health effects; an ADI includes safety factors to reflect variable susceptibility in the human population and other uncertainties. An ADI is much smaller than the theoretical threshold.

For regulatory purposes, the accepted way to establish safe levels of most chemicals in food and drinking water is to divide the NOAEL by safety (uncertainty) factors to derive a value for ADI. By international agreement an arbitrarily fixed uncertainty factor of 10 is used to compensate for different sensitivities among individuals, and an additional factor of 10 for extrapolating from an experimental animal to humans. This simplified approach is used in risk assessment by organizations like the World Health Organization (WHO) as well as by government regulatory agencies all over the world.[75] Its main justification for continued use lies largely in the fact that the ADIs have been used historically with no obvious negative consequences for human health.

The reference dose (RfD) is a contemporary surrogate used by the EPA instead of the ADI. The development of an RfD follows a somewhat stricter procedure than used for an ADI, sometimes resulting in lower values for acceptable intake. The standard procedures for establishing an RfD consist of the following steps[76]:

1. Select the most sensitive species for which adequate studies are available. This should always be done unless compelling evidence shows that the most sensitive species does not adequately reflect the situation in humans and a better animal model is available. When available, human data are always given priority, taking into account the limitations in epidemiological studies.
2. Select the principal or critical studies using the appropriate route of exposure. RfDs are route-specific (i.e., they refer specifically either to inhalation or to oral intake). Thus, a study based on intravenous injection would not be appropriate for the establishment of an RfD for oral intake, or for inhalation. Ideally, the critical study should be sufficiently well conducted to permit the establishment of a NOAEL with acceptable accuracy.
3. Select supporting studies. Investigations from a wide variety of sources may provide additional aid in interpreting the results from the critical study. Comparisons of metabolism and pharmacokinetics of the toxic chemical in the animal

used for the critical study with such data derived from humans are of particular value.

4. Identify the NOAEL, or if such data are not available, the LOAEL for the most sensitive end point. Chemical agents frequently have multiple toxicological effects. For example, arsenic induces a large number of chronic toxic effects upon ingestion: intestinal disturbances, anemia, neurological damage, peripheral vascular disorders, and others. However, based on epidemiological data, the induction of skin disorders and vascular complications is the most sensitive toxicological end point with a LOAEL of 14 μg arsenic per kg body weight per day [μg/(kg·day)].
5. The NOAEL for the most sensitive end point is adjusted downward by orders of magnitude to reflect uncertainty as follows:
   - Reduce the NOAEL found in humans by an uncertainty factor of 10 to account for variation in the general population, thus protecting the most sensitive subpopulations (e.g., children, elderly).
   - Reduce the NOAEL by an additional uncertainty factor of 10 when extrapolating from animals to humans.
   - Reduce the NOAEL by an additional uncertainty factor of 10 if the data are derived from a subchronic instead of a chronic study.
   - If the test data do not show a NOAEL (i.e., adverse effects were expressed at all dose levels), the LOAEL is selected and reduced by an additional factor of 10 to account for the uncertainty introduced by this extrapolation.

The uncertainty factors are applied sequentially. The EPA also applies a "modifying factor," which ranges from 1 to 10, to reflect a qualitative professional judgment of uncertainties which are not accounted for by the preceding uncertainty factors. The derivation of an RfD is illustrated in Example 5-4.

EXAMPLE 5-4. DERIVATION OF AN RfD. In a subchronic oral toxicity study in mice, a lowest observed adverse effect level (LOAEL) of 5 mg/(kg · day) was determined for a specific agent. The quality of the data is given a high rating by the expert evaluating the data. What is the RfD?

Solution

Area of uncertainty	Uncertainty factor
Variation within a population	10
Extrapolation from animals to humans	10
Extrapolation from subchronic to chronic	10
Extrapolation from LOAEL to NOAEL	10
Modifying factor	1

$$\text{RfD} = \frac{5\ \text{mg/(kg} \cdot \text{day)}}{10 \times 10 \times 10 \times 10 \times 1} = 0.5\ \mu\text{g/(kg} \cdot \text{day)}$$

Similar approaches are taken for the development of an RfD for other exposure durations (e.g., acute, chronic). Another example is that of "developmental exposures RfD." A developmental exposures RfD covers the special case when exposure of short duration, even a single event, prior to conception, during gestation,

or prior to sexual maturation, could result in adverse effects that would not occur in an adult. For example, investigations have revealed that precognitive development in young children may still be sensitive to the RfD of 0.0014 mg/(kg · day) for lead. Hence, the EPA at one time lowered the RfD for lead by a factor of 10. Today, the EPA restricts specific lead sources and attempts to quantify cumulative lead exposure, since there are so many environmental sources of lead and the neurotoxic threshold for young children is so low.

Apart from the arbitrary assignment of safety factors, ADIs, RfDs, and related values are also associated with the same uncertainty as those inherent in the NOAELs or LOAELs from which they are derived. Above all, the misuse of such data as some kind of physical constants is not recommended. The following note in the Integrated Risk Information System (IRIS) (to be identified in the next section) for arsenic amply illustrates this point[77]:

> Note: There was not a clear consensus among Agency scientists on the oral RfD. Applying the Agency's RfD methodology, strong scientific arguments can be made for various values within a factor of 2 or 3 of the currently recommended RfD value, i.e., 0.1 to 0.8 μg/(kg·day). **It should be noted, however, that the RfD methodology, by definition, yields a number with inherent uncertainty spanning perhaps an order of magnitude** [emphasis added]. . . . Risk managers should recognize the considerable flexibility afforded them in formulating regulatory decisions when uncertainty and lack of clear consensus are taken into account.

For most noncarcinogenic effects a considerable body of data exist that will permit adequate high-to-low dose extrapolation to be made on the basis of dose-response or dose-effect relationships. However, the establishment of such relationships with respect to immunological effects poses special problems. The characterization of such relationships for pulmonary (and other) sensitizers remains largely an unresolved issue, and constitutes one of the most enigmatic problems in the risk assessment of air pollutants.

Toxicological Databases

A number of databases exist that document RfDs, ADIs, and other toxicological data for both noncarcinogenic and carcinogenic substances. Some of these are accessible as computerized databases maintained by government agencies.

An example is IRIS, the EPA's preferred source for toxicity information for hazardous wastes typically encountered in site remediation projects. IRIS is updated monthly. The IRIS database may be accessed directly from the EPA's website (*www.epa.gov/iris/*). On accessing the database, the user is provided with several options, including the opportunity to review a list of recent updates to the IRIS database and a list of chemical substances currently included in the database. The substance list contains links to chemical summaries and toxicological profiles for each of the substances included in IRIS. These chemical summaries and toxicological profiles include information regarding both noncarcinogenic effects and carcinogenic effects. Details of the relevant toxicological studies are discussed, and toxicity indices {i.e., RfDs, reference concentrations (RfCs), slope factors (SFs) [carcinogen potency factor (CPFs)], and unit risk factors} are presented. In addition, each IRIS substance file contains bibliographic information and a summary of the revisions or updates that the EPA has made to the file.

Other databases and references include the following:

1. Health Effects Assessment Summary Tables (HEAST)
   - Updated quarterly
   - Includes interim RfD and CPF values
   - Prepared by the EPA's Environmental Criteria and Assessment Office (ECAO)
2. Toxicological profiles prepared by U.S. Agency for Toxic Substances and Disease Registry (ATSDR); available on CD-ROM
3. The International Programme on Chemical Safety (IPCS) Environmental Health Criteria documents published by the World Health Organization (WHO), Geneva, Switzerland
4. The Joint FAO/WHO Meeting on Pesticide Residues (JMPR) published by the Food and Agriculture Organization (FAO), Rome, Italy
5. Numerous database vendors that compile various other sources of data (e.g., National Library of Medicine, Toxnet, Syracuse Research Center)

One of the more comprehensive sources of information is the set of ATSDR toxicologic profiles listed as Item 2 above. U.S. federal law (SARA, 1986) mandated that the ATSDR prepare profiles for the toxic substances which are most commonly found at Superfund sites. Each profile can be as much as 100 pages in length and presents the following information:

1. Assessment of all relevant, peer-reviewed toxicological testing and information.
2. Determination of whether adequate information exists to define levels of exposure that present a significant risk to human health, and if it does not exist, the additional testing needed.
3. Presentation of the known acute, subacute, and chronic health effects at various levels of exposure in both experimental and epidemiological studies.
4. Description of the substance's relevant chemical and pathologic properties, such as pharmacokinetic data and environmental fate information.

Calculation Procedure for Assessment of Noncarcinogenic Risk

The preliminary assessment of noncarcinogenic risk of a hazardous waste site as recommended by the EPA is typically calculated in four major steps[78]:

1. Identify discrete exposure conditions:
   - Exposure route
   - Frequency
   - Duration
   - Administered dose
2. Derive appropriate RfDs for each discrete set of conditions.
3. Evaluate hazard for noncarcinogenic effects as a ratio of exposure dose to the recommended RfD.
4. Aggregate hazard for multiple chemical agents and exposure pathways as a hazard index, where appropriate.

Application of this four-step process is explained in detail in Chapter 14. The ratio referred to in the third step is utilized to quantify risk from noncarcinogens. As the fourth step indicates, the hazard index for individual chemicals may be summed for chemicals affecting a particular target organ or acting by a common mechanism in order to provide a final measure of noncarcinogenic toxic risk. If the sum of hazard indices is less than one, then the risk of adverse health effects is considered acceptable. This additive method is illustrated in Example 5-5.

EXAMPLE 5-5. THE USE OF RfDS TO EVALUATE NONCARCINOGENIC HAZARD. Because of contamination from hazardous waste originating from a metal plating facility, the water from a nearby community water supply well contained cyanide at a concentration of 30 μg/L, nickel at 120 μg/L, and chromium (III) at 12,400 μg/L. The daily water intake is assumed to be 2 L, and the body weight of an adult human 70 kg. Do these exposures indicate an unacceptable hazard?

Solution. The dose for cyanide is:

$$\frac{0.03 \text{ mg/L} \times 2 \text{ L/day}}{70 \text{ kg}} = 8.57 \times 10^{-4} \text{ mg/(kg} \cdot \text{day)}$$

The remaining calculations are shown in the following table, with RfD values from the IRIS database:

Substance	Concentration C, mg/L	Dose, mg/(kg · day)	RfD,[79] mg/(kg · day)	Hazard ratio, dose/RfD
Cyanide	0.03	8.57E-4	0.02	0.04
Nickel	0.12	3.43E-3	0.02	0.17
Chromium (III)	12.4	0.35	1.0	0.35
Total				0.56

The individual ratios and the total are below 1, and this preliminary evaluation does not indicate an unacceptable noncarcinogenic hazard.

Strictly speaking, additivity assumes similar type of action. It is justified, for example, when dealing with mixed exposure to cadmium, mercury, uranium, and chromium, all of which affect the kidneys. However, adding the hazard indices for separate compounds in the manner demonstrated may overestimate risk if they induce toxic effects by different mechanisms. Thus, for this example, the compounds present in the mixture do not affect a common target organ or act by a common mechanism and should be segregated by critical effect, and separate hazard indices calculated for each toxicological end point. One approach is to sum hazard indices on an organ-specific basis. This should be done only when the mechanisms of the noncarcinogenic toxicants are well known and the effects are virtually mutually exclusive.

5-5 CARCINOGENS*

Cancer is a disease marked by cellular abnormalities in which a normal cell becomes altered and divides irreversibly and uncontrollably. The definition of cancer favored by standard textbooks in pathology is:

*Section 5-5 was prepared by Drs. Bruce Molholt and Robert Nilsson.

> ... an abnormal mass of tissue, the growth of which exceeds and is uncoordinated with that of the normal tissue and persists in the same excessive manner after cessation of the stimuli which evoke the change.[80]

Cancer is characterized by tumors or neoplasms (meaning "new growth"); however, not all tumors are cancerous. Tumors can be divided into **benign** and **malignant** types, where the latter are commonly known as cancer. Solid malignant tumors derived from lymphocytes are lymphomas, from muscle are sarcomas, and those from epithelial cells, carcinomas. Ninety percent of all human cancers are carcinomas, mostly appearing in skin, where they are easily recognized and cured by surgery.

The most important difference between benign and malignant tumors is that a malignant tumor will invade surrounding structures and spread (metastasize) to distant sites; in contrast, a benign tumor will not. Malignant growths also tend to grow more rapidly than benign neoplasms and are composed of highly atypical cells. In a malignant tumor the cells have become altered in such a way that they no longer respond to regulatory signals that coordinate the rate of cell division to prevent invasion of surrounding tissues.

The multistep process that occurs between initial exposures to carcinogens and the frank development of malignancy is termed **carcinogenesis.** In humans, the process as induced by radiation or chemicals takes about 10 to 40 years on average, dependent on the nature of the carcinogenic agents involved and other factors such as heredity.

Within the last decade the multiple steps of carcinogenesis have begun to be understood. Current theory recognizes three stages[81]:

- *Initiation*—the alteration of a normal cell to a cancer cell through mutation.
- *Promotion*—multiplication and selection of initiated cells.
- *Progression*—expansion of the cancer cell line by invasion of local tissue and metastasis to distant organ sites.

These intermediate steps in carcinogenesis are summarized in Figure 5-19.

Initiation of Cancer

What alterations are required to convert a normal cell into a cancer cell? Because a tumor can arise from one single transformed cell, the expression of cancer is carried over from one cell generation to the next. Thus, the first change that must occur is genetic, that is, alteration of the structure and function of the DNA of a cell. Another clue indicating that DNA is the critical target of carcinogens is provided by the observation that patients with diseases like xeroderma pigmentosa or Bloom's syndrome have a predisposition for developing cancer. In these diseases DNA repair is deficient.[82]

In many cases, transformation to a cancer cell has been associated with changes in specific **oncogenes** into a form that has an altered function in comparison with its natural counterpart (often called a **protooncogene**).[83] One normal function of oncogenes appears to be growth control during development of the human embryo and fetus when proliferation of tissues must proceed at breakneck pace to make a fully formed human within just nine months of gestation. An oncogene can regulate

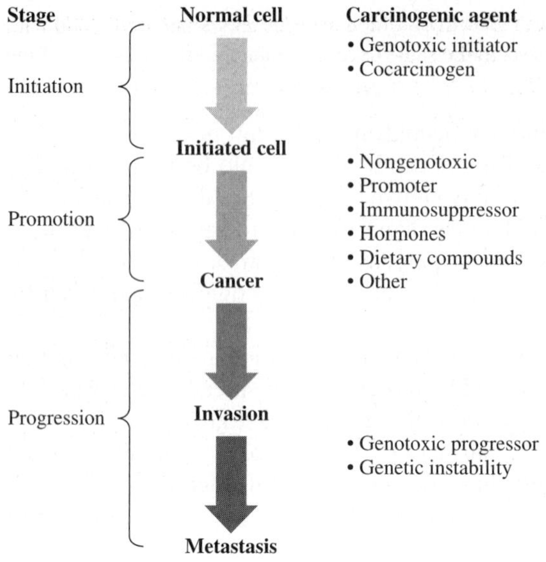

FIGURE 5-19
Three stages in carcinogenesis.

growth by controlling growth factors directly or by suppressing programmed cell death **(apoptosis).**

On birth, many oncogenes are repressed (turned off) and normally remain repressed permanently or are reactivated only transiently. However, mutation or other genetic transformation can fully activate an oncogene. There are evidently several ways in which an oncogene may be activated. As in human Burkitt's lymphoma, the change of position of a protooncogene by translocation (swapping of "arms" of chromosomes) to an active genetic region, in this case the most active gene of a lymphocyte that synthesizes immunoglobulin, may result in a greatly enhanced expression of the gene. Successive duplications (amplification) of a protooncogene may, likewise, give an enhanced expression. Another way of oncogene activation involves the change of individual bases in the DNA of a protooncogene (point mutation) that will result in the synthesis of a protein with a changed structure that is always active in accelerating cell growth and reproduction rather than controllable.

If an oncogene is activated, the affected cell of the child or adult individual is completely capable of autonomous growth and replication because it makes its own "hormone" and has a receptor for it. In such cases the mutated cells contain new antigens on their surface, which stimulate the production of antibodies that may destroy the mutated cells. However, if these cells are not recognized as foreign by the immune system, and continue to replicate abnormally, they will become the stem cell population for that malignant clone called cancer.

It is established, however, that a single mutated oncogene cannot cause cancer. At least two changes in the same cell are necessary (i.e., cancer can result only when the second copy undergoes mutation). A mutated oncogene is rarely inherited. If inherited, the person has a genetic predisposition to cancer because instead of two changes within the same cell, only an additional single change is necessary to initiate cancer. The resulting tumor would normally develop during childhood because much less time is required for a single oncogene change. However, most cancers develop

in adults and the incidence of cancer in children is low. This suggests that mutation of genes after birth causes cancer in most cases.

Only a rare subset of DNA mutations, or chromosomal rearrangements, activates oncogenes. Because there are roughly 70,000 human genes, out of which only a very minor portion are protooncogenes, the majority of DNA mutations have no effect on carcinogenesis. Furthermore, even within protooncogenes, the vast majority of mutations will inactivate rather than activate them to become an oncogene.

After these statements about oncogenes, it should be pointed out that in many aspects the role of oncogenes in the process of carcinogenesis remains obscure, largely because the mechanisms that regulate normal cell division and tissue differentiation are largely unknown. Further, for many sites, known activated oncogenes have been found only in a minor fraction of the tumors; thus, there are likely other routes to cancer, including as yet unidentified oncogenes. Regardless of what role oncogenes may play in carcinogenesis, the initiation of cancer is clearly genetic. Those agents that cause mutation of genes are referred to as **genotoxic.** Such agents consist of many carcinogenic chemicals and ionizing radiations.

Promotion of Cancer

The second stage of carcinogenesis is an increase in the number of initiated cells that may develop into "true" cancer cells.[84] This step is referred to as **promotion,** and it is triggered by promoting carcinogens (promoters). Promoters, in general, are **nongenotoxic.** Although they act in a wide variety of ways, basically they increase the replication rate of initiated cells, or they cause the cells to undergo differentiation that involves the switching on of new genes. Well-known examples in experimental animals are croton oil with respect to skin cancer, and phenobarbital for promotion of liver tumors. In humans, high-fat diets are considered to promote breast cancer in women and colon cancer in both sexes. A promoter is not usually the same carcinogen that initiated carcinogenesis.

Cells that have undergone two or more genetic changes necessary for initiation and that have been promoted still do not normally progress to frank malignancy. Because these transformed cells display foreign ("nonself") antigens on their outer surfaces, they may be recognized and rejected by the same components of the immune system (T cells) that reject invading parasites. However, the immune system can be inhibited in a number of ways (e.g., by the AIDS virus). Some toxic chemicals may also act to suppress the body's ability to recognize and reject promoted cells. Such chemicals are termed immunosuppressants, some of which are used clinically in organ transplantations.

Progression of Cancer

Following promotion, the final two steps necessary for malignancy consist of **invasion** of adjacent tissues and **metastasis** (i.e., movement) to distant organ sites. Invasion and metastasis are collectively known as oncogenic **progression.**[85]

For invasion to occur, localized clumps of cancer cells require enzymes not normally available to dissolve the sheath that surrounds the organ of origin. For metastasis, the localized tumor must next invade the circulatory system for move-

ment to another organ. At the distant organ site, these far flung cells must be able to grow in a foreign environment (i.e., without the support of complementary tissues from the organ of origin). These are not normal capabilities of cells and can only come about by further genetic change. Hence, the development of invasive and metastatic cancer requires additional mutations beyond that necessary to initiate carcinogenesis.

Molecular genetic examination of tumors has found progressive cancers to contain many more mutations than the two required for initiation. In addition, fully malignant cells often contain some number of chromosomes other than 46. The promotion and progression of a clone of initiated cells are accompanied by a cascade of genetic alterations, pointing to a genetic instability, the nature and significance of which remain unclear but involve loss of DNA repair and chromosome maintenance functions.[86] Nevertheless, further exposure to genotoxic agents seems necessary for the progression of initiated and promoted cells. This requirement constitutes the basis for the rapid decrease of cancer risk in smokers after cessation of smoking.

Role of Carcinogenic Chemicals

Most of our understanding of human carcinogenesis by exposure to carcinogenic chemicals comes from the workplace or from experiments in test animals. Certain types of cancer—like breast cancer in women, prostate cancer in men, and lung cancer in smokers—are common causes of death. To investigate the role of chemicals in the etiology of such cancers is difficult because epidemiological investigations require very large cohorts, unless the effect is very striking. The chances of detecting an effect improve if the chemical agent in question induces a specific type of cancer that is not common in the general population. Some of these specific cancers with known etiology are angiosarcoma of the liver caused by vinyl chloride exposure, oats cell carcinoma of the lung caused by exposure to bis(2-chloromethyl)ether, and cancer of the scrotum of chimney sweepers induced by exposure to polynuclear aromatic hydrocarbons in soot.

It is clear that exposure to carcinogenic chemicals can result in cancer via different mechanisms. Genotoxic carcinogens may prompt the initiating and progressing mutations necessary for the full-fledged development of human cancer, while nongenotoxic carcinogens speed up carcinogenesis through promotion or immunosuppression.

This distinction serves to divide carcinogenic chemicals into **genotoxic** and **nongenotoxic (epigenetic)** agents.[87] Since the implications with respect to hazard and risk assessment are profound, the distinction between the two groups serves a useful purpose. However, whereas government agencies in Germany, the Netherlands, and the United Kingdom have accepted the concept,[88] the EPA with one exception[89] has been reluctant to make use of this distinction. More exceptions will be allowed by the EPA if their 1996 draft revised cancer risk guidelines are finalized.

Genotoxic carcinogens

With respect to genotoxic carcinogens, toxicologists have found that chemical substances that are known to be active in carcinogenic initiation and progression also cause mutation in bacteria and other short-term test organisms. Because the

structure of DNA is the same for all organisms, these test organisms reveal some theoretical information about how specific chemicals prompt alterations of genetic material in humans.

Certain alkylating agents, like ethylene oxide and methyl methane sulfonate, react directly with DNA and can cause initiation. However, most initiating carcinogens are not active as such but require metabolic activation.[90] One example is benzo(a)pyrene, one of the major carcinogens of tobacco smoke and coke oven emissions. Initiation of carcinogenesis by benzo(a)pyrene first requires its metabolic activation to benzo(a)pyrene diol epoxide (Figure 5-20). The diol epoxide forms a covalently bonded adduct with guanine in DNA,[91] which appears to swivel the base about the glycosidic bond such that the guanine now appears as thymine during DNA replication or repair. In this manner, benzo(a)pyrene causes a so-called point mutation in DNA (Figure 5-21). It is of interest that the first human oncogene that was isolated and sequenced from the lethal bladder cancer of a heavy smoker contained a guanine-thymine transversion at the 35th base of the H-ras oncogene as its sole mutation.

A point mutation can include base-pair substitution or frameshift mutation. A base-pair substitution involves the replacement of one base pair by another. A frameshift mutation is the case of adding or deleting one or more base pairs of DNA. DNA altered in such a fashion mostly becomes dysfunctional. However, it may continue to function but in a grossly different manner than normal. Some mutagenic agents cause breakage of the backbone of the DNA molecule—events that may show up as chromosome breaks, translocations, or deletions. Also, certain genotoxic agents interfere with the replication of chromosomes and cause an uneven number to be distributed between the daughter cells.

Nongenotoxic carcinogens

Unlike carcinogenic initiation and progression, which are caused by genotoxic chemicals, carcinogenic promotion is caused by chemicals that lack genotoxic activity. Nongenotoxic (epigenetic) carcinogens act in a number of ways, including induction of cell lethality in regenerative tissues, stimulation of mitosis, inhibition of apoptosis, inhibition of gap junction contact with normal cells, and alteration of DNA methylation. As shown in Table 5-6, many well-known carcinogens such as

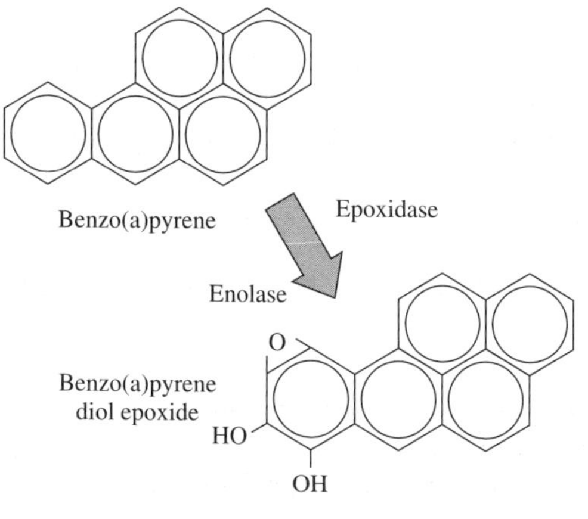

FIGURE 5-20
Metabolic activation of benzo(a)pyrene.

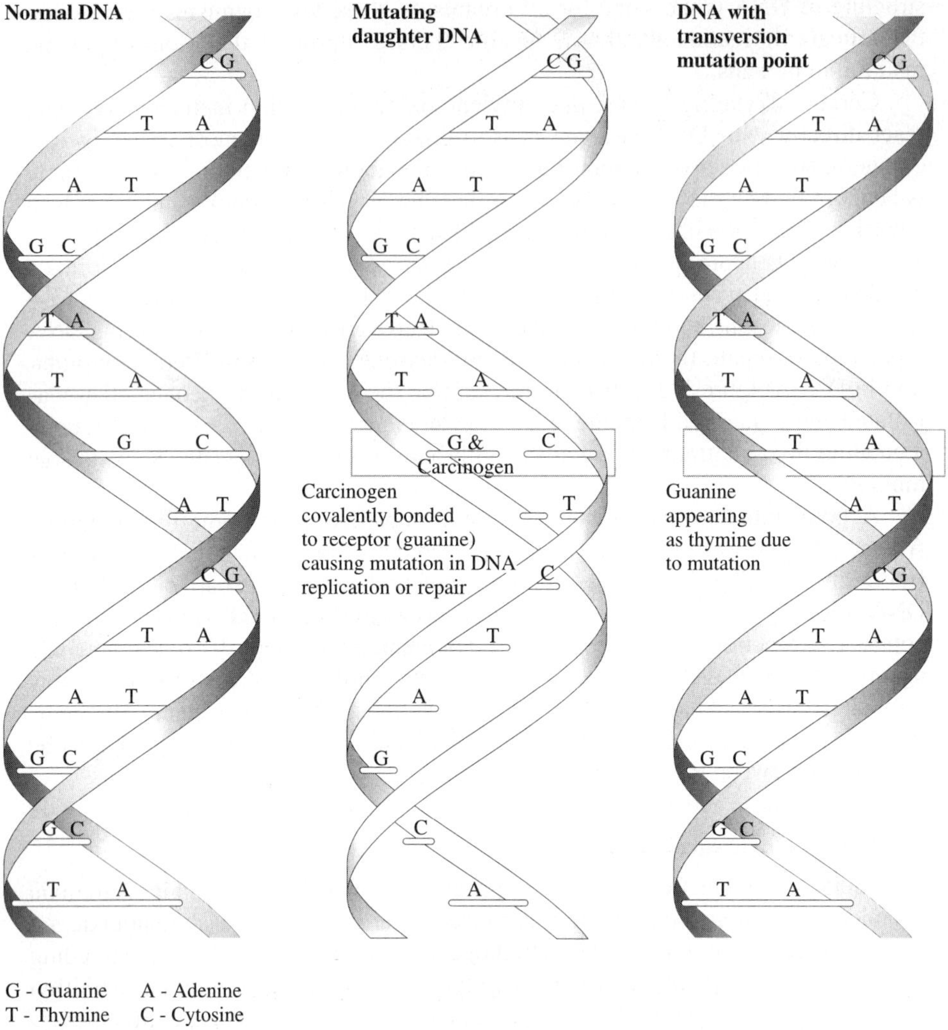

FIGURE 5-21
Example of a point mutation through base pair substitution.

chloroform, DDT, TCE, PCBs, and dioxins are nongenotoxic promoters. As noted above, this is important in that dose-response relationships for nongenotoxic carcinogenic promoters appear to contain threshold doses below which promotion fails, whereas genotoxic initiators may continue to mutate DNA at minute doses, albeit in direct proportion to dose.

Complete carcinogens

Some carcinogens can interact at more than one step in the carcinogenic process. For example, benzo(a)pyrene is a polycyclic aromatic hydrocarbon that can act both as an initiator and as a promoter of carcinogenesis. Carcinogens that both initiate and promote carcinogenesis are known as **complete carcinogens.** Complete

TABLE 5-6
Examples of genotoxic and nongenotoxic carcinogens

Directly genotoxic initiators:	*Nongenotoxic promoters:*
Bis(chloromethyl)ether	Alachlor
Chromium (Cr^{+6}, inhalation)	Aldrin
Di(chloroethyl)methylamine (and other nitrogen mustards)	Ammonium perchlorate
Diepoxybutane	Bis(2-ethylhexyl)phthalate
Ethylene oxide	Carbon tetrachloride
Formaldehyde (inhalation)	Chlordane
Iodine-131	Chloroform
N-nitrosodimethylamine (and other nitrosoamines)	DDT
Plutonium-239	1, 1-Dichloroethylene
Propylene oxide	Dichloromethane (methylene chloride)
Radium-226	Dieldrin
Radon-222	Diethylstilbestrol
	1, 4-Dioxane
	Heptachlor
	Hexachloroethane
Initiators with genotoxic metabolites:	Isophorone
Aflatoxin	Kepone
4-Aminobiphenyl	Lindane
Benzo(a)pyrene	Mirex
Benzidine	Nitrilotriacetic acid
1, 3-Butadiene	Pentachlorophenol
Dibenz(a,h)anthracene	Perchloroethylene (tetrachloroethylene)
Epichlorohydrin	Polychlorinated biphenyls
Ethylene dichloride	Saccharin
Hydrazine	2, 3, 7, 8-TCDD (and other polychlorinated dioxins and dibenzofurans)
2-Naphthylamine	Trichloroethylene
Propylene dichloride	Toxaphene
Vinyl chloride	

carcinogens exhibit wide variations in chemical structure. However, it has been possible to roughly classify most of these agents with respect to functional group, or other structural characteristics. In general, complete carcinogens exert their promoting effects through irritation or necrosis of target tissues. For their initiation activity, complete carcinogens may directly interact with DNA or have metabolites that are mutagenic. Examples of organic complete carcinogens that directly or indirectly initiate are shown as the first two groups of chemicals in Table 5-7.

Other Carcinogens

The mechanism of action of inorganic carcinogens is enigmatic. For arsenic[92] and possibly also for other metals, interference with DNA repair may be involved. Several polychlorinated dibenzo(p)dioxins have been found to induce tumors in rodents at low doses. However, they seem to lack genotoxic action, possibly acting as promoters and by immunosuppression. Because they are both powerful promoters and highly immunosuppressive, at significant doses dioxins appear to be potent carcinogens, especially when chlorinated at the 2, 3, 7, and 8 positions (see Figure 5-22).

TABLE 5-7
Further examples of organic and inorganic complete carcinogens

Chemical type	Examples
Directly alkylating organics	
Epoxides	Ethylene oxide Butadiene epoxide
Aldehydes	Formaldehyde
Ethylenimines	Ethylenimine Triethylene melamine
Alkyl halogenides	Methyl bromide
Halogenated ethers	Bis(chloromethyl)ether
Mustards	Bis(2-chloroethyl)sulfide Bis(2-chloroethyl)amine Cytoxan (endoxan)
Nitrosamides, nitrosoureas	N-methylnitrosourethane N-methylnitrosourea
Organics requiring metabolic activation	
Aromatic amines	2-Naphthylamine o-Toluidine 3,3′-Dichlorobenzidine
Nitrosamines	Dimethylnitrosamine Diethylnitrosamine Nitrosopiperidine
Azo dyes	*o*-Aminoazotoluene 4-Dimethylaminoazo-benzene Direct blue 6
Hydrazines	1,2-Dimethylhydrazine Cycasin (natural product)
Polycyclic aromatic hydrocarbons	Benzo(a)pyrene Methylcholanthrene Dibenz(a,h)acridine Dibenz(a,h)anthracene
Alkaloids (natural products)	Aflatoxins Pyrrolizidine alkaloids
Inorganic inhalation carcinogens	
Metals	Beryllium Nickel Chromium(VI) Cadmium
Metalloids	Arsenic
Solid-state carcinogens	Asbestos

Cocarcinogens are agents that enhance carcinogenesis when administered simultaneously with a carcinogen. Although sometimes acting in the same way as promoters, they may also function by affecting the metabolism of a carcinogen in

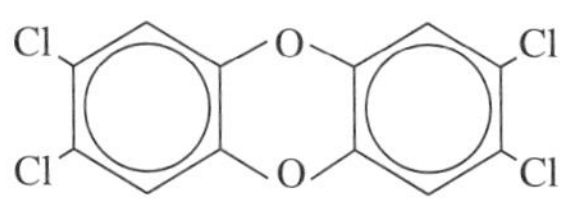

FIGURE 5-22
2,3,7,8-tetrachlorodibenzo(p)dioxin.

such a way as to increase the level of the active intermediate. Phenol is an example of a cocarcinogen that is not a promoter.

Affinity

An affinity between a carcinogen and a receptor is very important, as discussed in Section 5-2. An example of the importance of affinity is the case of polychlorinated biphenyls (PCBs) which were produced commercially by chlorinating benzene-containing precursors. During commercial production the rate and extent of precursor molecule chlorination are virtually impossible to control. Hence, these organochlorines were produced and now exist in the environment as complex mixtures of hundreds of individual **congeners.** Although each congener of these complex organochlorines has a similar molecular formula, each has its own toxicity properties when tested in experimental animals.

At one time a uniform carcinogenic potency was assigned for all PCBs, but over the past decade it has become clear that risk assessment for entire categories of these complex organochlorines is meaningless. A constructive approach was established first for dioxins (polychlorinated dibenzo(p)dioxins or PCDDs) where toxic equivalency factors (TEFs) were established for classes of polychlorination. Later, in line with international recommendations concerning dioxin toxicity, the EPA embraced a more useful approach to TEFs, which is dependent on whether chlorination sites include the 2, 3, 7, and 8 positions, inclusively, of PCDD molecules. According to the latest information available, a similar relationship holds among PCB congeners such that pentachlorobiphenyls (PeCBs) are the most toxic class of congeners followed by hexachlorobiphenyls (HxCBs) and tetrachlorobiphenyls (TeCBs). This information suggests the relative carcinogenic potencies shown in Table 5-8.

Classes of Carcinogens

How do we know which chemical substances are carcinogenic and just how potent they are? The preferred basis for making this decision is the case where sufficient hu-

TABLE 5-8
Relative carcinogenic potencies of PCB classes

PCB class	Number of chlorines	Relative potency
PePCBs	5	1.0
HxPCBs	6	0.5
TePCBs	4	0.1
>HxPCBs	>6	<0.1
<TePCBs	<4	<0.1

man epidemiologic data exist, mainly from studies of workers, to make a clear causal connection between the exposure to specific substances and subsequent induction of human cancers. Although human epidemiologic studies are severely limited, as discussed in Section 5-3, for some carcinogens, the causal relationship is clearly supported by strong statistical correlations. For example, fully half of the known cases of angiosarcoma of the liver have occurred in vinyl chloride workers. Similarly, fully 50 percent of those workers exposed 5 years or more to 2-naphthylamine dyes or who have worked in uranium mines have succumbed to bladder cancer or lung cancer, respectively.

The number of clear human epidemiologic studies is small. A total of only about 50 compounds (e.g., benzene, vinyl chloride) and complex exposures (e.g., aluminum production, tobacco smoke) have sufficient data available to permit their classification as human carcinogens.[93] The most potent human carcinogens known, the aflatoxins, are of natural origin. Their presence in food products through infestation by toxin-producing fungi constitute a serious problem in several tropical and subtropical countries.

The EPA utilizes the following description for Group A—human carcinogens: Compounds for which "there is sufficient evidence from epidemiologic studies to support a causal association between exposure to the agent and cancer." This designation is based on the weight of evidence. The designation as a human carcinogen represents a consensus opinion with an expert group rather than a scientific fact. The weight of evidence system is shown in Table 5-9. Of the approximately 50 human carcinogens presently known, 12 are contaminants found in the environment (see Table 5-10). The remainder are primarily pharmaceutical products (e.g., azathioprine, an immunosuppressant, and diethylstilbestrol, a synthetic estrogen).

Because of limited epidemiological data, most of our understanding about the potential for a toxic substance to cause cancer is based on animal testing. Substances not having an apparent epidemiological association with human cancer but which have induced cancer in test animals are referred to as **potential** human carcinogens. Several hundred substances for which there is inadequate human epidemiological data have been assigned as potential human carcinogens on the basis of data from laboratory animals, usually rats and mice. Many of these substances are commonly found in hazardous waste sites. In fact, of the more than 800 hazardous substances regulated under CERCLA, the EPA has identified 160 substances as "potential carcinogens or as substances having carcinogenic potential."[94]

TABLE 5-9
EPA weight-of-evidence classification system for carcinogenicity

Group	Description
A	Human carcinogen
B1	Probable human carcinogen, limited human data available
B2	Probable human carcinogen, sufficient evidence in animals and inadequate or no evidence in humans
C	Possible human carcinogen
D	Not classifiable as to human carcinogenicity
E	Evidence of noncarcinogenicity for humans

TABLE 5-10
Human environmental carcinogens

Carcinogen	Cancer	Genotoxin
Arsenic	Lung, skin	Yes
Asbestos	Lung, mesothelioma	No
Benzene	Leukemia	Rarely*
Benzidine	Bladder	Yes
Bis-chloromethyl-ether	Lung	Yes
Chromium, hexavalent	Lung	Yes
2-Naphthylamine	Bladder	Yes
Nickel	Lung	Yes
Plutonium-239	Lung	Yes
Radium-226	Lung, osteosarcoma	Yes
Radon-222	Lung	Yes
Vinyl chloride	Liver, angiosarcoma	Yes

*Only for cells derived from bone marrow.

Testing for Carcinogenicity

Testing of chemical carcinogens primarily involves animal experiments, particularly lifetime studies in rodents. In these tests, doses are generally administered to animals at levels at or below the maximum dose that neither reduces lifespan (other than via tumor induction) nor results in clinical symptoms of toxic effects [frequently referred to as maximum tolerated dose (MTD)]. Testing at these maximal doses is done for the purpose of increasing the likelihood of inducing cancer.

Such studies are very resource-demanding and can only be carried out for a limited number of economically important chemical products. Although such information is indispensable, many pitfalls exist with respect to its interpretation. The limitations are basically of three kinds: (1) statistical limitations posed by the limited size of the test groups used, (2) difficulties in interpreting the relevance of certain types of tumors for humans, and (3) uncertainties associated with extrapolation from high to low doses.

Limited size of test groups

The first issue has already been discussed under the section dealing with animal data in general (Table 5-4). In a human population, a 1 percent increase of a serious disease is of major concern. To ensure that an excess tumor incidence of 1 percent over a background spontaneous rate of 10 percent can be detected, theoretically, about 20,000 animals must be used. Actually, such a "mega mouse experiment" has indeed been performed for a cost of $15 million in 1980 using the potent experimental carcinogen 2-acetylaminofluorene.[95] The purpose was mainly to study the dose-response relationship in the low-dose region. In contrast to a 1 percent incidence, the common regulatory goal in the United States is 10^{-6}, which would necessitate testing for a 0.0001 percent excess cancer rate. But this is clearly impractical.

Relevance of tumors

Relative to the significance of certain types of tumors found in animals, there has been considerable controversy with respect to liver tumors found in the mouse,

kidney tumors in male rats, bladder tumors induced by nongenotoxic carcinogens in the bladders of rodents, and certain other experimental neoplasms. Primary liver cancer is rare in humans, whereas in the particular mouse strain mostly used in the United States for such tests (the B6C3F1 mouse) the spontaneous incidence is in the range 18 to 47 percent in males and 3 to 8 percent in females.[96] This incidence is significantly increased by a number of unspecific factors like hormonal treatment or by dietary changes. Thus, when the test animals are given chemicals near the maximum tolerated dose (MTD), which usually is several orders of magnitude above expected exposures for humans, there is a concern that it may trigger unphysiological processes (e.g., overloading of detoxification mechanisms) that do not occur at lower doses.[97,98]

Many if not most of the substances classified as potential carcinogens received such a classification solely because they induced liver tumors in these select mouse and rat strains. In contrast to the United States, three European regulatory agencies have discounted the finding of liver tumors at high doses in rodents (e.g., induced by the insecticide lindane) on grounds of a nongenotoxic mechanism of action which has no relevance to humans at realistic exposures.[99]

Bladder tumors in rodents illustrate another controversial situation. Although it is true that the background spontaneous incidence of bladder tumors in rats is low, chronic exposure to a number of agents that cause irritation or trauma to the bladder epithelium may stimulate cell division with the subsequent formation of tumors. Examples of such promoting agents are foreign solid bodies, alteration of urinary pH, and infectious agents. Thus, many compounds inducing such tumors in rodents, like the food additives cyclamate, saccharin, o-phenylphenol as well as butylated hydroxyanisole (BHA), are now recognized as promoters.[100,101]

The difference of opinion about interpretation of cancer studies has resulted in the unsatisfactory situation that a compound may be considered a carcinogen in one country but not in another. Such discrepancies of carcinogenic potential may result in unnecessary compliance with criteria or standards based on an overprotection of human health. At least with respect to judgment of the quality of an animal study as such, the **WHO International Agency for Research on Cancer (IARC)** is able to provide guidance that is universally accepted. These evaluations are published regularly in the series "IARC Monographs on the Evaluation of the Carcinogenic Risks to Humans."

Notwithstanding the issues discussed above, for many compounds rats and mice do seem to represent adequate models for humans. This is demonstrated by comparative dosimetry in animals and humans for those carcinogens that have adequate human data. In these cases, concordant animal data are remarkably similar (within an order of magnitude) in simulating the observed human carcinogenic response.

Other tests

As stated earlier, toxicologists have found that chemical substances that are known to be active in carcinogenic initiation also cause mutation in bacteria and other short-term test organisms. Because the structure of DNA is the same for all organisms, these tests provide a basis for prediction of carcinogenicity. However, the correlation between mutagenicity and carcinogenicity has been found to be relatively good only for initiators or complete carcinogens but not for compounds with a

promoter-like action. The following end points have been used for prediction of potential carcinogenic effects:

- Point mutations (base pair substitutions, frame shift mutations, small additions or deletions)
- Disturbances of DNA repair (unscheduled DNA synthesis)
- Clastogenic effects (chromosomal breaks resulting in fragments, losses, rearrangements of chromosomes)

The numerous test systems that now are available range from bacterial cells (the Ames test), yeasts, cultured mammalian cells, to whole animals (fruit flies, mice). Usually, results from representative test systems for all three categories are preferred before making an evaluation. Short-term *in vitro* tests are easy to perform and have resulted in the generation of a tremendous amount of data of varying quality. Pundits have suggested that the definition of a "nonmutagen" is a chemical that has not yet been tested in a sufficient number of test systems. When using this kind of data, it is important not to rely solely on the results from a single test, but to assess the weight of evidence obtained from several sources. Finally, a positive result from mutagenicity testing only gives an indication of a potential for carcinogenicity and should be used only as supporting information.[102]

Dose-Response Relationships for Carcinogens

The previous pages have explained how the assessment of human cancer risk based on animal tests requires two major considerations:

1. Is the tested chemical a carcinogen?
2. How are data from experimental animals applied to humans?

These interpretations are difficult, but far less difficult than deciding the potency of a carcinogen. This is an important concept because carcinogens vary greatly in regard to their potency. To predict the likely outcome from human exposure to carcinogenic chemicals requires the development of dose-response relationships. To develop these relationships, toxicologists ordinarily depend on animal data. In Figure 5-23, a typical dose-response curve in the high dose region is given for a hypothetical complete carcinogen.

The approach to developing dose-response relationships for carcinogens differs somewhat from the approach for noncarcinogens. For ease of comparisons, doses (the abscissa) are almost always plotted as the daily administered dose per unit of body weight [mg/(kg · day)]. Dose is plotted against response (the ordinate), which is the incidence of the total number of tested animals with tumors (dimensionless). The curve in Figure 5-23 represents the best fit of mean values. It should be noted that the ordinate is reported as **excess incidence of cancer,** or the incidence in the exposed animals minus the normal spontaneous (background) incidence in the control population.

When assessing lifetime cancer risk to humans, it is widely accepted that carcinogenesis works in a manner such that it is possible, however remote, that exposure to a single molecule of a genotoxic carcinogen could result in one of the two mutations

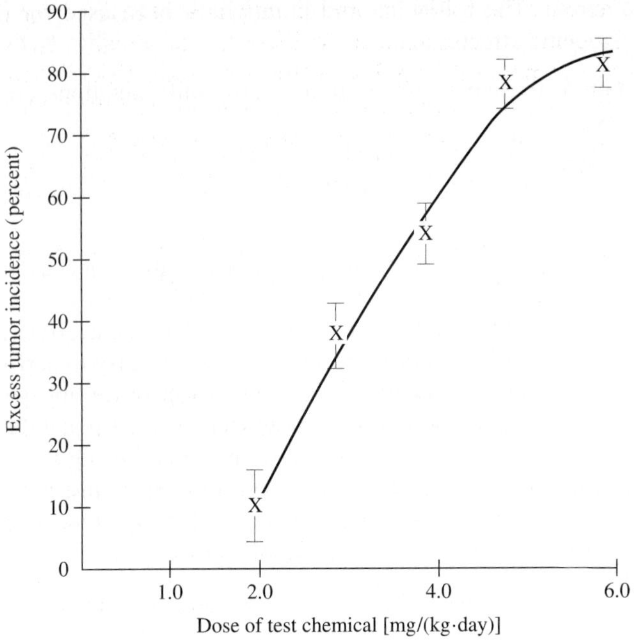

FIGURE 5-23
Hypothetical dose-response curve for a typical complete carcinogen.

necessary to initiate cancer (i.e., genotoxins do not exhibit thresholds). In theory, therefore, the dose-response curve is asymptotic to "zero" incidence. Because there is no threshold, there is no safe level, only "acceptable levels." The public has expressed concern with any level, and regulatory agencies tend to establish a regulatory goal of 1 in 1 million (1×10^{-6} or 0.0001 percent) excess lifetime cancer risk.

As stated before, it is not practical to test in the range of such a low incidence. In fact, laboratories conduct animal tests at high doses to increase the likelihood of inducing cancer in a portion of the test population. Thus, to apply the results of animal tests to human exposures, the data from the high doses used in the tests must be extrapolated to the low doses of public concern.

EXAMPLE 5-6. COMPARISON WITH 10^{-6} GOAL. For the dose-response relationship shown in Figure 5-23, what is the minimum dose tested? What is the mean incidence induced by this dose? How does this compare with the 10^{-6} goal?

Solution. The minimum dose tested is 2.0 mg/(kg · day), yielding a mean excess tumor incidence of about 10 percent. This incidence is 100,000 times (0.10/0.000001) greater than the 10^{-6} goal.

The calculation of carcinogenic risk involves the use of a **slope factor** (SF). An SF is basically the slope of the dose-response curve at very low exposures and was formerly referred to as the **carcinogen potency factor** (CPF). An SF is generally the 95 percent upper confidence limit of the slope of the dose-response curve at very low exposures. New approaches for defining the SF may be allowed by the EPA if their 1996 draft revised cancer risk guidelines are finalized. The dimensions of a

slope factor are expressed as the inverse of daily dose $[\text{mg}/(\text{kg} \cdot \text{day})]^{-1}$. Although the units of SF could be explained as "risk/dose," mathematically the stated units are necessary to compute the unitless estimated cancer risk. With a slope factor derived, the calculation of carcinogenic risk is straightforward. Quantification of carcinogenic risk of an exposure simply requires converting the dose to the appropriate terms $[\text{mg}/(\text{kg} \cdot \text{day})]$ and multiplying it by the slope factor as demonstrated in Example 5-7.

EXAMPLE 5-7. USE OF SLOPE FACTORS. What are the maximal number of excess lifetime cancer cases expected for a population of 100,000 adults with a daily intake of 0.14 mg of benzene?

Solution

1. The slope factor for benzene is 0.029 $[\text{mg}/(\text{kg} \cdot \text{day})]^{-1}$.
2. Assuming an adult weight of 70 kg, the lifetime cancer risk is calculated as:

$$\begin{aligned}
\text{Individual cancer risk} &= (0.14\ \text{mg/day})\,\frac{1}{70\ \text{kg}}\,\frac{0.029}{\text{mg}/(\text{kg}\cdot\text{day})} \\
&= 0.000052 \text{ or } 0.0052\% \\
\text{Maximal cases} &= \text{risk} \times \text{exposed population} \\
&= 0.000052 \times 100{,}000 \\
&= 5 \text{ excess lifetime cancer cases}
\end{aligned}$$

Low-Dose Extrapolation Models

To make the necessary extrapolations to low doses, toxicologists use various mathematical models. Current models are usually divided into two types: **tolerance distribution models** and **mechanistic models**.[103]

Tolerance distribution models

Tolerance distribution models assume the existence of a threshold as described in Section 5-4. Some of these models (e.g., the Mantel-Bryan model)[104] can be used for promoters, nongenotoxic carcinogens, and for certain tumor incidence data that exhibit threshold-like appearance for certain genotoxic carcinogens (e.g., the bladder tumor data for 2-acetylaminofluorene in the mega mouse experiment cited above).[105]

It should be kept in mind that most of the modeling carried out for low-dose extrapolation of promotive effects does not consider the following additive theory: The population is already subjected to a high level of promotive stress, to which an additional promotive insult may result in a dose-response relationship that lacks a dose threshold. However, such additivity presupposes a common mechanism of action.

Mechanistic models

In mechanistic models, it is assumed that a certain number of reactions, events, "hits" (concept derived from radiation biology), or transition stages, related to a critical target in the cell (DNA), are necessary to transform a normal cell to a cancer

cell. These models are important because there seems to be a general consensus that tolerance distribution models (e.g., Mantel-Bryan) should not be used. The exception is when available data are sufficiently accurate to exclude models involving a linear component in the low-dose region, or when a genotoxic mechanism might be involved.

The **multistage model,** first proposed by Armitage and Doll in 1954,[106] is an extension of the one-hit model developed for ionizing radiations. One generalized version of this model where at least one of the stages is assumed to be dose-related takes the form[107]:

$$P(D) = 1 - \exp\left[-\left(q_0 + q_1 D + q_2 D^2 + \cdots + q_k D^k\right)\right],\ k \geq 1 \qquad (5\text{-}3)$$

where $P(D) =$ the probability of cancer at dose D

$q_k =$ coefficients to best fit to the data

$D^k =$ the applied dose raised to the kth power

$k =$ the number of stages (usually set arbitrarily at the number of dose levels minus one)

The most likely estimate at very low doses becomes increasingly unstable with a small change in the response at experimental doses. Therefore, a further development in modeling was the replacement of the linear term in the polynomial function by its 95 percent confidence limit to achieve more stable estimates of risk above background than are obtained for the most likely estimates. This so-called **linearized multistage model** is currently used routinely by the EPA and has, therefore, become the most widely used model for estimation of cancer risk.[108,109] At low doses the function becomes essentially linear. The multistage model is very flexible in fitting data sets because it is a polynomial function of dose. The so-called carcinogen potency factors provided by the EPA have been based on this model and are calculated from animal data by using commercially available computer programs.[110]

The multistage model is favored by the EPA because it generally gives conservative risk estimates for low exposures, but the model has several weaknesses.[111,112] As already pointed out, the model is less well suited for promoters such as PCBs, greatly overestimating risk by not considering a threshold dose. For formaldehyde a much better fit to the experimental data was found by using a five-stage model rather than the EPA version with the conventional restrictions.[113] Equation 5-3 given earlier predicts that the probability of tumor induction will eventually approach unity when the dose is sufficiently high. Thus, a good fit will not be obtained for data sets where the dose-response function rises steeply and then reaches a plateau (strongly concave, Michaelis-Menten kinetics), e.g., that for vinyl chloride. To cope with such a situation, a regression procedure is used for computing the maximum likelihood polynomial function from a data set. Whenever the model does not fit the data sufficiently well, data at the highest dose are deleted, and the model is refitted to the rest of the data. This is continued until an acceptable fit is obtained. A more recent development is the **Moolgavkar-Knudson model,** which is based on cellular dynamics and transformations and incorporates time-to-tumor data.[114]

Extrapolation error

It is very possible that the extrapolation of data from the 10 to 90 percent carcinogenesis range of test animals to 0.0001 percent carcinogenesis yields results

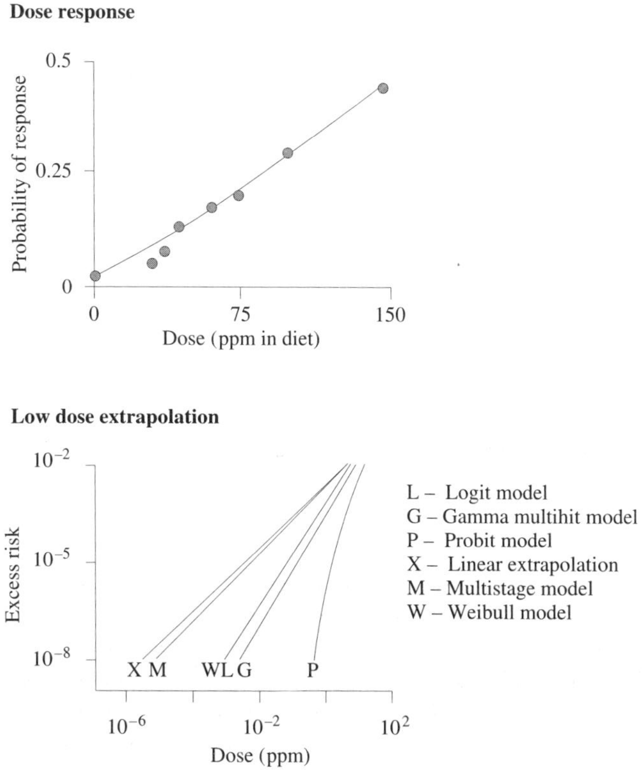

FIGURE 5-24
Extrapolation of dose-response relationship for liver tumors induced by 2-acetylaminofluorene in the low-dose range.[116]

that are conservative by several orders of magnitude for carcinogenic initiators and are incorrect for carcinogenic promoters and immunosuppressors. The range of error in experimental dose-response extrapolation for low carcinogenic frequencies in potential exposed human populations is indicated in Figure 5-24, which shows the results from an experiment with 2-acetylaminofluorene.

The EPA is well aware of the problems associated with overly conservative risk estimates, and has repeatedly stressed[115] that the unit cancer risk estimate only provides a plausible *upper* limit for a risk that can very well be much lower. The problem is that, in reality, official EPA unit risk estimates are widely used, more or less, as absolute standards. Draft revised cancer risk guidelines modified by the EPA in 1998 allow use of alternative models.

The adherence to a single model to cover all situations without taking due consideration of all relevant biological data (epigenetic/genotoxic action, promotive mechanisms, etc.), and sometimes even ignoring available data points in the high-dose part of the dose-response curve in trying to obtain a fit to the model used (e.g., the successive exclusion of high-dose data points in the application of the linearized multistage model), may have given mathematical modeling an undeserved bad reputation among biologists. In particular, the EPA's use of the linearized multistage model for compounds like chloroform, chlorinated dioxins, and ingested arsenic has

TABLE 5-11
Carcinogenic potencies depend upon exposure route

Carcinogen	Inhalation SF	Ingestion SF	Ingestion/inhalation ratio
Vinyl chloride	0.295	1.9	6.4
Arsenic	15	1.5	0.1

been criticized. Although the data sets derived from conventional carcinogenicity studies in animals are limited, the intelligent use of mathematical procedures can—at least in some cases—result in the rejection of a certain type of model for that particular situation, and thereby improve the confidence in low-level risk extrapolation.

Toxicological databases for carcinogens

Most of the toxicological databases discussed in Section 5-4 also apply to carcinogens and some also provide slope factors (SFs). IRIS is a prime example. Section 5-1 discussed the effect that the route of exposure can have on a substance's toxicity. This also holds true for some carcinogens. Thus, the selection of an SF from a database must consider the exposure route. For example, vinyl chloride is 6 times more potent by ingestion than inhalation, whereas with arsenic it is the other way around, the ingestion risk being only 10 percent of that by inhalation (see Table 5-11).

For reasons that are not apparent, most metallic lung carcinogens, such as cadmium, hexavalent chromium, and nickel, do not appear to be carcinogenic by any other route of administration than inhalation. Thus, care must be exercised when defining exposure and in selecting the appropriate slope factor.

Mixed exposure to carcinogens

The conservative nature of the EPA's linear-dose response model suggests that the government's approach is overprotective of public health. However, the approach is underprotective from another perspective. To explain, the data used by the model stem from testing of animals exposed to a single compound when, in fact, it is rare that a given human population will receive potential exposure to only one carcinogenic substance in the environment. The normal human is exposed to a complex mixture of substances, including carcinogens that are both initiators and promoters or immunosuppressors. Consequently, the EPA approach to considering multiple exposures has been considered underprotective by some scientists.

At present, the simplified risk assessment methodology for dealing with mixed exposure to carcinogens is that overall carcinogenic risk is equivalent to the sum of individual risks of each substance. This is referred to as an additive model, and is illustrated in the following hypothetical example for an individual exposed to three carcinogens:

Carcinogen	Lifetime risk for specified exposure
Benzene	0.0005
Cadmium	0.0008
Vinyl chloride	0.0025
Total	$0.0038 = 3.8 \times 10^{-3}$, excess cancer probability of approximately 4 in 1000

These classic additive models may not be appropriate for multiplicative or antagonistic interactions that may occur either within or as a result of exposure to complex mixtures. The dosimetry of potential human responses to mixtures of different types of carcinogens is more complex, as revealed from epidemiological studies of shipyard workers and uranium miners regarding the impact of cigarette smoking. For example, if the relative risk for lung cancer from tobacco smoking is 10 (ten times the lung cancer risk for nonsmokers), and the relative lung cancer risk from asbestos exposure is also 10, the classic additive model would conclude that the lung cancer risk from exposure to both tobacco smoke and asbestos would be 20 ($10+10$). However, it has been found that the lung cancer risk among asbestos workers who smoked was nearly 100 times that of workers in other industries who did not smoke.[117] This is a prototypic case of carcinogenic synergism in which the ultimate risk of exposure to an initiator and a promoter is the product ($10 \times 10 = 100$) of relative risks rather than the sum ($10 + 10 = 20$) of these relative risks.

Background and Lifestyle Carcinogenesis

Thirty-five percent of us will contract cancer in our lifetimes. Toxic substances in the environment do contribute to the carcinogenic burden on humans. It is widely accepted, however, that genetic predisposition, natural radiation, and lifestyle factors account for far more cancer deaths than toxic chemicals in the environment.

Background carcinogenesis

There is a background rate of carcinogenesis that has very little to do with human behavior, including where we live and what we eat or do. We live within a universe that is rather hostile to DNA, one in which we are constantly bombarded with cosmic radiation that is rather uniform over the Earth's surface, although slightly more pronounced at higher altitudes and over the poles. In addition there are naturally occurring radionuclides, including radon gas, that are associated with the composition of the Earth's crust and provide further sources of radiation exposure. This cosmic ray and radionuclide flux, at approximately 150 mrem per year at sea level in North America, is one background genotoxic influence on human cells that is responsible for tens of thousands of DNA repair events per day per cell—or an astronomical 10^{20} (100 quintillion) DNA repair events per person per year. Fortunately, the DNA repair process replaces the damaged DNA base with the same base. However, the DNA repair process can go awry, leading to a mutation. If this mutation occurs in or near an oncogene in a somatic cell, it can constitute one of the two mutations required for carcinogenic initiation, or it can be one mutation needed for progression.

Although each of us begins life at conception as a single cell, in the adult body there are approximately 100 trillion cells, each at risk of becoming cancerous. Every cancer begins as a clone of cells, all of which are derived from the same originally transformed single cell that has gained the ability to replicate faster than its neighboring cells. Although a very frequent disease of humans at the level of a cell, cancer initiation is an extremely rare event. Because cancers are clonal and one out of three humans, each with 100 trillion cells, develop a malignancy in a lifetime, the frequency of cancer initiation on a cellular basis may be calculated as 1 per 300 trillion cells, or 3.3×10^{-15}.

It is thought that the balance between faithful DNA repair and more mistake-prone DNA repair was a driving force in evolution, yet with minimal damage to our species. Less-than-perfect DNA repair enzymes allow for a low rate of accumulated mutations both in somatic cells (potentially cancer initiating) and germinal cells (leading to changes in the next generation, a minority of which may be more fit for a changing environment). In retrospect, had cancer been a lethal disease prior to puberty, chances are that this mechanism for DNA repair would not have survived in its present form. If so, we less well-adapted survivors of a less dynamic evolutionary lineage would carry a more perfect DNA repair system today. In that case, it would be doubtful if any of us could read these words, let alone reason—a dubious trade-off for a slightly errant DNA repair system.

Lifestyle and occupational carcinogenesis

In addition to persistent background mutagenesis, we encounter a wide variety of chemicals in our daily lives as part of our diet and lifestyle, having nothing to do with hazardous wastes. Many plants contain natural pesticides which are potentially carcinogenic. Some of us favor foods rampant with naturally formed carcinogens: mushrooms (hydrazines), celery (psoralens), peanut butter (aflatoxin), and charcoal-broiled steaks [benzo(a)pyrene]. Such known sources pale, however, in comparison with the widespread habit of wrapping shredded leaves of Mother Nature's produce into small bundles and smoking them. The inhalation of tobacco smoke has been a major pathway for delivery of carcinogens into *Homo sapiens,* and responsible for more lung, bladder, and possibly also pancreatic and breast cancer than all other artificial sources of carcinogenic insult combined.

In addition to background sources of carcinogenic exposure such as cosmic rays, natural radionuclides, and lifestyle-associated sources such as food and cigarettes, many workers are exposed to carcinogens occupationally. Obviously, an awareness of this source of exposure has driven many regulations by the U.S. Occupational Safety and Health Administration (OSHA), such that the workplace is much safer today than ever from the standpoint of potentially carcinogenic insult.

"Safe" levels of carcinogenic chemicals in hazardous wastes

The tools for predicting the potential carcinogenic burden for an exposed human population have become more precise over the past decade; still, there are many uncertainties that severely limit the process. Carcinogenic risk assessment is best at comparing risks and allowing a prioritization of risks rather than quantifying actual risks. As explained in this section, extrapolations from high doses in animals to low doses in humans are fraught with statistical uncertainty and conservative assumptions and are probably incorrect for nongenotoxic carcinogens. Nonetheless, a somewhat inaccurate estimation of risk may be better than no estimation at all, and errors at present are in the direction of human health protection.

In the future, it is expected that improved approaches will be developed such as the use of pharmacokinetic models to account for many of the variables in scaling animal test data to estimate human risk. Nevertheless, the toxicologists will be able to determine only the likelihood of adverse effects; definitive answers will remain elusive.

Given these scientific limitations, what levels of carcinogenic exposure from hazardous wastes (or other sources) should be considered acceptable by society?

At present, the EPA and many state agencies have an upper bound "safe" guideline of 1 in 10,000 (1×10^{-4}) lifetime cancer risk to persons who might be maximally exposed to carcinogens from hazardous wastes. With this upper bound, the EPA has defined acceptable risks for carcinogens as the range of 1 in 1 million to 1 in 10,000 (10^{-6} to 10^{-4}) lifetime cancer risk. The figure "1 in 1 million" (1×10^{-6}) excess lifetime cancer risk is used as a general regulatory goal (see Section 14-1).

5-6 ECOTOXICOLOGY*

As discussed in the preceding sections, the mechanisms by which toxic substances affect mammalian species (viz. humans) has historically been the focus of considerable research. In contrast, the assessment of chemical releases on other components of the biosphere commenced with the study of "fish kills" in the early twentieth century, and extensive research on hazards to nonhumans has recently gained prominence. Rachel Carson's documentation of ecological damage from agricultural pesticide application in *Silent Spring* is widely recognized as a driving force behind the heightened public awareness about "ecotoxicity" that emerged during the 1960s.[118] With this came the recognition of the need to protect and manage a multitude of feral (i.e., not domesticated) species, interacting as communities within complex ecosystems, thus providing an impetus for the science of ecotoxicology. As this book enters its second edition, ecotoxicology is receiving more regulatory attention than human toxicology at many hazardous waste sites. This renewed emphasis, in part, derives from a realization of many environmental chemicals as **endocrine disrupters** discussed in Section 5-2.

Ecotoxicology is a relatively young scientific discipline that extends the principles of toxicology to natural systems to evaluate the potential impact from releases of toxic materials. While the focus of traditional mammalian toxicology is the effects of chemical exposure on individual species, primarily humans, ecotoxicology addresses effects on groups of populations interacting with the physical environment. In other words, a chemical that kills a percentage of individuals in a population is a catastrophe to humans; however, a similar effect in an ecosystem may be of little ecological significance compared to chemicals that are not directly lethal to an individual, but decrease primary productivity or reduce energy transfer which, in turn, affect system function directly and affect system structure indirectly. Therefore, it is important to understand not only the direct effects of a chemical on an organism, but also how the chemical interacts with the organism's environment (e.g., physical alteration of habitat by a chemical spill; loss of a food source; biological or chemical degradation).[119] As examples of the latter, an accidental oil spill can produce a variety of effects including reduction of photosynthesis from decreased light penetration, lethality to aquatic organisms from exposure to lighter aromatic hydrocarbons, and the transfer of heavier hydrocarbons through the food chain from accumulation in tissue.

It is important to note that the principles of exposure, absorption, metabolism, and manifestation used for the study of humans are also germane to ecotoxicology.

*Section 5-6 was prepared by David J. Stout.

Although the specific steps can vary widely (e.g., absorption for nonpredaceous fish occurs predominantly in the gills), the pharmacokinetic fundamentals remain the same. The concepts of structural affinity and chemical activity are likewise important to ecotoxicology. Nevertheless, the application of these principles and fundamentals to diverse populations of feral species interacting as communities in natural systems presents difficult challenges for scientists and engineers involved in decisions regarding the management of hazardous waste. This section will highlight the general techniques used to evaluate toxic effects and how the results of these tests are used in the fields of aquatic and terrestrial toxicology.

Evaluation of Toxic Effects

Ecotoxicologists answer numerous questions regarding constituent toxicity by using a variety of test systems that differ in complexity, predictive capacity, and cost. A number of factors frustrate the applications of these tests and the use of test results.

The origin of laboratory toxicity testing traces to pre–World War II investigations of the cause of fish kills and the use of primitive test methods often referred to as "one fish in a jar." The subsequent pursuit of an optimal test organism, synonymous to the "white rat," led to the testing of a wide variety of species ranging from primary producers (e.g., algae) to invertebrates (e.g., water fleas) to ultimate food chain predators (e.g., largemouth bass). Although standard toxicity test methods have been established for numerous species, the lack of scientific agreement on the species most sensitive to a contaminant and/or representative of community structure continues to pace the search for the elusive "white rat."

Some of the ecotoxicology tests, in the sense that they are designed to monitor changes at the subcellular, cellular, and organism level, resemble those developed for estimating toxicity to humans. In many cases, however, evaluation at the organism and suborganism level does not reveal the extent of a toxic effect on an ecosystem. For example, measurement of contaminant levels in edible fish tissue represents an excellent indicator of individual exposure, but does not convey information about the possible loss of a food source in an aquatic food chain.

Ecotoxicity test methods attempt to bridge the information gaps between measurable subcellular effects and shifts in structure and function of an ecosystem. Such testing encounters increasing complexity at ascending levels of organization: population, community, and ecosystem. This is suggested by Figure 5-25, which depicts the relationship between responses at levels of biological organization and the time scales for these responses.

Testing and evaluating toxic effects are also frustrated by interactions among chemicals released to the environment. Environmental releases of an industrial nature seldom consist of an individual chemical compound; typically, they include a complex mixture of chemicals. In fact, the toxicity of relatively few of the multitude of chemicals released to the environment are understood, and minimal information exists to discern the effects of a complex melange of hazardous effluents and emissions released from active facilities and contaminated sites.

These and the other complicating factors frustrate the practice of ecotoxicology. Notwithstanding these complications, well-founded approaches have been de-

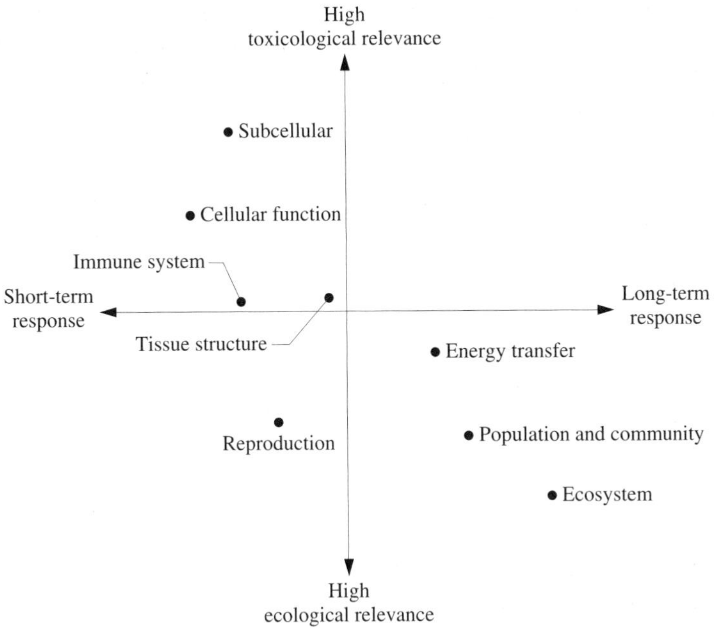

FIGURE 5-25
Toxicological/ecological relevance and time scale of toxic responses for levels of biological organization.[120]

veloped for evaluating toxic effects, providing reliable data for hazardous waste management decision making.

Toxic End Point

The selection and design of an appropriate test method first requires the specification of the toxic effect (**end point**) to be evaluated. This is often accomplished in ecotoxicity evaluations by using two distinct but complementary types of end points: assessment and measurement. **Assessment end points** are ecological characteristics, which, if found to be significantly affected, would indicate a need for mitigative action.[121] An assessment end point for an ecological risk assessment performed at a hazardous waste site could be the discernible adverse effect on populations of commercial or aesthetic value. For example, the potential effects of a discharge to surface water on the survival and abundance of a population of native trout presents an assessment end point of ecological and recreational value.

To be useful, an assessment end point must be measurable. Therefore, a **measurement end point** is selected that provides a quantitative expression (i.e., a measurable characteristic) related to the assessment end point.[122] For the example assessment end point, the measurement end point may quantify the frequency of mortality within this trout population by using either toxicity tests or field monitoring of changes in species abundance relative to the discharge location and dilution. Some

assessment and measurement end points used in ecological evaluations at hazardous waste sites are provided in Tables 5-12 and 5-13.

Bioassay

To evaluate the hazards associated with chemical exposure of nonhumans, ecotoxicologists have historically relied on the combination of short-term and long-term **bioassays.** A bioassay is a technique by which organisms, biological systems, or biological processes are used to measure effects of chemical exposure.

Results from short- and long-term toxicity tests are often used to establish water quality criteria, to estimate the killing or knock-down power of pesticides, to calculate permit limitations for effluent discharge, and to project the potential impacts of catastrophic spills and chemical accidents, and the actual and potential effects of hazardous waste disposal and cleanup. The integration of toxicity tests in waste

TABLE 5-12
Potential assessment end points[123]

Population	Ecosystem
Extinction	Productive capability
Abundance	
Yield/production	
Age/size class structure	
Massive mortality	
Community	**Human health concerns**
Market sport value	Contamination
Recreational quality	Gross mortality
Change to less useful desired type	

TABLE 5-13
Potential measurement end points[124]

Individual	Community
Death	Number of species
Growth	Species evenness/dominance
Fecundity	Species diversity
Overt symptomology	Pollution indices
Biomarkers	Community quality indices
Tissue concentrations	Community type
Behavior	
Population	**Ecosystem**
Occurrence	Biomass
Abundance	Productivity
Age/size class structure	Nutrient dynamics
Reproductive performance	
Yield/production	
Frequency of gross mortality	
Frequency of mass mortality	

management problems facing engineers can provide significant advantages in the decision-making process to achieve long-term solutions that are protective and cost-effective.

In the context of hazardous waste management, bioassays may be defined as the laboratory exposure of organisms to field-collected environmental samples for the purpose of identifying actual or potential toxic effects on resident species. Typically, such bioassays involve laboratory testing of soil, sediment, or water samples using a standard array of test organisms under controlled conditions.[125] A simple example is pumping contaminated surface water through a tank containing juvenile trout. Simulating exposure and interpreting effects on communities and ecosystems is far more demanding than measuring exposure and effects on populations of individuals exposed in laboratory systems.

Laboratory simulation of ecosystems has been attempted and frustrated by problems of scale. **Microcosms** have been used to imitate ecosystem structure and function on a reduced scale in the laboratory. For example, terrariums are common microcosms used as surrogates of simplified terrestrial ecosystems. To reduce difficulty presented by scale and lessen uncertainty in data extrapolation, macro model systems known as **mesocosms** have been developed as isolated portions or replications of natural settings. While microcosms and mesocosms are time- and cost-intensive and not commonly applied to hazardous waste management decision making, these tests are often used to evaluate the fate, transport, and effects of new chemicals seeking registrations.

The following discussions will examine the development and use of these toxicity tests, but an engineer should consult an ecotoxicologist and/or refer to guidance documents, as well as the scientific literature.[126,127]

Short-Term Toxicity Tests

Short-term toxicity tests, often referred to as **acute** toxicity tests, are used to evaluate severe effects from exposures spanning relatively short periods of time. While any response to a stressor represents a measurement end point, acute toxicity tests are most often used to evaluate percent affected by a chemical compound or perturbation. As in tests to predict potential human toxicity, the LD_{50} is used as an empirical estimate of the dose associated with a 50 percent lethal response in the test population (see Figure 5-11). Like an LD_{50}, the LC_{50} is used to measure lethality via inhalation or exposure in an aqueous solution. Short-term toxicity tests used to determine doses associated with sublethal responses provide an effective concentration, or EC_{50}. Figure 5-26 presents an idealized schematic of the effect of dose on organism response as measured by some physiological change, including percent death.

Acute toxicity tests are commonly designed to elicit a quantal response (i.e., an all-or-none effect: dead or alive). The relation between the concentration of the toxic substance and the percent of exposed organisms affected can be determined as a dose-response curve as discussed in Section 5-3. It is sometimes referred to as a dose-mortality curve.[128] The standard test for measuring acute lethality is designed to produce a 50 percent lethal response in a test organism after a 96-hour exposure duration. While responses for shorter exposure durations are often cited, the 96-hour LC_{50} is recognized as the standard.

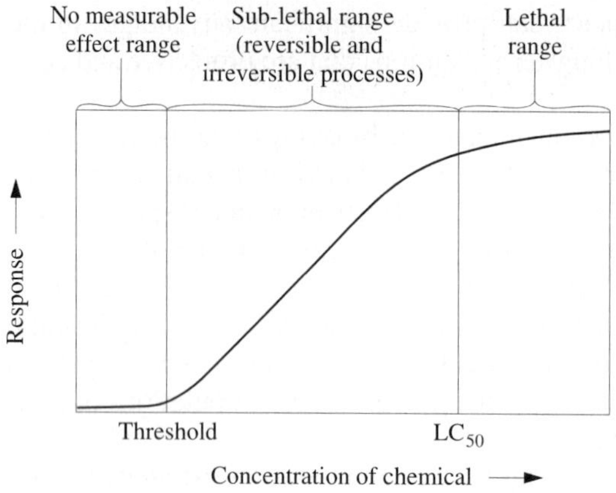

FIGURE 5-26
Idealized plot of dose-response relationship.[129]

Data for an acute toxicity test for sheepshead minnows exposed to sodium selenite are tabulated in Table 5-14 and the estimated LC_{50} values are presented in Figure 5-27.

An acute toxicity test can take many forms. For some organisms (e.g., invertebrates) death is not easily discernible; therefore, the 96-h EC_{50} is cited to reflect end points such as immobilization or loss of equilibrium which are considered surrogates of lethality. In an acute toxicity test, an organism may be exposed by one of four standard techniques: static, recirculation, renewal, or flow-through.[131] Essentially, these methods differ in regard to the introduction and maintenance of the test material in the exposure medium. Other water quality parameters (e.g., dissolved oxygen, hardness) may also be manipulated to evaluate their influence on toxicity.

Acute toxicity tests can be designed to answer numerous questions about waste management such as the toxicity of an individual chemical, mixture of chemicals, leachate, or other concentrated waste streams. Also, tests can be used to locate toxic "hot spots" at waste sites, and to evaluate the effectiveness of waste treatment technologies. Selecting the proper test organism, exposure system, and boundary conditions is important to producing meaningful results. Evaluation of the toxicity test results combines statistical procedures with interpretation based on experience.

TABLE 5-14
Test concentrations and mortality of sheepshead minnows exposed to sodium selenite[130]

Average measured concentration, selenium ion, mg/L	Percent mortality per exposure period			
	24 h	48 h	72 h	96 h
Control	0	0	0	0
2.0	0	0	0	0
3.6	0	5	10	20
7.1	0	25	35	45
14	10	30	60	75
27	40	100	100	100

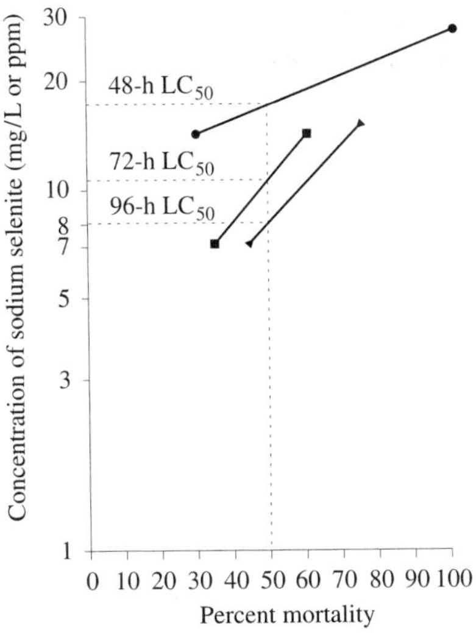

FIGURE 5-27
Graphical interpolation of time-dependent LC_{50} values for sodium selenite.[132]

Long-Term Toxicity Tests

Long-term toxicity tests, commonly referred to as **chronic** toxicity tests, are used to evaluate whether extended laboratory exposure of different life stages (e.g., embryos, juveniles, and adults) to a constituent will adversely affect individuals, species, or populations. These tests may measure effects on development, homeostasis, growth, and reproductive potential to establish benchmark concentrations for the onset of long-term effects. Such benchmark concentrations include the lowest observed effect level (LOEL), which represents the lowest concentration associated with a measurable effect, and the no observed effect level (NOEL), which denotes a concentration that did not produce a measurable effect, similar to human health benchmarks discussed in Section 5-4.

Since chemical concentrations associated with chronic effects are typically lower than those that produce acute responses, chronic toxicity usually provides a more sensitive measurement of toxicity. To develop acceptable exposure concentrations (e.g., water quality standards), ecotoxicologists use a subcategory of chronic toxicity tests referred to as **life-cycle** tests. These specific tests evaluate the responses of individuals from a species over a complete life cycle to evaluate effects on survival, growth, and reproduction.[133] For example, a life-cycle test for a freshwater fish may include exposures of embryos and juvenile fish at various stages of maturation, and adults continuing through reproduction and hatching of the next generation. An old tenet in aquatic toxicology is that the early life stages are the most sensitive to toxicants. An evaluation of data from 56 complete or partial "life-cycle" toxicity tests showed that embryo-larval and early juvenile life stages were the most sensitive in most cases.[134]

The problems associated with chronic toxicity tests are the time and costs for conducting such tests. While life-cycle tests provide useful information for decisionmaking, these tests are limited in that (1) not all species will reproduce in a laboratory setting, (2) they are unnecessary and incredibly expensive to conduct, (3) the toxicant is introduced at a constant rate for a fixed duration whereas in nature the

magnitude, frequency, and duration of exposure can vary considerably, and (4) only a small set of all toxic chemicals released to the aquatic environment can feasibly be evaluated.[135]

Biomarkers

The **biomarker** approach to identifying exposure or physiological effects differs from the more traditional laboratory methods of environmental testing (e.g., chemical measurements, toxicity tests) by using naturally occurring organisms as markers of environmental response.[136] The underlying concept for biomarkers is that selected measurement end points, typically composed of biochemical and physiological responses, can provide sensitive indices of exposure or effects other than death (nonlethal stress).[137] Confronted with the numerous complexities frustrating the measurement of environmental exposure and effects in dynamic, natural systems, there is increasing desire to evaluate exposure and effects of chemicals using biomarkers. These measurements of body fluids, cells, or tissues can indicate in cellular or subcellular terms the relative magnitude of chemical exposure or organism response.[138]

While extensive research continues to refine existing methods and to develop new markers of exposure and effect, experience has identified several biomarkers of value to hazardous waste management decisions. Measures of exposure to chemicals that bioaccumulate in tissue is easily accomplished by an analysis of tissue residue. In the absence of tissue residues, empirical constants often derived and used by ecotoxicologists to predict bioaccumulation include the **octanol-water partition coefficient** (K_{ow}), which approximates a chemical's behavior between the lipid and water phase, and the **bioconcentration factor** (BCF), which describes the relationship between the chemical concentration in an organism and the concentration in the exposure medium (typically water). These parameters were discussed in Section 3-2. The BCF is empirically expressed as[139]

$$C_{org} = (\text{BCF})(C_w) \qquad \text{or} \qquad \text{BCF} = C_{org}/C_w \tag{5-4}$$

where C_{org} = concentration in biota (mg/kg)
BCF = bioconcentration factor [(mg/kg in tissue)/(mg/L in water) or L/kg]
C_w = concentration in water (mg/L)

Table 5-15 provides physicochemical parameters for some typical compounds which bioaccumulate. Additional data may be found in Appendix A.

EXAMPLE 5-8. PREDICTING BIOCONCENTRATION. Routine monitoring of sediments and surface water downstream from a hazardous waste site indicate that reportable concentrations of the pesticide DDT are present. On the average, sediments and surface water contain 33 mg/kg and 0.25 μg/L, respectively. Considering that the U.S. Food and Drug Administration has identified levels of 5 ppm or higher of DDT as requiring action, is there a reason to be concerned about potential accumulation of DDT in fish tissue near this site?

Solution. Since the routine monitoring does not provide information regarding DDT residues in fish tissue, we refer to Table 5-15 and bioconcentration factor (BCF). Transforming the log BCF of 4.73, we estimate a BCF of 54,000 (mg/kg DDT in tissue per mg/L DDT in water, or L/kg).

TABLE 5-15
Selected physicochemical parameters and bioconcentration factors for typical compounds that bioaccumulate[140]

Compound	Log K_{ow}	Log BCF (fish)	Log K_{oc}	S, mg L^{-1}
Chlordane	3.32	4.15	5.15	0.56
DDT	6.19	4.73	5.39	0.0055
Dieldrin	3.50	3.68	3.23	0.195
Lindane	4.82	2.51	2.96	0.150
Chlorpyrifos	4.99	2.65	4.13	0.3
Cyhexatin	5.38	2.79	>3.64	<1.0
2,4-D	1.57	1.30	1.30	900
2,4,5-T	0.60	1.63	1.72	238
TCDD	6.15	4.73	5.67	0.0002

K_{ow} = octanol-water partition coefficient (unitless)
BCF = bioconcentration factor (L/kg)
K_{oc} = organic carbon partition coefficient (mL/g)
S = water solubility (mg/L)

Substituting our monitoring data and the empirical BCF for DDT in Equation 5-4 gives

$$C_{org} = [54{,}000\ (\text{mg/kg})/(\text{mg/L})] \times (0.00025\ \text{mg/L}) = 13.5\ \text{mg/kg}$$

We estimate a DDT tissue residue of 15 mg/kg. On the basis of this estimate, it is probable that DDT residues in tissue are an issue that should be evaluated.

A chemical that bioaccumulates and/or bioconcentrates can increase in tissue concentration as the chemical is transferred up through trophic levels in the food chain in a process known as **biomagnification.** Biomagnification in two simple aquatic food chains is depicted in Figure 5-28 for two organochlorine pesticides: DDD and toxaphene.

Tissue residues alone do not convey information about biological responses to chemical exposure. Furthermore, measuring tissue residues is not feasible with chemicals that do not readily bioaccumulate (compounds easily eliminated, for example) or with complex mixtures that require time- and cost-intensive analyses that may not identify all toxic chemicals. In such cases, other indirect measures may be preferred to indicate a biological response to exposure that is of toxicological significance.[142] A biomarker that provides information about both exposure and nonlethal stress to lead is the enzyme δ-aminolevulinc acid dehydratase (δ-ALA-D). This enzyme catalyzes a reaction involved in heme synthesis and is very sensitive to inhibition by inorganic lead.[144,145] δ-ALA-D provides an inexpensive measurement of lead exposure, which eschews the possible problem of lead contamination of environmental samples. At waste sites where lead is of particular interest, δ-ALA-D measurements may provide useful information about lead exposure and nonlethal organism response,[146] but its utility is reduced at hazardous waste sites contaminated by numerous trace metals. Another common biomarker for trace metal exposure is the metal-binding protein metallothionen. Metallothionen functions as a regulator for normal zinc and copper metabolism in addition to providing a mechanism for

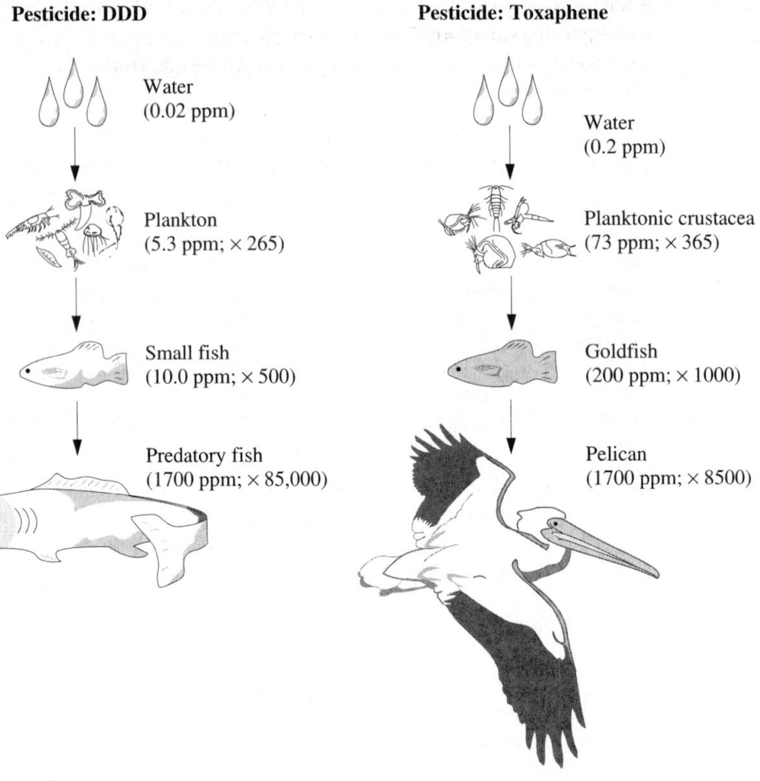

FIGURE 5-28
Biomagnification of pesticides in aquatic food chains.[143]

metal detoxification. Changes in metallothionen activity provide a sensitive marker of trace metal exposure.

Other biomarkers of environmental exposure and effects include induction of metabolic (detoxifying) enzymes such as mixed function oxidases (MFO) or cytochrome P450 by organic contaminants; inhibition of cholinesterase enzymes in the brain by insecticide compounds such as organophosphates; and nonspecific effects such as skeletal abnormalities. Additional information about the methods and uses of other specific biomarkers to evaluate exposure to hazardous waste is available in various references.[147,148]

Aquatic Toxicity of Trace Metals

Trace metals naturally cycle through and between the lithosphere, atmosphere, and hydrosphere. Enriched levels of trace metals may occur naturally because of the weathering of a geologic substrate or volcanic emissions, or from anthropogenic sources such as industrial effluent, or nonpoint sources such as urban storm runoff and atmospheric washout of particulate or gaseous emissions. Mining, fossil fuel combustion, smelting, plating, and other historical anthropogenic activities have

altered the biogeochemical cycling of trace metals and frequently enriched the levels of trace metals in the environment. The release from a hazardous waste site may represent a significant local source, but rarely the only source of exposure to local biota.

In aquatic systems, these trace metals may be present as both soluble and insoluble forms within the water column, sediments, and biotic tissues. Trace metal toxicity occurs when a receptor is exposed to an available form of a trace metal for a sufficient duration at a concentration that elicits an adverse response. The potential for trace metal toxicity to aquatic organisms depends on numerous factors, including the chemical and physical characteristics of the water and sediment, composition and health of the biological communities present, and the magnitude and availability of the trace metal. Some of the factors which influence trace metal toxicity in aquatic systems are outlined in Table 5-16.

Many trace metals are essential micronutrients for the maintenance of aquatic life, and toxic only when available to organisms at levels that exceed nutritional requirements. Trace metals such as copper, iron, zinc, manganese, cobalt, and selenium are essential to metabolism; yet, exposure of aquatic organisms to these same elements at elevated levels can adversely affect development and health, or cause death. Other trace metals such as lead, cadmium, and mercury do not play an essen-

TABLE 5-16
Factors influencing the toxicity of trace metals in the aquatic environment[141]

Form of metal in water	Inorganic Organic	Soluble	Ion Complex ion Chelated ion Molecule
		Particulate	Colloidal Precipitated Adsorbed
Presence of other metals or poisons	Joint action No interaction Antagonism		More than additive Additive Less than additive
Factors influencing physiology of organisms and possibly form of metal in water	Temperature pH Dissolved oxygen Light Salinity		
Condition of organism	Stage in life history (egg, larva, etc.) Changes in life cycle (e.g., moulting, reproduction) Age and size Sex Starvation Activity Additional protection (e.g., shell) Adaptation to metals Altered behavior		

tial role in daily maintenance, but adversely affect an organism if available at toxic levels.

Upon exposure, uptake occurs by three main processes: from the water column through respiratory surfaces (e.g., gill); adsorption from water onto body surfaces; and from ingested food, particles, or water through the digestive system.[150] After intake, as discussed in Section 5-1, organisms may absorb, transform, excrete, or store (bioaccumulate) metals. A schematic depicting contaminant exposure and uptake for a fish is presented in Figure 5-29.

Many free-swimming or mobile aquatic organisms (e.g., fish) actively excrete trace metals, particularly essential nutrients, through the gills, gut, feces, and urine, while less mobile or sessile organisms (e.g., plants and bivalves) are less capable of regulating internal metal concentrations. Once the processes that regulate metals have been saturated, bioaccumulation within tissue or other tolerance mechanisms may preclude an adverse effect. However, these protective mechanisms are often too transient or insufficient to preclude toxicity.

The mechanisms of metal toxicity are divided into three general categories: blocking of essential functional groups of biomolecules (e.g., protein and enzymes), displacing the essential metal ion in biomolecules, and modifying the active conformation of biomolecules.[151] These mechanisms of action may cause death at high-level exposures, but also may manifest nonlethal effects on structure (histology/morphology), function (growth, development), biochemistry (blood and enzymes), and reproduction.

Example of Aquatic Toxicity of Copper to Fish

To illustrate the toxic mechanisms and effects of trace metals on fish, the example of copper is selected and discussed thoroughly in the following pages. Copper, a micronutrient, is a necessary component of many enzymes and proteins. Cerulo-

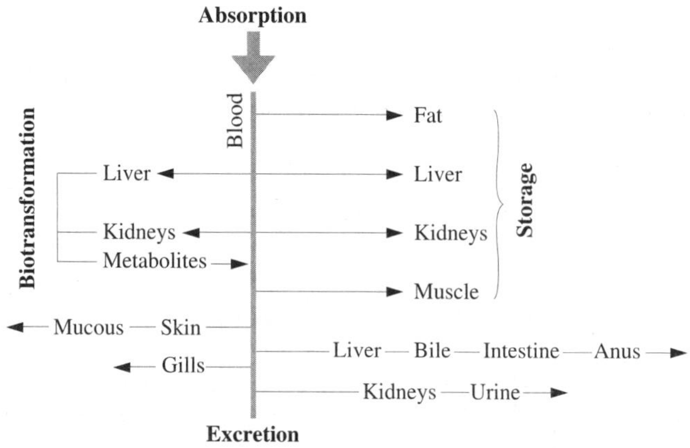

FIGURE 5-29
Possible movement and fate of a contaminant after absorption into the bloodstream of fish.[149]

plasmin, a protein located in blood serum, accounts for 95 percent of the circulating copper in mammals. Copper, mercury, and silver vary in the magnitude of their toxicity to aquatic plants, invertebrates, and fish, but as shown in Table 5-17 these metals are commonly the most toxic to aquatic biota. The sensitivity of aquatic organisms to copper is extreme for those species having a high surface-to-volume ratio, high respiratory flows, and an extensive, highly permeable gill surface that facilitates rapid uptake of large amounts of copper.[153]

The principal toxic copper species is cupric ion (Cu^{+2}), although two ionized copper hydroxides ($CuOH^+$ and $Cu_2OH_2^{+2}$) have also proven to be toxic.[154] The toxicity of copper in water is controlled primarily by alkalinity (which may be correlated with hardness) and pH. Copper is less toxic at higher alkalinities that accompany higher hardness values, which reduce bioavailability through formation of copper carbonate complexes.[155] Copper toxicity increases by reductions in alkalinity, dissolved oxygen, chelating agents, humic acids, pH, and suspended solids. For example, a low level of dissolved oxygen in the water column would require an increase in the ventilation rate of aquatic organisms. The elevated ventilation rate would then increase the flow of water containing copper, facilitating the uptake of copper and increasing the overall body burden.

On uptake and distribution, the maximum fish tissue copper concentrations are found in the liver, followed by the gill, kidney, and muscle. In studies that measured tissue concentrations, the liver was the principal accumulator of copper,[156–162] accumulating approximately 10 times the amount measured in muscle tissue.[163] The liver retains copper for the longest duration[164] and serves as the best indicator of copper exposure. The copper residing in the liver associates with metalloenzymes at concentrations that could potentially disrupt normal metabolic activity, hence producing deleterious effects in reproduction and development.[165] In Table 5-18, the distribution of copper in freshwater fish tissues is listed as measured by investigators in several studies.

Acute and chronic studies have been performed on copper for a variety of freshwater fish species at different life stages. In chronic studies encompassing a complete

TABLE 5-17

Range of acute toxicity (48- to 96-h LC_{50} or EC_{50} $\mu g/L$) data for commonly used freshwater test organisms[152]

Test organism	Copper	Mercury	Lead	Zinc
Arthropoda (crustaceans)	5–3000	0.02–40	—	30–9000
Annelida	6–900	—	—	—
Mollusca	40–9000	90–2000	—	500–20,000
Vertebrata (fish)				
Salmonidea	10–900	3–20,000	1000–500,000	50–7000
Centrachidae	700–10,000	—	20,000–400,000	1000–20,000
Cyprinidae	23–2000	3000–10,000	2000–500,000	400–50,000
Algae				
Chlorophyta	1–8000	<0.8–2000	500–1000	30–8000
Chrysophyta (diatoms)	5–800	—	—	—

life cycle, all species tested were found to be most sensitive to copper in their early life stages (larval and juvenile).[169–171] In the critical early stages copper did not produce embryo mortality, but it has been postulated that sublethal effects occurred in the embryo that ultimately influenced the success and survival of larvae and juveniles.[172]

In addition to the obvious effects on survival, copper also produces deleterious results in growth and reproduction. Inhibition of spawning, marked decreases in egg hatchability, and reductions in growth have been observed in a number of species over a wide span of copper concentrations.[173–178] Like the effects on survival the greatest reductions in growth occur in the early life stages.

At the lower levels of organization copper produces changes in blood chemistry, including increases in red blood cell count, hematocrit, hemoglobin, plasma glucose, and lactic dehydrogenase.[179] Although these changes in blood chemistry are good indicators of environmental stress their transient nature prevents their measurement as definitive biomarkers.

TABLE 5-18
Distribution of copper in fish tissues[166]

		Copper, ppm dry weight				
Species	**Cu source**	**Gills**	**Liver**	**Kidneys**	**Muscle**	**References**
Brook trout	Tank, Cu = 0.09 ppm	8.2	238	16.5	17	McKim and Benoit (1971)[156]
Brook trout	Tank, Cu = 0.003 ppm	5.0	239	19.7	15.3	McKim and Benoit (1971)[156]
Brown bullhead catfish	Tank, Cu = 0.027 ppm 30 days	9.4	33	10.0	—	Brungs et al. (1973)[157]
Brown bullhead catfish	Tank, control	5.7	24	8.0	—	Brungs et al. (1973)[157]
Brown bullhead catfish	Tank, Cu = 0.027 ppm 20 months	6.9	11	10.0	—	Brungs et al. (1973)[157]
Brown bullhead catfish	French River Cu = 2–9 ppm	1.8	30.3	2.5	1.3	Hutchinson and Fitchko (1974)[167]
Brown bullhead catfish	Tank, control	3.8	29	7.8	—	Brungs et al. (1973)[157]
Perch	D'Alene-Spokane River (polluted)	—	2.8–44.5	—	0.4–1.06	Funk et al. (1975)[168]
Perch	Tank, Cu = 0.044 ppm 20 days	19.0	191	—	—	Collvin (1984)[160]
Perch	Control	3.0	3.0	—	—	Collvin (1984)[160]
Bluegill sunfish	Tank, Cu = 0.040 ppm 24 months	5.0	61	13.0	—	Benoit (1975)[158]
Bluegill sunfish	Control, Cu = 0.003 ppm	3.0	7.0	22.0	—	Benoit (1975)[158]

Similar to combinations of many toxicants, additional effects could result from the association of copper with other contaminants. A combination of copper and other heavy metals, or xenobiotics, could produce synergistic, antagonistic, and additive effects in toxicity. For example, the combination of zinc and copper yielded an additive toxic response,[180,181] and a mixture of zinc, copper, and cadmium elicited an antagonistic response.[182]

Aquatic Toxicity of Hydrocarbons

The large universe of hydrocarbons naturally present and resulting from anthropogenic releases is addressed as two principal subgroups: aliphatic and aromatic. In the aquatic environment, the fate and persistence of a hydrocarbon are important parameters relative to potential aquatic exposure and associated toxicity. As discussed in Section 4-3, aromatic hydrocarbons, which tend to partition to organic matter, resist degradation, and persist in the aquatic environment, possess a greater potential to affect aquatic life than the aliphatic compounds which are generally less persistent because of greater volatility, solubility, and biodegradability. Further, hydrocarbons that are substituted with halogens, nitrogen, or other subgroups are generally more persistent and toxic than an unsubstituted, parent hydrocarbon. Table 5-19 presents aquatic toxicity data of some common monocyclic aromatic hydrocarbons for algae, invertebrates, and fish.

EXAMPLE 5-9. PATTERNS IN AQUATIC TOXICITY FOR ORGANIC COMPOUNDS. A chemical manufacturing facility produces chlorinated benzenes. As part of their effluent monitoring program they conduct acute toxicity tests. Assuming the toxicity data presented in Table 5-19 are representative of the results for their toxicity testing, can you identify a pattern relative to the toxicity of these chlorinated benzenes?

TABLE 5-19
Acute toxicity (96-h LC_{50}, mg/L) of monocyclic aromatic hydrocarbons to algae, invertebrates, and fish[184]

	Algae		Invertebrates		Fish	
Compound	**Fresh water**	**Salt water**	**Fresh water***	**Salt water**	**Fresh water**	**Salt water**
Benzene	525	—	200	39	22.5	33
Toluene	>433	>433	310	56	13	280–480
Chlorobenzene	228	342	86	16.4	16	10
1,2-Dichlorobenzene	95	44	2.4	2.0	5.6	9.7
1,3-Dichlorobenzene	64	52	28	2.9	5.0	7.8
1,4-Dichlorobenzene	97	57	11	2.0	4.3	7.4
1,2,4-Trichlorobenzene	36	8.9	—	0.5	3.4	21
1,2,3,5-Tetrachlorobenzene	17	0.7–7.1	9.7	0.3	6.4	37
1,2,4,5-Tetrachlorobenzene	50	7.3	—	1.5	1.6	0.8
Pentachlorobenzene	6.7	2.1	5.3	0.2	0.25	0.8
Ethylbenzene	>438	>438	75	88	150	280
Nitrobenzene	44	9.6	27	6.7	43	59
Dinitrotoluene	1.5	0.4	0.7	0.6	0.3	—
2,4,6-Trinitrotoluene	5	50	—	—	—	2.6

*48-h LC_{50}.

Solution. From the data presented in Table 5-19, one can compare the relative sensitivities of algae, invertebrates, and fish; identify potential differences between freshwater and saltwater species; and compare the relative toxicities of the mono-, di-, tri-, tetra-, and pentachlorinated benzenes. In regard to any differences in toxicity related to chlorine substitution, the concentration associated with acute toxicity in 20 of the 23 cases decreases for the following comparison: mono to di, di to tri, tri to tetra, and tetra to penta. This indicates that the acute toxicity of chlorinated benzenes increases as the chlorine content increases. This is no coincidence, as a common trend for the toxicity of halogenated hydrocarbons is that the acute toxicity increases as the degree of halogen substitution increases.

An important, pervasive class of hydrocarbons that persist in the aquatic environment is the polynuclear aromatic hydrocarbons (PAHs). Approximately 200,000 metric tons of PAHs are released annually to the aquatic environment from sources such as petroleum spillage, deposition of particulates from fossil fuel combustion, and other anthropogenic and natural processes. The hydrophobic character of PAHs leads to rapid association with particulate matter and partitioning to either sediments or tissue. PAHs are acutely toxic to aquatic organisms at concentrations ranging from 0.2 to 10 mg/L,[183] and acute toxicity tends to increase with increasing molecular weight.[184] However, higher molecular weight PAHs (i.e., larger than pyrene, which has a molecular weight of 202) are not considered acutely toxic to aquatic organisms, apparently because solubility is less than the concentration required to elicit a response.[186] Of greater interest are the potential effects resulting from nonlethal, chronic exposure. Although PAHs are hydrophobic, their metabolites are more water soluble and reactive. As discussed earlier for human toxicity, these reactive metabolites can bind to protein, DNA, and other macromolecules, leading to cell damage, mutagenesis, and possible cancer. While PAH-induced cancer-like growths or developmental anomalies have been described in species of aquatic organisms, the role of PAH in carcinogenesis in aquatic organisms is uncertain.[187]

Aquatic Toxicity of Pesticides

The expansion of the synthetic organic chemical industry after World War II led to rapid development of synthetic compounds to control species of insects, plants, rodents, and other "pests." These chemicals were intended to be persistent and to render an adverse effect on a **target** species of pest without damaging **nontarget** organisms. However, historical data for compounds such as DDT, Parathion, and other persistent pesticides provide clear evidence about deleterious effects on nontarget organisms of widespread pesticide use.

The effect of these organic compounds on aquatic organisms is related to several properties fundamental to their use as pesticides. First, pesticides are chemical compounds capable of affecting all types of biota, including nontarget species, to varying degrees, depending on physiological and ecological factors. Second, many pesticide compounds are, by design, resistant to degradation to enhance their persistence and effectiveness. Finally, pesticides are usually applied in bulk quantity and dispersed over large areas, which results in atmospheric transport to natural systems and runoff contamination of receiving surface water bodies.[188]

For our discussion of aquatic toxicity, the pesticides of interest are primarily insecticides and herbicides. Three important classes of insecticides are the organochlo-

rines (e.g., DDT, methoxychlor, toxaphene, aldrin, dieldrin), organophosphates (e.g., Parathion, Malathion, Diazinon), and the carbamates (e.g., Sevin, Temik). Most insecticides act by interfering with nervous system function. These neuroactive insecticides are believed to involve effects on nerve impulse transmission; the receptor for acetylcholine, a neurotransmitter; or acetylcholinesterase, an enzyme[189] (see Figure 5-10). Phenoxy acid herbicides (2,4-D and 2,4,5-T) effectively eradicate plants by growth disruption.

As shown in Table 5-20, the acute aquatic toxicity of organochlorine compounds is generally greater than organophosphates, and insecticides are relatively more toxic than herbicides.

Nonlethal effects of pesticide exposure are numerous and diverse, and related to a broad spectrum of metabolic and behavioral responses, including alterations in enzyme production, growth, reproduction, activity, production of tumors, and teratogenic effects.[190] Some sublethal effects of pesticide exposure on aquatic organisms are provided in Table 5-21.

TABLE 5-20
Relative toxicities (measured as induced mortalities) of pesticides to aquatic animals[185]

Pesticide type	Plankton	Shrimp	Crab	Oyster	Fish
Herbicide	1	1	1	1	1
Organochlorine compounds	×3	×300	×100	×100	×500
Organophosphorus compounds	×0.5	×1000	×800	×1	×2

TABLE 5-21
Sublethal effects of pesticides on aquatic organisms[198]

Test species	Test material	Observed effect	Effective concentration
American oyster	Chlordane	Reduced shell growth	6.2 μg/L
Four marine algae	Kepone	Effect on growth of cultures	350–600 μg/L
Fathead minnow	Kelthane	Alteration in normal schooling behavior after 24 h	218 μg/L
	Dursban		44 μg/L
	Disulfoton		413 μg/L
	Pydrin		3.6 μg/L
	Permethrin		7.2 μg/L
Yellow perch	Methoxychlor	Effect on respiration	5.0 μg/L
Juvenile blue crabs	Mirex	Paralysis	Mirex bait
Pinfish	Naled (Dibrom)	Brain acetylcholinesterase inhibition	84–99% inhibition at lethal concentrations
Brook trout	DDT	Prevents the establishment of a visual conditioned avoidance response	20–60 μg/L
Bluegill sunfish Largemouth bass	Parathion	Effect on locomotor orientation	10–25 μg/L
Grass shrimp	Methyl or ethyl Parathion	Impairment of antipredation behavior	0.1 μg/L
Sheepshead minnow	Kepone	Scoliosis (curvature of spine)	0.8 μg/L

Water Quality Criteria

To regulate the discharge of pollutants to surface water and to protect the use of such water by indigenous aquatic biota and humans, the Clean Water Act requires that the EPA and the states develop water quality criteria. These criteria are developed on the basis of the measured toxic effects on aquatic biota and humans of exposure to chemicals in surface water. For aquatic life, these criteria specify a pollutant concentration which, if not exceeded, should protect most, but not all, aquatic life.[191] To develop water quality criteria, the EPA requires data in four areas: acute toxicity to animals (eight data points), chronic toxicity to animals (three data points), toxicity to plants, and tissue residues.[192,193] The specific data requirements for each of the four areas, and the methods used to reduce and analyze the data are provided in the EPA guidance for deriving water quality standards.[194,195,196]

It is important to note that for several toxicants the influence of factors that mitigate toxicity are accounted for in the development of criteria. For example, the acute and chronic water quality criteria for several trace metals depend on the measured water hardness value. The chronic and acute criteria for copper recognize that freshwater aquatic organisms should not be adversely affected if the 4-day average concentration (in μg/L) of copper does not exceed the value given by $e^{0.8545[\ln\ (\text{hardness})]-1.465}$ more than once every 3 years on the average and if the 1-hour average concentration does not exceed the value given by $e^{0.9422[\ln\ (\text{hardness})]-1.464}$ more than once every 3 years on the average.[197] Therefore, at a water hardness level of 50 and 200 mg/L $CaCO_3$, the acceptable 4-day average concentrations are 6.5 and 21 μg/L, respectively, and the acceptable 1-hour concentrations are 9.2 and 34 μg/L. Additional equations to estimate freshwater aquatic life criteria for metals are provided in Table 5-22.

EXAMPLE 5-10. ECOTOXICOLOGY OF ZINC. Downstream of a metal-plating lagoon holding wastewater and sludge, surface water monitoring data indicate 4-day and 1-hour average zinc concentrations of 14.2 μg/L and 27 μg/L. An average water hardness value of 100 mg/L $CaCO_3$ was also measured for the stream. For the measured concentrations for zinc and water hardness, and the water quality criteria for zinc, is there reason to be concerned regarding the long-term protection of aquatic life?

Solution. From Table 5-22 the equation to estimate the zinc criterion for long-term (chronic) protection of aquatic life is as follows:

$$\mu\text{g/L Zn} = e^{0.8473[\ln\ (\text{hardness})]+0.7614}$$

Substituting the measured hardness value of 100 mg/L $CaCO_3$, we estimate that the criterion for the 4-day average zinc concentration is:

$$\begin{aligned} \text{Zn} &= e^{0.8473[\ln\ (100)]+0.7614}\,\mu\text{g/L}, \\ &= \underline{\underline{106\ \mu\text{g/L}}} \end{aligned}$$

Because the measured 4-day average zinc concentration is nearly an order of magnitude less than the criterion, there is no empirical reason for concern.

While water quality criteria continue to help ensure the maintenance of designated water body uses and the protection of water column aquatic life, these numerical criteria were not intended to protect organisms that inhabit the sediment. The evaluation and protection of sediment quality is an arena of aquatic toxicology

TABLE 5-22
Hardness-dependent aquatic toxicity criteria[199]

	Acute criterion = $\exp\{m_A[\ln(\text{hardness})] + b_A\}$		Chronic criterion = $\exp\{m_C[\ln(\text{hardness})] + b_C\}$	
	m_A	b_A	m_C	b_C
Cadmium	1.128	−3.828	0.7852	−3.490
Copper	0.9422	−1.464	0.8545	−1.465
Chromium (III)	0.8190	3.688	0.8190	1.561
Lead	1.273	−1.460	1.273	−4.705
Nickel	0.8460	3.3612	0.8460	1.1645
Silver	1.72	−6.52	—	—
Zinc	0.8473	0.8604	0.8473	0.7614

that is rapidly expanding, including the development of supplemental information pertaining to sediment toxicity and quality analyses.[200,201]

Early environmental statutes regulated the release of chemicals to water; therefore, the advances in and data available from measuring toxicity to aquatic ecosystems far exceeds the database for other branches of ecotoxicology. However, of the tens of thousands of hazardous materials produced and ultimately released to the environment, the EPA has developed water quality criteria to protect freshwater aquatic life for less than 30 pollutants.[202]

Terrestrial Toxicology

As discussed earlier, ecotoxicology initially focused on aquatic species. A second group of organisms to gain research attention were birds, particularly large birds of aesthetic and recreational value whose abundance visibly declined over a short period of time because of widespread use of several chemical substances, most notably DDT.

Only within the past 10 to 20 years has ecotoxicology begun to examine the effects of chemical releases on terrestrial species. With the exception of effects on sessile terrestrial plants, the effects of nonlethal chemical exposure on terrestrial organisms are not as easily measured because organisms are extremely mobile over large areas, exposed to numerous natural stressors such as disease, predation, and habitat loss, and demonstrate a penchant for avoidance, rather than confrontation.

Feral Plants and Mammals

The media attention afforded to the effects of acid deposition on forests in Germany, Canada, and the United States has increased public awareness regarding chemical effects on plants. Herbicide use in agriculture to inhibit "weed" growth, in Vietnam to eradicate jungle foliage, and along roadways and utility lines to regulate plant growth has also increased public interest about potential adverse effects on both target and nontarget receptors.

Another contemporary debate continues about the application of domestic sewage sludge ("biosolids") as a soil enhancement to land used for agriculture, recreation, and/or residential gardening. The land treatment of hazardous waste sludge is a related example. The concentrations of trace metals in sludge raises concern about potential plant phytotoxicity (inhibition of photosynthesis) and/or reductions in plant community productivity. The presence of the mobile lighter organic compounds in sludge could cause potential foliar damage such as chlorosis (loss or damage to chlorophyll), whereas heavier, less mobile organics may accumulate in plant tissue and be transferred through the food chain.

While many mammalian species have been evaluated in laboratory toxicity tests as surrogates for humans, information is limited regarding effects on feral species. An exception is the effects of nonlethal organohalogen exposure on reproduction. It is well established that PCBs can be found in wildlife in many areas, and studies have attempted to correlate the high tissue residues with breeding deficiencies in mammals.[203] While some disagreement remains among scientists regarding the mechanism of action, the diminished capacity of an organism to reproduce apparently is definitive. Recent evidence suggests that PCBs and other endocrine disruptors provide a double-edged reproductive inhibition, not only leading to eggshell thinning in females, but also to testicular feminization in males.

Avian Toxicology

The visible effects of chemical use on avian species of recreational and aesthetic value has increased research in avian toxicology. The database for avian effects of chemical exposure has expanded significantly because of regulatory requirements for avian toxicity testing as part of new chemical registration. As in the earlier discussion for aquatic organisms, most avian research has focused on the effects of several trace metals and the organohalogens, most notably pesticides and PCBs.

The intake of spent lead shot by waterfowl and the consumption of methylmercury by fish-eating birds are well-researched cases of avian poisoning though trace metal exposure. To maintain efficient grinding of food, most waterfowl ingest sand, small pebbles, and other forms of grit. It is during this routine event that many species ingest spent lead shot. Upon ingestion the lead shot is slowly eroded, becoming available for internal transport and distribution. In many cases it results in death from tissue hypoxia. While lead poisoning from spent shot presents an interesting case study, it is important to note that trace metals do not selectively affect waterfowl. Other species of insect-, seed-, and flesh-eating birds are also susceptible to trace metal poisoning from direct exposure to contaminated soil and water and indirect exposure via the food chain.

Avian toxicity research regarding the effects of hydrocarbon exposure has focused primarily on organochlorine and organophosphate pesticides and the PCBs. Although other hydrocarbons have not been as thoroughly scrutinized in toxicity tests, additional information regarding exposure via the food chain is available as tissue residues and physical impairment resulting from oil spills. Because pesticides generally accumulate in the food chain, species positioned at the top of a food chain are often susceptible to significant tissue burdens from long-term, low-dose

exposure. As energy demands increase during migration, reproduction, and other stressful activities, these materials are often liberated from fatty tissue deposits, and distributed to target organs where a variety of adverse responses occur. From these examples, it is clear that isolated contamination, transported by both physical and biological processes, can affect local receptors as well as migratory species.

DISCUSSION TOPICS AND PROBLEMS

5-1. Which are the main **routes of exposure** by which the organism can be exposed to chemical substances present in the environment?

5-2. The LD_{50} of a certain substance is appreciably higher by inhalation than by ingestion. What conclusions may be drawn from this observation?

5-3. What are the consequences with respect to uptake by the mammalian organism after inhalation of an aerosol with a mean aerodynamic particle diameter of

(*a*) 1.5μ
(*b*) 10μ?

5-4. Which properties favor accumulation of a substance in the mammalian body?

5-5. Identify substances that tend to accumulate in

(*a*) fatty tissues
(*b*) bone
(*c*) the kidney

5-6. What is the difference between dose expressed as administered (or inhaled) dose and target dose? Which is the more appropriate to use for determination of a dose-response relationship? Why?

5-7. How is an accumulating lipophilic compound handled by the mammalian organism to facilitate elimination in urine?

5-8. What are the differences between acute and chronic exposure, and how may these differences relate to acute and chronic health effects?

5-9. Give an example of a chemical compound that specifically induces damages to

(*a*) the kidney
(*b*) the liver
(*c*) the bone marrow
(*d*) the central nervous system
(*e*) the testicles

5-10. By which type of mechanism do neurotoxic organophosphorus compounds act?

5-11. What is the difference between a dose-response and a dose-effect relationship?

5-12. What is the fundamental difference between risk assessment of carcinogenic and mutagenic agents and the compounds that induce effects like malformations or neurotoxicity?

5-13. The NOEL for the herbicide atrazine was found to be 0.5 mg/(kg · day) based on induction of cardiotoxic effects (EKG alterations) in beagle dogs in a long-term feeding study.

(*a*) Assuming the quality of the study to be high, what is the ADI (RfD)?
(*b*) Can this ADI (RfD) be used to determine an acceptable occupational exposure level for farm workers when the intake occurs mostly by inhalation?

5-14. What type of disease could be induced by mutation in

(*a*) somatic cells
(*b*) sex (germinal) cells?

5-15. Why would a laboratory test that may identify a mutation in a single gene in a bacterium be of interest to human toxicology?

5-16. (*a*) List the different stages of carcinogenesis.
(*b*) What characterizes a cancer promoter?
(*c*) What is the difference between an epigenetic and a genotoxic carcinogen?

5-17. Which evidence is available to support the idea that the induction of malignant tumors involves changes in the hereditary material (i.e., DNA)?

5-18. What are the limitations of using laboratory animals in identifying human carcinogens?

5-19. What is the difference between a cohort study and a case control study, and what are the strengths and weaknesses inherent in their design, respectively?

5-20. In 1982 homes were constructed in the close vicinity of an abandoned hazardous waste site. In 1984 sampling indicated that the site was heavily contaminated with a human bladder carcinogen, beta-naphthylamine. According to a cross-sectional epidemiological investigation (completed by the end of 1986) of the area residents, an increased incidence of cancer of the prostate, bladder, lung, and rectum was found in comparison with a control population. The conclusion was that chemical contaminants from the hazardous waste caused the observed increase in cancer. Why was this conclusion unwarranted?

5-21. From epidemiological studies the carcinogenic potency factor (slope factor) for benzene was estimated by the EPA in 1988 to be 0.03 $[\text{mg}/(\text{kg} \cdot \text{day})]^{-1}$. On the basis of this estimate, what would be the hypothetical **yearly** increase in the cancer incidence of the U.S. *population* from benzene in gasoline due to exposure during filling of gasoline in motor vehicles? Use the following assumptions:

- 30 million car owners fill the gasoline tank twice a week at self-service stations.
- Filling averages 3 minutes.
- The concentration of benzene in the breathing zone is 0.5 ppm.
- The volume of inhaled air is 14 L/min.
- The degree of absorption of benzene in the lungs is 50 percent.
- The annual cancer incidence for the entire U.S. population can be estimated to be about 1 million cases.

5-22. What would be the closest equivalent for aquatic organisms to an oral LD_{50} in a rodent?

5-23. What is a life-cycle test?

5-24. Why is it not enough to investigate the acute toxic effects of chemicals in a specific species of fish to evaluate the impact on this species of the release of the chemical to the aquatic environment? What alternatives exist to test for ecotoxicologic effects in a few selected species of organisms?

5-25. Given the fact that MCPA [(4-chloro-2-methylphenoxy)acetic acid] and certain arsenic compounds are used as weed killers, and their toxicity to fish is relatively low, which segment of the aquatic ecosystem would you expect to be the potential main target of their adverse action?

5-26. (*a*) What physical property would you use to predict the BCF in a semiquantitative way?
(*b*) How would you define "biomagnification"?

5-27. Which terrestrial animal species would you use to test for the potential environmental impact of DDT?

5-28. Which regulatory standards in the United States have protection of aquatic life as one of their main goals? Are these standards designed to protect all forms of aquatic organisms?

5-29. Name one important environmental factor that will influence aquatic toxicity of several heavy metal ions to such an extent that it influences regulatory standards. Why?

5-30. The maximum permitted levels of pesticide residues in food products are set by tolerance levels. Assuming that the current tolerance level of Lindane of 4 ppm for hog fat would also apply to fish, would trout from a lake with a concentration of 0.1 ppm of Lindane in water (caused by accidental contamination) be considered fit for human consumption? Base your estimate on a log BCF value for Lindane of 2.51 mg/kg tissue per mg Lindane/L water.

5-31. You are a consultant who has been asked to evaluate the ecological impacts from a hazardous waste stream discharge into an adjacent estuary. Which of the following types of information would **not** be useful to your evaluation? Choose one and briefly explain.

(*a*) The physical and chemical properties of the waste stream constituents
(*b*) Characterization of the resident benthic populations
(*c*) Aquatic toxicity benchmarks
(*d*) The volume and timing of the waste stream discharge
(*e*) Surface water and sediment analytical data

REFERENCES

1. Keetan, William T.: *Biological Science*, 2nd ed., W. W. Norton, pp. 93, 167, 1972.
2. Gute, David M., and N. B. Haner: "An Applied Approach to Epidemiology and Toxicology for Engineers," Instructor's Resource Guide under review by U.S. Department of Health, Education and Welfare, Cincinnati, Ohio, 1991.
3. U.S. Department of Health and Human Services: "Toxicological Profile for Benzene," Public Health Service, Agency for Toxic Substances and Disease Registry, ATSDR/TP/-88/03, Atlanta, 1989.
4. Klaassen, C. D., and K. Rozman: "Absorption, Distribution, and Excretion of Toxicants," *Casarett and Doull's Toxicology—The Basic Science of Poisons*, M. O. Amdur, J. Doull, and C. D. Klaassen, eds., 4th ed., Pergamon Press, Elmsford, New York, p. 51, 1991.
5. Klaassen and Rozman, "Absorption, Distribution, and Excretion of Toxicants."
6. U.S. Department of Health and Human Services: "Toxicological Profile for Lead," Public Health Service, Agency for Toxic Substances and Disease Registry, ATSDR/TP-88/17, Atlanta, 1990.
7. International Programme on Chemical Safety (IPCS): "Environmental Health Criteria No. 24, Lindane," World Health Organization, Geneva, 1991.
8. Klaassen C. D., and D. L. Eaton: "Principles of Toxicology," *Casarett and Doull's Toxicology—The Basic Science of Poisons,* M. O. Amdur, J. Doull, and C. D. Klaassen, eds., 4th ed., Pergamon Press, Elmsford, New York, pp. 12–50, 1991.
9. Klaassen and Rozman, "Absorption, Distribution, and Excretion of Toxicants."
10. Ibid, p. 54.
11. U.S. Environmental Protection Agency: "Interim Methods for Development of Inhalation Reference Doses," Office of Health and Environmental Assessment, EPA/600/8-88/066F, Washington, D.C., August 1989.
12. Klaassen and Rozman, "Absorption, Distribution, and Excretion of Toxicants."
13. U.S. Environmental Protection Agency, "Interim Methods for Development of Inhalation Reference Doses."
14. Klaassen and Eaton, "Principles of Toxicology," pp. 12–49.
15. Clarkson, T. W.: "Effects—General Principles Underlying the Toxic Action of Metals," *Handbook on the Toxicology of Metals*, L. Friberg, G. F. Nordberg, and V. B. Vouk, eds., 2nd ed., Elsevier, Amsterdam, p. 135, 1986.

16. World Health Organization: "Recommended Health-Based Limits in Occupational Exposure to Heavy Metals," Report of a WHO Study Group, WHO, Geneva, 1980.
17. Manahan, Stanley E.: *Toxicological Chemistry and Guide to Toxic Substances in Chemistry,* Lewis Publishers, Chelsea, Michigan, 1989.
18. Sipes, I. G., and A. J. Gandolfi: "Biotransformation of Toxicants," *Casarett and Doull's Toxicology—The Basic Science of Poisons*, M. O. Amdur, J. Doull, and C. D. Klaassen, eds., 4th ed., Pergamon Press, Elmsford, New York, pp. 88–126, 1991.
19. U.S. Department of Health and Human Services: "Toxicological Profile for Benzene," Public Health Service, Agency for Toxic Substances and Disease Registry, ATSDR/TP/-88/03, Atlanta, 1989.
20. Ibid.
21. Klaassen and Eaton, "Principles of Toxicology."
22. Klaassen and Rozman, "Absorption, Distribution, and Excretion of Toxicants."
23. Clarkson, "Effects—General Principles Underlying the Toxic Action of Metals."
24. Gargas, M. L.: "Chemical-specific Constants for Physiologically-based Pharmacokinetic Models," *CIIT Activities*, vol. 11, no. 3, March 1991.
25. Klaassen and Rozman,"Absorption, Distribution, and Excretion of Toxicants," op. cit., p. 82.
26. Ecobichon, D.: "Toxic Effects of Pesticides," *Casarett and Doull's Toxicology—The Basic Science of Poisons*, M. O. Amdur, J. Doull, and C. D. Klaassen, eds., 4th ed., Pergamon Press, Elmsford, New York, p. 583, 1991.
27. Hathway, D. E.: *Molecular Aspects of Toxicology*, The Royal Society of Chemistry, Burlington House, London, 1984.
28. Kamrin, Stanley E.: *Toxicology,* Lewis Publishers, Chelsea, Michigan, 1988.
29. Bailey, James E., and David F. Ollis: *Biochemical Engineering Fundamentals*, 2nd ed., McGraw-Hill, New York, p. 94, 1986.
30. Kamrin, *Toxicology*.
31. Goldstein, J. A.: "Structure-activity Relationships for the Biochemical Effects and the Relationship to Toxicity," *Halogenated Biphenyls, Terphenyls, Naphthalenes, Dibenzodioxins and Related Products*, R. Kimbrough, ed., Elsevier/North Holland Biomedical Press, Amsterdam, p. 175, 1980.
32. Wold, S., W. J. Dunn, and S. Hellberg: "Toxicity Modeling and Prediction with Pattern Recognition," *Env. Health Perspectives*, vol. 61, pp. 257–268, 1985.
33. Goldstein, J. A.: "Structure-activity Relationships for the Biochemical Effects."
34. Schmidt, R. F., and G. Thews: *Human Physiology*, Springer Verlag, Berlin, 1983.
35. Ecobichon, "Toxic Effects of Pesticides."
36. Goyer, R. A.: "Toxic Effects of Metals," *Casarett and Doull's Toxicology—The Basic Science of Poisons*, M. O. Amdur, J. Doull, and C. D. Klaassen, eds., 4th ed., Pergamon Press, Elmsford, New York, pp. 643–645, 1991.
37. Ariëns, E. J., et al.: *Introduction to General Toxicology,* Academic Press, New York, p. 38, 1976.
38. Schmidt and Thews, *Human Physiology*.
39. Goyer, "Toxic Effects of Metals."
40. U.S. Department of Health and Human Services, "Toxicological Profile for Benzene."
41. Trevan, J. W.: "The Error of Determination of Toxicity," *Proc. Roy. Soc. London (Biol.)*, vol. 101, pp. 483–514, 1927.
42. Klaassen and Eaton, "Principles of Toxicology."
43. Merck & Co., Inc.: *The Merck Index*, Merck & Co., Rahway, New Jersey, 1983.
44. Ibid.
45. The British Crop Protection Council: *The Pesticide Manual—A World Compendium*, 9th ed., C. R. Worthing, ed., Unwin Brothers Ltd., Old Woking, Surrey, U.K.

46. Merck & Co., *The Merck Index.*
47. The British Crop Protection Council: *The Pesticide Manual—A World Compendium.*
48. Merck & Co., *The Merck Index.*
49. The British Crop Protection Council, *The Pesticide Manual—A World Compendium.*
50. Merck & Co., *The Merck Index.*
51. Ibid.
52. The British Crop Protection Council, *The Pesticide Manual—A World Compendium.*
53. Schwetz, B. A., J. M. Norris, G. L. Sparschu, V. K. Rowe, P. J. Gehring, J. L. Emerson, and C. G. Gerbig: "Toxicity of Chlorinated Dibenzo-p-dioxins," *Env. Health Perspectives*, vol. 5, pp. 87–99, 1973.
54. Pohjanvirta, R., and J. Tuomisto: "Han/Wistar Rats are Exceptionally Resistant to TCDD," II. *Arch. Toxicol. Suppl.*, vol. 11, pp. 344–347, 1987.
55. U.S. Department of Health and Human Services, "Toxicological Profile for Lead."
56. Roels, H., J.-P. Buchet, and R. Lauwerys, et al.: "Impact of Air Pollution by Lead on the Heme Biosynthesis Pathway in School-age Children," *Arch. Environm. Health*, vol. 31, pp. 310–316, 1976.
57. Foster, P. M., D. M. Creasy, J. R. Foster, L. V. Thomas, M. W. Cook, and S. D. Gangolli: "Testicular Toxicity of Ethylene Glycol Monomethyl and Monoethyl Ethers in the Rat," *Toxicol. Appl. Pharmacol.*, vol. 69, pp. 385–399, 1983.
58. Ibid.
59. U.S. Department of Health and Human Services, "Toxicological Profile for Lead."
60. U.S. Department of Health and Human Services: "Toxicological Profile for Benzene."
61. International Programme on Chemical Safety (IPCS): "Environmental Health Criteria No. 27, Guidelines on Studies in Environmental Epidemiology," World Health Organization, Geneva, 1983.
62. U.S. Department of Health and Human Services, "Toxicological Profile for Benzene."
63. Ibid.
64. OECD: "OECD Guidelines for Testing of Chemicals," OECD, Paris, 1987.
65. International Programme on Chemical Safety (IPCS): "Principles for the Toxicological Assessment of Pesticide Residues in Food," *Environmental Health Criteria*, vol. 104, WHO, Geneva, pp. 49–53, 1990.
66. OECD, "OECD Guidelines for Testing of Chemicals."
67. International Programme on Chemical Safety (IPCS), "Principles for the Toxicological Assessment."
68. Schwetz, et al., "Toxicity of Chlorinated Dibenzo-p-dioxins."
69. Ibid.
70. Pohjanvirta and Tuomisto, "Han/Wistar Rats Are Exceptionally Resistant to TCDD."
71. Ibid.
72. Henck, J. W., M. A. New, R. J. Kociba, and K. S. Rao: "2,3,7,8-Tetrachlorodibenzo-p-dioxin: Acute Oral Toxicity in Hamsters," *Toxicol. Appl. Pharmacol.*, vol. 59, pp. 405–407, 1981.
73. U.S. Department of Health and Human Services, "Toxicological Profile for Benzene."
74. Barnes, D. G., and M. Dourson: "Reference Dose (RfD): Description and Use in Health Risk Assessments," *Regulatory Toxicol. Pharmacol.*, vol. 8, pp. 471–486, 1988.
75. Hutt, P. B.: "Use of Quantitative Risk Assessment in Regulatory Decision-making under Federal Health and Safety Statutes," *Risk Quantitation and Regulatory Policy*, D. G. Hoel et al., eds., Banbury Report No. 19, Cold Spring Harbor Laboratory, New York, 1985.
76. Barnes and Dourson, "Reference Dose (RfD): "Description and Use in Health Risk Assessments."

77. IRIS: U.S. EPA Integrated Risk Information System: "Data File for Arsenic," updated January 10, 1991.
78. U.S. Environmental Protection Agency: "Superfund Public Health Evaluation Manual," Office of Emergency and Remedial Response, Washington, D.C., October 1986.
79. IRIS: "Data File for Cyanide, Nickel, and Chromium III."
80. Robbins, S. L., and R. S. Cotran: *Pathological Basis of Disease*, 2nd ed. W.B. Saunders, Philadelphia, 1979.
81. Weisburger, J. H., and G. M. Williams: "Chemical Carcinogens," *Casarett and Doull's Toxicology—The Basic Science of Poisons*, M. O. Amdur, J. Doull, and C. D. Klaassen, eds., 4th ed., Pergamon Press, Inc., Elmsford, New York, p. 129, 1991.
82. Ibid.
83. Ramel, C.: "Mutation Spectrum in Carcinogenicity," *Mechanisms of Environmental Mutagenesis-Carcinogenesis*, A. Kappas, ed., Plenum Press, New York, 1990.
84. Weisburger and Williams, "Chemical Carcinogens," p. 131.
85. Ibid., pp. 131–132.
86. Ramel, "Mutation Spectrum in Carcinogenicity."
87. Weisburger and Williams, "Chemical Carcinogens."
88. Nilsson, R., R. B. Jaeger, and M. Tasheva: "Why Different Regulatory Decisions When the Scientific Information Base Is the Same. I—Human Risk Assessment," *The Role of Science in Pesticide Management, Regulatory Toxicol. Pharmacol.*, Stockholm, May 20–22, 1992.
89. Paynter, O. E., G. J. Burin, R. B. Jaeger, and C. A. Gregorio: "Goitrogens and Thyroid Follicular Cell Neoplasia: Evidence for a Threshold Process," *Regulatory Toxicol. Pharmacol.*, vol. 8, pp. 102–119, 1988.
90. Weisburger and Williams, "Chemical Carcinogens," p. 129.
91. Gelboin, H. V., and P. O. P. Ts'o, eds.: *Policyclic Hydrocarbons and Cancer*, Academic Press Inc., New York, vol. 1, 1978.
92. Rossman, T. G., M. S. Meyn, and W. Troll: "Effects of Sodium Arsenite on the Survival of UV-irradiated *Escherichia coli:* Inhibition of a RecA-dependent Function," *Mut. Res.*, vol. 30, pp. 157–162, 1975.
93. IARC (WHO): *IARC Monographs on the Evaluation of the Carcinogenic Risks to Humans—Overall Evaluations of Carcinogenicity: An Updating of IARC Monographs*, Lyon, vols. 1–42, suppl. 7, 1987.
94. U.S. Environmental Protection Agency: "Emergency Planning and Community Right-to-Know," Section 313, Office of Pollution Prevention and Toxics, Washington, D.C., March 1998.
95. Littlefield, N. A., J. H. Farmer, P. W. Gaylor, and W. G. Sheldon: "Effects of Dose and Time in a Long-term, Low-dose Carcinogenesis Study," *J. Environm. Pathol. Toxicol.*, vol. 3, pp. 17–34, 1979.
96. Doull, J., B. A. Bridges, R. Kroes, L. Golberg, I. C. Munro, O. Paynter, H. C. Pitot, R. Squire, G. Williams, and W. Darby: "The Relevance of the Mouse Liver Hepatoma to Human Carcinogenic Risk," a report to the International Expert Advisory Committee to the Nutrition Foundation, Washington, D.C., 34 pp., 1983.
97. Weisburger and Williams, "Chemical Carcinogens."
98. Doull et al., "The Relevance of the Mouse Liver Hepatoma."
99. Nilsson et al., "Why Different Regulatory Decisions."
100. Weisburger and Williams, "Chemical Carcinogens."
101. Nilsson et al., "Why Different Regulatory Decisions."
102. *Federal Register*: Guidelines for Carcinogen Risk Assessment, U.S. Environmental Protection Agency, vol. 51, no. 185, pp. 33993–34003, September 24, 1986.
103. Johannsen, F. R.: "Risk Assessment of Carcinogenic and Non-Carcinogenic Chemicals," *Critical Reviews in Toxicology*, vol. 20, p. 351, 1990.

104. Mantel, N., and W. Bryan: "Safety Testing of Carcinogenic Agents," *J. Natl. Cancer Inst.,* vol. 27, pp. 455–470, 1961.
105. Littlefield, et al., "Effects of Dose and Time."
106. Armitage, P., and R. Doll: "The Age Distribution of Cancer and a Multistage Theory of Carcinogenesis," *Brit. J. Cancer*, vol. 8, pp. 1–12, 1954.
107. Crump, K.: "An Improved Procedure for Low-dose Carcinogenic Risk Assessment from Animal Data," *J. Env. Pathol. Toxicol.*, vol. 5, pp. 339–348, 1982.
108. *Federal Register*, "Guidelines for Carcinogen Risk Assessment."
109. Crump, "An Improved Procedure for Low-dose Carcinogenic Risk Assessment."
110. *Tox Risk*, K. S. Crump Division of Clement International Corp., Ruston, Louisiana.
111. Zeise, L., R. Wilson, and E. A. C. Crouch: "Dose-response Relationships for Carcinogens: A Review," *Environm. Health Persp.*, vol. 73, pp. 259–308, 1987.
112. Carlborg, F. W.: "Multistage Dose-response Models and Carcinogenesis," *Food Cosm. Toxicol.*, vol. 19, pp. 361–365, 1981.
113. Ibid.
114. Moolgavkar, S. H., and A. G. Knudson: "A Model for Human Carcinogenesis," *J. Natl. Cancer Inst.*, vol. 66, pp. 1037–1052, 1981.
115. *Federal Register*, "Guidelines for Carcinogen Risk Assessment."
116. Littlefield, et al., "Effects of Dose and Time."
117. Selikoff, I. J., E. C. Hammond, and J. Churg: "Asbestos Exposure, Smoking and Neoplasia," *J. Am. Med. Assoc.*, vol. 204, pp. 106–112, 1968.
118. Carson, Rachel: *Silent Spring*, Houghton Mifflin, New York, 1962.
119. Connell, D. W., and G. J. Miller: *Chemistry and Ecotoxicology of Pollution,* John Wiley & Sons, New York, 1984.
120. McCarthy, J. F., and L. R. Shugart: "Biomarkers of Environmental Contamination," *Biomarkers*, J. F. McCarthy and L. R. Shugart, eds., Lewis Publishers, CRC Press, Boca Raton, Florida, p. 10, 1990. (With permission.)
121. Suter, G.: "Ecological Endpoints," Ecological Assessment of Hazardous Waste, U.S. Environmental Protection Agency, Office of Research and Development, Environmental Research Laboratory, Corvallis, Oregon, EPA 600/3-89/013, 1989.
122. Suter, "Ecological Endpoints."
123. Ibid.
124. Ibid.
125. Athey, L. A., J. M. Thomas, J. R. Skalski, and W. E. Miller: "Role of Acute Toxicity Bioassays in the Remedial Action Process at Hazardous Waste Sites," U.S. Environmental Protection Agency, Office of Research and Development, Environmental Research Laboratory, Corvallis, Oregon, EPA 600/8-87/044, 1987.
126. Ibid.
127. American Society for Testing and Materials: *Standard Practice for Conducting Toxicity Tests with Fishes, Macroinvertebrates, and Amphibians.* Philadelphia, 1980.
128. Parrish, P. R.: "Acute Toxicity Tests," *Fundamentals of Aquatic Toxicology*, G. M. Rand and S. R. Petrocelli, eds., Hemisphere Publishing, Washington, D.C. 1985.
129. Heath, A. G.: *Water Pollution and Fish Physiology*, CRC Press, Boca Raton, Florida, p. 7, 1987.
130. Ibid., p. 49.
131. Ibid.
132. Ibid.
133. McKim, J. M.: "Early Life Stage Toxicity Tests," *Fundamentals of Aquatic Toxicology*, G. M. Rand and S. R. Petrocelli, eds., Hemisphere Publishing Corporation, Washington, D.C. 1985.
134. Ibid.

135. Heath, A. G., *Water Pollution and Fish Physiology*, CRC Press, Boca Raton, Florida, p. 7, 1987. (With permission.)
136. Adams, S. M., L. R. Shugart, and G. R. Southworth: "Application of Bioindicators in Assessing the Health of Fish Populations Experiencing Contaminant Stress," *Biological Markers of Environmental Contamination*, J. F. McCarthy and L. R. Shugart, eds., Lewis Publishers, CRC Press, Boca Raton, Florida, 1990.
137. DiGuilio, R. T.: "Biomarkers," *Ecological Assessment of Hazardous Waste,* U.S. Environmental Protection Agency, Office of Research and Development, Environmental Research Laboratory, Corvallis, Oregon, EPA 600/3-89/013, 1989.
138. McCarthy and Shugart: "Biomarkers of Environmental Contamination."
139. Connell, D. W.: *Bioaccumulation of Xenobiotic Compounds,* CRC Press, Boca Raton, Florida, 1990.
140. Ibid.
141. Connell, D. W., and G. J. Miller: *Chemistry and Ecotoxicology of Pollution,* John Wiley & Sons, New York, p. 319, 1984.
142. DiGuilio, "Biomarkers."
143. Connell, *Bioaccumulation of Xenobiotic Compounds*. Ibid., p. 169.
144. Goyer, R. A.: "Toxic Effects of Metals," *Casarett and Doull's Toxicology–The Basic Science of Poisons,* M. O. Amdur, J. Doull, and C. D. Klaassen, eds., 4th ed., Pergamon Press, Elmsford, New York, pp. 643–645, 1991.
145. DiGuilio, "Biomarkers."
146. Ibid.
147. Ibid.
148. McCarthy and Shugart, "Biomarkers of Environmental Contamination."
149. Heath, A. G.: *Water Pollution and Fish Physiology,* p. 62.
150. Ibid.
151. Connell and Miller, *Chemistry and Ecotoxicology of Pollution.*
152. Leland, H. V., and J. S. Kuwabara: "Trace Metals," *Fundamentals of Aquatic Toxicology,* G. M. Rand and S. R. Petrocelli, eds., Hemisphere Publishing Corporation, Washington, 1985.
153. Hodson, P. V., U. Borgmann, and H. Shear: "Toxicity of Copper to Aquatic Biota," *Copper in the Environment*, Part II. *Health Effects*, J. O Nriagu, ed., John Wiley and Sons, New York, 1979.
154. Howarth, R. S., and J. B. Sprague: "Copper Lethality to Rainbow Trout in Waters of Various Hardness and pH," *Water Research*, vol. 12, pp. 455–462, 1978.
155. Stiff, M. J.: "Copper/Bicarbonate Equilibria in Solutions of Biocarbonate Ion Similar to Those Found in Natural Water," *Water Research,* vol. 5, pp. 171–176, 1971.
156. McKim, J. M., and D. A. Benoit: "Effects of Long-Term Exposure to Copper on the Survival, Growth and Reproduction of Brook Trout (*Salvelinus fontinalis*)," *Journal of the Fisheries Resource Board of Canada*, vol. 28, pp. 655–662, 1971.
157. Brungs, W. A., E. A. Leonard, and J. M. McKim: "Acute and Long-Term Accumulation of Copper by the Brown Bullhead, *Ictalurus nebulosus*," *Journal of the Fisheries Resource Board of Canada,* vol. 30, pp. 583–586, 1973.
158. Benoit, D. A.: "Chronic Effects of Copper on Survival, Growth, and Reproduction of the Bluegill (*Lepomis macrochirus*)," *Transactions of the American Fisheries Society,* vol. 104, pp. 353–358, 1975.
159. Phillips, G. R., and R. C. Russo: "Metal Bioaccumulation in Fishes and Aquatic Invertebrates: A Literature Review," Ecological Research Series, EPA 600/3–78/103, 1978.
160. Collvin, L.: "Uptake of Copper in the Gills and Liver of the Perch, *Perca fluviatilis*," *Ecological Bulletins,* vol. 36, pp. 57–61, 1984.
161. Norris, R. H., and P. S. Lake: "Trace Metal Concentrations in Fish from the South Esk River, Northeastern Tasmania, Australia," *Bulletin of Environmental Contamination and Toxicology,* vol. 33, pp. 23–27, 1984.

162. Stout, D. J.: "Copper and Its Distribution in Fish Muscle Tissue in North Carolina," Master's Project. Duke University School of Forestry and Environmental Studies, Durham, North Carolina, 1986.
163. Leland, H. V., S. N. Luoma, and J. M. Fielden: "Bioaccumulation and Toxicity of Heavy Metal and Related Trace Elements," *Journal of Water Pollution Control Federation*, vol. 51, pp. 1592–1616, 1979.
164. Phillips and Russo, "Metal Bioaccumulation in Fishes and Aquatic Invertebrates."
165. Harrison, F. L.: "Effects of Copper Speciation on Aquatic Ecosystems," *Workshop Proceedings: The Effects of Trace Elements on Aquatic Ecosystems*, 1984.
166. Stout, "Copper and Its Distribution in Fish Muscle Tissue."
167. Hutchinson, T. C., and J. Fitchko: "Heavy Metal Concentrations and Distributions in River Mouth Sediments Around the Great Lakes," *Proceedings of the International Conference on Transport of Persistent Chemicals in Aquatic Ecosystems*, National Research Council of Canada, Ottawa, Canada, 1974.
168. Funk, W. H., et al.: *An Integrated Study on the Impact of the Metallic Trace Element Pollution in the Cuoer-D'Alene Spokane River Drainage System*, Washington State University and University Idaho Report to OWRR, B-044 Washington and B-015 Idaho, 1975.
169. Mount, D. I.: "Chronic Toxicity of Copper to Fathead Minnows (*Pimephales promelas, rafinesque*)," *Water Research,* vol. 2, pp. 215–223, 1968.
170. McKim and Benoit, "Effects of Long-Term Exposure to Copper."
171. Brungs, et al., "Acute and Long-Term Accumulation of Copper."
172. McKim, J. M., J. G. Eaton, and G. W. Holcombe: "Metal Toxicity to Embryos and Larvae of Eight Species of Freshwater Fish II: Copper," *Bulletin of Environmental Contamination and Toxicology,* vol. 19, pp. 608–616, 1978.
173. Mount, "Chronic Toxicity of Copper to Fathead Minnows."
174. McKim and Benoit: "Effects of Long-Term Exposure to Copper."
175. McKim, J. M., and D. A. Benoit: "Duration of Toxicity Tests for Establishing 'No Effect' Concentrations for Copper with Brook Trout (*Salvelinus fontinalis*)," *Journal of the Fisheries Resource Board of Canada*, vol. 31, pp. 449–452, 1974.
176. Brungs, et al., "Acute and Long-Term Accumulation of Copper."
177. Benoit, D. A.: "Chronic Effects of Copper on Survival, Growth, and Reproduction of the Bluegill (*Lepomis macrochirus*)," *Transaction of the American Fisheries Society,* vol. 104, pp. 353–358, 1975.
178. McKim, et al., "Metal Toxicity to Embryos and Larvae."
179. Nem'cok, J., and L. Borross: "Comparative Studies on the Sensitivity of Different Fish Species to Metal Pollution," *Acta Biologica Academie Scientifica Hungarie*, vol. 33, pp. 23–27, 1982.
180. Lloyd, R.: "The Toxicity of Mixture of Zinc and Copper Sulphate to Rainbow Trout (*Salmo gairdneri, Richardson*)," *Annals of Applied Biology*, vol. 49, pp. 535–538, 1961.
181. Sprague, J. B.: "Lethal Concentrations of Copper and Zinc for Young Atlantic Salmon," *Journal of the Fisheries Resource Board of Canada*, vol. 21, pp. 17–26, 1964.
182. Roch, M., R. N. Nordin, A. Austin, C. P. J. McKean, J. Densiger, R. D. Kathman, J. A. McCarter, and M. J. R. Clark: "The Effects of Heavy Metal Contamination on the Aquatic Biota of Buttle Lake and the Campbell River Drainage (Canada)," *Archives of Environmental Contamination and Toxicology*, vol. 14, pp. 347–362, 1985.
183. Neff, J. M.: "Polycyclic Aromatic Hydrocarbons," *Fundamentals of Aquatic Toxicology*, G. M. Rand and S. R. Petrocelli, eds., Hemisphere Publishing Corporation, Washington, D.C. 1985.
184. Moore, J. M., and S. Ramamoorthy: *Organic Chemicals in Natural Waters, Applied Monitoring and Impact Assessment,* Springer-Verlag, New York, 1984.
185. Connell, D. W., and G. J. Miller: *Chemistry and Ecotoxicology of Pollution*, John Wiley & Sons, New York, 1984.

186. Neff, "Polycyclic Aromatic Hydrocarbons."
187. Ibid.
188. Connell, *Chemistry and Ecotoxicology of Pollution.*
189. Ibid.
190. Ibid.
191. *Federal Register*: "Water Quality Criteria Documents; Availability," U.S. EPA, vol. 45, no. 231, pp. 79318–79379, 1980.
192. Ibid.
193. U.S. Environmental Protection Agency: "Quality Criteria for Water 1986," U.S. Environmental Protection Agency, Office of Water Regulations and Standards, EPA 440/5-86/001, Washington, D.C., 1986.
194. *Federal Register*, "Water Quality Criteria Documents; Availability."
195. U.S. Environmental Protection Agency: *Water Quality Standards Handbook*, U.S. Environmental Protection Agency, Office of Water Regulations and Standards, Washington, D.C., 1983.
196. U.S. Environmental Protection Agency, "Quality Criteria for Water 1986."
197. Ibid.
198. Nimmo, D. R.: "Pesticides," *Fundamentals of Aquatic Toxicology*. G. M. Rand and S. R. Petrocelli, eds., Hemisphere Publishing, Washington, D.C., p. 359, 1985.
199. *Federal Register*, vol. 56, no. 223, p. 58444, 1991.
200. Burton, G. A., ed.: *Sediment Toxicity Assessment*, Lewis Publishers, Chelsea, Michigan, 1992.
201. Baudo, R., J. Giesy, and H. Mantau: *Sediments: Chemistry and Toxicity of In-Place Pollutants,* Lewis Publishers, Chelsea, Michigan, 1990.
202. U.S. EPA, Amendments to the Water Quality Standards Regulation to Establish the Numeric Criteria for Priority Toxic Pollutants Necessary to Bring All States Into Compliance with Section 303-c)(2)(B). Proposed Rule 40 CFR Part 131.FR 56 (223): 581/20-58478, 1991.
203. Fuller, G. B., and W. C. Hobson: "Effects of PCBs on Reproduction in Mammals," *PCBs and the Environment*, vol. II, J. S. Waid, ed., CRC Press, Boca Raton, Florida, 1986.

ADDITIONAL READING

Ames, B. N.: "Dietary Carcinogens and Anticarcinogens," *Science*, vol. 221, pp. 1249–1264, 1983.
Annual Review of Pharmacology and Toxicology, Annual Reviews, Inc., Palo Alto, California.
Carson, R.: *Silent Spring*, Houghton Mifflin, New York, 1962.
Casarett and Doull's Toxicology–The Basic Science of Poison, M. O. Amdur, J. Doull, and C. D. Klaassen, eds., 4th ed., Pergamon Press, Elmsford, New York, 1991.
Clayton, G. D., and F. E. Clayton: *Patty's Industrial Hygiene and Toxicology*, 3rd ed., John Wiley and Sons, New York, 1982.
Congress of the United States, *Identifying and Regulating Carcinogens*, Office of Technology Assessment, U.S. Government Printing Office, Washington, D.C., 1987.
Doll, R., and R. Peto: "The Causes of Cancer: Quantitative Estimates of Avoidable Risks of Cancer in the United States Today," *J. National Cancer Institute*, vol. 66, no. 6, pp. 1192–1308, June 1981.
Efron, E.: *The Apocalyptics–Cancer and the Big Lie*, Simon and Schuster, New York, 1984.
Haley, T. J.: *Toxicology*, Hemisphere Publishing, Harper & Row, 1987.
Hamilton and Hardy's Industrial Toxicology, 4th ed., rev. by Asher Finkel, PSG Publishing, Littleton, Massachusetts, 1983.

Hayes, A. W., ed.: *Principles and Methods of Toxicology*, 2nd ed., Raven Press, New York, 1989.

Hayes, W. J., and E. R. Laws, eds.: *Handbook of Pesticide Toxicology*, Academic Press, New York, 1991.

Hynes, H. B. N.: *The Biology of Polluted Waters*, Liverpool University Press, Great Britain, 1971.

Johannesen, F. R.: "Risk Assessment of Carcinogenic and Noncarcinogenic Chemicals," *Critical Reviews in Toxicology*, CRC Press, Pearl River, New York, vol. 20, issue 5, pp. 341–367, 1990.

Johnson, D. W., and M. T. Finley: *Handbook of Acute Toxicity of Chemicals to Fish and Aquatic Invertebrates*, U.S. Department of Interior, Fish and Wildlife Section Resource Publication 137, 1980.

Lu, F. C.: *Basic Toxicology*, Hemisphere Publishing, Washington, D.C., 1985.

Maughan, J. T.: *Ecological Assessment of Hazardous Waste Sites*, Van Nostrand Reinhold, New York, 1993.

Nebert, D., and F. J. Gonzalez: "P450 Genes: Structure, Evolution and Regulation," *Annual Review of Biochemistry*, vol. 56, pp. 945–993, 1987.

Paustenbach, D. J., ed.: *The Risk Assessment of Environmental Hazards: A Textbook of Case Studies*, John Wiley and Sons, New York, 1989.

Rodricks, J. V.: *Calculated Risks: Understanding the Toxicity and Human Health Risks of Chemicals in Our Environment*, Cambridge University Press, Cambridge, U.K., 1992.

Sax, N. I., ed.: *Dangerous Properties of Industrial Materials*, 7th ed., Van Nostrand Reinhold, New York, 1989.

Suter, G. W., ed.: *Ecological Risk Assessment*, Lewis Publishers, Boca Raton, Florida, 1993.

Tannock, I. F., and R. P. Hill: *The Basic Science of Oncology*, Pergamon Books, 1987.

Tardiff, R. G., and J. V. Rodricks, eds.: *Toxic Substances and Human Risk—Principles of Data Interpretation*, Plenum Press, New York, 1987.

Zeise, L., R. Wilson, and E. A. C. Crouch: "Dose-Response Relationships for Carcinogens: A Review," *Environmental Health Perspectives*, vol. 73, pp. 259–308, 1987.

ACKNOWLEDGMENTS

The authors wish to acknowledge the following individuals who assisted with the preparation of this chapter for the first edition: Section 5-5, Carcinogens, was prepared by Dr. Bruce Molholt, Principal Toxicologist for *Environmental Resources Management (ERM),* and Dr. Robert Nilsson, Professor of Toxicology, *Royal University of Stockholm*. Dr. Nilsson contributed extensively to the entire chapter. Section 5-6, Ecotoxicology, was prepared by David Stout, manager of environmental science for *Blasland, Bouck and Lee, Inc.* The principal reviewer for ERM was Dr. Richard Shank.

Reviewers for the second edition were Dr. Bruce Molholt, B. C. Robison, Robin Streeter, Connie Faustini, and Randy Shuler, Ph.D., *ERM*.

CHAPTER 6

Environmental Audits

The greatest of faults, I should say, is to be conscious of none.
Carlyle

6-1 INTRODUCTION

As we enter the twenty-first century, there has been an explosion in the use of environmental auditing in corporate America. Auditing has become a standard management practice at large companies. In 1995, a survey on voluntary environmental auditing practices found that 75 percent of U.S. corporations had established some form of environmental auditing program.[1] The percentage is believed to have even further increased since 1995.[2]

Environmental audits are conducted for a wide range of goals and objectives. Traditionally, audits have been used as a preventative tool. An environmental audit derives its name from "financial audit," to which it bears some resemblance. A financial audit, developed largely in the 1800s as a means for detecting fraud, consists of an independent systematic examination of accounting records to verify that financial reports reflect generally accepted accounting principles. Similarly, an environmental audit is an independent, systematic method of verifying that environmental regulations, internal policies, and good operating practices are being followed.

The concept of an environmental audit had its beginnings in the late 1970s.[3] In comparison with a financial audit, an environmental audit has broader objectives. Today, environmental auditing can take many forms, each with its own set of objectives. Environmental auditing is recognized as a specialized field with dedicated staff and unique tools. It has become a standard operating practice at most corporations.

The EPA defines environmental auditing as a "systematic, documented, periodic, and objective review by a regulated entity of facility operations and practices related to meeting environmental requirements."[4] Such a definition stresses regulatory compliance, and verification of compliance was the driving force behind the initial developments in environmental auditing. Today, it still remains one of the main reasons for conducting audits. Audits with this objective are termed **compliance audits.**

The need for compliance audits is clear. The body of environmental law and regulations has increased in size and complexity over the past two decades, seemingly without bounds (see Figure 6-1). Violation of these requirements could result in imposition of criminal penalties, civil penalties, and injunctive relief. Additionally, enforcement by regulatory agencies is increasingly under great public scrutiny, with pressure to do even more (see Figure 6-2). In summary, the requirements are great, the price of noncompliance is high, and the chances and consequences of getting caught have increased. Thus, determining what specific regulatory requirements are imposed on an operation, finding out whether the operation is in compliance, and pinpointing possible violations in time to take proactive measures are the major goals of compliance auditing, but not the only useful purpose that an audit can serve.

A compliance audit is merely a "snapshot" of plant operations and procedures, identifying instances of either compliance with regulations or violation of them. A compliance audit is quantal and superficial. It is retrospective. More is needed. As companies have gained experience with audits, they have realized that the auditing process is very useful in extending beyond the mere identification of regulatory violations to an analysis of the root causes of any identified violations, and even the potential for future problems. This perspective, termed a **management audit,** determines whether an adequate compliance management "system" is established, implemented, and used correctly to integrate environmental compliance into everyday operating procedures.

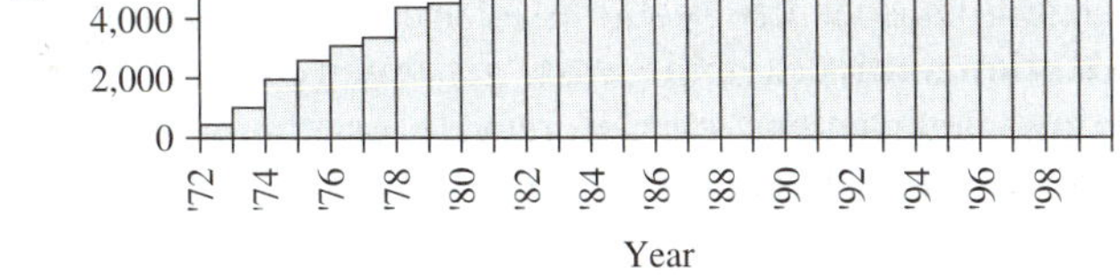

FIGURE 6-1
Growth of Federal environmental regulations. Pages in the Code of Federal Regulations totaled 16,400 in 1999—almost 10 percent growth in one year.[5]

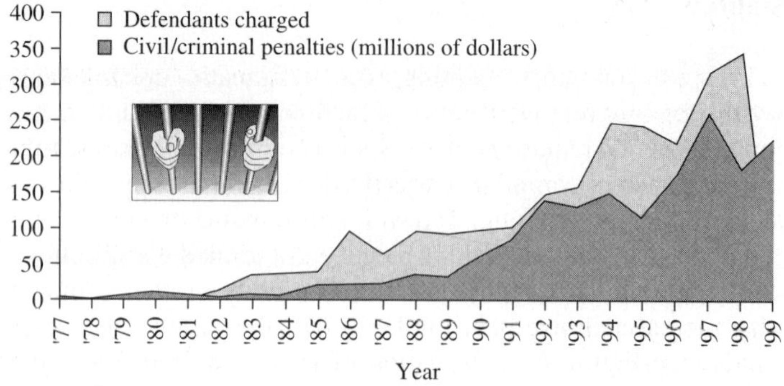

FIGURE 6-2
Trends in U.S. EPA enforcement.[6]

These types of audits are also used to evaluate a facility's adherence to either internal or external management system requirements that aim to develop, implement, maintain, and review existing environmental policies.[7] An example of this form of audit is conducted to evaluate adherence to ISO 14000 specifications and guidelines.

Such an audit examines cultural, management, and operational elements to include internal policies, human resources, training programs, budgeting and planning systems, monitoring and reporting systems, and information management systems. A management audit detects potential systematic breakdowns that could manifest themselves as an environmental problem.

Another type of environmental audit is a **liability definition audit.** These are typically done for prospective buyers of real estate and for proposed mergers and acquisitions. Such audits identify environmental problems that could reduce the value of a property or expose the buyer to liability. Not conducting such audits before acquiring property is done at the buyer's peril. The term "due diligence" implies gathering facts and information in a prudent manner and applying technical knowledge to identify where problems exist. The fact-finding of an audit helps to establish due diligence.

Other types of environmental audits are waste contractor, risk definition, and waste minimization. The **waste contractor audit** uses features of both a compliance audit and a liability definition audit to analyze commercial facilities used to store, treat, and dispose of hazardous waste. The **risk definition audit** analyzes the operations of facilities that handle hazardous materials and substances. These audits serve to help in obtaining insurance coverage and are required by some governments as part of catastrophe prevention planning. The results also provide environmental guidance for hazards not specifically addressed in current environmental regulations. **Waste minimization audits** examine waste generated by a facility with the objective of identifying viable actions to reuse, recycle, or otherwise reduce the quantity and toxicity of each waste stream (see Section 7-2).

Features of Effective Environmental Auditing

A review of mature auditing programs being conducted by both the public and the private sector has identified many common features deemed to be an important part

of an effective auditing program[8]:

- Explicit top management support for environmental auditing and commitment to follow up on audit findings.
- An environmental audit team separate from and independent of the persons and activities to be audited.
- Adequate team staffing and auditor training.
- Explicit audit program objectives, scope, resources, and frequency.
- A process that collects, analyzes, interprets, and documents information sufficient to achieve audit objectives.
- A process that includes specific procedures to promptly prepare candid, clear, and appropriate written reports on audit findings, corrective actions, and schedules for implementation.
- A process that includes quality assurance procedures to ensure the accuracy and thoroughness of environmental audits.

These features show that conducting an audit is a process rather than a mechanical application of a rigid, absolute methodology. Although an audit is systematic, as its procedures occur in a methodical manner, the procedures cannot be mechanical because much of the audit involves interpretation. The key findings of an audit typically spring from the ability of an auditor to probe in an open-ended manner, well beyond any checklist, in search of basic facts and specific patterns. In its best form, an audit is part of a continuous improvement program, with the findings of an audit triggering a number of follow-up actions, including subsequent audits.

An audit may seem to the casual observer to be an assessment. An environmental audit is more than an assessment: an audit requires clear goals, objectivity, systematic procedures, statistical verification, direct checking, and reproducibility.[9] An assessment implies an overview with less direct checking and less detail. Although both require professional judgment, an audit demands more. As an example, it is significant whether an audit collects its own samples or simply reviews existing data from previous sampling. Collecting new samples is essentially the only way to directly verify the existing sampling data. This point is so significant as to lead some practitioners to classify programs that do not sample as an assessment or review rather than true audits. Experience has shown that most environmental auditing programs do not collect samples and, therefore, these programs would be less comprehensive than formal audits.

Regardless of the type of audit being performed, it typically would unfold in four stages as follows:

1. Audit program planning (including objectives and scope).
2. Preaudit preparation (including previsit data collection).
3. On-site activities.
4. Evaluation of audit data and reporting of findings.

The next four sections of this chapter describe the approach taken in each stage of conducting a compliance audit of a plant, stressing those aspects that address hazardous wastes. As stated earlier, a compliance audit is just one type of audit, and its selection in this chapter for purposes of illustration does not diminish the importance of the other types of audits. Some other types (e.g., waste minimization and due diligence) are contrasted with compliance auditing in the closing section of this chapter.

6-2
PROGRAM PLANNING

A compliance audit program consists of many elements, including preaudit preparation and a visit to the plant. Prior to performing either of these activities, it is essential to plan the overall program. Planning would include obtaining commitment by management, defining the objectives and requirements, and developing an information management system. It is also important to organize the audit team and make sure that the program has the resources and tools necessary to address the issues that need to be investigated. In addition, one must consider the legal measures that may be appropriate to protect the results of the program. In summary, the auditing program must be tailored to meet the needs of the company's management and the facility being audited.

Commitment by Management

No compliance audit program can be successful without the clear commitment of senior management. This commitment must come before the start of the process in two ways. First, management must provide the program with the necessary resources and if necessary, direct plant personnel to provide access and to cooperate with the auditors during the process. Second, it is critical that management articulate an explicit written commitment to follow up an audit finding and correct the problems that the audit will uncover. Undertaking an audit program without a commitment to cure the ills that it may uncover is probably worse than not conducting an audit at all. Not correcting such faults suggests to employees that compliance is not a high-priority item. Further, leaving them uncorrected, perhaps with statements in a file that they were known, is viewed by a regulatory agency as flagrant disregard for the environment and will likely result in severe penalties. Under the EPA's environmental auditing policy on incentives for self-policing, finding a violation through voluntary efforts, correcting it, and self-reporting it to appropriate agencies eliminates any gravity-based (or punitive) penalties.[10]

The value of correcting faults is illustrated by the true case of two manufacturing plants with the same type of violation involving the Form R Chemical Release Report (a provision of SARA). One company had retained a consulting firm to conduct an audit, and the auditors discovered that the plant had filed incorrect reports for each of the 3 previous years; the reports had submitted miscalculated information for three chemicals and had omitted two chemicals that had substantially exceeded the reporting threshold. The plant self-reported the violations to the EPA and filed correct reports within 30 days of discovery. The plant received a fine of $2000. In contrast, a second plant operating in the same state was inspected by the EPA in the same year. The EPA inspector discovered that the information for one chemical had been substantially misreported, and the plant subsequently received a $60,000 fine. The difference derives from the fact that one plant self-reported and the other was caught by a regulatory agency.

Most companies recognize the risks of noncompliance and elect to incur the expense of an audit as part of doing business. Some small companies will elect not to

perform an audit. That is, they perceive that the risk of noncompliance does not justify the expense of conducting an audit. This decision may be made out of ignorance or believing that the chance of discovery is low or believing that any violation would be insignificant. Nevertheless, the record is replete with examples where the EPA conducted an inspection of a small company on a tip from a disgruntled employee/contractor.

Definition of Objectives and Scope

The process of performing an audit begins by selecting the overall objective of the audit (i.e., is the goal to ensure compliance, to minimize waste generation, or to look for liabilities that might be there from past practices?). After selecting the overall objective of the audit, the auditor must define the scope of the audit. Are all areas of environmental management to be covered or only one specific part (e.g., air quality or hazardous waste)? A compliance audit could cover all or just part of the various environmental management topics identified in Table 6-1. A comprehensive compliance audit would address all of these topics and perhaps related health, safety, and transportation requirements.

TABLE 6-1
Topics of an environmental compliance audit

Hazardous waste regulations	Identification Packaging Storage Waste minimization Emergency preparation Employee training Waste disposal (current) Record keeping and reporting
Past disposal of hazardous waste	Release notification procedures Remedial investigations and actions
Emergency planning requirements	Emergency response planning Emergency release reporting Hazardous chemical inventory Toxic chemical releases
Air pollution regulations	Permitting Record keeping and reporting
Water pollution regulations	Spill prevention and control plans Facility controls Discharge permitting Sampling/analysis
Underground storage tanks	Registrations, notifications Testing Inventory control Incidents/leakage
Solid waste regulations	Permitting Record keeping and reporting

After selecting the type of audit and its scope, other decisions need to be made: How will successful performance be measured over time? How frequently will the audits be done? How will the results be reported? Each of these is important to evaluate prior to designing the actual audit process. Defining these parameters in the beginning makes a major difference regarding the types of resources and techniques needed to conduct the actual audit.

Confidentiality

It may be appropriate to conduct an audit under legal arrangements to protect confidentiality. It is important to plan this carefully because such legal protections may withstand challenge only if they are built into the auditing process. Why should one consider holding an audit confidential? The reasons go beyond the direct connection of finding violations. For example, the audit report may describe proprietary business information. Also, there may be potential for an adverse community reaction to sensitive information if not communicated clearly in an appropriate forum. There is even the potential for self-incrimination if the disclosed violations carry criminal penalties. Many of these concerns will disappear by developing an effective strategy for correcting problems noted by the audit. Oral reporting will eliminate concern with disclosing highly sensitive issues. For information that clearly needs protection from public disclosure (e.g., proprietary data on processes), it is desirable to use legal protective methods.

One basis for protecting information from the requirement that it be disclosed in connection with litigation is the attorney-client privilege. Attorney-client privilege applies to communications between a client and his or her attorney. If an attorney is involved in the auditing process or if corporation counsel is available to the auditor, information may be communicated directly to the attorney. If it is anticipated that an interview may reveal sensitive information, it may be advisable to have an attorney conduct the interview.

A second basis for protecting information from discovery is the attorney work product privilege, which is a conditional privilege. At least three specific circumstances must be met to establish this privilege. First, legal counsel must have a direct involvement in conducting the audit. Second, the attorney work product is prepared in anticipation of litigation. This does not necessarily mean that litigation is ongoing or even imminent; however, litigation must be more than a remote possibility. The documentation that establishes this privilege should indicate a clear reason for the anticipation of litigation. Third, the distribution of documents prepared under attorney work product privilege must be very limited. The reports of the auditor must first go to the attorney for distribution on a limited basis. The marking of the documents as "Privileged and Confidential—Attorney Work Product" is critical, and all individuals in the process must be educated on the confidentiality and the clear requirement not to reveal the protected information to third parties. In fact, even accidentally revealing the contents of such a document may constitute a waiver of the confidentiality. The disclosure of the existence of an audit document does not constitute a waiver; however, the disclosure of any of its contents may.[11,12]

As a final point, confidentiality applies only to the reports and to the interpretations of the audit results as contained in the reports. Disclosure-limiting methods such as attorney-client privilege cannot protect the underlying facts from disclosure. For example, attorney-client privilege can protect an audit report from being disclosed and subpoenaed for use in litigation. On the other hand, detailed laboratory reports, faulty inspection records, faulty manifests, and similar items cannot be protected under attorney-client privilege. These are, in a sense, facts; they existed prior to the audit and are facts a regulatory agency has a right to see that cannot be protected under a disclosure-limiting method.

EXAMPLE 6-1. CONFIDENTIALITY. Company A is planning an initial public offering of its stock, and prior to preparing disclosure statements in connection with its offer to sell stock to the public, it conducts a compliance audit. Company B conducts a routine compliance audit program, reviewing every site every 2 years. Company C discovers a leaking underground storage tank as the result of an audit. Can any of these results be held confidential?

Solution. Company A may not be able to hold the results confidential. The audit may discover some environmental liabilities that would be required to be revealed by the rules of the United States Securities and Exchange Commission. While this is a complex area of law, in general a prospective buyer has a right to know about potential liabilities, including environmental liabilities.

Company B also may have some difficulty in holding the results confidential. As a matter of course, many companies attempt to protect the results of routine compliance audits from discovery by a regulatory agency or a party suing the company in the absence of specific statutory requirements to reveal the information.

The case of Company C is more straightforward because in certain states the company has a clear obligation to immediately report leaking underground storage tanks to the appropriate governmental authority; failure to do so could result in civil or even criminal sanctions.

Confidentiality and privilege are complex areas of the law, and legal counsel should be consulted when these questions arise.

Organization of Auditing Program

After the objectives and scope are defined for the audit program, the next step is to organize the program. There are many questions to be answered: What is the frequency of reaudits, if any? How to schedule an audit of multiple sites? Will the audit program be conducted by environmental consultants? Or will it be done by the corporate staff as part of their normal jobs? A third alternative is to have the audit managed by a small corporate oversight group with the audit function being conducted by staff from the operating divisions. Each of these three groups (consultants, corporate staff, and operating staff) has advantages and disadvantages in its approach to various issues. Mature, successful auditing programs tend to integrate participation by all three groups.

EXAMPLE 6-2. SOURCE OF AUDIT TEAM. What are the comparative capabilities and limitations of conducting an audit of a manufacturing plant with consultants, corporate staff, or operating staff?

Solution

Issue	Consultant	Corporate staff	Operating staff
Objectivity	High	Partial loss	Questionable
Auditing expertise	High	Varies	Typically nominal
Technical expertise	Broad and specialized	Industry-specific	Plant- or corporate-specific
Regulatory expertise	Broad and specialized	Industry-specific	Varies
Memory	None	Corporate memory	Corporate and plant memory
Familiarity with operations	Low	Moderate	High
Participation in solution	Nominal	Indirect	High and direct

Team of Auditing Specialists

Conducting a successful audit requires a team effort, directed by a lead auditor supported by auditor specialists. The lead auditor is responsible for efficient performance of the audit within the developed objectives and scope. The selection of the team has a major influence on the success of the audit. The size and makeup of the team depend on the size of the facility and the complexity of the issues that must be addressed.

Looking at the types of issues typically covered in an audit reveals that the area of knowledge required of the auditors is very wide. Auditors need to be "expert" in federal and state environmental regulations and sometimes local or county regulations. They need to know the specific regulations that apply to the facility's particular operations and also possess a knowledge of peer facilities. For a manufacturing facility, the team needs to know production (i.e., how a facility operates, how its manufacturing processes function, where "things" could be hidden, and how "things" are most likely to fail). Auditors need to have a knowledge of waste treatment and pollution control technology. The team needs to have a variety of scientific disciplines. They need a knowledge of information management systems and general management techniques for organizing staff and programs. In summary, a comprehensive audit team would offer specific experience in engineering, regulatory analysis, regulatory enforcement, chemistry, production, and perhaps others such as pollution control systems, geology, and community relations. An audit could function with a team of just two experienced individuals, a regulatory specialist and an engineer. More complex facilities require additional team members. For example, if specialized issues such as contamination of soil and groundwater are to be examined, additional auditors may be required.

In addition to knowledge in specific technical areas, the auditing team should have both training and experience in conducting audits. The auditors should be independent (i.e., free and capable of investigating the topic thoroughly and completely, probing appropriately through document reviews and interviews). Finally, the team

members should have sensitivity, with the ability to communicate accurately, verbally and in writing, with different levels of personnel.

6-3
PREAUDIT PREPARATION

After the overall structure of the audit program has been established and agreed on, actual preaudit preparation can begin. Preaudit preparation will include obtaining previsit information from facilities targeted for auditing, preparing the audit plan, developing the audit protocol, and establishing a schedule.

All these activities should be initiated well in advance of the anticipated date for the on-site audit activities. This will help ensure that the auditors are able to "bone up" on important regulatory issues. Providing time for modifications to the audit plan and protocol based on previsit information will also help ensure smoother on-site audit activities at the facility.

Previsit Data Collection

An audit usually requires facility staff to complete and return a previsit questionnaire that familiarizes the audit team with operations of the facility, its regulatory standing, and the environmental setting. Entering the facility without this information will result in wasting valuable on-site time instead of evaluating specific issues. It is essential to have received a completed previsit questionnaire before finalizing the audit team. The responses will identify whether special expertise may be needed on the audit team (e.g., air specialists, hydrogeologists) as well as special equipment and materials.

Alerting the auditors to the specific circumstances that they will encounter at the facility allows the auditors to focus their preparations on potential problem areas that deserve particular attention. The questionnaire also will help identify the type of information needed by the auditors, which the facility can begin to compile and organize as part of its preparations. From the responses to the completed previsit questionnaire, the auditing team will identify the key people to be interviewed and the key documents to be examined. The audit visit can then be scheduled and an agenda planned.

What type of information is typically solicited by the previsit questionnaire? It is the critical information necessary to plan the audit, including the following:

1. A management chart showing key personnel and their line and staff responsibilities.
2. A site layout showing the location of various operations, particularly the key items that should be examined (e.g., waste storage areas, sewer locations and drainage systems, underground storage tanks, and groundwater monitoring wells).
3. A brief description and block diagram of operating processes and any pollution control units, showing recent or planned modifications. Block diagrams are very

useful because they allow the audit team to anticipate the types of chemicals handled by the facility and the environmental emissions and wastes generated by the process. The team members can preview pertinent environmental regulations and compile data on specific chemicals, appropriate maintenance programs and other information that will help them to be more fully prepared for the on-site audit.

4. A list of major hazardous substances or all raw materials used at the facility.
5. A list of all types of waste generated and how each is managed.
6. A list of all environmental regulatory permits of the site, with copies. Permit information tells what type of reporting is required, and what limits are in place. The lack of permits for certain activities would alert the auditor to potential omissions on the basis of knowledge of similar operations.
7. All environmental plans and policy manuals. For example, spill prevention and control plans provide relevant information about the facility and the issues related to storage of waste and hazardous materials (e.g., inspection programs and structural and nonstructural controls that should be in place to prevent spills).
8. Information about noncompliance history of any sort. This is particularly useful because it allows the auditor to concentrate on areas that have attracted regulatory attention and would likely receive close regulatory scrutiny in the future.

The comprehensiveness of the questionnaire should conform to the size of the facility, the sophistication of facility management, and the type of support provided by the management for the auditing efforts. In any event, the questionnaire should be of moderate length (8 to 12 pages) and directed only at key factors. Examples of key questions under selected headings other than hazardous waste are as follows:

General issues

- What are the production processes at the plant?
- What is the acreage of the facility?
- How long has the facility been in operation?
- What is the past use of the site?
- Have you been inspected by any regulatory agency in the last year?
- Have you been cited for noncompliance in the last year?
- Who is responsible for environmental issues at the facility?

Hazardous materials

- What is the latest hazardous materials inventory?
- What quantity of petroleum products are stored on-site?
- Do any aboveground or underground tanks exist on-site?

Wastewater

- Where are wastewater discharge locations (point and nonpoint)?
- What permits are in place?
- Has there been any correspondence with the local sewer agency or the water board concerning compliance within the last 2 years?

Groundwater

- Does this plant have any required groundwater monitoring programs?

Air quality

- What air permits does this plant have? For what operations?

The plant should submit any available lists of items (e.g., underground storage tanks) with the completed questionnaire. Additionally, the plant staff should assemble backup information for the above items (e.g., environmental permits) and have it available when the auditors arrive.

A review of the previsit questionnaire completed by facility personnel will probably raise some questions that necessitate telephone interviews with facility management.

EXAMPLE 6-3. HAZARDOUS WASTE QUESTIONS. What are some examples of key questions and solicitations regarding hazardous waste that need to be included in a previsit questionnaire sent to a manufacturing plant?

Solution

- List monthly volumes and types of hazardous and nonhazardous solid waste generated at the facility.
- Please have available all manifests generated within last 3 years.
- List and briefly describe each waste storage area.
- How long are wastes stored?
- How are wastes stored?
- Are wastes treated or disposed of on-site?

Audit Protocol

One of the most critical audit tools is the audit protocol or the questionnaire used at the facility during the on-site audit. It is a guide to the types of questions that must be answered; it is a "road map." The use of an audit protocol ensures consistency and comprehensiveness in implementing the scope of the audit across different facilities. It forms a record of the audit procedures, becomes a basis for development of the audit findings, and also serves as a basis of any future critique of the audit. It is much more than a data inventory form.

The many different types of audit protocols range from general instructions and checklists to detailed yes/no guides that may even include scoring mechanisms. Table 6-2 describes the various types of protocols and the advantages and disadvantages of each. Figure 6-3 presents examples from three types of protocols.

All protocols are prepared with separate sections for air, water, waste, and whatever else is deemed important. Each type of protocol offers its own advantages and disadvantages. Regardless of the type, a protocol basically uses either of two approaches: **checklist** or **review and evaluation.** A checklist must be designed so that the question can be answered without leading or prompting the auditor or respondent. The checklist forces a rigid structure that may miss a subjective and critical fact. The review and evaluation approach is commonly used by more experienced auditors.

TABLE 6-2
Types of audit protocol

Type	Description	Advantages	Disadvantages
General instruction list	Divides audits into sequence of audit steps; very general in approach	Very flexible; gives broad freedom to auditors	Requires very experienced auditors; hard to score
Review and evaluation list	General topics "reminder" within a given area; details not included	Very flexible; gives broad freedom to auditors	Requires experienced auditors; hard to score
Detailed instruction checklist	Summarizes details of specific regulations and standards; aimed as method to walk auditor through steps	Consistency; fills in lack of knowledge	Bulky/long; too much detail; may miss critical decision points
Yes/no checklist	Organized summary of almost every regulation; common in companies doing first audits	Consistency; easy to use; easy synopsis	Long; rigid; tends to lead to "wrong" questions; poor on systems
Yes/no guide	More structured version of above	Consistency; very easy to use	Very long; requires interpretation; may miss critical decision points
Expanded analysis checklist	Usually added to "detailed instructions" or "yes/no" checklists to capture more detail	Allows extended response for questions that don't fit into checklist format	Long; rigid; may still miss critical decision points
Scored checklist	Add a weighted rating system to "yes/no" checklist	Allows comparison of similar facilities	Leads to "passing score" mentality
Verification checklist	Annotates checklists with instructions on verification techniques	Increases consistency and reliability of audit	Very long
Procedure-driven checklist	Addresses internal approaches/policies and systems in place to ensure compliance rather than get a snapshot of current compliance	Gives best analysis of overall potential for compliance	Requires experienced auditors; may not identify all current compliance issues

Type of audit protocol	Example
General instruction	Determine whether laboratory analytical procedures comply with 40 CFR, Part 136
Yes/no guide	Does the facility store oil or oil products above ground? *Yes* → Is total storage capacity greater than 660 gallons? *No* → No further analysis required Is total storage capacity greater than 660 gallons? *No* → No further analysis required *Yes* ↓
Yes/no checklist	Does the site have a written Spill Prevention Control and Countermeasure Plan? Does the plan include . . .

FIGURE 6-3
Examples of protocol entries.

This protocol is basically organized as a general topical outline with reminders of important points. It allows open, narrative answers with entries made to identify the basis for the determination (e.g., visual observation, interviewer's statements).

Audit protocols should not be viewed as static instruments to be placed in a file drawer after the performance of an audit. Established audit programs should periodically modify protocols to incorporate enhanced knowledge of specific facility operations and new regulatory requirements. This will reduce the potential that any staff changes in the audit team will adversely affect the quality of the audit.

Agenda and Schedule

After the preparation of a protocol, the site agenda can be prepared. The agenda lists the main issues that the audit will cover, the key personnel to be interviewed, a schedule for conducting the visit of the facility, and a list of documents and other information that the team wishes to review. Typical schedule items would start with the opening interview and the overview presentation, proceed through the site tour and each of the environmental medium-specific audits, and close with the exit interview. Excerpts from a typical audit schedule are as follows:

Day 1—Monday	
9:00–10:00 A.M.	Meet with host facility manager, environmental manager, and key operations staff to make introductions and review purpose of the audit. Refine schedules for interviews, tours, and special requirements. Receive overview presentation of facility operations.
10:00–11:30 A.M.	Tour the facility to provide audit team with orientation of overall facility. This is a brief "ride-through" of key areas of the facility.
11:30 A.M.–12:00 P.M.	Conduct briefing for facility manager.
1:00–5:00 P.M.	Complete tour of the facility. Refine audit schedule.
Day 2—Tuesday	
8:00 A.M.–5:00 P.M.	Conduct audit of hazardous waste management compliance. Physically inspect hazardous waste generation points (shops, process areas, storage yards, warehouses, treatment facilities). Inspect waste shipment manifests, waste analysis records, regulatory permits, groundwater monitoring, and scheduled inspection plans. Conduct interviews with key facility staff.
5:00–6:00 P.M.	Reconcile day's findings. Discuss follow-up needs (important to hold such meetings at end of each day).
Day 4—Thursday	
8:00 A.M.–2:00 P.M.	. . .
2:00–5:00 P.M.	Conduct audit of fuel/oil storage and spill response compliance. Physically inspect oil storage areas, tank farms, and drum storage areas. Review oil spill contingency plans, spill reports, and spill prevention and control plans. Conduct interviews with key staff.
Day 5—Friday	
8:00–10:00 A.M.	. . .
10:00–11:00 A.M.	Make an informal briefing to key environmental staff on preliminary findings of audit. Reconcile findings.
11:00 A.M.–12:00 P.M.	Conduct formal exit interview with facility manager.

A site agenda serves many purposes. It points out time conflicts among the members of the audit team. Sending the agenda to the facility in advance of the visit is imperative to reduce the disruption of operating activities and to ensure that the facility personnel are prepared for the visit (e.g., the facility can collect and organize the needed records in advance so that the audit team does not experience unnecessary delays while the information is being pulled from files).

6-4 ON-SITE AUDIT

After the acceptance of the site agenda by the facility, the team members visit the site. The primary purpose of the visit is to review documents maintained at the facility, to interview facility employees, and to inspect thoroughly all relevant operations

conducted at the facility. This is an intensive and thorough period of collecting data.

Orientation

The first step on arrival at the site is an orientation and initial kick-off meeting, typically with the facility manager. The environmental coordinator and, in some cases, many of the facility staff involved in the audit may also participate. It is important to gain early rapport with these individuals, to start on a positive note, and to ensure that the planned approach to the audit will make the most efficient use of everyone's time. Facility personnel should clearly understand that the audit is an improvement tool, not a policing effort.

The auditor should prepare handouts to describe the scope of the audit and should clearly explain that the efforts of facility staff will be needed to conduct the audit. The audit will likely disrupt the facility's schedule, and the audit team needs to acknowledge that this may represent an inconvenience, if not a problem, to the facility staff. Of course, it is important to point out the reasons for the audit and its benefits to them. This initial interview should encourage questions from the facility staff to assure that the staff members have a thorough understanding of the audit goals and the organized approach.

Several other issues must be discussed prior to proceeding with the audit. The first issue is whether the audit will be kept confidential as part of the attorney-client work product privilege. Similarly, it is important to clarify how the audit and its results will be revealed to employees and how this information will be used. Later, as the audit team proceeds through the audit of the facility, discussing issues and inspecting operations, employees will question the team as to what they are doing. The facility employees may assume, unless told otherwise, that an auditor is a regulator. Alternately, they may think an auditor is a company spy trying to check up on them. Such perceptions may evoke less than ideal responses as the audit team solicits information. Therefore, it is important to have cleared with the facility manager an agreement as to how this audit will be discussed and how information will be used. If the information will be used without attribution, facility personnel need to know this so they can be frank. On the other hand, if attribution will be made, personnel need to know this too. Also, it is important to decide with the facility manager whether personnel will receive any feedback about the results of the audit.

Safety measures and escort issues must be reviewed during orientation. In many circumstances, access to certain parts of a facility may require entry into restricted work areas or may require certain protective equipment (e.g., respirators, protective hats, and boots). This requirement must be understood prior to entering such areas and, in some cases, prior to arriving at the facility. It is important to understand what kind of emergency response signals may be given so that the audit team will respond appropriately and evacuate if necessary.

Having discussed the logistical and overview issues, the audit team should review the results of the previsit questionnaire. This is the time to expand the team's understanding of facility operations, particularly the operational setup and chemical inventory. Finally, the facility manager will need to define who will be available and

their schedules for interviews, inspections, and other elements of the audit. This may necessitate some adjustment to the previously prepared schedule.

Having completed the orientation, the audit team should take a quick tour of the facility. The goal is simply to obtain an overview of the facility, not to get into details. This tour may suggest further refinement of the previously planned schedule and logistics. After making these adjustments, the auditors should assign themselves specific responsibilities for interviews, document reviews, visual inspections, and other audit activities.

Interviewing Facility Employees

Much of the information gathered during the audit will emerge during discussions with facility personnel—as part of the protocol and as questions prompted by points discovered during visual inspections. These interviews are a particularly valuable method for obtaining subjective information. Interviews can be conducted informally in operating areas of the facility or formally in an office setting. In either case, the success of an interview, depending on the quantity and quality of the information obtained, depends largely on the techniques used by the auditor.

One necessary step in conducting a successful interview is to plan for the interview in advance. Before interviewing a person, the interviewer needs to organize thoughts, to identify issues to be discussed with the person, and to define what the desired outcome will be in terms of the types of information needed to draw conclusions.

When conducting a formal interview, the opening moments are critical to establishing both a relaxed atmosphere and a professional relationship with the person:

- Introduce yourself, your background, and your objectives.
- Ensure that the person has the appropriate amount of time available and is prepared to spend that time with you discussing the issues.
- Explain how the information will be used during the audit. Particularly, establish whether the information will be used anonymously or will be attributed. This may strongly affect the way that the person will respond.
- After the above explanation is completed, begin the information-gathering stage by requesting an overview of the interviewee's job. The response may be surprising! People will describe their job in many ways that are different from what one might expect from reading a job description or analyzing an organization chart.

As the interview progresses, it is important to continue to use the proper amount of respect. Do not prod; placing someone in a position of having to defend what they have done or just said will not reveal all the needed information. Instead, try to probe constructively with open types of questions. Do not ask questions that are just "yes/no." Instead, try to have open-ended but concrete questions that require the interviewee to express himself/herself so details will be given that otherwise could be overlooked by a yes/no checklist. Make sure that you are not using leading questions that would telegraph evaluations. Some key words in asking questions are "what," "who," and "how" (e.g., "how do you know. . . ," "how do you ensure. . ."). Silence is important in letting the person think through the response.

It is vital to provide feedback during the interview. Confirm that you have understood correctly what the person has told you by summarizing for them the information you are learning. Do not jump to conclusions; analysis and conclusions come later. When the interview is over, make sure you end on a positive note. Thank the person for the time and effort, and indicate that you will call if you have additional questions. An interview should ideally last about 20 to 30 minutes and never longer than 45 to 60 minutes. Clearly, this requires preparation and organization.

Document Review

Complying with environmental regulations, particularly those pertaining to hazardous waste, requires extensive record keeping and other documentation such as analytical results, permits, emergency plans, training and monitoring programs, and manifests. These documents contain information on the various characteristics of the facility and its operations, as well as the controls used to ensure compliance. As such, the documents warrant a thorough analysis as part of a compliance audit.

The most important documents to review are permits, because a major component of an audit is to compare operations against applicable permits. Thus, an auditor examines each permit for more information than simply that needed to ensure it has not expired. Nearly every regulatory program requires permitting. A major facility may have many permits for different activities—air quality, discharges of wastewater and storm water to natural water bodies and public sewers, currently generated and previously generated hazardous waste, other solid waste, underground tanks, groundwater, pesticides, and others. A permit covers a number of issues, including:

- A description of relevant facility equipment and operations
- The specific start and stop dates and the date for renewal
- The quantitative limitations on discharges, operations, or process inputs
- Monitoring, maintenance, and reporting requirements (e.g., quarterly reports on actual discharges compared to that permitted)

The review of a permit should include examining the application prepared by the facility prior to obtaining the permit. A permit application typically specifies equipment and procedures to be used by the facility. In many cases, these have changed since filing the original application. Not only may the current equipment and procedures not be correctly described in the application, but the facility may never have applied for a permit for the changes and thus would not have complied with the typical requirements of most regulations. It is important to determine whether the proper person from the facility, in terms of both level and authority, signed the permit application. In some circumstances, the person who did sign may not have had the proper authority.

A permit may specify limitations on discharges to the environment, the use of particular equipment, or some other activity. The permit requires periodic reports or at least some documentation of compliance with these limitations (e.g., monitoring data). An audit should ascertain what types of reports and record keeping are necessary to document compliance, whether the facility has met these documentation requirements, and whether the substance of the documentation shows that the facility is in compliance.

The need to use judgment when examining documented data, and not to take them at face value, is illustrated by an example of pH monitoring at an oil terminal. The facility had recorded daily for over a year that the pH of its discharge was always 7.0. The auditor asked the technician to take another measurement. The technician carefully obtained a representative sample in an especially clean bottle and used pH paper to measure the pH. The auditor was puzzled, however, when the technician entered 7.0 in the log book when it was obviously 6.2. It turned out that the technician was color blind and did not know it!

Some records such as manifests are not associated with a particular permit, yet are just as important. These too require auditing. In auditing any record, it is necessary not only to examine it for completeness or timely submittal but also to check the accuracy of the data. This may necessitate having to go back to original data (e.g., log books, laboratory results, strip chart, or whatever originated the record). A random sampling of the arithmetic calculations used to make the record should also be recomputed. The records for some facilities can be too extensive for manual review, necessitating the auditor to use information systems and database tracking.

During the document review process, it is important for auditors to compare environmental data recorded or reported by the facility across all regulatory program areas for consistency. All too often a facility may be reporting inconsistent information. For example, air emission estimates provided as part of air permitting activities may significantly differ from air emission estimates reported under SARA 313 for hazardous waste activities.

After analyzing the permits and record keeping, the next step is to review various types of documentation required as part of planning emergency responses. Typical of what one might find at a facility in the United States would be the Spill Prevention Control and Countermeasure (SPCC) Plan, a contingency plan for hazardous material and hazardous waste releases, a preparedness prevention plan, an OSHA 1910.120 emergency action plan, or a written plan for collecting inventory and emissions for Title III of SARA. Similarly, many regulations require written training programs and, in many cases, documented records to verify that the training has taken place at the proper time intervals. In many circumstances training programs have not been updated and thus the auditor may find that the equipment, techniques, and chemicals specified in the written training programs no longer suffice for existing operations at the facility.

The documents will list numerous items that are to be kept and maintained at the facility. For example, the contingency plan and the preparedness prevention plan will list specific items required for emergency response such as evacuation plans and certain types of emergency response equipment such as fire extinguishers, eye washes, and protective gear. The auditor should prepare a list of these items while reviewing the documentation. When conducting the physical inspection, the auditor should verify the presence of each of these items and the equipment maintenance schedules to determine whether they are in good repair.

EXAMPLE 6-4. STORAGE FACILITY. In conducting a compliance audit of a manufacturer, the previsit questionnaire indicated that spent solvents from a cleaning operation are accumulated and stored in an aboveground tank at the corner of the property. After a period of a few weeks, the waste is loaded into a rail tank car and shipped to a recycling facility. The waste is definitely hazardous because of its ignitability. You examine inventory records, manifests, inspection records, and other documents. All seems in order

until employees are unable to answer your questions regarding leak detection and fire precaution measures. What example Resource Conservation and Recovery Act (RCRA) requirements are at issue and what information as an auditor must you obtain and analyze to evaluate the issue?

Solution

RCRA issue: Have the employees received proper training? Training records must be checked and staff questioned as to their knowledge of pertinent regulatory requirements.

RCRA issue: Are leaks covered in the various contingency plans under the titles such as Preparedness Prevention and SPCC? These plans must be reviewed and the issue of secondary containment evaluated.

RCRA issue: Has the past response to any leaks complied with these plans? Inspection records must be thoroughly reviewed and the rail car loading areas carefully examined for stains.

RCRA issue: Is the storage in compliance with labeling requirements? The storage area must be inspected and markings on the tank (i.e., ignitability labels) examined.

RCRA issue: Is the tank properly located? Inspect the tank to determine if it is 50 feet from a property line (a rule for ignitable waste) and away from sparks or other ignition sources.

Note: If storage is greater than 90 days, the TSDF would be subject to permit requirements, making this example more complicated.

Data Sampling

In some cases, facility management may want all records reviewed for critical issues. In many other cases, proper environmental management can result in too many records for an auditor to check all of them. The auditor, therefore, must sometimes sample from the whole in a representative manner. The sample size and the sample itself can have a statistical basis or be based on the auditor's judgment. The method depends on the objective of the protocol.

If the objective is to define the percentage of a population which is not in compliance, the sample selection must be made in a statistically valid manner reflecting the potential for bias to enter the sampling process. For example, if the population is distributed randomly, a random number table can be used. However, if the population is stratified (e.g., a facility has three separate operations each of which files manifests), the sampling should be stratified accordingly.

If the objective is to discover any violation, the sampling may be concentrated away from subgroups expected to be in compliance. In any event, the auditor should collect the sample and not depend on the facility to do it. The sampling process and the sample must, of course, be thoroughly documented.

Visual Inspections

In conjunction with the detailed review of the various documents, the audit must conduct the detailed visual inspection of the constructed systems (e.g., treatment plant, containment devices) and the operating procedures (e.g., routine monitoring). The use of a detailed checklist of environmental inspection items is not recommended. Such a list is unwieldy and voluminous. Besides, it promotes a checklist mentality,

which experience has shown cannot capture every possible excursion from good management practices (see Sections 7-4 through 7-6).

A more fundamental approach, open to the full capabilities of human analysis and deduction, is needed. The approach starts by considering hazardous materials as well as hazardous waste. This is because the way hazardous materials are used can affect the classification of waste as hazardous. In addition, hazardous materials, if not properly managed, might lead to some hazard to employees or the environment. In fact, many hazardous wastes started out as mismanaged hazardous materials. In inspecting the use of hazardous materials, the auditor is essentially looking for good management practices, whether required by law or not, which prevent these materials from affecting human health or the environment. In many circumstances, complying with just the specific requirements of regulations will not prevent hazards. For example, a chlorinated solvent storage area is not governed by hazardous waste regulations; however, if the base of the storage area is constructed of asphalt, leaking solvents still could pass through to the soil. Over a period of time it is inevitable that leakage would occur, which would lead to soil contamination.

The safe facility is one that goes beyond regulations and institutes management practices appropriate to the hazardous material being handled, the operation that uses it, and the physical and environmental setting. Therefore, when visually inspecting a facility, the auditor must make two specific sets of considerations: whether the facility complies with regulations and whether the facility uses good management practices. Compliance inspections certainly are important because the potential for fines and other problems in not complying with regulations is obviously great. However, following good management practices in the absence of regulations can be just as important. Many Superfund sites started life as waste disposal operations that had complied with applicable regulatory requirements in effect at the time.

The art of conducting an inspection is a process of recognition and evaluation. The following is a simple summary for an audit that focuses on hazardous waste:

1. Look everywhere for hazardous materials and wastes (all wastes, not just hazardous wastes, because some waste may be improperly classified by the facility).
2. Every time a hazardous material or waste is recognized, ask:

   (*a*) How is it classified (raw material, rework, spent, waste)?
   (*b*) How is it handled?
   (*c*) Why (i.e., what is the basis)?
   (*d*) What could go wrong?
   (*e*) Are the necessary precautions in place to prevent a hazard?

Again, compliance audits typically encompass all media and not just hazardous waste. In such cases, inspections go well beyond hazardous waste and hazardous materials. The inspection will cover air pollution control devices, examining them for water discharges (from scrubbers) or solid waste generation (from baghouses). The inspection covers transformers, storm drain inlets, pollution control equipment operations, and a host of other features.

EXAMPLE 6-5. STAINING PROCESS. When conducting a compliance audit of a manufacturer of sporting equipment, you discover a staining process. Substances to be dyed are dipped into four tanks in the following sequence:

Tank 1	Methylene blue dye in alcohol
Tank 2	6N HCl in ethanol
Tank 3	Distilled water
Tank 4	Tap water

The spent dye bath is collected in 55-gallon drums, then taken to an accumulation area and shipped off site as a nonhazardous waste. The acid-ethanol mixture is drained to the municipal sewer system, as are the distilled water and tap water.

As an auditor, you must determine if the wastes are hazardous and, if so, is the method for managing the waste in compliance with hazardous waste regulations. What example RCRA requirements are at issue and what information as an auditor must you obtain and analyze to evaluate the issue?

Solution

RCRA issue: Is the waste properly classified? A review of Material Safety Data Sheets (MSDSs) and any test results to identify waste characteristics is required.

RCRA issue: Assuming that the waste is hazardous, is the accumulation in compliance with 90-day time limits and other requirements? Inspect the accumulation area and check inventory records.

RCRA issue: Is the waste disposed of improperly to the sewer system? Ask the employees to show the written permission from the local sewer authority for this discharge. Obtain a copy from the local authority of their industrial discharge requirements and compare their requirements (e.g., minimum pH) with the waste.

The spent acid-ethanol mixture would more than likely meet the characteristics of a corrosive waste (pH below 2) and thus would have to be managed as a hazardous waste. As such, action should be taken (e.g., neutralization, transport off-site to treatment facility) to prevent corroding the company's own sewer lines and causing other problems. Under the EPA's "mixing rule," mixing of the spent acid-ethanol mixture with the contents of tanks 3 and 4 would not relieve the company from having to manage the waste as hazardous.

EXAMPLE 6-6. CATALYST RECLAIMER. A small facility reclaims waste catalysts, most of which are listed as a hazardous waste under RCRA. In reviewing the accumulation records, some materials are noted that have been on-site for over 3 years without reclamation. When questioned, the manager says that because of depressed economic conditions, reclamation of these particular catalysts is not profitable. He intends to hold them until the market improves. What RCRA issue does this raise and how would you evaluate it?

Solution. Some Superfund sites are former recycling facilities that did not recycle but merely stored materials and ultimately went bankrupt. Because of this history, RCRA has specific provisions to preclude "speculative accumulation." Receipts, inventory records, and sales records must be examined to determine if the facility has met the criterion of 70 percent sales in a year.

Example Inspection—Storage Facilities

The improper storage of hazardous material and hazardous waste is one of the leading sources of environmental contamination. This section describes how storage operations should be inspected as an example of how all inspections can be approached.

A methodical inspection of a storage operation is needed to verify whether storage containers are leaking, containers are properly sealed, drip pans are used with

absorbent to immediately soak up spills, and containers are clearly marked so that the contents are quickly identified and characterized in case of an emergency or a spill. Other concerns include checking to see that aisle space is adequate, incompatible wastes are stored separately from each other, any flammables in a confined space have proper sprinkler protection, ignitable material is properly grounded, drains do not represent a ready pathway for release of leaking or spilled contents, and so forth.

Next, the auditor should evaluate the storage area itself: Is it properly secure? Are there signs posted to indicate that it is a hazardous waste or hazardous material storage area? Is the base impervious, with no cracks? Is the material used to construct the base compatible with the types of materials stored in the area? Is the emergency equipment operable and in its proper location? Is secondary containment necessary?

What are the most common sins in storing hazardous materials? The private inspections conducted by ERM during 1988–1990[13] found many examples of not complying with regulations or not employing good management practices. Table 6-3[14] summarizes these examples.

One common problem is the existence of storm drains in liquid storage and transfer areas or drains that connect to wastewater sewers. Almost 20 percent of the sites inspected had this problem. Another frequent problem is improper handling of empty drums (e.g., regarding drums with 2 to 3 inches of material on the bottom as empty, and allowing the contents to leak onto the ground). Containers for storage of hazardous waste were not always in good condition and suffered from bulges and corrosion, which can quickly lead to leakage. In many cases, drums containing ignitible waste were not grounded; sparks touched off during container filling and moving or due to static buildup might set the facility ablaze. Many inspected facilities had improper base material or otherwise inadequate containment for the wastes and hazardous materials stored in drums or tanks.

Containers are supposed to be stored completely closed. In many circumstances, employees have left funnels in the top of containers, believing that this would count as closed. However, unless funnels screw into the top and have self-closing mechanisms,

TABLE 6-3
Most common deficiencies in storing hazardous wastes and hazardous materials

Deficiency	Percent of audited facilities having this deficiency
Inadequate containment/drum/tank storage areas	68
Lack of integrity testing	65
Lack of or inadequate inspection program for storage	48
Improper labeling	44
Inadequate training program	40
Lack of or inadequate spill prevention, control, and contingency plan	34
Incomplete wastewater analysis	24
Lack of or inadequate hazardous waste contingency plan	18
Inadequate handling of materials containing PCB	18
Storm drains in liquid storage and transfer areas	18
Inadequate handling of empty drums	17

this is actually a violation of regulations and is dangerous, because if such a container were tipped, it would immediately leak. Drums are supposed to be stored stably with adequate aisle space. Because some facilities fail to do this, not all drums can be reached for inspection and leaks cannot be responded to quickly.

The storage of hazardous waste must comply with a fairly complete series of regulations. For example, an important and fundamental issue is whether the waste is properly classified as hazardous or nonhazardous. In many circumstances, facilities have misunderstood the classification system or do not correctly evaluate the waste. States may regulate some waste as hazardous that would otherwise be considered nonhazardous. Consequently, some facilities have put regulated hazardous waste into ordinary trash containers yet have classified some waste as hazardous that could have been properly disposed of as nonhazardous. The auditor should examine the "paper trail" prepared by the facility to record the decisions for classifying a waste and the basis for those decisions, whether tests, interpretations, or outside advice. The conscientiousness with which the facility made its determination is an important consideration of the auditor because the answer as to whether a waste is hazardous is not always clear.

Another problem with the storage of regulated hazardous waste is improper labeling. Hazardous waste containers require labels that clearly specify the contents and the start date for accumulation and storage. At least 40 percent of the facilities privately inspected have incomplete or improperly completed labels on some drums. One of the most common problems is an incorrect or missing start date for accumulation.

The lack of proper inspection programs for hazardous waste storage is another frequently encountered problem. Hazardous waste regulations require a fairly formalized self-inspection program of storage operations. Typical problems noted in inspections included no programs at all or lack of proper documentation of required inspections. It is incredible that some self-inspection programs have never recorded errors or any problems at all over a period of years. No storage operation is perfect, and it is highly unlikely that an adequate self-inspection program would go a year without finding at least some problem that would require an actual correction. In such cases an auditor should review the self-inspection program as documented and walk through the program with the facility employee responsible for the self-inspection, waiting for the employee to identify potential problems. In many cases, the auditor will find that the employee is overlooking important points and simply not doing an appropriate and proper inspection.

EXAMPLE 6-7. DRUM STORAGE FACILITY. One of the conditions listed on a permit for a hazardous waste storage facility states that the facility shall store no more than 2000 drums. What actions should be undertaken at the on-site audit to confirm compliance with this requirement?

Solution. The facility is expected to maintain an inventory of drums with weekly if not daily records. The inventory sheets need to be examined and the data analyzed for randomly selected days to ensure the total number of drums stored on those days is less than 2000. The employee who conducts the inventory needs to be questioned to ascertain if the techniques are adequate. The auditor should count or estimate the drums actually in storage on the day of the audit and compare this to the current inventory sheet.

6-5
EVALUATION AND PRESENTATION

The audit data are evaluated during three periods while still at the site. The first occurs in the form of immediate feedback as the auditor collects information. The second is during daily reconciliation meetings conducted on-site. The third occurs as the audit team performs a comprehensive analysis on completion of all physical inspections, interviews, and document reviews. The evaluation of data is such that the findings for the subsequent written report are developed before the audit team leaves the site.

Immediate Feedback

As an auditor gathers information, it is important to provide feedback to facility personnel. This involves communicating frequently with the personnel and identifying positive points and exceptions when found. Audits should not be conducted in an adversarial manner. The auditor should use the opportunity to explain the violation/exception to the person in charge of that situation so that he or she is able to learn the proper procedure and correct a problem. Even if facility personnel are able to correct a problem immediately, the auditor still should record the discrepancy as well as the response. An experienced auditor should consider internal procedures when reviewing documents, inspecting facilities, and interviewing personnel, specifically, whether the procedures lead to proper types of operations.

Although feedback should be provided routinely, it is important not to draw hasty conclusions. Frequently, as an auditor gathers more information, the initial conclusion may prove incorrect or there may be a reasonable explanation. This is particularly true for conclusions drawn only on the basis of a single interview. Facility personnel may contradict each other regarding procedures, thus requiring the auditor to sort out the differences. The auditor should not only acknowledge the rationale provided by a facility employee for an exception, but listen to it carefully.

Working Papers

The evaluation of the data collected as part of an audit is a major and complex effort that needs to be completed in a compressed period of time while the team is still at the site. To facilitate this evaluation, it is critical that the extensive notes from the interviews, document reviews, and inspections be organized carefully. Most auditors refer to the organization of these notes as **working papers.**

Working papers are a method for ensuring that all the steps have been taken and for organizing the information obtained from each source. They explain the rationale used for checking various items, and provide a basis for review by the audit team leader, or for others who may follow, as to the source of information and how supportive it is of the conclusions.

An auditor should include the following information in working papers: background information, all descriptive information gained from interviews and inspections, any analysis and supplementary information used to reach any conclusions.

The information is documented in the working papers as collected in the form of field notes. In recording field notes, several factors are critical:

1. Write legibly—transcribe your field notes in such a way that you can clearly identify what your thoughts were and what you gained from each interview or each step of your evaluation process.
2. Initial and date all pages.
3. Document the source of information.
4. Relate the information to the individual parts of the protocol that require such data.
5. Summarize findings and draw conclusions even if they are only on an interim basis.
6. Identify other questions prompted by an interview or inspection that require follow-up.

The purpose of these working papers is to provide an exhibit that supports all conclusions, whether positive or negative. Working papers are a factual written record that is made to support final conclusions. They should not contain hearsay statements (i.e., when the individual being interviewed provides information he or she has heard from others but does not know to be true from personal observation). When such statements raise serious questions, interviewing individuals who do have knowledge should follow up to obtain the facts to address the compliance issues.

After the working papers are completed, they should be read to make sure they would make sense out of context. At some point in the future, someone else may have to refer to them. Some large and complex facilities may have such extensive data to warrant a specialized database to track all information gathered.

Daily Conferences

In order to facilitate better communications during the performance of an audit, the audit team should hold a daily conference at the end of each day or the beginning of the following day. The purpose of these meetings is to review the interim audit findings, discuss data gaps to address, and plan the upcoming daily activities.

During the daily conferences, individual auditors should reconcile their interim findings with the findings identified by the other audit team members. These meetings should also be used to identify the necessary follow-up questions or activities required to eliminate data gaps or uncertainties in the audit findings.

Comprehensive Evaluation

The comprehensive evaluation is an integrated analysis of all information, in which all members of the audit team participate. The team as a whole reviews each step of the protocol, ensuring that each step is completed and that appropriate data-gathering techniques have been used. For example, if data gathering has required taking a sample (e.g., selecting and reviewing 100 of the 1000 manifests generated by the facility during a year), the team makes sure that the individual auditor selected an appropriate sample that correctly represents the whole.

Having reviewed each protocol step for completeness, the individual auditors should draw appropriate conclusions for their areas of the protocol and summarize each exception and observation supporting that conclusion. The supporting information should be referenced or clearly discussed. After pulling together all findings and exceptions, the team should critically review each of the findings: Did the auditor talk to the correct employees? Have all the equipment or other items that might apply been identified? Is the regulatory requirement fully understood by the auditor? Each finding must be carefully examined by the entire team to ensure that it is substantiated. The team should develop a complete list of audit exceptions and observations, then organize and summarize the list to group common exceptions and to look for patterns and trends. Team interaction is vital to stimulate the level of thinking necessary to establish patterns and to identify causes.

Exit Interview

Having completed the comprehensive evaluation, the audit team conducts an exit interview with the key employees of the site. One of the common pitfalls in an exit interview is focusing only on the deficiencies and failing to complete each step of the process. To prevent this, the audit team should prepare a discussion sheet covering all points to be addressed at the exit interview.

Typically, the exit interview will be conducted with the same group, if not an expanded group, that the audit team met on arrival at the facility. The exit interview must allow time for full discussion; it must address all findings no matter how minor. A report will eventually follow and it should not contain any surprises.

When conducting the exit interview, the audit team should set the stage. It is important to acknowledge the cooperation and the aid that the facility has provided. Rapport can be established by a willingness to discuss all the issues thoroughly and to listen to any explanation provided by facility personnel. The exit interview goes through the findings one at a time, making sure to avoid any personal attack, instead discussing methods that might be used to solve the problems. Nevertheless, a philosophy of zero tolerance for problems, however minor, should prevail. Frequently, conflicting opinions can occur. If this happens, the audit team and the facility personnel should talk the dispute out completely and document how the conflict has occurred and how to resolve it.

The final topic of the exit interview is describing the next step in the audit process—the report. The team defines how it will report the results of the audit and discusses the specific format of the report. The team should also explain how facility personnel will be able to challenge and to explain their responses to the findings of the audit team.

Audit Report

Preparation of the audit report can follow two basic approaches. First, the audit team can develop the report during the performance of the audit. The audit report can then be left with the facility personnel for their immediate review. Second, the audit

report can be prepared off-site after completion of the on-site audit activities. In this case, it is important to develop the report quickly so that the audit findings are fresh in the minds of the facility personnel.

In either case, the following guidelines should apply to the report format and style:

- Be clear and concise.
- Be sure to distinguish between isolated incidents and chronic problems.
- State the facts as discovered.
- Do not draw unsubstantiated conclusions: limit findings to the facts.
- State the nature of the problem clearly and exactly without using generalities.
- Give regulatory or management practice citations, very clearly citing where requirements have not been met; however, be very careful not to draw legal conclusions. **In particular, do not state that there is a violation.**

Two examples of an acceptable writing style are as follows:

1. A written waste analysis plan is required (40 CFR 265.13(b)(1–4)) for all hazardous wastes treated on site or stored on site. The facility's plan does not include the following two requirements listed in the regulation:
   a. Rationale for selection of parameters to be analyzed—parameters are listed but not justified.
   b. Frequency of review and repeat analysis—none is specified.

2. In reviewing a random sample of pH monitoring records for 15 days of the previous year for discharge no. 1, three excursions were identified that exceeded permit limits. We were not able to confirm that these incidents were properly reported.

An audit is an **improvement tool.** Therefore, the audit team should present recommendations in addition to a set of findings. It may be appropriate to present recommendations in a separate report so that they could be protected under the attorney work product privilege. It is helpful to present any recommendations in a "value-added format"—one that will provide the foundation for implementing corrective actions and ultimately designing an efficient, manageable facility compliance system. For each negative finding, the report should provide:

- A description of the specific requirements
- A discussion of the deficiency
- A recommended corrective action

Although the recommendations should clearly outline the type of action needed to resolve any negative findings, they should not represent complete work plans. A typical example would start with a finding, identify the requirements, estimate the potential liability, and present a recommendation as follows:

Finding: Incomplete RCRA employee training records

Requirements: RCRA requires that employees receive annual training for emergency procedures and the hazards associated with handling of hazardous wastes. RCRA requires the facility to maintain a copy of the facility's training plan and to maintain detailed training records for a minimum of 3 years.

Deficiency: Facility records were not available confirming that facility maintenance personnel who handle hazardous waste have been properly trained; a training outline or training plan had not been developed

Recommendation: Review facility's hazardous wastes and develop a training program that addresses potential hazards; incorporate facility's standard emergency response plan into a training document; designate a trainer; develop a record keeping system; update Material Safety Data Sheet (MSDS) system. Train maintenance personnel in hazardous waste handling requirements.

Finally, the audit report and its distribution must adhere to the disclosure requirements established early in the process (e.g., confidentiality or use of attorney work product privilege). Even if confidentiality is required, the auditor must assume that the report may be made public at any time because this does happen. An uninformed public can easily misinterpret results or take something out of context. A subsequent investigation may use the report as evidence. Therefore, it is important to be candid, but circumspect.

Action Plan

It can be seen that an audit program helps to troubleshoot potential violations of regulations and excursions from good management practices. However, without an action plan to address the problem and its cause, the next audit may find the same shortcoming. Therefore, after an audit report is completed, the facility (not the audit team) must develop an action plan—a written response to the audit that specifies a clear set of actions and tasks with deadlines. It also clearly identifies the persons responsible for implementing each action. A system of tracking deadlines to see that they are met should be established. Follow-through on audit findings is very important. If a regulatory agency discovers that a facility has conducted an audit and did not attempt to resolve the findings, the agency's response will probably be much harsher than if the audit had never taken place.

In summary, the goals of environmental compliance audits are:

- To obtain a clear understanding of the environmental status, risk, and liability of a facility.
- To allow timely, proactive correction of potential sources of problems within the budget process rather than having to respond to deficiencies uncovered by regulatory agencies or emergencies.
- To improve overall environmental performance of facility management systems.

The audit brings an increased attention to environmental issues, prompting facility personnel to pay more attention to these issues, and elevating the issues to the level of consciousness and concern that they deserve. With a good audit program, a facility should experience a decreasing pattern of serious discrepancies. This will provide the facility and corporate management increased confidence in the facility's control of environmental issues, including anticipating new issues as new regulations occur.

6-6
OTHER TYPES OF AUDITS

Other types of audits include management audits, waste management contractor audits, due diligence audits, and waste minimization audits.

Management Audits

The traditional audit evolved as a method to verify that a facility has instituted adequate procedures, dedicated sufficient resources, and installed the necessary systems to comply with environmental regulations. It is a mere "snapshot" of compliance. Yet, the overall environmental objective is compliance on a continuing, long-term basis. This is a function of organization, guidance, controls, and communications. Each of these functions is the subject of a management audit whose purpose is not the identification of specific violations but the detection of activities likely to experience problems and their root causes.

Specifically, a management audit addresses a number of interpretive questions under each of several broad issues as follows:

1. Clearly defined responsibilities—How are they established and communicated? How are assignments reinforced? Are any assignments overlapping, shared, or conflicting?
2. Adequate system of authority—How is it granted? Who grants exceptions? How are authorizations and exceptions recorded? Is there a real understanding of the issues?
3. Division of duties—What potential exists for conflict of interest?
4. Trained and experienced personnel—What methods are used to determine who needs training? What training have personnel had to help them understand and complete their duties?
5. Documentation—What records are developed? How are records maintained? Do type and amount of documentation correspond to the importance of the activity? How are exception reports developed? Are the records reliable? Accurate?
6. Internal verification—How does management review and evaluate environmental success? What is the accountability of managers for adherence to policies and procedures?
7. Protective measures—What engineered controls are in place to prevent noncompliance? What procedures?

A management audit detects areas likely to experience compliance problems and, more importantly, the root causes of these problems. A key component of any management audit is an analysis of how organizational resources could be used more effectively to assure compliance.

One increasingly common management audit is to evaluate conformance with the International Organization for Standardization (ISO) 14000 International Environmental Management Standards. These standards include provisions for environmental management systems and environmental performance evaluation.[15]

The ISO 14001 standard is the only one in the 14000 series that is written to be auditable. The core elements to be verifiable in the ISO 14001 standard include:

- An environmental policy signed by senior management including a commitment to compliance, the prevention of pollution, and continual improvement
- Objectives and targets developed by the facility for itself, based on its assessment of the significant environmental aspects of its operations and compliance requirements
- Implementation and control procedures and processes
- Performance monitoring and measurement with senior management review

An ISO 14001 management audit would use these core elements as benchmarks for evaluating the compliance with the standard. However, these core elements can also be used as guidance to focus the performance of any management audit.

EXAMPLE 6-8. TRAINING PROGRAMS. How would an auditor, in conducting a management audit of a California facility handling oil, hazardous waste, and hazardous materials, evaluate the adequacy of programs?

Solution. First, the auditor must review the regulations that apply to the facility and define their training requirements (see Figure 6-4). The auditor would then review the jobs and training records of selected employees who perform tasks covered by these regulations to ascertain data such as the following:

1. Besides meeting the basic requirements, does the level of training match the degree of hazard posed by the various jobs?
2. What system is used to identify employee training needs? Is this done on an annual basis? After changing positions?
3. How is employee knowledge tested?
4. Are employee records kept properly, and where are they located?

Waste Management Contractor Audits

Some generators ship a significant amount of nonhazardous and hazardous waste off-site for treatment and disposal by private contractors. A fundamental principle of waste law in the United States is that of joint and several liability for management of hazardous waste—the liability for remediation of problems created by a waste treatment and disposal contractor can rest with the generator (see Section 2-1).

When remediation is required, it can cost several orders of magnitude more than the originally contracted cost for the treatment and disposal of the waste. In addition, generators may also be financially responsible for damages to third parties derived from waste transport, treatment, and disposal operations. As a result of these potential liabilities, waste generators exercise great care in the selection of waste contractors. One tool to use in the selection of contractors, and in the verification that the selected contractor continues to operate in a prudent manner, is to audit the contractor's facilities.

A waste contractor audit follows closely the same approach of a facility compliance audit. It uses the same three phases (e.g., preaudit planning). The compliance issues are the same and it places the same emphasis on document review, interviews, and inspections. However, there are some key differences. A waste contractor audit extends beyond compliance to liabilities (e.g., potential for damages to third parties and potential for remediation). For example, the characteristics of the area surrounding the disposal facility (e.g., type of land use, presence of water wells) is much more important in an audit of a waste disposal facility. A waste contractor

Requirements	Regulations			
	CWA Spill prevention control and countermeasure plan	RCRA Contingency plan and emergency procedures	OSHA Hazard communications	California business plan requirements
Who must be trained?	Persons who operate or maintain equipment that may discharge oil, wastewater treatment operators	Persons who handle hazardous waste	Employees exposed to hazardous substances in their work area	Employees responsible for responding to hazardous emergencies
Who can do the training?	A designated person who is accountable for oil spill prevention	A person trained in hazardous waste management practices	A person familiar with workplace hazards and self-protection measures	A person with a higher level of training than the level being taught
What must be included?				
Proper hazardous materials handling procedures			✓	✓
Applicable laws, rules, regulations, and rights	✓	✓	✓	
Operation and maintenance of routine equipment	✓	✓		✓
Proper use of emergency equipment, procedures, systems, and plans	✓	✓	✓	✓
Proper waste management procedures		✓		
Methods for recognizing the release of hazardous chemicals			✓	✓
Physical and health hazards in the work area			✓	✓
Location and use of chemical and hazard information (labeling, MSDSs, plans, etc.)			✓	✓
When must training occur?	When person begins job; then periodically	Within 6 months after employment or new job; then annually	At time of initial assignment; then whenever new hazard is introduced	Annually; schedule specified in business plan
What records must be kept?	General documentation is required	Written job title and description; written description of the type and amount of training to be given; records documenting that training was completed	Written description of the type and amount of training that will be given; records documenting that training was completed	None required; however, documentation is a good management practice

FIGURE 6-4
Regulatory training requirements for Example 6-8.

audit also analyzes the financial strength of the company that owns the facility; a company possessing sufficient assets and insurance probably can absorb the costs of remediation without having to pass them to generators.

Another difference between compliance and waste contractor audits is the manner in which the on-site interviews and inspections are conducted—a waste contractor audit typically provides little ongoing feedback to facility personnel. Unlike compliance audits, improvement is not the primary objective. The objective is to identify risks, typically on a basis comparable with other facilities. Hence, a final difference between compliance and waste contractor audits is that the latter may tend to apply uniformly a detailed worksheet type of protocol in which the answers to a long checklist may possibly be scored.

Due Diligence Audits

Transactions involving real property or company acquisitions can convey the liabilities associated with contamination present in the soil or groundwater. Lenders, too, place their secured capital at risk if collateralized property is contaminated. Certain states require audits prior to the transfer of industrial property [e.g., New Jersey's Environmental Cleanup Responsibility Act of 1984 (ECRA)].

The liability for the costs of remediating a contaminated property has surprised many landowners, tenants, and lending institutions as illustrated in the following real examples:

1. A purchaser of land finds that the property was once used for a landfill; the purchaser must pay $6 million to clean up groundwater in the area.
2. A company moves from a site that they had leased for 6 years for an electronics manufacturing operation. The company is acquired 8 years later. Two years after that, groundwater contamination is discovered. The owner of the site successfully sues the acquiring company to pay for cleanup, claiming that the manufacturer was the "only tenant likely to have used a sufficient quantity of the chemical known to have caused the contamination."
3. A bank lends money to a company. When the business started to have financial difficulties, the bank stepped in and ran part of it. The bank ultimately foreclosed. Subsequently, the property was found to be contaminated. The bank was found to be liable for a major part of the remediation. The cost of remediation was 25 times the amount of the original loan.

Because the value of real property can be greatly diminished by the potential for having to remediate the site, the need to know the potential liabilities is crucial to both the buyer and seller of a property. In fact, the costs of removal of asbestos insulation or remediation of soil and groundwater contamination resulting from historic uses can greatly exceed the "pristine" value of a property. Lending institutions may lose the property's value as collateral and possibly become liable for the cleanup costs upon foreclosure. For these reasons, all prudent buyers of property or companies and virtually all commercial lending institutions require an evaluation of environmental risks before completing the mortgage transactions. In fact, due diligence audits have become a routine part of commercial real estate transactions. The purchaser of real property may establish an "innocent landowner defense" if it conducts appropriate

inquiries at the time of purchase that would establish that it had "no reason to know" that site contamination existed prior to the purchase.

The industry standard for conducting due diligence audits of real property is provided by the American Society for Testing and Materials (ASTM) Standard Practice E 1527-97. The overall goal of this standard is to identify "recognized environmental conditions" that could result in CERCLA liability at a site. According to the ASTM, the term *recognized environmental conditions* means "the presence or likely presence of any hazardous substances or petroleum products on a property under conditions that indicate an existing release, a past release, or a material threat of a release of any hazardous substances or petroleum products into structures on the property or into the ground, ground water, or surface water of the property."[16]

EXAMPLE 6-9. DUE DILIGENCE AUDIT. What parties may require a due diligence audit and why?

Solution. The buyer and seller need to know the liabilities in negotiating a price, and the lender must assure that the value of the real assets are free of major environmental liabilities. A tenant or landlord should evaluate liabilities, both on coming onto a property and on leaving it, to settle any future claims. Finally, a regulatory agency as part of its objective to protect the environment may bar the transfer of a property to its new owner until an audit is completed and the site remediated.

Elements of due diligence audit

A due diligence audit differs in two key ways from a compliance audit. It is usually conducted in a phased approach, and it focuses on historic practices and current operations that may have resulted in releases of hazardous substances. The largest liabilities often derive from historic rather than current practices, and defining liabilities encounters much uncertainty because of difficulties in detecting such problems. For example, unrecorded waste disposal operations, use of contaminated fill material, and underground storage tank leaks can easily have occurred decades ago at industrial properties. Such activities are much more difficult to discern with certainty than a determination of whether a facility has all necessary regulatory permits. Further, the emphasis to complete the transaction quickly makes time a critical factor, compressing the schedule to a matter of 2 to 4 weeks for larger transactions and often just a few days for smaller ones. The combination of tight schedules and difficult-to-define liabilities clearly reduces the certainty of the findings of a due diligence audit.

The decided advantage of having full knowledge of such liabilities during negotiations over the selling price will compel some parties to seek a thorough audit. The level of detail to which such an audit is conducted depends on the characteristics and history of the site and time constraints set by the transaction schedule. Underdeveloped or residential property obviously requires a lower level of effort than industrial facilities. For the latter, a thorough audit will include the following issues:

1. Soil/groundwater contamination.
2. Presence of hazardous waste and hazardous substances (including building construction materials).
3. Waste discharges to receiving waters/septic fields.
4. Proper permits and compliance.
5. Presence of underground tanks, waste storage areas, pits, sumps, and floor drains.
6. Special issues (e.g., asbestos, radon).

Emphasis on groundwater and soil contamination

Of these previously discussed issues, identifying whether the soil and groundwater are contaminated and defining the extent of such contamination will require the greatest effort. While a visual inspection of a property and its buildings, manufacturing processes, and hazardous material and waste handling procedures may give some indication of contamination, this must not be considered the only source of information. Instead, it is necessary to conduct an in-depth review of historical usage of the property to include specifically:

- Past chemical usage, particularly hazardous substances and materials and how they were stored and handled
- Past waste management and disposal practices
- Past spills, leaks, or pollution incidents
- Past wastewater sewer lines, lagoons, leach fields, underground tanks
- Past property development involving fill material

The buyers and sellers of property frequently have not disclosed the proposed transaction to employees at the time of an audit, severely restricting the use of employee interviews to obtain this information. Therefore, the auditors must compile and analyze the following types of data which typically are not addressed as part of a compliance audit:

1. Past aerial photographs
2. Plant and public engineering archives (e.g., construction drawings)
3. Historical fire insurance maps and street directories
4. Public records regarding enforcement and inspection activities on the site and adjacent properties by federal, state, or local environmental officials
5. Title searches for utility rights-of-way and previous owners
6. Review of federal and state lists and databases of contaminated sites reported for the property or nearby areas
7. Interviews with regulatory agency personnel with inspection authority
8. Any existing site studies
9. Public records of building permits and fire marshal permits

An audit should also address the surrounding properties if the possibility exists that their historical activities may have resulted in groundwater contamination.

Another difference between due diligence audits and compliance audits is that the former are usually conducted in a phased approach. The preceding steps are usually considered the first phase. Each successive phase, if needed, is done with increasing rigor to increase confidence that potential liabilities have been identified and evaluated as shown in Figure 6-5. The previously cited ASTM approach defines

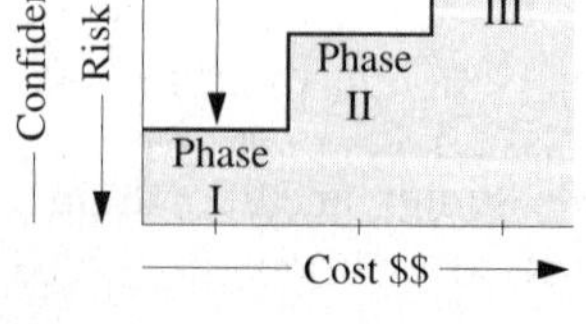

FIGURE 6-5
Phased property transfer audit.

the elements of a Phase I approach; however, it does not define subsequent phases. One can extrapolate from the ASTM definition to subsequent phases. In a simple example, the phases are illustrated as follows:

- Phase I investigates historic uses and examines current uses to conclude with a finding no more definitive than "we think we may have a problem" (i. e., Phase I is speculative).
- Phase II collects samples and conducts laboratory tests to reach a conclusion on the order of "we now know that we have a problem."
- Phase III conducts more sampling and performs engineering analysis to conclude something like "the problem is this big."

For example, the first phase may include only an on-site inspection and records review. If the results of the first phase show some indications of possible contamination, the audit would progress to a site investigation to include appropriate soil and groundwater sampling as discussed in Chapter 15. The results of the site investigation provide a basis for estimating probable remedial actions and their costs in subsequent phases.

Related Audits

The type of audit performed for property transfers is done for two other cases:

- Mergers, acquisitions, and divestitures
- Financial reporting

The full costs of environmental management are often not represented on the financial statements of target acquisitions. Acquired companies can saddle the parent with unexpected liabilities. The costs of site remediation, required pollution control systems, capital improvements, and fines can dramatically undermine the value of the acquisition if not factored into the transaction in the early stage. A due diligence audit addresses this concern by investigating the following:

- **Environmental compliance.** What are the liabilities of violations and the engineering, legal, and capital costs of bringing a facility into compliance?
- **Ability to expand production.** Are there environmental limits or constraints, such as local air pollution restrictions and permit requirements, that will prevent a company from expanding production levels?
- **Liabilities for remediation.** To what extent could remediation be required in the future?
- **Liabilities from past waste disposal practices.** To what extent could historical or current waste disposal practices result in future remediation liabilities?

Financial liabilities arising from environmental noncompliance and site remediation frequently represent material exposures requiring disclosure under public securities laws in the United States. The obligation for public companies to report such financial exposures may arise from specific environmental disclosure requirements adopted by the U.S. Securities and Exchange Commission, financial statement disclosures within the context of accounting for financial contingencies, and general accounting standards. The net effect is an obligation to disclose environment-related

exposures which may have a material effect upon the capital expenditures, earnings, and competitive position of the corporation or its subsidiaries. An audit will identify current, contingent, and future environmental exposures; anticipate the magnitude and timing of the costs; and provide a substantiated basis for adequate disclosure.

Waste Minimization Audits

Waste minimization is the reduction of any solid, liquid, or gaseous waste at its source of generation. It is a noble goal and certainly the preferred method for managing all waste. Accordingly, this book devotes the entire next chapter (Chapter 7) to the subject of pollution prevention. The first steps to successfully achieving the goal of waste minimization is an audit, consisting of planning the entire program and fully characterizing all wastes. From there, waste minimization options are identified, evaluated, selected, implemented, monitored, and communicated.

DISCUSSION TOPICS AND PROBLEMS

6-1. Prepare a protocol and conduct a compliance audit of one of the laboratories at your university.

6-2. You are auditing a refinery site to determine the likelihood of a release from its existing tanks and pipes. The site has more than 150 tanks, 75 miles of pipes, and more than 2000 valves. You have 1 day to inspect this part of the site. Devise and justify a sampling plan to determine the condition of the tanks, pipes, and valves.

6-3. Please correct or state what is wrong with the following examples of bad audit report writing:

(*a*) The audit team inspected 20 transformers and 10 capacitors of which 3 showed signs of leaks.

(*b*) Several of the drums at the hazardous waste area had no labels, as required by 40 CFR 262.

(*c*) The fire training area has been contaminated by waste solvents. The audit team recommends that the facility excavate the contaminated soils.

(*d*) In a review of the manifests generated by the site, 20 were found to be incorrectly completed.

6-4. You have been asked to audit PCB compliance at a very large site containing 65 PCB transformers or capacitors and 110 non-PCB transformers. They are spread out over the site. You wish to inspect to determine if the transformers are properly labeled as PCB.

(*a*) How would you select your sample?

(*b*) If three of the first five transformers you checked were not labeled (and should be), would you change your plan?

6-5. Review the conditions listed below for a hazardous waste storage permit and indicate the specific items you would expect to review and analyze to determine compliance with the permit. Please provide your methods by individual permit condition.

(*a*) The facility shall store no more than 5000 drums.

(*b*) The drums shall be segregated by waste type in a manner so that any spillage of incompatible materials will not mix.

(*c*) For the flammable materials stored in drums, proper fire protection methods will be used.
(*d*) No drum shall be stored longer than 45 days.
(*e*) Inspections for spillage, leakage, and poor drum quality will take place daily. Repairs will be recorded.
(*f*) No explosives or reactives may be stored for more than 10 days.

6-6. What is wrong with the following statements:

(*a*) "OK, we'll do an audit program, but we can't give a copy of the audit report to the plants!"
(*b*) "We don't need a spill plan. We don't have any spills!"
(*c*) "Well, how good did we do? What was my score?"
(*d*) "We send the solvent off to a recycler so they can reuse the hazardous waste. The sludge in the degreaser? Oh, that just goes in the dumpster."

6-7. Electrical transformers are often oil-filled, sometimes with PCB-containing oils. A common occurrence is the location of transformers near storm drains where a transformer leak or fire could result in spillage to the storm drain. What might the plant do to correct this deficiency?

6-8. Hazardous waste container storage areas must be designed to provide secondary containment. How should a facility address and handle rainwater which collects in the diked areas? One facility opened the valve and let it drain out. Is this permissible?

6-9. When asked if a facility had a copy of its air permits, a facility representative answered, "Yes," claiming that the facility was in full compliance. When the facility was asked if it had a copy of the air permit application referenced in the issued permit, a facility representative said no one had any knowledge of such a document. Why is it necessary for the plant to have the air permit application even when the facility has been issued a permit?

6-10. A facility that operates in New Jersey services vehicles and equipment, changes the oil, and repaints them. During the audit, the facility manager stated he checked the EPA regulations, and, on the basis of this information, none of the waste which was generated at the facility was hazardous. What is the deficiency in this approach, and what else should he do to arrive at a correct waste identification determination?

6-11. One facility had undergone some improvements in the process that it had used for 30 years to manufacture the same product. When asked about the air permits for the improvements, the facility manager stated, "The raw materials were the same, the products were the same; we just moved some equipment around. And besides, we're grandfathered." What is incorrect about this assumption?

6-12. A municipal sewage treatment plant produces a sludge containing toxic metals at concentrations in excess of TCLP limits. After dilution with stormwater runoff, the concentration is less than the TCLP limits. Is the sludge a hazardous waste?

6-13. A light manufacturing facility produces 125,000 gallons per week of sanitary sewage and uncontaminated wash water which are discharged to a municipal sewer system. A small parts rinsing operation flushes about 12 ounces per week of trichloroethylene down the sink. Does the facility have to manage their wastewater as a hazardous waste (see 40 CFR 261.3)?

6-14. Prepare a yes/no checklist for auditing a facility's contingency plan/preparedness prevention plan (see 40 CFR 265.50–265.56).

6-15. What would a waste management contractor audit cover in its scope that a compliance audit would not?

6-16. Describe what is meant in Example 6-2 when the consultant is described as having no memory.

REFERENCES

1. "The Voluntary Environmental Audit Survey of U.S. Business," Price Waterhouse, March 28, 1995.
2. Voluntary Environmental Audit Protection Act: Hearings Before the United States Congress, Senate Committee on the Judiciary, Subcommittee of Administration and Oversight of the Courts, One Hundred Fourth Congress, Second Session of Senate Bill 582, May 21, 1996.
3. Cahill, L. B., and R. W. Kane: *Environmental Audits,* Government Industries, Inc., Rockville, Maryland, October 1989.
4. *Federal Register:* "Environmental Protection Agency's Environmental Auditing Policy Statement," vol. 51, no. 131, July 9, 1986.
5. United States Government Printing Office Code of Federal Regulations, 1999.
6. U.S. Environmental Protection Agency Enforcement and Compliance Assurance FY99 Accomplishments Report.
7. ANSI/ISO 14001, Environmental Management Systems—Specifications with Guidance for Use, American Society for Testing Materials, 1996.
8. *Federal Register:* "Environmental Protection Agency's Environmental Auditing Policy Statement."
9. ANSI/ISO 14010, Guidelines for Environmental Auditing—General Principles, American Society for Testing Materials, 1996.
10. "USEPA Incentives for Self-Policing: Discovery, Disclosure, Correction, and Prevention Violations," *Federal Register,* vol. 60, no. 246, Friday, December 22, 1995.
11. Frost, E. B., and S. Seigel: "Environmental Audits: How to Protect Them from Disclosure," *Toxic Law Reporter,* p. 1211, February 17, 1991.
12. Reed, P. D.: "Environmental Audits and Confidentiality: Can What You Know Hurt You as Much as What You Don't Know?" Environmental Law Institute, p. 303, October 1983.
13. Roig, Randy, ERM-West, Inc: Personal communication, August 1990.
14. Source: Audits of private facilities conducted by the ERM Group.
15. EPA Standards Network Fact Sheet, ISO 14000: International Environmental Management Standards, USEPA, EPA/625/F-97/004, April 1998.
16. American Society for Testing and Materials (ASTM) Standard Practice E 1527-97, "Standard Practice for Environmental Site Assessments: Phase I Environmental Site Assessment Process."

ADDITIONAL READING

Blakeslee, W. H., and T. M. Grabowski: *A Practical Guide to Plant Environmental Audits,* Van Nostrand Reinhold, New York, 1989.

Friedman, F.: "Why Not to Be a Manager? Is This Job Really Worth It?" *Environmental Forum,* p. 20, May/June 1991.

Friedman, F. B.: *Practical Guide to Environmental Management,* Environmental Law Institute Monograph, 1990.

Frost, E. B.: "Voluntary Environmental Compliance Audits: A DOJ Policy Failure," *Toxic Law Reporter,* BNA, p. 499, September 18, 1991.

Greeno, J. L., et al.: *The Environmental Health and Safety Auditor's Handbook,* Arthur D. Little, 1988.

Moore, J., et al.: "Why Risk Criminal Charges by Performing Environmental Audits?" *Toxic Law Reporter,* BNA, p. 499, September 18, 1991.

U.S. Department of Justice: "Factors in Decisions on Criminal Prosecutions for Environmental Violations in the Context of Significant Voluntary Compliance or Disclosure Efforts by the Violator," Washington, D.C., July 1, 1991.

ACKNOWLEDGMENTS

The authors wish to acknowledge the following individuals who assisted with the preparation of this chapter for the first edition: Randy Roig, Ph.D., Brian Bennett and Ann Sirois, Esq., *ERM*. The principal reviewer for ERM was Howie Wiseman.

The reviewer for the second edition was Lance Traves, RHSP, CHMM, *ERM*.

CHAPTER 7

Pollution Prevention

Conspicuous consumption of valuable goods is a means of reputability to the gentleman of leisure.
The Theory of the Leisure Class, 1899,
Thorstein Veblen

In the previous chapter, methods were presented to identify environmental problems at the origin of waste generation. This chapter examines practices that reduce or eliminate hazardous waste at the point of generation.

7-1 GENERAL CONSIDERATIONS

Definitions

Many terms have been used to describe the activities that eliminate or reduce the generation of hazardous waste: waste minimization, waste reduction, source reduction, waste diversion, pollution prevention, recycling, and reuse. In a 1986 report to Congress, the U.S. Environmental Protection Agency (EPA) defined waste minimization as the reduction of volume or toxicity of waste:

> . . . the reduction, to the extent feasible, of hazardous waste that is generated or subsequently treated, stored, or disposed of. It includes any source reduction or recycling activity undertaken by a generator that results in either (1) the reduction in total volume or quantity of hazardous waste, or (2) the reduction of toxicity of hazardous waste, or both, so long as the reduction is consistent with the goal of minimizing present and future threats to human health and the environment.[1]

In a 1989 policy statement,[2] the EPA suggested the following hierarchy for management of wastes:

- Source reduction
- Recycling
- Treatment
- Disposal

In this chapter the term **waste minimization** applies to any management technique or process modification that ultimately reduces the mass or toxicity of waste sent on to treatment and disposal facilities. The term does not include any action that merely concentrates the hazardous contents of a waste to reduce waste volume or dilutes it to reduce degree of hazard. The somewhat broader term **pollution prevention** has an evolving definition that includes:

- Managing chemicals to reduce risk
- Identifying and estimating all releases
- Waste minimization

An overview of pollution prevention techniques is illustrated in Figure 7-1. This chart provides a clear establishment of priorities in hazardous waste management, ranging from changing products to ultimate disposal of hazardous wastes that cannot be avoided.

Historical Overview

A common perception is that pollution prevention efforts began in the 1980s. Actually there are numerous examples from the nineteenth century of industrial practices that reduced waste generation. At the time it was characterized as simply good en-

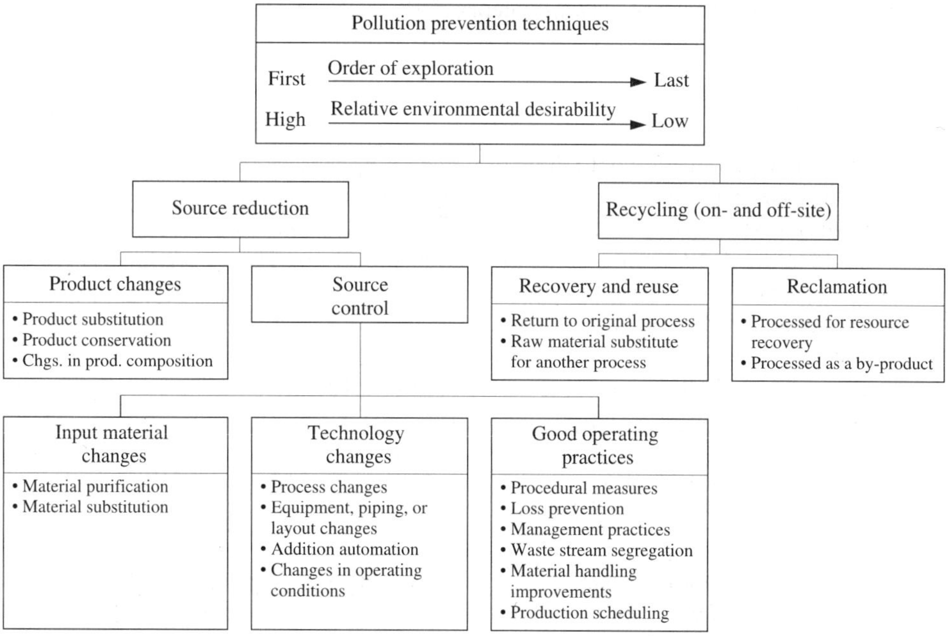

FIGURE 7-1
Pollutant prevention techniques.

gineering economics to create a salable product rather than a waste or to reduce the cost of raw materials. The early twentieth century contains numerous examples of waste reduction efforts.[3] A book on waste utilization was published in London in 1915.[4] The technical literature of the mid-twentieth century includes many articles on waste reduction, such as those by Trembler[5] and Besselievre[6] to cite two examples. In a 1964 paper on industry's approach to pollution abatement, Rocheleau and Taylor of E. I. DuPont de Nemours & Co. wrote:

> Emphasis has been placed on in-plant control measures, because operation of a wastewater treatment works is only part, often a small part, of industrial control problems. . . . It is now standard policy with nearly all chemical companies to apply effective wastewater controls as soon as possible in the development of a new or improved process. If a potential waste problem can be recognized early enough, it can often be prevented from becoming a problem, or at least it can be minimized.[7]

It is somewhat ironic that industries that were early pioneers in waste reduction are now being penalized for being too forward thinking. As an example, "save-alls" were originally installed in paper mills in the 1940s as a way of recovering fibers, and preventing them from becoming a source of pollution. They became so effective and widespread that they are now considered part of the manufacturing process, and no credit would be given today to this device as a waste minimization technique, because it comes as a standard part of paper manufacturing equipment and is thought of as standard operating procedure.

What is new today are the dramatic increases in the cost of disposal, legal, and regulatory incentives and public demands that industry simply stop producing hazardous wastes. In particular, the "cradle-to-grave" concept of hazardous waste regulation and associated liabilities causes all industries to develop means and measures for reducing hazardous waste generation.

Economic Incentives

It is common practice today to think of wastes simply as resources out of place. It is often the case that a second look at the manufacturing process can identify additional ways that wastes can be reduced, sometimes at little or no cost. A frequently cited example is the 3M Company's 3P (Pollution Prevention Pays) program. A recent article reports the following amounts of waste eliminated as a result of the company's source reduction programs:

Waste type	Eliminated	United States only
Air pollutants, tons	121,000	110,000
Water pollutants, tons	14,600	13,500
Sludge/solid waste, tons	314,000	303,000
Wastewater, 10^6 gallons	1,600	1,000

Since its inception in 1975, "The processes and product changes made to achieve these reductions have yielded cumulative first year savings of $420 million (U.S. savings: $350 million)." These savings were achieved in such diverse ways as savings on capital costs of unneeded pollution control facilities and sales from new products from former wastes.[8]

Several changes in the economic environment justify a second look at the way a manufacturer handles waste. The most dramatic is the sharp increase in the cost of hazardous waste disposal. A 1988 report from DuPont suggests that the cost of managing a pound of hazardous waste is escalating at a rate of 20 to 30 percent annually.[9] Recycling proposals that were economically infeasible only a few years ago now look very cost-effective. In fact, with costs of hazardous waste disposal ranging between \$200 and \$1000 per 55-gallon drum, depending on the chosen method of disposal, recycling and waste minimization can be critical parts of increasing profitability.

An important aspect in evaluating the economics of environmentally desirable proposals is to examine the "big picture" and not simply focus on microeconomics at the plant level. This is illustrated in Figure 7-2, which examines the competitive advantages that can accrue to corporations from waste minimization activities. This big picture approach is discussed in detail in Section 7-3, Life Cycle Analysis.

Legal and Regulatory Incentives

A second motivation for conducting waste minimization evaluations in the United States is that they are required by law. In the 1984 amendments to RCRA, known as the Solid and Hazardous Waste Amendments, Congress defined as national policy that "... the generation of hazardous waste be reduced or eliminated as expeditiously as possible." In order to ship hazardous waste to a disposal facility, generators are now required to certify that they have a waste minimization program in place. While this requirement lacks a strong method of enforcement, new legislation, including Pollution Prevention and Air Toxics legislation and RCRA reauthorization, include regulatory "hammers" that will force recalcitrant industries into waste minimization programs.

Perhaps the most effective legislation does not deal with waste minimization at all. Under the Emergency Planning and Community Right-to-Know (EPCRA) provisions of SARA (see Section 2-4), industries are required to furnish information

Competitive advantage from:	Product design	Raw materials	Manufacture, sales, and distribution	Packaging	Use	Final disposal
	• Chemical usage • Ease of recycling • Ease of reuse • Ease of disassembly (e.g., modular design) • Size	• Use of renewable resources • Reducing impact from transport	• Pollution prevention, waste minimization • Process hazards management	• Reduced packaging • Vegetable-based inks • Concentrating product	• Aerosol-free cleaning agents • Low-dosage pesticides	• Recycling • PET versus HDPE
	Management and organization		Cross-functional management strategies, integrating environmental expertise into decision-making processes			
	Technologies		Creating new technologies that are environmentally sound			

FIGURE 7-2
Adding value from an environmental standpoint: competitive advantage.

on toxic waste releases to the local community, providing data that were not generally known before. Not surprisingly, many communities have reacted negatively to the knowledge that a local plant is releasing toxic compounds into the air that they breathe.

The 1990 Pollution Prevention Act in the United States was designed to reduce the generation of waste by a range of institutional measures including:

- Establishing a national pollution prevention strategy
- Providing grants to states to establish programs to promote source reduction
- Establishing a source reduction clearinghouse

This Act requires industries to provide detailed annual reports on the effectiveness of source reduction for all chemicals that they must already report under the community right-to-know provisions of SARA. These reports are known as Annual Waste Minimization reports. Because these reports will become public information, this Act provides another strong incentive to develop an effective source reduction program.

In February 1991, the EPA instituted a voluntary 33/50 program which had an overall national goal of 50 percent reduction in the generation of 17 high-priority toxic chemicals by 1995. The program had an interim goal of 33 percent reduction by the end of 1992. The 17 chemicals listed in Table 7-1 accounted for 1.4 billion pounds of releases and transfers within the United States in 1988.

In 1995, the 50 percent goal was realized when 750 million pounds of the 17 chemicals were reported, down from the nearly 1.5 billion pounds reported in 1988. The total reduction of these 17 chemicals between 1988 and 1996 was 896 million pounds, a decrease of 60 percent.[11] The program, having met its goals, was eliminated in 1996.

TABLE 7-1
Target chemicals covered by the EPA's 33/50 program[10]

Chemical	Quantity released in 1988, million pounds
Benzene	33.1
Cadmium	2.0
Carbon tetrachloride	5.0
Chloroform	26.9
Chromium	56.9
Cyanide	13.8
Dichloromethane	153.4
Lead	58.7
Mercury	0.3
Methyl ethyl ketone	159.1
Methyl isobutyl ketone	43.7
Nickel	19.4
Tetrachloroethylene	37.5
Toluene	344.6
1,1,1-Trichloroethane	190.5
Trichloroethylene	55.4
Xylene	201.6

Another legal inducement is the potential future liability for any hazardous wastes produced. As pointed out in Chapter 2, an industry that generates a hazardous waste is responsible for that waste forever. Because these future liabilities are of unknown magnitude, many industries do not want to ship any wastes to what may become future Superfund sites. Finally, the cost of insurance, when it is available at all, has risen dramatically.

Public Demands

Another important incentive is that the public will no longer tolerate the continued production of hazardous waste. This intolerance manifests itself in many forms, including the new legislation cited previously and consumer boycotts of products from companies perceived not to be environmentally responsible. As will be seen in Section 8-5, it has become virtually impossible to site new hazardous waste disposal facilities and many existing facilities have closed. Much of the opposition to proposed new facilities is generated by genuine, even if mistaken, fears of excessive health risks. Another major source of opposition is the perception that industry has no need to produce these wastes in the first place and the proposed facility is simply not necessary. One of the more effective groups at preventing the siting of new facilities notes in its newsletter:

> By slowing and stopping the siting process, communities are forcing industry and government to look hard at why people are opposing existing treatment and disposal methods. . . . After all, what makes more sense than to avoid generating waste in the first place.[12]

It is a *sine qua non* with such groups that all hazardous waste generation can and should be eliminated. While most environmental engineers and industrial managers would consider this a clearly unachievable objective, the minimization of waste is unquestionably an appropriate goal. To illustrate how waste minimization has been effective, the president of a leading commercial hazardous waste company stated: "Before waste minimization was enacted there were 350 million tons of waste generated annually; now that number is 250 million tons a year."[13]

7-2 MANAGEMENT STRATEGIES

Overview

Given the goal of minimizing the generation of hazardous waste at a manufacturing facility, how does one proceed? The following outline is suggested:

1. Planning and organization
2. Characterization of waste and losses
3. Development of waste minimization options
4. Technical, regulatory, and economic feasibility
5. Implementation (including training)
6. Monitoring and optimization
7. Continued and ongoing evaluations of reaching a "zero" generation status.

The first step includes obtaining a commitment from management and establishing a clear set of goals. Typically this step would also include identifying individual responsibilities and organizing the group of individuals who will establish the Waste Minimization Plan.

The second step is to understand the types and sources of wastes—an outcome of the audits discussed below. Steps 3 and 4 analyze this information in a framework that includes a range of perspectives: environmental, economic, legal, political, and technical. The end result of this analysis should be a comprehensive Waste Minimization Plan. Implementation of this plan must concern virtually all aspects of the company including research and development, marketing, production, training, maintenance, raw materials procurement. Finally, these efforts should be examined to ensure that they are accomplishing the objectives in an optimal way. The EPA has divided these activities into four phases as illustrated in Figure 7-3.

Audits

The steps for successfully achieving the goal of waste minimization begin with an audit following the methods outlined in Chapter 6. The waste minimization audit

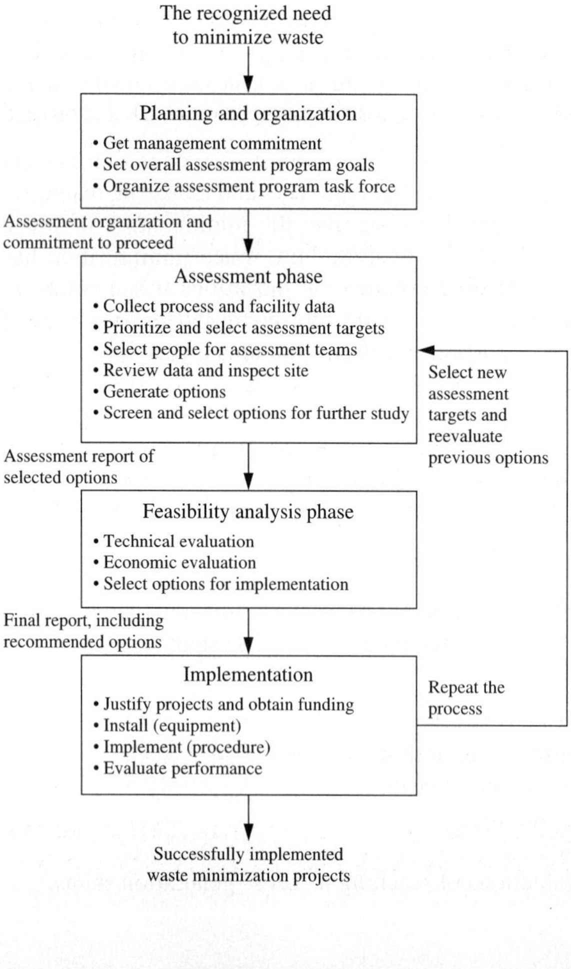

FIGURE 7-3
EPA's waste minimization assessment procedure.[14]

consists of planning the entire program and fully characterizing all wastes. From there, waste minimization options are identified, evaluated, selected, implemented, monitored, and communicated.

The planning of a waste minimization program is critical to its success. Planning starts with obtaining commitment of top management, setting goals, establishing incentives, and budgeting financial and human resources. Planning also includes communicating the policy and goals to facility employees, designating a person to lead the program, and organizing a team to support it. It also includes preparing schedules, worksheets, and other necessary tools.

The primary functions of a waste minimization audit are to characterize waste generation and to identify obstacles to source reduction. The formality of a waste minimization audit depends not only on the complexity of the facility and its wastes but also on the stage of waste reduction that the facility has already implemented. Hirshorn and Oldenburg suggest four stages[15]:

1. *Commonsense waste reduction*—immediate opportunities without having to change production processes. Examples are reducing the amount of water used in cleaning equipment and ceasing the comingling of hazardous and nonhazardous waste.
2. *Information-driven waste reduction*—relatively easy opportunities resulting from simple changes in production prompted by information on successful cases in similar industries. Examples are changing raw materials (e.g., replacing an organic solvent with a water-based product, or installing a closed-loop recycling unit).
3. *Audit-dependent waste reduction*—opportunities requiring capital investment and production changes that seem to have a questionable economic payback without the detailed information of a formal audit.
4. *Research-and-development-based waste reduction*—opportunities facing severe technical obstacles that require extensive study, if not research, to overcome.

The first and second stages can be implemented with a fairly simple audit using a checklist. The third stage requires a formal audit. A formal waste minimization audit surveys each operation that generates waste to determine how the waste is generated, what are its characteristics, how it is managed, and what are the costs. This provides baseline conditions for evaluating progress toward meeting waste minimization goals. It also provides the necessary data to prioritize waste streams and to identify alternative options for minimizing the high-priority wastes.

The waste minimization audit includes a thorough survey to provide the following data for each hazardous waste stream:

- Specific sources
- Generation rate
- Physical characteristics
- Chemical characteristics
- Production rate of process/operation generating the waste
- Current management methods from generation to final disposition
- Costs of each management method

The first step in the waste survey is to assemble and organize existing information on all waste streams, not just hazardous waste. This can come from a variety of

sources to include mandatory reports required on a one-time-only or periodic basis by regulatory agencies (e.g., annual summaries, quarterly reports, waste shipment manifests, contracts for disposal and tolling services, permits and permit applications, laboratory data, and other similar sources).

The assembled sources of information provide a basis for identification and initial characterization of each waste stream. The second task is to relate each generated waste directly to the specific operation or process responsible for its generation. Necessary sources of information to complete this include:

- Process flow diagrams
- Material and heat balances (both design and actual)
- Operating manuals and process descriptions
- Equipment lists
- Equipment specification and data sheets
- Piping and instrument diagrams
- Plot and elevation plans
- Equipment layouts and work flow diagrams

The characteristics of a waste depend not only on the process design but how the process is operated and what raw materials are used as inputs to the process. Therefore, other supporting sources of information include[16]:

- Product composition and batch sheets
- Material application diagrams
- Material safety data sheets
- Material purchase records
- Product and raw material inventory records
- Operator data logs
- Operating procedures
- Production schedules
- Production records
- Maintenance procedures
- Maintenance records

Methods

Perhaps the simplest waste minimization efforts involve housekeeping changes. Such items as segregating hazardous waste, sweeping floors prior to washing, and training employees to be sensitive to the implications of their actions can dramatically reduce waste generation. Only after these relatively inexpensive measures are undertaken should production changes be contemplated.

EXAMPLE 7-1. OFFICE BUILDING. A downtown office building had large transformers located below the street level. Since air cooling was not feasible, groundwater from a well was pumped and treated with chlorine gas to cool the transformers. The groundwater was then treated to remove the chlorine so it could be discharged to the river through a permitted NPDES discharge. For years, environmental engineers processed the necessary extraction and discharge permits and facility personnel maintained the chlorination and dechlorination systems, at an annual cost of approximately $140,000. Chlorine cylinders were located in a small room below the street that was vented to the street level, near

Where capital expenditures are contemplated, it must be recognized that pollution prevention projects must compete with other proposed capital projects for limited company funds. Capital costs for new waste minimization equipment are often considered a drain on a company's profitability. In order to fully appreciate the actual cost of such equipment, it is necessary to look at the savings resulting from reduced cost of waste disposal. Economic methods applied to such decisions include estimates of annualized cost savings, internal rate of return, and present worth analysis.

A theoretical econometric model developed by Anderson[18] incorporates the analysis of energy and material flows with waste generation to provide a method of evaluating approaches to waste reduction (see Figure 7-4).

The total cost of production is estimated by:

$$C = C_1 R + C_2 E + C_3 W_m + C_4 W_e + C_5 L \tag{7-1}$$

where C_i = appropriate unit costs
R = amount of raw material
E = energy usage
W_m = waste from material input
W_e = waste from energy usage
L = labor usage

Anderson provides generalized production curves, based on mass and energy balances, showing the relationship between mass of raw material (R, abscissa), agent input (e.g., capital, labor, etc.) (K, ordinate) and the mass of product output (Q, curve parameter). The optimization is illustrated in Figure 7-5. The equation for the curves is:

$$Q = \frac{\nu R}{m} \tag{7-2}$$

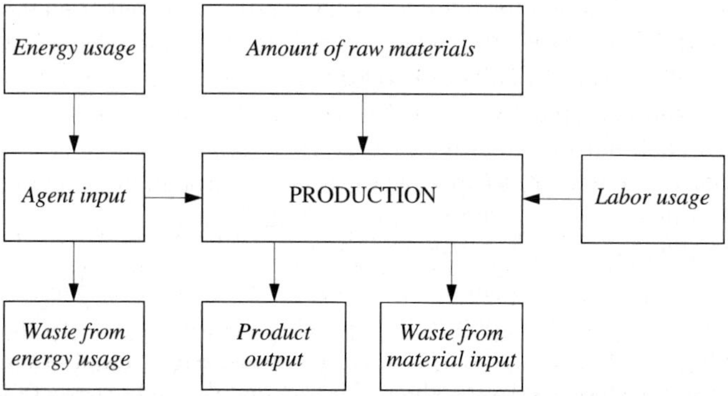

FIGURE 7-4
An economic production model.

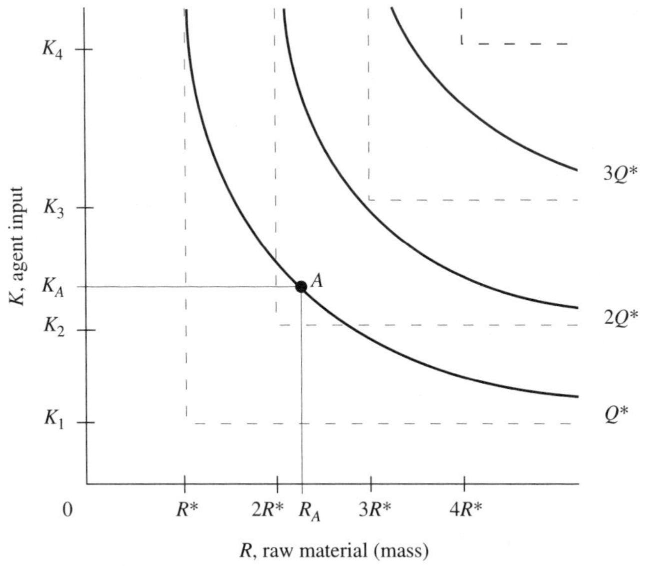

FIGURE 7-5
Optimization of economic production model.[19]

where $Q =$ maximum amount of output that could be produced given R units of input and a mass per unit of output of m

$\nu =$ fraction of input mass R that becomes product

Agent input is an economic term to account for the nonmaterial input to production. It includes capital, labor, and the associated energy. The cost of K would include the cost of the agent's services plus the energy costs.

Using this terminology, waste production is estimated as $(1 - \nu)R$, implying that waste production is proportional to the material input to the process. If a given production level is selected (e.g., Q^*), moving to the left along the curve reduces W_m, but increases W_e. At point A in the figure, Q^* units of product are produced, using R_A units of material and K_A units of agent input. The curve implies that it is not possible to produce Q^* units of product with less than R^* mass of raw material, no matter how much K is increased. The waste production at point A is $W_m = R_A - \nu R^*$, which can be divided into by-product waste $[(1 - \nu)R_A]$ and economic waste $[\nu(R_A - R^*)]$.

Anderson's conclusions regarding policy are:

> W_e is an inevitable flow from any production process so that policy directed at the elimination of this flow is doomed to failure. Incentives are instead required to induce firms to use agent-energy combinations which have high thermal efficiencies . . . and/or result in flows of a less harmful nature . . . **for by-product wastes, waste reduction can only be achieved by a reduction of the output produced (assuming the nature of the output and material input remain unchanged).** [Emphasis added.] Finally, while policies aimed at reducing economic waste may be effective in doing so, they are likely to result in increased flows of W_e if output levels are to be maintained.[20]

While the above theoretical analysis is somewhat pessimistic regarding the ability to reduce wastes without curtailing production, this chapter will examine techniques to change the nature of the output and material input, while still maintaining productive manufacturing facilities. To start the process, microeconomic calculations used to evaluate minimization alternatives will be presented.

Present worth analysis is used to bring all future costs to an equivalent amount of money which, if invested today, would produce the needed future sums. For example, if 10 percent interest is assumed, one dollar today is effectively worth \$1.10 in 1 year and \$1.00(1.1)(1.1) = \$1.21 in 2 years. In general:

$$\mathrm{PW} = \frac{F}{(1+i)^n} \tag{7-3}$$

where PW = present worth
F = future sum
i = interest rate (10%: $i = 0.1$)
n = number of years

Present worth analysis typically considers all future operating costs as a uniform series. The present worth for a uniform series of annual payments is given by:

$$\mathrm{PW} = A \times \mathrm{PWF} = A \times \frac{[(1+i)^n - 1]}{i(1+i)^n} \tag{7-4}$$

where A = annual cost and PWF = present worth factor.

The process is illustrated by Example 7-2. This example does not take into account changes in annual costs due to inflation and other factors. Because it is typical for waste treatment costs to rise much faster than other manufacturing costs, the above analysis is conservative. The selection of an interest rate is normally based on the present cost of money for the industry in question. Design life would be based on the expected life of the equipment in question, perhaps modified by considerations of potential changes in regulations or other factors that may influence how long the new process will remain in place. Example 7-3 illustrates how the effect of inflation can be incorporated into the calculation.

EXAMPLE 7-2. PRESENT WORTH ANALYSIS. A manufacturing process has present operating costs of \$10.9 million annually and waste disposal costs of \$3.3 million. The plant manager wishes to upgrade the process by installing new equipment at a capital cost of \$23.3 million. This will reduce the annual operating and waste disposal costs to \$7.1 and \$1.4 million respectively. Compare the two processes on a present worth basis, assuming no salvage value. Use 12 percent interest for 10 years as a basis for calculation.

Solution. The present worth factor is:

$$\mathrm{PWF} = \frac{[(1+i)^n - 1]}{i(1+i)^n} = \frac{[(1.12)^{10} - 1]}{0.12(1.12)^{10}} = 5.650$$

For the present process:

$$\mathrm{PW} = (10.9 + 3.3)(5.650) = \$80.233 \text{ million}$$

For the new process:

$$PW = 23.3 + (7.1 + 1.4)(5.650) = \$71.327 \text{ million}$$

On a present worth basis, the new process saves $8.9 million.

It should be noted that this answer does not consider other important factors such as income taxes, investment tax credits, depreciation, and inflation. The last is addressed in Example 7-3. These various factors can be combined together in an ROI (return on investment) approach. Because a company has limited capital funds, the typical procedure is to compare in an annual budgeting process the ROIs of alternative capital outlays and select those providing the higher returns. As a rule of thumb, a capital expenditure should pay for itself in 2 to 4 years. In this example, the annual savings of $5.7 million repay the capital expenditure in 4 years, signifying a marginal opportunity. What may drive a decision to make the outlay is whether the upgraded process will reduce the long-term liability represented by the disposal component.

EXAMPLE 7-3. INFLATION ANALYSIS. Revise Example 7-2 to account for a 5 percent estimated inflation rate for manufacturing costs and a 12 percent rate for waste disposal costs. Use 12 percent interest for 10 years as a basis for calculation.

Solution. Calculations are shown in Table 7-2.

These calculations can also be performed by applying Equation 7-3 to each future cost. For example the cost of the present process in year 2 is $16.16 million. The present worth of this figure at 12 percent is:

$$\frac{16.16}{(1.12)^2} = \frac{16.16}{1.25} = 12.93$$

TABLE 7-2
Inflation calculations for Example 7-3 (in millions of dollars)

	Present process			New process			Difference	PW
Year n	Manufacturing*	Waste disposal†	Total T_p	Manufacturing*	Waste disposal†	Total T_n	$\Delta = T_p - T_n$	$\Delta/(1+i)^n$
0	10.90	3.30		23.30		23.30	−23.30	−23.30
1	11.45	3.70	15.14	7.10	1.40	8.50	6.64	5.93
2	12.02	4.14	16.16	7.46	1.57	9.02	7.13	5.69
3	12.62	4.64	17.25	7.83	1.76	9.58	7.67	5.46
4	13.25	5.19	18.44	8.22	1.97	10.19	8.26	5.25
5	13.91	5.82	19.73	8.63	2.20	10.83	8.89	5.05
6	14.61	6.51	21.12	9.06	2.47	11.53	9.59	4.86
7	15.34	7.30	22.63	9.51	2.76	12.28	10.35	4.68
8	16.10	8.17	24.27	9.99	3.09	13.09	11.19	4.52
9	16.91	9.15	26.06	10.49	3.47	13.96	12.10	4.36
10	17.75	10.25	28.00	11.01	3.88	14.90	13.11	4.22

Total savings = $\sum$ PW = $26.72 million

*increase at 5%
†increase at 12%
PW = present worth at $i = 0.12$ (12%)

For the present process:

$$\begin{aligned}\text{PW} &= (15.14/1.12) + (16.16/1.25) + (17.25/1.40) + (18.44/1.57)\\ &\quad + (19.73/1.76) + (21.12/1.97) + (22.63/2.21) + (24.47/2.48)\\ &\quad + (26.06/2.77) + (28.00/3.11)\\ &= 110.75\end{aligned}$$

For the new process:

$$\begin{aligned}\text{PW} &= \text{capital cost } + \sum \text{ (years 1–10)}\\ &= (23.30) + (8.50/1.12) + (9.02/1.25) + (9.58/1.40)\\ &\quad + (10.19/1.57) + (10.83/1.76) + (11.53/1.97) + (12.28/2.21)\\ &\quad + (13.09/2.48) + (13.96/2.77) + (14.90/3.11)\\ &= 84.03\end{aligned}$$

New process savings = 110.75 − 84.03 = $26.72 million

Equations 7-3 and 7-4 may be inverted to determine a future value or annual payment corresponding to a specified present worth. A final compound interest equation determines the annual payment necessary to produce a specified future amount:

$$A = \frac{F \times i}{[(1+i)n - 1]} \tag{7-5}$$

While present worth analysis provides a simple mechanism for comparing costs, it does not take into account many factors that should influence waste minimization decisions. One modification of the analysis is to use a **benefit/cost analysis.** While this is often thought of as another form of economic analysis, the values used are usually arrived at by present worth analysis. In addition to the costs identified above, benefits might include the value of goods produced and reduction in paperwork and other transactional costs associated with waste disposal. Additionally one could estimate the value of benefits such as reduction in future liabilities and other less tangible items such as corporate public relations and potential public benefits. Sometimes negative benefits are termed **disbenefits** (i.e., hardships created by the alternative). This leads to two common equations used to calculate the **benefit cost ratio** (B/C):

$$\frac{(B - D)}{C} \tag{7-6}$$

$$\frac{B}{(C + D)} \tag{7-7}$$

where B = benefits
C = costs
D = disbenefits

As illustrated in Example 7-4, the selection of the B/C equation may provide slightly different values of the ratio.

EXAMPLE 7-4. BENEFIT/COST ANALYSIS. Calculate the B/C ratio given the following present worth data about a waste minimization alternative:

Benefits (million \$):	
Reduction in future liabilities	\$6.0
Elimination of potential fines estimated at	\$2.0
Savings in waste treatment costs estimated at	\$4.5
Total benefits:	\$12.5
Disbenefits: Increased production costs \$7.5	
Cost of new equipment \$4.97	

Equation 7-6: $(B - D)/C = (12.5 - 7.5)/4.97 = \underline{\underline{1.01}}$
Equation 7-7: $B/(C + D) = 12.5/(4.97 + 7.5) = \underline{\underline{1.00}}$

A modification of benefit/cost analysis is termed **cost-effectiveness analysis.** In this case, no attempt is made to put a price on intangibles. Instead all monetary costs are evaluated for various alternatives by present worth analysis. Other factors are evaluated on an effectiveness scale. Effectiveness might include safety considerations, potential effects on production (delays, product quality, etc., that cannot easily be quantified), hidden costs (paperwork, monitoring), public relations considerations, ease of implementation and future liabilities. Cost-effectiveness is a particularly useful technique in evaluating factors for which it is simply inappropriate to assign dollar values. For example society deems it indelicate to assign a dollar value to human life or an additional cancer case, even when the objective is to save lives or reduce the incidence of cancer. In the latter cases, risk is used as the measure of effectiveness (see Figure 17-8).

A plot of effectiveness versus cost provides decision makers with a graphical illustration of the decision to be made. The process is illustrated in Example 7-5. It must be emphasized that this is simply a way of presenting information to decision makers. It is of value only to the extent that the selection of factors, weights, and relative effectiveness scores is credible to management. The most appropriate use of cost-effectiveness analysis is to screen a large number of potential pollution prevention alternatives to determine which ones are appropriate candidates for detailed feasibility studies.

EXAMPLE 7-5. COST-EFFECTIVENESS ANALYSIS. A manufacturing plant is faced with the loss of a hazardous waste disposal site. It is evaluating three alternative strategies:

(*a*) Ship waste to a site in a distant state.
(*b*) Construct major in-plant changes to reduce the production of hazardous waste.
(*c*) Rebuild the manufacturing facility using an entirely new process that does not produce any hazardous waste.

Solution. To begin an analysis, management identifies six factors that it believes cover the intangibles (see Table 7-3). The factors are assigned weights that represent management's value judgment as to the relative importance of each factor. A further subjective evaluation assigns a relative effectiveness score to each alternative for each factor. The raw score is simply the weighted sum of the individual values, and the effectiveness is obtained by subtracting the raw score of the present alternative (the do-nothing alternative provides a baseline on which to judge the others).

From the results in the table, it is seen that alternative B is twice as effective at achieving objectives, but adds only about 30 percent to the cost. Alternative C is actually less effective (see Figure 7-6).

TABLE 7-3
Cost-effectiveness data for Example 7-5

Factor	Weight	Effectiveness			
		Present	A	B	C
Short-term liability	2	−3	−2	−1	−1
Long-term liability	3	−5	−3	−2	−1
Compliance	5	−2	+1	+2	+2
Production impact	5	0	−2	−1	−3
Public relations	1	0	+2	+2	+4
Regulatory impact	1	−3	−2	−1	0
Raw score		−34	−18	−2	−6
Effectiveness		0	16	32	28
Cost, million $ *		0	8.6	10.7	52.4

*Present worth of all identifiable capital and operating costs less savings compared to present operation.

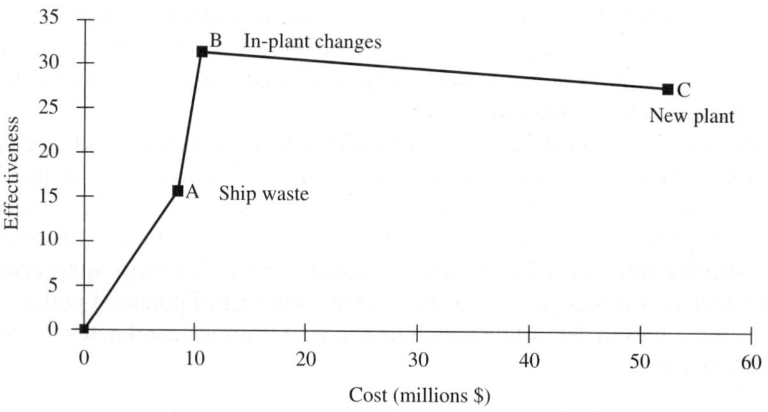

FIGURE 7-6
Cost-effectiveness plot.

It is possible to use this type of analysis simply to deal with intangibles, in advance of any attempts at estimating cost. For example, an industry with dozens of process lines may wish to prioritize them to determine which processes should be evaluated first.

Table 7-4 was used by Intel Corp. to prioritize waste streams. For toxicity information the lowest available median lethal dose (LD_{50}) for human oral toxicity was used. Where no human data existed, animal data were used, and in the absence of oral data from animal studies, other routes (e.g., dermal) were utilized.

For waste streams with multiple components, the higher rating is selected. Total ratings are determined for all waste streams and used as a basis for selecting which waste streams to study first.

TABLE 7-4
Waste stream prioritization[21]

	Prioritize waste streams using the following parameters:	
A.	Monthly amount produced	
	Amount > 5000 lb	=4
	5000 lb ≥ Amount > 1000 lb	=3
	1000 lb ≥ Amount > 500 lb	=2
	500 lb ≥ Amount > 50 lb	=1
	Amount < 50 lb	=0
B.	Toxicity rating (LD_{50})	
	LD_{50} < 10 mg/kg	=3
	10 mg/kg ≤ LD_{50} < 100 mg/kg	=2
	100 mg/kg ≤ LD_{50} ≤ 1000 mg/kg	=1
	LD_{50} > 1000 mg/kg	=0
C.	Land disposal ban applicability (LD)	
	Yes	=4
	No	=0
D.	Disposal cost (DC)	
	Cost > $1500/ton ($6.25/gallon)	=3
	$1500/ton ≥ Cost > $1000/ton ($4.17/gallon)	=2
	$1000/ton ≥ Cost ≥ $500/ton ($2.08/gallon)	=1
	Cost < $500/ton	=0
E.	Flash point (FP)	
	Flash point < 20°F	=4
	20°F ≤ flash point < 80°F	=3
	80°F ≤ flash point < 140°F	=2
	140°F ≤ flash point ≤ 200°F	=1
	Flash point > 200°F	=0
F.	Reactivity (R)	
	Yes	=1
	No	=0
G.	Carcinogenicity, mutagenicity, and/or teratogenicity (CMT)	
	Yes	=4
	No	=0
	Total rating = sum of seven factors	

Financing

Major pollution prevention programs may require substantial capital investment. Potential sources of capital include:

- Corporate resources
- Commercial bank loans
- Business development corporations
- Trade associations
- State and federal governments

Where corporate and commercial bank funding is not available, business development corporations (BDCs) may be able to assist. These institutions were established to assist the private sector in retaining and expanding employment. BDCs exist in most states and were specifically set up to assist businesses to secure funding that would not normally be available through established lending institutions.

Trade associations may be able to secure funding for a group of industries with a common problem and centralized solution. In New Jersey, metal finishers established a nonprofit research foundation that examined the feasibility of a centralized waste treatment facility:

> Ninety members of the Master's Association of Metal Finishers each paid $2,500, and additional assistance was obtained through grants ... a conventional treatment system would cost each company approximately $200,000 to purchase and install at their own plants ... the centralized system would save companies at least $14,000 a year.[22]

In order to encourage industries to implement waste reduction programs, state and federal governments have instituted programs to provide assistance. This assistance ranges from technical assistance on how to accomplish waste minimization to financial assistance in paying for the changes. Illinois's program included the development of an interactive computerized database that provides industry with access to the technical literature on waste minimization.[23] Technical assistance programs are operated by many states as part of their economic development efforts. These programs provide information on regulatory compliance and technologies for achieving waste reduction. Small business development centers and state-funded university research programs are other ways government is providing direct technical assistance to industry. In general, these programs are aimed at small businesses, which are unlikely to have the required specialized knowledge in-house. Most state programs, such as that of South Carolina's Department of Health and Environmental Control, are nonregulatory in nature and offer assistance from individuals with many years of industry experience.

Funding for pollution prevention efforts is also available from many states. Since the requirements and level of funding vary from state to state, it is important to obtain this information before analyzing a proposal. Even if a state program merely provides low interest loans for waste minimization projects, the difference in the interest rate may be sufficient to change the way corporate management will view the proposal.

Obstacles

While the previous introductory sections would lead to the conclusion that every industry should adopt wide-ranging waste minimization programs, there are a number of impediments to such action. Hirschorn and Oldenburg characterize this as a serious national problem:

> ... nearly every part of American society is mentally locked into the established, institutionalized pollution control culture, a paradigm that defines environmental protection in terms of what is done to wastes and pollution *once they are produced*. ... The switch from pollution control to pollution prevention is a classic example of a paradigm change—a truly profound change in the way people think about something. Changing to a new paradigm takes time.[24]

In addition to a change in mindset, there are a number of institutional impediments to waste reduction. These include some governmental policies, "short-term" management strategies, information and educational needs, and economic disincentives.

Certain federal regulations have created some disincentives for waste reduction programs. For example, RCRA specifically prohibits issuance of tax-free industrial bonds for funding pollution control equipment to achieve compliance. Additionally, some waste reduction efforts may require a permit under state or federal regulations (e.g., certain types of recycling activities will require a "Part B" permit under RCRA). The added cost of permit preparation and the possibility that the permit might not be granted, may be sufficient to stifle some waste minimization initiatives.

As noted above, the two most obvious sources for these funds are corporate resources or banks. Unfortunately, lending institutions are reluctant to provide commercial loans for such projects, unless it can be shown that they will also improve profitability.[25]

A similar impediment in some corporations is the inability of management to take a long-term view of corporate decisions. Often the payback period for waste reduction efforts is many years. If the plant manager is evaluated solely on the next quarter's bottom line, purchase of equipment that does not add to production is undertaken at personal cost.

An effective pollution prevention effort must include research and development, process design, production management, and other corporate functions in order to incorporate meaningful changes into the manufacturing process. Information needs range from a real understanding of the costs (long-term as well as short-term) of waste management practices to newly developed technologies that will enable a plant to dramatically reduce waste generation. Although the information is available, corporate management needs strong incentives to devote the necessary resources to acquiring it. Government-applied incentives are often reduced to more forms to fill out to demonstrate pollution prevention efforts, which may result in less prevention.

Because many waste minimization procedures involve changes in operation rather than equipment, it is important that a training component be included in any waste minimization plan. Workers need to know how their actions affect both productivity and the environment.

7-3 LIFE CYCLE ANALYSIS*

The environmental impact of all phases of industrial activity, from raw materials acquisition and R&D to the final disposition of a product and its packaging, is having a far-reaching effect on environmental quality and on public health. As a result, industry, environmental groups, and governments are attempting to identify a systematic means of evaluating and minimizing the environmental impact of products and processes. One of the more promising systematic approaches for identifying and evaluating opportunities to improve the environmental performance of industrial activity is termed life cycle analysis (LCA) (see Figure 7-7). LCA provides an analytical framework for investigating the entire range of environmental impacts (e.g., air emissions, wastewater, solid and hazardous waste, renewable resources, and energy utilization).

*Section 7-3 prepared by Marc Wendell, ERM.

By employing LCA methods, corporations are identifying opportunities to reduce environmental impacts, while maintaining or gaining competitive advantage. The use of life cycle methods at the design stage of a product or process is particularly noteworthy. This is true whether the concern is manufacturing operations, product use, or disposal. For example, the typical car contains about 180 pounds of plastic. However, when a car is disposed of, the different plastics in the car cannot be easily sorted. To counter this life cycle problem, BMW introduced a new series of small cars in the United States in June 1992. Plastic parts are stamped by type during fabrication so they can be sorted quickly and accurately. In addition, the cars are designed for disassembly. Design for disassembly is one of the many life cycle–oriented "design for" techniques being embraced by the auto and other industries. Design for ease of assembly, design for field repair, and design for transportability are other examples.[26]

The remainder of this section provides a brief overview of life cycle analysis, focusing on its use in addressing hazardous waste issues.

Background

Although LCA has been used in various forms over the past several decades in Europe and in the United States, it remains a relatively new concept, the methodology of which is still in the developmental stages. In the United States the origin of LCA can be traced back, prior to its environmental usage, to military procurement. Contrary to its perception as wasteful, the defense industry has implemented systems to improve efficiency. For example, defense systems analysts used life cycle methodology to

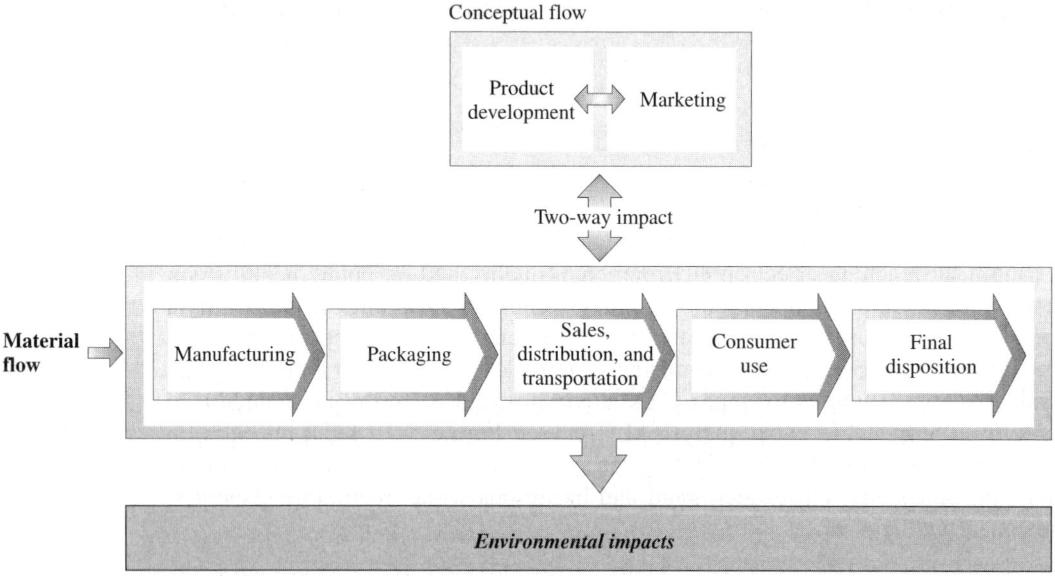

FIGURE 7-7
Product life cycle.

account for the previously neglected operating and maintenance costs associated with owning equipment and systems. The differences in life cycle costs among systems were factored into the analysts' procurement decision-making processes. This allowed low-maintenance systems with high capital cost to compete with lower capital cost systems that required extensive maintenance.[27]

Paralleling the concern of defense systems analysts, the environmentalists, policy makers, and corporations of today are employing LCA in an attempt to take into account the previously neglected environmental costs (and benefits) of products, processes, and other business activities.

Performing LCA within Industry

Private corporations are the most likely group to be performing LCA in the marketplace today. These companies are using LCA methods to compare design variations for a single product or to compare new or proposed products to their existing lines. From this vantage point, LCAs are valuable tools to:

- Provide a means of identifying tradeoffs associated with product modifications and design alternatives (such as reducing the weight of packaging material and increasing the use of recycled material), raw material sourcing (e.g., agricultural versus fossil fuel), technology changes (in process or materials), and various waste management options.
- Serve as a baseline for comparison with future modifications.
- Quantify and monitor a product's energy consumption and emissions in an effort to identify potential resource and waste reduction opportunities associated with specific manufacturing operations (e.g., raw material extraction, fabrication, and transportation).

Some common characteristics exist in all industrial LCA programs regardless of the products manufactured. In particular, LCA methodologies generally involve some variation of the following three-phase process:

$$\text{Inventory} \Rightarrow \text{impact (or interpretation)} \Rightarrow \text{improvement}$$

One common approach being used in industry to perform LCA is outlined in Table 7-5.

Inventory Phase

Organize teams

A project coordinator is usually designated to manage an LCA. The coordinator then assembles one or more teams, each of which assesses a limited number of targeted products or activities.

The teams include those personnel who have a significant stake in the outcome of process or product changes. They are cross-functional groups with broad practical experience, including personnel from areas such as operations and maintenance, manufacturing, product/design engineering, finance and accounting, and corporate environmental health and safety (EHS). By bringing together a diverse team that

TABLE 7-5
The life cycle analysis model

Phase 1, inventory	Phase 2, impact analysis	Phase 3, improvement analysis
• Organize LCA teams	• Assess compliance with company requirements	• Assess alternatives for improvement
• Determine priorities		
• Target products for LCA	• Assess risk, where appropriate	
• Diagram the LCA		
• Determine usage and generation rates (of raw materials, wastes, and energy)		

includes all parties with a stake in the outcome, the team can overcome resistance to change, facilitate the sharing of ideas, and address conflicts among departments.

Target products/processes for analysis

Even though LCA was conceived for the purpose of investigating the full range of impacts across the entire life cycle of a product, in practice it is financially impractical for an organization to perform detailed life cycle analyses on all its activities. Instead, it is usually necessary for the corporation to target those issues requiring detailed life cycle analysis. For example, a specific product may be targeted because of concern over its usage of toxic materials, air emissions, etc.

The targeting process usually requires a preliminary assessment of products, processes, chemical usage, waste generation, and emissions by experienced company personnel, particularly those with a broad base of experience with product and environmental issues. It is not necessary to compile exacting quantitative information during targeting. Instead, it is better to develop an initial, qualitative understanding of the environmental concerns of the company (e.g., its products, processes, and activities). Such targeting can be performed with readily available sources, such as operating logs, and from the practical knowledge of experienced production and process design personnel. Outside consultants with broad regulatory experience and an understanding of regulatory trends can also be a useful source of information, providing an outsider's "objective" insight.

Diagram the life cycle

After an activity has been targeted, the life cycle can be diagrammed as simple box flowcharts (see Figure 7-8). The scope of the diagram should be thought through carefully before it is developed, since the diagram will set the limits of the investigation. The analysis might begin with the acquisition of raw materials and end with disposal to the municipal waste stream, or it might extend to the production of raw materials and the decomposition products of the waste (see Figure 7-8).

The actual scope, or boundary conditions, of a particular analysis depends on the needs of the investigation, which in turn depends on the issues of concern. For example, if waste generation from the manufacture of a product is of concern, it might

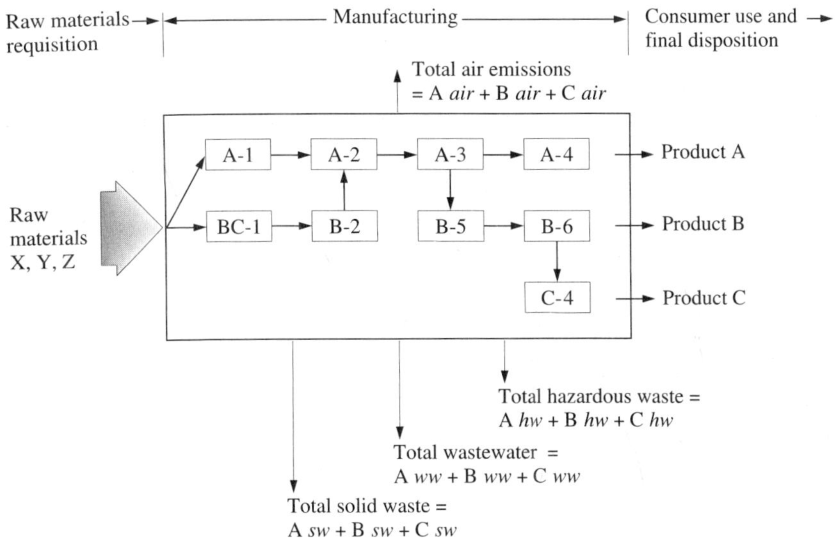

FIGURE 7-8
A manufacturing plant with products A, B, and C (note several stages of the product life cycle are not shown in this particular diagram).

be possible to limit attention to key manufacturing operations, and the treatment, storage, disposal, and recycling (TSDR) facility that ultimately receives the waste.

In contrast, if the entire product life cycle is of concern, the "boundary conditions" of the analysis might need to be expanded to include all of the following life cycle stages: marketing, research and development, raw material acquisition, material manufacture, final product fabrication, packaging, distribution, consumer use, and disposal.

Identify input and output

Following the completion of the diagram, the location and amounts of materials and wastes entering and leaving each stage of the life cycle are identified. Raw materials and intermediates and the emissions, wastes, and products should be identified for each stage of the life cycle. The practical knowledge of experienced personnel can be used to accelerate this evaluation.

If the main concern of the LCA is the reduction of hazardous waste [e.g., hazardous waste and hazardous air pollutants (HAPs)], emphasis should be placed on inventorying these wastes at each stage of the life cycle. For example, not only are hazardous wastes generated as a direct result of the production stage, they also result indirectly, from the maintenance activities associated with raw materials processing, manufacturing operations, and transportation equipment. In addition, returned goods can be a significant source of waste from outdated or damaged goods. All possible generation points need to be evaluated for possible inclusion in the final life cycle diagram.

The final diagrams and accompanying documents should contain the following information:

- **Targets** (e.g., product A, process B, and/or activity C)
- The **location** of waste emission or energy or material usage (e.g., a process vent, wastewater stream, and/or equipment cleanout)
- The type of **medium** transfer (e.g., air emissions, wastewater, and solid or hazardous waste)
- The **amount** of material or waste transferred (and/or energy utilized).

Readily available sources of information for estimating the rates of material usage and waste generation include sales figures, material safety data sheets (MSDSs), operating and maintenance records and manuals, permit information, vendor information, and facility (e.g., purchasing) records. An example of a process assessment worksheet is shown in Figure 7-9.

The diagramming activity is designed to promote improvement by providing a visual framework for systematic analysis of the life cycle, including the identi-

Process Assessment

Process: Chemical Manufacturing Work sheet #: 1 of 1

General mass balance equation: **Input + Generation = Output + Destruction**

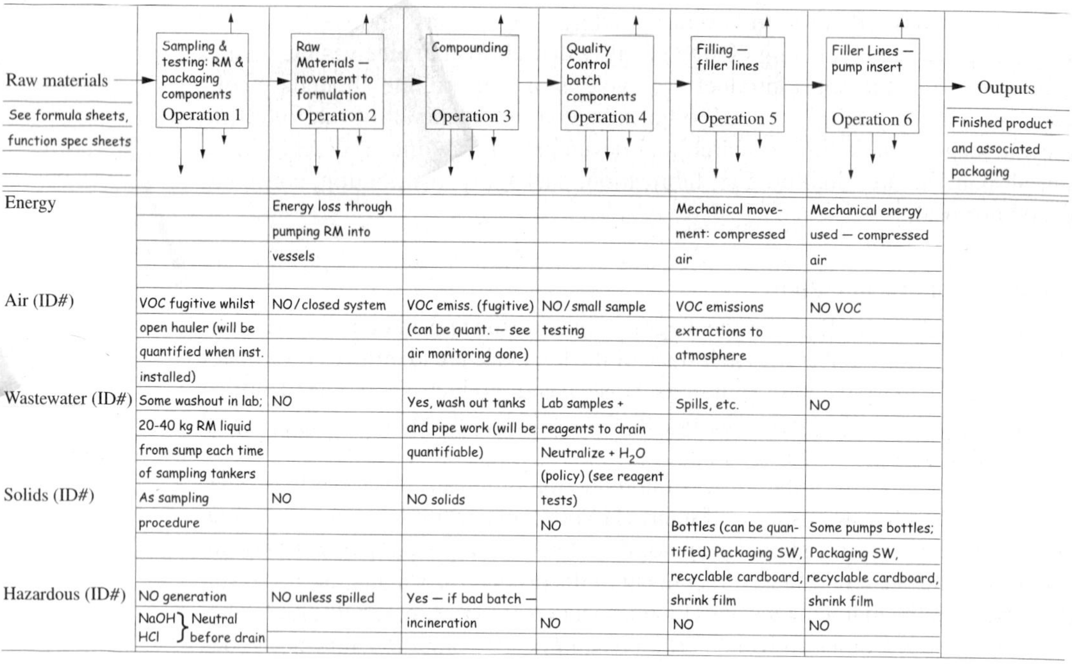

	Operation 1	Operation 2	Operation 3	Operation 4	Operation 5	Operation 6
Energy		Energy loss through pumping RM into vessels			Mechanical movement: compressed air	Mechanical energy used — compressed air
Air (ID#)	VOC fugitive whilst open hauler (will be quantified when inst. installed)	NO/closed system	VOC emiss. (fugitive) (can be quant. — see air monitoring done)	NO/small sample testing	VOC emissions extractions to atmosphere	NO VOC
Wastewater (ID#)	Some washout in lab; 20-40 kg RM liquid from sump each time of sampling tankers	NO	Yes, wash out tanks and pipe work (will be quantifiable)	Lab samples + reagents to drain Neutralize + H_2O (policy) (see reagent tests)	Spills, etc.	NO
Solids (ID#)	As sampling procedure	NO	NO solids	NO	Bottles (can be quantified) Packaging SW, recyclable cardboard, shrink film	Some pumps bottles; Packaging SW, recyclable cardboard, shrink film
Hazardous (ID#)	NO generation; NaOH, HCl } Neutral before drain	NO unless spilled	Yes — if bad batch — incineration	NO	NO	NO

FIGURE 7-9
Process assessment worksheet.

fication and tracking of root causes of waste. For example, the worksheets aid in "normalizing" waste generation rates to the production rates of finished goods (e.g., pounds of hazard waste per thousand "widgets" produced). The normalized waste factors help to distinguish between real reductions in waste and the effects that accompany fluctuations in production rates. In doing so they provide a solid basis for setting environmental goals and tracking performance.

Impact (Interpretation) Phase

After the inventory has been developed, it is necessary to interpret the impacts of waste emissions and energy usage. There are a number of alternative strategies for interpreting life cycle impacts, most of which are still in the developmental stages (e.g., regulatory, hazard based, energy utilization, risk).

Regulatory strategy

After an inventory has been developed, a review of regulated chemicals, wastes, emission, etc., should be performed to determine whether (1) any of the life cycle stages are regulated or (2) further characterization of a material, waste, air emission, or process is necessary to assess regulatory applicability. The resultant "hit list" of regulated materials, wastes, and/or processes forms one important basis for investigating improvement opportunities.

Nonregulated targets

Corporations can promote the improvement of EHS performance by setting priorities beyond compliance, either expanding efforts beyond regulated materials, wastes, and processes or increasing the force behind specific regulatory items. For example, many corporations are accelerating their rate of chlorofluorocarbon (CFC) replacement beyond that mandated by law.

The Clean Production Program of the EPA's Office of Toxic Substances (OTS) uses three primary criteria to set (nonregulatory) priorities for addressing particular processes:

- Potential for reducing risk
- The significance of the process or use to multiple program initiatives across the EPA, both regulatory and nonregulatory
- The importance of the process or chemical to agency voluntary pollution prevention efforts, including the 33/50 program (see Table 7-1)

While the above example draws from the EPA's prioritization strategy, the same basic strategy can be used by industry. Corporations can set priorities as to which products, processes, and/or activities receive in-depth life cycle analysis.

As the above example suggests, life cycle inventory information provides a basis for assessing compliance with regulatory, company, and customer requirements and developing a strategy to meet these requirements. If, for example, a facility is found to be nearing the large-quantity generator status for hazardous waste, an array of regulatory requirements may have to be met (e.g., record keeping, reporting, and training requirements), entailing significant costs. The company may, however,

decide instead to implement a pollution prevention program to reduce the potential regulatory burden, opting to remain a small-quantity generator, or become a conditionally exempt small-quantity generator. Again, it is the life cycle analysis information that can provide support for strategic decision making.

Finally, LCA inventory information may trigger the need for further in-depth investigation (e.g., risk assessment) of materials, processes, and management practices.

Improvement Phase

The improvement analysis generally consists of two parts—the identification of improvement options (to address each area of concern identified during the impact analysis), followed by a comparative evaluation of a set of options. For example, the reduction of hazardous waste generation from maintenance activities might be accomplished by the redesign of manufacturing equipment, changes in maintenance procedures, or a combination of the two. The process of identifying the best options usually begins with the applications of quality management tools (e.g., Pareto and cause-effect diagrams).

The comparative evaluation requires a comparison of the technical, economic, and regulatory feasibility of each option and is beyond the scope of this brief section. Instead, the remainder of this section will present some examples of improvement strategies that are currently being employed within industry.

Use Clusters

For example, the new Clean Production Program developed by the EPA's Office of Toxic Substances (OTS) employs the idea of "use clusters"—groups of chemicals used for the same or similar purposes, such as paint strippers—for purposes of targeting. Examining use clusters, rather than single chemicals, OTS has evaluated the potential risks and identified safer substitutes and cleaner technologies for the chemical substances and technologies associated with a cluster.

Compliance alternatives

Compliance with even the most stringent regulatory or corporate requirements can often be achieved by using different strategies and technologies. Similarly, environmental risks can be managed by using alternative approaches. From a company's viewpoint, the aim should be to minimize risk to meet the company's requirements, ensure compliance, and do so at a minimum cost. LCA can provide the information needed for such decision making.

As an example, the 1990 Clean Air Act Amendments have cost industry billions of dollars and will fundamentally impact how manufacturing facilities can operate in the future. However, there are opportunities, through emissions reduction, chemical substitution, and other process changes, to avoid or delay regulation under the Clean Air Act. LCA can be employed to gain competitive advantage from such regulations by providing a framework for the strategic evaluation of compliance alternatives.

Case Study 1: Evaluating packaging designs

One of the most important uses of LCA is to provide companies with a method of ensuring that new product formulations or packaging options result in meaningful environmental improvements. In the following simplified example (Table 7-6), an LCA was conducted on various packaging alternatives for a liquid fabric softener. Only four of the several possible alternatives were studied. The data are based on 1000 liters of a single-strength product. The table indicates *percent decreases* for each energy and emission category as compared to a nonrecycled, 64-ounce bottle made of virgin high-density polyethylene (HDPE).

The results suggest that the *source reductions* associated with *product concentration* offer greater reductions in life cycle energy requirements and emissions than either incorporation of 25 percent recycled HDPE (from other sources) or a 25 percent recycling rate for the bottle itself. Product concentration *plus the reuse of existing containers* provides a margin of additional benefits.

Relative to the nonrecycled virgin bottles, each of the four packaging options studied offered a reduction in energy requirements and solid waste emissions. Note that the emissions associated with each packaging option differ in composition, and this may require a more detailed examination for meaningful interpretation. This case study illustrates the merits of LCA in (1) providing a baseline as a reference for improvement, (2) guiding product development efforts, and (3) lending better understanding of the unique environmental attributes of different strategies.

Case Study 2: Evaluating chemical designs[29]

The traditional approach to synthesizing chemicals is to produce the greatest yield at the least cost. However, many of the synthetic steps, or synthetic pathways,

TABLE 7-6
Strategies for packaging improvement[28]

	Percent decrease in energy use			Percent decrease in emissions		
Strategy	**Process**	**Transport**	**Feed stock**	**Solid**	**Waterborne**	**Airborne**
1. Incorporate 25% recycled HDPI into the virgin HDPE bottle	3	0	9	9	(+4)	4
2. Assume 25% consumer recycling of the virgin HDPE bottle	3	2	11	11	(+4)	5
3. Use a concentrated (3×) product in smaller, virgin HDPE bottles	55	53	56	55	54	55
4. Use a concentrated (3×) product in virgin paper carton to refill HDPE bottle	53	58	94	91	40	62

that produce high yields also generate toxic by-products or use high-risk solvents and catalysts.

These toxins could be reduced or "designed out" of the manufacturing process if alternative synthetic sequences were found. This is a relatively new life cycle design concept that the EPA's Office of Toxic Substances is encouraging among researchers in the field of synthetic organic chemistry.

The following sections explore additional improvement strategies in more detail, illustrating specific methods to reduce the volume and toxicity of waste and to eliminate waste generation altogether. Sections 7-4 through 7-6 are not mutually exclusive but rather provide a convenient method of organizing the information. There is a significant amount of overlap because many methods effect a reduction in both volume and toxicity of waste.

7-4 VOLUME REDUCTION

An appropriate place to initiate waste minimization investigations is to examine ways to reduce the volume of hazardous waste. This can be accomplished by a number of methods including modifying production processes, segregation, and reuse. It should be noted that under some regulatory schemes, simply reducing the volume of the waste without an accompanying reduction in toxicity would not be considered "waste minimization." For example, under California's Hazardous Waste and Management Review Act of 1989,[30] "actions that merely concentrate the constituents of a hazardous waste to reduce its volume" are explicitly excluded from the definition of "source reduction."[31]

Process Modifications

Process modifications include changes in:

- Raw materials
- Equipment
- Operating procedures
- Materials storage
- End products

Raw material substitution includes such simple items as cleaning materials. In the printing industry, the common practice of using organic solvents for cleaning presses has been replaced with water-based cleaners. Merck, Sharp & Dohme reports replacing organic solvents with inorganic acids and bases, resulting in the reduction in the emission of 300,000 gallons of hexane.[32] Sometimes conversion to a higher-quality raw material can eliminate the generation of hazardous waste where the compound causing the waste to be considered hazardous is due to contamination of the raw material. Another type of substitution is Safety Kleen's Premium Gold solvent. This solvent is a nonhazardous solution and does not have to be managed as hazardous waste.

In a typical metal plating process, a metal piece is submerged in a bath, then removed and allowed to drain prior to rinsing. The piece must be moved away from

the bath during the drain step to allow the next piece to be submerged. This results in drag-out of plating bath. Simple equipment modifications such as the drain boards and drip tanks shown in Figure 7-10 can significantly reduce waste.

A major source of drag-out occurs in the rinse tanks used in most plating operations to remove plating solution from the part being plated. A significant reduction in drag-out to the rinse tank can be achieved through the use of a double drag-out tank not connected to the final flowing rinse tank (see Figure 7-11). In addition to significantly reducing the volume of wastewater, electroplaters have experienced up to 50 percent reduction in use of process chemicals.

Hunt[33] suggests the following additional measures to reduce drag-out:

- Increase drain time
- Air knife (low pressure air to blow solution off parts)
- Spray rinsing over process tank
- Minimize concentration of metal in plating bath
- Rack parts to maximize drainage
- Use drip bars to hold racked parts over plating tanks
- Maximum hole size on barrels
- Rotate barrels above plating tank

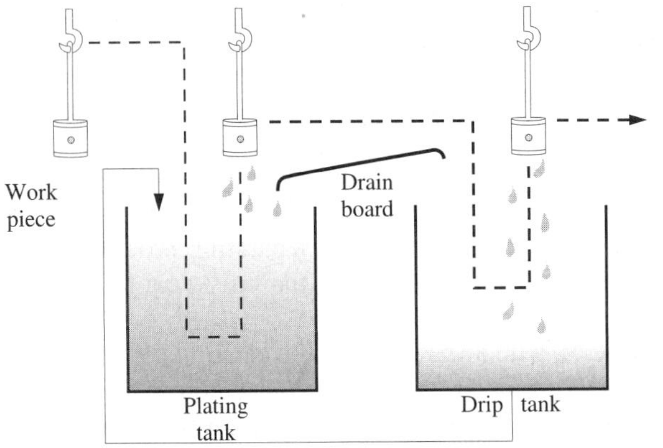

FIGURE 7-10
Drag-out recovery devices. *(Adapted from Higgins[34] with permission.)*

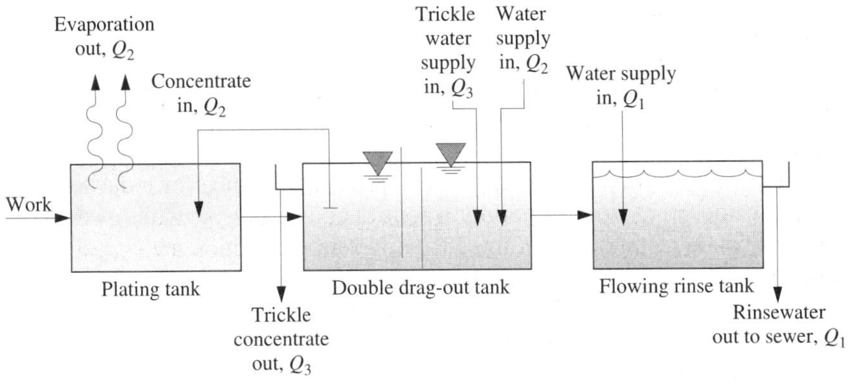

FIGURE 7-11
Providence method of double drag-out tank in plating operations.[35]

A final important aspect of any process change is the need to coordinate with production management. They must be an integral part of the planning, design, and implementation of any waste minimization efforts. Production supervisors must be committed not only to the changes in equipment, but to training staff as to the reasons for the changes. Creatively designed waste minimization efforts can fail because of a lack of understanding on the part of production staff as to the need for the changes.

Segregation

A primary tenet of source reduction is to avoid mixing wastes. As noted in Chapter 2, a mixture of a small amount of "listed" hazardous waste, pursuant to 40 CFR 261 Subpart D, with a larger amount of nonhazardous waste creates a large amount of material that must be treated as a hazardous waste! Another basic rule—don't make it a liquid if it is dry! Housekeeping operations as simple as sweeping prior to washing floors can substantially reduce waste volumes.

Keeping wastes segregated greatly facilitates any required treatment. Proper labeling of all lines and containers will greatly increase the likelihood that plant personnel will follow any changes in practices intended to enhance segregation of wastes. Where hazardous waste treatment is required, segregation permits treating smaller quantities of waste.

Another area where substantial volume reduction can be achieved is keeping noncontact cooling water separate from waste streams. Many manufacturing facilities are in buildings that were constructed long before waste management was a major concern and there is often a single sewer system for such buildings. In such buildings, all aqueous wastes drain to a common sewer, and separating process wastes through the installation of new piping systems is often the only way to properly handle wastes.

In order to design such a system, a detailed study of the existing sewer system, including dye studies to see which drains are connected, is required. Samples of wastes should be taken and a mass balance performed either on water or a specific chemical parameter to ascertain the validity of the sewer map.

Experience with a large number of industrial facilities has revealed two general rules regarding drainage diagrams:

1. No plant has an accurate sewer map—there is always unknown information.
2. After a detailed (and expensive!) survey of sewers, rule 1 still applies.

Some solid waste streams can be segregated effectively through minor changes in equipment. A major source of hazardous waste at a number of industries is baghouse dust emanating from air pollution control equipment. As with sewer systems, common dust collectors were utilized for different production areas, resulting in a mixing of different types of dust and precluding recycling. Lewis[36] reports that FMC Corporation was able to modify dust collectors as illustrated in Figure 7-12, so that each compartment served a single source.

Major impediments to waste segregation programs are materials that go into plant trash but don't belong there. Examples include laboratory samples, which

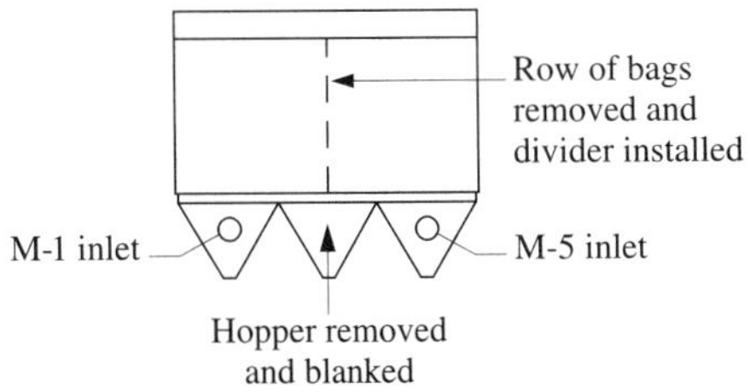

FIGURE 7-12
Modification of baghouse to segregate dust. *(Reprinted with permission from Air and Waste Management Association.)*

must be disposed of as hazardous wastes if they have any of the four hazardous waste characteristics (corrosive, ignitable, reactive, or fail the TCLP test). Other materials include solvents and pigments, for which special receptacles must be provided.

Reuse

Many materials that are disposed of as hazardous waste have a potential to be reclaimed for another application. In some instances, contaminated materials may be of adequate quality to serve as a solvent or cleaning material for a less sensitive application. In the printing industry, toluene is often used as both a cleaning agent for presses and a thinning agent for the inks, prior to their being placed on the presses. If the toluene used for cleaning is limited to cleaning one specific color ink, it may be possible to reuse that toluene to thin the same color ink.

The acid recovery process used by the steel industry provides an example of reuse of a liquid waste stream. Iron scale is removed by an acid cleaning process referred to as "pickling." The spent pickle liquor is a hazardous waste that must be neutralized and the resulting sludges are still considered hazardous and require further treatment or secure land disposal. Some example methods for acid recovery of spent pickle liquor are discussed in Chapter 8, Section 8-1. As another example, the recovery process shown in Figure 7-13 is reported[38] to reduce virgin sulfuric acid use by as much as 40 percent. The capital cost of recovery facilities exceeds the cost of neutralizing the waste. However, when the savings associated with reduction in purchase of acid and elimination of waste treatment are included, a 100,000 ton/year facility realizes a net annual savings of $350,000.

Another source of hazardous wastes is raw materials stored past their expiration date. Such dates are often established conservatively and it may be possible to reassay and requalify the material for use in production. A recent report by the U.S. Navy suggests that a total evaluation of specified shelf lives may be appropriate:

> Shelf life term is specified by Item Managers who are not usually chemists and, therefore, technically unqualified to make such an assessment. As a result, the suppliers are relied upon to specify a "safe" shelf term. The manufacturers have found that shelf term is inversely proportional to Navy sales and, accordingly, often recommend shelf lives that are unrealistically short. . . . It is possible that some goods may not even require specification of a shelf life.[39]

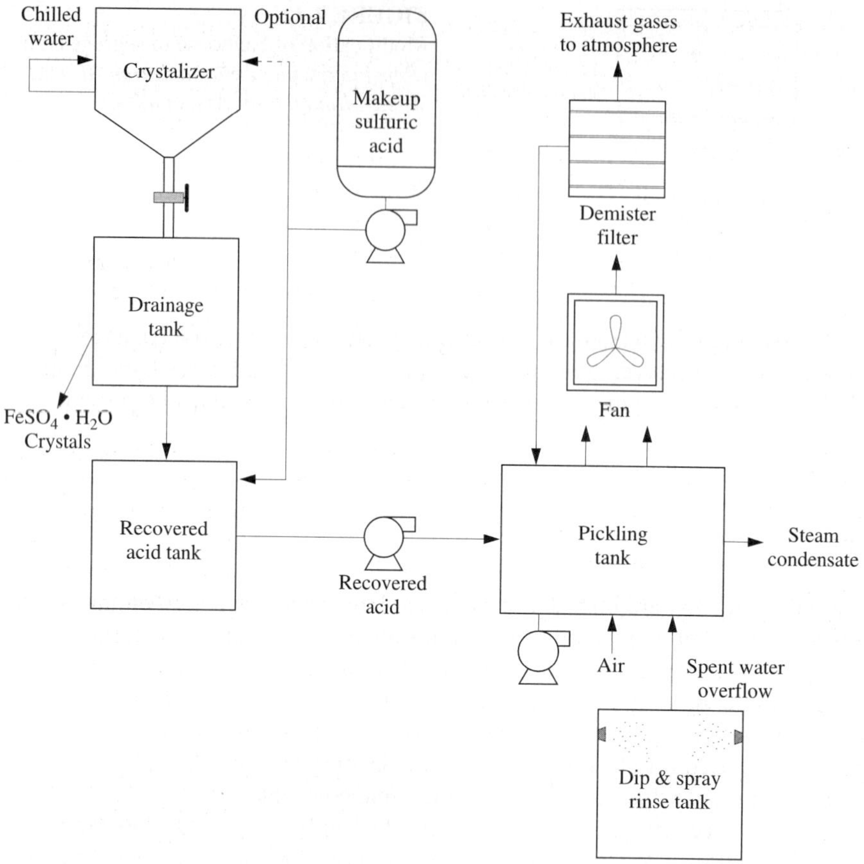

FIGURE 7-13
Acid recovery process for steel pickle liquor.[37]

7-5
TOXICITY REDUCTION

A number of waste minimization techniques reduce the concentration of contaminant in a liquid or solid waste stream, without necessarily diminishing the volume of wastes produced. It is often possible to lessen the toxic characteristics sufficiently so that the remaining waste is no longer considered a hazardous waste.

Process Modification

Process modifications were discussed in Section 7-4. Some specific examples of modifications that result in a less-toxic waste stream include dry powder painting and solvent recovery methods. Safety Kleen's Premium Gold solvent solution, as discussed in Section 7-4, is another example.

Powder coating technology has reduced the need for solvent-based paints in many industrial painting applications. The process consists of spraying a heat-fusible

powder on a metal surface. An electrode in the spray gun charges the powder particles. The metal to be coated carries an opposing charge and the particles are attracted to the metal surface. Applying heat to bind the paint to the metal completes the process. Overspray is collected and recycled. In addition to eliminating a hazardous waste stream, this process reduces material, labor, and energy costs compared to conventional solvent-based painting.[40] It also reduces air emissions. Where solvents must be used, they should be recycled (see Section 7-6).

EXAMPLE 7-6. PARTS MANUFACTURER. A manufacturer produced small parts through a complex series of steps. After initial processing, a copper cyanide solution was used to electroplate copper on a portion of the parts as a mask for subsequent heat treating. After initial heat treating, the copper was then removed from the parts by a cyanide-based stripping solution. The facility incurred significant environmental expenditures for wastewater treatment and remediation of releases from old cyanide lagoons.

Solution. What can be done to reduce costs and the risk of future remediation expenditures? This is another "true story." Facility product and process engineers developed an alternative heat-treating process that eliminated the need to plate and strip the parts. The new process utilized a more controllable and reliable process, resulting in an increase in product quality. In addition, the elimination of the plating and stripping processes and implemention of the new heat-treating system reduced the total cost to manufacture the parts by a substantial 30 percent.

Equipment Modification

As in the discussion of volume reduction, many pieces of equipment can be modified to reduce waste characteristics. Fairly innocuous waste streams can become hazardous wastes through spills or intentional batch dumps of concentrated liquids. Simple devices such as alarms on plating baths can avoid batch dumps and ensure that spent plating baths are transferred to the proper containers for reclamation of useful metals.

Where small quantities of powdered chemicals are used, it is common practice to purchase these materials in 50-pound bags. Initially an operator would simply cut the bags open and dump the powder into a hopper. Health and safety concerns subsequently required that the operating area where the bags are cut and dumped be properly ventilated to protect workers. Then air pollution control equipment in the form of a wet venturi scrubber was added to the ventilating system to avoid violating air pollution laws. This resulted in a liquid hazardous waste (the spent scrubber solution) being produced. The existing system was not designed as a whole, but rather, parts were added to meet changing regulatory requirements. The installation of a small packaged bag dumping station, where bags are placed in a sealed contained box and then cut open, eliminates a major waste source and a lot of unneeded equipment. What is required is a new look at the equipment.

Housekeeping Practices

Perhaps the simplest and most inexpensive method of reducing hazardous wastes at the source is through revision of housekeeping and maintenance procedures. Table 7-7 provides a list of items to look for in revising such procedures.

TABLE 7-7
Good operating practices*

Practice	Program elements
Waste minimization assessments	Form a team of qualified individuals Establish practical short-term and long-term goals Allocate resources and budget Establish assessment targets Identify and select options to minimize waste Periodically monitor the program's effectiveness
Environmental audits/reviews (see Chapter 6)	Assemble pertinent documents Conduct environmental process reviews Carry out a site inspection Report on and follow up on the findings
Loss prevention programs	Establish Spill Prevention, Control, and Countermeasures (SPCC) plans Conduct hazard assessment in the design and operation phases
Waste segregation	Prevent mixing of hazardous wastes with nonhazardous wastes Isolate hazardous wastes by contaminant Isolate liquid wastes from solid wastes
Preventive maintenance	Use equipment data cards on equipment location, characteristics, and maintenance Master preventive maintenance (PM) schedule Deferred PM reports on equipment Maintain equipment breakdown reports Keep vendor maintenance manuals handy Maintain computerized file
Training and awareness building	Provide training for: Safe operation of the equipment Proper materials handling Economic and environmental ramifications of hazardous waste generation and disposal Detecting releases of hazardous materials Emergency procedures Use of safety gear
Employee participation	"Quality circles" (free forums between employees and supervisors) to identify ways to reduce waste Solicit employee suggestions for waste reduction ideas
Production scheduling/planning	Maximize batch size Dedicate equipment to a single product Batch sequencing to minimize cleaning frequency (e.g., light-to-dark batch sequence) Schedule production to minimize cleaning frequency
Cost accounting/allocation	Cost accounting done for all waste streams leaving the facilities Allocate waste treatment and disposal costs to the operations that generate the waste

*Adapted from EPA.[41]

Materials handling procedures provide a good place to initiate the search for opportunities to reduce waste. The design of storage areas should allow for easy inspection of potential sources of leaks. For example, in a drum storage area, aisles should be wide enough to permit access for routine inspection. Specifying the inspections and not permitting the inspector to see potential problems is counterproductive. Adequate lighting must be provided in the design of storage areas. Other practices include keeping aisles clear of debris and keeping different chemicals separate to preclude contamination or reaction. Drums that remain on concrete pads often corrode due to "concrete sweating." Use of pallets avoids this.

Training must be a major element of housekeeping practices. It is essential to have an ongoing education program to account for turnover of production staff. The best designed waste minimization plan will not work without adequate instruction of the personnel involved. Employees should be taught the implications of their actions, and their thoughts on how to reduce problems should be solicited. For example, while the disposal of spent solvent in a cooling water sump may be clearly inappropriate to an environmental engineer, operating personnel will not know this unless someone takes the time to tell them. They will not take seriously rules about where to dump waste, unless the reasons are explained. Soliciting waste minimization suggestions from operating staff often identifies worthwhile ideas. Some industries provide cash awards for waste minimization suggestions.

Labeling is a simple housekeeping task that can markedly reduce the unnecessary generation of hazardous waste. A specific example is the use of solvents. Plants where both chlorinated and nonchlorinated solvents are used have found that clearly marking disposal storage tanks and sensitizing staff to the need to keep these materials separate has greatly reduced the generation of hazardous waste.

Finally, once a new housekeeping program is under way, it should be monitored to determine that the objectives are being met. A check on housekeeping practices should be part of routine inspections of the operating units. Employees should be encouraged to suggest additional practices that will further reduce waste generation.

Material Substitution

Most manufacturing processes were designed with product quality and profitability as the primary design criteria. The changing economics of waste management are forcing a new look at production formulations. Among the many material changes that have proved to be cost-effective and reduce wastes are the use of noncyanide baths in metal plating operations and the use of less toxic solvents in cleaning operations. Industries producing machine parts use large quantities of lubricating oils that become contaminated with metals and must be disposed of as a hazardous waste. Synthetic lube oils cost more but last longer than petroleum-based materials, and can be returned to the manufacturer for recycling. The analysis to evaluate the efficacy of such a proposal is to compare a low cost–high volume material against a high cost–low volume substitute.

Polaroid Corporation has a Toxic Use and Waste Reduction Program that groups chemicals into five categories:

I and II	Carcinogens, highly toxic materials, etc. (Category I chemicals are those selected for special priority in reduction)
III	Suspected carcinogens, moderately toxic materials, eye irritants, corrosives, etc.
IV	All other chemicals
V	All other wastes

Example 7-7 indicates how Polaroid applied its waste reduction program to the analysis of a new dye.

EXAMPLE 7-7. POLAROID CORPORATION. A NEW DYE. "In 1987 Polaroid's research division began evaluating new dyes to replace an existing material used for many years. After completing the research stage, several candidate dyes had been identified that delivered the desired product improvement (a spectrum that better matched the requirements of the photographic use). Each candidate also had a higher molar absorptivity than the existing material, so less dye per picture was required.

"The candidate dyes were evaluated by the process research and development group for 'manufacturability.' Due to considerations of ease of production, the choices were reduced to two. Serious process evaluation on the finalists began. One of the key process steps of each new candidate dye, as well as of the existing product, was an oxidation step using Cr(VI) in aqueous pyridine solvent. Cr(VI) is a Category II material in Polaroid's system, and the workup of the oxidized product was complicated by the mixed solvent, which was also difficult to recycle.

"A review of the literature led us to investigate possibilities of catalyzed air oxidations. Conditions were ultimately found to successfully implement this technology, completely eliminating the Category II Cr(VI) compound, and to link it to readily recoverable solvents. [Figure 7-14] plots the impact of this new dye on toxic use and by-product generation per film pack. The combination of increased molar absorptivity and greatly reduced process waste results in an 80 percent reduction in the waste being generated using the new dye, compared with the old. In addition, the wastes that are generated are more readily treated, so disposal costs are lower per pound of waste generated. Overall, this substitution is saving Polaroid about $1 million a year. For most chemical

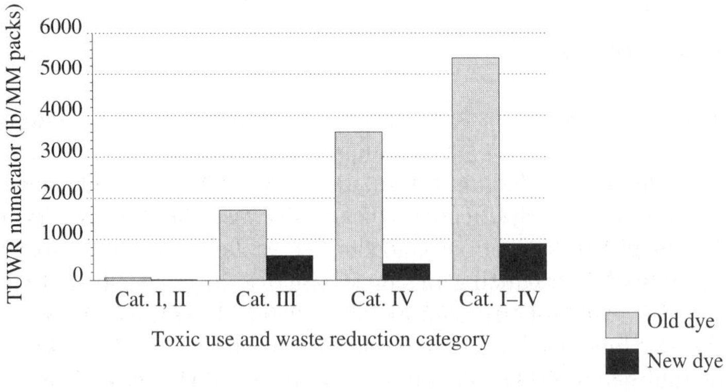

FIGURE 7-14
Toxic use and waste reduction—old dye compared to new dye. TUWR equals pounds of Cr(VI) and other toxic constituents in dye divided by millions of film packs—see Example 7-7. *(Reprinted with permission from Pollution Prevention Review, all rights reserved.)*

processes, our experience is that the minimum waste process is usually the minimum net cost process."[42]

7-6 RECYCLING

When it is not possible to reduce the volume or toxicity of a waste, it may be possible to recycle it to another process or another plant. Other factors being equal, on-site recycling is preferable because shipping hazardous waste off-site, even for recycling, carries the liability that the waste might be mishandled.

Water

Water is in most instances the easiest material to recycle, and this is the first place one should look when evaluating prospects for recycling. In terms of hazardous waste, aqueous waste streams that meet the characteristics of hazardous waste (e.g., toxic, corrosive, reactive) must be treated as a hazardous waste. If these waste streams can be recycled to some useful purpose, a hazardous waste is effectively eliminated.

In the instance of an extremely acidic line, reuse for waste treatment in neutralization of caustic wastes is an obvious solution. Rarely does waste acid exactly balance with alkali, but the total cost of chemicals can be cut by maximizing the recycle of high and low pH lines.

It is often possible to recycle process water to a less sensitive use. Often a high-quality water will become degraded in the manufacturing process and no longer suitable for use in that process, but may be adequate for another purpose. Exxon Corp., at its Baytown, Texas, refinery, recycles part of its wastewater to wet gas scrubbers.

Another potential source of recycle water is noncontact cooling water. Often such water is simply discharged as a heated but uncontaminated waste stream. If this source could be recycled, it would reduce the need for raw water makeup in some other part of the plant and perhaps some of the energy in the heated water could be conserved as well.

Solvents

Solvent recycling is common practice in many industries, and a wide range of solvents are currently recycled. Some of the more common solvents currently being recycled are shown in Table 7-8. Azeotropic distillation is commonly used to enhance solvent recovery. An azeotrope or azeotropic mixture is a mixture of liquids that behave as a single substance when boiled (i.e., both the vapor and the liquid have the same composition). Azeotropic distillation consists of adding a substance (typically steam) to form an azeotrope with the solvent to be recycled. The azeotrope will have a lower boiling point than the original mixture and recovery of the solvent will be facilitated. Azeotropic boiling points of compounds not listed in Table 7-8 may be found in standard references.[43,44]

Packaged solvent distillation units are available from a variety of manufacturers and usually consist of simple distillation units applicable to low boiling point

TABLE 7-8
Commonly recycled solvents*

	Boiling point, °F	
Solvent	**Atmospheric**	**Azeotropic**
Aliphatic hydrocarbon		
Hexane	157.0	142.9
Heptane	209.0	174.8
Aromatic hydrocarbon		
Benzene	176.0	157.0
Aromatic naphtha		
Toluene	232.0	185.0
Xylene	261–318	202.1
Turpentine		
Chlorinated hydrocarbon		
Trichloroethylene	189	163.8
Perchloroethylene	249	189.7
Methylene chloride	104	101.2
Alcohols		
Ethyl	173	172.7
Isopropyl	180.1	176.5
Ketones		
MEK	175	164.1
MIBK	241	190.2
Esters		
Ethyl acetate	170.9	158.7
Butyl acetate	259.2	194.4
Mixtures of solvents		
Toluene/xylene		
Ketones		
Alcohols		
Phenols		

*Adapted from Freeman[45] and Higgins.[46]

solvents. More sophisticated units are now becoming available that utilize vacuum distillation to handle solvents with higher boiling points (above 400°F). While package units are available at sizes as small as 3 gallons, for very small operations it may be more appropriate to return spent solvents to the original vendor or to a commercial recycle operation.

Another effective way to recycle solvents is **cascade reuse,** where a solvent is used once for high-quality cleaning and then used again for other cleaning operations that can perform effectively with a less-pure solvent. The subsequent solvent cleaning operation may occur at the same plant or involve shipping the used solvent to another company. An example would be solvents used in highly demanding pharmaceutical operations that are reused directly in metal product manufacturing.

Higgins[47] presents a case study illustrating the use of distillation to recover deionized (DI) water and spent solvent from electronics manufacturing. N-methyl-2-pyrrolidone (NMP) is used to clean semiconductors. NMP becomes contaminated with tin, lead, oil, and grease. The cleaning process is followed by a DI water rinse, and the water is degraded with NMP and its contaminants. A wide range

of purification processes were evaluated and laboratory studies demonstrated that vacuum distillation provided the required recovery of both NMP and DI water. Such a recovery system can be highly complex as illustrated in Figure 7-15.

Oil

Oil that has become contaminated with hazardous materials may require disposal as a hazardous waste. Often this material can be recycled rather than shipped for off-site disposal.

Adolph Coors Co. in Golden, Colorado, reports a 40 percent increase in oil reclamation by using ultrafiltration (see Section 9-6) compared to dissolved air flotation.[49]

The energy content of spent oil can be reclaimed in many instances, through burning in industrial boilers or furnaces. This is a common practice at power plants in California where state law permits the mixing of oil recovered from wastewater treatment with fuel oil for burning in boilers. Similarly, most refineries recycle slop oil from wastewater treatment plants.

Solids

A wide range of solid materials, including paper, metals, and plastics, are amenable to recycling. The recycling of paper and paperboard has become routine at most industries. Scrap metal such as swarf from machining operations is also a common candidate for recycling. The recycling of plastics is not as widespread. However, as the following example illustrates, it is a feasible alternative to incineration or land disposal of these materials.

Polystyrene (PS), $(C_6H_5CHCH_2)_n$, represents a major component in plastic waste. Kampouris, et al.[50] present laboratory studies of a solvent method of recovering PS scrap. The method consists of five steps:

1. Dissolution with solvents
2. Filtering
3. Reprecipitation of PS
4. Washing of polymer grains
5. Drying

The dissolution step employs miscible solvent/nonsolvent mixtures. Solvents utilized included MEK and *p*-xylene, and nonsolvents included methanol and *n*-heptane. These investigations also evaluated the separation of solvents for reuse.

The effectiveness of recycling was determined on the basis of three criteria:

- Rheological behavior of the solvent-PS solution
- Yield of usable PS polymer
- Adequate solvent separation

The authors report several advantages of solvent systems to recover plastics:

- Volume of plastic refuse is reduced.
- A range of plastics is amenable to this type of recycling.

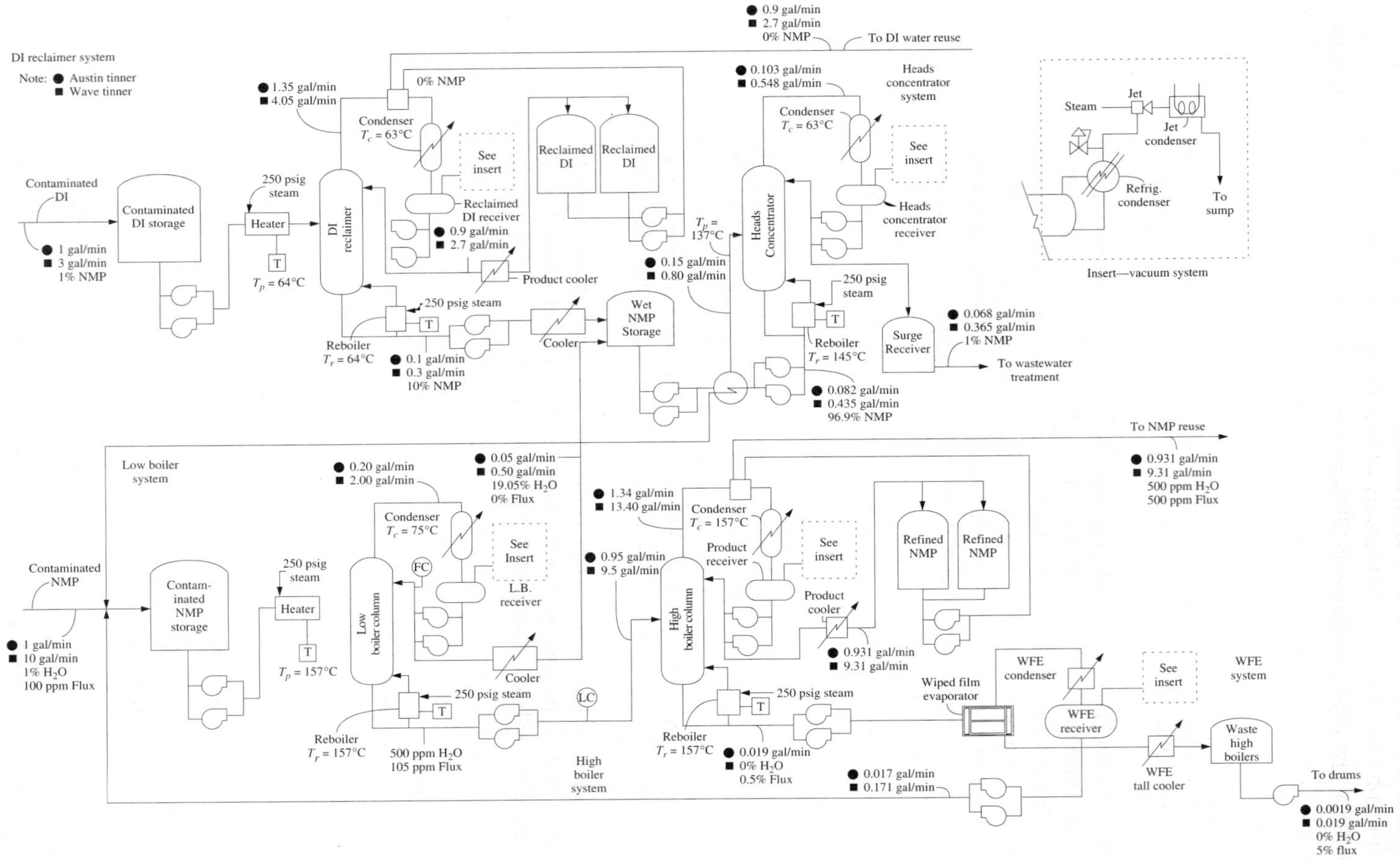

FIGURE 7-15
Vacuum distillation system.[48]

Phenol Formaldehyde Resin
NE:A05-0242
Phenol formaldehyde resin from vat coating of fiberglass. Solution contains 54% phenolic resin in 23% ethanol with 1% formaldehyde by net weight. Currently 5 drums available. Thereafter 2500 lbs/qtr. No restrictions on amounts. Sample Available. MD

FIGURE 7-16
Waste exchange listing.[53]

- The process removes impurities in the nonsolvent phase.
- The reprecipitated polymer can be produced in many forms (e.g., granules, fibrils, or powders).
- The end product is of high quality.

This study concludes that

> . . . the recovery of PS from foam waste by means of a solvent technique is feasible and even in the case of feeding with dirty stock, removal of contaminants can be ensured. Obviously, several parameters, such as waste transportation, solvent losses and solvents separation energy, have to be taken into account for an optimized operation of the process.[51]

Waste Exchanges

While reusing a waste product inside a manufacturing facility is the most desirable form of recycling, it is not always possible to find another department or process that can effectively utilize the waste. An alternative is to locate another company that can make use of your waste. A **waste exchange**[52] is a regional clearinghouse for such transactions. Waste exchanges maintain computer databases and/or publish periodic lists of wastes available or materials sought by various industries. A typical listing is illustrated in Figure 7-16. The waste exchange information base would include:

- Company ID code
- Category (acid, solvent, etc.)
- Primary usable constituents
- Contaminants
- Physical state
- Quantity
- Geographic area
- Packaging

DISCUSSION TOPICS AND PROBLEMS

7-1. Discuss how you would develop a waste minimization program for a small metal plating operation.

7-2. Discuss how you would identify potential changes in production methods that might be appropriate for the management of a large chemical plant to consider.

7-3. Discuss some of the impediments to implementing an effective waste minimization program.

7-4. Write a report identifying housekeeping and other activities that might be appropriate in minimizing the hazardous waste from a shop or laboratory identified by your instructor. Identify institutional impediments to implementing your solution.

7-5. Contact a waste exchange in your region and identify the types of wastes that have been successfully exchanged.

7-6. Identify institutional impediments to implementing a waste minimization program at a small manufacturing plant.

7-7. Identify institutional impediments to implementing a waste minimization program in your state.

7-8. Identify institutional impediments to implementing a waste minimization program for a manufacturing facility in a third world country.

7-9. Derive Equation 7-5 from 7-4 and 7-3.

7-10. A firm needs to expand its manufacturing capacity. The present facility could be expanded to meet the needs at a cost of $1.2 million. Waste treatment costs for the expanded facility would be about $184,000 per year. An entirely new facility that would reduce waste production could be built for approximately $3.35 million. The new facility would have an estimated annual waste treatment cost of $56,000. Assume a 20-year useful life for the facilities and a zero salvage value for both. If the prevailing interest rate is 10 percent, which alternative is preferable?

7-11. A company is faced with the decision of selecting either hazardous waste minimization process A or B. Process A involves purchasing machinery that costs $1.2 million with annual maintenance expenses of $150,000 for the first 10 years and $180,000 for the following 10 years. Process B calls for buying machinery that costs $1.5 million and has maintenance charges of $100,000 for the 20-year period. Both sets of machinery have zero salvage value. Which process is more economical for the company at a 15 percent rate of interest?

7-12. An industry is considering the construction of a hazardous waste treatment facility of 21 MGD (million gal/day) with the following annual costs and disbenefits:

Capital cost	$20.2 million
Annual O&M cost	$19,000/MGD
Transportation cost	$375,000 per year
Loss of agricultural land	$0.8 million per year

Benefits for the facility are:

Health benefits	$0.64 million per year
Environment and ecosystem	$2.25 million per year
Reuse of chemicals	$0.6 million per year

Compute the B/C ratio. Calculate the present worth using interest rates of 10 percent and 15 percent and compare the results. The expected life for the facility is 10 years.

7-13. The following information is provided for two alternative liquid hazardous waste treatment facilities:

Plant	Flow rate, 10^6 gal/day	Capital costs, $\$10^6$	Operating costs, $\$10^6$/year
X	4.5	10.7	0.85
Y	9	16.8	1.63

Assuming a constant inflation rate of 9 percent over the 20-year life of each plant, determine which of the two alternatives is more economically feasible. Use an internal rate of return of 7 percent.

7-14. The benefits and disbenefits accruing from a proposed waste minimization project that would involve an initial cost of $7 million are the following:

Benefits:	
Reduced air pollution	$50,000 per year
Reduced future fees	$0.3 million
Increased public image	$0.4 million
Disbenefits:	
Increased maintenance	$1.5 million

Compute the B/C ratio of this project based on a 10 percent rate of interest for a 20-year project life.

7-15. Redo Problem 7-11 assuming an inflation rate of 5 percent for process A and 8 percent for process B. How do your results differ from the previous computation?

7-16. A paint manufacturer has modified production scheduling for its processes to greatly reduce waste generation and associated environmental costs. The process is a basic batch mixing process in large (500 to 1000 gallon) stainless steel mix tanks. Product is drained from the tanks into drums. On Monday, light-colored paints are produced, with darker colors produced through the week to minimize the amount of thorough cleaning required. Between batches, tanks may be rinsed with water that is then drained to a trench and a settling tank. On Fridays, employees clean the mix tanks, which can entail having to enter the tanks to scrape out residuals. This is followed by water rinses, with the rinse water again discharged to the trench and settling tank. After the settling tank, the wastewater is discharged to the sanitary sewer. Twice a year, plant employees rent a backhoe and dig up the accumulated sludge in the trench settling tank. The nonhazardous sludge is placed into drums and absorbent is added to the drums to allow the material to be disposed at the local landfill. Plant personnel feel that there is no merit in looking at further waste minimization opportunities since their environmental costs associated with the process are so low. The plant represented its total environmental costs as:

Sludge disposal (landfill charge)	$10,000
Labor	7,560
Backhoe rental	720
Total	$18,280

Question 1: What cost items are not included in the plant numbers and what would be reasonable estimates of these costs?

Question 2: What are potential pollution prevention opportunities to reduce these costs? Try to identify opportunities for the following options:

Improved waste management
Waste collection and recycling
Waste capture and reuse in the process
Elimination of any hazardous constituents in the waste
Elimination of the waste
Elimination of the waste generating process

7-17. A management review of the information in Table 7-3 indicates that another major factor, safety, should be included in the analysis. Management rates this as important as compliance or production impact. This assessment reveals that alternative B should be rated −3 for safety. In the process of this assessment, the rating for production impact for alternative B is lowered from −1 to −3. Revise Example 7-5 to account for these changes. Indicate how this might affect the waste minimization decision.

For Problems 7-18 through 7-27, use the appropriate manufacturing flowsheet from *Shreve's Chemical Process Industries* (5th edition, McGraw-Hill, 1984), and develop an outline of a waste minimization plan for the manufacturing process specified.

7-18. By-product coking (Shreve's: Fig. 5.3, p. 76).
7-19. Hydrogen production (Fig. 7.3, p. 109).
7-20. China clay (Fig. 9.1, p. 152).
7-21. Portland cement (Fig. 10.1, p. 175).
7-22. Superphosphate (Fig. 16.3, p. 275).
7-23. Sulfuric acid (Fig. 19.3, p. 330).
7-24. Nitrocellulose (Fig. 22.2, p. 397).
7-25. Photographic film (Fig. 23.3, p. 419).
7-26. Water-based paint (Fig. 24.4, p. 432).
7-27. Styrene (Fig. 36.3, p. 696).

7-28. Prepare a one-page summary of the article by G. J. Hollod and R. F. McCartney.[54]
7-29. Prepare a one-page summary of the article by Y. Isooka et al.[55]
7-30. Prepare a one-page summary of the article by S. T. Johnnie.[56]
7-31. Prepare a one-page summary of the article by L. R. Martin.[57]
7-32. Prepare a one-page summary of the article by L. L. Zelnio.[58]
7-33. Visit the plant manager of a local manufacturing facility suggested by your instructor. Prepare a brief summary of its current waste minimization program.
7-34. Contact your state environmental agency and determine what programs your state has to encourage hazardous waste minimization.

REFERENCES

1. U.S. Environmental Protection Agency: *Report to Congress: Minimization of Hazardous Waste,* Volumes I and II, EPA/530-SW-86-033A, Office of Solid Waste, p. ii, October 1986.
2. *Federal Register:* "Pollution Prevention Policy Statement," U.S. EPA, vol. 54, Jan. 26, 1989.
3. Colten, C. E.: "Historical Development of Waste Minimization," *The Environmental Professional,* vol. 11, pp. 94–99, 1989.
4. Koller, T.: *The Utilization of Waste Products,* Scott Greenwood & Son, London, 1915.
5. Trembler, H. A.: "Waste Savings by Improvements in Milk Plant Equipment," in *Proceedings of the Industrial Waste Conference,* Purdue University, pp. 6–21, November 1944.
6. Besselievre, E. B.: "Industries Recover Valuable Water and By-Products from Their Wastes," *Wastes Engr.,* vol. 30, p. 760, 1959.
7. Rocheleau, R. F., and E. F. Taylor: "An Industry Approach to Pollution Abatement," *Journal Water Pollution Control Federation,* vol. 36, pp. 1185–1194, October 1964.
8. Bringer, R. P., and D. M. Benforado: "Pollution Prevention as Corporate Policy: A Look at the 3M Experience," *The Environmental Professional,* vol. 11, pp. 117–126, 1989.
9. Hollod, G. J., and R. F. McCartney: "Hazardous Waste Minimization: Part I. Waste Reduction in the Chemical Industry—DuPont's Approach," *J. Air Pollut. Control Assoc.,* vol. 38, no. 2, pp. 174–179, 1988.
10. U.S. EPA *Pollution Prevention News,* p. 8, January/February 1991.
11. U.S. EPA, Internet website, June 23, 1998.
12. Lester, S. U.: "EPA Says: 'Build a New Superfund Site or Lose Money to Clean up the Ones You've Got,' " *Everyone's Backyard,* vol. 6, no. 1, Citizens Clearinghouse for Hazardous Waste, Inc., Washington, D.C., pp. 6–8, Spring 1988.
13. *The State Newspaper,* Columbia, S.C.

14. U.S. Environmental Protection Agency: *Waste Minimization Opportunity Assessment Manual,* EPA/625/7-88/003, p. 4, 1988.
15. Hirschorn, J. S., and K. U. Oldenburg: "Waste Reduction Audit: Matching Types to Stages" in *Hazardous and Industrial Solid Waste Minimization Practices,* ASTM, STP 1043 (R. A. Conway et al., eds.), ASTM, Philadelphia, pp. 41–47, 1989.
16. U.S. Environmental Protection Agency, *Waste Minimization Opportunity Assessment Manual.*
17. Ghassemi, M.: "Waste Reduction: An Overview," *The Environmental Professional,* vol. 11, pp. 100–116, 1989.
18. Anderson, C. L.: "The Production Process: Inputs and Wastes," *J. Environ. Econ. Mgmt.,* vol. 14, no. 1, p. 11, 1987.
19. Ibid.
20. Ibid.
21. Smith, Dennis: "Intel Corp. Waste Minimization Program," University of California at Davis, 1989.
22. *Hazardous Waste Minimization Manual for Small Quantity Generators,* Center for Hazardous Materials Research, University of Pittsburgh, p. 10-2, October 1989.
23. Thomas, D. L., et al.: "A Technical Assistance and Waste Reduction Program for Illinois," *Waste Minimization–Proceedings of A&WMA Specialty Conference,* SP67, Air & Waste Management Assoc., pp. 128–141, Pittsburgh, 1988.
24. Hirschorn, J. S., and K. U. Oldenburg: "The Obstacles to Waste Reduction," *Chem. Engr. Progress,* pp. 31–35, June 1989.
25. *Hazardous Waste Minimization Manual for Small Quantity Generators.*
26. Greenwood, Mark A.: "Design for the Environment: DfE Concepts Are Being Integrated into OTS Activities, Chemicals in Progress Bulletin, Office of Toxic Substances, U.S. EPA, vol. 12, no. 4, December 1991.
27. Bailey, Paul E.: "Life-Cycle Costing and Pollution Prevention," *Pollution Prevention Review,* vol. 1, no. 1, Winter 1990/91.
28. Data compiled by Franklin Associates, Ltd., and provided courtesy of the Proctor & Gamble Company.
29. Ibid.
30. Calif. Health and Safety Code, Div. 20, Chapter 6.5, Article 11.9.
31. Calif. Code of Regulations (CCR) Title 22, Chapter 30, Article 6.1.
32. U.S. Congress: *Serious Reduction of Hazardous Waste: For Pollution Prevention and Industrial Efficiency,* OTA-ITE-317, Office of Technology Assessment, Washington, D.C., September 1986.
33. Hunt, G. E.: "Waste Reduction in the Metal Finishing Industry," *J. Air Poll. Control Assn.,* vol. 38, pp. 672–680, 1988.
34. Higgins, T. E.: *Hazardous Waste Minimization Handbook,* Lewis Publishers, Chelsea, Mich., p. 81, 1989.
35. Thibault, J. E.: "The Costs and Benefits of Source Reduction in Metal Finishing," *Massachusetts Hazardous Waste Source Reduction Conference Proceedings,* Massachusetts Dept. of Environmental Management, October 1983.
36. Lewis, D. A.: "Waste Minimization in the Pesticide Formulation Industry," *J. Air Poll. Control Assn.,* vol. 38, pp. 1293–1296, 1988.
37. U.S. Environmental Protection Agency: "Recovery of Spent Sulfuric Acid from Steel Pickling Operations," U.S. EPA, 625/2-78-017, 1978.
38. Ibid.
39. Roberts, R. M., J. L. Koff, and L. A. Karr: "Hazardous Waste Minimization-Initiation Decision Report," Tech. Memorandum TM 71-86-03, Naval Civil Engineering Laboratory, Port Hueneme, Calif., pp. 2–13, January 1988.
40. Higgins, *Hazardous Waste Minimization Handbook,* pp. 119ff.

41. U.S. Environmental Protection Agency, *Waste Minimization Opportunity Assessment Manual,* p. E-4.
42. Ahearn, J., H. Fatkin, and W. Schwalm: "Case Study: Polaroid Corporation's Systematic Approach to Waste Minimization," *Pollution Prevention Review,* pp. 257ff., Summer 1991.
43. R. C. Weast, ed.: *CRC Handbook of Chemistry & Physics,* 64th edition, CRC Press, Boca Raton, Fla., pp. D-9 to D-23, 1984.
44. Perry, R. H.: *Chemical Engineers' Handbook,* 6th edition, McGraw-Hill, New York, pp. 13–57ff, 1984.
45. Freeman, H. M.: *Hazardous Waste Minimization,* McGraw-Hill, New York, p. 328, 1990.
46. Higgins, *Hazardous Waste Minimization Handbook,* p. 56.
47. Ibid., pp. 65ff.
48. Ibid., p. 69.
49. McTeer, P., W. Word, and K. Lewandowski: "A New Application for Ultrafiltration Wastewater Treatment in the Industrial Environment," *Proceedings of 39th Industrial Waste Conference,* Purdue University, 1985.
50. Kampouris, E. M., C. D. Papaspyrides, and C. N. Lekakou: "A Model Recovery Process for Scrap Polystyrene Foam by Means of Solvent Systems," *Conservation & Recycling,* vol. 10, no. 4, pp. 315–319, 1987.
51. Ibid.
52. Slough, W.: "Waste Exchanges: Catalysts for Innovation in Waste Minimization," *J. Air Poll. Control Assn.,* vol. 38, p. 744, 1988.
53. Listings Catalog, Northeast Industrial Waste Exchange, Inc., Issue No. 36, p. 20, May 1990.
54. Hollod, G. J., and R. F. McCartney: "Hazardous Waste Minimization: Part I. Waste Reduction in the Chemical Industry—DuPont's Approach," *J. Air Pollut. Control Assn.,* vol. 38, no. 2, pp. 174–179, 1988.
55. Isooka, Y., Y. Imamura, and Y. Sakamoto: "Recovery and Reuse of Organic Solvent Solutions," *Metal Finishing,* vol. 82, no. 6, pp. 113–118, June 1984.
56. Johnnie, Susan T.: "Waste Reduction in the Hewlett-Packard, Colorado Springs Division, Printed Circuit Board Manufacturing Shop," *Hazardous Waste Hazardous Mater.,* vol. 4, no. 1, pp. 9–22, 1987.
57. Martin, L. R.: "Demanding Waste Reduction: The Roles of Public Interest Organizations in Promoting the Institutionalization of Waste and Toxics Reduction," *The Environmental Professional,* vol. 11, pp. 132–141, 1989.
58. Zelnio, L. L.: "Reclamation of Metalworking Fluids from Individual Machine Tools," *Lubrication Engineering,* January 1987.

ADDITIONAL READING

Bridges, J. S., et al.: "Results from a Cooperative Federal, State, and Trade Association Waste Minimization Research Program," *Hazardous Waste Hazardous Mater.,* vol. 6, no. 17, 1989.

Brinkman, D. W.: "Technologies for Re-Refining Used Lubrication Oil," *Lubrication Engineering,* vol. 43, no. 5, pp. 324–328, May 1987.

Caldart, C. C., and C. W. Ryan: "Waste Generation Reduction: A First Step toward Developing a Regulatory Policy to Encourage Hazardous Substance Management Through Production Process Change," *Hazardous Waste Hazardous Mater.,* vol. 2, no. 3, pp. 309–331, 1985.

Cranford, B.: "Federally Sponsored Waste Minimization Research and Development for Hazardous and Non-Hazardous Wastes," *J. Air Pollut. Control Assn.,* vol. 39, no. 34, 1989.

Davies, R., R. Curtis, and R. Laughton: "Recycling Metal Stamping Plant Wastes," *Water and Pollution Control,* vol. 123, no. 5, pp. 26–27, September/October 1985.

deYoung, R.: "Exploring the Difference Between Recyclers and Non-Recyclers: The Role of Information," *J. Environ. Sys.,* vol. 6, p. 341, 1989.

Freeman, H. M.: *Hazardous Waste Minimization—Industrial Overviews,* JAPCA Reprint Series RS14, Air & Waste Management Assoc., Pittsburgh, Pennsylvania, 1989.

Johnson, J. C., et al.: "Metal Cleaning by Vapor Degreasing," *Metal Finishing*, September 1983.

Kalnes, T. N., et al.: "Recycling Waste Tube Oils for Profit (UOP Direct Hydrogeneration Process)," *Hazardous Waste Hazardous Mater.,* vol. 6, no. 51, 1989.

Kenson, R. E.: "Recovery and Reuse of Solvents from VOC Air Emissions," *Environmental Progress,* vol. 4, no. 3, pp. 161–165, August 1985.

Kikendall, T. R.: "Converting from Conventional to Compliance Coating Systems," *Industrial Finishing,* vol. 58, no. 6, pp. 40–42, June 1982.

Kolaczkowski, S. T., and B. D. Crittenden: *Management of Hazardous and Toxic Wastes in the Process Industries,* Elsevier Applied Science, 1987.

Loehr, R.: *Reducing Hazardous Waste Generation,* National Academy of Sciences, National Academy Press, 1985.

Nunno, T. J., and H. M. Freeman: "Minimizing Plating Bath Wastes in the Electronics Products Industry," *J. Air Poll. Control Assn.,* vol. 37, pp. 723–729, 1987.

Overby, C.: "Design for the Entire Life Cycle: A New Paradigm?" in *Proceedings of the International Conference on Pollution Prevention* (EPA/600/9-90/039), June 1990.

Overcash, M. R.: *Techniques for Industrial Pollution Prevention,* Lewis Publishers, Chelsea, Mich., 1986.

Sarokin, D., W. R. Muir, C. G. Miller, and S. Sperber: *Cutting Chemical Wastes,* Inform, Inc., New York, 1986.

Theodore, L., and McGuinn, Y. C., *Pollution Prevention,* Van Nostrand-Reinhold, New York, 1992.

Tsai, E. C., and R. Nixon: "Simple Techniques for Source Reduction of Wastes from Metal Plating Operations," *Hazard. Waste Hazard. Mater.,* vol. 6, no. 67, 1989.

U.S. Environmental Protection Agency, *Waste Minimization Issues and Options,* vol. I, II, and III, EPA/530-SW-86-041, Office of Solid Waste, Washington, D.C. (NTIS:PB87-114351, PB87-114369 and PB87-114377), October 1986.

U.S. Environmental Protection Agency: *Proceedings of the International Conference on Pollution Prevention,* EPA/600/9-90/039, June 1990.

U.S. Environmental Protection Agency: *Report to Congress: Minimization of Hazardous Waste,* vol. I and II, EPA/530-SW-86-033A, Office of Solid Waste, Washington, D.C. (Available from NTIS: PB87-114336 and PB87-114344), October 1986.

U.S. Environmental Protection Agency: *Seminar Publication—Solvent Waste Reduction Alternatives,* EPA/625/4-89/021, September 1989.

Waste Minimization—Proceedings of A&WMA Specialty Conference, SP67, Air & Waste Management Assoc., Pittsburgh, 1988.

ACKNOWLEDGMENTS

The authors wish to acknowledge the following individuals who assisted with the preparation of this chapter for the first edition: Marc Wendell, Ph.D.; Brian Bennett, P.E.; Donald Grogan; and Donald Landeck, *ERM;* and Sharmistha Law, *Bucknell University.* The principal reviewer for ERM was Amir Metry, P.E., Ph.D.

The authors also wish to acknowledge Daniel Reinke, *ERM*, who assisted with the preparation of this chapter for the second edition. The principal reviewer for the second edition was Ken Johnson, *ERM*.

CHAPTER 8

Facility Development and Operations

When society requires to be rebuilt, there is no use attempting to rebuild it on the old plan.

John Stuart Mill

A facility is defined as the contiguous land, structures, and other improvements and appurtenances used for storing, recovering, recycling, treating, or disposing of hazardous waste. This chapter covers the steps in the development and operation of hazardous waste management facilities.

This chapter begins with an explanation of facility types and their operations. Next, the chapter briefly highlights the step of assessing the need for facilities, thus providing a lead to the controversial subject of siting. This is followed by a review of how the public can participate in siting, including a case study of a successful project to site a major facility. Although this review of public participation addresses just the subject of siting, the general concepts of public communication presented are important to virtually all aspects of hazardous waste management. The chapter concludes with a discussion of permitting.

8-1 FACILITY TYPES

The concept of a facility dedicated to the management of hazardous waste is not new. Long before the enactment of U.S. hazardous waste laws, many generators recognized the need for specialized treatment and disposal of these wastes. Many generators constructed and operated their own captive facilities referred to as **on-site facilities.** For example, Dow Chemical constructed a rotary kiln incinerator at its Midland, Michigan, plant in 1954 for the primary purpose of thermal destruction of organic waste from the production of a great number of organic chemicals.

Hundreds of plants developed simple, inexpensive surface impoundments (many of them unlined). Thousands of industrial wastewater treatment units were installed to neutralize acids and precipitate metals.

Other generators, not having a suitable site or not generating a sufficiently large volume of waste to justify the investment in an on-site facility, transported their waste off-site to specialized facilities for treatment and disposal. Such facilities are referred to as commercial **off-site facilities.** The commercial hazardous waste management industry began developing off-site facilities in the late 1960s. Some of these early facilities, such as the Rollins Environmental Services facility built in Bridgeport, New Jersey, in 1971, are still operating today, at the same sites but with vastly modernized equipment and operations.

Just as there are many types of hazardous waste, there are many ways in which hazardous wastes can be managed. In fact, there are at least 50 commercially proved technologies for the recovery and treatment of hazardous waste. A hazardous waste facility may function with just one technology, or it may combine multiple technologies, particularly if it is a commercial facility serving a number of generators.

The predominant types of facilities, other than storage facilities, are depicted in Figure 8-1 under the following major categories:

- **Recovery/recycling facilities:** Recover material as a salable product (typically solvents, oils, acids, or metals); some use the waste for its energy value (e.g., cement kilns).
- **Treatment facilities:** Change the physical or chemical characteristics of a waste, or degrade or destroy waste constituents, using any of a wide variety of physical, chemical, thermal, or biological methods.
- **Land disposal facilities:** Permanent emplacement of waste on or below land surface, typically after the waste has been treated to meet defined standards called land disposal restrictions (LDR).

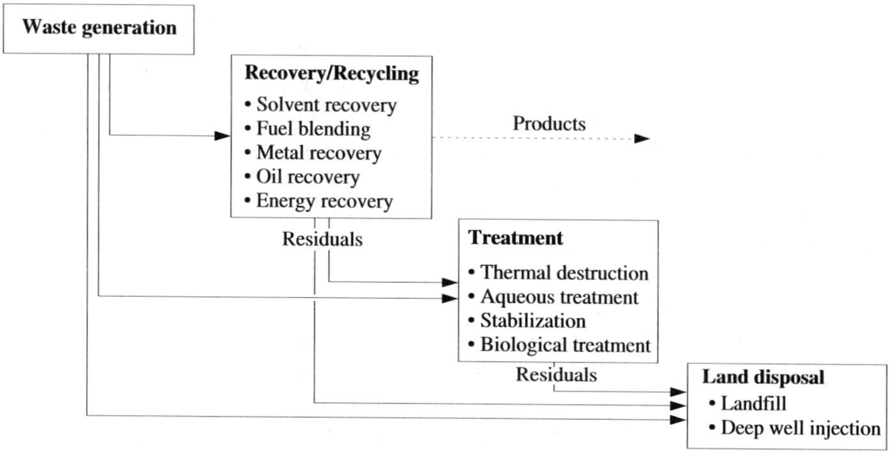

FIGURE 8-1
Hazardous waste recovery, treatment, and disposal facilities.

There are some differences between a commercial off-site facility and a captive on-site facility. Public reaction to their siting contrasts remarkably. This derives in part from the fact that an off-site facility accepts waste from outside the community, while an on-site facility handles only that waste generated by what could be a long-standing and important economic activity in the community. From a technical perspective, the off-site facility generally handles a wider range of waste types, is larger and more complex, and is responsible for waste transport to the facility. Ignoring this, however, the risks posed by off-site and on-site facilities are comparable, and depend far more on types of waste received, design, operation, and other site-specific factors. The two must meet the same hazardous waste regulations. Thus, no further technical distinction is made in this chapter.

Numerous references are available that describe in detail the various types of facilities.[1] In fact, some books may be dedicated to a single type.[2] The major types of facilities are discussed in Chapters 9 through 13 of this textbook. The following summaries are developed from a report intended to introduce the reader to basic categories of facilities.[3] Within each category, many subcategories exist, depending on a particular technology employed.

Recovery Facilities

Solvent recovery

These facilities separate contaminants from waste solvents and thus restore the solvent to its original quality or to a lower-grade product (e.g., lacquer thinner). This practice is fairly widespread for halogenated solvents. Distillation (batch, continuous, or steam) is used by most commercial solvent processors, and typically recovers about 75 percent of the waste solvent (see Section 7-6). The residue, known as still bottoms, can be a liquid or a sludge, depending on a number of conditions, and typically requires management as a hazardous waste. Other separation technologies used by solvent processors include filtration, simple evaporation, centrifugation, and stripping.

Oil recovery

Used lubricating oils can be recovered to a quality essentially equal to virgin lubricating oils. This process is typically referred to as oil re-refining. The term oil recovery excludes the larger practice of fuel blending, which does not recover lubricating oils. Two methods may be employed in waste oil re-refining: the acid/clay method or the capital-intensive distillation method. Because the former produces a large amount of spent acidic oily clay residue, distillation is preferred.

Acid regeneration

Acid recovery usually involves the separation of unreacted acid from an acid waste such as spent pickle liquor generated by the steel industry. One method, used in the steel industry, is to cool sulfuric acid to precipitate ferrous compounds. In another method, acid can be regenerated by injecting it into a spray roaster. In the spray roaster, spent pickle liquor is heated to form iron oxide and gaseous hydrochloric acid (HCl). On exiting the top, HCl is separated from other gases for reuse at a pickling line.[4]

Metals recovery

The technologies for recovering metals can be grouped as pyrometallurgical or hydrometallurgical. Pyrometallurgy uses the differences in melting and boiling properties to separate metals at high temperatures. The heat is typically provided by roasting or smelting. The hydrometallurgy technology extracts and concentrates metals from liquid waste using any of a variety of processes such as ion exchange, electrodialysis, reverse osmosis, membrane filtration, adsorption, sludge leaching, electrowinning, solvent stripping, and precipitation. Nonliquid wastes first require dissolution.

Fuels blending

Waste-derived fuels can be made from waste oils, solvents, and distillation bottoms by either blending different wastes that have high BTU values or by mixing wastes with coal or fuel oils. Wastes, particularly oils, normally need pretreatment to remove bottoms, sediments, and water. This is achieved through separation and dehydration. Fuels blending is a simple process that is well developed.

Coincineration in industrial kilns/furnaces

The manufacturers of cement, light aggregate, and some other products use a special type of rotary kiln. The destruction of hazardous waste by coincineration with primary fuels and raw materials in such industrial kilns has been practiced for a long time. Cement kilns operate at high temperatures (2600 to 3000°F) and provide very long residence times, which make possible very high destruction of principal organic hazardous constituents (POHCs, see Section 12-2). The waste thus represents a supplementary fuel. The waste may be blended at the site where the waste is burned or blended at a separate facility and shipped to the site of the industrial kiln.

An example of a recovery facility is shown in Figure 8-2. This particular facility has the capability to recover waste solvents and oils.

Treatment Facilities

Thermal destruction

Exposing waste to high temperatures in the presence of oxygen results in either partial or complete destruction of organic waste materials. The most common application is an incinerator, which is an "enclosed device using controlled flame combustion, the primary purpose of which is to thermally break down hazardous waste."[5] Existing federal regulations specify that incineration must achieve a destruction and removal efficiency (DRE) of at least 99.99 percent. Chapter 12 describes several technologies for achieving thermal destruction.

Aqueous treatment

An aqueous waste treatment system removes and/or detoxifies hazardous constituents that are dissolved or suspended in water. The selection and sequence of unit processes are determined by the characteristics of incoming wastes and the required effluent quality. Figure 8-3 shows the unit process diagram for destruction of aqueous waste containing cyanide. (Cyanide destruction is discussed in Section 9-4.)

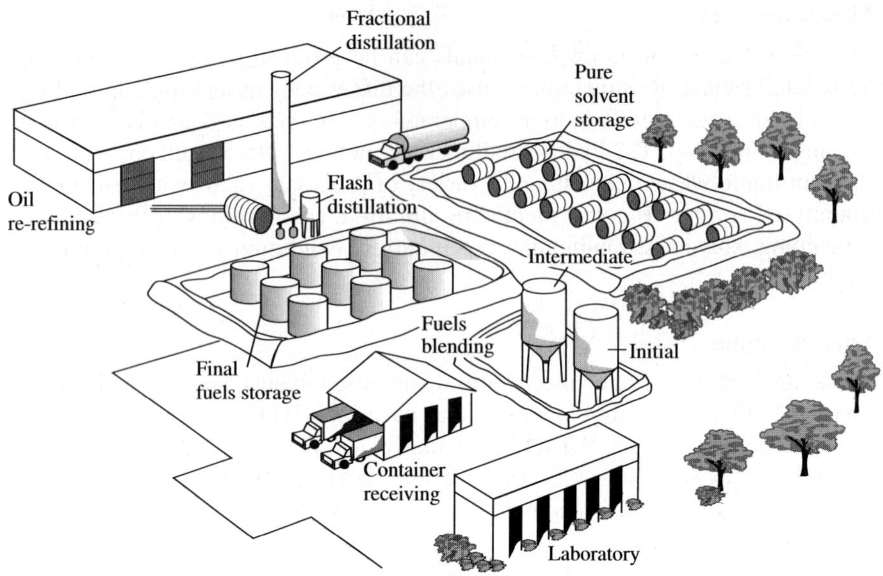

FIGURE 8-2
Liquid organics recovery facility.[6]

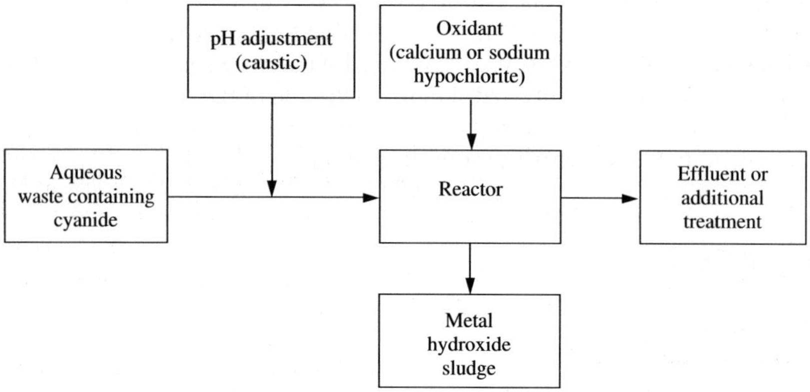

FIGURE 8-3
Cyanide destruction process diagram.

An example of a full-service aqueous treatment facility is shown in Figure 8-4. Typically, it would feature most, if not all, of the following unit processes and could treat essentially any aqueous waste:

- Cyanide destruction
- Chrome reduction
- Two-stage metal precipitation
- pH adjustment (neutralization)
- Solids filtration

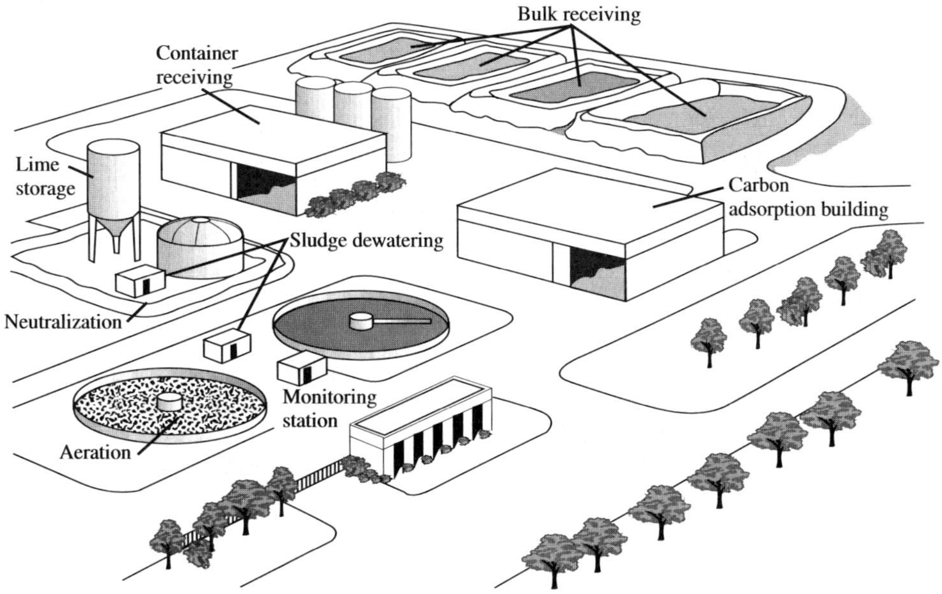

FIGURE 8-4
Full-service aqueous treatment facility.[7]

- Biological treatment
- Carbon adsorption
- Sludge dewatering

Stabilization

Stabilization involves admixing materials with wastes to improve the handling and physical characteristics of the wastes and to reduce mobility of contaminants in a landfill environment. Stabilization may either induce a chemical change to produce an essentially insoluble form or chemically bind the waste in a matrix of crystal lattice having a very low solubility. Several types of stabilization processes have been developed, and these are described in Chapter 11.

Biological treatment

Biological treatment systems utilize microorganisms to degrade organic wastes. Organic substances are either mineralized or transformed into compounds having a lower molecular weight. The primary process variables are the method of contacting the waste with biomass (microbes), the moisture content of the waste (e.g., liquid, slurry, or solid), and the method and degree of aeration. Biological methods are explained in Chapter 10.

Land Disposal Facilities

Landfill

A landfill involves permanent emplacement of hazardous waste. In many instances the waste will undergo some form of pretreatment (e.g., stabilization) prior

to land disposal. Regulatory requirements (LDRs) define the level of treatment required. Emplacement would occur either in relatively shallow trenches or in mounded cells or vaults that extend mostly, if not totally, above grade. A landfill facility has several protective elements (e.g., liners), both design and operational, to contain emplaced wastes, to minimize the generation of leachate, and to remove leachate as it is generated. It is important to note that landfills should be operated as containment facilities with necessary provisions for virtually perpetual maintenance and monitoring after closure. Landfills are described in Chapter 13.

Deep well injection

This disposal method features injection of liquid waste into a deep, porous subsurface geological formation containing salt water. The receiving formation is well below potable aquifers and should have an impermeable confining layer above it. The injection well consists of the injection pipe surrounded by a series of concentric casings with the annuli containing cement and noncorrosive fluids to detect and prevent leakage into shallow formations.

Other

Surface impoundments are used for the retention of waste liquids and sludges. They require liners and many of the other protective elements associated with a landfill, particularly if they are closed with some waste left in place. A land farm is a biological treatment method in which soil is the treatment medium; however, some waste constituents are not degraded but incorporated into the soil; the method thus represents a form of disposal.

Fully Integrated Facility

Some large commercial facilities employ aqueous treatment, incineration, land disposal, and possibly other components to form a fully integrated facility as depicted in Figures 8-5, 8-6, and 8-7.

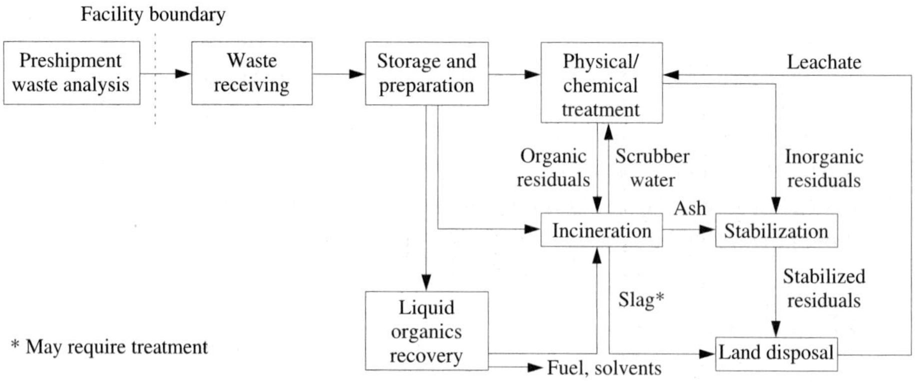

FIGURE 8-5
Waste management flow diagram for fully integrated hazardous waste management facility.

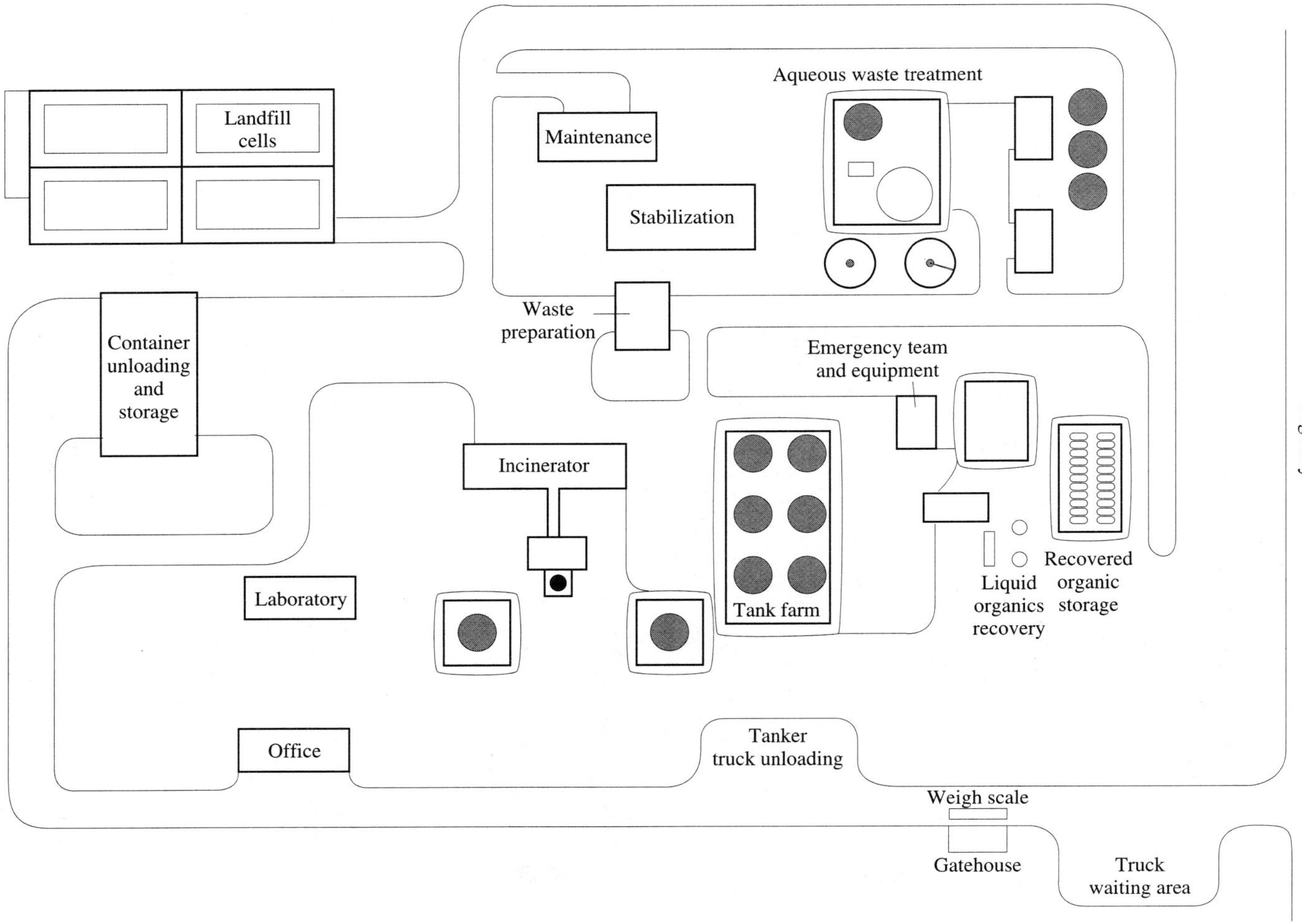

FIGURE 8-6
Site plan of fully integrated hazardous waste management facility.

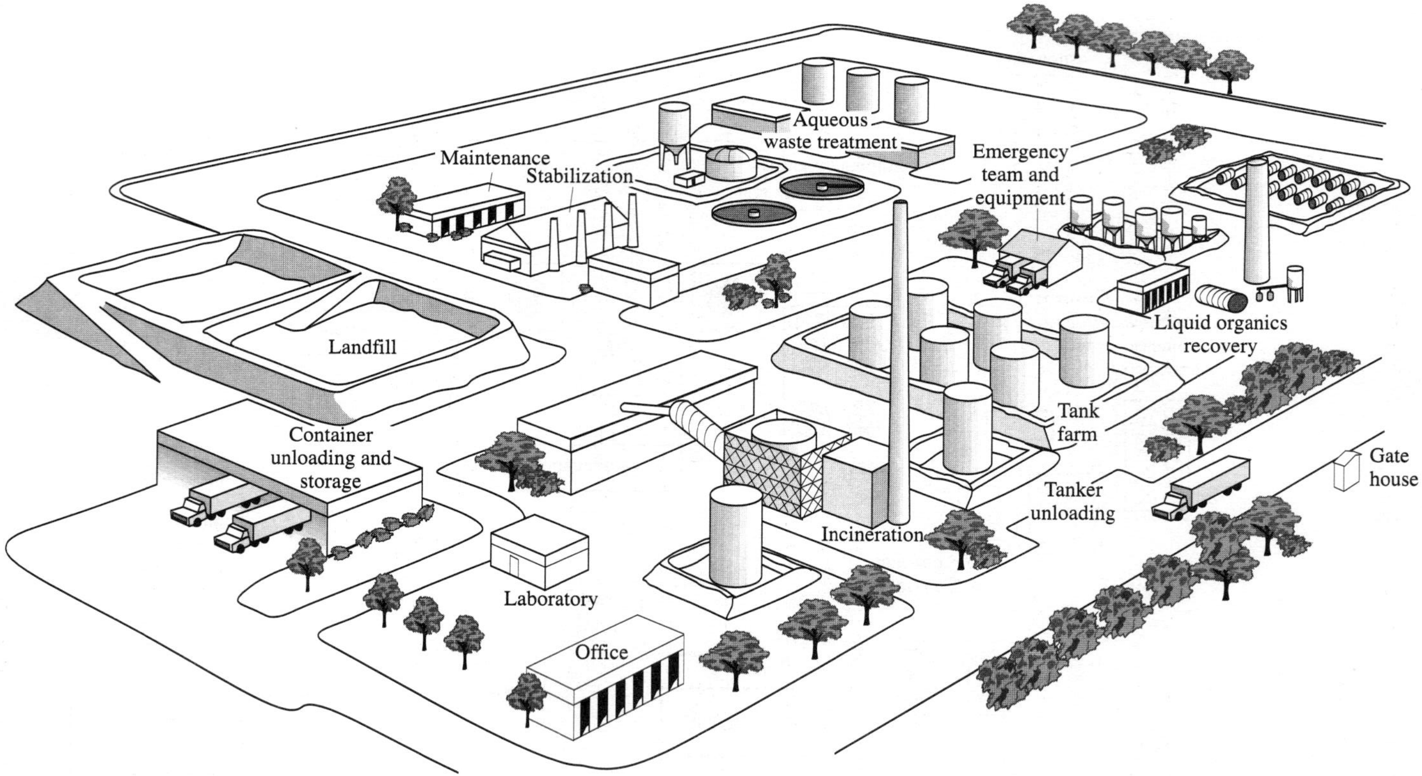

FIGURE 8-7
A hypothetical fully integrated hazardous waste management facility.[8]

8-2 FACILITY OPERATIONS

The operations at a hazardous waste facility consist of five subsystems:

1. Preshipment waste analysis
2. Waste receiving
3. Waste storage and preparation
4. Waste treatment
5. Residuals management

Figure 8-8 matches these subsystems with the components of the fully integrated facility depicted in the preceding figures. It is important to note that all components operate under an umbrella of a number of special measures. These special precautionary measures include security, inspections, maintenance, training, incident prevention, emergency planning, safety, monitoring, record keeping, auditing, and reporting.

Preshipment Waste Analysis

A waste analysis plan is a critical document for any hazardous waste facility. The plan specifies the parameters for which each waste will be analyzed, the sampling and analytical methods to be used, and the frequency of analysis. Before a facility treats, stores, or disposes of a waste, it must profile that waste, including a detailed chemical and physical analysis of a representative sample of the waste (see Figure 8-9). Commercial facilities require this full characterization prior to shipment by the generator. Representative sampling of a waste shipment is conducted on arrival at the facility to verify that the composition of the shipped waste matches the fully characterized waste. The purpose of the full characterization before shipment is to satisfy the following requirements:

	Operations subsystems				
Facility components	Preshipment waste analysis	Waste receiving	Waste storage and preparation	Waste treatment	Residuals management
Analytical laboratory	✓	✓			
Truck waiting area		✓			
Gate house		✓			
Weigh scale		✓			
Drum unloading and storage		✓	✓		
Tank farm		✓	✓		
Bulk waste storage		✓	✓		
Physical/chemical treatment				✓	✓
Incinerator				✓	✓
Stabilization plant				✓	✓
Landfill cells					✓

FIGURE 8-8
Operations subsystems and their components.

Waste Profile Form

Generator Name ____
Address ____

City, State, Zip ____
Technical Contact ____ (Name) (Title)
Area Code ( ) Telephone # ____
Facility EPA ID # ____

Billing Address ____

City, State, Zip ____
Technical Contact ____ (Name) (Title)
Area Code ( ) Telephone # ____
Business SIC Code ____
Duns # ____

Common Name of Waste ____
Generating Process ____

Chemical Composition (Totals must add up to 100%)

	%
	%
	%
	%
	%
	%
	%
	%
	%
	%
	%

EPA Haz Waste # ____
(Quan) (Units) PERIN (Time Interval)
Rate of Generation ____
(Quan) (Units) (Container)
Volume in Storage ____
Is Waste DOT Hazardous ☐ (Yes) ☐ (No)
Proper DOT Shipping Name ____
Hazard Class ____ ID # ____
Transportation Equipment ____
Placarding ____

Metals (mg/l or ppm)

	Total	Leachate		Total	Leachate
Arsenic (As)			Selenium (Se)		
Barium (Ba)			Silver (Ag)		
Cadmium (Cd)			Copper (Cu)		
Chromium (Cr)			Nickel (Ni)		
Chromium, Hex			Zinc (Zn)		
Lead (Pb)					
Mercury (Hg)					

Inorganics (mg/l or ppm)

Total CN			Bromide		
Free CN			Iodide		
Sulfide			Asbestos		
Bisulfite					
Sulfite					
Sulfate					
Phosphate					
Fluoride					
Chloride					

Organics (mg/l or ppm)

Endrin			BOD		
Lindane			COD		
Methoxychlor			Organohalide		
Toxaphene			Organo-sulfur		
2,4-D			Mercaptans		
2,4,5-T					
Phenol (ics)					
PCB					
TOC					

Physical Description
Physical State ☐ Liquid ☐ Semi Solid ☐ Solid
Phases/Layering ☐ None ☐ Bilayered ☐ Multilayered
Total Solids (wt %) ____ Suspended Solids (wt %) ____
Type of Solids ☐ Organic ☐ Inorganic ☐ Mixed
Specific Gravity ____
Viscosity ☐ High ☐ Medium ☐ Low
Flash Point (°F) ____ Type ____
Boiling Point (°C) ____ Freezing Point (°C) ____
Vapor Pressure (mm Hg @ 25°C) ____
BTU/lb ____ % Ash Content ____
pH (Avg) ____ (Range) ____ To ____
Total Alkalinity/Acidity (%) ____
Odor ____
Color ____

Hazardous Properties ☐ None
☐ Ignitable ☐ Corrodes Steel ☐ Toxic Vapor
☐ Reactive ☐ Pyrophoric ☐ Shock Sensitive
☐ Explosive ☐ Water Reactive ☐ Radioactive
☐ Biological ☐ Pathogen ☐ Etiological
☐ Pesticide Residuals ☐ Other ____

NFPA Hazard Identification
Toxicity: Oral ____ Dermal ____ Inhalation ____

Please attach all material safety data sheets, handling precautions, additional hazard information, support data & comments.

I believe that the above information is true, accurate and complete to the best of my knowledge and, that all known and suspected hazards have been disclosed.

____ By ____ ____ ____
Date Name Title Signature

FIGURE 8-9
Waste profile form.

1. Determine if the waste may be accepted by the facility in terms of both its permit and its capability to treat or dispose of the waste.
2. Identify the inherent hazards of the waste so that the facility can take appropriate precautions to prevent incidents.
3. Determine the physical characteristics and chemical constituents of the waste to allow selection of effective waste processing and disposal methods.
4. Select the verification parameters to be tested on arrival at the facility to ensure that each shipment of waste is the same type as the fully characterized waste.
5. Develop an estimate of the cost of treatment or disposal.

Waste Receiving

Waste shipments typically arrive by truck at a facility's gatehouse. On accepting the waste, the facility signs the manifest and sends a copy to the generator. At that point, the facility shares liability with the generator and the transporter. Thus, it is critical that the preshipment waste analysis has already been completed and the shipment scheduled. Without prior scheduling of the incoming shipment or if the shipment is improperly documented, the gatehouse may refuse entry of the truck. This can lead to concerns over "illegal dumping" of these wastes.

Scheduled and properly documented shipments are directed to the receiving station where any packaging is checked, the loaded truck is weighed, and representative samples are collected for testing the verification parameters. The waste may arrive as bulk liquids in a tank truck, containerized liquids or sludges in drums, bulk shipments of contaminated soil in dump trucks, or by a number of other methods. Collecting a representative sample can pose a difficult task considering that a waste may be in multiple phases and states or have pockets of high contamination. The receiving station must use previously established procedures for each situation to ensure the collection of a representative sample.

On collection of the sample, the laboratory analyzes a portion for the verification parameters and retains the remainder for subsequent testing of treatability parameters. On verification of the waste shipment, the truck is directed to an unloading area where it is emptied and then reweighed before it leaves the facility.

The mere "emptying" of a truck can pose a difficult challenge if the waste has stratified, a container has leaked, or a solidification reaction has occurred. It is important that facilities have planned procedures and are prepared with special equipment to resolve such problems. The truck may need to be decontaminated to remove any trace residues, with the rinse water being handled as additional hazardous wastes.

Waste Storage and Preparation

On unloading, the wastes are moved into storage. This can consist of tanks or impoundments for bulk liquids, hoppers for solids and sludges, or pads and warehouses for containers. The objectives of storage and preparation are fourfold: (1) store the waste safely before introduction as feed into the system of unit treatment and disposal processes, (2) provide adequate accumulation time during periods when treatment

and disposal process systems are out of service, (3) facilitate mixing, blending, and repackaging of waste as deemed necessary, and (4) allow staged input of various wastes with reagents to the subsequent unit treatment processes.

Fire prevention and protection are obviously important safety considerations. The storage of certain types of hazardous waste requires automatic alarms and possibly sprinklers. The facility must provide adequate water supply for extinguishing fires plus the capability to collect and store fire water runoff. The storage or treatment of any water-reactive waste necessitates an alternative type of fire protection system.

A key issue in providing safe storage is compatibility. This has two independent considerations:

1. The compatibility of the waste with the material used to construct the container, tank, or liner in contact with the waste (e.g., certain solvents should not be stored in plastic containers).
2. The compatibility of the waste with other wastes stored together (e.g., containers of cyanide waste should not be located near acid waste).

The key is to segregate incompatible wastes by placing them in separate areas constructed of suitable materials. If stored together, incidents such as leaks could result in mixing of incompatible wastes. Different chemical reactions could occur. Some reactions could produce excessive pressure, thus posing fire or explosion hazards. Others could produce toxic fumes or gases. Figure 8-10 presents a compatibility chart and indicates that careful planning must be given to chemical storage. Such requirements apply to hazardous waste generators as well as to treatment and disposal facilities.

EXAMPLE 8-1. WASTE COMPATIBILITY. Given the following list of wastes stored in containers in a common area, what could happen in the event of a fire in which high-pressure water was used to extinguish it? To provide safe segregation, what is the minimum number of separate storage areas required? Does this list provide sufficient information to make this determination?

Waste A—Spent pickle liquor from a steel manufacturer
Waste B—Metal plating rinsewater from cyanide bath
Waste C—Cadmium-bearing sludges from a metal plater
Waste D—Spent degreaser from an electroplater
Waste E—Solvent-based paint sludges from a metal finishing operation
Waste F—Distillation bottoms from a phenolics manufacturer
Waste G—Laboratory packs from a university
Waste H—Pesticides with an expired shelf life
Waste I—Soil contaminated with a spill of diesel fuel
Waste J—Alkali cleaners
Waste K—Off-spec sodium product
Waste L—Arsenical wastewater treatment sludges from veterinarian product manufacturer
Waste M—Drum-reconditioning waste

Solution. These limited descriptions do not characterize the wastes sufficiently for the purposes of storage. For example, drum reconditioning wastes could include acid and alkaline rinses, spent solvents, and paint sludges. Laboratory packs could include any of innumerable chemicals used in research laboratories. Additional information may be obtained from the waste profiles needed for facilities to accept these wastes. In this example, it is assumed that the waste profiles would identify the following compatibility

KEY

Reactivity code	Consequences
H	Heat generation
F	Fire
G	Innocuous and nonflammable gas generation
GT	Toxic gas generation
GF	Flammable gas generation
E	Explosion
P	Violent polymerization
S	Solubilization of toxic substances
U	May be hazardous but unknown

Example:

H, F, GT	Heat generation, fire, and toxic gas generation

Reactivity group

No.	Name	1	2	3	4	5	6	7	8	9	10	11	12	13	14	15	16	17	18	19	20	21	22	23	24	25	26	27	28	29	30	31	32	33	34	101	102	103	104	105	106	107
1	Acids, minerals, non-oxidizing	1																																								
2	Acids, minerals, oxidizing		2																																							
3	Acids, organic		G, H	3																																						
4	Alcohols & glycols	H	H, F	H, P	4																																					
5	Aldehydes	H, P	H, F	H, P		5																																				
6	Amides	H	H, GT				6																																			
7	Amines, aliphatic & aromatic	H	H, GT	H		H		7																																		
8	Azo compounds, diazo comp. & hydrazines	H, G	H, GT	H, G	H, G	H			8																																	
9	Carbamates	H, G	H, GT						H, G	9																																
10	Caustics	H	H	H		H				H, G	10																															
11	Cyanides	GT, GF	GT, GF	GT, GF					G			11																														
12	Dithiocarbamates	H, GF, F	H, GF, F	H, GF, GT		GF, GT		U	H, G				12																													
13	Esters	H	H, F						H, G		H			13																												
14	Ethers	H	H, F												14																											
15	Fluorides, inorganic	GT	GT	GT												15																										
16	Hydrocarbons, aromatic		H, F														16																									
17	Halogenated organics	H, GT	H, F, GT					H, GT	H, G		H, GF	H						17																								
18	Isocyanates	H, G	H, F, GT	H, G	H, P			H, P	H, G		H, P, G	H, G	U						18																							
19	Ketones	H	H, F						H, G		H	H								19																						
20	Mercaptans & other organic sulfides	GT, GF	H, F, GT						H, G									H	H	H	20																					
21	Metals, alkali & alkaline earth, elemental	GF, H, F	GF, H, F	GF, H, F	GF, H, F	GF, H, F	GF, H	GF, H	GF, H	GF, H	GF, H	GF, H	GF, GT, H	GF, H				H, E	GF, H	GF, H	GF, H	21																				
22	Metals, other elemental & alloys as powders, vapors or sponges	GF, H, F	GF, H, F	GF					E, F, GT	U	GF, H							H, E	GF, H		H, GF, F		22																			
23	Metals, other elemental & alloys as sheets, rods, drops, moldings, etc.	GF, H, F	GF, H, F						H, F, G									H, F						23																		
24	Metals & metal compounds, toxic	S	S	S			S	S			S														24																	
25	Nitrides	GF, H, F	H, F, E	H, GF	GF, H, E	GF, H			U	H, G	U	GF, H	GF, H	GF, H				GF, H	U	GF, H	GF, H	E				25																
26	Nitrites	H, GT, GF	H, F, GT	H							U											H, P			S	GF, H	26															
27	Nitro compounds, organic		H, F, GT			H					H, E											H, GF, E				H, GF, E		27														
28	Hydrocarbons, aliphatic, unsaturated	H	H, F			H																	H, E						28													
29	Hydrocarbons, aliphatic, saturated		H, F																											29												
30	Peroxides & hydroperoxides, organic	H, G	H, E		H, F	H, G		H, GT	H, F, E	H, F, GT		H, E, GT	H, F, GT					H, E	H	E	H, F, GT	H, E	P, G		H, G	H, GF, E	H, P, GT		H, P		30											
31	Phenols & cresols	H	H, F						H, G										H, P			GF, H				GF, H					H	31										
32	Organophosphates, phosphothioates, phosphodithioates	H, GT	H, GT						U		H, E											H									U		32									
33	Sulfides, inorganic	GT, GF	HF, GT	GT		H			E										H												H, GT			33								
34	Epoxides	H, P	H, P	H, P	H, P	U		H, P	H, P		H, P	H, P	U								H, P	H, P	H, P		H, P	H, P					H, P	H, P	U	H, P	34							
101	Combustible & flammable materials, misc.	H, G	H, F, GT																			H, G, F				H, GF, F					H, F, GT					101						
102	Explosives	H, E	H, E	H, E					H, E		H, E			H, E								H, E	H, E	H, E	E	E					H, E	H, E		H, E	H, E	H, E	102					
103	Polymerizable compounds	P, H	P, H	P, H					P, H		P, H	P, H	U									P, H	P, H	P, H	P, H	P, H					P, H	P, H		P, H			H, E	103				
104	Oxidizing agents, strong	H, GT		H, GT	H, F	H, F	H, F, GT	H, F, GT	H, E	H, F, GT		H, E, GT	H, F, GT	H, F	H, F		H, F	H, GT	H, F, GT	H, F	H, F, GT	H, F, E	H, F, E	H, F		H, F, E	H, F, GT	H, E	H, F	H, F	H, G	H, F	H, F, GT	H, F, GT	H, F, G	H, F, G	H, E	H, F, GT	104			
105	Reducing agents, strong	H, GF	H, F, GT	H, GF	H, GF, F	GF, H, F	GF, H	H, GF	H, G				H, GT	H, F				H, E	GF, H	GF, H	GF, H						H, GF	H, E			H, E	GF, H	GT, GF, H		H	GF, H	H, E	H, P, GF	H, F, E	105		
106	Water & mixtures containing water	H	H						G										H, G			GF, H	GF, H		S	GF, H								GT, GF						GF, GT	106	
107	Water reactive substances	Extremely reactive! Do not mix with any chemical or waste material!																																								107

FIGURE 8-10
Compatibility chart for storage of hazardous waste.[9]

groups for purposes of storage:

Waste	Reactivity group	Name
A	2, 24	Oxidizing mineral acids, toxic metal compounds
B	11	Cyanides
C	24	Toxic metal compounds
D	17	Halogenated organics
E	16	Aromatic hydrocarbons
F	31	Phenols and cresols
G	16	Aromatic hydrocarbons
H	32	Organophosphates
I	29	Saturated aliphatic hydrocarbons
J	10	Caustics
K	107	Water reactive substances
L	24	Toxic metal compounds
M	10	Caustics

A number of events could occur if a high-pressure water hose were used in an attempt to extinguish a fire among the above wastes stored together in individual containers. For example, the force of the water stream could tip over the containers of mineral acids and cyanides, thereby creating highly toxic hydrogen cyanide gas. If the containers leaked, the elemental sodium could come into contact with water, liberating hydrogen gas, which spontaneously ignites. Similarly, acids could react violently with alkalis, releasing heat. Violent reactions in surrounding containers could release the nonhalogenated solvents, which would form an explosive mixture with air.

From the information in Figure 8-10, a total of four separate areas are needed for the compatible storage of these wastes as follows:

Area	Wastes
1	A, C, L
2	B, J, M
3	D, E, F, G, H, I
4	K

Certain types of wastes will require prior mixing. For example, the concentration of waste constituents can vary considerably because of large differences in incoming waste strength. This is particularly true at most commercial treatment facilities. Mixing can control such variations to a range that will not upset performance of the subsequent unit treatment processes. Some waste must undergo other forms of preparation before feeding to the waste treatment processes. For example, contaminated soil can contain rocks, bricks, and reinforcing rods that could damage material handling equipment if not removed.

Waste Treatment

While the waste is maintained in storage, a treatment schedule is developed that will identify the waste to be treated, its storage location, any necessary preparations, the

method of treatment, and the rate at which the waste is fed. On commencement of waste treatment operations, the waste is typically fed by bulk material handling systems such as pipelines or conveyors to the equipment used to perform the prescribed treatment steps.

Treatment operations may be carried out on a batch or continuous basis. A facility must monitor operations carefully to assure that its performance attains the desired results. Operational monitoring is done with instrumentation, direct human observation, and chemical analysis. This typically involves extensive record keeping using a combination of computers, chart recorders, and manually entered paper logs.

Hazardous waste can be treated by any of a large number of commercially proved unit processes. Many of these methods are discussed in Chapters 9 through 12 of this textbook. The treatment methods fall into four categories:

- Phase separation (e.g., sedimentation, steam stripping)
- Component separation (e.g., ion exchange, electrodialysis)
- Chemical transformation (e.g., chemical oxidation, incineration)
- Biological transformation (e.g., fixed film aerobic treatment)

The selected method of treatment depends not only on the type of waste but also on the waste's individual physical and chemical characteristics and the specifications for the treated waste. Figure 8-11 presents a diagram showing how basic characteristics (e.g., percent organics and percent moisture) can influence the selection of

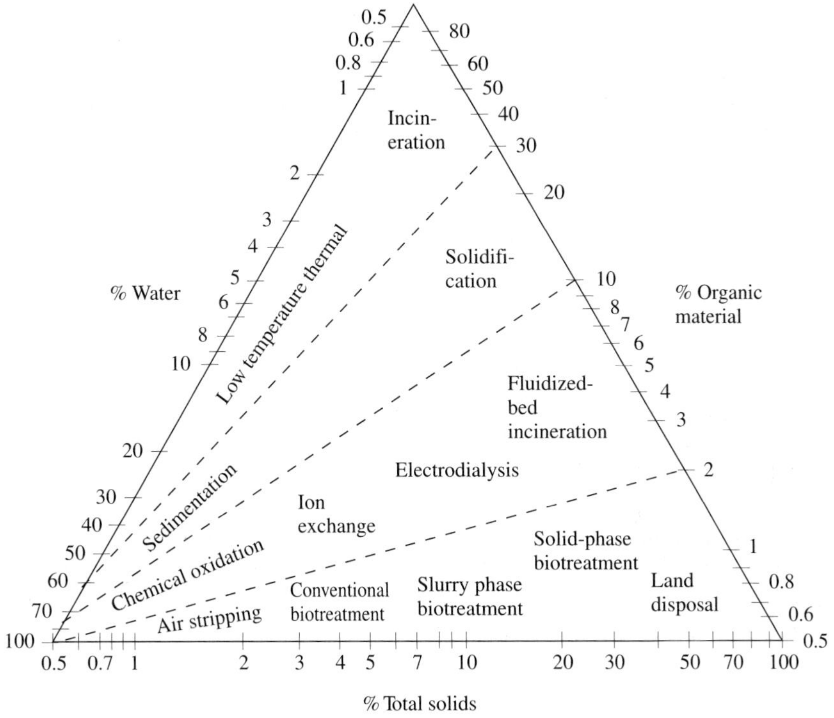

FIGURE 8-11
General suitability of various treatment methods versus physical and chemical characteristics of waste.

particular treatment methods. This figure is for illustrative purposes. The actual selection of a treatment method requires consideration of many more variables than just the concentration of moisture and total organic carbon.

Whatever treatment is deemed suitable, the treatment of a hazardous waste typically involves a multiple-step system. Interconnection of unit treatment processes in an integrated system can attain more efficient and effective treatment. This is illustrated in the example shown in Figure 8-12. This example is very involved and represents a system that could handle essentially any hazardous waste; most treatment facilities would employ only a few of the processes illustrated.

Residuals Management

Each waste treatment process produces gaseous emissions, wastewater effluents, or residuals requiring subsequent management if not additional treatment. An incinerator, for example, produces combustion gases that require scrubbing that in turn produces an acidic washwater requiring wastewater treatment. Incineration also produces fly ash and bottom ash requiring disposal, if not treatment. The unit treatment processes are not the only operations that generate residuals. Spillage and runoff from storage areas require treatment. The opening of containers may need to be done under negative atmospheric pressure with the fumes collected and treated. A full-service facility can usually provide all necessary treatment of residuals. Specialized facilities may have to store the residuals and transport them as hazardous waste to another facility capable of treating them.

Special Measures

A hazardous waste facility needs to take a number of special precautionary measures for all of its day-to-day operations. These measures can be quite complex for full-service, commercial facilities and much less involved for single-purpose, on-site facilities. Nevertheless, both types of facilities need all of these measures to prevent incidents. In fact, the facility's permit application must provide detailed plans and procedures on how the facility will implement precautionary measures. These measures can be listed as follows:

- Security
- Inspection and maintenance
- Incident prevention
- Emergency planning
- Employee training
- Safety
- Monitoring
- Reporting
- Record keeping
- Audits

A facility needs to provide strong **security** to prevent vandals and inadvertent intruders who could become exposed from contact with waste, damage the equipment, or cause illicit dumping. Most facilities use a combination of security guards, total

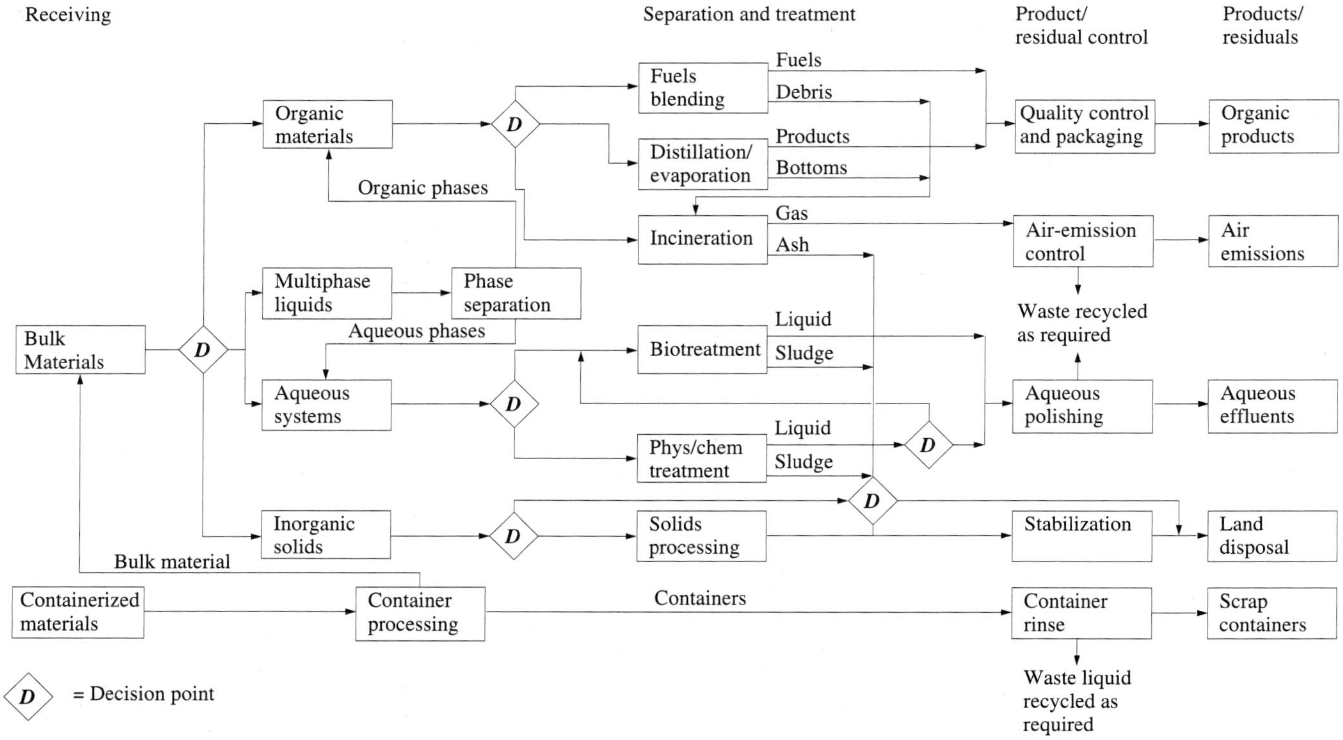

FIGURE 8-12
Integrated hazardous waste treatment processes.[10]

enclosure (usually with fences), controlled entry points, adequate lighting, proper warning signs, and 24-hour surveillance. The guards operate the gatehouse where they prevent entry of unscheduled trucks and monitor entry of visitors.

A facility needs regular and thorough **inspection** of its equipment, structures, and other appurtenances with preventive and corrective **maintenance** taken. This requires preparing a list of items to be inspected, a schedule, and the typical problems that may be encountered. The inspection should examine process equipment, storage areas, emergency equipment, monitoring equipment, and security devices. Basically, the inspection should check for equipment malfunctions, structural deterioration, operator errors, and discharges that could lead to the release of hazardous waste constituents. Some critical components of a facility require daily inspection. For example, the operator should inspect storage areas, looking not only for leaks but for signs of corrosion and mismanagement (e.g., loose lid on a drum). The unloading area is critical to inspect because of the higher potential for spills.

EXAMPLE 8-2. INSPECTION OF UNLOADING AREA. What potential problems could an inspector find in an unloading area?

Solution. The spill containment structure may have signs of deterioration. There may be spillage outside the spill containment structure. The loading dock equipment may be inoperable or show signs of corrosion. The eyewash may not have sufficient water pressure. Absorbent material used to control spills may be low if not out of stock. Access for fire protection equipment may be obstructed. Incoming trucks may be leaking waste.

It is important to implement **incident prevention** measures to minimize the possibility of a fire, explosion, spill, or any other unplanned, sudden release of hazardous waste constituents. Such a program begins by conducting a risk analysis to identify hazards and potential incidents. The risk analysis would indicate the routine activities and waste processing operations that have a higher probability of initiating an incident. In this way, preventive measures can be instituted and appropriate responses planned in the event one of the more probable incidents were to occur. Fire, for example, is a major concern. Having alarms, fire control equipment, water of adequate pressure, and personnel trained to fight fires is important, but so is maintaining adequate aisle space to allow unobstructed movement of fire protection equipment. Of course, the inspection and compatibility considerations discussed earlier in this section are important to prevent incidents.

If an incident occurs, the facility should follow a set of procedures prepared as part of an **emergency plan** that describes the actions that facility personnel should take in response to fires, explosions, or other incidents. The plan should provide for interaction with local police departments, fire departments, hospitals, and other appropriate emergency and community services. The plan should also describe evacuation procedures for facility personnel. Practice drills are important. One facility employee should be appointed as emergency coordinator and take leadership responsibility for implementing the plan.

It is important that the facility train its employees to perform their duties effectively and safely and to know how to respond to an emergency. The **training** usually consists of both classroom instruction and supervised on-the-job training. Each employment position should have a description of its responsibilities and duties, the content of the necessary introductory and continuing training, and a frequency for review sessions (usually annually). The facility must decide if its own personnel are

to respond to on-site emergencies (and train them accordingly), or if it will rely on outside assistance.

One of the most important measures is **environmental monitoring**—collecting samples of the environmental media and testing for the presence of hazardous substances that may have been released by the facility. The objective is to detect potential problems before they impact human health and the environment. Early detection should allow sufficient time for adequate warning of potentially impacted individuals and effective implementation of remedial measures. Important monitoring points are groundwater wells for storage and land disposal facilities (see Section 15-7) and air monitoring stations at critical locations around the facility. Monitoring could also include surface water, employees (e.g., blood samples), and flora and fauna.

Regulatory compliance could be thought of as a separate function, yet this is essentially the driving force for the entire facility. The market for facilities is heavily influenced by dynamic regulatory programs that continue to undergo significant change. The operation of a facility is thus geared around managing waste in a manner that meets, if not exceeds, environmental regulations. This is no easy matter. For example, the **record keeping** requirements for documents and reports alone represent a major undertaking; nearly 200 U.S. Environmental Protection Agency (EPA) regulations specify retention periods.[11]

Audits are an excellent tool for ensuring that the facility operating staff is complying with all standard operating procedures, best management practices, and regulatory requirements. Actually, diverse groups from outside the facility conduct periodic inspections and audits from varied perspectives. Federal and state agencies inspect the facility to determine compliance with regulatory requirements. Many clients (i.e., generators of waste) audit facilities to ensure that the facility's practices for managing wastes do not pose future liabilities for the generators. Some facilities may even allow local citizen groups to tour the facility to review particular operations important to the community (e.g., truck traffic and monitoring data). The corporate parent of the facility may also conduct periodic audits. For these reasons, some facilities conduct their own internal audits. Chapter 6 is devoted to a discussion of audits.

Closure

Some of the nation's Superfund sites were originally hazardous waste management facilities that were not closed properly when operations ceased. A closure plan is now a necessary part of a permit application. This plan provides a clear and orderly set of steps and methods to be followed on cessation of all operations at a facility. The steps are designed to ensure that the closed facility (1) poses a minimal risk to the environment and human health and (2) requires minimal postclosure maintenance.

A closure plan requires assurance that funds are available to close the facility even if the facility owner enters bankruptcy proceedings. This assurance can be in the form of a bond, corporate guarantee, or some other financial instrument. The monetary amount is determined from a cost estimate prepared as part of the closure plan. To satisfy the EPA regulations, the cost estimate must equal the maximum costs of closing all waste management units ever activated at the facility.

Closure of a storage or treatment facility requires removal of all remaining waste to another facility. All equipment and structures that had been in contact with waste must be decontaminated. This may entail removal of concrete pads used to hold waste containers as well as contaminated soil where leaks had occurred.

EXAMPLE 8-3. ESTIMATE OF CLOSURE COSTS. What key activities should be included in the closure cost estimate for a treatment facility?

Solution. A treatment facility usually consists of waste storage and an integrated system of various types of treatment equipment (see Figure 8-12 for an involved example). EPA regulations specify that the following key elements must be included in the closure cost estimate for a treatment facility: on-site treatment of the maximum possible inventory of waste, removal and transportation to an off-site facility of all wastes that cannot be treated on site, decontamination of all equipment and other facility components, monitoring and security during closure, and certification of closure. This listing ignores the cost of remediation of groundwater if such were to occur.

The closure of land disposal facilities, where waste has been emplaced, differs from closure of a treatment facility. In the former case, waste is not to be removed but contained, necessitating installation of cover systems plus implementation of a long-term postclosure plan providing for leachate management, monitoring, maintenance, security, and other measures typically for a period of 30 years, if not longer.

8-3 NEEDS ASSESSMENT

The first step in developing a facility, even prior to selecting a site, is to decide what type of facility is needed and its capacity. It is readily apparent that the ideal characteristics for a site depend on the type and scale of facility to be located there. The process for making such a decision is termed a **needs assessment.**

In its logical sense, a needs assessment involves an analysis of the following:

1. The current and future types and quantities of hazardous wastes generated that would be managed at the proposed facility.
2. The current methods for managing these wastes.
3. The methods expected or desired in the future for managing the generated waste.
4. The capacity of existing facilities providing such methods.

The need for new facilities is thus determined by comparing future waste generation (no. 1) with anticipated capacity (nos. 3 and 4). A shortfall in capacity indicates a need for new facilities. Economists would equate this with a comparison of "supply" and "demand."

Need is a concept that has different dimensions, depending on one's perspectives. A private sector developer would examine need on a market basis (i.e., "supply" and "demand"). From this perspective, waste might be diverted from facilities now in operation to new types of facilities. The capacity shortfall and the potential economic return must be sufficiently large to justify the financial risk of the investment. A state government may perceive need as sufficient facility capacity to manage all waste generated within the state, thus avoiding having to depend on facilities in other states. With this view, possible diversion from existing in-state facilities may be excluded from the calculation. An environmental group may perceive need as

that capacity needed only after implementation of extensive waste minimization and source reduction programs by waste generators. California's siting program requires each county to complete a needs assessment.

EXAMPLE 8-4. DIFFERENT ASSESSMENTS OF NEED. Which needs assessment made by the following entities would likely indicate the greatest need for new facility capacity: a private developer considering market conditions, a state government seeking independence from other states, or an environmental group emphasizing waste minimization? Which is correct?

Solution. The state government would probably show a greater need for new facility capacity. A state, by seeking independence, would ignore the capacity of facilities in other states. Yet, the state must recognize that the capacity of commercial facilities within its own borders is not totally available because interstate commerce provisions allow the facilities to serve generators in other states. In contrast, a private developer would consider a new facility as having to operate in an interstate market; facilities in other states could compete with the new facility. A developer, thus, may show less need for additional capacity unless the developer projects that waste now going to competing out-of-state facilities might be sent instead to a new facility. In any event, the perspective of an environmental group emphasizing waste minimization could likely be that no additional facilities are needed. All three might be "correct."

Most needs assessments have focused on off-site (commercial) hazardous waste management facility development and their capacity to meet defined "need." Many such assessments have failed to adequately account for existing and potential new on-site treatment systems, and especially to comprehend the inherent contradictions between anticipated demand for new off-site treatment or disposal facilities and potential expansion by hazardous waste generators of their own waste reduction/source reduction efforts. In response to the rapid rise in prices during the late 1980s and 1990s for off-site hazardous waste treatment and disposal, many generators initiated programs to reduce the volume of hazardous wastes they were generating. These waste reductions, in turn, diminished the need for new off-site facilities.

8-4 SITE SELECTION

A primary objective of a site selection process is to assure that new facilities are located at intrinsically superior sites that, by virtue of their natural features and land use setting, provide a high degree of protection to public health and the environment. In this concept, the land use setting and natural features function as an additional line of defense if the facility operations do not perform as planned. Proximity to sources of waste generation is another important criterion.

A site selection process usually proceeds through a phased approach, as depicted in Figure 8-13. It begins with the use of regional screening techniques to reduce a large study area, such as an entire state or region, to a manageable number of discrete search areas. Computerized geographic information systems (GIS) are available to assist in this task (see Section 15-8). Because of this screening, the search areas should have higher probabilities of containing suitable sites. The next phase evaluates the discrete search areas in more detail and identifies candidate sites within them. These candidate sites are then evaluated in even more detail at a site-specific level

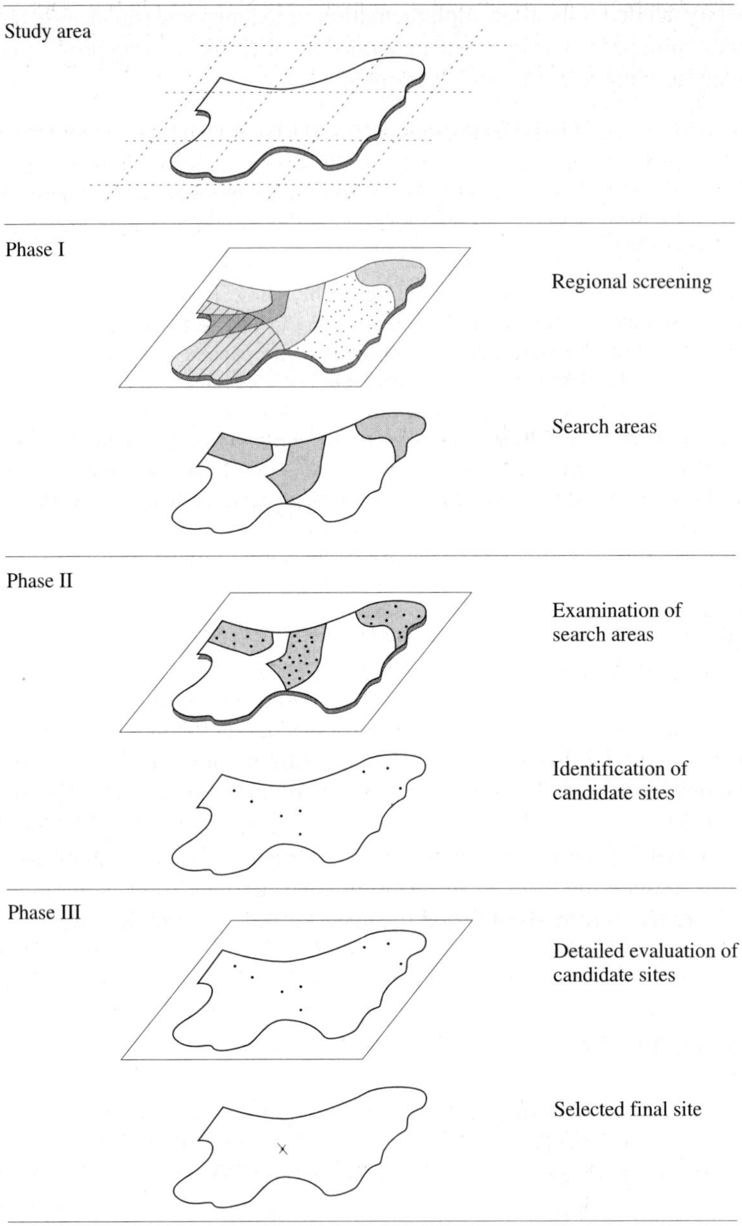

FIGURE 8-13
Phased site selection process.

of analysis to provide the basis for selecting a site for the facility. "The overall site selection process is thus one of increasingly intensive analysis of progressively smaller areas."[12]

A phased approach is usually employed by a governmental agency when the selection process is under public scrutiny. Another approach to site selection, fre-

quently implemented by the private sector, is to preempt a phased approach and simply select from an inventory of a few specific sites. The inventory, for example, may consist of those sites advertised as "for sale" or sites already owned by the facility developer. Defenders of such an approach contend that if the characteristics of a site selected in such a manner meet all regulatory requirements, the site is suitable. Thus, it should still be considered suitable regardless of whether a phased approach, starting with regional screening, may have located an environmentally superior site. Even so, sites selected in a preemptive approach may necessitate more expensive design features to mitigate less than ideal characteristics.

Site Screening Methods

There are alternative methods used to integrate and evaluate data in the phased approach of progressing from a study area to the selection of the final site. The following list contains four examples of screening methods (i.e., methods for data synthesis).

- Intuitive
- Stepped-down exclusion
- Scaling
- Criteria combinations

The **intuitive** method is one in which the decision maker examines all data as a whole and judges whether the site is acceptable. The argument for using this method is that the site characteristics are too functionally interdependent with each other and with the design and operation of the proposed facility to support an assessment derived from an examination of individual factors. This is akin to Gestalt psychology.[13]

The **stepped-down exclusion** method examines each siting factor sequentially, determines a criterion (definitive threshold of acceptability) for the factor, and applies the criterion to eliminate areas from further consideration. A criterion may be thought of as an environmental, technical, social, economic, or political constraint to siting a facility. The process continues by selecting another factor and applying a criterion for it. The method progresses through the entire sequence of siting factors to identify "not unacceptable areas." The process may stop short of applying all criteria if application of the previous factors had accomplished the siting objective.

The **scaling** method is one of weighting the site data. That is, the data are modified from their original form by replacing them with numerical surrogates having a common scale. For example, permeability of the formation below a site is a typical consideration in selecting a site for a land disposal facility. Clay is an indicator of low permeability, and a site underlain by 10 feet of clay may have a value of 10, while one with sandy silt may be assigned a value of 3. Similar values are assigned for other data describing site suitability. The next step in scaling is to assign weights to indicate the relative importance of the factors (e.g., permeability and depth to groundwater may be judged to have a weight of 8 and 4, respectively). After assigning weights, all factors are combined in a multiplication-summation step to yield a numerical score for each site. The selected site would be the one with the best score.

The **criteria combination** method can use either the stepped-down exclusion or scaling technique. Instead of working with a single set of criteria or scaling values, alternatives are developed. This involves selecting a particular combination of criteria and identifying sites satisfying them. The process is then repeated for another combination. For example, one combination may stress that the facility be located over thick deposits of clay, and it may give less consideration to factors such as proximity to population. Another combination of criteria may relax the requirement for clay but require that the facility be located on remote, public-owned land. The criteria combination method results in the selection of a set of starkly different sites representing different perspectives about what constitutes a better site.

EXAMPLE 8-5. COMPARISON OF SITE-SCREENING METHODS. What are the capabilities and limitations for the four site-screening methods listed previously?

Solution

Method	Capabilities	Limitations
Intuitive	Possibly the only way to consider public input and subjective information when selecting from otherwise technically superior sites.	Difficult to justify decisions, especially at regional level.
Stepped-down exclusion	Works well in regional screening and in application of mandatory criteria.	Not appropriate for preferential criteria because it may unnecessarily exclude a site that could rate "high" in subsequent evaluations.
Scaling	Works well if the number of sites is small and the differences among sites are broad.	May introduce biases. Incorrectly treats each factor as wholly independent. Significance of the quantitative output is not readily apparent.
Criteria combinations	Only method that takes into account (1) the interdependence among factors and (2) differences in professional opinion. Excellent for encouraging public participation.	Does not rank sites. Data management difficult for a large number of sites.

Siting Criteria

Most screening methods require siting criteria. Siting criteria stem from siting **factors** associated with a specific consideration important in judging the suitability of a site. For example, the potential for leachate to migrate to groundwater is a consideration with land disposal facilities. Accordingly, it is desirable that a land disposal site have one or more of the following characteristics: the permeability of the uppermost formation is low, the depth to groundwater is high, and the thickness of any clay deposits is high. Thus, the pertinent siting factors are permeability, depth to groundwater, and thickness of clay. The factors applicable to a treatment facility may be very different, including industrial land use zoning and proximity to major sources of waste generation.

A **criterion** is a definitive threshold of acceptability for a siting factor. For example, a criterion for permeability (hydraulic conductivity) could be "maximum of 1×10^{-7} cm/s." Sites having a value higher than this would be deemed unsuitable. Similarly, a criterion for thickness of clay could be "minimum of 10 feet."

A great number of siting factors can be applied when selecting the site for a hazardous waste facility. Table 8-1 lists some example factors frequently used in such endeavors.

Siting criteria can be divided into two types: mandatory and discretionary. **Mandatory** criteria represent a legal requirement, a regulatory standard, or some other aspect deemed so important that it cannot be violated under any circumstance. Most siting criteria are **discretionary**—they represent desirable but not mandatory features. Discretionary criteria reflect preferences and value judgments.

The development of siting criteria acceptable to the public requires much input from the public. An example of a process to develop siting criteria is one used by the state of New Jersey. A state-appointed advisory council, representing a broad range of interests, used a six-step process:

1. Make basic assumptions.
2. Identify types of facilities.
3. Analyze facility failure possibilities—types and consequences.
4. Establish objectives of siting criteria.
5. Select siting factors.
6. Determine siting criteria.

The first two steps resemble a needs assessment. On the basis of prior studies, the advisory council assumed that: "(1) New Jersey needs new hazardous waste facilities, (2) one or more of these new facilities will be sited within the State's borders, and (3) all facilities engender some degree of risk and, therefore, siting criteria must be developed to help minimize the risk to public health and the environment."[15] Any approach to the development of siting criteria must consider what types of facilities are needed because the possible failure events vary according to type of facility. An analysis of failure events indicates potential consequences from which a set of siting objectives can be established. For example, the site selected for a hazardous waste land disposal facility should have intrinsic features that satisfy the following objectives:

TABLE 8-1
Example hazardous waste facility siting factors[14]

Category	Example siting factors
Surface water	Flood hazard areas, drinking water supplies, reservoirs
Groundwater	Hydraulic conductivity, depth to groundwater, thickness of clay deposits, aquicludes/aquitards, recharge areas, proximity to wells, karst area, groundwater flow direction
Environmentally sensitive lands	Wetlands, habitat for endangered species, parks
Population	Proximity to dwelling units, proximity to schools, population density
Industry/waste generators	Proximity to major waste generators

- Provide structural stability
- Protect surface water quality
- Protect groundwater quality
- Protect environmentally sensitive areas

Meeting each objective requires identifying pertinent siting factors and then determining a criterion for each factor. An example follows:

- Objective: provide structural stability
- Siting factor: riverine flood hazard areas
- Siting criterion: all facilities prohibited in flood hazard areas

EXAMPLE 8-6. NARRATIVE AND NUMERICAL SITING CRITERIA. What are examples of narrative and numerical siting criteria for protecting groundwater from the possible subsurface migration of leachate from a hazardous waste landfill?

Solution. The protection of groundwater is a hydrogeologic issue. The relevant siting factors, as discussed in Chapter 4, could include:

- Permeability
- Groundwater flow velocity
- Depth to bedrock
- Karst topography
- Time of travel

The criteria must be carefully selected to reflect the geology of the study area. A narrative criterion for permeability and time of travel could read as follows:

Allowed only in areas having a sufficient thickness of naturally occurring geologic materials of low permeability (e.g., fine-grained geologic deposits, dense bedrock) to assure that effective remedial action can be taken.[16]

If a numerical format is desired, the criterion could specify a minimal thickness (e.g., 20 feet) of regionally continuous clay having a permeability no greater than 1×10^{-7} cm/s. The EPA has determined that an appropriate criterion is that the time for groundwater to travel 100 feet along the flow line originating at the base of a hazardous waste unit is to be greater than 100 years.[17]

EXAMPLE 8-7. SITING CRITERIA SENSITIVE TO PUBLIC'S CONCERN. The types of criteria produced in Example 8-6 were presented and discussed in great detail in a series of well-attended public meetings in New Jersey in 1982. The public objected to the narrative criteria because qualified descriptors such as "sufficient" give too much discretion and flexibility, opening the door to finesse and compromise. The public also eschewed arbitrary numbers such as 100 years, 20 feet, and 10^{-7} cm/s. Of most concern, the public noted that criteria stressing permeability would drive the site search toward regional clay deposits that happen to overlie New Jersey's major aquifers used by a large percentage of its population for drinking water. The failure of a facility in such a location could have major consequences. What are examples of siting criteria sensitive to the public's concerns listed above?

Solution. This task begins by examining risk. Risk involves two independent considerations—**likelihood** and **consequential.** That is, risk is composed of the potential for an event to occur and the severity of the consequences if it does occur. Theoretically, risk is the product of multiplying the two. In this instance, the risk to groundwater can be reduced by applying criteria that address either the likelihood of contaminant release or the consequences if such a release were to occur.

Factors such as permeability provide for containment. Containment is a measure of the potential for release; it ignores the consequences of such a release. The public did not accept the position that clay prevents release of toxic pollutants; therefore, a criterion requiring clay does not meet the public's concerns. The public in New Jersey asked for criteria that would protect the groundwater even if the facility operation and clay did not function as planned. The criteria, therefore, should address consequences of failure, however remote.

One option is to prohibit the siting of facilities except above groundwater not presently used for water supplies and so severely contaminated from past practices or natural salinity as to make it highly unlikely that the groundwater would ever be remediated. The consequences of failure would be nominal, as additional contamination would not affect current or future users. The public again expressed disapproval, claiming that (1) this would represent a license to pollute and (2) this "writes off" major aquifers when the policy eventually should be to remediate the past contamination and allow use of the groundwater.

Developing criteria sensitive to the public's concerns requires consideration of the entire groundwater flow regime: The criteria should eliminate recharge areas of water supply aquifers, particularly areas where the uppermost aquifer is used for drinking water supplies down-gradient from a facility. Also, the site conditions should foster the early detection of leakage, if it occurs, and its remediation. The following groundwater flow regime would meet these requirements: The groundwater under the site should flow either horizontally (parallel to the water table) or upward toward the water table (discharge area) until discharging to surface water. In such an area the recharge to deep, regional aquifers would be nominal, possibly nonexistent. In areas with downward flow, in contrast, any migrating pollutants could bypass monitoring wells, and remediation would be more difficult.

Thus, the above requirements can be captured in the following criterion: "The flow of groundwater in the uppermost saturated unit is predominantly parallel to or upward toward the water table and the predominant groundwater flow direction is toward the nearby surface water bodies without any intermediate withdrawals from the uppermost saturated zone for public or private water supply."[18] This criterion is effective only in conjunction with other criteria concerned with the complexity of the geology, groundwater travel time between the waste and property boundaries, and uses made of the surface water into which the groundwater discharges. Such requires a site-specific assessment, with some flexibility to assure the spirit of the criteria are met without eliminating the entire study area.

Value Judgments and Tradeoffs

A site selection process will involve value judgments because:

1. Criteria are based on an "acceptable risk" concept (there are *no* circumstances for which risks, however remote, are zero).
2. The line between "acceptable" and "unacceptable" is typically unclear, and often varies from one point of view to the next.

More significant than these two reasons, value judgments are required because the ideal site with optimum conditions for all siting considerations (e.g., hydrogeology, land use, and transportation) does not exist. The siting process must, therefore, examine tradeoffs in its determination of what constitutes "adequate" or "better" conditions, and this will inevitably encounter differences in judgments, even among leading authorities in the same field.

Individuals responsible for siting facilities often make value judgments too early—when establishing criteria rather than when the criteria are applied. Establishing inflexible or narrow criteria could preempt important decisions that should be deferred until after the application of the criteria and the collection of site-specific data. Example 8-7 illustrates the limitation of requiring that a facility be located in areas having regional deposits of clay. As another example, a decision to require a facility be located only in remote areas would immediately eliminate areas in or near cities where suitable sites may exist, especially for treatment facilities. In fact, on detailed examination, urban sites may prove to be technically superior in key ways. For example, risks from accidents involving the transportation of hazardous waste might be reduced at urban sites because transportation distances are shorter. Such sites may also be more politically viable because most generators are located near cities. As a result, "equity-based" siting criteria can be applied. Decisions to consider only sites in remote areas preempts application of the stringent criteria to urban sites.

Site selection processes that start with inflexible and narrow criteria typically conclude with a tradeoff analysis of nothing other than trivial differences among sites. It is important to provide a wide choice and keep many options open as part of the final site selection. This facilitates a tradeoff analysis of stark differences that should encourage participation by affected parties. Such an objective can be met by establishing different combinations of criteria and selecting a set of alternative sites for each combination. This progressively screens the study area without reducing value judgment flexibility until the end when different sites epitomizing different opinions can be contrasted and more data can be obtained.

8-5
PUBLIC PARTICIPATION

The process of selecting sites for new facilities and obtaining permits for them is commonly referred to as "siting." It can cost tens of millions of dollars and take a decade or more of effort before a definitive decision is received from the permitting agency. That decision is frequently a denial. The Province of Ontario started a process in the early 1980s to site a full-service facility. In 1990, after the selection of a site, public hearings commenced that were expected to continue for another $3\frac{1}{2}$ years.[19] The cost in Ontario of siting a facility may exceed the cost of constructing the facility, that is, if one is ever built. In a private sector proposal, Clean Harbors, Inc., spent \$12 million in a now-terminated attempt to site an incinerator in Massachusetts.[20]

Even if a permit is granted, subsequent litigation can delay placing the facility into operation and add to the mounting costs. Almost all efforts to develop new, modern facilities have failed because of an obstacle much greater than market forces: public opposition. While every aspect of developing and operating a facility is conducted under close scrutiny, nothing seems to raise as much opposition as siting a new facility. The situation has given rise to the acronym, NIMBY (not in my back yard).

Root Causes of Public Opposition

Why do hazardous waste facilities arouse so much public opposition? Why can the opposition become so vehement? The answers stem from many causes including risk perception, public mistrust, inequities in risk sharing, and other factors.

As discussed in Chapter 1, typical lay people perceive a facility in their community as (1) imposed on the community (involuntarily encountered), (2) having no real benefit, and (3) representing an unknown but substantial risk. To a public already cautioned by the label "hazardous," all three perceptions can prompt severe public response. While public concern with hazardous waste spurred regulatory initiatives that created and drove the expanding market for hazardous waste management services, the same concern prompts opposition to new facilities for treating and disposing of waste.

Compounding these problems is the general observation in society today that there is not enough public trust on which to promote understanding and acceptance. The engineering and science professions believe they have developed the capabilities required to construct, operate, and monitor safe waste management facilities within technically acceptable risk limitations. However, the public does not necessarily share the same belief. The link between risk and technology is perceived in a climate of public distrust of both industry and government. This is part of the legacy of Love Canal and similar episodes that were deemed at their time to provide adequate disposal. Underdesigned, undercapitalized, and underregulated, many of the early hazardous waste facilities have, in fact, ceased to operate and eventually became Superfund sites. Although government is the watchdog to ensure that these episodes will not be repeated, the public lacks confidence in the adequacy of the laws and regulations and the capabilities and willingness of regulatory agencies to enforce them.

Once formed, public opposition is not known to lessen measurably. A number of different countervailing programs have been attempted, without success, to "overcome" opposition. As if public mistrust and the heightened risk perceptions were not enough, new factors can emerge to harden opposition. One of these factors is inequity in risk sharing. There is no real link between waste management at a facility in one community and waste generation in another. The host community bears the risk of waste treatment and disposal while other communities benefit from the jobs, local purchases, and other economic activities associated with the industries that generate the waste. California, for one, has attempted to counter NIMBY-based siting resistance with equity-based efforts termed "yes in many back yards" (YIMBY).[21] Each county was expected to provide for sufficient capacity (in commercial facilities) to manage all the hazardous wastes being generated by industry in that county. Local rejection (under NIMBY pressure) of appropriate facilities proposed for development could be appealed to a new statewide siting board empowered (under certain limited conditions) to override such parochial local decisions. Unfortunately, weak state agency implementation of this law left the appealing concept largely untested.

Another factor behind public opposition is the perceived loss of a community's sacred right to shape its own future.[22] Opposition stems in part from how the public first becomes aware of a proposal to site a facility. Most of the public, even in a process that has been conducted openly, will learn of the project after a site has

been identified in its community and publicized. The *post facto* revelation of a siting decision as a *fait accompli*, particularly without proper explanation, can prompt an intense feeling that citizens have no way of influencing the decision other than to oppose it, a decision that will affect their well-being.[23] Getting the public involved early is a difficult undertaking. When the public has been given an opportunity to participate in "front-end" planning, the public usually fails to perceive its stake and typically will not participate. Only after sites are identified do citizens in the potential communities get involved, marked by a rapid "yawn to yell" transition. On reaching this stage, discussions between the facility developer and the public make little progress toward finding common ground. How can there be when the developer, having to site a facility somewhere, will not consider any option that involves siting the facility elsewhere, and the public will not consider any option that involves siting the facility in its community? Satirists refer to these discussions as a "dialogue of the deaf."

Another compounding factor is that some well-organized political activist groups exist that oppose all new hazardous waste facilities in general. These groups will support local communities in their opposition to a proposed facility. Support can consist of an information clearinghouse and expert witnesses. The positions of these groups usually disagree sharply with those of facility proponents, thus adding a degree of confusion for laypersons struggling to understand a technical issue.

Public Involvement

With all these obstacles, facilities have been successfully sited where success can be measured as (1) selecting a technically suitable site, (2) identifying public concerns, and (3) addressing them. Some facility developers have even gained public acceptance (see Case Study of Siting a Hazardous Waste Facility). If one analyzes the procedure that was followed for the successful permitting of these new facilities, it is clear that in each case the community participated extensively in the process and had an influential and constructive effect. Simply stated, public involvement was the key to success, fostering trust on which understanding and acceptance could build. It is not sufficient to merely profess to have adequate technology and expect public trust. For the public to have confidence, it must participate in the process of making decisions on issues perceived as having great effect on its future.

Public involvement is a political art, possibly even foreign to the professional scientist and engineer who typically respond only with technical information. Some such response is not the answer. The public is skeptical of technical experts. In fact, technical information may even be counterproductive; observers of opposition to nuclear power facilities have noted an inverse relationship between knowledge and public support.

The triad of the facility developer, the local community, and the regulatory agency is closely interrelated in the issue of siting new facilities (see Figure 8-14). State governments have assumed active, nontraditional roles. One role may be to ensure that local authority does not veto what is otherwise a suitable, needed site. Some state governments, however, are actually leading the search for new sites, a role that governments do not necessarily relish. Other state governments are taking a less active but supportive role, endeavoring to use institutional measures to promote

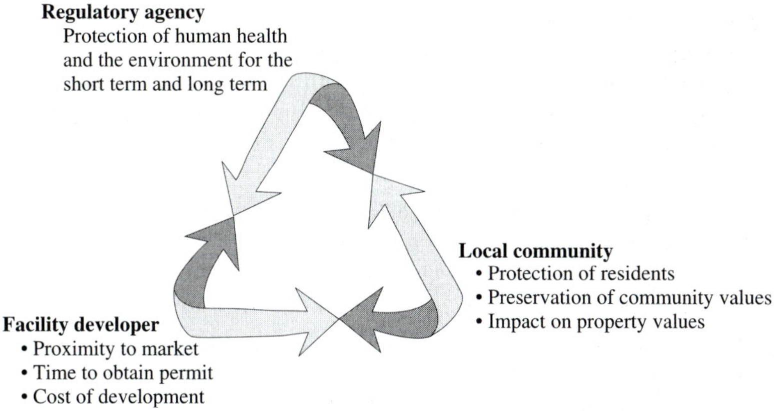

FIGURE 8-14
Key issues among triad of participants in siting a facility.

negotiation between the community and facility developer in the common interest of all its citizens.

In the supportive role, government initiatives include a credible waste management plan justifying the need for a new facility, assistance grants to proposed host communities, extensive public information, public participation in selection of siting criteria if not the site itself or at least the type of facility developed at the site, and a negotiated siting agreement between the facility developer and the local community. To the facility developer, this level of public involvement may seem excessive, if not impossible, to deal with. However, without it, experience shows that an expensive and long siting process and the high probability of regulatory denial are inevitable.

Environmental Justice

In the 1990s, a new term emerged that names another siting concern, **environmental justice.** This concern derived from an awareness that minority populations and/or low-income populations were bearing a disproportionate amount of exposure to health and environmental risks. Within the field of hazardous waste management, a number of waste treatment and disposal facilities that later became Superfund sites had been located in low-income communities or communities with minority populations.

This led President Clinton to issue Executive Order 12898 in 1994, focusing federal agency attention on the issue of environmental justice. The EPA responded by developing a strategy to address these concerns. The EPA defines environmental justice as the "fair treatment for people of all races, cultures, and incomes, regarding the development of environmental laws, regulations, and policies." Accordingly, the EPA has developed recommendations for enhancing public involvement in siting and permitting of facilities. The intent is to promote "early, direct, and meaningful community involvement" especially at sites with environmental justice concerns. This includes going beyond the minimum public participation requirements of the EPA's media permitting programs (see *www.epa.gov/swerosps/ej/index.html#ejhist*).

Case Study of Siting a Hazardous Waste Facility

It is reasonable to estimate that siting a new facility has a success rate of less than 1 percent. Almost all failures derive from public opposition, with many proposals ending in ugly incidents: facility proponents wearing bulletproof vests before an alarmed public, or police using stun guns on protesters.[24] Contrary to this trend, one headline captures the spirit underlying one of the very few success stories of siting a new hazardous waste facility.

> Small Town in Alberta Embraces What Most Reject: Toxic Waste[25]

The Province of Alberta in Canada sponsored a siting process that eventually prompted two separate communities to vote overwhelmingly to accept a large, full-service waste management facility. Because only one facility would be constructed, the two communities competed actively for the facility and even lobbied Parliament to be selected.

Why did the Alberta process achieve success when almost all others have failed? The case is documented in several references[26,27,28] and presented here as an example of the cornerstone effect of an open public participation program that educates the public and involves citizens fully in the decision-making process.

As background, a private proposal to site a facility in Alberta had failed in 1979 amid strong public opposition.[29] This prompted the Alberta government to develop a special (i.e., hazardous) waste management plan. The plan included a needs assessment showing the extent of existing and future hazardous waste generation in the province, especially by Alberta's large oil and gas industry. One element was to form a committee of representatives from government and the public to develop and implement a site selection process consisting of two parts:

1. The development and scientific application of technical siting criteria to ensure the intrinsic suitability of the selected site.
2. The resolution of sociological issues such as to foster *acceptance* by the host community.

The application of siting criteria followed a sequential progression, as discussed earlier in this chapter, to yield large regions of land free of technical constraints (e.g., steep slopes, significant groundwater recharge). The application of such criteria is a fundamental land use planning technique. It is a necessary part of siting facilities; however, it is not the key to success. In fact, it is reasonable to state that many failed siting proposals had undertaken a similar screening of constraints and identified technically suitable if not superior sites. The Alberta success hinged on the approach taken after the screening—soliciting volunteer communities rather than naming sites and threatening the use of eminent domain. That is, Alberta did not plan to preempt or override the traditional preeminence of local government in land use decisions, even though the facility was needed for the benefit of the whole province. In fact, Alberta's approach was termed the "invitational process"[30] because the Alberta government had to be invited by the local government.

In 1981, after preparing the regional screening maps but before naming specific sites, the Alberta task force met with counties, municipalities, and other local governments having constraint-free land within their borders. The meetings provided an open forum to explain the need for a facility, to discuss the planned siting process,

and to receive local input. After these introductory meetings, the local governments had to decide whether to request further consideration to site the facility in their jurisdiction or to withdraw from the site selection process. Opting for further consideration did not forfeit their eventual right to reject the facility. Nearly 75 percent of the local governments took the next step by inviting the Alberta task force to conduct an environmental analysis of potential sites in their area.[31]

The next step in the process included more local meetings and workshops to educate the community. Local review committees were formed. This public communication program had two objectives:

1. Engender a sense of individual responsibility for overall industrial, municipal, and household waste problem of which hazardous waste is one small part.
2. Demonstrate that the risks of a facility were no greater than those posed by other industries.

The Alberta task force did not identify the potential sites. Instead, the local governments reviewed the constraint maps and delineated specific sites within the constraint-free areas. Following that, the task force conducted environmental analysis of the sites, including geologic drilling to verify the constraint mapping.

Some communities were eliminated because of environmental constraints revealed by the environmental analysis. For communities having sites that satisfied the environmental constraints, local government again could elect to be removed from further consideration. Many did drop out because of public opposition. Many communities, however, perceived that a facility would benefit their local economy and that the risks posed by a facility were manageable. About 25 percent of the communities studied decided to continue with the process.[32]

Several factors obviously had much to do with gaining such a high degree of public acceptance. For example, the institutional arrangement that the province would own the site was important to ensure perpetual care. Clearly, the local government retaining the sole right to accept or reject a facility without intervention by a higher government fostered a high level of trust and respect. The communication program built on this trust and was able to educate the public on the highly emotive issue without forming and hardening opposition. This has proved very difficult to achieve in most other siting cases.

The communication program worked with the media but also had a number of other communication vehicles:

- Materials to assist at public meetings
- Displays and participation at local conferences and conventions
- A print campaign of brochures and handout materials
- A literature review on progress made in other jurisdictions
- A publication and newspaper-clipping inventory
- Preparation of key officials for on-the-spot radio, television, and newspaper interviews
- An educational program for schools
- Development of the mapping program for presentation in public meetings
- A newsletter outlining the issues, activities, and progress[33]

In 1983 the province had narrowed the list of candidate sites for the facility to five communities. This marked the beginning of detailed site investigations and

a series of intensive educational seminars. Community opinion leaders, including facility opponents, were sent on informational tours of active facilities in Europe and the United States where they visited and talked with the host communities about their experiences.

A necessary requirement for siting a facility was that the community supported it. To gauge support, each community held a referendum in 1984. Seventy-nine percent of the voters in a community named Swan Hills voted in favor of the facility. Another community had a 77 percent favorable vote.[34]

Soon after the plebiscites, the provincial cabinet selected a site near the community of Swan Hills as the location for the facility. The town had a population of about 2500, with oil and gas production representing the dominant industry. Many residents worked in the oil and gas fields and were already familiar with hazardous operations. The community leadership was concerned with its dependence on the oil industry, and it envisioned a facility as a progressive and needed step toward diversifying its economic base.

It should be noted that no "host fees" were paid to Swan Hills.[35] Host fees are typically proposed in the United States, not so much as an inducement but as a means to compensate a community for accepting a facility that will benefit many other communities. Although frequently offered by facility developers, host fees can backfire and harden opposition if not carefully presented; they can be easily misconstrued as a bribe instead of compensation for accepting an unwanted facility.

The facility opened in 1987, employing 94 workers,[36] and featuring an incinerator, a landfill, a deep well, and physical-chemical treatment units[37] (see Figure 8-15). The facility manager has operated the facility in the manner in which it was sited—including an "open door" policy with full disclosure of information. A key feature of the operation is that the public has ready access to facility personnel to seek additional information.

FIGURE 8-15
Alberta Special Waste Treatment Centre at Swan Hills. (*Alberta Special Waste Management Corporation.*)[39]

An \$85 million expansion was completed in 1992 that quadrupled its capacity. The facility now has two rotary kiln incinerators with a combined capacity of 43,000 metric tons. All waste is treated. The treated liquid residue is discharged to a deep injection well, and solid residue is stabilized and disposed of in an on-site landfill. A furnace to recover metals from PCB-contaminated transformers was also added. In the year 2000, the facility employed 125 people. There have been two incidents resulting in fines to Bovar Waste Management, Inc., the Alberta-based operating company. The causes of both incidents were readily corrected.

Two major changes from how the facility was originally operated demonstrate that it continues to receive widespread support from the Swan Hills community. The facility originally was organized as a public-private sector partnership with the Alberta government owning a majority share; such an institutional arrangement was deemed important to gain community support. That changed in 1996 when Bovar acquired 100 percent of the shares. Also, the facility originally was limited to accepting only that waste generated in Alberta. With a proved record of performance, that policy was changed later to all of Canada. In 2000, the waste importation policy was widened further to include waste from the United States and outside the continent.[38]

Public Participation in Other Activities

This section has focused on public participation in the site selection process. However, the public is keenly interested not just in siting but in all aspects of facility development and operations. In particular, the public has expressed concern with the permitting of facilities, the operation and closure of facilities, and the transportation of waste. Activities other than the development of facilities also encounter public concern. For example, the remediation of contaminated sites can alarm members of the surrounding community, who perhaps believe that the planned remediation will stop short of what is necessary or that the proposed remediation techniques will place short-term human health at risk. The same fundamental concepts discussed in this section generally apply to the other areas (e.g., getting the public involved as early as possible).

8-6 PERMITTING

Permitting is the action of regulatory agencies to authorize construction and operation of a facility. Facilities cannot operate without holding a valid permit, issued on finding that the facility site, design, and operations do not violate health and environmental standards.

A hazardous waste treatment and disposal facility is probably the most closely regulated nonnuclear or nonmedical activity in the United States today. A new term, "Part B permit," has entered the language, and the RCRA permit application alone can require hundreds of pages of documentation. This permitting requirement may be just one of a host of federal, state, and, frequently, local requirements that must be met. In one instance, for example, a commercial hazardous waste treatment company was required to obtain 24 separate permits to open an integrated treatment and landfill facility. Only one permit addressed hazardous waste regulations; the balance of permits dealt with transportation, stormwater, zoning, and a host of other concerns.

By any measure, the EPA permit application process for a hazardous waste facility is lengthy, expensive, and burdensome. In many states, a facility developer must also submit a separate permit application to satisfy the state's particular requirements. Preparing the permit application takes considerable effort and time. Federal regulations (primarily 40 CFR Parts 264 and 270) specify the technical requirements of a permit application. The major elements of a typical application are listed in Table 8-2.

Extensive detail is required for any element. For example, in "Process information," a facility having tanks must provide the following information for each tank:

- Dimensions and capacity
- Assessment of tank system integrity
- External control protection
- Tank system installation and testing plans and procedures
- Description of feed systems, safety cutoff, bypass systems, and pressure controls
- Diagram of piping, instrumentation, and process flow
- Design, construction, and operation of secondary containment system
- Leak detection requirements
- External liner, vault, double-walled tank, or equivalent device
- Containment and leak detection requirements for ancillary equipment

TABLE 8-2
Elements of a permit application for a hazardous waste facility

Major element	Example subjects
Facility description	Location and configuration of structures, improvements and appurtenances; traffic patterns
Waste characteristics	Chemical and physical analyses of all waste types, waste analysis plan (e.g., sampling and testing methods)
Process information	Container management practices, control of run-on, tank system integrity, piping and instrumentation diagrams, spill prevention and control, liner system compatibility, process engineering analyses, waste compatibility
Groundwater monitoring	Demonstration of migration potential, description of wells, sampling and analysis plan, corrective action program
Procedures to prevent hazards	Security, inspection schedule, communication and emergency equipment, preventive procedures, management of reactive and ignitable wastes
Contingency plan	Responses to fires, explosions, or other unplanned release of hazardous waste
Personnel training	Programs to prepare personnel to operate and maintain facility
Closure and postclosure requirements	Waste inventory, equipment decontamination, final cover, long-term monitoring, inspection and maintenance schedule, financial assurance, liability requirements
Other federal laws	Demonstration of compliance with other applicable federal laws (e.g., Endangered Species Act)

There are extensive requirements for technical information for unit waste treatment or disposal operations. Some types of treatment or disposal have to meet **performance standards.** For example, an incineration facility must conduct a trial burn to demonstrate 99.99 percent destruction and reduction efficiency (DRE) for principal organic hazardous constituents (POHCs) (see Chapter 12). DREs are higher for a select few compounds, such as PCBs. Concern with products of incomplete combustion (PICs) from incineration has led some states to establish a 99.9 percent combustion efficiency as an additional requirement. Some states further require three-dimensional dispersion modeling of air emissions to calculate concentration of contaminants at the location of receptors (see Chapter 4). The calculated concentrations are compared with toxicology data to assess health risks. Applicants may need to prepare detailed risk assessment documents.

Preparing a permit application has many pitfalls that may have adverse consequences for years to come if not addressed effectively. For example, it is important that the applicant provide the details that allow the permit writer (i.e., the regulatory agency) to extract specific terms and conditions. Permits written in general terms are subject to a variety of interpretations and may result in unexpected and unwarranted enforcement actions that may threaten the facility's viability. On the other hand, a commercial facility designed to accept waste from industry at large will need the flexibility to operate in an efficient and cost-effective manner and to accept a wide range of waste types.

The technical requirements are fairly clear but still leave open many decisions that, in aggregate, constitute a judgment as to what is an "acceptable" level of risk. However minutely small the calculated risk is, it is not zero. This contributes to public concern and becomes a controversial point of conflict between the public and the facility developer. Lengthy delays on the part of regulatory agencies have occurred as they endeavor to judge the "bottom line" acceptability of a facility—without definitive standards to guide their decision but with the pressure of a constituency bitterly opposing the facility.

The public's concern with hazardous waste facilities has prompted some state governments to involve citizens more fully in the review of a permit application. Table 8-3 shows what one state, Pennsylvania, has proposed as a schedule of review steps in its permitting procedure. This schedule assumes no Notice of Deficiency (a finding that the information in the application is either incomplete or identifies an operational or design feature not in compliance with regulatory requirements) on the review for completeness, no Notice of Deficiency on the technical review, no new issues as the result of the public hearing on the draft permit, only 5 days for the public hearing, and a technical review in 180 days. Experience shows that each assumption is highly unlikely. A Notice of Deficiency would extend the permitting period for several months or a year, if not longer. Additionally, changes in facility design, configuration, or even ownership will delay the process even longer. The permitting process itself includes several points where additional public hearings can be convened and decisions appealed. As a result, it is not unusual to take 5 years or more to issue a permit.

Outside of the permitting process, facilities can face challenges in the form of local zoning restrictions, litigation by affected parties, or simply political intervention. These actions string out the construction and operation of the facility and add to the mounting costs. The developer must "carry" the costs of facility development

TABLE 8-3
Proposed schedule for review of permit application in Pennsylvania

Step	Duration
Preapplication conference	Optional
Site investigation by state	Optional
Notify host community application has been submitted (within 10 days)	
Publish notice in state bulletin and twice in local paper	15 days
State reviews application for completeness	30 days
State publishes notice of start of technical review	15 days
Public review and comment period	60 days
Technical review by state	180 days
If requested, state holds fact-finding meeting with the public	45 days
State prepares draft permit and fact sheet	45 days
EPA review of draft permit and fact sheet	60 days
State publishes notice that draft permit has been prepared	15 days
Public review and comment period	45 days
If requested, public hearing is held	60 days
State can extend the public review and comment period if new issues are raised	?
State prepares a responsiveness summary on the permit and publishes comments	45 days
State prepares and issues the permit	30 days
	645 days (21.5 months)

without the prospect of revenue generation on the horizon, making the project less attractive to potential investors. These delays and cost increases can be enough to make undercapitalized ventures withdraw applications for permits and cease when investors realize that the risk of denial makes competitive investments more attractive.

The facts speak for themselves. The typical schedule *intended* by state government for reviewing and approving a permit application may take from 6 months to 2 years. Actual experience shows that only from one-third to one-half the number of permit applications received during a 7-year period (1980–1986) were completely processed. Of the applications submitted, only 53 facilities were permitted, of which 46 were existing facilities undergoing some form of expansion or modernization. About half were commercial facilities, the remainder being on-site operations.[40]

One critic of the particular procedure for reviewing a permit application claims that it "reverses George Orwell's *1984*" and it is "a prescription for governmental paralysis."[41] While the schedule may be lengthy, it does respond to a need and a growing outcry to involve the public more in siting. The question remains of how to make this involvement meaningful and productive.

DISCUSSION TOPICS AND PROBLEMS

8-1. Which types of facilities do not generate residual waste?

8-2. What types of facilities could recover or treat the residual wastes from a solvent recovery facility?

8-3. Briefly describe 10 measures for ensuring the safe operation of a hazardous waste facility.

8-4. Define five criteria for siting a hazardous waste incinerator? Should any of these be mandatory?
8-5. How could a computer prove useful in siting hazardous waste facilities?
8-6. Name a positive and negative way in which public concern with hazardous waste impacts the development of facilities to treat or dispose of the waste.
8-7. Name five root causes of the general reaction of the public to oppose the siting of hazardous waste facilities in its community.
8-8. What did the Province of Alberta do to address these root causes? Which was/were the most important?
8-9. Would one expect public opposition to differ depending on the type of facility proposed?
8-10. Identify the representation one would expect to find on a public advisory committee to a statewide site selection process.
8-11. What are Part A and Part B permit applications, and what are their main elements?
8-12. What are the implications of the term "acceptable risk" as it applies to the siting and permitting of hazardous waste facilities?
8-13. Why is a "needs assessment" important to the development of a facility?
8-14. Determine the minimum number of separate storage areas necessary to provide safe segregation of the following wastes: plating sludges containing Cd; cleaning agents containing caustics; chemical wastes (aromatic hydrocarbons); waste cresol from a wood-treating operation; sludge from plating operation containing cyanides.
8-15. Under what conditions can a surface impoundment be considered a land disposal facility?
8-16. What are two fundamental measures for preventing a deep injection well from contaminating groundwater?
8-17. What is the difference between a waste profile and a waste analysis plan?
8-18. Name five media/receptors that could be monitored for release of hazardous waste constituents from an incineration facility.
8-19. Which is more important in selecting a site for a new hazardous waste facility: applying siting criteria in a scientific fashion to select a technically superior site or involving the public in site selection?
8-20. What is the process difference between an aqueous waste treatment facility and a municipal sewage treatment plant?

REFERENCES

1. Freeman, H. M., ed.: *Standard Handbook of Hazardous Waste Treatment and Disposal,* McGraw-Hill, New York, 1988.
2. *Incinerating Hazardous Wastes,* edited by H. M. Freeman, Technomic Publishing Co., Lancaster, Pa., 1988.
3. ERM, Inc.: "Hazardous Waste Management Technologies—An Assessment," report to Pennsylvania Department of Environmental Resources, February 22, 1983.
4. Schriefer, John: "Reaping the Value from Dust and Slag," *Iron Age New Steel,* February 1997.
5. 25 PA Code §75.260.
6. Clark-McGlennon Associates, Inc.: "An Introduction to Facilities for Hazardous Waste Management," report to the New England Regional Commission, Boston, November 1980.
7. Ibid.
8. Ibid.
9. Hatayama, H. K., et al.: "A Method for Determining the Compatibility of Hazardous Waste," EPA-600/2-80-076, U.S. Environmental Protection Agency, Cincinnati, Ohio, April 1980.

10. Freeman, H. M., ed.: *Standard Handbook of Hazardous Waste Treatment of Disposal,* p. 11.10, 1988. McGraw-Hill, New York.
11. Ackerman, S. A.: "Do I Need to Keep All These Records?" *Environment Today,* newsletter by Montgomery McCracken, Walker, and Rhoads, Philadelphia, Summer 1991.
12. Buckingham, P. L.: "Regional Screening Approaches to Selecting Sites for New Secure Landfills," *Proceedings of Fourth Annual Madison Conference of Applied Research & Practice on Municipal & Industrial Waste,* University of Wisconsin, Madison, September 28–30, 1981.
13. Rogers, Golden, and Halpern, Inc.: "Regional Screening and Conflict Resolution," report prepared for Maryland Department of Natural Resources, 1977.
14. ERM–New England, Inc.: "Development and Application of Criteria for Siting Hazardous Waste Facilities in Vermont," report prepared for Vermont Agency of Natural Resources, February 1991.
15. Buckingham, P. L.: "Protecting Ground Water with Siting Criteria," *Proceedings of Hazardous Materials Management Conference,* Philadelphia, July 14, 1983.
16. New Jersey Department of Environmental Protection: "Basis and Background Document for N.J. Major Commercial Hazardous Facility Siting Criteria," Trenton, N.J., February 1983.
17. McCoy and Associates: "Locational Criteria for Siting TSD Facilities May Have Broad Impact," *The Hazardous Waste Consultant,* November/December 1986.
18. New Jersey Department of Environmental Protection, "Basis and Background Document."
19. ERM–New England, Inc., "Development and Application of Criteria."
20. McGlennon, J. A. S.: "Siting New Facilities: Problems and Possibilities," presented to the American Institute of Chemical Engineers, August 25, 1986.
21. Morell, David: "Siting and the Politics of Equity," *Hazardous Waste,* vol. 1 (1984), pp. 555–571, Mary Ann Liebert, Inc., publishers. Reprinted in Robert W. Lake, ed., *Resolving Locational Conflict,* Rutgers University Center for Urban Policy Research, New Brunswick, N.J., 1987, pp. 117–136.
22. Ibid.
23. Morell, D. L., and C. Magorian: *Siting Hazardous Waste Facilities: Local Opposition and the Myth of Preemption,* Ballinger, Cambridge, Mass., 1982.
24. Tomsho, Robert: "Small Town in Alberta Embraces What Most Reject: Toxic Waste," *The Wall Street Journal,* vol. CXXV, no. 126, p. A1, December 27, 1991.
25. Ibid.
26. Ibid.
27. McQuaid-Cook, J., and K. S. Simpson: "Siting a Fully Integrated Waste Management Facility," *Journal of the Air Pollution Control Association,* vol. 36, no. 9, p. 1031, September 1986.
28. Kupchanko, E. E.: "Case Study of a Successful Facility Siting," *Standard Handbook of Hazardous Waste Treatment and Disposal,* H. M. Freeman, ed., McGraw-Hill, New York, p. 3.39, 1988.
29. McQuaid-Cook and Simpson, "Siting a Fully Integrated Waste Management Facility."
30. Ibid.
31. Ibid.
32. Ibid.
33. Kupchanko, E. E., "Case Study of a Successful Facility Siting."
34. McQuaid-Cook and Simpson, "Siting a Fully Integrated Waste Management Facility."
35. Tomsho, "Small Town in Alberta Embraces."
36. McQuaid-Cook and Simpson, "Siting a Fully Integrated Waste Management Facility."
37. Ibid.
38. Personal communication between Phillip Buckingham and Tom Kinderwater, Plant Manager, Bovar Waste Management, Inc., February 15, 2000.

39. Alberta Special Waste Management Corporation, Edmonton, Alberta, Canada.
40. National Governors Association: "Siting New Treatment and Disposal Facilities," Washington, D.C.
41. Kury, F. L.: comments on draft report of "The Proposed Public Participation Program for Hazardous Waste Management in Pennsylvania," submitted to Pennsylvania Department of Environmental Resources by Rogers, Golden, and Halpern, February 8.

ADDITIONAL READING

Andrews, R. N. L., and F. M. Lynn: "Siting of Hazardous Waste Facilities," *Standard Handbook of Hazardous Waste Treatment and Disposal,* H. M. Freeman, ed., McGraw-Hill, New York, pp. 3.3–3.16, 1989.

Andrews, R. N. L., and T. K. Pierson: "Hazardous Waste Facility Siting Processes: Experience from Seven States," *Hazardous Waste,* vol. 1, no. 3, pp. 377–386, 1984.

Bacow, L. S., and J. F. Milkey: "Overcoming Local Opposition to Hazardous Waste Facilities: The Massachusetts Approach," *Harvard Environmental Law Review,* vol. 6, pp. 265–305, 1982.

Baker, R. D., et al.: "Management of Hazardous Waste in the United States," *Hazardous Waste and Hazardous Materials,* vol. 9, pp. 37–59, 1992.

Bingham, G., and T. Mealey: *Negotiating Hazardous Waste Facility Siting and Permitting Agreements,* The Conservation Foundation, Washington, D.C., 1988.

Clark-McGlennon Associates: *A Decision Guide for Siting Acceptable Hazardous Waste Management Facilities in New England,* New England Regional Commission, Boston, Mass., 1980.

Daley, Peter: "Comprehensive Hazardous Waste Treatment Facilities," *Standard Handbook of Hazardous Waste Treatment and Disposal,* H. M. Freeman, ed., McGraw-Hill, New York, pp. 11.1–11.18, 1989.

Dawson, G. W., and B. W. Mercer: *Hazardous Waste Management,* John Wiley and Sons, New York, 1986.

Freeman, H. M. ed.: *Standard Handbook of Hazardous Waste Treatment and Disposal,* McGraw-Hill, New York, 1989.

Greenberg, M. R., R. F. Anderson, and K. Rosenberger: "Social and Economic Effects of Hazardous Waste Management Sites," *Hazardous Waste,* vol. 1, no. 3, pp. 387–396, 1984.

Hadden, S. G., J. Veillette, and T. Brandt: "State Roles in Siting Hazardous Waste Disposal Facilities: From State Preemption to Local Veto," Chapter 10 in J. P. Lester and A. O'M. Bowman, eds., *The Politics of Hazardous Waste Management,* Duke University Press, Durham, N.C., pp. 196–211, 1983.

Hirschhorn, J. S.: "Siting Hazardous Waste Facilities," *Hazardous Waste,* vol. 1, no. 3, pp. 423–429, 1984.

Kupchanko, E. E.: "Case Study of a Successful Facility Siting," *Standard Handbook of Hazardous Waste Treatment and Disposal,* H. M. Freeman, ed., McGraw-Hill, New York, pp. 3.39–3.49, 1989.

National Audubon Society: *Siting Hazardous Waste Facilities: A Handbook,* National Audubon Society, Tavernier, Fla., 1983.

O'Hare, M., L. S. Bacow, and D. R. Sanderson: *Facility Siting and Public Opposition,* Van Nostrand Reinhold, New York, 1983.

Seley, J. E.: *The Politics of Public Facility Siting,* Lexington Books, Lexington, Mass., 1983.

Smith, M. A., F. M. Lynn, and R. N. L. Andrews: "Economic Impacts of Hazardous Waste Facilities," *Hazardous Waste,* vol. 3, no. 2, 1986.

Smith, M. A., F. M. Lynn, R. N. L. Andrews, R. Olin, and C. Maurer: *Costs and Benefits to Local Government Due to Presence of a Hazardous Waste Management Facility and Related Compensation Issues,* Governor's Waste Management Board, Raleigh, N.C., pp. 195–204, 1985.
Susskind, L. E.: "The Siting Puzzle: Balancing Economic and Environmental Gains and Losses," *Environmental Impact Assessment Review,* vol. 5, pp. 157–163, 1985.
U.S. EPA: "Hazardous Waste Treatment, Storage and Disposal Facility Report for 1987," U.S. Environmental Protection Agency, Washington, D.C., 1987.

ACKNOWLEDGMENTS

The authors wish to acknowledge the following individuals who assisted with the preparation of this chapter for the first edition: Alan S. MacGregor, Ruth Baker, and David Morell, *ERM*. The principal reviewer for ERM was John A. S. McGlennon, *ERM*.

The reviewer for the second edition was David Morell, *ERM*.

CHAPTER 9

Physicochemical Processes

'If seven maids with seven mops, Swept it for half a year,
Do you suppose', the Walrus said, 'That they could get it clear?'
'I doubt it' said the Carpenter, and shed a bitter tear.
Lewis Carroll

The physicochemical processes described in this chapter include a selection of technologies that can be used for hazardous waste treatment and recycling, as well as for groundwater or soil remediation. Each section presents a description of a technology, the theory behind its application, and the procedure for designing the physical facility to apply the technology. This chapter presents a selection of available technologies, as space limitations preclude in-depth treatment of all technologies currently available. Additional processes, which are predominantly *in situ*, are discussed in Chapter 16.

9-1 AIR STRIPPING

Air stripping is a mass transfer process that enhances the volatilization of compounds from a liquid phase (water) into a vapor phase (air), by actively passing air through water. This improves the transfer between the air and water phases. Air stripping has been shown to be a cost-effective technique for removing volatile organic compounds (VOCs) from contaminated water, and the process is one of the most commonly used for remediating VOC-impacted groundwater, being most effective for low concentrations (< 200 mg/L).

Air stripping can be performed by using stripping basins or stripping towers. Stripping basins consist of diffused aeration or mechanical aeration (aeration basins); stripping towers include packed towers, trayed towers, and spray systems.

Aeration basins are simple and relatively inexpensive, and may be used for groundwater containing highly volatile compounds, if high efficiency is not required. An aeration basin is a continuous-flow tank with bubble diffusers located at the bottom through which compressed air is diffused, thereby providing an increased gas-phase surface area for the mass transfer of VOCs. Such basins provide roughly 90 percent removal of VOCs and operate as a completely mixed reactor. The mass transfer driving force is low because of the dilution. However, performance can be improved with baffles placed in the basin, since they promote plug flow and therefore increase the driving force for mass transfer.

Trayed, spray, and cross-flow towers increase the surface area of the groundwater available for mass transfer. The water cascades down a tower of slats or structured packing material while air is drawn through the tower by a fan. Although the air does not truly flow countercurrently to the water, less airflow is required than for aeration basins. The treatment efficiency is moderate because of plug flow of the water and because the cascading droplets of water provide an additional surface area for mass transfer. The energy requirement is low because the pressure drop through the tower is low. However, mist generated by the towers may be a nuisance surrounding the facility, and this treatment process has limited application when air temperatures are low (i.e., below the freezing point of water).

This section focuses on packed air stripping towers with countercurrent flow, which are most frequently used for remediating groundwater, although the theory can be applied to other types of air stripping systems.

Process Description

The process consists of countercurrent flow of air and contaminated groundwater through a vertical column packed with spherically shaped pieces, which improve the effective surface area for mass transfer and ensure homogeneity of the mass transfer system. The process is illustrated in Figure 9-1. A contaminated water stream is brought in at the top of the stripper and distributed evenly over the packing, while the air stream is brought in through the bottom of the column. The packing can be composed of individual pieces, or it can be a structure stacked in the column. Typically, the packing materials consist of plastic shapes that have a high surface-to-volume ratio and provide the necessary transfer surface to allow volatile components to move from the liquid stream to the air stream.

The air stream leaves the column at the top; the water stream at the bottom. A mass balance equation for the column may be written by setting the mass of contaminant removed from the water equal to the mass added to the air:

$$Q_W\,[C_{\text{in}} - C_{\text{out}}] = Q_A\,[A_{\text{out}} - A_{\text{in}}] \tag{9-1}$$

where Q_W = water flow rate (m^3/s)

C = concentration of contaminant in water ($kmol/m^3$)

Q_A = airflow rate (m^3/s)

A = concentration of contaminant in air ($kmol/m^3$)

Assuming that the influent air contains none of the contaminant ($A_{\text{in}} = 0$) and the effluent water also is free of contamination ($C_{\text{out}} = 0$), the stripper will

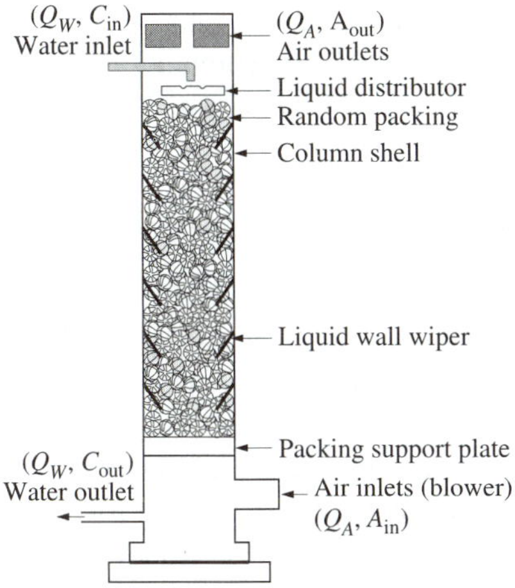

FIGURE 9-1
Packed tower air stripper.

work at 100 percent efficiency. Finally, the concentration in the effluent air, A_{out}, is determined from Henry's law (see Section 3-2) by using the dimensionless Henry's law constant H', and assuming equilibrium:

$$A_{out} = H'C_{in}$$

$$Q_W\,[C_{in}] = Q_A\left[H'C_{in}\right]$$

$$Q_W = Q_A\left[H'\right] \tag{9-2}$$

or

$$H'\,(Q_A/Q_W) = 1 \tag{9-3}$$

This theoretical relationship is true for ideal equilibrium and perfect stripping. The expression $R = H'(Q_A/Q_W)$ is termed the **stripping factor.** Values of this factor must be greater than 1 for stripping to occur. Q_A/Q_W is termed the air-to-water ratio.

Air stripping systems are generally limited to VOC concentrations of less than a few hundred milligrams per liter. Nonvolatile compounds (i.e., those with $H' \leq 0.01$) will not be removed effectively. However, this process is applicable to a wide range of VOCs and is capable of removing contaminants to extremely low levels. If designed correctly, packed columns can achieve removal efficiency greater than 99.9 percent.

Because the flows are countercurrent, the air and water approach equilibrium, and the required volume of air is minimized. To reach true equilibrium, a column of infinite height would be required. Because the capital cost for column construction and the operating costs for the blower increase with column height, the packed tower height is somewhat limited. However, the surface area of contaminated groundwater can be increased by spray formation, with nozzles or mechanical agitators.

Stripping Theory

Two-phase resistance or two-film theory is generally used for description of mass transfer in the gas-liquid exchange processes. The overall resistance to mass transfer

is considered to be the sum of two resistances, a liquid-phase resistance and a gas-phase resistance (the Onda correlations are generally used for estimating the mass transfer coefficients). According to this two-film theory, the transfer of VOCs from water to air (Figure 9-2) consists of transfer from:

- Bulk liquid to liquid film
- Liquid film to air film
- Air film to bulk air

The rate at which a contaminant is transferred from water to air, for low solubility contaminants, is represented by an overall transfer rate constant $K_L a(\text{s}^{-1})$ which is the product of two variables: K_L, the liquid mass transfer coefficient $\{(\text{kg} \cdot \text{mol})/[\text{s} \cdot \text{m}^2)(\text{kg} \cdot \text{mol})/\text{m}^3]\}$ or (m/s), and a, the area-to-volume ratio of the packing (m^2/m^3). For design purposes, $K_L a$ should be determined experimentally. However, for dilute solutions, $K_L a$ can be estimated from the Sherwood and Holloway[1] equation:

$$K_L a = \alpha \times D_L \times \left(0.305 \, \frac{L}{\mu}\right)^{1-n} \left(\frac{\mu}{\rho D_L}\right)^{0.5} \tag{9-4}$$

where $D_L =$ liquid diffusion coefficient (m^2/s)
$L =$ liquid mass loading rate [$\text{kg}/(\text{m}^2 \cdot \text{s})$]
$\mu =$ liquid viscosity (1.002×10^{-3} Pa · s at 20°C)
$\rho =$ density of water (998.2 kg/m^3 at 20°C)
$\alpha, n =$ constants from Table 9-1

Diffusion coefficients may be estimated by the Wilke-Chang method (see Section 3-2).

Mathematical correlations developed by Onda[3,4] also provide values of $K_L a$ that are valid over a range of contaminants and are preferred for preliminary design. The Onda correlations are[5]:

$$k_L \left(\frac{\rho_L}{\mu_L g}\right)^{1/3} = 0.0051 \left(\frac{L}{a_w \mu_L}\right)^{2/3} \left(\frac{\mu_L}{\rho_L D_L}\right)^{-0.5} (a_t d_p)^{0.4} \tag{9-5}$$

and

$$\frac{a_w}{a_t} = 1 - \exp\left[-1.45 \left(\frac{\sigma_c}{\sigma}\right)^{0.75} \left(\frac{L}{a_t \mu_L}\right)^{0.1} \left(\frac{L^2 a_t}{\rho_L^2 g}\right)^{-0.05} \left(\frac{L^2}{\rho_L \sigma a_t}\right)^{0.2}\right] \tag{9-6}$$

$$\frac{k_G}{a_t D_G} = 5.23 \left(\frac{G}{a_t \mu_G}\right)^{0.7} \left(\frac{\mu_G}{\rho_G D_G}\right)^{1/3} (a_t d_p)^{-2} \tag{9-7}$$

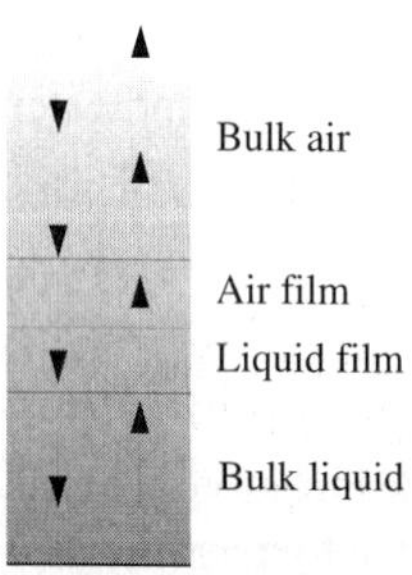

FIGURE 9-2
Two-film theory.

TABLE 9-1
Constants for Equation 9-4[2]

Packing	Size, mm	α	n
Raschig rings	12	920	0.35
	25	330	0.22
	38	295	0.22
	50	260	0.22
Berl saddles	12	490	0.28
	25	560	0.28
	38	525	0.28
Tile	75	360	0.28

Note: millimeters × 0.0394 = inches

The overall mass transfer coefficient may be found by:

$$\frac{1}{K_L a} = \frac{1}{H' k_G a} + \frac{1}{k_L a} \tag{9-8}$$

Example units for the Onda correlations are:

k_L	liquid-phase mass transfer (m/s)
ρ_L	liquid density (kg/m^3) = 998.2 at 20°C
μ_L	viscosity of water [kg/(m · s) = 1.002×10^{-3} at 20°C
μ_G	viscosity of gas [kg/(m · s)] = 1.1×10^{-5} at 20°C
g	gravity (m/s^2) = 9.81
L	liquid mass loading rate [kg/(m^2 · s)]
a	area/volume ratio; use a_w
a_w	wetted packing area (m^2/m^3) (used as equal to a in $K_L a$)
D_L	liquid diffusivity (m^2/s) (see Section 3-2)
a_t	total packing area (m^2/m^3) (see Table 9-2)
d_p	nominal packing diameter (m)
σ_c	critical surface tension of packing material (N/m) = 0.033 for polyethylene, 0.040 for PVC
σ	surface tension of water (N/m) = 0.073 at 20°C
k_G	gas-phase mass transfer (m/s)
D_G	gas diffusivity (m^2/s) (see Appendix B and Section 3-2)
G	gas mass loading rate [kg/(m^2 · s)]
ρ_G	density of air (kg/m^3) = 1.205 at sea level at 20°C
$K_L a$	overall mass transfer (s^{-1})
H'	Henry's constant (dimensionless) (see Appendix B and Section 3-2)

EXAMPLE 9-1. ONDA CORRELATIONS. Given the following characteristics of an air stripping column to remove chloroform ($CHCl_3$) from groundwater: flow = 170 gal/min; column diameter = 3 ft; packing = 2-in polyethylene Tri-paks®; $D_L = 6 \times 10^{-6}$ cm^2/s; air-to-water ratio = 100:1. Determine the overall mass transfer coefficient $K_L a$ using the Onda correlations. Assume 20°C (293.2 K).

TABLE 9-2
Packing characteristics

Size, mm	Intalox saddle		Raschig ring		Pall ring		Berl saddle		Tri-pak	

	F	***a***	***F***	***a***	***F***	***a***	***F***	***a***	***F***	***a***
13	200	623	580	364	—	—	240	466	—	—
25	98	256	155	190	52	206	110	249	28	279
50	40	118	65	92	25	102	45	105	16	157
75	22	92	37	62	—	—	—	—		
88	—	—	—	—	16	85	—	—	12	125

F = packing factor (dimensionless); a = area/volume ratio (m^2/m^3). Data for ceramic shapes except Pall ring and Tri-pak® data for plastic shape.

Solution

$$\frac{\sigma_c}{\sigma} = \frac{0.033}{0.073} = 0.45$$

Diameter = 3 ft = 0.91 m; area = $\pi(0.91)^2/4 = 0.650\ m^2$
Unit weight of water = 8.34 lb/gal

$$L = \frac{170\frac{\text{gal}}{\text{min}} \times \frac{1\ \text{min}}{60\ \text{s}} \times 8.34\frac{\text{lb}}{\text{gal}} \times \frac{1\ \text{kg}}{2.204\ \text{lb}}}{0.650\ \text{m}^2} = 16.49\ \text{kg}/(\text{m}^2 \cdot \text{s})$$

$$a_t = 157\ \text{m}^{-1} \quad \text{(Table 9-2)}$$

$$D_G = 0.09404\ \text{cm}^2/\text{s (see Appendix B)} \times 10^{-4}\ \text{m}^2/\text{cm}^2 = 9.404 \times 10^{-6}\ \text{m}^2/\text{s}$$

$$Q_W = 170\frac{\text{gal}}{\text{min}} \times \frac{\text{ft}^3}{7.48\ \text{gal}} \times \frac{\text{min}}{60\ \text{s}} = 0.379\ \text{ft}^3/\text{s}$$

$$Q_A = 100 \times Q_W = 37.9\ \text{ft}^3/\text{s}$$

$$G = \frac{37.9\frac{\text{ft}^3}{\text{sec}} \times 0.02832\frac{\text{m}^3}{\text{ft}^3} \times 1.205\frac{\text{kg}}{\text{m}^3}}{0.650\ \text{m}^2} = 1.99\ \text{kg}/(\text{m}^2 \cdot \text{s})$$

$$D_L = 6 \times 10^{-6}\frac{\text{cm}^2}{\text{s}} \times \frac{\text{m}^2}{10^4\ \text{cm}^2} = 6 \times 10^{-10}\ \text{m}^2/\text{s}$$

$$d_p = 2\ \text{in} \times \frac{\text{m}}{39.37\ \text{in}} = 0.051\ \text{m}$$

From Appendix B for chloroform: $A = 9.84$; $B = 4.61 \times 10^3$.

$$\begin{aligned} H &= \exp\left[9.84 - (4.61 \times 10^3/T)\right] \\ &= \exp\left[9.84 - (4.61 \times 10^3/293.2)\right] \\ &= 2.786 \times 10^{-3}\ (\text{atm} \cdot \text{m}^3)/\text{mol} \end{aligned}$$

From Equation 3-7:

$$H' = \frac{H}{RT} = \frac{2.786 \times 10^{-3}\ (\text{atm} \cdot \text{m}^3)/\text{mol}}{\left[8.205 \times 10^{-5}\ (\text{atm} \cdot \text{m}^3)/(\text{mol} \cdot \text{K})\right](293.2\ \text{K})}$$
$$= 0.116\ (\text{dimensionless})$$

Reynolds number:

$$\frac{L}{a_t \mu_L} = \frac{16.48\ \text{kg}/(\text{m}^2 \cdot \text{s})}{157\ \text{m}^{-1} \times 1.002 \times 10^{-3}\ \text{kg}/(\text{m} \cdot \text{s})} = 104.76\ (\text{dimensionless})$$

Froude number:

$$\frac{L^2 a_t}{\rho_L^2 g} = \frac{(16.48^2)(157)}{(998.2^2)(9.81)} = 4.36 \times 10^{-3}\ (\text{dimensionless})$$

Weber number:

$$\frac{L^2}{\rho_L \sigma a_t} = \frac{16.48^2}{(998.2)(0.073)(157)} = 0.0237\ (\text{dimensionless})$$

From Equation 9-6:

$$a_w = 157\left\{1 - e\left[-1.45\,(0.45)^{0.75}\,(104.76)^{0.1}\left(4.36 \times 10^{-3}\right)^{-0.05}(0.0237)^{0.2}\right]\right\}$$
$$= 85.6\ \text{m}^2/\text{m}^3$$

From Equation 9-5:

$$k_L\left[\frac{998.2}{\left(1.002 \times 10^{-3}\right)(9.81)}\right]^{1/3} = 0.0051\left[\frac{16.48}{(85.6)\left(1.002 \times 10^{-3}\right)}\right]^{2/3}$$
$$\times\left[\frac{1.002 \times 10^{-3}}{(998.2)\left(6 \times 10^{-10}\right)}\right]^{-0.5}\left[(157)(0.051)\right]^{0.4}$$

$$k_L = 2.00 \times 10^{-4}\ \text{m/s}$$

From Equation 9-7:

$$\frac{k_G}{(157)\left(9.404 \times 10^{-6}\right)}$$
$$= 5.23\left[\frac{1.99}{(157)\left(1.81 \times 10^{-5}\right)}\right]^{0.7}\left[\frac{1.81 \times 10^{-5}}{(1.205)\left(9.404 \times 10^{-6}\right)}\right]^{1/3}\left[(157)(0.051)\right]^{-2}$$

$$k_G = 1.374 \times 10^{-2}\ \text{m/s}$$

$$\frac{1}{K_L a} = \frac{1}{H' k_G a} + \frac{1}{k_L a}$$
$$= \frac{1}{(0.116)\left(1.374 \times 10^{-2}\right)(85.6)} + \frac{1}{\left(2.00 \times 10^{-4}\right)(85.6)} = 64.32$$

$$\underline{\underline{K_L a = 0.0155\ \text{s}^{-1}}}$$

In order to develop a design equation, a section of the stripping tower is selected (see Figure 9-3). The section has a superficial surface area B, and a differential thickness dz.

The rate of mass transfer per unit volume of the tower is:

$$M = (Q_W\,dC)/(B\,dz) = \left(Q_W/B\right)(dC/dz) \tag{9-9}$$

where $M =$ mass transfer rate [kmol/(s · m^3)]
$Q_w =$ liquid flow rate (m^3/s)
$C =$ contaminant concentration (kmol/m^3)
$B =$ surface area (m^2)
$z =$ depth in column (m)
$dC/dz =$ concentration gradient (kmol/m^4)

The rate of mass transfer can also be defined by examination of the concentration gradient across the air/water interface:

$$M = K_L a[C - C_{eq}] \tag{9-10}$$

where $C_{eq} =$ concentration in water in equilibrium with the air at a specific point
$[C - C_{eq}] =$ the degree to which the system is out of equilibrium

From the definition of Henry's law (Equation 3-6):

$$C_{eq} = A/H' \tag{9-11}$$

where $A =$ concentration in air (kmol/m^3)
$H' =$ dimensionless Henry's constant (see Section 3-2).

The concentration imbalance $[C - C_{eq}]$ is the driving force for transfer. Setting Equation 9-10 equal to Equation 9-9, and converting to molar units by substituting L/M_w for Q_w/B:

$$(L/M_w)(dC/dz) = K_L a[C - C_{eq}] \tag{9-12}$$

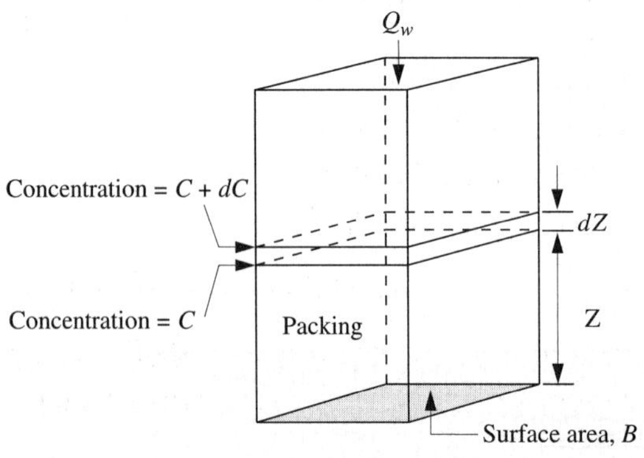

FIGURE 9-3
Differential section.

where $L =$ liquid molar loading rate [kmol/(s · m^2)]
$M_W =$ molar density of water (55.6 kmol/m^3) = 3.47 lbmol/ft^3

Equation 9-12 is a first-order linear differential equation. Separating the variables gives

$$dz = \frac{L}{M_W K_L a} \times \frac{dC}{[C - C_{eq}]} \tag{9-13}$$

The first term on the right is independent of C, and is designated HTU, the height of a transfer unit:

$$\text{HTU} = \frac{L}{M_W K_L a} \tag{9-14}$$

Integrating from the bottom of the column to the top, we obtain the required depth of packing in the column:

$$Z = \text{HTU} \times \text{NTU} \tag{9-15}$$

where NTU (the number of transfer units) is:

$$\text{NTU} = \int_{C=C_{out}}^{C=C_{in}} \frac{dC}{[C - C_{eq}]} \tag{9-16}$$

The equilibrium concentration C_{eq} may be determined from a mass balance from the bottom of the column up to the differential section dz. Let A and C be the concentrations in the air and water at the section, respectively. The mass gained by the air must equal the mass lost by the water:

$$Q_A(A - A_{in}) = Q_w(C - C_{out}) \tag{9-17}$$

From Equation 9-11: $A = C_{eq}H'$ and $A_{in} = 0$; therefore

$$Q_A C_{eq} H' = Q_w(C - C_{out}) \tag{9-18}$$

Solving for the equilibrium concentration gives

$$C_{eq} = Q_w(C - C_{out})/(Q_A H') = (C - C_{out})/R \tag{9-19}$$

Substituting Equation 9-19 into Equation 9-16 and integrating produces:

$$\text{NTU} = \left(\frac{R}{R-1}\right) \ln\left(\frac{(C_{in}/C_{out})(R-1)+1}{R}\right) \tag{9-20}$$

where $R =$ stripping factor $H'(Q_A/Q_W)$ (dimensionless)
$C_{in} =$ concentration in the influent water (kmol/m^3)
$C_{out} =$ concentration in the effluent water (kmol/m^3)

Since the units for concentration cancel, C_{in} and C_{out} may be expressed in mass concentration units (mg/L or μg/L).

Practical values of R range from 2 to 10 or greater. The stripping factor is one of the primary design variables in the system. Realistic values of the air-to-water ratio (Q_A/Q_W) range from 5 to several hundred.

It should be noted that in the chemical engineering literature the stripping factor is typically defined as[6]

$$R = H'(G/L) \tag{9-21}$$

where H' = Henry's law constant (dimensionless)
G = molar flow rate of air (kmol/h)
L = molar flow rate of water (kmol/h)

At the low concentrations encountered in environmental work, G/L and Q_A/Q_w produce approximately the same values.

EXAMPLE 9-2. PRELIMINARY DESIGN OF AIR STRIPPING COLUMN. A groundwater supply has been contaminated with ethylbenzene. The maximum level of ethylbenzene in the groundwater is 1 mg/L, and this must be reduced to 35 μg/L by an air stripping column. The following data are provided:

$K_La = 0.016\ \text{s}^{-1}$; $Q_W = 7.13$ L/s; temperature $T = 20°\text{C} = 293.3$ K

$H = 6.44 \times 10^{-3}(\text{atm} \cdot \text{m}^3)/\text{gmol}$

Select: Column diameter = 2.0 ft (0.61 m); air-to-water ratio $(Q_A/Q_W) = 20$

Determine: Liquid loading rate L
Stripping factor R
HTU, NTU, and height of packing in column

Solution. From Equation 3-6: $H' = H/RT = 6.44 \times 10^{-3}/(8.205 \times 10^{-5} \times 293.3) = 0.27$.

1. Liquid loading rate:
   Cross-sectional area of column = 0.292 m²
   Mass rate = 1.0 kg/L × 7.13 L/s = 7.13 kg/s
   Mass loading = 7.13/0.292 = 24.4 kg/(s · m²)
   $L = [24.4\ \text{kg/(s} \cdot \text{m}^2)](10^3\ \text{g/kg})(1/18\ \text{mol/g}) = \underline{\underline{1360\ \text{mol/(s} \cdot \text{m}^2)}}$

2. Stripping factor: $R = H'(Q_A/Q_W) = 0.27(20) = \underline{\underline{5.4\ \text{(dimensionless)}}}$

3. Height of transfer unit: $$\text{HTU} = \frac{L}{M_W K_L a} = \frac{1360}{(55{,}600)(.016)} = \underline{\underline{1.53\ \text{m}}}$$

4. Number of transfer units: $$\text{NTU} = \left(\frac{R}{R-1}\right) \ln\left[\frac{(C_{\text{in}}/C_{\text{out}})(R-1)+1}{R}\right]$$

$$\text{NTU} = (1.23) \ln\left[\frac{(1000/35)(5.4-1)+1}{5.4}\right]$$

$$= \underline{\underline{3.88\ \text{transfer units}}}$$

5. Height of packing in column $Z = \text{NTU} \times \text{HTU} = (3.88)(1.53)$
   $= \underline{\underline{5.93\ \text{m} = 19.45\ \text{ft}}}$

In an actual design, a safety factor of 20 percent would be added and the height of packing raised to the next whole number:

$$Z = 19.45(1.2) = 23.34 \text{ ft}$$

Use $\underline{\underline{Z = 24 \text{ ft}}}$.

Design Considerations

In designing a stripping system such as a packed tower, the design parameters include column diameter, height, air-to-water ratio, and type and size of packing. In practice, stripping towers typically have diameters of 0.5 to 3 m and heights of 1 to 15 m. The air-to-water ratio ranges from as low as 5 to several hundred and is controlled primarily by flooding and pressure drop considerations. The stripping system must be designed to meet all of the performance constraints, such as the required removal amounts, allowable pressure drop, and operating and construction costs. The removal amounts are determined from the results of the risk assessment and required cleanup goals. The construction costs are closely related to the tower dimensions, with capital cost being a function of tower volume; operation costs increase with column height.

The pressure drop in a tower should be between 200 and 400 N/m^2 per meter of tower height (0.25 to 0.5 inH_2O/ft) to avoid flooding.[7] The pressure drop appears to be the parameter that most directly affects the operating cost of the tower.

As the airflow in a tower is increased, it will ultimately hold back the free downward flow of water and cause flooding of the tower. The smaller-size packings are more susceptible to flooding. Channeling occurs when water flows down the tower wall rather than through the packing. Distribution plates should be placed approximately every 5 diameters to avoid this. Using small-size packing will also reduce the tendency of flow to channel.

Other design considerations include mist control, algal growth control (to avoid column fouling), chemical precipitation (ferrous ion, calcium carbonate) control, and pretreatment (for chemical stabilization). Vapor treatment may be necessary to treat the air stripper air emissions prior to release to the atmosphere (usually required if mass emissions exceed 1 lb/day, or if risk assessment so indicates). This is often accomplished by passing the air stream through an activated carbon unit (see Figure 9-4).

Flooding and pressure drop calculations are simplified by the use of Figure 9-5. The notation for the figure is:

$G =$ air loading rate [lb/(ft$^2 \cdot$ s)]

$L =$ liquid loading rate [lb/(ft$^2 \cdot$ s)][= kg/(s $\cdot$ m^2) $\times$ 0.2048]

$F =$ packing factor (dimensionless, see Table 9-2)

$\rho_A =$ air density (lb$_f$/ft^3)(= kg/m^3 $\times$ 0.06243)

$\rho_W =$ water density (lb$_f$/ft^3)

$g =$ gravitational constant (32.17 ft/s^2)

Because the ordinate axis of Figure 9-5 is not dimensionless, the English units shown must be used. Figure 9-5 may be used to estimate the pressure drop for a

FIGURE 9-4
An air stripping unit.

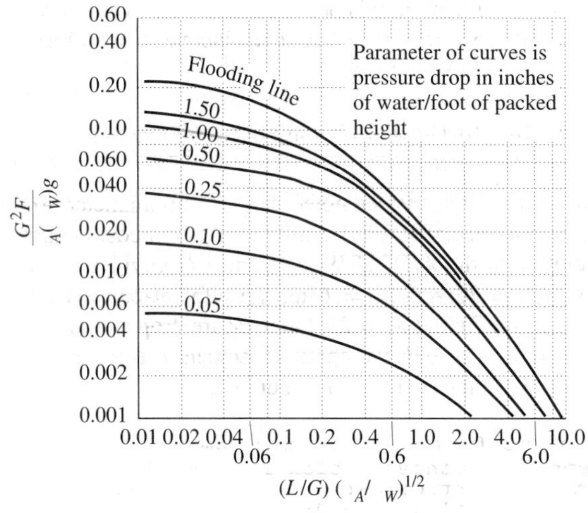

FIGURE 9-5
Generalized pressure drop correlations.[8]

specified design. Additionally, it may be used to modify a design by moving to a different acceptable pressure drop and determining a new value for G.

Characteristics of several typical packing shapes are shown in Table 9-2. Additional information is available in several references.[9,10,11]

EXAMPLE 9-3. DESIGN OF AIR STRIPPING TOWER. Using the data from Example 9-2, determine the pressure drop through the tower. Examine the impact on effluent quality of varying the air-to-water ratio (A/W) and the packing height.

Solution

1. Select 3-in (75-mm) ceramic Raschig rings as packing.

2. Calculate pressure drop. The parameters for Figure 9-5:

$$L = 24.4\frac{\text{kg}}{\text{s} \cdot \text{m}^2} \times 0.2048 = 4.997\frac{\text{lb}}{\text{s} \cdot \text{ft}^2}$$

$$Q_A = Q_W(A/W) = 7.13(20) = 142.6\ \text{L/s} = 0.1426\ \text{m}^3/\text{s}$$

$$\rho_A = \text{air density} = 1.205\ \text{kg/m}^3 = 0.075\ \text{lb/ft}^3\ (@20°\text{C})$$

$$\rho_W = \text{water density} = 62.3\ \text{lb/ft}^3\ (@20°\text{C})$$

$$G = \frac{(0.1426\ \text{m}^3/\text{s})(1.205\ \text{kg/m}^3)}{(0.292\ \text{m}^2)} = 0.5885\frac{\text{kg}}{\text{s} \cdot \text{m}^2}$$

$$= 0.5885\frac{\text{kg}}{\text{s} \cdot \text{m}^2} \times (0.2048) = 0.1205\frac{\text{lb}}{\text{s} \cdot \text{ft}^2}$$

$$F = 37\ (\text{Table 9-2})$$

$$\text{Ordinate} = G^2F/\rho_A(\rho_W)g = (0.1205)^2(37)/(0.075)(62.3)(32.17) = \underline{\underline{0.0036}}$$

$$\text{Abscissa} = (L/G)(\rho_A/\rho_W)^{.5} = (4.997/0.1205)(0.075/62.3)^{.5} = \underline{\underline{1.44}}$$

This point intersects the curves of Figure 9-5 at $dP = 0.1$ inches of water per foot of packing depth (82 Pa/m). This pressure drop is below the recommended range of 0.25 to 0.5 inH_2O/ft, indicating that the airflow rate could be increased significantly without causing flooding.

$$\text{Pressure drop} = (24\ \text{ft})(0.1\ \text{in/ft}) = \underline{\underline{2.4\ \text{in H}_2\text{O} = 600\ \text{Pa}}}$$

3. Evaluate impact on effluent quality. It should be noted that column diameter, *A/W*, and packing shape are design parameters to be selected to provide a cost-effective design. The following output from the AIRSTRIP program[12,13] provides a range of designs. Note that at $A/W = 20$ and $Z = 6$ m, the effluent concentration approximates that of Example 9-2. For $A/W = 20$, the pressure drop is 106 Pa/m. For 6 m this yields a pressure drop of 636 Pa, about 25 percent greater than the calculation shown above, but within the accuracy of Figure 9-5.

```
                  Sample output from AIRSTRIP program
Contaminant: Ethylbenzene           Concentration In: 1000.0 μg/L
C-Raschig Rings: 76.2 mm            Atmospheric Pressure: 101.3 kPa
Temperature: 20.0°C                 Liquid Loading Rate: 24.4 kg/m²sec
Min Packing Depth: 2.0 meter        Minimum A/W Ratio: 10.0
Max Packing Depth: 6.0 meter        Maximum A/W Ratio: 50.0
```

Effluent Concentration (μg/L)

Packing Depth (meter)	A/W = 10	A/W = 20	A/W = 30	A/W = 40	A/W = 50
2.0	359.1	310.6	292.5	282.9	276.8
3.0	231.1	180.2	162.7	153.7	148.2
4.0	152.1	105.8	91.2	84.0	79.7
5.0	101.5	62.5	51.3	46.1	30.0
6.0	68.4	37.1	29.0	25.3	23.2
R	2.7	5.4	8.1	10.8	13.5
dP (Pa/m)	<50	106	208	371	612

Limitations

Although widely employed, air stripping is not without its limitations. First, air stripping is simply the transfer of contaminants from one medium (water) to another (air). In some cases, the reduction in the contaminant concentrations in the groundwater provide the necessary protection to the public health while the increase of contaminants in the atmosphere may be inconsequential. In other cases, the air may require treatment by techniques such as activated carbon (see Section 9-2). Also, as discussed in the preceding portions of this section, air stripping is effective only for volatile and semivolatile compounds. Finally, stripping towers may be inclined to foul because of iron oxidation; growth of algae, bacteria, or fungi; and fine particulate matter in the water. This may necessitate pretreatment of the water or frequent cleaning of the tower.

9-2 CARBON ADSORPTION

Sorption is a process by which a component accumulates at a phase interface. The adsorbent most widely used in environmental applications is carbon that has been processed to significantly increase the internal surface area (activated carbon). Use of different raw materials and processing techniques results in a range of carbon types with different adsorption characteristics. Activated carbon is available in both powdered and granular form. Granular activated carbon (GAC) is most commonly used for removal of a wide range of toxic organic compounds from groundwater and industrial waste streams and is discussed in this section. Powdered activated carbon is more often used in biological treatment systems and is briefly discussed in Section 10-6.

Process Description

A typical activated carbon contactor is illustrated in Figures 9-6 and 9-7. Carbon is held in place by a plenum plate. Contaminated water enters the top of the column, is contacted with the carbon, and exits through an underdrain system at the bottom. Typical applications require facilities for air scouring and backwashing the carbon to avoid the buildup of head loss due to the accumulation of solid particles present in the influent. Additionally, the system must allow for the removal of spent carbon for regeneration and the addition of new carbon.

Carbon adsorption systems are typically continuous-flow columns, set up in series so that the final column in the system is in effect a polishing unit (see Figure 9-8). Because the downflow beds also act as filtration units, they must be periodically backwashed.

Other modifications of the process include downflow columns in parallel, upflow expanded-bed systems in series, and moving bed systems where new carbon is continually added to the system. The moving bed is operated in a counterflow mode wherein regenerated carbon is continuously added at the top and spent carbon is continuously removed at the bottom for regeneration, while the contaminated water enters at the bottom and exits at the top.

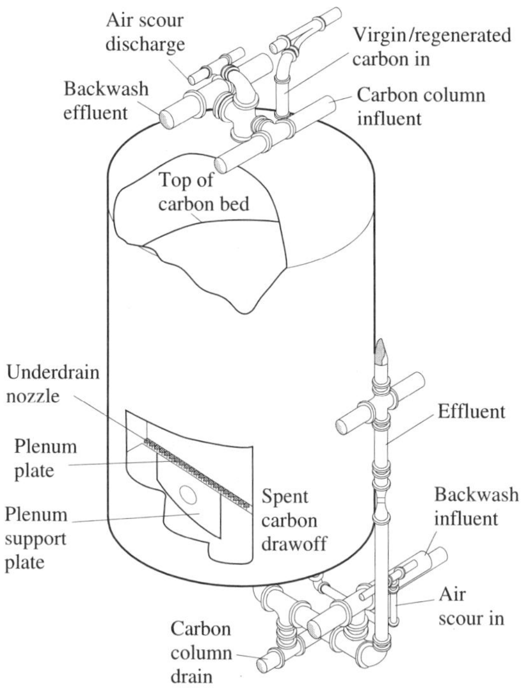

FIGURE 9-6
Schematic of carbon contactor.[14]

FIGURE 9-7
Carbon contactors. *(Courtesy of Calgon Carbon Corp.)*

Upflow expanded beds are used when the influent contains significant amounts of suspended matter. Use of an upflow mode eliminates the filtering of solids within the carbon system, which can be more efficiently removed with sand filters rather than carbon.

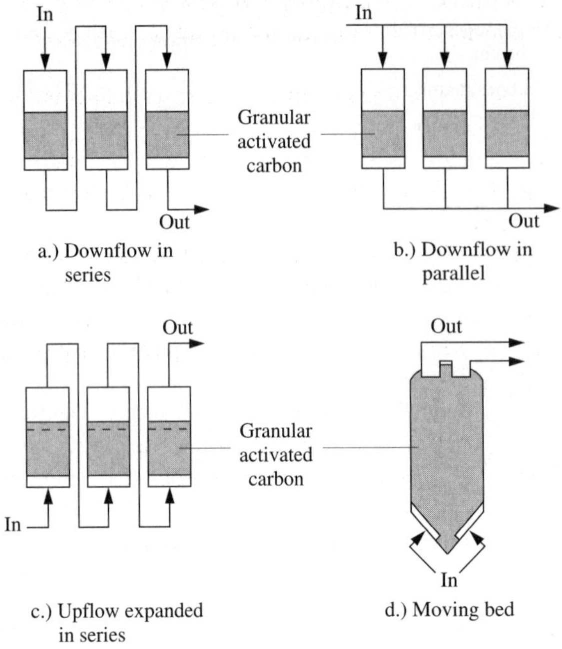

FIGURE 9-8
Schematic of activated carbon adsorption systems.[15]

Activated carbon adsorption is a well developed technology that is capable of effectively removing a wide range of soluble organic compounds. It is able to produce a very high-quality effluent, and is utilized in drinking water systems as well as for pollution control.

Adsorption Theory

Sorption is the process by which a component moves from one phase to another across some boundary. In adsorption, the process takes place at a surface. Examination of a cross section of an activated carbon particle, reveals a porous structure with a large internal surface area. As illustrated in Figure 9-9, movement of an or-

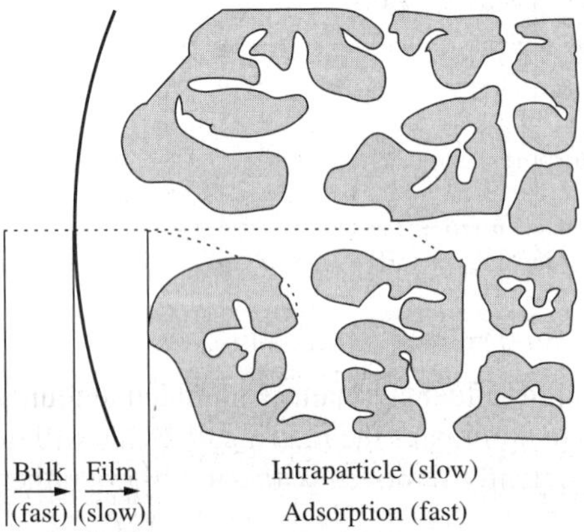

FIGURE 9-9
Transport processes.[16]

ganic and/or inorganic molecule to a surface site requires four separate phenomena: bulk fluid transport, film transport, intraparticle (pore and/or surface diffusion), and physical attachment.

Film and intraparticle transport are slow compared to other rate mechanisms and therefore are the rate-limiting steps that control design. The rate of diffusion tends to increase with increasing solute concentration and temperature. Where intraparticle diffusion controls, the rate decreases with increasing molecular weight of the solute and decreasing pore size of the carbon.

The driving forces that control adsorption include electrical attraction, a chemical affinity of the particular organic molecule for the adsorbent, van der Waal's forces (weak attractive forces acting between molecules), and the hydrophobic nature of the organic. A summary of some of the factors that affect the adsorption capacity of carbon for organic compounds is listed in Table 9-3.

A plot of the amount of contaminant adsorbed per unit mass of carbon q against the concentration of contaminant in the bulk fluid at constant temperature is referred to as an adsorption isotherm. There are several different mathematical forms of isotherms.

Where the sorbent surface is homogeneous, isotherm data may be approximated by the Langmuir equation:

$$q = abC/(1 + bC) \tag{9-22}$$

where a, b = constants. The Langmuir equation is developed by assuming that a fixed number of adsorption sites are available, and that the adsorption is reversible. The Langmuir equation can be linearized by plotting the inverse of q against the inverse of C.

The Freundlich isotherm (Equation 4-39) is an empirical model which can be expressed mathematically:

$$q = KC_f^{1/n} \tag{9-23}$$

TABLE 9-3
Factors affecting carbon adsorption

Factor	Effect
Solubility	Less soluble compounds are adsorbed to a greater extent than more soluble compounds.
Molecular structure	Branch-chain organics are more easily adsorbed than straight-chain organics.
Molecular weight	Larger molecules are generally more easily adsorbed than smaller molecules. However, when pore diffusion governs the adsorption process, the adsorption rate decreases with molecular weights above values specific to each type of carbon and within a given class of organics.
Polarity	Less polar (or weakly ionized) organics are more easily adsorbed than polar (or strongly ionized) organics.
Hydrocarbon saturation	Double- or triple-carbon bond (unsaturated) organics are adsorbed more easily than single-carbon bond (saturated) organics.

$$q = \text{mass of contaminant adsorbed (mg)} = \frac{(C_i - C_f) \times V}{\text{mass of carbon (mg)}} \tag{9-24}$$

where $V =$ volume of solution (L)
$C_i =$ initial concentration of contaminant in solution (mg/L)
$C_f =$ final concentration of contaminant in solution (mg/L)
$K, n =$ empirical constants (see Appendix B)

K and n can be determined from laboratory studies by a logarithmic plot, as illustrated in Example 9-4.

EXAMPLE 9-4. DETERMINING ISOTHERM CONSTANTS. 100 mL of a 600-mg/L solution of xylenes is placed in each of five containers with different amounts of activated carbon and shaken for 48 hours. The samples are filtered and the concentration of xylene measured, yielding the following analyses:

Container:	1	2	3	4	5
Carbon, mg:	600	400	300	200	50
Xylenes, mg/L:	25	99	212	310	510

Determine the Freundlich constants, K and n, and plot the isotherm. Initial contaminant $C_i = 600$ mg/L; Volume $V = 0.1$ L.

Solution

M, g	C_f, mg/L	$X = (C_i - C_f)V$, mg	q, mg/g	$\log_{10} q$	$\log_{10} C_f$
0.6	25	57.5	95.8	1.98	1.40
0.4	99	50.1	125.3	2.10	2.00
0.3	212	38.8	129.3	2.11	2.33
0.2	310	29.0	145.0	2.16	2.49
0.05	510	9.0	180.0	2.26	2.71

Freundlich constants are determined by linear regression on the last two columns: constant = 1.71; slope = 0.187.

$$K = 10^{1.71} = \underline{\underline{51.3 \text{ mg/g}}}$$

$$1/n = \underline{\underline{0.187}}$$

$$q = 51.3C_f^{0.187}$$

A plot of the adsorption isotherm for these data is shown here:

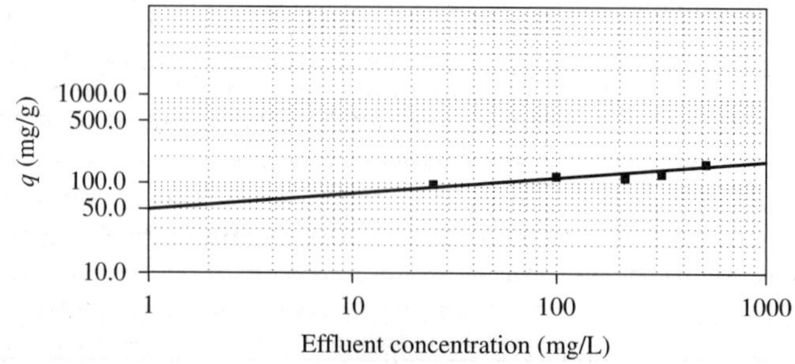

EXAMPLE 9-5. USE OF ADSORPTION ISOTHERM DATA. Use the isotherm data from Example 9-4 to determine the amount of carbon required to treat 10,000 gal/day of water contaminated with 600 mg/L xylenes. Assume a required effluent of 10 mg/L and that the facility will operate at the same temperature and pH for which the isotherm was developed. Determine daily carbon usage for a batch reactor.

Solution. After the required contact time, the carbon and solution should reach equilibrium with a concentration of xylenes of 10 mg/L. From the Freundlich equation developed in Example 9-4:

$$q = 51.3\, C_f^{0.187}$$

If $C_f = 10$ mg/L,

$$q = 78.9 \text{ mg xylenes/g carbon}$$

Carbon required = 0.0789 g/g = 0.0789 lb/lb
Xylenes removed = 10,000 gal/day × (600 − 10) mg/L × 10^{-6} L/mg × 8.34 lb/gal
= 49.21 lb xylenes per day
Carbon required:
(49.21 lb xylenes/day)/(0.0789 lb xylenes/lb carbon) = 624 lb/day

Design Considerations

While full-scale adsorbers will reach equilibrium at least near the inlet, they are not allowed to reach equilibrium along their entire bed length. This is because the adsorption zone (or mass transfer zone) must be contained in the column to maintain effluent quality. Therefore, the column is not fully exhausted at breakthrough. With carbon beds in series, the adsorption zone can be allowed to pass completely through the first bed in the series, thus more fully saturating the entire bed, and more closely approaching the conditions of an isotherm test.

The design of adsorption units requires column tests that simulate the actual operation of full-scale units. In the laboratory, small-diameter (e.g., 2-in-diameter) columns are filled with the carbon to be tested and the contaminated groundwater or wastewater is run through the columns. The column effluent is monitored for the contaminants of interest.

The volume within the carbon bed where adsorption takes place is referred to as the **adsorption zone** (also known as a mass transfer zone). As the carbon becomes exhausted (i.e., all of the adsorption sites are filled), the adsorption zone moves down the column. This is illustrated in Figure 9-10 where the effluent concentration from a column is plotted against time (or volume of water treated, V). At $V = V_1$, the effluent concentration is very small. A small portion of the carbon at the top is totally exhausted (dark shading) and the adsorption zone (light shading) is near the top of the column. At breakthrough, V_b, the adsorption zone reaches the bottom of the column. Breakthrough is typically defined as the point where a specified amount of the influent is detected in the effluent (usually 5 or 10 percent). After breakthrough the effluent concentration begins to rise rapidly, but the carbon in the adsorption zone at breakthrough is not yet exhausted. Operating columns in series allows complete exhaustion of the first carbon column without releasing significant amounts of contaminants in the effluent.

The results of laboratory column studies with three 5-cm- (2-in-) diameter columns in series, each containing a carbon depth of 2.3 m (7.5 ft), are plotted

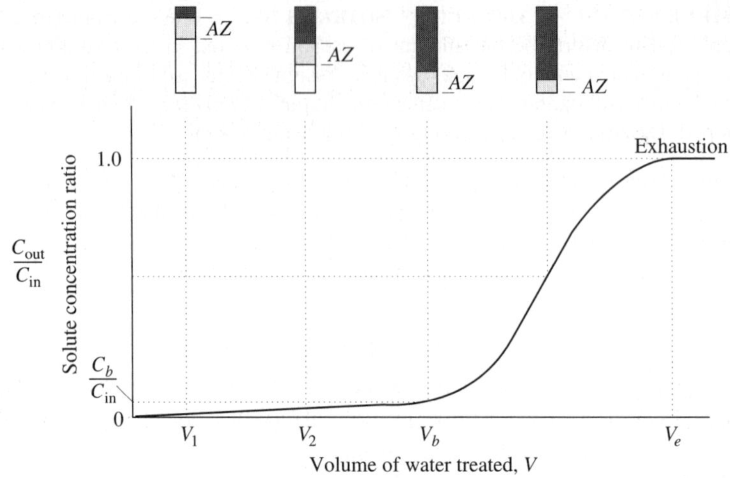

FIGURE 9-10
Carbon adsorption breakthrough curve showing movement of an adsorption zone.[17]

in Figure 9-11. The effluent concentration C_{out} is divided by the influent concentration C_{in} to provide the fractional breakthrough of contaminant versus time for each column. Breakthrough does not occur in the second column until the carbon in column 1 is almost fully exhausted.

Hutchins[18] presented a simplified way of examining such data by evaluating the bed-depth service time (BDST). A horizontal line is drawn through Figure 9-11 at $C_{out}/C_{in} = 0.9$, representing the time in each column where breakthrough is at 90 percent of the influent concentration. The line intersects the curve for column 1 at time = 22 days. Selected data for the three columns are presented in Table 9-4.

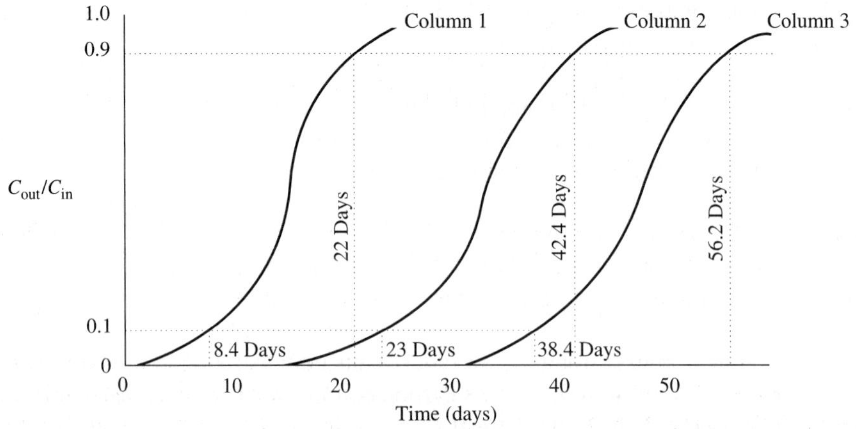

FIGURE 9-11
Laboratory breakthrough curves.

TABLE 9-4
Bed-depth service time

Column number	Cumulative depth, m	Service time (breakthrough), days, for percent of contaminant remaining 10%	60%	70%	80%	90%
1	2.3	8.4	12.3	13.8	14.5	22.0
2	4.6	23.0	27.7	29.3	30.8	42.4
3	6.9	38.4	41.2	43.3	44.8	56.2

Bed-depth service time curves are plotted as Figure 9-12. The lower line represents 90 percent removal of organics. This line can be represented by the Bohart-Adams[19] equation.

$$t = aX + b \tag{9-25}$$

where X = depth in column (m).

$$a = \frac{F_1 N}{C_{\text{in}} V} \tag{9-26}$$

where a = slope (h/m)

F_1 = conversion factor for units = 10^3 for metric units shown

= 1998 if N has units of lb/ft^3 and V has units of (gal/min)/ft^2

N = adsorptive capacity of carbon (mass of contaminant removed per volume of carbon in the column, kg/m^3)

C_{in} = influent contaminant concentration (mg/L)

V = superficial velocity through column (m/h) (m^3/h per m^2 of column)

$$b = (F_2/KC_{\text{in}}) \times \ln\left[(C_{\text{in}}/C_{\text{out}}) - 1\right] \tag{9-27}$$

where b = intercept (h)

F_2 = conversion factor = 10^3 for the units shown

= 16,025 if K has units of ft^3/(lb · h)

K = adsorption rate constant required to move an adsorption zone through the critical depth [m^3/(kg · h)] (m^3 of liquid treated per kg impurity fed per hour)

C_{out} = contaminant concentration at breakthrough (mg/L)

This technique provides carbon dosage based on a single fixed bed. The actual carbon usage rate will be lower for beds in series. At $t = 0$:

$$X(0) = (F_2/F_1) \times (V/NK) \times \ln\left[(C_{\text{in}}/C_{\text{out}}) - 1\right] \tag{9-28}$$

The abscissa intercept, $X(0)$, is the critical bed depth (i.e., the minimum to obtain satisfactory effluent at time zero). The ordinate intercept b measures the time required for an adsorption zone to pass through the critical depth. The slope term a provides a measure of the velocity of the adsorption zone (i.e., the speed at which carbon is exhausted). The velocity of the adsorption zone is $1/a$.

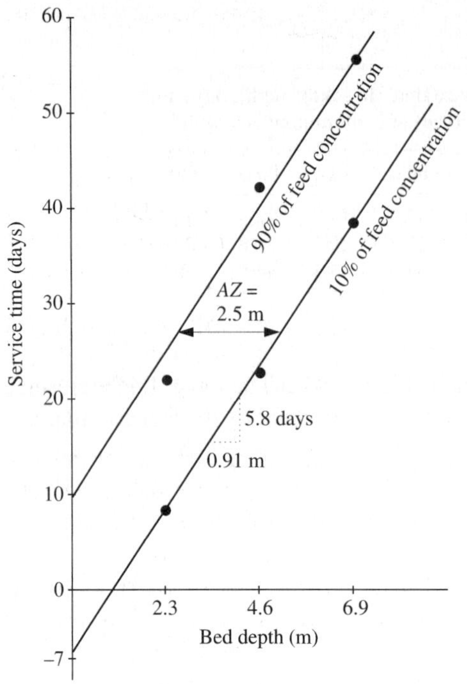

FIGURE 9-12
Bed-depth service time curves corresponding to Table 9-4.

The rate of carbon utilization may be determined by using this velocity:

$$\text{Carbon utilization} = \text{area} \times (1/a) \times \text{unit weight}$$

After a BDST curve has been established, Equation 9-25 can be modified to account for changes in flow rate:

$$t = a'X + b \tag{9-29}$$

$$a' = a(V/V') \tag{9-30}$$

where V' = the new flow rate (m^3/h per m^2 of column).

Similarly, the BDST equation is easily modified for changes in the feed concentration:

$$a' = a \times C_{\text{in}}/C'_{\text{in}} \tag{9-31}$$

$$b' = b \times \frac{C_{\text{in}}}{C'_{\text{in}}} \times \frac{\ln[(C'_{\text{in}}/C'_{\text{out}}) - 1]}{\ln[(C_{\text{in}}/C_{\text{out}}) - 1]} \tag{9-32}$$

where C'_{in} and C'_{out} represent the influent and effluent concentrations under the new operating conditions.

The horizontal distance between the curve representing exhaustion (90 percent in Figure 9-11) and breakthrough (10 percent in Figure 9-11) represents the height of an adsorption zone. Where a number of columns in series are used, the total number of columns is related to the height of an adsorption zone by:

$$n = (\text{AZ}/d) + 1 \tag{9-33}$$

where n = number of columns in series (rounded up to a whole number)
AZ = height of an adsorption zone
d = height of a single column

EXAMPLE 9-6. PRELIMINARY DESIGN OF CARBON ADSORPTION SYSTEM. A groundwater flowing at a rate of 0.145 m^3/min (55,000 gal/day) requires treatment to reduce organic concentration from 89 mg/L to 8.9 mg/L (90 percent removal). Laboratory studies are run in columns 2.3 m (7.5 ft) long by 0.051 m (2 in) diameter at a flow rate of 0.5 L/min (190.2 gal/day). The results are shown in Figures 9-11 and 9-12. Assume a unit weight of carbon of 481 kg/m^3 (30 lb/ft^3).

Determine: the height of the adsorption zone, number and size of columns required, and the carbon usage rate.

Solution

1. Height of adsorption zone, AZ = 2.5 m (8.25 ft) (from Figure 9-12)
2. Number and size of units: $n = (\mathrm{AZ}/d) + 1 = (2.5/2.3) + 1 = 2.09$; therefore, number of units = 3 columns.

   Area of lab columns = 2.043×10^{-3} m^2 (3.14 in^2)
   Loading rate in laboratory columns:

$$V = (5.0 \times 10^{-4}\ \mathrm{m^3/min})/2.043 \times 10^{-3}\ \mathrm{m^2}$$
$$= 0.245\ \mathrm{m^3/(m^2 \cdot min)}\ [6.14\ \mathrm{gal/(ft^2 \cdot min)}]$$

   Using the same loading rate for the full scale unit yields:

$$\text{Area} = Q/V = \frac{0.145\ \mathrm{m^3/min}}{0.245\ \mathrm{m^3/(m^2 \cdot min)}} = 0.59\ \mathrm{m^2}\ (6.32\ \mathrm{ft^2})$$

   Diameter = 0.867 m (2.84 ft)
3. BDST equation for 90 percent removal (use curve for 10 percent of feed concentration in Figure 9-11)
   Slope a = 5.8 days/0.91 m = 6.34 days/m (1.93 days/ft)
   Intercept b: −7 days
   Equation of line: $t = 6.34X - 7$
   Velocity of adsorption zone = $1/a$ = 0.158 m/day (0.52 ft/day)
   Carbon utilization = area × $(1/a)$ × unit weight = 0.59 m^2 × 0.158 m/day × 481 kg/m^3 = 44.8 kg/day (98.7 lb/day)

Adsorption Kinetics

To predict the shape and velocity of the active adsorption zone, the kinetics of adsorption must be characterized. As previously discussed, the rate of adsorption is controlled by film and intraparticle mass transfer, and this is why the active adsorption zone is also called the mass transfer zone (MTZ). Mass transfer occurs by two diffusion steps in series: (1) diffusion from the bulk solution across a hydrodynamic boundary layer ("film transport") to the particle surface and (2) intraparticle diffusion from the surface toward the center of the particle. Intraparticle diffusion may be modeled as pore diffusion, surface diffusion, or both. Pore diffusion refers to the diffusion of dissolved solute in the adsorbent pore space, while surface diffusion

refers to diffusion of adsorbed solute along the pore surface. The rate of diffusion depends on the magnitude of the diffusion coefficient and on the spatial scale over which diffusion occurs.

The shape of the mass-transfer zone is related to these diffusion mechanisms, and dictates the minimum bed length required to maintain a satisfactory effluent concentration. The MTZ moves through the reactor at a velocity related to the mass loading and the equilibrium uptake of the adsorbent. Knowledge of the MTZ length, velocity, and reactor empty-bed contact time is needed to make estimates of adsorbent use rates and cost.

Adsorption rate parameters include the film mass transfer coefficient, which incorporates the free-liquid diffusivity, the hydrodynamic boundary layer thickness, and the intraparticle diffusion coefficient. These parameters can be obtained from breakthrough curve experiments described previously. The resulting "breakthrough curve" can be analyzed in several ways; one such way, using the Bohart-Adams approach, has been described in detail. Breakthrough curve data directly yield information about the MTZ length and reactor efficiency for the loading and flow conditions of the experiment. By appropriate design of the laboratory reactor using scaling equations, the breakthrough curve (plotted on dimensionless coordinates) will represent the same reactor performance as a full-scale reactor operated under the same mass loading and flow conditions. In addition, breakthrough curve data measured in the laboratory will be fitted with a mass-transfer model to estimate film and intraparticle diffusion coefficients. Mass-transfer-based mathematical models calibrated in such a manner have been shown to have wide applicability in describing and predicting the behavior of sorption processes, and the calibrated model can be used to investigate alternative design and process configurations to optimize adsorbent use.

Estimation of Adsorption Rate Parameters

The film mass transfer coefficient k_f can be independently estimated by using correlations for packed beds and can be measured in the laboratory by using the short-bed adsorber technique of Weber and Liu.[20,21] The value of k_f depends on adsorbate free-liquid diffusivity and the thickness of the hydrodynamic boundary layer. The boundary layer thickness is governed, in turn, by the particle size, the relative velocity between the particle and the bulk fluid, and the fluid viscosity. The four parameters that govern k_f are generally incorporated in two dimensionless numbers, the Reynolds number N_{Re}, which is used to characterize hydrodynamic conditions, and the Schmidt number N_{Sc}, which is used to characterize the relative importance of mass and momentum diffusivity. The Reynolds number is computed from:

$$N_{\text{Re}} = \frac{vd}{\upsilon} \tag{9-34}$$

where v = relative velocity between the particle and the bulk fluid (interstitial velocity)

d = particle diameter

υ = kinematic viscosity

The Schmidt number is the ratio of momentum to mass diffusivity and is defined as:

$$N_{\mathrm{Sc}} = \frac{v}{D_L} \tag{9-35}$$

Computation of N_{Sc} requires estimates of D_L, the free-liquid diffusivity. The Gnielinski correlation was recommended by Roberts et al. (1985) to estimate k_f:

$$\frac{k_f d}{D_L} = \psi \left(2 + 0.644 N_{\mathrm{Re}}^{1/2} N_{\mathrm{Sc}}^{1/3}\right)[1 + 1.5(1 - \varepsilon)] \tag{9-36}$$

where ε is the bed porosity and ψ is an experimentally determined shape factor on the order of 2. Such correlations may be calibrated by using film mass-transfer coefficients determined directly from laboratory measurements by the short-bed adsorber technique.

The short-bed adsorber technique involves performing a breakthrough test using a bed length of adsorbent shorter than the mass-transfer zone length. Under such conditions, the reactor will exhibit immediate breakthrough, and the initial effluent concentration C_{out} will be related to the product of the film mass transfer coefficient and the interfacial area available for mass transfer:

$$k_f a = -\frac{Q\rho_b}{M} \ln \frac{C_{\mathrm{out}}}{C_{\mathrm{in}}} \tag{9-37}$$

where $a \cong$ particle surface area per unit volume ($\mathrm{m}^2/\mathrm{m}^3$)
$Q =$ flow rate
$\rho_b =$ bed density
$M =$ mass of adsorbent

Using this technique, k_f can be determined for a range of particle sizes and flow rates.

In contrast to the film mass transfer coefficient, intraparticle diffusion coefficients are intrinsic characteristics of the adsorbate and the adsorbent, and do not depend on the hydrodynamic conditions at the particle surface. Because pore size restrictions and chemical interactions between adsorbate and adsorbent may affect the measured intraparticle diffusion rate, reliable correlations for estimating intraparticle diffusivities are not available. Therefore, estimates for these parameters must be made from measured breakthrough curves.

To estimate intraparticle diffusion coefficients, breakthrough curve data are fit with a mathematical model. One such mass transfer model is written as follows. The liquid phase mass balance is:

$$\frac{\partial C}{\partial t} = -v\frac{\partial C}{\partial x} + D_h \frac{\partial^2 C}{\partial x^2} - \frac{3(1 - \varepsilon)k_f}{\varepsilon R}(C - C_s) \tag{9-38}$$

where $C =$ bulk liquid phase concentration, ML^{-3}
$v =$ liquid velocity, LT^{-1}
$D_h =$ hydrodynamic dispersion coefficient, $\mathrm{L}^2\mathrm{T}^{-1}$
$\rho_s =$ solid phase density

Note that the rate of film transport, which appears as a sink in the liquid-phase mass

balance, must equal the rate of uptake by the solid phase:

$$\frac{3(1-\varepsilon)k_f}{\varepsilon R}(C - C_s) = \frac{\rho_s(1-\varepsilon)}{\varepsilon}\frac{\partial q_{\text{ave}}}{\partial t} \tag{9-39}$$

where R = carbon particle radius, L
C_s = liquid-phase concentration at the carbon surface, ML^{-3}
q_{ave} = average solid phase concentration

The isotherm is used to relate C_s and q_{ave}, and thus is an integral component of a mass transfer model. The solid phase mass balance is written:

$$\varepsilon_s\frac{\partial C_p}{\partial t} + \rho_s\frac{\partial q_{\text{ave}}}{\partial t} = \frac{D_p\varepsilon_s}{r^2}\frac{\partial}{\partial r}\left[r^2\frac{\partial C_p}{\partial r}\right] + \frac{D_s\rho_s}{r^2}\frac{\partial}{\partial r}\left[r^2\frac{\partial q_{\text{ave}}}{\partial r}\right] \tag{9-40}$$

where C_p = liquid phase concentration in the carbon pores, ML^{-3}
ε_s = solid phase porosity
D_p = pore diffusion coefficient, L^2T^{-1}
D_s = surface diffusion coefficient, L^2T^{-1}

The solid-phase mass balance, as written, includes both surface and pore diffusion. When data are fit with a pore diffusion model and the diffusion coefficient required to fit the data is greater than the free-liquid diffusivity, it is generally assumed that surface diffusion makes a significant contribution to intraparticle flux. In practice, one of these mechanisms is usually neglected. It is often found that the pore diffusion model (PDM) and the homogeneous surface diffusion model (HSDM) fit breakthrough curves equally well. The coupled partial differential equations describing mass transfer controlled adsorption must be solved by numerical methods. Simplified analytical techniques have been developed, and computer packages to solve these equations are available commercially.

9-3 STEAM STRIPPING*

Process Description

Steam stripping is utilized for the removal of volatile and sometimes semivolatile compounds from groundwater or wastewater. This process is capable of reducing volatile organic compounds (VOCs) in water to very low concentrations. Table 9-5 shows the low treatment limits that the U.S. EPA established for the Organic Chemicals, Plastics, and Synthetic Fibers (OCPSF) category, based on the use of steam stripping as the model technology.[22,23] In some cases, discharge VOC concentrations from steam strippers can approach or be below analytical detection limits.

Both steam and air strippers are based on the transfer of organics from the liquid phase to the gas. However, as discussed in this section, the higher concentration of organics in a steam stripper requires more complex process design techniques

*Section 9-3 was prepared by David Potter, P.E., Science Applications International Corp.

TABLE 9-5
Best Available Technology Economically Achievable (BATEA) effluent limitations set for OCPSF products and production based on steam stripping technology (40 CFR Part 414.101)

Contaminant	Maximum concentration for any one day, μg/L	Maximum concentration for monthly average, μg/L
Benzene	134	57
Carbon tetrachloride	380	142
Chlorobenzene	380	142
1,2,4-Trichlorobenzene	794	196
Hexachlorobenzene	794	196
1,2-Dichloroethane	574	180
1,1,1-Trichloroethane	59	22
Hexachloroethane	794	196
1,1-Dichloroethane	59	22
1,1,2-Trichloroethane	127	32
Chloroethane	295	110
Chloroform	325	111
1,2-Dichlorobenzene	794	196
1,3-Dichlorobenzene	380	142
1,4-Dichlorobenzene	380	142
1,1-Dichloroethylene	60	22
1,2-trans-dichloroethylene	66	25
1,2-Dichloropropane	794	196
1,3-Dichloropropylene	794	196
Ethylbenzene	380	142
Methylene chloride	170	36
Methyl chloride	295	110
Hexachlorobutadiene	380	142
Tetrachloroethylene	164	52
Toluene	74	28
Trichloroethylene	69	26
Vinyl chloride	172	97

than for an air stripper. The functional differences between steam stripping and air stripping are:

- Steam, rather than air, is used as the stripping gas.
- The stripping gas, steam, is infinitely soluble in the liquid phase, water.
- Steam strippers operate at much higher temperatures than air strippers.
- The organics in the water may be recovered as a separate liquid phase.

A schematic diagram of a steam stripper operating at atmospheric pressure is presented in Figure 9-13, and a schematic diagram of a steam stripper operating under vacuum is presented in Figure 9-14.

In both Figures 9-13 and 9-14, the bottom portion of the steam stripper is stripping wastewater in a manner similar to an air stripper. This is known as the stripping section of the column. The top section of the column, above the feed point, is known as the rectifying section. This section of the column enriches the organic content of the steam to a point where a separate organic phase can be achieved in the overhead decanter. The steam is enriched because it is (theoretically) in equilibrium with the saturated liquid that is fed to the top of the column. The combination of the strip-

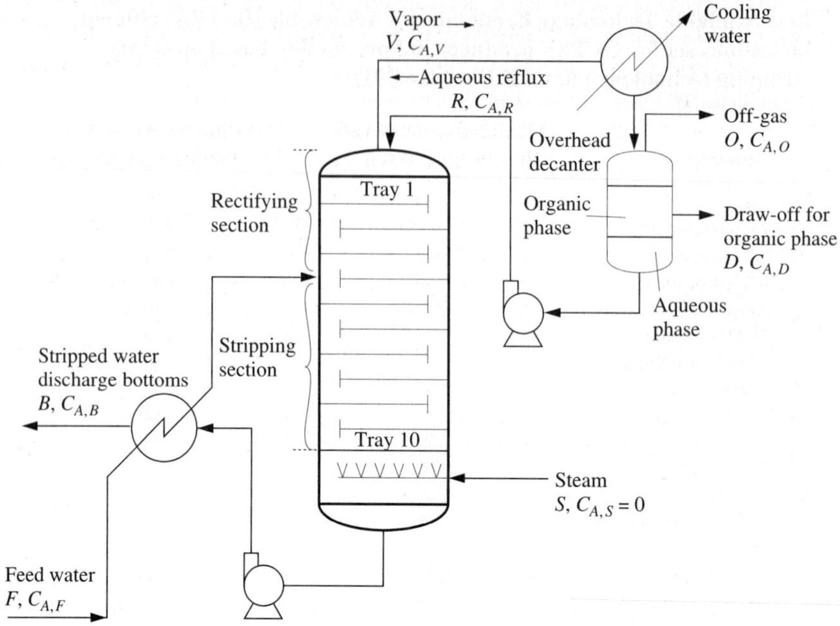

FIGURE 9-13
Atmospheric pressure steam stripping column.

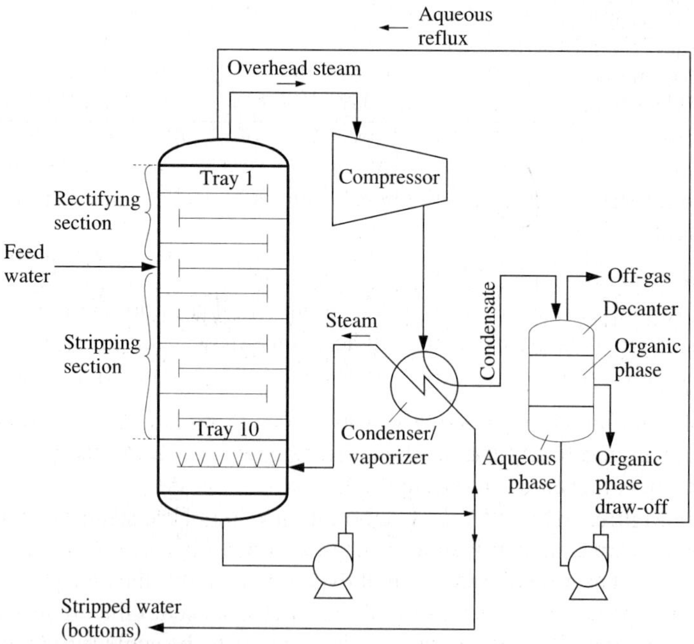

FIGURE 9-14
Vacuum steam stripper.

ping and rectifying sections is a mass transfer process known as distillation in the chemical process industry. Distillation has many other production applications in the chemical industry in addition to its specialized application to wastewater treatment as steam stripping.

In Figure 9-13, the contaminated water feed to the steam stripper is preheated to near-boiling temperature by exchanging heat with the stripped water exiting from the bottom of the stripping column. The contaminated water enters the column at the feed point, and the water flows downward through the stripping section of the column. Steam passes countercurrently up through thc column. The column operates at a temperature that is slightly higher than the normal (1 atmosphere) boiling point of water, usually in the range of 215° to 220°F. The temperature difference between the top and the bottom of the column is modest, generally on the order of a few degrees Fahrenheit.

At the elevated temperature inside the column, the volatile organics in the water exert a higher vapor pressure than at ambient conditions. As the organics vaporize in the column, they are transferred from the liquid phase into the gas phase. As the steam travels up the column, the concentration of organics increases in the stripping steam.

The steam exits the top of the column where it undergoes a phase change to a liquid in the overhead condenser. (Condenser is a term describing a heat exchanger where a gas is cooled to effect a change to the liquid phase.) This liquid is supersaturated with the organics, so a separate organic layer forms in the decanter. This organic phase can have a density greater or less than the aqueous phase in the decanter. In this phase, the organics will either float or sink in the decanter and can be separated. In Figure 9-13 this is shown as a floating "organic phase."

The aqueous phase from the decanter is returned to the top of the column where it flows down the column. This aqueous phase is saturated with the organics because the organic concentration in the aqueous phase is in equilibrium with the separate organic phase in the decanter. An analysis of phase-equilibria thermodynamics and a mass balance will show that, because this feed to the top of the column is saturated with the organics, the steam leaving the top of the column will contain a high enough concentration of organics to form a separate "organic" phase when the steam is condensed to a liquid phase (see Example 9-7).

Figure 9-14 shows a steam stripper operating under vacuum conditions in the column. A vacuum stripper has a stripping section and a rectifying section and many similarities to an atmospheric stripper. However, because it is operating at reduced pressure, the boiling point of water is reduced. Consequently, the entire vacuum stripping column operates at a lower temperature than an atmospheric stripping column. Typical operating values are 140° to 180°F and 0.2 to 0.5 atmosphere.

In a vacuum steam stripper, the overhead steam is mechanically compressed to an elevated pressure that will cause the gas to condense to a liquid. The mechanical compression of the gas also creates a vacuum on the compressor inlet side that, in turn, creates vacuum conditions in the steam stripper. The mechanical compression of a gas gives off a great deal of heat that is recovered by vaporizing part of the bottoms flow to create the stripping steam. The overhead mixture is decanted into separate organic and aqueous phases as in an atmospheric pressure stripper, and the aqueous phase is returned to the top of the stripping column as in an atmospheric stripper.

Steam strippers can also operate at elevated pressures. The operating temperature of the stripper will be higher than the normal boiling point of water because the column is under pressure. Operation of a pressurized steam stripper is relatively common in refineries where hydrogen sulfide and ammonia are stripped from the wastewater before it is discharged to a biological wastewater treatment plant. However, use of pressurized steam stripping is not a normal practice outside the petroleum refining industry because the higher pressure complicates the operation of a steam stripper.

A steam stripper could be applied to water containing volatile organic compounds that are highly soluble in water (e.g., acetone, ethanol). In such an application, a separate organic phase would not form in the overhead decanter. The purpose of such an application would be to increase the concentration of VOCs in the water and thereby reduce the volume of water for a subsequent treatment unit. Although this is theoretically sound, as a practical matter the extra capital and O&M costs of a steam stripper outweigh the benefits gained by reducing the volumetric rate to downstream unit operations.

Design Theory

Development of column design theory occupies several chapters in mass transfer textbooks.[24,25] Consequently, the detailed process design of steam strippers is a complicated mass transfer problem that requires engineers who are experienced in column design. However, an understanding of process design fundamentals can be applied to engineering feasibility studies without the benefit of detailed process design experience. In this section, the basic design theory underlying design of distillation columns is presented on a conceptual level, including:

- Thermodynamics
- Material balance
- Mass transfer
- Process design selection

These subsections are discussed on a conceptual basis to provide familiarity with the process design principles and practices involved in column design. A list of fundamental chemical engineering references is provided at the end of this chapter for more detailed discussion of these topics.[26,27,28,29,30]

Thermodynamics

The conceptual principles of the complex liquid-vapor equilibrium and phase separation processes involved in the design of a steam stripper are presented in this section. The fundamental differences and similarities between liquid-vapor equilibrium design for steam strippers and air strippers should then be readily apparent.

Up to this point, the equilibrium between an organic in the vapor phase and an organic in the liquid phase has been represented by a simple Henry law model (Equation 3-5):

$$p_g = HC_l \tag{9-41}$$

where p_g = partial pressure of gas
H = Henry's constant
C_l = concentration in liquid

However, Henry's constant is applicable to a limited temperature range (usually 0 to 35°C), to simple binary systems (usually water and a single organic), and to very dilute systems. These simplifications are not valid within a steam stripper. In an air stripper, the feedwater concentration is generally less than 100 mg/L, and the concentration of organics decreases in the liquid phase as the water is processed through the air stripper. In contrast, the feedwater to a steam stripper generally is 100 mg/L or more, and the concentration of organics in the liquid phase increases above the feed point. Consequently, the assumption of a very dilute system is not valid in a steam stripping column. This means the interaction between various organics in the system must be considered in estimating liquid-vapor equilibrium.

Activity coefficient models were developed to consider such interactions among components in a mixture. In fact, Equation 3-5 is a generally accepted simplification of the more complex activity coefficient model of liquid-vapor equilibrium:

$$p_{g,A} = \frac{a_{A,B}}{\delta_{A,B}} X_A P_{vp,A} \tag{9-42}$$

where $p_{g,A}$ = partial pressure of component A above the specified system of A and B
$a_{A,B}$ = activity coefficient of component A in component B (unitless)
$\delta_{A,B}$ = fugacity coefficient of gas A at the system total pressure (unitless)
X_A = mole fraction of component A in the mixture of A and B
$P_{vp,A}$ = saturation vapor pressure of component A at system temperature

The fugacity coefficient and the activity coefficient in Equation 9-41 represent non-ideal gas and liquid behavior, respectively. In a steam stripper, the modest operating pressures usually result in a fugacity coefficient near 1.

Several models exist for estimating the activity coefficient for a given system. Two examples are the nonrandom two liquid (NRTL) model and the universal quasi-chemical (UNIQUAC) model, both of which are generally available on computerized thermodynamics packages.[31] To develop activity coefficient models, the mutual solubility of two or more components is experimentally measured. These data are then reduced to models that utilize other thermodynamic properties, mole fractions, and absolute temperature.

When more than two components are present in a mixture, the binary activity coefficients between all components are calculated by using published data and models such at NRTL or UNIQUAC. An additional set of correlations is utilized to account for the interactions between the various components in the mixtures that can increase or decrease the binary activity coefficients. Finally, a special case activity coefficient is utilized to estimate the activity coefficient at infinite dilution. This infinite dilution activity coefficient is not applicable in the overhead decanter, where a separate organic phase is formed. In the overhead decanter, a separate set of liquid-liquid activity coefficients must be estimated to calculate the expected phase separation.

There is much fundamental data available to determine activity coefficients for binary organic/water mixtures as well as data for ternary and quaternary water/organic mixtures. The generally accepted published source for this data is the Dechema series.[32,33,34] In addition, many companies and vendors in the chemical process industries have their own internal databases developed for materials they process. These databases are developed from empirically measured parameters, but they are not published because they represent proprietary process information. Models to account for multicomponent interactions and to estimate activity coefficients are available on the open market, and there are a number of in-house modifications of these models that are also considered proprietary information in the chemical process industry.

An activity coefficient model can also be used to predict the separation of the "organic phase" and the "aqueous phase" in the overhead decanter. This would, however, be a liquid-liquid equilibrium model. Ideally, the composition of the organic and aqueous phases would be a simple phase-equilibrium calculation based on the solubility of organics in water. However, in practice there is some water in the organic phase. Cosolvent effects can increase the solubility of a given organic in water, and a strong hydrophobic organic can extract highly water soluble organics from the aqueous phase.[35] The proper way to model this separation is very similar to the above methods discussed for liquid-vapor equilibria. This includes a combination of measured water-organic and organic-organic activity coefficients[36] and computer modeling of the component interactions in a multicomponent two-phase liquid mixture. The NRTL and UNIQUAC models are often used for steam stripper design because these methods support estimation of both liquid-vapor and liquid-liquid equilibria.

Material Balance

A schematic diagram of an atmospheric steam stripper is presented in Figure 9-13, with notation for a material balance. For purposes of illustration only, this discussion will assume that the feed wastewater is a simple binary water-organic mixture. As previously noted, this is not the case for most applications of steam strippers. The student can develop a more complex material balance, as needed, using these same principles.

The material balance consists of two balances—one for the organic and one for water. Because there are no chemical reactions taking place in the steam stripper, the accumulation and generation terms are zero for both balances. Using the generalized material balance approach for nonreacting steady-state processes presented in Equation 3-48, the material balance for the organic component A is expressed as:

$$\text{Input} = \text{output}$$
$$FC_{A,F} = BC_{A,B} + DC_{A,D} + OC_{A,O} \tag{9-43}$$

where F = feedwater mass rate (lb/h)

$C_{A,_}$ = concentration of component A in the various streams (weight percent)

B = bottoms discharge mass rate (lb/h)

D = overhead organic phase withdrawal mass rate (lb/h)

O = offgas (from overhead decanter) mass rate (lb/h)

Equation 9-43 can be simplified by the recognition that the concentration of component A out through the bottoms discharge and the mass rate of stream O are almost always negligible terms in this mass balance:

$$\begin{aligned} FC_{A,F} &= B\underset{\text{neg.}}{\not{C}_{A,B}} + DC_{A,D} + \underset{\text{neg.}}{\not{O}}C_{A,O} \\ &= DC_{A,D} \end{aligned} \tag{9-44}$$

Example 9-7 shows the basis for considering the $C_{A,B}$ and O terms to be negligible. Although this is almost always true, the assumptions presented in Example 9-7 should be confirmed before using the simplified form of Equation 9-42. It is also worth noting that the overhead phase withdrawn from the decanter is not assumed to be a pure organic phase ($C_{A,D} \neq 100.0$). In practice, this organic phase can contain a very high dissolved water fraction, up to 50 percent.

EXAMPLE 9-7. Demonstrate the basis for the negligible terms in Equation 9-44. Discuss limitations of this simplified water mass balance.

Given:
1. Nonreactive system described in Figure 9-13
2. Feed organics = 230 mg/L
3. Organics removal efficiency = 99.4 percent
4. Decanter temperature = 35°C

Solution. The generalized water mass balance on the nonreactive system described in Figure 9-13:

$$\text{Input} = \text{output}$$

$$F(1 - C_{A,F}) + S(1 - C_{A,S}) = D(1 - C_{A,D}) + O(1 - C_{A,O}) + B(1 - C_{A,B})$$

terms which may be considered negligible are:

$C_{A,F}$ is 230 mg/L or 2.30×10^{-4} weight fraction. So $1 - C_{A,F} = 0.999770$. Within the accuracy of these types of engineering calculations, at most three figures are significant. Therefore, 0.999770 is rounded to 1.00 and $C_{A,F}$ is negligible in terms of the water mass balance calculations.

$C_{A,S}$ is theoretically zero. In practice, plant steam will contain less than 10 mg/L of water treatment chemicals. By the same logic applied to $C_{A,F}$, $C_{A,S}$ may then be considered negligible.

$C_{A,B}$ may be calculated as (1 − removal efficiency/100)$C_{A,F}$ = (1 − 0.994)(230 mg/L) = 1.38 mg/L = $C_{A,B}$. $C_{A,B}$ may then be considered negligible since as a weight fraction $C_{A,B} << 1$. (This is the same logic as used for $C_{A,F}$.)

O is the overhead gas rate. This gas is almost entirely composed of noncondensible gases (which are oxygen and nitrogen). The noncondensibles come from air dissolved in the feedwater. In Example 9-8 it is shown that the solubility of air in water is 0.0022 wt% at 25°C and atmospheric conditions. Therefore, the first estimate of O is based on noncondensible flow only:

$$O_{\text{NC}} = (0.0022/100)F = 2.2 \times 10^{-5} F$$

The noncondensible gases flow over the liquid surface in the decanter and then to the atmosphere. This liquid surface can be the organic layer or the aqueous layer, depending on the relative densities of the aqueous and organic layers.

The highest water content in the noncondensible gas flow occurs if the top layer is a pure water layer. This case is assumed for the rest of this discussion. At 35°C, pure water exerts a vapor pressure of 91 mmHg or 0.12 atmosphere. The total flow of water

and noncondensible gas is:

$$O_{\text{tot}} = O_{\text{water}} + O_{\text{NC}}$$

and in terms of pressure:

$$\begin{aligned} P_{\text{tot}} &= 1 \text{ atm} = P_{\text{water}} + P_{\text{NC}} \\ &= 0.12 \text{ atm} + P_{\text{NC}} \\ P_{\text{NC}} &= 0.88 \text{ atm} \end{aligned}$$

Therefore, since

$$\frac{O_{\text{NC}}}{P_{\text{NC}}} = \frac{O_{\text{water}}}{P_{\text{water}}}$$

then

$$\begin{aligned} O_{\text{water}} &= O_{\text{NC}}(0.12/0.88) \\ &= 0.136 O_{\text{NC}} \\ O_{\text{tot}} &= O_{\text{water}} + O_{\text{NC}} = 0.136 O_{\text{NC}} + O_{\text{NC}} \\ &= 1.14 O_{\text{NC}} \end{aligned}$$

As discussed above, $O_{\text{NC}} = 2.2 \times 10^{-5} F$. So

$$\begin{aligned} O_{\text{tot}} &= 1.14\,(2.2 \times 10^{-5} F) \\ &= 2.51 \times 10^{-5} F \end{aligned}$$

Since $O_{\text{tot}} << F$, O_{tot} may be treated as negligible in the water mass balance calculations.

D is the flow of the organic layer, which often contains a significant water fraction. For purposes of this calculation, assume a very high water fraction in the organic layer, e.g., 50 percent water by weight.

The organic layer total mass flow is composed of an organic flow and a water flow:

$$D_{\text{tot}} = D_{\text{water}} + D_{\text{organic}}$$

The pure organic portion is the outlet term of an organic mass balance. For simplification, assume 100 percent of the organics in the feedwater go into the organic decant layer:

$$\begin{aligned} D_{\text{organic}} &= (C_{A,F})F \\ &= (2.30 \times 10^{-4})F \\ D_{\text{water}} &= (0.50/0.50)\,D_{\text{organic}} \\ &= D_{\text{organic}} \\ D_{\text{total}} &= D_{\text{water}} + D_{\text{organic}} \\ &= D_{\text{organic}} + D_{\text{organic}} \\ &= 2D_{\text{organic}} \\ &= 2(2.30 \times 10^{-4})F \\ &= 4.60 \times 10^{-4} F \end{aligned}$$

Conclusion: $D_{\text{total}} << F$ and therefore may be treated as negligible in the overall water mass balance.

Using the generalized material balance approach presented in Equation 3-24, the material balance for water is expressed as:

$$F\left(1 - C_{A,F}\right) + S\left(1 - C_{A,S}\right) = D\left(1 - C_{A,D}\right) + B\left(1 - C_{A,B}\right) + O\left(1 - C_{A,O}\right) \tag{9-45}$$

where F = feedwater mass rate (lb/h)

$1 - C_{A,_}$ = concentration of water in the various streams (weight percent)

$S =$ steam mass rate (lb/h)
$D =$ overhead organic phase withdrawal mass rate (lb/h)
$B =$ bottoms discharge mass rate (lb/h)
$O =$ offgas (from overhead decanter) mass rate (lb/h)

This generalized material balance can be simplified by recognizing that the concentration of water is near unity for the steam, feed, and bottoms streams, and that the water mass rate of the organic phase withdrawal and the offgas is a negligible term in the water mass balance (demonstrated in Example 9-7):

$$F(1 - \underset{\text{neg.}}{\not{C}_{A,F}}) + S(1 - \underset{\text{neg.}}{\not{C}_{A,S}}) = \underset{\text{neg.}}{\not{D}}(1 - C_{A,D}) + B(1 - \underset{\text{neg.}}{\not{C}_{A,B}}) + \underset{\text{neg.}}{\not{O}}(1 - C_{A,O})$$

$$F + S = B \tag{9-46}$$

As can be confirmed from this balance, essentially all of the condensed water vapor from the steam used in the stripping column ends up in the bottoms discharge flow. This can easily result in a net increase of 10 percent in the water flow rate after water is treated in a stripping column. This must be considered in sizing downstream equipment or subsequent treatment processes.

The above simplified material balances should be used for purposes of estimating material balances around a steam stripper. The exact form of these equations is used for purposes of detailed process design. The concentration of organic A at various points in the process is often negligible in terms of the overall material balance, but these concentration terms are essential for calculation of mass transfer rates and in the detailed process design of a steam stripper.

EXAMPLE 9-8. AIR SOLUBILITY. Calculate the weight percent of air that is soluble in water at 25°C, 1 atm, and equilibrium conditions. Assume (1) ideal gas behavior; (2) ideal liquid behavior.

Solution. Ideal gas and liquid behavior is assumed, so the Henry law expression is applicable:

$$P = HX$$

where $P =$ partial pressure (atm)
$H =$ Henry's constant (atm)
$X =$ mole fraction solute in solvent

According to Perry's *Chemical Engineers Handbook*, 5th edition, pp. 3–96: for air/water, at 25°C, $H = 7.20 \times 10^4$ atm.

$$\begin{aligned} X &= P/H \\ &= 1 \text{ atm}/7.20 \times 10^4 \text{ atm} \\ &= 1.39 \times 10^{-5} \text{ moles air/mole water} \end{aligned}$$

$$\begin{aligned} \text{wt \% air} &= (\text{mole fraction air/water}) \left(\frac{\text{MW air}}{\text{MW water}} \right) \times 100\% \\ &= (1.39 \times 10^{-5}) \left(\frac{\text{moles air}}{\text{moles water}} \right) \left(\frac{28.9 \text{ g air/mole air}}{18 \text{ g water/mole water}} \right) \times 100\% \\ &= 0.0022\% \end{aligned}$$

Conclusion: Air is nearly insoluble in water.

Mass Transfer

Mass transfer design of a steam stripper is fundamentally different from the mass transfer design of an air stripper. These different design approaches are a result of the different organic concentrations in an air stripper and a steam stripper as well as the different solubilities of the stripping gases in water.

In the case of an air stripper, air is so sparingly soluble in water that air is treated as a water-insoluble gas for mass transfer purposes. In addition, the ratio of the airflow to liquid flow is so high that there is no significant approach to equilibrium conditions between the organic concentrations in the gas phase (air) and the liquid phase (water). The design of an air stripper is consequently based on an estimation of the resistances to mass transfer (e.g., the two-film theory).

In a steam stripper, the stripping gas is entirely soluble in water (steam is water vapor!), and the concentration of organics at the top (rectifying) section of a steam stripper can easily be several percent. This is 4 to 5 orders of magnitude higher than the liquid concentrations in an air stripper. Consequently, operating gas-liquid conditions do significantly approach liquid-vapor equilibrium conditions. This applies to both the water-steam and the organic components. As a result, a steam stripper is designed as an equilibrium stage process; mass transfer resistances are not directly considered in this design approach.

The equilibrium stage approach is illustrated in Figure 9-15. As the gas travels up the column, it passes through each tray and through the liquid that is on top of the tray (say tray $n+1$). The gas then exits the liquid, picking up organics from the liquid on the tray $(n+1)$, and proceeds upward to the next tray $(n+2)$. Under ideal conditions, the gas leaving the top of the tray $(n+1)$ would be in equilibrium with the liquid on the tray $(n+1)$. In practice, however, the mass transfer only approaches equilibrium conditions.

The mass transfer at tray $n+1$ may be described by the following mass balance and equilibrium equations:

$$\text{Input} = \text{output}$$
$$GY_{A,n} + LX_{A,n+2} = GY_{A,n+1} + LX_{A,n+1} \tag{9-47}$$

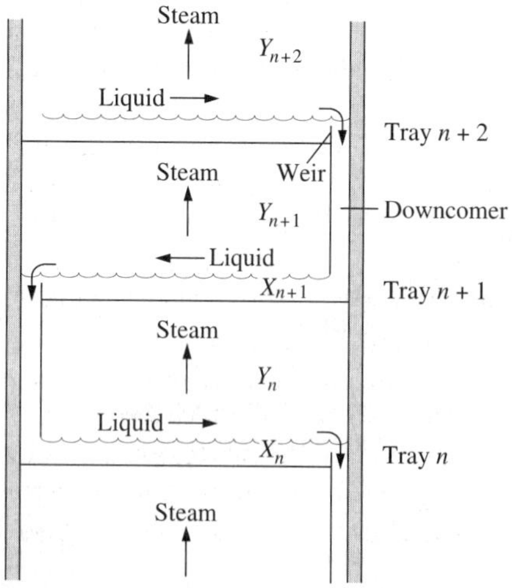

FIGURE 9-15
Mass transfer across trays.

where $G =$ steam molar rate (lb-mole/h)
$L =$ water molar rate (lb-mole/h)
$Y_A =$ mole fraction of organic component A in the gas above the designated tray number (lb-moles A/lb-moles gas)
$X_A =$ mole fraction of organic component A in the liquid on the designated tray number (lb-moles A/lb-moles water)

The equilibrium conditions between the liquid at and the gas above each tray are determined from the thermodynamic liquid-vapor equilibrium considerations previously discussed. A tray efficiency is used to estimate the approach to these equilibrium conditions.

$$E = \frac{\text{actual vapor pressure of component } A}{\text{equilibrium vapor pressure of component } A} = \frac{(Y_{A,n})\,(P_{T,n})}{(a_{A,n}/\delta_{A,n})\,(X_{A,n})\,(P_{vp,A})} \tag{9-48}$$

or

$$(Y_{A,n}) = E\,(a_{A,n}/\delta_{A,n})\,(X_{A,n})\,(P_{vp,A})\,/\,(P_{T,n}) \tag{9-49}$$

where $E =$ tray efficiency (unitless fraction)
$Y_{A,n} =$ mole fraction of organic component A in the gas above tray n (lb-moles A/lb-moles gas)
$a_{A,n} =$ activity coefficient of organic component A in liquid on tray n (unitless)
$\delta_{A,n} =$ fugacity coefficient of organic component A in gas above tray n (unitless)
$X_{A,n} =$ mole fraction of organic component A in gas above tray n (lb-moles A/lb-moles liquid)
$P_{vp,A} =$ vapor pressure of organic component A at conditions on tray n
$P_{T,n} =$ total pressure of system above tray n

Obviously, a tray with a higher efficiency is more desirable for mass transfer purposes. Tray efficiency is a function of both the mechanical design of the tray and the process conditions for operating the tray. Tray efficiencies of 50 to 65 percent are common for steam strippers.

Packed column steam strippers are actually designed as equilibrium stage (or tray) columns. Mechanical and process considerations are then utilized to estimate the height equivalent to a theoretical plate (HETP). HETP is expressed in feet of packing depth, which is equivalent to one theoretical equilibrium stage. The calculation of HETP is also treated as a proprietary trade secret in the chemical process industry, although there are correlations available in published literature.

The calculation of the number of theoretical trays required to meet a given set of process requirements is now routinely performed on computers. However, a review of an older technique, the McCabe-Thiele diagram, is useful for conceptually understanding the design of a steam stripper. In Figures 9-16 and 9-17, a McCabe-Thiele diagram is presented for a simple binary water/organic mixture. The diagram in Figure 9-16 consists of:

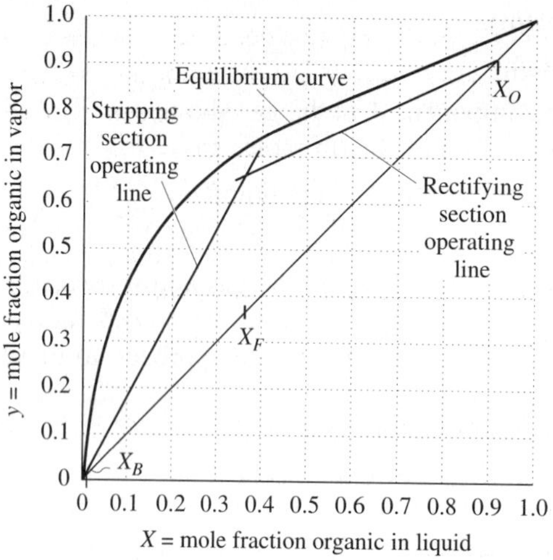

FIGURE 9-16
Construction of McCabe-Thiele diagram.

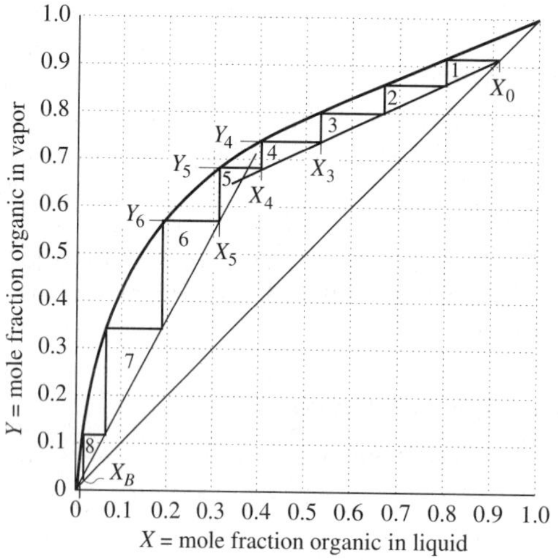

FIGURE 9-17
Completed McCabe-Thiele diagram.

- An equilibrium curve
- The rectifying section operating line
- The stripping section operating line
- The bottom (X_B), feed (X_F), and overhead vapor (X_O) organic mole fractions

In Figure 9-17, the number of theoretical equilibrium stages (also called theoretical trays or theoretical plates) is stepped off from X_O to X_B. The underlying principle in this approach is that the liquid leaving a theoretical tray is in equilibrium with the vapor leaving that same tray. Also, the slopes of the rectifying section and stripping section operating lines are functions of the ratio of the liquid-to-gas (steam) mass flowrates (L/G).

For example, in Figure 9-17, set tray n equal to tray 5, tray $n+1$ equal to tray 6, and tray $n+2$ equal to tray 7. The equilibrium conditions on tray 6 are the point, in Figure 9-17, on the equilibrium curve where step 6 is indicated. The liquid then

flows down from tray 6 to tray 7 on the straight vertical line. Similarly, the vapor from tray 6 steps off against the operating line to tray 5. At step 5, the operating lines for the rectifying section and stripping section intersect. Consequently, in this example, tray 5 is the feed tray. The rectifying section is located above tray 5 and the stripping section is located below tray 5. The preceding discussion is presented only as an example (for purposes of explaining the equilibrium stage design concept); McCabe-Thiele diagrams are generally not used for the design of steam strippers. The limitations of the McCabe-Thiele diagram in a steam stripper application are:

- The method is applicable to pure binary mixtures (not multicomponent mixtures).
- The method does not graphically recognize the proper mass balances for the top and bottom trays in a steam stripper.

Current process design methods generally utilize a computerized process simulator package, as discussed later in this section. For more information on the McCabe-Thiele design method, the reader is referred to Treybal,[24] McCabe, Smith, and Harriott,[25] and Wankat.[37]

In practice, the design of a column is much more complicated than the simple graphical method presented in Figures 9-16 and 9-17. However, the concept of steps and equilibrium stages is an important one. In actual process design, process simulation software is used to calculate the complex equilibrium relationships between vapor and liquid at a given tray along with mass and energy balances. The vapor then travels up to a higher tray, the liquid flows down to the next lower tray, and another set of complex equilibrium calculations must be performed for each tray. Although these calculations are too complex to perform by hand, the equilibrium stage concept is still fundamental to the design of distillation columns such as a steam stripper.

Curve of Constant Bottoms Concentration

For a given water flow rate and a given set of feed and bottoms concentrations, there are a large number of potential column designs that will satisfy the process requirements. By increasing the steam rate, a smaller number of theoretical trays would be required to meet the specified design performance requirements. This would be reflected on the McCabe-Thiele diagram by a lower slope on the operating lines in the stripping and rectifying sections of the column. A number of these solutions can be prepared to develop a curve of constant bottoms concentration, shown in Figure 9-18. The curve presented in Figure 9-18, based on actual pilot plant data, shows a number of solutions to a set of design constraints in terms of number of theoretical trays versus steam flow rate.

The prudent design of a steam stripper will be to the left of the inflection point on this curve. In this manner, if process conditions shift the design curve, a small change in the steam rate will bring the operation back onto the design curve. Contrast this to the portion of the design curve to the right of the tangent point. In this area, a shift in the design curve will require a large change in steam flow rate in order to return the design curve. Such a large increase in steam rate may be beyond the steam supply capabilities of a given installation or exceed the capacity of a given tray hydraulic design, besides causing a large increase in steam use and operating costs.

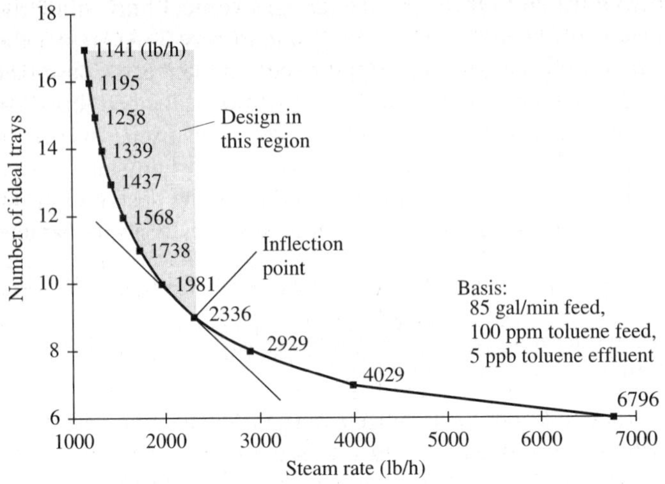

FIGURE 9-18
Design curve—constant bottoms concentration.

Design Considerations

The initial analysis of the feasibility of a steam stripper must consider two key factors:

- The strippability of the organics
- Whether the organics will form a separate organic phase in the overhead decanter

For a single organic in wastewater, this is a relatively easy analysis. However, for water containing a mixture of organics, this often requires the use of a computerized process simulator to assess the thermodynamics of the interactions between the various organics in the wastewater. However, there are some general considerations that can be examined before utilizing a process simulator. These types of considerations can be used to eliminate obviously poor candidates for steam stripping.

As a general rule of thumb, any priority pollutant that is analyzed by direct injection on a gas chromatograph can be considered a good candidate for high-efficiency steam stripping. This can easily be determined by a detailed reading of the analytical reports for a water sample. Another rule of thumb is that any compound with a boiling point below 150°C is a potential candidate for steam stripping,[38] and the lower the boiling point the more easily a compound would be expected to be steam stripped. A key factor not included in this rule of thumb is what effects the presence of other organics will have on the activity coefficient of a given organic. A large increase in the activity coefficient will effect an increase in the strippability of a given organic compound.

The second feasibility factor is whether a separate organic phase can be formed in the overhead condenser. This analysis involves looking at the solubility limits of the various organics in water. If even one organic compound has a low solubility limit (e.g., less than 1 percent), then there is a basis to expect a separate phase to form in the decanter. Some organic compounds are infinitely soluble in water, and

will not form a separate organic phase, by themselves, in the overhead decanter. Additional treatment of the aqueous phase would be necessary if a separate organic phase is not formed. Treatment options other than steam stripping are then, usually, more economical. Some examples include simple alcohols such as methanol and ethanol. Generally speaking, the infinitely soluble organic compounds are simple low molecular weight organics.

Two example water/organic mixtures are shown in Table 9-6. Mixture A contains organic compounds that are all infinitely soluble in water. Consequently, this groundwater is not a candidate for steam stripping. Mixture B contains the same organic compounds along with toluene and xylene, both of which are sparingly soluble in water. The Mixture B groundwater was actually being treated by a steam stripper. When the toluene forms an insoluble phase in the overhead decanter, the other organics partition to the organic phase to a high degree (e.g., in excess of 90 percent). The point of this illustration is that only one sparingly soluble organic need be present to make steam stripping feasible. As a general rule, the organics will preferentially partition to the organic phase created by the single sparingly soluble organic.

After an initial desktop feasibility analysis has been performed, further analysis of promising candidates should be considered. Such an analysis would be performed using a process simulator. A positive result from a process simulator would then normally be followed by pilot-scale tests using samples of the candidate wastewater or groundwater. These pilot-scale tests may be a simple two- to three-day lab program using a single water sample. Or the pilot-scale tests may involve a field pilot plant that is installed for several weeks.

The key factor in determining the appropriate pilot program is the variability of the water composition, including the presence of inorganic salts that will also have an effect on activity coefficients. A properly designed steam stripper can handle water composition variations of less than 1 order of magnitude. Consequently, simple lab-scale programs are appropriate for these types of candidate waters. This typically includes groundwaters and wastewaters from continuously operated manufacturing processes that produce a single product.

Where large day-to-day variations in composition are expected or a number of batch or campaign production units (switching products every few weeks or months) are utilized, a field pilot plant may be advisable. The purpose of such a field pilot plant would be to ascertain the variability in the activity coefficients over

TABLE 9-6
Example water mixtures for steam stripping feasibility analysis

Mixture A	37 mg/L methanol
	194 mg/L ethanol
	114 mg/L *n*-butanol
Mixture B	37 mg/L methanol
	194 mg/L ethanol
	114 mg/L *n*-butanol
	110 mg/L toluene
	14 mg/L xylene

a sustained period of time. On the basis of the range of activity coefficients, a proper stripping column could be designed along with an agitated equalization tank that would dampen composition variations to a level that is acceptable for operation of a steam stripper. This type of variability in composition is usually present only in industrial wastewater from large complexes or from batch production plants.

Field pilot plant trials may also be appropriate where fouling or plugging of the trays is a concern. An example would be a wastewater containing monomers that polymerize at elevated temperature.

Mechanical Design

The design of a steam stripper can utilize any of several types of internals to provide the necessary gas-liquid contact time for effective mass transfer of the organics. For steam strippers, the practical choice is usually one of three types of internals:

- Random packing
- Valve trays
- Sieve trays

An important concept in the design of tray towers is the **turndown ratio.** This is the ratio of the maximum rate of flow of either the gas or liquid to the corresponding minimum rate of flow. This is particularly important in environmental applications because of the possibility of significant variability in the feed flow rate. As an example, if a steam stripper with a turndown ratio of 3 were designed to handle contaminated groundwater at a maximum flow of 100 gal/min, it could not function properly if the flow dropped below 33 gal/min.

Random packing (see Section 9-1), has the advantage of providing the highest mass transfer capability for a given column height. Random packing also provides for a very high turndown ratio of steam rates—about 10. However, random packings are limited in their ability to pass solids in the groundwater. Valve trays (Figure 9-19) have a high turndown ratio that approaches random packing performance and are capable of passing modest suspended solids in the water. Sieve trays (Figure 9-19) have the best solids-passing capabilities, but the steam turndown ratio is about 4.

In addition to turndown ratios and solids-passing capabilities, a number of site-specific items need to be considered, such as inspection, maintenance, and the availability of packings in the more exotic materials of construction (see the following section, Other Design Considerations). Concerns over unacceptable downtime can drive internals selection to a less efficient but more easily maintained or repaired

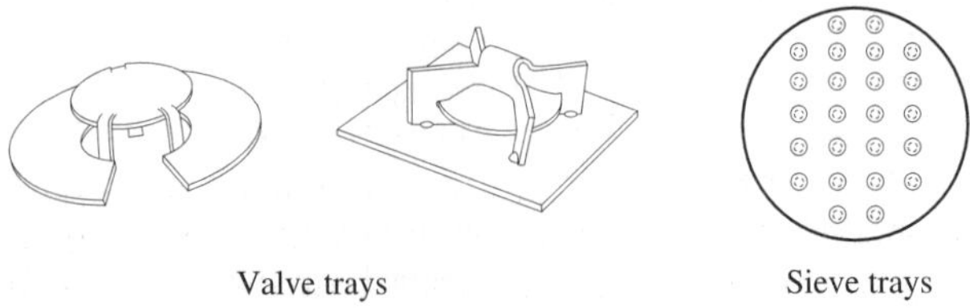

FIGURE 9-19
Plate types.

design. Unacceptable downtime can shut down a production process or lead to high fines from regulating authorities because the authorities expect virtually 100 percent compliance with discharge limitations.

In addition to the selection of internals, a basic decision must be made regarding atmospheric operation or vacuum operation of a steam stripping column. The choice of a vacuum stripper may be determined by a limited or nonexistent steam supply at a proposed site. Other factors include the availability of operators who are trained in mass transfer processes and the need to maintain a mechanical vapor compressor.

All of the above factors are site-specific and should not be considered an exhaustive list of factors that can affect the mechanical design of a steam stripping column.

Other Design Considerations

Groundwaters containing high concentrations of iron are commonly found in many parts of the United States. Iron will be oxidized rapidly in the high-temperature environment of a steam stripping column, and a ferric hydroxide precipitate will form. This can easily overwhelm the solids-passing capabilities of even the best mechanical design of column internals. Consequently, it is sometimes necessary to add a pretreatment step to precipitate iron and other ions that can cause excessive scaling or precipitation in a steam stripping operation.

The elevated temperature of a steam stripping operation often requires exotic materials of construction. The presence of free chlorides and elevated temperatures will attack the common austenitic stainless steels (e.g., 304, 316), so the use of Monel™, titanium, or other specialty alloys is often indicated for wetted parts where high temperature is present. Other exotic materials of construction could include a carbon steel column lined or coated with corrosion-resistant materials. The exact materials of construction are usually determined on a case-by-case basis.

9-4 CHEMICAL OXIDATION

In general, the objective of chemical oxidation is to detoxify waste by adding an oxidizing agent to chemically transform waste components. For example, an organic molecule can be converted to carbon dioxide and water or to an intermediate product that may be less toxic than the original. This intermediate molecule might be amenable to further treatment by the biological methods presented in the next chapter.

Chemical oxidation of wastes is a well-established technology that is capable of destroying a wide range of organic molecules, including chlorinated VOCs, mercaptans, and phenols, and inorganics such as cyanide. This section focuses on the oxidizing agents most commonly used for hazardous waste treatment: ozone, hydrogen peroxide, and chlorine. Ultraviolet (UV) light is usually added along with ozone and/or hydrogen peroxide to accelerate the oxidation of chlorinated VOCs.

Oxidation and reduction reactions occur in pairs to make up an overall redox reaction. In chemical oxidation for hazardous waste treatment, an oxidizing agent is added to oxidize the waste components of concern, which serve as the reducing agents. Oxidizing agents are nonspecific and will react with any reducing agents

present in the waste stream. Therefore, these processes are most economical when organics other than the ones of concern are in low concentration. For example, in the oxidation of cyanides, the presence of large quantities of other organic molecules will require excessive amounts of oxidizing agent to be used. Additionally, reactions between some oxidizing agents (e.g., chlorine) and some organics (e.g., hydrocarbons) may result in chlorine substitution rather than destruction of the organic. The resulting chlorinated hydrocarbon may be more toxic than the original waste.

Although chemical oxidation is typically applied to liquid hazardous wastes and contaminated groundwater, soils may also be amenable to these processes. Contaminated soils can be excavated and treated in a slurry form in reaction vessels. However, because excavation is an expensive process, the tendency in soil cleanup technology is to use *in situ* processes. The feasibility of *in situ* chemical oxidation depends on the ability to get reagents in contact with contaminants, which is a function more of the soil properties such as permeability than of the process chemistry. In remediation technologies employing addition of ozone or hydrogen peroxide (two of the most common oxidizing agents in hazardous waste treatment) to soils *in situ,* enhanced volatilization of contaminants and biological activity of native microorganisms, rather than chemical oxidation, are the primary mechanisms of remediation. Therefore, *in situ* biological treatment and soil vapor extraction have been more widely developed and applied for site remediation. These two technologies are discussed in Section 10-7 and Section 16-8, respectively.

Process Description

Chemical oxidation is typically conducted in completely mixed tanks or plug flow reactors. The contaminated water is introduced at one side of the tank and the treated water exits at the other side. The oxidizing agent is either injected into the contaminated water just before it enters the tank or is dosed directly into the tank. Complete mixing of the water with the oxidizing agent must be provided, either by mechanical agitation, pressure drop, or bubbling into the tank. Complete mixing, which prevents short-circuiting in the tank, is necessary to ensure contact of the contaminants with the oxidizing agent for a minimum period of time, and, thus, reduce the chemical dosage required to obtain a specific effluent concentration.

Ozone is produced from atmospheric oxygen by using electrical energy to split the oxygen into two oxygen radicals ($O\cdot$), which readily combine with other oxygen molecules (O_2) to form ozone (O_3). Ozone is unstable under normal environmental conditions and readily decomposes back to oxygen. Ozone is added to the liquid waste as a gas, either through porous diffusers at the bottom of a tank, or through an injector where the pressure drop produced draws ozone into the injector and mixes it with the liquid.

Hydrogen peroxide (H_2O_2) and UV light are generally used together. A typical arrangement is shown in Figure 9-20. Hydrogen peroxide is stored on site as a 35 to 50% solution. The high solubility of H_2O_2 in water obviates the need for mixing equipment, reducing the number of moving parts in the system. The size of the combined H_2O_2/UV system can be significantly reduced through the use of high-energy UV lamps, capable of imparting 500 watts/liter to the water.[39]

The last oxidizing agent discussed in this section, chlorine (Cl_2), is a dense gas that readily liquefies under pressure. Liquid chlorine is commercially available in

150-lb and 1-ton cylinders. Chlorine is also available in solid form such as calcium hypochlorite [$Ca(OCl)_2$]. For aqueous waste treatment applications, liquid chlorine is evaporated to a gas and then mixed with water in a solution feeder. This concentrated solution is then added to the liquid waste. The dry calcium hypochlorite is also dissolved in water prior to application to the liquid waste.

Some common oxidation and reduction reactions are listed in Table 9-7.

EXAMPLE 9-9. CHEMICAL QUANTITY DETERMINATION. A waste stream of 20,000 gal/day contains 270 mg/L of cyanide as NaCN. Determine the theoretical (stoichiometric) amount of chlorine required daily to destroy this waste.

Solution. Atomic weights: Na = 23, C = 12, N = 14, Cl = 35.45.

From Table 9-7: $2NaCN + 5Cl_2 + 12NaOH \rightarrow N_2 + Na_2CO_3 + 10NaCl + 6H_2O$. This implies that 2.5 moles of Cl_2 are required for each mole of NaCN. The total mass of NaCN discharged daily is:

$$(20{,}000 \text{ gal/day})(8.34 \text{ lb/gal})(270 \text{ parts}/10^6 \text{ parts}) = 45.04 \text{ lb/day NaCN}$$

Pound-moles of NaCN discharged per day:

$$(45.04 \text{ lb/day})/(49 \text{ lb/lb-mol}) = 0.92 \text{ lb-mol/day}$$

$$Cl_2 \text{ required daily} = 2.5\,(0.92) = 2.3 \text{ lb-mol } Cl_2$$
$$= 2.3 \text{ lb-mol } Cl_2\ (70.9 \text{ lb/lb-mol})$$
$$= \underline{\underline{163 \text{ lb/day chlorine required}}} \text{ (stoichiometric)}$$

Because of short-circuiting and/or incomplete reaction time, more than stoichiometric amounts of the chemical are required in practice.

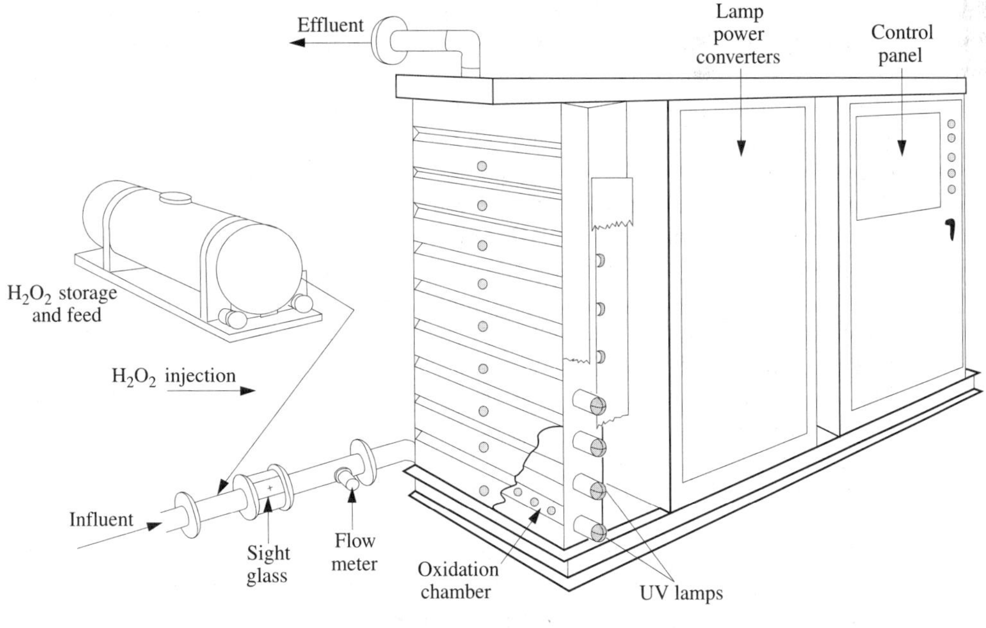

FIGURE 9-20
Equipment arrangement for a typical UV/H_2O_2 system.[40]

TABLE 9-7
Some typical oxidation and reduction reactions

Reagent	Description	Reaction	Comments	Reference
		Oxidation		
Ozone	Cyanide oxidation	$NaCN + O_3 \rightarrow NaCNO + O_2$		41
Ozone	Phenol oxidation	(See Fig. 9-23)		42, 43
Hydrogen peroxide	Cyanide oxidation	$NaCN + H_2O_2 \rightarrow NaCNO + H_2O$	pH 9.5–10.5	44
	Sulfide oxidation	$H_2O_2 + H_2S \rightarrow 2H_2O + S$ $4H_2O_2 + S^{=} \rightarrow SO_4^{=} + 4H_2O$		45
Chlorine	Ferrous to ferric	$2Fe^{++} + HOCl + 5H_2O \rightarrow$ $2Fe(OH)_3 + Cl^- + 5H^+$		46
Chlorine	Cyanide to cyanate	$NaCN + Cl_2 \rightarrow CNCl + NaCl$ $CNCl + 2NaOH \rightarrow$ $NaCNO + H_2O + NaCl$	CNCl is a toxic gas	47
	Cyanate to CO_2 and N_2	$2NaCNO + 3Cl_2 + 4NaOH \rightarrow$ $N_2 + 2CO_2 + 6NaCl + 2H_2O$		
Chlorine dioxide	Manganous ion to manganese dioxide	$Cl_2 + 2NaClO_2 \rightarrow 2ClO_2 + 2NaCl$ $Mn(NO_3)_2 + 2e^- \rightarrow MnO_2 + 2NO_2^-$	MnO_2 insoluble	48, 49
Potassium permanganate	Sulfide	$4\ KMnO_4 + 3H_2S \rightarrow$ $2K_2SO_4 + S + 3MnO + MnO_2 + 3H_2O$		
Ozone/UV	Acetaldehyde	$CH_3CHO + O_3 \rightarrow CH_3COOH + O_2$	Forms H_2O_2	
UV/peroxide	Dichloromethane (CH_2Cl_2)	$CH_2Cl_2 + 2H_2O_2 \rightarrow CO_2 + 2H_2O + 2HCl$		50
Oxygen	Formaldehyde (CH_2O)	$CH_2O + 1/2\ O_2 \rightarrow CO_2 + H_2O$		
Oxygen	Cyanide	$2\ CN^- + O_2 \rightarrow 2CNO^-$ $CNO^- + 2H_2O + 2H^+ \rightarrow CO_2 + NH_4^+ + H_2O$	Requires activated C and Cu catalyst	51
		Reduction		
Sulfur dioxide	Chromium(VI)	$3SO_2 + 3H_2O \rightarrow 3H_2SO_3$ $2CrO_3 + 3H_2SO_3 \rightarrow Cr_2(SO_4)_3 + 3H_2O$	Cr(III) is then precipitated as $Cr(OH)_3 \downarrow$ by adding lime	52 53
Ferrous sulfate	Chromium(VI)	$2CrO_3 + 6FeSO_4 + 6H_2SO_4 \rightarrow$ $3Fe_2(SO_4)_3 + Cr_2(SO_4)_3 + 6H_2O$		
Sodium borohydride	Copper	$NaBH_4 + 8Cu^+ + 2H_2O \rightarrow$ $8Cu + NaBO_2 + 8H^+$		54, 55

The environment under which redox reactions take place greatly influences their effectiveness. Parameters that can be used to quantify effectiveness for use in hazardous waste treatment are the free energy of a reaction and the oxidation reduction potential (ORP). Factors such as temperature and pH, the presence of catalysts, and concentration of other reactants will influence selection of the oxidizing agent and will determine whether cost-effective treatment is feasible.

The power of an oxidizing or reducing agent is measured by its electrode potential. An indication of how a reaction will proceed can be determined from the free energy considerations (see Section 3-3). The free energy of a reaction, ΔG^0, is related to the standard electrode potential E^0, and the equilibrium constant K by:

$$\Delta G^0 = -nFE^0 = -RT \ln K \tag{9-50}$$

where ΔG^0 = free energy of a reaction (kcal/gmol)
n = number of electrons exchanged during the reaction per gmol
F = Faraday constant 23.062 kcal/(volt-equiv.)
E^0 = standard electrode potential (volts)
R = gas constant = 1.98×10^{-3} kcal/(gmol · K)
T = temperature (K)

E^0 may be determined from tabulated half reactions (see Table 9-8). To assemble a complete reaction from two half-reactions, an oxidation half-reaction is added to a reduction half-reaction. The process is illustrated by Example 9-10.

To balance redox reactions occurring in acidic solutions:

- Balance atoms other than H and O
- Balance O by adding H_2O
- Balance H by adding H^+
- Balance net charge by adding e^-
- Add half reactions after equalizing e^- gain and loss
- Cancel similar terms on both sides of equation

It is possible to measure the ORP directly by means of a galvanic cell made up of a gold or platinum anode and a reference electrode (the cathode). This galvanic cell instrument is capable of measuring only those potentials from species that react with the electrode and, thus, does not provide a direct measure of E^0. However, for some reactions of interest in waste treatment, the ORP value does provide a useful measurement, as shown in Figure 9-24 for hydrogen peroxide.

The **Nernst equation** may be derived by substituting the definition of K from Chapter 3. This equation allows E to be determined at conditions other than standard:

$$E = E^0 - (RT/nF) \ln Q \tag{9-51}$$

$$Q = \frac{\Pi\ (\text{products})^x}{\Pi\ (\text{reactants})^y} \tag{9-52}$$

where x, y represent the coefficients of the products and reactants in the original equation and Q is termed the reaction quotient. Note that at equilibrium, $Q = K$.

For the reaction $aA + bB \rightarrow cC + dD$:

$$Q = \frac{[C]^c\,[D]^d}{[A]^a\,[B]^b} \tag{9-53}$$

TABLE 9-8
Standard electrode potentials at 25°C[56–59]

Reaction	E^0, volts
$Cl_{2(g)} + 2e^- \rightarrow 2Cl^-$	+1.36
$HOCl + H^+ + 2e^- \rightarrow Cl^- + H_2O$	+1.49
$ClO^- + H_2O + 2e^- \rightarrow Cl^- + 2OH^-$	+0.90
$NH_2Cl + H_2O + 2e^- \rightarrow Cl^- + NH_3 + OH^-$	+0.75
$NHCl_2 + 2H_2O + 4e^- \rightarrow 2Cl^- + NH_3 + 2OH^-$	+0.79
$NH_3Cl^+ + H^+ + 2e^- \rightarrow Cl^- + NH_4^+$	+1.40
$NHCl_2 + 3H^+ + 4e^- \rightarrow 2Cl^- + NH_4^+$	+1.34
$O_3 + 2H^+ + 2e^- \rightarrow O_2 + H_2O$	+2.07
$O_3 + H_2O + 2e^- \rightarrow O_2 + 2OH^-$	+1.24
$H_2O_2 + 2H^+ + 2e^- \rightarrow 2H_2O$	+1.78
$HO_2^- + 2e + H_2O \rightarrow 3OH^-$	+0.85
$ClO_2 + 2H_2O + 5e^- \rightarrow Cl^- + 4OH^-$	+1.71
$MnO_4^- + 4H^+ + 3e^- \rightarrow MnO_2 + 2H_2O$	+1.68
$MnO_4^- + 8H^+ + 5e^- \rightarrow Mn^{2+} + 4H_2O$	+1.49
$MnO_4^- + 2H_2O + 3e^- \rightarrow MnO_2 + 4OH^-$	+0.58
$O_{2(g)} + 4H^+ + 4e^- \rightarrow 2H_2O$	+1.23
$O_2 + 2H_2O + 4e^- \rightarrow 4OH^-$	+0.40
$HBrO + H^+ + 2e^- \rightarrow Br^- + H_2O$	+1.33
$MnO_2 + 4H^+ + 2e^- \rightarrow Mn^{2+} + 2H_2O$	+1.23
$ClO_2 + e^- \rightarrow ClO_2^-$	+1.15
$Fe(OH)_3 + e^- + 3H^+ \rightarrow Fe^{2+} + 3H_2O$	+1.06
$Fe^{3+} + e^- \rightarrow Fe^{2+}$	+0.77
$ClO_2 + 2H_2O + 5e^- \rightarrow Cl^- + 4OH^-$	+0.76
$ClO_3 + H_2O + 2e^- \rightarrow ClO_2 + 2OH^-$	+0.35
$S_{(s)} + 2H^+ + 2e^- \rightarrow H_2S_{(g)}$	+0.17
$NO_3^- + H_2O + e^- \rightarrow NO_2^- + 2OH^-$	+0.01
$\frac{1}{4}CO_{2(g)} + H^+ + e^- \rightarrow \frac{1}{24}(\text{glucose}) + \frac{1}{4}H_2O$	−0.20
$H^+ + e^- \rightarrow \frac{1}{2}H_{2(g)}$	0
$Na^+ + e^- \rightarrow Na_{(s)}$	−2.72
$Mg^{2+} + 2e^- \rightarrow Mg_{(s)}$	−2.37
$Cr_2O_7^{2-} + 14H^+ + 6e^- \rightarrow 2Cr^{3+} + 7H_2O$	+1.33
$Cr^{3+} + e^- \rightarrow Cr^{2+}$	−0.41
$Fe^{2+} + 2e^- \rightarrow Fe_{(s)}$	−0.44
$Cu^{2+} + e^- \rightarrow Cu^+$	+0.16
$Cu^{2+} + 2e^- \rightarrow Cu_{(s)}$	+0.34
$Ag^{2+} + e^- \rightarrow Ag^+$	+2.0
$Ag^+ + e^- \rightarrow Ag_{(s)}$	+0.8
$AgCl_{(s)} + e^- \rightarrow Ag_{(s)} + Cl^-$	+0.22
$Au^{3+} + 3e^- \rightarrow Au_{(s)}$	+1.5
$Zn^{2+} + 2e^- \rightarrow Zn_{(s)}$	−0.76
$Cd^{2+} + 2e^- \rightarrow Cd_{(s)}$	−0.40
$Hg_2Cl_{2(s)} + 2e^- \rightarrow 2Hg_{(l)} + 2Cl^-$	+0.27
$Mn(OH)_3 + e \rightarrow Mn(OH)_2 + OH^-$	+0.15
$Mn(OH)_3 + 2e^- \rightarrow Mn + 2OH^-$	−1.56
$2Hg^{2+} + 2e^- \rightarrow Hg_2{}^{2+}$	+0.91
$Al^{3+} + 3e^- \rightarrow Al_{(s)}$	−1.68
$Sn^{2+} + 2e^- \rightarrow Sn_{(s)}$	−0.14
$PbO_{2(s)} + 4H^+ + SO_4^{2-} + 2e^- \rightarrow PbSO_{4(s)} + 2H_2O$	+1.68
$Pb_2^+ + 2e^- \rightarrow Pb_{(s)}$	−0.13
$NO_3^- + 2H^+ + 2e^- \rightarrow NO_2{}^- + H_2O$	+0.84

TABLE 9-8
Standard electrode potentials at 25°C *(continued)*

Reaction	E^0, volts
$NO_3^0 + 10H^+ + 8e^- \rightarrow NH_4^+ + 3H_2O + NO_3^- \rightarrow NH_4^+$	+0.88
$N_{2(g)} + 8H^+ + 6e^- \rightarrow 2NH_4{}^+$	+0.28
$NO_2^- + 8H^+ + 6e^- \rightarrow NH_4{}^+ + 2H_2O$	+0.89
$2NO_3^- + 12H^+ + 10e^- \rightarrow N_{2(g)} + 6H_2O$	+1.24
$O_{2(aq)} + 4H^+ + 4e^- \rightarrow 2H_2O$	+1.27
$SO_4^{2-} + 2H^+ + 2e^- \rightarrow SO_3^{2-} + H_2O$	−0.04
$S_4O_6^{2-} + 2e^- \rightarrow 2S_2O_3^{2-}$	+0.18
$S_{(s)} + 2H^+ + 2e^- \rightarrow H_2S_{(g)}$	+0.17
$SO_4^{2-} + 8H^+ + 6e^- \rightarrow S_{(s)} + 4H_2O$	+0.35
$SO_4^{2-} + 10H^+ + 8e^- \rightarrow H_2S_{(g)} + 4H_2O$	+0.34
$2HOCl + 2H^+ + 2e^- \rightarrow Cl_{2(aq)} + 2H_2O$	+1.60
$Cl_{2(aq)} + 2e^- \rightarrow 2Cl^-$	+1.39
$2HOBr + 2H^+ + 2e^- \rightarrow Br_{2(l)} + 2H_2O$	+1.59
$Br_2 + 2e^- \rightarrow 2Br^-$	+1.09
$2HOI + 2H^+ + 2e^- \rightarrow I_{2(s)} + 2H_2O$	+1.45
$I_{2(aq)} + 2e^- \rightarrow 2I^-$	+0.62
$I_3^- + 2e^- \rightarrow 3I^-$	+0.54
$ClO_2 + e^- \rightarrow ClO_2^-$	+1.15
$CO_{2(g)} + 8H^+ + 8e^- \rightarrow CH_{4(g)} + 2H_2O$	+0.17
$6CO_{2(g)} + 24H^+ + 24e^- \rightarrow C_6H_{12}O_6\text{(glucose)} + 6H_2O$	−0.01
$CO_{2(g)} + H^+ + 2e^- \rightarrow HCOO^-$ (formate)	−0.31
$H_2BO_1 + 5H_2O + 8e^- \rightarrow BH_4^- + 8OH^-$	−1.24
$H_2BO_1 + H_2O + 3e^- \rightarrow B + 4OH^-$	−1.79
$H_2BO_1 + 3H^+ + 3e^- \rightarrow B + 3H_2O$	−0.8698
$Ba(OH)_2 + 2e^- \rightarrow Ba + 2OH^-$	−2.99
$Cd^{2+} + 2e^- \rightarrow Cd$	−0.3521
$Cd(OH)_2 + 2e^- \rightarrow Cd^{++} + 2OH^-$	−0.809
$Fe^{1+} + 3e^- \rightarrow Fe$	−0.037
$Ni(OH)_2 + 2e^- \rightarrow Ni + 2OH^-$	−0.72

where $[C] =$ molar concentration of component C
$c =$ the coefficient of component C in the reaction

In simplified form, at 25°C, Equation 9-51 reduces to

$$E = E^0 - (0.059/n)\ \log_{10} Q \tag{9-54}$$

EXAMPLE 9-10. CALCULATION OF FREE ENERGY. Determine the free energy for oxidation of Fe^{2+} by MnO_2 in acid solution:

Solution. The half-reactions from Table 9-8 are given below. (Note that $2H_2O + 4H^+$ appears on both sides when the two half-reactions are combined.)

Oxidation:

$$Fe^{2+} \rightarrow Fe^{3+} + e^- \qquad E^0 = -0.77\text{ V}$$

Reduction:

$$MnO_2 + 4H^+ + 2e^- \rightarrow Mn^{2+} + 2H_2O \qquad E^0 = +1.23\text{ V}$$

To balance electron charges, multiply the iron half-reaction by 2:

$$2Fe^{2+} \rightarrow 2Fe^{3+} + 2e^-$$

and add the two half-reactions (e^- cancels out):

$$2Fe^{2+} + MnO_2 + 4H^+ \rightarrow 2Fe^{3+} + Mn^{2+} + 2H_2O$$

Total electron potential:

$$E^0 = -0.77\ V + 1.23\ V = +0.46\ V$$

From Equation 9-50:

$$\Delta G^0_{rxn} = -(2\ \text{equiv. gmol})(23.062\ \text{kcal/V-equiv.})(+0.46\ V) = -21.2\ \text{kcal/gmol}$$

The equilibrium constant may also be determined from Equation 9-49:

$$\ln K = -\Delta G^0/RT = \frac{(21.2\ \text{kcal/gmol})}{[1.98 \times 10^{-3}\ \text{kcal/(gmol} \cdot \text{K)}](298\ \text{K})} = 36.0$$

$$K = 4.1 \times 10^{15}$$

Design Considerations

Ozone

Ozone has long been used as a disinfectant in water and wastewater applications. Ozone is a blue gas at normal temperatures and pressures and has a recognizable pungent odor. It is irritating to mucous membranes and is a significant surface air pollutant in many areas. Ozone is a powerful oxidant and readily reacts with most toxic organics. Most of the reaction products are less toxic and many are biodegradable.

Reactions between ozone and organic molecules include the insertion of oxygen into a benzene ring, the breaking of double bonds, and the oxidation of alcohols. Ozonolysis of the C=C bond produces aldehydes and ketones:[60]

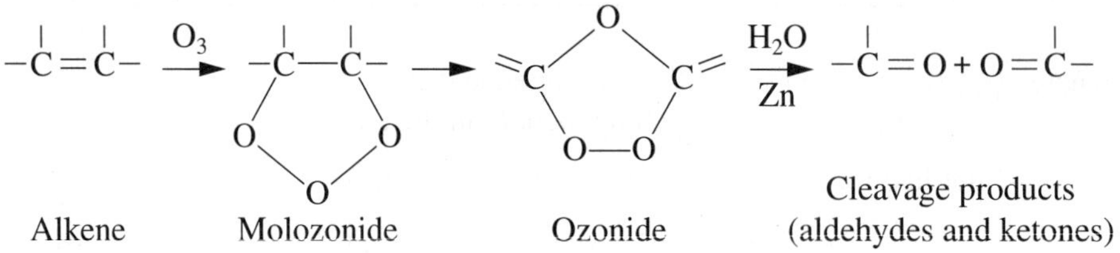

(9-55)

Alcohols react with ozone to form organic acids:

$$3RCH_2OH + 2O_3 \rightarrow 3RCOOH + 3H_2O \tag{9-56}$$

Aromatic compounds may be converted to toxic phenolics at pH greater than 9 and in the presence of redox salts such as iron or manganese.[61] Ozone reacts with a wide range of organic compounds as shown in Figure 9-21. Note the dramatic reduction in the time required by using ultraviolet light with ozone.

The primary advantage of ozone as an oxidizing agent is that it has a very high free energy, which indicates that the oxidation reaction may proceed to completion. Glaze[63] reports a value of $\Delta G^0 = -400$ kJ/mol for the reaction of ozone to form water and oxygen:

$$O_3 + 2H^+ + 2e^- \rightarrow H_2O + O_2 \tag{9-57}$$

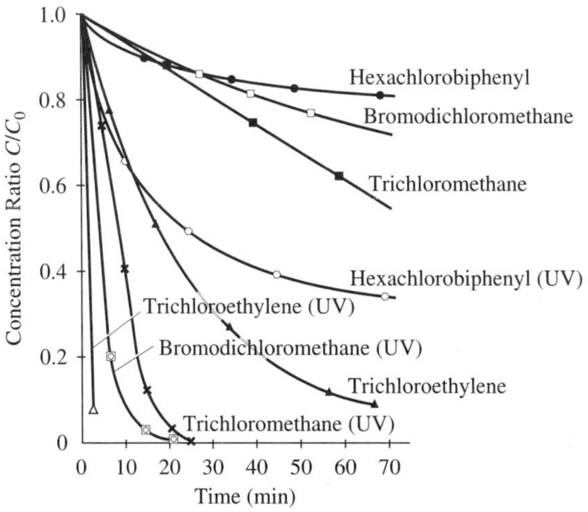

FIGURE 9-21
Destruction of chlorinated organics by ozone with and without UV radiation.[62]
Note: At pH 6–7, ozone dose rate = 1.0–1.4 mg/L min; UV frequency is 254 nm from a low-pressure mercury lamp, with a flux of 0.42 W/L.

However, kinetics indicate that the reaction rates for ozone oxidation are not always high. A disadvantage of ozone is its rapid dissociation to oxygen in the atmosphere. This dissociation can compete with any chemical oxidation reactions during waste treatment, and therefore, ozone must be generated and immediately applied at the treatment location.

The transformation of compounds in aqueous solutions by ozone via its decomposition in water is due to a chain reaction involving the ·OH (hydroxyl) radical.[64] The reaction rate constant for destruction of organics by ·OH is typically 10^7 to 10^9, many orders of magnitude greater than for O_3 alone.[65] A schematic indicating the various types of O_3 reactions is presented in Figure 9-22.

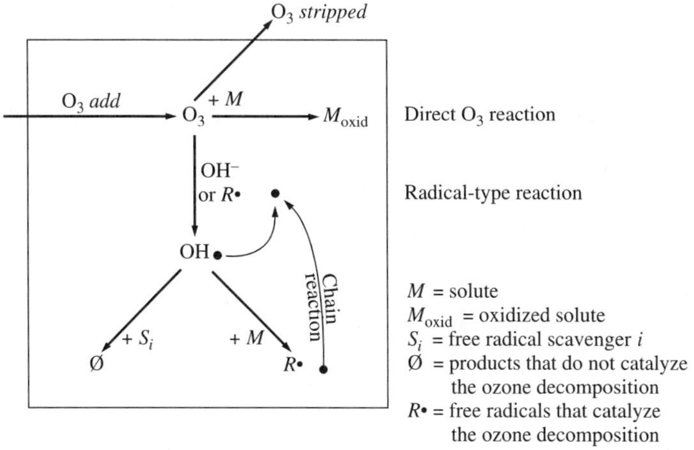

FIGURE 9-22
Reaction of ozone in aqueous solution.[66]

Phenol is readily degraded by ozone. Early work by Eisenhauer[67] indicated that the destruction of phenol by ozone follows the pathways indicated in Figure 9-23. Because simple organic acids such as acetic and oxalic acids do not readily degrade unless catalyzed by UV light or aided by H_2O_2,[68] recent applications of O_3 to oxidation of toxic wastes have typically included either UV light or H_2O_2.

The principal disadvantages of using ozone are its cost and inability to inject the gas efficiently. Posselt and Weber[70] indicate that only about 150 grams of O_3 can be produced per kilowatt-hour (kWh) of power input, compared to a theoretical value of 1058 g/kWh. Ozone generation costs are about $2400/kg per day for plants producing 900 kg/day, and are dramatically higher for small systems.[71] Ozonation of cyanide is reported to cost $0.14 to $0.64 per pound of cyanide destroyed.[72]

Historically, ozone was introduced into the subsurface through conventional injection or monitoring wells. In the presence of compounds susceptible to oxidation, the rapid reaction rate with ozone limits ozone transport over large distances. This results in a limited radius of influence around the injection point. To provide the volume of ozone adequate to destroy a given contaminant mass, the cost of the delivery system and required ozone volume tended to be high.

Laboratory research and field tests have generally concluded that the effectiveness of ozone treatment is related to the design of the delivery system. Simply speaking, the more ozone that can be effectively delivered to the contaminant mass, the more successful the treatment. This seems to be a rather obvious statement, but reconsider the application of ozone through conventional injection wells. In order to effectively treat a contaminant mass, the oxidizable mass immediately adjacent to the well must be destroyed before the ozone can diffuse further away from the well. Because the ozone bubbles will have a tendency to rise vertically or migrate through preferential pathways, similar to the behavior observed during air sparging, the treatment efficiency will tend to decrease with distance from the injection well.

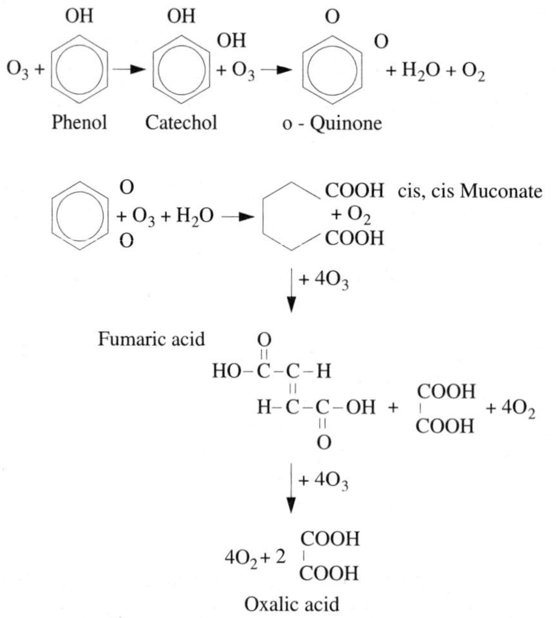

FIGURE 9-23
Degradation of phenol by ozone.[69]

This means more injection wells are required to achieve mass destruction. Recent advancements in technology have resulted in the application of ozone through horizontal wells to achieve an increased effective radius of influence and efficiency of mass destruction, recirculation wells, and the use of smaller bubbles under the influence of magnetic fields to increase diffusion.

EXAMPLE 9-11. OZONE DESTRUCTION OF TRICHLOROETHYLENE. Determine the amount of ozone required to treat 760 L/min of groundwater contaminated with trichloroethylene at 150 μg/L. Assume 90 percent removal is required, and the data of Figure 9-21 apply.

Solution. The figure indicates an ozone dose rate of 1.0 to 1.4 mg/(L · min). Select the highest value [1.4 mg/(L · min)]. From the figure, 90 percent removal requires approximately 1 hour of contact at 1.4 mg/(L · min).

$$\text{Required dose} = 1.4 \text{ mg/(L} \cdot \text{min)} \times 60 \text{ min} = 84 \text{ mg/L}$$

Mass of ozone required daily:

$$(760 \text{ L/min})(1440 \text{ min/day})(84 \text{ mg/L}) = 9.193 \times 10^7 \text{ mg/day} = \underline{\underline{91.9 \text{ kg/day}}}$$

Note that using ozone combined with ultraviolet light at 0.42 W/L the contact time is reduced to approximately 1 minute.

Hydrogen Peroxide

Hydrogen peroxide functions like ozone in that, in the presence of an iron catalyst (Fenton's reagent), it produces the hydroxyl radical (·OH). This radical reacts with organic compounds to produce a reactive organic radical (·R)

$$\cdot OH + RHFe^{2+} \rightarrow \cdot R + H_2O \tag{9-58}$$

The organic radical can again react with peroxide to produce an additional hydroxyl radical:

$$\cdot R + H_2O_2Fe^{2+} \rightarrow \cdot OH + ROH \tag{9-59}$$

Hydrogen peroxide has been shown to be effective in oxidizing organics in soil through *in situ* treatment. Watts et al.[73] report 99.9 percent degradation of pentachlorophenol (PCP) contaminated soils, noting that reduction in total organic carbon closely followed the removal of PCP. They further note:

> The phenomenon that organic carbon was removed rapidly after PCP degradation suggests that hydroxyl radical attack on the products is more rapid than on the parent compound. High rates of product degradation may be explained by the lower oxidation state of the ring as it is hydroxylated [*i.e., hydroxyl added*] and the increased water solubility of the products.[74]

Schmitt[75] reported on the use of hydrogen peroxide to treat a paint stripping liquid waste containing the following contaminants:

Methylene chloride:	50–1000 mg/L
Phenol:	500–6000 mg/L
Chromium (VI):	15–400 mg/L

Initial pilot studies showed that 99 percent of the phenol could be destroyed following Cr^{6+} reduction and precipitation, at a dosage of 2.5 g H_2O_2 per g phenol,

with an iron catalyst level of 0.05 mg Fe^{2+}/mg H_2O_2. However, a considerable amount of color remained. Color could be removed at a peroxide/phenol dosage of 7:1, but this released O_2 gas and raised the temperature of the waste more than 60°F.

Further studies demonstrated that H_2O_2 treatment should precede Cr^{6+} reduction. In order to ensure phenol destruction and avoid overdosing, an oxidation-reduction potential (ORP) control system following the data in Figure 9-24 was recommended. Not only was the phenol effectively oxidized, but the reduction and subsequent precipitation of chromium was enhanced.

As illustrated in Figure 9-25, the effectiveness of H_2O_2 may be greatly enhanced by UV light. The application of H_2O_2 + UV light is a well-established technology with operating costs ranging from $1 to $10 per thousand gallons treated.[77] When UV is used to catalyze peroxide reactions, the ability of the water to transmit light becomes of consequence, and information must be obtained on light transmissibility. Pretreatment to remove color, iron, or suspended solids may be required.

The effectiveness of treatment by H_2O_2 + UV depends on which organic compounds are to be destroyed. This is illustrated for halogenated aliphatics in Figure 9-26.

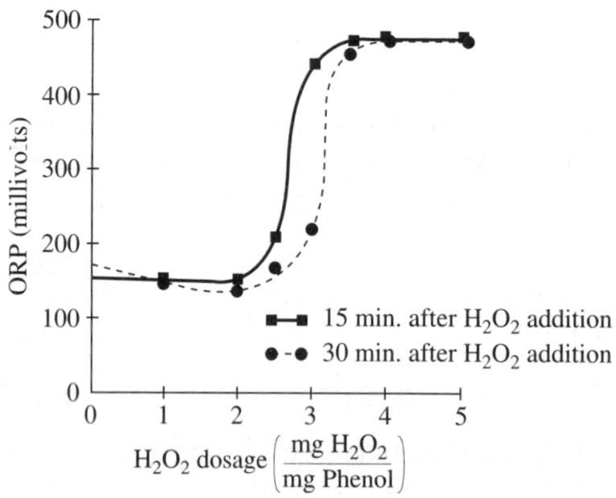

FIGURE 9-24
Control of H_2O_2 dosage by ORP.[76]

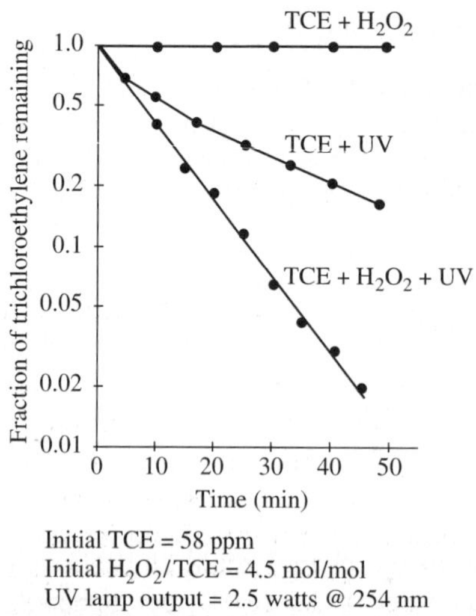

FIGURE 9-25
Decomposition of trichlorethylene with H_2O_2, UV and H_2O_2+UV.[78]

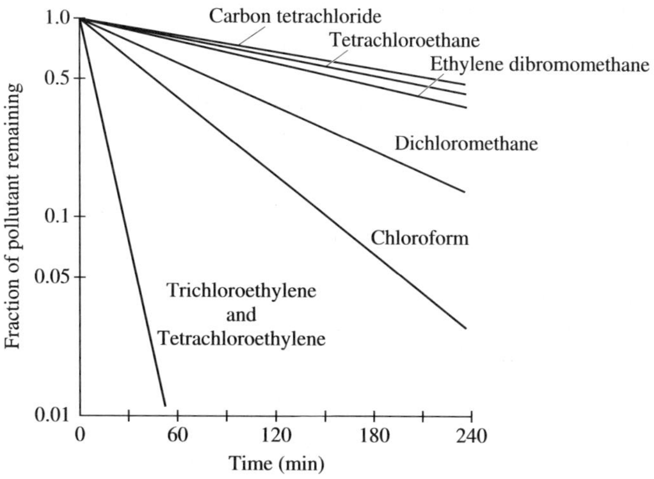

FIGURE 9-26
Destruction of halogenated aliphatics by $H_2O_2 + UV$ at 20°C.[79]

EXAMPLE 9-12. REMOVAL OF PHENOL WITH H_2O_2 AND IRON CATALYST. Calculate the amount of H_2O_2 and iron catalyst (as Fe^{2+}) required to remove 99 percent of the phenol in a 10,000 L/day (2640 gal/day) waste stream containing 6000 mg/L of phenol. Use the dosages found by Schmitt[80] for oxidation of phenol after hexavalent chromium reduction.

Solution. The required dosages for 99 percent removal are 2.5 g H_2O_2/g phenol and 0.05 mg Fe^{2+}/mg H_2O_2.

Mass of phenol to be treated per day: 10,000 L/day × 6000 mg/L = 6×10^7 mg/day = 60 kg/day

Mass of H_2O_2 required daily = 60 kg/day × 2.5 g H_2O_2/g phenol = 150 kg/day (330 lb/day)

Mass of Fe^{2+} required daily = 150 kg/day × 0.05 mg Fe^{2+}/mg H_2O_2 = 7.5 kg/day (16.5 lb/day)

Chlorine

Chlorine and chlorine compounds such as chlorine dioxide are frequently utilized as chemical oxidizing agents in waste treatment applications. For aqueous waste treatment applications, chlorine is evaporated to a gas and mixed with water to provide a hypochlorous acid (HOCl) solution:

$$Cl_2(aq) + H_2O \rightarrow HOCl + H^+ + Cl^- \qquad (9\text{-}60)$$

Hypochlorous acid ionizes as follows:

$$HOCl \rightarrow H^+ + OCl^- \qquad (9\text{-}61)$$

Equation 9-61 is very pH dependent, with the dissociation constant of hypochlorous acid being approximately 7.6 at ambient temperatures.[81] The oxidizing power of chlorine tends to increase with pH. At pH values higher than 7.5, hypochlorite (OCl^-) is the predominant species in Equation 9-61.

A classical application of oxidation with chlorine is the destruction of cyanide.[82] Under alkaline conditions cyanide is converted to the less toxic cyanate:

$$CN^- + OCl^- \rightarrow CNO^- + Cl^- \tag{9-62}$$

Caution must be exercised to maintain pH above 10 for reaction 9-62 to prevent cyanogen chloride (CNCl), a toxic gas, from being produced:

$$NaCN + Cl_2 \rightarrow CNCl + NaCl \tag{9-63}$$

Under caustic conditions, cyanogen chloride is converted to sodium cyanate:

$$CNCl + 2NaOH \rightarrow NaCNO + H_2O + NaCl \tag{9-64}$$

Cyanate is destroyed by further alkaline reaction with chlorine:

$$2NaCNO + 3Cl_2 + 4NaOH \rightarrow N_2 + 2CO_2 + 6NaCl + 2H_2O \tag{9-65}$$

In practice, excess chlorine must be added to ensure that cyanogen chloride is not produced.

The above reactions provide information to determine the stoichiometric amount of chlorine needed. However, the actual amounts required for real treatment systems would depend on other constituents in the waste, such as metals and reducing agents that create a chemical demand for chlorine. Additionally, chlorine is not stable in aqueous solutions and some will be lost to decomposition. When the cyanide is in the form of iron or nickel complexes, destruction of cyanide by chlorine is more difficult. Ferrocyanide ($[Fe(CN)_6]^{4-}$) is converted to ferricyanide ($[Fe(CN)_6]^{3-}$) and additional destruction by chlorine is ineffective. For Ni, an excess dosage of approximately 20 percent is recommended.[83]

EXAMPLE 9-13. CHEMICAL OXIDATION OF CYANIDE. A waste stream contains 85 kg of CN^- daily. Determine the stoichiometric amounts of chlorine (Cl_2) and NaOH required to: (1) oxidize the cyanide to cyanate and (2) oxidize the cyanide to N_2. Disregard the NaOH required to maintain the pH = 10.0 to prevent reaction 9-63.

Molecular weights: CN^-: 26; Cl_2: 70.9; NaOH: 40.

Solution. The conversion of CN^- to N_2 occurs through Equations 9-62 and 9-65.

1. Oxidation of cyanide to cyanate.
   From Equation 9-62, the required molar ratio $OCl^-/CN^- = 1$. From Equation 9-60, $Cl_2/OCl^- = 1$. Therefore, the molar ratio $Cl_2/CN^- = 1$ and the mass ratio $Cl_2/CN^- = 1 \times 70.9/26 = 2.73$ kg/kg. The mass of Cl_2 required daily is 85 kg/day × 2.73 kg/kg = 232.1 kg/day (510.6 lb/day).
   Mass of NaOH required daily = none according to Equation 9-62. (*Note:* pH must be well above 7.6 to ensure that OCl^- is present rather than Cl_2.)
2. Oxidation of cyanide to N_2.
   Step 1—As calculated above, 232.1 kg/day of Cl_2 are required.
   Step 2—From Equation 9-65, the molar ratio $Cl_2/CNO^- = 3/2 = 1.5$, and the mass ratio is $1.5 \times 70.9/26 = 4.09$ kg/kg. The mass of Cl_2 required daily is 85 kg/day × 4.09 kg/kg = 347.7 kg/day (764.9 lb/day).
   Total mass of Cl_2 required = 232.1 + 347.7 = 579.8 kg/day (1,275.6 lb/day).
   Mass of NaOH required daily
   Step 1—None

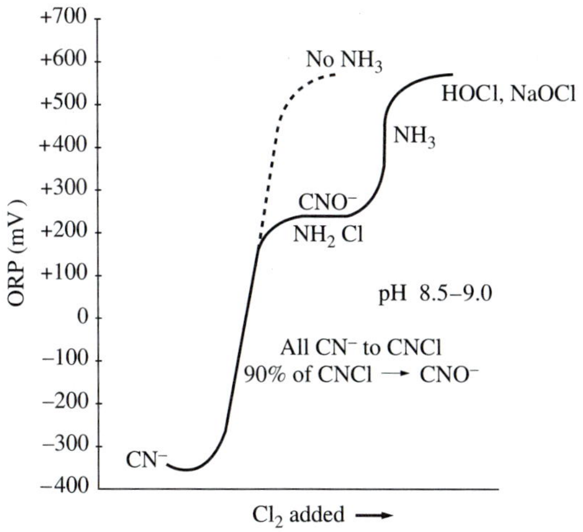

FIGURE 9-27
Control of cyanide destruction by ORP.[84]

Step 2—From Equation 9-65, the molar ratio $NaOH/CNO^- = 4/2 = 2$ and, from Equation 9-62, the molar ratio $CNO^-/CN^- = 1$. Therefore, the molar ratio $NaOH/CN^- = 2$ and the mass ratio $NaOH/CN^- = 2 \times 40/26 = 3.08$ kg/kg. The mass of NaOH required daily is 85 kg/day × 3.08 kg/kg = 261.8 kg/day (576 lb/day).

Total mass of NaOH required= 261.8 kg/day (576 lb/day).

In a continuous treatment system, destruction of cyanide by chlorine can be controlled by oxidation reduction potential as illustrated in Figure 9-27.

9-5 SUPERCRITICAL FLUIDS

Supercritical fluids (SCF) extraction and supercritical water oxidation (SCWO) represent emerging technologies in hazardous waste management. Supercritical fluids are materials at elevated temperature and pressure that have properties between those of a gas and a liquid. In SCF extraction, organics in soils, sediments, or water are dissolved in the fluid at elevated temperature and pressure conditions and released from the SCF at lower temperatures and pressures. In supercritical water oxidation, air and contaminated water are brought together above the critical point of water where complete oxidation of organic components occurs rapidly.

Process Description

In SCF extraction, the contaminated stream (liquid or solid) is introduced into an extraction vessel into which the extraction fluid, pressurized and heated to the critical point in a compressor, is continuously loaded (see Figure 9-28).[85] The organic contaminant in the contaminated stream dissolves in the SCF. The SCF is then expanded by passing it through a pressure reduction valve. The expansion of the SCF

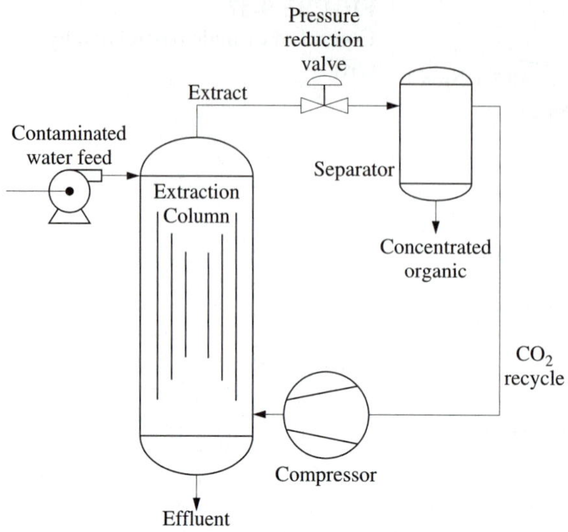

FIGURE 9-28
Schematic diagram of an SCF process for extracting organic compounds from water using CO_2.[86]

lowers the solubility of the organic contaminant in the SCF and, thus, results in the separation of the organic contaminant from the extracting fluid. The SCF is then recompressed and recycled to the extraction vessel.

Alternatively, instead of reducing the pressure of the exiting SCF, the temperature can be reduced at a constant pressure by passing the SCF through a heat exchanger. The end result is also a reduction in the solubility of the organic contaminant in the SCF and the separation of both components. In this case, a blower is used instead of a compressor, and the separated extracting fluid entering the process vessel is heated in the same heat exchanger used to cool the exiting SCF.

In supercritical water oxidation, pressurized and heated water (at supercritical conditions) containing the organic contaminants is mixed with compressed air in a reactor (oxidizer). Figure 9-29 shows a schematic of the process.[87] The oxidation of many organics in the reactor takes place in a short time [residence times of less than 1 minute at 600 to 650°C (1112 to 1200°F) have been reported to produce removal efficiencies of 99.9999 percent].[88] From the oxidizer, the mixture passes through a solids separator (the solubility of inorganic compounds in water decreases by about 3 to 4 orders of magnitude when the water is at supercritical conditions) and through high- and low-pressure vapor separators to separate the gases (mainly CO_2 and N_2) from the effluent water. A portion of the energy recovered from the high-pressure separator is used to compress the influent air.

Theory of Supercritical Fluid Applications

Fluids are normally divided into two phases: liquids and gases. However, at elevated temperatures and pressures, a point is reached where this distinction is no longer discernible. This is termed the **critical point.** An SCF is a single-phase fluid at temperature and pressure above the critical point. An SCF has properties between those of the liquid and gas phases (i.e., densities approaching those of the liquid phase and diffusivities and viscosities approaching those of the gas phase). As a

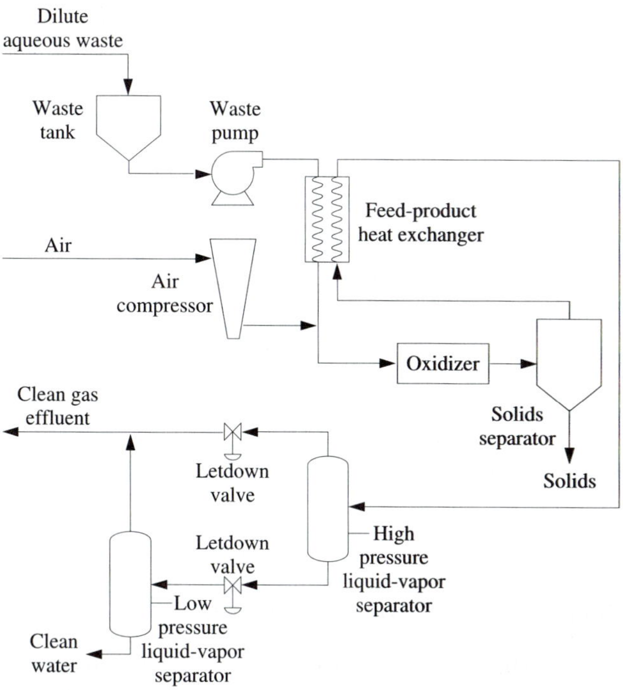

FIGURE 9-29
Process schematic diagram for SCF oxidation of an aqueous waste.[89]

result of these properties, organic compounds are highly soluble in SCFs and can easily transfer to the SCF from their original medium. Critical conditions for a number of SCF solvents are given in Table 9-9.

Phase diagrams like the pressure-temperature diagram for water shown in Figure 9-30 aid in illustrating the relationships between a compound's properties. Similar diagrams can be constructed for other pure compounds. Measurements represent the pressure at which the compound is present in the solid phase (from point *A* up to the triple point) and the pressure at which the compound is present in the liquid

TABLE 9-9
Critical constants[90]

Solvent	Temperature, °C	Pressure, atm	Density, g/cm^3
Carbon dioxide	31.1	73.0	0.460
Water	374.15	218.4	0.323
Ammonia	132.4	111.5	0.235
Benzene	288.5	47.7	0.304
Toluene	320.6	41.6	0.292
Cyclohexane	281.0	40.4	0.270

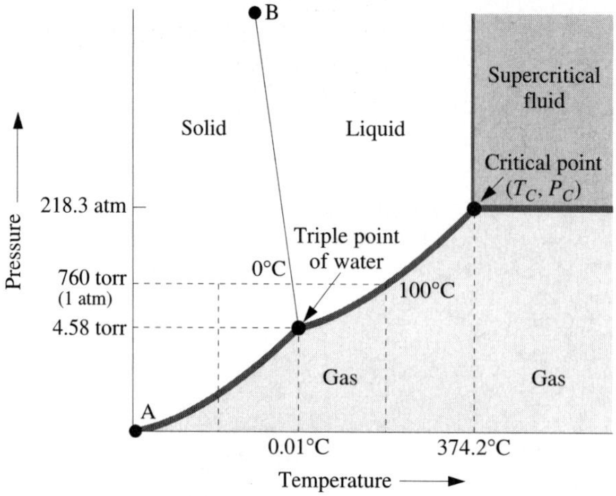

FIGURE 9-30
Pressure temperature phase diagram for water.[92]

phase (from the triple point up to the critical point). The solid-liquid equilibrium relationship is illustrated by the line from the triple point through point B. At the **triple point,** all three phases coexist in equilibrium.

EXAMPLE 9-14. DEGREES OF FREEDOM. Use the phase rule to determine the degrees of freedom (i.e., number of variables that can be selected independently) at the triple point of water.

The phase rule (Equation 3-28) is:

$F = C + 2 - P$
$C = 1$ component (i.e., water)
$P = 3$ phases (solid/liquid/vapor)

Therefore, $F = 0$ independent variables, and the triple point is considered invariant.

As shown in Figure 9-30, the **critical point** (T_C, P_C) represents the highest values of T and P where vapor/liquid equilibrium can exist. The material in the area labeled "Supercritical fluid" is neither a liquid nor a gas. For example, a liquid can be vaporized (i.e., instantaneously converted to a vapor) if the pressure is reduced at constant temperature. An SCF undergoes a gradual change in properties from those of a liquid to properties of a vapor with no instantaneous phase change.

Properties are commonly expressed in terms of reduced temperature and pressure (i.e., the actual value divided by the critical value). The basis for this is the **theorem of corresponding states:**

> All gases when compared at the same reduced temperature and reduced pressure, have approximately the same compressibility factor and all deviate from ideal gas behavior to about the same degree.[91]

The compressibility factor Z is simply the inverse of the density ρ.

Figure 9-31 shows isotherms of reduced temperature (T/T_C) on a plot of reduced density (ρ/ρ_C) (ordinate) versus reduced pressure (P/P_C) (abscissa). All three variables are termed "reduced" because values of T, ρ, and P have been divided by their critical values.

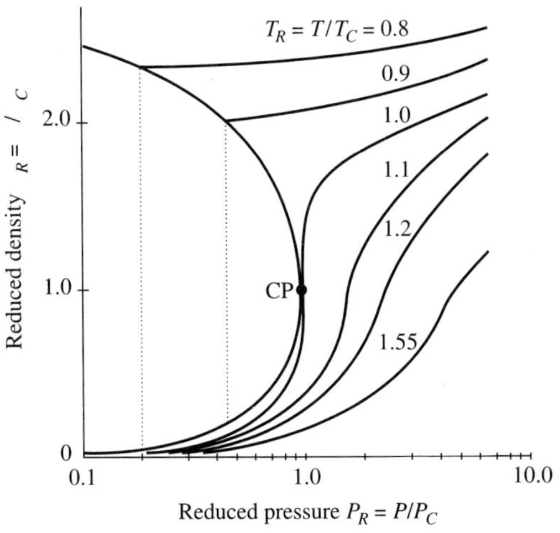

FIGURE 9-31
Variation in reduced density (ρ_R) near the critical point for a pure compound.[93]

Figure 9-31 illustrates one of the properties of SCFs. Near the critical point (CP), a small change in pressure (from $P_R = 1$ to $P_R = 1.1$) can create a large change in density (from $\rho_R = 1$ to $\rho_R = 1.6$) at constant temperature ($T_R = 1$). At the higher density the SCF acts as a more effective solvent.

In most instances, the viscosity changes rapidly near the critical point. Curves of generalized reduced viscosity as a function of temperature and pressure are illustrated in Figure 9-32.

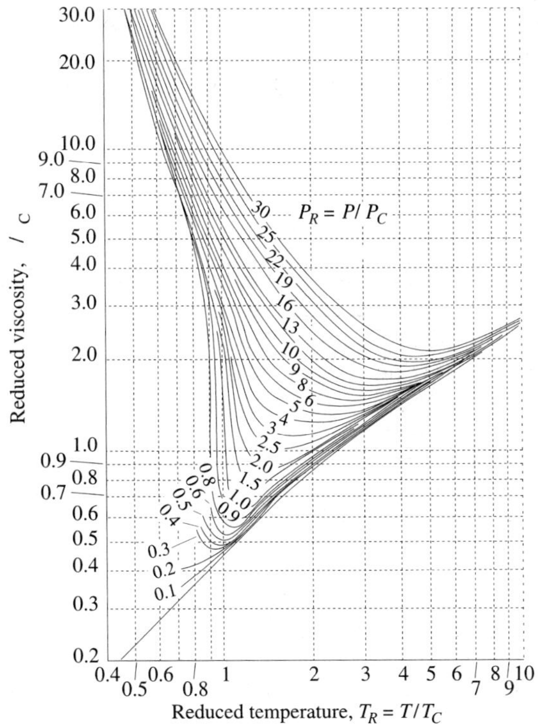

FIGURE 9-32
Generalized reduced viscosities.[94]

For CO_2, the viscosity at 250 atm and 37°C is about 0.09 cP, an order of magnitude below typical organic solvents. Diffusivities of organic constituents in supercritical fluids are typically orders of magnitude higher than is usual for organic liquids at normal conditions. This lower solvent viscosity and higher diffusivity permits much higher mass transfer rates of organic contaminants into SCFs.

Properties of supercritical water are illustrated in Figure 9-33. As the temperature is raised to near the critical point, the density changes from that of a liquid to that of a vapor, without going through an abrupt phase change. The high dielectric constant of water at ambient conditions is due to strong hydrogen bonds, which provide the very high solubility for inorganic compounds. At supercritical conditions, the dielectric constant can drop to 2.5. As an example of the impact on solubility, NaCl has a solubility in water at ambient conditions of 37 percent by weight (370,000 mg/L). At supercritical conditions, this drops by almost 4 orders of magnitude to about 100 mg/L. Many organic compounds, on the other hand, become essentially miscible in supercritical water. For example, benzene is miscible in supercritical water, but has a solubility of only 1700 mg/L at ambient temperature and pressure (Appendix B).

Design Considerations

Among the many decisions to be made in designing an SCF extraction system is the selection of solvent. Treybal[96] suggests the factors shown in Table 9-10 as appropriate for selecting solvents for solvent extraction.

It is essential to be able to reuse the solvent in order for these systems to be cost-effective. Separation of the SCF from the feed stream depends on the difference in density of the two fluids. For SCF extraction, reactions between the solvent and solute would be undesirable. However, for oxidation of organics in supercritical water, reactivity is essential. Concern with solvent toxicity has led to increased use of CO_2 as an SCF. Potential hazards with SCF extraction technology would include

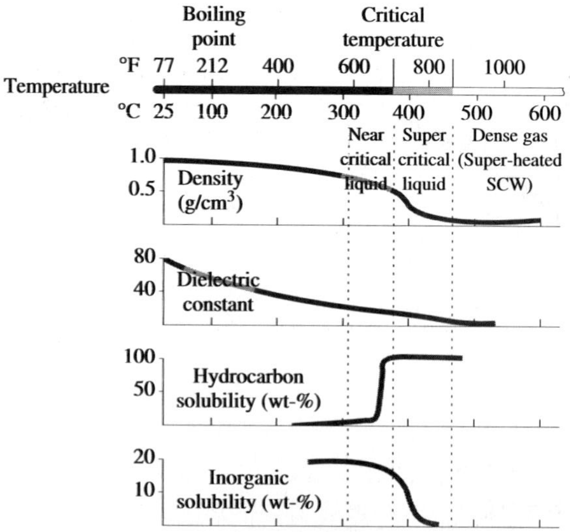

FIGURE 9-33
Properties of water at 250 atm.[95]

TABLE 9-10
Factors for selection of solvents

Distribution coefficient	Recoverability
Density	Critical temperature and pressure
Toxicity	Chemical reactivity
Interfacial tension	Cost
Hazard	

corrosive characteristics and flammability. The lower the critical temperature and pressure of the SCF, the less expensive the process is.

The distribution coefficient (partition coefficient) of a compound between two phases can be calculated by using Equation 9-66. In this equation, the extract is the extraction solvent leaving the reactor, and the raffinate is the effluent.

$$K = C_e/C_r \tag{9-66}$$

where K = distribution coefficient (dimensionless)

C_e = concentration of organic in the extract at equilibrium (weight %)

C_r = concentration of organic in the raffinate (effluent) at equilibrium (weight %)

The distribution coefficient is used to determine the *minimum* quantity of extraction solvent needed to extract a compound from a feed. For example, a value of $K = 4$ indicates that *at least* 1 kg of extraction solvent would be required to extract all of the compound from 4 kg of feed.[97] In actual application, the quantity of extraction solvent would be a function of the removal required and substantially above the minimum.

In supercritical water oxidation (SCWO), organic compounds are destroyed rather than removed. Reactions occur in a single homogeneous phase, characteristic of SCFs, and produce the end products shown in Table 9-11.

The process is reported to be cost-effective for aqueous wastes with organic concentrations in the range of 1 to 20 percent by weight. In this range the heat of combustion raises the SCF temperature from the critical temperature of water (374°C) to 600–650°C.[98]

Additional design considerations include: (1) the ability of the waste to be pressurized, (2) the potential to form char, and (3) the removal of the solids produced.[99] If the waste is a solid, slurry, or suspension, the solid particles must be reduced

TABLE 9-11
SCWO end products

Initial chemical	Resulting end product
Organic compounds	Carbon dioxide
Chlorine	Chlorides
Nitrogen compounds	Nitrates
Sulfur	Sulfates
Phosphorous	Phosphates

to less than 100 μm in diameter, which is the particle size limit for high-pressure pumps. Charring may occur when organics are heated slowly to the supercritical water conditions. The formation of char is undesirable, as the char is oxidized more slowly than solubilized organics, resulting in the need for longer detention times in the oxidizer.

The separation of inorganic salts precipitated during the process may be a problem. These solids may adhere to the reactor walls, decreasing the effective volume available for the reaction and, thus, the residence time in the reactor. Also, the solids may adhere to the pipelines downstream of the reactor, eventually plugging the lines. Solids problems can be avoided by periodically washing the reactor and pipelines. The precipitated solids present a removal and disposal problem because of the potential for either handling a concentrated, corrosive brine or a dilute brine that has to be further treated for removal of toxic metals prior to disposal.

Another major design consideration for both SCF extraction and supercritical water oxidation is that the materials of construction must be able to withstand the extreme pressures and temperatures required by SCF systems and the corrosive characteristics of the supercritical streams. SCF extraction vessels are typically constructed of stainless steel or glass, while supercritical water oxidation has been carried out in vessels made of Hastelloy C-276 and Inconel 625 (high-nickel alloys).

TABLE 9-12
Results of bench-scale evaluation tests of SCF oxidation*

Compound	Temp., °C	Time, min	Destruction efficiency, %	Reference
Aliphatic hydrocarbons				
Cyclohexane	445	7.	99.97	101
Aromatic hydrocarbons				
Biphenyl	450	7.	99.97	102
o-Xylene	495	3.6	99.93	103
Halogenated aliphatics				
1,1,1-Trichloroethane	495	3.6	99.99	104
1,2-Ethylene dichloride	495	3.6	99.99	105
1,1,2,2-Tetrachloroethylene	495	3.6	99.99	106
Halogenated aromatics				
O-Chlorotoluene	495	3.6	99.99	107
Hexachlorocyclopentadiene	488	3.5	99.99	108
1,2,4-Trichlorobenzene	495	3.6	99.99	109
4,4-Dichlorobiphenyl	500	4.4	99.993	110
DDT	505	3.7	99.997	111
PCB 1234	510	3.7	99.99	112
PCB 1254	510	3.7	99.99	113
Oxygenated compounds				
Methyl ethyl ketone	460	3.2	99.96	114
Methyl ethyl ketone	505	3.7	99.993	115
Dextrose	440	7.	99.6	116
Organic nitrogen compounds				
2,4-Dinitrotoluene	457	0.5	99.7	117
2,4-Dinitrotoluene	513	0.5	99.992	118
2,4-Dinitrotoluene	574	0.5	99.9998	119

*Source: Modell.[120]

As noted in the introduction, SCF processes are emerging technologies, and few full-scale applications exist. Generally accepted design procedures do not as yet exist and current practice is to perform bench-scale studies in conjunction with SCF equipment manufacturers to ensure that the process is both feasible and cost-effective for a specific application. Table 9-12, from Modell,[100] provides the results of such tests for a range of organic compounds, indicating the potentially wide applicability of SCF processes.

9-6 MEMBRANE PROCESSES

The application of membranes to separate water from contaminants is a well-established technology in industry. For example, a membrane process (reverse osmosis) is used to produce industrial water of extremely high quality in the manufacture of semiconductors, for which the mineral content of ordinary drinking water is too high. The term **membrane** usually refers to a barrier to flow that will allow the passage of water, ions, or small molecules (**semipermeable membrane**). However, membrane processes do not operate as a conventional filtration process. In most applications, the solution flows in parallel to the membrane and the transfer of solute or solvent occurs through application of direct electrical current (electrodialysis) or high pressures (reverse osmosis and ultrafiltration) to the solution. The membrane may be made of a solid matrix or swollen gel.[121]

The membranes used in industrial processes are subject to fouling and degradation, which limit their applicability in many waste treatment situations. The application of these processes to hazardous waste management is generally limited to those instances where extremely toxic materials cannot be removed by more cost-effective technologies. Research is being conducted to produce membranes resistant to fouling and new cleaning procedures to prevent fouling.[122]

This section discusses three membrane processes of interest in managing hazardous wastes: **electrodialysis, reverse osmosis,** and **ultrafiltration.**

Process Description

Electrodialysis, reverse osmosis, and ultrafiltration have different modes of operation and, thus, use different membrane types. Each of these processes is described in the following paragraphs.

Electrodialysis

Electrodialysis consists of the separation of ionic species from water by applying a direct-current electrical field. The ion-selective membranes used in electrodialysis (cation and anion exchange membranes) permit the passage of cations through the cation exchange membrane and of anions through the anion exchange membrane. By alternating cation and anion exchange membrane between two electrodes, alternate dilute and concentrated cells are created (see Figure 9-34). This results in the separation of the ionic species from water. Newer cell design incorporates the periodic reversal of current to reduce fouling.

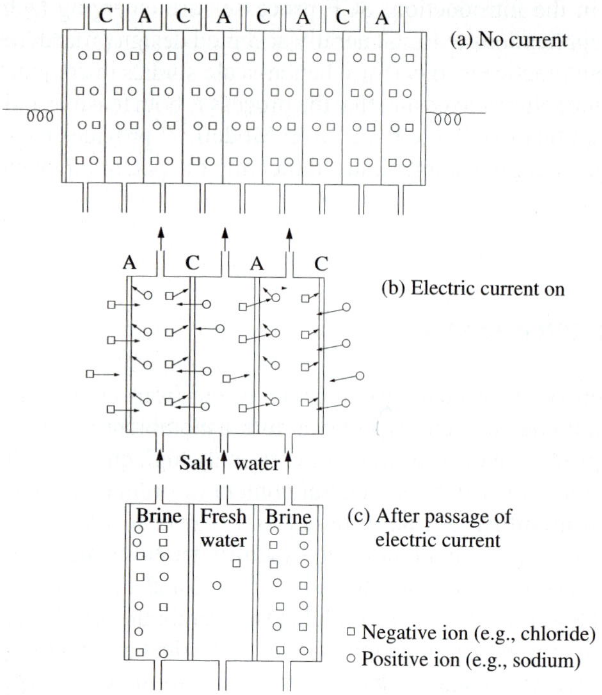

FIGURE 9-34
Operation of electrodialysis systems.[123]

Electrodialysis is used in the electroplating industry for the recycling of metals and purified water. Because a single pass through an electrodialysis cell usually removes only 30 to 60 percent of the metal, recycling of the produced brine to the feed inlet is necessary until the desired metal concentration is achieved.[124] This brine is returned to the plating bath, while the treated water is returned to the rinse tanks.

As shown in Figure 9-35, hundreds of individual cells can be combined between two electrodes.[125] The membranes are about 0.5 mm thick and the spacers that distribute the liquid over the surface of the membrane are 1 mm thick. The electrodes must be continually rinsed to prevent buildup of hydrogen gas at the cathode and oxygen or chlorine at the anode.

Typical electrodialysis units operate at pressures in the order of 276 to 414 kPa (40 to 60 psi). Generally, 90 percent of the feed is transformed to product water, while approximately 10 percent remains as the concentrate.[126]

A modification of the technology using a three-compartment cell[130] is shown in Figure 9-36. A salt solution MX (cation M, anion X), enters the central compartment and is separated into acid (HX) and basic (MOH) streams by electrical energy. The bipolar membrane is a composite of an anion selective membrane and a cation-selective membrane with an interface. Water in the interface is dissociated electrically to form H^+ and OH^-.

This three-compartment cell has been applied to the regeneration of spent pickle liquor. The process starts with pretreatment consisting of neutralization with KOH

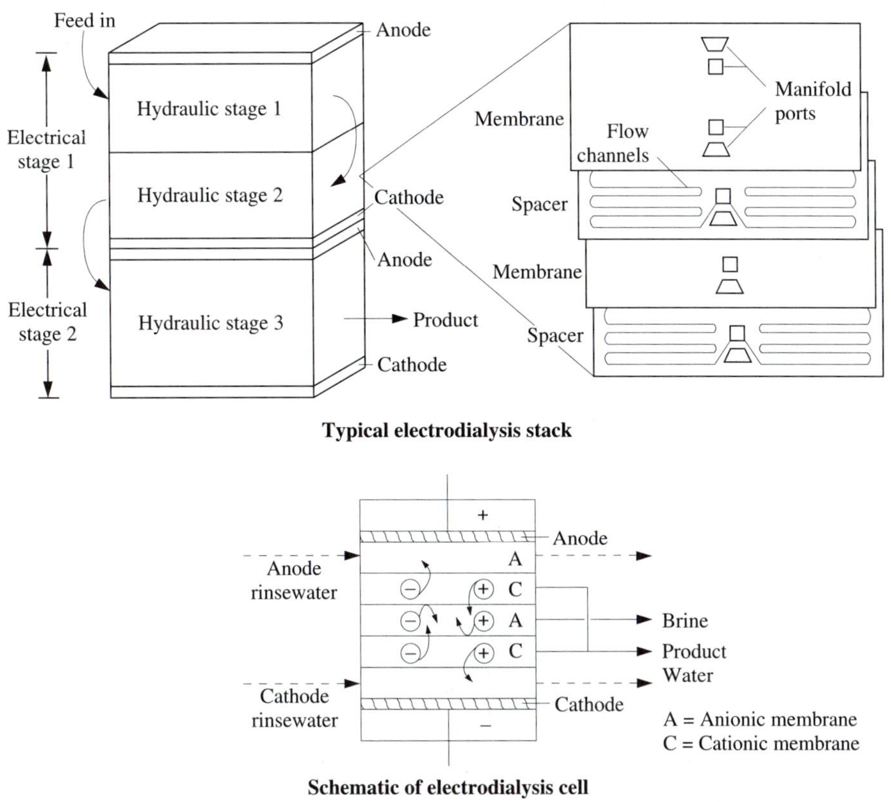

FIGURE 9-35
Schematic of electrodialysis system.[127–129]

to form a KF/KNO_3 solution, which is then filtered by diatomaceous earth. An electrodialysis cell then regenerates hydrofluoric and nitric acids, which are returned to the pickling process. A conventional electrodialysis unit is used to recover water from the filter cake wash. The process is illustrated in Figure 9-37.

Reverse Osmosis

In reverse osmosis, a solvent is separated from a solution by applying a pressure greater than the osmotic pressure, thus forcing the solvent through a semipermeable membrane.[132]

The driving force for osmosis, the transport of solvent across a semipermeable membrane, is the gradient of the chemical potential of the solvent in the direction of flow. The process is illustrated schematically in Figure 9-38. In the left diagram, water flows from the pure water side to the salt side because the concentration of salt is higher on the right and the system tends to move toward an osmotic equilibrium state. The semipermeable membrane will allow water molecules to pass, but only a small fraction of salt molecules. This results in an increase in pressure in the salt solution until the osmotic pressure is reached (center diagram). At this point, equilibrium exists and the net transport of solvent is zero. **Reverse osmosis** (RO) occurs if mechanical pressure greatly in excess of the osmotic pressure is applied to

FIGURE 9-36
Three-compartment electrodialysis cell.

FIGURE 9-37
Pickle liquor recovery process.[131]

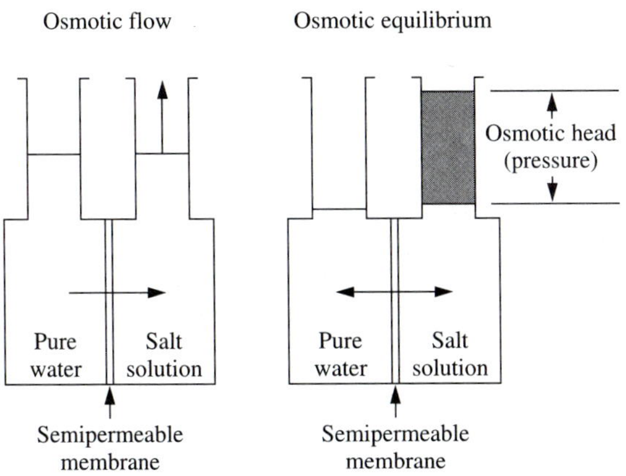

FIGURE 9-38
Pictorial explanation of reverse osmosis.[133]

the salt solution. This results in the transfer of water from the salt solution through the membrane into the pure water side (right diagram).

A reverse osmosis unit consists of several modules to which the feed is added through a high-pressure pump. The feed must be pumped at a pressure high enough as to maintain the design pressure difference in the last module.[134] The difference between the system and osmotic pressure is usually around 1000 kPa (145 psi) with a maximum system pressure of 5500 kPa (800 psi).

Reverse osmosis is used in the electroplating industry to recover both metals and water. Figure 9-39 presents a flow diagram of the process. The rinsewater from the first tank is passed through a reverse osmosis unit to recover a concentrated metal solution that is recycled to the plating bath and a purified water stream that is recycled to the third rinse tank. Figure 9-40 shows a photograph of a typical unit.

Reverse osmosis has been applied to both organic and inorganic hazardous waste problems. Martin et al.[136] report on EPA research where a reverse osmosis unit was operated at 0.69 to 2.07 MPa (100 to 300 psi) to remove organics. Reported rejection rates were 97 percent for dimethylphthalate, and 98 percent for naphthalene. Higgins[137] reports that reverse osmosis can concentrate divalent metals in plating rinse water to a 10 to 20 percent solution, greatly facilitating the recycling of these materials.

An important RO design consideration is the need to dispose, recycle, or reuse both the permeate and concentrate streams, since complete separation of solute from solvent is not possible.

Ultrafiltration

The third membrane process to be discussed in this chapter, ultrafiltration, separates solutes from a solvent on the basis of molecular size and shape by passing the solution through a membrane module, where a pressure difference is maintained across the membrane. Water and small molecules move through the membrane to the lower-pressure side, while the larger molecules are retained by the membrane.

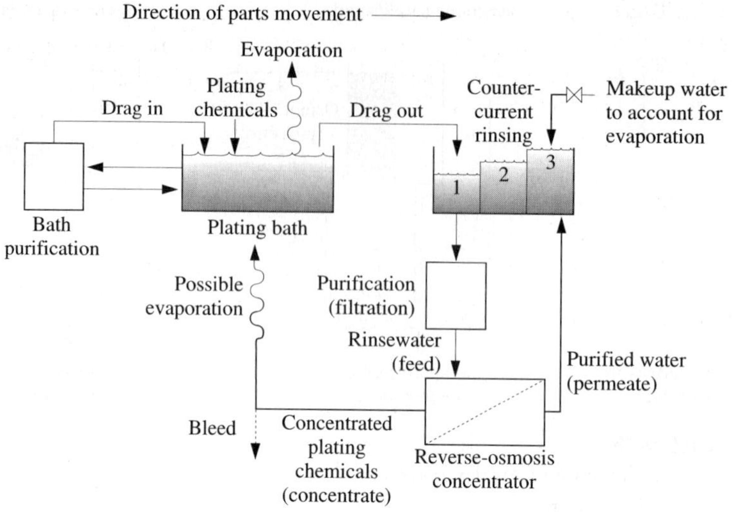

FIGURE 9-39
Flow diagram for a reverse osmosis system treating plating rinsewater.[135]

FIGURE 9-40
Reverse osmosis unit. *(Courtesy of Process Science, Inc.)*

To prevent fouling, the solution is passed at a high velocity over the membrane, resulting in small separation efficiencies. To improve the separation efficiency, the feed is recycled through the ultrafiltration unit several times or passed through several modules in series.[138]

In ultrafiltration, solutes of molecular weight greater than about 500 and less than 500,000 can be separated from a solution. The lower limit is related to the pore size of commercially available membranes. Solutes above the upper limit of molecular sizes are no longer separated from solution by ultrafiltration, but by conventional

filtration.[139] Ultrafiltration membranes generally allow the passage of water and most low molecular weight solutes (including inorganic ions) while retaining molecules having molecular weights greater than the stated molecular weight cutoff. Because of the lower osmotic pressures of the high molecular weight solutes separated by ultrafiltration membranes, the operating pressure differences are lower than those used in reverse osmosis, in the order of 35 to 690 kPa (5 to 100 psi).[140]

Ultrafiltration systems operate at lower pressures than reverse osmosis because the osmotic pressures of the larger molecular weight solutes are generally lower, but more important, because the pores are larger and solvent transport is primarily advective (or convective); e.g., the volumetric flux (m/s) of solution (J_w), described by Hagan-Poiseulle flow in cylindrical pores, is:

$$J_w = \frac{\varepsilon r^2 \Delta P}{8 \mu \delta} \tag{9-67}$$

where $\varepsilon =$ membrane porosity
$r =$ membrane pore size
$\Delta P =$ net transmembrane pressure
$\mu =$ viscosity
$\delta =$ membrane thickness

Ultrafiltration can be operated as a batch, single-stage continuous or feed-and-bleed, and multistage continuous operation. Figure 9-41 depicts the batch and continuous operation modes. In all operation modes, the feed is passed through ultrafiltration membrane modules several times to improve the solute separation efficiency. Ultrafiltration units have been used in the recycle of electrophoretic paints, the removal of starches and enzymes from wastewaters, the recycling of polyvinyl alcohol in the textile industry, and the separation of oil-water emulsions.[141] In most applications, both the water and the concentrated solute are reused.

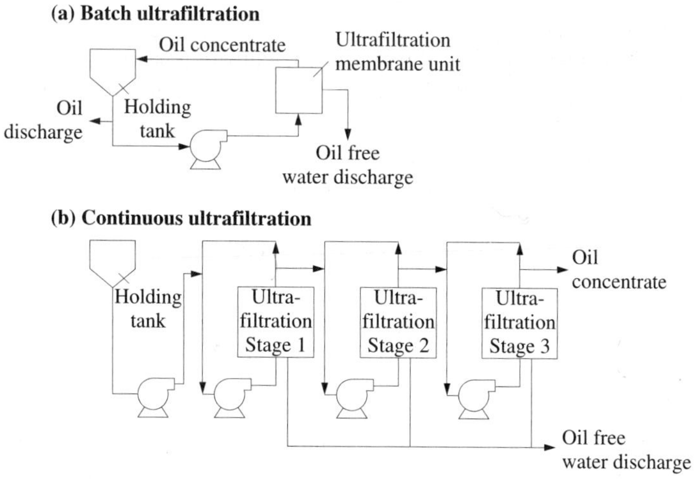

FIGURE 9-41
Batch and continuous ultrafiltration operations.[143]

Theory

Electrodialysis

The current required to operate an electrodialysis unit can be calculated by using Faraday's law, which indicates that to transfer 1 gram equivalent of a substance by electrical force, a single cell will require 1 faraday of electrical current (96,487 coulombs). In n cells, the required current is[142]:

$$I = (FQN/n) \times (E_1/E_2) \tag{9-68}$$

where I = current (amperes)
F = Faraday's constant = 96,487 C/g-equivalent [or (A · s)/g-equivalent]
Q = flow rate (L/s) (gal/min × 0.06308 = L/s)
N = normality of solution (g-equivalents/L)
n = number of cells between electrodes
E_1 = removal efficiency (fraction)
E_2 = current efficiency (fraction)

The current efficiency E_2 represents the number of gram-equivalents of product actually produced per faraday applied. Current efficiency is adversely affected by incfficient selectivity of the electrodialysis membranes and diffusional losses due to concentration gradients.

The voltage may be determined from Ohm's law:

$$E = IR \tag{9-69}$$

where E = voltage requirement (volts)
R = resistance through the unit (ohms)

The required power is:

$$P = I^2R \tag{9-70}$$

where P = power (watts).

The **current density** (CD) is defined as the current passing through a unit area of membrane (A/m^2). The ability of the solution to carry current is a function of the normality N of the solution. The ratio CD/N provides an important characterization of the system. A large value can result in too much current for the ions available in solution. This can lead to concentration polarization (i.e., regions of low ion concentration), which will result in an increase in resistance and a decrease in system efficiency, as described in the section Design Considerations.

EXAMPLE 9-15. ELECTRODIALYSIS OF NICKEL-BEARING WASTE. An electroplating plant has 1700 m^3/day of a nickel-bearing waste. The nickel concentration is 15,400 mg/L as $NiSO_4$. Assume the following characteristics of the system:

Resistance through unit: 10.5 Ω; current efficiency = 90 percent; maximum CD/N ratio = 6000 (A/m^2)/(g-equivalent/L); membrane area = 1.0 m^2. Provide a preliminary design of a system to produce 90 percent removal of Ni. Determine the number of membranes, power required, and the annual electrical cost at \$0.07/kWh. (Molecular weight of $NiSO_4$ = 154.7 g/gmol.)

Solution

1. Normality:

$$\text{Equivalent weight of NiSO}_4 = \frac{\text{molecular weight}}{\text{g-equivalent/gmol}}$$

$$= \frac{154.7\ \text{g/gmol}}{2\ \text{g-equivalent/gmol}}$$

$$= 77.4\ \text{g/g-equivalent}$$

Concentration = 15,400 mg/L = 15.4 g/L
Normality = 15.4/77.4 = $\underline{\underline{0.2N}}$

2. Maximum current density = 6000(0.2) = 1200 A/m^2
Current = 1200 A/m^2 × 1.0 m^2 = $\underline{\underline{1200\ \text{A}}}$

3. Number of cells: $I = (FQN/n) \times (E_1/E_2)$; but $E_1 = E_2 = 0.9$. Therefore,

$$n = (FQN/I)$$

$$= \frac{96{,}487\left(\frac{\text{A}\cdot\text{s}}{\text{g-equivalent}}\right) \times 1.7 \times 10^3 \left(\frac{\text{m}^3}{\text{day}}\right) \times 10^3 \left(\frac{\text{L}}{\text{m}^3}\right)\left(\frac{0.2\ \text{g-equivalent}}{\text{L}}\right)}{1{,}200\ \text{A} \times 86{,}400\ \text{s/day}}$$

$$= 316.4\ \text{cells}$$

Number of membranes = 2 × 316.4 = 633 ≅ $\underline{\underline{640\ \text{membranes in 320 cells}}}$

4. Power = I^2R = (1200 A)^2 × 10.5 Ω = 1.51 × 10^7 W = $\underline{\underline{1.51 \times 10^4\ \text{kW}}}$

5. Annual electrical cost = 1.51 × 10^4 kW ×(365 × 24) h × \$0.05/kWh = $\underline{\underline{\$9.23\ \text{million}}}$

Reverse Osmosis

In reverse osmosis systems, the osmotic pressure, solute rejection, and water and solute fluxes are of interest. The flow across the membrane exerts an osmotic pressure π, determined by the **van't Hoff equation**[144]:

$$\pi = \phi_c N C_s RT \tag{9-71}$$

where π = osmotic pressure (atm)
ϕ_c = osmotic pressure coefficient
N = number of ions in each molecule
C_s = concentration (gmol/L)
R = universal gas constant = 0.082 (atm · L)/(gmol · K)
T = absolute temperature (K)

EXAMPLE 9-16. OSMOTIC PRESSURE. Determine the osmotic pressure for a sodium chloride brine solution with concentration = 23,400 mg/L at 25°C, given $\phi_c = 0.92$.

Solution

$$\pi = \phi_c N C_s RT$$

$N = 2(\mathrm{NaCl} = \mathrm{Na}^+ + \mathrm{Cl}^-)$; molecular weight = 58.45 g/gmol

C_s = 23,400 mg/L = (23.4 g/L)/(58.45 g/gmol) = 0.4 gmol/L

$$\pi = 0.92 \times 2 \times (0.4\ \text{gmol/L}) \times [0.082\ (\text{atm} \cdot \text{L})/(\text{gmol} \cdot \text{K})] \times (298.2\ \text{K})$$

$$= \underline{\underline{18.01\ \text{atm}}} \qquad (264.7\ \text{psi})$$

Membrane rejection of contaminants is defined in terms of influent and effluent concentrations:

$$R = 100 \times (C_{\text{in}} - C_{\text{out}})/C_{\text{in}} \tag{9-72}$$

where R = rejection (%)
C_{in} = concentration in feed (mg/L or gmol/cm^3)
C_{out} = concentration in permeate (mg/L or gmol/cm^3)

The flux of water through a semipermeable membrane is proportional to the difference in the operating pressure and the osmotic pressure.[145]

$$J_w = (D_w C_w V_m / RT\ \Delta Z)(\Delta P - \Delta\pi) \tag{9-73}$$

where J_w = water flux through the membrane [gmol/(cm^2 · s)]
D_w = diffusivity of water in membrane (cm^2/s)
C_w = concentration of water in the membrane (gmol/cm^3)
V_m = partial molar volume of water ≅ 0.018 L/gmol = 18 cm^3/gmol
R = universal gas constant = 82.057 (atm · cm^3)/(gmol · K)
T = absolute temperature (K)
ΔZ = membrane thickness (cm)
ΔP = pressure differential across membrane = $P_{\text{in}} - P_{\text{out}}$ (atm)
$\Delta\pi$ = osmotic pressure differential across membrane $\pi_{\text{in}} - \pi_{\text{out}}$(atm)

Equation 9-73 applies only if D_{w}, C_{w}, and V_{m} are independent of pressure. This is a valid assumption at pressures of up to 14,200 kPa (2050 psi). The term $(D_w C_w V_m / RT\ \Delta Z)$ is referred to as the coefficient of water permeation, W_p, and is a characteristic of the particular membrane utilized with units of gmol/(cm^2 · s · atm):

$$J_w = W_p \times (\Delta P - \Delta\pi) \tag{9-74}$$

Similarly, the salt flux through the membrane is driven by the concentration gradient:

$$J_s = (D_s K_s / \Delta Z)(C_{\text{in}} - C_{\text{out}}) \tag{9-75}$$

where J_s = salt flux through the membrane [gmol/(cm^2 · s)]
D_s = diffusivity of the solute in the membrane (cm^2/s)
K_s = solute distribution coefficient (dimensionless)
C = concentration (gmol/cm^3)

The concentration gradient governing solute transport is the difference between the *local* concentration at the membrane surface and the concentration in the permeate solution.

The term $D_s K_s/\Delta Z$ is referred to as the solute transport parameter K_p and is often defined empirically on the basis of bulk concentrations at the inlet and outlet of a membrane module, $(C_{\text{in}} - C_{\text{out}})$. When defined in this manner, the solute transport parameter depends not only on the solute and the membrane material, but on the design and hydrodynamic conditions within the module.

The coefficient of permeability is also a characteristic of the particular membrane and has units of cm/s:

$$J_s = K_p \times (C_{\text{in}} - C_{\text{out}}) \tag{9-76}$$

EXAMPLE 9-17. SELECTION OF REVERSE OSMOSIS SYSTEM. A wastewater stream at 25°C is to be treated by a reverse osmosis system. The wastewater has a flow rate of 50 gal/min and a concentration of 5000 mg/L of sodium chloride. A 75 percent recovery rate is required.

A reverse osmosis vendor has recommended a membrane with the following characteristics:

$$W_p = 1.9 \times 10^{-6} \frac{\text{gmol}}{\text{cm}^2 \cdot \text{s} \cdot \text{atm}}$$

Area for a bundle = 300 ft^2
50 percent recovery rate
Optimal pressure differential across the membrane = 500 psi
95 percent rejection rate of the salt

Determine:
1. The osmotic pressure of the solution.
2. The water flow through the membrane.
3. Number of units required for 75 percent recovery.
4. The final product water quantity and quality.

Solution

1. The osmotic pressure of the solution using the van't Hoff equation.

$$\pi = \phi_c N C_s RT$$

$$\phi_c = 0.92, \text{ from Example 9-16}$$

$$N = 2$$

$$C_s = 5000 \text{ mg/L} = \frac{5 \text{ g/L}}{58.45 \text{ g/gmol}} = 0.09 \text{ gmol/L}$$

$$\pi = (0.92)(2)(0.09)(0.082)(298.3) = 4.05 \text{ atm} = \underline{\underline{59 \text{ psi}}}$$

2. The water flow through the membrane.

$$J_w = W_P(\Delta P - \Delta\pi)$$

$$= 1.9 \times 10^{-6} \frac{\text{gmol}}{\text{cm}^2 \cdot \text{s} \cdot \text{atm}} (500 \text{ psi} - 59 \text{ psi}) \left(\frac{1 \text{ atm}}{14.7 \text{ psi}}\right)$$

$$= \underline{\underline{5.7 \times 10^{-5}}} \text{ gmol/(cm}^2 \cdot \text{s)}$$

3. To treat 50 gal/min of water:

$$50 \frac{\text{gal}}{\text{min}} \times 3.79 \frac{\text{L}}{\text{gal}} \times \frac{1 \text{ min}}{60 \text{ s}} \times \frac{1000 \text{ grams}}{\text{L}} \times \frac{\text{gmol}}{18 \text{ g}} \times \frac{\text{cm}^2 \cdot \text{s}}{5.7 \times 10^{-5} \text{ gmol}}$$

$$= 3.1 \times 10^6 \text{ cm}^2 \text{ of membrane required}$$

Each bundle of reverse osmosis membrane has 300 ft^2 of area. The number of bundles for the first phase is

$$3.1 \times 10^6 \text{ cm}^2 \times \frac{\text{bundle}}{300 \text{ ft}^2} \times \frac{(3.28 \text{ ft})^2}{\text{m}^2} \times \frac{1 \text{ m}^2}{(100 \text{ cm})^2} = \underline{\underline{11.1}}$$

Therefore, 12 bundles are necessary for the first phase.

Each pass has a 50 percent recovery rate. In order to obtain a 75 percent recovery rate, the reject stream must be passed through a second reverse osmosis unit. A 95 percent rejection rate of the salt is expected. Therefore, the concentration in the reject can be calculated:

$$\text{Concentration reject} = 0.95 \times 5000 \text{ mg/L} \times 2 = 9500 \text{ mg/L}$$

Determine the osmotic pressure of the reject stream of the first reverse osmosis unit using the van't Hoff equation.

$$\pi = \phi_c N C_s RT$$

$$C_s = 9500 \text{ mg/L} = \frac{9.5\text{g/L}}{58.45 \text{ g/gmol}} = 0.16 \text{ gmol/L}$$

$$\pi = 0.92 \times 2 \times 0.16 \frac{\text{gmol}}{\text{L}} \times 0.082 \frac{\text{atm} \cdot \text{L}}{\text{gmolK}} \times 298.2 \text{ K} = 7.2 \text{ atm}$$

$$7.2 \text{ atm} \left(\frac{14.7 \text{ psi}}{\text{atm}} \right) = \underline{\underline{106 \text{ psi}}}$$

The water flow through the membrane is calculated by

$$\begin{aligned} J_w &= w_p(\Delta P - \Delta\pi) \\ &= 1.9 \times 10^{-6} \frac{\text{gmol}}{\text{cm}^2 \cdot \text{s} \cdot \text{atm}} (500 \text{ psi} - 106 \text{ psi}) \left(\frac{1 \text{ atm}}{14.7 \text{ psi}} \right) \\ &= \underline{\underline{5.1 \times 10^{-5}}} \text{ gmol/(cm}^2 \cdot \text{s)} \end{aligned}$$

To treat the 25 gal/min reject stream, the number of bundles required is found by the following calculations:

$$25 \frac{\text{gal}}{\text{min}} \times 3.79 \frac{\text{L}}{\text{gal}} \times \frac{1 \text{ min}}{60 \text{ s}} \times \frac{1000 \text{ g}}{\text{L}} \times \frac{\text{gmol}}{18 \text{ g}} \times \frac{\text{cm}^2 \cdot \text{s}}{5.7 \times 10^{-5} \text{gmol}}$$

$$= 1.54 \times 10^6 \text{ cm}^2 \text{of membrane required}$$

$$1.54 \times 10^6 \text{ cm}^2 \times \frac{\text{bundle}}{300 \text{ ft}^2} \times \frac{(3.28 \text{ ft})^2}{\text{m}^2} \times \frac{1 \text{ m}^2}{(100 \text{ cm})^2} = \underline{\underline{5.52}}$$

Therefore, 6 bundles are necessary for the second pass.

4. Calculate the final product water quantity and quality.

a. For cell 1

$$\begin{aligned} \text{Quantity} &= \text{recovery rate} \times \text{feed rate} \\ &= 0.50 \times 50 \text{ gal/min} = 25 \text{ gal/min} \\ \text{Quality} &= (1 - \text{rejection rate}) \times \text{feed salt concentration} \\ &= (1 - 0.95) \times 5000 \text{ mg/L salt} \\ &= 250 \text{ mg/L salt} \end{aligned}$$

b. For cell 2

$$\begin{aligned} \text{Quantity} &= \text{recover rate} \times \text{feed rate} \\ &= 0.50 \times 25 \text{ gal/min} = 12.5 \text{ gal/min} \end{aligned}$$

$$\text{Quality} = (1 - \text{rejection rate}) \times \text{feed salt concentration}$$

$$\text{Feed salt concentration} = \text{reject salt concentration from cell 1}$$
$$= 9500 \text{ mg/L (see step 3 above)}$$

$$\text{Quality} = (1 - 0.95) \times 9500 \text{ mg/L salt}$$
$$= 475 \text{ mg/L salt}$$

c. Final quantity and quality

$$\text{Quantity} = \text{recovery cell 1} + \text{recovery cell 2}$$
$$= 25 + 12.5 \text{ gal/min}$$
$$= 37.5 \text{ gal/min}$$

Quality is calculated by mass balance as follows:

$$C_f = (Q_1 \times C_1 + Q_2 \times C_2)/Q_f$$
$$= (25 \times 250 + 12.5 \times 475)/37.5$$
$$= 325 \text{ mg/L salt}$$

Ultrafiltration

In ultrafiltration, which is designed for the separation of macromolecular solutes, solute rejection and accumulation at the membrane surface may lead to the formation of a gel layer, which in effect constitutes a secondary membrane. The solvent flux may be expressed as:[146]

$$J_w = \frac{(\Delta P - \Delta\pi)}{(R_g + R_m)} \tag{9-77}$$

where J_w = water flux through the membrane [gmol/(cm^2 · s)]
R_g = resistance due to the gel layer (g · cm)/(gmol · s)
R_m = membrane resistance (g · cm)/(gmol · s)

and ΔP, $\Delta\pi$ have units dynes/cm^2.

Osmotic pressure can usually be neglected and the equation reduced to:

$$J_w = \Delta P/(R_g + R_m) \tag{9-78}$$

For dilute solutions of large solutes, R_g tends to be much less than R_m and may be neglected. When this is not the case, however, concentration polarization can occur and R_g will control (see Design Considerations for a discussion of concentration polarization).

As mentioned previously, ultrafiltration systems can be operated in different distinct modes, and each mode has different operating characteristics. In the batch concentration mode (also called ultrafiltration), the feed solution reservoir is generally open to the atmosphere, and an initial volume of solution, V_0, is concentrated to a final retentate volume, V_r. The permeate volume leaving the system, V_p, is then $V_0 - V_r$. In this mode of operation, the concentrations of all molecules having permeation factors less than unity increase over time, in both the feed solution reservoir and in the permeate. In the semibatch or diafiltration mode, an initial volume of solution, V_0, is maintained constant by replacing the volume lost as permeate with solute-free buffer solution. Because the volume of feed solution in the system is constant, the concentration of retained molecules remains constant. In contrast, the

concentration of permeating molecules decreases over time in both the reservoir and the permeate. The ultrafiltration system is operated until a chosen filtration volume has permeated the membrane, which corresponds to a desired separation efficiency or degree of treatment.

Model Development

Mass distribution within the reactor and the transport of both solute and solvent across the membrane must be characterized to model the separation of solutes having different molecular sizes. The scale over which the mass distribution within the reactor system is characterized establishes a distinction between "microscopic" and "macroscopic" models. In this section, a simple macroscopic model will be described. Macroscopic models have the advantage that they are simple to calibrate and apply, because no attempt is made to characterize the distribution of mass as a function of position within the membrane module. Rather, transport is characterized by using effective model coefficients based on readily measured macroscopic concentrations: reactor inlet concentration and permeate concentration. The advantages of macroscopic modeling is that fewer model coefficients are required, and the resulting mass-balance equations can be solved analytically. However, the transport parameters measured depend on operating conditions, and must be considered phenomenological. As before, solute rejection is defined as:

$$R = 1 - \frac{C_p}{C_f} \tag{9-79}$$

where C_f is the bulk feed concentration and C_p is the permeate concentration. The permeate concentration is given by the ratio of the solute flux J_s to the volumetric solution flux J_w:

$$R = 1 - \frac{J_s}{J_w C_f} \tag{9-80}$$

Therefore, solute flux is given by:

$$J_s = (1 - R) J_w C_f \tag{9-81}$$

Equations describing solute transport in batch ultrafiltration systems can be developed by incorporating the solute flux expression into a system mass balance relationship. The recycle configuration of the semibatch system configuration is employed to achieve high crossflow velocities designed to minimize secondary membrane formation and resistance. The high recycle rates also allow operational flexibility to obtain backmixing conditions that approximate completely mixed flow, thereby simplifying the system mass balance equations. In any particular application, backmixing conditions should be verified by residence time distribution analysis. It should be noted that, while the macroscopic system is modeled as a completely mixed reactor, the microscopic hydrodynamic conditions at the membrane surface are not characterized. Equating mass accumulation within the system to the rate of mass permeation through the membrane, the general differential mass balance can be written:

$$V_s \frac{dC_f}{dt} + C_f \frac{dV_s}{dt} = J_s A = J_v A C_p = J_v A (1 - R) C_f \tag{9-82}$$

where A is the membrane surface area, V_s is the volume of solution in the system, and other terms are as defined previously. It is assumed that both J_w and membrane

characteristics remain constant, which neglects potential concentration polarization effects. It is further assumed that the solute does not undergo transformation and that adsorption to the membrane surface is negligible.

In the diafiltration mode, the system volume is maintained constant and equal to V_0; therefore, the volume derivative in Equation (9-82) is equal to zero. Integrating the mass balance equation with the initial condition $C_f = C_{f0}$ when $t = 0$ and making the substitution $J_v At = V_p$ yields:

$$C_f = C_{f0} \exp\left((1 - R)\frac{V_p}{V_0}\right) \tag{9-83}$$

If the ultrafiltration system is operated in the concentration mode, the system volume V_s is a function of time or permeate volume throughput V_p:

$$V_s = V_0 - J_v At = V_0 - V_p \tag{9-84}$$

and the time derivative of the system volume is equal to $J_v A$. Integrating the mass balance equation with the initial condition $C_f = C_{f0}$ when $t = 0$ yields:

$$C_f = C_{f0}\left(\frac{V_0}{V_0 - V_p}\right)^R \tag{9-85}$$

The rejection may be determined from the log-linearized forms of the integrated mass balance equations, or from nonlinear regression. For the diafiltration mode, a plot of $\ln(C_f/C_{f0})$ versus V_p/V_0 yields a slope equal to $(R - 1)$, and for the concentration mode, a plot of $\ln(C_f/C_{f0})$ versus $\ln[V_0/(V_0 - V_p)]$ yields a slope equal to R. Alternative linearization forms that allow the determination of C_{f0} from the intercept of the linearized plot can also be employed.

Solute Transport Model Verification

The permeation of natural organic matter (molecular weight approximately 3000 Da) from a reservoir will serve as an illustration of the modeling approach. The ultrafiltration system employed a 1-kDa polysulfone hollow fiber module. Representative results are shown in Figure 9-42.

Macromolecule Solution Fractionation

The permeation model can be used to design alternative fractionation procedures for recovering surfactants, separating oil-water emulsions, etc. For a given

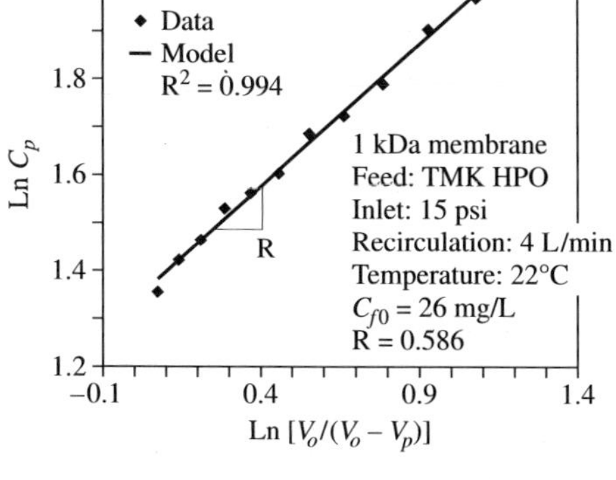

FIGURE 9-42
Verification of the solute permeation model for Tomhannock natural organic matter (NOM).

initial volume, the mass throughput depends on the initial mass present, $C_{f0}V_0$, membrane rejection, and the permeate volume throughput V_p. Model calculations demonstrate that the most efficient mass throughput is achieved (1) at high feed concentrations and (2) in the concentration mode. However, an important objective of any fractionation procedure is to minimize the effects of concentration polarization, solute accumulation on the membrane surface, and variations in membrane rejection. Therefore, attempting to achieve separation by using very high concentration factors is not a practical strategy.

The diafiltration mode has several advantages over the concentration mode from an operational point of view. The original solution volume is kept constant by the continuous addition of buffer solution containing the same inorganic matrix as the feed. The effects of concentration polarization are minimized, therefore, because the concentration of retained molecules is kept constant, while the concentrations of permeating molecules decrease as a function of permeate volume throughput. Further, as a matter of operational convenience, the diafiltration mode can be operated continuously. Model calculations indicated that the efficiency of the concentration mode and the operational benefits of the diafiltration mode could be combined by employing a sequential mode of operation. Under this scenario, the sample is first concentrated to enhance mass throughput efficiency and reduce sample volume. The concentration mode is followed by a period of operation in the diafiltration mode; the smaller sample volume maximizes the value of V_p/V_0 for a given total throughput volume. Figure 9-43 illustrates model calculations of the feed concentration and mass throughput for this combined operation. It can be seen that maximizing mass throughput (i.e., separation of molecular weight components) requires fairly large V_p/V_0 values.

Design Considerations

Concentration polarization is a phenomenon that occurs in all three membrane processes described in this section. In general, concentration polarization results in an increased power requirement without an appreciable performance enhancement.

In electrodialysis, there is a value of current that produces a depletion of ions on the feed side of the membrane because the ions are being transferred through the

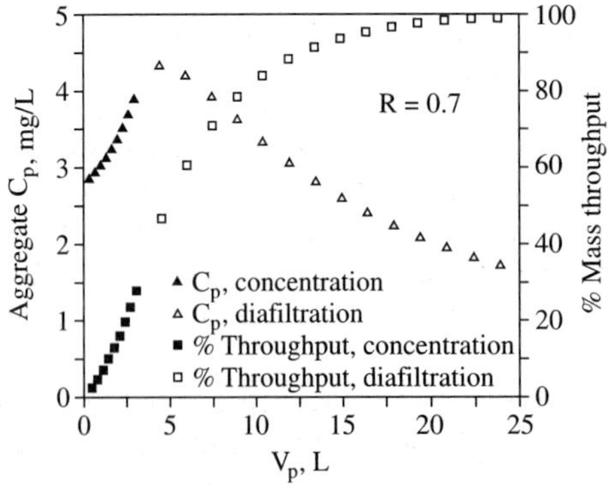

FIGURE 9-43
The effect of operation protocol on aggregate permeate concentration and mass throughput (molecular weight separation).

membrane faster than they can be transferred from the membrane to the solution. An increase in resistance occurs not only because there are no ions to transfer the current at the membrane boundary, but also because the excess current is used to ionize water ("water splitting") into H^+ and OH^- ions, and the increased OH^- concentration (increased pH) on the concentrating side of the membrane results in scaling by precipitation of pH-sensitive substances, such as calcium carbonate.[147]

In reverse osmosis, concentration polarization occurs because the solute builds up on the feed side of the membrane, thus exceeding the bulk concentration. As a result, the local osmotic pressure increases, necessitating higher system pressure to produce the same water recovery, and the transfer of solute through the membrane increases, producing a less pure water. In addition, the membrane may deteriorate faster if certain substances accumulate on the membrane surface and the water recovery may decrease if precipitation of sparingly soluble salts occurs on the membrane surface.[148]

Finally, concentration polarization occurs in ultrafiltration because of the buildup of macrosolutes on the membrane surface. These macrosolutes have very low diffusion coefficients, and thus their diffusion back into the bulk liquid is slow. Eventually, a gel layer develops on the feed side of the membrane and the process becomes controlled by mass transfer and independent of the applied pressure.[149]

The methods to control concentration polarization include the use of a high feed velocity or turbulent flow (all membrane processes), lower current density to feed concentration ratio (electrodialysis), lower recovery (reverse osmosis), and higher feed temperature (ultrafiltration).[150]

Fouling of membranes can be reduced by periodic cleaning with alkalis or acids and, in the case of electrodialysis and reverse osmosis, reversal of flow. Pretreatment of the feed is also useful to reduce membrane fouling and may include filtration, softening, adsorption, pH control, and dispersant or surfactant addition to prevent scaling or wetting of membranes by oil emulsions, respectively.

Conlon[151] notes several design restrictions for electrodialysis systems:

- The ion concentration ratio (concentrate versus product water) should be below 150.
- The voltage applied to the membrane stack should be below 80 percent of the limiting voltage (i.e., where excess current will travel through adjacent membranes and generate enough heat to damage membranes).
- The maximum current density should be 70 percent of the limiting current density.
- Reversal of the direct current on a timed cycle (30 seconds every 20 to 30 minutes) will reduce fouling of the membranes.

Other requirements for electrodialysis systems include the use of membranes and electrodes that function adequately in both current flow directions. In addition, because the size of the unit and the energy consumption depend on the total dissolved solids (TDS) concentration, electrodialysis systems are generally used for low TDS concentration feeds (less than 5000 mg/L), unless the recovered product is valuable enough to offset the costs.[152]

Additional design considerations for reverse osmosis systems include the feed temperature and pH. High feed temperatures and pH values may result in membrane

hydrolysis and, thus, reduced performance. However, within the operating range of temperature and pH values, increased temperature results in an increased water flux through the membrane with a constant solute rejection.[153]

For effective separation in ultrafiltration, the solute molecule should be at least 1 order of magnitude larger than the solvent molecule. Ultrafiltration membranes are specified in terms of a molecular weight cutoff (MWC), a value representing the molecular weight that results in 90 percent of the solute of that size in a very dilute solution being retained by the membrane.[154] However, the MWC is simply a theoretical value useful to rank membranes. Other solute characteristics (i.e., shape and affinity for the membrane) and the operating conditions (i.e., whether a layer of solute has formed at the membrane surface or fouling has occurred) also play a role in the retention capability of the membrane.[155]

DISCUSSION TOPICS AND PROBLEMS

9-1. Derive Equation 9-20 starting with Equation 9-16.

9-2. Provide a preliminary design of an air stripping column to remove toluene from groundwater. Levels of toluene range from 0.1 to 2.1 mg/L and this must be reduced to 50 μg/L. A hydrogeologic study of the area indicates that a flow rate of 110 gal/min is required to ensure that contamination not spread. Laboratory investigations have determined the overall transfer constant, $K_L a = 0.020\ \text{s}^{-1}$. Use a column diameter of 2.0 feet and an air-to-water ratio of 15. Specifically determine: liquid loading rate, stripping factor, and height of the tower. Provide a sketch of the unit indicating all required appurtenances.

9-3. Using the data in Problem 9-2, determine the pressure drop through the tower.

9-4. Using the data in Problem 9-2, determine the impact on effluent quality by varying the air-to-water ratio and the packing height.

9-5. Determine the overall mass transfer coefficient $K_L a$, given the following data from an air stripping column: contaminant = tetrachloroethene (C_2Cl_4); liquid mass loading rate = 10.2 kg/(m^2 · s); packing = 1-in polyethylene Tri-paks®; air mass loading rate = 1.5 kg/(m^2 · s); temperature = 6°C; diffusion coefficient in water = 1.1×10^{-6} cm^2/s.

9-6. Recalculate $K_L a$ from Example 9-1 incorporating the following changes: H' is given as 0.0704 (dimensionless) at a temperature of 6°C; $a_t = 138\ \text{m}^2/\text{m}^3$.

9-7. Design an air stripping column to remove TCE from water. The initial concentration of TCE is 1.3 mg/L and this must be reduced to 75 μg/L. Use the following criteria: water flow rate: 350 gal/min; water temperature: 16°C; air temperature: 23°C; packing: use 2-in Intalox saddles.

9-8. Demonstrate that the stripping factor used in this book is numerically similar to that in common use by chemical engineers:

$R = H'G/L$

H' = Henry's constant (dimensionless)

G = gas loading rate (kmol/h)

L = liquid loading rate (kmol/h)

9-9. 100 mL of a solution with a TOC (total organic carbon) concentration of 0.5 percent is placed in each of five containers with activated carbon and shaken for 48 hours. The samples are filtered and the concentration of TOC measured, yielding the following analyses:

Container:	1	2	3	4	5
Carbon, g:	10	8	6	4	2
TOC, mg/L:	42	53	85	129	267

Determine the Freundlich constants K and n, and plot the isotherm.

9-10. 200 mL of a solution with a *para*-xylene concentration of 500 mg/L is placed in each of six containers with activated carbon and shaken for 24 hours. The samples are filtered and the concentration of p-xylene measured, yielding the following analyses:

Container:	1	2	3	4	5	6
Carbon, g:	24	20	16	12	8	4
p-xylene(mg/L):	10.7	14.6	23	29	48	107

Determine the Freundlich constants K and n, and plot the isotherm.

9-11. Estimate the daily carbon utilization to remove chlorobenzene from 1.0 million gal/day of groundwater saturated with chlorobenzene. Assume a chlorobenzene concentration of 5 mg/L is acceptable for discharge to a POTW.

9-12. Provide a preliminary design of a carbon adsorption system for removal of 2,4,6 trichlorophenol from 250,000 gal/day of water. The following data are provided: Bohart-Adams model: $a = 2.3$ days/ft; $b = -10$ days in laboratory tests where trichlorophenol concentration was reduced from 395 mg/L to 10 mg/L at a loading of 4.0 gal/(ft^2 · min). The adsorption zone was 19.0 feet.

9-13. A groundwater aquifer is contaminated with benzene. Contrast the relative effectiveness of air stripping and activated carbon adsorption by providing a preliminary design of each process.

9-14. Using the information in Example 9-6, determine the Bohart-Adams coefficients N and K.

9-15. Revise Example 9-6 for a flow rate of 250,000 gallons per day.

9-16. Revise Example 9-6 for a contaminant concentration in groundwater of 239 mg/L.

9-17. Derive the conversion factors F_1 and F_2 in the Bohart-Adams model.

9-18. Derive a simplified Nernst equation (Equation 9-54) from Equation 9-51.

9-19. A plating waste with a flow of 50 L/min contains 274 mg/L of NaCN. Determine the amounts of chlorine as $Ca(OCl)_2$ and NaOH required to:

(*a*) Oxidize the cyanide to cyanate
(*b*) Oxidize the cyanide to CO_2 and H_2O.

9-20. A plating waste with a flow of 40,000 gal/day contains 274 mg/L of NaCN. Determine the amounts of chlorine as Cl_2 and NaOH required to oxidize the cyanide to CO_2 and N_2.

9-21. Demonstrate that reactions 9-63 and 9-64 are equivalent to reaction 9-62.

9-22. An industrial waste stream contains 650 mg/L of hexavalent chromium at a flow rate of 35 L/min. Determine the amount of sulfur dioxide required to reduce the chromium to the less toxic trivalent form.

9-23. A waste contains 150 mg/L of copper at a flow rate of 75 L/min. Determine the amount of sodium borohydride required to treat this waste.

9-24. A waste contains 250 mg/L of cadmium and 450 mg/L lead at a flow rate of 45,000 gal/day. Determine the amount of sodium borohydride required to treat this waste.

9-25. Provide a preliminary design of an electrodialysis unit to treat a plating waste with the following characteristics: Ni concentrations = 2.1% as $NiSO_4$; flow rate = 20 gal/min. Determine the number of membranes, power required, and annual electrical cost at \$0.07/kWh. The effluent Ni concentration must be less than 500 mg/L.

9-26. Determine if the following mixtures of organics in wastewater are good condidates for steam stripping. State reasons. Apply the OCPSF discharge limitations as criteria for the steam stripper bottoms concentrations.

Mixture A	Mixture B	Mixture C
93 mg/L benzene	133 mg/L phenol	56 mg/L ethanol
127 mg/L acetone	27 mg/L nitrotoluene	182 mg/L acetone
	42 mg/L methanol	158 mg/L methyl ethyl ketone

9-27. Show, through material balances, that the ratio of liquid to gas (L/G) in the stripping section of a steam stripper is higher than in the rectifying section. State any necessary assumptions or givens.

9-28. Using the equilibrium curve in Figure 9-16, and given the following equations for operating lines:

Stripping section: $Y = 1.38X$

Rectifying section: $Y = 0.65X + 0.39$

Use the McCabe-Thiele method to estimate the number of theoretical trays. Assume: $X_B = 0.05$; $X_F = 0.35$; $X_O = 0.9$

REFERENCES

1. Perry, R. H., and D. W. Green, eds.: *Perry's Chemical Engineers' Handbook,* 6th ed., McGraw-Hill, New York, pp. 18–32, 1984.
2. Ibid., pp. 18–33.
3. Ibid., pp. 18–34.
4. Onda, K., H. Takeuchi, and Y. Okumoto: "Mass Transfer Coefficients Between Gas and Liquid Phases in Packed Columns," *Journal Chemical Engineering of Japan,* vol. 1, p. 58, 1968.
5. Cornwell, D. A.: "Air Stripping and Aeration," in *Water Quality and Treatment* (F. W. Pontious, ed.), McGraw-Hill, 1990.
6. McCabe, W. L., J. C. Smith, and P. Harriot: *Unit Operations of Chemical Engineering,* 4th ed., McGraw-Hill, New York, 1985, p. 466.
7. Ibid., p. 622.
8. Eckert, J. S.: "Selecting the Proper Distillation Column Packing," *Chem. Engr. Prog.,* vol. 66, pp. 39–44, 1970.
9. Haarhoff, J., and D. Schoeller: "Theory and Design of Countercurrent Packed Aeration Towers," Airstrip, Ames, Iowa, 1988.
10. Perry, R. H., and D. W. Green, eds.: *Perry's Chemical Engineers' Handbook,* pp. 18-21 to 18-23.
11. Treybal, R. E.: *Mass Transfer Operations,* McGraw-Hill, New York, pp. 196ff, 1980.
12. Haarhoff, J., and D. Schoeller, "Theory and Design of Countercurrent Packed Aeration Towers"
13. Haarhoff, J., and J. L. Cleasby: "Evaluation of Air Stripping for the Removal of Organic Drinking Water Contaminants," *Water South Africa,* vol. 16, pp. 13–22, 1990.
14. Metcalf and Eddy, Inc.: *Wastewater Engineering,* McGraw-Hill, New York, p. 316, 1990.
15. Anon.: "Activated Carbon Application Bulletin 20-4b," Calgon Corp., 1973.
16. Weber, Jr., W. J.: "Evolution of a Technology," *Journal of Environmental Engineering,* ASCE, vol. 110, no. 5, pp. 899–917, 1984.
17. Metcalf & Eddy, Inc., *Wastewater Engineering,* p. 321.

18. Hutchins, R. A.: "New Method Simplifies Design of Activated Carbon Systems," *Chem. Engr.,* vol. 80, pp. 133–138, August 23, 1974.
19. Bohart, G. D., and E. Q. Adams: "Some Aspects of the Behavior of Charcoal with Respect to Chlorine," *J. Amer. Chem. Soc.,* vol. 42, pp. 523–544, 1920.
20. Weber, Walter J., and Liu, K. T: "Determination of Mass Transport Parameters for Fixed-bed Adsorbers," *Chemical Engineering Communications,* vol. 6, p. 49. 1980.
21. Liu, K. T., and Weber, Walter J: "Characterization of Mass Transfer Parameters for Adsorber Modeling and Design," *Journal of the Water Pollution Control Federation,* vol. 53, no. 10, pp. 1541–1550, 1981.
22. U.S. EPA: "Development Document for Effluent Guidelines and Standards for the OCPSF Source Category, vol. III (Best Available Technology)," EPA 440/1-83/009b.
23. U.S. EPA: "Organic Chemicals and Plastics and Synthetic Fibers Category Effluent Limitations Guidelines, Pretreatment Standards, and New Source Performance Standards," 40 CFR Parts 414 and 416, *Federal Register,* vol. 52, no. 214 pp. 42522–42584, November 5, 1987.
24. Treybal, R. E.: *Mass Transfer Operations,* 2nd ed., McGraw Hill, New York, 1968.
25. McCabe, W. L., J. C. Smith, and P. Harriott: *Unit Operations of Chemical Engineering,* 4th ed., McGraw-Hill, New York, 1985.
26. King, C. J.: *Separation Processes,* McGraw-Hill, New York, 1980.
27. Perry & Green: *Perry's Chemical Engineers' Handbook.*
28. Smith, J. M., and H. C. Van Ness: *Introduction to Chemical Engineering Thermodynamics,* 4th ed., McGraw-Hill, New York, 1987.
29. Balzhiser, Samuels, and Eliassen: *Chemical Engineering Thermodynamics,* Prentice-Hall, Engelwood Cliffs, NJ, 1972.
30. Reid, R. C., Prausnitz, and Poling: *The Properties of Gases and Liquids,* 4th ed., McGraw-Hill, New York, 1987.
31. Ibid, p. 254 ff.
32. DECHEMA Chemistry Data Series: *Vapor-Liquid Equilibrium Data Collection,* vol. 1, parts 1 and 1a, DECHEMA, Frankfurt, 1981.
33. DECHEMA Chemistry Data Series: *Vapor-Liquid Equilibrium Data Collection,* vol. I, part 1b, DECHEMA, Frankfurt, 1986.
34. DECHEMA Chemistry Data Series: *Activity Coefficients at Infinite Dilution,* vol. IX, parts 1 and 2, DECHEMA, Frankfurt, 1986.
35. Pinal, R., Suresh, Rao, Lee, Cline, and Yalkowsky, "Cosolvency of Partially Miscible Organic Solvents on the Solubility of Hydrophobic Organic Chemicals," *Environmental Science and Technology,* vol. 24, no. 5 (1990), pp. 639–647.
36. DECHEMA Chemistry Data Series: *Liquid-Liquid Equilibrium Data Collection,* vol. V, part 1, DECHEMA, Frankfurt, 1979.
37. Wankat, Phillip C.: *Equilibrium Staged Separations,* Elsevier, New York, pp. 157–198.
38. U.S. Environmental Protection Agency: "Technical Resource Document—Treatment Technologies for Halogenated Organic Containing Waste," U.S. EPA PB88-131271, December 1987.
39. Bernardin, F. E., and E. Froelich: "Chemical Oxidation of Dissolved Organics Using Ultraviolet-Catalyzed Hydrogen Peroxide," *Superfund '90,* Haz. Materials Control Research Instit., Washington, D.C., pp. 768–771, 1990.
40. Ibid., p. 770.
41. Fochtman, E. G.: "Chemical Oxidation and Reduction," *Standard Handbook of Hazardous Waste Treatment and Disposal,* Harry M. Freeman, ed., McGraw-Hill, New York, pp. 7.41ff, 1989.
42. Glaze, W. H.: "Chemical Oxidation in American Water Works Association," *Water Quality and Treatment,* 4th ed., F. W. Pontius, ed., McGraw-Hill, New York, pp. 747ff, 1990.

43. Eisenhauer, H. R.: "The Ozonation of Phenolic Wastes," *J. WPCF,* vol. 40, pp. 1887–1889, 1968.
44. Fochtman, E. G., "Chemical Oxidation and Reduction."
45. Eckenfelder, W. W.: *Industrial Water Pollution Control,* McGraw-Hill, New York, p. 301, 1989.
46. Glaze, W. H.: "Chemical Oxidation in American Water Works Association."
47. Eckenfelder, W. W.: *Industrial Water Pollution Control,* op. cit., p. 306.
48. Posselt, H. S., and W. J. Weber: "Chemical Oxidation," in W. J. Weber, *Physicochemical Processes for Water Quality Control,* Wiley-Interscience, New York, pp. 398, 391, 1972.
49. Patterson, J. W.: *Industrial Wastewater Treatment Technology.*
50. Sundstrom, D. W., et al.: "Destruction of Halogenated Aliphatics by Ultraviolet Catalyzed Oxidation with Hydrogen Peroxide," *Hazardous Waste & Hazardous Materials,* vol. 3, pp. 101–110, 1986.
51. Weber, W. J., and O. Corapcioglu: "Catalytic Oxidation of Cyanides," *Proceedings of the 36th Industrial Waste Conference,* Purdue University, Ann Arbor Sci. Publ., Ann Arbor, Mich., 1982.
52. Fochtman, E. G.: "Chemical Oxidation and Reduction."
53. Ibid, p. 7.43.
54. Lindsay, M. J., M. E. Hackman, and R. J. Tremblay: "Sodium Borohydride Reduces Hazardous Waste," *Proceedings of the 40th Industrial Waste Conference,* Purdue University, Ann Arbor Sci. Publ., Ann Arbor, Mich., p. 479, 1985.
55. Fochtman, E. G.: "Chemical Oxidation and Reduction," p. 7.49.
56. Weast, R. C., ed.: *Handbook of Chemistry and Physics,* CRC Press, p. D-156ff, 1984.
57. Snoeyink, V., and D. Jenkins: *Water Chemistry,* Wiley, pp. 328–329, 1980.
58. Posselt, H. S., and W. J. Weber: "Chemical Oxidation," in W. J. Weber, *Physicochemical Processes for Water Quality Control,* Wiley-Interscience, New York, pp. 398, 391, 1972.
59. Weber, W. J., and O. Corapcioglu: "Catalytic Oxidation of Cyanides," *Proceedings of the 36th Industrial Waste Conference,* Purdue University, Ann Arbor Sci. Publ., Ann Arbor, Mich., 1982.
60. Morrison, R. T., and R. N. Boyd: *Organic Chemistry,* 3rd ed., Allyn & Bacon, p. 218, 1973.
61. Eckenfelder, W. W., *Industrial Water Pollution Control,* op. cit., p. 301.
62. Glaze, W. H.: "Drinking Water Treatment with Ozone," *Environ. Sci. Tech.,* vol. 21, p. 227, 1987.
63. Ibid.
64. Staehelin, J., and J. Hoigne: "Decomposition of Ozone in Water in the Presence of Organic Solutes Acting as Promoters and Inhibitors of Radical Chain Reaction," *Environ. Sci. Tech.,* vol. 19, p. 1206, 1985.
65. Glaze, W. H.: "Drinking Water Treatment with Ozone."
66. Hoigne, J., and H. Bader: "Rate Constants of Reactions of Ozone with Organic and Inorganic Compounds in Water," *Water Research,* vol. 17, pp. 173–183, 1983.
67. Eisenhauer, H. R.: "The Ozonation of Phenolic Wastes," *J. WPCF,* vol. 40, pp. 1887–1889, 1968.
68. Miller, G. W., et al.: *An Assessment of Ozone and Chlorine Dioxide Technologies for Treatment of Municipal Water Supplies,* EPA 600/2-78-147 (NTIS No. PB-285972), 1978.
69. Eisenhauer, H. R.: "The Ozonation of Phenolic Wastes."
70. Posselt, H. S., and W. J. Weber: "Chemical Oxidation," op. cit., p. 368.
71. Glaze, "Drinking Water Treatment with Ozone."
72. Patterson, J. W., *Industrial Wastewater Treatment Technology,* p. 14, Butterworths, Stoneham, Mass., 1985.

73. Watts, R. J., M. D. Udell, and P. A. Rauch: "Treatment of Pentachlorophenol-Contaminated Soils Using Fenton's Reagent," *Hazardous Waste & Hazardous Materials,* vol. 7, pp. 335–345, 1990.
74. Ibid.
75. Schmitt, R. J.: "Combining Chromium Treatment with Hydrogen Peroxide Oxidation of Phenolics," *Proceedings of the 36th Industrial Waste Conference,* Purdue University, Ann Arbor Sci. Publ., Ann Arbor, Mich., pp. 375–382, 1982.
76. Ibid, p. 379.
77. Barich, J. T., and J. D. Zeff: "A Review of Ultrox, Ultraviolet Oxidation Technology," *Superfund '89,* Haz. Materials Control Research Instit., Washington, D.C., pp. 264–266, 1989.
78. Sundstrom, D. W., et al., "Destruction of Halogenated Aliphatics by Ultraviolet Catalyzed Oxidation with Hydrogen Peroxide," *Hazardous Waste & Hazardous Materials,* vol. 3, pp. 101–110, 1986.
79. Eckenfelder, W. W., *Industrial Water Pollution Control,* op. cit., p. 308.
80. Schmitt, R. J.: "Combining Chromium Treatment with Hydrogen Peroxide Oxidation of Phenolics."
81. Haas, C. N.: "Disinfection," in American Water Works Association, *Water Quality and Treatment,* 4th ed., F.W. Pontius, ed., McGraw-Hill, New York, p. 883, 1990.
82. Milne, D.: "Chemistry of Waste Cyanide Treatment," *Sewage & Industrial Wastes,* vol. 23, p. 174, 1951.
83. Posselt, H. S., and W. J. Weber: "Chemical Oxidation," op. cit., p. 396.
84. Eckenfelder, W. W., *Industrial Water Pollution Control,* op. cit., p. 310.
85. Thomason, T. B., and M. Modell: "Supercritical Water Destruction of Aqueous Wastes," *Hazardous Wastes,* vol. 1, p. 459, 1984.
86. McHugh, M. A., and V. J. Krukonis: *Supercritical Fluid Extraction,* Butterworths, Stoneham, Mass., p. 131, 1986.
87. Modell, M.: "Supercritical-Water Oxidation," *Standard Handbook of Hazardous Waste Treatment and Disposal,* Freeman, H. M., editor in chief, McGraw-Hill, New York, pp. 8.153–8.168, 1989.
88. Thomason, T. B., and M. Modell: "Supercritical Water Destruction of Aqueous Wastes."
89. Modell, M.: "Supercritical-Water Oxidation," p. 8.159.
90. Perry, R. H., and D. W. Green, eds.: *Perry's Chemical Engineers' Handbook,* op. cit., p. 3–111.
91. Smith, J. M., and H. C. Van Ness: *Introduction to Chemical Engineering Thermodynamics,* 4th ed., McGraw-Hill, New York, p. 86, 1987.
92. Josephson, J.: "Supercritical Fluids," *Environ. Sci. & Tech.,* vol. 16, p. 549A, 1982.
93. McHugh, M. A., and V. J. Krukonis: *Supercritical Fluid Extraction,* Butterworths, Stoneham, Mass., p. 5, 1986.
94. Reid, R. C., J. M. Prausnitz, and T. K. Sherwood: *The Properties of Gases and Liquids,* 3rd ed., McGraw-Hill, New York, p. 422, 1977.
95. Thomason, T. B., and Modell, M., "Supercritical Water Destruction of Aqueous Wastes," p. 459.
96. Treybal, R. E.: *Mass Transfer Operations,* 3rd ed., p. 488, McGraw-Hill, New York.
97. Perry, R. H., and D. W. Green, eds.: *Perry's Chemical Engineers' Handbook,* op. cit., p. 15-5.
98. Thomason, T. B., and M. Modell, "Supercritical Water Destruction of Aqueous Wastes," p. 462.
99. Modell, M., op. cit., p. 154.
100. Modell, M.: "Supercritical-Water Oxidation," *Standard Handbook of Hazardous Waste Treatment and Disposal,* H. M. Freeman, editor-in-chief, McGraw-Hill, New York, pp. 8.153–8.168, 1989.

101. "Detoxification and Disposal of Hazardous Organics by Processing in Supercritical Water," Final Report, U.S. Army Contract No. DAMD 17-800-C-0078, prepared by MODAR, Inc., Houston, Texas, 1987.
102. Ibid.
103–113. Modell, M., *Processing Methods for the Oxidation of Organics in Supercritical Water,* U.S. Patent 4,543,190, September 1985.
114. "Detoxification and Disposal of Hazardous Organics by Processing in Supercritical Water," Final Report, U.S. Army Contract No. DAMD 17-800-C-0078, prepared by MODAR, Inc., Houston, Texas, 1987.
115. Modell, 1985.
116. "Detoxification and Disposal of Hazardous Organics by Processing in Supercritical Water," Final Report, U.S. Army Contract No. DAMD 17-800-C-0078, prepared by MODAR, Inc., Houston, Texas, 1987.
117. Thomason, T. B., and M. Modell, "Supercritical Water Destruction of Aqueous Wastes," *Hazardous Waste,* vol. 1, no. 4, p. 453, 1984.
118. Ibid.
119. Ibid.
120. Modell, M.: "Supercritical Water Oxidation," op. cit., p. 8.163.
121. Gregor, H. P., and C. D. Gregor: "Synthetic Membrane Technology," *Scientific American,* pp. 112–128, July 1978.
122. Palmer, S. A. K., M. A. Breton, T. J. Nunno, D. M. Sullivan, and N. F. Surprenant: "Technical Resource Document: Treatment Technologies for Metal/Cyanide Containing Wastes," vol. III, U.S. Environmental Protection Agency, NTIS PB88-143896, pp. 6–9, 1987.
123. Perry, R. H., and D. W. Green, eds.: *Perry's Chemical Engineers' Handbook,* pp. 17–37.
124. MacNeil, J., and D. E. McCoy: "Membrane Separation Technologies," in *Standard Handbook of Hazardous Waste Treatment and Disposal,* H. M. Freeman, ed., McGraw-Hill, New York, pp. 6–103, 1989.
125. "Membrane Separation Technology: Applications to Waste Reduction and Recycling," *The Hazardous Waste Consultant,* vol. 3, no. 3, McCoy & Assoc., Lakewood, Colo., p. 4–21, 1985.
126. MacNeil, J., and D. E. McCoy, "Membrane Separation Technologies."
127. "Membrane Separation Technology: Applications to Waste Reduction and Recycling," *The Hazardous Waste Consultant.*
128. Applegate, Lynn E., "Membrane Separation Processes," *Chemical Engineering,* June 11, 1984, pp. 64–89.
129. Katz, William E., "State-of-the-Art of the Electrodialysis Reversal (EDR) Process," *Industrial Water Engineering,* vol. 21, no. 1, pp. 12–20.
130. Anon.: "Aquatech Acid Recovery System," *Chemical Engineering,* December, 1989.
131. Mani, K. N.: "Electrodialysis Water Splitting Technology," Aquatech Systems, 1985.
132. Eykamp, W., and Steen, J.: "Ultrafiltration and Reverse Osmosis," in *Handbook of Separation Process Technology,* R. W. Rousseau, (ed.). Wiley-Interscience, New York, p. 836, 1987.
133. Conlon, W. J.: "Membrane Processes," *Water Quality and Treatment,* 4th ed., American Water Works Assoc., p. 714, 1990.
134. Eykamp, W., and J. Steen, "Ultrafiltration and Reverse Osmosis," p. 838.
135. MacNeil, J., and D. E. McCoy, "Membrane Separation Technologies," p. 6–97.
136. Martin, J. F., M. J. Stutsman, and E. Grossman: "Hazardous Waste Volume Reduction Research in the U.S. E.P.A.," *Hazardous and Industrial Solid Waste Minimization Practices,* ASTM STP1043, Amer. Soc. for Testing & Materials, pp. 163–171, 1989.
137. Higgins, T.: *Hazardous Waste Minimization Handbook,* Lewis Publishing, Chelsea, Mich., p. 99, 1989.

138. Perry, R. H., and D. W. Green, eds.: *Perry's Chemical Engineers' Handbook,* pp. 17–32.
139. Weber, W. J.: *Physicochemical Processes for Water Quality Control,* John Wiley & Sons, New York, p. 330, 1972.
140. Perry, R. H., and D. W. Green, eds.: *Perry's Chemical Engineers' Handbook,* pp. 17–28.
141. Ibid. pp. 17–32.
142. Rich, G. L.: *Unit Processes of Sanitary Engineering,* John Wiley & Sons, New York, p. 130, 1963.
143. Perry, R. H., and D. W. Green, eds. *Perry's Chemical Engineers' Handbook.*
144. Perry, R. H., and D. W. Green, eds.: *Perry's Chemical Engineers' Handbook,* op. cit., pp. 17–23.
145. Weber, W. J., and E. H. Smith: "Removing Dissolved Organic Contaminants from Water," *Env. Sci. & Tech.,* vol. 20, pp. 970–979, 1986.
146. Perry, R. H., and D. W. Green, eds.: *Perry's Chemical Engineers' Handbook,* op. cit., pp. 17–29.
147. Klein, E., R. A. Ward, and R. E. Lacey: "Membrane Processes—Dialysis and Electrodialysis," in R. W. Rousseau, ed., *Handbook of Separation Process Technology,* Wiley-Interscience, New York, pp. 972–975, 1987.
148. Weber, W. J.: *Physicochemical Processes for Water Quality Control,* pp. 322–323.
149. Eykamp, W., and J. Steen, "Membrane Separation Technologies," p. 827.
150. Weber, W. J.: *Physicochemical Processes for Water Quality Control,* pp. 342, 323, 334.
151. Conlon, W. J.: "Membrane Processes," *Water Quality and Treatment,* 4th ed., American Water Works Assoc., pp. 735ff, 1990.
152. MacNeil, J., and D. E. McCoy: "Membrane Separation Technologies," pp. 6.102, 6.104.
153. Weber, W. J.: *Physicochemical Processes for Water Quality Control,* pp. 320, 319.
154. Eykamp, W., and J. Steen, "Membrane Separation Technologies," p. 829.
155. Ibid.

ADDITIONAL READING

Angell, K. G.: "In Situ Remedial Methods: Air Sparging," *The National Environmental Journal,* pp. 20–23, January/February 1992.

Applegate, Lynn E.: "Membrane Separation Processes," *Chemical Engineering,* June 11, 1984, pp. 64–89.

Baker, E. G., et al.: "Catalytic Destruction of Hazardous Organics in Aqueous Wastes: Continuous Reactor System Experiments," *Hazard. Waste Hazard. Mater.,* vol. 6, no. 1, p. 87, 1989.

Branscome, M. C., et al.: "Field Assessment of Steam Stripping Volatile Organics from Aqueous Waste Streams," *Proceedings of the Thirteenth Annual Research Symposium; Land Disposal, Remedial Action, Incineration and Treatment of Hazardous Wastes,* EPA 600/9-87/015, Cincinnati, Ohio, 1987.

Breton, M., et al.: "Technical Resource Document—Treatment Technologies for Solvent Containing Wastes," EPA/600/2-86/095, October 1986.

"Chemical Treatment," vol. 2 in *Innovative Site Remediation Technology,* W. C. Anderson, ed., American Academy of Environmental Engineers, 1994.

deFillippi, R. P., and M. E. Chung: "Laboratory Evaluation of Critical Fluid Extractions for Environmental Applications," EPA/600/2-85/045 (NTIS No. PB 85-189843), April 1985.

Dobbs, R. A., and J. M. Cohen: "Carbon Adsorption Isotherms for Toxic Organics," EPA-600/8-80-023, 1980.

Fair, G. E., and F. E. Dryden: "Comparison of Air Stripping Versus Stream Stripping for Treatment of Volatile Organic Compounds in Contaminated Ground Water," *Superfund '89,* Haz. Materials Control Research Instit., Washington, D.C., pp. 558–561, 1989.

Freeman, Harry M., ed.: *Innovative Hazardous Waste Treatment Technology Series,* vol. 2, *Physical/Chemical Processes,* Technomic Publishing, Lancaster, Pa., 1990.

Freeman, Harry M., ed.: *Standard Handbook of Hazardous Waste Treatment and Disposal,* McGraw-Hill, New York, 1989.

Helling, R. K., and J. W. Tester: "Oxidation of Simple Compounds and Mixtures in Supercritical Water: Carbon Monoxide, Ammonia, and Ethanol," *Environmental Science & Technology,* vol. 22, no. 11, pp. 1319–1324, 1988.

Lyandres, S., et al.: "Evaluation of Membrane Processes for the Reduction of Trace Organic Contaminants," *Environ. Prog.*, vol. 8, p. 239, 1989.

Lyman, Warren J., William F. Reehl, and David H. Rosenblatt: *Handbook of Chemical Property Estimation Methods,* Environmental Behavior of Organic Compounds, McGraw-Hill, New York, 1982.

McCabe, W. L., J. C. Smith, and P. Harriot: *Unit Operations of Chemical Engineering,* 4th ed., McGraw-Hill, New York, 1985.

Niramalakhandan, N., Y. H. Lee, and R. E. Speece: "Designing a Cost-Efficient Air-Stripping Process," *Journal AWWA,* vol. 79, no. 1, p. 56, 1987.

Raghaven, R., et al.: "Cleaning Excavated Soil Using Extraction Agents: A State-of-the-Art Review," Rep. No. EPA/600/2-89/034, 1989.

Shanableh, A., and E. F. Gloyna: "Supercritical Water Oxidation—Wastewaters and Sludges," *Water Science & Technology,* vol. 23, pp. 389–398, 1991.

"Soil Vapor Extraction," vol. 8 in *Innovative Site Remediation Technology,* W. C. Anderson, ed., American Academy of Environmental Engineers, 1994.

"Soil Washing/Soil Flushing," vol. 3 in *Innovative Site Remediation Technology,* W. C. Anderson, ed., American Academy of Environmental Engineers, 1994.

"Solvent/Chemical Extraction," vol. 4 in *Innovative Site Remediation Technology,* W. C. Anderson, ed., American Academy of Environmental Engineers, 1994.

Streebin, L. E., H. M. Schornick, and A. M. Wachinski: "Ozone Oxidation of Concentrated Cyanide Wastewater from Electroplating Operations," Proceedings of the 35th Industrial Waste Conference, Purdue Univ., Ann Arbor Sci. Publ., Ann Arbor, Mich., 1983.

Verschueren, Karel: *Handbook of Environmental Data on Organic Chemicals,* 2nd ed., Van Nostrand Reinhold, New York, 1983.

Yu, X., et al.: "Supercritical Fluid Extraction of Coal Tar Contaminated Soil," *Environmental Science & Technology,* vol. 24, no. 11, pp. 1732–1738, November 1990.

ACKNOWLEDGMENTS

The authors wish to acknowledge the following individuals who assisted with the preparation of this chapter for the first edition. Section 9-3, Steam Stripping, was prepared by David Potter, *Science Applications International Corp.* Review assistance was provided by Karen Winegardner, Deborah Watkins, William Murphy, Richard Dulcey, Jeffrey Case, John Haasbeek, Ruth Baker, and Amy Wong, *ERM*; Profs. W. C. King and M. Prince and Steven Galloway, *Bucknell University*; and Wayne Schuliger, *Calgon Carbon Corporation.* The principal reviewers for ERM for the first edition were James Fitzgerald and Dr. Elsie Millano, *ERM.*

Reviewers for the second edition were Mark Bordelon, Catherine Chomat, Michael Eversman, Kenneth Schroeder, and Stanislaus Zagula, Ph.D., *ERM*; and Prof. James Kilduff, Ph.D., *Rensselaer Polytechnic Institute.*

CHAPTER 10

BIOLOGICAL METHODS

It is probably not unscientific to suggest that somewhere or other some microorganism exists which can, under suitable conditions, oxidize any substance which is theoretically capable of being oxidized.

E. F. Gale, *The Chemical Activities of Bacteria* (1952)[1]

For thousands of years, human civilization has benefited from biologically mediated processes. For example, the production of beer, wine, bread, cheese, yogurt, and many other food products relies on microbially catalyzed reactions. Beginning with penicillin in 1928, the pharmaceutical industry has developed and commercialized many microbially produced antibiotics and other useful drugs. The use of microorganisms to systematically improve human health and the environment is also a recent phenomenon. The biological treatment of municipal sewage in the United States began in the early 1900s with the goal of improving sanitation. By the 1950s, the expansion of the synthetic chemicals industry had sparked the widespread use of biological treatment of industrial process waters.

Over the past 40 years, the realization dawned that many past waste management practices by manufacturers and users of synthetic chemicals actually resulted in the contamination of groundwater and soil with **xenobiotic** (synthetic or unnatural) compounds. A number of these xenobiotics are classified as hazardous wastes. A vast majority of organic hazardous wastes can be treated biologically because virtually all organic chemicals can be degraded if the proper microbial communities are established, maintained, and controlled. With a sound understanding of microbiology and many of the same engineering principles that have governed the treatment of industrial wastewater, environmental scientists and engineers can design and operate treatment systems that create the environment necessary for the effective treatment of hazardous wastes.

This chapter provides the reader with an overview of the biological technologies used to treat contaminated soils and groundwater. Successful biological treatment projects require a basic understanding of the microbiology involved as well

as the engineering factors required for design, implementation, and operation. With this in mind, this chapter starts with a basic introduction to microbiology, focusing on microbial metabolism. That is followed in Section 10-2 by a discussion of the biodegradation of xenobiotics (including the topic of compound biodegradability, the role of aerobic and anaerobic conditions, and the importance of microbial communities and adaptation). Section 10-2 summarizes the growth requirements of microbial populations necessary to metabolize or otherwise degrade xenobiotics. Section 10-3 presents the classical mathematical equations that describe the kinetics of microbial growth, and provides a mathematical approximation for the biological treatment of hazardous waste.

The central part of this chapter describes the various hazardous waste biological treatment methods, organized as to liquids (contaminated groundwater, industrial process wastewater, and landfill leachate), slurries (sludges and contaminated soils mixed with clean or contaminated waters), solids (contaminated soils), and vapors. A major part of this chapter covers *in situ* methods (e.g., treating contaminated groundwater and subsurface contaminants where they are found without excavating the overlying soil or extracting the groundwater for treatment in aboveground engineered reactors).

The chapter concludes with a discussion of the major classes of xenobiotics, to include their biodegradability and applicable treatment methods. Brief descriptions of treatability studies and emerging methods are also presented.

10-1 BASICS OF MICROBIAL METABOLISM

Microorganisms

Biological treatment systems typically rely on the interactions of different types of living organisms, including bacteria, fungi, algae, and protozoa. Although all of these organisms are "microorganisms," each is biologically distinct, with its own properties and characteristics. One characteristic is the basic cellular structure. Other characteristics are determined from the reactions and various substrates from which the organisms derive their energy and organic carbon. Bacteria, fungi, algae, and protozoa are briefly described below. More detailed information can be found in the reference texts listed in the Additional Reading section at the end of this chapter.

Cells are the smallest individual units of life. Each cell contains genetic information in the form of DNA, typically present as **chromosomes** consisting of DNA and bound proteins. Certain cells may also contain extrachromosomal pieces of DNA known as **plasmids.** All cells have membrane boundaries composed of lipids and proteins, and plant cells have structural walls in addition to the membrane. Cell membranes are semipermeable structures, able to selectively allow molecules to enter or leave the cell. Confined by the cell membrane is the liquid cytoplasm, which contains the majority of the cell's RNA and proteins.

The **procaryote** is the simplest type of living cell, consisting of a single chromosome, cytoplasm, cell membrane, and possibly one or more plasmids. Bacteria and blue-green algae are procaryotes. The blue-green algae are also known as blue-green bacteria or cyanobacteria. A more complex type of cell is the **eucaryote,** which

contains a nucleus made up of a semipermeable cell, within which is held one or more chromosomes. The nucleus effectively separates the chromosomes from the cytoplasm. Fungi, protozoa, all plant and animal cells, and most algae are eucaryotes.

All living organisms require organic carbon for use as building blocks of cellular components, as well as a source of energy to power the biochemical reactions taking place within cells. **Autotrophic** organisms use carbon dioxide as a source of organic carbon for the manufacture of cellular components. If light is used as a source of energy, these organisms are referred to as **photoautotrophs.** If oxidation-reduction reactions of inorganic compounds (such as ammonia) are used as a source of energy, these organisms are referred to as **chemoautotrophs. Heterotrophic** organisms use organic compounds, not carbon dioxide, to supply the organic carbon needed by the cell. If light is used as a source of energy, these organisms are referred to as **photoheterotrophs.** If oxidation-reduction reactions of organic compounds are used as a source of energy, these organisms are referred to as **chemoheterotrophs.** Chemoheterotrophic bacteria can use a variety of organic compounds, including xenobiotic compounds, as a source of both carbon and energy during metabolism. Chemoheterotrophs are the most important class of microorganisms used in biological treatment systems. Table 10-1 summarizes the classification of microorganisms according to energy and carbon source.

Of the many different organisms involved in biological treatment of hazardous waste, the most commonly relied on are heterotrophic bacteria. Nevertheless, fungi and higher living forms can be effective in specific applications. The latter as predators can affect biotreatment results in the mixed biological populations that occur naturally outside the laboratory.

Bacteria are widely distributed throughout the biosphere. Individual bacterial species are able to exist under almost any conditions found on the planet. Bacteria have three principal shapes: spherical cocci, cylindrical rods, and helical spirilla. Sizes range from 0.5 to 1.0 μm in diameter for cocci, 0.1 to 1 μm in width by 1.5 to 3.0 μm in length for rods, and 0.5 to 5 μm in width by 6 to 15 μm in length for spirilla. These unicellular organisms multiply by binary fission. Microbial reaction rates increase with increasing temperature, and approximately double with every 10°C increase up to some limiting temperature. Optimal growth temperatures are 12 to 18°C for psychrophilic organisms, 25 to 40°C for mesophilic organisms, and 55 to 65°C (or higher) for thermophilic organisms. Many bacteria are motile, using one

TABLE 10-1
Types of microorganisms

Classifications of microorganisms	Energy source	Carbon source (substrate)
Autotrophic		
Photoautotroph	Light	Carbon dioxide
Chemoautotroph	Oxidation-reduction reactions of inorganic compounds	Carbon dioxide
Heterotrophic		
Photoheterotroph	Light	Organic carbon
Chemoheterotroph	Oxidation-reduction reactions of organic compounds	Organic carbon

or more flagella to propel themselves through water. Many bacteria also have the ability to form spores under unfavorable growth conditions. Spores develop when the cell's genetic information is encapsulated in a tough polysaccharide coating. Spores can survive under extreme physical and chemical conditions. Once environmental conditions improve, spores germinate and the resulting active cell resumes its normal existence.

Fungi are heterotrophic organisms that commonly inhabit soils, though some species prefer water environments. Most fungi consist of long filaments of cells (hyphae) which together form mycelium. Fungi are classified into four distinct groups according to their mode of reproduction (sexual or asexual), encompassing fission, budding, and spore formation. Fungi are able to survive over a wide pH range (2 to as high as 10), and can also survive under low moisture and nitrogen levels. Fungi are involved in the decay of plant and animal matter. Some fungal species are especially effective in the degradation of hazardous chemicals.

Algae are unicellular or multicellular photoautotrophs. When exposed to light, the cells produce oxygen from water through photosynthesis. At night without light, oxygen is used during respiration. Algae are typically present in natural aquatic environments and can also be found in holding basins for wastewater treatment systems. Algae typically play only a limited role in the treatment of hazardous waste compounds.

Protozoa are usually motile single cells, the majority of which are aerobic heterotrophs. Protozoa are much larger than bacteria, and often graze on bacteria as a source of carbon and energy. Protozoa help to maintain bacterial population levels in liquid-phase biological treatment systems, and are believed to play a similar regulating role in soil-based ecosystems.

Cellular Metabolism

Microbial behavior has two principal goals: to maintain the living state, and to grow/reproduce where conditions are favorable. To accomplish these goals, complex substrates are enzymatically degraded into simpler molecules, providing the cell with organic carbon and energy in the form of **adenosine triphosphate (ATP)** in a process known as **catabolism.** At the same time, the cell is synthesizing complex molecules and polymers needed for life from these simpler molecules using energy provided by the ATP in a process known as **anabolism.** The anabolic processes produce protoplasm, which is composed of proteins, carbohydrates, DNA, and other cellular components. The typical metabolic steps involved in the biodegradation of organic substrates (to include organic wastes) are shown in Figure 10-1.

The sum total of all the biochemical reactions taking place within a cell is collectively known as **cellular metabolism.** An organic waste, being a substrate, is thus degraded by the metabolic action of microorganisms. Degradation alters the molecular structure of organic compounds. The degree of alteration determines whether **biotransformation** or **mineralization** has occurred. Biotransformation refers to the simplification of an organic compound to a daughter compound. Mineralization under aerobic conditions results in the complete breakdown of organic molecules into cellular mass, carbon dioxide, water, and inert inorganic residuals. Biotransformation is partial degradation and mineralization is complete degradation.

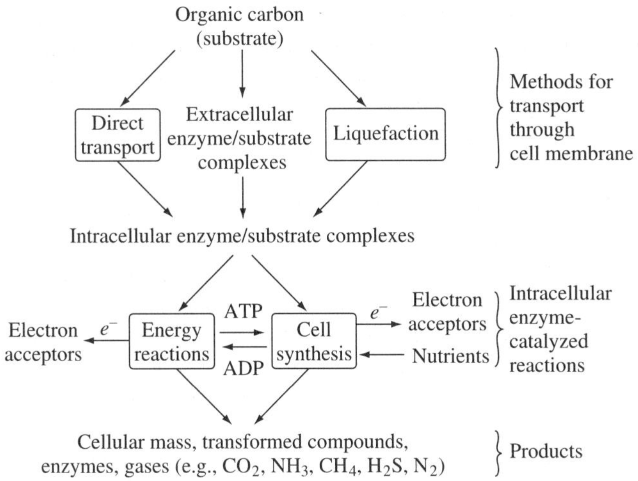

FIGURE 10-1
Metabolic steps in the degradation of organic wastes.

Aerobic, Anaerobic, and Anoxic Conditions

Catabolic reactions involve a transfer of electrons from the source of organic carbon to an electron acceptor, and the biological process by which this occurs is termed **respiration.** In **aerobic respiration,** bacteria utilize oxygen as the terminal acceptor of electrons removed from oxidized organic compounds in the generation of ATP. Energy is stored in the ATP molecule when a high energy bond is formed linking a phosphate group to ADP, adenosine diphosphate. The mass of oxygen required for aerobic biodegradation can be calculated by stoichiometric or laboratory determinations.

Anaerobic respiration occurs in the absence of free oxygen. Under these conditions, some microorganisms have the ability to use inorganic oxygen-bearing compounds (e.g., sulfate) as terminal electron acceptors in the generation of ATP. An anoxic environment can be considered a transition zone between strict anaerobic and aerobic conditions (no free oxygen and excess free oxygen, respectively). The term **anoxic** is commonly used to describe conditions under which free oxygen is present only at very low concentrations (so low that aerobic metabolism is limited) or when oxygen is absent and nitrite or nitrate is present. However, by strict definition an anoxic environment is one in which the redox potential (E_H) of paired reactions is greater than zero. Anoxic reactions can involve facultative and obligate anaerobes. Using a corresponding definition, an anaerobic environment is one in which free oxygen is absent and the redox potential of paired reactions is less than zero. Under these highly reduced conditions, only obligate anaerobes can conduct sulfate reduction, acetogenesis/fermentation, and methanogenesis processes. A comparison of redox pairs, redox potential, microbiological processes, and environmental conditions is shown in Table 10-2.[2]

Fermentation also occurs under anaerobic conditions when ATP is generated as a consequence of oxidizing one organic compound while reducing another organic compound. Some organisms can grow only in the presence of free oxygen, and are known as **strict aerobes.** Organisms that can grow only in the complete absence of

TABLE 10-2
Redox pairs important in anoxic/anaerobic environments

Redox pair	E_H, volts	Microbiological process	Environmental conditions
CO_2/CH_4	−0.42	Methanogenesis	Highly reduced
CO_2/acetate	−0.28	Acetogenesis/fermentation	Highly reduced
SO_4^{-2}/H_2S	−0.22	Sulfate reduction	Highly reduced
NO_3^-/NO_2^-	+0.42	Nitrate reduction	Moderately reduced
NO_3^-/N_2	+0.74	Denitrification	Anoxic/oxidized
Fe^{+3}/Fe^{+2}	+0.76	Iron reduction	Anoxic/oxidized
O_2/H_2O	+0.82	Aerobic respiration	Oxidized

free oxygen are known as **strict anaerobes. Facultative anaerobes** can grow either in the presence or in the absence of free oxygen.

Special Forms of Metabolism

Microbial metabolism relies on the biochemical breakdown of primary substrates in order to provide the cell with organic carbon and energy. As substrate concentrations decrease, the amount of carbon and energy derived from their utilization may be less than that required by a cell to maintain growth. This threshold concentration (minimal substrate level) at which microbial growth just balances microbial decay from endogenous metabolism is known as S_{min} (the mathematical relationship between S_{min} and substrate concentration is derived in Section 10-4). Many environmentally significant chemicals are present in the environment at concentrations below S_{min} levels; therefore, their concentration would not ordinarily sustain biodegradation. Specialized forms of microbial metabolism, such as **cometabolism** and **secondary substrate utilization,** may allow these low concentrations of xenobiotics to be biologically treated.

The phenomenon of **cometabolism** was first discovered in the late 1950s when obligate methane-utilizing *Pseudomonas* species were being characterized.[3–5] When fed a mixture of air, methane, ethane, propane, and butane, only methane was used by the cell to provide organic carbon and energy for the cell. However, a nonspecific monooxygenase enzyme produced by the cultures also converted ethane into acetic acid and acetaldehyde, propane into propionic acid and acetone, and butane into *n*-butyric acid and 2-butanone. This phenomenon of nonspecific enzyme activity has been successfully exploited by the pollution control field.

Although many similar definitions of cometabolism and fortuitous metabolism can be found in the technical literature, the following definitions will be used herein[6]:

- **Cometabolism** is the transformation of a nongrowth substrate (known as the "secondary substrate") in the obligate presence of a growth substrate or another transformable compound. A nongrowth substrate is a compound that will not support cell division, but may be transported into the cell with a concomitant increase in biomass. Although cometabolism is sometimes also called fortuitous metabolism, the terms have subtly different meanings.

- **Fortuitous metabolism** is defined as the transformation of a nongrowth substrate without the presence of another usable substrate.

Cometabolism plays a significant role in microbial communities treating hazardous waste. The daughter products of the secondary substrates may be used as primary substrates and/or secondary substrates by other microbial species present in the microbial community.

Cometabolism is involved in the biotransformation of many xenobiotic compounds, including 2,3,6-trichlorobenzoate,[7] alkyl benzene sulfonate,[8] cyclohexane,[9] parathion,[10] 2,4-D,[11] and 2-chlorophenol.[12] Cometabolism is a potential alternative for compounds that otherwise would degrade very slowly, such as trichloroethylene (TCE). It has been well documented that functioning methanotrophs (a group of heterotrophic bacteria that use methane as their source of carbon and energy) will indirectly oxidize TCE and other chlorinated aliphatics as a secondary substrate in the presence of oxygen. Methanotrophs produce the enzyme methane monooxygenase (MMO) when metabolizing methane, the primary substrate. MMO has broad substrate enzymatic specificity and can transform TCE to TCE epoxide. The TCE epoxide readily decomposes in water.[13,14]

In addition to cometabolism, the biodegradation of low concentrations of xenobiotic compounds may be accomplished by secondary substrate utilization. **Secondary substrate utilization** is a form of simultaneous substrate utilization in which one compound is present at a concentration above its $S_{\min}$ level (the primary substrate), while a second compound is present below its $S_{\min}$ level (the secondary substrate).[15,16] Microbial growth occurs at the expense of the primary substrate, and the secondary substrate is also degraded. Secondary substrate utilization was demonstrated in a pure microbial culture with acetate as the primary substrate and methylene chloride as the secondary substrate.[17]

Enzymatic Control of Cellular Metabolism

Both catabolic and anabolic reactions are catalyzed through the action of enzymes, which are proteins highly specific for each task they perform. Biodegradation requires the presence of adequate numbers of suitable enzymes to effectively transform the contaminants of concern. Since the synthesis of enzymes requires organic carbon and energy (ATP), enzyme production must be controlled in order for a cell to regulate its energy and carbon utilization. Several mechanisms allow the cell to do this, ensuring that growth will occur in the most efficient manner for any given set of environmental conditions. This is especially important under unfavorable growth conditions.

In general, cells possess two types of enzyme systems: those always expressed (constitutive) and those regulated by the presence of an effector molecule (inducer or repressor). Constitutive enzymes generally act on substrates that are readily available and provide the cell with abundant quantities of carbon and energy. These substrates typically include various proteins, fats, and carbohydrates that are found throughout the biosphere. Inducible enzymes, in contrast, generally act on substrates that are only periodically available. Many xenobiotic compounds fall into this latter category.

High intracellular energy levels indicate to the cell that it is meeting all of its carbon and energy requirements through the action of enzymes already actively

functioning. If these are constitutive enzymes, the synthesis of inducible enzymes is repressed. The control mechanism preventing synthesis of inducible enzymes on the basis of intracellular energy levels is termed catabolite repression. Catabolite repression was first observed by the French microbiologist Monod when a pure batch culture of bacteria sequentially utilized a mixture of two sugars. In this culture, the constitutively degradable substrate (glucose) provided all of the energy needed by the cells, and thus the inducible enzymes needed for degradation of the second substrate (lactose) were not synthesized until all of the first substrate was depleted.

Catabolite repression has been extensively studied by using the inducible lactose degrading enzyme system. These studies have shown that the number of enzymes present in a cell is governed by the rate of enzyme synthesis, and that there are both positive and negative factors that influence the rate of inducible enzyme synthesis. High intracellular energy levels cause a decrease in the concentration of an effector molecule necessary for high rates of enzyme synthesis. At the same time, high rates of enzyme synthesis require adequate concentrations of a suitable inducer to be present. Thus, the degree of catabolite repression occurring in the cell reflects both the efficiency of cellular metabolism (reflected in high energy levels) and the concentration of inducer present.

The specificity of constitutively produced enzymes differs from microbial species to species. The substrate best able to support microbial growth is always metabolized first. Hence, there is no single substrate able to cause catabolite repression in all bacteria.

Although catabolite repression–enzyme induction offers a convenient method of predicting the number of substrates that may be used at a given time by a cell, growth under low nutrient conditions (cell growth rates less than 0.01 h^{-1}) may involve several other factors that influence cellular metabolism. For example, a cell may synthesize totally different enzymes that are more efficient at scavenging carbon. In other cases, enzymes may be modified, or greater quantities may be produced. Cells may also increase their metabolic efficiency by increasing the transfer rate of substrate across the cell membrane; increasing the rate of substrate metabolism within the cell, causing a greater driving gradient to be formed between the environment and the cell; decreasing the synthesis of cellular components requiring the substrate; and coordinating the synthesis of all cellular components differently.

Departures from Monod's growth rate–substrate concentration relationship may also occur. The maintenance energy of the cell, that is, the energy required just to keep the cell alive, may become more significant, resulting in a lower growth rate for the concentration of substrate provided. The concentration of usable substrate may be increased with cell lysis products, or growth may continue on cell lysis products alone—a process known as cryptic growth. Starving cells have been known to degrade nonessential cellular components, or they may uncouple energy production from growth, in which case artificially low specific growth rates would be observed. Metabolism of a substrate may not be initiated below a threshold concentration. Growth under low nutrient conditions may have to be considered as a special case of microbial metabolic reaction to the environment. Biological interactions occurring between species also influence the growth rate and survival of individual organisms.[18]

Another important metabolic control mechanism exerts its influence on the activity of preexisting enzymes, rather than on enzyme synthesis. It is possible for a

chemical to bind to an enzyme, altering its specificity or ability to function. When a readily degradable substrate effects this change, the process is called **catabolite inhibition** or **competitive inhibition.** Catabolite inhibition is an almost instantaneous process, and removal of the effecting molecule allows growth on the initial substrate to continue without a lag.

It is evident that microbial metabolism is a complex process, controlled primarily by catabolite repression–enzyme induction. Under favorable growth conditions, the cell may be able to attain sufficient carbon and energy by using its constitutive enzymes, and only a single substrate at a time will be metabolized. Under environmental conditions inadequate to supply the cell with sufficient energy (intermediate intracellular energy levels), catabolite repression is reduced and induction is possible, allowing for the simultaneous utilization of two substrates. At low cellular energy levels, simultaneous utilization of mixed substrates can occur. The biodegradation of any substrate, including xenobiotics, is subject to these metabolic controls.

10-2 BIODEGRADATION OF XENOBIOTICS

The biodegradation of xenobiotic chemicals relies on the fundamental properties of microbial growth and metabolism already discussed. Experience and research have shown that most synthetic organics are biodegradable, making biological treatment a technically feasible alternative for many environmental problems. Nevertheless, the literature is replete with cases where specific compounds have either resisted biodegradation, or their degradation occurs so slowly as to make biological treatment ineffective.

Compound Biodegradability

Compounds resistant to biodegradation are commonly referred to as recalcitrant or refractory. **Recalcitrance** is the inherent resistance of a chemical to undergo any degree of biotransformation or biodegradation. To properly label a compound recalcitrant, the chemical must be resistant to biotransformation under all conditions (for example, by all species of microorganisms, all microbial communities, and under all combinations of environmental conditions). Since it is impossible to evaluate all of these situations, true recalcitrance cannot be proved. Compounds not known to be biodegraded are more properly said to be persistent. **Persistence** is defined as the resistance of a chemical to undergo biodegradation under a defined set of conditions, such as those created in a laboratory environment or those found at a hazardous waste site. Although persistence is also commonly used to describe compounds that are only slowly biodegraded, this use of the term is incorrect and frequently misleading. Compounds slowly biodegraded should be described as such.

Even with persistent compounds, continuing research has often been successful in identifying some microorganisms, or consortium of microorganisms, that will degrade them. Nevertheless some noteworthy compounds that have been subjected to considerable research still remain labeled as persistent, such as TCDD (dioxin), highly chlorinated PCBs, and highly chlorinated pesticides.

What makes one organic chemical persistent, and another not? The inherent biodegradability of a chemical depends to a large extent on its molecular structure. Unsubstituted hydrocarbons, except for multiple-ring polycyclic aromatic hydrocarbons (PAHs), generally degrade rapidly. Yet certain substituents will resist biodegradation. Small changes in the configuration of a biodegradable molecule may result in persistence, as shown in Figure 10-2.

Attempts to correlate degradability with various structural parameters indicate that the following conditions, either individually or in combination, are frequently associated with persistent or slowly biodegradable compounds[19]:

- Halogenation
- Large number of halogens
- Highly branched
- Low solubility in water
- Atomic charge difference

Some general conclusions regarding the ability of microorganisms to degrade specific classes of organic chemicals have been reported by researchers.[20–23] These can be summarized as follows:

- Typically, straight-chain aliphatic compounds are easily aerobically degraded. However, introduction of extensive branching of alkane substituents results in sterically hindered configurations that often are resistant to degradation. Also, unsaturated aliphatics are less readily transformed than the saturated analogs.
- Simple aromatic compounds are usually degradable by several mechanisms of ring cleavage. The introduction of halogens will result in decreased biodegradability because halogens stabilize the ring. The degree of decreased biodegradability usually is a function of the number of halogens incorporated because biodegradation will require dehalogenation at some stage. However, microorganisms capable of utilizing haloaromatic as well as haloaliphatic compounds as the sole source of carbon and energy have been isolated from natural environments.
- Even highly chlorinated aromatic compounds like PCBs can be degraded, albeit slowly. Highly substituted PCB isomers are degraded much more slowly than less halogenated derivatives, and catabolic enzymes preferentially attack and cleave the ring carrying the lowest number of chlorines. For halogenated substances, degradation of isomeric substrates may require distinctly different pathways. In addition, many of these compounds are preferentially or initially degraded under anaerobic conditions. Sequential anaerobic/aerobic treatment may be required for mineralization.

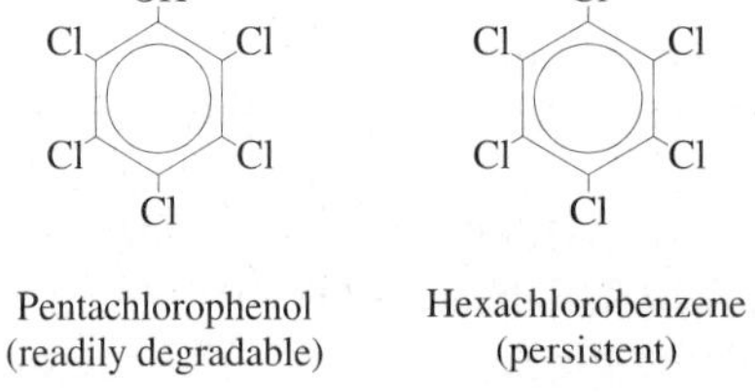

FIGURE 10-2
Similarly structured compounds having different biodegradability.

- Biodegradation of nitrogen- and sulfur-containing compounds is often linked to their utilization as nutrient sources. Branched chain alkyl or aryl sulfonates are often slowly degraded.
- Polymeric materials are among the most resistant to microbial attack. Nevertheless, microorganisms have been isolated that will utilize nylon polymer or, to a more limited extent, polystyrene, as the sole source of carbon and energy.

Many exceptions exist to these guidelines. An environmental professional considering the use of biodegradation for the treatment of a hazardous substance may find useful information in the technical literature describing previous biotreatability studies. For substances whose biodegradability has not been reported in the literature, bench-scale or pilot-scale treatability testing may be required (see Section 10-10). Many standardized tests exist to determine biotreatability for wastewater treatment applications, including 5-day biochemical oxygen demand (BOD_5), ultimate oxygen demand, respirometry studies, and batch or continuous flow laboratory- or pilot-scale treatability studies. These tests are usually conducted in a tiered sequence, starting with simple laboratory measures and advancing to more involved studies only if the screening tests prove positive.

Aerobic versus Anaerobic Treatment

Much of the early research and field work on bioremediation focused exclusively on aerobic processes, relying on oxygen as the terminal electron acceptor. As the knowledge base of environmental microbiology increased, a new understanding and appreciation of anaerobic biological processes was realized. As discussed in Section 10-1, strict anaerobic conditions require the complete absence of free oxygen. For the treatment of xenobiotics at hazardous waste sites, several wastes have been found that undergo anaerobic transformation more efficiently than under aerobic conditions. Of particular interest is the transformation of various halogenated compounds. A chemical such as tetrachloroethene (a dry cleaning fluid also known as perchloroethene or PCE) is too highly oxidized for efficient degradation by aerobic processes. Yet, reductive dechlorination of this compound occurs in a subsurface, anaerobic environment. The concept of reductive dehalogenation involves the cleaving of a chlorine atom or other halogen from an organic compound by oxidation-reduction. This is illustrated in Figure 10-3 for PCE. In this example, a reduced organic compound loses an electron and thus becomes oxidized. A mediator transfers the electron to the halogenated compound, thus cleaving the halogen from the compound.

Many contaminated subsurface environments are anoxic, presumably because of aerobic biological oxygen depletion. These sites may also have high concentrations of reduced metals or alternative electron acceptors like nitrate and sulfate. Thus they are perfect candidates for utilizing anoxic/anaerobic biological processes. Anoxic/anaerobic treatment eliminates the need (and cost) for artificially creating aerobic subsurface conditions in order to utilize biodegradation. Since anaerobic processes produce less energy from metabolism, less biomass is produced and lower levels of macronutrients and micronutrients are needed. The benefits of anoxic/anaerobic processes can also be used in part or whole in the aboveground treatment of soils or aqueous wastes.

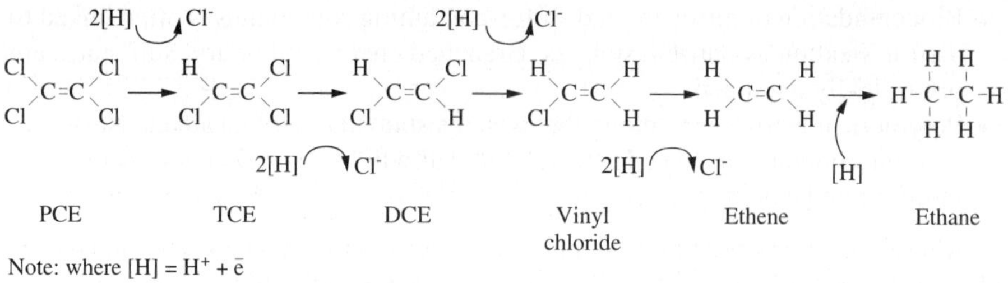

FIGURE 10-3
Anaerobic reductive dehalogenation of perchloroethene (PCE) to ethene and ethane.

Importance of Microbial Communities in Biodegradation

It is extremely rare to find any ecological niche in the biosphere inhabited by only a single species of microorganism. Many microbial species are typically found coexisting together in a given environment. Microorganisms are adaptable to widely changing environmental conditions, a characteristic that is made possible by several species-species interactions. The seven types of interactions most commonly cited as occurring between two species are shown in Table 10-3,[24] and may simply be considered beneficial, neutral, or detrimental.

Microbial communities can also be classified on the basis of broader considerations, rather than by the specific interactions taking place. One such classification system is shown in Table 10-4.[25] Of major importance from an environmental engineering point of view are the cometabolic (Class 5) and unknown mixed culture reactions (Class 7) communities. The major reason these communities are so important stems from the way biodegradation of xenobiotic chemicals is achieved.

The successful biodegradation of a xenobiotic chemical requires: (1) the existence of microbial enzymes able to effect a change in the molecule, (2) the availability of organisms possessing such enzymes, and (3) the presence of environmental conditions favorable for microbial growth. The foreign chemical structure of xenobiotics may require the combined metabolic action of microbial communities to effect total biodegradation because of the difficulty of meeting the three requirements with a single species. Class 7 communities have long been implicated in the biodegradation of xenobiotic chemicals because of the failure to isolate single species able to degrade

TABLE 10-3
Microbial interaction between species

Interaction	Definition
Neutralism	Lack of interaction
Commensalism	One member benefits while the other is unaffected
Mutualism	Each member benefits from the other
Competition	A race for nutrients and space
Amensalism	One adversely changes the environment for the other
Parasitism	One organism steals from the other
Predation	One organism ingests the other

TABLE 10-4
Classes of microbial communities

Class	Structure based on:
1	Provision of specific nutrients by different members of the community
2	The alleviation of growth inhibition, including removal of metabolites that are inhibitory to the producer species (see also hydrogen transfer communities)
3	Interactions that result in the alteration of individual population growth parameters, thereby producing more competitive and/or efficient community performance
4	A combined metabolic activity not expressed by individual populations alone
5	Cometabolism
6	Hydrogen transfer reactions
7	The interaction of several primary species

some of them. For example, activated sludge, composed of bacteria, yeast, fungi, and protozoa, is particularly adept at metabolizing many compounds persistent to single species.

A microbial consortium, in which each species serves a vital link in a chain if not web of transformation and mineralization steps, may be necessary for the degradation of many organic compounds. For example, a mixed microbial community of three species of anaerobic bacteria was found to be able to degrade 3-chlorobenzoate. However, none of the species could, individually, fully utilize the chemical. The initial step as shown in Figure 10-4 is the reductive dechlorination by a specific bacterium to yield benzoate. This dechlorinating bacterium receives no carbon source from this step. Another species ferments the benzoate to yield acetate and dissolved hydrogen gas (H_2). This second species, termed the benzoate oxidizer, is not known to use any other chemical for substrate. Because of the production of H_2 and the need to keep its levels below a threshold to enable benzoate oxidation, a third bacterial species is needed. This is a methanogen that uses the H_2 with CO_2 to produce methane. A portion of the H_2 is returned to the dechlorinating bacterium for the reduction step. Presumably, the dechlorinating bacterium receives some form of

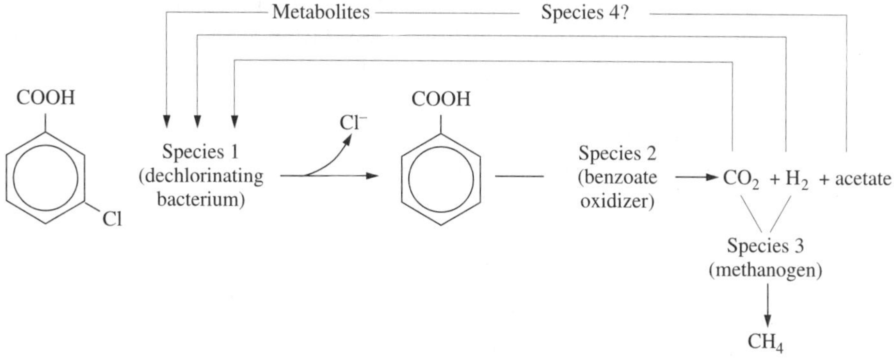

FIGURE 10-4
Mixed culture of three microbial species involved in the anaerobic degradation of chlorobenzoate.

carbon from one or more of the metabolites produced in the subsequent oxidation of acetate, meaning that still another species of bacterium is possibly required.[26]

Adaptation

Many microorganisms are unable to utilize a hazardous compound on initial exposure. However, when microorganisms are exposed to organic contaminants for various lengths of time they tend to develop, or adapt to, an increased ability to degrade these substrates. The rate at which microbial adaptation occurs, to a great extent, determines how quickly the concentration of contaminants diminishes in impure water and soils.

Adaptation is the culmination of natural selection, producing a microbial culture with new or modified enzymes with beneficial enzymatic activity. These new or modified enzymes allow for the biochemical transformation of a formerly persistent compound. Such adapted strains may arise by selection among strains already present, and by mutation and gene transfer (in the latter case, e.g., via plasmids).

Since the composition of an enzyme, and hence the structure and function, is specified by the DNA of a microorganism, changes to enzymes can arise only through preceding changes or mutations in an organism's DNA. From an environmental perspective, many mutations that occur are desirable because they increase the number of specific chemical compounds that can be acted on enzymatically. In this way novel enzymatic activities may appear in a population. It is also possible that the gene encoding for such enzymes was initially present in a nonexpressed form, and an appropriate mutation now allows the gene to be fully expressed on a constitutive basis.

Apart from modifying existing enzymes, a microorganism may gain the ability to produce entirely new enzymes by acquiring new DNA. Small extrachromosomal pieces of DNA known as plasmids exist in many species of bacteria, and are able to replicate autonomously. Under appropriate conditions plasmids can be exchanged between compatible cells, even those of different species. Plasmid DNA frequently codes for the production of proteins involved in antibiotic resistance, heavy metal resistance, and the biodegradation of many organic compounds.

Since adaptation may require long periods of time under natural conditions, one approach for enhancing biodegradation **(bioaugmentation)** at a site is to introduce microorganisms that have already been successfully adapted in the laboratory or at a similarly contaminated site. Such commercially available cultures can be added in large numbers to contaminated soils and liquids. However, the continued viability of such inoculated cultures is frequently subject to debate.

Microbial Growth Requirements

Biological treatment processes must promote and maintain a microbial population (biomass) that metabolizes the target wastes. A number of factors influence the effectiveness of biological treatment, including the type and concentration of carbon sources (primary or secondary substrates), electron acceptors, moisture, temperature, pH, inorganic nutrients, and the presence of toxic or inhibitory growth conditions. These factors are described in greater detail below.

Electron acceptors

The choice of electron acceptors depends on whether aerobic or anaerobic biological activity is desired. For aerobic biodegradation, free oxygen is needed to serve as the terminal electron acceptor. Under anaerobic conditions, inorganic compounds (ferric iron, nitrate, sulfate, and carbon dioxide) and certain organic compounds (during fermentation) can be used as terminal electron acceptors.

The mass of oxygen required by aerobic systems can be calculated from stoichiometry or laboratory determinations. For example, the following stoichiometric equation for the theoretical mineralization of a branched alkene in catabolic reactions skips the intermediate steps of oxidation to an alcohol and then to an acid:

$$C_7H_{12} + 10O_2 \rightarrow 7CO_2 + 6H_2O$$

The oxygen required to completely mineralize one milligram of alkene ($MW = 12 \times 7 + 12 \times 1 = 96$) is calculated as follows:

$$\frac{10 \text{ mmol } O_2}{\text{mmol alkene}} \times \frac{32 \text{ mg } O_2}{\text{mmol } O_2} \times \frac{1 \text{ mmol alkene}}{96 \text{ mg alkene}} = 3.33 \text{ mg } O_2/\text{mg alkene}$$

EXAMPLE 10-1. OXYGEN REQUIREMENTS. (a) What are the oxygen requirements for the anabolic metabolism of alkene to cellular material? (b) What are the oxygen requirements if this cellular material then undergoes endogenous respiration?

Solution

(a): The stoichiometric equation for cellular material is $C_5H_7O_2N$. Under anabolic conditions, the oxygen required in the conversion of alkene to cellular material is supplied from water, not diatomic oxygen (O_2), as represented by the following equations:

$$C_7H_{12} + 14H_2O \rightarrow 7CO_2 + 40e^- + 40H^+$$

$$5O_2 + 20H^+ + 20e^- \rightarrow 10H_2O$$

$$5CO_2 + NH_3 + 20H^+ + 20e^- \rightarrow C_5H_7O_2N + 8H_2O$$

$$C_7H_{12} + NH_3 + 5O_2 \rightarrow C_5H_7O_2N + 2CO_2 + 4H_2O$$

Even though oxygen is utilized in the process, diatomic oxygen from the environment is not required, and the anabolic reactions do not contribute to the biochemical oxygen demand.

(b): The summary chemical equation for the anabolic conversion of alkene to biomass is:

$$C_7H_{12} + 5O_2 + NH_3 \rightarrow C_5H_7O_2N + 2CO_2 + 4H_2O$$

The new cellular material, if mineralized in a process termed endogenous respiration, also exerts oxygen demand as follows:

$$C_5H_7O_2N + 5O_2 \rightarrow 2H_2O + NH_3 + 5CO_2$$

Summing these two equations yields:

$$C_7H_{12} + 10O_2 \rightarrow 7CO_2 + 6H_2O$$

which is the same equation as for alkene catabolism. Catabolism required 3.33 mg O_2 per mg of alkene, but since the 10 oxygen atoms used in anabolism came from water, only 1.67 mg O_2 are required for endogenous respiration.

Moisture

Biodegradation requires adequate moisture levels for two reasons. First, since cellular tissue is 75 to 80 percent water, moisture is needed for microbial growth. Second, water is used as a conveyance medium for transport of organic/inorganic nutrients to cells, the transport of biodegradation waste products away from cells, and the movement of cells themselves within the environment.

Biodegradation in soil systems can occur at moisture levels well below saturation. In the technical literature, the required minimum soil moisture level reported varies widely, depending on the source consulted. For example, one source states that bacteria fail to grow if the water content of the medium falls below 92 percent relative humidity.[27] However, it is generally accepted that the minimum moisture content necessary for treatment of contaminated soil is 40 percent of saturation.

Temperature

Temperature has a major influence on the microbial growth rate. Cellular activity, particularly enzyme systems, responds to heat so that the rate of cell growth increases sharply with increasing temperature until the optimum growth temperature is reached. Figure 10-5 shows for a particular microorganism how cellular activity observed below the optimum temperature can more than double with a rise of about 10°C. Increases in temperature just a few degrees above an organism's optimum growth temperature can slow growth dramatically. Continued exposure to high temperature may result in cell death.

Unlike high temperatures, low growth temperatures are usually not lethal. Instead, many cells have the ability to become dormant. Metabolic activity decreases at temperatures below the optimum because of reduced enzyme activity and a loss of the fluidity of the cell membranes. These changes may restrict the transport of substrate molecules.[28] A rapid change to low-temperature conditions produces a much greater reduction in cell activity than a corresponding gradual decrease to the same temperature. Gradual decreases may enable the microorganisms to acclimate.

The effect of temperature on cellular activity, and thus the rate of biological degradation, can be expressed simply as follows:

$$\frac{r_t}{r_{20}} = \theta^{(T-20)} \tag{10-1}$$

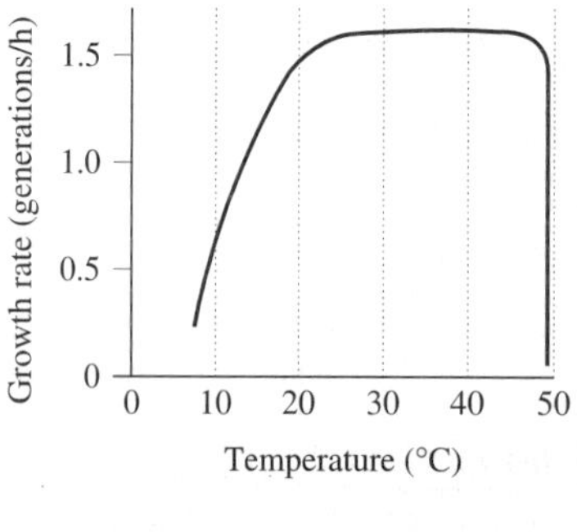

FIGURE 10-5
Effect of temperature on the growth rate of a particular species of mesophilic bacteria.

where r_t = activity rate at T°C
r_{20} = activity rate at 20°C
θ = temperature-activity coefficient
T = temperature (°C)

In conventional wastewater treatment systems, values for θ can range from 1 to 1.14.[29]

Each species of microorganism has a distinct optimal growth temperature, and microorganisms have been isolated that are able to survive and grow across a wide range of temperatures. Viable species of bacteria have been found in thermal springs and glaciers. Optimal growth temperature provides a basis for classification of the typical ranges in which bacteria function. Within this range there exists an optimum for which the maximum cell growth rate occurs, all else being equal. This classification system is shown in Table 10-5.

pH

Enzyme activity frequently depends upon pH conditions in the surrounding environment. The mass of a particular enzyme that is in a catalytically active state changes as pH changes, with the maximum activity occurring at a pH optimal to that specific enzyme. Accordingly, the growth of bacteria depends on pH. Most bacteria grow best in a relatively narrow range around neutrality (i.e., in a range of pH of 6 to 8). Die-off typically occurs below a pH of 4 to 5 and above a pH of 9 to 9.5. Certain microbial strains have pH optimums outside these limits.

Microbial activity can alter the pH of the surrounding environment. Examples include anaerobic fermentation, which converts organic compounds to organic acids, depressing pH. Nitrification also lowers pH, as does the carbon dioxide produced by aerobic degradation. The breakdown of organonitrogen compounds can raise pH by releasing NH_4^+. If these pH changes are significant and not buffered, the altered pH can create a microbial growth environment that is inhibitory or toxic to the microbial populations present.

Inorganic nutrients

Cell mass is composed of water, organic compounds, and inorganic materials. The inorganic cellular material contains numerous elements including phosphorus, sodium, sulfur, potassium, calcium, magnesium, iron, and various trace metals. Metabolism requires these elements as nutrients in addition to organic carbon as substrate. Phosphorus and nitrogen are referred to as **macronutrients** because the synthesis of cellular tissue requires much more of these than other nutrients. Frequently, nitrogen and phosphorus are not available in sufficient amounts in hazardous waste and must be added, usually as ammonia and orthophosphate.

TABLE 10-5
Classification of bacteria based on temperature

Classification	Functional range	Optimum
Psychrophilic	−2 to 30°C	12 to 18°C
Mesophilic	20 to 45°C	25 to 40°C
Thermophilic	45 to 75°C	55 to 65°C

An empirical formula for the organic portion of bacterial protoplasm, $C_{60}H_{87}O_{23}N_{12}P$ (or simply as $C_5H_7O_2N$), provides a basis for calculating the theoretical amount of nitrogen and phosphorus required for synthesis of cellular mass from waste.[30] The total organic carbon to nitrogen to phosphorus ratio (TOC:N:P ratio) for anabolic reactions is therefore approximately 100:23:4.3.

On the basis of general experience, environmental engineers commonly use a rule of thumb TOC:N:P ratio of 20:5:1. This ratio is less than the theoretical ratio because not all of the organic carbon metabolized by bacteria is synthesized to new cellular material. As much as half, if not more, of the metabolized carbon may be oxidized by catabolic reactions to carbon dioxide without the need for corresponding nutrients.

EXAMPLE 10-2. NUTRIENT REQUIREMENTS. What is the ratio of the nutrient requirements per unit of biochemical oxygen demand (BOD) for the aerobic treatment of alkene, assuming no endogenous respiration takes place?

Solution. In the absence of experimental data, environmental engineers customarily assume that half of a hazardous organic substrate is converted to cellular mass (anabolism) and the other half oxidized for energy (catabolism). Assuming a starting alkene mass of 1 mg, 0.5 mg of the mass will undergo catabolism while the remaining 0.5 mg will undergo anabolism. The formula for alkene is C_7H_{12} and the formula for biomass is $C_5H_7O_2N$.

As shown previously in the text, the catabolism of 1 mg alkene requires 3.33 mg of oxygen; therefore 0.5 mg alkene will require only 1.67 mg of oxygen. From Example 10-1, it follows that no diatomic oxygen is required for the remaining 0.5 mg of alkene to be anabolized. Since endogenous respiration is excluded from consideration, the BOD of 1 mg alkene under these conditions is 1.67 mg of oxygen.

The actual mass of carbon in 0.5 mg of alkene to be converted to cellular carbon is calculated as follows:

$$\text{Carbon into cellular mass} = (\text{mass of alkene})(\text{fraction of carbon in alkene})$$
$$= (0.5)[84/(84+12)] = (0.5)(0.875) = 0.4375 \text{ mg}$$

The empirical TOC:N:P ratio of 100:23:4.3 can be used to calculate the mass of nitrogen and phosphorus needed. Using 0.4375 mg of carbon as the TOC:

$$\text{N needed} = (0.4375 \text{ mg})(23/100) = 0.100 \text{ mg}$$
$$\text{P needed} = (0.4375 \text{ mg})(4.3/100) = 0.019 \text{ mg}$$

Therefore, the BOD:N:P ratio for treatment of alkene = (1.67):(0.100):(0.019) = 100:6:1.1

Environmental engineers have commonly used a rule of thumb BOD:N:P ratio of 100:5:1, based on general experience. The rule of thumb approximates the empirical ratio for alkene, and the true ratio would depend on the split between catabolic and anabolic reactions, the degree of endogenous respiration, and other factors.

Rather than calculate theoretical requirements or use a general rule of thumb, one can determine the need for nutrients as part of laboratory treatability tests (see Section 10-10). The specific inorganic nutrient requirements are very much dependent on the waste, the availability of nutrients in the waste, the yield of biomass, and the design of the treatment system. The general rule of thumb provides a basis for calculating ranges to be tested or for conducting economic evaluations. The objective is to ensure that organic carbon (the target substrates) is the growth-limiting nutrient.

Micronutrients

In addition to nitrogen and phosphorus, many other inorganic nutrients are needed at lower concentrations to ensure unhindered metabolism. These micronutrients include sulfur, potassium, calcium, magnesium, iron, nickel, copper, zinc, various vitamins, and others. The micronutrients should have a minimum concentration of 1 to 100 μg/L.[31] In most cases, all micronutrients can be obtained by bacteria from the environment and do not have to be added, particularly if the waste is a contaminated soil or one that has been in contact with soil (e.g., groundwater). However, the biological treatment of an industrial process wastewater may require the addition of micronutrients.

Inhibitory and/or toxic conditions

Inhibitory or toxic conditions can occur in the microbial growth environment because of many factors, including:

- Concentrations of one or more substrates beyond that tolerated by the microorganisms
- Concentrations of inorganic compounds, such as metals, beyond that tolerated
- The presence of antibiotics produced by competing microorganisms
- The presence of predator microorganisms such as protozoa

Concentration effects of substrates on the microbial growth rate are discussed further in the next section.

10-3
GROWTH KINETICS

The performance of biological waste treatment can be measured by the rate at which microorganisms metabolize the waste which, in turn, is usually related to their rate of growth. Bacteria reproduce by binary fission, and the time required to divide is referred to as generation or doubling time, which can be as little as 15 to 20 minutes under optimal growth conditions. Although individual cells reproduce at an approximately constant linear rate in optimum conditions, the total population of biomass grows at exponential rates.

Batch Culture

A classic laboratory study is the batch liquid culture, which starts with a small inoculation of cells from a single bacterial species into a solution containing a high concentration of a single substrate. The pure culture is then incubated under optimum conditions. The pattern for population growth of such a culture follows five distinct phases as shown in Figure 10-6 and described below:

- **Lag phase:** The cells require a period to adapt to the new environment before beginning their growth (this is a period of enzyme synthesis); preadapted cells experience a much shorter lag phase.
- **Exponential-growth phase:** The biomass grows exponentially as cells divide at a constant rate that exceeds the rate at which the cells are dying; generally

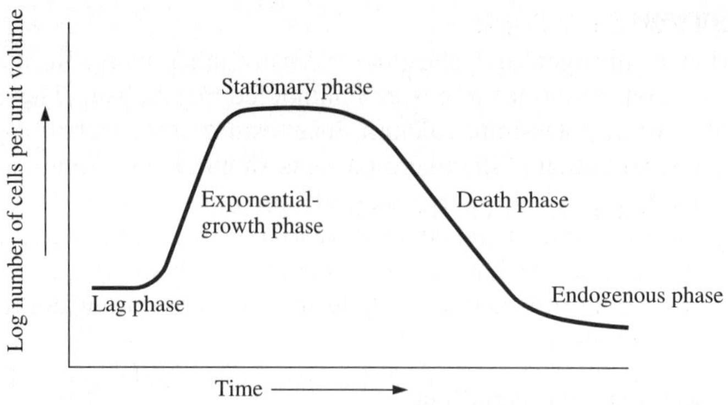

FIGURE 10-6
Growth of bacteria in batch culture.

this is the maximum growth rate for the established environmental conditions.

- **Stationary phase:** On approaching exhaustion of the substrate or limiting nutrient, the number of dying cells offsets the number of new cells.
- **Death phase:** Without the addition of substrate, dying cells outnumber new cells. The death rate can be as great as the "inverse" of the exponential growth phase.
- **Endogenous phase:** In some cases, the rate of population decline slows as the number of dead cells becomes adequate to supply the nutrients needed for new cells.

The buildup in the microbial population, or biomass, corresponds to the removal of substrate (degradation of organic contaminants) as shown in Fig. 10-7.

Kinetics of Batch Culture

The rate at which cells divide is the **specific growth rate** μ. It is measured as generations per unit time, or the inverse of the doubling time. The term provides a starting point for deriving a mathematical formulation of growth kinetics. Without die-off, the bacterial cells accumulate in the batch culture at an exponential rate,

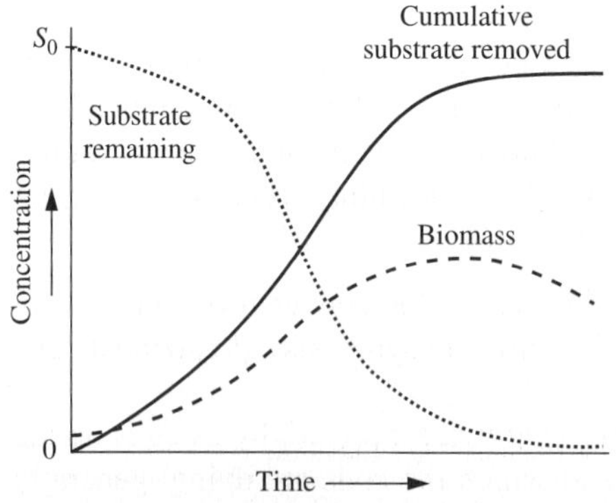

FIGURE 10-7
Degradation of organics and corresponding increase in biomass.

which can be defined as follows:

$$\frac{dX}{dt} = \mu X \tag{10-2}$$

where $X =$ concentration of biomass (mass/unit volume)
$\mu =$ specific growth rate (time^{-1})
$t =$ time

In the exponential-growth phase, cell division is unrestrained and occurs at a constant specific growth rate, which is the maximum observed for batch conditions. As shown in Figure 10-6, the growth in biomass slows as the stationary phase is approached because the specific growth rate begins to decrease in response to the diminishing concentration of substrate. A classic relationship exists between bacterial growth rate and substrate concentration when all other conditions are kept constant. This is depicted in Figure 10-8 for a hypothetical example. The specific growth rate at lower concentrations of substrate is almost linearly dependent on substrate concentration. As the concentration of the substrate increases, the specific growth rate eventually reaches a maximum that is maintained independently of higher concentrations. The actual maximum specific growth rate depends on the nature of the substrate and other factors, but not on substrate concentration.

Environmental engineers and scientists have traditionally formulated the relationship between specific growth rate and concentration according to the equation first proposed in 1942 by Monod[32]:

$$\mu = \mu_{max} \frac{S}{K_S + S} \tag{10-3}$$

where $\mu =$ specific growth rate (time^{-1})
$\mu_{max} =$ maximum specific growth rate (time^{-1})
$S =$ concentration of substrate in solution (mass/unit volume)
$K_S =$ half-velocity constant (i.e., substrate concentration at which the specific growth rate is one-half μ_{max}) (mass/unit volume)

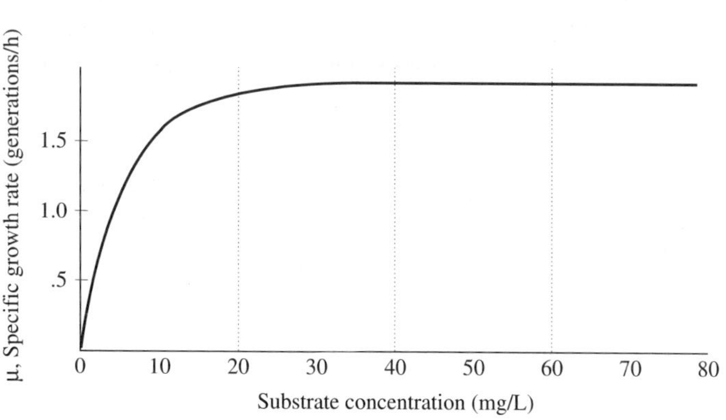

FIGURE 10-8
Effect of substrate concentration on bacterial growth rate (hypothetical example).

Although an oversimplification, the Monod equation reasonably depicts the general observation that the specific growth rate is almost linearly dependent on substrate concentration at the lower range of concentrations and approximately independent at higher concentrations. The **half-velocity constant** K_S represents an approximate division between the two ranges as shown in Figure 10-9.

The Monod equation implies that some growth occurs even at extremely low concentrations of substrate, a condition not only unproved but also probably untrue. The equation also may not describe reasonably well the conditions that yield very rapid growth.[33] Nevertheless, the equation represents an acceptable starting point for describing kinetics of growth.

Substituting Equation 10-3 in Equation 10-2 yields the following expression:

$$\frac{dX}{dt} = \frac{\mu_{\max} SX}{K_S + S} \tag{10-4}$$

It is generally accepted that the mass of new cells synthesized per unit mass of substrate removed is constant for a given substrate and bacterial species. The ratio of the increase in cellular mass to the decrease in mass of substrate is termed the **growth yield coefficient** Y and can be expressed as follows:

$$Y = \frac{dX/dt}{dS/dt} \tag{10-5}$$

Substituting Equation 10-5 in Equation 10-4 results in the following relationship:

$$\frac{dS}{dt} = \frac{\mu_{\max} SX}{Y(K_S + S)} \tag{10-6}$$

The term $\mu_{\max}/Y$ can be replaced by the term k, which is referred to as the **degradation rate constant** or maximum rate of substrate removal per unit weight of biomass. This results in the following expression for the rate of substrate degradation:

$$\frac{dS}{dt} = \frac{kSX}{K_S + S} \tag{10-7}$$

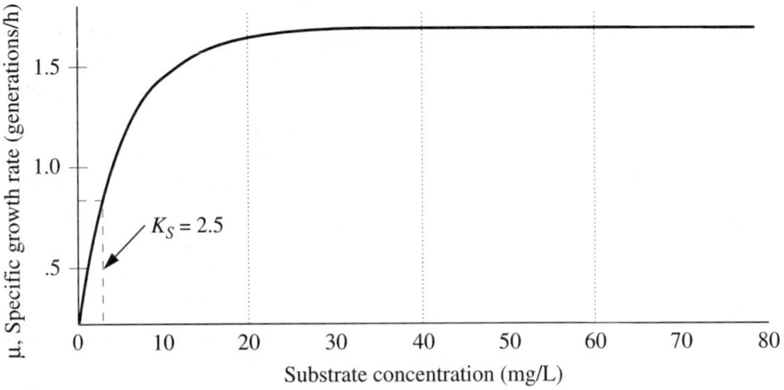

FIGURE 10-9
Monod half-velocity constant (hypothetical example).

Similarly, an expression for the rate of biomass formation is as follows:

$$\frac{dX}{dt} = \frac{YkSX}{K_S + S} \tag{10-8}$$

Actually, in any biological treatment system, a portion of the cells are not proliferating but rather are in a maintenance if not death phase. The net yield in biomass is thus the hypothetical amount of new cell formation represented by Equation 10-8 less any reduction due to death and maintenance. The reduction can be grouped and defined as **endogenous decay.** Inclusion of decay in Equation 10-8 yields the following:

$$\frac{dX}{dt} = \frac{YkSX}{K_S + S} - bX \tag{10-9}$$

where b = endogenous decay constant (time^{-1}).

These basic equations form the foundation for any detailed analysis of the kinetics of batch aqueous biological treatment systems. The four kinetic parameters (k, K_S, Y, and b) are determined through treatability tests. Table 10-6 summarizes the definitions and significance of these parameters.

Treatment of Toxic or Inhibitory Compounds

Many hazardous wastes can inhibit their own degradation at increased concentrations, as shown in Figure 10-10. Two cases are shown: one for an organic chemical with no toxic inhibition and the other for an organic chemical whose toxic nature inhibits growth at higher concentrations. In both cases, the specific growth rate at lower concentrations increases in a nearly linear relationship. However, as the concentration increases further, the specific growth rate of the toxic substance slows,

TABLE 10-6
Important kinetic parameters

Term	Definition	Significance
k	Degradation rate constant (time^{-1})	Indication of degradability of an organic compound; for biological treatment to be effective, high values are preferred.
K_S	Half-velocity constant (mass/unit volume)	Indication of the efficiency with which degradation will occur; to degrade organic compounds to low concentrations, low values of K_S are necessary. Determining K_S is very difficult at sites with complex contamination (e.g., variable concentrations of multiple compounds).
Y	Yield of biomass generated per unit mass of substrate removed (mass/unit mass)	Indication of utilization of substrate for the production of biomass.
b	Endogenous decay constant (time^{-1})	Indication of the rate of loss of biomass due to natural death and decay.

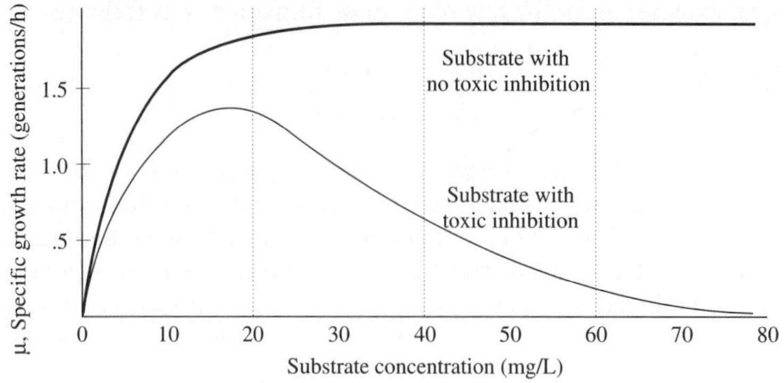

FIGURE 10-10
Effect of toxic inhibition on the bacterial growth rate (hypothetical example).

eventually reaching a threshold concentration above which growth ceases. Concentrations greater than those shown could result in microbial toxicity and die-off.

If the concentration of an aqueous waste with a toxic nature is low enough that the effect of inhibition is negligible, the equations that have been derived from Monod kinctics are still applicable. However, for higher concentrations, the equations must be modified. The effect of toxic inhibition, as demonstrated in Figure 10-10, can be described by Andrews kinetics (also referred to as the Haldane equation[34]) as follows:

$$\frac{dS}{dt} = \frac{kSX}{K_S + S + \dfrac{S^2}{K_i}} \tag{10-10}$$

where K_i = inhibition coefficient (mass/unit volume). From this expression, the biokinetics for various liquid-phase treatment reactors can be derived. The expressions become particularly complicated in dealing with mixtures of wastes in which one waste may inhibit the degradation of another.

For any organic chemical exhibiting toxic inhibition, concentration is critical. The maximum μ, or μ^*, for such a substance occurs at concentration S^* (see Figure 10-11). If the concentration in the reactor exceeds S^*, inhibition or die-off can commence on a continuing, irreversible basis, resulting in loss of the reactor's treatment capability. Because of the great number of potentially toxic substances and the synergism among them, determination of the concentration that results in toxic inhibition requires specific testing. Even when the aqueous waste proves toxic to the biomass, certain operational measures such as acclimation of the microbes, introduction of hardier species, and dilution of the waste can possibly mitigate toxic inhibition.

Minimum Substrate Concentration

Many cases of soil and groundwater contamination involve dilute concentrations of hazardous waste constituents (e.g., low parts per million or parts per billion).

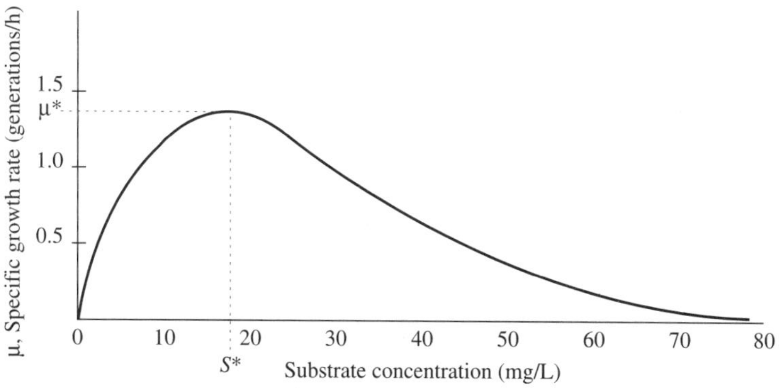

FIGURE 10-11
Critical concentration (S^*) for an inhibitory substance.

Although such a low waste concentration may still pose a harmful risk to receptors, it may not support a viable biomass needed for treatment. The rate of growth decreases as substrate concentrations diminish, eventually reaching a point where growth ceases. The biomass may still survive, but not grow, at low concentrations if the diffusion of substrate molecules is sufficient to meet the energy needs for maintenance. At even lower concentrations, a threshold will be reached where not even energy needs are met, resulting in loss of the biomass.

Explanations for this threshold include the following: The flux of energy from degradation of insufficient substrate cannot sustain the biomass; the concentration may not be sufficient to induce enzyme production; and spores may form that cause microorganisms to become dormant.[35,36] This theoretical concentration behavior is noted universally in laboratory cultures, in full-scale treatment systems, and in natural water. It may explain the persistence of low levels of otherwise biodegradable substances in natural water.[37]

The kinetic expressions provide a basis for examining the minimum substrate concentration required to sustain a viable biomass. Equation 10-9 shows that the minimum sufficient substrate would occur when formation of new biomass equals loss by endogenous decay, or as expressed in the following equation[38]:

$$\frac{YkS_{\min}X}{K_S + S_{\min}} = bX \tag{10-11}$$

Solving for $S_{\min}$ gives

$$S_{\min} = \frac{bK_S}{Yk - b} \tag{10-12}$$

Researchers have reported typical $S_{\min}$ concentrations for aerobic aqueous treatment systems in the range of 0.1 to 1.0 mg/L. In contrast, for the specific conditions of many contaminated sites, some toxic organics are considered to pose an unsafe carcinogenic risk at lower concentrations, as low as 0.01 mg/L, if not lower. When concentrations of toxic organics do not exceed their $S_{\min}$, it is necessary to induce fortuitous metabolism or cometabolism by adding a primary substrate. Otherwise, effective biodegradation may not occur.

It should be noted that this mathematical derivation pushes the Monod equation perhaps beyond its limit. Approached from another perspective, the concept of a minimum substrate concentration is refuted mathematically by examining Equation 10-7 for the case when S approaches zero. Equation 10-7 reduces to the following as $S << K_S$:

$$\frac{dS}{dt} = \frac{k}{K_S} XS \tag{10-13}$$

Researchers have noted bacteria with very low K_S values in the open ocean and very low microbial concentrations (as low as 0.001 mg/L) in the subsurface.[39] Both cases represent situations in which low concentrations of organics are present. This observation and Equation 10-13 suggest that organic substrates may indeed degrade at trace concentrations, albeit at extremely low rates perhaps measured in months if not years. The theoretical S_{min} is also the minimum effluent concentration that can eventually be achieved when treating any substrates.

Mathematical Approximation for Hazardous Waste Cases

The application of the classic wastewater treatment equations in the field of hazardous waste is sometimes limited because of multiple complexities, including the need to use nonconventional treatment methods, the toxic inhibition effects of some wastes, the competitive nature of some mixed wastes, the dilute concentrations of substrate, the difficulty in accounting for abiotic losses (which can be significant), and concepts relatively new to the environmental engineering field such as cometabolism along with primary and secondary substrates. In the field of hazardous waste treatment, the classical equations are applicable primarily to the treatment of aqueous wastes.

Research continues to be conducted to formulate appropriate degradation kinetics specific for hazardous waste treatment in soil and subsurface environments. In many situations a simple **first-order approximation** has been used with reasonable accuracy to describe the degradation of organic wastes. This approximation stems from Equation 10-7 and is based on two key assumptions concerning the concentrations of substrate and biomass. For the typical hazardous waste application, in which the target organic chemical is at a very low concentration, the numerical value of S is much smaller than K_S. Therefore, assuming S is insignificant by comparison, Equation 10-7 reduces to Equation 10-13 as explained previously:

$$\frac{dS}{dt} = \frac{k}{K_S} XS \tag{10-13}$$

As a second assumption, if the biomass concentration X changes little with time, it can be considered a constant. Integration of Equation 10-13 yields:

$$\ln\left(\frac{S}{S_0}\right) = -\frac{k}{K_S} Xt \tag{10-14}$$

where t = duration of treatment (time)

S_0 = initial substrate concentration (mass/unit volume)

The constant X can be grouped with the two degradation coefficients as follows:

$$k' = \frac{k}{K_S} X \tag{10-15}$$

where k' = first-order degradation rate constant (time^{-1}).

Substituting Equation 10-15 in Equation 10-14 and solving for S yields the following expression for determination of contaminant concentration after treatment for the time interval t:

$$S = S_0 e^{-k't} \tag{10-16}$$

In most cases where investigators have reported hazardous waste degradation rates in the literature, they have reported it as this first-order degradation rate coefficient or in terms of half-life. Half-life can be determined from Equation 10-16 by setting $S/S_0 = 0.5$, yielding:

$$k' = \frac{0.693}{t_{1/2}} \tag{10-17}$$

where $t_{1/2}$ = half-life (days).

EXAMPLE 10-3. HALF-LIFE CALCULATION. What is the half-life of a toxic compound for which a laboratory treatability test yielded a first-order degradation rate constant of 0.02 day^{-1}?

Solution. Substituting values of 0.5 for S/S_0 and 0.02 day^{-1} for k' in Equation 10-16 and solving for t yields:

$$t = \frac{\ln 0.5}{0.02\ \text{day}^{-1}}$$
$$= 35 \text{ days}$$

10-4
EX SITU SYSTEMS—TREATMENT OF LIQUIDS

Liquid hazardous wastes amenable to biological treatment include contaminated groundwater and landfill leachate. Other examples include industrial process wastewater if it is of a primarily aqueous nature (e.g., rinse water).

Basics of Liquid-Phase Treatment

Many different bioreactor designs have been used in the treatment of liquid waste streams, including batch reactors, standard activated sludge systems and their variations, fluidized beds, and membrane systems. Many of these technologies, or variants, were originally developed for the treatment of industrial wastewater. The typical hazardous wastes treated by these methods include contaminated groundwater, industrial process wastewater, and landfill leachate containing toxic organic substances.

Liquid-phase treatment consists of passing the contaminated liquid through a reactor containing either suspended or attached biomass of highly active and acclimated microorganisms. Flow can be continuous or batch, and the reactor can be

operated under aerobic or anaerobic conditions. For aerobic conditions, oxygen is added by diffused or surface aeration. More detailed discussions of conventional wastewater treatment systems can be found in wastewater engineering texts.[40,41]

A generalized process flow diagram for aqueous treatment is shown in Figure 10-12. Liquid waste streams may receive complementary physical/chemical treatment before and after the biological treatment process.

Pretreatment, if needed, could consist of several steps, depending on the types of waste:

- **Equalization—**to dampen/modulate hydraulic surges and variable organic loadings in continuous flow systems.
- **Chemical treatment—**typically to precipitate toxic metals, if present, but could involve other steps such as breaking of emulsions.
- **Physical separation—**sedimentation of metallic precipitates, removal of floating material, etc.
- **Conditioning—**typically to supply nutrients and to adjust pH to optimum range.

After pretreatment, the liquid waste flows into the bioreactor where the dissolved organics are metabolized by the biomass with a resulting growth of cellular mass. This can be achieved only by biodegradation. The actual yield of biomass depends on numerous factors. As stated earlier, not all utilization of substrate results in synthesis of cells: Some substrate is oxidized to produce energy.

The removal of a portion of the organic waste may occur by methods other than biodegradation (i.e., abiotic losses). Volatilization can result in significant removal of some organics, especially in aerated systems. Some organics may not be metabolized but adsorbed with colloidal contaminants onto the biomass. In sum, abiotic losses can account for a significant amount of the organic waste removed with no reduction of the toxic nature of the waste. The chemical constituents have merely been transferred to other media, which may require their own particular treatment.

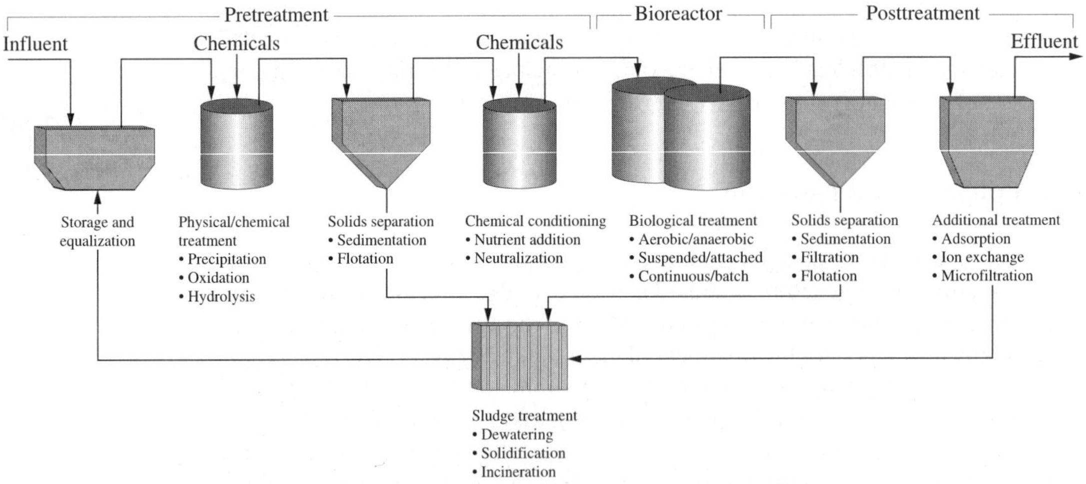

FIGURE 10-12
Conventional liquid-phase treatment.

Some of the major factors to consider in the use of liquid-phase systems include:

- **Mixing regime:** Complete mixing provides internal equalization to compensate for variations in concentration of waste constituents. The opposite regime is termed plug flow, and it provides for more efficient removal of waste constituents if the influent flow and concentration have nominal variation. Another factor is that the reactor must provide sufficient mixing for the waste to come into contact with the biomass. This holds obvious importance in treating waste having a great amount of inert solid material (e.g., contaminated soil).
- **Hydraulic retention time:** The hydraulic retention time of a reactor is equal to its volume divided by the rate of flow. Typically, treatment costs decrease as hydraulic retention time decreases. Minimizing the hydraulic retention time is constrained by the fact that as it is reduced, the ability to handle shock loads decreases. Also, as the hydraulic retention time is decreased the biomass concentration must increase to achieve effective biodegradation. Experience shows that extremely high biomass levels are unacceptable in liquid treatment systems because of potential problems with overloading the subsequent clarification process, as well as the inability to transfer adequate amounts of oxygen in a limited liquid volume.
- **Solids retention time:** The population in a reactor of waste-degrading microorganisms, measured as biomass, is important because the rate of degradation shows a linear dependence on the concentration of biomass. The concentration of biomass can be increased by recycling, a process in which the microorganisms (or biological solids) generated in the bioreactor are withdrawn with the effluent and allowed to settle in a clarifier, with the sludge returned to the bioreactor. Related to this concept is the term solids retention time (SRT), also referred to as sludge age, or mean cell residence time, or θ_c. In designing biological reactors, SRT is a key factor. The implication is simple: Within limits, a longer value of θ_c results in more efficient degradation, smaller reactor size, and lower cost. If the SRT drops below the cell regeneration time, biomass will wash out faster than it forms new cells.

  The SRT can be calculated as follows:

$$\theta_c = \frac{XV}{\text{rate of biomass wastage}} \tag{10-18}$$

  where θ_c = solids retention time (time)
  V = volume of reactor (volume)

  Thus, the SRT can exceed the hydraulic retention time. An obvious relationship exists that the greater the SRT, the greater is the concentration of biomass. It follows that removal efficiency increases as SRT increases. A key advantage of recycling other than to increase the concentration of biomass is that it serves to retain the specific "degraders" that have acclimated to the particular organic chemicals being treated. If the SRT is too short (i.e., less than the time to generate new cells), the biomass will diminish and wash out.
- **Total dissolved solids:** The concentration of dissolved solids can affect liquid-based biological treatment. If the concentration (as measured by total dissolved solids or TDS) is too high, microbes die because of osmotic rupture of the cell membrane. If the TDS varies widely, the activity of the microbial population

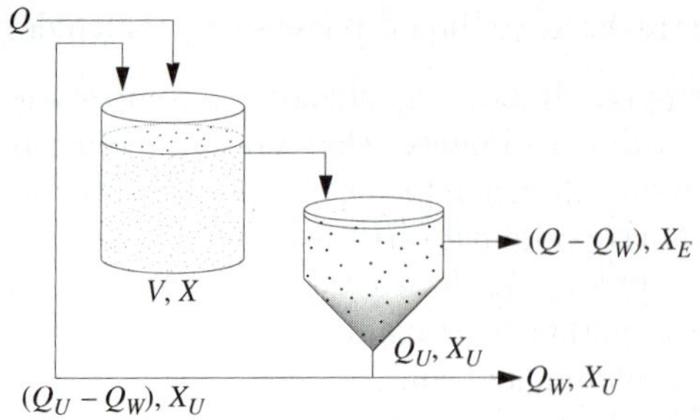

FIGURE 10-13
Flow and solids balance for system with recycle.

can decline. Environmental engineers have developed practical rules of thumb based on experience: TDS should not exceed 40,000 mg/L, and TDS should not vary by more than a factor of 2 over a period of a few days.

- **Other factors:** Other factors can influence waste treatment efficiency but not to the degree represented by the key factors previously discussed. Examples of these other factors include alternative carbon sources, a concept in which a waste contains multiple organic compounds capable of serving as sources of carbon for the biomass. Research shows that because of preferential differences, the biomass may degrade the compounds sequentially, possibly delaying the degradation of a target compound. However, further research shows that in systems with a long SRT, the compounds are removed simultaneously, each at a different rate.[42]

By controlling these factors, biodegradation can be enhanced to reduce contaminant concentrations to required levels. Optimization of these variables will reduce economic costs and increase system reliability.

EXAMPLE 10-4. SOLIDS RETENTION TIME. A 5000-gal bioreactor operates at a biomass concentration of 2000 mg/L, measured as mixed liquor volatile suspended solids (MLVSS) and treats 10,000 gallons per day of liquid waste containing 1000 mg/L of total organic carbon (TOC). The suspended solids are separated in a clarifier following the bioreactor with recycle of separated sludge. The recycle flow rate is 5000 gals/day. Each day, 300 gal of recycle are wasted. The effluent from the clarifier contains 40 mg/L MLVSS. What is the solids retention time? How does that compare with the rule of thumb? If too short, how can the SRT be increased?

Solution. The conceptual flow and solids balance for a system with recycle are shown as Figure 10-13. SRT is calculated for a recycle system as depicted in Figure 10-13 as follows:

$$\theta_c = \frac{XV}{(Q - Q_W)X_E + Q_W X_U} \qquad (10\text{-}19)$$

For this example, all variables are known except for X_U. To determine X_U, a solids balance must be done around the clarifier. For conservation of matter (assuming no accumulation in the clarifier),

$$(Q + Q_U - Q_W)X = (Q - Q_W)X_E + Q_U X_U$$

Solving for the underflow sludge concentration, X_U,

$$X_U = \frac{(Q + Q_U - Q_W)X - (Q - Q_W)X_E}{Q_U}$$

$$= \frac{(10{,}000 + 5000 - 300)(2000) - (10{,}000 - 300)(40)}{5000}$$

$$= 5800 \text{ mg/L}$$

Now solving for θ_c,

$$\theta_c = \frac{(2000)(5000)}{(10{,}000 - 300)(40) + (300)(5800)}$$

$$= \underline{\underline{4.7 \text{ days}}}$$

This is a very short SRT, probably too short for efficient treatment. It needs to be increased. Not wasting any recycle (i.e., $Q_W = 0$) would yield an SRT as follows:

$$\theta_c = \frac{(2000 \text{ mg/L})(5000 \text{ gal})}{(40 \text{ mg/L})(10{,}000 \text{ gal/day})}$$

$$= \underline{\underline{25 \text{ days}}}$$

To increase the SRT even more would require improving clarification in a manner that reduces the solids in the effluent from the clarifier to below 40 mg/L.

Aerobic Batch Systems

Batch treatment using suspended biomass is the simplest operational bioreactor design. In a true **batch process,** an empty vessel is initially filled with contaminated water and inoculated with a suitable microbial population. The reactor is allowed to operate with mixing under either aerobic or anaerobic conditions. At the completion of the treatment period, the contents of the tank are completely removed and the process is repeated with new materials as necessary.

A demonstration project treating groundwater contaminated with PAHs from wood-preserving wastes employed a two-stage batch reactor system. The first tank operated with a 4-day treatment period, and the second tank operated for 8 days. The first tank received inocula of bacteria specialized for treating low and medium molecular weight PAHs, and the second tank was prepared with bacteria acclimated for the treatment of high molecular weight PAHs. The percent removal attained for low, medium, and high molecular weight PAHs was > 99, > 98, and 98, respectively. The system reduced pentachlorophenol concentrations by 88 percent.[43]

A variation of the batch process combines equalization, biotreatment, and sedimentation in a single tank. The three steps are conducted sequentially (see Figure 10-14); hence the term **sequencing batch reactor (SBR)** is used to describe these systems. The biomass can be maintained by discharging only the clarified effluent after the sedimentation step. The time allocated to each step can be adjusted within the constraint of the incoming flow. This provides much greater flexibility than continuous flow systems, allowing SBRs to achieve high performance when treating

variable wastewaters. The simplicity and flexibility of the SBR approach makes it especially appropriate for small-scale processes.

If the waste stream requiring treatment flows continuously, at least two SBRs are needed: one to fill while the other sequences through the treatment steps. Placement of SBRs in series helps to ensure that the biomass in each tank is acclimated to the transformed compounds being discharged from the previous tank. This can speed the sequential transformation of waste. A commercial hazardous waste treatment facility has used SBRs to remove 75 percent of the organic contamination in leachate and other aqueous wastes.[44]

A European research project showed the potential capability of SBRs to foster genetic exchange in the microbial community of a bioreactor. This is important because bacteria adapted in a laboratory to a specific waste frequently do not survive in the field bioreactor because of predatory activity or because of washout due to failure to flocculate or attach. The experiment compared rapid and slow fill rates for an SBR operation, under the following conditions:

Waste influent:	660 mg/L of the pesticide 2,4-D and 270 mg/L of fructose
Fill time:	Fast fill: 20 minutes
	Slow fill: 6 to 24 hours
Reaction time:	Constant
Drain time:	30 min
Volume exchange per cycle:	50%
Inoculation:	Donor bacteria with laboratory-developed capability of degrading 2,4-D

Under "fast fill" conditions, transconjugants (i.e., microorganisms that have acquired genetic material from another microbe) acquired the capability to degrade 2,4-D from the inoculated donors and increased significantly in population. The "feast or famine" conditions created by the fast fill cycle exerted selective pressure on the transconjugants to preserve their capability for degrading 2,4-D. In the slow fill cycle, they "abandoned" this capability. Longer reaction times also resulted in the loss of this capability.[45]

Anaerobic Batch Systems

Batch reactors operating under anaerobic conditions also play a major role in the waste treatment field. The treatment of municipal sewage has long depended on anaerobic bioreactors to digest excess sludge to methane and carbon dioxide in an

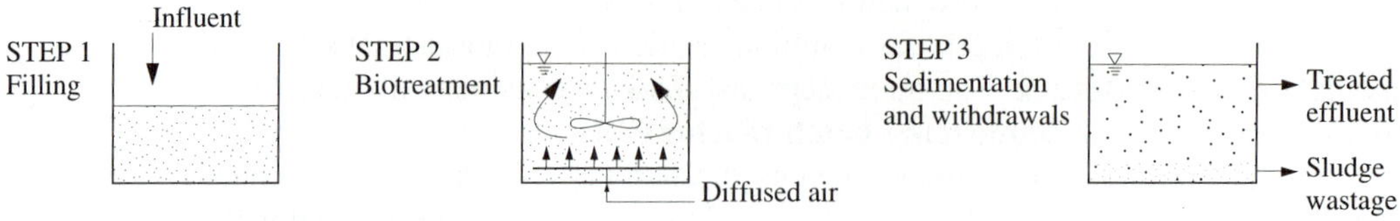

FIGURE 10-14
Sequencing batch reactor operation.

oxygen-free environment. Industries with high-strength, readily degradable wastewater (e.g., food products, pharmaceutical, pulp and paper, and chemicals) have increasingly used anaerobic systems because of the savings realized with energy conservation and greatly reduced sludge production. The specific systems used in these applications typically feature attached growth processes similar to that shown in Figure 10-15 with modifications to maintain anaerobic conditions.

Continuous Flow and Aerobic Suspended Growth Systems

Continuous flow systems utilize either suspended growth or attached growth biomass. Figure 10-16 depicts a typical suspended growth system in which the bioreactor is mixed continuously, thereby facilitating contact between the substrate (i.e., organics in the liquid waste) and the suspended biomass. The effluent washes the suspended biomass out of the bioreactor to a step where the biomass is separated from the treated effluent, typically by sedimentation or clarification. Subsequent steps (e.g., filtration) may also be utilized, depending on the regulatory limitations on the quality of the effluent.

As discussed earlier in this section, a major portion of the settled biomass (i.e., sludge) is returned to the bioreactor to maintain the proper solids retention time (i.e., sludge age) and the proper ratio between substrate and acclimated microorganisms (i.e., food:biomass ratio). Recycling increases the solids retention time beyond the simple hydraulic retention time. This speeds degradation and reduces the size of the needed reactor. Excess sludge must be removed for treatment and disposal. The typical sludge age of an effective system is about 20 to 30 days, and no system should have a sludge age outside the range of 10 to 50 days. A lower sludge age would tend to result in system washout, instability or filamentous organism growth. A higher sludge age could produce nonsettling, pinpoint floc. An additional consideration is

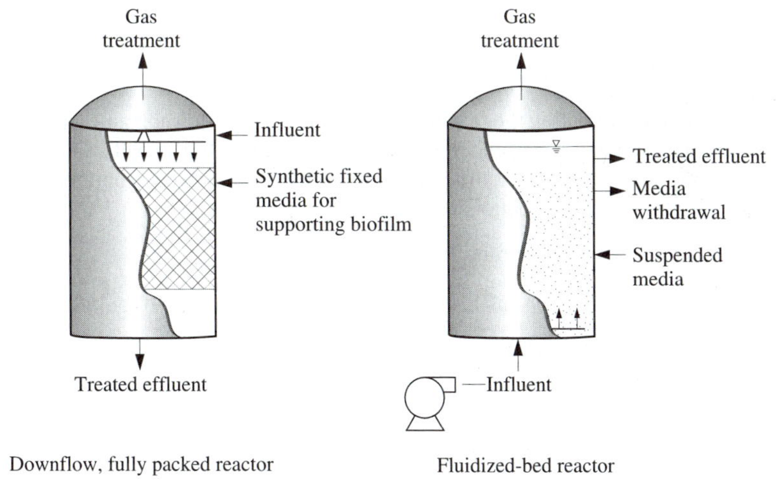

FIGURE 10-15
Examples of attached growth systems

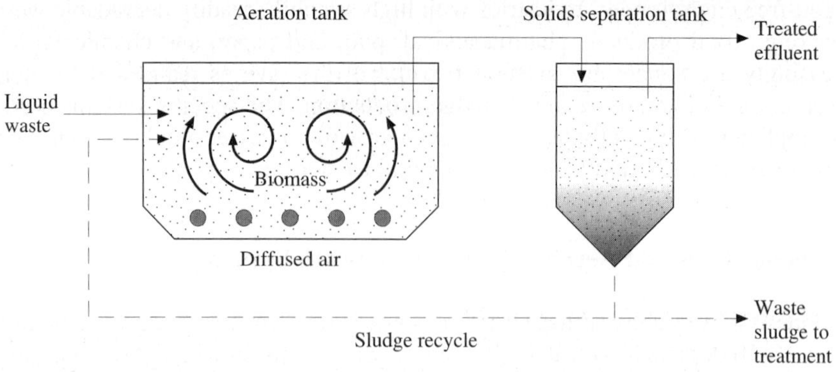

FIGURE 10-16
Completely mixed, suspended growth treatment system.

that the biomass mixed liquor suspended solids (MLSS) concentration should not exceed 5000 mg/L; a higher concentration could result in the clarifier failing.

EXAMPLE 10-5. BIOMASS CALCULATIONS. An industrial plant generates a process wastewater stream that has averaged 440 gal/min and 4800 mg/L of COD over the past 3 years. A regulatory agency has stated that the organic chemical constituents are hazardous and must be reduced to 100 mg/L measured as COD. The storm-water runoff from the site also contains the organic chemicals of concern, and all runoff from an annual average rainfall of 60 inches must also be treated.

The industrial plant conducted some studies and found the following:

- Rainfall/runoff measurements over 3 months reported a total of 12 inches of rainfall over that period, which yielded 3,300,000 gal of runoff containing 300 mg/L of COD.
- Treatability tests of the process wastewater reported the following data:
    - BOD:COD:1:2
    - Biomass production rate $Y = 0.7$ mg/mg BOD removed
    - Endogenous decay rate $b = 0.05\ \text{day}^{-1}$

The plant is evaluating the feasibility of treating the wastewater and runoff in a full-scale completely mixed, suspended growth system with solids recycle. On the basis of the above information, could a sufficient biomass concentration be maintained to prevent settling problems in a conventional system?

Solution. Calculate average runoff:

$$\text{Runoff} = (3.3 \times 10^6\ \text{gal/12 in}) \times (60\ \text{in/year}) = 16.5 \times 10^6\ \text{gal/year}$$
$$= \underline{\underline{45{,}200\ \text{gal/day}}}$$

Calculate average daily flow:

$$\text{Flow} = (440\ \text{gal/min} \times 1440\ \text{min/day}) + 45{,}200\ \text{gal/day}$$
$$= 633{,}600\ \text{gal/day} + 45{,}200\ \text{gal/day}$$
$$= 678{,}000\ \text{gal/day}$$

Calculate average BOD of influent:

$$\text{Ave. COD} = \frac{(633{,}600 \times 4800 \text{ mg/L}) + (45{,}200 \times 300 \text{ mg/L})}{678{,}800}$$

$$= 4500 \text{ mg/L}$$

$$\text{Ave. BOD} = 0.5 \times 4500 \text{ mg/L}$$

$$= \underline{\underline{2250 \text{ mg/L}}}$$

Estimate suspended biomass (MLSS):

The steady-state equations include a number of interdependent variables. Use of these equations may necessitate assuming values for select variables. These assumed values represent starting points from which refinements may be necessary. For this example, it is first necessary to assume a sludge age θ_c and hydraulic detention time θ_h as follow:

$$\theta_c = 50 \text{ days}$$

$$\theta_h = 7 \text{ days}$$

With this starting point, the suspended biomass can be calculated for this type of system from Equation 10-20:

$$X = \frac{Y\,(S_0 - S)}{(1 + b\theta_c)}\,\frac{\theta_c}{\theta}$$

$$= \frac{0.7\left[2250 - (0.5 \times 100)\right]}{\left[1 + 0.05\,(50)\right]}\,\frac{50}{7} \qquad (10\text{-}20)$$

$$= \underline{\underline{3100 \text{ mg/L}}}$$

This is a reasonable level for the suspended solids in a conventional reactor. It is within the range defined by a minimum of 1000 mg/L and a maximum of 5000 mg/L as discussed earlier.

The suspended growth process described above is a conventional system, and many variations to this basic process design exist. The incorporation of hydraulic design measures can create plug flow conditions, unlike the completely mixed conditions of the conventional system. Plug flow systems theoretically can be more efficient than continuous flow systems.[46] However, they are more easily upset by rapid changes in loadings, particularly of toxic compounds. Completely mixed suspended growth systems withstand shock loadings of nondegradable toxic compounds better than most other systems, because influent concentrations are diluted almost immediately throughout the reactor.

PACT process

In another variation, powdered activated carbon is added to the bioreactor to remove organic compounds not metabolized by the biomass. The activated carbon also adsorbs organics that otherwise could be toxic to the biomass. A discussion of this treatment process can be found in a current book on hazardous waste technologies.[47] This powdered activated carbon treatment (PACT) process has been patented, and is depicted in Figure 10-17.

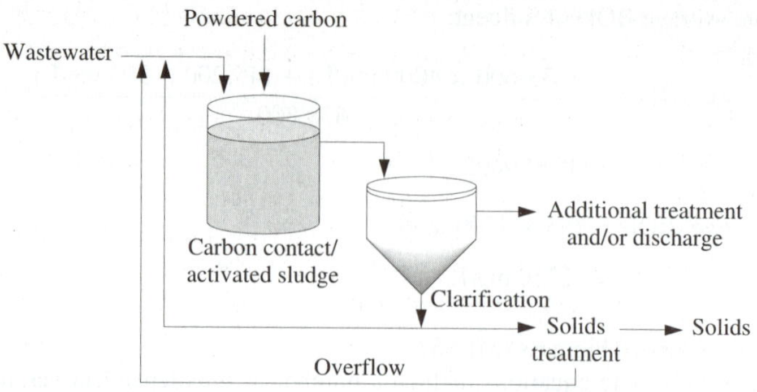

FIGURE 10-17
Zimpro PACT™ process.

Membrane bioreactors

The term **membrane bioreactor** is used to describe a suspended growth biological treatment system coupled with a membrane filtration clarification system. In this concept, clarification utilizes either external membranes with cross-flow ultrafiltration, or internal membranes with microfiltration. Cross-flow ultrafiltration is more resistant to biological fouling, and can tolerate suspended biomass concentrations of up to 30,000 mg/L; microfiltration systems are limited to biomass concentrations of up to 15,000 mg/L.[48]

The major advantages of these systems include the ability to:

- Maintain very high suspended biomass concentrations in the bioreactor
- Easily implement advanced bioprocess control within the reactor (such as the introduction of specific substrates for use in cometabolic biodegradation)
- Efficiently remove biomass regardless of the physiological settling characteristics of the bacteria
- Quickly separate biomass without the need for large clarification basins, quiescent conditions, and extended detention times.

Membrane bioreactor systems have been used in the treatment of oily wastes,[49] alcohols,[50] and food manufacturing wastes.[51]

Attached Growth Systems

Conventional suspended growth systems are best suited to wastes containing moderate to high concentrations of organics, as high as 5000 mg/L of TOC. Low concentrations of organics do not yield enough growth to provide a properly flocculated biomass. This produces poor final settling of biomass such that new cells cannot replace those lost through washout and endogenous decay. Fixed-film processes may be better suited to the treatment of low-concentration wastewaters.[52]

Attached growth (or fixed-film) systems rely on the ability of microorganisms to attach to surfaces of inert media (see Figure 10-15). Contaminated water is passed through a bioreactor, which houses the media. The resulting microbial growth at-

taches to the media and forms a film. The biomass remains in the reactor except that which sloughs off the supporting media. Part of the effluent and biomass may be recycled.

Attached growth systems can develop high concentrations of biomass in relatively small reactors because relatively little biomass is lost with the effluent. This results in a low food-to-microorganism ratio and the long solids retention time necessary to foster slow-growing microorganisms. Both enhance degradation of hazardous contaminants, particularly wastes with low concentrations of organics. Attached growth systems require that the microbes have the ability to attach to the appropriate media; such ability varies among strains of microbes and must be examined during the design phase.

The plug flow nature of attached growth systems would seemingly make them more vulnerable to shock loadings. It has been shown, however, that significant biodegradation can occur in attached growth systems at substrate concentrations that would inhibit suspended growth reactors.[53] It is believed that the outer layers of the biofilm protect the inner layers from toxic effects.

The field of municipal sewage treatment historically has relied extensively on **trickling filters,** a form of attached growth treatment using rock as the inert medium. For a period, numerous rotating biological contactor systems were constructed for biomass attainment. These rotating contactors consisted of closely spaced vertical disks mounted to a rotating shaft suspended horizontally just above the surface of the liquid undergoing treatment. The circular disks provide a large surface area for supporting the biofilm. By rotating the disks, the biomass is aerated and treatment occurs. A pilot-scale unit of such a system was reported to have reduced an already low concentration of chlorinated phenolic compounds by 36 to 68 percent.[54]

Other applications feature **packed towers** with synthetic porous media having a high area-to-volume ratio as shown in Figure 10-15. The influent is normally sparged with air before moving as plug flow through the columns. These units are usually constructed in modular form to provide the necessary treatment volume (i.e., treatment time). Such a system was used to treat groundwater contaminated with ethylene dichloride (EDC).[55] The system reduced concentrations of EDC from 560 mg/L in the influent to about 1 mg/L in the effluent. A check of volatilization indicated that 99 percent of the removal of EDC was due to degradation. Batch tests indicated an initial degradation rate of 60 mg/kg/h, depending of course on the substrate.

Submerged fixed-film reactor

A hybrid reactor design particularly useful in the treatment of contaminated liquids is the submerged fixed-film reactor. This reactor utilizes a trickling filter medium as a support for biomass growth, but the packing medium is maintained in a submerged state within the reactor shell. Oxygen to the system is supplied by air sparging. "Growth" and "decay" forms of this reactor design have been used successfully in the field. In the growth reactor configuration, the concentration of hazardous organics in the water is sufficient to maintain growth of the biofilm. The contaminants present are used as primary substrates for microbial growth. This design has been successfully used to treat ground and industrial process waters contaminated with methyl ethyl ketone, xylene, ethylbenzene, cyclohexanone, cyclohexanol, acetone, tetrahydrofuran, 2-butanol, and methyl propyl ketone[56] as well as isomeric toluic

acids present in an industrial landfill leachate.[57] In the decay form of the reactor, a healthy biofilm is initially established within the reactor by operating in a water recycle mode and by adding supplemental substrates. Then, continuous flow to the reactor is begun and the biofilm effectively scavenges low-concentration organics (total concentration of 50 mg/L or less) from the water. These low influent organic levels are insufficient to sustain a growing biofilm. Effluent concentrations in the low part-per-billion range are possible for specific organics. Over time the biofilm deteriorates, causing a decrease in contaminant removal effectiveness. At this point flow to the reactor is stopped and the biofilm is reestablished through water recycling and supplemental feeding. Decay-type reactors are suitable to the treatment of BTEX (benzene, toluene, ethylbenzene, and xylene) and other organics present in contaminated groundwaters.[58]

Fluidized-bed reactor

Another design is the fluidized-bed reactor (Figure 10-15). A fluidized bed is achieved by growing the biofilm not on a fixed medium but on particles of sand, activated carbon, or other inert medium. The water in the reactor can be modified as needed for pH, inorganic nutrients, and oxygen levels. The reactor can be operated under aerobic or anaerobic conditions. The influent is introduced from below the bed in an upflow reactor; the upflow velocity must be sufficiently high to keep the particles suspended. Withdrawal of treated effluent from the top of the reactor maintains the upflow velocity. A portion of the inert medium is withdrawn periodically as the bed expands. The virtually complete mixing reduces the susceptibility to shock loadings. Further, these systems can remove more organic loading per unit reactor volume than other fixed-film processes.[59]

Fluidized-bed reactors have been used to successfully treat contaminated groundwater and industrial wastewater. A 250-gal/min system treating groundwater containing a total analine and nitrobenzene concentration of approximately 80 mg/L has attained a consistent removal efficiency of greater than 99.5 percent. Another system, operating under anaerobic conditions, has been used to treat a 4000 gal/min groundwater stream containing approximately 8 mg/L of ammonium perchlorate. Effluent perchlorate levels from this system have been as low as 4 μg/L.[60]

10-5 *EX SITU* SYSTEMS—SLURRY-PHASE TREATMENT

Slurry-phase systems involve the treatment of contaminated soils/sludges mixed with clean or contaminated liquids. Slurry-phase treatment can take place in aboveground reactors or within existing/constructed lagoons. Only aboveground bioreactors will be discussed in this section; lagoon treatment systems are covered in Section 10-8 as part of *in situ* treatment technologies.

Slurry-phase reactors received attention in the late 1980s and early 1990s as a potential process for biodegrading difficult-to-treat soils, especially those contaminated with high levels of heavy oils, PAHs, pesticides, and chlorophenols. It was believed that only aggressive mixing and treatment under slurry conditions would offer the potential for success. Because of the high costs associated with excavation, materials handling, and residuals handling, the *ex situ* form of this biological treatment process was rarely used and rapidly fell out of favor. In contrast, *in situ*

anaerobic slurry treatment in lagoons has been gaining acceptance and use for the treatment of nitroaromatic compounds (explosives) and other persistent compounds.

The same basic slurry treatment principles and concerns apply to both *ex situ* and *in situ* treatment approaches. Slurry treatment typically requires initial mixing or agitation. The agitation not only homogenizes the slurry but also promotes the following actions:

- Breakdown of solid particles
- Desorption of waste from solid particles
- Contact between organic waste and microorganisms
- Oxygenation of the slurry by aeration
- Volatilization of contaminants

Slurry-phase treatment is similar to conventional suspended growth biological treatment except that the biomass is not recycled and the suspended solids can frequently be inert. The fundamental treatment steps are mixing/aeration, desorption, and biodegradation. These can occur in an in-place reactor or in a multiple-stage system, and flow can be continuous or batch. A multiple-stage system includes pretreatment steps, such as passing the soil through a sieve or adding surfactants, to enhance desorption and fractionate the waste. The treated soil and other inert solids typically can be left on the site. Figure 10-18 shows the steps typically involved in the use of bioslurry reactors.[61]

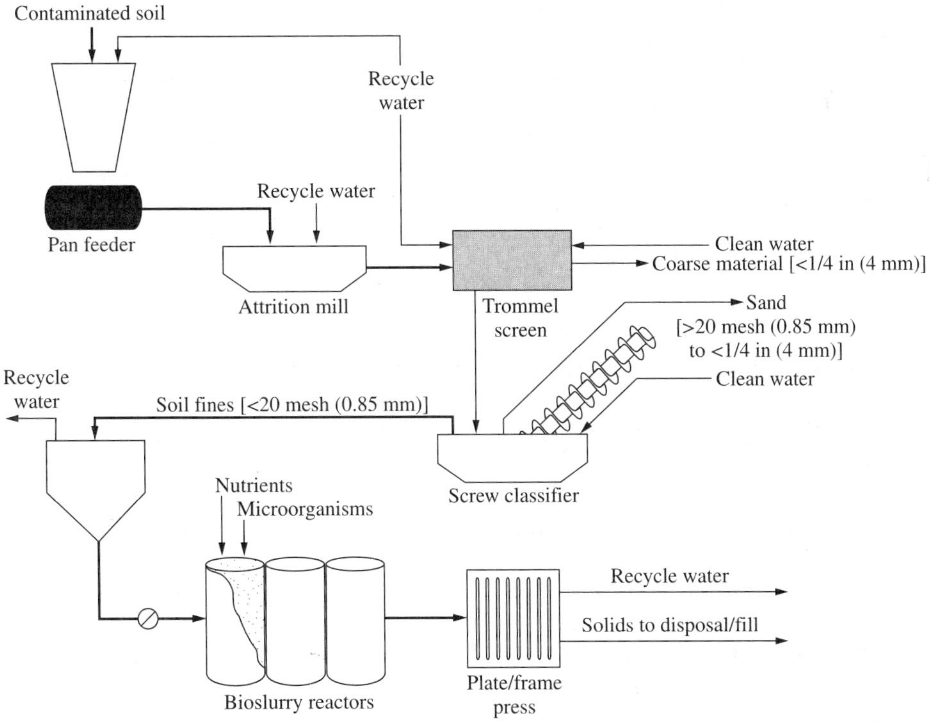

FIGURE 10-18
Multiple-stage slurry-phase treatment system.[61]

The effectiveness of slurry-phase treatment depends on a number of theoretical factors including pretreatment, desorption, solids concentration, mixer design, and retention time.

Pretreatment

A waste can be pretreated in a number of ways to increase the effectiveness and efficiency of the overall system. Pretreatment methods include those that enhance desorption (e.g., adding surfactants and reducing soil particle size) and those that concentrate the waste to be treated (e.g., fractionation of soil by particle size). Fractionation holds much merit for treating soil because contaminants preferentially adsorb to finer soil particles. Thus, fractionation can (1) eliminate the heavier particles, which require higher energy to keep suspended, and (2) concentrate the contaminants prior to degradation. A research project fractionated a soil contaminated with polynuclear aromatic version (PAHs) and demonstrated that the fines, which could easily be kept suspended in a slurry, represented 37 percent of the dry mass of the soil but contained 94 percent of the contamination[62] (see Figure 10-19).

Desorption

The contaminant usually must be desorbed from the soil particles before biodegradation can occur. One exception is the case where waste exists as droplets to which microorganisms have attached. Desorption is particularly important with the soil-water suspensions characteristic of slurry-phase treatment.

The rate at which the pollutant desorbs from the soil particles is influenced by the physical and chemical properties of both the soil and the contaminant. Essentially, the same factors control desorption and sorption. A portion of each pollutant is bound in a stable position along micropores within soil aggregates and particles. The relatively large size of microorganisms excludes them from the micropores, preventing their direct access to the sorbed contaminant. However, some microorganisms do attach to the outer surfaces of the soil particles. To be accessible, the contaminant must desorb to the micropore water and then move by diffusion to the outer surface of the soil particle where the solute is either degraded by attached microorganisms or passes to the bulk aqueous phase. Local equilibria may develop within the soil particle and act to slow further desorption. The high affinity of many compounds for soil (e.g., PAHs have a high affinity for clay minerals) will also inhibit desorption.

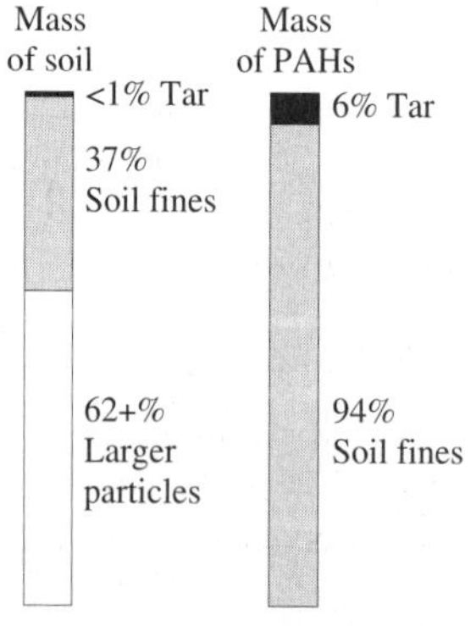

FIGURE 10-19
Distribution of mass of soil and contaminants by soil fraction.

Investigations in the Netherlands and United States claim that desorption-diffusion from the sorbed phase to the solute phase is the rate limiting factor for slurry phase treatment.[63–65] This conclusion derives from observations in laboratory tests that: (1) the rate of degradation is a function of the concentration of the contaminant in solution rather than the total sorbed mass and (2) microbial populations grow at a linear rather than logarithmic rate. Another investigator observes that this is true for lower molecular weight PAH compounds, even for those with fairly low solubility; however, for higher molecular weight PAH compounds, the degradation rate is the limiting factor.[66] With regard to the higher, unsubstituted PAHs, their solubility may be too low to support significant biological activity, again identifying desorption as the rate-limiting factor.

It was noted in the Dutch research projects that in experimenting with different mechanical mixing techniques and the same pollutant, the mixing technique that reduced the soil particle size to 30 μm resulted in degradation rates several times greater than the technique that yielded a particle size of 60 μm.[67] Thus, an important consideration in pretreatment and mixing is the reduction of particle size. Similarly, washing the soil with a surfactant or other suitable chemical can facilitate transfer of contaminants into the aqueous phase.

Concentration of solids in reactor

The reactor volume for a given throughput can be reduced by maximizing the solids content. Various applications have reported satisfactory degradation for solids content as low as 5 percent and as high as 50 percent (dry weight basis). The maximum solids content depends on the type of soil and the reactor design. The physical factors involved in keeping a slurry in suspension may establish an upper constraint of 30 to 40 percent solids.[68] However, the practical limit may be less if a high solids concentration proves toxic to the biomass.

Mixer design

Mixing and particle suspension are critical to desorption and biological degradation. Keeping large concentrations of sludge and heavy soil particles suspended in a reactor requires high-energy mixers such as turbine mixers or surface aerators. The design of the mixers and reactor is perhaps the most critical factor in slurry-phase treatment. For example, baffle systems are required to improve mixing and prevent settling. Mixing does supply oxygen but may also promote volatilization, possibly necessitating the use of enclosed tanks.

The importance of the mechanical aspects of the reactor has prompted advanced designs. One concept applies technology developed in the mineral processing industry to improve energy efficiency[69] (see Figure 10-20). The dual drive design of the reactor suspends independently the finer and coarser particles found in contaminated soil. The axial flow impeller rotates at 20 to 30 rpm to agitate the slurry. Other rake arms turn at 2 rpm and push the coarser particles to a central airlift that pumps them to a removal system. Diffusers mounted on the rake arms release fine bubbles of air, which keep the finer particles suspended. The manufacturer claims the system reduces operating costs to only 20 percent of conventional reactors primarily because of energy savings.

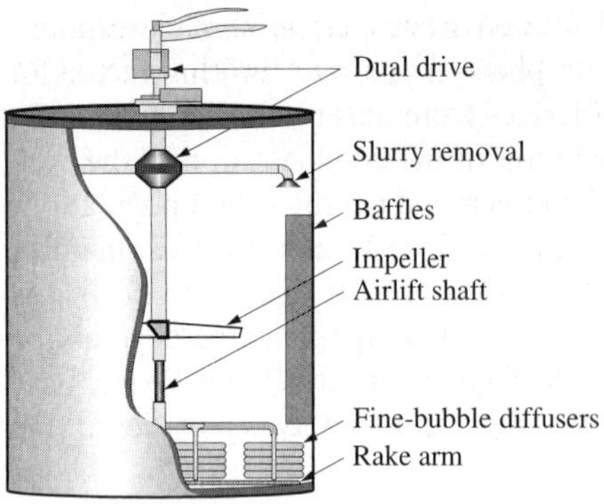

FIGURE 10-20
EIMCO Biolift™ slurry-phase bioreactor.[70]

Retention times

As stated before, the biomass grows on the surface of the soil particles. Clearly, it represents a very small portion of the overall solids content of the slurry, and separation and recycling of the biomass has not proved practical. Therefore, compared to liquid treatment methods employing much larger concentrations of biomass, longer hydraulic retention times in the slurry bioreactor are necessary. For a continuous flow process, the rate at which the soil is fed and removed from the bioreactor is critical. If the removal rate is higher than the bacterial generation rate, the biomass will diminish.

Applications—Wood-Preserving Wastes

Wood-preserving wastes contain high concentrations of creosote oil, pentachlorophenol (PCP), and PAHs (including potentially carcinogenic PAHs).

EXAMPLE 10-6. TWO-STEP SLURRY-PHASE TREATMENT DEGRADATION RATES. A laboratory test evaluated a two-step batch process to treat creosote waste: (1) suspending creosote waste with surfactants for 7 days at 20 percent solids in a roughing vessel and (2) transferring the supernatant to a polishing reactor for 14 days of additional biological treatment.[71] The reported reductions in concentrations of selected PAHs are shown in Table 10-7.

Assuming the above bench-scale data are representative of a larger operation and starting at the same initial concentrations, how many days of treatment in a pilot-scale operation would be necessary to degrade each of the constituents to 100 ppm or less?

TABLE 10-7
Reported reduction in polycyclic aromatics

Constituent	Initial concentration, ppm	Final concentration, ppm
Phenanthrene	13,000	<100
Fluorene	7,000	<100
Fluoranthrene	8,000	<500
Pyrene	6,000	<500
Benzo(*a*)pyrene	9,000	<100
Benzo(*b*)fluoranthrene	13,000	<500
Benz(*a*)anthracene	11,000	<100

Solution. The analysis begins by identifying the constituent most resistant to treatment, which is pyrene. Assuming first-order kinetics and accepting the limitations of Equation 10-16, the degradation rate can be determined:

$$S = S_0 e^{-k't}$$

$$e^{-k'(21\text{ days})} = \frac{500}{6000}$$

$$k' = 0.12/\text{day}$$

Again using Equation 10-16, calculate t for $S = 100$ ppm:

$$S = S_0 e^{-k't}$$

$$100 = 6000 e^{-(0.12/\text{day})t}$$

$$t = \underline{\underline{34\text{ days}}}$$

The pilot-scale test should be designed with a longer test period than 34 days because it involves two extrapolations (i.e., to a larger scale operation and to a final concentration not previously achieved). Practical experience with biological treatment has shown that such extrapolations can fall short of expectations.

Applications—Contaminated Soil

The use of slurry-phase treatment for contaminated soil has attracted the interest of researchers in Europe and the United States. A laboratory-scale study in the Netherlands indicated that this method biomineralizes soil contaminated with hexachlorocyclohexane (HCH) (400 ppm of α-HCH and 230 ppm of β-HCH) at rates much greater than achieved for solid-phase treatment methods.[72] The mixing of soil slurries in a batch reactor containing 5 to 25 percent solids readily degraded α-HCH at an initial rate of 32 mg/kg of soil per day, much higher than the 2.5 mg/kg per day measured in a land treatment experiment of the same waste. The tests showed that temperature affected the initial degradation rates; the rate at 30°C was about 33 percent higher than the rate at 20°C, which was about 6 times higher than that at 10°C. However, given sufficient time, the same degree of degradation was achieved at all three temperatures. The difference observed between degradation at the low and high temperatures is believed to be due less to changes in biological activity and more to the sensitivity of solubility, and hence desorption, to temperature.

The laboratory experiments described above were not able to degrade the β-isomer of HCH. This is a result of 2 factors. First, this isomer has a lower solubility, which would reduce the rate of desorption. Second, this isomer is believed to be recalcitrant because of its spatial chloride arrangement.

EXAMPLE 10-7. SLURRY-PHASE TREATMENT OF CONTAMINATED SOIL. For the laboratory-scale study described above, a degradation rate of 32 mg/kg of soil per day was reported for α-HCH. Convert this rate to a first-order rate.

Solution

$$k' = (32\text{ mg/kg of soil/day})(1\text{ kg soil/400 mg})$$
$$= 0.08/\text{day}$$

Another Dutch experiment[73] treated PAH-contaminated soil in a laboratory reactor under the following conditions:

- Treatment as a slurry (dry solids content of 40 to 50 percent)
- Continuous feed
- Continuous mixing
- Room temperature

For a soil sample containing a low concentration of total PAHs (6 mg/kg) and no fuel oil, no degradation occurred. However, for another soil sample containing a significant amount of fuel oil and a much higher concentration of total PAHs (3900 mg/kg), degradation occurred, but at a linear rate without exhibiting logarithmic growth kinetics typically associated with biological treatment.

The lower molecular weight PAH compounds (those having two and three rings) showed rapid degradation with essentially complete mineralization after 43 days or less. Although the higher molecular weight PAHs had slower rates, the experiment was able to degrade the higher PAHs present to less than half of their initial concentration.

10-6
EX SITU SYSTEMS—SOLID-PHASE AND VAPOR-PHASE TREATMENT

Solid-phase treatment includes a broad class of biological methods used to treat contaminated soils and sludges containing minimal free moisture. There are three major engineering/process approaches to solid-phase treatment:

- **Land treatment:** Biodegradation of wastes incorporated into a defined soil layer.
- **Composting:** Codegradation of hazardous waste with larger amounts of supplementary organics mixed together in windrows, piles, or vessels.
- **Soil piles:** Static aeration of oversized piles of contaminated soil.

With each treatment process, both biotic and abiotic contaminant removal is possible. Common abiotic removal mechanisms include volatilization, leaching, and immobilization.

It is worth reiterating that successful treatment of soils and sludges depends on:

- The presence of metabolically suitable microorganisms
- Creating and maintaining favorable microbial growth conditions
- Eliminating or reducing the toxicity of the contaminants
- Providing adequate contaminant bioavailability
- Providing adequate treatment time.

Detailed design information about solid-phase remediation systems can be found elsewhere.[74] The remainder of this section will describe the general characteristics of land treatment, composting, and soil pile systems.

Land Treatment

The term **land treatment** was originally used to describe a process whereby liquid wastes (such as oils or oily sludges) were applied to open land at controlled rates.

After application, the contaminated materials were mixed into the shallow surface soils. Naturally occurring physical, chemical, and biological processes at work in the contaminated soils degraded and immobilized the waste constituents. A number of operational steps are used to stimulate bacterial growth (e.g., aeration, pH adjustment, nutrient addition, moisture control, and mixing). Since the practice has a resemblance to standard agriculture principles, it was commonly referred to as **land farming.** However, a hazardous waste land treatment site is not farmed in an agricultural sense; no agricultural crops are grown because of concern with crop uptake of toxic constituents. In comparison with slurry-phase treatment, land treatment is usually less expensive.

Process description

Wastes undergoing land treatment are applied uniformly to the surface of an engineered site by spreading them over the surface or injecting them just below the surface. Some type of cultivation equipment, such as a disk harrow or rototiller, is then used to incorporate the wastes into the soils (see Figure 10-21).

The cultivation step tends to equalize the concentration of waste constituents. However, it serves two more important processes: (1) It mixes the soil with waste, thereby maximizing contact among the waste, indigenous microbes, and chemical additives such as nutrients. (2) It provides the aeration necessary for biological degradation. The depth to which the waste is cultivated, typically about 10 to 30 cm, represents the zone of incorporation (ZOI) (or the "plow zone," in operators' terms).

The soil constitutes the treatment medium. The area where treatment occurs is referred to as the zone of treatment, and it may extend as much as 1.5 m below the soil surface, well below the zone of incorporation. The physical, chemical, and biological properties of both the soil and the waste interact with external variables such as climate to determine the fate of contaminants. As a result, contaminants can be degraded or immobilized. Figure 10-22 shows the possible biological, chemical, and physical mechanisms that affect the fate of various waste constituents after land treatment.

Soil naturally contains a large and diverse microbial community. This community can degrade applied organics, basically as an aerobic treatment process in which

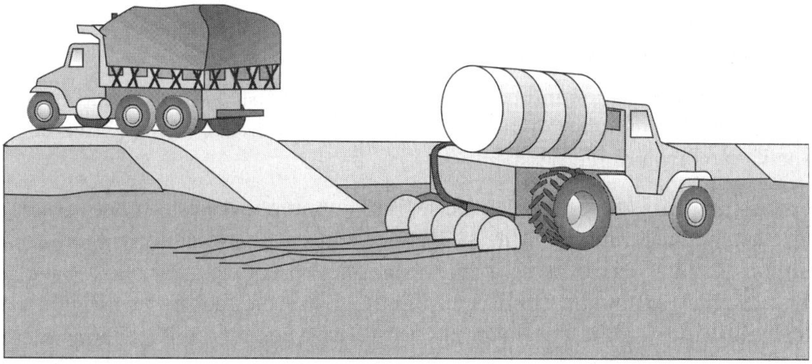

FIGURE 10-21
Land treatment system and equipment.

	Precipitation	Ion exchange	Neutralization	Volatilization	Migration	Fixation	Adsorption	Biological or chemical degradation	Plant uptake
Organics	●	●	●	●	●		●	●	●
Oils				●	●			●	
Metals	●	●			●	●	●	●	●
Inorganic acids, bases, and salts	●	●	●		●		●		

FIGURE 10-22
Fate of constituents in land treatment.

virtually all biological degradation occurs within the zone of incorporation. Besides bacteria and fungi (procaryotes), other eucaryotic biological forms such as protozoa, mites, and earthworms act to degrade organics.

Land treatment also removes waste constituents other than organics. Suspended solids are removed by physical settling and filtration on the soil. Heavy metals are removed from solution by adsorption onto soil particles and by precipitation and ion exchange in the soil. Certain soils have a high absorptive capacity for metals and exhibit high levels of removal.

No particular type of soil is ideal. The more permeable soils tend to aerate and drain better, allowing larger applications of organics than the fine, low-permeability soils. The latter, however, generally have a higher adsorptive capacity for metals. The specific site conditions must be evaluated with regard to the waste characteristics and the operational practices. Land treatment systems are routinely operated with an organic concentration as high as 5 to 10 percent (total mass of waste as a percent of mass soil in the ZOI). Land treatment systems can also be operated for wastes having low concentrations of organics (e.g., contaminated soil and sediment). This latter case may not require cultivation into fresh soil (i.e., the contaminated soil itself can be the treatment medium).

Assimilative capacity

The key to successful land treatment is to apply wastes at a rate and frequency that do not exceed the assimilative capacity of the site. The assimilative capacity must be viewed from three perspectives[75]:

- **Capacity limiting—**addresses conservative, immobile waste constituents (e.g., metals) that accumulate in the soil with successive applications, their concentration eventually increasing to a threshold level requiring the site to be closed. The cumulative amount applied during the life of the facility is critical.
- **Rate limiting—**addresses nonconservative waste constituents that are degraded in the soil over a period of weeks, months, or even years (e.g., oil). The cumulative amount applied in comparison with the cumulative amount degraded is critical.

- **Application limiting—**addresses mobile waste constituents that may readily migrate from the site (e.g., volatile organics). It also addresses the maximum hydraulic loading that will infiltrate through the site surface without runoff. The amount applied at any one time is critical.

Metals pose a special problem to the continued operation of a land farm. Proper control of soil pH will facilitate their immobilization in the soil. The soil pH can be adjusted to the optimum range (typically no less than pH 6 to 7) and maintained within such range by application of lime and/or other soil amendments. The presence of amphoteric metals (e.g., Se and Sb) would necessitate a narrow acceptable pH range because their solubilities increase with increasing pH. The immobilization of metals within the upper portion of the treatment zone will result in ever-higher concentrations of metals with each application of waste. Eventually the concentrations rise to levels that are toxic to the microbial community. At that point, biodegradation will slow or stop, requiring closure of the site or replacement of the soil with uncontaminated soil. If the soil is replaced, the removed soil must be disposed of in a manner consistent with its waste characteristics. Even if the concentrations of metals are not toxic to the microorganisms in the soil, they may still reach phytotoxic levels and inhibit establishment of any planned postclosure vegetative cover. Whether phytotoxic or biotoxic, the buildup of metals typically represents the limiting constraint on the life of the land treatment facility.

Waste migration pathways

The major drawback to land farming operations is the inability to control contaminant migration from the incorporation zone. Volatilization of organics and wind-blown dust represent air pathways for migration of waste constituents. Runoff can carry contaminated stormwater to surface streams. Infiltration of rainfall can leach waste constituents into groundwater, a major environmental concern. A study of several land treatment sites after their closure showed that the concentrations of copper, lead, and zinc in the soils below the zone of incorporation were about the same as observed in background soils. However, concentrations of chromium and nickel appeared greater than background concentrations, indicating that they had migrated.[76]

With proper site selection, development, and operation, the potential for adverse environmental impacts can be minimized. Adequate construction and maintenance of control structures (e.g., berms, runoff channels) can virtually eliminate the potential for surface runoff from the treatment site. Collected runoff can be treated and discharged or reapplied to the soil during dry weather. Volatilization of organics can be controlled by keeping waste application rates in safe ranges considering waste characteristics and site conditions. Even under dry conditions, the oil in refinery waste will control wind-blown dust. For nonoily wastes, wind-blown dust can be prevented by adding water to maintain proper soil moisture. Of particular importance is the selection of sites with the hydrogeologic characteristics that offer protection against migration into groundwater.

Mathematical models of land treatment

Within the zone of treatment, a contaminant moves among four phases—soil grains, pore water, oil or any other immiscible liquid in the pores, and unsaturated pore space—in response to a variety of physical and chemical influences including

advection, dispersion, adsorption/desorption, and volatilization. A contaminant may also degrade biologically. Each of these actions can be modeled to mathematically describe the fate and behavior of the contaminant. Such a model can predict the total amount of a pollutant leached below the treatment zone or lost as vapor under varying design and operating practices.

Different models exist for predicting the fate and behavior of waste constituents in a land treatment facility. These models can be highly involved because of the complexity of vadose zone processes. Most of the framework of any model addresses physical and chemical processes of mobilization. Biological degradation is typically approximated by the first-order kinetics of Equation 10-17.

Land treatment—applicable wastes

The widest use of land treatment occurs in the petroleum industry, where it has been used at more than 100 locations in the United States, primarily to treat American Petroleum Institute (API) Separator sludges, slop oil emulsion solids, induced and dissolved air flotation float, and leaded and unleaded tank bottoms.[77] These sludges are biodegradable to varying extents and can be treated by land application techniques, with half-lives ranging from 200 to 500 days. The limiting parameter is usually the oil loading rate, although the cumulative loading of certain metals contained in the waste (e.g., lead) may be limiting. Land treatment is also used in other industries, for example, in the wood products industry for treatment of waste containing wood preservatives, primarily creosote and pentachlorophenol.

Land Treatment Case Study

The BP Oil Alliance Refinery near Belle Chasse, Louisiana, conducted a pilot land treatment project for 1 year on two 50- × 120-ft plots to demonstrate that the process does degrade or immobilize its refinery waste.[78] The project also provided data for designing a full-scale facility.

Construction details

The soil at the test plots is fine grained with a high clay content (Unified Soil Classification System designation of CH and CL). In constructing the plots, a subsurface drainage system, consisting of 4-in drainage laterals spaced at 20-ft intervals, was placed on a low-permeability clay layer, which naturally occurs at an average depth of 5 to 6 ft below the existing grade. Soil from outside the plots was added to elevate the plots 3 ft above the existing grade. This additional soil helped to provide a treatment zone of about 4 to 4.5 ft in depth plus a buffer of 3 ft between the treatment zone and the seasonal high water table. To prevent run-on and contain runoff, a dike was constructed 2 ft above the land treatment surface around each plot.

Operations

Thickened and filtered sludges from the refinery's API Separator units, API skim treatment tank, dissolved air flotation units, and miscellaneous tank bottoms were applied in quarterly batches at an oil loading rate of 827 barrels per acre per year on one plot and 513 barrels per acre per year on the other. The zone of incorporation was tilled weekly, with soil moisture, pH, and nutrient balance monitored periodically.

TABLE 10-8
Land treatment of petroleum refinery sludge

Time of application, days after initial application	Amount of applied waste, tons	Oil and grease concentration of applied waste, %
0	20.1	16.5
110	30.6	13.1
229	30.6	16.8
298	24.7	18.5

The date of application and the amount of waste applied to the plot receiving waste at the rate of 827 barrels per acre per year are shown in Table 10-8.

Sampling and monitoring

The demonstration project included quarterly sampling of lysimeters (devices to collect soil water from vadose zone soils), soil cores at various depths, and the subsurface drainage system water. Analysis of samples for all hazardous constituents in the waste materials applied to the plots revealed the following:

1. The applied waste contained six volatile organics, nine semivolatile organics, and 12 metals.
2. Second quarter sampling of soils showed applied organic waste constituents in the 0- to 1-ft interval but none at deeper intervals. Metal concentrations showed a buildup in the 0- to 1-ft interval.
3. Waste was not applied between the second and third quarter; therefore, the third quarter soil coring showed a decrease in organic constituents, with metal concentrations exhibiting substantial variations.
4. Fourth quarter soil core sampling gave results consistent with the previous results.
5. No organic hazardous constituent present in the applied waste was detected in the soil core liquid.
6. Five metals were detected in the lysimeter samples, and all except for barium were at very low levels.
7. None of the hazardous constituents in the waste were detected in the subsurface drainage system water.

EXAMPLE 10-8. GENERAL ASSESSMENT OF TREATMENT EFFECTIVENESS. On the basis of the sampling results in Table 10-8, what overall conclusion can be made about the effectiveness of the project? What important information is not provided by the above results?

Solution. The project demonstrated that the operation could treat the waste at the applied rates without migration via groundwater during the 1-year period of the test. However, the results do not address oil and grease degradation and how that factor limits the rate at which waste can be applied over the long term.

Degradation kinetics

To support a determination of the rate at which waste can be applied in a full-scale facility over a longer period, it is necessary to calculate oil degradation kinetics. The oil and grease concentrations exhibited significant variation across each plot.

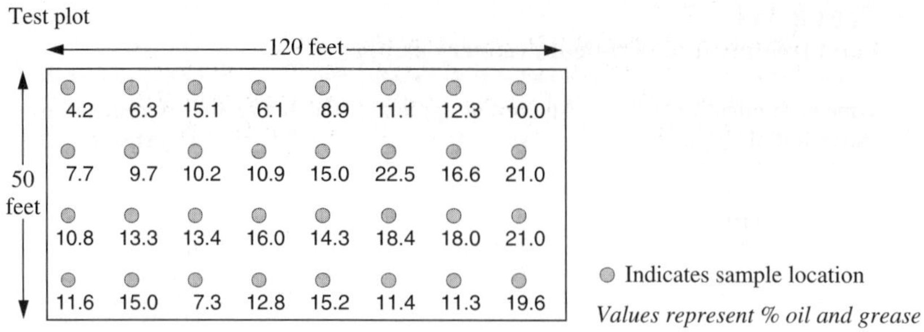

FIGURE 10-23
Final oil and grease concentrations at the test plot.

This variation was deemed natural and not due to improper operating procedures. To provide a better definition, 32 grab samples were taken of the upper 4 to 6 in of the surface soil, which is the zone of incorporation (ZOI), and analyzed for oil and grease (mass basis). The grab sampling occurred just over 1 year (368 days) after the initial application; results are shown in Figure 10-23.

EXAMPLE 10-9. DEGRADATION RATE OF LAND TREATMENT OF PETROLEUM REFINERY SLUDGE. On the basis of the results in Table 10-8 and Figure 10-23 in the case study, what are the half-life and degradation rate constants for the oil phase? Assume a soil density of 96 lb/ft^3.

Solution

1. Calculate oil loadings. The initial application increased the oil concentration in the ZOI by the following incremental amount:

$$\text{Incremental increase} = \frac{\text{loading of oil}}{\text{(area of plot)(depth of sampling)(density of ZOI)}}$$

$$= \frac{(16.5\%)(20.1 \text{ tons})(2000 \text{ lb/ton})}{(50 \text{ ft})(120 \text{ ft})(4 \text{ in})(1 \text{ ft}/12 \text{ in})(96 \text{ lb/ft}^3)}$$

$$= \frac{6600 \text{ lb}}{192{,}000 \text{ lb}}$$

$$= 3.4\%$$

Similarly, the incremental loadings represented by all applications are shown in Table 10-9.

2. Calculate residual oil. From Figure 10-23, the following statistics are calculated:

Mean	12.7 (% oil and grease)
Standard deviation	5.0
Coefficient of variation (standard deviation/mean × 100)	39.4%

3. Calculate degradation rate. From the above, residual oil at day 368 is 13.0 percent. Therefore, from Equation 10-16, the first-order kinetics degradation rate is as

TABLE 10-9
Incremental loadings to land treatment plot

Days after initial application	Applied oil (lb)	Incremental increase to ZOI oil concentration, %
0	6,600	3.4
110	8,000	4.2
229	10,000	5.2
298	9,200	4.8

follows:

$$13.0 = \sum S_0 e^{-k'(368-t_0)} \text{ for } t_0 = 0, 110, 229, 298 \text{ days}$$
$$= 3.4e^{-368k'} + 4.2e^{-258k'} + 5.2e^{-139k'} + 4.8e^{-70k'}$$
$$k' = \underline{\underline{0.0016/\text{day}}}$$

4. Calculate half-life. By definition, half-life is $t_{1/2}$, where $S_t/S_0 = 0.5$. Rearranging Equation 10-16 gives

$$0.5 = e^{-k't_{1/2}}$$
$$t_{1/2} = \frac{-\ln 0.5}{0.0016/\text{day}}$$
$$= 433 \text{ days}$$

Improved version of land treatment

Because of the concern with off-site migration of contaminants and the steady accumulation of metals and persistent compounds (such as high molecular weight PAHs) in treatment zone soils, the widespread use of this process is in decline. In many circumstances it has been replaced with a similar approach that incorporates controls for contaminant migration. The improved version with engineering controls may be referred to as solid-phase bioremediation, and is commonly used to treat soils contaminated with various organic contaminants, including pentachlorophenol.[79] Treatment takes place in constructed cells, which include the following contaminant migration controls:

- An impermeable barrier between the bottom of the treatment cell and the groundwater table
- Leachate collection system below the zone of incorporation
- Surface water runoff collection systems
- Aboveground enclosure of the cell for air pollution control (optional)

The concern with groundwater contamination has prompted the use of a synthetic liner below the treatment zone, somewhat similar to a landfill liner. Drainage pipes above the liner allow any generated leachate to be removed. The leachate pipes are covered with a buffer zone of clean permeable soil. As an alternative to the synthetic membrane and leachate collection system, it may be acceptable to construct the bottom of the cell with several feet of compacted clay. The relatively impermeable clay later will resist water infiltration and form a barrier to groundwater contamination.

Contaminated soils of a depth conducive to mechanical mixing are layered on top of the buffer soil/clay layer for treatment. Just as in land farming operations previously discussed, the contaminated soil can be adjusted for nutrient levels, pH, and moisture content. Operation, monitoring, and sampling are performed in a similar manner. Once the soil layer meets regulatory standards, it is removed from the cell. A new layer of contaminated soils is then put in place, and the process is repeated as necessary.

Many of the operational problems associated with the original land farming method do not apply to the improved version. For example, the continued accumulation of metals or persistent compounds within the treatment zone does not occur because each contaminated soil lift is treated once and then removed. The moisture content of the treatment zone soils is easier to control with the leachate collection system, which essentially eliminates overwatering concerns.

Depending on the volatility, concentration, and environmental effects of the contaminants present, an airtight enclosure of the treatment cell may be required. Temporary buildings or greenhouses can be used to provide enclosure. In addition to providing control of volatile emissions, the enclosure prevents excessive infiltration of rainwater and may also help to elevate the soil temperature in colder regions. Contaminated air from within the enclosure can be treated with vapor-phase activated carbon or other suitable air pollution control technologies.

Composting

Composting can successfully convert complex organic solids to simpler organics and humic materials under aerobic and anaerobic conditions. The degradation typically occurs at elevated temperatures, usually in the thermophilic range. Composting has a fairly long history in treating municipal solid waste and municipal sewage sludge. When contaminated soils contain high concentrations of organics, it may be possible to treat the contaminants by a process similar to that used for composting municipal wastes.

Most contaminated soils do not contain a sufficiently high concentration of organics to sustain composting. However, these soils can be composted by mixing a small percentage of the soils with other readily biodegradable solids, bulking agents, and a microbial inoculum. Bulking agents, such as wood chips, create void spaces for the passage of air needed to supply the necessary oxygen. Biological solids (such as straw, hay, and vegetable wastes) are readily biodegraded, generating heat in the process, which enhances biological activity. This increase in temperature is necessary for aerobic thermophilic (55 to 65°C) microorganisms to thrive. The microbial inoculum typically used is animal manure; pig, cow, and chicken manures have been used successfully.

The composting temperature is important for municipal solid waste and sewage sludge because the higher operating temperature of thermophilic bacteria can sterilize the waste of pathogenic bacteria. Such sterilization is not important with hazardous waste treatment and, therefore, mesophilic bacteria may suffice, depending on their ability to degrade the waste. If mesophilic conditions are desired, the excess heat produced by composting will require removal. Heat removal is typically accomplished by drawing air through the pile: cool air enters while heated air is withdrawn.

Composting—aeration techniques

Composting of waste will occur readily given the presence of appropriate microorganisms, adequate aeration, optimum temperature, necessary nutrients, pH, and moisture. Aeration is critical, and several techniques have been developed in municipal waste applications to enhance aeration and thereby promote decomposition. Basically there are three types and each can be applied to hazardous waste–organic mixtures as follows (see Figures 10-24 and 10-25):

- **Windrows:** In this technique the mixture of waste and organics is piled in long rows and turned once or twice per week, if not daily. The turning aerates the mixture and releases excess heat. Turning also increases the release of volatiles.
- **Static piles:** This is similar to windrowing except that the piles are not turned. Rather, by building the piles on top of a grid of perforated pipes, air can be mechanically drawn or forced through the piles by a vacuum or forced air system. A negative aeration system created by a vacuum at the base of the pile enables process air to be treated before it exhausts to the atmosphere if needed.
- **Enclosed reactor system:** By composting within an enclosed reactor, the operation can be optimized to complete the process in as little as 3 days compared with as much as 30 days for windrows.[80] It also facilitates control of volatile emissions by collecting process air in the headspace above the compost, by a diffuser system below the compost, or by an exhaust system within the com-

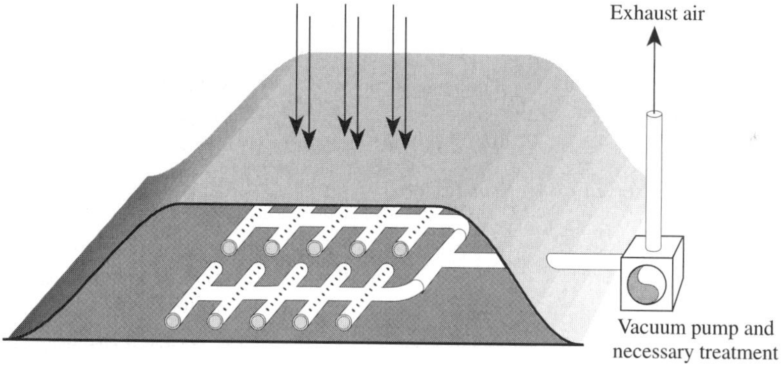

FIGURE 10-24
Composting methods—static piles.

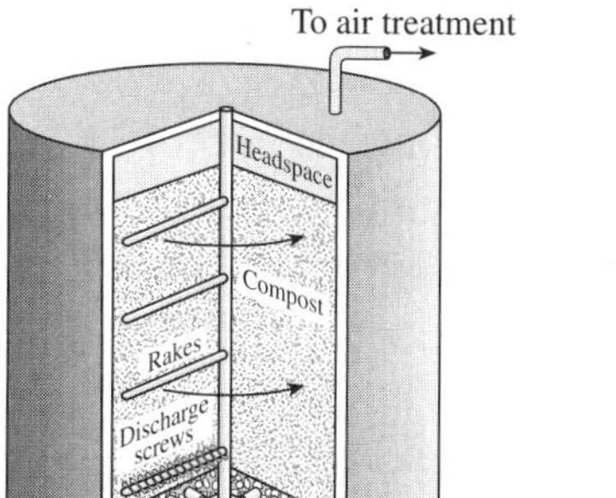

FIGURE 10-25
Composting methods—enclosed reactor.[85]

post. A reactor can be operated in a batch or continuous mode, with or without agitation.

Windrow or static pile composting is usually conducted outdoors and exposed to the elements, which presents the potential for release of contaminants to the soils, groundwater, and the atmosphere. Some of the contaminant migration controls required for land farming operations previously discussed might also be applicable or required for composting projects. Composting can also be conducted on slab within buildings or covered structures. Under these conditions, many of the problems of contaminant migration are mitigated.

Aerobic composting applications

Bench-scale tests and field demonstrations have been reported on explosive and propellant manufacturing wastes,[81] soils contaminated with coal tar,[82] and wastes containing ethylene glycol.[83] The test variables and results are shown in Table 10-10.

TABLE 10-10
Test results of composting hazardous waste

Item	Sediments and soil containing explosive and propellant manufacturing waste	Soil contaminated with coal tar	Waste containing ethylene glycol
Waste constituents	76,000 mg/kg total explosives (primarily TNT)	20,000 mg/kg coal tar	1900 mg/kg ethylene glycol 120,000 mg/kg TOC
Aeration methods	Static pile	Static pile	Static pile
Compost pile volume	34 yd^3	4 yd^3	32 gal
Compost pile height		4 ft	
Bulking agent	None (thermal source served as bulking agent)	Not disclosed	Wood chips
Additives	Alfalfa, straw, manure, horse feed, and fertilizer	Not disclosed	Dry molasses, sewage sludge, soil, ammonium nitrate, 10-10-10 fertilizer lime
Mix ratio of waste to additives	1% (vol.) and 24% (mass)	Not disclosed	2.5%, 5%, 10%, 20% (vol.)
Air supply	Vacuum	Forced air using blower; offgases recycled through the compost	Forced air using blower; adjusted to maintain minimum of 10% O_2 in exhaust gas
Moisture	Not reported, other than 400 gal water added initially	Not reported	40-60%
Operating temperature	Approximately 55°C	65–85°F (18–29°C)	Exceeded 120°F (49°C)
Test period	22 weeks	80 days	30 days
Reduction of waste constituents	99% of total explosives	94% of total hydrocarbons 95% of 2-ring PAHs 94% of 3-ring PAHs 89% of 4-ring PAHs	>95% of ethylene glycol
Comment	Split between mineralization and transformation to nonextractable compound was not determined	87% of total hydrocarbon removed in 35 days. 5-ring PAH did not degrade.	Test did not account for abiotic losses

The test with soil contaminated with coal tar yielded some interesting observations. Although the five-ring PAHs did not degrade, the data suggest that they became immobilized. The recycling of offgases proved effective: Only 1 percent of the volatiles were trapped by carbon filters, with the remainder assumed to have been degraded or entrapped in the compost matrix. Although the residue was not tested, the researchers believe it is suitable for cover in landfills and as industrial fill.

Windrow composting has been used in various forms to treat contaminated soils. For example, aerobic thermophilic composting has been used to mineralize trinitrotoluene (TNT), dinitrotoluene (DNT), RDX, and HMX. The 6.5 cubic yard (yd^3) field pilot test used compost consisting of contaminated soil (20 to 30 percent by volume), food wastes, alfalfa, cow and chicken manure, and wood chips. Water mixed with molasses was used to keep the pile humidified and to provide a readily biodegradable food source for the microorganisms. After approximately 40 days of treatment, the target explosives concentrations were below detection limits.[84] In another study, approximately 2840 yd^3 of TNT- and RDX-contaminated soil were used to construct five windrow piles, each with a slightly different mixture of hay, wood chips, manure, and potatoes/potato wastes. Thermophilic conditions were attained in the piles, and acceptable explosives concentrations were reached in only 28 days of treatment.[86]

EXAMPLE 10-10. COMPOSTING DEGRADATION RATE. The owner of a property contaminated with petroleum must remediate the site completely in 1 year. Approximately 40,000 yd^3 of soil contaminated with a mean total petroleum hydrocarbon (TPH) concentration of 10,000 mg/kg, ranging from "not detected" to over 98,000 mg/kg, must be remediated to a mean TPH of not greater than 100 mg/kg, with no sample greater than 500 mg/kg, if it is to be left on-site. As an alternative to the costly option of incineration and off-site disposal, the owner has conducted laboratory tests of composting that show the reductions in TPH listed in Table 10-11. How many days will it take to meet the mean 100 mg/kg standard? How many days will it take to meet the absolute 500 mg/kg standard?

Solution. The short-term tests show rapid reduction of TPH on a continuing basis, suggesting that the contamination is readily degradable via composting. The question remains if the objectives of 500 mg/kg as a maximum and 100 mg/kg as a mean value can be achieved, because neither has been achieved in the short-term tests. This question can be addressed by analyzing the degradation kinetics. Accepting the limitations of Equation 10-16, it can be used to determine the degradation rate constants. For Sample

TABLE 10-11
Composting of contaminated soil

	Concentration of TPH, mg/kg					
Day	**Sample 1**	**Sample 2**	**Sample 3**	**Sample 4**	**Sample 5**	**Sample 6**
0	18,600	9100	4500	98,300	8100	6900
3	16,400	5500	4200	87,700	5800	2600
6	13,200	2100	3000	55,800	5100	2500
10	11,200	1900	1600	43,600	3200	2100

1, the constant is calculated as follows:

$$\frac{S}{S_0} = e^{-k't}$$

$$k' = \frac{-\ln(11{,}200/18{,}600)}{10 \text{ days}}$$

$$k' = 0.05/\text{day}$$

The mean degradation rate constant for the six sampler is 0.10/day. Again using Equation 10-16, the number of days to reduce the mean TPH concentration from 10,000 mg/kg to 100 mg/kg is calculated as follows:

$$\frac{S}{S_0} = e^{-k't}$$

$$k' = \frac{-\ln(100/10{,}000)}{0.10/\text{day}}$$

$$t = 46 \text{ days}$$

Similarly, reducing 98,000 mg/kg to 500 mg/kg would require the following time:

$$\frac{S}{S_0} = e^{-k't}$$

$$k' = \frac{-\ln(500/98{,}000)}{0.10/\text{day}}$$

$$t = 53 \text{ days}$$

Both objectives can be met. However, scaling up from bench-scale tests and extrapolating to a longer time period will most likely encounter much slower rates of degradation. Pilot testing is necessary.

Anaerobic/aerobic composting

Many difficult-to-degrade compounds, such as nitroaromatics and pesticides, are more amenable to anaerobic treatment than aerobic biodegradation. This had led to bench-scale, pilot-scale, and full-scale projects in which conditions within the compost pile are sequentially changed from anaerobic to aerobic through the use of organic amendments and controlled aeration. If required, the anaerobic/aerobic cycles can be repeated as required. This approach is expected to continue to be accepted and utilized by the waste treatment field.

Soil Piles

Although soil piles may be considered a form of static pile composting, they form a distinct treatment process. Soil piles are an effective method for treating large volumes of contaminated soil, particularly if available land area is limited. Soil piles typically differ from composting piles in the following ways:

- Contaminated soils contain low levels of organic contaminants.
- Indigenous microorganisms are used for biodegradation.

- Supplemental organics are not added.
- Aerobic biodegradation only is utilized.
- Air distribution system piping can be installed throughout the pile at various depths as needed.
- The piles can be taller, minimizing the treatment space required.
- The process is conducted in the mesophilic temperature range.
- The piles are not mechanically mixed during the treatment process.

Soil pile treatment involves stockpiling contaminated soil in large mounds, typically over an air piping system and an impermeable liner. Rather than aerate the soil by cultivating it as in land treatment, the piping system pulls a vacuum through the heap or blows air through it. The former is preferred as it enables treatment of volatile emissions if necessary. The piping system can also be installed at various depths within the pile, as needed depending on overall pile height and anticipated airflow characteristics. Inert bulking agents such as wood chips may be added if needed to ensure adequate vapor and liquid flow through the soils. Inorganic nutrients can be added either as a water solution or as a powder during pile construction. Water is added to the pile as needed to maintain adequate moisture levels. The mounds can be several meters in height and may be covered if necessary. By using a bottom liner, leachate can be collected and recirculated to the pile or treated as necessary. The operation is monitored for pH, nutrient balance, and moisture, with adjustments made as required.

Soil piles degrade waste at a somewhat slower rate than land treatment, because of the low concentrations of organics and the lack of frequent cultivation. For example, one project reported remediation times from 4 to 18 months.[87] Reported cost data suggest this process is less expensive than land treatment for large volumes of contaminated soil because, in part, of the reduced need for land area. This treatment method, albeit slower than land treatment, offers the advantage that migration by volatilization and leaching, two primary concerns with land treatment, can be virtually eliminated. The emission of volatile organics is controlled by covering the heap and maintaining a vacuum, with the exhaust from the vacuum passed through a vapor treatment system. Leaching is controlled by covering the pile to prevent excess water infiltration and by removing water collecting on the liner.

Soil pile technology has been used extensively for soils contaminated with easy-to-degrade petroleum hydrocarbons such as gasoline, jet fuel, and diesel. The use of soil piles is diminishing because of the preference for using *in situ* treatment technologies at many sites.

Vapor-Phase Systems

Biological processes can also be used to treat air contaminated with VOCs and hazardous air pollutants (HAPs), as well as odor-causing compounds. Typically, the vapor-phase contaminants being treated are emissions from *ex situ* systems used to treat liquid or solid hazardous wastes or from *in situ* remediation systems. Vapor-phase bioreactor designs include biofilters, biotrickling filters, and bioscrubbers. The term **biofiltration** is commonly used to describe any such biological vapor treatment process.

Biofilters are the most widely used and accepted form of vapor-phase biological treatment system and have been described in detail elsewhere.[88–90] In a **biofilter system,** the gas to be treated typically first enters a preconditioning unit in which the moisture level and gas temperature are adjusted and particulates are removed. The gas is then conveyed to the biofilter air distribution system, from which it passes through a biologically active filter bed made of packing media. The biomass is attached to the various surfaces of the packing materials. The packing media usually consist of a mixture of bulking agents (such as wood chips), compost, peat, and various amendments (such as inorganic nutrients, activated carbon, pH buffering compounds, and possibly a supplemental organic carbon source). As the contaminated air passes through the media, contaminants and nutrients are transferred to the biomass, where cellular metabolism degrades the chemicals into simpler compounds and metabolic by-products.

Biofilters are most effective at treating air streams containing a total organic load of 1500 ppmv (a measurement of gas concentration by volume) or less. Typical vapor contact times required for odor treatment range from 10 to 40 seconds. The treatment of higher concentrations of VOCs and HAPs requires a longer contact time: 20 to 90 seconds is common. Biofilters have been successfully used to treat a wide range of gaseous compounds, including:

- Aliphatic hydrocarbons (hexane)
- Aromatic hydrocarbons (benzene, toluene, ethylbenzene, xylenes, and styrene)
- Alcohols (methanol, ethanol, and propanol)
- Aldehydes (formaldehyde and acetaldehyde)
- Ketones (acetone, methyl ethyl ketone, and methyl isobutyl ketone)
- Carbonic acids (butyric acid)
- Carbonic acid esters (ethylaldehyde and methyl methacrylate)
- Phenols
- Sulfides (hydrogen sulfide and dimethyl sulfide)
- Mercaptans (methyl mercaptan)
- Nitriles (acetonitrile and acrylonitrile)
- Ammonia

Biofilter reactor designs are characterized by the lack of a free-phase liquid, and the filter bed thickness is usually limited to a height of 1 m or less. Biofilter reactor designs include open soil beds, open single beds, open multiple beds, closed single beds, and closed multiple beds. Depending on the reactor design, the contaminated air can be passed upflow or downflow through the beds. The choice of which biofilter reactor design to use is dependent on the waste stream to be treated and the specific project constraints imposed.

Biotrickling filters consist of contaminant-degrading biofilms attached to some sort of plastic or ceramic packing material. Free water is recirculated over the packing to aid in contaminant mass transfer and maintenance of favorable microbial growth conditions. The water recirculation system also allows for finer pH control. Since a synthetic packing is used, biotrickling filters can be designed as packed columns with correspondingly smaller footprints. The contaminated air stream can be introduced cocurrent or countercurrent to the recirculating liquid.

Bioscrubbers are tower-like reactors in which vapor-phase contaminants are initially absorbed into a recirculating liquid. This liquid can then undergo subsequent biological treatment in the same chamber or in a separate reactor.

Biofilters have been extensively used for odor-treatment applications, especially the treatment of odors generated at municipal wastewater treatment plants. Small modular biofilters have been marketed by several firms for the treatment of hydrocarbon-contaminated vapors generated as part of soil vapor extraction projects. The number of such systems is limited. Several large-scale (10,000 ft^3/min of air or greater) biofilter systems have been installed for the treatment of industrial VOC/HAP waste streams, especially within the wood products industries. Widespread acceptance of such systems is not anticipated in the immediate future.

Biotrickling filters may eventually overcome some of the operational difficulties associated with biofilters, but the acceptance and commercial use of this biosystem design has also been limited.

10-7 *IN SITU* BIOLOGICAL TREATMENT SYSTEMS—MONITORED NATURAL ATTENUATION

In situ is a Latin phrase meaning "in its original place." *In situ* bioremediation is the method of treating subsurface contaminants where they are found without excavating the overlying soil or extracting the groundwater for treatment in aboveground, engineered reactors. A variety of *in situ* processes are available to treat contaminated soils in the vadose zone as well as contaminated soils and groundwater present in the saturated zone. *In situ* treatment also refers to the in-place treatment of solids and liquids present in lagoons or other natural impoundments.

Background

In order to understand and appreciate *in situ* biological treatment systems, a brief review of subsurface contamination and the factors involved in its remediation are in order.

As discussed in Section 4-1, organic contaminants migrate into the subsurface from many sources, including surface product spills, leaking underground tanks, unlined lagoons, buried wastes, and leaking disposal facilities. Organic contaminants typically exist in four phases:

- Free product
- Adsorbed or otherwise bound to soil particles and interstitial openings in the geologic matrix (sorbed phase)
- Dissolved in the groundwater (solute phase)
- Volatilized compounds (present in interstitial pore spaces in the vadose zone)

The amount in any phase varies, depending on the total volume of product released, the characteristics of the contaminants and the subsurface, and factors such as the soil-water partition and adsorption coefficients (see Sections 3-2 and 4-3). In most cases, the total mass of the contaminant in the solute phase is very small compared with the total mass present in either the sorbed or free phase. However, the solute phase impacts a much greater volume of subsurface materials because of the mobility of groundwater.

Any free product in the subsurface continues to serve as a source of soil and groundwater contamination. Even if free product (LNAPL or DNAPL) (see Section 4-2) recovery efforts at a site are successful, a significant contaminant mass will typically remain adsorbed to vadose and saturated zone soils. This represents a long-term source of groundwater contamination due to desorption by infiltrating precipitation and the seasonal horizontal and vertical movement of groundwater.

EXAMPLE 10-11. PHASE DISTRIBUTION OF FUEL IN SUBSURFACE. A slow leak from a diesel fuel tank has contaminated a site, both the soil below the tank and the underlying aquifer. After removal of free product, testing shows the following information:
Extent of heavy contamination of soil at source (sorbed phase):

$$10 \text{ m} \times 25 \text{ m} \times 3 \text{ m}$$

Arial extent of groundwater contamination (solute phase):

$$100 \text{ m (maximum width)} \times 300 \text{ m (length) at 5 m depth}$$

Average concentration of diesel fuel:
Dissolved in groundwater: 6 mg/L
Absorbed in soil at source: 4000 mg/kg
Soil characteristics:
Type: silty sand
Density: 1800 kg/m^3
Aquifer characteristics:

Average depth contaminated with fuel	5 m
Effective porosity	30%
Hydraulic gradient	0.0042 ft/ft
Hydraulic conductivity	0.02 cm/s

On the basis of this information and assuming that the groundwater contamination plume follows a truncated oblate ellipsoid, what is the phase distribution of fuel?

Solution. The volume of contaminated groundwater can be approximated as a pyramid on its side with base 5 m × 100 m and a height of 300 m. Thirty percent of this volume is void space that can contain groundwater.

$$\text{Volume of contaminated ground water} = 1/3(100 \text{ m} \times 300 \text{ m} \times 5 \text{ m})(0.30) = 15{,}000 \text{ m}^3$$

$$\text{Mass of fuel in solute phase} = 15{,}000 \text{ m}^3 \times 6 \text{ mg/L} \times 103 \text{ L/m}^3 \times 10^{-6} \text{ kg/mg} = 90 \text{ kg}$$

$$\text{Mass of contaminated soil at source} = (10 \text{ m} \times 25 \text{ m} \times 3 \text{ m})(1800 \text{ kg/m}^3) = 1.35 \times 10^6 \text{ kg}$$

$$\text{Mass of fuel in sorbed phase} = 1.35 \times 10^6 \text{ kg} \times 4000 \text{ mg/kg} = 5400 \text{ kg}$$

Thus, the mass of contaminants in the sorbed phase is 60 times greater than that dissolved in the groundwater.

Early projects to remediate contaminated groundwater usually involved the use of "pump and treat" technology (see Section 16-7). "Pump and treat" processes used extraction wells to pump contaminated groundwater to the surface, where it underwent in engineered aboveground reactors the *ex situ* treatment necessary to remove

organic and/or inorganic contaminants (see Section 10-4). The time scale for pump and treat operations to reduce residual soil contamination to acceptable levels can be enormously long since many contaminants desorb very slowly from soils and/or have only limited solubility in water. There are numerous accounts that document the long time scale involved with desorption. For example, a study of gasoline-contaminated soils showed that infiltration of 46 pore volumes of water (which would take years in some aquifers) removed only 1.6 percent of the absorbed fraction.[91] An analysis by the EPA showed that pump and treat operations rapidly lowered groundwater contamination to 10 to 50 percent of its initial concentration, but contamination quickly reached a residual concentration that stayed fairly constant with continuous pumping for a long period of time.[92] Groundwater scientists have predicted that to reduce groundwater concentrations by a factor of 100 could require 100 to 200 years of pumping for ideal conditions,[93] and thousands of years for contaminants with very low solubility.[94] Thus, although "pump and treat" processes continue to play a valuable role in hydraulic plume control, they are generally considered ineffective as a treatment technology to achieve mass removal.

The limitations of pump and treat, plus its high economic costs, served as driving forces for the development of new remediation technologies. Various physical, chemical, and biological *in situ* processes have been proposed and commercialized to meet these needs.

The concept of *in situ* bioremediation was first developed in the 1970s by a subsidiary of Sun Oil Company for the treatment of gasoline, diesel fuels, and other mineral oil hydrocarbons. In the 1990s, *in situ* bioremediation gained rapid acceptance as further advances were made in the technology. The concept is a simple one; it relies on the indigenous microorganisms present in the subsurface to degrade the contaminants present.

Successful *in situ* treatment requires careful consideration of a complicated set of factors involving not only microbiology, but also hydrogeology, geochemistry, and engineering. *In situ* biodegradation follows the same principles as other biological treatment methods except that the contaminated subsurface, rather than an engineered vessel, functions as the bioreactor. Although it is easy to control the microbial growth conditions within an engineered vessel, these conditions are more difficult to control in the field.

This section addresses the *in situ* degradation processes that occur naturally in the subsurface without enhancement. The subsequent section covers the methods used to enhance, stimulate, and manage the naturally occurring processes.

Definition and Mechanisms of Monitored Natural Attenuation

One of the wonders of the environment is its natural capability to eventually remediate itself of most contamination. Monitored natural attenuation (MNA) is a remediation approach that involves quantifying, evaluating, and monitoring the natural processes that work to reduce mass and concentration of contaminants in soil and groundwater. Monitored natural attenuation is an appropriate approach when natural processes will retain or destroy the contaminants to levels that are protective of human health and the environment. For some sites, MNA may be as effective or even more effective than active remediation.

Natural attenuation is being used at sites with a wide range of contaminants, including chlorinated solvents; benzene, toluene, ethyl benzene, and xylene (BTEX) compounds; polycyclic aromatic hydrocarbons; phenols; polychlorinated biphenyls; pesticides; and inorganics. The use of natural attenuation at underground storage tank sites with petroleum releases increased significantly in the early 1990s. As of 1995, natural attenuation is the second most popular option for sites with contaminated soils. It is being used at roughly 29,000 sites, equivalent to 28 percent of all active sites dealing with soils remediation. Natural attenuation is the most common treatment option at sites with contaminated groundwater. It is being used at 17,000 sites, or approximately 47 percent of the active sites dealing with groundwater contamination.[95] Most of these sites involve contamination by fuel hydrocarbons. Table 10-12 summarizes advantages and disadvantages of implementing MNA as compared to other remediation options using intrusive methods (e.g., excavation of contaminated soils).

In their technical directive on the use of MNA, the EPA stated:

> The term "monitored natural attenuation," as used in this Directive, refers to the reliance on natural attenuation processes (within the context of a carefully controlled and monitored clean-up approach) to achieve site-specific remedial objectives within a time frame that is reasonable compared to other methods.[96]

Natural attenuation in groundwater systems results from the integration of several subsurface attenuation mechanisms that are classified as either destructive or nondestructive. Biodegradation and abiotic degradation processes, such as hydrolysis, are destructive mechanisms. Nondestructive attenuation mechanisms include sorption, dispersion, dilution from recharge, and volatilization, among others.

Biodegradation

Biodegradation is the most important destructive natural attenuation mechanism. As discussed earlier in this chapter, certain microorganisms can digest hazardous substances such as fuel hydrocarbons or solvents that are hazardous to humans. In most instances, degradation is merely a complex reduction/oxidation (redox)

TABLE 10-12
Advantages and disadvantages of monitored natural attenuation

Advantages	Disadvantages
Less generation of remediation waste Lower potential of cross-media contamination Potentially lower overall remediation costs Potential for application to all or part of given site Can be used as a follow-up to active remediation Potential application alone or in combination with other remediation technologies Lower risk of human exposure to contaminated media	Longer remediation time Institutional controls may be necessary More comprehensive and more costly site characterization Potential for contaminant migration Contingency plan may be required Toxicity of transformation products may exceed that of the parent compound Monitoring will be required for as long as contamination levels remain above cleanup goals Potential for renewed mobility of previously stabilized contaminants Extensive education efforts may be necessary for regulators and the public to accept MNA

reaction. The electrons or reducing equivalents (hydrogen or electron-transferring molecules) produced must be transferred to a terminal electron acceptor. Table 10-13 lists redox reactions that can be responsible for biodegradation of benzene and toluene in different redox environments.

The existence of biological activity can be confirmed and evaluated through investigation of the concentrations of the chemical species or "geochemical parameters" that are either consumed or produced during biodegradation. Table 10-14

TABLE 10-13
Coupled benzene and toluene reduction/oxidation reactions

Degradation type	Electron donor	Reaction
		Benzene
Aerobic	O_2	$C_6H_6 + 7.5O_2 \Rightarrow 6CO_2 + 3H_2O$
Denitrification	NO_3^-	$C_6H_6 + H^+ + 6NO_3^- \Rightarrow 6CO_2 + 3N_2 + 6H_2O$
Manganese reduction	MnO_2	$C_6H_6 + 15Mn^{+4} + 12H_2O \Rightarrow 6CO_2 + 30H^+ + 15Mn^{+2}$
Iron reduction	Fe^{+3}	$C_6H_6 + 30Fe^{+3} + 12H_2O \Rightarrow 6CO_2 + 30H^+ + 30Fe^{+2}$
Sulfate reduction	SO_4^{-2}	$C_6H_6 + 7.5H^+ + 3.75SO_4^{-2} \Rightarrow 6CO_2 + 3.75H_2S + 3H_2O$
Methanogenesis	CO_2	$C_6H_6 + 4.5H_2O \Rightarrow 2.25CO_2 + 3.75CH_4$
		Toluene
Aerobic	O_2	$C_6H_5\text{–}CH_3 + 9O_2 \Rightarrow 7CO_2 + 4H_2O$
Denitrification	NO_3^-	$C_6H_5\text{–}CH_3 + 7.2NO_3^- + 7.2H^+ \Rightarrow 7CO_2 + 3.6N_2 + 7.6H_2O$
Manganese reduction	MnO_2	$C_6H_5\text{–}CH_3 + 18MnO_2 + 36H^+ \Rightarrow 7CO_2 + 18Mn^{+2} + 22H_2O$
Iron reduction	Fe^{+3}	$C_6H_5\text{–}CH_3 + 36Fe^{+3} + 14H_2O \Rightarrow 7CO_2 + 36Fe^{+2}$
Sulfate reduction	SO_4^{-2}	$C_6H_5\text{–}CH_3 + 4.5SO_4^{-2} + 3H_2O \Rightarrow 2.25H_2S + 2.25HS^- + 7HCO_3^- + 0.25H^+$
Methanogenesis	CO_2	$C_6H_5\text{–}CH_3 + 5H_2O \Rightarrow 2.5CO_2 + 4.5CH_4$

TABLE 10-14
Geochemical parameters, role in biodegradation, and indication of biodegradation at a site

Geochemical parameters	Role in biodegradation	Indication of biodegradation
Oxygen	Determines aerobic/anaerobic/anoxic conditions Consumed during respiration	Lower than background
Nitrate	Consumed during respiration	Lower than background
Sulfate	Consumed during respiration	Lower than background
Reduction oxidation potential	Determine reductive or oxidative conditions	Lower than background
Iron (II)	Produced during respiration	Higher than background
Manganese	Produced during respiration	Higher than background
Methane	Produced during respiration	Higher than background
Chloride	Produced from dechlorination of chlorinated organics	Higher than background
Alkalinity	Increases as a result of respiration (denitrification)	Higher than background
CO_2 or dissolved inorganic carbon	Produced during respiration	Higher than background
Reduced transformation products	Produced from dechlorination of chlorinated organics	Present
Ethane or ethene	Produced from dechlorination of chlorinated organics	Present

lists potential geochemical parameters and their significance to the evaluation of biodegradation and natural attenuation. In theory, in any environment in which microbial activity occurs, there is a progression from aerobic to anaerobic (methanogenic) conditions. Microbial activity can be expected to progress rapidly through these conditions. There is a sequence of electron acceptors used in this progression through distinctly different redox states: aerobic respiration, denitrification, manganese reduction, iron reduction, sulfate reduction, and methanogenic (fermentative). Different redox states will be observed throughout a plume. Figure 10-26 shows the change in concentration of electron donors caused by degradation of an organic contaminant plume as a function of distance.

Evaluation and Implementation Approaches

Several organizations have published technical approaches for evaluation and implementation of natural attenuation.[97–100] The protocols generally consist of a stepwise, but iterative, process of evaluation that develops "lines of evidence" for the occurrence of natural attenuation. The three lines of evidence to support natural attenuation as a remedy for contamination in soil and groundwater are:

- Documented loss of contaminants
- Geochemical and analytical data
- Direct microbiological evidence

In general, the EPA favors the first line of evidence, and has indicated it may accept the second one. The third line of evidence is the EPA's least preferred method for demonstrating the occurrence of natural attenuation.

Documented loss of contaminants

This simply involves the documentation of the reduction of the total mass or concentration of contaminants by natural processes through demonstration of: a reduction in concentrations in a monitoring well with time, a reduction in concentration in different wells with distance, or a reduction in the mass of contaminants between two monitoring well transects.

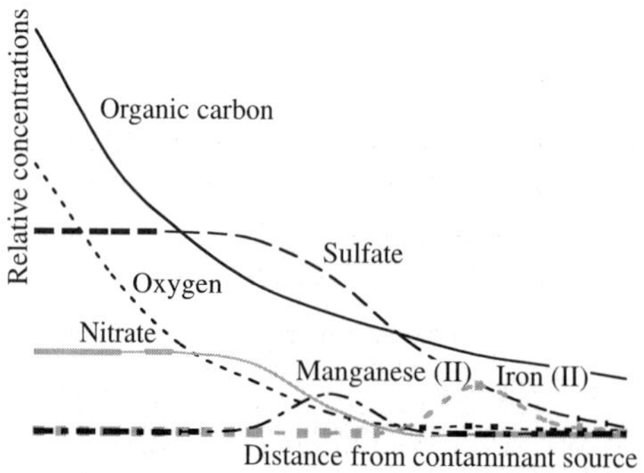

FIGURE 10-26
Relative electron acceptor concentrations versus distance in a theoretical plume.

Geochemical analytical data

Contaminant biodegradation can be confirmed by directly correlating decreases in contaminant concentrations with increases in metabolic by-product concentrations using mass balance calculations or changes in geochemical indicators. Geochemical data can also be used to show that concentrations of electron donors or acceptors in groundwater are sufficient to facilitate degradation of the dissolved contaminants. Solute fate and transport models may be used as part of the mass balance calculations.

Direct microbiological evidence

The potential for biodegradation of the contaminants of interest can be indicated by using studies of microcosms. Samples of aquifer material are taken into the laboratory where the microbial activity is then evaluated.

Implementation Steps

The technical approaches to implementing MNA typically include the following steps:

- Review available site and regional historical case data to evaluate whether natural attenuation is a possibility. The EPA has published a scoring procedure for initial site screening for natural attenuation.[101]
- Develop a preliminary site conceptual model and assess the potential for natural attenuation.
- If needed, perform additional site characterization to support natural attenuation.
- Update the conceptual model.
- Simulate long-term site behavior (with explicit treatment of uncertainty). This step includes estimation of site-specific attenuation rates.
- Perform an exposure pathway analysis identifying all potential receptors and exposure pathways. This step may also include solute transport modeling to evaluate whether receptors will be affected.
- Evaluate whether the time for remediation by MNA is reasonable compared to the time required if different remedial actions are taken.
- If natural attenuation is recommended, prepare a long-term monitoring and contingency plan, if needed.

Calculation of Site-Specific Attenuation Rates

Estimation of site-specific attenuation rates is a fundamental part of documenting the reduction of the total mass or concentration of contaminants by natural processes (first line of evidence). In addition, site-specific attenuation rates can be used to simulate long-term plume behavior and predict the time necessary to meet cleanup objectives in the on-site groundwater.

The overall attenuation rate (i.e., the change in a solute's concentration in groundwater over time) includes all processes acting to reduce the concentrations of the contaminants, such as advection, dispersion, dilution from recharge, sorption,

and biodegradation. The overall attenuation rate can often be described by a first-order rate constant. In one dimension, first-order decay is described by the following ordinary differential equation:

$$\frac{dC}{dt} = -kC \tag{10-21}$$

where $C =$ concentration in groundwater at time t (mg/L)

$k =$ overall attenuation rate (first-order rate constant) (day^{-1})

The solution for this equation is:

$$C = C_0 e^{-kt} \tag{10-22}$$

where $C_0 =$ concentration at time zero.

This is the same as Equation 10-16 presented in Section 10-3, with C substituted for S. With this equation, the overall attenuation rate can be estimated on the basis of the changes over time in contaminant concentration in groundwater samples collected from individual monitoring wells at a site. The procedure consists of plotting the natural log of the ratio C/C_0 versus time, and calculating the slope of the best-fitting line.

Since biodegradation is the most important destructive attenuation mechanism, estimating the site-specific biodegradation rate is critical to documenting loss of contaminants. To determine the portion of the overall attenuation rate that can be attributed to biodegradation, the effects of abiotic processes such as advection, dispersion, dilution from recharge, and sorption must be subtracted from the total attenuation rate. The abiotic effects can be subtracted using either data analysis or biodegradation rates measured during microcosm studies. The EPA states in its technical protocol that microcosms should be used to estimate biodegradation rates only when it is impossible to estimate the rate of attenuation from data obtained by monitoring wells in the plume.[102]

The EPA's technical protocol presents two methods of calculating biodegradation from data from monitoring wells in the plume.[103] One method involves the use of a normalized data set (i.e., a set of data that has been corrected to exclude the effects of abiotic processes) to compute a decay rate. The second method was derived by Buscheck and Alcantar and is valid for steady-state plumes.[104] Others have suggested the use of screening solute transport models such as BIOSCREEN, BIOCHLOR, and BIOPLUME III to iteratively fit plume characteristics as a means of calculating biodegradation rates.[105,106]

Normalized data set

A normalized data set[107] can be developed by using tracers. These are compounds or elements associated with the contaminant plume that are relatively unaffected or predictably affected by biologic processes occurring within the aquifer. Trimethylbenzene isomers, which can be biologically recalcitrant, may be used as a tracer for fuel hydrocarbon plumes. Chloride or the carbon nucleus of the chlorinated compound can also be used as a tracer. The corrected concentration of a compound is the concentration that would be expected at one point (B) located down-gradient from another point (A) if the processes of dispersion and dilution had not been occurring between points A and B.

$$C_{B,\text{corr}} = C_B(T_A/T_B) \tag{10-23}$$

where $C_{B,\text{corr}}$ = corrected contaminant concentration at a down-gradient point B
C_B = measured contaminant concentration at point B
T_A = tracer concentration at a point A up-gradient
T_B = tracer concentration at point B down-gradient

Then, the biodegradation rate can be estimated between any two points (A and B) of a normalized data set as follows:

$$C_{B,\text{corr}} = C_A e^{-\lambda t} \tag{10-24}$$

where $C_{B,\text{corr}}$ = normalized contaminant concentration at down-gradient point B
C_A = contaminant concentration at up-gradient point A; if point A is the first point in the normalized data set, then $C_A = C_{A,\text{corr}}$
λ = first-order biological decay rate (first-order rate constant) (day^{-1})
t = time of contaminant travel between points A and B (day)

This relationship can be used to calculate the first-order biological decay rate constant between two points by solving Equation 10-24 for λ:

$$\lambda = -\ln\left(\frac{C_{B,\text{corr}}/C_A}{x/v_c}\right) \tag{10-25}$$

where x = distance between two points (m)
v_c = retarded solute velocity (m/day)

Method of Buscheck and Alcantar

This method[108] uses regression techniques and analytical solutions to calculate an apparent first-order biodecay (biodegradation) rate for steady-state plumes. The slope of the line formed by making a natural log-linear plot of contaminant concentration versus distance down-gradient along the flow path is used in the following equation:

$$\lambda = (v_c/4\alpha_x)\left\{[1 + 2\alpha_x(k/v_x)]^2 - 1\right\} \tag{10-26}$$

where λ = first-order biological rate constant for a steady-state plume (day^{-1})
v_c = retarded contaminant velocity in the x direction (m/day)
α_x = dispersivity (m)
k/v_x = slope of line formed by making a natural log-linear plot of contaminant concentration versus distance down-gradient along flow path

Some researchers advise that rates calculated using the Buscheck and Alcantar method should always be substantiated, since the analysis may yield misleading results, especially if an insufficient number of wells is included in the analysis or if the plume is not at steady state.[109]

EXAMPLE 10-12. A site has the following concentrations of an organic compound. The retarded contaminant velocity in the down-gradient direction is 30 m/year and the

corresponding dispersivity is 15 m. Determine the attenuation rate using the Buscheck and Alcantar method.

Distance, m	0	11	20	30	40	51	60	70	80
Organic compound concentration, mg/L	300	201	150	112	90	74	60	52	43

The Buscheck and Alcantar method uses the following equation to calculate the degradation rate:

$$\lambda = (V_c/4\alpha_x)\left\{[1 + 2\alpha_x(k/V_x)]^2 - 1\right\}$$

Data: $V_c = 30$ m/year; $\alpha_x = 15$ m.

Solution. A plot of the concentrations with distance is shown below. From this plot, the value of k/V_x (the slope of plot of concentration vs. distance) is:

$$k/V_x = -0.00235/\text{m}$$

With this value, λ can be calculated as:

$$\lambda = [30/4(15)]\{[1 + (2)(15)(-0.00235)]^2 - 1\}(1\ \text{year}/365\ \text{days}) = 0.0019/\text{day}$$

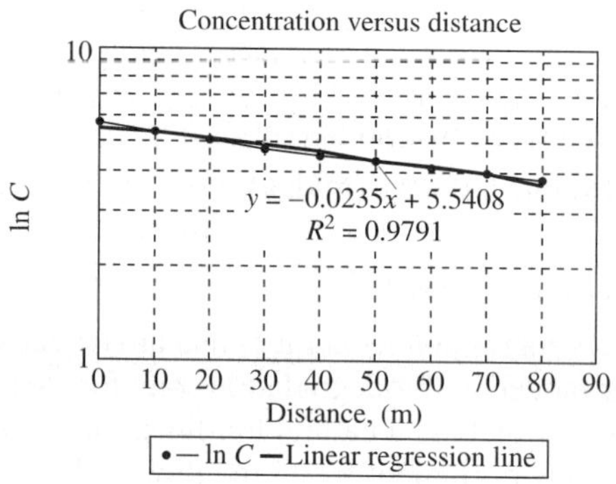

Modeling

Screening solute transport models such as BIOSCREEN,[110] BIOCHLOR,[111] and BIOPLUME III[112] can be used to iteratively fit plume characteristics as a means of calculating biodegradation rates. Descriptions of these models can be found in the indicated references.

Evaluation of Time Required for Remediation

The same models mentioned for calculation of attenuation rates can be used to model the future behavior of the plume to estimate the time required for MNA to reduce the contaminant concentrations to the selected cleanup objectives. In addition, a number of other solute transport models, such as RT3D[113] and BioRedox[114] can also be used. The EPA has indicated that time frames for MNA are considered “reasonable” (and,

thus, MNA may be used for remediation of the site) if they are no more than 2 to 3 times longer than the time required by other remedial actions.

Long-Term Monitoring

Long-term monitoring is an essential component of a natural attenuation as a remedy. The location, number, and other pertinent parameters regarding monitoring wells for both the evaluation of MNA and long-term monitoring of MNA should be determined on a site-specific basis. Monitoring well networks will typically include wells located in the source area, wells located down-gradient along the centerline of the flow path, and wells located on transects across the contaminant plume that measure and monitor the width of the plume (see Figure 10-27). In addition, in most cases a site will require one or more down-gradient wells between the plume edge and the receptors, where concentration of contaminants in groundwater samples either cannot be detected or are at levels lower than their cleanup objectives. These wells are considered "sentinel wells," because they provide a means of detecting the migration of the plume front.

Case Studies

The EPA and other researchers have widely published the results of the analysis of natural attenuation at several fuel hydrocarbon and chlorinated ethene release sites. Fuel hydrocarbon release sites include Hill Air Force Base, Utah[115,116]; Patrick Air Force Base in Florida[117]; U.S. Coast Guard Air Station, Traverse City, Michigan[118]; and a gas condensate–contaminated aquifer.[119] Chlorinated ethene release sites include the St. Joseph Michigan Superfund Site[120,121]; Aberdeen Proving Grounds in Maryland[122]; Plattsburg Air Force Base in New York[123]; and Dover Air Force Base

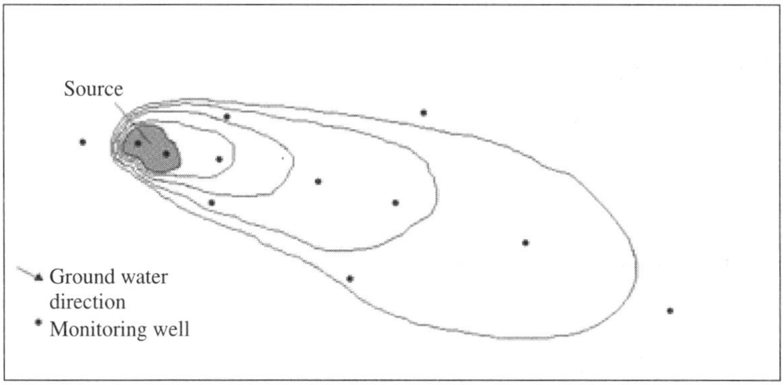

FIGURE 10-27
Example plume-monitoring well network.

in Delaware.[124,125] These case studies describe site conditions, type of contaminants, lines of evidence used, and design of the MNA remediation system.

Costs

The most attractive advantage to implementing MNA rather than an intrusive remediation is the potential for lower project costs. Table 10-15 lists several Superfund sites where MNA was considered as an alternative or included as part of an alternative.[126] The table shows the estimated capital and O&M costs and time to complete remediation for each alternative remediation approach presented in the site's Record of Decision. Although implementing MNA typically includes collection of more samples more frequently, at more locations, and for more analyses, there is still considerable cost advantage over groundwater extraction and treatment. Despite the significant cost advantage, it is not an acceptable alternative at sites where the groundwater plume presents unacceptable risks or where the longer remediation time is not acceptable.

EXAMPLE 10-13. Assume a groundwater plume containing tetrachloroethylene is being naturally attenuated at a rate of 0.009/day. The distance from the source (where the tetrachloroethylene concentration is 10 mg/L) to the receptor, where the concentration cannot exceed 0.005 mg/L, is 300 m. (a) If the groundwater velocity is 0.4 m/day, can MNA be used at the site? (b) What would you consider to be a safe distance to the nearest receptor to be able to use MNA under these conditions?

Data:

Attenuation rate = 0.009/day
Distance to receptor = 300 m
Attenuation rate = 0.009/day
Concentration at the source = 10 mg/L
Maximum concentration at the receptor = 0.005 mg/L (this is the maximum contaminant limit for drinking water)
Velocity of the contaminant in groundwater = 0.4 m/day

Solution

$$\begin{aligned}\text{Transport time} &= 300\text{ m}/0.4\text{ m/day}\\ &= 750\text{ days}\end{aligned}$$

Calculated concentration at the receptor:

$$\ln C/C_0 = -kt$$

or

$$\ln C = \ln C_0 - kt = \ln 10 - 0.009(750)$$

Therefore, $C = 0.0117$ mg/L and MNA is not acceptable.

For an acceptable case, the distance to the receptor would have to be such that C would be at least 1 order of magnitude less than the target concentration (MCL):

$$\begin{aligned}t &= -(\ln C/C_0)(1/k)\\ &= -(\ln 10/0.0005)(1/0.009)\\ &= 1100\text{ days}\\ \text{Safe distance} &= (1100\text{ days})(0.4\text{ m/day})\\ &= 440\text{ m}\end{aligned}$$

TABLE 10-15

Comparison of remedial alternative costs from Records of Decision at several sites considering MNA

Site	Source type	General hydrogeology	Groundwater contaminants	Alternatives*	Capital cost†	Annual O&M†	Time to remediate, years
Prestolite Battery, Vincennes, Indiana	Lead-acid battery manufacturing	Sands/gravels (0–22 ft bgs), organic silty clay aquitard (0–15 ft thick), sand/gravel unit (~18 ft)	Metals including antimony and beryllium and VOCs including PCE and TCE	MNA and institutional controls (deed restrictions and prohibition of water supply well installation in plume area)	$30,000	$30,000	20
				✓GW extraction, air stripping, and discharge to publicly owned treatment works and institutional controls as above	$460,000	$150,000	10
				GW extraction, air stripping, and discharge to publicly owned treatment works and institutional controls as above	$469,000	$171,000	10
Hanford 1100 Area, Benton County, Washington	Department of Defense facilities including former battery acid, paint and solvent, and antifreeze and degreaser pits and a landfill	Glacially deposited gravels and sands (16–44 ft thick); silt aquitard (~30 ft thick); and silt, gravel, and sand layers underlying	Large (>1 mile long) TCE groundwater plume. Various source areas with VOCs, TPH, SVOC, PCBs, pesticides, and metals	✓Source control, MNA, evaluation of further needs, installation of additional monitoring wells, and institutional controls including prohibition of water supply wells	$685,000	$24,300	30
				Source control and GW extraction at 100 gal/min with air stripping	$1,536,000	$256,300	17
				Source control and GW extraction at 300 gal/min with air stripping	$3,557,000	$505,000	13
				Source control and GW extraction at 100 gal/min with UV/chemical oxidation treatment	$2,072,000	$262,000	17
				Source control and GW extraction at 300 gal/min with UV/chemical oxidation treatment	$4,228,000	$538,000	13

TABLE 10-15
Continued

Site	Source type	General hydrogeology	Groundwater contaminants	Alternatives*	Capital cost†	Annual O&M†	Time to remediate, years
Juncos Landfill (OU2—Groundwater)‡ Juncos, Puerto Rico	Municipal and industrial landfill	Deeply weathered rock (saprolite) overlaying fractured bedrock (granodiorite)	Metals, carbon disulfide, chloroform, and pesticides	No action—natural attenuation without monitoring	$0	$0	13
				✓MNA and institutional controls including GW use restrictions	$51,624	$42,250	13
				GW extraction, metals removal and air stripping discharge to surface water, and institutional controls during the remediation including GW use restrictions	$867,802	$490,071	<13
PSC Resources Site, Palmer, Massachusetts	Abandoned waste oil and solvent reclamation facility	River valley flood plain composed of thick sequences of highly permeable alluvial sand and gravel deposits	Chlorinated solvents, ketones, BTEX, phthalates, phenols, and PAHs	✓Source control with MNA including long-term monitoring (30 years) of groundwater, wetland sediment, and surface water (Quaboag River) and sediment	$0	$38,000	4–11
				Source control with GW extraction via an interceptor/barrier recovery trench and treatment (metals and suspended solids removal and air stripping) and discharge to river	$793,634	$69,000	3–7
				Source control with GW extraction via extraction wells and treatment (metals and suspended solids removal and air stripping) and discharge to river	$425,130	$141,000	3–7

*The alternative with a check mark is the selected alternative for the site.

†Costs shown are for the groundwater portion of the remedial alternative only and do not include costs for source control.

‡The published Record of Decision is for the groundwater operable unit only. Source control is considered a separate operable unit and is described in a separate Record of Decision.

Key:

MNA = monitored natural attenuation; PCE = tetrachloroethene; TCE = trichloroethene; ft = feet; bgs = below ground surface; GW = groundwater; BTEX = benzene, ethyl benzene, toluene, and xylenes; VOCs = volatile organic compounds; TPH = total petroleum hydrocarbons; SVOC = semivolatile organic compounds; PCBs = polychlorinated biphenyls; PAHs = polynuclear aromatic hydrocarbons.

10-8 *IN SITU* BIOLOGICAL TREATMENT SYSTEMS—ENHANCED DEGRADATION

The previous section has shown that natural biodegradation can provide an effective method for remediation of subsurface contamination. Its major disadvantage is a long remediation time. Remediation can be accelerated by various *in situ* technologies for enhancing, stimulating, and managing the actions of subsurface microbiological communities.

Most of the enhanced degradation methods involve the conveyance of nutrient materials to and metabolic by-products away from indigenous microorganisms. This must be accomplished by using either a liquid or a gas as carrier fluids. Materials dissolved in water (such as organic nutrients, oxygen, and organic carbon) can be introduced to subsurface soils by gravity infiltration from the surface or by injection wells. Oxygen, organic nutrients (such as propane or butane), and certain inorganic nutrients (such as nitrogen in the form of ammonia) can be introduced to the subsurface as gases. Engineered systems can be designed to direct the flow of amended water or gas through the subsurface, and are described in this section. These processes help to optimize subsurface microbial growth conditions.

Management of Oxygen and Nutrients in the Subsurface

Groundwater naturally contains low concentrations of oxygen because of the minimal reaeration resulting from its laminar flow below the surface. Thus, even moderate biological activity readily depletes the oxygen in the groundwater. For example, the oxidation of 1 mg of gasoline theoretically requires 2.5 mg of oxygen.[127] Depending on the target hydrocarbon considered, even higher oxygen-to-hydrocarbon ratios (up to 3.5 to 1) are common.[128]

EXAMPLE 10-14. OXYGEN DEMAND FOR *IN SITU* TREATMENT. For the contaminated groundwater case described in Example 10-11, what is the mass of oxygen required to mineralize the gasoline?

Solution. From the stoichiometry of the degradation of simple hydrocarbons discussed earlier in this chapter (see Section 10-2, Microbial Growth Requirements), it can be seen that their mineralization requires about 2.5 g oxygen (O_2) per gram of hydrocarbon (HC).

$$\begin{aligned}\text{Oxygen demand} = {} & (2.5\text{ g O}_2/\text{g HC})(1\text{ kg O}_2/1000\text{ g O}_2)(5400\text{ kg HC} + 90\text{ kg HC}) \\ & \times (1000\text{ g HC}/1\text{ kg HC}) \\ = {} & 13{,}725\text{ kg O}_2\end{aligned}$$

The saturated zone is also typically low in the inorganic nutrients (primarily nitrogen and phosphorus) required by microorganisms for growth with high substrate concentrations. Because of these limitations, the concept of aerobic *in situ* bioremediation originally centered in the 1970s to the mid-1980s on the delivery of oxygen and nutrients to the saturated zone, typically by withdrawing groundwater, adding oxygen and nutrients, and reinjecting the enriched water. The injected water

with nutrients and oxygen then moved through the aquifer, stimulating the growth of native microorganisms and resulting in the degradation of the contaminants. Few applications of *in situ* bioremediation as practiced today had been made until the mid-1980s, when technical advancements and heightened interest prompted a much greater level of activity.

Since oxygen was quickly depleted in the subsurface, many efforts were made to develop practical methods of increasing groundwater oxygen levels. Rather than remove groundwater for aeration and reinjection, air sparging within wells was used to increase groundwater dissolved oxygen levels. However, this process was limited by the relatively low maximum dissolved oxygen concentration in water that is in contact with air (8 to 12 mg/L). One approach to overcome this solubility limitation was to add liquid hydrogen peroxide to wells, resulting in an increase in groundwater oxygen levels. However, concentrated hydrogen peroxide solution was expensive, and the peroxide readily decomposed in contact with soils containing various metals and other readily oxidizable materials. Chemical stabilizers were added to peroxide solutions to help minimize excessive decomposition, but peroxide treatment remained an expensive treatment option.

In the 1980s, soil vapor extraction (SVE) systems used the movement of air through subsurface soils to volatilize susceptible chemicals. This low-cost remediation technology also offered the benefit of providing a continuous source of oxygen to the subsurface soils, enhancing *in situ* biodegradation in the vadose zone. When SVE is optimized for biological treatment rather than volatilization, it is known as bioventing. The extension of SVE to the sparging of saturated zone groundwater and soils to volatilize contaminants also resulted in the addition of oxygen to these materials. In a like manner, when groundwater sparging is optimized for biological treatment it is known as biosparging. In recent years, proprietary magnesium peroxide compounds in solid or slurry form have been added to groundwater wells or subsurface soils to slowly release free oxygen over an extended period of time.[129] Aerobic cometabolism has also been exploited with the addition of oxygen and organic cosubstrates to the subsurface environment.

The initial reliance on aerobic *in situ* biotreatment technologies alone reflected the limited knowledge of anaerobic microbial ecology at the time, especially as it related to the biodegradation of hazardous organics. As this knowledge and experience base increased, the importance of anaerobic subsurface microbiology became apparent. Instead of biodegradation stopping when subsurface oxygen was depleted, anaerobic processes continued the beneficial treatment of many xenobiotic compounds. In some cases, anaerobic processes alone were required for the treatment of certain compounds, such as the anaerobic dechlorination of perchloroethene. Throughout the 1990s, much attention focused on exploiting anaerobic biological treatment processes. For example, a proprietary hydrogen release compound is now commercially available that slowly releases lactic acid when exposed to water. The lactic acid can be fermented by microorganisms, releasing hydrogen that is then used by other microorganisms to reductively dechlorinate various chlorinated aliphatic compounds.[130]

Listing of *in Situ* Biotreatment Methods

A number of biotreatment technologies have become standard practice in the hazardous waste treatment field. The predominant vadose zone biotreatment methods

are bioventing, natural attenuation, and phytoremediation, while the predominant saturated zone biotreatment methods include biosparging, natural attenuation, biobarriers, and possibly phytoremediation. Natural attenuation was discussed in the previous section. The other methods are described below. The *in situ* treatment of lagoon/impoundment materials is discussed separately at the end of this section.

It should be noted that many of the treatment methods listed above may be applied in either an aerobic or anaerobic mode. In addition, many actions short of these full processes can be implemented at a site to improve aerobic or anaerobic treatment conditions. For example, a nutrient solution can be applied to surface soils or within a trench for infiltration by gravity into subsurface soils. Anaerobic conditions can be created by flooding soil pores with water, effectively reducing or eliminating oxygen diffusion into the subsurface. Many other such simple treatment techniques are available for use.

Bioventing

General description

Bioventing is an *in situ* remediation technology that provides oxygen to indigenous microbial populations in order to stimulate the biodegradation of contaminants in the vadose zone. Bioventing is not used to treat contaminants in the capillary fringe or in the saturated zone. Because of effective performance as well as low capital and operating costs, bioventing has established itself as the preferred biotreatment method for the vadose zone.

There are two main configurations for bioventing: air injection and air extraction. Both configurations successfully provide oxygen to the vadose zone in a cost-effective manner. The air extraction mode operates like soil vapor extraction (SVE) (see Section 16-8). Figure 10-28 shows two typical bioventing system arrangements.

Extraction bioventing systems operate at much lower airflow rates than corresponding SVE systems. This reduces volatization of contaminants, allowing for increased biodegradation of the contaminants.

Bioventing has been most successfully used at sites with petroleum hydrocarbon contamination. Contaminants treated include BTEX (benzene, toluene, ethylbenzene, and xylenes), PAHs (polyaromatic hydrocarbons), phenols, and some chlorinated aliphatic compounds. High molecular weight and/or less volatile hydrocarbons such as diesel fuel and kerosene are better suited to treatment by bioventing since volatilization through soil vapor extraction alone has limited effectiveness.

Design considerations

A number of texts discuss the design of bioventing systems in detail.[131–134] The major factors affecting microbial processes in bioventing include[135]:

- Availability and type of electron acceptors
- Moisture content
- Soil temperature
- Nutrient supply
- Contamination concentration
- Bioavailability
- Biodegradation rates

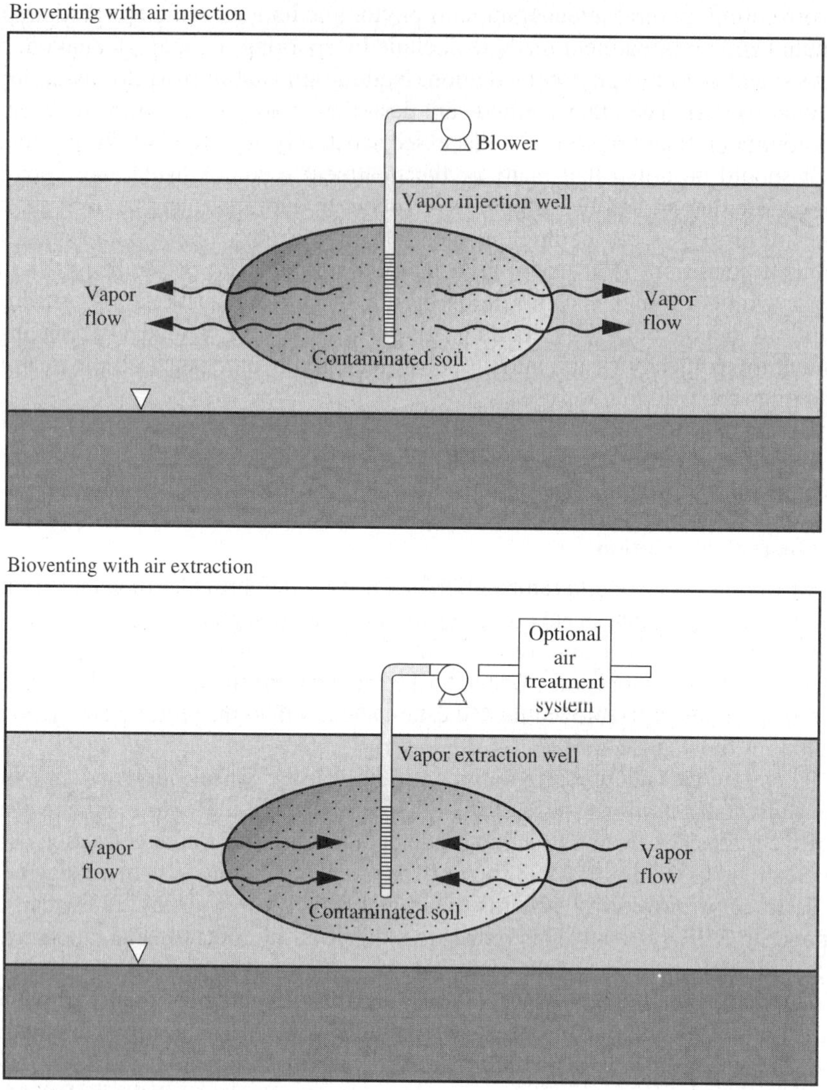

FIGURE 10-28
Drawing of typical bioventing systems.

Bioventing success depends on a proper site characterization and clear understanding of the nature of the contaminants. At a minimum, the site characterization needs to assess the site geology, hydrology, soil permeability, soil chemistry, and temperature. The selection of bioventing as a remedial alternative requires an understanding of the solubility, volatility, and biodegradability of each contaminant present.

Oxygen utilization rate. For aerobic biodegradation, oxygen is used as the terminal electron acceptor. For the treatment of petroleum hydrocarbons, a general

guideline is that aerobic biodegradation requires between 3.0 and 3.5 pounds of oxygen for every pound of hydrocarbon present. The rate at which air (and oxygen) is added to the subsurface in a bioventing program must fall between a determined range to be effective. The minimum amount of air usually meets that volume of oxygen utilized by indigenous microorganisms to enhance the aerobic degradation process. This volume, or **oxygen utilization rate,** is commonly determined from data obtained during pilot or full-scale bioventing tests, and it is calculated as the percent change in O_2 over time. To estimate biodegradation rates from the oxygen utilization rate, a stoichiometric relationship of the oxidation of the target hydrocarbon needs to be determined and used. Commonly, hexane is used as a representative hydrocarbon, and its stoichiometric relationship (ignoring any use of hexane for cell synthesis) is:

$$C_6H_{14} + 9.5O_2 \rightarrow 6CO_2 + 7H_2O$$

From the utilization rates (change of oxygen in percent per day) the biodegradation rate in terms of milligrams of hexane-equivalent units per kilogram of soil per day can be estimated by using Equation 10-27:

$$K_{\text{bio}} = K_{\text{ox}} V D_{\text{ox}} M_{\text{H:O}}/100 \tag{10-27}$$

where K_{bio} = biodegradation rate (mg/kg per day)
K_{ox} = oxygen utilization rate (percent per day)
V = volume of air per unit mass of soil (L/kg)
D_{ox} = density of oxygen in air assuming a 21.5% oxygen content (mg/L)
$M_{\text{H:O}}$ = mass ratio of hydrocarbon to oxygen required for mineralization

Assumptions generally used in completing Equation 10-27 are:

Porosity of soil: 0.15 to 0.35 for the vadose zone (varies based on soil type and moisture content)
Soil bulk density: 1440 kg/m^3 (varies with actual soil composition)
Density of oxygen in air: 1330 mg/L (varies with temperature and atmospheric pressure)
Mass ratio of hydrocarbon to oxygen: 1/3.5 based on hexane stoichiometry (varies with mineralization stoichiometry of specific hydrocarbon used)

Table 10-16 presents oxygen uptake rates for various hydrocarbon contaminants.[136]

As an alternative to calculating biodegradation rates from oxygen consumption, carbon dioxide generation rates have also been used. Since carbon dioxide can also be released abiotically from the subsurface, biodegradation rates based on carbon dioxide generation must be corrected for the abiotic contribution. Alkaline soils also absorb carbon dioxide, making generation rates more difficult to interpret.

Air permeability. A practical consideration in injecting the air is to assess how much is lost to dispersion or flow along preferential pathways. An *in situ* air permeability test is a companion to the respirometry test. The air permeability test can vary in scale from using a simple injection and monitoring point to a complex monitoring network incorporating an inert gas. The ultimate goal of the air permeability test,

TABLE 10-16
Oxygen uptake rates

Contaminant	*In situ* O_2 respiration, % O_2/day	Estimated biodegradation rate, mg hexane equivalents/kg/day
JP-4 jet fuel	9.6	7.5
JP-5 jet fuel	3.1	2.4
JP-5 jet fuel	6.0	4.7
JP-4 jet fuel	13.2	10.6
Mixed fuels	16.3	13.0
Solvents (nonchlorinated)	4.6	3.6
PAHs	9.1	7.0
Crude oil	21.1	16.8

however, is to assess subsurface airflow, and define preferential pathways, anisotropy, and vapor movement with increased pressure. The maximum amount of air to be injected to the subsurface is limited by safety considerations. Air should not be introduced at pressures and flows that would promote vapor migration away from the treatment area, possibly toward sensitive receptors.

The air permeability equations used in this section (Equations 10-28, 10-29, and 10-30) were developed by the U.S. Air Force as part of a technical protocol for bioventing.[137] The dynamic method of calculating air permeability requires that air be extracted or injected at a constant rate from a single well while measuring the pressure changes at several soil gas monitoring points. Under these conditions, the permeability to vapor flow can be calculated when Q and m are known and the calculated slope of P' versus $\ln t$ is used:

$$k = \frac{Q\mu}{4A\pi m} \tag{10-28}$$

where A = slope of the straight line P' versus $\ln t$, where P' is the gauge pressure [g/(cm · s^2)] measured at a distance from the vapor extraction and time is t in seconds
k = soil permeability to airflow (cm^2)
m = stratum thickness, generally the well screen interval (cm)
r = radial distance from monitoring point to well (cm)
μ = viscosity of air [1.8×10^{-4} g/(cm · s) at 18°C]
t = time from the start of the test (s)
Q = volumetric flow rate from the extraction well (cm^3/s)

Air permeability can also be calculated using a steady-state testing method when the dynamic method cannot be used. For wells operating under vacuum, the dynamic air permeability is calculated from Equation 10-29:

$$k_{sgp} = \frac{Q\mu \ln(R_w/R_i)}{H\pi P_w[1 - (P_{atm}/P_w)]} \tag{10-29}$$

where k_{sgp} = soil gas permeability (cm^2)
μ = dynamic viscosity of air [1.8×10^{-4} g/(cm · s) at 18°C]
Q = volumetric flow rate from/to the well (cm^3/s)

P_{atm} = ambient pressure [at sea level 1.013×10^6 g/(cm · s^2)]
H = length of screen (cm)
R_i = maximum radius of influence at steady state (cm)
P_w = absolute pressure at the vent well [g/(cm · s^2)]
R_w = radius of the vent well (cm)

For wells operating under injection pressure, the dynamic air permeability is calculated from Equation 10-30:

$$k_{sgp} = \frac{Q\mu \ln(R_w/R_i)}{H\pi P_w[1 - (P_w/P_{atm})^2]} \tag{10-30}$$

When air injection pressure and flow are in balance to the target area, offgas treatment is avoided and lower operating costs are realized. Bioventing is commonly used to treat volatile organic compounds (VOCs) and semivolatile organic compounds (SVOCs). However, it can also be used to treat subsurface contaminants that are biodegradable but nonvolatile (e.g., some phenols and phthalates).

Design practices

The following points highlight common testing and assessment work for bioventing projects:

- Conduct respirometry test
- Conduct air permeability test
- Assess preferential pathways
- Calculate the radius of influence and effects of preferential pathway to identify well spacing and number of wells
- Calculate the time to pulse aeration events (correlate to oxygen utilization and dispersion)
- Assess injection air pressure/flow relationship to avoid pushing soil vapors to potential receptors
- Install pressure and vapor monitoring probes (in stratified lithology use nested probes)
- Conduct bioassay and biodegradation test

Limitations

Although bioventing has gained rapid acceptance, the process has certain limitations. The most important of these limitations include:

- The water table cannot be within several feet of the surface.
- Saturated soil lenses and/or low permeability soils reduce performance.
- Vapors can build up in basements within the radius of influence of air injection wells. This problem can be alleviated by extracting air near any structure of concern.
- Extremely low soil moisture content may limit biodegradation.
- Monitoring of the off gases at the soil surface may be required.
- Low soil temperatures may slow biodegradation.
- Aerobic biodegradation of many polychlorinated compounds may not be effective unless there is a primary substrate present. As an alternative, an anaerobic treatment period may be needed.

EXAMPLE 10-15. Bioventing is being considered for use at a site with gasoline-contaminated soils. As a prelude to the final selection of the technology, air pressure, oxygen influence, and *in situ* respirometry tests were performed. Air was injected into the subsurface at three different rates: 100 L/min, 250 L/min, and 520 L/min. Helium gas was also injected as part of the respirometry test. Monitoring points were positioned 2, 4, 6, 8, 12, and 14 m from the air injection point. Data generated from the three tests are summarized in the tables below.

Pressure test (mbars above 1 atm)

Well	Distance, m	Initial Pressure, mbar	100 L/min at 5 mbar	250 L/min at 16 mbar	520 L/min at 27 mbar
BW1	2	0	1	2	2.8
BW2	4	0	0.7	0.3	0.6
BW3	6	0	0.4	0.7	0.9
BW4	8	0	0.28	0.4	0.49
BW5	12	0	0.12	0.14	0.16
BW6	14	0	0	0	0

Oxygen influence test

Well	Distance, m	Initial O_2 reading	100 L/min at 5 mbar	250 L/min at 16 mbar	520 L/min at 27 mbar
BW1	2	2	19.6	19.8	20
BW2	4	5	18.4	18.8	19
BW3	6	3	17.2	17.6	17.8
BW4	8	3	16	16.6	16.8
BW5	12	5	7	9.2	10
BW6	14	4	4	5.0	5.4

***In situ* respirometry test**

Lapse time, h	% O_2	% CO_2	% He
0	16.0	0	1.50
2	15.6	0.4	1.45
4	14.8	0.6	1.45
6	14.0	1.0	1.45
12	12.0	1.2	1.40
14	11.6	1.4	1.35
16	10.8	1.6	1.35
24	8.6	1.8	1.25
26	7.8	1.9	1.25
28	7.0	2.0	1.2
34	5.2	2.2	1.15
36	4.6	2.4	1.15
42	2.6	2.8	1.1
52	2.0	3.0	1.1

Soil gases from the contaminated zone showed the presence of gasoline. The soil gas oxygen levels in the contaminated zone ranged from 2 to 5 percent, while the soil gas oxygen levels in uncontaminated soils ranged from 19 to 20.1 percent (data not shown in the table).

Using the information given, answer the following questions:

(a) Is bioremediation a viable treatment technology?
Results from the *in situ* respirometry test can be used to answer this question. If aerobic respiration occurs, a decrease in the percent oxygen level would be expected along with an increase in the percent carbon dioxide level. These results are observed over the 52-h test.

The helium tracer gas was injected with air at the start of the respirometry test. Detection of helium during the test indicates that the subsurface gas sampled is the same gas that was initially injected. This infers that changes to the composition of the gas are the result of subsurface biological activity in the test zone and not the influx of new subsurface gases. Qualitatively, the small change in helium levels observed (1.5 to 1.1 percent) over the 52-hour period supports the idea that the oxygen decline is a function primarily of microbial oxygen utilization.

According to this information, aerobic bioremediation via bioventing is a viable vadose zone treatment technology.

(b) What is the radius of oxygen influence?
Results from the oxygen influence test can be used to answer this question. The radius of oxygen influence is defined as the maximum distance from the air injection well at which subsurface oxygen levels are at the minimum concentration needed for aerobic biodegradation. On a practical basis, the radius of oxygen influence can also be defined as the distance from the air injection well that oxygen levels can be noticeably increased above a given background concentration.

From the problem information given above, the soil gas oxygen levels in the contaminated zone ranged from 2 to 5 percent. Therefore, monitoring point subsurface oxygen levels must be increased above 5 percent to be considered within the radius of influence.

From the table, at an airflow injection rate of 100 L/min, subsurface oxygen levels were greater than 5 percent up to a distance of 12 m from the injection well. For the 250 L/min and 520 L/min injection rates, subsurface oxygen levels were greater than 5 percent at a distance of 14 m from the injection point.

According to this information, the oxygen radius of influence is 12 to 14 m, depending on the injected airflow rate and injection pressure. (*Note:* From this analysis, the spacing between bioventing wells would be 24 m, assuming preferential flow does not exist.)

(c) What is the oxygen utilization rate?
The *in situ* respirometry test data are used to answer this question. The oxygen utilization rate is typically defined as the percent of oxygen consumed per day (in a 24-h period). From the respirometry test table, at the beginning of the test (zero lapsed hours) the oxygen concentration was 16 percent. Twenty-four hours later, the oxygen concentration was reduced to 8.6 percent. By difference, the oxygen utilization rate is 7.4 percent per day. (The difference in the percent oxygen level between 4 and 28 h is 7.8 percent, while the difference between 12 and 36 h is 7.4 percent).

Thus the oxygen utilization rate is approximately 7.5 percent per day.

(d) What is the biodegradation rate?
Equation 10-27 can be used to calculate the biodegradation rate in terms of milligrams of hexane-equivalent units per kilogram of soil per day:

$$K_{bio} = (K_{ox})(V)(D_{ox})(M_{H:O})/100$$

Assuming a soil porosity of 20 percent, a density of oxygen in air of 1330 mg/L, a mass ratio of hydrocarbon (hexane) to oxygen of 1 to 3.5, and a soil density of 1600 kg/m^3, the values used in Equation 10-27 are:

$$K_{ox} = 7.5/\text{day}$$

A cubic meter contains 1000 L, so a 20 percent soil porosity corresponds to 200 L/m^3. The soil density is 1600 kg/m^3. Since V is defined as the liters of gas per kilogram of soil,

$$V = (200\ \text{L/m}^3)/(1600\ \text{kg/m}^3) = (0.125\ \text{L/kg})$$

$$D_0 = 1330\ \text{mg/L}$$

$$M_{H:O} = 1/3.5 = \text{ approximately } 0.3$$

So,

$$K_{bio} = (7.5/\text{day})(0.125\ \text{L/kg})(1330\ \text{mg/L})(0.3)/100$$

The biodegradation rate is approximately 3.75 mg/kg/day.

(e) What should the operating flow rate be?
The pressure and oxygen radius-of-influence tests both suggest that an effective radius of influence can be achieved by any of the flow/pressure combinations used. However, to reduce the possibility of pushing vapors away from the treatment zone, the lowest flow, 100 L/min, would be the safest choice. In addition, a fivefold increase in flow from 100 L/min to 520 L/min would only add an additional 2 m of influence.

(f) What is the system cycling time based on oxygen utilization?
Assume an injection cycle can achieve 20 percent oxygen saturation and that 5 percent oxygen saturation is the minimum threshold for aerobic conditions. The oxygen saturation can be reduced from 20 percent to 5 percent before it will have to be increased via renewed air injection, a difference of 15 percent. Since approximately 7.5 percent oxygen is consumed over a 24-h period (see part c), oxygen would need to be replenished:

$$(20\% - 5\%)/(7.5\%/\text{day}) = 2\ \text{days}$$

Biosparging

General description

Biosparging is an *in situ* remediation technology that involves the injection of air under pressure into the saturated zone to promote biodegradation in both the saturated and unsaturated zones. Volatilization of the contaminant is a secondary objective. Because of upward and lateral movement of the volatized contaminants, biosparging is used in conjunction with other technologies, usually SVE systems, to remove the volatized contaminant from the unsaturated zone. Offgas treatment may be required for the extracted vapors, depending on the site conditions and system design. A drawing of a typical biosparging system is shown in Figure 10-29. Biosparging systems have been described in detail in several texts.[138–140]

Design considerations

The maximum amount of air injected in a biosparging system is generally limited to the saturation limit of dissolved oxygen in groundwater. On a practical basis,

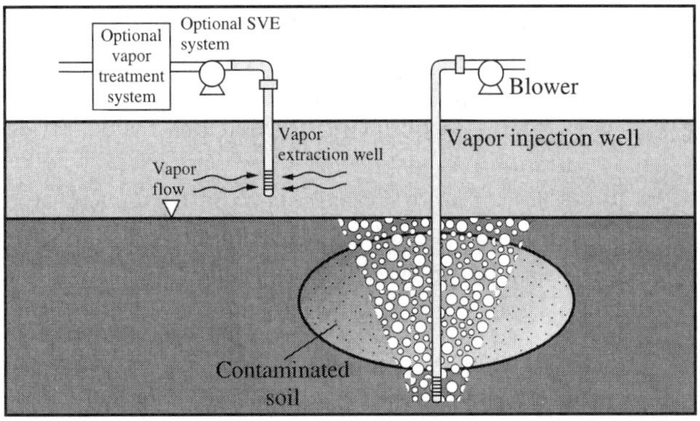

FIGURE 10-29
Biosparging system.

this is usually between 5 and 6 mg/L. Injecting a greater volume of air than the dissolved oxygen saturation capability of groundwater would make it necessary to consider the effect of increased volatilization. In addition, the volume of air as a function of pressure and flow must be considered as part of the safe operation of the system. Air should not be introduced at pressures and flows that would push dissolved-phase hydrocarbons away from the treatment zone or generate vapors in areas of sensitive receptors. Also, because of the compressibility of air and reduced hydrostatic pressure created during sparging, a site's geotechnical stability should be assessed. It is important to consider that 98 mbar of pressure is required to overcome 1 m of groundwater submergence before airflow will occur in the saturated zone.

As with bioventing, biosparging needs to be assessed for loss by dispersion of flow along preferential pathways. The scale of this depends on the complexity of the geology, hydrology, and subsurface structures such as utility corridors. The ultimate goal of this test is to assess subsurface airflow and to define preferential pathways, anisotropy, and hydrocarbon movement (dissolved and vapor phase) with increased pressure and flow. The ideal goal of a biosparging program is to inject air for only that time required to increase the dissolved oxygen level to saturation, then turn off the system and allow biodegradation to occur for the period until aeration is needed again.

Design practices

The major steps in designing a biosparging system are:

- Calculate breakout pressure (the minimum pressure required to overcome hydrostatic pressure).
- Assess injection airflow versus pressure relationship to avoid pushing dissolved-phase hydrocarbons and minimizing volatilization.
- Conduct a soil vapor extraction test for recovery of offgas due to volatilization of VOCs. This is principally a safety feature since the typical SVE safety design factor is 3 times the airflow of a sparged air system. An SVE system design would be needed for companion operation with the biosparge system.
- Conduct a sparge test and calculate the radius of influence and preferential pathways to define well spacing. The radius of influence is defined by dissolved

oxygen concentration changes but should also consider groundwater mounding, pressure, and how VOC concentrations change with distance from the sparge well.

- Install pressure and vapor monitoring probes (use nested probes in stratified lithology).
- Calculate time for aeration periods (correlate to time to reach saturation or target dissolved oxygen level and account for effect of dispersion).
- If necessary, conduct bioassay and/or biodegradation tests.

Limitations

The most important limitations to biosparging include:

- Contaminants that form complexes with the soil matrix have decreased volatization rates.
- Low-permeability soils prevent the free flow of air through groundwater and in vadose zone soils.
- Low-permeability layers overlying the aquifer prevent upward movement of volatized vapors.
- Channeling of air through heterogeneous soils.
- Sites where contaminated groundwater is less than 5 ft below ground surface.
- Possible clogging or plugging of soils due to increased iron precipitation or accumulated biomass.

Despite these limitations, biosparging has been proved effective for many saturated zone bioremediation projects.

Phytoremediation

General description

Since the mid-1990s a new *in situ* process involving the use of plants has been evolving as a cost-effective means to remediate hazardous constituents at certain disposal and spill sites. Actually, interest in the use of plants to treat wastes was developing as early as 1967 out of investigations of a plant's ability to uptake and translocate contaminants.[141] Phytoremediation is attracting increased attention by research groups, technology vendors, industry, and the EPA.

Phytoremediation is most applicable for the treatment of vadose zone soils. Depending on the plant or tree root depth, the process may also be suitable for the treatment of shallow saturated zone soils and groundwater. Phytoremediation is considered an *in situ* biotreatment process because it does not require the excavation of vadose/saturated zone soils to effect subsurface treatment. However, to the phytoremediation field, the term *in situ* refers to plant processes that take place below the ground surface, while *ex situ* treatment is used to describe plant processes that take place above ground surface.

As discussed here, **phytoremediation** is the use of plants to facilitate, either directly or indirectly, the attenuation of hazardous constituents present in soil. There are a number of phytoremedial approaches that vary dependent on a number of site-specific factors including (but not limited to) contaminant type, site and contaminant suitability for phytoremediation, geographic location, and topography. Phytoremediation has some economic advantages; however, the regulatory and technical obstacles

can be formidable. The success of a phytoremediation project requires a multidisciplinary approach including, at a minimum, botany, plant physiology, soil chemistry, soil microbiology, ecotoxicology, and possibly special laboratory techniques.

Some plants possess unique properties that can be used to reduce the toxicity or mobility of waste constituents. Phytoremediation has been at least partially effective in the cleanup of metals, pesticides, solvents, explosives, crude oil, polycyclic aromatic hydrocarbons, and landfill leachates at carefully selected sites.[142] Plant species used in phytoremediation may be selected based on their ability to draw constituents of concern up into the plant by the root system where they can be subsequently transported into plant stems and leaves, and ultimately harvested and removed from the site. In an alternative technique, chemicals exuded by plants or associated microbes may act on chemicals and alter/immobilize constituents either in the root zone *(in situ)* or within the plant stem and leaves *(ex situ)*.

Applicability of phytoremediation

In order to remediate contaminants successfully by phytoremediation, a number of factors need to be considered, including the contaminants of concern, existing vegetation, and potential risk to human health and wildlife. If phytoremediation is found to be a reasonable remediation method, the type of plants used to remediate a site will depend, in part, on the contaminants of concern. Different plant species are appropriate for different types of contaminants.

Remediation of metals. Metal contaminants may be treated by phytoextraction /phytoaccumulation (P/A)[143] or phytostabilization.[144] **Phytoextraction/phytoaccumulation** is the uptake and translocation of metal contaminants from the soil into the plant matter via plant roots.[145] The plants act as filters or traps for the metals and remove them from contaminated soils. Once these plants have performed their function they are harvested and incinerated with waste ash disposed of at a hazardous waste landfill. The harvested plants may also be recycled for metal extraction. The plants are replanted for as many cycles as needed to decrease risk to acceptable levels. Plants that are being researched for potential use in heavy metal treatment include Indian mustard *(Brassica juncea)*, oats *(Avena sativa)*, barley *(Hordeum vulgare)*, and alfalfa *(Medicago sativa)*.[146] **Phytostabilization** is the interaction between plants and contaminants that retards the rate of contaminants leaching to groundwater. This can be achieved through a variety of ways (e.g., chemical processes such as pH control, physical means such as increased sorption, or reduction of rainfall infiltration through increased evapotranspiration). For these processes to be successful in remediating metals it may be necessary to add nutrients such as phosphates or lime to the soil.[147]

Remediation of organics. Techniques for the phytoremediation of organic compounds differ significantly from those used for inorganic (metal) contaminant cleanup. Organic compounds are metabolized to simpler molecules and/or elemental constituents, while biological action on metals changes their oxidation state (and hence their mobility). Treatment of organics can occur through phytovolatilization, plant-assisted bioremediation (also called phytostimulation or enhanced rhizospheric biodegredation), and phytotransformation/phytodegradation.

Phytovolatilization occurs when water and contaminants are transported from the ground, through the plant, and into the atmosphere by evapotranspiration. Phyto-

volatilization therefore is not a contaminant destruction process, but rather a cross-media transfer of contaminants.

Plant-assisted bioremediation occurs when roots exude chemicals into the rhizosphere (the soil containing roots and their associated bacteria and fungi) that assist the natural microbial breakdown of contaminants. Exudates include enzymes, aliphatic and aromatic compounds, amino acids, sugars, and low molecular weight carbohydrates, as well as mucigel.[148,149] These exudates create conditions that are favorable to fungi and microbial growth and colonization, whose actions degrade organic compounds. Plants typically used in plant-assisted phytoremediation include mulberries *(Morus sp.)*, apple trees *(Malus sp.)*, osage orange *(Maclura pomifera)*, and grasses with fibrous roots, including rye *(Elymus sp.)*, fescue *(Festuca sp.)*, and Bermuda *(Cynodon dactylon)*.[150,151]

Phytotransformation occurs when contaminants are broken down and taken into the plants via metabolic processes and incorporated into the plant tissue as nutrients.[152] The process of drawing contaminants into the roots, stems, and leaves of the plant where they are converted to other chemical forms is considered *ex situ* remediation within the phytoremediation field. Species commonly used in phytotransformation include grasses (rye, Bermuda, and fescue), phreatophyte trees (poplar, willow, cottonwood, and aspen), and herbaceous species (legumes, clover, and alfalfa).

The determination of which phytoremediation technique to use partly depends on the ability of the contaminant to be affected by plant activity. Prediction of a plant's ability to treat contaminants is often related to water solubility. The K_{OW}, octanol water partition coefficient, of a contaminant is a measurement of the contaminant's affinity hydrophobicity (see Section 4-3). The higher the K_{OW} of the contaminant the higher its ability to be absorbed by the plant. The K_{OC} of the plant indicates the contaminant's affinity to sorb to organic carbon (present in soil particles). The higher the K_{OC} the more strongly the contaminant is sorbed to the soil and not available for uptake into the plant. Generally, the likelihood of phytoremedial success is high for contaminants soluble in water. The K_{OW} and K_{OC} can be calculated by using the Briggs equation as follows:

$$\log\,(\text{water solubility}) = 0.84 - 1.18 \log K_{OW} \text{ (for liquids)} \tag{10-31}$$

The K_{OC} can then be calculated[153]:

$$\log K_{OC} = 0.52 \log K_{OW} + 0.62 \tag{10-32}$$

The proper phytoremediation technique is selected not only on the basis of the constituent of concern but also on the basis of the media contaminated. Different techniques are used in different media. Of the techniques listed above, phytoextraction, phytostabilization, phytovolatilization, phytotransformation, and plant-assisted bioremediation are all useful techniques in the remediation of contaminated soils. **Rhizofiltration** is useful in the remediation of water. For example, sunflowers have been used to treat radioactive waters from the Chernobyl nuclear plant in Ukraine. Constructed wetlands have been used to treat waters from ammunition plant wastes.[154]

Design considerations

Plants that are the most useful in the remediation of organic compounds have a high amount of biomass, have a high tolerance for concentrated contaminants,

produce exudates, have long growing seasons, grow quickly, and have roots that extend to the depth of the contamination. Research on appropriate plant species is just beginning to identify viable species-contaminant combinations. Using native species is preferential to using nonnatives because of their proved ability to establish in the site-specific ecosystem.

For phytoremediation purposes, useful biomass consists of the parts of the plant that are available for evapotranspiration. Biomass includes both the leaf surface area and the plant root system. Trees may be more effective in remediation compared to grasses, in part because of their large biomass. The biomass of a plant may also be directly related to a decrease in contaminant leaching and a high rate of evapotranspiration. These side benefits of phytoremediation are themselves a viable site treatment method when the primary objective is merely to control the water-induced leaching of contaminants to groundwater.

Trees are also used in phytoremediation because of their depth of root penetration. Roots that can reach to the depth of contaminants are more beneficial to the remediation goal than are short/shallow root systems. Deep tree roots may also be able to reach a shallow saturated zone.

The success of a phytoremediation project can be shown by documenting reductions in soil contaminant levels and/or by observing the accumulation of contaminants in plant tissues. For example, a comparison of soil contaminant concentrations in two similar plots of contaminated land, one treated with plants and the other void of plants, can be made. If treatment is successful, the plant-treated soils will have lower contaminant concentrations than the untreated soils.

If project success is determined by the accumulation of contaminants in plant tissues, a different control is used for comparison. In this case, the concentrations of metals in tissue from plants grown on contaminated soil plots are compared with the metal concentrations in tissue from plants grown on a similar, uncontaminated soil plot. Only the tissue from plants grown on contaminated soils should show elevated metals concentrations.

To better ensure project success, bench or pilot studies may be required before field-scale phytoremediation efforts are implemented.

Design factors

Typical phytoremediation design factors include:

- Topography (grading)
- Appropriateness of planting mixed stands or single species (monoculture) stands
- Determining synergistic or adversarial effects in mixed-stands
- Appropriate spacing for planting (allowing for plant growth)
- Planting depth
- Degree of plant root penetration
- Growth rates of plants in various levels of contaminants
- Ability of plants to control water infiltration
- Potential impact of natural plant succession (the evolution through pioneer species from grasses to scrubs to trees)

For example, too much slope results in increased water runoff that creates a plant water deficit and plant mortality, while too little slope may cause ponding, which can adversely affect many species of plants and increase plant mortality.

Phytoremediation projects that include mixtures of both tree and grass species are less likely to be damaged by insect infestation and disease impacts. Grass species, in seed form, can be planted in order to allow for quick moisture control and effective cover over a large portion of soil. Nursery-grown trees could then be planted in either the winter or early spring, prior to fruiting, after the grasses have established. Finally, shade-tolerant grasses can be planted after tree establishment.

Monitoring plan. A monitoring plan is necessary to measure the success level of the designed phytoremediation project. Monitoring throughout the project can indicate the need for design modification, thus allowing the plan to evolve with the project, if necessary. The complexity of the monitoring plan depends on a number of factors, especially the site-specific risk posed by the presence of contaminants.

The monitoring plan should include erosion control, evapotranspiration, the effectiveness of degradation (contaminant reduction), and the process of succession. Erosion control can be measured by the presence or absence of airborne particulate emissions, and by the quality of water runoff. Evapotranspiration can be measured by assessing the water balance of the site (locally measure precipitation compared to the stormwater runoff rate), field observation data to estimate plant biomass (amount of surface area available for evapotranspiration), and the volatile organic compound emissions from the plants and direct measurements of soil moisture content. To determine the effectiveness of contaminant degradation, soil nutrients, soil oxygen content, root development, and the measured levels of contaminants throughout various stages of the project can be measured.

A monitoring plan also addresses the natural change of site vegetation over time (succession). After planting the ideal species for the site, pioneer species, or naturally occurring ones, which are quick to invade recently disturbed areas, will recruit to the project area. There are three ways to manage succession at a phytoremediation site. The pioneer species can be eradicated or ignored, or favorable species can be introduced to the area prior to pioneer recruitment, thus decreasing the amount of surface area available for pioneer establishment.

Economics. The cost of a phytoremediation project depends on the type of contaminants present on-site, the plant species selected for planting, and the complexity of the monitoring plan. Phytoremediation has proved to be less expensive than excavation and either on- or off-site treatment of large volumes of contaminated media. Even though some forms of phytoremediation would include the harvesting, removal, and incineration of aboveground plant materials, the volume of this plant waste may be only 10 percent of the volume of subsurface soils that would require excavation and treatment/disposal. This significant volume reduction translates into lower remediation costs.[155]

Limitations

Phytoremediation can introduce a new exposure pathway on-site. That is, the presence of hazardous constituents in plants may be bioavailable to the food chain (especially via leaves) at an unacceptable concentration. This pathway should be evaluated by considering soil constituent concentrations, plant uptake concentrations (either measured or calculated), receptor species, routes of exposure, and ultimately determining estimated resultant absorbed dosages. These dosages are then compared

to thresholds such as the "no observed adverse effect level" (NOAEL) to determine if any unacceptable conditions could potentially result (see Chapter 14). Though more conservative approaches have been mandated in some cases, it may be prudent to limit receptor species to those capable of roaming outside the test area, and to allow a home range that is unaffected by contamination to be considered in the risk calculation.

Phytoremediation is not generally a "quick fix," sometimes taking years to reduce contaminant concentrations to required levels, even at carefully selected sites with mostly favorable characteristics. It should be noted, however, that some contaminants in groundwater, such as chlorinated solvents, may be treated significantly faster. Nevertheless, if the contaminant concentrations at a site currently are immediately harmful to humans or wildlife, a faster remediation alternative may be necessary.

Biobarriers

Biobarriers are fixed-place zones of enhanced biological activity created in saturated soils. Groundwater contaminated with organic/inorganic pollutants is actively or passively funneled through the biologically active zone to receive treatment. The indigenous subsurface microbial activity can be enhanced through the addition of various materials (primary or cosubstrates, inorganic nutrients, surfactants, and appropriate electron acceptors) and by creating aerobic or anaerobic conditions as appropriate. Complementary subsurface metabolic capabilities can possibly be created through the addition of specialized microbial cultures.

Aerobic and anaerobic biobarriers differ widely in their degree of sophistication. For example, a simple barrier can consist of a series of saturated zone wells in which air only is sparged to create aerobic subsurface conditions. It is also possible to create an aerobic zone by injecting slow-release oxygen compounds into the wells. The basic aerobic treatment zone can be enhanced by the addition of inorganic nutrients and/or through bioaugmentation with specific microbial populations.

As an example of the use of a biobarrier with microbial augmentation, a special methyl *tert*-butyl ether (MTBE)-degrading mixed culture was used in a field pilot test to remediate MTBE-contaminated groundwater.[156] One portion of the site served as a control treatment area and was not modified. A second portion of the site used well injection points to aerate the plume. The third portion of the site inoculated subsurface soils with the MTBE-degrading culture and was also aerated with subsurface injection. The portion of the site receiving the specialized culture showed significant reductions in MTBE levels within 1 month. The aerated-only portion of the site showed essentially no reduction in MTBE levels over the initial 4-month period; afterward, MTBE levels declined somewhat, though not as significantly as the levels observed with the augmented culture. The control portion of the site did not demonstrate enhanced MTBE removal.

Anaerobic biobarriers can be used alone or in sequence with aerobic barriers. Barriers with anaerobic biological activity can be used to reductively dechlorinate halogenated aliphatics (such as PCE and TCE), treat other organics preferentially degraded under anaerobic conditions, and reduce the valence state of metals (e.g., iron) such that they can now catalytically dechlorinate compounds.

Lagoons/Impoundments

In the past, many waste disposal practices involved depositing contaminated sludges in lagoons or other impoundments. *In situ* biological treatment processes focus on resuspending and remediating these materials in place. Aerobic and anaerobic processes can be employed for this purpose. This is the *in situ* form of slurry-phase treatment presented in Section 10-5.

Figure 10-30 shows how a sludge storage lagoon can be modified by the addition of surface aerators to form an *in situ* aerobic slurry bioreactor. The mixing action resuspends the settled sludge, making it more bioavailable, and increases oxygenation of the water. Adding other amendments, such as inorganic nutrients or pH-adjusting chemicals, is easy under these conditions. It is also possible to inoculate the slurry with specialized microbial cultures without much difficulty.

The effectiveness of slurry-phase treatment depends on a number of theoretical factors including pretreatment, desorption, solids concentration, mixer design, and retention time. The contents of the lagoon can also be mixed by using pumps or dredges as necessary. On completion of biological treatment, the solids are settled, possibly to be left in place, and the liquid is decanted for further treatment as necessary. Although treatment in lagoons has some economic merits, such applications may result in excessive losses to the atmosphere if high concentrations of volatile compounds are present.

In situ lagoon processes have been used successfully to treat soils and sludges contaminated with petroleum hydrocarbons as well as wood-preserving wastes containing pentachlorophenol (PCP), creosote, and PAHs. More recently, anaerobic lagoon treatment has been used to process soils and sludges contaminated with nitroaromatic compounds generated from munitions manufacturing. Initially, anaerobic conditions are created in the slurry by adding a mass of readily biodegradable substrate. Mixing is discontinued once the substrate has been dispersed through-

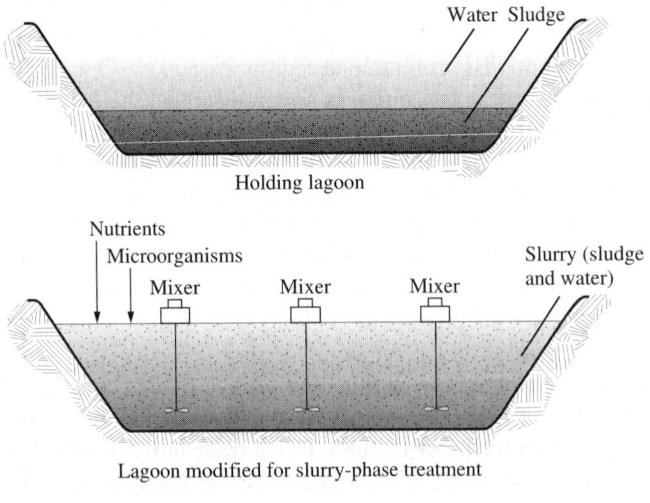

FIGURE 10-30
In situ lagoon treatment.

out the slurry. On depletion of the dissolved oxygen in the settled solids, anaerobic biodegradation of the nitroaromatic compounds occurs. A follow-up aerobic biodegradation treatment period with mixing and oxygenation can be necessary.

10-9 MAJOR CLASSES OF XENOBIOTIC COMPOUNDS AMENABLE TO BIOLOGICAL TREATMENT

Many xenobiotic compounds are amenable to biological treatment. Several texts that discuss the biodegradation of xenobiotic compounds in detail served as the basis for this section.[157,158] The Additional Reading section at the end of the chapter lists other useful texts. The reader is also advised to consult the technical literature for detailed information about biochemical degradation pathways for specific compounds.

Although it is beyond the scope of this chapter to provide an extensive review of the biochemical breakdown of all such compounds, generalizations can be made about the aerobic and anaerobic treatment of major categories of xenobiotic compounds. These categories include:

- Petroleum hydrocarbons (aliphatics, aromatics, PAHs, nonchlorinated solvents, and oxygenates)
- Halogenated hydrocarbons (aliphatics and aromatics)
- Nitroaromatics and explosives compounds
- Pesticides
- Metals/inorganics

Representative constituents of each category, along with information about the aerobic and anoxic/anaerobic biological treatment of each category, are discussed below.

Petroleum Hydrocarbons

Most petroleum products are actually complex mixtures of chemicals. For example, gasoline, jet fuel, diesel fuel, fuel oils, and tars may each contain several hundred individual chemicals. Each of these chemicals has individual physical, chemical, and biological properties.

Petroleum hydrocarbons can be classified into several major categories, such as aliphatic compounds, alicyclic compounds, aromatic compounds, and polycyclic aromatic hydrocarbons, as shown in Table 10-17. Chlorinated versions of these compounds, as well as simple nonchlorinated solvents will be discussed separately.

Aliphatic hydrocarbons

Aliphatics are made of chains of carbon atoms of various lengths, with the atoms arranged in linear or branched forms. **Alkane** compounds contain single carbon bonds only, while **alkenes** have at least one double carbon bond, and **alkynes** have at least one triple carbon bond. **Alicyclic** compounds consist of carbon rings of various sizes, and are classified as to the presence of single, double, or triple carbon bonds (such as cyclohexane, cyclohexene, and cyclohexyne, respectively). Alicyclic compounds can also exist in various branched or substituted forms.

TABLE 10-17
Petroleum hydrocarbon categories and examples

Aliphatic compounds (straight and branched carbon chains)
Alkanes (single bonds): propane, hexane, octane, decane Alkenes (double bonds): ethylene, 2-methylpropene Alkynes (triple bond): ethyne, 2-butyne
Alicyclic compounds (carbon rings of various sizes)
Cycloalkane (single bonds): cyclopentane, cyclohexane Cycloalkenes (double bonds): cyclopentene, cyclohexene Cycloalkynes (triple bond): cyclopentyne, cyclohexyne
Single-ring aromatic compounds (based on benzene ring)
Unsubstituted: benzene Substituted: toluene, xylenes, ethylbenzene, benzoate
Polycyclic aromatic compounds (two or more fused benzene rings)
Unsubstituted: naphthalene (2 rings), anthracene (3 rings), phenanthrene (3 rings), fluorene (3 rings), pyrene (4 rings), benzo(*a*)anthracene (4 rings), benzo(*a*)pyrene (5 rings) Substituted: 1-methylnaphthalene (2 rings), 1-methyl-phenanthrene (3 rings), 1-methylfluorene (3 rings)

Aliphatic hydrocarbons are typically treated by aerobic biological processes. Alkanes are the easiest compounds to degrade, followed by alkenes and alkynes. Straight-chain compounds are easier to degrade than branched-chain structures. With a process known as **β-oxidation,** oxygen is added to the terminal end of a straight-chain alkane to form a carboxylic acid, and the two terminal carbon atoms are then cleaved from the molecule. With long-chain hydrocarbons, this process can repeat itself as necessary. β-oxidation is stopped when branches are encountered in a hydrocarbon molecule.

The biodegradation of alkanes works best for compounds at least 10 carbon atoms in length. Shorter structures tend to exhibit increased toxicity toward many microorganisms. Very short alkanes (such as methane, ethane, propane, and butane) are present as gases, and can be used either as primary substrates or cometabolically degraded. Other enzymatic processes are available in certain microorganisms to degrade branched or substituted molecules, alicyclic compounds, alkenes/cycloalkenes, and alkynes/cycloalkynes; however, these reactions occur less frequently and at slower reaction rates. The anaerobic treatment of aliphatic compounds is of limited significance and typically does not play a role in bioremediation projects.

Aromatic hydrocarbons

Aromatics are based on the benzene molecule, in which six carbon atoms are arranged in a ring with alternating single and double carbon bonds. The ring can be substituted with various functional groups. Benzene, ethylbenzene, toluene, and xylene (ortho-, meta-, and para-substituted isomers), commonly referred to as BTEX, compose the major aromatic fraction of gasoline. The BTEX constituents frequently have their own individual cleanup levels at site remediation projects, with benzene typically being regulated most stringently.

Aromatic hydrocarbons can be degraded by either aerobic or anaerobic biological processes. In aerobic treatment, many different metabolic pathways are involved. Typically the first step involves the addition of oxygen to the benzene ring, eventually forming catechol. The catechol molecule then undergoes ring cleavage by ortho- or metacleavage pathways. The end products of this process are then fed into the cell's anabolic pathways.

The anaerobic biodegradation of aromatic hydrocarbons has been intensively investigated from the mid-1980s, and typically involves the action of microbial communities or consortia. With certain aromatic compounds such as benzoates and phenols, the double bonds of the benzene ring are initially reduced by the addition of hydrogen to the molecule. The ring is then cleaved and further modified to form saturated fatty acids or dicarboxylic acids. Phenols can also be degraded through the addition of a carboxyl group to the ring, which is further reduced to cyclohexanone and subjected to ring cleavage. Aromatics with substituents (such as cresols and toluene with arylmethyl groups) can have the substituents initially oxidized to carboxyl groups (the oxygen added comes from water), with subsequent ring cleavage. Benzene is the monoaromatic compound slowest to degrade under all anaerobic conditions.

The anoxic/anaerobic biodegradation of aromatic hydrocarbons has been shown to occur under denitrifying, sulfate reducing, and methanogenic conditions. In addition, ferric iron and manganese oxide can also serve as alternative electron acceptors under appropriate anoxic conditions. These processes play a key role in the success of monitored natural attenuation (see Section 10-7).

Polycyclic aromatic hydrocarbons (PAHs)

PAHs, also known as polynuclear aromatic hydrocarbons (PNAs), consist of two or more fused aromatic rings. As the number of rings increases, the solubility of the compound decreases and the persistence of the compound increases. Persistence may be linked to the decrease in a compound's bioavailability. PAHs can be biodegraded aerobically and anaerobically.

Aerobic treatment is most pronounced in treating PAHs with two or three rings (such as naphthalene, anthracene, and phenanthrene). As with benzene and its derivatives, oxygen is added directly to one of the rings, which then undergoes cleavage. This process repeats itself as necessary. In addition to aerobic biodegradation by bacteria, certain fungi and algae also have the ability to degrade PAHs. The white rot fungus *Phanarocheate chrysosporium,* for example, produces lignin-degrading enzymes able to degrade two- to five-ring PAHs. The combined metabolic action of microbial communities is especially important in the degradation of compounds with higher ring numbers.

Three- to five-ring PAHs are important from an environmental perspective because many of them are confirmed or suspected human carcinogens, and as such must be treated to low levels in contaminated soils and groundwater. Anaerobic processes play a significant role in the degradation of these compounds. As with monoaromatic ring compounds, the attack on the PAHs can involve initial ring reduction by hydrogen addition or by the addition of oxygen (from water) to an aromatic ring. The biodegradation rates of large molecular weight PAHs are very slow, and extended periods of time may be required for their treatment. It is not uncommon for these compounds to have biochemical half-lives on the order of 6 months or more.

The use of surfactants may increase their solubility and hence their biodegradation rate.

Nonchlorinated hydrocarbon solvents

These solvents and their chemical derivatives are commonly used as pure compounds in manufacturing and processing applications. For example, alkanes (hexane), alcohols (methanol and ethanol), ketones [methyl ethyl ketone (MEK) and methyl isobutyl ketone (MIBK)], and aldehydes (formaldehyde) are common solvents, and are readily biodegradable under aerobic conditions. Many of these compounds are also readily biodegradable under anaerobic conditions.

Oxygenates

Oxygenates are a category of hydrocarbon-based derivatives attracting increased attention. One such class of derivatives, the ethers, has been extensively used as oxygenates in reformulated gasolines. These oxygenates include methyl *tert*-butyl ether (MTBE), ethyl *tert*-butyl ether (ETBE), and *tert*-amyl methyl ether (TAME). The oxygenates pose a formidable environmental problem because they are highly soluble, their movement is only minimally retarded in soils, they produce taste and odor problems in drinking water at very low concentrations, their long-term human health effects are unknown, and they appear to be relatively resistant to aerobic and anaerobic biodegradation under *in situ* conditions.

MTBE

Much research into the biodegradation of MTBE and the other oxygenates was conducted during the 1990s and is continuing today. Although these compounds were once thought to be recalcitrant or generally persistent, laboratory research and limited field activities have shown that they can be degraded by pure or mixed cultures under aerobic or anaerobic conditions. Because the knowledge base for treating these compounds is so new and limited, conflicting information is sometimes reported in the technical literature.[159,160] The paragraphs below summarize some of the laboratory findings and field reports as of 1999.

MTBE can serve as a primary substrate (food and energy source) for at least two different aerobic pure cultures.[161] Aerobic treatment of MTBE has also been shown to be a cometabolic process in at least some bacteria and fungi. Cometabolic primary metabolites discovered to date include propane, *n*-butane, *iso*-butane, pentane, heptane, benzene, and xylene.[162,163,164] Both pure and mixed cultures able to biodegrade MTBE have been isolated and studied.

MTBE biodegrades to *tert*-butyl formate (TBF), which quickly degrades to *tert*-butyl alcohol (TBA). TBA appears to be a more stable daughter product, but is itself biodegraded. TBF and TBA may be used as indicator compounds for the biodegradation of MTBE in both laboratory and field studies.

Conflicting information exists whether MTBE and the aromatic gasoline fraction [benzene, ethylbenzene, toluene, and xylenes (BTEX)] can be degraded simultaneously in contaminated aquifers.[165,166] Some field studies have reported that significant MTBE degradation occurs once the BTEX fraction has been treated in a plume under aerobic conditions. This finding is supported by laboratory work that indicates that MTBE degradation is competitively inhibited by the presence of pentane, heptane, benzene, and xylene.[167] Other field reports indicate that MTBE and

BTEX plume concentrations can decrease together in an aerated plume.[168] Field studies of MTBE-contaminated sites indicate that natural biodegradation of MTBE does occur, but the rates of degradation are much slower than for BTEX at the same sites.

One field pilot study investigated the treatment of an MTBE-only groundwater plume with a special MTBE-degrading mixed culture.[169] One portion of the site was undisturbed and served as a control treatment area. A second portion of the site used well injection points to aerate the plume. The third portion of the site used inoculated subsurface soils (a process referred to as bioaugmentation) with the MTBE-degrading culture and was also aerated with subsurface injection. The portion of the site receiving the specialized culture showed significant reductions in MTBE levels within 1 month. The aerated only portion of the site showed essentially no reduction in MTBE levels over an initial 4-month period, and then MTBE levels showed a decline (though not as significant as that observed with the augmented culture). This finding indicates that perhaps indigenous MTBE-degrading organisms need an acclimation or adaptation period before they can effectively treat the compound. The control portion of the site did not demonstrate enhanced MTBE removal.

A review of field data has also concluded that MTBE can be biodegraded under anaerobic conditions, including denitrifying conditions, iron reduction, sulfate reduction, and methanogenesis.[170,171,172] In some situations, TBA appeared as a daughter product, and was slowly degraded. In at least one case MTBE levels were shown to decrease while BTEX concentrations in a plume remained stable.[173] Biodegradation modeling indicated that MTBE plume sizes were less than that expected to occur without bioattenuation, implying that biodegradation was taking place.

Halogenated Hydrocarbons

Halogenated hydrocarbons include both aliphatic and aromatic species. Such chlorinated compounds make up some of the most common environmental pollutants, as shown in Table 10-18.

Halogenated aliphatics

Such aliphatics were considered resistant to biological treatment for many years, but research in the 1980s and 1990s showed that this was not the case. Many of the chlorinated aliphatics were shown to be cometabolized under aerobic conditions. Cosubstrates discovered include methane (used by methanotrophs), phenol and toluene (used by *Pseudomonas* species), and ammonia (used by *Nitrosomonas* species). Nonspecific oxygenases initially form an epoxide with the target aliphatic. Under aerobic conditions, the chlorinated aliphatics with the lowest degree of chlorination showed the highest degree of destruction. Vinyl chloride was a good target compound, while perchloroethene was resistant to aerobic treatment.

Under anaerobic conditions, reductive dehalogenation of chlorinated aliphatics takes place under cometabolic conditions. In reductive dechlorination, halogen atoms on the target compound are sequentially removed and replaced by hydrogen atoms. The chlorinated aliphatic is actually used as an electron acceptor, not a substrate. Another carbon source must be supplied for microbial energy. An anaer-

TABLE 10-18
Halogenated hydrocarbon categories and examples

Aliphatic compounds
Tetrachloroethene (perchloroethene or PCE)
Trichloroethene (TCE)
cis-1,2-Dichloroethene (cis-1,2-DCE)
trans-1,2-Dichloroethene (trans-1,2-DCE)
1,1-Dichloroethene (1,1-DCE)
Chloroethylene (vinyl chloride or VC)
1,1,1-Trichloroethane (1,1,1-TCA)
1,1,2-Trichloroethane (1,1,2-TCA)
1,1-Dichloroethane (1,1-DCA)
1,2-Dichloroethane (1,2-DCA)
Chloroethane
Tetrachloromethane (carbon tetrachloride or CT)
Trichloroethane (chloroform or CF)
Dichloromethane (methylene chloride or MC)
Chloromethane
Aromatic compounds
Chlorobenzenes (various forms)
Chlorophenols (various forms)
Chlorobenzoates (various forms)
Chlorocatechols (various forms)

obic (methanogenic) mixed culture using acetate as a food source was first used to demonstrate that trichloroethene (TCE), 1,1,1-trichloroethane (1,1,1-TCA), and carbon tetrachloride (CT) could be transformed into carbon dioxide and methane. Since that time many other chlorinated aliphatic compounds have been shown to be treatable under anaerobic conditions, including perchloroethene (PCE). In contrast to the aerobic treatment of chlorinated aliphatics, the efficiency of anaerobic treatment increases with the degree of halogenation of a compound. Thus PCE is very effectively treated with reductive dehalogenation, while vinyl chloride is not. An increasing number of halorespiring microorganisms have also been isolated that are able to link dechlorination to growth.

Halogenated aromatics

Such aromatics include chlorophenols, chlorobenzenes, and polychlorinated biphenyls (PCBs). The halogens in order of increasing difficulty to treat are bromine, iodine, chlorine, and fluorine. The halogenated aromatics utilize biodegradation pathways similar to those involved with nonchlorinated aromatics. Under aerobic biodegradation, for example, chlorobenzenes are oxidized to chlorocatechols, which undergo ring cleavage and subsequent dechlorination of the ring cleavage products. In some circumstances dechlorination can occur before ring cleavage. Microbial consortia are usually involved in mediating these processes.

Halogenated aromatics can also undergo reductive chlorination. Once again, the more highly substituted aromatics are easier to dechlorinate than monohalogenated

compounds. The efficiency of dechlorination may also vary with the position that the halogen occupies on the aromatic ring. Ortho- and meta-substituted sites are good candidates for dechlorination, while ortho-substituted sites may be much more resistant to dechlorination. This phenomenon has been repeatedly demonstrated with high molecular weight PCBs, for example.

To date, the reductive dechlorination of aromatics has been most extensively studied under methanogenic conditions. The interest and research into dechlorination under sulfate reducing conditions is increasing.

Explosives Compounds

Most explosives are organonitro compounds, and include nitroaromatic groups, nitrate esters, and nitroamine groups. The most common explosives encountered in hazardous waste treatment include:

- TNT (2,4,6-trinitrotoluene)
- RDX (hexahydro-1,3,5-trinitro-1,3,5-triazine)
- HMX (octahydro-1,3,5,7-tetrazocine)

The dinitro aromatic compounds 2,4-DNT (2,4-dinitrotoluene) and 2,6-DNT (2,6-dinitrotoluene) are also frequently found with TNT, since TNT is formed from the successive nitration of toluene.

Nitroaromatic contaminants are commonly found at Department of Defense ordnance sites as well as ammunition manufacturing plants. The contaminated wastewater generated during munitions production (also called "pink water" because of its distinctive color) was typically placed into unlined lagoons without further treatment. Water leaching from these lagoons contaminated both the underlying soils as well as groundwater in many cases. The remediation of these sites usually involves the treatment of large volumes of soil and groundwater.

Much of our understanding of the biological processes involved in explosives treatment was generated during the 1990s from laboratory, field pilot, and limited full-scale treatment projects. Both aerobic and anaerobic processes have been used to treat nitroaromatic compounds.[174]

Anaerobic treatment

The anaerobic treatment of nitroaromatic compounds may be easier to accomplish since many reports of its success do not mention the need for specialized or acclimated cultures. It appears that the microbial capability to degrade these compounds may be widely distributed in nature (either in soils or easily added in the form of various animal manures), since beneficial anaerobic biological treatment typically follows once anaerobic conditions in soil and/or groundwater are created. However, there have been concerns that anaerobic treatment may produce undesirable and possibly carcinogenic daughter products. In contrast, several reports commenting on successful aerobic treatment of explosives compounds employ specialized microbial populations. Fewer and/or potentially less harmful daughter products are reportedly produced during aerobic biodegradation. The possibility of using sequential anaerobic and aerobic biological processes to reduce or eliminate daughter product concerns has also been explored.[175,176]

Bioprocess designs used to treat explosives include slurry-phase aerobic reactors, sequential anaerobic/aerobic processing of soils and liquids, and windrow composting. These processes have been successfully used to remediate TNT, DNT, RDX, HMX, nitrobenzene, and nitroglycerine. Several of these pilot and full-scale applications are described in greater detail below.

Contaminated Groundwater

A pilot-scale fluidized-bed reactor with a preacclimated aerobic culture was used to treat various mono-, di-, and trinitrotoluenes present in groundwater. Different contaminant loading rates and operating conditions were evaluated, and the fluidized-bed process was found to be effective in removing mono- and dinitrotoluenes from the water.[177] A 250-L pilot soil slurry reactor was used to evaluate the effectiveness of treating high concentrations of RDX (approximately 10,000 mg/kg dry soil) and HMX (approximately 4800 mg/kg dry soil) under anaerobic conditions with a sewage sludge inoculum. RDX concentrations were reduced to less than 500 mg/kg dry soil in approximately 45 days. RDX concentration reductions were not observed until HMX removal was complete, and then required an additional 35 days.[178]

Lagoons

On a larger scale, existing or constructed lagoons on site can be used as slurry-phase reactors to treat contaminated soils. Soils can be processed and slurried with water before being placed in the lagoon or bioreactor. Nutrients and a readily biodegradable carbon source are then added, and indigenous aerobic biological activity quickly depletes the free oxygen and added carbon source present in the water. Once anaerobic conditions are created, the soils can be allowed to settle, and anaerobic biological activity is used to treat the nitroaromatics present.[179]

Contaminated soils

Aboveground soil piles treated sequentially under anaerobic and aerobic conditions have also shown success. Anaerobic conditions are initially created in the contaminated soils by adding electron-donating substrates such as starch or sucrose, along with a microbial inoculum (cow manure and/or anaerobic digester sludge), and water. The pile is kept under anaerobic conditions until the nitroaromatics are reduced to their corresponding amines. The aerobic part of the process consists of making an aerated soil pile by mixing the treated soils with composted yard waste. Oxygen is supplied to the soils by actively blowing or drawing air through the pile. Alternatively, the pile can be passively aerated by allowing air to permeate into soils from buried piles. Residual amine compounds undergo aerobic biodegradation and/or can be polymerized into the soil organic matter through biological activity.[180] However, the toxicity of bound residues was not examined.

Lastly, windrow composting has been used in various forms. Aerobic thermophilic composting can be used to reductively denitrify and mineralize TNT, dinitrotoluene (DNT), RDX, and HMX. The composition of one such compost in a 6.5-yd^3 field pilot test included contaminated soil (20 to 30 percent by volume), food wastes, alfalfa, cow and chicken manure, and wood chips. Water mixed with molasses was used to keep the pile humidified. Explosives concentrations were below detection limits after approximately 40 days of treatment.[181] Approximately 2840 yd^3 of TNT- and RDX-contaminated soil were used in another study to construct five windrow piles, each with a slightly different mixture of hay, wood chips,

manure, and potatoes/potato wastes. Thermophilic conditions were reached in the piles, and acceptable explosives concentrations were reached in 28 days.[182]

Pesticides

The "pesticide" category also includes organic chemicals with insecticidal and herbicidal properties. Pesticides have a wide variety of chemical structures, ranging from simple aliphatics to complex chlorinated ring compounds. Most pesticides can be classified into one of nine major categories based on their chemical structures, as shown in Table 10-19.[183] The reader is advised to review the primary literature for detailed information about the biodegradation of specific pesticides; a thorough general discussion of pesticide biodegradation can be found elsewhere.[184]

Because of their widely different structures, it is difficult to make accurate generalizations about the biodegradability of pesticides. Some pesticides are readily biodegradable, and must be reapplied to soils frequently to maintain their toxicological properties. Other pesticides are relatively resistant to biodegradation, and have been reported to persist in the environment for tens of years.

Pesticides are amenable to both aerobic and anaerobic biological treatment. Aerobic processes provide the best form of treatment for certain compounds, although others are best treated using anaerobic means. Sequential anaerobic/aerobic treatment strategies are rapidly gaining acceptance for the treatment of mixtures of pesticides.[185]

Cometabolic processes and/or the action of mixed microbial communities are frequently associated with aerobic treatment. Reductive dehalogenation processes are associated with anaerobic biodegradation. The biodegradation of pesticides can be limited by their low solubility, making treatment a bioavailability issue. In addition, the enzymatic capability to degrade these compounds does not appear to be widely distributed in nature. Therefore, indigenous microbial populations may only contain low levels of pesticide-degrading microorganisms. Low population densities of pesticide-degrading microorganisms result in low field biodegradation rates. Although microbial metabolism can adapt to certain pesticides, this may be a lengthy process.

A brief review of the biodegradability of the nine major classes of pesticides is provided below[186]:

- **Halogenated aliphatic** pesticides include various fumigants. Cometabolism under aerobic conditions and reductive dehalogenation under anaerobic conditions have been shown to occur with these compounds.
- **Chlorinated cyclic aliphatic** pesticides generally react poorly to aerobic treatment; anaerobic processes are preferred.
- **Halogenated aromatic** pesticides are susceptible to aerobic treatment, but treatment effectiveness decreases as the degree of halogenation increases. Anaerobic processes or combinations of anaerobic/aerobic processes may be best, especially for the treatment of DDT.
- **Chlorinated phenoyalkanoate** pesticides include many herbicides, and are generally readily degraded under both aerobic and anaerobic conditions.
- **Halogenated analine-based** pesticides include many herbicides. Some of these compounds are readily biodegraded or mineralized, although others are resistant

TABLE 10-19
Pesticide categories and examples

Halogenated aliphatics
DBCP (1,2-Dibromo-3-chloropropane) Dalapon (2,2-Dichloropropionic acid)
Chlorinated cyclic aliphatics
Lindane (gamma-hexachlorocyclohexane)
Halogenated aromatics
DDT (dichlorodiphenyltrichloroethane)
Chlorinated phenoyalkanoates
2,4-D (2,4-Dichlorophenoxy acetic acid) 2,4,5-T (2,4,5-Trichlorophenoxy acetic acid) MCPA (4-chloro-2-methylphenoxy-acetic acid) Silvex
Halogenated aniline-based
Propanil Alachlor Trifluralin
Cyclodienes
Heptachlor Chlordane Dieldrin Aldrin
Carbamates
Aldicarb Carbofuran
Pyrethroids
Permethrin Fenvalerate Fluvalinate
Organophosphonates
Glyphosate Parathion

to treatment and may evidence only minor chemical modifications. Aerobic or anaerobic processes are suitable.

- **Cyclodiene** pesticides include many popular insecticides. Many of these compounds are persistent, and require either cometabolic or anaerobic treatment processes.
- **Carbamates** include insecticides and herbicides. Both aerobic and anaerobic processes are suitable for use, and some of these compounds may be easily biodegraded.

- **Pyrethroid** pesticides are less toxic than many previous compounds, and their use is increasing. Both aerobic and anaerobic processes have been used successfully in treating these compounds.
- **Organophosphonate** pesticides include insecticides. Glyphosate is rapidly metabolized, while parathion is very toxic and requires cometabolism and/or mixed microbial cultures to be treated successfully.

Metals/Inorganics

Although the biological treatment of organic compounds receives the most attention, biological processes are beginning to play an increasing role in the treatment of metals and other inorganic environmental contaminants. Both microorganisms and plants have roles to play in these processes. The role of plants in treating organic and inorganic contaminants is discussed under the topic of phytoremediation in Section 10-8. Some of the most important uses of microorganisms to treat metals and inorganics include:

- Changing the valence state of metals, reducing their toxicity and/or solubility
- Removing heavy metals and radionuclides from water by adsorption
- Detoxifying cyanide
- Removing excess nitrogen compounds (ammonia and nitrate) from soils/groundwater through nitrification
- Changing the structure and properties of certain metals through methylation.

Microorganisms can be used in aboveground systems and *in situ* to change the valence state of metals. For example, hexavalent chromium, a human carcinogen, can be reduced to trivalent chromium, a benign form of the metal.[187] Uranium in batch experiments has been reduced to highly insoluble uraninite following the addition of ethanol as a carbon source and trimetaphosphate as an inorganic nutrient.[188] Laboratory work with metal-reducing bacteria has shown that in addition to chromium reduction, iron(III) can be reduced to iron(II). Iron(II) is able to catalytically dechlorinate organic compounds (such as trichloroethene) in much the same way as zero iron dechlorination.[189] Other work has shown that sulfate-reducing bacteria can be used to convert dissolved sulfate to sulfide, precipitating metals out as insoluble metal sulfides in the process. This treatment mechanism is potentially suitable for use with cobalt, cadmium, nickel, lead, and zinc.[190] Using indigenous microorganisms to modify subsurface environments for metals treatment adds a new dimension to *in situ* biological treatment.

Cyanide is susceptible to biological treatment with properly acclimated microorganisms. Cyanide-tolerant cultures can be developed in the laboratory. These cultures have the ability to successfully tolerate increasing concentrations of cyanide. One such culture was used in subsequent pilot-scale tests.[191]

Many industrial and agricultural sites have groundwater contaminated with high concentrations of nitrogen-containing compounds such as ammonia, nitrite, and nitrate. High nitrate levels render water unsafe for human consumption. An excess of ammonia, nitrite, and nitrate in industrial wastewater streams has been successfully treated by biological nitrification and denitrification. Nitrification converts ammonia to nitrate. The complementary denitrification process then converts nitrate to

nitrite and eventually nitrogen gas. *In situ* denitrification is now being explored for the treatment of contaminated soils and groundwater. Several recent studies have reported on successful laboratory and field pilot denitrification projects. Various added carbon substrates and nutrient sources have been added to these systems, including vegetable oil with added phosphate,[192] sodium acetate with trimetaphosphate,[193] and dissolved sugar with corn meal pellets.[194]

Lastly, certain microorganisms have the ability to methylate metals such as selenium and mercury. The methylated metals can then be volatilized from the surrounding environment. Although the methylated metals may be removed from the immediate microbial environment, this form of the metals may actually be more hazardous to humans.

10-10
BIOLOGICAL TREATMENT—TREATABILITY STUDIES AND OTHER CONSIDERATIONS

Biotreatability Studies

Treatability studies can be used to investigate the feasibility of various biological processes and/or to investigate the usefulness of different bioengineering designs. Treatability studies provide information:

- Unavailable from the technical literature
- When data from existing pilot- or full-scale treatment systems is inappropriate for use (for example, different compounds to be treated, different reaction conditions, etc.)
- About site-specific processes and/or engineering information

Treatability testing typically progresses through three steps as follows:

- Laboratory screening
- Bench-scale testing
- Pilot-scale testing

Laboratory screening is a quick test to establish whether biological treatment has the potential to treat a waste. A biological treatment technology is usually considered a reasonable option if the test results in greater than a 20 percent reduction in the mass or concentration of the target compound, excluding abiotic losses. These screening studies are used to establish treatment feasibility, that is, whether an aerobic and/or anaerobic biological process can be used to treat one or more target organics. Since screening studies are frequently conducted under optimal environmental conditions (such as excess inorganic nutrients, favorable temperature, good mixing, and excess oxygen for aerobic processes), favorable laboratory results do not automatically imply that the process can be successfully scaled up and used in the field.

Bench-scale treatability tests are moderately quick tests for the purpose of selecting a process for further evaluation and for identifying the key parameters that have the greatest influence on biodegradation. With aqueous waste streams, bench-scale tests are usually conducted in batch reactors with adjustments made in loading rates, chemical additives (e.g., nutrients), physical parameters (e.g., aeration and mixing rates), and other variables to study their relative impact on the rate of degra-

dation. The data provided by bench testing should lead to the selection of one or two processes for further testing at the pilot-scale level.

The pilot-scale test is done not only to verify the process but to provide kinetic degradation data for final design. A pilot-scale test may require weeks to months to complete, but it provides data from real-world conditions most directly suitable for scale-up and design.

Treatability Studies for Hazardous Wastes

Properly designed and conducted treatability studies are useful because they:

- Provide a relatively low-cost method of answering treatability questions
- Allow evaluation of new technologies or combinations of technologies
- Provide technology support information to local, state, and federal regulatory agencies
- Result in significantly reduced project capital and operating expenses when successfully used in technology selection.

Organic and inorganic contaminants present in soils, sludges, liquids, and vapors are suitable for use in treatability tests. In many cases, biological treatment is evaluated as a complementary technology for use with other physical/chemical remediation processes. In these situations, biological processes can be used as a pre- or post-treatment technology in combination with other processes.

Detailed discussions of biotreatability studies and protocols can be found in several reference texts and the primary technical literature.[195,196] The most important components to be considered in treatability testing include:

- Problem definition
- Experimental design
- Analytical procedures
- Time constraints
- Financial constraints

Treatability studies are designed to answer specific technical questions; hence no single testing protocol exists that is applicable to all situations. The study **problem definition** is critically important because it determines what questions will be asked and hopefully answered. Typical problem definition questions include:

- Are the target compounds amenable to aerobic or anaerobic biological treatment?
- Will indigenous microorganisms be appropriate for use?
- What concentrations of the chemical are treatable?
- What final contaminant concentration is attainable?
- What conditions are needed for successful treatment?
- How long will the process take?
- Will any hazardous daughter products be produced?
- What reactor design is best for this application?

Once the problem definition is completed, the experiments must be designed to answer these specific questions. At a minimum, the **experimental design** must take into account:

- Whether the study will be pragmatic or more esoteric in nature (for example, using contaminated soils from the site versus spiking a known mass of contaminant into a surrogate matrix such as sand)
- What controls (positive and negative) will be employed
- The frequency of sampling to be used
- The analyses to be performed
- Duration of the study
- Whether a contaminant mass balance will be needed
- The acceptable variability and statistical validity required
- Quality assurance/quality control (QA/QC) issues to be met

It is also critical to control (or account for) the loss of target compounds due to abiotic factors such as volatilization or adsorption. This is especially true in situations where the starting concentration of target compounds is low (low parts per million or parts per billion levels).

There are few widely accepted standardized treatability protocols. In 1988 the EPA published a series of aerobic and anaerobic test protocols in the Code of Federal Regulations[197] (these tests are also summarized elsewhere[198]). The test procedures were primarily designed for evaluating industrial wastewater, but some soil protocols are included. In 1991, the EPA released an interim guide for conducting aerobic treatability screening studies for contaminants found at Superfund sites.[199] These procedures require the use of control and reactive test conditions. At least three or four sampling points over time are used with both the control and reactive conditions. It is recommended that duplicate or triplicate samples be collected at each sampling event. Data from the control samples are used to adjust for abiotic losses occurring during the test. Biological treatment is considered a potential remediation technology if the average final contaminant concentration, corrected for abiotic loss, is at least 20 percent less than the average contaminant concentration observed at the beginning of the experiment. Many other treatability protocols are based on academic test procedures described in the primary technical literature. These aerobic and anaerobic protocols can be modified as necessary for use in specific studies.

The **analytical procedures** used must consider the quantitation of the target contaminants only or a combination of target contaminants and indirect indicators of biological treatment (such as changes in pH, chloride levels with the treatment of chlorinated compounds, biochemical oxygen demand, or chemical oxygen demand). Treatability data to be submitted to regulatory agencies may require the use of standardized analytical procedures performed by certified laboratories. It is important to know the final detection level of the target compounds required to ensure that samples of sufficient mass or volume are supplied to the laboratory. It may also be possible (or desirable) to monitor the production of any daughter products of microbial metabolism, as well as to note whether these products are transitory or persistent in nature.

Time constraints imposed on a project must be considered to determine if they will interfere with the integrity of the study. Treatment times that are inadequate (too short) may result in artificially negative treatability results. Although many aromatic and aliphatic hydrocarbons undergo rapid biodegradation under favorable test conditions, highly branched chain aliphatics or polycyclic aromatic hydrocarbons (PAHs) do not. A treatability study designed to investigate the aerobic biodegrada-

tion of gasoline in groundwater may require only several weeks to complete, while a similar study to evaluate the biodegradation of PAHs may require a minimum of several months in order to generate valid results.

Financial constraints imposed on a treatability study also need to be considered in terms of interfering with the integrity of a study. Highly customized studies frequently have high costs because the treatability laboratory cannot accomplish economies of scale. Each study is a one-of-a-kind test requiring senior level technical support at least for design and data interpretation. Analytical expenses frequently make up 50 to 75 percent of the total project costs, depending on the type of analyses performed, the number of samples analyzed, and the laboratory turnaround time required. In many situations private parties do not fully understand the costs associated with properly designed and conducted studies, and hence the budgets for such work are seriously underfunded. This situation typically results in reducing the number of samples taken by eliminating entire sampling points or by collecting only single rather than duplicate or triplicate samples. It is also common to see a reduction in the number of positive or negative controls performed. Such shortcuts may result in the generation of ambiguous or misleading data with limited useful value.

Treatability Studies for Industrial Wastewaters

The key treatability components discussed above also hold for the evaluation of industrial wastewaters. Many standardized treatability tests have been developed for evaluating municipal or industrial wastewaters, and are described in the technical literature as well as wastewater engineering texts. These tests frequently focus on monitoring aggregate values for changes in organic carbon (such as BOD, COD, or TOC), or are concerned with the oxygen demand posed by aerobic biotreatment processes. A complete evaluation of biological treatment variables can be a labor/cost-intensive undertaking involving much scientific and engineering judgment.

Further innovation is needed in the approach and tools used to evaluate and design biological treatment processes. One tool for reducing the effort required is that of evaluating biodegradation rates based on automated respirometry. Respirometry, or the measurement of the rate of oxygen uptake, has been largely automated and integrated with ancillary computer and modeling analysis to provide instantaneous data and interpretations.[200] It has long been employed for many applications in the pollution control field, for the simple reason that oxygen uptake rates provide a rapid indicator of microbial activity. One application is to model the rate of biodegradation. To do this, one must be able to utilize the oxygen uptake data to compute cell growth or substrate utilization rates.[201] A major effort showed that respirometric data can be employed to determine cell growth rates which can be transformed into equivalent growth or substrate utilization data.[202] This is accomplished by using equations which are derived based on the principles of COD and energy balances for aerobic systems.

Several research and field application efforts validated the use of respirometry as a means to obtain values of the biokinetic constants for model calibration.[203] Experience shows that the model predictions that were calibrated with respirometric data accurately described the field performance of aerobic biological treatment systems.

This approach saves times and provides kinetically meaningful data for designing and operating biological treatment systems. For example, proactive adjustments can be made as a treatability test progresses, enabling rapid turnaround times (as soon as 30 hours from receipt of a sample for a bench-scale test).

Limitations of Treatability Tests

Many important pollutants may be found biodegradable under laboratory conditions. However, biodegradation under field conditions is a different matter. It is often observed that metabolic rates are much slower in the field for many reasons. Environmental conditions such as temperature, pH, and oxygen may not be optimal. Microbial cultures isolated in the laboratory may not thrive or even survive in competition with the vast population of microbes that naturally occur in the field. Also, field conditions may favor the production of intermediates not encountered in the laboratory.

Another possible reason for slower rates is that the composition and concentration of target organics may differ significantly between field and laboratory conditions. The spatial and temporal concentrations of a target compound at a waste site typically exhibit great variability, a condition that may not be addressed in the laboratory. The waste may also contain multiple contaminants other than the target contaminant. Either condition may inhibit degradation of the target compound. The degradation of some persistent compounds may require a sequence of events that may not be easily duplicated in a laboratory setting at a reasonable cost. Control of abiotic losses of the compound (e.g., via volatilization) is another consideration.

Because of these many obstacles, the design of a full-scale treatment system may be difficult. It involves a series of distinct tasks, each of which is crucial to the development of a successful operation. The tasks start with literature reviews and desktop evaluations, and frequently progress through a series of laboratory and field treatability tests before a suitable full-scale engineering design can be successfully completed.

Modeling of Biological Processes

A number of specialized computer models are available as aids in selecting, evaluating, and predicting the performance of *in situ* biological treatment processes. These models include:

- BIOF&T 2-D/3-D (biodegradation, flow, and transport in saturated and unsaturated zone soils)
- BIOMOD 3-D (bioremediation, fate, and transport model for MODFLOW)
- BIOPLUME III (transport of dissolved hydrocarbons under the influence of oxygen-limited biodegradation)[204]
- BioTrends (evaluation of natural attenuation trends)
- BIOSCREEN (fate and transport model)[205]
- BIOCHLOR (fate and transport model)[206]

With valid input data and when used properly, many models can provide valuable information about biodegradation rates and project duration.

The biological waste treatment field made great progress during the 1990s. The understanding, acceptance, and use of bioprocesses to treat hazardous wastes have increased at a rate long predicted but previously little realized. However, significant improvement is still possible and desirable. Listed below are several areas of interest that will continue to attract attention:

- **Anaerobic treatment:** Emphasis continues on understanding and exploiting the biological processes that take place under anaerobic conditions, especially those associated with monitored natural attenuation.
- **Anaerobic/aerobic cycling:** With an improved understanding of anaerobic processes comes the opportunity to combine them with aerobic processes. The benefit of such an approach is becoming increasingly apparent in the treatment of soils contaminated with nitroaromatics, pesticides, and polychlorinated solvents.
- **Aboveground liquid-phase reactors:** The development of novel reactor systems employing specialized microbial populations and operating conditions will continue to extend the types of contaminants treatable in ground and industrial process waters.
- **MTBE:** The extensive groundwater contamination problems associated with the use of this gasoline oxygenate are becoming increasingly evident. Efforts will be directed to the development of cost-effective processes to treat groundwater both aboveground and *in situ*. It is especially important to determine whether the metabolic capability for degrading MTBE is widely distributed in the biosphere and, if so, under what conditions it can be encouraged. If this capability is not widely distributed, under what conditions can it be introduced and maintained in the subsurface?
- **Genetically engineered microorganisms (GEMs):** The use of the first genetically engineered microorganism approved for field trial in the hazardous waste treatment field (*Pseudomonas flourescens* HK 44) was first reported in the technical literature in March 2000.[207] Although genetically engineered organisms hold great promise for the waste treatment field, their widespread acceptance by regulatory agencies and the general public in the near future is questionable. Although work will continue to advance, the use of GEMs may have a more immediate role in aboveground reactor systems.
- **Development of specialized cultures to treat resistant compounds:** Work will continue to develop specialized cultures able to treat chemicals that have remained difficult to biodegrade in the past. These compounds include highly halogenated chemicals.

The bioremediation field will continue to advance in basic knowledge and applications experience. The widespread acceptance of monitored natural attenuation provides proof that sound science and engineering can result in cost-effective waste treatment strategies.

DISCUSSION TOPICS AND PROBLEMS

10-1. Name three advantages and three disadvantages of biological treatment in comparison with physical-chemical treatment.

10-2. Why are chemoheterotrophic bacteria the most important class of microorganisms involved in bioremediation?

10-3. What are the differences between catabolism and anabolism? Between biotransformation and mineralization?

10-4. How are electron acceptors used in microbial metabolism? Name the different aerobic and anaerobic electron acceptors and their corresponding transformation products.

10-5. How are cometabolism, fortuitous metabolism, and secondary substrate utilization different from each other? Why are they important from a bioremediation perspective?

10-6. Explain the difference between recalcitrance and persistence.

10-7. How does the chemical structure of a compound affect its biodegradability?

10-8. Name several advantages and disadvantages associated with aerobic biological treatment. With anaerobic biological treatment.

10-9. Name several classes of environmentally significant chemicals. Give examples of each.

10-10. Why are microbial communities important in biodegradation?

10-11. How do microorganisms adapt to new environments or substrates?

10-12. Explain reductive dehalogenation. In order for contaminant destruction to occur, why must an additional carbon source be present?

10-13. How are the major microbial growth requirements (temperature, moisture, pH, organic substrates, inorganic nutrients, and electron acceptors) controlled in *in situ* and *ex situ* treatment systems?

10-14. Explain the growth response of microorganisms to increasing concentrations of inhibitory and noninhibitory substrates.

10-15. Name three advantages and disadvantages of conducting biotreatability studies.

10-16. Why is it typically not possible to directly scale-up biotreatability study results to the design of full-scale treatment systems? Can these limitations be overcome?

10-17. What are the advantages and disadvantages associated with suspended growth and fixed-film bioreactor designs?

10-18. What are the similarities and differences among land treatment, composting, and soil pile treatment systems?

10-19. What are the advantages of anaerobic/aerobic cycling in composting systems? What types of contaminants are especially amenable to this form of treatment?

10-20. Name the three types of vapor-phase bioreactors. How do these reactors differ from each other in terms of design and operation?

10-21. Why are *in situ* treatment processes typically preferred for subsurface remediation over pump-and-treat processes?

10-22. How is the physical delivery/removal of materials controlled in vadose and saturated zone materials?

10-23. Explain the difference between soil vapor extraction (SVE) and bioventing.

10-24. Explain how biosparging systems can also encourage aerobic biological activity in the vadose zone.

10-25. In addition to biodegradation, what physical/chemical processes are involved in natural attenuation?

10-26. Monitored natural attenuation is usually associated with the treatment of saturated zone soils and groundwater. Does natural attenuation also occur in vadose zone soils?

10-27. Why are anaerobic biodegradation processes so important in monitored natural attenuation projects?

10-28. Name the key geochemical parameters monitored during natural attenuation projects. What are their roles in biodegradation, and how do they indicate that biodegradation is occurring?

10-29. Name the three lines of evidence used to support natural attenuation. What information is needed for each line of evidence?

10-30. How are phytoextraction, phytoaccumulation, phytostabilization, phytovolatilization, phytostimulation, phytotransformation, phytodegradation, and rhizofiltration different from each other?

10-31. Explain the concept of biobarriers. How can aerobic and anaerobic conditions be created and maintained in biobarriers?

REFERENCES

1. Gale, E. F.: *The Chemical Activities of Bacteria,* Academic Press, London, 1952.
2. Liss, Steven N., and Katherine T. Baker: "Anoxic/Anaerobic Bioremediation," *Bioremediation,* edited by K. T. Baker and D. S. Herson, McGraw-Hill, New York, pp. 297–342, 1994.
3. Leadbetter, E. R., and J. W. Foster: "Bacterial Oxidation of Gaseous Alkanes," *Archiv für Mikrobiologie,* vol. 35, pp. 92–104, 1960.
4. Leadbetter, E. R., and J. W. Foster: "Oxidation Products Formed from Gaseous Alkanes by the Bacterium *Pseudomonas methanica,*" *Archives of Biochemistry and Biophysics,* vol. 82, pp. 491–492, 1959.
5. Leadbetter, E. R., and J. W. Foster: "Studies on Some Methane-Utilizing Bacteria," *Archiv für Mikrobiologie,* vol. 30, pp. 91–118, 1958.
6. Dalton, H., and D. I. Stirling: "Cometabolism," *Philosophical Transactions of the Royal Society of London B,* vol. 297, pp. 481–496, 1982.
7. Horvath, R. S.: "Cometabolism of the Herbicide 2,3,6-Trichlorbenzoate by Natural Microbial Populations," *Bulletin of Environmental Contamination and Toxicity,* vol. 7, pp. 273–276, 1972.
8. Horvath, R. S., and B. W. Koft: "Degradation of Alkyl Benzene Sulfonate by *Pseudomonas* Species," *Applied Microbiology,* vol. 23, pp. 407–414, 1972.
9. Beam, H. W., and J. J. Perry: "Microbial Degradation of Cycloparaffinic Hydrocarbons via Cometabolism and Commensalism," *Journal of General Microbiology,* vol. 82, pp. 163–169, 1974.
10. Daughton, C. G., and D. P. H. Hsieh: "Parathion Utilization by Bacterial Symbionts in a Chemostat," *Applied and Environmental Microbiology,* vol. 34, pp. 175–184, 1977.
11. Kunc, F., and J. Rybarova: "Effect of Glucose on the Amount of Bacteria Mineralizing 2,4-Dichlorophenoxyacetic Acid in Soil," *Folia Microbiology,* vol. 28, pp. 54–56, 1983.
12. Pal, H. S., T. Murphy, A. C. Carter, and S. W. Drew: "Rapid Assays for Microbial Degradation of 2-Chlorophenol," Bulletin 128, Virginia Water Resources Research Center, Virginia Polytechnic Institute and State University, Blacksburg, Va., 1980.
13. Leahy, M. C., et al.: "Biodegradation of Clorinated Aliphatics by a Methanotrophic Consortium in a Biological Reactor," *Proceedings of the 2nd National Conference of Biotreatment: The Use of Microorganisms in the Treatment of Hazardous Materials and Hazardous Waste,* HMCRI, Washington, D.C., pp. 3–9, November 27–29, 1989.
14. Roberts, P. V., et al.: "In Situ Aquifer Restoration of Chlorinated Aliphatics by Methanotrophic Bacteria," Technical Report No. 310, Department of Civil Engineering, Stanford University, June 1989.
15. McCarty, P. L., B. E. Rittmann, and M. Reinhard: "Trace Organics in Groundwater," *Environmental Science & Technology,* vol. 15, pp. 40–51, 1981.
16. Rittmann, B. E., P. L. McCarty, and P. V. Roberts: "Trace Organics Biodegradation in Aquifer Recharge," *Ground Water,* vol. 18, pp. 236–243, 1980.
17. Lapat-Polasko, L. T., P. L. McCarty, and A. J. B. Zehnder: "Secondary Substrate Utilization of Methylene Chloride by an Isolated *Pseudomonas* Sp.," *Applied and Environmental Microbiology,* vol. 47, pp. 825–830, 1984.

18. Skladany, George J.: "Characterization of a Mixed Microbial Population Able to Degrade 2-Chlorophenol and Nitrilotriacetic Acid," Master of Science Thesis, Clemson University, Clemson, S.C., 1984.
19. Grady, C. P. Leslie, Jr.: "Biodegradation of Hazardous Wastes by Conventional Biological Treatment," *Hazardous Waste & Hazardous Materials,* vol. 3, no. 4, pp. 333–365, 1986.
20. Niemi, G. J., et al.: "Structural Features Associated with Degradable and Persistent Compounds," *Environmental Toxicology and Chemistry,* vol. 6, Pergammon Journals, Elmsford, N.Y., pp. 515–527, 1987.
21. Rochkind, M. L., J. W. Blackburn, and G. S. Saylor: "Microbial Decomposition of Chlorinated Aromatic Compounds," EPA/600/2-86/090, Hazardous Waste Engineering Laboratory, U.S. Environmental Protection Agency, Cincinnati, Ohio, September 1986.
22. Nilsson, Robert (professor, University of Stockholm): Personal communication, 1990.
23. Vogel, T. M., C. S. Criddle, and P. L. McCarty: "Transformations of Halogenated Aliphatic Compounds," *Environmental Science and Technology,* vol. 21, no. 8, pp. 722–736, 1987.
24. Bungay, H. R., III, and M. L. Bungay: "Microbial Interactions in Continuous Culture," *Advances in Applied Microbiology,* vol. 10, pp. 269–290, 1968.
25. Bull, A. T., and J. H. Slater: "Microbial Interactions and Community Structure," *Microbial Interactions and Communities,* vol. 1, edited by A. T. Bull and J. H. Slater, Academic Press, New York, pp. 13–44, 1982.
26. Tiedje, J. M., and T. O. Stevens: "The Ecology of an Anaerobic Dechlorinating Consortium," *Environmental Biotechnology,* edited by G. S. Omenn, Plenum Press, New York, pp. 3–14, 1990.
27. Singleton, Paul, and Diana Sainsbury: *Dictionary of Microbiology,* John Wiley & Sons, New York, 1978.
28. Irvine, R. L., and P. A. Wilderer: "Aerobic Processes," *Standard Handbook of Hazardous Waste Treatment and Disposal,* McGraw-Hill, New York, 1988.
29. Metcalf & Eddy, Inc.: *Wastewater Engineering. Treatment, Disposal, Reuse,* 3rd ed., McGraw-Hill, New York, 1991.
30. Ibid.
31. United States Environmental Protection Agency: *Corrective Action: Technologies and Applications,* EPA/625/4-89/020, Cincinnati, Ohio, September 1989.
32. Monod, J.: *Recherches sur La Croissance des Cultures Bacteriennes,* Herman et cie., Paris, 1942.
33. Bailey, J. E., and D. F. Ollis: *Biochemical Engineering Fundamentals,* 2nd ed., McGraw-Hill, New York, 1986.
34. Grady, C. P. L.: "Biodegradation of Toxic Organics: Status and Potential," *Journal of Environmental Engineering,* vol. 116, no. 5, pp. 805–828, September/October 1990.
35. Kobayashi, Hester, and Bruce E. Rittman: "Microbial Removal of Hazardous Organic Compounds," *Environmental Science and Technology,* vol. 16, no. 3, pp. A170-A183, 1982.
36. Srinivasan, P. N., and J. W. Mercer: "Simulation of Biodegradation and Sorption Processes in Ground Water," *Ground Water,* vol. 26, no. 4, pp. 475–487, July 1988.
37. Alexander, Martin: "Biodegradation of Organic Chemicals," *Environmental Science and Technology,* vol. 19, no. 2, pp. 106–111, 1985.
38. Bhattacharya, S. K.: "Innovative Biological Processes for Treatment of Hazardous Wastes," *Proceedings of the 11th National Conference Superfund '90,* Hazardous Materials Control Research Institute, Washington, D.C., pp. 847–852, November 26-28, 1990.
39. Bouwer, E. J., and P. L. McCarty: "Utilization Rates of Trace Halogenated Organic Compounds in Acetate-Grown Biofilms," *Biotechnology and Bioengineering,* vol. 27, pp. 1564–1571, 1985.

40. Metcalf & Eddy, Inc.: *Wastewater Engineering. Treatment, Disposal, Reuse,* 3rd ed.
41. Grady, C. P. L., Jr., G. T. Diagger, and H. C. Lim: *Biological Wastewater Treatment,* 2nd ed., Marcel Dekker, New York, 1998.
42. Grady, C. P. L.: "Biodegradation of Toxic Organics: Status and Potential."
43. SBP Technologies, Inc.: "Bioremediation of Polycyclic Aromatic Hydrocarbons & Pentachlorophenol," Stone Mountain, Ga., 1988.
44. Irvine, R. L., and P. A. Wilderer: "Aerobic Processes."
45. Wilder, P. A., and M. A. Rubio: "Sequencing Batch Reactors: A Versatile Technology for the Biological Treatment of Hazardous Leachates," *Contaminated Soil '88,* Kluwer Academic Publishers, Boston, 1988.
46. Metcalf & Eddy, Inc.: *Wastewater Engineering. Treatment, Disposal, Reuse,* 3rd ed.
47. Irvine, R. L., and P. A. Wilderer: "Aerobic Processes."
48. Personal Communication with Dr. A. Paul Togna, Envirogen, Inc., Lawrenceville, N.J., February 29, 2000.
49. Knoblock, M. D., P. M. Sutton, P. N. Mishra, and A. Janson: "Membrane Biological Reactor System for Treatment of Oily Wastewater: Pilot to Full Scale Results," *Water Environment Research,* vol. 66, pp. 133ff, 1994.
50. Pitre, Michael P., David N. Enegess, and Ronald Unterman: "The New Wave in Wastewater Treatment," *Environmental Protection,* pp. 30–32, 1999.
51. Cantor, J., P. M. Sutton, R. Steinheber, and M. Myronyk: "Membrane Filtration," *Industrial Wastewater,* pp. 18–22, 1999.
52. Skladany, George J., and Evan K. Nyer: "Decay Theory Biological Treatment for Low-Level Organic Contaminated Ground Water," *Proceedings of the Second National Outdoor Conference on Aquifer Restoration, Ground Water Monitoring and Geophysical Methods* (Las Vegas, Nev.), National Water Well Association, Dublin, Ohio, 1988.
53. Grady, C. P. L.: "Biodegradation of Toxic Organics: Status and Potential."
54. Tokuz, R. Y.: "Biotreatment of Hazardous Organic Wastes Using Rotating Biological Contactors," *Environmental Progress,* vol. 10, no. 3, pp. 198–204, 1991.
55. Bouwer, E. J., and P. L. McCarty: "Utilization Rates of Trace Halogenated Organic Compounds in Acetate-Grown Biofilms," *Biotechnology and Bioengineering,* vol. 27, pp. 1564–1571, 1985.
56. Skladany, G. J., J. M. Thomas, G. Fisher, and R. Ramachandran: "The Design, Economics, and Operation of a Biological Treatment System for Ketone Contaminated Ground and Solvent Recovery Process Waters," *Proceedings of the 42nd Annual Purdue Industrial Waste Conference,* Lewis Publishers, Chelsea, Mich., pp. 53–58, 1987.
57. Skladany, George J.: "Onsite Biological Treatment of an Industrial Landfill Leachate: Microbiological and Engineering Considerations," *Hazardous Waste and Hazardous Materials,* vol. 6, pp. 213–222, 1989.
58. Skladany, George J.: "The Use of Submerged Fixed-Film Reactors to Treat Aqueous Wastes," *Bioremediation: Field Experience,* edited by P. E. Flathman, D. E. Jerger, and J. H. Exner, Lewis Publishers, Boca Raton, Fla., pp. 237–258, 1993.
59. Bhattacharya, S. K.: "Innovative Biological Processes for Treatment of Hazardous Wastes," *Proceedings of the 11th National Conference Superfund '90,* Hazardous Material Control Research Institute, Washington, D.C., November 26–28, 1990.
60. Pitre, Michael P., David N. Enegess, and Ronald Unterman: "The New Wave in Wastewater Treatment."
61. EIMCO Process Equipment Company: "EIMCO Biolift Reactor," Salt Lake City, Utah, 1988.
62. Ahlert, R. C., and D. S. Kosson: "Aerobic Mineralization of Organic Contaminants Bound on Soil Fines," *3rd International Conference on New Frontiers for Hazardous Waste Management,* Pittsburgh, September 1989.
63. Bachmann, A., and A. J. B. Zehnder: "Engineering Significance of Fundamental Concepts in Xenobiotics Biodegradation in Soil," *Contaminated Soil '88,* Kluwer Academic Publishers, Boston, pp. 671–678, 1988.

64. Dekreuk, J. F., and G. J. Annokkee: "Applied Biotechnology for Decontamination of Polluted Soils: Possibilities and Problems," *Contaminated Soil '88,* Kluwer Academic Publishers, Boston, pp. 679–686, 1988.
65. Alvarcz, L. M., et al.: "The Effects of Sorption on the Biotransformation Rate of TCE by Methanotrophs," *62nd Annual Water Pollution Control Conference,* October 16, 1989, San Francisco.
66. Smith, J. R., J. Lynch, and G. R. Brubaker: "Overview of Selected Bioremediation Technologies," *Hazardous Waste and Hazardous Materials Conference,* April 12–14, 1989, New Orleans.
67. Bachmann, A., and A. J. B. Zehnder: "Engineering Significance of Fundamental Concepts in Xenobiotics Biodegradation in Soil."
68. Stroo, H. F.: "Bioremediation of Hydrocarbon-Contaminated Soil Solids Using Liquid/Solids Contact Reactors," *Proceedings of the 10th National Conference Superfund '89,* Hazardous Materials Control Research Institute, Washington, D.C., pp. 332–337, November 27, 1989.
69. Brox, G. H.: "A New Solid/Liquid Contact Bioslurry Reactor Making Bio-Remediation More Cost Competitive," *Proceedings of the 10th National Conference Superfund '89,* Hazardous Materials Control Research Institute, Washington, D.C., November 27, 1989.
70. EIMCO Process Equipment Company: "EIMCO Biolift Reactor."
71. Portier, R. J.: "Examination of Site Data and Discussion of Microbial Physiology with Regard to Site Remediation," *Proceedings of the 10th National Conference Superfund '89,* Hazardous Materials Control Research Institute, Washington, D.C., pp. 351–361, November 27, 1989.
72. Bachmann, A., and A. J. B. Zehnder: "Engineering Significance of Fundamental Concepts in Xenobiotics Biodegradation in Soil."
73. Dekreuk, J. F., and G. J. Annokkee: "Applied Biotechnology for Decontamination of Polluted Soils: Possibilities and Problems."
74. Cookson, John T.: *Bioremediation Engineering: Design and Application,* McGraw-Hill, New York, 1995.
75. Brown, K. W., G. B. Evans, Jr., and B. D. Frentrup: *Hazardous Waste Land Treatment,* Butterworth, Woburn, Mass., 1983.
76. Loehr, R. C., et al.: *Mobility and Degradation of Residues at Hazardous Waste Land Treatment Sites at Closure,* U.S. Environmental Protection Agency, EPA/600/S2-90/018, Ada, Okla., July 1990.
77. Smith, J. R., J. Lynch, and G. R. Brubaker: "Overview of Selected Bioremediation Technologies," Hazardous Waste and Hazardous Materials Conference, April 12–14, 1989, New Orleans.
78. "Land Treatment Demonstration," BP Oil Alliance Refinery, report prepared by ERM, Metairie, La., May 24, 1988.
79. Cookson, John T.: *Bioremediation Engineering: Design and Application.*
80. Smith, J. R., J. Lynch, and G. R. Brubaker: "Overview of Selected Bioremediation Technologies."
81. Williams, R. T., P. S. Ziegenfuss, G. B. Mohrman, and W. E. Sisk: "Composting of Explosives and Propellant Contaminated Sediments," *Proceedings of the 21st Mid-Atlantic Industrial Waste Conference,* Pennsylvania State University, pp. 599–611, June 25–27, 1989.
82. Taddeo, A., M. Findlay, M. Dooley-Danna, and S. Fogel: "Field Demonstration of a Forced Aeration Composting Treatment for Coal Tar," *Proceedings of the Second National Conference of Biotreatment: The Use of Microorganisms in the Treatment of Hazardous Materials and Hazardous Waste,* Washington, D.C., pp. 57–62, November 27–29, 1989.

83. Mays, M. K., L. J. Sikoa, Ph.D., and J. W. Hatton: "Composting as a Method for Hazardous Waste Treatment," *Proceedings of the 10th National Conference Superfund '88,* Hazardous Materials Control Research Institute, Washington, D.C., pp. 298–300, November 29, 1989.
84. Gwinn, Rosa E: "Composting Treatment of Explosives-Contaminated Soil, Tooele, Utah," *Bioremediation of Nitroaromatic and Haloaromatic Compounds,* edited by B. C. Alleman and A. Leeson, Battelle Press, Columbus, Ohio, pp. 51–56, 1999.
85. United States Environmental Protection Agency: "In-Vessel Composting of Municipal Wastewater Sludge," EPA/625/8-89/016. Center for Environmental Research Information, Cincinnati, OH, 1989.
86. Brenner, Richard, and Dennis Potter: "Cost-Effective Explosive Compounds Bioremediation by Windrow Composting, Hawthorne Army Depot," *Bioremediation of Nitroaromatic and Haloaromatic Compounds,* edited by B. C. Alleman and A. Leeson, Battelle Press, Columbus, Ohio, pp. 57–63, 1999.
87. Hater, G. R.: "Bioremediation of In-Situ and Ex Situ Contaminated Soils," *Proceedings of the 4th Annual Hazardous Materials Management Conference,* Tower Conference Management Company, 1988.
88. Leson, Gero, and A. Winer: "Biofiltration: An Innovative Pollution Control Technology for VOC Emissions," *Journal of the Air & Waste Management Association,* vol. 41, pp. 1045–1054, 1991.
89. Skladany, G. J., M. A. DeShusses, J. S. Devinny, A. P. Togna, and T. S. Webster: "Biological Systems: Biofilters," *Odor and VOC Control Handbook,* edited by H. J. Rafson, McGraw-Hill, New York, pp. 8.150–8.191, 1998.
90. Devinny, J. S., M. A. DeShusses, and T. S. Webster: *Biofiltration for Air Pollution Control,* Lewis Publishers, New York, 1998.
91. Wilson, S. B.: "In Situ Remediation of Ground Water and Soils," *Proceedings of the 10th National Conference Superfund '89,* Hazardous Materials Control Research Institute, Washington, D.C., pp. 227–240, November 27–29, 1989.
92. Cannon, J. Z.: "Considerations in Ground Water Remediation at Superfund Sites," memorandum from acting assistant administrator of U.S. Environmental Protection Agency, October 18, 1989.
93. Mackay, D. M., and J. A. Cherry, *Environmental Science and Technology,* vol. 23, no. 6, June 1989.
94. Hall, C. W.: "Practical Limits to Pump and Treat Technology for Aquifer Remediation," *Preconference Workshop,* Annual Conference of Water Pollution Control Federation, Dallas, 1989.
95. United States Environmental Protection Agency: "Summary of the August 1996 Meeting of the Bioremediation Action Committee," Office of Solid Waste and Emergency Response Technology Innovation Office, Washington, D.C., August 6–7, 1996.
96. United States Environmental Protection Agency: *Use of Monitored Natural Attenuation at Superfund, RCRA Corrective Action, and Underground Storage Tank Sites,* Office of Solid Waste and Emergency Response, Directive 9200.4-17, Washington, D.C., 1997.
97. United States Environmental Protection Agency: *Technical Protocol for Evaluating Natural Attenuation of Chlorinated Solvents in Ground Water,* Office of Research and Development, Washington, D.C., EPA/600/R-98/128, September 1998.
98. American Society for Testing and Materials: "Guidance for Remediation by Natural Attenuation at Petroleum Release Sites," Remediation by Natural Attenuation Task Group, Philadelphia, 1999.
99. Wisconsin Department of Natural Resources: "Interim Guidance on Natural Attenuation for Petroleum Releases Bureau for Remediation and Redevelopment," Pub-RR-614, October 1999.

100. Wiedemeier, T. H., J. T. Wilson, D. H. Kampbell, R. N. Miller, and J. E. Hansen: "Technical Protocol for Implementing Intrinsic Remediation with Long-Term Monitoring for Natural Attenuation of Fuel Contamination Dissolved in Groundwater," U.S. Air Force Center for Environmental Excellence, San Antonio, Tex., 1995.
101. United States Environmental Protection Agency: *Technical Protocol for Evaluating Natural Attenuation of Chlorinated Solvents in Ground Water.*
102. Ibid.
103. Ibid.
104. Buscheck, T. E., and C. M. Alcantar: "Regression Techniques and Analytical Solutions to Demonstrate Intrinsic Bioremediation," edited by R. E. Hinchee, J. T. Wilson, and D. C. Downey, Battelle Press, Columbus, Ohio, pp. 109–116, 1995.
105. United States Environmental Protection Agency: *Seminars—Monitored Natural Attenuation for Ground Water,* Office of Research and Development, EPA/625/K-98/001, September 1998.
106. Brady, P. V., M. V. Brady, and D. J. Borns: *Natural Attenuation—CERCLA , RBCAs and the Future of Environmental Remediation,* Lewis Publishers, Boca Raton, Fla., 1998.
107. United States Environmental Protection Agency: *Technical Protocol for Evaluating Natural Attenuation of Chlorinated Solvents in Ground Water.*
108. Buscheck, T. E., and C. M. Alcantar: "Regression Techniques and Analytical Solutions to Demonstrate Intrinsic Bioremediation."
109. McNab, W. W., Jr., and R. D. Brendan: "A Critique of a Steady-State Analytical Method for Estimating Contaminant Degradation Rates," *Ground Water,* vol. 36, pp. 983–987, 1998.
110. Newell, C. J., R. K. McLeod, and J. R. Gonzales: *BIOSCREEN: Natural Attenuation Decision Support System User's Manual,* Version 1.3, EPA/600/R-96/087, 1996.
111. Aziz, C. E., C. J, Newell, J. R. Gonzales, P. Haas, T. P. Clement, and Y. Sun: *BIOCHLOR Natural Attenuation Decision Support System Beta Version 1.0, User's Manual,* U.S. EPA, National Risk Management Research Laboratory, Subsurface Protection and Remediation Division, March 1999.
112. Rifai, H. S., C. J. Newell, P. Bedient, J. R. Gonzales, S. Dendrou, L. Kennedy, and J. T. Wilson: *BIOPLUME III—Natural Attenuation Decision Support System User's Manual,* Version 1.0, EPA/600/R-98/010, January 1998.
113. Clement, T. P.: "RT3D (Version 1.0), a Modular Computer Code for Simulating Reactive Multi-Species Transport in 3-Dimensional Groundwater Systems," The Battelle Pacific Northwest National Laboratory, PNNL-SA-11720, 1997.
114. Carey, G. R., P. J. Van Geel, and J. R. Murphy: *BIOREDOX V1.0: a Coupled Biodegradation-Redox Model for Simulating Natural and Enhanced Bioremediation of Organic Pollutants—Verification Manual,* Conestoga-Rovers & Associates, Waterloo, Ontario, Canada, 1998.
115. Lu, Guoping, T. P. Clement, C. Zheng, and T. Wiedemeier: "Natural Attenuation of BTEX Compounds: Model Development and Field-Scale Application," *Ground Water,* vol. 37, pp. 707–769, 1999.
116. United States Environmental Protection Agency: *Bioremediation Field Initiative Site Profile: Hill Air Force Base Superfund Site,* Office of Research and Office of Solid Waste and Environmental Protection Development Emergency Response, EPA/540/F-95/506C, September 1995.
117. Wiedemeier, T. H., J. T. Wilson, D. H. Kampbell, R. N. Miller, and J. E. Hansen: "Technical Protocol for Implementing Intrinsic Remediation with Long-Term Monitoring for Natural Attenuation of Fuel Contamination Dissolved in Groundwater."
118. Wilson, B. H., J. T. Wilson, D. H. Kampbell, and B. E. Bledsoe: "Traverse City: Geochemistry and Intrinsic Bioremediation of BTX Compounds," *Proceedings of the Sym-*

posium on Intrinsic Bioremediation of Ground Water, U.S. Environmental Protection Agency, pp. 94–102, 1994.

119. Gieg, L. M., R. V. Kolhatkar, M. J. McInterney, R. S. Tanner, S. H. Harris, K. L. Sublette, and J. M. Suflita: "Intrinsic Bioremediation of Petroleum Hydrocarbons in a Gas Condensate-Contaminated Aquifer," *Environmental Science and Technology,* vol. 33, pp. 2550–2560, 1999.
120. Weaver, J. W., J. T. Wilson, and D. H. Kampbell: "Case Study of Natural Attenuation of Trichloroethene at St. Joseph, Michigan," *Proceedings of the Symposium on Natural Attenuation of Chlorinated Organics in Ground Water,* Dallas, September 11–13, 1996, EPA/540/R-96/509, 1996.
121. Weaver, J. W., J. T. Wilson, and D. H. Kampbell: "Extraction of Degradation Rate Constants from the St. Joseph, Michigan Trichloroethene Site," *Proceedings of the Symposium on Natural Attenuation of Chlorinated Organics in Ground Water,* Dallas, September 11–13, 1996, EPA/540/R-96/509, 1996.
122. William, B.: "Natural Attenuation of Chlorinated Volatile Organic Compounds in a Freshwater Tidal Wetland, Aberdeen Proving Ground, Maryland," U.S. Geological Survey, Water-Resources Investigations Report 97-4171, 1997.
123. Wiedemeier, T. H., J. T. Wilson, and D. H. Campbell: "Natural Attenuation of Chlorinated Aliphatic Hydrocarbons at Plattsburgh Air Force Base, New York," *Proceedings of the Symposium on Natural Attenuation of Chlorinated Organics in Ground Water,* Dallas, September 11–13, 1996, EPA/540/R-96/509, 1996.
124. Ellis, D. E., E. J. Lutz, G. M. Klecka, D. L. Pardieck, J. J. Salvo, M. A. Heitkamp, D. J. Gannon, C. C. Mikula, C. M. Vogel, G. D. Sayles, D. H. Kampbell, J. T. Wilson, and D. T. Maiers: "Remediation Technology Development Forum Intrinsic Remediation Project at Dover Air Force Base, Delaware," *Proceedings of the Symposium on Natural Attenuation of Chlorinated Organics in Ground Water,* Dallas, September 11–13, 1996, EPA/540/R-96/509, 1996.
125. Clement, P., C. D. Johnson, Y. Sun, G. M. Klecka, and C. Bartlett: "Modeling Natural Attenuation of Chlorinated Solvent Plumes at Dover Air Force Base Area-6 Site," *Natural Attenuation of Chlorinated Solvents, Petroleum Hydrocarbons and Other Organic Compounds,* edited by Bruce C. Alleman and Andrea Leeson, Battelle Press, Columbus, Ohio, vol. 5, no. 1, p. 29(6), 1999.
126. United States Environmental Protection Agency: RODScan database.
127. Verheul, J. H. A. M., R. Van Den Berg, and D. H. Eikelboom: "In-Situ Biorestoration of a Subsurface Contaminated with Gasoline," *Contaminated Soil '88,* Kluwer Academic Publishers, Boston, pp. 705–715, 1988.
128. Cookson, John T.: *Bioremediation Engineering: Design and Application.*
129. Koenigsberg, Stephen S., and Robert D. Norris, editors: *Accelerated Bioremediation Using Slow Release Compounds, Selected Battelle Conference Papers 1993–1999,* Regenesis Bioremediation Products, San Clemente, Calif., 1999.
130. Ibid.
131. Hinchee, R. E., R. N. Miller, and P. C. Johnson: *In Situ Aeration, Bioventing, and Related Remediation Processes,* Battelle Press, Columbus, Ohio, 1995.
132. United States Environmental Protection Agency: *Bioventing Principles and Practices,* vol. 1: *Bioventing Principles*, EPA/540/R-95/534A, Washington, D.C., 1995.
133. United States Environmental Protection Agency: *Bioventing Principles and Practices,* vol. 2: *Bioventing Design,* EPA/540/R-95/534B, Washington, D.C., 1995.
134. United States Army Corps of Engineers: *Engineering and Design: Soil Vapor Extraction and Bioventing,* EM 1110-1-4001, 1995.
135. United States Environmental Protection Agency: *Bioventing Principles and Practices,* vol. 1: *Bioventing Principles.*

136. Hinchee, R. E., and S. E. Ong: "A Rapid In-Situ Respiration Test for Measuring Aerobic Biodegradation Rates of Hydrocarbons in Soil," *Journal of the Air & Waste Management Association,* vol. 42, pp. 1305–1312, 1992.
137. Hinchee, R. E., S. K. Ong, R. N. Miller, D. C. Downey, and R. Frandt: *Test Plan and Technical Protocol for a Field Treatability Test for Bioventing,* US Air Force Center for Environmental Excellence, San Antonio, Tex., 1992.
138. Hinchee, R. E.: *Air Sparging for Site Remediation,* Battelle Press, Battelle Memorial Institute, Columbus, Ohio, 1994.
139. Miller, R.: *Technology Overview Report: Air Sparging,* Ground-Water Remediation Technologies Analysis Center (GWRTAC), Pittsburgh, 1996.
140. American Academy of Environmental Engineers: *Innovative Site Remediation Technology: Design and Application,* vol. 7: *Vacuum Extraction and Air Sparging,* Annapolis, Md., 1998.
141. Edwards, N. T.: "Polycyclic Aromatic Hydrocarbons (PAHs) in the Terrestrial Environment," *Journal of Environmental Quality,* vol. 12, pp. 427–441, 1983.
142. Hossner, L. R., R. H. Loeppert, R. J. Newton, and P. J. Szaniszlo: Literature Review: Phytoaccumulation of Chromium, Uranium, and Plutonium in Plant Systems, Amarillo National Resource Center for Plutonium, Amarillo, Tex., May 1998.
143. Ibid.
144. United States Environmental Protection Agency: *Phytoremediation Resource Guide,* U.S. Environmental Protection Agency, Office of Solid Waste and Emergency Response, EPA/B-99/003, June 1999.
145. Ibid.
146. Ebbs, S. D.: "Phytoextraction of Zinc by Oat, Barley, and Indian Mustard," *Environmental Science & Technology,* vol. 32, pp. 802–805, 1998.
147. Schnoor, J. L.: *Phytoremediation,* Ground-Water Remediation Technologies Analysis Center (GWRTAC), Publication TE-98-01, Pittsburgh, 1997.
148. Cunningham, S. D., T. A. Anderson, A. P. Schwab, and F. C. Hsu: "Phytoremediation of Soils Contaminated with Organic Pollutants," *Advances in Agronomy,* vol. 56, pp. 55–114, 1996.
149. Fletcher, J. S., P. K. Donnelly, and R. S. Hedge: "Bioaccumulation of PCB-Degrading Bacteria by Compounds Released from Plant Roots," *Bioremediation of Recalcitrant Organics,* edited by R. E. Hinchee, R. E. Hoeppel, and D. B. Anderson, Battelle Press, Columbus, Ohio, pp. 131–136, 1995.
150. Hedge, R. S., and J. S. Fletcher: "Influence of Plant Growth Stage and Season on the Release of Root Phenolics by Mulberry as Related to Development of Phytoremediation Technology," *Chemosphere,* vol. 12, pp. 2471–2479, 1996.
151. Schnoor, J. L.: *Phytoremediation.*
152. United States Environmental Protection Agency: *Phytoremediation Resource Guide.*
153. Trapp, S., and J. C. McFarlane, editors: *Plant Contamination—Modeling and Simulation of Organic Chemical Processes,* Lewis Publishers, Boca Raton, Fla., 1994.
154. Schnoor, J. L.: *Phytoremediation.*
155. United States Environmental Protection Agency: *A Citizen's Guide to Phytoremediation,* U.S. Environmental Protection Agency, Office of Solid Waste and Emergency Response, Technology Innovation Office, EPA 542-F-96-014, September 1996.
156. Salanitro, J. P., G. E. Spinnler, C. C. Neaville, P. M. Maner, S. M. Stearns, P. C. Johnson, and C. Bruce: "Demonstration of the Enhanced MTBE Bioremediation (EMB) In Situ Process," *In Situ Bioremediation of Petroleum Hydrocarbon and Other Organic Compounds,* Battelle Press, Columbus, Ohio, vol. 5, no. 3, p. 37(10), 1999.
157. Baker, Katherine T., and Diane S. Herson: "Microbiology and Biodegradation," *Bioremediation,* edited by K. T. Baker and D. S. Herson, McGraw-Hill, New York, pp. 9–60, 1994.

158. Cookson, John T.: *Bioremediation Engineering: Design and Application.*
159. Alleman, Bruce C., and Andrea Leason, editors: *Natural Attenuation of Chlorinated Solvents, Petroleum Hydrocarbons, and Other Organic Compounds,* Battelle Press, Columbus, Ohio, 1999.
160. Alleman, Bruce C., and Andrea Leason, editors: *In Situ Bioremediation of Petroleum Hydrocarbons and Other Organic Compounds,* Battelle Press, Columbus, Ohio, 1999.
161. Le Roux, F., G. Hernandez, and J. P. Vandecastelle: "Mineralization of Ethyl *t*-Butyl Ether by Defined Mixed Cultures," *In Situ Bioremediation of Petroleum Hydrocarbons and Other Organic Compounds,* edited by B. C. Alleman and A. Leeson, Battelle Press, Columbus, Ohio, pp. 25–30, 1999.
162. Hyman, Michael, and Kirk O'Reilly: "Physiological and Enzymatic Features of MTBE-Degrading Bacteria," *In Situ Bioremediation of Petroleum Hydrocarbons and Other Organic Compounds,* edited by B. C. Alleman and A. Leeson, Battelle Press, Columbus, Ohio, pp. 7–12, 1999.
163. Koenigsberg, S., C. Sandefur, W. Mahaffey, M. DeShusses, and N. Fortin: "Peroxygen Mediated Bioremediation of MTBE," *In Situ Bioremediation of Petroleum Hydrocarbons and Other Organic Compounds,* edited by B. C. Alleman and A. Leeson, Battelle Press, Columbus, Ohio, pp. 13–18, 1999.
164. Garnier, Patrice, R. Auria, M. Magana, and S. Revah: "Cometabolic Biodegradation of Methyl T-Butyl Ether by a Soil Consortium," *In Situ Bioremediation of Petroleum Hydrocarbons and Other Organic Compounds,* edited by B. C. Alleman and A. Leeson, Battelle Press, Columbus, Ohio, pp. 31–36, 1999.
165. Reid, J. B., H. J. Reisinger, P. G. Bartholomae, J. C. Gray, and A. S. Hullman: "A Comparative Assessment of the Long-Term Behavior of MTBE and Benzene Plumes in Florida, USA," *Natural Attenuation of Chlorinated Solvents, Petroleum Hydrocarbons, and Other Organic Compounds,* edited by B. C. Alleman and A. Leeson, Battelle Press, Columbus, Ohio, pp. 97–102, 1999.
166. Hurt, K. H., J. T. Wilson, F. P. Beck, and J. S. Cho: "Anaerobic Biodegradation of MTBE in a Contaminated Aquifer," *Natural Attenuation of Chlorinated Solvents, Petroleum Hydrocarbons, and Other Organic Compounds,* edited by B. C. Alleman and A. Leeson, Battelle Press, Columbus, Ohio, pp. 103–108, 1999.
167. Koenigsberg, S., C. Sandefur, W. Mahaffey, M. Deshusses, and N. Fortin: "Peroxygen Mediated Bioremediation of MTBE," *In Situ Bioremediation of Petroleum Hydrocarbons and Other Organic Compounds,* edited by B. C. Alleman and A. Leeson, Battelle Press, Columbus, Ohio, pp. 13–18, 1999.
168. Defibaugh, Stephen T., and Daniel S. Fischman: "Biodegradation of MTBE Utilizing a Magnesium Peroxide Compound: A Case Study," *In Situ Bioremediation of Petroleum Hydrocarbons and Other Organic Compounds,* edited by B. C. Alleman and A. Leeson, Battelle Press, Columbus, Ohio, pp. 1–6, 1999.
169. Salanitro, J. P., G. E. Spinnler, C. C. Neaville, P. M. Maner, S. M. Stearns, P. C. Johnson, and C. Bruce: "Demonstration of the Enhanced MTBE Bioremediation (EMB) In Situ Process."
170. Hurt, K. H., J. T. Wilson, F. P. Beck, and J. S. Cho: "Anaerobic Biodegradation of MTBE in a Contaminated Aquifer."
171. Wilson, B. H., H. Shen, J. Cho, and J. Vardy: "Use of BIOSCREEN to Evaluate Natural Attenuation of MTBE," *Natural Attenuation of Chlorinated Solvents, Petroleum Hydrocarbons, and Other Organic Compounds,* edited by B. C. Alleman and A. Leeson, Battelle Press, Columbus, Ohio, pp. 115–120, 1999.
172. Anthony, J. W., B. M. Henry, T. H. Wiedemeier, E. K. Gordon, J. B. Bidgood, R. C. Hinchee, and J. E. Hansen: "Methodology to Evaluate Natural Attenuation of Methyl Tertiary-Butyl Ether," *Natural Attenuation of Chlorinated Solvents, Petroleum Hydrocarbons, and Other Organic Compounds,* edited by B. C. Alleman and A. Leeson, Battelle Press, Columbus, Ohio, pp. 121–134, 1999.

173. Hurt, K. H., J. T. Wilson, F. P. Beck, and J. S. Cho: "Anaerobic Biodegradation of MTBE in a Contaminated Aquifer."
174. Alleman, Bruce C., and Andrea Leeson, editors: *Bioremediation of Nitroaromatic and Haloaromatic Compounds,* Battelle Press, Columbus, Ohio, 1999.
175. Autenrieth, R. L., M. D. Jankowski, J. S. Bonner, and M. Kodikanti: "Optimizing the Biotransformation of RDX and HMX," *Bioremediation of Nitroaromatic and Haloaromatic Compounds,* edited by B. C. Alleman and A. Leeson, Battelle Press, Columbus, Ohio, pp. 21–26, 1999.
176. Riefler, R. G.: "Modeling the Mineralization of Trinitrotoluene and Dinitrotoluenes in an Aerobic/Anoxic Biofilm," *Bioremediation of Nitroaromatic and Haloaromatic Compounds,* edited by B. C. Alleman and A. Leeson, Battelle Press, Columbus, Ohio, pp. 27–32, 1999.
177. Spain, J. C., S. F. Nishino, M. R. Greene, J. E. Forbort, N. A. Nogalski, R. Unterman, W. M. Riznychok, S. E. Thompson, P. M. Sleeper, and M. A. Boxwell: "Field Demonstration of FBR for Treatment of Nitrotoluenes in Groundwater," *Bioremediation of Nitroaromatic and Haloaromatic Compounds,* edited by B. C. Alleman and A. Leeson, Battelle Press, Columbus, Ohio, pp. 7–14, 1999.
178. Guiot, S. R., C. F. Shen, L. Paquet, J. Breton, J. Hawari, S. Thiboutot, and G. Ampleman: "Pilot-Scale Anaerobic Bioslurry Remediation of RDX- and HMX-Contaminated Soils," *Bioremediation of Nitroaromatic and Haloaromatic Compounds,* edited by B. C. Alleman and A. Leeson, Battelle Press, Columbus, Ohio, pp. 15–20, 1999.
179. Jerger, Douglas E., and Patrick Woodhill: "Technology Development for Biological Treatment of Explosives-Contaminated Soils," *Bioremediation of Nitroaromatic and Haloaromatic Compounds,* edited by B. C. Alleman and A. Leeson, Battelle Press, Columbus, Ohio, pp. 33–38, 1999.
180. Green, Roger B., Paul W. Burns, and Gary R. Hater: "Pilot Demonstration of the Sequential Anaerobic-Aerobic Bioremediation of Explosives-Contaminated Soils," *Bioremediation of Nitroaromatic and Haloaromatic Compounds,* edited by B. C. Alleman and A. Leeson, Battelle Press, Columbus, Ohio, pp. 45–50, 1999.
181. Gwinn, Rosa E: "Composting Treatment of Explosives-Contaminated Soil, Tooele, Utah," *Bioremediation of Nitroaromatic and Haloaromatic Compounds,* edited by B. C. Alleman and A. Leeson, Battelle Press, Columbus, Ohio, pp. 51–56, 1999.
182. Brenner, Richard, and Dennis Potter: "Cost-Effective Explosive Compounds Bioremediation by Windrow Composting, Hawthorne Army Depot," *Bioremediation of Nitroaromatic and Haloaromatic Compounds,* edited by B. C. Alleman and A. Leeson, Battelle Press, Columbus, Ohio, pp. 57–63, 1999.
183. Cookson, John T.: *Bioremediation Engineering: Design and Application.*
184. Ibid.
185. Gray, N. C. C., P. R. Cline, G. P. Moser, L. E. Moser, H. A. Guiler, A. L. Gray, and D. J. Gannon: "Full-Scale Bioremediation of Chlorinated Pesticides," *Bioremediation of Nitroaromatic and Haloaromatic Compounds,* edited by B. C. Alleman and A. Leeson, Battelle Press, Columbus, Ohio, pp. 125–130, 1999.
186. Cookson, John T.: *Bioremediation Engineering: Design and Application.*
187. McLean, J. S., T. J. Beveridge, and D. Phipps: "Chromate Removal from Contaminated Groundwater Using Indigenous Bacteria," *Bioremediation of Metals and Inorganic Compounds,* edited by B. C. Alleman and A. Leeson, Battelle Press, Columbus, Ohio, pp. 121–126, 1999.
188. Abdelouas, Abdesselam: "Long-Term Stability of Uraninite Following In Situ Remediation of U(VI)," *Bioremediation of Metals and Inorganic Compounds,* edited by B. C. Alleman and A. Leeson, Battelle Press, Columbus, Ohio, pp. 115–120, 1999.

189. Gerlach, Robin, Al Cunningham, and Frank Caccavo, Jr.: "Chromium Elimination with Microbially Reduced Iron–Redox-Reactive Biobarriers," *Bioremediation of Metals and Inorganic Compounds,* edited by B. C. Alleman and A. Leeson, Battelle Press, Columbus, Ohio, pp. 13–18, 1999.
190. McGregor, R., D. Blowes, R. Ludwig, E. Pringle, and M. Pomeroy: "Remediation of a Heavy Metal Plume Using a Reactive Wall," *Bioremediation of Metals and Inorganic Compounds,* edited by B. C. Alleman and A. Leeson, Battelle Press, Columbus, Ohio, pp. 19–24, 1999.
191. Nordwick, S., M. Canty, R. Hiebert, and L. Thompson: "Microbiological Treatment of Cyanide, Metals, and Nitrates in Mining Byproducts," *Bioremediation of Metals and Inorganic Compounds,* edited by B. C. Alleman and A. Leeson, Battelle Press, Columbus, Ohio, pp. 31–36, 1999.
192. Hunter, W. J.: "Use of Vegetable Oil in a Pilot Scale Denitrifying Biobarrier," *Bioremediation of Metals and Inorganic Compounds,* edited by B. C. Alleman and A. Leeson, Battelle Press, Columbus, Ohio, pp. 47–51, 1999.
193. Nuttall, H. Eric, Lijun Deng, Abdesselam Abdelouas, and Werner Lutze: "In Situ Denitrification: A Field Demonstration," *Bioremediation of Metals and Inorganic Compounds,* edited by B. C. Alleman and A. Leeson, Battelle Press, Columbus, Ohio, pp. 59–64, 1999.
194. Nelson, Sheldon N., Kirk T. O'Reilly, Miriam L. Barbonio, and Gerald Vogt: "Laboratory and Field Testing of Engineered Denitrification," *Bioremediation of Metals and Inorganic Compounds,* edited by B. C. Alleman and A. Leeson, Battelle Press, Columbus, Ohio, pp. 91–96, 1999.
195. Skladany, George J., and Katherine H. Baker: "Laboratory Treatability Studies," *Bioremediation,* edited by Katherine T. Baker and Diane S. Herson, McGraw-Hill, New York, pp. 97–172, 1994.
196. Cookson, John T.: *Bioremediation Engineering: Design and Application.*
197. United States Environmental Protection Agency: Code of Federal Regulations (40 CFR Ch. 1, 7-1-99 Edition), U. S. Government Printing Office, Washington, D.C., pp. 563–631, 1988.
198. Skladany, George J., and Katherine H. Baker: "Laboratory Treatability Studies."
199. United States Environmental Protection Agency: *Guide for Conducting Treatability Studies under CERCLA: Biodegradation Remedy Selection, Second and Final Draft,* U.S. EPA Publication, Washington, D.C., March 1993.
200. Rozich, A. F., and A. F. Gaudy, Jr.: *Design and Operation of Activated Sludge Processes Using Respirometry,* Lewis Publishers, Chelsea, Mich., 1992.
201. Gaudy, A. F. Jr., A. Ekambaram, A F. Rozich, and R. J. Colvin: "Comparison of Respirometric Methods for Determination of Biokinetic Constants for Toxic and Nontoxic Wastes," *Proceedings of the 44th Purdue Industrial Waste Conference,* Lewis Publishers, Chelsea, Mich., pp. 393–403, 1990.
202. Gaudy, A. F., Jr., A. F. Rozich, N. R. Moran, S. T. Garniewski, and A. Ekambaram: "Methodology for Utilizing Respirometric Data to Assess Biodegradation Kinetics," *Proceedings of the 42nd Purdue Industrial Waste Conference,* Lewis Publishers, Chelsea, Mich., pp. 573–584, 1988.
203. Rozich, A. F., and R. J. Colvin: "Formulating Strategies for Activated Sludge Systems," *Waste Engineering & Management,* vol. 137, no. 10, pp. 39–41, 1990.
204. Rifai, H. S., C. J. Newell, P. Bedient, J. R. Gonzales, S. Dendrou, L. Kennedy, and J. T. Wilson: *BIOPLUME III—Natural Attenuation Decision Support System User's Manual,* Version 1.0.
205. Newell, C. J., R. K. McLeod, and J. R. Gonzales: *BIOSCREEN: Natural Attenuation Decision Support System User's Manual,* Version 1.3.

206. Aziz, C. E., C. J, Newell, J. R. Gonzales, P. Haas, T. P. Clement, and Y. Sun: *BIOCHLOR Natural Attenuation Decision Support System Beta Version 1.0, User's Manual.*
207. Ripp, S., D. E. Nivens, Y. Ahn, C. Werner, J. Jarrell IV, J. P. Easter, C. D. Cox, R. S. Burlage, and G. S. Sayler: "Controlled Field Release of a Bioluminescent Genetically Engineered Microorganism for Bioremediation Process Monitoring and Control," *Environmental Science and Technology,* vol. 34, pp. 846–853, 2000.

ADDITIONAL READING

Microbiology, Environmental Microbiology, and Microbial Ecology

Alef, Kassem, and Paolo Nannipieri: *Methods in Soil Microbiology and Biochemistry,* Academic Press, San Diego, 1995.
Atlas, R. M., and R. Bartha: *Microbial Ecology: Fundamentals and Applications,* 4th ed., Addison-Wesley, New York, 1997.
Atlas, Ronald A.: *Principles of Microbiology,* 2nd ed., Mosby Yearbook, New York, 1996.
Lim, D. V.: *Microbiology,* 2nd ed., McGraw-Hill, New York, 1997.
Maier, R. M., I. L. Pepper, and C. N. Gerba: *Environmental Microbiology,* Academic Press, San Diego, CA, 1999.
Mitchell, R.: *Environmental Microbiology,* Wiley-Liss, New York, 1992.
Nester, E. W., C. E. Roberts, N. N. Pearsall, and D. G. Anderson: *Microbiology: A Human Perspective,* McGraw Hill, New York, 1998.
Paul, E. A., and F. E. Clark: *Soil Microbiology and Biochemistry,* Academic Press, San Diego, 1989.
Paul, Eldor A., and Francis E. Clark: *Soil Microbiology and Biochemistry,* 2nd ed., Academic Press, San Diego, 1996.

Biochemistry, Metabolism, and Bioremediation

Alexander, Martin: *Biodegradation and Bioremediation,* 2nd ed., Academic Press, San Diego, 1999.
Baker, Katherine T., and Diane S. Herson, (editors): *Bioremediation,* McGraw-Hill, New York, 1994.
Cookson, John T.: *Bioremediation Engineering: Design and Application,* McGraw-Hill, New York, 1995.
Garrett, R. H., and C. M. Grisham: *Biochemistry,* 2nd ed., Saunders College Publishing, New York, 1999.
Ingraham, J. L., O. Maaloe, and F. C. Neidhardt: *Growth of the Bacterial Cell,* Sinauer Associates, Sunderland, Mass., 1983.
Lehninger, A. L., D. L. Nelson, and M. M. Cox: *Principles of Biochemistry,* W. H. Freeman, New York, 2000.
Metzler, D. E.: *Biochemistry: Chemical Reactions of Living Cells,* 2nd ed., Academic Press, New York, 2000.
Silver, S., A. M. Chakrabarty, B. Iglewski, and S. Kaplan: *Pseudomonas Biotransformations, Pathogenesis, and Evolving Biotechnology,* American Society for Microbiology, Washington, D.C., 1990.
Sokatch, J. R., and L. N. Ornston, (editors): *The Bacteria,* vol. X. *The Biology of Pseudomonas,* Academic Press, New York, 1986.
Stryer, L. L.: *Biochemistry,* 4th ed., W. H. Freeman, New York, 1996.

United States Environmental Protection Agency: *Microbial Decomposition of Chlorinated Aromatic Compounds,* Office of Research and Development, EPA/600/2-86/090, Cincinnati, September 1986.
Voet, D., and J. G. Voet: *Biochemistry,* 2nd ed., John Wiley & Sons, New York, 1995.

Biological Wastewater Treatment Systems

Celenza, Gaetano J.: *Industrial Waste Treatment Process Engineering,* vol. 2: *Biological Processes,* Technomic Publishing Company, Lancaster, Pa., 1999.
Grady, C. P. L. G., Jr., G. T. Diagger, and H. C. Lim: *Biological Wastewater Treatment,* 2nd ed., Marcel Dekker, New York, 1998.
Metcalf & Eddy, Inc.: *Wastewater Engineering—Treatment, Disposal, Reuse,* 3rd ed., McGraw-Hill, New York, 1991.

Biological Hazardous Waste Treatment Systems, Processes, and Applications

Adriano, D. C., J.-M. Bollag, W. T. Frankenberger, Jr., and R. C. Sims: *Bioremediation of Contaminated Soils* (Monograph 37), American Society of Agronomy, Crop Science Society of America, and the Soil Science Society of America, Madison, Wis., 1999.
Alleman, Bruce C., and Andrea Leeson, (editors): *Bioreactor and Ex Situ Biological Treatment Technologies,* Battelle Press, Columbus, Ohio, 1999.
Alleman, Bruce C., and Andrea Leeson, (editors): *Bioremediation of Nitroaromatic and Haloaromatic Compounds,* Battelle Press, Columbus, Ohio, 1999.
Alleman, Bruce C., and Andrea Leeson, (editors): *In Situ Bioremediation of Petroleum Hydrocarbon and Other Organic Compounds,* Battelle Press, Columbus, Ohio, 1999.
Alleman, Bruce C., and Andrea Leeson, (editors): *Natural Attenuation of Chlorinated Solvents, Petroleum Hydrocarbons, and Other Organic Compounds,* Battelle Press, Columbus, Ohio, 1999.
Alleman, Bruce C., and Andrea Leeson, (symposium chairs): *In Situ and On-Site Bioremediation:* vols. 1–5, Battelle Press, Columbus, Ohio, 1997.
American Academy of Environmental Engineers: Innovative Site Remediation Technology: *Design and Application,* vol. 7: *Vacuum Extraction and Air Sparging,* Annapolis, Md., 1998.
American Society for Testing and Materials: *Guidance for Remediation by Natural Attenuation at Petroleum Release Sites,* Remediation by Natural Attenuation Task Group, Philadelphia, 1999.
Bajpai, Rakesh K., and Mark E. Zappi, (editors): *Bioremediation of Surface and Subsurface Contamination,* The New York Academy of Sciences, New York, 1997.
Baker, Katherine T., and Diane S. Herson, (editors): *Bioremediation,* McGraw-Hill, New York, 1994.
Brady, P. V., M. V. Brady, and D. J. Borns: *Natural Attenuation—CERCLA , RBCA's and the Future of Environmental Remediation,* Lewis Publishers, Boca Raton, Fla., 1998.
Cookson, John T.: *Bioremediation Engineering: Design and Application,* McGraw-Hill, New York, 1995.
Devinny, J. S., M. A. Deshusses, and T. S. Webster: *Biofiltration for Air Pollution Control,* Lewis Publishers, New York, 1998.
Flathman, P. E., D. E. Jerger, and J. H. Exner, (editors): *Bioremediation—Field Experience,* Lewis Publishers, Boca Raton, Fla., 1994.
Hinchee, R. E., A. Leeson, and L. Semprini, (editors): *Bioremediation of Chlorinated Solvents,* Battelle Press, Columbus, Ohio, 1995.

Hinchee, R. E., A. Leeson, L. Semprini, and S. K. Ong, (editors): *Bioremediation of Chlorinated and Polycyclic Hydrocarbon Compounds,* Battelle Press, Columbus, Ohio, 1994.
Hinchee, R. E., and Robert F. Olfenbuttel, (editors): *In Situ Bioreclamation—Applications and Investigations for Hydrocarbon and Contaminated Site Remediation,* Butterworth-Heinemann, Boston, 1991.
Hinchee, R. E., and Robert F. Olfenbuttel, (editors): *On-Site Bioreclamation—Processes for Xenobiotic and Hydrocarbon Treatment,* Butterworth-Heinemann, Boston, 1991.
Hinchee, R. E., B. C. Alleman, R. E. Hoeppel, and R. N. Miller, (editors): *Hydrocarbon Bioremediation,* Battelle Press, Columbus, Ohio, 1994.
Hinchee, R. E., C. M. Vogel, and F. J. Brockman, (editors): *Microbial Processes for Bioremediation,* Battelle Press, Columbus, Ohio, 1995.
Hinchee, R. E., D. B. Anderson, F. B. Metting, Jr., and G. D. Sayles, (editors): *Applied Biotechnology for Site Remediation,* Battelle Press, Columbus, Ohio, 1994.
Hinchee, R. E., (editor): *Air Spaging for Site Remediation,* Battelle Press, Columbus, Ohio, 1994.
Hinchee, R. E., G. D. Sayles, and R. S. Skeen, (editors): *Biological Unit Processes for Hazardous Waste Treatment,* Battelle Press, Columbus, Ohio, 1995.
Hinchee, R. E., G. S. Douglas, and S. K. Ong, (editors): *Monitoring and Verification of Bioremediation,* Battelle Press, Columbus, Ohio, 1995.
Hinchee, R. E., J. A. Kittel, and H. J. Reisinger, (editors): *Applied Bioremediation of Petroleum Hydrocarbons,* Battelle Press, Columbus, Ohio, 1995.
Hinchee, R. E., J. Fredrickson, and B. C. Alleman, (editors): *Bioaugmentation for Site Remediation,* Battelle Press, Columbus, Ohio, 1995.
Hinchee, R. E., J. L. Means, and D. R. Burris, (editors): *Bioremediation of Inorganics,* Battelle Press, Columbus, Ohio, 1995.
Hinchee, R. E., J. T. Wilson, and D. C. Downey, (editors): *Intrinsic Bioremediation,* Battelle Press, Columbus, Ohio, 1995.
Hinchee, R. E., R. E. Hoeppel, and D. B. Anderson, (editors): *Bioremediation of Recalcitrant Organics,* Battelle Press, Columbus, Ohio, 1995.
Hinchee, R. E., R. N. Miller, and P. C. Johnson, (editors): *In Situ Aeration: Air Sparging, Bioventing, and Related Remediation Processes,* Battelle Press, Columbus, Ohio, 1995.
Hinchee, R. E., S. K. Ong, R. N. Miller, D. C. Downey, and R. Frandt: *Test Plan and Technical Protocol for a Field Treatability Test for Bioventing,* US Air Force, Center for Environmental Excellence, 1992.
Leeson, Andrea, and Bruce C. Alleman, (editors): *Bioremediation of Metals and Inorganic Compounds,* Battelle Press, Columbus, Ohio, 1999.
Leeson, Andrea, and Bruce C. Alleman, (editors): *Bioremediation Technologies for Polycyclic Aromatic Hydrocarbon Compounds,* Battelle Press, Columbus, Ohio, 1999.
Leeson, Andrea, and Bruce C. Alleman, (editors): *Engineered Approaches for In Situ Bioremediation of Chlorinated Solvent Contamination,* Battelle Press, Columbus, Ohio, 1999.
Leeson, Andrea, and Bruce C. Alleman, (editors): *Phytoremediation and Innovative Strategies for Specialized Remedial Applications,* Battelle Press, Columbus, Ohio, 1999.
Means, J. L. and R. E. Hinchee, (editors): *Emerging Technology for the Bioremediation of Metals,* Battelle Press, Columbus, Ohio, 1994.
Miller, R.: Technology Overview Report: *Air Sparging,* Ground-Water Remediation Technologies Analysis Center (GWRTAC), Pittsburgh, October 1998.
National Research Council: *In Situ Bioremediation,* National Academy Press, Washington, D.C., 1993.
Norris, R. D., et al.: *Handbook of Bioremediation,* CRC Press, Inc., Lewis Publishers, Boca Raton, Fla., 1994.
Nyer, E. K., et al.: *In Situ Treatment Technology,* CRC Press, Boca Raton, Fla., 1996.

Schnoor, J. L.: *Phytoremediation,* Publication TE-98-01, Ground-Water Remediation Technologies Analysis Center, Pittsburgh, October 1997.

United States Army Corps of Engineers: *Engineering and Design: Soil Vapor Extraction and BioVenting,* EM 1110-1-4001, 1995.

United States Environmental Protection Agency: *Bioventing Principles and Practices,* vol. 1: *Bioventing Principles,* EPA/540/R-95/534a, Washington, D.C., 1995.

United States Environmental Protection Agency: *Bioventing Principles and Practices,* vol. 2: *Bioventing Design,* EPA/540/R-95/534b, Washington, D.C., 1995.

United States Environmental Protection Agency: *Citizen's Guide to Phytoremediation,* Office of Solid Waste and Emergency Response, Technology Innovation Office, EPA/542/F-96/014, September 1996.

United States Environmental Protection Agency: *Phytoremediation Resource Guide,* Office of Solid Waste and Emergency Response, EPA/B-99/003, June 1999.

United States Environmental Protection Agency: *Remediation Case Studies: Bioremediation,* EPA/542/R-95/002, Washington, D.C., 1995.

United States Environmental Protection Agency: *Seminars—Monitored Natural Attenuation for Ground Water,* Office of Research and Development, EPA/625/K-98/001, September 1998.

United States Environmental Protection Agency: *Technical Protocol for Evaluating Natural Attenuation of Chlorinated Solvents in Ground Water,* Office of Research and Development, Washington, D.C., EPA/600/R-98/128, September 1998.

Ward, C. H., J. A. Cherry, and M. R. Scalf, (editors): *Subsurface Restoration,* Ann Arbor Press, Chelsea, Mich., 1997.

Wickramanayake, Godage B., and Robert E. Hinchee, (editors): *Bioremediation and Phytoremediation: Chlorinated and Recalcitrant Compounds,* Battelle Press, Columbus, Ohio, 1998.

Wickramanayake, Godage B., and Robert E. Hinchee, (editors): *Designing and Applying Treatment Technologies: Chlorinated and Recalcitrant Compounds,* Battelle Press, Columbus, Ohio, 1998.

Wickramanayake, Godage B., and Robert E. Hinchee, (editors): *Natural Attenuation: Chlorinated and Recalcitrant Compounds,* Battelle Press, Columbus, Ohio, 1998.

Wisconsin Department of Natural Resources: *Interim Guidance on Natural Attenuation for Petroleum Releases,* Bureau for Remediation and Redevelopment, PUB-RR-614, October 1999.

Wise, Donald L., and Debra J. Trantolo, (editors): *Remediation of Hazardous Waste Contaminated Soils,* Marcel Dekker, New York, 1994.

Journals

Applied and Environmental Microbiology, American Society for Microbiology, Washington, D.C.

Bioremediation Journal, CRC Press, Boca Raton, Fla.

ACKNOWLEDGMENTS

The authors wish to acknowledge the following individuals who assisted with the preparation of the first edition chapter: Derek Ross, Ph.D.; Al Rozich, Ph.D., P.E.; Jack Harrison, P.E.; Nicholas Hollingshad, *ERM*; and Michael Goldman, *Buchart-Horn Engineers*. The principal reviewer for ERM was Brian Flynn.

This second edition chapter was co-authored by George Skladany, Ph.D., *ERM*. In addition to updating and revising the existing text, he edited the new material that now makes up approximately half of the chapter. New text and example problems were contributed by ERM associates Daniel Yamashiro and Andrew McIntosh (bioventing and biosparging); Victoria Oster, P.E., and Elsie Millano, Ph.D., P.E. (monitored natural attenuation); Beth Baker and John Beath, P.E. (phytoremediation); and George Skladany (various facets of microbial metabolism, vapor-phase treatment, classes of xenobiotics amenable to biological treatment, and hazardous waste treatability studies). Jim Lute, Ph.D., of ERM reviewed portions of the document and provided valuable input as the manuscript was being written.

CHAPTER 11

Stabilization and Solidification

There is nothing stable in the world.
John Keats

11-1 INTRODUCTION

Stabilization and solidification have been widely applied in the management of hazardous wastes. The technologies are being applied to (1) the treatment of industrial wastes, (2) the treatment of wastes prior to secure landfill disposal, and (3) the treatment of contaminated land where large quantities of soil containing contaminants are encountered.

In general terms, stabilization is a process where additives are mixed with waste to minimize the rate of contaminant migration from the waste and to reduce the toxicity of the waste. Thus, stabilization may be described as a process by which contaminants are fully or partially bound by the addition of supporting media, binders, or other modifiers.[1] Likewise, solidification is a process employing additives by which the physical nature of the waste (as measured by the engineering properties of strength, compressibility, and/or permeability) is altered during the process.[2] Thus, objectives of stabilization and solidification would encompass both the reduction in waste toxicity and mobility as well as an improvement in the engineering properties of the stabilized material.[3]

Although stabilization and solidification are physicochemical processes, they are discussed separately from those in Chapter 9. This allows for more detailed examination of these widely employed hazardous waste management techniques, including presentation of stabilization mechanisms, techniques, and design considerations along with the tests used to evaluate the treatment effectiveness.

Definitions

Stabilization is a process employing additives (reagents) to reduce the hazardous nature of a waste by converting the waste and its hazardous constituents into a form that minimizes the rate of contaminant migration into the environment, or reduces the level of toxicity.

Fixation is often used synonymously with stabilization. Stabilization is accomplished through the addition of reagents that[4]:

- Improve the handling and physical characteristics of the waste
- Decrease the surface area across which transfer or loss of contaminants can occur
- Limit the solubility of any pollutants contained in the waste
- Reduce the toxicity of the contaminants

In contrast, **solidification** is described as a process by which sufficient quantities of solidifying material, including solids, are added to the hazardous materials to result in a solidified mass of material.[5] Solidifying the mass is accomplished through the addition of reagents that increase the strength, decrease the compressibility, and decrease the permeability of the waste.

In most cases, the processes of stabilization and solidification are combined in the treatment of hazardous wastes. Therefore, throughout this chapter the term stabilization will be utilized to mean both stabilization and solidification except where specifically noted otherwise.

The potential for contaminant loss from a stabilized mass is usually determined by leaching tests. **Leaching** is the process by which contaminants are transferred from a stabilized matrix to a liquid medium such as water.

During stabilization, certain contaminants may be destroyed by, for example, the dechlorination of chlorinated hydrocarbons. Other organics may "disappear" as a result of volatilization. However, the stabilization of inorganic contaminants that are already in their atomic form, such as cadmium, lead, and other metals, should mimic nature. By studying the form in which they occur in nature, we can learn much about the optimum stabilization method. For example, zinc phosphate ore bodies provide the source of zinc for metallurgical processes, and a stabilization technique that precipitates the zinc as zinc phosphate is likely to be the optimum technique in terms of environmental effectiveness. Stabilization must be considered as a waste treatment process that reduces, to an acceptable or *geologically slow* rate, the movement of contaminants into the environment.

Regulations

The cleanup of hazardous waste sites that endanger the public health and welfare and the environment requires the application of remedial technologies that are both effective and permanent. In this regard, stabilization offers technological and health and safety advantages over alternative remedial options such as landfilling that manage waste by transfering it from one location to another. Stabilization may also be utilized in situations where other treatment methods, such as bioremediation or incineration, are not technically applicable or cost-effective. For long-term solutions

it is important to consider more permanent technologies such as stabilization. As a result, stabilization is specifically identified in the U.S. National Contingency Plan.[6]

It has been recognized that the disposal of liquid wastes in landfills is an environmentally unsound practice. Liquids migrate downward through the landfill, assist in the generation of leachate, and potentially migrate through defects in the liner system. As a consequence of the high mobility of liquid hazardous wastes, their disposal in landfills has been prohibited in the United States by RCRA as amended by the Hazardous and Solid Waste Amendments of 1984.[7] This ban has resulted in the increased use of stabilization in order to preclude disposal of liquid wastes. In this application, the stabilization agents must chemically bond the hazardous chemicals in liquid wastes. The agents cannot act simply as absorbents (physically containing the liquids much the same way in which a sponge holds water) and must pass the Liquids Release Test[8] discussed later in the chapter.

The U.S. regulations have thus brought into focus two distinctly different technological objectives for stabilization: (1) stabilization of materials prior to land disposal and (2) the stabilization of contaminants to be left in place as part of a site cleanup. Stabilization prior to land disposal may require a less stringent performance measure (eliminating free liquids). Alternatively, for those situations where the treated material is left in place, the stabilization process may require considerably greater performance (reducing leachability).

Applications

The three major areas of application for stabilization technologies are:

- *Land disposal*—the stabilization of wastes prior to secure landfill disposal
- *Site remediation*—the remediation of contaminated sites
- *Solidification of industrial wastes*—the solidification of nonhazardous, unstable wastes, such as sludges

Land disposal

The land disposal of liquid waste increases the likelihood of contaminant migration (and is currently banned under U.S. regulations). Liquid waste, along with wet sludges, must be stabilized prior to landfilling. To effectively stabilize liquids, the stabilization agents cannot be absorbents (such as sawdust). The liquids absorbed by agents could be easily released (desorbed) in the landfill when compressed under additional loads. That is, as more material is landfilled, the weight of the overlying material would squeeze the liquids out of the underlying material. Therefore, the liquids must be chemically and physically bound by the stabilization reagents so that they are not expelled by the consolidation stresses or leached out by the downward percolation of precipitation. The Liquids Release Test[8] simulates these consolidation stresses. Note that under U.S. regulations, incineration is the preferred treatment method over land disposal for certain organic wastes. However, there are provisions for demonstrating alternative technologies in the land disposal restrictions.

Site remediation

The remediation of contaminated sites having organic wastes, inorganic wastes, and/or contaminated soils may be accomplished by employing stabilization tech-

nology. For site remediation, stabilization is used to (1) improve the handling and physical characteristics of the wastes, (2) decrease the rate of contaminant migration by decreasing the surface area across which the transfer of pollutants can occur and by limiting the solubility of pollutants, and (3) reduce the toxicity of certain contaminants. Stabilization may be considered to be a more permanent remedial solution than other alternatives such as containment. Stabilization is particularly well suited for sites where the hazard involves large quantities of soils contaminated at low levels. In many instances it may not be environmentally sound nor cost-effective to excavate, transport, and landfill or incinerate soils contaminated with low levels of pollutants because of (1) the additional air pollution caused by excavation equipment, trucks, and the exposure of buried contaminated soils to the air, which enhances the volatilization of organics, and (2) the increased risks as a result of traffic accidents.

EXAMPLE 11-1. EVALUATION OF LANDFILLING VERSUS STABILIZATION. The total quantity of contaminated soils at the Pepper's Steel and Alloy site was 120,000 tons.[9] Evaluate the transportation cost of the landfilling option versus stabilization for the management of the hazardous waste at this site. Each truck can carry 31,500 pounds to the nearest suitable landfill site at Emile, Alabama, at a distance of 850 miles. The trucking cost per mile was \$2.00 and the total stabilization cost was \$67 per ton. Identify advantages and disadvantages of landfilling for this site.

Solution. Stabilization cost estimate:

$$120{,}000 \text{ tons} \times \$67/\text{ton} = \$8{,}040{,}000$$

Trucking cost estimate:

$$(120{,}000 \text{ tons} \times 2000 \text{ pounds/ton})/31{,}500 \text{ pounds/truck} = 7619 \text{ trucks}$$

$$7619 \text{ trucks} \times 850 \text{ miles/truck} \times \$2.00/\text{mile} = \$12{,}952{,}381$$

Because trucking costs alone are almost \$13 million, stabilization is certainly more cost-effective.

Advantages of landfilling: Landfilling permits the actual reclamation of the contaminated land at the Pepper's Steel and Alloy site and eliminates one site from the national inventory of contaminated sites.

Disadvantages of landfilling: Valuable landfill space is being used by soils amenable to other treatment, precluding the use of that space by wastes better suited for landfilling.

Has site remediation employing stabilization been widely used? The answer for the United States is yes. Table 11-1 contains a listing of Superfund waste sites where stabilization was selected and implemented as the hazardous waste management technology of choice. The table also presents information regarding the type and quantity of treated contaminants. Early studies using four different types of artificially contaminated soils representative of those found at U.S. Superfund sites have demonstrated significant reductions in contaminant leachability.[11] In fact, the selection of stabilization for remediation of Superfund sites increased to a peak in 1992 but has since declined, as the data plotted on Figure 11-1 show.[12]

It should be noted that the data plotted reflect the choice of stabilization in the ROD or action memo. In some cases, even though laboratory testing demonstrated the suitability of the technology,[13] subsequent field tests have led to the

TABLE 11-1
Summary of full-scale stabilization sites[10]

Site	Contaminant	Physical form	Binder	Percentage binder added	Treatment (batch/ continuous *in situ*)
Independent Nail, S.C.	Zn, Cr, Cd, Ni	Solid/soils	Portland cement	20%	Batch plant
Midwest U.S. Plating Company	Cu, Cr, Ni	Sludge	Portland cement	20%	*In situ*
Unnamed	Pb/soil 2–100 ppm	Solid/soils	Portland cement and proprietary ingredient	Cement (15–20%) proprietary (5%)	*In situ*
Marathon Steel, Phoenix, Ariz.	Ph, Cd	Dry landfill	Portland cement and silicates	Varied 7–15% (cement)	Concrete batch plant
Alaska refinery	Oil/oil sludges	Sludges, variable	Portland cement and proprietary ingredient	Varied 50+%	Concrete batch plant
Unnamed, Kentucky	Vinyl chloride Ethylene dichloride	Sludges, variable	Portland cement and proprietary ingredient	Varied 25+%	*In situ*
Northeast refinery	Oil sludges, Pb, Cr, As	Sludges, variable	Kiln dust (high CaO content)	Varied, 15–30%	*In situ*
Velsicol Chemical	Pesticides and organics (resins etc.) up to 45% organic	Sludges, variable	Portland cement and kiln dust, proprietary ingredient	Varied (cement 5–15%)	*In situ*
Amoco Wood River	Oil/solids Cd, Cr, Pb	Sludges	Proprietary ingredient	NA, proprietary	Continuous flow (proprietary)

TABLE 11-1
Continued

Site	Contaminant	Physical form	Binder	Percentage binder added	Treatment (batch/continuous *in situ*)
Pepper Steel & Alloy, Miami	Oil-saturated soil Pb—1000 ppm PCBs—200 ppm As—1–200 ppm	Soils	Pozzolanic and proprietary ingredient	~30%	Continuous feed (mixer proprietary design)
Vickery, Ohio	Waste acid PCBs (<500 ppm) Dioxins	Sludges (viscous)	Lime and kiln dust	~15% CaO ~5% kiln dust	*In situ*
Wood treating, Savannah	Creosote wastes	Sludges	Kiln dust	20%	*In situ*
Chemical refinery, Texas	Combined metals, sulfur, oil sludges, etc.	Sludges (synthetic oil sludges)	Portland cement and proprietary ingredient	NA	Continuous flow
API Sep. Sludge, Puerto Rico	API separator sludges	Sludges	Portland cement and proprietary ingredient	50% cement ~4% proprietary	Concrete batch plant
Metaplating, Wisconsin	Al—9500 ppm Ni—750 ppm Cr—220 ppm Cu—2000 ppm	Sludges	Lime	10–25%	*In situ*

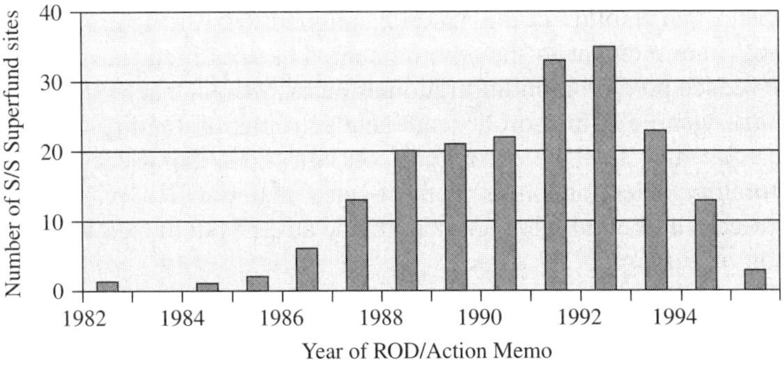

FIGURE 11-1[12]

implementation of an alternative remedy (e.g., McColl Site, Fullerton, California). Alternatively, as early as 1994 stabilization/solidification (S/S) had been implemented and the site identified for removal from the Superfund National Priorities List (NPL).[14] In 1995 and 1996, three and four additional sites, respectively, were proposed for deletion from the NPL after S/S processes were implemented.[15,16]

Solidification of industrial wastes

As a legacy from our past waste management practices, a wide variety of organic and inorganic industrial wastes can be found in pits, ponds, and lagoons. Solidification (improving of physical properties such as strength and compressibility) improves the engineering properties and may reduce the rate at which contaminants migrate into the environment. Although many of the nonhazardous waste materials may not endanger the public health or the environment, the materials are frequently structurally unstable, aesthetically unsuitable, and their condition precludes other uses of the site area. Thus, the primary goal of solidification is the improvement of the structural integrity of the material. For such solidification projects, the effectiveness of the solidification process can often be evaluated by measurement of the material's strength.[17]

Figure 11-2[18] depicts a waste solidification project where cement kiln dust is added to neutralize and stabilize an oily sludge. In a project such as this, improving

FIGURE 11-2
Typical waste solidification project. *(Photo provided by Enreco, Inc.)*

the appearance and stability of the material, coupled with the neutralization, results in an overall improvement to the environment. The solidified mass could then be covered to reduce precipitation infiltration as described in Chapter 13.

In another example, the soil beneath an electroplating facility was remediated by stabilization such that the chromium concentrations from leaching tests were below laboratory detection limits in the treated material.[19] Mixing was with an electric-powered unit employing a mixing blade auger system (see Section 11-5 for a discussion of mixing).

11-2 MECHANISMS

An understanding of the fundamental physical and chemical mechanisms that control the effectiveness of the stabilization reagents is essential to the correct implementation of stabilization as a hazardous waste management technology. Further, as new reagents are developed or existing reagents are modified and adapted to new and different wastes, fundamental stabilization mechanisms should be considered to evaluate the potential for success or failure of such developments. Successful stabilization employs one or more of the following mechanisms:

- Macroencapsulation
- Microencapsulation
- Absorption
- Adsorption
- Precipitation
- Detoxification

Macroencapsulation

Macroencapsulation is the mechanism by which hazardous waste constituents are physically entrapped in a larger structural matrix[20]; that is, the hazardous waste constituents are held in discontinuous pores within the stabilizing materials. On physical degradation (breakdown) of the stabilized material, even to relatively large particle sizes, the entrapped materials are free to migrate. The stabilized mass may break down over time (as measured on a geologic time scale) because of imposed environmental stresses. These stresses include such things as repeated cycles of wetting and drying or freezing and thawing, introduction of percolating fluids, and physical loading stresses. Thus, contaminants stabilized by only macroencapsulation may find their way into the environment if the integrity of the mass is not maintained.

Figure 11-3 is an example of an oily waste that is stabilized, in part, as a result of macroencapsulation. In the stabilized sample shown in this photograph, note the dark nodules of oily waste encapsulated within the light gray matrix of fly ash, lime, and cement binder. The waste in these pores is bound only by physical entrapment in the stabilizing matrix. Should this matrix degrade as a result of cycles of freezing and thawing or wetting and drying, the stabilized oily waste would be free to migrate into the environment.

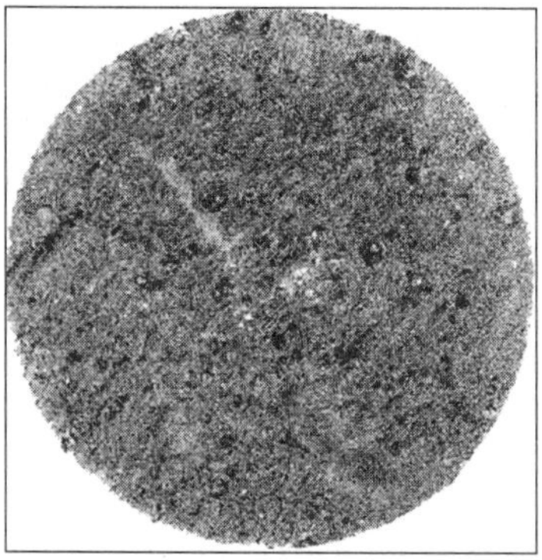

FIGURE 11-3
Organic waste encapsulated in binder matrix. *(Photo courtesy of Debra Cook, Bucknell University.)*[21]

The degree of macroencapsulation is enhanced by the type and energy per unit mass of mixing. Laboratory-prepared samples are generally well blended with a high energy per unit mass compared to field mixing. This gives rise to a greater degree of macroencapsulation in the laboratory (a greater number of particles to be encapsulated) than in the field (fewer, but larger, particles to be encapsulated). Therefore, laboratory studies must be conducted using field mixing conditions to the extent practicable.

Microencapsulation

In **microencapsulation,** hazardous waste constituents are entrapped within the crystalline structure of the solidified matrix at a **microscopic** level. As a result, even if the stabilized materials degrade into relatively small particle sizes, most of the stabilized hazardous wastes remains entrapped. However, as with macroencapsulation, because the waste is not chemically altered or bound, the rates of contaminant release from the stabilized mass may increase as the particle size decreases and more surface area is exposed.

Figure 11-3 also provides an example of an oily waste that is stabilized, in part, as a result of microencapsulation. The oily waste is encapsulated within the small pores of the light gray matrix of fly ash, lime, and cement binder. The waste in these very small pores is bound only by physical entrapment in the stabilized mass, as in macroencapsulation. If the stabilized materials degrade to very small particles over the long term, the stabilized oily waste would be released from the matrix.

As previously described in the macroencapsulation section, laboratory-prepared samples are generally well blended with a high energy per unit mass as compared with field mixing. This gives rise to a greater degree of microencapsulation in the laboratory than in the field, and this must be considered when deciding whether to use solidification or not.

Absorption

As described in Section 3-2, absorption is the process by which contaminants are taken into the sorbent in very much the same way a sponge takes on water. As applied

in stabilization, absorption requires the addition of solid material (sorbent) to soak up or **absorb** the free liquids in the waste. The process is primarily employed to remove free liquid to improve the waste-handling characteristics, that is, to solidify the waste. For many common absorbents, liquids are free to squeeze out of the material should the mass be subjected to consolidating stresses. Thus, the use of absorption is considered only a temporary measure to improve handling characteristics.

The most common absorbents include:

- Soil
- Fly ash
- Cement kiln dust
- Lime kiln dust
- Clay minerals including bentonite, kaolinite, vermiculite, and zeolite
- Sawdust
- Hay and straw

Some of the absorbents previously listed, such as cement kiln dust, have additional benefits due to their pozzolanic (self-cementing) characteristics. The use of these materials as stabilization reagents is discussed later in this chapter.

Adsorption

In addition to the physical entrapment within the stabilized and solidified mass described above, electrochemical interactions may occur.[22] **Adsorption** is the phenomenon by which contaminants are electrochemically bonded to stabilizing agents within the matrix. These are typically considered surface phenomena, and the nature of the bonding may be through van der Waal's or hydrogen bonding. Further discussion of adsorption mechanisms is contained in Section 3-2. Contaminants that are chemically adsorbed (fixed) within the stabilized matrix are less likely to be released into the environment than those that are not fixed. Unlike microencapsulation and macroencapsulation, where simple particle breakdown may enhance the rate of contaminant migration, additional physicochemical stress is necessary to desorb the material from their adsorbing surfaces. As a result, the treatment is considered more permanent.

The stabilization of organic wastes with organically modified clays illustrates how adsorption can be used in waste stabilization. Organically modified clays are clays that have been altered by replacing exchangeable inorganic cations adsorbed on the clay surfaces with long-chain organic cations, rendering the clays **organophilic.** Organophilic clays have an affinity for organic molecules. Without modification, naturally occurring clays are generally organophobic. Organic waste molecules are then adsorbed to the clay as shown in Figure 11-4. The adsorption bond strength must then be overcome if the organic waste molecules are to be released to migrate into the environment (see Organically Modified Clays in Section 11-3). For metal contaminents, sorption is strongly pH dependent[23] as is metals precipitation, described below.

Precipitation

Certain stabilization processes will precipitate contaminants from the waste, resulting in a more stable form of the constituents within the waste.[24] Precipitates such as

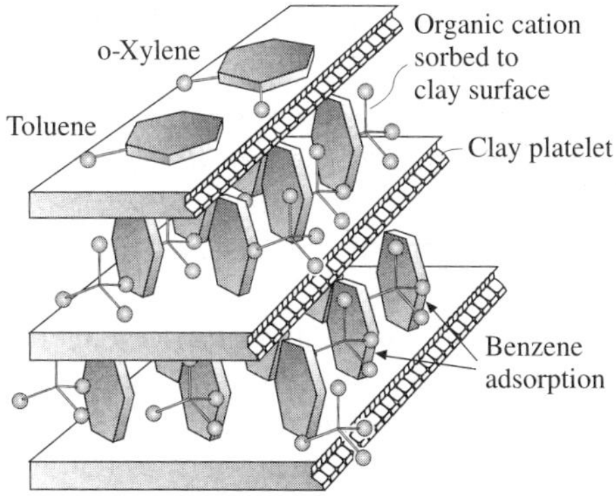

FIGURE 11-4
Organic waste adsorbed to an organophilic clay (modified from Reference 25).

hydroxides, sulfides, silicates, carbonates, and phosphates are then contained within the stabilized mass as part of the material structure. This phenomenon is applicable to the stabilization of inorganic wastes such as metal hydroxide sludges. For example, metal carbonates are typically less soluble than metal hydroxides. At high pH, the reaction to form a metal carbonate from a metal hydroxide sludge, is as follows:[26]

$$Me(OH)_2 + H_2CO_3 \rightarrow MeCO_3(s) + 2H_2O \qquad (11\text{-}1)$$

where Me represents a metallic cation.

The permanence of the fixation of a metal as a metallic carbonate depends on, among other things, the pH. In this example, the metallic carbonate is quite stable except at low pH. Under strongly acidic conditions, the metal may be redissolved and may then be free to migrate as a solute into the environment. Thus, the assessment of the effectiveness of stabilization must properly account for present and future environmental conditions.

Detoxification

Certain chemical reactions taking place during the stabilization process may result in a waste with reduced toxicity. Detoxification is any mechanism that changes a chemical constituent into another constituent (or form of the same constituent) that is either less toxic or nontoxic.

An example of this is the reduction of chromium in the +6 valence state to chromium in the +3 valence state during stabilization with cement-based materials. The trivalent chromium has a lower solubility and toxicity than hexavalent chromium. A number of fixation systems can be used to detoxify chromium by reducing its valence state; ferrous sulfate and a combination of ferrous sulfate and sodium sulfate have been found to be effective.[27] Leaching of the reduced chromium poses a lesser threat to the environment than the leaching of the original hexavalent chromium.

11-3 TECHNOLOGY

A detailed description of stabilization technology can best be compartmentalized by a discussion of the various types of additives (reagents) employed. Although it

is recognized that this classification into categories is somewhat arbitrary because more than one reagent may be employed for any given waste stabilization project, this approach allows examination of the mechanisms by which each of the reagents act. Where agents are used in conjunction with each other in a synergistic fashion, it will be noted in the text. The term **binder** is used to denote a reagent that contributes to the strength gain associated with stabilization. The term **sorbent** is used to denote a reagent that primarily contributes to retaining contaminants in the stabilized matrix.

As a first approximation, the effectiveness of each of the stabilization reagents upon selected waste types is shown in Table 11-2. The application of Table 11-2 depends on several factors, such as the concentration of the contaminant, the quantity of the reagent, and the synergistic effects of multiple contaminants and reagents to name a few. Nonetheless, the information of Table 11-2 is useful for preliminary screening and the development of bench scale studies.

There is an extensive range of sorbents and binders available in the worldwide marketplace, including several proprietary reagents. This text describes those that are not proprietary to enable the student to appreciate the complexity of the chemistry involved and to begin to understand the nature of the stabilization process.

Cement

Stabilization of hazardous wastes frequently employs cement as the principal reagent. What is cement? The most common cement is **portland cement,** which is made by firing a mixture of limestone and clay (or other silicate) in a kiln at high temperatures. The kiln produces a clinker, which is ground to a powder that is a mixture of calcium, silicate, aluminum, and iron oxides. The main constituents are tri- and dicalcium silicates.[28,29] For cement-based stabilization, waste materials are mixed with cement followed by the addition of water for hydration, if necessary because the waste does not have enough water. The hydration of the cement forms a crystalline structure, consisting of calcium aluminosilicate. This results in a rock-like, monolithic, hardened mass. **Concrete,** by definition, is a particulate composite consisting of hydrated cement and aggregate.

In its simplest form, the reaction of the tricalcium silicate can be expressed by the reaction equation:

$$2(3CaO \bullet SiO_2) + 6H_2O \rightarrow 3CaO \bullet 2SiO_2 \bullet 3H_2O + 3Ca(OH)_2 \qquad (11\text{-}2)$$

and the reaction of dicalcium silicate by:

$$2(2CaO \bullet SiO_2) + 4H_2O \rightarrow 3CaO \bullet 2SiO_2 \bullet 3H_2O + Ca(OH)_2 \qquad (11\text{-}3)$$

Additional reactions occur during the hydration of portland cement as shown in Figure 11-5. The resulting silicate gel is termed tobomorite. These reactions are quite slow, contributing to the lengthy curing time associated with portland cement concretes.

The most rapid reaction in portland cement is:

$$3CaO \bullet Al_2O_3 + 6H_2O \rightarrow 3CaO \bullet Al_2O_3 \bullet 6H_2O + \text{ heat} \qquad (11\text{-}4)$$

TABLE 11-2
Reagent applicability for waste stabilization[30]

Waste component	Cement-based	Pozzolan-based	Thermoplastic	Organic polymer
Nonpolar organics as: oil and grease, aromatic hydrocarbons, halogenated hydrocarbons, PCBs	May impede setting. Decreases durability over a long time period. Volatiles may escape on mixing. Demonstrated effectiveness under certain conditions.	May impede setting. Decreases durability over a long time period. Volatiles may escape on mixing. Demonstrated effectiveness under certain conditions.	Organics may vaporize on heating. Demonstrated effectiveness under certain conditions.	May impede setting. Demonstrated effectiveness under certain conditions.
Polar organics as: alcohols, phenols, organic acids, glycols	Phenol will significantly retard setting and will decrease durability in the short run. Decreases durability over a long time period.	Phenol will significantly retard setting and will decrease durability in the short run. Alcohols may retard setting. Decreases durability over a long time period.	Organics may vaporize on heating.	No significant effect on setting.
Acids as: hydrochloric acid, hydrofluoric acid	No significant effect on setting. Cement will neutralize acids. Types II and IV portland cement demonstrate better durability characteristics than Type I. Demonstrated effectiveness.	No significant effect on setting. Compatible, will neutralize acids. Demonstrated effectiveness.	Can be neutralized before incorporation.	Can be neutralized before incorporation. Ureaformaldehyde demonstrated to be effective.
Oxidizers as: sodium hypochlorate, potassium permanganate, nitric acid, potassium dichromate	Compatible.	Compatible.	May cause matrix breakdown, fire.	May cause matrix breakdown, fire.
Salts as: sulfates, halides, nitrates, cyanides	Increases setting times. Decreases durability. Sulfates may retard setting and cause spalling unless special cement is used. Sulfates accelerate other reactions.	Halides are easily leached and retard setting. Halides may retard setting, most are easily leached. Sulfates can retard or accelerate reactions.	Sulfates and halides may dehydrate and rehydrate, causing splitting.	Compatible.
Heavy metals as: lead, chromium, cadmium, arsenic, mercury	Compatible. Can increase set time. Demonstrated effectiveness under certain conditions.	Compatible. Demonstrated effectiveness on certain species (lead, cadmium, chromium).	Compatible. Demonstrated effectiveness on certain species (copper, arsenic, chromium).	Compatible. Demonstrated effectiveness with arsenic.
Radioactive materials	Compatible.	Compatible.	Compatible.	Compatible.

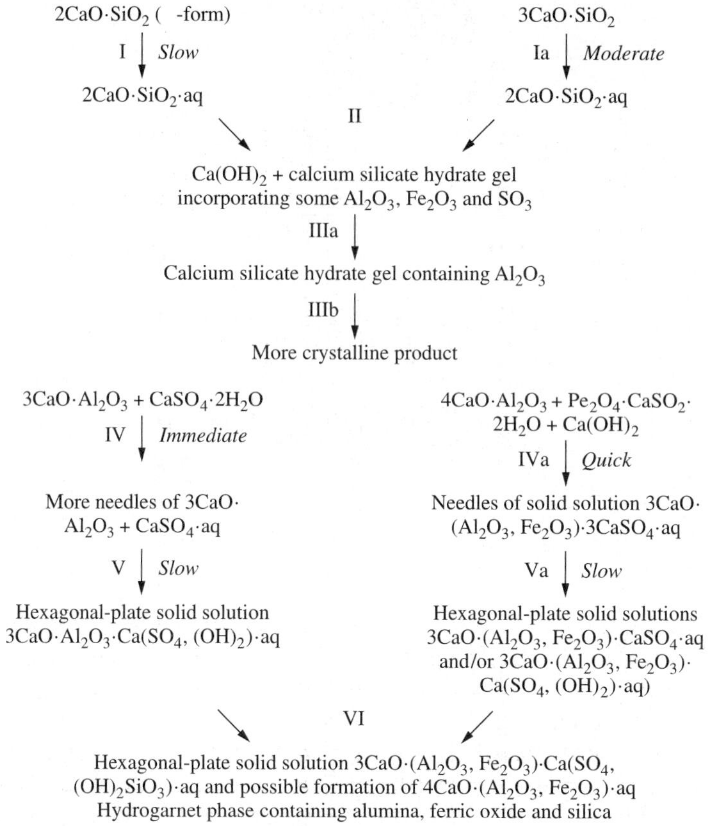

FIGURE 11-5
Portland cement reactions.[31]

This reaction provides portland cement with its initial set. A schematic representation of the tricalcium and dicalcium cement hydration reactions is shown in Figure 11-6.

Cement-based stabilization is best suited for inorganic wastes, especially those containing heavy metals. As a result of the high pH of the cement, the metals are retained in the form of insoluble hydroxide or carbonate salts within the hardened structure.[33,34] Studies have shown that lead, copper, zinc, tin, and cadmium are likely bound in the matrix by chemical fixation, forming insoluble compounds while mercury is predominantly held by physical microencapsulation.[35]

On the other hand, organic contaminants interfere with the hydration process, reduce the final strength, and are not easily stabilized.[36–42] They may also reduce the crystalline structure formation resulting in a more amorphous material.[43] Other additives may be incorporated into the stabilizing mix with the cement to reduce organic contaminant interference with cement hydration and enhance stabilization. These additives may include organically modified and natural clays, vermiculite, and soluble sodium silicates.

It is difficult to determine precisely where a specific contaminant will react within the stabilized cement matrix. In a study of the binding chemistry of hazardous waste

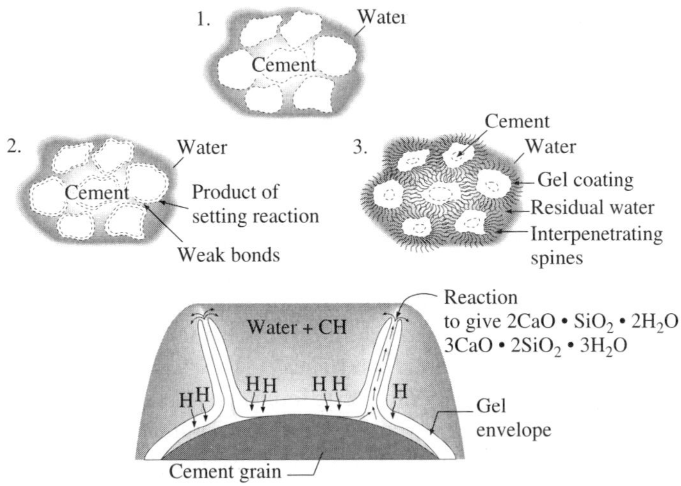

FIGURE 11-6
Schematic representation of cement hydration.[32]

in cement, the locations of the heavy metals lead and chromium were studied.[44] The results indicated that the lead precipitated on the outer surface of the hydrated cement particles whereas chromium was widely and more uniformly dispersed through the hydrated cement particles.

Stabilization with cement has been used for the fixation of inorganic wastes such as metal hydroxide sludges and metal-contaminated soils as shown in the first four sites listed on Table 11-1. The wide use of stabilization of inorganics derives from (1) the lack of better alternatives (for example, metals do not biodegrade and do not change in atomic structure when incinerated) and (2) available and understood physiochemical mechanisms such as precipitation and adsorption.

There are a number of advantages of cement-based stabilization. The technology of cement is well known, including handling, mixing, setting, and hardening. Cement is widely employed in the construction field, and, as a result, the material costs are relatively low and the equipment and personnel readily available. Extensive dewatering of wet sludges and waste is not typically necessary, as water is required for cement hydration. In fact, cement-based stabilization may proceed with sufficient water so as to make the material pumpable.[45] The system can handle variations in the chemistry of the waste.[46] Finally, the alkalinity of the cement can neutralize acidic wastes. In one case, a cement-based stabilization process was used to immobilize a mixed cocktail of organic and inorganic contaminants and the site was subsequently developed into a football field.[47] The principal disadvantage is the sensitivity of cement to the presence of certain contaminants that could retard or prohibit proper hydration and the resulting setting and hardening of the material.

Pozzolans

A **pozzolan** is a material that can react with lime in the presence of water to produce a cementitious material. The reaction of aluminosilicious material, lime, and water

results in the formation of a concrete-like product termed **pozzolanic concrete.** Pozzolanic materials include fly ash, ground blast furnace slag, and cement kiln dust. The resulting pozzolanic structures are termed amorphous aluminosilicates.[48] Fly ash, the most commonly employed pozzolan, is typically about 45% SiO_2, 25% Al_2O_3, 15% Fe_2O_3, 10% CaO, 1% MgO, 1% K_2O, 1% Na_2O, and 1% SO_3. In addition, the amount of unburned carbon may vary from source to source.

Like cement-based stabilization, most stabilization applications involving the use of pozzolans are for inorganic materials. The high pH environment is well suited for waste contaminated with heavy metals. An example in the literature shows that the untreated sludge from an Imhoff digester tank was treated with fly ash and lime reagents to stabilize elevated levels of cadmium, chromium, copper, iron, lead, magnesium, manganese, selenium, silver, and zinc.[49] Although the stabilized material had a soil-like consistency, the leaching tests indicated that the stabilization process significantly slowed leaching of several of these components. Unburned carbon in fly ash may adsorb organics from the waste. As a result, a pozzolan such as fly ash may have beneficial effects in the stabilization of both organic and inorganic species.

It should be noted that when fly ash is used, quicklime (CaO) is needed to achieve satisfactory stabilization. One problem identified with the use of fly ash is the growth of the mineral ettringite in the presence of sulfates. Since ettringite swells in contact with water, destabilization of the matrix may occur. It has however been found that minimum amounts of barium hydroxide can eliminate the formation of ettringite.[50]

As shown in Table 11-1, pozzolans (such as unspecified pozzolans and kiln dust) have been used on a number of Superfund sites.

EXAMPLE 11-2. EVALUATION OF STABILIZATION EFFECTIVENESS. The total mass of contaminants leached in a column leaching test from the unstabilized and pozzolanic reagent stabilized Imhoff sludge example presented above is as follows[51]:

Parameter	Untreated sludge	Treated sludge
Arsenic	0.724	0.049
Cadmium	3.3	0.126
Chromium	0.566	1.16
Copper	3.86	3.2
Iron	48.4	1.57
Lead	0.238	0.019
Manganese	11.5	1.48
Zinc	120	3.02

(*a*) Calculate the percent reduction in the mass of the leached constituents.
(*b*) What do you consider the average reduction in leached constituent concentrations?
(*c*) Which chemical parameter requires further consideration in evaluating the effectiveness of stabilization and why?

Solution

(*a*) Reduction (%) = (100) × (Untreated sludge − treated sludge)/(untreated sludge). For example, $R = [(0.724 - .049)/0.724] \times (100) = \underline{\underline{93.2\%}}$

Continuing in this manner yields:

Parameter	Untreated sludge	Treated sludge	Reduction, %
Arsenic	0.724	0.049	93.2
Cadmium	3.3	0.126	96.2
Chromium	0.566	1.16	−104.9
Copper	3.86	3.2	17.1
Iron	48.4	1.57	96.8
Lead	0.238	0.019	92.0
Manganese	11.5	1.48	87.0
Zinc	120	3.02	97.5

(*b*) Because each constituent has a different toxicity and mobility, it is not reasonable to average the value and report the average reduction in contaminant leachability. Each value must be assessed individually; the average is meaningless at best, misleading at worst.

(*c*) The chromium concentrations actually increased in the leachate of the stabilized sludge over the untreated sludge. The high chromium concentrations suggest a high mobility and therefore a +6 valence state. The stabilization process should consider first reducing the chromium to a +3 valence state and then precipitation as chromium hydroxide.

Lime

Stabilization of sludges[52] is frequently accomplished through the addition of calcium hydroxide $Ca(OH)_2$, or **lime** (also called hydrated lime). The resulting reaction with materials included in the waste may result in hydrates of calcium silicate, calcium alumina, or calcium aluminosilicate. These materials are formed from the reaction of calcium in the lime and aluminosilicates in the waste. As with all additives, additional stabilization can be accomplished through the use of other ingredients in smaller quantities. Note that lime may also be added to raise the pH of acidic sludges with other reagents that provide the main stabilization reactions, such as fly ash. Lime-based stabilization is typically best suited for inorganic contaminants and has been widely employed for metal sludges.

Soluble Silicates

The use of silicious materials in metals stabilization has been long applied in full-scale commercial processes. In one such process, silica reagents are acidified to form a monosilic acid solution to which metal-bearing wastes are added.[53] In another, a combination of liquid soluble silicates and cement form the basis of a process that has been demonstrated to be effective in stabilizing soils contaminated with high concentrations of lead, copper, and zinc.[54]

Organically Modified Clays

In the past, the use of stabilization and solidification for organic wastes has been limited because the ability of traditional stabilization additives to hold the organics

in the stabilized matrix was limited. In fact, as pointed out earlier, organics interfere with the cement hydration process. Organically modified clays have been recently employed in conjunction with other stabilization reagents in order to entrap the organic portion of the waste to be stabilized.[55–61]

Organically modified clays are produced when natural clays are organically modified to become organophilic.[62,63] This characteristic is in contrast to their original organophobic nature. The modification process is accomplished through the replacement of inorganic cations within the clay crystalline mineralogical structure with organic cations, typically quaternary ammonium ions. After this replacement process, organic molecules are adsorbed within the crystalline structure of the clay that then swells in the presence of organic contaminants. The clay structure of a montmorillonite, shown in Figure 11-7, depicts the clay as a tetrahedral layer on each side of an octahedral layer. The components of this two-to-one structure are bound tightly by covalent bonds. However, as a result of isomorphous (same form) substitutions, the clay has a net negative charge that is balanced by exchangeable cations in the intercrystalline region (see Figure 11-7).[64]

In the production of organically modified clay, the exchangeable inorganic cations shown in the intercrystalline region are replaced with organic cations. The result is a nearly identical clay structure where organic cations occupy the intercrystalline region, as shown on Figure 11-8. In this figure, the organic cations are quaternary ammonium ions. These organic ions attached to the clay readily adsorb other organic species. The effectiveness of organically modified clays in the stabilization of wastes containing organics is due to the adsorption of organic contaminants to the clay which in turn can be encapsulated by cement or other binders.

Organophilic clays typically are added to the waste first and allowed to interact with the organic components.[65,66] Additional agents are added to provide shear strength and solidify the material into a monolithic mass.

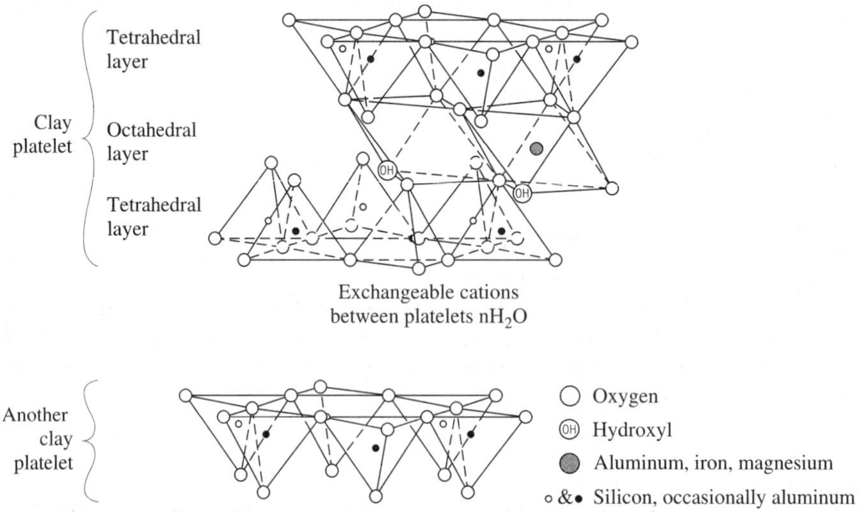

FIGURE 11-7
Montmorillonitic clay structure.[64]

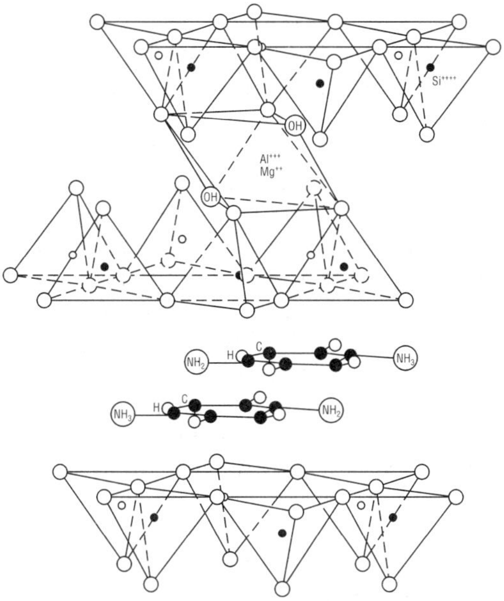

FIGURE 11-8
Organically modified clay structure.[69]

The effectiveness of organically modified clays compared to other sorbents in waste stabilization is illustrated on Figure 11-9. To stabilize an acidic petroleum sludge, a finely ground cement (MC-500) was used as a binder, and both unmodified and modified clays were used as sorbents. By weight, the sludge:sorbent:binder ratio was 1.0:0.4:0.25 for this series.

As shown on Figure 11-9, the total organic carbon in the leachate from the organically modified clay mixes was one-third less than the bentonite and attapulgite mixes. The unconfined compressive strength for samples mixed with organically modified clays was nearly an order of magnitude stronger than the unmodified clay mixes.

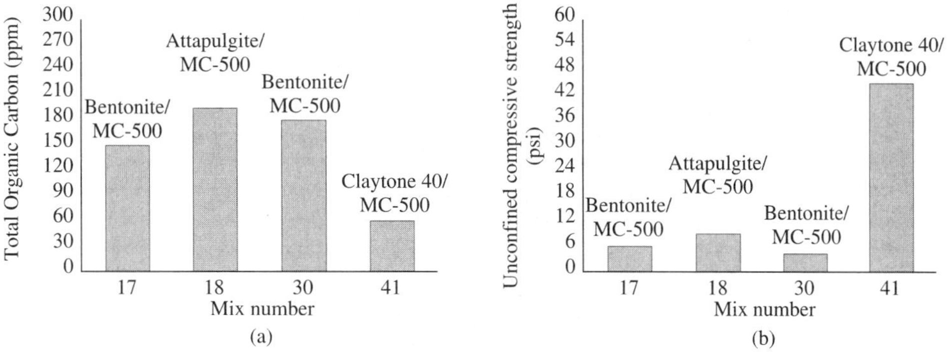

FIGURE 11-9
Stabilization effectiveness of organically modified clay.[71] Claytone 40 is an organically modified clay from Southern Clay Products Inc. Bentonite and attapulgite are processed, naturally occurring clays. Mix 30 has a larger bentonite content than mix 17.

The impact of organically modified clays on the stabilization of organics is further illustrated in Figure 11-10. In this figure investigating the effect of sorbent-to-waste ratio, the phenol concentration in the leachate decreased as the sorbent quantity increased for two different organophilic clays. Correspondingly, the strength of the stabilized materials increased with increasing organophilic clay content. Detailed studies measuring the partitioning coefficients show that the adsorption of phenols by organophilic clays from water increases with chlorine addition (i.e., phenol < chlorophenol < dichlorophenol < trichlorophenol).[67] In another study, the use of an organophilic clay for adsorption of phenolic compounds was demonstrated to be a successful presolidification adsorbent for cement-based stabilization.[68]

Both the type of quaternary ammonium ion used and the nature of the base clay affect the ability of the organically modified clay to adsorb organics.[70] Figure 11-11 shows the relationship between the trichloroethene concentration in the soil pore water and the adsorption on the organically modified clay. An important fundamental is illustrated in this figure. Utilizing different quaternary ammonium ions with the

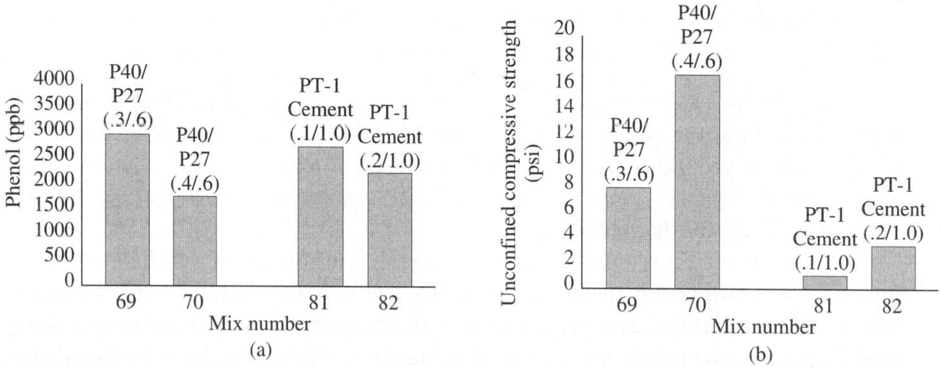

FIGURE 11-10
Effect of organically modified clay.[72] PT-1 and P40 are organically modified clay from Bentec Inc.

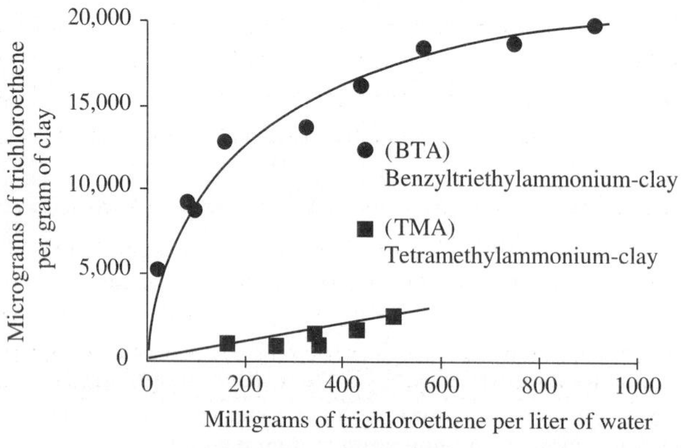

FIGURE 11-11
Effect of quaternary amine on organophilic clay.[73]

identical base clay produces two very different organophilic clays as measured by their adsorptive capacities.

Modified Lime

Organically modified lime products have been specifically developed for the stabilization of organic wastes. The lime-based process has been developed to convert toxic organic wastes with significant percentages of hydrocarbon to an inert mass. One system using hydrophobized lime is patented and has been used since 1976 in Germany.[74] The process employs a calcium oxide–based reagent modified with stearic acid. Heat and water-as-steam are given off during the reaction (along with volatile organic contaminants). The waste is converted to a dry water-repellent (hydrophobic) powder. The material becomes quite strong and relatively impermeable to water. The resulting material is often considered suitable as a construction material for road and embankment fill.

In this regard, quicklime (CaO) has been modified to form a group of patented technologies for S/S of oily sludges, oil-contaminated soils, acidic oil sludges, and heavy metals in the resulting calcium hydroxide [$Ca(OH)_2$] and calcium carbonate ($CaCO_3$) matrices. This dispersion by chemical reaction (DRC) process has been found to be successful in stabilizing liquid contaminants and heavy metals in contaminated soils, although the chemical properties of solid organics (asphalt tar and pesticides) were not generally improved.[75]

Thermosetting Organic Polymers

Hazardous wastes may be stabilized through an organic polymer process that involves mixing of a monomer, such as urea-formaldehyde, that acts as a catalyst to form a polymeric material.[76] A sponge-like mass is thereby formed, trapping solid particles of hazardous waste within the matrix (microencapsulation). This process does, however, leave some wastes untrapped, particularly liquid wastes. Because liquids generally remain after thermosetting organic polymer treatment, the final waste product would need to be dried and containerized prior to ultimate disposal.

The principal advantage of this process is that it generally results in a low-density material relative to other fixation techniques. Further, only small quantities of additives are required to solidify wastes. Thus, this particular technique is most applicable in the solidification of liquid, nonvolatile, organic hazardous wastes. This method has also been evaluated to solidify low-level radioactive waste prior to ultimate disposal.

This technology is generally not applicable to site remediation processes because of its cost, fire hazard, production of water from the waste matrix, and volatilization of organics. However, in one study for the remediation of sediments in the James River near Richmond, Virginia, the sediments were contaminated with the organic pesticide Kepone.[77] Bench-scale studies determined that an organic epoxy grout was the most effective stabilization agent in treating the Kepone-contaminated sediment.

Thermosetting organic polymers may ultimately prove useful for remedial action, but this technology has not yet been applied to a full-scale site remediation.

Thermoplastic Materials

Hazardous waste may be stabilized by blending molten thermoplastic materials with wastes at high temperatures.[78] Molten thermoplastic materials include asphalt, paraffin, bitumen, polyethylene, polypropylene, and sulfur. When cooled, the solidified material is characterized as a thermoplastically coated waste and typically containerized (put in drums) for ultimate disposal. When bitumen has been employed, the typical waste-to-bitumen ratio has been between 1:1 and 2:1.[79] This technique has been predominantly employed for radioactive waste applications because of its cost. Further, organics are volatilized at the operating temperatures, requiring control of air emissions during the process.

The use of thermoplastic stabilization has received attention for mixed waste, that is, waste that is both hazardous and radioactive.[80] The waste is stabilized so that it is no longer considered hazardous, and then it is disposed of in accordance with the requirements for radioactive waste.

A number of research studies have been conducted, indicating the potential for this technique.[81–83] The primary limitations include the presence of materials within the matrix that could deteriorate the thermoplastic material and the presence of organic chemicals that act as solvents to the stabilizing thermoplastic materials. Thus, a potential for long-term degradation exists if these detrimental components are in the waste product. Thermoplastically stabilized waste, however, is quite resistant to leaching and biodegradation.

In recent efforts to reuse contaminated material, soils contaminated with mineral oils were stabilized with bitumen and added to road paving materials.[84]

Vitrification

Vitrification does not require the addition of reagents but is a stabilization technique and, as such, is included under this technology section. Archaeologists estimate that humans have been making glass for more than 7000 years. Glassmaking, or *vitrification,* involves melting and fusion of materials at temperatures normally in excess of 1600°C followed by rapid cooling into a noncrystalline, amorphous form. The application of vitrification to the management of hazardous wastes has been explored as both an *in situ* technique and an in-plant technique. Vitrification is considered a solidification and stabilization procedure because it renders the waste more structurally stable with a reduced potential for contaminant migration into the environment. At the time of this writing (1998), a number of field-scale demonstration projects have been completed, along with the successful remediation of the Wasatch Superfund site.

In situ vitrification

In situ vitrification is considered an emerging technology, with one vendor (GeoSafe) providing commercial systems. In the *in situ* process, the high tempera-

tures needed to melt the soil into a molten mass are achieved through the application of an electric current. As the current flows through the soil, heat builds up, eventually causing the soil to melt. On melting, the soils become more conductive and the molten mass becomes a heat transfer medium, allowing the molten mass to grow. The entire process is begun using a surficial layer of graphite and glass frit to serve as a starter path. Typically, electrodes spaced a maximum of 26 feet apart in a rectangular pattern are employed. Once initiated, the electronically conductive molten mass grows downward and outward to the desired treatment depth.[85] Schematically, the process stages are shown in Figure 11-12. The resulting residue bears a strong resemblance to obsidian glass, which is formed by nature's own vitrification process.

The soil is melted at a rate of 3 to 6 tons per hour, giving a rate of melt advance of 1 to 2 inches per hour. But what exactly happens to the contaminants during this process? As the temperature increases, organic materials first vaporize and then pyrolyze (break down in the absence of oxygen) into their elemental components. The gases move slowly through the molten mass (which is quite viscous) toward the surface, and those gases that are combustible burn in the presence of oxygen on offgassing. All gases are collected (a hood is constructed over the area being processed) and treated to ensure air emission standards are met. Because of the high temperature of the melt (1600 to 2000°C), residual contamination remains in the residue (the glass).

Shown in Figure 11-13 is a schematic of the system used by GeoMelt to implement the process shown in Figure 11-12. A hood is employed to capture gases from the process. These captured gases then require treatment. The exact nature of the offgas treatment system depends on the site contaminants.

Inorganic contaminants behave in a similar manner—some decompose while others are dissolved in or react with the melt. For example, nitrates yield gaseous nitrogen (N_2) and oxygen (O_2).

The porosity of soils can vary over a large range, but volume reduction is commonly in the range of 25 to 50 percent. Soil in its *in situ* state is a three-phase material with solids, liquids, and gaseous components. The liquids and gases make up the volume of voids. The molten material during the vitrification process is entirely liquid phase, made up of melted solids as the original liquids and gases are offgassed.

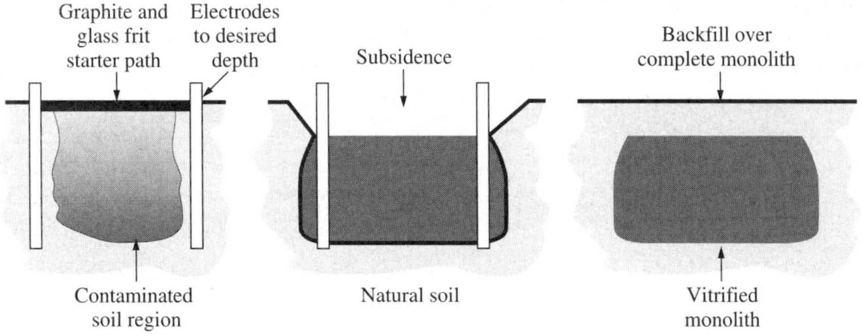

FIGURE 11-12
In situ vitrification.

Thus, on cooling, there is a net volume reduction in the vitrification process. This is in contrast with the previously discussed techniques where the addition of reagents causes a net volume increase.

In situ vitrification has been used in a number of demonstration projects, including a Superfund site in Washington where 3100 tons of contaminated soil and debris were treated. Results of *in situ* tests performed on buried wastes showed that vitrification can successfully immobilize or destroy both hazardous and radioactive contaminants and achieve up to 75 percent volume reduction.[86]

In addition to *in situ* vitrification, the GeoSafe Corporation has developed variations on the original *in situ* methods termed *Staged In Situ Vitrification, Stationary Batch,* and *Continuous* (http://www.geomelt.com).

EXAMPLE 11-3. *IN SITU* VITRIFICATION. A sandy soil with a porosity of 0.3 is contaminated with a wide range of organic and inorganic contaminants to a depth of 3 meters. The site is a candidate for *in situ* vitrification. Calculate the surface subsidence expected at the completion of vitrification.

Solution. Basis of calculation: 1 square meter of surface.

$$V_T = 3 \text{ m}^3/\text{m}^2$$

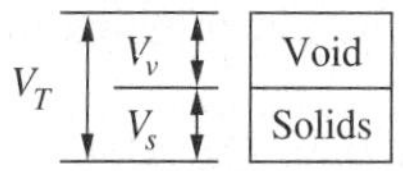

$$\text{Porosity } n = \frac{V_{\text{voids}}}{V_{\text{total}}} = 0.3$$

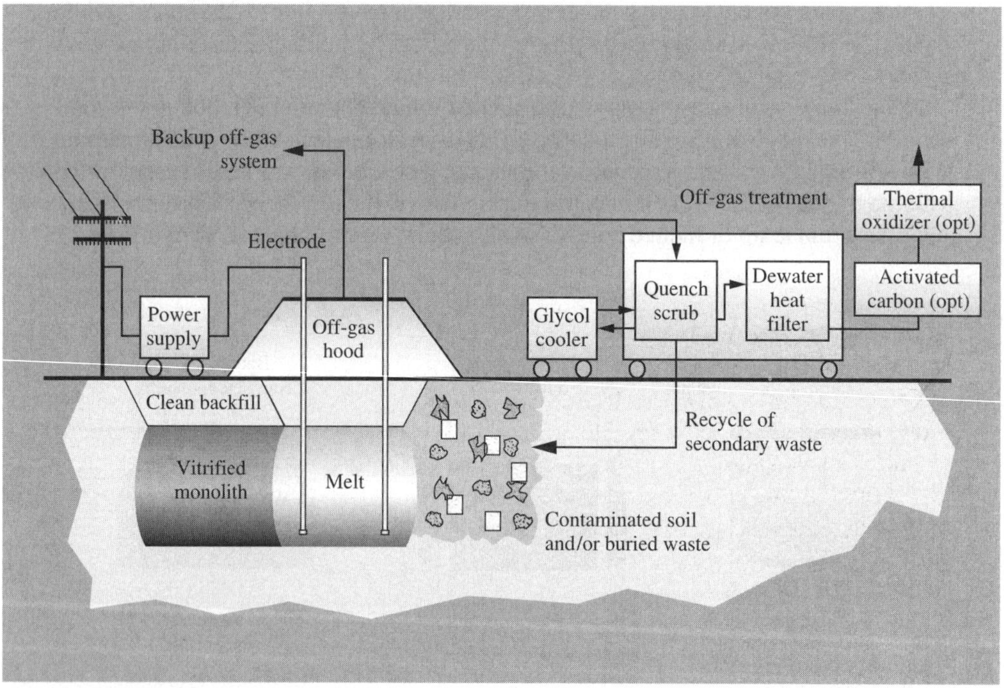

FIGURE 11-13
GeoMelt/SV system schematic.[83]

Prior to vitrification:

$$V_V = n \cdot V_T = 0.3 \times 3 = 0.9 \text{ m}^3/\text{m}^2$$
$$V_S = V_T - V_0 \Rightarrow V_S = 2.1 \text{ m}^3/\text{m}^2$$

After vitrification:

$$n = 0 \text{ (porosity is effective after vitrification)}$$
$$V_T = V_S = 2.1$$
$$\Delta V = 3 - 2.1 = 0.9 \text{ m}^3/\text{m}^2$$
$$\Delta V = 1 \text{ m}^2 = \Delta H = 0.9 \text{ m} = \text{ surface subsidence}$$

In-plant vitrification

The use of contaminated soils as feedstock for glassmaking (vitrification) offers several advantages as a hazardous waste management technique. The technology has the potential both to treat hazardous waste, particularly contaminated soil, and to produce a product that may be usable (e.g., as a road aggregate). The technology employs a glassmaking furnace operating at 1600°C. A starter mix of recycled glass, fly ash, and limestone is used to initiate the process. The contaminated soil is then introduced into the furnace for the melting and fusion stages that last a minimum of 5 hours.[87] At the time of this writing, the use of glassmaking technologies and facilities for the management of hazardous wastes is not employed on a commercial scale although it has been demonstrated on a pilot scale under the U.S. EPA SITE (Superfund Innovative Technology Evaluation) program.[88,89]

Additional considerations for this vitrification technology include air emissions and leachability of the residue (the glass). Air emissions can be captured and treated as necessary for environmental protection by traditional air treatment technologies including ammonia for NO_x, lime for SO_x, and condensation of volatiles in a fractionating pebble bed.[90] As for the *in situ* vitrification product, leaching tests have shown the glass to be inert. Despite the test results, the heightened environmental awareness may dictate that the product be used in applications that limit its distribution to the general public. For example, significant quantities of clean fill are required for daily cover, intermediate berms, and drainage layers in landfill construction and operation. The use of the vitrified product in these applications would provide an added margin of safety to the process and preserve our mineral resources for other future uses.

At least one company (Seiler Pollution Control Systems) has constructed a high-temperature vitrification plant designed to recycle hazardous waste into a nonhazardous glass/ceramic product.[91] Products resulting from the process include abrasive blast media, roof shingle grit, and castings and meet criteria for nonhazardous material established by the U.S. EPA and the State of California. Similarly, the Savannah River Technology Center has produced more than 500 tons of vitrified waste from a variety of industrial environments.[92]

11-4 TESTING

What does it mean to say that a hazardous waste has been stabilized and solidified? How can the effectiveness of the stabilization and solidification process be

measured? How strong does the solidified material need to be? How fast can contaminants leach from the stabilized material? The answers to these questions are complex. The assessment of the effectiveness of stabilization requires the measurement of physical, engineering, and chemical properties of the stabilized material. The measured property depends on the measurement technique; that is, different tests for the same property may yield different measures of that property. To further complicate the evaluation and design process, it is difficult, if not impossible, to predict the long-term environmental stresses to which these stabilized materials might be subjected—freezing and thawing, wetting and drying, percolation of precipitation, and pressure from overburden load. Thus, if the range of environmental stresses cannot be predicted, these stresses cannot be simulated in the laboratory. Even if the long-range environmental stresses could be predicted, how can the predicted environmental stresses be simulated in the laboratory?

As an added complication, the hardening (hydration or curing, if you prefer) process is time dependent and may take many months. Civil engineers traditionally test samples of concrete 28 days after mixing, and this procedure has carried over into the evaluation of stabilized materials. However, since the hydration process goes on long after 28 days, the leaching test results from 28-day samples may overestimate the leachability of a fully cured sample. In one study, reductions in moisture content on samples obtained 9 and 18 months after curing demonstrated this ongoing, long-term nature of the process.[93]

As a natural outcome of these complexities, a large number of laboratory tests are utilized to evaluate the effectiveness of stabilization. No one test is ideal for all wastes and all applications. Each, when properly interpreted, provides partial insight into the effectiveness of a particular stabilization technique and reagent mix for a specific waste. This section discusses the available tests, how they are conducted, the type of information these tests provide, and the circumstances for which the tests are most applicable. Because it is inevitable that tests and the interpretation of the test results will change with time, the fundamental aspects of the tests are emphasized.

The selection of the appropriate tests and the interpretation of the test results depend on the objectives of the stabilization program. For example, site remediation was identified earlier in this chapter as a major application area of stabilization technology. The selection of the type and amount of reagents used in stabilization depends on specific goals for the site remediation program. As described in Section 17-6, risk assessment may be used to establish these goals. For example, the calculated risk from groundwater might be a function of the amount of arsenic that is estimated to leach daily from a proposed stabilization alternative. In this case, the selected tests must estimate the amount of leachable arsenic.

Extraction and Leaching Tests

In most cases, a reduction in the rate at which contaminants can migrate into the environment is the first and foremost reason for selecting stabilization and solidification as a hazardous waste management technique. As precipitation infiltrates the stabilized waste, contaminants migrate from the stabilized mass into the water (the transport medium) and continue their trek into the environment. Of course, in

the short term (geologically speaking) efforts are usually taken to minimize precipitation infiltration and to collect the leachate prior to release into the surrounding environment (see Sections 13-4 and 13-5).

Leaching tests are listed in Table 11-3. The fluid to which the contaminants are leached is called the **leachant.** After the leachant has become contaminated it is termed the **leachate.** The terms **extraction** and **leaching** are used interchangeably and are, as defined earlier, the process by which contaminants are transferred from a solid or stabilized matrix to the leachant. Finally, the overall ability of a stabilized material to leach contaminants is termed **leachability.**

Do leaching tests have a single purpose? If that were so, it would be much easier to define a single acceptable leaching test protocol. However, there are many reasons to conduct a leaching test. The leaching test may be conducted as a **regulatory test** run to provide the basis for a consistent and uniform decision-making process. In this case a highly reproducible test protocol is needed and the results can be compared to predetermined standards. The stabilized material either passes or fails. Alternatively, a leaching test may be a **predictive test** run to generate data that can be used to model real-world contaminant migration. In this case, the data would be employed in subsequent fate and transport analysis (Chapter 4) and risk assessments (Chapter 14). **Investigatory tests** may also be run to study the basic binding mechanisms, interference factors, and the underlying principles of stabilization technology. Because each of these test types has different goals, different test methods have evolved. As the test methods are discussed, consider which purpose is best served by the method described.

How is material leached from a stabilized mass? As the leachant passes through and around, the sample contaminants are transferred from the stabilized mass to the leachant. This may occur as contaminants are dissolved into the leachant, washed from the surfaces of the stabilized material, or as contaminants diffuse from within the stabilized mass to the leachant. Thus, the leachability depends on the physical and chemical properties of both the stabilized material and the leachant. Studies have shown that the main factors affecting leachability are the alkalinity of the stabilized product; the surface-to-volume ratio of the waste; and the tortuosity, a measure of the path length for diffusion.[94] The mechanisms of leaching must be considered in the selection and evaluation of leaching test methods as described below.

The test method affects the leachability of the sample. Specifically, there are a number of test variables that affect the contaminant concentrations in the leachate:

TABLE 11-3
Leaching test methods

Paint filter test
Liquids release test
Extraction procedure toxicity test (EPTox)
Synthetic precipitation leaching procedure (SPLP)
Toxicity characteristic leaching procedure (TCLP)
Modified uniform leach procedure (ANS 16.1)
Equilibrium leach test
Dynamic leach test
Sequential leach test (sequential chemical extraction)
Multiple extraction procedure

- Leachant-to-waste ratio
- Surface area of the waste (e.g., area may be increased by grinding the stabilized mass into small particles)
- Type of leachant (e.g., distilled water, acetic acid, simulated acid rain)
- pH of the leachant
- Contact time
- Extent of agitation
- Number of replacements of fresh leachant
- Extraction vessel
- Temperature

The impact of these variables should be self-evident. For example, the greater the ratio of leachant to waste, the greater the amount of contaminant leached from a stabilized mass. As each test method is described, consider the previously listed variables and their impact on the test result.

Paint Filter Test

The placement of liquid hazardous wastes in landfills in the United States is banned. The paint filter test evaluates the absence or presence of free liquids in bulk and containerized hazardous wastes.[95] This test is rapid, economical, easy to conduct, and easy to evaluate. In this test, wastes are placed in a standard paint filter; if liquid is drained by gravity through the filter within 5 minutes, the hazardous waste is considered to contain free liquids and must be treated prior to landfilling. The test may also be used after stabilization to determine if the stabilization process has been effective in eliminating free liquids from the hazardous waste.

In the strictest of interpretations, the paint filter test is not a leaching test as leachant is not added to the system. The test may, however, yield a leachate. The paint filter test provides results that trigger minimal technology requirements for the "stabilization" of the hazardous waste. The test is adequate to evaluate the presence of free liquids in materials prior to disposal in a landfill. The paint filter test, however, is not adequate to evaluate the effectiveness of stabilization for other applications such as site remediation.

The stresses causing liquids to flow are very small and do not simulate the long-term stresses expected in a landfill. An alternative test, the "liquids release test" (see below), has been developed to better account for the overburden stress in a landfill.

Liquids Release Test

The liquids release test[96] determines if a solidified liquid waste is suitable for landfilling. The procedures were developed to preclude the use of sorbents such as sawdust to "solidify" liquid wastes. In this test a "consolidation" stress is applied to test how readily leachate can be squeezed from a solidified mass.

Extraction Procedure Toxicity Test

The EP toxicity test, an older regulatory test now phased out as the method to classify wastes by toxicity in the United States, can be used to generate a liquid extract from solid wastes.[97] A particular waste is considered EP toxic if its extract has concentrations of any of the eight metals and six pesticides that are greater than the "standards" previously specified in the federal regulations. In this test, the solidified monolithic block is crushed to pass a 9.5-mm sieve. A 0.04M acetic acid (pH = 5) leaching solution is used at a liquid-to-solid ratio of 16:1. The extraction takes place over a period of 24 hours with agitation. The liquid extract is analyzed for specific chemical constituents. For a stabilized mass, the extraction liquid (leachant) is a mixture of deionized water and acetic acid so that the specified pH is achieved. Chemical analyses of the filtered extract are then conducted to determine the concentration of specified organic and inorganic constituents.

As discussed in Chapter 2, the EP toxicity test has been extensively used in the past to classify materials as hazardous or nonhazardous. In this sense it is considered a regulatory test. It is not a design test, as there is no realistic way to apply the results to any sort of transport, fate, or risk analysis. The measured chemical concentrations are simply artifacts of the test procedure. For example, if the pH is changed, the chemical concentrations will change. Similarly, if the liquids-to-solids ratio is changed, the chemical concentrations will change. The test procedure allows the volatile organic contaminants to be released into the atmosphere during the 24-hour extraction period. As a result, a test of this type is not well suited to evaluating wastes with a significant volatile organic component. Most importantly, the TCLP test discussed below has now widely replaced this test as a means to evaluate the effectiveness of stabilization.

Toxicity Characteristic Leaching Procedure (TCLP)

The toxicity characteristic leaching procedure (TCLP) was adopted on November 7, 1986, by the U.S. EPA under the Hazardous and Solid Waste Amendments of 1984.[99] This test, a regulatory test, was adopted as a replacement for the EP toxicity test to determine if a particular waste meets the applicable technology-based treatment standards to be land-disposed. The TCLP is also widely used to evaluate the effectiveness of stabilization and later (1990) replaced the EP toxicity test for classification of materials as hazardous or nonhazardous.[100] In this test method,[101] the stabilized material is crushed to a particle size smaller than 9.5 mm. The extraction fluid is made with water and acetic acid to a pH of 2.88 ± 0.05, selected to mimic conditions in a municipal landfill. The crushed material is mixed with a weak acetic acid extraction liquid, in a liquid-to-solid weight ratio of 20:1, and agitated in a rotary extractor for a period of 18 hours at 30 rpm and 22°C. After 18 hours of agitation, the sample is filtered through a 0.6 to 0.8 μm glass fiber filter, and the filtrate is defined as the TCLP extract. This TCLP extract is analyzed for a wide variety of hazardous waste constituents including volatile and semivolatile organics, metals, and pesticides as shown in Table 11-4. The results of the extract analysis are compared to the regulatory level shown on this table to determine the nonhazardous classification.

TABLE 11-4
Maximum concentration of contaminants for the toxicity characteristics (40 CFR 261.24)[98]

Waste code	Contaminant	Regulatory level, mg/L
D004	Arsenic	5.0
D005	Barium	100.0
D018	Benzene	0.5
D006	Cadmium	1.0
D019	Carbon tetrachloride	0.5
D020	Chlordane	0.03
D021	Chlorobenzene	100.0
D022	Chloroform	6.0
D007	Chromium	5.0
D023	*o*-Cresol	200.0*
D024	*m*-Cresol	200.0*
D025	*p*-Cresol	200.0*
D026	Cresol	200.0*
D016	2,4-D	10.0
D027	1,4-Dichlorobenzene	7.5
D028	1,2-Dichloroethane	0.5
D029	1,1-Dichloroethylene	0.7
D030	2,4-Dinitrotoluene	0.13†
D012	Endrine	0.02
D031	Heptachlor (and its epoxide)	0.008
D032	Hexachlorobenzene	0.13†
D033	Hexachlorobutadiene	0.5
D034	Hexachloroethane	3.0
D008	Lead	5.0
D013	Lindane	0.4
D009	Mercury	0.2
D014	Methoxychlor	10.0
D035	Methyl ethyl ketone	200.0
D036	Nitrobenzene	2.0
D037	Pentachlorophenol	100.0
D038	Pyridine	5.0†
D010	Selenium	1.0
D011	Silver	5.0
D039	Tetrachloroethylene	0.7
D015	Toxaphene	0.5
D040	Trichloroethylene	0.5
D041	2,4,5-Trichlorophenol	400.0
D042	2,4,6-Trichlorophenol	2.0
D017	2,4,5-TP (Silvex)	1.0
D043	Vinyl chloride	0.2

*If *o*-, *m*-, and *p*-cresol concentrations cannot be differentiated, the total cresol (D026) concentration is used. The regulatory level of total cresol is 200 mg/L.

†The quantification limit is greater than the calculated regulatory level. The quantification limit therefore becomes the regulatory level.

Special procedures are used to test the sample for the presence of volatile organic compounds within the extract. These special procedures use a zero-head-space extraction vessel, a cylindrical container with an airtight seal at each end. Extraction in the zero-head-space extractor maintains the volatile organics within the extract, thus allowing reliable measurement of the leachability of volatile organics.

The TCLP is subject to criticism with respect to its use for the evaluation of stabilization effectiveness for several reasons. First, a solidified monolithic mass is broken down to pass the 9.5-mm sieve, reducing beneficial effects of macroencapsulation and microencapsulation. It has been shown that as the particle size reduces, the leachability increases.[102,103] Further, the low-pH environment during extraction may not be representative of the real-world field conditions although it may be representative in a landfill containing domestic garbage. As a result, proponents argue that those severe steps yield conservative results. However, a stabilized material with a high alkalinity, such as cement-based stabilization, may rapidly increase the pH of the leachant and, in effect, result in leaching under basic rather than acidic conditions. Despite these criticisms, the TCLP is particularly useful in comparing the effectiveness of one treatment technology with another or one stabilization mix or process with another.[104]

It is important to note that the EP toxicity test and the TCLP test yield concentrations for specific constituents. These data are sometimes compared to 100 times the drinking water standard to determine if a sample has "passed" or "failed" the TCLP. The factor of 100 is used to account for dilution into the environment. Dragun et al.[105] concluded that the TCLP and EP toxicity methods may alter the soil mineralogy and render predictions of metals migration of little value. These tests do not yield data on the time rate at which contaminants will be leached into the environment. They simply yield a concentration that can be compared to standards or to leachate concentrations obtained using alternative stabilizing techniques and mixes.

The synthetic precipitation leaching procedure[106] is a method to evaluate the mobility (leachability) of both organic and inorganic contaminants from liquids, solids, and wastes with a leachant intended to be representative of the precipitation of the region of the site. The test is similar to the TCLP but the leachant is less aggressive. For soil samples east of the Mississippi (and for waste samples), the extraction fluid is a 60/40 mix of sulfuric and nitric acid combined with water to a pH of 4.2 ± 0.05. For soil samples west of the Mississippi, the extraction fluid is a 60/40 mix of sulfuric and nitric acid mixed with water to a pH of 5.0 ± 0.05. As in the TCLP, the liquid-to-solid ratio is 20:1 by weight and the particle size of the material must be less than 9.5 mm. It is believed that the SPLP is better suited than the TCLP to evaluate the effectiveness of *in situ* stabilization, as the extraction fluid is more realistic than the acetic acid fluid used in the TCLP.

American Nuclear Society Leach Test (ANS 16.1)

The ANS 16.1 leach test,[107] also called the uniform leach test (similar to the International Atomic Energy Agency standard leach test), was developed to determine the leachability of solidified radioactive wastes. This test was designed for use with a solid block of material and, as a result, may not be applicable for soil-like products. The test requires an accurate calculation of surface area so that diffusivity can be determined and then the rate of loss of contaminants for larger amounts of the waste can be predicted. In the ANS 161 test, a solidified test specimen is rinsed in a leaching medium of aerated demineralized water. Following the 30-second rinse, the specimen is left unagitated in a nonreactive container (usually glass) for specific time

intervals. The leaching medium is removed and replaced at various time intervals throughout the 14-day test period. The test yields a number termed the **leachability index** that can be related to the diffusion coefficient. The values of diffusivity should be used with caution, and transport analysis must take into account the amount of disposed waste, the annual net infiltration of water or movement of groundwater through the landfill, the factors affecting the leaching of the waste, the possible interactions between the waste and the leachate, and the fate of the landfill leachate after it has passed from the waste.

Equilibrium Leach Test

The equilibrium leach test[108] is a batch extraction process that utilizes distilled water as the extraction fluid. In this test, a dried sample, ground to pass through an ASTM No. 100 sieve, is mixed with 4 times its mass of distilled water (4:1 liquid-to-solid ratio). The mixture is agitated for 7 days, and the filtered extract is analyzed for total dissolved solids. Specified chemical analyses are also performed on the filtered extract.

Dynamic Leach Test

The dynamic leach test provides data on the rate of leachability, as does the multiple extraction procedure described later.[109,110] The concentrations obtained from all of the leaching tests previously mentioned except the ANS 16.1 leach test, are "equilibrium" or "static" values, and therefore they do not provide any information regarding the rate of release or its dependence on time. To overcome this deficiency, dynamic leach tests are performed. The dynamic leach test, adapted from the American Nuclear Society test (ANS) 16.1, is a true leaching test used to determine the mobility of contaminants through a material matrix, by measuring the rate of leachability. In this test, a monolithic solidified specimen is immersed in distilled water at a specific volume-to-surface-area ratio. The specimen is placed in a nylon mesh harness, rinsed, and suspended in containers filled with distilled water. Fresh leaching agents are used after certain liquid-solid contact times have elapsed, and the solid and leachate are separated.[111] Over a period of 9 days, the leachant is renewed at intervals calculated according to a diffusion model.[112] The samples may be analyzed for organics and/or inorganics, and the pH of the leachant is measured after each transfer. The amount leached in each interval, relative to the total amount of contaminant initially present, is determined, and an apparent diffusion coefficient can be calculated. With this type of data available, time-related behavior, or diffusive transport, can be evaluated and used for the long-term prediction of leaching rate.

Sequential Leach Test (Sequential Chemical Extraction)[113]

This test attempts to evaluate the leachability of metals from a solid material. Five successive chemical extractions of increasing aggressiveness (pH varying from neutral to very acidic) are performed to separate the contaminants into five fractions: the ion exchangeable fraction, the surface oxide and carbonate bound metal ions,

the metal ions bound to iron and manganese oxides, the metal ions bound to organic matter and sulfides, and the residual metal ions.[114] The first three fractions are classified as being "available for short- and medium-term leaching," and the last two fractions are "unavailable for leaching." In this test, a sample is dried in an oven at 60°C and ground to pass through an ASTM No. 325 sieve (45-μm opening size). Then, a 0.5-gram sample is placed in a polysulfone centrifuge tube and subjected to a series of five successive extraction procedures, each suitable for extracting a specific fraction of the metals. In each separate extraction, the specific extraction fluid is added and the mixture is agitated and/or heated for a specific time period before being centrifuged to separate the solid and liquid portions. Chemical analyses are performed on the liquid portion, and the solid portion is rinsed with distilled water, centrifuged, and isolated for use in the next extraction procedure. The procedure is slightly modified for performing a total extraction, in which a new sample is used for each extraction instead of the material from the previous extraction.

Multiple Extraction Procedure

The U.S. EPA multiple extraction procedure appeared in the *Federal Register* on November 22, 1982, and was required by the EPA for temporary exclusion or delisting petitions. It has subsequently been revised.[115] The test attempts to duplicate leaching conditions that a waste would be subjected to from repetitive contact with acid rain at an improperly designed sanitary landfill. The test is not applicable for treated wastes that will be placed in a hazardous waste landfill, but it does provide severe conditions for the evaluation of the effectiveness of a particular treatment technique.

At the start of this test, the procedures are the same as for the EP toxicity test. However, the remaining solid portion is reextracted nine additional times with a synthetic acid rain extraction fluid, consisting of a 60/40 mixture of sulfuric and nitric acids with a pH of 3.0. In each of the nine additional extractions, the mixture is agitated for 24 hours and separated. The extract analysis is the same as that specified in the EP toxicity test. If the contaminant concentrations in the extracts have increased over the seventh and eighth extractions, the extractions are continued, past the ninth repetition, until the concentrations in the extract do not increase.

As in the case of the other tests, these results should not be directly applied to actual field conditions. These results were determined for contaminants released under controlled laboratory conditions that are not directly comparable with conditions at a specific site.

Chemical Tests

The chemical analysis of leachate is a complex task. The type of analysis and the analytical procedures are frequently specified in the leaching procedure by citing standard methods or U.S. EPA protocols. A discussion of analytical methods is beyond the scope of this chapter. Nonetheless, Table 11-5 contains a list of chemical tests performed on the leachate or soil frequently employed to evaluate the effectiveness of stabilization.

TABLE 11-5
Chemical test methods[116]

Total waste analysis
Metals by inductively coupled plasma spectrometry or atomic absorption
Organics by gas chromatography (GC) and mass spectrometry (MS)
Total organic carbon
Loss on ignition
pH
Fourier transform infrared spectra (FTIR)
Differential scanning calorimetry and thermal gravimetric analysis

Much of the focus of chemical testing of stabilized materials relates to the analysis of the extracted fluid or the identification of contaminants in the waste. Because our application of stabilization as a hazardous waste management technology relies on the fixing of the waste within the matrix, tests that help in our evaluation of that process can be useful. One such test is Fourier transform infrared spectra (FTIR), used to evaluate the presence and type of chemical bonding between the organic contaminant and the binding reagents by interpretation of the shift in the infrared frequencies. Differential scanning calorimetry data are used to determine the excess energy needed to release the organic contaminant from the stabilized matrix. Finally, gas chromatography and mass spectrometry (GC/MS) techniques can be used to identify products in the stabilized matrix that may result from degradation of the original organic contaminants. Taken as a group these three tests can give insight into the nature and extent of the chemical bonding of organic contaminants in the stabilized matrix.[117]

Physical and Engineering Property Tests

A number of other tests have been adapted from the civil engineering area to evaluate the physical integrity and engineering properties (strength, compressibility, and permeability) of stabilized materials. These tests are described below and are summarized in Table 11-6.

Moisture content

The moisture content test[118] is used to determine the amount of water (or liquid) in a given quantity of material. For untreated materials, the moisture content aids in the determination of water needs for hydrating reagents. For solidified materials, the moisture content is needed to calculate the dry unit weight (or dry density) of solids given measurements of bulk density. For soil-like materials, the moisture content influences the compactability of the materials. For hardened, concrete-like materials, the moisture content is less significant.

The wet weight of a small (few-gram) sample is first determined. The sample is dried at 105°C for 24 hours and reweighed. The moisture content can then be calculated. Water content on a dry weight basis is presented as a percentage and is calculated as the weight of the water divided by the weight of the dry mass. This is numerically different from the moisture content typically used by environmental engineers, which is expressed on a wet weight basis (weight of water/total weight).

Some wastes contain more liquid than dry solids, and it is possible to have moisture content on a dry weight basis exceeding 100 percent. This is a straightforward test, but misleading results may arise when the organic contaminant level is high because organics will volatilize during the drying period.

Wet and dry bulk unit weight (wet and dry bulk density)

The bulking (increase in volume) associated with stabilization is evaluated by using bulk unit weights that provide information for weight-volume relationships and for calculating density and void volume. These data are used to determine space requirements for ultimate disposal and to compute earth pressures in stability calculations.

The bulk unit weight is determined by measuring and weighing the stabilized samples. The bulk unit weight is calculated as the weight of the solids plus pore fluid per unit volume of material. The dry unit weight is determined from the weight of oven-dried solids per unit volume. Typical units are gm/cm^3 or lb/ft^3.

Specific gravity

Specific gravity and density are measures of the mass-to-volume ratio of the individual solid component of the three-phase material. The density of solids divided by the density of water is the specific gravity. This parameter is needed to calculate certain physical properties such as porosity and unit weight. Specific gravity is numerically equal to the density of solids in SI units as water has a density of 1.0 g/cm^3.

The procedure to determine specific gravity is difficult to execute with precision and is not generally amenable to a solidified material that contains a number of different materials, each having a different specific gravity. At best, the measured specific gravity represents a composite of the individual components.

TABLE 11-6
Physical, engineering, and durability property tests

Property	Purpose
Physical properties	
Moisture content	Phase calculations (saturation, void ratio)
Wet and dry bulk density (unit weight)	Stress and volume calculations
Specific gravity	Phase calculations (saturation, void ratio)
Particle size distribution	Classification descriptor (e.g., distinguishes sand, silt, or clays)
Pocket penetrometer	As-mixed strength
Microstructural examination by:	
X-ray diffraction	Crystallization
Optical microscopy	
SEM (scanning electron microscopy)	
Supernatant formation during curing	Curing time
Rate of setting	Excess liquid in mixture
Engineering properties	
Strength	Stability analysis
Compressibility (consolidation)	Settlement analysis
Hydraulic conductivity	Flow and transport calculations
Durability properties	
Wet/dry	Long-term integrity
Freeze/thaw	Long-term integrity

Particle size distribution

The grain-size distribution[119] of the waste or soil breaks the total mass of soil into various fractions by particle size. This particle size distribution is important to the design of stabilization alternatives. Fine-grained wastes are subject to wind dispersion, are difficult to handle during construction, and present problems in producing a high-strength solidified waste. On the other extreme, wastes with a significant coarse fraction may require prescreening to enable mixing equipment to operate and to preclude particle segregation.

To determine the particle size distribution, sieves are employed for the coarse fractions. Gravitational methods employing a hydrometer are used to separate the fine fraction particles with sizes ranging from 0.074 mm to micrometer size. In accordance with Stokes' law, larger particles settle more quickly through water than smaller particles, and a hydrometer allows measurement of the differences in particle size by determining the change in the fluid density as a function of time. Because the stabilized mass is often a monolith, particle size distribution is not generally applicable to stabilized materials.

Pocket penetrometer

The pocket penetrometer provides information on the strength of the material being tested (i.e., provides an assessment of the as-mixed properties of the stabilized sludge prior to curing). A pocket penetrometer is a standardized cylinder that is pushed into the material to measure the penetration resistance (force), estimating the unconfined compressive strength in tons/ft^2 (see Figure 11-14). This test is also used in field applications.

Microstructural examination

The stabilized mass can be examined by X-ray diffractometry, optical microscopy, scanning electron microscopy, and energy dispersive microscopy to better understand the nature of the stabilization processes at work. The microstructural analyses are part of basic research into stabilization mechanisms. These techniques allow improved understanding of the mechanisms by which specific contaminants are bound in the matrix. For example, in a laboratory study the microstructure, made of a stabilized heavy metal sludge, was examined. The original sludge contained chromium, nickel, cadmium, and mercury and had been stabilized using a combination of ordinary portland cement and fly ash.[120] The microstructural examination

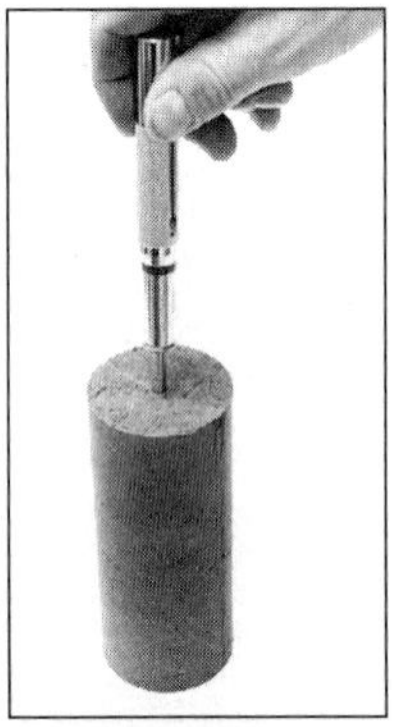

FIGURE 11-14
Pocket penetrometer evaluating a stabilized waste.
(Photo courtesy of Debra Cook, Bucknell University.)

revealed that there was a wide variability in the sludge composition and a corresponding wide variability in the chemical composition of the stabilized material. Variability on a microscopic level, may be related to variability of properties measured on the macroscopic level such as leaching, strength, and durability. At this point in time, more research is needed to answer this and other questions regarding the details of stabilization for hazardous waste management.

Supernatant formation during curing and rate of setting

These test methods characterize the performance of treated wastes in the first few hours after mixing. They are particularly suitable for concrete-like (high-slump) materials and not suited for soil-like materials. These tests give early information regarding the suitability of the mix.

To determine the supernatant formation, a freshly mixed sample is placed in a 1-L graduated cylinder, and as the solids settle and supernatant rises, the quantities of each are recorded. Generally speaking, the formation of any supernatant is undesirable.

The set time represents the time required for the fluid-like material to achieve some initial, yet minimal, strength. The set time can be measured with a vicat needle.[121] In this test, a standard "needle" penetration is measured at various times. Set is defined as the time for less than $\frac{1}{4}$-inch penetration.

Unconfined compressive strength test

The unconfined compressive strength test[122,123] is conducted to determine the strength of cohesive materials. Cohesive materials include materials ranging from soft clays to concrete. In this test, a cylindrical specimen typically 2.8 inches in diameter by 5.6 inches tall is axially loaded to failure as shown in Figure 11-15. The axial load and corresponding deformation are measured. From these measurements, the applied stress and resulting strain can be calculated and the stress-strain relationship plotted as shown in Example 11-4. The unconfined compressive strength is then the maximum applied stress in the test.

FIGURE 11-15
Unconfined compression strength test. *(Photo courtesy of Debra Cook, Bucknell University.)*

EXAMPLE 11-4. UNCONFINED COMPRESSIVE STRENGTH TEST. A sample of sludge was stabilized with an attapulgite clay sorbent and binders of fly ash, lime, and cement. The sample data measured are:

Diameter (d):	7.01 cm (2.70 in)
Height (H_0):	14.55 cm (5.73 in)
Weight (M):	829.6 g (1.83 lb)

The following data were recorded during the test:

Load P, lb	Deflection, dial reading, in
0.0	0.057
231.0	0.068
405.8	0.079
620.5	0.091
861.9	0.102
1131.4	0.113
1408.3	0.124
1651.1	0.136
1814.0	0.147
1905.8	0.158
1969.5	0.170
1634.9	0.181

Compute sample area, volume, wet bulk density, stress, and axial strain. Plot the stress-strain curve. What is the unconfined compressive strength?

Solution. The initial area A_0 of the cylindrical test specimen is:

$$A_0 = \frac{\pi d^2}{4} = 38.6 \text{ cm}^2 \ (5.98 \text{ in}^2)$$

where $d =$ sample diameter.

The volume V of the cylindrical test specimen is:

$$V = A_0 H_0 = 38.6 \times 14.55 = 561.7 \text{ cm}^3 \ (34.27 \text{ in}^3)$$

where $H_0 =$ initial height of sample.

The density ρ is:

$$\rho = M/V = 829.6 \text{ g}/561.7 \text{ cm}^3 = 1.48 \text{ g/cm}^3 \ (92.2 \text{ lb/ft}^3)$$

As shown on the data table, at the start of the test, the applied load P is zero. The axial deformation δ must also be zero, although there is an initial dial reading of 0.057 in corresponding to an axial deformation of zero.

Computing the axial strain ϵ_a for the first load and deflection dial readings yields:

$$\epsilon_a = \frac{\delta}{H_0} \times 100 = \frac{0.068 - .057}{5.73} = .002$$

or in percent

$$\epsilon_a(\%) = 0.2\%$$

The stress σ can now be calculated as force per unit area:

$$\sigma = \frac{P}{A_c} = \frac{P}{\frac{A_o}{1-\epsilon_a}} = \frac{231}{\frac{5.98}{1-0.002}} = 38.5 \text{ psi}$$

Note that the area is corrected, A_c, to account for the sample getting "fatter" as it gets "shorter"; remember, there is no net volume change, so as the sample is strained, the area increases. Continuing to calculate σ and ϵ_a shows:

Load *P*, lb	**Deflection, in**	**strain ϵ_a, %**	**Corrected area, in^2**	**Stress, psi**
0.0	0.057	0.00	5.982	0.0
231.0	0.068	0.20	5.994	38.5
405.8	0.079	0.40	6.006	67.6
620.5	0.091	0.59	6.018	103.1
861.9	0.102	0.79	6.03	142.9
1131.4	0.113	0.99	6.042	187.3
1408.3	0.124	1.19	6.054	232.6
1651.1	0.136	1.38	6.066	272.2
1814.0	0.147	1.58	6.078	298.5
1905.8	0.158	1.78	6.09	312.9
1969.5	0.170	1.97	6.103	322.7
1634.9	0.181	2.17	6.115	267.4

The unconfined compressive (UC) strength is the peak compressive strength. For these data, the UC strength is 322.7 psi.

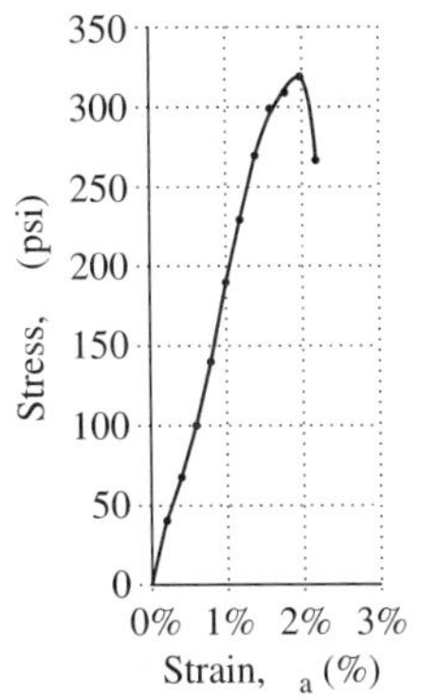

The use of the unconfined compressive strength test to evaluate the effectiveness of stabilization and solidification has evolved into widespread application as an international standard for the industry. Stabilized and solidified materials must have adequate strength to enable them to support the loads of materials placed over them. Further, the test results would be expected to correlate with the effectiveness of stabilization of inorganic wastes as the inorganics are tied up in the hydrating matrix. This test also offers several other advantages for evaluation of the effectiveness of stabilization. The test is relatively quick (and thus inexpensive), and the required testing equipment is widely available. The test also provides an indication of the ductility/brittleness of the stabilized material. The test can be conducted over time

to provide a measure of the improvement of the stabilization process as a function of time. In general, for any given stabilization reagent, the stronger the stabilized hazardous waste, the more effective the stabilization process, particularly for inorganic contaminants. Thus, the unconfined compressive strength is often used as an indicator parameter of the effectiveness of stabilization.

It is generally accepted that an unconfined compression strength of 50 psi represents a suitable strength measure for stabilized materials on the basis of the calculated typical overburden pressures in a landfill. It is not necessary, however, from a stability standpoint, to achieve a specific unconfined compressive strength to achieve stability in a landfill. Engineering materials can be quite stable and yet may exhibit an unconfined compressive strength of zero (e.g., sand). Alternatively, the strength of engineering materials may be characterized by their frictional properties, termed the angle of internal friction. This is analogous to the sliding block on an inclined plane problem examined in introductory physics. The block does not begin to slide down the inclined plane until the angle of inclination increases to the point that the down plane component of gravity exceeds the frictional force resisting the movement. The frictional force is equal to the normal force times the coefficient of friction between the plane and the block. Similarly, the strength of engineering materials may depend on the coefficient of friction of the material sliding against itself, thus the term **angle of internal friction.** For example, the strength of sand and gravel is characterized by an angle of internal friction generally greater than 30°, but the unconfined compressive strength of that material is zero. Structures, dams, sand dunes, and other examples of geotechnical stability can be identified for granular materials with unconfined compressive strengths of zero.

In the long term, it appears that the unconfined compressive strength may not be a very reliable indicator for the effectiveness of stabilization,[124] and it is likely that this test will demonstrate its limited usefulness as the science and art of stabilization evolve. Specifically, the effectiveness of stabilization as determined with chemical analysis of leachate, particularly with organic sludges, does not necessarily correlate with unconfined compressive strength, as shown in Figure 11-16. Further, the

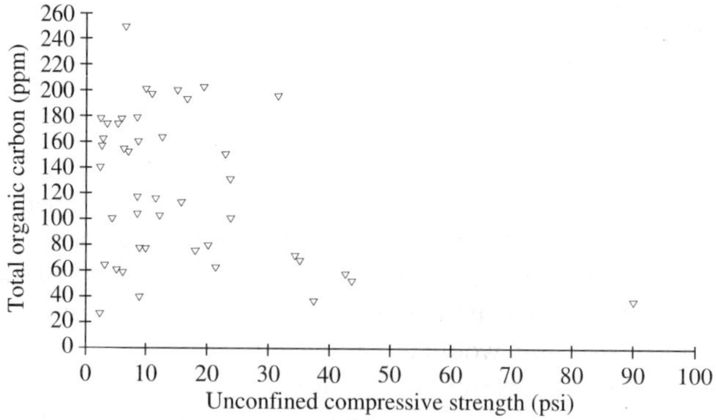

FIGURE 11-16
Unconfined compression and organic waste stabilization. *(From Evans and Pancoski, 1989).*[125]

strength properties may change with time, and the unconfined compressive strength measured at 28 days may not reliably predict long-term strength behavior without due consideration of product durability.

Consolidation

It is recognized that some form of consolidation test may be well suited for the evaluation of the settlement of stabilized waste. The consolidation test measures the time rate of one-dimensional compression as well as the relationship between applied vertical stress and total deformation. During a consolidation test, an axial load is applied to the material, which is restrained from lateral deformation (see Figure 11-17). The time rate of deformation (consolidation) is measured for each load applied. The resulting vertical deformation, in terms of axial strain or void ratio, is plotted versus the applied load (see Figure 11-18). For soft compressible materials, liquids will be squeezed out of the pores as a result of the applied load. The figure shows axial compression with increasing stress as the material is loaded through the recompression range and the virgin compression range. On unloading, the material rebounds. The liquid that is squeezed out of the pores can be chemically analyzed and used in an assessment of the rate of contaminant migration into the environment.

For well-stabilized, rigid materials the compression will be primarily elastic without the expulsion of fluids from the pores of the stabilized mass. Thus, this test can be used to simulate the field condition of stabilized materials in a large landfill. In the field, stabilized material is loaded by the overlying fill and cover materials. As a result of lateral drainage and anisotropy (i.e., difference in properties such as permeability based on whether flow is horizontal or vertical) due to placement in

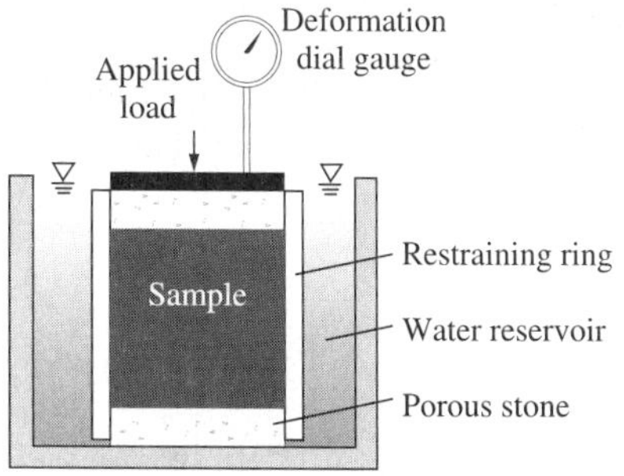

FIGURE 11-17
Consolidation test schematic.

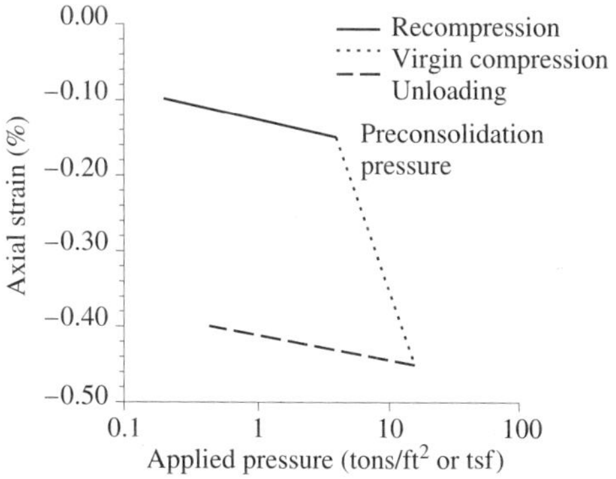

FIGURE 11-18
Idealized consolidation behavior.

layers or sections in the field, it is likely that the time rate of consolidation predictions provided by the laboratory consolidation test data will overestimate the time required for consolidation in the field condition.

Shown in Figure 11-19[126] is the consolidation behavior for three samples of sludge, each stabilized with different proportions of stabilization reagents. The data show that the compressibility of sample A is about 4 times greater than the compressibility of sample B at an applied pressure of 10 tons per square foot. The data also show that sample B exhibits a substantial preconsolidation pressure as compared with sample A. Consolidation results such as these are used in evaluations of the suitability of the stabilization process in terms of allowable settlement (deformation) of the stabilized materials.

Hydraulic conductivity

Hydraulic conductivity, sometimes called permeability, is the ability of a material to conduct or discharge water in response to an applied hydraulic gradient (see Chapter 4). This property is valuable in that it indicates the ease with which water will pass through the material. Hydraulic conductivity is used to estimate the flow rate of fluid through a porous material. As a result, hydraulic conductivity is used to calculate information regarding the rate at which chemicals in the treated waste may migrate into the environment. Some of the factors affecting the hydraulic conductivity of a material include density, degree of saturation, temperature, type of permeating fluid, hydraulic gradient, and particle size distribution. The contaminant concentrations in the effluent can also be measured and values of effective diffusivity determined.

The triaxial permeability test[128,129] is frequently employed to evaluate the hydraulic conductivity of stabilized materials. This test is also called the flexible wall permeability test because the sample is encapsulated in a flexible membrane that permits the sample to shrink or swell under the applied confining stresses. A schematic of the triaxial permeability test is shown on Figure 11-20. Flow is induced by elevating the hydraulic head on one end of the sample with respect to the other. The *in situ*

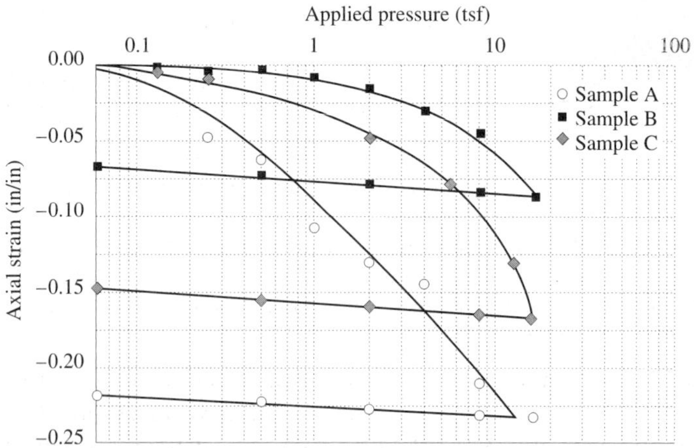

FIGURE 11-19
Consolidation test results.[127]

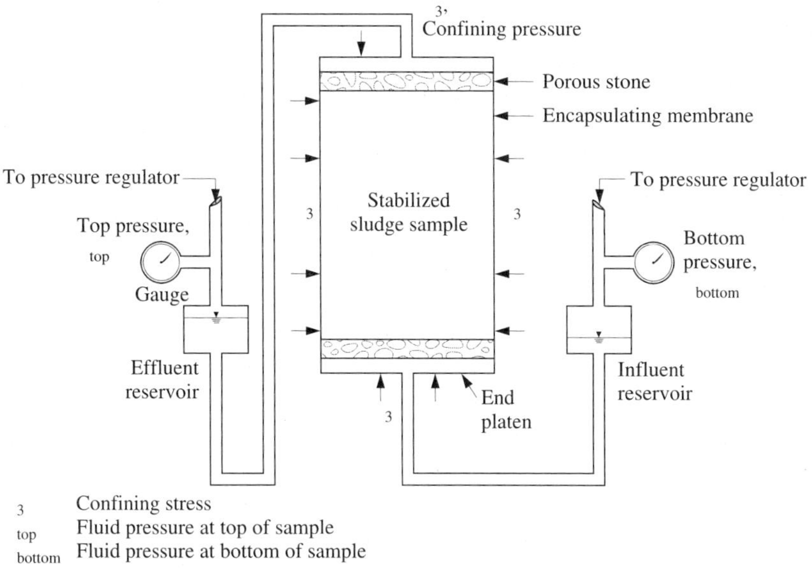

FIGURE 11-20
Triaxial permeability test schematic (modified from Evans and Fang, 1988).[130]

stress can be simulated by applying pressure to the triaxial cell water surrounding the membrane-encapsulated sample.

EXAMPLE 11-5. EFFECTIVE STRESS FOR HYDRAULIC CONDUCTIVITY TESTING. Stabilized sludge will ultimately be disposed in a 5-acre landfill to a maximum depth of 35 feet. The total unit weight of the stabilized sludge is 110 lb/ft^3. What confining stress would you select for a triaxial permeability test?

Solution. The confining stress for a triaxial permeability test should simulate the *in situ* stresses. For the above conditions, the maximum *in situ* vertical stress would be:

$$\sigma_v(\text{max}) = 35 \times 110 = \underline{\underline{3850}}\ \text{lb/ft}^2$$

The minimum stress would be at the top of the fill, immediately beneath the cover. Assuming a cover 5 feet in thickness (see Chapter 13) and a unit weight of 110 lb/ft^3, the minimum vertical stress would be:

$$\sigma_v(\text{min}) = 5 \times 110 = \underline{\underline{550}}\ \text{lb/ft}^2$$

Because the hydraulic conductivity is stress dependent, it is recommended that hydraulic conductivity tests be conducted over a range of stress levels to a maximum of 3850 lb/ft^2. For example, for a three-test series, confining stresses of 500, 2175, and 3850 lb/ft^2 could be used.

By monitoring the influent and the effluent quality and quantity, the effectiveness of stabilization can be evaluated along with the effective diffusivity of the contaminant in the stabilized mass. Typical test results for permeation of an acidic oil sludge stabilized with a combination of cement, fly ash, lime, and attapulgite are shown in Figure 11-21. Note that pore volume displacement is a dimensionless parameter

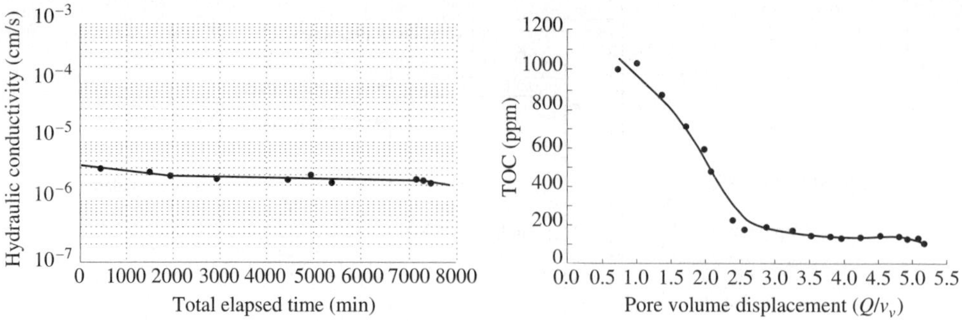

FIGURE 11-21
Triaxial permeability test results.[131]

calculated as the cumulative volume of flow (Q) divided by the volume of voids (V_v) in the sample.

EXAMPLE 11-6. HYDRAULIC CONDUCTIVITY TEST. The stabilized sludge described in Example 11-5 was tested according to your recommendation. The sample is 14.2 cm long with a diameter of 7.1 cm. The cylindrical sample was tested in a triaxial cell, shown schematically in Figure 11-20, using a bottom pressure σ_{bottom} on the influent reservoir of 55 psi and a top pressure $\sigma_{\text{top}} = 50$ psi. A total flow of 1.76 cm^3 was measured over an elapsed time of 3 hours. Calculate the hydraulic conductivity.

Solution. Darcy's law, $q = KiA$, can be rewritten as

$$K = \frac{q}{iA}$$

where K = hydraulic conductivity (cm/s)
i = hydraulic gradient (dimensionless)
A = cross-sectional area in a direction normal to flow (cm^2)
q = flow rate (cm^3/s)

The area A is:

$$A = \frac{\pi d^2}{4} = \pi \times \frac{7.1^2}{4} = 39.6 \text{ cm}^2$$

The gradient is

$$i = \frac{\Delta h}{L}$$

where Δh = change in hydraulic head
L = length of flow path

The hydraulic head on the bottom of the sample is:

$$h_{\text{bottom}} = \frac{\sigma_{\text{bottom}}}{\gamma_w}$$

where γ_w = unit weight of water.

$$h_{\text{bottom}} = \frac{55 \text{ lb/in}^2}{62.4 \text{ lb/ft}^3} \times \frac{144 \text{ in}^2}{\text{ft}^2} \times \frac{12 \text{ in}}{\text{ft}} \times \frac{(2.54 \text{ cm})}{\text{in}}$$

$$= 3869 \text{ cm}$$

The hydraulic head on the top of the sample is:

$$h_{\text{top}} = \frac{\sigma_{\text{top}}}{\gamma_w} = \frac{50 \text{ lb/in}^2}{62.4 \text{ lb/ft}^3} \times \frac{144 \text{ in}^2}{\text{ft}^2} \times \frac{12 \text{ in}}{\text{ft}} \times \frac{2.54 \text{ cm}}{\text{in}}$$

$$= 3517 \text{ cm}$$

The gradient is

$$i = \frac{\Delta h}{L} = \frac{3869 \text{ cm} - 3517 \text{ cm}}{14.2 \text{ cm}} = 24.8$$

The flow rate q is:

$$q = \frac{1.76 \text{ cm}^3}{3 \text{ h} \times 60 \text{ min/h} \times 60 \text{ s/min}} = 1.63 \times 10^{-4} \text{ cm}^3/\text{s}$$

The hydraulic conductivity may now be computed as:

$$K = \frac{q}{iA} = \frac{1.63 \times 10^{-4} \text{ cm}^3}{24.8 \times 39.6 \text{ cm}^2 \cdot \text{s}}$$

$$= 1.7 \times 10^{-7} \text{ cm/s}$$

Durability test methods

Durability test methods are needed to assess the long-term performance of the stabilized mass. Durability testing is used to evaluate the ability of a material to resist repeated cycles of weathering.[132] A successfully solidified waste material will remain intact throughout these tests, so that the area of the waste directly exposed to leaching is not greatly increased. Experience has shown that most specimens that withstand freeze/thaw cycles also withstand wet/dry cycles, but the reverse is not necessarily true.[133] As a result, the freeze/thaw test, discussed in the following section, was found to be the more severe of the two durability tests.

Wet/dry durability (ASTM D4843)[134]

The wet/dry durability test evaluates the resistance of stabilized materials to the natural weathering stresses of repeated wetting and drying cycles. Cured test specimens are subjected to twelve test cycles. Each cycle consists of a period of 5 hours submerged under water and 42 hours in an oven under low-temperature drying conditions (160°F). The change in volume, moisture content, and weight is determined after each cycle. After the 12 cycles, the total sample weight loss is determined or the number of cycles to cause disintegration is determined if the sample is not intact after the 12 cycles are completed.

Freeze/thaw durability (ASTM D4842)[135]

The resistance of a material to the natural weathering stresses of freezing and thawing are evaluated in the freeze/thaw durability test. Cured test specimens are

exposed to twelve cycles, each consisting of a 24-hour freezing period at −20°C and a 24-hour thawing period in water. As in the wet/dry test, performance is evaluated by calculating the weight loss after twelve cycles or the number of cycles that cause disintegration, whichever occurs first. In these durability tests, the weight loss after each cycle can be compared to that of a control specimen.

11-5 FIELD IMPLEMENTATION

The selection of the type of stabilization process depends on factors such as waste characteristics, materials handling and processing, stabilization objectives, regulatory requirements, and economics.[136] Selection of alternatives for site remediation is discussed in Chapter 17. This section discusses alternative methods for implementing stabilization technologies in the field.

In-Drum Mixing Alternatives

To improve the physical characteristics of drummed hazardous wastes prior to land disposal, containers of toxic and hazardous liquids and sludges can be stabilized by in-drum solidification procedures. The drum serves as both the mixing vessel and shipment container for the solidified hazardous waste. However, no credit is taken for the long-term integrity of the drum, assuming the drum will ultimately disintegrate in a landfill.

Reagents are added to the drum containing the waste to be stabilized. Because this may include new drums from a presently operating manufacturing process or old drums from an on-site remediation, it is necessary to evaluate the condition of the drum to determine if it has suitable integrity for the mixing and shipment. Generally, the process involves the addition of the chemical reagents and mixing by a top-entering mixing blade. This method of solidification has been used extensively and may be performed remotely by robots.

In Situ Mixing Alternatives

The most widely employed method for implementing the stabilization and solidification alternative is *in situ* mixing. A photograph of typical *in situ* mixing is shown on Figure 11-2. *In situ* mixing frequently uses commonly available construction equipment to accomplish the mixing process. In the typical case of the treatment of a large lagoon, this method is often the most economical because it minimizes the handling of wastes. This process was used for the stabilization of a site originally used for oil recovery.[137] A blend of 100 parts sludge to 15 parts lime to 5 parts cement kiln dust was employed. The reagents were added to the lagoons by cranes with 100-foot-long booms and mixed with backhoes by using a kneading action. The completion of fixation was defined according to criteria as follows:

- Visual inspection indicating the material to be solid and earth-like
- No free liquid

- Possible to excavate with material sitting in bucket without overflowing
- Ratio of reagent to fixed material of 1 to 6

Contractor innovations have improved the thoroughness of mixing compared to that achieved with a backhoe. The backhoe bucket can be replaced with vibratory injectors that simplify the process of adding reagents and increase the mixing energy to obtain a more uniform mix.

For sites requiring relatively deep *in situ* stabilization or more thorough mixing, modified augers with mixing blades have been employed. Figure 11-22 shows a large-diameter single auger. Figure 11-23 shows a schematic of the entire process. There are, however, always questions as to the thoroughness of an *in situ* mixing process.

FIGURE 11-22
In situ mixing process with modified augers. *(Photo courtesy of Geo-Con.)*[138]

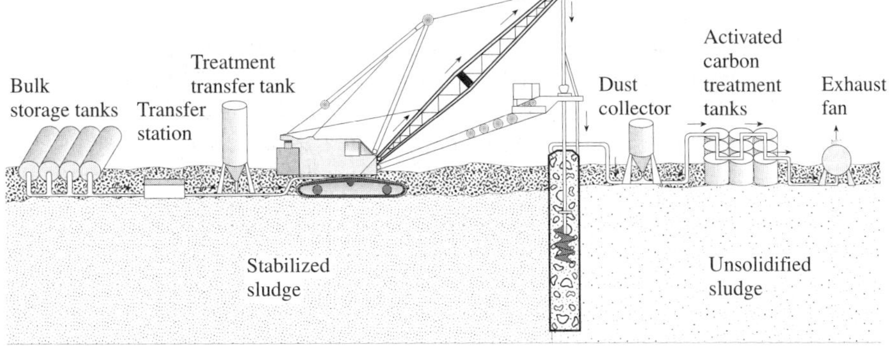

FIGURE 11-23
Schematic of an *in situ* mixing process with modified augers.[139]

Another *in situ* mixing method is known as **jet grouting.** This method of mixing subsurface materials *in situ* has been used for underpinning of foundations, excavation support, groundwater control, and, of interest here, stabilization. For stabilization, jet grouting is a method by which a column of material is mixed in the subsurface by a high-velocity jet. The technique uses a rotating high-pressure jet to excavate and mix *in situ* materials with the injection fluid. There are variations in the system such as the number of rods used, as shown on Figure 11-24. First, drilling to the desired maximum depth and using a consistent and uniform rotation and lifting from the bottom up forms the mixed column. For stabilization, the grout would be designed for the *in situ* contaminants and the desired performance for the stabilized material. The principal advantages of jet grouting over other *in situ* methods for stabilization are the smaller equipment with the ability to work near buildings and buried utilities and the relatively quiet, low-vibration equipment. Later, this method will be shown to be useful for the installation of subsurface vertical barriers (see Chapter 16).

Plant Mixing Alternatives

A more thorough way to blend stabilization reagents with the hazardous wastes employs a mechanical mixer using either batch or continuous processes. With this alternative, the hazardous waste is supplied to a continuous or batch mixing process where agents are added and blended. The treated materials are then placed in their ultimate disposal area. A batch process adds fixed amounts of waste and reagents for blending for a fixed contact time (see Figure 11-25). A continuous process provides for ongoing input of waste and reagents and provides the necessary contact time by design of the feed and mixing equipment. These processes can provide significantly improved control over the mix proportions and the thoroughness of blending. Equipment used can include pug mills or extruders. The blended material is then transported for disposal. It is typically placed back on site, compacted, and allowed to cure in place. These processes enable better control over the reagent-to-waste ratios

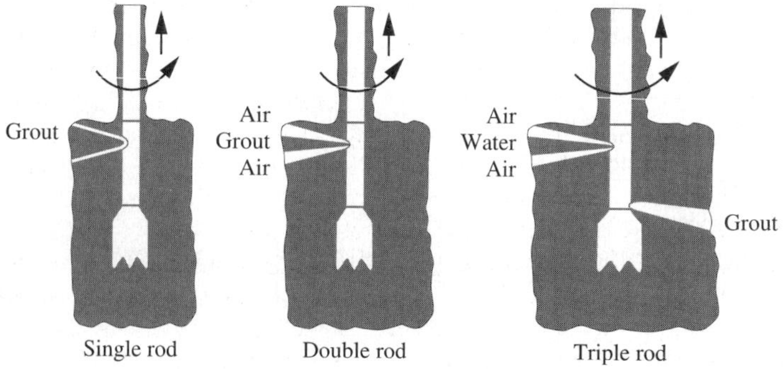

FIGURE 11-24
Typical jet grouting systems.[140]

FIGURE 11-25
Remote process batch mixing. *(Photo courtesy of Enreco, Inc.)*[142]

and ensure better and more uniform mixing than *in situ* techniques. However, even with batch mixing, it may be difficult to ensure the treated materials are adequately blended with the reagents and no unmixed lumps remain in the batch. Plant mixing is typically implemented on-site by employing transportable equipment.

EXAMPLE 11-7. At a secure landfill, wastes with free liquids (failing the paint filter test) are stabilized with fly ash.[141] The alkaline fly ash is derived from power plants burning western coal. The on-site mixing alternative is employed, using a batch process. The wastes are mixed with fly ash by a front-end loader working on a concrete mixing pad. The front-end loader measures the quantity of fly ash, usually two or three buckets, and places the fly ash on the mixing pad for each batch. Wastes are then added directly from the tank trucks and roll-off containers. Additional fly ash is then added, and each batch mixed by using the front-end loader to turn and spread the material. The blended material is then transported to the landfill and spread in uniform lifts with a bulldozer.

An oily waste was stabilized by the process described above. The waste, a brown oil-water emulsion, was analyzed and found to contain 14 percent oil and grease, 2.2 percent total organic carbon (TOC), and 72 percent volatile solids including 180 mg/kg 1,1,1-trichloroethane, 71 mg/kg butylbenzylpthalate, 15 mg/kg cadmium, 150 mg/kg copper, 20 mg/kg lead, and 15 mg/kg zinc. From visual observations, the waste required 1.2 parts fly ash to 1 part waste by weight for stabilization. After stabilization, testing showed an unconfined compressive strength of 1758 kg/m^2 (0.18 tsf) at the time of mixing and 48,828 kg/m^2 (5.0 tsf) after 6 days of curing. After stabilization, the material was analyzed and found to contain 89 percent total solids and 14 percent volatile solids, including 1.5 mg/kg 1,1,1-trichloroethane, 1.1 mg/kg 4-methyl-2-penatone, 12 mg/kg toluene, 14 mg/kg ethyl benzene, and 25 mg/kg total xylenes.

(a) Calculate the reduction in contaminant concentrations due to the dilution only.
(b) Identify other processes that could account for further reductions in contaminant concentrations.

Solution

(*a*) $C_0 \times W_w = C_f(W_w + W_{FA})$

where $C_0 =$ original contaminant concentration

$W_w =$ weight of waste

$C_f =$ final contaminant concentration

$W_{FA} =$ weight of fly ash

$$\frac{C_0}{C_f} = \frac{W_w + W_{FA}}{W_w} = \frac{1.0 + 1.2}{1.0} = \frac{2.2}{1} = \underline{\underline{2.2}}$$

$$\frac{C_f}{C_0} = \frac{W_w}{W_w + W_{FA}} = \frac{1.0}{1.0 + 1.2} = \frac{1}{2.2} = \underline{\underline{0.45}}$$

The contaminant concentrations are 45 percent of original as a result of dilution only; that is a reduction of 55 percent, assuming no change in density.

(*b*) The metals could react in the pozzolanic reactions of the fly ash.
The volatiles are likely released through volatilization during the extensive handling.
The heat of any pozzolanic reaction enhances the rate of volatilization.
Fly ash is a pozzolan, and hydration reactions are time-dependent.

11-6 DESIGN

The selection and design of a stabilization process necessitate characterization of both the waste and the site. A flowchart for these evaluations[143] is presented in Figure 11-26. The most important lesson regarding design of stabilization projects is that often a poor correlation exists between bench-scale studies and full-scale field performance. The variations are frequently related to the thoroughness with which the waste and reagents are mixed; in the laboratory the mix is well blended whereas in the field, such control may be lacking. A full-scale field demonstration (50 to 100 m^3) should be a critical part of the design process.

The selection and design of the appropriate stabilization formulation and technique begin with treatability studies. Treatability studies are conducted to select and develop the appropriate type of binder to eliminate or reduce the hazard of the hazardous waste. Should treatability studies indicate the potential for stabilization, further studies are needed to demonstrate the suitability and to optimize the mix formulation.

The development of the stabilization formulation is an art, not a science. How then does one start the process? The selection of the appropriate formulations is based on the nature of the waste coupled with judgment incorporating published literature and corporate experience. In addition to consideration of the mechanisms of stabilization discussed in Section 11-2, Conner[144] suggests the following additional factors to consider:

- Curing time—field and laboratory
- Volume increase/decrease
- Cost—ingredients, labor, and equipment
- Schedule—project duration, weather restrictions
- Type of mixing—batch or continuous

- Heat generation—heat of hydration and other chemical reactions
- Gas generation—loss of volatiles, material porosity, possible toxicity
- Mixing time—field and laboratory, maximum and minimum
- Other restrictions—flash point, explosion potential, fire potential
- Handling characteristics—waste and reagents

Treatability tests are conducted by mixing different types and quantities of reagent with small (often about 500-g) samples of the waste. The studies are conducted in a phased manner, progressively applying what has been learned to arrive at a suitable solution. The first phase typically includes bench-scale mixing and testing of leachability and strength. For example, in one study four variables (sand/cement ratio, water cement ratio, wet sludge quantity, and additive dosage) were varied in a statistically designed experiment to determine the variable effect on the strength and leachability.[145] As the mix proportions and ingredients are refined, additional

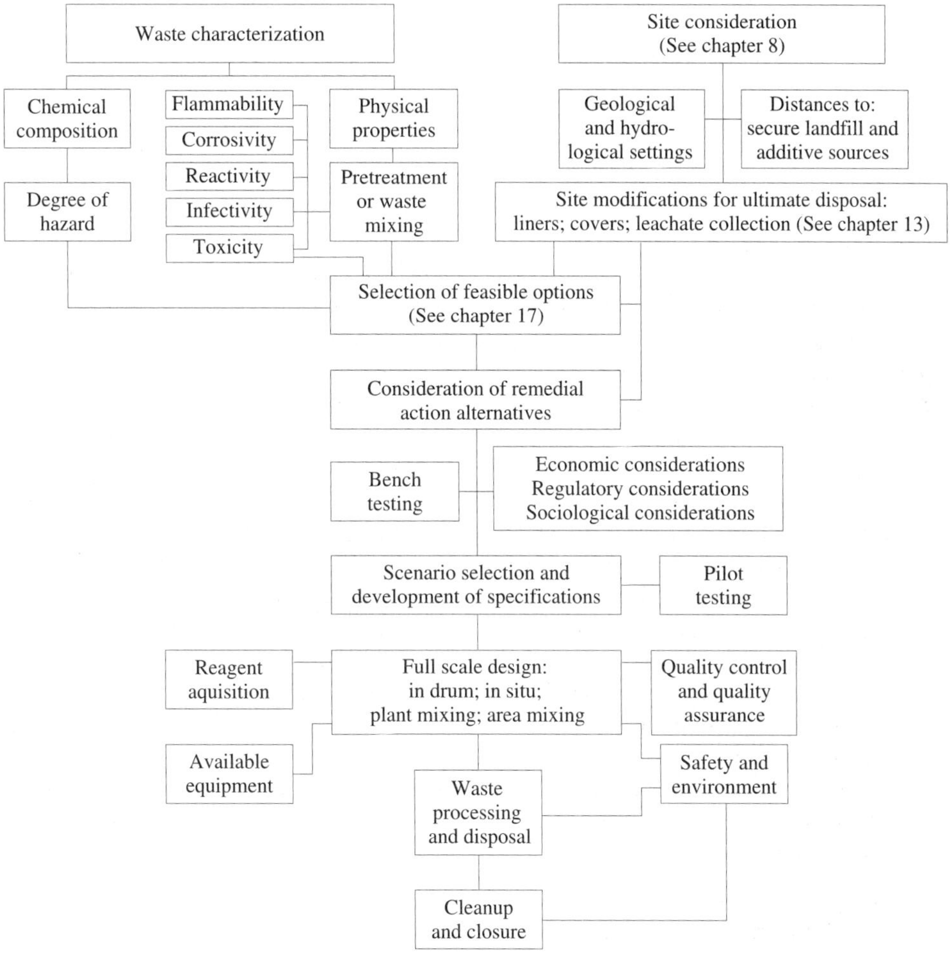

FIGURE 11-26
Flowchart for stabilization process design.[143]

bench scale testing typically includes more extensive engineering property testing such as hydraulic conductivity and weathering durability. It has been suggested that the mix proportions and test results could be plotted to define an "area" of acceptable test results; in essence, a quality control chart[146] (e.g., a plot of a specific organic of concern in the leachate versus the proportion of cement in the mix). Finally, field pilot-scale studies are undertaken to confirm that the laboratory tests can be replicated under field conditions.

EXAMPLE 11-8. STABILIZATION COST ESTIMATE. Stabilization of sludge from a wastewater treatment plant is being considered. The plant treats 0.3×10^6 gal/day (0.3 mgd) with a suspended solids concentration of 220 mg/L (ppm) in the influent wastewater. The treatment plant has a suspended solids removal efficiency of 90 percent. From bench-scale studies, adequate stabilization can be achieved by adding 8 percent lime by weight. Lime stabilization costs are $80.00 per ton including materials, labor, and equipment.

What is the annual cost? How much lime is needed per year?

Solution

$$\text{Solids in the sludge} = (0.3 \times 10^6 \text{ gal/day}) \times \left(220 \frac{\text{parts}}{10^6 \text{ parts}}\right) \times (0.9) \times (8.24 \text{ lb/gal})$$

$$= 495.4 \text{ lb/day}$$

$$\text{Annual sludge production} = 495.4 \text{ lb/day} \times 365.25 \text{ days per year}$$

$$= 180{,}944 \text{ lb/year} = 90.5 \text{ ton/year}$$

$$\text{Annual cost} = 90.5 \text{ ton/year} \times \$80/\text{ton} = \underline{\underline{\$7238/\text{year}}}$$

$$\text{Lime usage} = 90.5 \text{ ton/year} \times 0.08 = \underline{\underline{7.2 \text{ ton/year}}}$$

11-7
CASE STUDIES

Case Study 1

Under the U.S. EPA's SITE (Superfund Innovative Technology Evaluation) program, an *in situ* soil stabilization system for PCB-contaminated soil was evaluated.[147,148] The system employed a proprietary additive incorporating organically modified clay and deep soil mixing. Initial results were promising as the unconfined compressive strength was high, the permeability low, and the material stable to cycles of wetting and drying. However, many samples subjected to cycles of freezing and thawing showed degradation. The effect of stabilization on the PCB leachate concentration was inconclusive, since all analytical results (untreated and treated samples) were near detection limits. The *in situ* test equipment operated satisfactorily, resulting in a dense and homogeneous structure of low porosity. The 1989 cost was estimated at $110 per ton. A follow-up study 1 year later showed improvement in the physical properties although the freeze/thaw durability weight loss results were unsatisfactorily high.[149]

Case Study 2

It must be emphasized that although theory and bench-scale testing can provide tremendous insight into and confidence in stabilization processes, the work in the field must be adequately controlled to yield comparable results. A recent report by Harwell Laboratories[150] described the findings of an investigation of a waste stabilization site that has been operating since 1974. The Leigh facility employs a process which, according to company literature, encapsulates a wide range of organic and inorganic wastes in a "synthetic rock which is safe, impermeable, non-polluting and non-leaching." Testing revealed much of the material was weaker than required and could not support drilling equipment for sampling in one pit. Measured values of hydraulic conductivity in the pit with the lowest strength were generally 3 orders of magnitude higher than those measured elsewhere. Free liquids were present in one pit with phenol concentrations up to 140 mg/L. The problems were attributed to a decision in the early 1980s to include organically contaminated wastes and use less cement in the stabilization process. This case points to the need to continually test, evaluate, monitor, and change the stabilization reagents and/or mix proportions as the nature of the waste changes.

Case Study 3

A steel manufacturer's chromium wastewater treatment plant generates a filter cake currently classified as hazardous (EPA code F006). The listed constituents are cadmium, hexavalent chromium, nickel, and cyanide (complexed). A stabilization process employing lime and fly ash in a pozzolanic reaction has been shown successful in significantly reducing the mobility of the hazardous constituents. The contaminants are fixed in the calcium-aluminosilicate structure formed during the hydration process. For the process to succeed, the hexavalent chromium is first reduced to trivalent chromium. Based upon the results of both TCLP and EP toxicity testing, a petition to delist the stabilized waste has been filed.[151]

Case Study 4

Stabilization of sandy soils contaminated primarily with cadmium were stabilized at an industrial site in the State of New York.[152] The total cadmium concentrations in the relatively dense sand and gravel soil were approximately 130 mg/L, and tests revealed TCLP concentrations of 2.2 mg/L exceeding the regulatory limit of 1.0 mg/L. Prior to construction, treatability studies evaluated numerous reagents including portland cement, blast furnace slag, fly ash, ferrous sulfate, and potassium permanganate along with several water-to-reagent ratios. These laboratory studies evaluated strength, leachability, grout characteristics, and volumetric expansion. From these studies and considerations of materials availability, a grout mixture of 25 percent cement with 1 percent ferrous sulfate was selected.

For this site, jet grouting was chosen as the method of *in situ* mixing. Jet grouting, the field method to implement the stabilization alternatives, was selected because of the site's restrictive access and the need to work around underground utilities

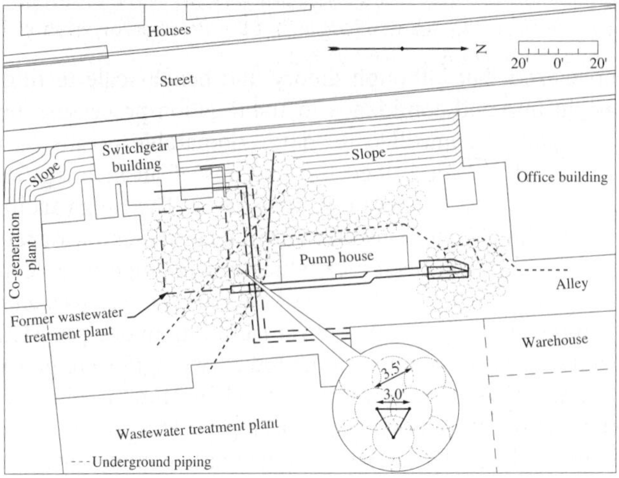

FIGURE 11-27
Layout of jet grout columns.[152]

with minimal disturbance. Shown on Figure 11-27 are the existing structures, roads, slopes, and parking lot, as well as the layout of the jet grout columns.

Construction verification testing included 39 field-molded samples, all of which showed cadmium not detected (detection limits were 0.048 or 0.0014 mg/L) as tested in TCLP. Seventeen field samples obtained by drilling and SPT sampling (see Chapter 15 for the SPT) revealed four samples having cadmium above the regulatory limit of 1.0 mg/L in the TCLP. The failed areas were remediated by using a denser pattern of jet grouting columns in these areas. The project was considered successful because it came in on schedule and below budget, and the remediation received state approval.

DISCUSSION TOPICS AND PROBLEMS

11-1. Define the following terms: (a) solidification, (b) stabilization, (c) fixation, and (d) hydration.

11-2. Define and provide an example of the following mechanisms: (a) macroencapsulation, (b) microencapsulation, (c) adsorption, and (d) absorption.

11-3. Differentiate between an engineering property and a physical property.

11-4. Describe how portland cement can stabilize metallic wastes.

11-5. Identify similarities and differences between the following tests: (a) TCLP and EP tox, (b) paint filter test and liquids release test, and (c) unconfined compression test and one-dimensional consolidation test.

11-6. For each leaching test discussed, identify whether the test is principally a regulatory test, a predictive test, or an investigatory test.

11-7. Construct a table listing leach test variables and each leaching test discussed. Using the EP tox test as a standard, identify whether the contaminant concentrations are expected to be higher or lower as a result of differences in each of the variables.

11-8. Identify three "problems" associated with a stabilization project which might go undetected in the laboratory bench scale studies.

11-9. The data in Table P11-9 were obtained during an unconfined compression test conducted on a stabilized petroleum sludge. Plot the stress-strain relationship and determine the unconfined compressive strength.

TABLE P11-9

Load, kg	Axial deflection, cm
0	0
508	0.028
893	0.056
1364	0.086
1896	0.114
2488	0.142
3098	0.170
3632	0.200
3991	0.229
4191	0.257
4332	0.287
3597	0.315

Original sample length = 15 cm
Original sample diameter = 7.1 cm

11-10. Identify the advantages and disadvantages of the unconfined compression test as a means of evaluating stabilized waste.

11-11. Where did the value of 2175 lb/ft^2 come from in Example 11-5?

11-12. Prove that the specific gravity in English units is equal to the bulk density in SI units.

11-13. Differentiate between batch and continuous processes in waste stabilization.

11-14. The subsurface at the site of a toy manufacturing site consists of a well-graded mixture of gravel and sand with 5 percent silt. Contaminants found in the soil include benzene, toluene, xylene and vinyl chloride to depths of 3 m. The water table is at a depth of 4 m. Select the appropriate stabilization reagents and mixing technique; justify your selection. Suggest other remedial alternatives which might be feasible.

11-15. A lagoon 100 m by 50 m by 3 m deep is found to contain a PCB- and metal-contaminated sludge at a 10 percent solids content by volume. Sludge dewatering can be used to increase the volumetric solids content to 40 percent at a cost of \$34 per cubic meter of sludge processed. Stabilization of the sludge costs \$110 per metric ton of sludge treated. Leachate can be treated at a cost of \$20 per cubic meter. By calculating the volume of sludge and leachate generated by dewatering, compare the cost of two alternatives: stabilization of the entire sludge or dewatering and leachate treatment with stabilization of the dewatered sludge. Comment on the overall environmental effectiveness of the two approaches.

11-16. Differentiate landfilling versus stabilization in terms of long-term environmental considerations for the Pepper's Steel and Alloy site in Example 11-1.

11-17. Repeat Example 11-3 for a saturated soil porosity of 30 percent, a wet density of 2 g/cm^3, and a trichloroethylene concentration of 300 mg/L.

11-18. Calculate the mean effective stress, gradient, and hydraulic conductivity for a sample described as follows: $L = 7.1$ cm, diameter $= 7.1$ cm, $\sigma_{top} = 350$ kPa, $\sigma_{bottom} = 400$ kPa, $\sigma_{cell} = 500$ kPa, and $Q = 3\ cm^3$ after 6 hours.

11-19. For a site 100 m × 50 m × 3 m deep, calculate the cost for *in situ* vitrification assuming 900 kWh/tonne of soil processed. Assume the site consists of sandy soils having a total density of 1.9 g/cm^3 at a porosity of 0.4. Also, calculate the expected surface settlement. Calculate the initial moisture content and comment on its suitability with respect to *in situ* vitrification.

REFERENCES

1. Conner, J. R.: *Chemical Fixation and Solidification of Hazardous Wastes*, Van Nostrand Reinhold, New York, 692 pp., 1990.
2. Pojasek, R. B. (ed.): *Toxic and Hazardous Waste Disposal*, Ann Arbor Science Publishers, Ann Arbor, Mich., 1979.
3. Cullinane, M. J., L. W. Jones, and P. G. Malone: *Handbook for Stabilization/Solidification of Hazardous Waste*, U.S. Environmental Protection Agency Hazardous Waste Engineering Research Laboratory (HWERL), EPA/540/2-86/001, 1986.
4. Malone, P. G., and L. W. Jones: *Survey of Solidification/Stabilization Technology for Hazardous Industrial Wastes*, EPA 600/2-79-056, 41 pp., 1979.
5. Truett, J. B., R. L. Holberger, K. W. Barrett: "Feasibility of *in situ* Solidification/Stabilization of Landfilled Hazardous Wastes." EPA-600/2-83-088, September 1983.
6. *Federal Register*: National Oil and Hazardous Substances Pollution Contingency Plan, Final Rule, vol. 55, p. 8666, March 8, 1990.
7. Environmental Laboratory: "Guide to the Disposal of Chemically Stabilized and Solidified Wastes," EPA-IAG-D4-0569, September 1982.
8. *Federal Register*, U.S. EPA: "Liquids Release Test," 40 FR, vol. 51, no. 247, p. 46824, December 1986.
9. Dole, L.: "In Situ Immobilization of PCBs at the Pepper's Steel and Alloys Site: A Success Story," Company Report, Qualtec, Inc., 1989.
10. Freeman, H. M., and E. F. Harris (eds.): *Remediation: Innovative Treatment Technologies*, Technomic Publishing, Lancaster, Pa., 1996.
11. U.S. Environmental Protection Agency: *Stabilization/Solidification of CERCLA and RCRA Wastes: Physical Tests, Chemical Testing Procedures, Technology Screening, and Field Activities*, EPA/625/6-89/022, 1989.
12. U.S. Environmental Protection Agency, Innovative Treatment Technologies Annual Status Report Database Version 2.0, EPA 542-C-96-002 (1996).
13. Zarlinski, S. J., N. W. Kingham, and J. Blevins, "Solidification Treatment of Thiopene and BTEX Contaminated Soils." *Proceedings of the 16th Superfund Conference*, Washington, D.C., November 6–8, 1995, vol. 2, pp. 899-908.
14. U.S. EPA (1994). National Oil and Hazardous Substances Pollution Contingency Plan; National Priorities List, Notice of Intent to Delete the Dupont County Road X23 Site, Lee County, Iowa from the National Priorities List: Request for Comments, *Federal Register*, vol. 59, p. 44689.
15. Millano, E. F. "Hazardous Waste: Storage, Disposal, Remediation, and Closure," *Water Environment Federation*, vol. 68, no. 4, pp. 586–607, 1996.
16. Millano, E. F. "Hazardous Waste: Storage, Disposal, Remediation, and Closure," *Water Environment Federation*, vol. 69, no. 4, pp. 689–719, 1997.
17. Hwang, Daekyoo: "Geotechnical Performance Evaluation of Solidified Wastes," in *Hazardous and Industrial Waste—Proceedings of the 20th Mid-Atlantic Industrial Waste Conference*, Howard University, June 19–21, 1988.
18. Enreco, Inc. (now Williams Environmental Services, Inc.), Amarillo, Texas.
19. Graetis, T. A., W. Werntgas, and D. Hockenberry: "In Situ Stabilization/Fixation of Chromium-Contaminated Soil," *Proceedings of 17th Superfund Hazardous Waste Conference*, Washington, D.C., October 15–17, 1996, pp. 745–752.
20. Somerville, R. B.: "Encapsulation Solidification of Hazardous Wastes," in *1986 AICHE Symposium—Hazardous Waste Management*, Overland Park, Kansas, November 14, 1986.
21. Debra Cook, Bucknell University, Lewisburg, Pa.
22. Bricka, R. M., and D. O. Hill: "Metal Immobilization by Solidification of Hydroxide and Xanthate Sludges," in *Proc. 4th International Hazardous Waste Symposium on*

Environmental Aspects of Stabilization/Solidification of Hazardous and Radioactive Waste, May 3–6, 1987.
23. Cheng, K. Y., and P. L. Bishop: "Sorption, Important in Stabilized/Solidified Waste Forms," *Hazardous Waste and Hazardous Materials*, vol. 9, p. 289, 1994.
24. Camp, Dresser, and McKee Inc.: "Mobile Treatment Technologies for Superfund Wastes," EPA 540/2-86/003, 1986.
25. Cadena, F.: "Use of Tailored Bentonite for Selective Removal of Organic Pollutants," *J. Environ. Eng. Proc. Am. Soc. Civ. Eng.*, vol. 115, p. 756, 1989.
26. Conner, *Chemical Fixation and Solidification*, op. cit.
27. Carpenter, C. J.: "Ferrous Sulfate/Sodium Sulfide Chromium Reduction Metals Precipitation," *Proceedings of the 5th National Conference on Hazardous Waste and Hazardous Materials*, Las Vegas, Nev., pp. 52–56, 1988.
28. Chou, A. C., H. C. Eaton, F. W. Cartledge, and M. E. Tittlebaum: "A TEM Study of the Interactions Between Certain Organic Hazardous Wastes and Type I Portland Cement," Hazardous Waste Research Center, Louisiana State University, Baton Rouge, La.
29. Walsh, M. B., H. C. Eaton, M. E. Tittlebaum, F. K. Cartledge, and D. Chalasani: "The Effect of Two Organic Compounds on a Portland Cement-Based Stabilization Matrix," *Hazardous Waste and Hazardous Materials*, vol. 3, no. 1, pp. 111–123, 1986.
30. U.S. EPA, *Stabilization/Solidification of CERCLA and RCRA Wastes.*
31. Lea, F. M.: *The Chemistry of Cement and Concrete*, Chemical Publishing Company, New York, p. 727, 1971.
32. Ashby M. F., and D. R. H. Jones: *Engineering Materials 2: An Introduction to Microstructures, Processing and Design*, Pergamon Press, Oxford, U.K., 368 pp, 1986.
33. Bishop, P., D. Gress, and J. Olafsson: "Cement Stabilization of Heavy Metals: Leaching Rate Assessment," in *Industrial Wastes—Proceedings of the 14th Mid-Atlantic Industrial Waste Conference*, College Park, Md., 1982.
34. Shively, W., P. Bishop, D. Gress, and T. Brown: "Leaching Tests of Heavy Metals Stabilized with Portland Cement," *J. WPCF*, vol. 38, no. 3, pp. 234–241, 1986.
35. U.S. Army Construction Engineering Research Laboratory: "Critical Review of Cement Based Stabilization/Solidification Techniques for the Disposal of Hazardous Wastes," ADA-184 427, Ill, 1986.
36. Walsh, et al.: "The Effect of Two Organic Compounds on a Portland Cement-Based Stabilization Matrix," op. cit.
37. Skipper, D. G., H. C. Eaton, F. K. Cartledge, and M. E. Tittlebaum: "The Microscopic Fracture Morphology of Hardened Type I Portland Cement Paste Containing Parachlorophenol," *Hazardous Waste and Hazardous Materials*, vol. 4, no. 4, pp. 389–402, 1987.
38. Cullinane, M. J., R. M. Bricka, and N. R. Francingues, Jr.: "An Assessment of Materials that Interfere with Stabilization/Solidification Processes," in *Proc. 13th Annual Research Symposium*, Cincinnati, Ohio, pp. 64–71, 1987.
39. Chalasani, D., F. K. Cartledge, H. C. Eaton, M. E. Tittlebaum, and M. B. Walsh: "The Effects of Ethylene Glycol on a Cement-Based Solidification Process," *Hazardous Wastes and Hazardous Materials*. vol. 3, no. 2, 1986.
40. Cartledge, F. K., H. C. Eaton, and M. E. Tittlebaum: *The Morphology and Microchemistry of Solidified/Stabilized Hazardous Waste*, EPA/600/S2/056, January 1990.
41. Jones, L. W.: *Interference Mechanisms in Waste Stabilization/Solidification Processes*, EPA/600/S2-89/067, April 1990.
42. Montgomery, D. M., C. J. Sollars, R. Perry, S. E. Tarling, P. Barnes, and E. Henderson: "Treatment of Organic-Contaminated Industrial Wastes Using Cement-based Stabilization/Solidification—I. Microstructural Analysis of Cement-Organic Interactions," *Waste Management & Research*, vol. 9, pp. 103–111, 1991.
43. Chou, A. C., H. C. Eaton, F. W. Cartledge, and M. E. Tittlebaum: "A Transmission Electron Microscopic Study of Solidified/Stabilized Organics," *Hazardous Waste and Hazardous Materials*, vol. 5, no. 2, pp. 145–154, 1988.

44. Cocke, D. L.: "The Binding Chemistry and Leaching Mechanisms of Hazardous Substances in Cementitious Solidification/Stabilization Systems," *Proceedings of the 2nd Annual Symposium of the Gulf Coast Hazardous Substance Research Center*, Lamar University, Beaumont, Texas, pp. 81–102, February 15–16, 1990.
45. Sams, T. L., and T. M. Gilliam: "Systematic Approach for the Design of Pumpable, Cement-Based Grouts for Immobilization of Hazardous Wastes," *Environmental Aspects of Stabilization and Solidification of Hazardous and Radioactive Wastes*, ASTM STP 1033, P. L. Cote and T. M. Gilliam (eds.), American Society for Testing and Materials, Philadelphia, pp. 15–20, 1989.
46. Daniali, S.: "Solidification/Stabilization Technology for Hazardous Industrial Wastes," *Journal of Hazardous Materials*, vol. 24, no. 2–3, pp. 225–230, 1990.
47. Lee, C., and W. L. Haas: "Site Recycling: From Brownfield to Football Field," *Environ. Prot.*, vol. 6, no. 7, p. 24, 1995.
48. Rademaker, P. D., and R. B. Wiegers: "The Processing of Industrial Waste for Immobilization and/or Recycling Applying Pozzolanic Reactions," presented at *Environmental Technology—2nd European Conference on Environmental Technology*, pp. 411–421, 1987.
49. Malone, P. G., and L. W. Jones: *Solidification/Stabilization of Sludge and Ash from Wastewater Treatment Plants*, EPA/600/S2-85-058, 1985.
50. Dermata, D., and Z. Meng: "Utilization of Fly Ash for Stabilization/Solidification of Heavy Metal Contaminated Soils," *Proceedings of 2nd International Conference on Energy and the Environment*, Prague, December 1995, pp. 563–581.
51. Malone, P.G. and L. W. Jones: Op. cit.
52. National Lime Association, "Biosolids Stabilization: Which Class A Stabilization Is Most Economical," National Lime Association, 42 pp., 1996.
53. Ezell, D., and P. Suppa: "The Soliroc Process in North America: A Stabilization/Solidification Technology for the Treatment of Metal-Bearing Wastes with Reference to Extraction Procedure Testing," *Environmental Aspects of Stabilization and Solidification of Hazardous and Radioactive Wastes*, ASTM STP 1033, P. L. Cote and T. M. Gilliam, (eds.), American Society for Testing and Materials, Philadelphia, pp. 7–14, 1989.
54. U.S. Environmental Protection Agency: "Chemfix Stabilization/Solidification Process, Clackamas, Oregon," *Superfund Innovative Technology Evaluation*, EPA/540/S5-89/011, December 1990.
55. Beall, Gary W.: "Method of Immobilizing Organic Contaminants to Form Non-Flowable Matrix Therefrom," U.S. Patent Number 4,650,590, March 17, 1987.
56. Evans, J. C., S. E. Pancoski, and G. Alther: "Organic Waste Treatment With Organically Modified Clays." *3rd International Conference on New Frontiers for Hazardous Waste Management*, pp. 48–57, Pittsburgh, EPA 600/9-89/072 September 10–13, 1989.
57. Boyd, S. A., J. F. Lee, and M. M. Mortland: "Attenuating Organic Contaminant Mobility by Soil Modification." *Nature*, vol. 3, pp. 345–347, 1988.
58. Boyd, S. A., M. M. Mortland, and T. J. Pinnavaia: "Use of Modified Clays for Adsorption and Catalytic Destruction of Contaminants," in *Proc. 13th Annual Research Symposium*, Cincinnati, 1987.
59. LaGrega, M. D., J. C. Evans, C. O. Acuna, S. J. Zarlinski, and D. S. Hall: "Stabilization of Acidic Refinery Sludges," *Journal of Hazardous Materials*, vol. 24, pp. 169–187, 1990.
60. Montgomery, D. M., C. J. Sollars, R. Perry, S. E. Tarling, P. Barnes, and E. Henderson: "Treatment of Organic-Contaminated Industrial Wastes Using Cement-based Stabilization/Solidification—II. Microstructural Analysis of the Organophilic Clay as a Pre-Solidification Adsorbent," *Waste Management & Research,* vol. 9, pp. 113–125, 1991.

61. Faschan, A., M. Tittlebaum, F. Cartledge, and R. Portier: "The Use of Organioclays to Reduce Organic Waste Volatization during Solidification," *Hazardous Waste & Hazardous Material,* vol. 9, no. 4, 1992.
62. Boyd, S. A., M. M. Mortland, and C. T. Chiou: "Sorption Characteristics of Organic Compounds on Hexadecyltrimethylammonium-smectite," *Soil Sci. Soc. Am. J.*, vol. 52, pp. 652–657, 1988.
63. Boyd, S. A., S. Shaobai, J. Lee, and M. M. Mortland: "Pentachlorophenol Sorption by Organo Clays," *Clays and Clay Minerals*, vol. 36, no. 2, pp. 125–130, 1988.
64. Grim, R. E.: *Clay Mineralogy*, McGraw-Hill, New York, 1968.
65. Alther, G. R., J. C. Evans, and S. E. Pancoski: "Organically Modified Clays for Stabilization of Organic Hazardous Waste," *Superfund '88—Proceedings of the 9th National Conference*, Washington, D.C., November 28–30, 1988.
66. Sheriff, T. S., C. J. Sollars, D. Montgomery, and R. Perry: "The Use of Activated Charcoal and Tetra-Alkylammonium-Substituted Clays in Cement-Based Stabilization/Solidification of Phenols and Chlorinated Phenols," *Environmental Aspects of Stabilization and Solidification of Hazardous and Radioactive Wastes*, ASTM STP 1033, P. L. Cote and T. M. Gilliam (eds.), American Society for Testing and Materials, Philadelphia, pp. 273–286, 1989.
67. Mortland, M. M., S. Shaobai, and S. A. Boyd: "Clay-Organic Complexes as Adsorbents for Phenol and Chlorophenols," *Clays and Clay Minerals*, vol. 34, no. 5, pp. 581–585, 1986.
68. Lo, I. M. C., "Solidification/Stabilization of Phenolic Waste Using Organo-Clay Complex," *Journal of Environmental Engineering*, ASCE, vol. 122, no. 9, pp. 850–855, September 1996.
69. Grim, *Clay Mineralogy.*
70. Evans, J. C. and S. E. Pancoski: "Organically Modified Clays," *Transportation Research Record* 1219, Geotechnical Engineering 1989, Transportation Research Board, National Research Council, pp. 160–168, 1989.
71. Evans, J. C., and G. Alther: "Hazardous Waste Stabilization Using Organically Modified Clays," Geotechnical Engineering Congress 1991, ASCE, Geotechnical Engineering Special Technical Publication No. 27, pp. 1149–1162, June 1991.
72. Ibid.
73. Smith, J. A., P. R. Jaffe, and C. T. Chiou: "Effect of Ten Quaternary Ammonium Cations on Tetrachlormethane Sorption to Clay from Water," *Environmental Science and Technology*, vol. 24, no. 8, pp. 1167–1772, 1990.
74. Company literature: "Monitoring of Potential Leachate Contamination from Contaminated and DCR-Treated Material," Miller Environmental Ltd., United Kingdom, undated.
75. Marion, G. M., Payne, J. R., and Brar, G. S., *Site Remediation via Dispersion by Chemical Reaction (DCR)*, Special report of the Cold Regions and Engineering Lab, Hanover, N.H., report no. AD-A-330681/8/XAB, August 1997.
76. Environmental Laboratory: "Guide to the Disposal of Chemically Stabilized and Solidified Wastes." op. cit.
77. McNeese, J. A., et al.: "Laboratory Studies of Fixation of KEPONE® Contaminated Sediments," in *Toxic and Hazardous Waste Disposal*, Robert B. Pojasek (ed.), vol. II, Ann Arbor Science Publishers, Ann Arbor, Mich., pp. 219–220, 1979.
78. Tittlebaum, M. E., R. K. Seals, F. K. Cartledge, and S. Engles: "State of the Art on Stabilization of Hazardous Organic Liquid Wastes and Sludges," *CRC Critical Reviews in Environmental Control*, vol. 15, no. 2, pp. 179–211, 1985.
79. Doyle, R.: "Use of Extruder/Evaporator to Stabilize and Solidify Hazardous Waste," in *Toxic and Hazardous Waste Disposal*, R. Pojasek (ed.), Ann Arbor Science Publishers, Ann Arbor, Mich., vol. 1, pp. 65–91, 1979.

80. Swindlehurst, D. P., R. D. Doyle, and A. J. Mattus: "The Use of Bitumen in the Stabilization of Mixed Wastes," *Environmental Aspects of Stabilization and Solidification of Hazardous and Radioactive Wastes*, ASTM STP 1033, P. L. Cote and T. M. Gilliam (eds.), American Society for Testing and Materials, Philadelphia, pp. 21–27, 1989.
81. Mattus, A. J., M. M. Kaczmarsky, and C. K. Cofer: "Leaching and Comprehensive Regulatory Performance Testing of an Extruded Bitumen Containing a Surrogate Sodium Nitrate-Based, Low-Level Waste," *Environmental Aspects of Stabilization and Solidification of Hazardous and Radioactive Wastes*, ASTM STP 1033, P. L. Cote and T. M. Gilliam (eds.), American Society for Testing and Materials, Philadelphia, pp. 28–39, 1989.
82. Unger, S. L., R. W. Telles, and H. R. Lubowitz: "Surface Encapsulation Process for Stabilizing Intractable Contaminants," *Environmental Aspects of Stabilization and Solidification of Hazardous and Radioactive Wastes*, ASTM STP 1033, P. L. Cote and T. M. Gilliam (eds.), American Society for Testing and Materials, Philadelphia, pp. 40–52, 1989.
83. Modified from Heiser, J. H., E.-M. Franz, and P. Colombo: "A Process for Solidifying Sodium Nitrate Waste in Polyethylene," *Environmental Aspects of Stabilization and Solidification of Hazardous and Radioactive Wastes*, ASTM STP 1033, P. L. Cote and T. M. Gilliam (eds.), American Society for Testing and Materials, Philadelphia, PA, 53–62, 1989.
84. Fall, C., M. LaRoche, M. A. LaForte, and G. Poulin. "Treatment of Contaminated Soils by Stabilization with Bitumenfoam," *Proceedings of the 6th Symposium and Exhibition on Groundwater and Soil Remediation*, Montreal, March 18–21, 1997, p. 617.
85. Campbell, B. E., J. E. Hansen, and C. L. Timmerman: "In Situ Vitrification (ISV): An Evaluation of the Disposition of Contaminant Species during Thermal Processing," *15th Intl. Conf. on Incineration and Thermal Treatment Technologies*, May 6–10, 1996, Savannah, Ga.
86. Kindle, C. H., and L. E. Thompson: "Buried Waste Remediation: A New Application for In-Situ Vitrification," NTIS DE91012773, U.S. Dept. of Energy, Washington, D.C., 1991.
87. MacNeill, K. R.: "Recycling Waste Streams Using Glass Making Activities," *Proceedings of Contaminated Land: Policy, Regulation, and Technology*, IBC Technical Services, London, February 1991.
88. U.S. Environmental Protection Agency: "The Babcock and Wilcox Cyclone Furnace Soil Vitrification Technology," EPA/540/F-92/010, September 1992.
89. U.S. EPA (1992Y). Technology Demonstration Summary, DuPont/Oberlin Microfiltration System, Palmerton, Pa., April-May 1992. NTIS PB92-224351. Center for Environmental Research Information, Cincinnati.
90. MacNeill, "Recycling Waste Streams Using Glass Making Activities."
91. Richards, R. J. [Rrichards@ee.net]. *Seiler Pollution Control Systems* [www.1seiler.com], April 15, 1998.
92. *SRTC Expertise in Vitrification Technology* [www.srs.gov/general/sci-tech/technologies/vitrification/expertise.html].
93. dePercin, P. R., and S. Sawyer: "Long-term monitoring of the Hazcon stabilization process at the Douglassville, Pennsylvania Superfund Site," *Journal of Air and Waste Management Assoc.*, vol. 41, no. 763, 1991.
94. van der Sloot, H. A., G. J. de Groot, and J. Wijkstra: "Leaching Characteristics of Construction Materials and Stabilization Products Containing Waste Materials," *Environmental Aspects of Stabilization and Solidification of Hazardous and Radioactive Wastes*, ASTM STP 1033, P. L. Cote and T. M. Gilliam (eds.), American Society for Testing and Materials, Philadelphia, pp. 125–149, 1989.

95. U.S. Environmental Protection Agency, "Paint Filter Liquids Test," SW-846, Method 9095A, December 1996.
96. U.S. Environmental Protection Agency, "Liquids Release Test (LRT) Procedure," SW-846, Method 9096, September 1994.
97. U.S. Environmental Protection Agency: "Extraction Procedure (EP) Toxicity Test," *Code of Federal Regulations*, 40 CFR, part 261, subpart C, sec. 261-24, July 1, 1982.
98. U.S. EPA, "Identification and Listing of Hazardous Wastes, Toxicity Characteristics," 40 CFR 261.24, *Federal Register*, vol. 18, no. 261, pp. 55–56, July 1, 1999.
99. *Federal Register*, U.S. EPA, "Toxicity Characteristic Leaching Procedure (TCLP)," 40 CFR, vol. 51, no. 286, app. 2, part 268, p. 40643, November 7, 1986.
100. U.S. EPA: "Hazardous Waste Management System; Identification and Listing of Hazardous Waste; Toxicity Characteristics Revisions." Final Rule, *Federal Register*, vol. 55, p. 11797, 1990.
101. U.S. Environmental Protection Agency, "Toxicity Characteristic Leaching Procedure," SW-846, Method 1311, July 1992.
102. Brown, T., and P. Bishop: "The Effect of Particle Size on the Leaching of Heavy Metals from Stabilized/Solidified Waste," in *Proc. International Conference on New Frontiers in Hazardous Waste Management*, 1985.
103. Bishop, P. L.: "Leaching of Inorganic Hazardous Constituents from Stabilized/Solidified Hazardous Wastes," *Hazardous Wastes and Hazardous Materials*, vol. 5, no. 2, pp. 129–143, Spring 1988.
104. Thurnau, R. C. and M. P. Esposito: "TCLP as a Measure of Treatment Effectiveness," *Hazardous Wastes and Hazardous Materials*, vol. 6, no. 4, pp. 347–361, November 4, 1989.
105. Dragun, J., et. al., "Misapplication of the EP-Tox, TCLP, and CAM-WET Tests to Derive Data on Migration Potential of Metals in Soil Systems." In *Petroleum Contaminated Soils*, vol. III, P. T. Kostecki and E. J. Calabrese (eds.), Lewis Publishers, Chelsea, Mich., p. 75, 1990.
106. U.S. Environmental Protection Agency, "Synthetic Precipitation Leaching Procedure," SW-86, Method 1312, September 1994.
107. American Nuclear Society (ANS): "ANSI/ANS 16.1-1986, American National Standard Measurement of Solidified Low-Level Radioactive Wastes by a Short-Term Test Procedure," American Nuclear Society, LaGrange Park, Ill., April 14, 1986.
108. Environment Canada and Alberta Environmental Center, "Test Methods for Solidified Waste Characterization," Alberta, Canada, 1986.
109. Cote, P. L. and D. P. Hamilton: "Evaluation of Pollutant Release from Solidified Aqueous Wastes Using a Dynamic Leaching Test," in *Proc. Hazardous Wastes and Environmental Emergencies Conference*, pp. 302–308, 1984.
110. Cote, P. L., and D. Isabel: "Applications of a Static Leaching Test to Solidified Hazardous Waste," presented at the *ASTM International Symposium on Industrial and Hazardous Solid Wastes*, Philadelphia, 1983.
111. Hannak P., A. J. Liem, and P. Cote: "Methods for Evaluation Solidified Waste," Alberta Environmental Centre, Vegreville, Alberta, Canada, and Wastewater Technology Centre, Environmental Protection Service, Environment Canada, Burlington, Ontario, Canada, presented to the *2nd Annual Symposium on Solid Waste Testing and Quality Assurance Organized by U.S. Environmental Protection Agency*, Washington, D.C., 14 pp., July 15–18, 1986.
112. Stegemann, Julia: "Results of a Cooperative Study of Twelve Test Methods for Solidified Waste Characterization," Environment Canada, Wastewater Technology Centre, Burlington, Ontario, presented at the *34th Ontario Industrial Waste Conference*, Toronto, June 1987.

113. Ibid.
114. Shively, W., and S. Dethloff: "Hazardous Waste Treatment Demonstration Guide—Test Methods for Solidification/Stabilization of Hazardous Wastes," EPA seOS8801/44/1, 82 pp, September 14, 1987.
115. U.S. Environmental Protection Agency, "Multiple Extraction Procedure," SW-846, Method 1320, September 1986.
116. Weitzman, L., and L. E. Hamel: "Evaluation of Solidification/Stabilization as a Best Demonstrated Available Technology for Contaminated Soils," op. cit.
117. Soundararajan, R., E. F. Barth, and J. J. Gibbons: "Using an Organophilic Clay to Chemically Stabilize Waste Containing Organic Compounds," *Hazardous Materials Control*, vol. 3, no. 1, pp. 42–45, January–February 1990.
118. American Society for Testing and Materials (ASTM), "Laboratory Determination of Water (Moisture) Content of Soil and Rock," D2216-98, *Annual Book of Standards*, 1998.
119. American Society for Testing and Materials (ASTM) "Particle Size Analysis of Soils," D422-63, *Annual Book of Standards*, 1998.
120. Roy, A., H. C. Eaton, F. K. Cartledge, and M. E. Tittlebaum: "Solidification/Stabilization of a Heavy Metal Sludge by a Portland Cement/Fly Ash Binding Mixture," *Hazardous Waste and Hazardous Materials*, vol. 8, no. 1, pp. 33–41, 1991.
121. American Society for Testing and Materials (ASTM), "Standard Test Method for Time of Setting of Hydraulic Cement Vicat Needle," ASTMC191-92, *Annual Book of Standards*, 1992.
122. American Society for Testing and Materials (ASTM), "Standard Test Method for Compressive Strength of Molded Soil-Cement Cylinders" D1633-96, *Annual Book of Standards*, vol. 4.08, 1996.
123. American Society for Testing and Materials (ASTM), "Standard Test Method for Unconfined Compressive Strength of Cohesive Soil," D2166-98a, Annual Book of Standards, 1998.
124. Evans, J. C., and S. E. Pancoski: "Stabilization of Petroleum Sludges," *Superfund '89—Proceedings of the 10th National Conference*, HMCRI, Washington, D.C., November 27–29, 1989.
125. Ibid.
126. Zarlinski, S. J., and J. C. Evans: "Weathering Resistance of Stabilized Petroleum Sludges," *Proceedings of the 11th National Conference of the Management of Uncontrolled Hazardous Waste Sites,* Washington, D.C., pp. 712–715, November 1990.
127. Ibid.
128. American Society for Testing and Materials (ASTM), "Standard Test Method for Permeability of Granular Soils (Constant Head)," D2434-68 (1998), *Annual Book of Standards*, vol. 4.08, 1998.
129. U.S. Environmental Protection Agency, "Saturated Hydraulic Conductivity, Saturated Leachate Conductivity, and Intrinsic Permeability," SW-846, Method 9100, September 1986.
130. Evans, J. C., and H. Y. Fang: "Triaxial Permeability and Strength Testing of Contaminated Soils," *Advanced Triaxial Testing of Soil and Rock*, ASTM STP 977, American Society for Testing and Materials, Philadelphia, pp. 387–404, 1988.
131. Zarlinski and Evans, "Weathering Resistance of Stabilized Petroleum Sludges," op. cit.
132. Yang, G. C. C.: Durability Study of a Solidified Mercury-containing Sludge. *J. Hazard. Mater.*, vol. 34, p. 217, 1993.
133. Cote, P. L. and D. P. Hamilton, "Evaluation of Pollutant Release from Solidified Aqueous Wastes Using a Dynamic Leaching Test," op. cit.
134. American Society for Testing and Materials (ASTM), "Standard Test Method for Wetting and Drying Test of Solid Wastes," D4843-88 (1999), *Annual Book of Standards*, 1998.

135. American Society for Testing and Materials (ASTM), "Standard Test Method for Determining the Resistance of Solid Wastes to Freezing and Thawing," D4842-90, ASTM, W. Conshohocken, Pa.
136. Wiles, C. C., et al.: "Status of Solidification/Stabilization in the United States and Factors Affecting Its Use," EPA/600-D-89/159, U.S. EPA, 1989.
137. Curry, M. F. R.: "Fixation/Solidification of Hazardous Waste at Chemical Waste Management's Vickery, Ohio Facility," *Superfund '89—Proceedings of the 10th National Conference*, November 27-29, 1989.
138. Geo-Con, Monroeville, Pa., U.S.A.
139. U.S. Environmental Protection Agency: "Technology Evaluation Report: Site Program Demonstration Test, Soliditech, Inc. Solidification/Stabilization Process," vol. 1, EPA/540/5-89/005a, February 1990.
140. Welsh, J. P., and G. K. Burke: "Vertical Cutoffs and Bottom Sealing by Jet Grouting," *ASCE Specialty Conference on Characterization, Containment, Remediation and Performance in Environmental Geotechnics, The Geoenvironment 2000*, ASCE Geotechnical Special Publication No. 46, February, 1995, pp. 1207–1221.
141. Shively, W. E., and M. A. Crawford: "Extraction Procedure Toxicity and Toxicity Characteristic Leaching Procedure Extractions of Industrial and Solidified Hazardous Waste," *Environmental Aspects of Stabilization and Solidification of Hazardous and Radioactive Wastes*, ASTM STP 1033, P. L. Cote and T. M. Gilliam (eds.), American Society for Testing and Materials, Philadelphia, pp. 150–1692, 1989.
142. Enreco, Inc. (now Williams Environmental Services, Inc.), Amarillo, Texas.
143. Cullinane, M. J., and L. Jones: "Solidification and Stabilization of Hazardous Wastes," *HMCRI Journal of Hazardous Materials Control*, vol. 2, no. 2, p. 22, March–April, 1989.
144. Conner, *Chemical Fixation and Solidification of Hazardous Wastes*, op. cit.
145. Shin, H-K, N-R Her, and J-K Koo: "Design Optimization for Solidification of Hazardous Wastes," *Hazardous Waste and Hazardous Materials*, vol. 5, no. 3, pp. 239–250, 1988.
146. Rushbrook, P. E., Baldwin, and C. B. Dent: "A Quality-Assurance Procedure for Use at Treatment Plants to Predict the Long-Term Suitability of Cement-Based Solidified Hazardous Wastes Deposited in Landfill Sites," *Environmental Aspects of Stabilization and Solidification of Hazardous and Radioactive Wastes*, ASTM STP 1033, P. L. Cote and T. M. Gilliam (eds.), American Society for Testing and Materials, Philadelphia, pp. 93–113, 1989.
147. U.S. Environmental Protection Agency: "Superfund Innovative Technology Evaluation," EPA/540/8-89/010, November 1989.
148. U.S. Environmental Protection Agency: "Superfund Innovative Technology Evaluation, Technology Demonstration Summary, International Waste Technologies In Situ Stabilization/Solidification, Hialeah, FL," EPA/540/S5-89/004, June 1989.
149. U.S. Environmental Protection Agency: "International Waste Technologies/Geo-Con In situ Stabilization/Solidification Update Report," *Superfund Innovative Technology Evaluation*, EPA/540/S5-89/004a, January 1991.
150. Harwell Laboratories: ENDS Report 173, June 1989.
151. Stephanatos, B. N., and A. MacGregor: "Stabilization and Delisting of Hazardous Wastes: An Effective Approach for Reducing High Sludge Disposal Costs," *Proceedings of the 12th National Conference on the Management of Uncontrolled Hazardous Waste Sites*, Washington, D.C., 1991.
152. Day, S. R., S. J. Zarlinski, and P. Jacobson: "Stabilization of Cadmium-Impacted Soils Using Jet-Grouting Techniques," *ASCE Specialty Conference on In Situ Remediation of the Geoenvironment,* J. C. Evans (ed.), ASCE Geotechnical Special Publication No. 71, October 1997, pp. 388–402.

ADDITIONAL READING

Conner, J. R.: *Chemical Fixation and Solidification of Hazardous Wastes*, Van Nostrand Reinhold, New York, 692 pp., 1990.

Cullinane, M. J., and L. Jones: "Solidification and Stabilization of Hazardous Wastes, Part 1," *Hazard. Mater. Control,* vol. 2, p. 9, 1989.

Cullinane, M. J., and L. W. Jones: *Stabilization/Solidification of Hazardous Waste,* U.S. Environmental Protection Agency Hazardous Waste Engineering Research Laboratory (HWERL), EPA/600/D-86/028, Cincinnati, Ohio, 1986.

Cullinane, M. J., L. W. Jones, and P. G. Malone: *Handbook for Stabilization/Solidification of Hazardous Waste*, U.S. Environmental Protection Agency (U.S. EPA) Hazardous Waste Engineering Research Laboratory (HWERL), EPA/540/2-86/001, 1986.

Means, J. L., et al.: *The Application of Solidification/Stabilization to Waste Materials*, Lewis Publishers, CRC Press, Inc., Boca Raton, Fla., 334 pp., 1995.

ACKNOWLEDGMENTS

The authors wish to acknowledge the following individuals who assisted with the preparation of this chapter for the first edition: Ed Van Keuren, Steve Montagna, James Archer, Daekyoo Hwang, Nick Hollingshad, *ERM*; Charles Johnson, Eric Smalstig, Marc Smugar, John Floyd, Brian Swanson, Jason Fralick, *Bucknell University*; Dennis Contois, Eric Mortenson, *Pennsylvania State University–Harrisburg*; Adam Faschan, *Louisiana State University*.

The principal reviewer for the *ERM* for the first and second editions was Elsie Milano, Ph.D.

CHAPTER 12

Thermal Methods*

Lasciate ogni speranza voi ch'entrate!
Abandon all hope, ye who enter here!
Dante

12-1 INTRODUCTION

Incineration systems are designed to destroy only organic components of waste; however, most hazardous waste is nonexclusive in its content and therefore will contain both combustible organics and noncombustible inorganics. By destroying the organic fraction and converting it to carbon dioxide and water vapor, incineration reduces the waste volume and, to the extent that the organic components include toxic components, its threat to the environment.

Hazardous wastes come in all physical forms: liquid, solid, and somewhere in between. Commercial hazardous waste incineration systems have to handle the gamut of waste streams (see Section 12-6). Nevertheless, many hazardous waste incinerators are concerned only with liquid wastes. Section 12-5 addresses this special type of system. While gaseous waste streams are not regulated as hazardous waste, the combustion of such wastes is certainly a part of the overall incineration process and is addressed in Section 12-4. The theory and equipment design for each type of waste is similar but different, and their combination into a single system is often required.

Regulations governing hazardous waste incineration were essentially nonexistent until 1980. The Clean Air Act was the only regulatory constraint on incineration

*Chapter 12 was prepared by Richard D. Ross and Joseph Santoleri of Four Nines, Inc.

of any type before that date. Table 12-1 shows the historical perspective of the regulatory process and some of the process developments along the way.

Incineration is a rather simple chemical process when viewed superficially. Organic compounds are confined to those containing carbon, hydrogen, and sometimes oxygen, with the possible addition of metals and nonmetals such as halogens and nitrogen. This should make the chemistry of combustion easy to decipher. Unfortunately, this chemistry applies only when reactions are complete, and in combustion they often are not.

Good combustion is good oxidation of the organic components—carbon and hydrogen. To achieve this, air, which contains only 21 percent oxygen by volume, must be thoroughly mixed with the carbon and hydrogen of the fuel (waste) to produce a stoichiometric product of carbon dioxide and water. Unfortunately, air also contains 79 percent nitrogen, which is inert and gets in the way of the combustion process. Given a completely homogeneous system (a well-stirred reactor), which requires both **T**ime and **T**urbulence, the complete oxidation of carbon and hydrogen should occur at some prescribed **T**emperature. Thus, the "three T's" of combustion affect the reaction. Decrease one of these factors and the other two must be increased to achieve the same degree of combustion efficiency.

For this reason, few combustion reactions are complete at their minimum theoretical temperature or with exactly the theoretical amount of air to provide stoichiometric amounts of oxygen for the carbon and hydrogen in the fuel. Over a finite

TABLE 12-1
Historical perspective

Significant legislation	Year	Waste types	Technology
	1955	Solids, drummed waste	Open burning
		Liquids	Sewer or stream
		Industrial solids	Grate incinerator
	1960	Sewage sludge	Multiple hearth
		Waste gases	Fume incineration
		Nonhalogenated solvents	Liquid injection
		Heavy tars	
Federal Water Quality Act	1965	Aqueous wastes	Liquid injection
Clean Air Act	1970	Sludges, drummed waste	Rotary kiln
		All wastes	Incinerator APC
		Solids and sludges	Fluid bed
		HCl recovery	Liquid injection
		Halogenated liquids	Liquid injection heat recovery
	1975	Drummed wastes	Rotary hearth
Resource Conservation and Recovery Act	1976		
Clean Air Act Amendments	1977		
RCRA Regulation for Hazardous Waste Incinerators	1980	Halogenated liquids	Molten salt process
		All wastes	Required destruction and removal efficiency
	1982	Hazardous wastes	Performance standards
		Contaminated soils	Huber AER
RCRA Amendments	1985		

range, as the turbulence (mixing) of the reactor (burner/incinerator) improves and the time allowed for the reaction to take place increases, the amount of excess air (oxygen) needed in the reaction decreases. If combustion is viewed as an isosceles triangle with time, temperature, and turbulence as the three sides, then a decrease in the length of one side requires a compensatory increase in the other two. This is combustion!

Typical combustion process reactions are

$$C + 2(O_2 + N_2) \rightarrow CO_2 + 2N_2 + O_2 + \text{HEAT} \tag{12-1}$$

$$H_2 + 2(O_2 + N_2) \rightarrow H_2O + 2N_2 + 3O_2 + \text{HEAT} \tag{12-2}$$

$$CH_4 + 3(O_2 + N_2) \rightarrow CO_2 + 2H_2O + 3N_2 + O_2 + \text{HEAT} \tag{12-3}$$

EXAMPLE 12-1. COMBUSTION CALCULATION. A waste mixture (by weight) of 30 percent toluene, 65 percent acetone, and 5 percent water is to be burned in a liquid injection type incinerator at a rate of 1000 lb/h with 20 percent excess air.

Compound	Formula	Heating value, Btu/lb
Toluene	$C_6H_5CH_3$	18,252
Acetone	CH_3COCH_3	13,120
Water	H_2O	0

The ambient temperature is 60°F and the incinerator is lined with 9 in of refractory having a thermal conductivity of 13 (Btu · ft^2 · °F)/in at 1800°F.

1. What is the total heat release in the incinerator?
2. What is the percent by volume of each component in the flue gas?

Solution

1.

Toluene heat release	=	$0.30 \times 18{,}252$	=	5,476 Btu/lb
Acetone heat release	=	$0.65 \times 13{,}120$	=	8,528 Btu/lb
Water heat release			=	0 Btu/lb
Heat release per pound of mixture			=	14,004 Btu/lb
Heat release in the incinerator	=	$1000\ \text{lb/h} \times 14{,}004\ \text{Btu/lb}$	=	14,004,000 Btu/h

2. $C_6H_5CH_3 + 9\ O_2 \rightarrow 7\ CO_2 + 4\ H_2O$
The molecular weight of toluene is 92
The weight of toluene is $0.30 \times 1000\ \text{lb/h} = 300\ \text{lb/h}$
$300/92 = 3.26$ lb-mol/h of toluene

Constituent	MW	lb-mol/h
Toluene	92	3.26
O_2	32	29.34
CO_2	44	19.56
H_2O	18	13.04

The oxygen in the table above is the stoichiometric amount (i.e., that needed to just react with the toluene). An excess amount of 20 percent has been specified and this excess

will be in the flue gas as O_2.

$$0.2 \times 29.34 = 5.87 \text{ mol/h of } O_2 \text{ in addition to that above.}$$

But air, not oxygen, is used in the incinerator, and air is 79 percent nitrogen. Therefore, all of the N_2 will be in the flue gas.

$$29.34 + 5.87 = 35.21 \text{ lb-mol of } O_2$$

are required, or

$$79/21 \times 35.21 = 132.46 \text{ lb-mol of } N_2$$

Using the same approach for acetone gives

$$CH_3COCH_3 + 4O_2 \rightarrow 3CO_2 + 3H_2O$$

The molecular weight of acetone is 58. The weight of acetone is 0.65×1000 lb/h = 650 lb

$$650/58 = 11.21 \text{ lb-mol/h of acetone}$$

Constituent	**MW**	**lb-mol/h**
Acetone	58	11.21
O_2	32	44.84
CO_2	44	33.63
H_2O	18	33.63

$$\text{Excess } O_2 = 0.2 \times 44.84 = 8.99 \text{ lb-mol/h}$$

$$\text{Total } O_2 = 44.84 + 8.99 = 53.83 \text{ lb-mol/h}$$

Therefore the nitrogen is $79/21 \times 53.83 = 202.5$ lb-mol/h.
In addition to the above there is 5 percent water in the waste, or 50 lb/h, or $50/18 =$ 2.78 lb-mol.
By adding the moles of flue gas generated from the three waste components the total moles of each can be determined.

	CO_2		**H_2O**	**O_2**	**N_2**
Toluene	19.56		13.04	5.22	117.75
Acetone	33.63		33.63	8.99	202.50
Water			2.78		
Totals	53.19		49.45	14.21	320.25
Total lb-mol = 437.10					
Mole % = volume %					
Flue gas is therefore	12.17%	CO_2			
	11.30%	H_2O			
	3.25%	O_2			
	73.27%	N_2			

12-2 REGULATIONS

The design and evaluation of incineration systems, as with most aspects of hazardous waste management, is driven by laws and regulations enacted to assure environmental and human health protection. In the United States the law governing incineration systems is the Resource Conservation and Recovery Act (RCRA)[1] as amended by the Hazardous and Solid Waste Amendments of 1984 (HSWA).[2] This statute was discussed in Section 2-2. As noted in Chapter 2, this statute provided far more technical details than other environmental legislation. For example, in contrast to the usual practice of leaving technical details to the EPA, HSWA required specific destruction and removal efficiencies:

> ... for each incinerator which receives a permit under section 3005(c) after the date of enactment of the Hazardous and Solid Waste Amendments of 1984, the attainment of the minimum destruction and removal efficiencies required by regulations in effect on June 24, 1982[3]

Incineration facilities are regulated by rules promulgated under RCRA. The specific requirements for an operating permit[4] mandate a trial burn and a detailed report including a complete waste analysis quantifying the hazardous constituents in the wastes to be burned, a detailed engineering description of the incinerator, etc.

The Trial Burn Plan, if approved, will identify *Principal Organic Hazardous Constituents* (POHCs) that must be monitored during the *Trial Burn* operation. POHCs are specific organic compounds selected on the basis of their difficulty to incinerate. Facility standards[5] include general requirements for waste analysis, performance standards, operating requirements, etc.

The following federal performance standards apply to hazardous waste incinerators in the United States[6]

- Emissions of particulates—0.08 grains/dry standard cubic foot (dscf) (180 mg/dscm) corrected to 7% O_2 (see Example 12-11) in the flue gas. This is a federal regulation, and state regulations may be more stringent.
- Emissions of HCl—4 lb/h or 99% control. RCRA regulations will probably change to risk based limits for HCl and chlorine.
- Carbon monoxide emissions—100 parts per million by volume (ppmv) as a 60-minute rolling average corrected to 7% oxygen measured on a dry basis.
- Metals emissions—Emissions of 10 priority metals must meet risk based guidelines.
- Destruction and removal efficiency—The incinerator must demonstrate its capability to achieve a 99.99% DRE on one or more selected Principal Organic Hazardous Constituents during a supervised Trial Burn.

DRE is defined as:

$$\text{DRE} = \frac{W_{\text{in}} - W_{\text{out}}}{W_{\text{in}}} \times 100 \tag{12-4}$$

where DRE = destruction and removal efficiency (%)

W_{in} = mass feed rate of a particular POHC (kg/h or lb/h)

W_{out} = mass emission rate of the same POHC (kg/h or lb/h)

EXAMPLE 12-2. PERFORMANCE STANDARDS. The mixture described below is being incinerated at 2000°F with 50% excess air and a residence time of 2.1 seconds. Principal Organic Hazardous Constituents (POHCs) for this waste are benzene, chlorobenzene, and toluene. The flow rate of gas from the incinerator is measured at 12,500 dscfm (dry standard cubic feet per minute). The O_2 concentration in the flue gas is 7.0%. The inlet and outlet mass flow rates of various chemical compounds measured during the burn are as follows:

Compound	Formula	MW	Inlet, lb/h	Outlet, lb/h
Benzene	C_6H_6	78.11	2015	0.537
Chlorobenzene	C_6H_5Cl	112.5	1150	0.109
Ethylbenzene	C_8H_{10}	106.17	2230	0.757
Toluene	C_7H_8	92.10	637	0.022
Xylenes	C_8H_{10}	106.17	3040	1.25
Hydrochloric acid	HCl	36.45	0	10.7
Particulates				23.4

1. Calculate DRE for all of the organic compounds.
2. Determine if this emission meets requirements for:
   (*a*) POHCs
   (*b*) Particulates
   (*c*) HCl
3. Comment on the results.

Solution

1. For benzene, from Equation 12-4, DRE $= [(2015 - 0.537)/2015] \times 100 = 99.9733$. DREs for the other compounds are shown in the table below.

Compound	Formula	MW	Inlet, lb/h	Outlet, lb/h	DRE, %	Number of chlorines	Molar feed rate, lb-mol/h	Moles HCl, lb-mol/h	Weight HCl, lb/h
Benzene	C_6H_6	78.11	2015	0.537	99.9733	0	25.80	0.00	0.0
Chlorobenzene	C_6H_5Cl	112.5	1150	0.109	99.9905	1	10.22	10.22	372.6
Ethylbenzene	C_8H_{10}	106.17	2230	0.757	99.9661	0	21.00	0.00	0.0
Toluene	C_7H_8	92.1	637	0.022	99.9965	0	6.92	0.00	0.0
Xylenes	C_8H_{10}	106.17	3040	1.250	99.9589	0	28.63	0.00	0.0
HCl		36.45		10.7	97.12829*				
Particulates				23.4			Total inlet HCl (lb/h)		372.6
Particulates out 0.218 grains/dscf									

*Removal efficiency, %.

2. (*a*) The DRE for benzene is less than 99.99; therefore, this emission does not meet the regulatory requirement for DRE.
   (*b*) The concentration of particulates leaving the system may be calculated by dividing the outlet mass rate by the stack flow rate:

$$\frac{(23.4\ \text{lb/h}) \times (7000\ \text{grains/lb})}{(12{,}500\ \text{dscfm}) \times (60\ \text{min/h})} = 0.218\ \text{grains/dscf}$$

   This is above the limit of 0.08 grains/dscf.

(*c*) Only one compound (chlorobenzene) contains chlorine. The inlet HCl can be determined by multiplying the inlet feed rate (1150 lb/h) by the molar ratio of hydrochloric acid to chlorobenzene (36.45/112.5) times the number of chlorine atoms in chlorobenzene (1)

$$\text{Inlet HCl} = (1150 \text{ lb/h}) \times (36.45/112.5) \times (1) = 372.6 \text{ lb/h}$$

The removal efficiency for HCl is:

$$[(372.6 - 10.7)/372.6] \times 100 = 97.128\%$$

The HCl emissions fail the requirements on both bases: The total mass rate exceeds 4 lb/h and the removal efficiency is less than the required 99%.

3. The emissions from this unit fail to meet any of the three regulations listed, and if these were the results of a test burn, this unit clearly would not pass. It is also interesting to note that the three POHCs listed in the problem do not include xylenes, which have the largest mass rate leaving the unit.

12-3 COMBUSTION

Introduction

Combustion of hazardous wastes does not differ greatly from the combustion of conventional fuel as discussed in the previous section except that the wastes may contain many different organic compounds. Every organic compound and every waste has a measurable heating value (expressed in Btu/lb) which can be determined experimentally by bomb calorimetry; however, the heating values of most common waste chemicals and mixtures are available from the literature[7] (see Table 12-2 and Appendix C).

Excess Air

When organic wastes are burned with a stoichiometric amount of air (oxygen), the products of complete combustion should not include any oxygen. This is known as "perfect combustion," which is not possible in commercial burners or incinerators. Perfect combustion is an infinitely narrow line between excess air combustion on the one side and a deficiency of air (starved combustion or pyrolysis) on the other (see Figure 12-1). Incinerators must always utilize excess air to achieve combustion; however, they may accomplish this in two stages—the first stage operating under starved air (pyrolitically) and the second with an excess of air. Excess air is also used in incinerators for temperature control because the excess air absorbs heat generated during the combustion reaction.

Fuels

Organic wastes often have heating values high enough to support combustion. In such systems auxiliary fuel is needed only for ignition of the waste. In many hazardous

TABLE 12-2
Miscellaneous waste data

Waste	Btu value/lb (as fired)	Weight, lb/ft^3 (loose)	Weight, lb/ft^3	Content by weight: Ash, %	Content by weight: Moisture, %
Kerosene	18,900		50	0.5	0
Benzene	18,210		55	0.5	0
Toluene	18,440		52	0.5	0
Hydrogen	61,000		.0053	0	0
Acetic acid	6,280		65.8	0.5	0
Methyl alcohol	10,250		49.6	0	0
Ethyl alcohol	13,325		49.3	0	0
Turpentine	17,000		53.6	0	0
Naphtha	15,000		41.6	0	0
Newspaper	7,975	7		1.5	6
Brown paper	7,250	7		1.0	6
Magazines	5,250	35		22.5	5
Corrugated paper	7,040	7		5.0	5
Plastic coated paper	7,340	7		2.6	5
Coated paper milk cartons	11,330	5		1.0	3.5
Citrus rinds	1,700	40		0.75	75
Shoe leather	7,240	20		21.0	7.5
Butyl sole composition	10,900	25		30.0	1
Polyethylene	20,000	40–60	60	0	0
Polyurethane (foamed)	13,000	2	2	0	0
Latex	10,000	45	45	0	0
Rubber waste	9,000–11,000	62–125		20–30	
Carbon	14,093		138	0	0
Wax paraffin	18,621		54–57	0	0
$\frac{1}{3}$ wax–$\frac{2}{3}$ paper	11,500	7–10		3	1
Tar or asphalt	17,000	60		1	0
$\frac{1}{3}$ tar–$\frac{2}{3}$ paper	11,000	10–20		2	1
Wood sawdust (pine)	9,600	10–12		3	10
Wood sawdust	7,800–8,500	10–12		3	10
Wood bark (fir)	9,500	12–20		3	10
Wood bark	8,000–9,000	12–20		3	10
Corn cobs	8,000	10–15		3	5
Rags (silk or wool)	8,400–8,900	10–15		2	5
Rags (linen or cotton)	7,200	10–15		2	5
Animal fats	17,000	50–60			0
Cotton seed hulls	8,600	25–30		2	10
Coffee grounds	10,000	25–30		2	20
Linoleum scrap	11,000	70–100		20–30	1

The above chart shows the various Btu values of materials commonly encountered in incinerator designs. The values given are approximate and may vary, depending on their exact characteristics or moisture content.
Source: Incinerator Institute of America, New York, N.Y.

waste incineration applications the waste heating value is low, so conventional fuels must be used to bring the waste to a temperature where rapid oxidation of the organic fraction of the waste can occur.

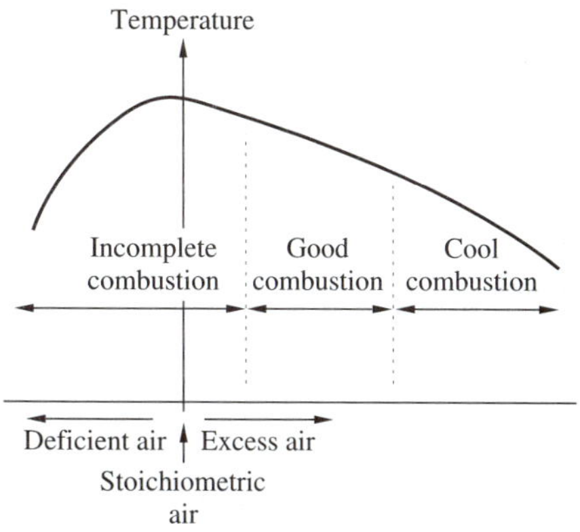

FIGURE 12-1
Excess air temperature curve.

The fuels used in an incineration system to provide auxiliary heat may be any commercially available fuel such as natural gas (methane), propane (LPG), light or heavy fuel oil, or possibly a waste fuel such as a spent solvent mixture, which may not be a hazardous waste. Combustion characteristics of some commercial fuels are shown in Table 12-3. Combustion constants for a range of substances are shown in Table 12-4.

There are hundreds of hydrocarbon waste mixtures that have sufficient heating value to be considered a fuel. Primary among these are solvent mixtures that have been used for process washdown or cleanup. Often these mixtures have heating values approaching those of commercial fuels so they may be used in a hazardous waste incinerator as a fuel. For example: consider a waste solvent mixture of two commercial solvents such as toluene and acetone. Toluene has a heating value of 18,440 Btu/lb, and acetone has a heating value of 13,119 Btu/lb. In a 50/50 weight mixture the combined heating value would be 15,686 Btu/lb, which makes an excellent fuel.

Wastes Containing Sulfur, Halogens, Nitrogen, and Inorganics

Hydrocarbon combustion produces CO_2, water vapor, and possibly some CO. The combustion of wastes containing sulfur produces SO_2 and possibly SO_3. Halogen-containing wastes (i.e., chlorine, fluorine, and bromine) produce the acid halogen

TABLE 12-3
Analysis for common fuels[8]

	Weight fraction, lb element/lb fuel		
Element	**Residual fuel oil (no. 6)**	**Distillate fuel oil (no. 2)**	**Natural gas**
C	0.866	0.872	0.693
H	0.102	0.123	0.227
N	—	—	0.08
S	0.03	0.005	—
Net heating value	19,500 Btu/lb	18,600 Btu/lb	21,301 Btu/lb

TABLE 12-4

Combustion constants*

No.	Substance	Formula	Molecular weight[a]	lb/ft³ [b]	ft³/lb[b]	Sp. gr., air = 1.000[b]	Heat of combustion[c] Btu/ft³ Gross	Btu/ft³ Net[d]	Btu/lb Gross	Btu/lb Net[d]
1	Carbon	C	12.01	—	—	—	—	—	14.093[g]	14.093[f]
2	Hydrogen	H_2	2.016	0.005327	187.723	0.06959	325.0	275.0	61.100	51.623
3	Oxygen	O_2	32.000	0.08461	11.819	1.1053	—	—	—	—
4	Nitrogen (atm)	N_2	28.016	0.07439[e]	13.443[e]	0.9718[e]	—	—	—	—
5	Carbon monoxide	CO	28.01	0.07404	13.506	0.9672	321.8	321.8	4.347	4.347
6	Carbon dioxide	CO_2	44.01	0.1170	8 .548	1.5282	—	—	—	—
Paraffin series C_nH_{2n+2}										
7	Methane	CH_4	16.041	0.04243	23.565	0.5543	1013.2	913.1	23,879	21,520
8	Ethane	C_2H_6	30.067	0.08029[e]	12.455[e]	1.04882[e]	1792	1641	22,320	20,432
9	Propane	C_3H_8	44.092	0.1196[e]	3.365[e]	1.5617[e]	2590	2385	21,661	19,944
10	*n*-Butane	C_4H_{10}	58.118	0.1582[e]	6.321[e]	2.06654[e]	3370	3113	21,308	19,680
11	Isobutane	C_4H_{10}	58.118	0.1582[e]	6.321[e]	2.06654[e]	3363	3105	21,257	19,629
12	*n*-Pentane	C_5H_{12}	72.144	0.1904[e]	5.252[e]	2.4872[e]	4016	3709	21,091	19,517
13	Isopentane	C_5H_{12}	72.144	0.1904[e]	5.252[e]	2.4872[e]	4008	3716	21,052	19,478
14	Neopentane	C_5H_{12}	72.144	0.1904[e]	5.252[e]	2.4872[e]	3993	3693	20,970	19,396
15	*n*-Hexane	C_6H_{14}	86.169	0.2274[e]	4.398[e]	2.9704[e]	4762	4412	20,940	19,403
Olefin series C_nH_{2n}										
16	Ethylene	C_2H_4	28.051	0.07456	13.4 12	0.9740	1613.8	1513.2	21,644	20,295
17	Propylene	C_3H_6	42.077	0.1110[e]	9.007[e]	1.4504[e]	2336	2186	21,041	19,691
18	*n*-Bulene (Butylene)	C_4H_8	56.102	0.1480[e]	6.756[e]	1.9336[e]	3084	2885	20,840	19,496
19	Isobulene	C_4H_8	56.102	0.1480[e]	6.756[e]	1.9336e	3068	2869	20,730	19,382
20	*n*-Pentene	C_5H_{10}	70.128	0.1852[e]	5.400[e]	2.4190[e]	3836	3586	20,712	19,363
Aromatic series C_nH_{2n-6}										
21	Benzene	C_6H_6	78.107	0.2060[e]	4.852[e]	2.6920[e]	3751	3601	18,210	17,480
22	Toluene	C_7H_8	92.132	0.2431[e]	4.113[e]	3.1760[e]	4484	4284	18,440	17,620
23	Xylene	C_8H_{10}	106.158	0.2803[e]	3.567[e]	3.6618[e]	5230	4980	18,650	17,760
Miscellaneous gases										
24	Acetylene	C_2H_2	26.036	0.06971	14.344	0.9107	1499	1448	21,500	20,776
25	Naphthalene	$C_{10}H_8$	128.162	0.3384[e]	2.955[e]	4.4208[e]	5854[f]	5654[f]	17,298[f]	16,708[f]
26	Methyl alcohol	CH_3OH	32.041	0.0846[e]	11.820[e]	1.1052[e]	867.9	768.0	10,259	9,078
27	Ethyl alcohol	C_2H_5OH	46.067	0.1216[e]	8.221[e]	1.5890[e]	1600.3	1450.5	13,161	11,929
28	Ammonia	NH_3	17.031	0.0456[e]	21.914[e]	0.5961[e]	441.1	365.1	9,668	8,001
29	Sulfur	S	32.06	—	—	—	—	—	3,983	3,983
30	Hydrogen sulfide	H_2S	34.076	0.09109[e]	10.979[e]	1.1898[e]	647	596	7,100	6,545
31	Sulfur dioxide	SO_2	64.06	0.1733	5.770	2.264	—	—	—	—
32	Water vapor	H_2O	18.016	0.04758[e]	21.017[e]	0.6215[e]	—	—	—	—
33	Air	—	28.9	0.07655	13.063	1.0000	—	—	—	—

All gas volumes corrected to 60°F and 30 inHg dry. For gases saturated with water at 60°F, 1.73% of the Btu value must be deducted.

[a] Calculated from atomic weights given in *Journal of the American Chemical Society,* February 1937.

[b] Densities calculated from values given in grams per liter at 0°C and 760 mm in the International Critical Tables allowing for the known deviations from the gas laws. Where the coefficient of expansion was not available, the assumed value was taken as 0.0037 per °C. Compare this with 0.003662, which is the coefficient for a perfect gas. Where no densities were available, the volume of the mol was taken as 22.4115 L.

[c] Converted to mean Btu per lb (1/180 of the heat per lb of water from 32°F to 212°F) from data by Frederick D. Rossini, National Bureau of Standards, letter of April 10, 1937, except as noted.

[d] Deduction from gross to net heating value determined by deducting 18.919 Btu per pound mol of water in the products of combustion. Osborne, Stimson, and Ginnings, *Mechanical Engineering,* p. 163, March 1935, and Osborne, Stimson, and Flock, National Bureau of Standards Research Paper 209.

TABLE 12-4

Combustion constants (*Continued*)

	ft³/ft³ of combustible						lb/lb of combustible						
	Required for combustion			Flue products			Required for combustion			Flue products			Experimental error in heat of combustion, % ±
	O_2	N_2	Air	CO_2	H_2O	N_2	O_2	N_2	Air	CO_2	H_2O	N_2	
1	—	—	—	—	—	—	2.664	8.863	11.527	3.664	—	8.863	0.012
2	0.5	1.882	2.382	—	1.0	1.882	7.937	26.407	34.344	—	8.937	26.407	0.015
3	—	—	—	—	—	—	—	—	—	—	—	—	—
4	—	—	—	—	—	—	—	—	—	—	—	—	—
5	0.5	1.882	2.382	1.0	—	1.882	0.571	1.900	2.471	1.571	—	1.900	0.045
6	—	—	—	—	—	—	—	—	—	—	—	—	—
7	2.0	7.528	9.528	1.0	2.0	7.528	3.990	13.275	17.265	2.744	2.246	13.275	0.033
8	3.5	13.175	16.675	2.0	3.0	13.175	3.725	12.394	16.119	2.927	1.798	12.394	0.030
9	5.0	18.821	23.821	3.0	4.0	18.821	3.629	12.074	15.703	2.994	1.634	12.074	0.023
10	6.5	24.467	30.967	4.0	5.0	24.467	3.579	11.908	15.487	3.029	1.550	11.908	0.022
11	6.5	24.467	30.967	4.0	5.0	24.467	3.579	11.908	15.487	3.029	1.550	11.908	0.019
12	8.0	30.114	38.114	5.0	6.0	30.114	3.548	11.805	15.353	3.050	1.498	11.805	0.025
13	8.0	30.114	38.114	5.0	6.0	30.114	3.548	11.805	15.353	3.050	1.498	11.805	0.071
14	8.0	30.114	38.114	5.0	6.0	30.114	3.548	11.805	15.353	3.050	1.498	11.805	0.11
15	9.5	35.760	45.260	6.0	7.0	35.760	3.528	11.738	15.266	3.064	1.464	11.738	0.05
16	3.0	11.293	14.293	2.0	2.0	11.293	3.422	11.385	14.807	3.138	1.285	11.385	0.021
17	4.5	16.939	21.439	3.0	3.0	16.939	3.422	11.385	14.807	3.138	1.285	11.385	0.031
18	6.0	22.585	28.585	4.0	4.0	22.585	3.422	11.385	14.807	3.138	1.285	11.385	0.031
19	6.0	22.585	28.585	4.0	4.0	22.585	3.422	11.385	14.807	3.138	1.285	11.385	0.031
20	7.5	28.232	35.732	5.0	5.0	28.232	3.422	11.385	14.807	3.138	1.285	11.385	0.037
21	7.5	28.232	35.732	6.0	3.0	28.232	3.073	10.224	13.297	3.381	0.692	10.224	0.12
22	9.0	33.878	42.878	7.0	4.0	33.878	3.126	10.401	13.527	3.344	0.782	10.401	0.21
23	10.5	39.524	50.024	8.0	5.0	39.524	3.165	10.530	13.695	3.317	0.849	10.530	0.36
24	2.5	9.411	11.911	2.0	1.0	9.411	3.073	10.224	13.297	3.381	0.692	10.224	0.16
25	12.0	45.170	57.170	10.0	4.0	45.170	2.996	9.968	12.964	3.434	0.562	9.968	—
26	1.5	5.646	7.146	1.0	2.0	5.646	1.498	4.984	6.482	1.374	1.125	4.984	0.027
27	3.0	11.293	14.293	2.0	3.0	11.293	2.084	6.934	9.018	1.922	1.170	6.934	0.030
28	0.75	2.823	3:573	—	1.5	3.323	1.409	4.688	6.097	—	1.587	5.511	0.080
29	—	—	—	—	—	—	0.998	3.287	4.285	1.998	—	SO_2 3.287	0.071
30	1.5	5.646	7.146	1.0	1.0	SO_2 5.646	1.409	4.688	6.097	1.880	0.529	SO_2 4.688	0.30
31	—	—	—	—	—	—	—	—	—	—	—	—	—
32	—	—	—	—	—	—	—	—	—	—	—	—	—
33	—	—	—	—	—	—	—	—	—	—	—	—	—

[e] Denotes that either the density or the coefficient of expansion has been assumed. Some of the materials cannot exist as gases at 60°F and 30 inHg pressure, in which case the values are theoretical ones given for ease of calculation of gas problems. Under the actual concentrations in which these materials are present, their partial pressure is low enough to keep them as gases.

[f] From *Combustion,* 3rd ed.

[g] National Bureau of Standards, RP 1141.

**Source: B & W Steam Book.* Based on *Fuel Flue Gases,* 1941 edition, courtesy of American Gas Association.

gas of each in a combustion reaction—HCl, HF, and HBr. Each acid halogen gas is formed according to the equilibrium conditions present at the time of the combustion reaction. Typical reactions are:

$$CS_2 + 3O_2 \rightarrow CO_2 + 2SO_2 + \text{heat} \tag{12-5}$$

$$CHCl_3 + O_2 \rightarrow CO_2 + HCl + Cl_2 + \text{ heat} \tag{12-6}$$

From the heating value of carbon disulfide (6236 Btu/lb) it can be readily seen that it will burn rapidly without the need for auxiliary fuel; however, chloroform ($CHCl_3$) has a lower heating (1350 Btu/lb) value and will require auxiliary fuel to complete the reaction. In actual combustion situations, some chlorine (Cl_2) will always be formed.

For the reaction

$$Cl_2 + H_2O \leftrightarrow 2HCl + \tfrac{1}{2}O_2$$

the equilibrium constant is defined by:

$$K = \frac{[P_{HCl}]^2\left[P_{O_2}\right]^{1/2}}{\left[P_{H_2O}\right]\left[P_{Cl_2}\right]} \tag{12-7}$$

P denotes the partial pressures (mole fractions) of each reactant. As shown in Figure 12-2, the value of K is approximately 1 at 670°C and rises to 100 at 1700°C.

EXAMPLE 12-3. HCl/Cl$_2$ EQUILIBRIUM. Chlorinated organics, when burned in air, will produce HCl gas and some chlorine gas in the combustion products. The amount of chlorine in the flue gas is proportional to the incineration temperature and the amount of excess air in the combustion process. It is desirable to maximize the HCl and minimize the Cl_2 in the flue gas because HCl is much more soluble in water than Cl_2. If much chlorine is formed, it will have to be removed with a caustic solution.

At 1200°C the K_p for the reaction is 30 while at 1100°C it is 20 (Figure 12-2). In burning 200 lb/h of monochlorobenzene in a liquid injection incinerator with 10 percent excess air, how much more chlorine will be in the flue gas at the lower temperature?

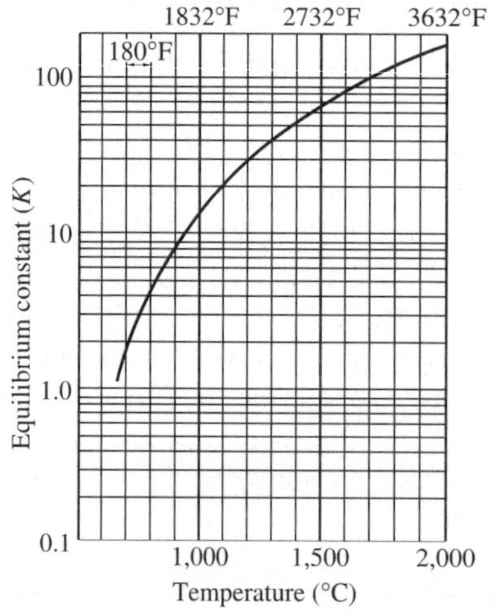

FIGURE 12-2
HCl/Cl$_2$ equilibrium.[9]

Solution. Monochlorobenzene: Formula: C_6H_5Cl MW = 112.5
Reaction: $C_6H_5Cl + 7\,O_2 = 6\,CO_2 + 2\,H_2O + HCl$

$$\frac{200 \text{ lb/h}}{112.5 \text{ lb/lb-mol}} = 1.78 \text{ lb-mol of monochlorobenzene/h}$$

The theoretical oxygen required is $7 \times 1.78 = 12.46$ mol and the total oxygen (including excess) is $12.46 \times 1.1 = 13.7$ lb-mol, so the O_2 in the flue gas will be $13.7 - 12.46 = 1.24$ lb-mol.

The molar flue gas composition will then be as follows:

Component			lb-mol/h	Mole %
Oxygen			1.24	1.8
Nitrogen	$13.7 \times 79/21$	=	51.50	74.9
CO_2	6×1.78	=	10.68	15.5
H_2O	2×1.78	=	3.56	5.2
HCl	1×1.78	=	1.78	2.6
Total			68.76	100.00

At 1 atmosphere the mole fraction is numerically the partial pressure, so if $K_p = 20$ and

$$P_{HCl} = 0.026 \qquad (0.026)^2 = 0.000676$$

$$P_{O_2} = 0.018 \qquad (0.018)^{1/2} = 0.134$$

$$P_{H_2O} = 0.52$$

$$P_{Cl_2} = ?$$

$$20 = \frac{(0.000676)(0.134)}{(0.52)(P_{Cl_2})}$$

$$P_{Cl_2} = 0.0000087 \text{ or } 8.7 \text{ ppm at } 1000°\text{C}$$

If $K_p = 30$, then

$$P_{Cl_2} = 0.0000058 \text{ or } 5.8 \text{ ppm at } 1200°\text{C}$$

The difference between operation at 1000°C and 1200°C is 2.9 ppm.

Note: This solution ignores the consumption of O_2 to convert HCl to Cl_2; this will have a small effect but needs to be stated.

Nitrogen compounds in the fuel (fuel-bound nitrogen) present a complication because they form various oxides of nitrogen during combustion, especially when excess oxygen is present. The oxides of primary concern are NO and NO_2. These two compounds are also formed in the combustion of fuels that do not contain nitrogen by a process called "nitrogen fixation." Nitrogen in the air used for combustion is "fixed" as a nitrogen oxide in the combustion process. "Fuel-bound nitrogen" is more easily formed into nitrogen oxides (NO_x) than oxides formed by fixation. NO_2 is not generally present in flue gases above 1200°F. While other oxides of nitrogen are present in the flue gas, NO predominates. Burning ammonia in air is a typical example of the combustion of a nitrogen compound:

$$NH_3 + \tfrac{1}{2}O_2 \rightarrow NH_2 + OH^- \tag{12-8}$$

The NH_2 then oxidizes or reduces as follows:
Oxidation:

$$NH_2 + O_2 \rightarrow HNO + OH^- \tag{12-9}$$

or reduction:

$$NH_2 + NO \rightarrow N_2 + H_2O \tag{12-10}$$

The oxidation/reduction ratio is 4/1 when oxygen is present in the mixture, so the products of Equation 12-9 are most likely.

In the oxidation reaction, the HNO reacts with oxygen as follows:

$$HNO + \tfrac{1}{2}O_2 \rightarrow NO + OH^- \tag{12-11}$$

In practice ammonia is burned with an excess of natural gas or steam to provide a reducing atmosphere to reduce the formation of NO (see Equation 12-10).

Metals

Inorganic components of wastes fed to an incinerator cannot be destroyed, only oxidized. Most of the inorganic materials are chemically classified as metals, and enter the combustion process as a component of a waste. Generally, these metals will exit the combustion process as oxides of the metal that enters.

If the metal enters the process as a metal salt that has a boiling point lower than the incinerator temperature, it may vaporize and not oxidize and therefore be present in the flue gas. For example, lead chloride has a vaporization temperature of 950°C (1747°F).[10] Lead that enters an incinerator as a chloride may show up as lead in the flue gas unless it is condensed in the air pollution control equipment. Lead oxide, however, is not volatile and would stay in the bottom ash. Most metal compounds will remain in the incinerator ash, but the volatility of certain metals such as arsenic, antimony, cadmium, and mercury can create problems in the flue gas. High metal content wastes are not good candidates for incineration, but appropriate air pollution control equipment can usually remove metals to acceptable flue gas levels for discharge to the atmosphere.

Combustion Calculations

To perform basic combustion calculations it is necessary to apply the chemistry of the reaction (e.g., Equations 12-5 and 12-6). If the waste chemistry cannot be determined exactly, then a waste fuel analysis similar to that shown on Table 12-5 is required.

Good combustion requires that excess air be present. The percent of excess air may be as low as 5 percent in a good combustor or burner and as high as 100 percent in a rotary kiln incinerator to provide good mixing. Excess air may also be used to control the temperature when burning a high heating value waste (see Figures 12-3, 12-4, and 12-5). Figure 12-3 demonstrates that the adiabatic combustion temperature for a fuel of any given heating value decreases as the excess air increases. Figure 12-4 similarly shows this effect for natural gas and also the effect of preheating combustion

TABLE 12-5
Waste fuel analysis

Type of waste	Heating Btu/lb	Volatiles* %	Moisture %	Ash %	Flash point COC†, °F	Fire point COC, °F	Sulfur %	Dry combus-tible %	Density, lb/ft^3
1. Coated fabric–rubber	10,996	81.20	1.04	21.20	265	270	0.79	78.80	23.9
2. Royalite	20,299	81.90	0.37	9.62	270	280	0.04	90.38	24.3
3. Coated felt–vinyl	11,054	80.87	1.50	11.39	165	170	0.80	88.61	10.7
4. Coated fabric–vinyl	8,899	81.06	1.48	6.33	155	175	0.02	93.67	10.1
5. Ensolite–expanded	10,216	88.90	0.35	9.99	265	300	0.32	90.01	5.7
6. Missile–rubber scrap	12,238	71.36	1.69	24.94	340	360	1.17	75.06	28.9
7. Fuel cell spray booth	12,325	79.10	1.74	20.67	125	130	0.25	79.33	9.5
8. Missile–rubber dust	9,761	62.36	0.87	36.42	250	260	1.06	63.58	9.9
9. Banbury–rubber scrap	13,242	60.51	1.74	4.18	145	180	0.53	95.82	34.9
10. Polyethylene film	19,161	99.02	0.15	1.49	180	200	0	98.51	5.7
11. Foam–cloth backed	10,185	55.06	0.37	31.87	215	235	1.44	68.13	11.3
12. Uppers–cloth	7,301	70.73	1.25	24.08	295	300	0.40	75.92	13.5
13. Foam–scrap	12,283	75.73	9.72	25.30	185	240	1.41	74.70	9.1
14. Tape–resin covered glass	7,907	15.08	0.51	56.73	300	330	0.02	43.27	9.5
15. Fabric–nylon	13,202	100.00	1.72	0.13	625	640	0	99.87	6.4
16. Fuel cell bladder and tire cord	15,227	87.35	1.24	3.57	270	290	0.55	96.43	19.5
17. Vinyl scrap	11,428	75.06	0.56	4.56	155	165	0.02	95.44	23.4
18. Liquid waste	13,140	100.00	3.2	1.04	68	68	0.07	95.76	53.0

*Test for volatiles was run independently. The difficulty of getting two similar samples accounts for the higher percentage of volatiles.

†Abbreviation for Cleveland open cup, a standard technique of flash point determination.

Source: Four Nines, Inc.

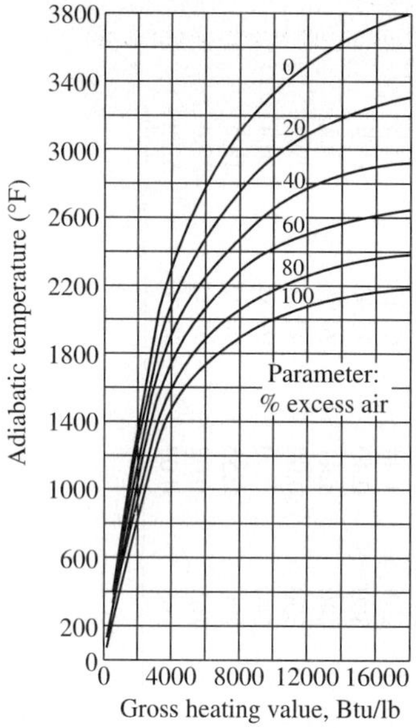

FIGURE 12-3
Adiabatic temperature versus heating value at various percentages of excess air.

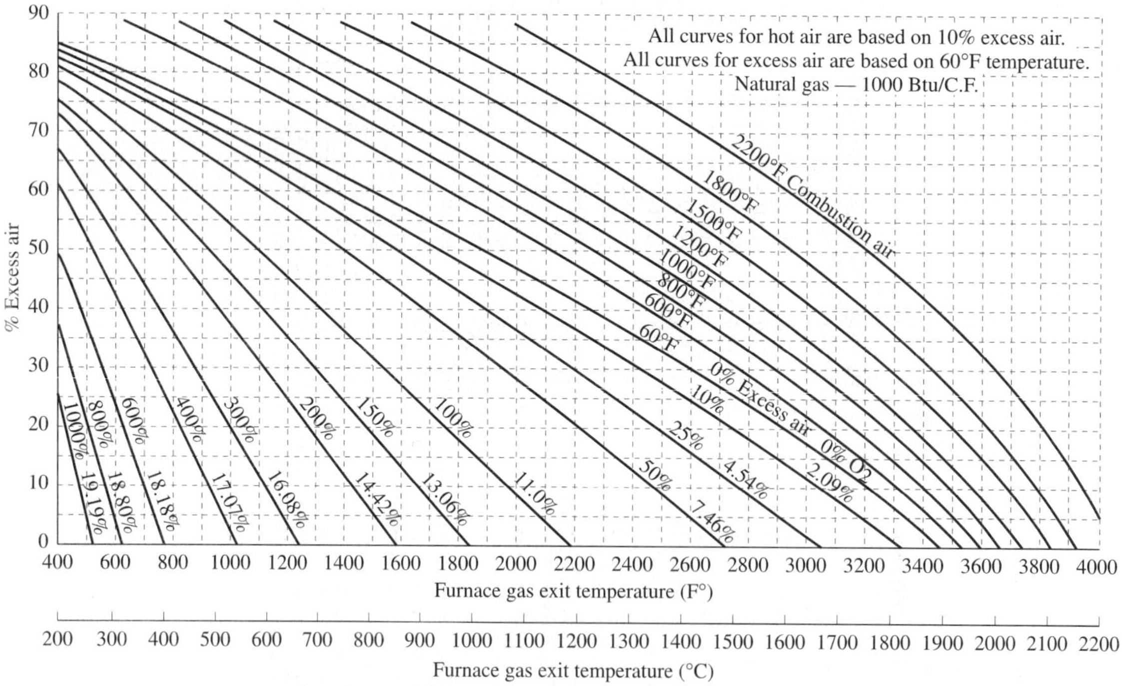

FIGURE 12-4
Excess air versus temperature.[11]

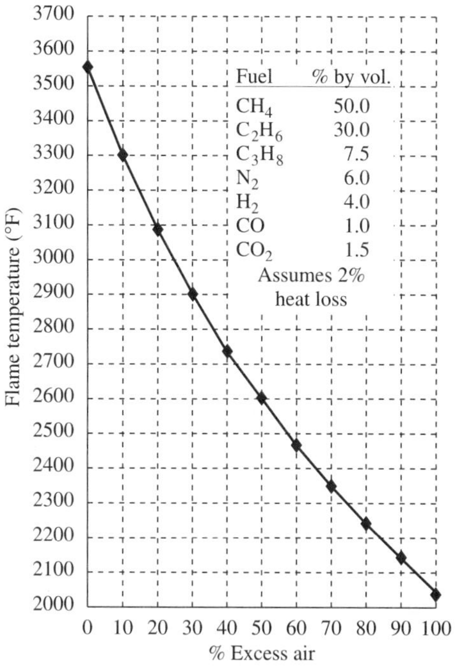

FIGURE 12-5
Flame temperature versus excess air.

air. Figure 12-5 illustrates flame temperatures that can be achieved on various fuels at excess air percentages from 0 to 100.

Combustion calculations are often based on the change in the amount of heat content between the reactants and the products of combustion. This heat content is known as enthalpy. (See Section 3-3.) Table 12-6 shows the enthalpy of common flue gases and air.

A typical combustion problem is solved in Example 12-4.

EXAMPLE 12-4. ENTHALPY—QUENCH. Many hazardous waste incinerators are followed by an "adiabatic quench" where the flue gases are cooled directly by water sprays to the adiabatic saturation temperature end point (i.e., the lowest temperature that can be achieved in single-stage cooling). Cooling is due to the evaporation of water, usually in a spray tower.

For simplicity, assume that the flue gases are essentially air at 2000°F and contain 20 percent water by weight. The total gas flow is 100,000 lb/h.

1. What is the adiabatic saturation temperature?
2. What is the cooling water requirement in gallons per minute?

Solution. The enthalpy of dry air at 2000°F from Table 12-6 is 513 Btu/lb. The enthalpy of water vapor at the same temperature is 2060 Btu/lb.

$$\text{Heat in the dry air} = 80{,}000 \text{ lb/h} \times 513 \text{ Btu/lb} \quad = 41.04 \times 10^6 \text{ Btu/h}$$

$$\text{Heat in the water vapor} = 20{,}000 \text{ lb/h} \times 2060 \text{ Btu/lb} \quad = \underline{41.20 \times 10^6 \text{ Btu/h}}$$

$$\text{Total heat in the flue gas} \quad = 82.24 \times 10^6 \text{ Btu/h}$$

TABLE 12-6

Enthalpy of combustion gases, Btu/lb

Temp., °F	O_2	N_2	Air	CO	CO_2	SO_2	H_2	CH_4	H_2O*	HCl	HF	HBr	Temp., °F
60	0.00	0.00	0.00	0.00	0.00	0.00	0.00	0.00	1060.00	0.00	0.00	0.00	60
100	8.80	9.90	9.60	10.00	8.00	5.90	137.00	21.00	1077.70	7.60	13.20	3.40	100
200	30.90	34.80	33.60	54.90	29.30	21.40	484.00	76.10	1122.40	26.70	45.20	12.00	200
300	53.30	59.90	57.70	59.90	52.00	37.50	832.00	136.40	1168.00	45.80	79.50	20.70	300
400	76.20	85.00	81.80	85.00	75.30	54.40	1182.00	202.10	1213.00	64.90	112.70	29.30	400
500	99.40	110.30	106.00	110.60	99.80	71.80	1532.00	272.60	1260.00	84.10	153.00	38.00	500
600	123.10	136.10	130.20	136.30	125.10	89.80	1882.00	347.80	1307.00	103.00	188.00	46.80	600
800	171.70	187.70	178.90	188.70	177.80	127.00	2584.00	511.20	1404.00	142.00	258.00	64.50	800
1000	221.70	240.70	235.00	242.70	233.60	165.50	3291.00	691.10	1504.00	182.00	329.00	82.90	1000
1200	272.50	294.70	288.50	297.80	290.90	205.10	4007.00	886.20	1608.00	222.00	400.00	101.00	1200
1400	324.30	350.80	343.00	354.30	349.70	245.40	4729.00	1094.00	1715.00	264.00	465.00	119.00	1400
1600	377.30	407.30	398.00	407.50	416.30	286.40	5460.00	1313.00	1827.00	302.00	538.00	138.00	1600
1800	430.70	465.00	455.00	465.30	470.90	327.80	6198.00	1542.00	1947.00	349.00	613.00	157.00	1800
2000	484.00	523.80	513.00	523.80	532.80	369.10	6952.00		2060.00	388.00	688.00	177.00	2000
2200	539.30	583.20	570.70	583.30	596.10	411.10	7717.00		2186.00	433.00	764.00	198.00	2200
2400	594.40	642.30	628.50	643.00	659.20	452.70	8490.00		2303.00	477.00	841.00	218.00	2400
2600	649.00	702.80	687.30	703.20	723.20	495.20	9272.00		2430.00	523.00	920.00	239.00	2600
2800	702.80	763.10	746.60	771.30	787.40	537.50	10060.00		2570.00	568.00	998.00	260.00	2800
3000	758.60	824.10	806.30	832.60	852.00	580.00	10870.00		2700.00	615.00	1085.00	283.00	3000
3200	816.40	885.80	866.00	894.00	916.70	622.50	11680.00		2830.00	665.00	1167.00	304.00	3200
3400	873.40	947.60	925.90	956.00	981.60	665.00	12510.00		2960.00	712.00	1249.00	326.00	3400
3600	931.00	1010.00	986.10	1018.00	1047.00	707.50	13330.00		3100.00	759.00	1331.00	347.00	3600
3800	988.00	1070.00	1046.00	1081.00	1112.00	760.00			3240.00	806.00	1409.00	369.00	3800
4000	1045.00	1132.00	1102.00	1138.00	1177.00	804.00			3380.00	854.00	1499.00	391.00	4000

*Water vapor enthalpy includes latent heat of vaporization.

$$\frac{82{,}240{,}000 \text{ Btu/h}}{80{,}000 \text{ lb/h}} = 1028 \text{ Btu/lb of dry flue gas}$$

This is the enthalpy of the saturated mixture per pound of dry flue gas.

1. Using Table C-7 (Appendix C), which gives the enthalpy of a saturated air-water vapor mixture in Btu/lb of dry air, find the temperature that most closely corresponds to 1028 Btu/lb. It is 186°F. This is the adiabatic saturation temperature for the system.
2. The latent heat of vaporization is the difference between the enthalpy of the gas and that of the saturated liquid. From Table C-8 (Appendix C):

	Enthalpy, Btu/lb		Latent heat of vaporization
Temp.	**Vapor**	**Liquid**	**Btu/lb**
180°F	147.99	1138.2	990.2
190°F	158.03	1142.1	984.1

Thus, the latent heat of water at 186°F is 986.6 Btu/lb. The heat available to evaporate water in the quench is that contained in the dry flue gas, or 41 million Btu/h.

$$\frac{41{,}040{,}000 \text{ Btu/h}}{986.6 \text{ Btu/lb water}} = 41{,}597 \text{ lb/h} = \underline{\underline{83 \text{ gal/min of water}}}$$

12-4 GASES AND VAPORS

Even though waste gases and vapors are not covered by federal RCRA regulations because RCRA essentially regulates liquid and solid hazardous wastes, there are a number of thermal processes that are of general interest and should be mentioned.

Waste gases usually consist of a hydrocarbon or a mixture of hydrocarbons in air. They may be in such high concentrations that ignition is not possible until more air is added. Such mixtures are said to be above the higher explosive limit (HEL). Mixtures of hydrocarbon vapors in air that can be ignited are said to be within the explosive range. Mixtures that have such low hydrocarbon concentrations that they cannot be ignited are said to be below the lower explosive limit (LEL). Most hydrocarbon air mixtures are of this type (see Table 12-7).

EXAMPLE 12-5. EXCESS AIR CALCULATIONS. A liquid injection incinerator has a stack gas that contains 7 percent oxygen by volume on a wet basis at standard conditions. The incinerator is burning toluene at a rate of 184 lb/h with air.

(*a*) What percent excess air is required?
(*b*) If the stack gas had been on a dry basis, what would the excess air be?
(*c*) What is the combustion efficiency of the system if the CO content of the flue gas is 500 ppmv?

Solution

(*a*) There is no other source of oxygen in the system except the combustion air. Toluene, $C_6H_5CH_3$ has a molecular weight of 92 and a heating value of 18,252 Btu/lb.

$$C_6H_5CH_3 + 9O_2 \rightarrow 7CO_2 + 4H_2O$$

TABLE 12-7
Combustibility characteristics of pure gases and vapors in air

Gas or vapor	Lower limit % by volume	Upper limit % by volume	Closed cup flash point, °F
Acetaldehyde	4.0	57	−17
Acetone	2.5	12.8	0
Acetylene	2.5	80	—
Allyl alcohol	2.5	—	70
Ammonia	15.5	26.6	—
Amyl acetate	1.0	7.5	77
Amylene	1.6	7.7	—
Benzene (benzol)	1.3	6.8	12
Benzyl chloride	1.1	—	140
Butene	1.8	8.4	—
Butyl acetate	1.4	15.0	84
Butyl alcohol	1.7	—	—
Butyl cellosolve	—	—	141
Carbon disulfide	1.2	50	−22
Carbon monoxide	12.5	74.2	—
Chlorobenzene	1.3	7.1	90
Cottonseed oil	—	—	486
Cresol *m- or p-*	1.1	—	202
Crotonaldehyde	2.1	15.5	55
Cyclohexane	1.3	8.4	1
Cyclohexanone	1.1	—	111
Cyclopropane	2.4	10.5	—
Cymene	0.7	—	117
Dichlorobenzene	2.2	9.2	151
Dichloroethylene (1,2)	9.7	12.8	57
Diethyl selenide	2.5	—	57
Dimethyl formamide	2.2	—	136
Dioxane	2.0	22.2	54
Ethane	3.1	15.5	—
Ether (diethyl)	1.8	36.5	−49
Ethyl acetate	2.2	11.5	28
Ethyl alcohol	3.3	19.0	54
Ethyl bromide	6.7	11.3	—
Ethyl cellosolve	2.6	15.7	104
Ethyl chloride	4.0	14.8	−58
Ethyl ether	1.9	48	−49
Ethyl lactate	1.5	—	115
Ethylene	2.7	28.6	—
Ethylene dichloride	6.2	15.9	56
Ethyl formate	2.7	16.5	−4
Ethyl nitrite	3.0	50	−31
Ethylene oxide	3.0	80	—
Furfural	2.1	—	140
Gasoline (variable)	1.4–1.5	7.4–7.6	−50
Heptane	1.0	6.0	25
Hexane	1.2	6.9	−15
Hydrogen cyanide	5.6	40.0	—
Hydrogen	4.0	74.2	—
Hydrogen sulfide	4.3	45.5	—
Illuminating gas (coal gas)	5.3	33.0	—
Isobutyl alcohol	1.7	—	82
Isopentane	1.3	—	—

TABLE 12-7
Combustibility characteristics of pure gases and vapors in air (*Continued*)

Gas or vapor	Lower limit % by volume	Upper limit % by volume	Closed cup flash point, °F
Isopropyl acetate	1.8	7.8	43
Isopropyl alcohol	2.0	—	53
Kerosene	0.7	5	100
Linseed oil	—	—	432
Methane	5.0	15.0	—
Methyl acetate	3.1	15.5	14
Methyl alcohol	6.7	36.5	52
Methyl bromide	13.5	14.5	—
Methyl butyl ketone	1.2	8.0	—
Methyl chloride	8.2	18.7	—
Methyl cyclohexane	1.1	—	25
Methyl ether	3.4	18	—
Methyl ethyl ether	2.0	10.1	−35
Methyl ether ketone	1.8	9.5	30
Methyl formate	5.0	22.7	−2
Methyl propyl ketone	1.5	8.2	—
Mineral spirits No. 10	0.8	—	104
Naphthalene	0.9	—	176
Nitrobenzene	1.8	—	190
Nitroethane	4.0	—	87
Nitromethane	7.3	—	95
Nonane	0.83	2.9	88
Octane	0.95	3.2	56
Paraldehyde	1.3	—	—
Paraffin oil	—	—	444
Pentane	1.4	7.8	—
Propane	2.1	10.1	—
Propyl acetate	1.8	8.0	58
Propyl alcohol	2.1	13.5	59
Propylene	2.0	11.1	—
Propylene dichloride	3.4	14.5	60
Propylene oxide	2.0	22.0	—
Pyridine	1.8	12.4	74
Rosin oil	—	—	266
Toluene (toluol)	1.3	7.0	40
Turpentine	0.8	—	95
Vinyl ether	1.7	27.0	—
Vinyl chloride	4.0	21.7	—
Water gas (variable)	6.0	70	—
Xylene (xylol)	1.0	6.0	63

$$\frac{184\ \text{lb/h}}{92\ \text{lb/lb-mol}} = 2\ \text{lb-mol toluene/per hour}$$

This requires 18 moles of O_2 or $100/21 \times 18 = 85.71$ moles air on an hourly basis. This is the stoichiometric amount of air required to burn the toluene. Two moles of toluene will produce 14 mol CO_2 and 8 mol H_2O plus $79/21 \times 18 = 67.7$ mol N_2. The total moles of flue gas at theoretical air requirements = 14 mol + 8 mol + 67.7 mol = 89.7 mol.

In burning with an excess of theoretical air, more nitrogen and some oxygen will be present in the flue gas. Oxygen is present at 7 percent by volume = 7 mol %

The total flue gas/h = 89.7 + mol N_2 + mol O_2
Let X be the moles of O_2 in the flue gas:

$$X = 0.07(89.7 + \text{ mol N}_2 + X)$$
$$\text{Moles of N}_2 = 79/21 \times X$$

Therefore:

$$X = 0.07(89.7 + 79/21X + X)$$
$$= 9.42 \text{ mol O}_2/\text{h}$$

Additional moles of N_2 added with the excess air/h = 79/21 ×9.42 = 35.5
Additional moles of air/h = 100/21 × 9.42 = 44.86
Total air in system/h = 85.71 + 44.86 = 130.57

$$\text{Excess air} = \text{excess/theoretical} = \frac{44.86}{85.71} \times 100 = \underline{\underline{52.34\%}}$$

(*b*) From part (*a*):
On a dry basis the theoretical moles of flue gas must be 89.7 − 8 mol water = 81.7 mol/h. If X = moles of O_2 in the flue gas, then:

$$X = 0.07 \times (81.7 + 79/21X + X)$$
$$= 8.59 \text{ mol O}_2/\text{h}$$

Mol N^2/h = 79/21 × 8.59 = 32.3 moles
Mol air/h = 100/21 × 8.59 = 40.9

$$\% \text{ excess air} = 40.9/85.71 \times 100 = \underline{\underline{47.7\%}}$$

The total flue gas will now be 81.7 lb-mol/h.
(*c*) 500 ppm is equivalent to 0.05% by volume CO.

$$\text{Combustion efficiency } \% = \frac{\text{CO}_2 - \text{CO}}{\text{CO}_2} \times 100$$
$$= \frac{14.05 - 0.05}{14.05} \times 100 = \underline{\underline{99.64\%}}$$

EXAMPLE 12-6. EXPLOSIVE LIMITS. A film drying system uses hot air to evaporate acetone. The acetone-air vapor mixture is transported in a duct to a thermal type VOC incinerator. Acetone has the following physical and chemical properties:

Molecular weight	54
Formula	CH_3COCH_3
Boiling point	56°C
Heating value	13,120 Btu/lb
Auto ignition temperature	537°C
Lower explosive limit	2.5% by volume
Higher explosive limit	12.8% by volume

The flow from the dryer is 1000 actual ft^3/min (acfm) at 150°F and 216 lb/h of acetone is evaporated. The molal volume = 379 ft^3/mol at 60°F.

(*a*) Determine the concentration of the acetone in the duct.
(*b*) Is this a dangerous situation? Why?
(*c*) Insurance regulations require that flammable vapor concentration in ducts be maintained at 25% of the LEL or less. How can the problem be solved?

Solution

(*a*) $\dfrac{216\text{lb/h}}{54\text{ lb/mol}} = 4\text{ lb-mol/h acetone}$

The molal volume at 150° is:

$$379 \times \frac{460 + 150}{460 + 60} = 445\text{ ft}^3/\text{mol at }150°\text{F}$$

or,

$$4\text{ lb-mol/h} \times 445\text{ ft}^3/\text{lb-mol} = 1780\text{ ft}^3/\text{h of acetone}$$

The acetone is a constituent in 1000 acfm or 60,000 acfh total flow; therefore the concentration of acetone in the flow stream from the dryer is:

$$\frac{1780}{60{,}000} = 0.0297 \text{ or } \underline{\underline{2.97\%\text{ by volume}}}$$

(*b*) It is dangerous because the acetone concentration in air is between the lower explosive limit (LEL) and the higher explosive limit (HEL). This means that the concentration is in the explosive range and any source of ignition could cause a fire or explosion in the system.
(*c*) The insurance limit on the acetone concentration is 0.25 × 2.5% or 0.625%, or the acetone concentration in the flow stream is 2.97/0.625 = 4.75 times higher than the insurance limit; therefore: 4.75 times the flow is needed to reduce the acetone concentration to this level, or 4750 acfm. This can be accomplished by adding 3750 acfm of air after the dryer.

This type of VOC emission is normally extremely dilute. The VOCs are mixed with air or occasionally a carrier gas such as nitrogen. For example, VOCs could come from coating and painting operations that employ volatile solvents as a vehicle to carry the coating, or may also result from the venting of storage tanks containing organic solvents or from leaks in pump valves and flanges. High VOC concentrations (over 10,000 ppm in air, or 1 percent by volume) are rare. Normal emission levels of VOCS from most sources are low. VOC emissions, or gaseous waste streams with high concentrations of VOCs that will support combustion, are usually burned in burners or flares or boilers. Waste streams with low VOC concentrations may be recovered by condensation or carbon adsorption or burned in a VOC incinerator.

Flares

Flares are used for waste gases that are above the HEL and may be mixed with air, ignited, and burned cleanly. Unfortunately, few waste gases will burn cleanly under these conditions so a steam injection method was conceived to provide for clean burning by providing mixing energy at the flare tip (see Figure 12-6). Flares can be elevated, i.e., burning from the tip of a flare stack, as commonly seen in a refinery,

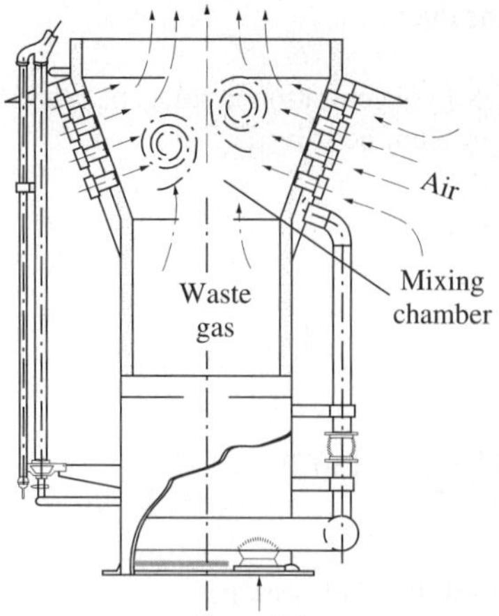

FIGURE 12-6
Flare tip used for combustion of waste gases that are above HEL.[12]

or can be enclosed at ground level and burned within the enclosure. Flares are used primarily to dispose of combustible gases during a process upset condition. Regular emissions of such combustible waste gases could be utilized to recover heat.

Catalytic VOC Incinerators

Low concentrations of VOC in air can be effectively removed by catalytic incineration. The VOC-air stream is directly heated with products of combustion from a fuel burner to a temperature at which the VOC will begin to burn on the surface of the catalyst. This temperature is determined by the type of VOC, the catalyst employed, and the VOC concentration in the air. The catalyst enhances the combustion reaction and causes oxidation of VOCs on the catalyst surface with very little heat loss. Therefore, catalytic incineration of VOC-air mixtures usually occurs at lower temperatures than thermal incineration and less auxiliary fuel is used.

The catalyst used for combustion of VOCs is usually a noble metal, e.g., platinum or palladium, although other catalysts are being employed today. Alumina is the most common catalyst support material because it will withstand high temperatures and is easily formed into a variety of configurations. The most popular configuration for catalyst support in incinerators is the honeycomb. The noble metal in a soluble form, often the nitrate, is dried onto the support surface to provide the maximum catalyst surface area with a lower airflow pressure drop.

Catalytic incinerators provide excellent destruction of most VOCs; however, care must be exercised in their application. A sudden increase in the VOC content of the VOC-air mixture to the incinerator may cause a temperature rise that will destroy the catalyst and support; therefore, it is best applied to waste gas streams where the VOC content is fairly constant.

Catalysts are also subject to poisoning (which lowers their effectiveness) by a variety of elements and chemical compounds such as chlorine, sulfur, and vanadium.

Heat recovery is a normal part of most catalytic systems. The incoming VOC-air mixture is preheated by the gases exiting the catalyst.

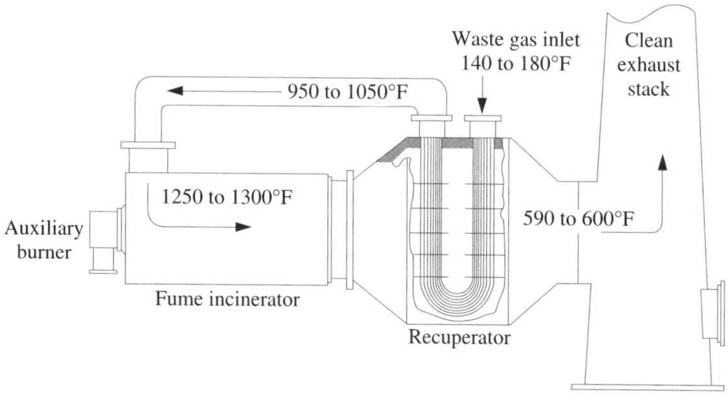

FIGURE 12-7
Thermal VOC incinerator.[13]

Thermal VOC Incinerators

The thermal incinerator for VOC-air mixtures is a catalytic incinerator without the catalyst. This usually means that the incoming mixture must be heated to a temperature where the VOC will be oxidized. Most VOCs are reduced to desirable levels at exit gas temperatures of 815°C (1500°F).

The thermal incinerator consists of a fuel burner firing into a chamber where the VOC-air mixture is adequately mixed with the burner combustion products and achieve the desired mixture temperature for at least $\frac{1}{2}$ second. Heat recovery in these systems is usually by preheating the cold incoming VOC-air stream with the hot effluent gas (see Figure 12-7).

EXAMPLE 12-7. VOC INCINERATOR WITH HEAT RECOVERY. A steel strip coater/drier emits 1500 ppmv of toluene vapor in a total air volume of 10,000 acfm at 230°F. Design a VOC thermal oxidizer with a heat recuperator schematically similar to the system shown in Figure 12-7.

In the process, the fumes enter the tubular recuperator at 230°F, where they are preheated to 900°F by the products of combustion from the incinerator. Natural gas is used as the auxiliary fuel and the preheated fumes are used as combustion air. The thermal oxidizer must operate at an outlet temperature of 1600°F to achieve the required destruction efficiency of toluene. Assume that the natural gas is methane (CH_4) with a heating value of 1000 Btu/ft^3 or 23,875 Btu/lb. Toluene ($C_6H_5CH_3$) has a heating value of 18,252 Btu/lb and a molecular weight of 92 lb/mol.

1. How much natural gas is required assuming a 5 percent heat loss?
2. What is the stack temperature?
3. To provide a $\frac{1}{2}$-second residence time in the incineration chamber, what should be its dimensions assuming a gas velocity of 20 ft/s?

Solution

1. Determine the volumetric flow rate and mass flow rate:
   1500 ppmv $= 0.15\%$ by volume
   $0.0015 \times 10{,}000 = 15$ ft^3 of toluene in the mixture at 230°F.

At 60°F the volumetric flow rate is:

$$15 \times \frac{460 + 60}{460 + 230} = 11.3 \text{ ft}^3\text{/minute}$$

The molal volume at 60°F is 379 ft^3/mol.

$$\text{Molar flow rate of toluene} = \frac{11.3 \times 60}{379} = 1.79 \text{ lb-mol/h}$$

$$\text{Mass flow rate of toluene} = 1.79 \times 92 = 164.7 \text{ lb/h}$$

$$\text{Heat release from toluene} = 164.7 \text{ lb/h} \times 18{,}252 \text{ Btu/lb}$$

$$= 3{,}006{,}000 \text{ Btu/h}$$

The total air flow to the system is

$$10{,}000 \text{ acfm} \times \frac{460 + 60}{460 + 230} = 7536 \text{ scfm}$$

From Table 12-3, the density of air is 0.076 lb/ft^3.

$$7536 \text{ scfm} \times 0.076 \text{ lb/ft}^3 \times 60 \text{ min/h} = 34{,}360 \text{ lb/h}$$

$$\frac{34{,}360 \text{ lb/h}}{29 \text{ lb/mol}} = 1185 \text{ mol/h of air}$$

In order to heat air from 900°F to 1600°F, the heat input must be equal to the difference in enthalpy values for air at these two temperatures:

At 1600°F the enthalpy of air = 398 Btu/lb (Table 12-6)

At 900°F the enthalpy of air = 207 Btu/lb

Enthalpy difference = 191 Btu/lb

Or, the total heat required is 191 Btu/lb × 34,360 lb/h = 6,564,000 Btu/h. Since there is a heat loss of 5 percent and the toluene contributes 3,006,000 Btu/h, the natural gas requirement is:

$$1.05(6{,}564{,}000) - 3{,}006{,}000 = 3{,}885{,}000 \text{ Btu/h}$$

$$\frac{3{,}885{,}000 \text{ Btu/h}}{1000 \text{ Btu/ft}^3} = \underline{\underline{3885 \text{ ft}^3\text{/h of natural gas}}}$$

$$\frac{3{,}885{,}000 \text{ Btu/h}}{23{,}875 \text{ Btu/lb}} = \underline{\underline{162.72 \text{ lb/h of natural gas}}}$$

2. The incinerator will burn both natural gas and toluene.
Natural gas as CH_4: $CH_4 + 2O_2 \rightarrow CO_2 + 2H_2O$

$$\frac{162.72 \text{ lb/h}}{16 \text{ lb/mol}} = 10.17 \text{ mol natural gas/h}$$

The combustion products per hour will be

$$CO_2 = 10.17 \text{ mol}$$

$$H_2O = 20.34 \text{ mol}$$

$$N_2 = 76.52 \text{ mol}$$

Toluene: $C_6H_5CH_3 + 9O_2 \rightarrow 7CO_2 + 4H_2O$

$$\frac{164.7 \text{ lb/h}}{92 \text{ lb/mol}} = 1.79 \text{ mol toluene/h}$$

The combustion products per hour will be:

$$CO_2 = 12.53 \text{ mol}$$
$$H_2O = 7.16 \text{ mol}$$
$$N_2 = 60.60 \text{ mol}$$

The total moles of O_2 used for both natural gas and toluene are

$$(10.17 \times 2) + (1.79 \times 9) = 36.45 \text{ mol/h}$$

This is equivalent to $36.45 \times 100/21 = 173.6$ mol air/h.

Since the inlet to the incinerator was 1185 mol air and 173.6 mol was used in the combustion process, this leaves 1011.4 mol air at the outlet plus the sum of the products above from the combustion of the natural gas and toluene. The outlet gas from the incinerator is made up as follows:

Air	1011.4 mol	× 29 lb/mol	= 29,331 lb
N_2	137.12 mol	× 28 lb/mol	= 3,839 lb
CO_2	22.70 mol	× 44 lb/mol	= 999 lb
H_2O	27.50 mol	× 18 lb/mol	= 495 lb
	1160.48 mol		34,664 lb

The heat needed in the recuperator is the difference in enthalpy between 900°F and 230°F or (207 − 41) = 166 Btu/lb. This must equal the heat given up by the flue gas entering the recuperator at 1600°F. The weight fraction of each component in the flue gas times its enthalpy at 1600°F, from Table 12-6, will give the flue gas enthalpy:

Component	**Weight fraction**	**Enthalpy, Btu/lb**	**Weighted enthalpy**
Air	0.846	398	337
N_2	0.111	407.3	45
CO_2	0.029	416.3	12
H_2O	0.014	1827.0	26
Enthalpy of flue gas =			420 Btu/lb

Let the heat in the stack gas be X.
If heat in = heat out in the recuperator then

$$(420 \times 34{,}664) - X = (166.12 \times 34{,}364)$$
$$X = 8{,}850{,}000 \text{ Btu into the stack}$$

$$\frac{8{,}850{,}000 \text{ Btu}}{34{,}664 \text{ lb/h}} = 255 \text{ Btu/lb of flue gas in stack}$$

By trial and error the temperature of the stack can be estimated. Assume a stack temperature of 1000°F and use the enthalpies from Table 12-6.

Component	Weight fraction	Enthalpy, Btu/lb	Weighted enthalpy
Air	0.846	235	199
N_2	0.111	241	27
CO_2	0.029	243	7
H_2O	0.014	1504	21
Enthalpy of flue gas =			254 Btu/lb

Since this almost matches the 255 Btu/lb calculated above, the stack temperature is approximately 1000°F.

3. The products of combustion from the incinerator total 1160.5 mol/h, which is approximately:

$$1160.5 \text{ mol/h } \times 379 \text{ ft}^3/\text{mol } \times 1 \text{ h/60 min } = 7330 \text{ scfm}$$

The average temperature in the incinerator is

$$\frac{1600 + 900}{2} = 1250°\text{F}$$

The average gas flow at 1250°F is:

$$\frac{460 + 1250}{460 + 60} \times 7330 = 24{,}104 \text{ acfm}$$

The velocity in the incinerator is:

$$20 \text{ ft/s } = 1200 \text{ ft/min}$$

The area of a chamber is:

$$24{,}104 \text{ ft}^3/\text{min} \div 1200 \text{ ft/min} = 20.75 \text{ ft}^2$$

A cross-sectional area of 20.75 has a diameter of 5.14 ft, say 5-ft diameter. The length of the chamber is:

$$20 \text{ ft/s} \times 0.5 \text{ s} = 10 \text{ ft}$$

Overall dimensions: 5.0 ft diameter × 10 ft length.

Regenerative Thermal VOC Systems

Regenerative thermal units offer higher fuel efficiencies (heat exchange efficiencies) than either the capital catalytic or recuperative thermal systems but they are larger and usually more expensive in capital cost. These units employ a "heat sink"—a chamber filled with heat-absorbing and heat-retaining ceramic materials. The VOC-air mixture passes into a thermal incineration chamber, mixing with the products of combustion from a fuel burner, which raises the mixture to 1500°F. The hot effluent gases from the incineration chamber pass through the ceramic-filled chamber on their way to the stack, preheating this chamber to a high temperature. Valves in the VOC-air inlet line then redirect the heated mixture through the heated ceramic chamber before it reaches the combustion chamber effecting significant heat

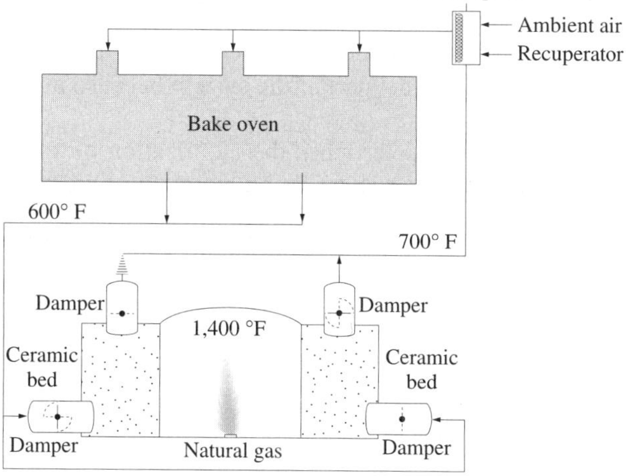

FIGURE 12-8
Regenerative VOC incinerator.[14]

recovery. Regenerative units can achieve 90 percent heat recovery. These systems are sensitive to high VOC content in the incoming mixture but can solve this problem in bypassing some of the inlet gases directly to the combustion chamber (see Figure 12-8).

12-5 LIQUID INJECTION INCINERATORS

Introduction

There are more liquid injection (LI) incinerators than all other types combined for the incineration of hazardous waste. The greatest proportion of hazardous waste incinerators in operation today are the LI type. With the inclusion of these systems by the Boiler and Industrial Furnace (BIF) regulations, the predominance is even greater. Various types of incinerators burn liquid wastes along with solids, sludges, and slurries. This section discusses the liquid injection type incinerator, which as the name implies is applicable, almost exclusively, to pumpable liquid waste. The waste is burned directly in a burner (combustor) or injected by atomizing nozzles into the flame zone or combustion zone of the incinerator chamber (furnace). The heating value of the waste is the primary determining factor for the location of the injection point.

Liquid injection incinerators are usually refractory-lined combustion chambers (horizontal flow or vertical flow either up or down), generally cylindrical in cross section, and equipped with a primary burner (waste and/or auxiliary fuel fired). Often secondary combustors or injection nozzles are required for incineration of flow heating value materials such as dilute aqueous-organic wastes.

Liquid incinerators operate at temperatures ranging from 1000°C (1832°F) to 1700°C (3092°F). Residence time for the combustion products in the incinerator may vary from milliseconds to as much as 2.5 seconds. The atomizing nozzle in the burner or the incinerator is a critical part of the system because it divides the liquid waste into fine droplets.

The viscosity of the waste determines the atomization of a liquid. Two-fluid atomizers, using compressed air or steam as an atomizing fluid, are capable of atomizing liquids with viscosities up to 70 centistokes (2.7 ft^2/h).

When considering the use of a liquid injection incinerator to burn a waste, it is necessary for the entire system and all of its auxiliary components to meet the federal and state regulations.

Determination of Waste Properties

The physical, chemical, and thermodynamic properties of the waste must be considered in the basic design of any incinerator system.

Any commercial facility receiving hazardous waste requires a complete analytical laboratory on site. Most commercial operators require a sample of the waste before they will provide a treatment cost to the generator. This sample is then analyzed for the parameters shown below, a treatment and handling regimen is developed and a price for treatment and unit disposal is provided to the generator. If the generator accepts this price then a contract is prepared that sets forth the responsibilities of both parties and imposes limits on physical and chemical properties for all waste sent to the disposal facility under that contract.

When each shipment of the waste material is received at the facility, the contractor will perform a cursory analysis known as a "fingerprint" analysis to determine that the waste meets the criteria of the contract. For example: If an initial sample of a liquid waste to be treated by incineration indicates a high heating value, say 10,000 Btu/lb, the contractor might offer the generator an attractive disposal price because it will reduce the amount of auxiliary fuel needed in the incinerator. If the actual waste received for disposal has a heating value of only 2000 Btu/lb then the generator has misrepresented the waste and the contractor needs to charge a higher disposal cost. This "fingerprint" is a heating value which can be determined by bomb calorimetry.

Today's commercial laboratories include sophisticated analytical instrumentation such as gas chromatographs coupled with mass spectrometers (GC/MS), which can quickly analyze for thousands of compounds, atomic absorption spectrophotometers for the analysis of metals, and many other devices for rapid characterization of waste.

The following information on the waste to be burned (destroyed) in the incinerator is required to design and properly engineer the total system and the auxiliary components.

1. Chemical composition
2. Heat of combustion
3. Viscosity
4. Corrosivity
5. Reactivity

6. Potential for polymerization

7. Ash content

8. Ash fusion temperature

The atomizer design in the burner or incinerator usually dictates the pressure requirements of the transport system. A hydraulic (pressure-type) atomizing nozzle requires high pressures and a pump design with close clearances between the impeller and the housing. High-viscosity materials and/or liquids containing solids can erode the pump surface, causing excessive wear and rapid deterioration of pump pressure. Reduction in pressure results in a reduction of flow as well as poor atomization. Thus, waste transport is critical in the proper design of any incinerator.

Some liquid wastes and sludges must be maintained at temperatures above ambient to remain pumpable. When cooled, these wastes can solidify in the pipeline. Conversely, if these wastes are heated too much, they may polymerize. Once polymerized, it may be impossible to melt the material. Further, with heated liquids, pump bearings may become overheated or centrifugal pumps may cavitate, so pump selection is critical.

Atomizers

The method of injecting liquid into the burner or incinerator furnace is one of the most important features of a well-designed system. The reasons for injecting the liquid as a fine spray are:

1. To disperse the liquid as a fine mist mixed with air for efficient combustion

2. To develop the desired pattern for the liquid droplets in the combustion zone with sufficient penetration and kinetic energy

3. To control the rate of flow of the liquid discharged to the combustion system

Organic and aqueous liquids pass through three stages before oxidation takes place. (Hydrocarbons ignite at temperatures as low as 20°C and as high as 650°C.) The liquid droplets are heated, vaporized, and superheated to ignition temperature. Therefore, these droplets receive heat by radiation and convection from the furnace. At the same time, they must also be in intimate contact with oxygen from the combustion air. If the liquid droplet diameter is large, fewer droplets will be produced per unit flow of waste, and the total surface area available for heat transfer is small. In a good atomizer the liquid droplet size is small, providing a larger surface area and resulting in rapid vaporization. Note the burning time for a 300-μm droplet is 150 ms compared to 30 ms for a 125-μm droplet (Figure 12-9).

There are three basic types of atomizers for liquid wastes. The first is the mechanical or pressure atomizer that is a small orifice through which high-pressure liquid expands as it goes from a high pressure to the low pressure of the combustion chamber. A high liquid pressure at the nozzle results in better atomization. Droplet size at full flow is usually on the order of 100 to 150 μm, and the atomization deteriorates as the flow is reduced. The mechanical atomizing nozzle is simple in construction, but is subject to fouling and plugging by small particles, and has poor turndown characteristics (i.e., atomization deteriorates rapidly below 50 percent flow capacity).

The second type of atomizer is the two-fluid internal mix atomizer where steam or compressed air acts as the second fluid and the waste or fuel as the first. The two

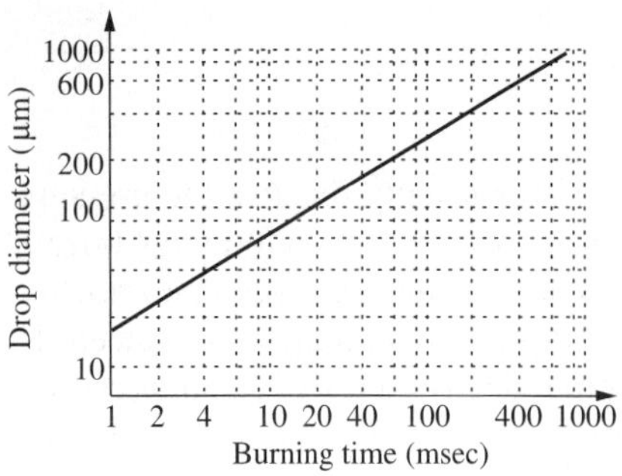

FIGURE 12-9
Burning time for atomized droplets.

fluids are mixed within the atomizer, and the energy of the steam or compressed air is used for atomization. Droplets down to 50 μm can result from good internal mix nozzle design. This nozzle has small orifices so it should not be selected for a "dirty" liquid waste. It is an excellent atomizer with good turndown (see Figure 12-10).

Because most wastes contain solid particles, the two fluid external mix atomizing nozzle is most widely used. Here the atomizing fluid and the waste mix outside the inner nozzle. The atomization may not be as good as in an internal mix nozzle, but because there are no small openings it does not foul or plug easily. In fact, particles as large as $\frac{1}{4}$ inch may be tolerated. It is significantly better than the mechanical nozzle in atomization ability (see Figure 12-11).

With high-viscosity materials, mechanical nozzle orifices have to bc made larger to minimize pressure drop, erosion, and blockage. Atomizers must be designed to cause a shearing action of the liquid to break it into many smaller diameter particles. Most atomizers for viscous liquids use a two-fluid nozzle.

Combustion Air Systems

Proper mixing of combustion air with the atomized liquid droplets is very important. As the liquid is vaporized and superheated to ignition temperature, oxygen reacts with the hydrocarbon vapor to combust and produce heat. As this occurs, there is a sudden

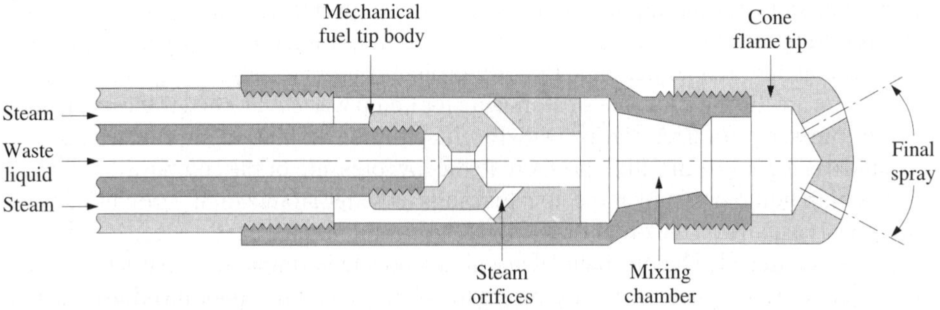

FIGURE 12-10
Internal mix nozzle.[15]

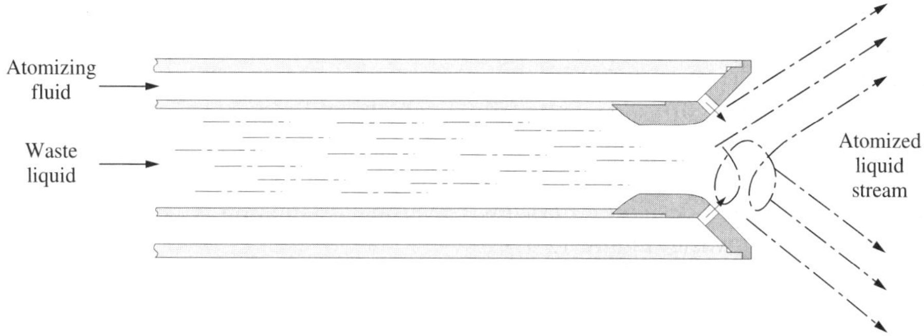

FIGURE 12-11
External mix atomizer.[16]

rise in temperature, increasing the velocity of the gases in the area surrounding the droplet, which causes greater mixing and completes the reaction.

As the viscosity of the liquid increases, droplet size increases. Hence, to completely vaporize and superheat the droplet, more time is needed. Increased turbulence created by high-intensity burners (Figure 12-12) speeds the reaction time. Also greater energy, imparted to the combustion air by a fan or blower, provides higher velocities in the combustion zone, improving the mixing of the air and the fuel droplets. In many burners this turbulence causes internal recycling of the hot products of combustion, which provides better heat transfer to the atomized droplets, achieving the ignition point more rapidly.

If the vaporized liquid contains solids, the incinerator design must allow the particles to be carried into the gas stream without agglomeration. A high swirl or a cyclonic design may cause the solids to be reagglomerated into larger particles that are more difficult to burn. Proper design of the air mixing device and the nozzle location are therefore important. Proper oxygen concentration at the surface of these solids is needed to ensure that gradual oxidation will occur. For solids, the particles first burn at the surface, and heat is transferred inward to the core of the particle. Sufficient time must be provided to permit complete burnout of the solid particles in suspension. It is for this reason that coal is finely pulverized before combustion in a fluid system.

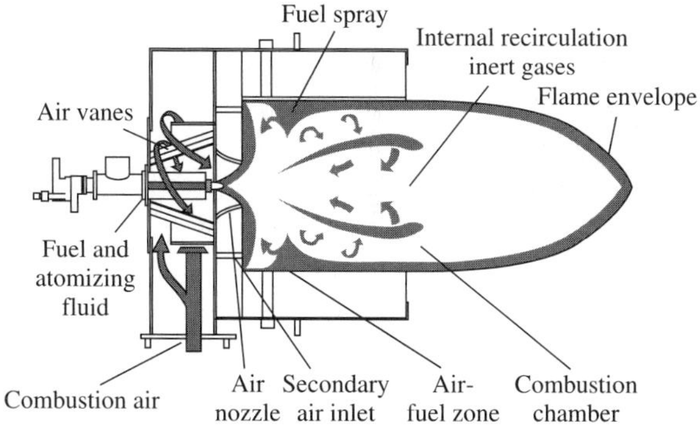

FIGURE 12-12
High intensity burner.[16]

Inorganic compounds are carried with the liquid into the gas stream as particulates. Depending on the type of atomizer, the composition of the solid, and the temperature of the primary oxidizer chamber, many particles will be reduced to submicrometer size and diffused into the gas stream. The heavier particulates may melt and agglomerate into a molten ash. The combustor must be designed to collect the molten ash without plugging the flow passages of the incinerator and quench system. A slanted horizontal design or a downward vertical orientation is most often used for these types of waste (Figure 12-13).

Combustors (Burners)

Primary and secondary combustion units are utilized in the liquid injection type incinerator systems. Primary units are used to burn wastes which have sufficient heating value to burn without the need for auxiliary fuel. With good burner design, heating values from approximately 2500 kcal/kg (4500 Btu/lb) and above can be burned without the use of auxiliary fuel. The burner design (e.g., air mixing, turbulence) determines the minimum heating value waste that can be burned without need for auxiliary fuel.

Figure 12-3, gross heating value versus adiabatic temperature, provides a guide to the ability of a combustor to burn a particular waste. Low-intensity or laminar flame burners, where the air pressure drop across the burner varies from 50 mm to 150 mm [2 in water column (w.c.) to 6 in w.c.], utilize high excess air levels (25 to 60 percent) to provide the mixing of air with the waste fuel. These burners usually operate

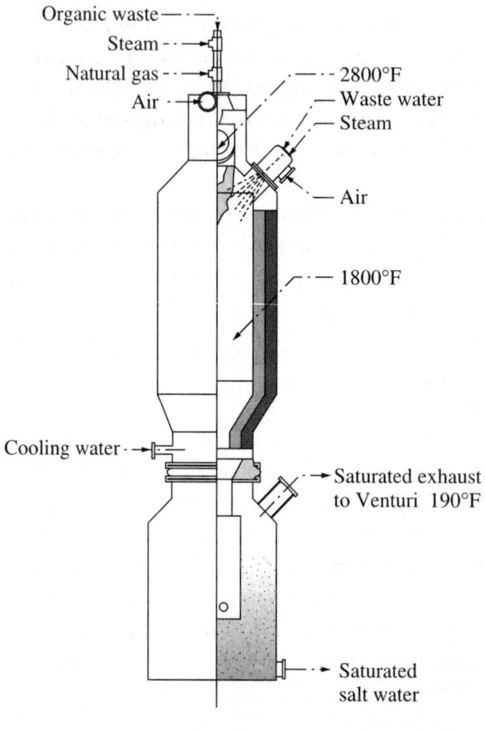

FIGURE 12-13
Down-fired combustor.[16]

with low turbulence. High-intensity burners operate with combustion air pressures of 200 to 500 mm w.c. (8 to 20 in w.c.), which results in increased turbulence because of the energy imparted by the high pressure air. This increased turbulence allows efficient combustion at much lower excess air levels (0 to 20 percent). The minimum temperature for combusting a fuel is approximately 1200 to 1315°C (2200 to 2400°F). Figure 12-3 indicates that a high-intensity burner would be able to oxidize a 2500 kcal/kg (4500 Btu/lb) waste material at 2200°F with 10 percent excess air. A low-intensity burner operating at 40 percent excess air and 2200°F would be limited to a minimum heating value of 3220 kcal/kg (5800 Btu/lb). This burner would require auxiliary fuel to properly burn a 2500 kcal/kg (4500 Btu/lb) waste.

Other Design Considerations

In the process industries, numerous liquid wastes are generated that are hard to segregate or categorize. These wastes can be defined as organic, aqueous-organic, or aqueous-organic-inorganic.

Wastes that are defined as organic, in most cases, will sustain combustion and may be injected into the primary combustor. Many organic wastes are utilized as a source of fuel in boilers, process furnaces, and cement, lime, and aggregate kilns.

Aqueous wastes that contain ash are more difficult to handle than wastes without ash. Non-ash-bearing wastes are easily handled in either a horizontally or vertically fired liquid incinerator. The waste is usually injected downstream from the burner and into the flame of the main burner. Care must be taken in the sizing, location, number, and direction of the aqueous waste nozzles.

Because a highly aqueous waste stream is seldom viscous, it may be atomized by hydraulic pressure in a mechanical atomizing nozzle (or in a dual-fluid nozzle by steam or compressed air). Use of compressed air for these nozzles has been very successful. The air serves as combustion air and reduces the heat load on the system that is required with steam. The design must control the proper ratio of liquid to atomizing fluid flow to ensure the best spray pattern and droplet size through the entire range of flow.

Atomizing nozzles must also be protected from excessive heat in the event of loss of waste flow. This can be achieved by maintaining atomizing fluid flow. In many units, atomizers are designed to be physically removed from the burner or incinerator chamber when not in use to prevent burnout of the nozzle tip.

Sufficient air must be provided at all times in the incinerator to oxidize the organics contained in the waste stream. Incinerators can produce soot when burning aqueous organic wastes unless air is well mixed at the nozzle.

Aqueous wastes containing organic and inorganic compounds are produced by industries that manufacture herbicides, pesticides, etc. The pharmaceutical industry also produces a number of high water content wastes, some containing phosphorus, chlorine, and sulfur. Aqueous solutions are not ignitable when the water content is greater than 75 percent. If the waste contains ash (salts), consideration must be given to the maximum operating temperature to minimize refractory spalling problems. Aqueous wastes are normally injected into the incinerator downstream from the primary combustor in a lower temperature zone, rather than directly into the high-temperature zone.

Incinerators designed for the oxidation of high-ash-containing wastes are usually arranged for vertical downward flow (see Figure 12-13). This provides for gravity discharge of molten ash from the waste into the quench zone. Critical to the design and operation of this type of incinerator is the refractory selection. Reactions often occur at elevated incineration temperatures between the metallic oxides formed from oxidation of the waste and the refractory lining (typically alumina and silica). Severe spalling and high maintenance result from improper selection of refractories. Pretesting of various refractories with the slag that results from the incineration of such wastes at various temperatures can be achieved in a muffle furnace. This often eliminates subsequent refractory problems in a full-scale system.

In burning aqueous wastes and using organic wastes as an auxiliary fuel, the primary purpose of the incineration process is to completely oxidize the organics present in the waste. Therefore, conditions must be established to provide the necessary atomization, mixing, temperature, and time for the various types of wastes to be incinerated. Certain conditions of waste composition may limit the operating temperature, e.g., aqueous wastes with salts. Therefore, it is important to determine how and where each waste may be introduced into the incinerator. Completely organic wastes are usually introduced into the primary combustor to provide the necessary heat input and temperature needed for the low heating value waste streams. If the aqueous waste stream will not cause a problem with refractories at high temperatures, it may also be introduced with organic waste in the primary combustor as long as it does not lower the primary combustor temperature below 2200°F.

Acid-Gas-Producing Liquid Wastes

If the waste contains sulfur, phosphorus, chlorine, or another halogen it will produce an acid gas on combustion. The primary concerns in the hazardous waste industry are with sulfur and chlorine because these are the most frequently encountered acid producers. As outlined in Section 12-3, the quantity of acid gases produced will vary, depending on their equilibrium constants at the operating temperature, but they must be reduced to acceptable emission levels before they can be discharged to the atmosphere. Acid gases are generated from many types of hazardous wastes, but acid gas removal methods are essentially the same (see Section 12-9).

12-6 SOLID WASTE INCINERATION

Introduction

Solid waste combustion occurs in suspension, on a grate, or on a solid hearth. Over the years many different incinerator designs have been employed for burning solid wastes. Suspension systems rely on a relatively uniform feed sizing. For example, lumber and woodworking operations burn sawdust suspended in a high-velocity air stream. The fluid-bed incinerator consists of a bed of sand or alumina fluidized with air, into which the waste is injected while the bed is in suspension.

Grate-type systems are suitable for large and irregular wastes that can be supported on a stationary or moving alloy grate that allows combustion air to pass through the grate into the waste. Most municipal solid waste incinerators are of the grate type. Hearth-type incinerators include the two-chamber, fixed-hearth; the rotary kiln; and the multiple-hearth.

Hazardous waste incineration is currently performed in all types of incinerators but mainly in suspension and hearth-type systems. The rotary kiln incinerator accounts for 75 percent or more of all the hazardous waste incinerators in the United States that handle liquid and solid waste. The two-chamber, fixed-hearth incinerator probably accounts for about 15 percent, and the multiple-hearth and fluidized bed incinerators combined account for the other 10 percent. All of these systems also can burn liquid hazardous wastes.

Grate-Type Incinerators

Stationary grate-type incinerators, which burn the waste on metal grates, provide air circulating below, above, and through the waste. They are not generally suitable for hazardous waste because the higher temperatures required for waste destruction may destroy the grates. Obviously, this type of incinerator depends on the waste to be of such character that it will be supported on the grate and will not fall through to the ash pit until it is burned. Burning cellulosic wastes (wood and paper) is the most practical use for most grate-type incinerators.

Traveling grate or belt incinerators have also been used for municipal trash, but seldom for process waste. One exception is the infrared incinerator, which employs a woven alloy belt or a segmented belt made of alloy castings moving through a radiant electric furnace. This primary furnace is followed by a secondary chamber where additional air and fuel are added to complete combustion of the vapors from the primary furnace. Only a few systems have been employed for soil decontamination, and with mixed results.

Hearth-Type Incinerators

Most hazardous waste is burned in hearth-type systems, of which there are several basic types:

- Rotary kiln
- "Controlled-air" or "two-chamber fixed-hearth" incinerator
- Multiple-hearth incinerator
- Monohearth incinerator (seldomly used)

The rotary kiln shown in Figure 12-14 consists of a refractory-lined cylindrical chamber that sits on trunnions and rotates slowly (0.5 to 2 rpm) on its longitudinal axis. The kiln is sloped 1 to 2° from the feed end to the ash discharge end so that the waste moves horizontally as well as radially through the cylinder. Waste burns as it moves toward the ash discharge end. While the ash is discharged at the low end, the flue gases from the kiln pass into a secondary combustion chamber and are heated to a higher temperature for complete destruction.

Rotary kiln advocates have consistently debated the subject of "slagging" versus "nonslagging" kiln operation. The slagging technology, of European design and preference, operates the kiln at temperatures where inorganic components of the waste, including steel drums, will be discharged to the ash quench chamber in molten form. Systems in the United States have long employed the nonslagging ash discharge where the ash temperature is kept below its melting point. Both systems are viable.

An advantage of the slagging system is that it melts everything and discharges a "liquid" ash. However, the refractory must withstand high temperatures. Although the operation requires closer temperature control, it does allow for more types of waste feed. The nonslagging approach doesn't require rigorous temperature control and doesn't provide the feed flexibility. Also, larger noncombustible materials must be removed from the ash discharge.

Real operating difficulty exists between the two operating temperature conditions because a system operating as neither nonslagging nor slagging is in a plastic condition where the waste can build a semisolid ash that cannot be discharged either dry or as a liquid. To avoid this condition, temperature control is required at all times.

The fixed-hearth incinerator shown in Figure 12-15 consists of a primary chamber having either a single-level hearth or a stepped hearth. In smaller units, waste is intermittently charged to the primary chamber, but ash is not removed until the accumulated quantity interferes with normal operation. In larger units a mechanical ram pushes the charge through the incinerator, and the ash is continually removed. Both types have secondary combustion chambers where the flue gases from the first chamber are heated to destroy organic compounds present in the flue gas.

Combinations of liquid and solid wastes are often burned in the same incinerator, and there is nothing to prevent a waste gas from also being burned. Judicious use of the energy from one waste to help burn another waste is good energy stewardship. Because liquid- or gas-burning equipment is usually employed in a solid waste incinerator to burn auxiliary fuel, it can be employed in much the same way to burn a waste liquid or gas.

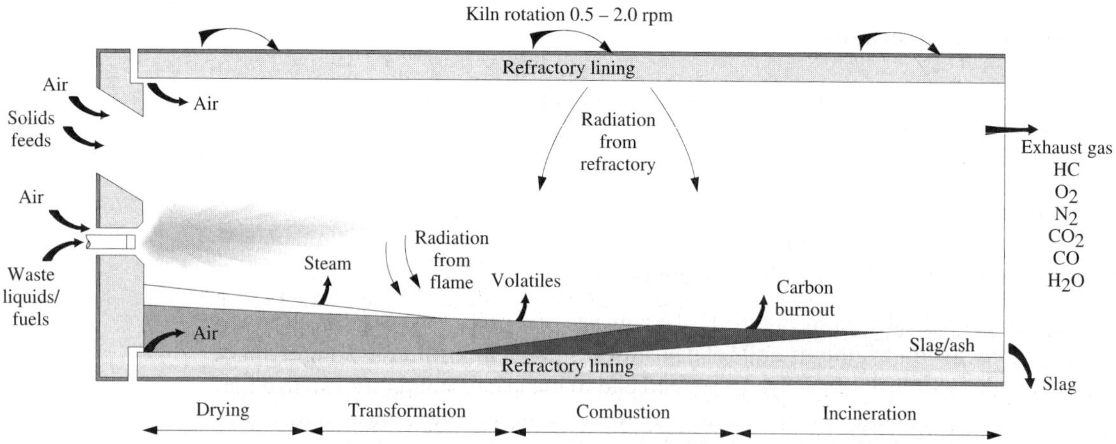

FIGURE 12-14
Rotary kiln incinerator.

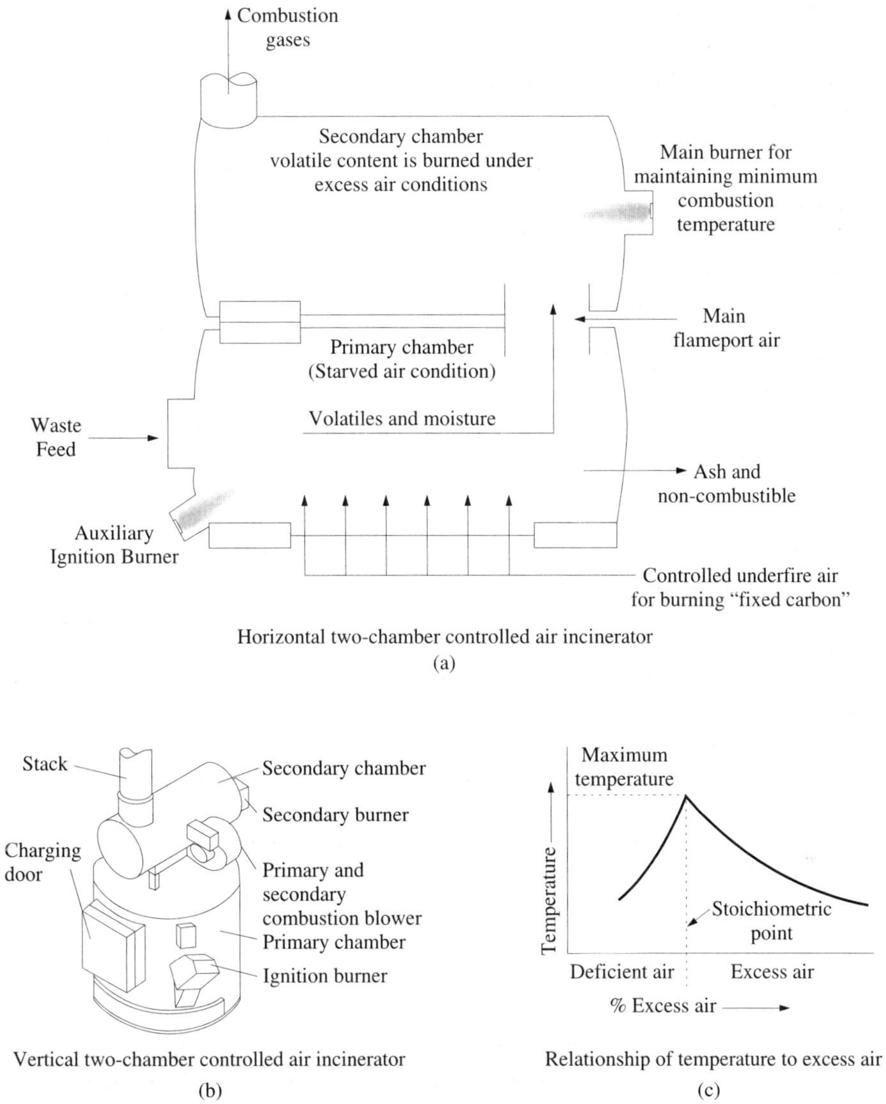

FIGURE 12-15
Fixed hearth incinerator.

The multiple-hearth incinerator shown in Figure 12-16 was originally utilized in an incineration mode for sewage sludge, although multiple-hearth systems had been used in the mining industry for years as ore roasters. It is a complicated, highly mechanical system used to burn sludges. It has seen limited use in hazardous waste incineration because the temperatures required to provide reasonable destruction efficiencies are not compatible with long equipment life. The system consists of two to six horizontal hearths in a vertical array. The waste is fed to the top hearth and ignited by fuel burners. An air-cooled central shaft moves rabble arms over the surface of each hearth in a carousel fashion. The ploughs on the arms move the

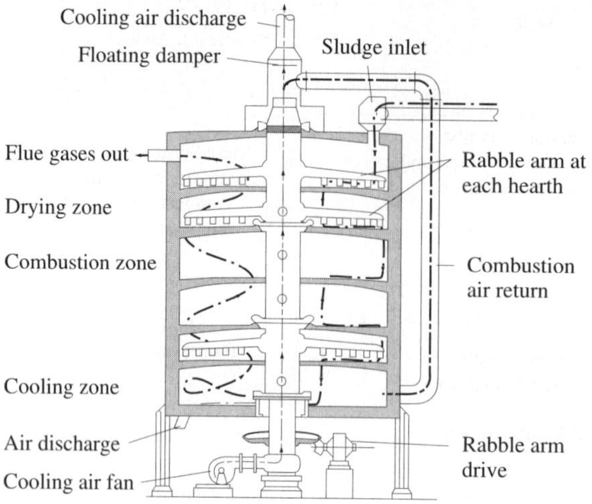

FIGURE 12-16
Multiple-hearth incinerator.

waste across the surface of the hearth until it reaches a pie-shaped opening. The waste drops through the opening to the next lower hearth and continues to burn, is cooled, and discharged as an ash. The flue gases move vertically upward within the outer shell and into the air pollution control equipment. A variation of this design, often called a "monohearth," employs a single hearth that revolves beneath fixed rabble arms, which move and level the waste. The monohearth also requires a secondary combustion chamber for complete combustion.

Fluidized-Bed Incinerators

The fluidized-bed incinerator shown in Figure 12-17 utilizes a "fluid hearth" consisting of sand or alumina on which and in which combustion occurs. Imagine a bed of sand sitting on a porous surface. Airflow from below with sufficient pressure will "fluidize" the bed of sand (hold it in suspension) as long as the velocity of the air is not so great that it transports the sand out of the system. This is a fluid bed in which the particles of the bed are in suspension, but do not flow out of the system. Waste is injected into this fluidized bed, either as a liquid, sludge, or uniformly sized solid. The air used to fluidize the bed is heated to at least the ignition temperature of the waste, and the waste begins to burn (oxidize) within the bed. Most of the ash remains in the bed while some exits the incinerator into the air pollution control equipment. Exiting hot flue gases can be used in a boiler or to preheat combustion air.

There are a number of variations for fluidized-bed systems. One is the circulating fluid-bed combustor where the gas velocities are above those required for mere fluidization. The variations in technology are illustrated in Figure 12-18.

Oxidation versus Pyrolysis

There are two distinct operating modes for hazardous waste incinerator systems: oxidizing (excess air) and pyrolytic (starved air). These modes apply to the first

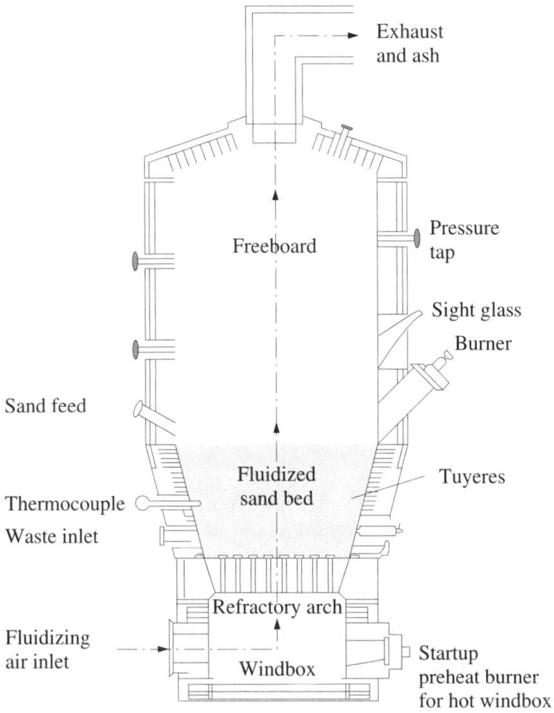

FIGURE 12-17
Fluidized incinerator.

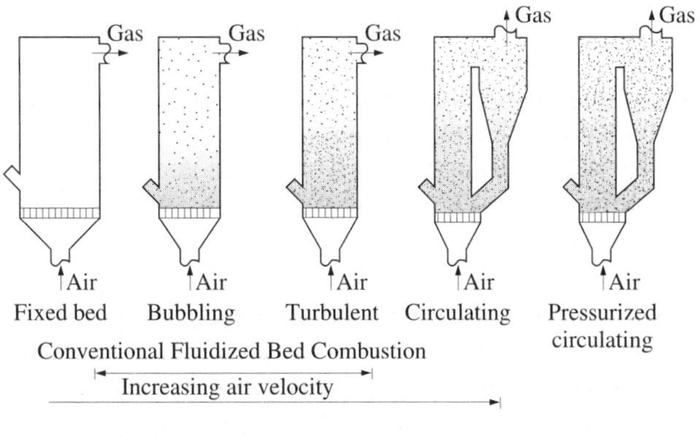

FIGURE 12-18
Fluid-bed incinerator types.

combustion stage only, because complete oxidation of the flue gases is required to make them acceptable for release to the atmosphere.

The two-chamber, controlled-air type hearth incinerator is most often operated in a pyrolytic (oxygen-starved) mode. While there are rotary kiln systems that claim pyrolytic operation, the actual operation is almost always in excess air (oxidizing). The two-chamber system is designed to be pyrolytic in the first stage to reduce gas velocity and, therefore, to reduce particulate carryover. This approach was cogent when it was possible to meet particulate emission regulations without the addition

of particulate capture equipment. Today's particulate emission levels make that impossible in most cases, so pyrolytic operation has few advantages. The partially combusted gases from the primary chamber, consisting of CO and some H_2, are heated and oxidized in the second chamber to CO_2 and water vapor. The ash, however, from many of these units often contained unacceptable levels of hazardous constituents.

Excess air systems that employ an oxidizing atmosphere in both a primary and secondary combustor, are more often used for hazardous waste incineration. The rotary kiln is such a device. Because the kiln is under negative pressure and adequate seals have not been devised to prevent the in-leakage of air, it is almost always in an excess air (oxidizing) mode.

Secondary Combustion Chamber

The secondary combustion chamber (SCC) is necessary in virtually every hazardous waste incinerator where solid wastes are burned because the primary chamber, whether it is a kiln, fixed-hearth, or multiple-hearth design, does not provide enough time, turbulence, or temperature to destroy the organic components of the waste to the required DRE. Low temperatures in the primary chamber, 705 to 815°C (1300 to 1500°F), often provide better operation than higher temperatures because they are below the slagging (melting) temperatures of inorganics in the waste. This is especially true in incinerators for soil decontamination. The primary chamber's function is to volatilize the organic fraction of the waste. The secondary chamber's function is to heat the vaporized organics to a temperature where they will be completely oxidized. This oxidation process occurs at 980 to 1200°C (1800 to 2200°F). Because commercial fuel is used as the heat source in operating most SCCs, lowering the temperature at which the DRE can be achieved is most economical. Again, the time, temperature, and turbulence triangle is at work. Regulatory agencies may require a flue gas residence time in the SCC of at least 2 seconds. The destruction of organics in the flue gas is a function of not one, but all three, factors. The use of a secondary combustion chamber is not usually required for incinerating liquid or gaseous wastes because these "three T's" can be achieved in a single chamber.

The SCC may be either a horizontal or vertical refractory-lined vessel with its own liquid or gas burners of sufficient capacity to heat the flue gases from the primary chamber to the desired operating temperature where the required destruction of hydrocarbons can be achieved. Vertical orientation is preferred over horizontal when the particulate loading from the primary combustor is expected to be substantial. This is because gravity will cause particulate deposits on the bottom of the chamber, which cause a removal problem. Design gas velocities in the SCC are usually 15 to 25 ft/s at operating temperatures of 1800°F or above. Figure 12-19 shows a rotary kiln incinerator with a vertical secondary combustion chamber.

EXAMPLE 12-8. ROTARY KILN—BATCH FEED. A rotary kiln designed for a nominal heat release of 15,000 Btu/(h · ft^3) has an inside diameter of 8 ft and is 30 ft long.

1. Determine the design heat release in Btu/h.
2. Normally a waste having a heating value of 750 Btu/lb is burned in the kiln. It is fed continuously through an auger. Occasionally a waste consisting of polyethylene pellets is "batch" fed to the kiln in 30-gallon fiber containers. The pellets have a

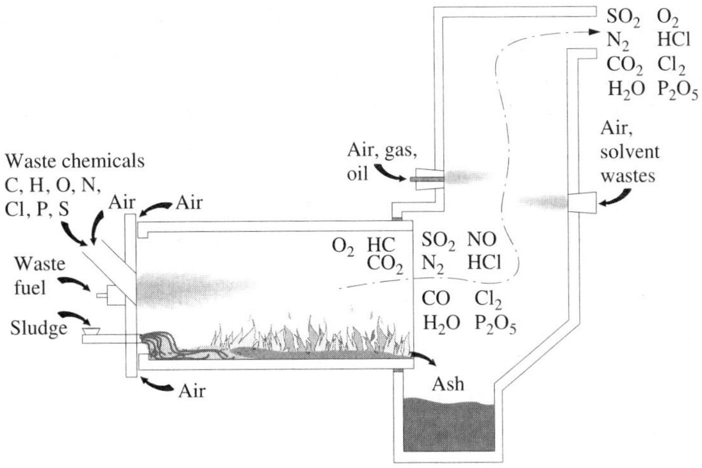

FIGURE 12-19
Rotary kiln with vertical secondary combustion chamber.

bulk density of 50 lb/ft^3 and a heating value of 18,350 Btu/lb. A single container is consumed in 6.5 minutes.

(*a*) Will the kiln operate within its design parameters?
(*b*) What kind of combustion would you expect?
(*c*) How would you correct the problem if you were the operator?

Solution

1.

$$\text{Kiln volume} = 30 \times \frac{(8)^2\pi}{4} = 1508 \text{ ft}^3$$

The heat release is:

$$1507 \text{ ft}^3 \times 15{,}000 \text{ Btu/(h} \cdot \text{ft}^3) = \underline{\underline{22{,}620{,}000 \text{ Btu/h}}}$$

2. (*a*) $30 \text{ gal} \times 8.35 \text{ lb/gal} \times \dfrac{50}{62.4} = 201 \text{ lb in each container.}$
201 lb × 18,350 Btu/lb = 3,688,000 Btu heat release.
But this will occur in 6.5 minutes; therefore the equivalent hourly heat release is:

$$\frac{60 \text{ min/h}}{6.5 \text{ min}} \times 3{,}688{,}000 \text{ Btu} = 34 \text{ million Btu/h}$$

This is about 1.5 times the design heat release of 22.6 million Btu/h, so the kiln will *not* operate within its design parameters.

(*b*) There will not be sufficient air in the system to burn the polyethylene waste at that rate, so the kiln will produce a dense black smoke.

(*c*) To correct the situation, package the polyethylene in smaller batches. The batch size will have to be smaller by at least the same factor (1.5). Therefore:

$$201/1.5 = 134 \text{ lb/batch}$$

$$134 \times 18{,}350 = 2{,}458{,}900 \text{ Btu heat release}$$

$$60/6.5 \times 2{,}458{,}900 = 22.7 \text{ million Btu/h}$$

The Btu/h is almost the same as the design value. Good practice would suggest that a safety margin of at least 10 percent be used, so about 120 lb/batch is a good size. Also, the containers must be fed at a rate no greater than one every 6.5 minutes, say 9 per hour.

Thermal Relief Vent

Most hazardous waste incinerator systems, except liquid incinerators, operate under a negative pressure. If the pressure in the system becomes positive because of failure of the induced draft fan, it is usually because of power failure. If this happens, the points that allow air to leak in may, because of the positive pressure, allow flue gases to leak out. This is undesirable. In a rotary kiln system this occurs at the kiln seals. In addition, many components of the air pollution system are designed to operate at close-to-ambient temperatures (e.g., their materials of construction are plastics that cannot withstand elevated temperatures). If power fails or water flow decreases, process sensors are designed to detect these problems immediately, but it is necessary to prevent either a positive discharge of flue gases through the kiln seals or an overtemperature in the air pollution control system. This is accomplished by installing a thermal relief vent (TRV) at the top of the SCC or between the SCC and the waste heat boiler or quench.

The TRV is a bypass stack (see Figure 12-20) that opens automatically when the emergency condition occurs. It provides sufficient draft to maintain a negative

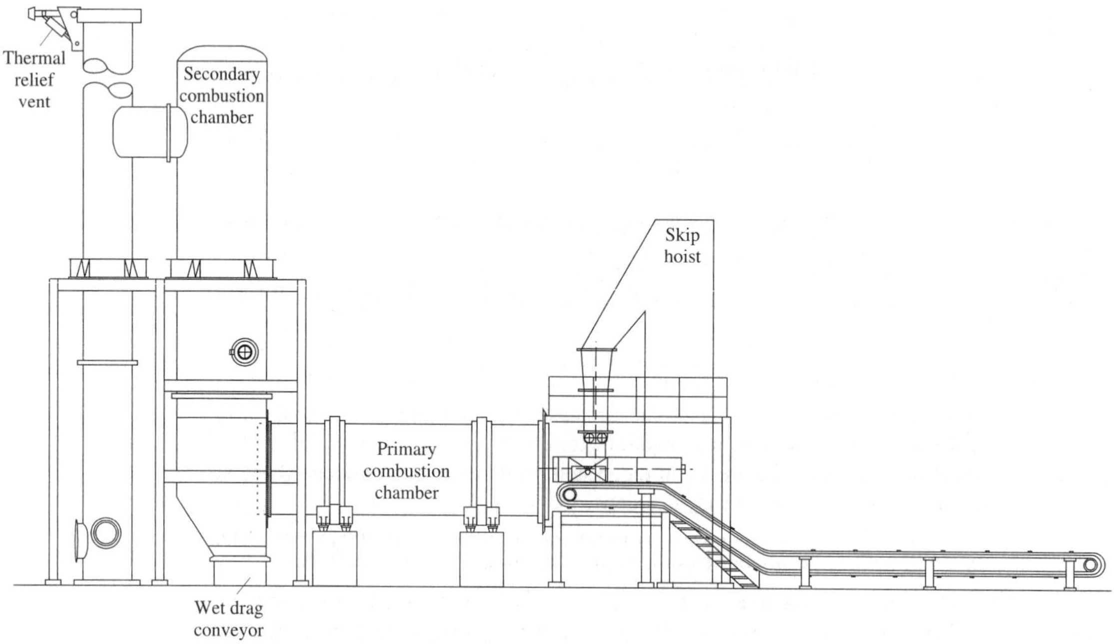

FIGURE 12-20
Partial elevation of an incinerator.

pressure condition in the kiln and prevents hot flue gas flow into the air pollution control system. It is downstream from the SCC so the reduced flow from the kiln at the high SCC operating temperature results in high organic compound destruction efficiencies. The low-flow condition entrains few particulates. Therefore, the threat of environmental pollution with particulate contamination is low. While the TRV is not an ideal solution, it is the safest and most practical approach to emergency conditions.

12-7 STORAGE AND FEED SYSTEMS

Liquid and Sludge Storage and Feed Systems

To provide a uniform feed from the storage tanks, good mixing is required. Layering or phase separation often occurs in large tanks, especially where a number of liquid wastes are introduced from different sources. Mixing may be accomplished by internal mixers, external recycle pumps, and air or steam spargers. Proper venting of storage and mixing tanks is also necessary to prevent pressure or vacuum conditions in the vapor space of the tank. Vent gases from storage tanks may be exhausted to the incinerator or controlled by other means such as condensation by refrigeration or carbon adsorption.

The preferred physical design for hazardous waste liquid and sludge storage tanks is a vertical vessel supported on legs and with a maximum capacity of 20,000 gallons. There are three important reasons for selecting this configuration:

- Leak detection from a leg-supported (elevated) tank is obviously easier than leak detection from a tank that has a flat bottom sitting on a concrete foundation.
- 20,000-gallon vertical tanks are about the largest size that can be reasonably agitated or mixed by normal methods.
- Hazardous waste tank containment volume must be equal to the largest tank within the containment.

Materials of construction of all components in the storage and feed system—tanks, pumps, mixers, piping, valves, and nozzles—must be compatible with the chemical composition of the waste to prevent corrosion and erosion. If the waste contains suspended solids that are not combustible, filtration of the waste may be required to prevent solids carryover into the incinerator and air pollution control system. For organic solids, filtration is not required and the design must provide for their transport into the incinerator.

High-viscosity wastes (slurries and sludges) and liquids containing solids present material handling problems. Various generic pump types are available, and the selection and the use of each pump depends on the properties of the waste stream. Centrifugal pumps often have too great a capacity for incineration systems when waste flows are low. Progressive cavity pumps have been used successfully when the construction materials have been selected properly to prevent corrosion and erosion. Gear pumps, which provide low flows at high pressures with low-to-medium viscosity wastes having some lubricating quality, are also widely utilized. Diaphragm pumps, operated by compressed air, are good for pumping heavy sludges.

In material handling (feed) systems where a variety of waste streams must be pumped into the incinerator, it is difficult to select the proper feed system to use with each waste. In some cases, pumps may not be practical because of the high maintenance costs and shutdowns that result. Pressurizing a storage or feed tank with an inert gas (nitrogen) or air where the waste is not highly flammable may be the safest and most economical method.

Solid Waste Storage and Feed Systems

The most difficult design problem in any hazardous waste incinerator is the material handling system and primarily the solid waste feed system. Storage of solid wastes is relatively simple. Solids may be stored either in small containers such as drums and boxes, large containers such as 20 cubic yard (or larger) roll-off boxes, or in bulk in some type of solid waste containment. Bulk storage, in other than containers, is not usual in hazardous waste incinerator facilities because it is difficult to contain and difficult to control air emissions. Therefore, small containers, steel drums, fiber and plastic drums, and similar containers are used for most hazardous waste other than heavy sludges or contaminated soil. The latter are usually shipped to the incinerator in roll-off boxes and stored in these containers until processed into the feed system.

Covered, dry storage areas are certainly required for roll-off containers and also for the drum storage and processing facility. What happens to the waste after receipt and storage is a function of the incinerator type. Only the rotary kiln has been utilized widely enough to provide a basis for a discussion of solid feed processing systems: rotary kiln systems constitute the majority of commercial hazardous waste incinerators handling solid wastes.

The solid waste feed system presents the most difficult design problem in any incinerator project, and its proper design is vitally important to the success of the incineration system. Commercial hazardous waste incinerators must provide for feeding:

1. Containerized waste
2. Bulk waste
3. Contaminated soils
4. Oversized items
5. Problem materials

Containers (i.e., steel, fiber, or plastic drums) or cardboard boxes can be fed directly into a rotary kiln through an airlock door by a hydraulic ram feeder (see Figure 12-21). Whole steel drums are fed only to slagging-type kilns, but the fiber or plastic drums will burn completely in nonslagging systems except for metal rings or closures. Problems with this type of feed system arise when the container contains a waste that is volatile and has significant heating value.

For example, if a 50-gallon drum of liquid toluene (HHV = 18,252 Btu/lb and 7.2 lb/gal) is fed into a hot rotary kiln, the potential heat release from that drum is over 6.5 million Btu. It has a vaporization temperature of only 230°F, so virtually all of the toluene in the drum will be vapor within a few seconds in a kiln operating at 1500°F. If vaporization took 15 seconds, the hourly heat release rate would be

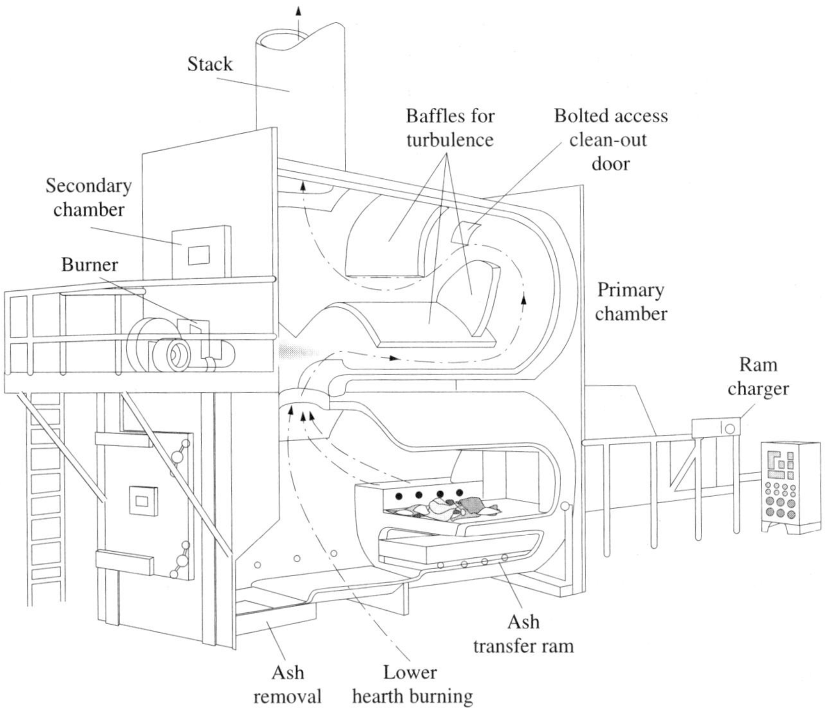

FIGURE 12-21
Hydraulic ram feeder.

$4 \times 60 \times 6.5$ million = 1560 million Btu/h. It is important to see this as a *rate* of heat release and not an actual release. Most rotary kilns are designed for heat release rates of 20,000 Btu/(h · ft^3). This would mean that the kiln would need to have a volume of 78,000 ft^3 to be within good design practice. Since a typical rotary kiln is on the order of 12 ft diameter × 30 ft long (3400 ft^3), this is obviously impractical.

It is even more important to recognize the combustion consequences of such a single waste charge. There would not be enough combustion air to satisfy the high volume of hydrocarbon vapor in the kiln, so smoke or soot would instantaneously appear. The oxygen in the kiln would drop to zero, and the CO concentration would rise to a level that would exceed regulatory limits.

This example is an extreme case for illustrative purposes only, but batch feeding always presents this problem. The higher the heat release in a batch charge of any size, the more difficult is the control of the combustion process.

Containers of all types, containing liquids or sludges, are usually preprocessed by pumping and bulking the waste into storage tanks for continuous feed through lances and atomizers. The containers, either empty or with some residual waste material, can then be batch-fed to the incinerator or shredded and fed continuously.

Bulk waste can also be fed by hydraulic ram. It presents the same batch charge problem as containers. It also can be fed by gravity through a chute. Continuous feed is still difficult if not impossible unless there is some "sizing" procedure, such as shredding ahead of the feed system.

The normal heating value for contaminated soils is low, but may occasionally have significant caloric levels as high as 4000 Btu/lb. If they are first processed through a sizing system such as a bar screen or grizzly, they can be fed best by a screw conveyor. This continuous type of feeder provides uniform feeding with variable feed rates depending on heating value.

Oversized items, regardless of their heating value, need to be reduced in size by shredding to be accepted by any feed system.

Special wastes, those that are extremely volatile, pyrophoric, or potentially explosive, must be fed through a special chute (often called a bottle drop chute) in quantities small enough to obviate any problem.

The best feed system, from a combustion standpoint, is one that will provide a steady, continuous feed. This is impossible to achieve with types 1, 2, 4, and 5 above unless these types are reduced to a relatively uniform size. This can be accomplished by shredding the waste before it is fed to the incinerator after first removing liquids and pumpable sludges from containers.

12-8 FLUE GAS TEMPERATURE REDUCTION

Introduction

When the flue gases leave the secondary combustion chamber, they are at temperatures usually exceeding 1800°F and at times as high as 2200°F. They contain particulate matter and may contain acid gases such as HCl, Cl_2, and other halogens; SO_2 and possibly P_2O_5 which must be removed; and heat, which also must be removed or recovered. There are two options: (1) recover the heat or (2) cool or quench the gases without recovering any heat.

Heat Recovery

Heat recovery is a major economic consideration because heat recovery equipment is expensive and there must be a practical use for the recovered heat. In Europe, where fuel cost is high, heat recovery is more widely employed than in the United States. Heat recovery from hazardous waste incinerators in the United States is virtually nonexistent in commercial operations because there is little use for steam on-site, the potential of selling steam to off-site customers is minimal (usually because hazardous waste incinerators are not located close to potential users), and the economics of steam-electric power generation are poor. In-plant hazardous waste incinerators may have entirely different economics, and this is where heat recovery often is viable.

The options for heat recovery are the use of waste heat boilers to generate steam or hot water, and recuperators to preheat combustion air with the incinerator flue gases or to reheat stack gases with flue gases to eliminate the stack water vapor plume after a wet scrubber. Boilers have been widely used in municipal solid waste (MSW) incinerators to produce steam for electric power generation. Here the waste heating value is reasonably constant. This is not always true with hazardous

waste incinerators. Also, the size of the MSW incinerator is large compared to the hazardous waste incinerator, making power generation economical.

The waste heat boiler (see Figure 12-22) is a special design for most hazardous waste incinerators. It should be capable of handling a high-temperature, high-particulate, high-acid content flue gas. The particulate in the flue gas from the SCC may be essentially "plastic," i.e., containing inorganics at temperatures above their melting point, or even their vaporization temperature which, on contacting cold boiler tubes, will deposit a solid residue on the tubes. For this reason, radiant sections of tubes with wide tube spacing are used in the inlet end of the boiler. Some manufacturers have even provided for rapping or shaking of the tubes to release the solid buildup into hoppers below. After the temperature is adequately reduced to eliminate this problem, the remaining boiler tubes can serve as a convective heat transfer device.

Most waste heat boilers on hazardous waste facilities that burn solids, or wastes containing solids, utilize a specially designed boiler of the "water tube" type. Liquid incinerators that burn low-ash liquids may use a "fire tube" type boiler. The generated steam pressure is usually nominal, below 100 psig, because there is no application for high-pressure steam in the plant and no customers for the steam external to the plant.

Design considerations for burning wastes that produce acid gases, especially halogen gases, involve operating restrictions more than design restrictions. A cold boiler fired on a waste which will produce HCl on combustion is a boiler with a short life because hydrochloric acid will condense on the tubes and rapid corrosion will result. Therefore, any boiler handling acid gases must be preheated on a fossil or nonacid fuel to at least its normal operating temperature before acid gas–producing fuel is burned. Similarly, on shutdown, the acid gas–producing fuel must be stopped and replaced with fossil fuel at least 10 minutes before the boiler fuel is cut off.

Flue Gas Quench

The other option for cooling flue gases is to use the direct injection of air or water. Water injection is more efficient because of its higher specific heat compared to air.

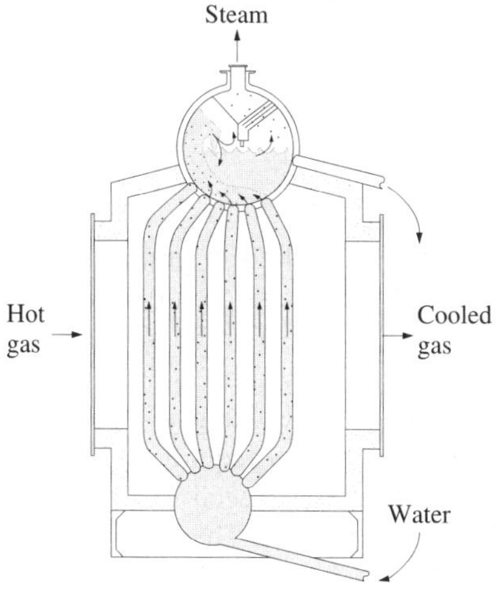

FIGURE 12-22
Waste heat boiler.

This also means that the net volume of flue gases cooled with water will be lower at any given temperature. Most hazardous waste incineration systems that do not employ a waste heat boiler, and many that do use a waste heat boiler, cool the flue gases either finally or initially by direct injection of water.

Cooling hot flue gases is a tricky problem. It is one of distribution and mixing. There is no such thing as a "dry bottom quench" (i.e., a quench process that evaporates all the water at adiabatic saturation temperatures). The vagaries of cooling a hot gas stream involve both heat and mass transfer between the cooling medium and the flue gases if the coolant is water. This means fine water droplets and good distribution throughout the flue gas stream. The higher the flue gas volume the more difficult this is to achieve. Usually, multiple, two-fluid, atomizing nozzles are employed. The number and location of the atomizers depend on the size of the quench chamber and the gas flow. Many quench systems are concurrent-flow (i.e., flue gas and water flow in the same direction) because the SCC is vertical and the hot gas flow is downward, but this does not exclude a counterflow arrangement where the equipment design is amenable to such an arrangement. The design must provide cooling of the gases to satisfy inlet temperature requirements to a spray dryer 260°C to 371°C (500°F to 700°F) or to an adiabatic saturation temperature of about 82°C (180°F) for low-temperature downstream equipment.

12-9 AIR POLLUTION CONTROL

Air pollution control systems associated with hazardous waste incinerators are unique in their combination of elements, but the elements themselves (scrubbers, baghouses, cyclones, etc.) are not unique.

Most hazardous waste incinerator air pollution systems require two functional elements: (1) a system or equipment item that will remove particulates from the flue gas stream and (2) a system or item of equipment for the removal of acid gases. These two functions may be achieved with wet systems or dry systems or with a combination of both.

The particulate removal system must specifically address the removal of the U.S. EPA's list of 10 priority metals which have regulated emission levels. These 10 metals are: arsenic, beryllium, cadmium, chromium, antimony, barium, lead, mercury, silver, and thallium.

Particulate Removal

Particle density in the flue gas from an incinerator is measured in grains/dscf or mg/dscm. The particle density in the gas varies widely, depending on two factors: gas velocity in the incinerator and actual particle size. The smaller the particle, the more easily it is carried by the flue gas at a low gas velocity. The higher the gas velocity, the larger the particle capable of being carried in the flue gas stream and the greater the number of particles of all sizes carried. Current federal regulations specify that particulate emissions from an incinerator shall not exceed 180 mg/dscm (0.08 grains/dscf) corrected to 7 percent oxygen in the flue gas. These regulations

will surely become more stringent, and some states now require levels as low as 0.015 grains/dscf.

Particle removal from gas streams can be effected by gravity separation, interference such as impacting the particles on another object, centrifugal separation, filtration through a filter media, electrostatic separation, or by sufficiently wetting a particle with water so that it is "scrubbed" from the gas stream. Particulate removal can be accomplished by completely dry removal methods, a combination of wet and dry methods, or completely wet methods.

The dry particulate removal methods include:

- *Impaction*—baffles and screens
- *Centrifugal separation*—cyclone separators
- *Filtration*—fabric filters
- *Electrostatic*—precipitators

Wet methods that employ water as a medium include:

- *Impaction*—packed and tray columns
- *Centrifugal separation*—wet cyclones
- *Particle wetting*—venturis and similar units
- *Particle conditioning and wetting*—collision scrubber
- *Electrostatic*—wet ionizer/precipitator

Acid Gas Removal

In an incineration system, especially a hazardous waste incineration system, there is a need to remove acid gases from the flue gas before discharge to the atmosphere. The normal acid gases encountered are HCl, SO_2, and HF. Occasionally HBr will be present. HCl and HF are very soluble in water and HBr slightly less. SO_2 is not as soluble so its removal depends on using a base such as lime, NaOH, or soda ash for neutralization. The same solution can be used to remove HCl and HF. Acid gas removal can be accomplished in either a wet or dry system. The efficiencies of acid gas removal vary between the wet and dry system. The wet system generally has a higher efficiency and uses more nearly stoichiometric amounts of base than the dry system.

The dry removal systems are:

- *Dry lime injection*—a variety of systems
- *Dry scrubber*—spray dryer

Dry removal systems are usually followed by a fabric filter to collect the neutralized acid gas as a salt.

The wet systems are:

- *Absorption/reaction*—packed and tray columns
- *Wetting contactors*—venturis/Calvert/Hydro-Sonic

In a wet system, the effluent is a salt solution of the acid, such as $CaCl_2$, NaCl, $CaSO_4$, or Na_2SO_4. These salt solutions are normally removed from the scrubber continuously and treated in on-site wastewater treatment facilities.

Control System Options

For a hazardous waste facility, air pollution control requires a system designed to remove particulates, including metals and acid gases. For example, hazardous waste incineration air pollution control systems today are being designed to eliminate the need for handling wet effluent; therefore, the system is designed to produce only a dry waste product. The flue gases are cooled to about 600°F in a boiler or by direct contact with water or air. They are then exposed to a neutralizing agent such as lime, which is pneumatically conveyed into the flue gas. The finer the lime particles, the better the neutralization efficiency. This is known as a dry-dry injection system (see Figure 12-23). The lime could also be fed into the hot gas stream as an atomized slurry so that the heat from the flue gas evaporates the water and the acid gases are neutralized by the lime. This is usually accomplished in a spray dryer, sometimes called a dry scrubber. The lime-sulfite mixture leaves the spray dryer as a solid, either at the bottom of the unit or entrained in the flue gases. This is known as a wet-dry system (see Figure 12-24). Sodium hydroxide or soda ash also might be used, but both are more expensive than lime and, because sodium salts are more hygroscopic, lime is preferred.

Both types of dry reaction systems are followed by particle removal devices such as an electrostatic precipitator or fabric filter (baghouse). The operating temperatures

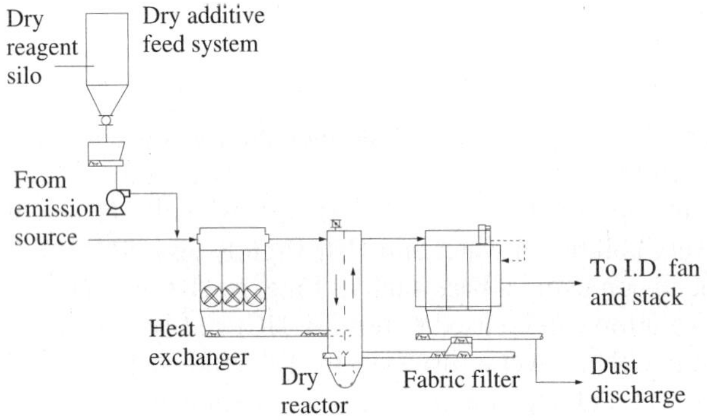

FIGURE 12-23
Dry/dry APC.

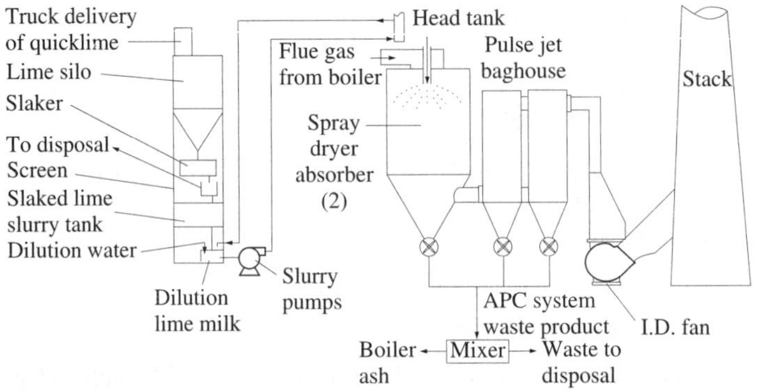

FIGURE 12-24
Wet/dry APC.

for these units are normally about 350 to 450°F. Baghouses are the preferred removal device in most systems because the acid-base reaction that begins upstream continues on the surface of the bags, increasing acid gas removal efficiency. Both the acid gases and the particulate from the incinerator are removed in the dry scrubber and baghouse. Because the reaction between lime and acid gases is slow and perfect mixing is a problem, the lime-to-acid gas ratio must be 1.5 to 2.0 times the stoichiometric reaction quantity.

Completely wet systems for particulate and acid gas removal are usually employed after an adiabatic quench where the flue gases have been saturated with water to the adiabatic saturation temperature. Then acid gases can be easily removed in a packed column utilizing a caustic or soda ash solution. Unlike the dry neutralization systems, the wet systems are quite efficient and are controlled by maintaining the scrubbing solution pH. The particulates can be removed from the flue gas before the acid gases or at the same time, depending on the device employed. Venturi scrubbers are one possibility for particulate removal. They are simple and rugged but require a high pressure drop to remove submicrometer particles to meet required emission levels and are therefore not very energy-efficient. They often are equipped with an adjustable "throat" to provide constant pressure drop over a range of gas flow rate conditions (see Figure 12-25).

The Calvert "collision scrubber" (see Figure 12-26) has a better efficiency than a venturi at any given pressure drop because it employs a conditioning tower where small particles are provided time to grow larger, thus requiring less energy for removal. The Hydro-Sonics scrubber (see Figure 12-27) is also capable of removing fine particles and at pressure drops between venturi and collision scrubbers.

The wet electrostatic precipitator and the ionizing wet scrubber impart an electrostatic charge to the particles in the gas stream, and after the particles are attracted to the collection plates, they are washed off by water. These devices, like the dry electrostatic precipitator, are load-sensitive to the amount of particulate (grains/scf) in the gas stream. If the loading increases for any given design or configuration, the emissions will also increase. By placing a number of units in series, each with the same removal efficiency, the particulate emission level can be controlled for almost any scenario as long as the inlet loading does not exceed the design loading (see Figure 12-28).

Packed towers are the most widely used wet contact devices for acid gas removal. In essence, they transfer pollutants from a gas stream to a liquid stream. This liquid is

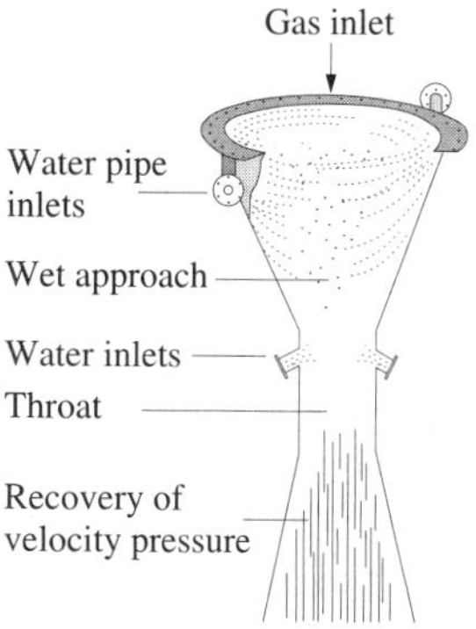

FIGURE 12-25
Venturi scrubber.

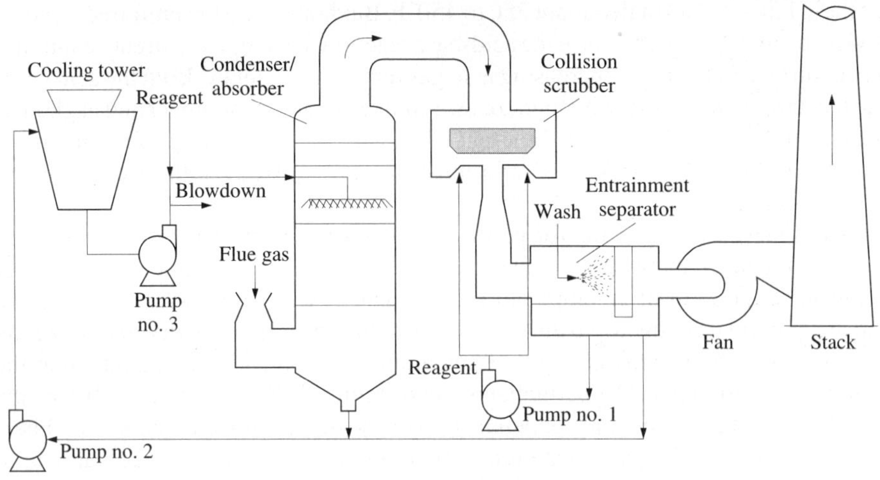

FIGURE 12-26
Calvert collision scrubber.

a waste that requires further treatment. Packed towers consist of a simple cylindrical column containing a ceramic or plastic material that provides an efficient medium for gas/liquid contacting (see Figure 9-1). Ceramic packing is used for higher operating temperatures. Plastic packing is the most widely used for acid gas removal because it is light in weight and inexpensive, and incinerator flue gases are normally cooled sufficiently to a lower temperature compatible with the plastic. PVC and polypropylene are the most common plastics utilized, but others are available. Packed columns are not suitable for particulate removal and are not recommended for use in flue gas streams where the particulate loading is high. Packed towers applied to transferring contaminants from water to air were discussed in Section 9-1.

Tray and bubble cap columns are also suitable for acid gas removal but have little advantage over the packed column in efficiency and no advantage in construction materials.

Fans and Blowers

These components are of prime importance to a hazardous waste incinerator because they move air into the incinerator and flue gas out of the system. The combustion air for most incinerators is supplied by either a forced-draft blower or an induced-draft fan or both. Hazardous waste incinerators are ideally operated under a slight negative pressure so that all leakage will be inward, but many liquid injection type incinerators do operate under a positive pressure.

Fans may be driven by electric motors, steam turbines, or hydraulic drives. The speed of the fan may be variable or fixed, depending on the service. Variable speed may be achieved through a variable speed motor or a hydraulic drive. For most incineration systems the fans and blowers are driven with fixed speed motors with a V-belt drive. Airflow on these systems is controlled with a mechanical damper at the inlet to the fan.

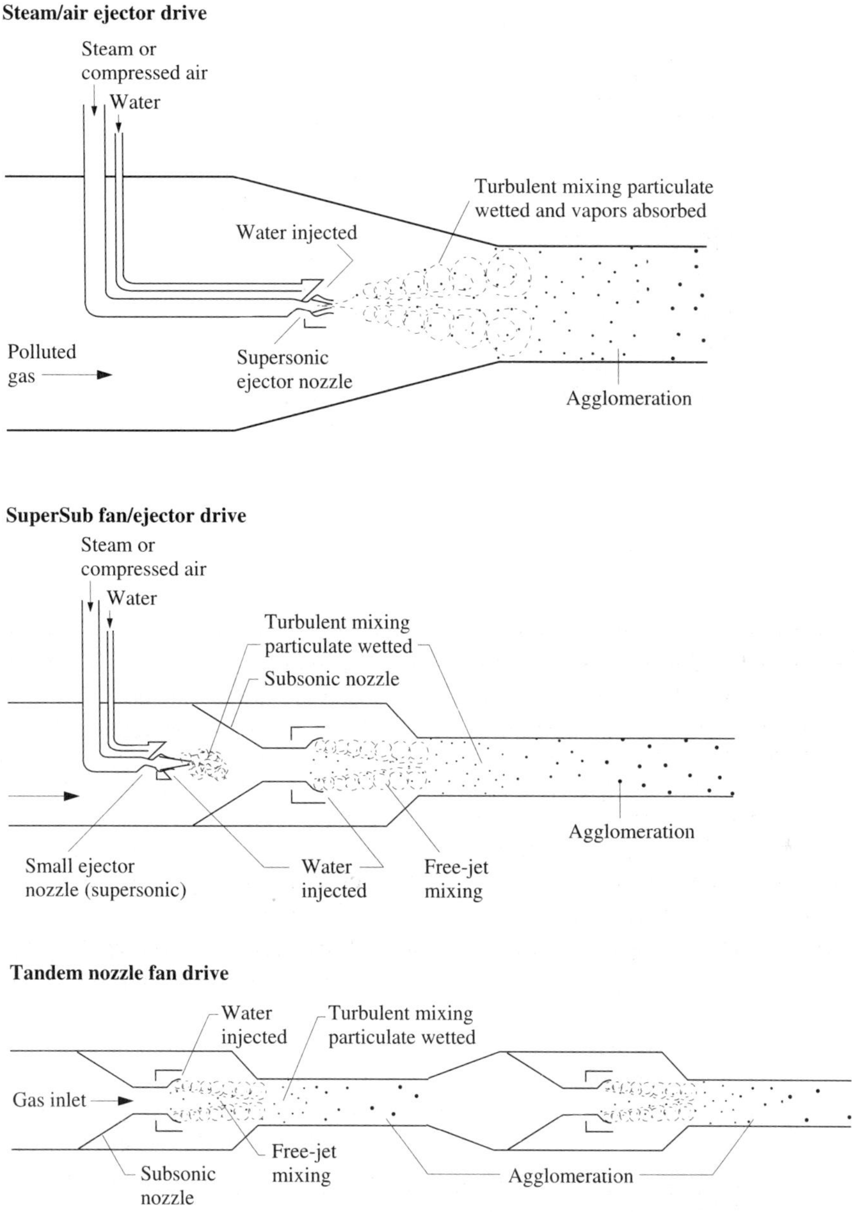

FIGURE 12-27
Hydro-Sonic gas cleaning equipment.

Coolers and Condensers

Flue gas at adiabatic saturation temperature that has passed through a mist eliminator still remains saturated with water at a temperature higher than the ambient air. When this flue gas exits the stack and is cooled by the ambient air, a dense steam plume is formed. Often this plume is conceived to be particulate air pollution by the public. To

FIGURE 12-28
Ionizing wet scrubber. (*Courtesy of Ceilcote Co.*)

overcome this steam plume, many incinerator designs employ a condensing step in the process to reduce the dew point of the stack gas and reduce, if not eliminate, the plume formation. The condensation is achieved with a direct contact system using cooling towers or with a tubular condenser employing either air or water to cool the flue gas. In a geographical area where water is scarce, there is an additional benefit because the condensate may be reused in the process.

Mist Eliminators

Flue gas from an incinerator, especially when it has been adiabatically cooled in a direct water contact system, contains a significant amount of water. As this gas is further cooled by heat loss in the duct, fine water droplets begin to form in the air stream. To remove these fine droplets and keep them from entering the stack, a mist eliminator or entrainment separator is used. Mist eliminators are generally impaction-type separation devices, although sometimes centrifugal separators are used. The impaction separator is either a wire or plastic mesh pad or a series of baffle plates that intercept the droplets and allow them to coalesce so that water may be drained from the system (see Figure 12-29).

Stacks

At the end of every incinerator system there must be a stack to release and disperse the flue gases to atmosphere. For hazardous waste incinerators, the natural draft created by the stack is not required because the fan forces the flue gas up the stack. Therefore, the critical design aspect of the stack is not its ability to produce a natural draft but rather its ability to disperse the flue gases. The proper height of the stack depends on the surrounding terrain, nearby buildings, and prevailing wind direction. Before

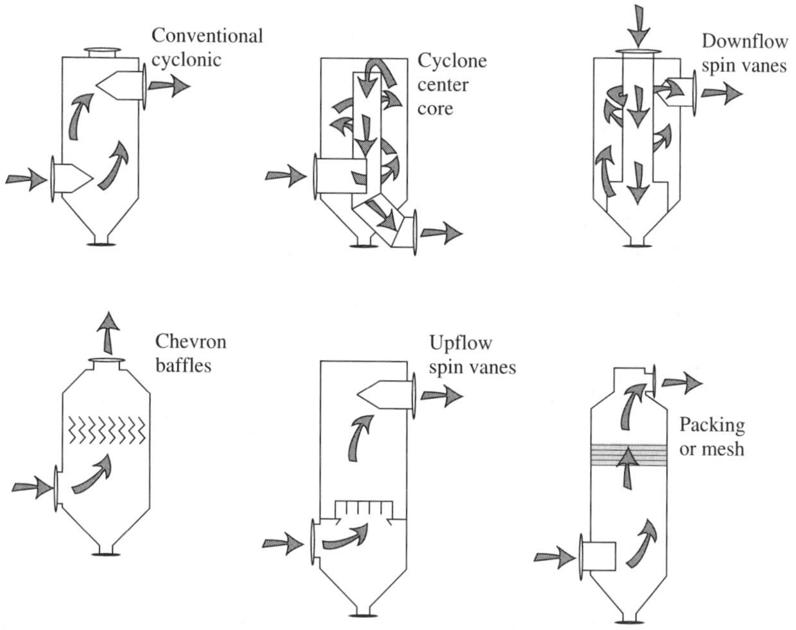

FIGURE 12-29
Entrainment separators.

deciding on the stack location and height, a dispersion analysis of the plume is usually performed. The dispersion analysis estimates the ground-level concentrations and location of pollutants present in flue gas so that a health risk analysis can be performed (see Section 4-4).

The stack must be structurally capable of withstanding the maximum possible wind velocity at its location and its design should provide a sampling platform and sampling ports at a location to meet the conditions for isokinetic sampling of the flue gas. Ideally this is 8 or more stack diameters above the gas inlet and at least 5 stack diameters from the discharge point. At this sampling location, two or more 4-in circular sampling ports located at 90° to each other are required.

EXAMPLE 12-9. PARTICULATE REMOVAL. A rotary kiln incinerator and secondary chamber is followed by an adiabatic quench to cool the gases to approximately 185°F. There are no acid gases present in the flue gas, so the only problem is particulate removal. Because the gas has been cooled to saturation, only wet removal systems are practical. This suggests that a venturi scrubber is a reasonable solution.

1. The flue gas has been examined by an impactor and the results show that 5 wt % of the particles are less than 0.5-μm size while the remaining 95 wt % are larger. The venturi scrubber operates with a 50-in w.c. gas pressure drop at which it will remove 99 percent of all particles larger than 0.5 μm.

   The stack gas emission limitations are 180 mg/dscm corrected to 7% O_2 in the flue gas. Assuming that the flue gas contains 7% O_2, what is the allowable particle concentration at the inlet to the venturi?

It must be assumed that the venturi scrubber will capture only those particles larger than 0.5 μm and none of those smaller, although in fact it may capture some small particles.

2. For small gas flows, the venturi is economical, especially because it is a low capital cost item. But for large flows the horsepower used by the fan would be a high operating cost. An ionizing wet scrubber would require only 4-in w.c. pressure drop per stage. Each stage is 86 percent efficient in particulate removal and the ionizing wet scrubber efficiency is independent of particle size. At the same inlet loading, how many ionizing stages are required to meet regulations?
3. Another option for particulate removal is a fabric filter, or baghouse; however, this cannot be used at saturation temperatures because of certain condensation on the filter bags, which would make bag cleaning impossible. Therefore, a partial water quench would be utilized to bring the flue gas temperature down to 350°F. Assuming a gas flow of 10,000 scfm at 350°F, and a pulse jet baghouse with a 99 percent removal efficiency, what filter bag area would you specify? Air to cloth ratios, i.e., ft^3/ft^2, are usually about 4/1 for a pulse jet baghouse.
4. If each filter bag is 6 in in diameter and 10 ft long, and the baghouse is divided into four chambers, three of which are "on line" and one is "off line" in the cleaning mode, how many bags are required per section?

Solution

1. Let X = weight of the total particulate into the scrubber.
Capture $= 0.99 \times 0.95X = 0.94X$ or 94 percent of the total.
Therefore:

$$\text{Allowable inlet loading } \frac{180 \text{ mg/dscm}}{1.0 - 0.94} = 3000 \text{ mg/dscm}$$

2. If each stage is 86 percent efficient then: $1 - 0.86 = 0.14$ or 14 percent of the inlet concentration would pass through the scrubber, so it would require two stages to reduce 3000 mg/dscm to less than 180 mg/dscm.
3.

$$\text{Gas flow} = 10{,}000 \times \frac{350 + 460}{60 + 460} = 15{,}576 \text{ acfm}$$

If $A/C = 4$ then: $15{,}576/4 = 3894 \text{ ft}^2$ of bag area

4.

$$\text{Area per bag } = 15.71 \text{ ft}^2$$

"On-line" number of bags $= 3894 \text{ ft}^2/15.71 \text{ ft}^2 = 248$ bags
If three compartments are on-line, then $248/3 = 83$ bags each plus an off-line compartment of 83 bags.
From a practical standpoint, baghouses of this size are usually modular in design, so a baghouse of this size or larger would be chosen for this application.

EXAMPLE 12-10. ACID GAS REMOVAL. A hazardous waste incinerator, equipped with a waste heat boiler, delivers 30,000 scfm of flue gas at 600°F. The flue gas is contacted with a lime slurry in a dry scrubber where acid gases are partially neutralized and the gases cooled to 320°F. They then pass through a baghouse for particulate removal and finally are cooled to adiabatic saturation and further neutralized by a 10 percent caustic solution in a conterflow packed tower.

The flue gas contains 700 lb/h of SO_2 and 500 lb/h of HCl. The dry scrubber lime feed rate is 1.6× stochiometric, and it is 80 percent efficient in removing SO_2 and 90 percent efficient in HCl removal.

1. What is the lime feed rate in lb/h?
2. How many lb/h of SO_2 and HCl will be in the flue gas entering the caustic scrubber?

Assume that CO_2 in the flue gas does not react with the lime.
Assuming the lime as CaO (calcium oxide),
$CaO + SO_2 \rightarrow CaSO_3$
$CaO + 2HCl \rightarrow CaCl_2 + H_2O$
Molecular weights: CaO = 56; SO_2 = 64; HCl = 36.5.

Solution

(700 lb/h)/(64 lb/mol) = 10.9 mol/h SO_2 in flue gas
(500 lb/h)/(36.5 lb/mol) = 6.85 mol/h HCl in flue gas

$$SO_2 \text{ requires } 10.9 \text{ mol/h CaO} \times 1.6 = 17.44 \text{ mol/h}$$

$$\text{HCl requires } 6.85 \text{ mol/h CaO} \times 1.6 = 10.96 \text{ mol/h}$$

$$\text{Total lime as CaO} = 28.4 \text{ mol/h}$$

1. Total lime usage = 28.4 mol/h × 56 lb/mol = 1590.4 lb/h
2. 0.2 × 700 lb/h = 140 lb/h SO_2 in flue gas
   0.1 × 500 lb/h = 50 lb/h HCl in flue gas

Note that 50 lb/h is well above the RCRA limit of 4 lb/h.

12-10 INSTRUMENTATION

Introduction

The basic control concepts utilized in a hazardous waste incineration system are essentially the same as those employed in any combination process. These can be quite simple and basic, electronic or pneumatic, but because they are utilized on a hazardous waste incinerator they must be reliable. They must be able to measure all of the system variables and to shut down the system *if* there is any indication of malfunction that might cause a discharge of the hazardous chemicals to the environment. These specific monitors are designed as automatic hazardous waste feed cutoff devices for the incinerator system. The basic operating parameters measured are temperature, flow, pressure, differential pressure, pH, and level.

The control strategy for most incinerators is straightforward. The primary combustion chamber exit gas temperature is controlled automatically by the primary combustion air damper, which in turn is linked with the burner fuel control valve to maintain temperature. The secondary combustion chamber exit temperature is automatically maintained by modulation of the combustion airflow rate and the burner fuel control valve. In a wet APC system the variable venturi throat is modulated to maintain constant pressure drop across the venturi. Draft is controlled by a damper on the induced draft (ID) fan inlet. Minor variations occur from plant to plant. For example, some control room operators prefer to control the draft manually while leaving the primary burner fuel and air on automatic control, the secondary burner fuel on automatic control, and the secondary burner combustion air on manual control. These variations occur as a result of control and incinerator system dynamics.

Temperature

Temperatures are measured by thermocouples installed in standard industrial protection tubes or thermowells. These are to be installed in the combustion gas stream and away from burners to obtain accurate temperature measurement.

Thermocouples are one of the more difficult sensing devices to maintain. Waste streams often create slagging conditions in the incinerator that affect thermocouple readings in both the primary and secondary chambers. Duplicate or dual element thermocouples should be provided in both chambers to indicate major deviations. Often, a variation in readings is noted before a shutdown occurs and replacement can be made upon shutdown. Manual logging of data on an hourly or once-per-shift basis can alert operators to instrument problems by comparing readings to normal expected levels and/or previous readings.

A poor thermocouple location may also cause problems. In some incinerators, primary chamber exit gas thermocouples are located too close to air seals. An increase in leakage created by kiln seal wear or worn airlocks will result in a low temperature at the thermocouple.

An irregular burner flame in the secondary combustion chamber due to a refractory spalling or a liquid atomizer failure may also affect the thermocouple reading. A daily review of this data will often spot the problem before a serious condition occurs.

Flow

Flow measurements are important for all waste feeds, combustion air, scrubbing solutions in the APC system, and flue gas flow rates.

The feed rate of the waste to the incinerator is monitored by mass flow devices, weigh belt scales for soils, and similar instruments. The readout in the control room gives instantaneous feed rate plus integrated totals.

Gas flow rates are measured by differential pressure across an orifice, pitot tubes, annubars, and other methods. These measurements are also important and require regular calibration of the instrumentation.

Pressure and Differential Pressure

Pressures are indicated on standard industrial pressure and vacuum gauges for low pressures and draft and on industrial Bourdon tube gauges for high pressures. In most waste incinerator systems it is important to maintain a negative pressure throughout. Pressure controllers are used to modulate dampers at the inlet of constant speed ID fans or to vary the speed of variable speed fans to maintain a constant pressure at the inlet to the fan.

Differential pressure transmitters and controllers are employed to modulate throat valves in venturi-type scrubbers or to set atomizing steam or air pressures at a constant differential pressure above the waste liquid pressure.

pH

Control of pH in wet scrubbers is important to ensure that the neutralizing agent for acid gas removal is added in sufficient quantities. By maintaining a set pH value, the maximum acid gas concentration of the existing flue gas is controlled effectively.

Level

Level control in wet scrubbing systems, liquid feed tanks, and storage tanks is obviously important. It is more difficult with sludges and corrosive liquids than with clean, noncorrosive wastes, so the selection of the proper instrumentation is important. The options include differential pressure systems such as diaphragms and bubblers, sonic and nuclear sensing units, conductivity sensors, floats, and gauges.

Table 12-8 lists the principal process variables that are monitored in the operation of an incinerator.

Automatic Waste Feed Cutoff

RCRA regulations require that every incinerator provide an automatic cutoff of waste feed streams under conditions that might lead to a discharge of the hazardous waste to the environment.

Instrumentation is provided to monitor process conditions, to provide data for ensuring compliance with regulatory requirements, and to ensure appropriate process response, control, and operational flexibility. The safety interlocks and shutdown features constitute a major portion of the control system. Periodically, the system should be checked to ensure that waste feeds are cut off when one of the process shutdown limits has been exceeded. Most operating permits require that the waste feed cutoffs be tested monthly and the information logged for review by the regulatory agency. The operating conditions under which the automatic waste feed shutoff

TABLE 12-8
Principal process variables monitored

Process variable	Measurements
Feed rate of waste*	Feed inlet
Temperatures	PCC exit gas*
	SCC exit gas*
	Stack gas
	Venturi scrubber inlet gas or baghouse inlet gas
Pressures	PCC feed end draft
	Venturi scrubber pressure drop or baghouse pressure drop*
Flows	Venturi scrubber water flow rate*
	Stack gas velocity*
Stack gas composition	CO*
	O_2*

*Recorded continuously via strip chart and/or digital data logger.

TABLE 12-9
Typical events triggering emergency control systems

Item	Control action
High stack temperature	ES
Loss of fan draft	ES
Loss of water flow to wet scrubber	ES
Power failure	ES
High secondary outlet temperature	ES
Low draft in kiln for > 15 seconds	AWFSO
Low absorber pH	AWFSO
Low secondary chamber temperature	AWFSO
Low oxygen (<3%) in the stack	AWFSO
Low SCC temperature	AWFSO
High CO, > 100 ppm, corr. 7% O_2, 1-h rolling average	AWFSO
High gas flow velocity in stack	AWFSO
High waste feed rate	AWFSO

Notes:
1. ES = emergency shutdown; when ES occurs, fuel and waste feed are stopped.
2. AWFSO = automatic waste feed shutoff
3. In addition to automatic operation, both the AWFSO and ES can be activated manually at the operator's discretion.

(AWFSO) system and emergency shutdown (ES) generally occur are presented in Table 12-9.

12-11 CONTINUOUS EMISSION MONITORS

Parameters most important to incinerator operation are the combustion chamber temperature, stack carbon monoxide, oxygen, and the combustion gas flow rate. They are used as indicators to ensure the proper destruction and removal of the toxic components. For a continuously operated system, such as a commercial disposal facility, redundant instrumentation is necessary to minimize emissions of hazardous chemicals and eliminate or minimize waste feed cutoffs and downtime. The installation, location, calibration, and maintenance of these instruments are important to the continued performance of the system.

RCRA regulations were designed to ensure that incineration systems that treat hazardous wastes operate in an environmentally responsible manner. To ensure that the performance standards are met, a trial burn is conducted. The trial burn identifies the operating conditions necessary to ensure the incinerator's ability to meet or exceed the performance standards throughout the life of the permit.

Currently, monitoring technologies are available to ensure incinerator operation at high combustion efficiency (> 99.9 percent) by continuous measurement of carbon monoxide (CO) and carbon dioxide (CO_2). Monitors measuring waste stream flow rates, combustion airflow, stack gas flow, and oxygen level in the stack also indicate that the total heat release (Btu/h) from the system is within design and permit conditions.

The following monitors are needed to ensure that DRE is maintained at the levels achieved in the trial burn:

- Temperature in the primary and secondary chambers
- Oxygen in the stack
- Carbon monoxide in stack corrected to 7 percent O_2 dry volume
- Stack gas volume flow rate

Instruments that analyze flue gases for CO, CO_2, SO_2, and HCl continuously in the stack are referred to as continuous emission monitors (CEMs). The U.S. EPA prefers extractive systems, which extract a sample of the flue gas from the stack, and condense water vapor before sending the gas to the analyzer. *In situ* CEMs use sensors/analyzers in the stack and are less costly but are typically used for process feedback rather than regulatory compliance. Redundant CEM systems are frequently used to allow maintenance, calibration, and/or replacement of the monitor without shutting down the incinerator.

Carbon Monoxide

The EPA has established a level of 100 ppm CO, rolling average corrected to 7 percent oxygen on a dry basis, as an acceptable level to ensure proper combustion achieving DRE greater than 99.99 percent. This has been determined from many trial burns to be conservative. Accurate CO readings require regularly scheduled calibration of each monitor. The calibration must be done by a qualified instrument technician, trained in this field, and certified calibration gases must be used.

Carbon Dioxide

Carbon dioxide monitors have been added to many incinerators to assess combustion efficiency. The combustion efficiency is calculated as follows:

$$\mathrm{CE} = \frac{\mathrm{CO_2}}{\mathrm{CO_2} + \mathrm{CO}} \times 100\% \tag{12-12}$$

where CO_2 = volume concentration (dry) of CO_2 (parts per million, volume, ppmv)
CO = volume concentration (dry) of CO (ppmv)

In some installations, this is continuously recorded. If the combustion efficiency drops below 99.9 percent, it indicates that CO levels are probably increasing, which alerts the operator to take corrective action before the high CO triggers automatic waste feed cutoff.

Oxygen

Oxygen analysis, when combined with stack gas volume flow, can be used as a means of determining total heat input to the system. An increase in oxygen at the maximum stack gas flow rate indicates a reduction in the maximum Btu/h heat release. This may be traced to kiln seal wear or leakage to the APC system. Incinerator

systems that use ionizing wet scrubbers for particulate removal in the APC system have this air leakage. Also air in-leakage through the ID fan shaft seal (which occurs at high-draft conditions needed with a venturi scrubber) will result in a higher O_2 level in the stack than in the SCC outlet. If the O_2 level drops at the maximum stack flow rate, total heat release may be increasing, because of a failure of the waste flow controllers, a "hot" charge included in the solid waste feed, or failure of an atomizing nozzle. All these events provide more fuel into the incinerator, which leads to the increase in heat release.

EXAMPLE 12-11. OXYGEN CORRECTION. Current hazardous waste regulations require that emissions be corrected to 7 percent oxygen in the stack on a dry basis.

The carbon monoxide (CO) concentration in the stack from a hazardous waste incinerator is measured at 20 ppm by volume at a temperature of 175°F. The oxygen concentration in the stack is measured to be 12 percent by volume on a wet basis. In another test the particulate concentration in the stack is measured at 20 mg/dscf at a stack oxygen concentration of 12 percent measured on a dry basis at 70°F. The water content of the stack gas is 10 mol %.

1. What is the corrected CO content in the stack gas?
2. What is the corrected particulate concentration?
3. Would this stack meet federal regulations (180 mg/dcsm corrected to 7% O_2)?

Solution

O_2	12%
H_2O	10%
$N_2 + CO_2$	78% (by difference)

On a dry basis the oxygen content is $12/(1 - 0.10) = 13.33\%$

1. The correction formula provided in the regulations (see 40 CFR Part 264.343) is as follows:

$$P_c = P_m \times \frac{14}{21 - Y}$$

where P_c = corrected concentration
P_m = measured concentration
Y = O_2 concentration in stack (dry basis)

Therefore:

$$P_c = 20 \times \frac{14}{21 - 13.33} = \underline{\underline{36.5 \text{ ppm of CO}}}$$

2. Since the O_2 in the second test is measured on a dry basis, no correction for water content is required.

$$P_c = 20 \times \frac{14}{21 - 12} = \underline{\underline{31.1 \text{ mg/dscf}}}$$

3. $31.1 \text{ mg/ft}^3 \times 35.3 \text{ ft}^3/\text{m}^3 = \underline{\underline{1098 \text{ mg/dscm}}}$
No, it would be well in excess of regulations.

Hydrogen Chloride

In the United States RCRA regulations require 99 percent removal (or 4 lb/h maximum emissions) of HCl. The new guidelines[17] for HCl and Cl_2 emissions limits

are risk-based which may result in lower emission limits. Most incineration systems do not include a continuous HCl monitor. The HCl is sampled and analyzed during the trial burn according to the maximum chlorine content in the feed. APC systems which use wet scrubbers for acid gas control employ pH control of the scrubbing solution to ensure that the HCl is removed.

For the dry APC systems, proper HCl removal efficiencies are likely provided: (1) combustion gas flow, (2) neutralizing agent flow, (3) baghouse/electrostatic precipitator (ESP) gas pressure drop are monitored and within the proper limits and (4) the chlorine level in the feed is not higher than that tested in the trial burn.

HCl monitors have been developed in the past few years. They are used routinely in Western European countries for monitoring controlled and uncontrolled HCl emissions from municipal and hazardous waste incinerators. The German government has established HCl standards and relies on the CEM data to demonstrate compliance. Experience in the United States with HCl monitors, however, is limited.

Total Hydrocarbons

There is currently no U.S. federal limit to total hydrocarbon (THC) emissions from hazardous waste incinerators, but some EPA regions and some states are willing to allow operators to use THC as an indicator of proper operation in place of the CO rolling average, especially with batch-fed rotary kilns. When the operator cannot comply with the 100 ppm 60-minute rolling average, a maximum 20 ppmv THC limit has been acceptable.

Combustion Gas Velocity

RCRA requires that flue gas velocity be monitored and recorded. The velocity is an indirect measure of total heat release, and high gas velocities usually mean higher particulate carryover in the exit gas.

Measuring combustion gas velocity is not a simple task. The usual direct methods are pitot tubes, annubars (which are pitot tubes that average velocity pressure across a duct), and vortex-shedding flow meters. These all work well in clean gas at ambient temperature but not as well at high temperatures, in corrosive conditions, or in dirty gas streams where they plug and foul. They are also sensitive to disturbances in flow patterns upstream of the probe.

The simplest and most reliable method for measuring combustion gas flow is by the induced draft (ID) fan motor amperage. During stack tests, four or more EPA Method 1[18] velocity traverses are made, each at a different ID fan damper setting, and hence, different gas flow rate. The data are then plotted and correlated with the motor amperage. This method works well for the majority of incinerator installations which use a constant-speed centrifugal fan and an inlet VANE-type damper to control flow. A system using a variable-speed fan would require a more complex algorithm incorporating fan rpm, pressure drop, and amperage to calculate volumetric gas flow rate.

Noncontinuous Parameters

At present, the RCRA standard for particulate emissions from incinerators is 0.08 grains/dscf corrected to 7 percent O_2. It is well recognized that particulate emissions may contain trace quantities of toxic metals. Concern about health hazards associated with metal emissions, both to the atmosphere and by ash leaching into the groundwater, has led to regulatory "guidelines."[19]

There are no CEMs that adequately provide information related to the quantitative particulate content of a flue gas. Opacity meters are used in the power industry for coal and oil-fired boilers, but are reliable only for dry stack gases. There are no federal requirements for continuous opacity measurement on RCRA incinerators. Some incinerators have installed opacity meters as an additional source of operating information.

12-12 TRIAL BURNS

Trial burns must be run on all hazardous waste incinerators in accordance with 40 CFR Part 270.62. The trial burn is run to verify that the incinerator is capable of operating in compliance with performance standards outlined in 40 CFR Part 264.343.

An analysis of the flue gas from the operating unit must be made to demonstrate a destruction and removal efficiency of 99.99 percent or greater for principal organic hazardous constituents contained in the waste.

The flue gas must also be sampled and analyzed to determine the amount of particulate present in terms of weight units per unit of dry volume corrected to 7 percent oxygen. The federal standard is currently 0.08 grains/dscf. Many states have more stringent emission standards for particulate.

Hazardous waste incinerators must also conform to HCl emission limits of no more than 4 lb/h or 99 percent removal of all HCl in the flue gas, but recent guideline changes will place these emissions on a risk basis and include an HCl/Cl_2 total flue gas content.

Carbon monoxide emissions may not exceed 100 ppmv on the basis of a 60-minute rolling average corrected to 7 percent oxygen. An exception to this regulation is made when the operator is willing to perform a risk assessment and monitor and maintain total hydrocarbons below 20 ppmv.

Flue gas analysis for oxygen is necessary to provide a correction for the CO and particulate analysis. Flue gas analysis for CO_2 is required for TSCA trial burns (i.e., substances such as PCBs that come under the Toxic Substances Control Act) to determine combustion efficiency.

Emission limits on 10 priority metals[20] have been listed in the EPA guidelines. Permissible feed rates and emission limits have been established for these metals and may not be exceeded unless a dispersion analysis and risk assessment is made for all metals exceeding these levels. Individual state air regulations may require analysis for other metals also.

Sampling and analysis for other pollutants such as sulfur oxides and NO_x may also be required during the trial burn if they are expected to be present in any

appreciable quantity. In some trial burns, the EPA or the state has required sampling and analysis for dioxins (PCDDs) and furans (PCDFs).

Sampling for all of these parameters is done in the stack of the incinerator. Samples are taken, and analyses are subsequently made in the laboratory except for O_2, CO, and CO_2, which must be measured in the field.

12-13
MOBILE SYSTEMS

The idea for a transportable or mobile incinerator developed during the Superfund program (see Section 2-4). Incineration is a reasonable method of contaminated soil remediation; however, it is no simple task to dig up an entire site and transport it to one of the few commercial hazardous waste incinerators in the United States. A site not close to any commercial disposal facility might spend more for transportation than for any other aspect of the cleanup program.

By placing the incinerator on wheels (mobile system), or making it transportable in modules, the incinerator can be transported to the Superfund site rather than moving the contaminated soil from the site to the incinerator. The idea caught on quickly, and a number of companies entered the field with a variety of portable and mobile units.

Today, available transportable systems include rotary kiln systems, fluid-bed incinerators, and infrared units. Most completely mobile units are small rotary kilns. These systems can be set up on-site and ready for operation in a time period within 1 week to several months.

The technology has developed into three basic areas:

- Incineration where the soils are heated to a temperature of more than 1200°F and the subsequent vapors burned in a secondary chamber (see Figure 12-30).
- Incineration of non-RCRA wastes such as contaminated soil from underground storage tanks (USTs) that contain fuel oil or gasoline. Many of these systems merely devolatilize the soils at lower temperatures and burn the flue gas as a VOC.
- Low-temperature devolatilizers that indirectly heat the waste to vaporize the organics and then condense the hydrocarbon vapors for reclamation.

Mobile incineration has been widely used for site cleanup. For example, a wood chemical treatment facility in Mississippi where telephone and electric poles had been treated with creosote was found to have a significant amount of creosote-contaminated soil. A mobile/transportable rotary kiln incinerator was used to process over 9000 tons of the contaminated soil, destroying the creosote and other organic contaminants and providing clean soil that could be landfilled on site.

A mobile incinerator was used to remediate soils and sediment contaminated with VOCs, PAHs, pesticides, and other organic compounds at the Baird and McGuire site in Massachusetts. A successful trial burn was held in January 1995 and full-scale operation of the incinerator began in June of the same year. By the time the project was completed 2 years later, a total of 248,000 tons of contaminated soil and sediment had been incinerated.

Many other mobile thermal technologies have been proposed for industrial and commercial use. Several of these are still attempting to find a niche in the market-

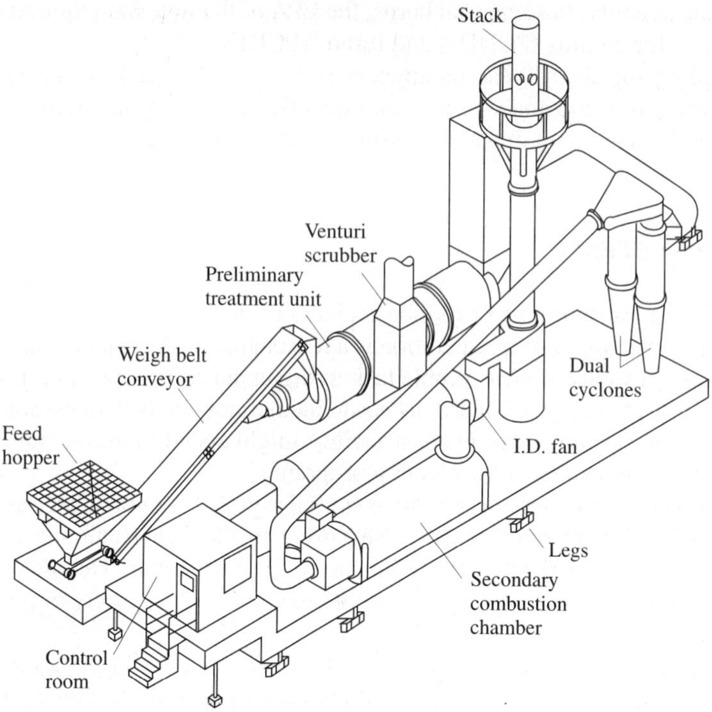

FIGURE 12-30
Mobile (transportable) incinerator.

place but their widespread use has been limited, due in part to economics or capacity restrictions. The U.S. EPA SITE program has provided some of the more promising technologies an opportunity to prove their commercial viability. For example, a mobile thermal desorption system using rotary kiln technology was tested at a demonstration project and DRE of 99.99 percent was demonstrated. After the SITE demonstration, the technology was used on at least two more sites.[21]

Some incineration technologies have become commercially viable. There are literally reams of literature and operating data available to the interested scholar concerning these "emerging" systems but their practical application has been limited.

DISCUSSION TOPICS AND PROBLEMS

12-1. 1500 lb/h of a waste mixture of 40 percent benzene, 50 percent toluene, and 10 percent water is burned with 25 percent excess air. Determine the total heat release and the percent by volume of each component in the flue gas.

12-2. The mixture in the table below is being incinerated at 2200°F with 50 percent excess air, a residence time of 2.3 seconds, and a stack flow rate = 37,500 dscfm. Principal organic hazardous constituents (POHCs) for this waste are benzene, tetrachlorophenol, and toluene. Determine if this emission meets requirements and comment on the results. The following table provides incinerator emissions. Oxygen in the flue gas is 7.0 percent.

Compound	Formula	MW	Inlet lb/h	Outlet lb/h
Benzene	C_6H_6	78.11	1025	0.087
Chlorobenzene	C_6H_5Cl	112.5	278	0.034
Ethylbenzene	C_8H_{10}	106.17	780	0.089
Tetrachlorophenol	C_6HOHCl_4	231.9	760	0.056
Toluene	C_7H_8	92.10	756	0.024
Xylene	C_8H_{10}	106.17	168	0.204
HCl		36.45		4.3
Particulates				20.3

12-3. Using the flue gas generated in Example 12-1 calculate the flame temperature in the incinerator if there is a 10 percent heat loss.

12-4. Good incinerator design provides for a flue gas residence time of 2 seconds in a liquid incinerator and a gas velocity of 20 ft/s. Using the information provided in Example 12-1, determine the inside diameter and length of the incinerator. Assume an incinerator temperature of 2679°F and a gas velocity of 20 ft/s.

12-5. Given the following stack data: CO = 17 ppmv at 190°F; O_2 = 10% by volume (wet basis); particulates are 11 mg/dscf. Determine the CO and particulate concentration corrected to 7% O_2. Assume the moisture content of flue gas is 10%.

12-6. Estimate the heat released from a rotary kiln incinerator 12 ft inside diameter × 45 ft long.

12-7. An incinerator that normally burns waste with a heating value of 700 to 1000 Btu/lb and is designed to handle 2×10^7 Btu/h must burn a rubber waste with a heating value of 12,000 Btu/lb. Estimate the maximum size of a waste batch for this system to perform effectively. Assume the residence time in the incinerator is 6 min.

12-8. Determine how much caustic will be required to reduce SO_2 emissions from the incinerator in Example 12-10 to 25 tons/year at 98 percent efficiency in the packed column scrubber if the operating year is 7200 hours.

12-9. How much additional caustic is required to reduce the HCl emissions from the incinerator in Example 12-10 to less than 4 lb/h?

12-10. A VOC thermal incinerator operates at 1500°F. How much natural gas at 1000 Btu/ft^3 will be required to incinerate the emissions from the film dryer described in Example 12-6. Assume the diluted mixture of acetone vapor and air is at 80°F.

12-11. Determine the adiabatic saturation temperature and cooling water flow for a spray tower. Assume the flue gas is essentially air at 1800°F and contains 12 percent water by weight. The total gas flow is 80,000 acfm. The specific weight of air at 60°F is 0.0763 lb/ft^3.

12-12. An incinerator burns 150 lb/h chlorobenzene and has a stack gas containing 5 percent oxygen by volume on a dry basis. Determine the percent excess air.

12-13. Discuss briefly the types of particulate removal devices that might be appropriate to a commercial hazardous waste incinerator using a rotary kiln. What factors would control the selection of air pollution control equipment?

12-14. An incinerator releases 8000 scfm of flue gas containing particulates that must be removed by fabric filtration. Assume an air-to-cloth ratio of 4.0 and a baghouse containing six modular units with one always off-line. Determine the number of bags per module.

REFERENCES

1. The Resource Conservation and Recovery Act of 1976, 42 USC 6901 et seq., 1976.
2. The Hazardous and Solid Waste Amendments of 1984 (HSWA) (P.L. 98-616), November 9, 1984.

3. HSWA, op. cit., Sect. 3004(o)(1)(B).
4. *Code of Federal Regulations* (40 CFR 270.19).
5. *Code of Federal Regulations* (40 CFR 264, Subpart O).
6. *Code of Federal Regulations* (40 CFR 264.343).
7. Perry, R. H., and D. W. Green, eds.: *Perry's Chemical Engineers' Handbook,* 6th ed., McGraw-Hill, New York, 1984, pp. 4–53.
8. Theodore, L., and J. Reynolds: *Introduction to Hazardous Waste Incineration,* John Wiley & Sons, New York, 1987.
9. Nittetu Chemical Engineering, Ltd., Tokyo.
10. West, R. C., ed., *Handbook of Chemistry & Physics,* 64th ed., CRC Press, 1983.
11. *North American Combustion Handbook,* North American Manufacturing Co., Cleveland, Ohio, 1979.
12. Flaregas, Inc.
13. Ross, R. D.: "Incineration of Solvent-Air Mixtures," *Chemical Engineering Progress,* August 1992.
14. Reeco, Morris Plains, N.J.
15. National Air Oil Co., Philadelphia.
16. T-Thermal Co., Conshohocken, Pa.
17. *Guidance on Metals and Hydrogen Chloride Controls for Hazardous Waste Incinerators,* vol. IV, U.S. EPA 1989.
18. *Code of Federal Regulations,* Title 40, Parts 60 and 61.
19. *Guidance on Metals and Hydrogen Chloride Controls for Hazardous Waste Incinerators,* vol. IV, U.S. EPA, 1989.
20. Ibid.
21. U.S. EPA: *Technology Profiles,* 10th ed., Vol. 1, *Demonstration Program: Superfund Innovative Technology Evaluation Program,* PEA/540/R-99/500a, February 1999.

ADDITIONAL READING

Bayer, Johan E.: "Incinerator Operations," *Journal of Hazardous Materials,* vol. 22, no. 2, pp. 243–247, November 1989.

Brunner, Calvin: "Industrial Sludge Waste Incineration," *Environmental Progress,* vol. 8, no. 3, pp. 163–166, August 1989.

Carnes, Richard A.: "RCRA Trial Burns: Adventures at Rollins," *Journal of Hazardous Materials,* vol. 22, no. 2, pp. 151–160, November 1989.

Cooper, Eliot D.: "Incineration Permitting: The Long and Winding Road," *Hazardous Materials & Waste Management,* January–February 1986, pp. 10–13.

Cundy, V. A., T. W. Lester, C. Leger, G. Miller, A. N. Montestruc, S. Acharya, A. M. Sterling, D. W. Pershing, J. S. Lighty, G. D. Silcox, and W. D. Owens: "Rotary Kiln Incineration—Combustion Chamber Dynamics," *Journal of Hazardous Materials,* vol. 22, no. 2, pp. 195–219, Nov. 1989.

Cundy, V. A., T. W. Lester, A. M. Sterling, J. S. Morse, A. N. Montestruc, Christopher B. Leger, and S. Acharya: "Rotary Kiln Injection I. An In Depth Study—Liquid Injection," *Journal of the Air Pollution Control Association,* vol. 39, no. 1, pp. 63–75, January 1989.

Freeman, Harry: "Recent Advances in the Thermal Treatment of Hazardous Wastes," *Management of Hazardous and Toxic Wastes in the Process Industries,* vol. 19, no. 4, pp. 34–36, 1987.

Leuser, R. M., L. A. Valazquez, and A. Cohen: "Remediation of PCB Soil Contamination by On-Site Incineration," *Journal of Hazardous Materials,* vol. 25, pp. 375–385, December 1990.

McFee, John, George Ramussen, and Craig Young: "The Design and Demonstration of a Fluidized Bed Incinerator for the Destruction of Hazardous Organic Materials in Soils," *Journal of Hazardous Materials,* vol. 12, no. 2, pp. 129–142, November 1985.
McGowan, T., and R. Ross: "Hazardous Waste Incineration Is Going Mobile," *Chemical Engineering,* vol. 98, pp. 114–123, October 1991.
Munoz, Henry, Frank Cross, and Joseph Tessitore: "Comparisons Between Fluidized Bed and Rotary Kiln Incinerators for Decontamination of PCB Soils/Sediments at CERCLA Sites," *Proceedings of the National Conference on Hazardous Wastes and Hazardous Materials,* pp. 424–245, March 4–6, 1986.
Santoleri, Joseph J.: "Design and Operation of Thermal Treatment Systems for Contaminated Soils," *Proceedings—A&WMA Annual Meeting,* Air & Waste Management Association, pp. 79–93, 1989.
Santoleri, Joseph J.: "Rotary-Kiln Incineration Systems: Operating Techniques for Improved Performance," *Hazardous and Industrial Wastes—Proceedings of the Mid-Atlantic Industrial Waste Conference,* Technomic Publishing Company, Inc., pp. 743–758, 1990.
Stoddart, Terry L., and Jeffrey J. Short: "Dioxin Surrogate Trial Burn: Full Scale Rotary-Kiln Incinerator for Decontamination Soils Containing 2,3,7,8-Tetrachlorodibenzo-P-Dioxin," *Chemosphere,* vol. 18, no. 1–6, pp. 355–361, 1989.
Takeshita, R., and Y. Akimoto: "Control of PCDD and PCDF Formation in Fluidized Bed Incinerators," *Chemosphere,* vol. 19, no. 1–6, pp. 345–352, 1989.
Theodore, L., and J. Reynolds: *Introduction to Hazardous Waste Incineration,* John Wiley & Sons, New York, 1987.
Tilly, Jean, Gary Dietrich, and Dan Pyne: "Exposure and Risk Assessment for a Proposed Hazardous Waste Incinerator," *Environmental Progress,* vol. 8, no. 3, pp. 207–211, August 1989.
U.S. Environmental Protection Agency, Office of Solid Waste: *Permitting Hazardous Waste Incinerators.* EPA/530-SW-88-024, Washington, D.C., April 1988.
Wiley, S. K.: "Incinerate Your Hazardous Waste," *Hydrocarbon Processing,* vol. 66, no. 6, pp. 51–54, June 1987.
Wood, R. W., R. E. Bastian, et al.: "Rotary Kiln Incinerators: The Right Regime," *Mechanical Engineering,* vol. III, no. 9, pp. 78–81, September 1989.

ACKNOWLEDGMENTS

Chapter 12 was prepared by Richard D. Ross and Joseph J. Santoleri of Four Nines, Inc. The authors wish to acknowledge the following individuals who assisted with the preparation of this chapter for the first edition: Jack Riggenbach, *ERM,* and Debra J. Brovero, *Bucknell University*. The principal reviewer for the first edition was Paul H. Woodruff, P.E., *ERM.*

The principal reviewer for the second edition was Arbind Prasad, *ERM.*

CHAPTER 13

Land Disposal

Now I would give a thousand furlongs of sea for an acre of barren ground.
Shakespeare

13-1 INTRODUCTION

Landfills will remain a significant part of our hazardous waste management practice. A landfill is defined as that system designed and constructed to contain discarded waste so as to minimize releases of contaminants to the environment. Landfills are necessary because (1) other hazardous waste management technologies such as source reduction, recycling, and waste minimization cannot totally eliminate the waste generated and (2) hazardous waste treatment technologies such as incineration and biological treatment produce residues. Thus the generation of hazardous waste cannot be reduced to zero in the foreseeable future. It is with this inevitability in mind that the topic of land disposal is addressed.

Historically, many land disposal facilities did not provide adequate environmental protection. As a result, the total environmental and societal costs of landfilling were not considered. Society is now paying these costs in the form of cleaning up many old sites, such as those identified by the Superfund program in the United States and by contaminated land inventories in the European Community. This has heightened the public sensitivity to the construction of new storage and land disposal facilities. With this historical perspective, what are the responsibilities of modern managers of hazardous waste? It is important to promote both minimization of the amount of hazardous waste generated at the source and treatment of the irreducible minimum. For unavoidable hazardous waste and hazardous residue from waste treat-

ment processes, it is necessary to design and construct safe land disposal facilities. Safe facilities must employ the best available technologies, incorporating state-of-the-art design and analysis methods.

When establishing guidelines for the design, construction, and operation of safe disposal facilities, the designers must anticipate imperfection and build physical and institutional redundancies to accommodate inevitable leakage. The goal must be a facility that minimizes releases of contaminants into the environment to an acceptable, perhaps nondetectable, rate. This contaminant release rate could be considered a geologically slow rate. Remember, many contaminants, particularly toxic metals, originated in our natural environment. At low concentrations many of these metals are not only nontoxic but are necessary for the sustenance of life.

Land disposal, as a *principal* hazardous waste management technique, is a vestige of the past, since the present emphasis on landfilling is reduced. However, land disposal remains a necessary part of any hazardous waste management system for residuals and is an attractive alternative as a component of remediation of uncontrolled hazardous waste sites. Professional waste management requires a decision on the final destination of these materials. Confining our choices to planet Earth, we have only three options: air, water, or land. These three media are interrelated (e.g., what is initially disposed of on the land may end up in the water). Nonetheless, our decisions should be made in full recognition of our options: air, water, or land?

EXAMPLE 13-1. JUSTIFICATION OF LAND DISPOSAL. Select a hazardous waste treatment technology described in the previous chapters and identify a potential need for use of land disposal facilities in conjunction with the selected treatment technology.

Solution. Incineration: The incineration of hazardous waste creates a number of secondary waste products, ash the most obvious. In addition, depending on the incinerator design and the wastes being burned, residues from air emission controls may require landfilling.

It is useful to differentiate between land disposal and land treatment, often known as land farming. The concept behind **land disposal** is that landfilling represents the ultimate fate of the waste. Land disposal also includes deep well injection, a topic only briefly covered later in this chapter. **Land farming** is a treatment technique wherein biologically degradable wastes are placed onto the land in low concentrations so that microbes can degrade the materials.

It is important to note the difference between land disposal facilities and storage facilities. Land disposal facilities represent a hazardous waste management technique that constitutes a final placement of the waste. In contrast, **storage facilities** represent a temporary management technique where the waste has not yet reached its final destination. Storage facilities hold waste prior to its shipment for treatment such as incineration, physical-chemical treatment, or reuse and recycling.

Examples of storage facilities include pits, ponds, lagoons, tanks, piles, and vaults. Storage facilities are designed such that waste can be retrieved whereas land disposal facilities are designed to function in perpetuity. Storage facilities and disposal facilities employ many of the same precautions in design and construction. This chapter focuses on land disposal; it should be noted that the approach employed for land disposal facilities (to minimize the rate of contaminant migration into the environment) is often the same approach employed for the design and construction of storage facilities.

U.S. Federal regulations under Subtitle D of RCRA define a municipal solid waste landfill as "a discrete area of land or an excavation that receives household waste," and regulations have been developed to keep regulated hazardous waste out of municipal waste landfills.[1]

The design of municipal solid waste landfills will not be explicitly covered in the textbook, even though municipal solid waste landfills are used worldwide. However, because municipal solid waste landfills receive significant quantities of household hazardous waste, many of the design concepts and materials employed for hazardous waste land disposal facilities are also employed for municipal solid waste facilities. Note that municipal solid waste landfills can be (but typically are not) designed and operated as bioreactors because much of the waste is decomposable.[2] The overall design of secure land disposal facilities includes (1) control of the top to minimize air emissions and infiltration of precipitation and (2) control of the bottom to maximize the collection of leachate and minimize contaminant transport through the bottom.

Although many design configurations are possible, a simplified cross section of a hazardous waste landfill, presented on Figure 13-1, depicts the major components

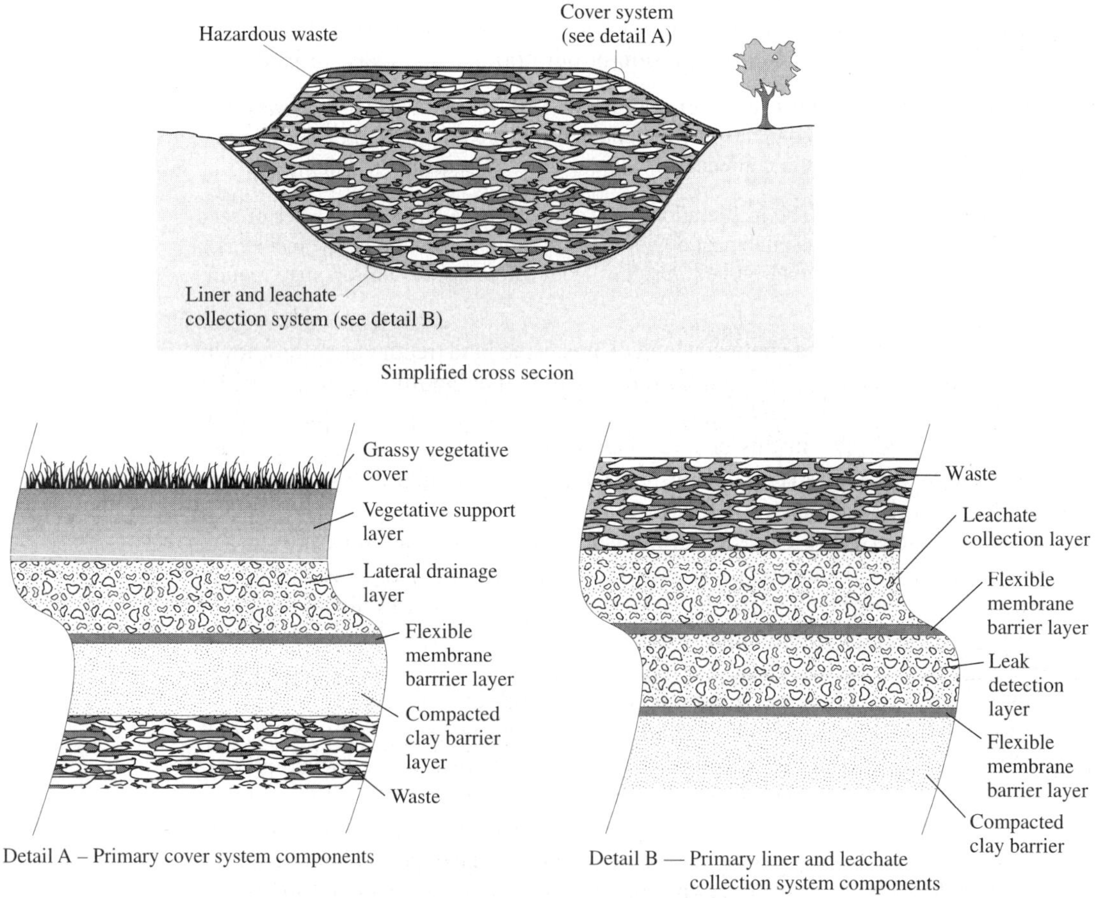

FIGURE 13-1
Simplified cross section of a hazardous waste landfill with details A and B.

of the systems engineered to protect the environment from the landfill wastes. The basis of design and analysis for the components of a landfill system are presented in this chapter. Discussed will be both (1) the components of storage and disposal facilities and their function and (2) the materials and their properties. Land disposal regulations and their history are covered in Sections 2-2 and 2-3; therefore, the focus of this chapter is only on the technologies employed in the design and construction of land disposal facilities.

Quality control (QC) and quality assurance (QA) are necessary for the construction of a land disposal facility that is protective of human health and the environment. QC is an inspection plan used to directly monitor the quality of construction (CQC) or manufacturing (MQC). QA is a plan of activities such as audits, which are independent of QC inspections, and serves to assess whether the construction quality assurance (CQA) or manufacturing quality assurance (MQA) is resulting in a product meeting the plans and specifications. While QC and QA are essential parts of the process to construct a land disposal, details are provided elsewhere.[3]

13-2
LANDFILL OPERATIONS

The management of hazardous waste in land disposal facilities requires the **tracking** of the waste, that is, the recording of the journey of the waste from the time it is generated to its ultimate disposal site. This tracking extends to the recording of the waste's location within the final disposal site—in smaller mappable units called **cells**. The cells are laid out in a grid pattern. Records are maintained, including who provided the waste, the nature of the waste, and where and when the waste was landfilled.

The principal reason for the careful tracking and disposal in cells is to ensure **waste compatibility**. Many wastes may react with each other, resulting in the potential for heat, combustion, and/or toxic fumes. By tracking the type of waste and its location within the land disposal facility, a waste compatibility check can be made to provide safe operating conditions for landfill operators as well as a safe long-term disposal to minimize or eliminate potentially harmful chemical reactions. A detailed compatibility chart is presented as Figure 8-10. Additional purposes of this detailed tracking include future resource recovery and/or future recovery for alternative treatment, as well as the ability to pinpoint defects in the system should specific contaminants show up in the environment at some time in the future (i.e., identify the source of the leak).

The wastes disposed in a landfill are in both bulk and containerized forms. Drummed wastes are typically aligned and covered with other wastes, such as contaminated soils or sludges, using care not to damage the drums. Drums may be stacked or may be placed in single lifts. In all cases the wastes are in solid, not liquid, form.

EXAMPLE 13-2. LAND DISPOSAL CELLS. A shipment of petroleum sludge, generated by a process that employs sulfuric acid to refine crude oils, arrives at Landfills-R-Us for disposal. The operator decides that it would be useful to dispose of this in the area containing sludge from an old storage lagoon from an electroplating wastewater treatment plant. What do you think?

Solution. Not a good idea because electroplating sludge would be expected to contain, among other things, cyanide. Acidifying the electroplating waste would generate hydrogen cyanide (HCN). HCN, a gas at slightly above room temperature, can enter the body by inhalation and skin absorption. It is highly toxic—lethal concentrations have been reported as low as 10 ppm.

Operational considerations may include the application of a **daily cover** placed at the close of each operational day. A daily cover typically consists of soil 0.3 m (1 foot) thick. The purpose is to minimize odor, airborne transport of contaminants, and potential for direct contact, and maximize aesthetics. Although the advantages of a daily cover are clear, it is also clear that daily cover, that is, placement of uncontaminated clean soil on a daily basis in a hazardous waste landfill, is both expensive and uses up valuable landfill space. Further, daily cover creates a highly anisotropic environment in the landfill and, as a result, seepage often exits along the relatively permeable daily cover layers and through the side slopes. Thus, the concept of daily cover for hazardous waste facilities is debatable in balancing the advantages with the disadvantages.

EXAMPLE 13-3. LAND DISPOSAL DAILY COVER. Landfills-R-Us asks: How much material do we need for daily cover for our proposed hazardous waste landfill? How much of our total volume (air space) will be used by the daily cover during the life of the landfill? How much longer could the landfill be used if daily cover were not employed? The landfill is 500 m by 250 m in average plan and 15 m high. Daily filling rates are 15 m by 10 m by 2 m and the daily cover is 0.3 m.

Solution. The total landfill air space is:

$$V_1 = 500 \times 250 \times 15 = 1.88 \times 10^6 \text{ m}^3$$

Each daily layer (lift) has a volume of:

$$V_2 = 15 \times 10 \times 2 = 300 \text{ m}^3$$

The daily cover has a volume of:

$$V_3 = 15 \times 10 \times 0.3 = 45 \text{ m}^3$$

The volume of air space (in percent) used by daily cover is:

$$V_{pc} = (V_3/V_1) \times 100 = \frac{45}{300} \times 100 = 15\%$$

The life of the landfill with daily cover is:

$$t_1 = V_1/(V_2 + V_3) = 5449 \text{ days}$$
$$= 21 \text{ years at 5 days per week}$$

The life of the landfill without daily cover is:

$$t_2 = V_1/V_2 = 6267 \text{ days}$$
$$= 24.1 \text{ years}$$

Therefore the landfill life will be extended by:

$$t_3 = t_2 - t_1 = 818 \text{ days or more than 2 years}$$

Operating land disposal facilities requires the use of heavy equipment, with the proper protective health and safety gear for the operators. The technology for land disposal facility operations is based on the routine technologies employed for construction. Typically, bulldozers, compactors, and trucks developed for heavy construction are used for land disposal operations. Often, modifications are made with respect to equipment weight and blade design to accommodate their use in a hazardous waste landfill. Forklifts and barrel snatchers may be used to handle containerized waste. In all cases, modifications are made to accommodate health and safety requirements for the operators.

As a consequence of precipitation while a cell is being filled and of infiltration after a cell is closed, leachate is generated in landfills. This necessitates the collection and treatment of leachate as an integral part of any hazardous waste land disposal facility. The unit processes utilized in leachate treatment are the same as those described in Chapters 9 and 10 for other liquid waste streams. Typically, the leachate characteristics and flow rates are definable and have characteristics to other industrial and hazardous wastewater. The predictability of the leachate characteristics increases for cells designated for specific wastes. Experience has shown that even a mixed waste cell will generate a more predictable leachate than those liquids from uncontrolled hazardous waste sites.

One of the guidelines affecting the design, construction, and operation of land disposal facilities is that long-term monitoring data are needed to provide effective environmental protection. Land disposal facilities include groundwater and air quality monitoring along with monitoring of leachate quality and quantity. The monitoring programs are designed and implemented to establish the performance of the landfill with respect to leachate generation and contaminant migration. The monitoring begins prior to landfill construction, and continues during its operation and long after its closure.

13-3
SITE SELECTION

The selection of a hazardous waste disposal site is a highly charged political topic, as discussed in Section 8-4. This chapter will briefly focus on some of the technical aspects of storage and disposal facilities. From a technical standpoint, the site selection should incorporate considerations of geology, seismology, hydrogeology, hydrology, population density, environmental receptors, and transportation. The natural environment of the site should exhibit characteristics that provide **redundancy** to the engineered systems designed to protect the public health and the environment. For example, an area exhibiting an upward gradient from deeper aquifers would, in the case of a spill or unforeseen leakage, carry contaminants away from the aquifer as opposed to into the aquifer. Locating the landfill within a clay formation or the presence of aquitards immediately beneath the site offers a level of natural protection.

The intrinsic value of the site environment also needs to be considered. For example, there are many areas in the United States where the groundwater is unsuitable for use because of background levels of naturally occurring chemical constituents. Sites such as these constitute better sites than those where the underlying groundwater aquifers are of high quality. Likewise, brownfield sites may make more environmental sense than greenfield sites.

In addition to the subsurface hydrology and geology, there are surface water hydrology considerations. Undoubtedly the construction of an engineered final cover will increase surface water runoff. Both the capacity and sensitivity of the receiving waters to surface water runoff and to leachate discharge after treatment require consideration. Streams of high value need special consideration, such as those that constitute drinking water supplies and/or those used for sport fishing.

In addition to these issues, the logistical issue of transportation (i.e., road capacity and nearness to regional markets) ought to be considered during the site selection. Additional discussion of site selection, including such considerations as adjoining land use, floodplains, wetlands, habitats for endangered species, breeding or stopping areas for migratory birds, prime agricultural land and proximity to human habitats (population densities, schools, hospitals, nursing homes and the like), for hazardous waste facilities is included in Section 8-4.

As siting becomes increasingly difficult, engineers are using the double-lined composite liner technologies developed for hazardous waste applications for municipal solid waste landfills. An example is the use of an abandoned limestone-dolomite quarry by the City of Bristol, Virginia.[4] This quarry averaged a depth of 350 feet with near-vertical sidewalls. The bottom liner system included, from top to bottom, an 18-inch-thick leachate collection system, a 60-mil high-density polyethylene (HDPE) geomembrane liner, a 3-foot compacted clay liner, a 12-inch-thick "witness" zone, a 2-foot-thick compacted clay liner, and a filter geotextile. Sidewalls were lined with a double-sided geocomposite and a 60-mil low-density polyethylene (LDPE) geomembrane. The following section of this chapter presents the liner and leachate collection systems and their function, and describes the materials typically used for each of these layers.

13-4
LINER AND LEACHATE COLLECTION SYSTEMS

It can be confidently said that more energy, effort, and resources have gone into the development of liner systems for hazardous waste disposal facilities than any other technical aspect of land disposal facilities. The purpose of a **liner** is to provide a barrier to minimize migration of contaminants. A liner with 100 percent effectiveness will prevent chemical constituents from migrating across the liner system into the environment. Because no liner is 100 percent effective, the engineered liner systems discussed below offer redundancy to accommodate imperfections. Because some leachate will inevitably be generated, leachate collection is required. Therefore, the landfill bottom consists of (1) alternating layers of materials to provide barriers to contaminants attempting to migrate from the landfill and (2) layers providing collection of these contaminants through collection systems (see Figure 13-1).

The complexity of liner systems has evolved as new materials are developed and as preceding generations of liner systems have proved to be inadequate. At the present time, landfill bottom systems have evolved in the United States to the multiple layer system shown in Figure 13-2.

The concepts behind the performance of the system as shown in Figure 13-2 are described and the terminology is made a bit more precise in the following paragraphs.

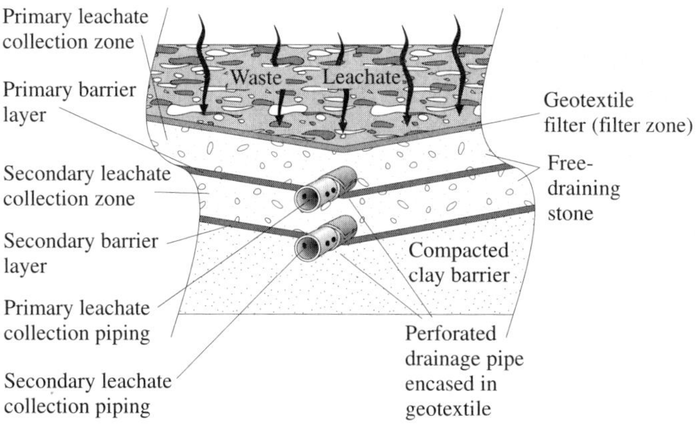

FIGURE 13-2
Schematic of liner and leachate collection systems for a hazardous waste disposal facility.

The infiltration of precipitation will be inevitable during the operational period and will continue, although to a greatly reduced degree, after landfill closure.

Leachate results from a combination of direct precipitation infiltration through landfill waste materials and any liquids squeezed out of landfill waste materials as a result of consolidation. Leachate seeps, or percolates, downward to the base of the landfill because of gravitational forces. Accompanying seepage forces may cause the leachate to carry particulate material in suspension (suspended solids) along with the dissolved constituents. Generally, the landfill environment is considered to be partially saturated with the exception of the very bottom.

The transport of contaminants by leachate through the landfill bottom is by diffusion and advective transport, as described in Section 4-2. At the base of the land disposal facility, the leachate first passes through a **filter zone** (see Figure 13-2). The filter zone may be a **geotextile,** and/or a well-graded sand and gravel through which the particulates are filtered out. A **geogrid** may also be included to provide additional structural stability. The filter separates the waste from a relatively free draining zone around the **primary leachate collection piping**. In this primary leachate collection zone the leachate is free to flow to the piping system for removal for treatment. Since flow through the liner is proportional to hydraulic head, it is desirable to minimize this head. The hydraulic head on the liner is minimized through proper design of the leachate collection system including pipe spacing, pipe size, drainage material, and slope.

The leachate collection system normally uses a granular material having a high hydraulic conductivity and is therefore free draining. Such materials, sand and gravel, must be designed to meet **filter criteria;** that is, they must be permeable enough to allow leachate to freely flow, while remaining fine enough to preclude the transport of solids. A geotextile layer also may be used to serve as a filter between the overlying waste and the leachate collection system. The material also provides structural support to the leachate collection piping. The piping must be designed to avoid collapse (crushing), and the granular material contributes to the pipe stability. One of the major concerns with the leachate collection system is clogging. Clog-

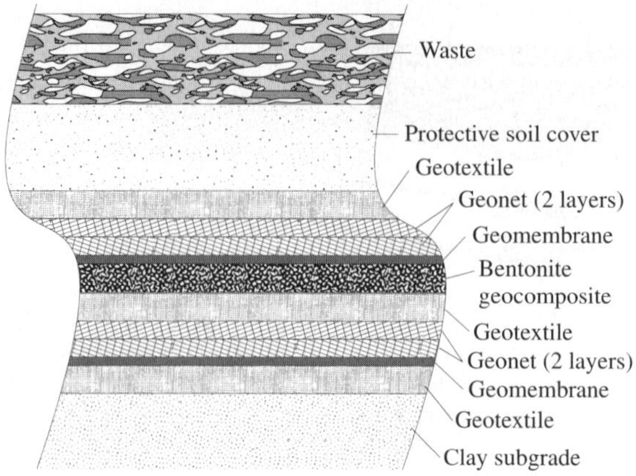

FIGURE 13-3
Schematic of geosynthetic liner and leachate collection systems.[8]

ging can occur from the migration of fines into the system (thus the need for filter considerations). Clogging can also occur from chemical and biological processes. Microorganisms in the leachate can establish communities in the granular material and/or piping, resulting in a reduced capacity to discharge leachate. Chemical processes such as precipitation can also contribute to clogging of the system. While the traditional leachate collection systems are constructed with granular materials, alternatives have been considered and used (particularly for municipal solid waste landfills as opposed to hazardous waste landfills). For example, **geonets** are grid-like geosynthetics that are designed to maximize their in-plane transmissibility and are used with geotextiles to prevent clogging from the infiltration of fine materials into the net matrix. Geonets have been used (see Figure 13-3) for leachate collection in landfills, waste piles, and heap leach pads and for leak detection between double liners in landfills and surface impoundments.[5] The performance of shredded scrap tires has been studied in the laboratory to evaluate their effectiveness as leachate drainage materials. One particular study demonstrated that, although the hydraulic conductivity decreased 2 to 3 orders of magnitude because of compression under the waste loading, the shredded tires proved to be an effective drainage material.[6] Not only does ground rubber meet the hydraulic need of the leachate collection system, but it has been shown to be an effective sorbent of organic contaminants such as benzene.[7] It remains an opportunity/responsibility of the profession to continue to develop and evaluate alternatives to the conventional approaches.

The **primary barrier layer** underlines the entirety of the primary leachate collection zone. In the United States, the primary barrier layer must be a synthetic material known as a **geomembrane** or **flexible membrane liner** (FML). Both geotextiles and geomembranes (as well as other products developed to function in lieu of previously employed naturally occurring materials) are known as **geosynthetics.**

EXAMPLE 13-4. LEACHATE GENERATION. As a first approximation, how much leachate (liters per minute) would be generated during the operation of a Pennsylvania landfill 1000 m by 1000 m in plan? The average annual rainfall is 100 cm/year. Comment on the answer.

Solution. The total rate of infiltration into the landfill can be conservatively assumed to be equal to the average annual precipitation:

$$
\begin{aligned}
L &= 1000 \times 1000 \times 1 = 1 \times 10^6\ \mathrm{m^3/year} \\
&= (1 \times 10^6\ \mathrm{m^3/yr})/(365.25\ \mathrm{day/year}) = 2738\ \mathrm{m^3/day} \\
&= (1 \times 10^6\ \mathrm{m^3/day})/(24\ \mathrm{h/day}) = 114\ \mathrm{m^3/h} \\
&= (1 \times 10^6\ \mathrm{m^3/h})/(60\ \mathrm{min/h}) = 1.9\ \mathrm{m^3/min} \\
&= (1 \times 10^6\ \mathrm{m^3/min})/(1 \times 10^3\ \mathrm{L/m^3}) = 1900\ \mathrm{L/min} \\
&= 700{,}000\ \mathrm{gal/day}
\end{aligned}
$$

Comment. This flow rate is calculated assuming that the waste has no capacity to absorb water (that is, the waste is at field capacity), that no evaporation or surface runoff is taking place, and that the entire landfill area is uncovered. The high rate of leachate generation demonstrates the importance of (1) maintenance of a cover over areas not actively being filled, (2) a low placement moisture content for the waste, and (3) timely placement of intermediate and final cover.

The **secondary leachate collection zone** functions much the same as the primary leachate collection system and underlies the primary barrier layer. This second system was originally known as the **leak detection system.** Now it is known as the secondary leachate collection system, in recognition of the inevitability of liquids containing contaminants entering this second layer. Given the effectiveness of the primary liner, the secondary leachate collection system handles a considerably reduced quantity of leachate. Beneath the secondary leachate collection system is yet another geomembrane, the **secondary barrier layer,** which acts in the same manner as the primary barrier layer, serving as a hydraulic barrier preventing downward flow of contaminants and allowing the secondary collection system to collect the leachate.

In a perfect system, the leachate is stopped by the primary barrier layer and is collected there for subsequent treatment. However, it has been shown that all liners leak—detailed studies of landfill liner systems built with the best available technology and construction quality control have revealed that a rate of leakage of about 22 gal/acre/day through the primary liner can be expected.[9] At one time, the EPA identified an action leakage rate of about 5 gal/acre/day, below which no corrective action is required.[10] This action rate of leakage of 5 gal/acre/day is much less than the leakage rate of 22 gal/acre/day that can be readily achieved. Currently, the EPA defines the action leakage rate as the maximum design flow rate that the leak detection system can remove without the fluid head on the bottom liner exceeding 30 cm (1 ft). If the action leakage rate is exceeded, response actions must be taken.[11,12]

Underlying the entire system of primary and secondary leachate collection and geomembrane barriers is a third barrier, a layer normally composed of a "natural material," compacted clay, or clay admixed into the natural underlying soils. The third barrier layer is required to control contaminants that may have passed through the secondary leachate collection system. Conceptually, the redundant barrier layers progressively reduce the rate at which contaminants exit the land disposal facility through hydraulic transport.

Is this the only arrangement for the liner and leachate collection system? No, the combinations and variations are endless. The liner system in Figure 13-3 is in service at a municipal solid waste and industrial waste landfill that employs exclusively geosynthetics for both drainage and barrier layers. The filters are geo-

textiles, drainage layers are geonets, and barrier layers are geomembranes and geocomposites. One such geocomposite, or **geosynthetic clay liner,** is a sodium montmorillonitic clay attached to a geotextile to create a thin clay sheet of low permeability.

A European design emphasizes natural materials and minimizes the use of geosynthetics. Shown in Figure 13-4 is an example of a composite liner employing a geomembrane directly on a clay (mineral sealing) layer. In employing such a design approach, it is recommended that a risk assessment be carried out "to define the degree of security attainable in relation to the various elements of the design and their interaction. Attention should be given to potential contaminant flow paths [to receptors]."[13]

Composite systems description

As illustrated by the examples presented above, the emphasis in landfill liner design has been on the resistance to contaminant transport by hydraulic means (advective contaminant transport). Attention has been drawn to the use of composite liner systems that limit both advective and diffusive contaminant transport.[14]

As described in Section 4-2, the rate of contaminant migration is affected by a number of factors including hydrolysis, sorption, cosolvation, ionization, precipitation, and biotransformation.[15] A liner designed to minimize the rate of contaminant migration into the environment must (1) minimize hydraulic conductivity, (2) minimize molecular diffusion rate and (3) maximize retardation. **Composite liners** are made of dissimilar materials, each employed to achieve one or more of these goals. Composite liner materials could include: bentonite, zeolite (suitable for the inor-

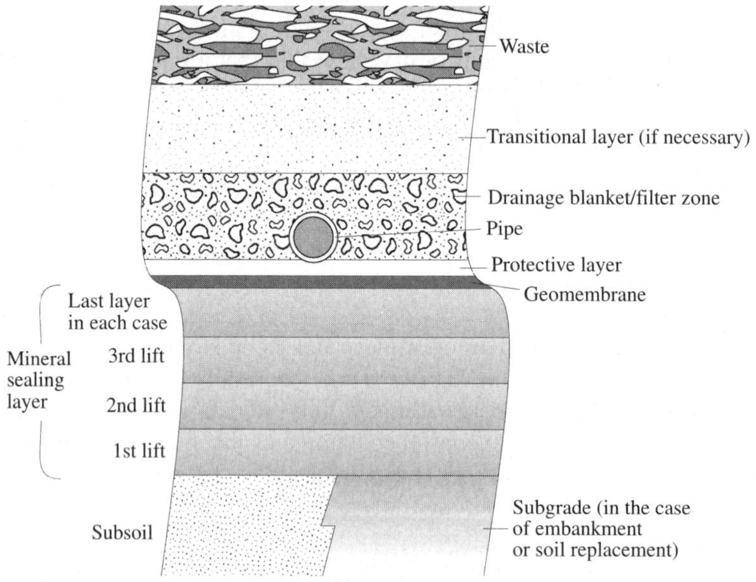

FIGURE 13-4
European liner and leachate collection systems.

ganic components of the leachate), organically modified clay, high-carbon fly ash (suitable for organic constituents), and geosynthetics.

The use of materials with high adsorptive capacities to attenuate contaminant migration is well established.[16] Earlier sanitary landfills for municipal waste were designed as attenuating landfills assuming leakage at a low rate. Attenuating landfills function on the premise that the undesirable constituents within the landfill leachate are attenuated by the underlying soils through adsorption, biological uptake, cation and anion exchange, dilution, and precipitation.[17]

The concept of designing liner systems that release contaminants only by molecular diffusion has been examined.[18] In this approach, hydraulic transport is eliminated by hydraulic barrier redundancy and the remaining transport mechanism, diffusion, is the basis for design. This approach can be taken one step further by introducing a composite liner system that seeks to minimize the diffusive transport.

Shown in Figure 13-5[19] is an example composite liner system that incorporates both the barrier and leachate collection zones in addition to materials that reduce the rate of diffusion through the system. Materials that delay the transport of inorganic contaminants include calcium bentonite, sodium bentonite, and zeolite (shown here). Materials that attenuate organic contaminants include high-carbon fly ash and organically modified clay (shown here). Organically modified clay is most effective in retaining higher molecular weight organics whereas the carbon in the fly ash is effective for low molecular weight organics. Studies have also shown that additives such as bentonite can enhance the ability of soil liners to retard off-site migration of radionuclides.[20]

The thickness of the attenuating layers depends on the allowable breakthrough time and steady-state transport rate which in turn are governed by hydraulic conductivity, hydraulic head, chemical potential, effective diffusivity, and retardation factor. The method of calculation can be based on the analytical solution to the advective-dispersive equation for solute transport (see Section 4-2).

The adsorbing barrier materials can be admixed with clay or attached to the geomembranes. At present, admixed soil liners are commonly constructed of native soil with a bentonite admixture to reduce hydraulic conductivity. Admixing attenuating materials can result in material of relatively low hydraulic conductivity ($< 1 \times 10^{-7}$ cm/s)[21,22] thereby reducing both advective and diffusive transport.

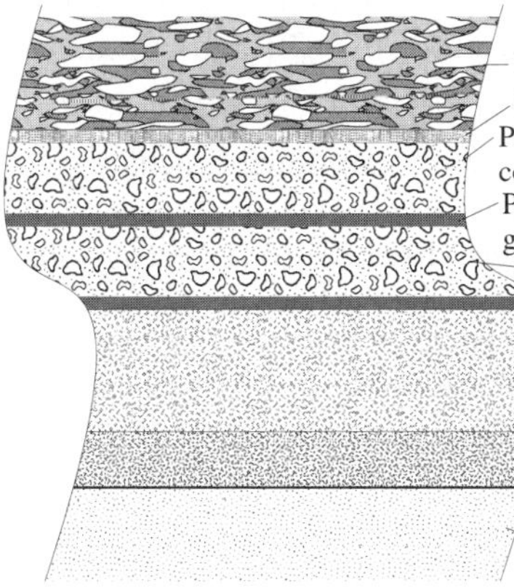

FIGURE 13-5
Schematic of composite liner system for a hazardous waste disposal facility. (*From Evans, et al., 1990.*)[21]

Attaching clays to geomembrane barriers is a well-developed manufacturing technique.[23] The use of these materials individually for adsorption is well known. Zeolite, carbon, and organically modified clays (see Figure 11-8) have been used in wastewater treatment processes.[24]

Adsorptive layers should be located on the underside of the secondary geomembrane liner because the geomembranes function as the primary barrier against contaminant transport by acting as a hydraulic barrier. In this way the contaminant loading to the adsorptive layers is decreased and their useful lives are extended. Contaminants that migrate through the geomembrane liners (either through diffusion or advection through liner defects) are sorbed by the underlying attenuating materials.

A composite liner that considers both diffusive and advective transport is more effective in minimizing the rate of contaminant transport in comparison to liners that consider only advective transport. The retardation and reduced diffusion rates result from adsorption and cation exchange of inorganics within the zeolite and/or bentonite and natural clays, and adsorption of organics by the organically modified clays and/or carbon-rich fly ash. Further retardation can be achieved by employing electrokinetics (described briefly in Section 16-10) to reduce the contaminant mobility.[25]

13-5 COVER SYSTEMS

A significant source of leachate generation at hazardous waste disposal facilities is the infiltration of water through the top of the landfill, by either direct precipitation or stormwater run-on. The following section describes the controls available to minimize precipitation infiltration and to minimize surface water entering the landfill.

General Description of Cover Systems

Much attention has been given to the design of the final cover for landfills.[26] The final cover design considers health and safety, aesthetics, and site usage after closure, coupled with engineering requirements for permeability, compressibility, and strength. Because the final cover is expected to remain in service for as long as the waste is present, a thorough design assesses potential failure mechanisms both immediately after construction and throughout the service life (30 years or more), and provides for the following:

- Controlling water movement into the landfill system to minimize leachate generation.
- Controlling animals and vectors that can introduce disease into the ecosystem.
- Protecting the public from the results of direct contact with waste.
- Controlling gas movement to avoid a decline in air quality.
- Minimizing fire potential to avoid air emissions and damage to landfill components.
- Ensuring overall stability of the cover on the landfill slopes where slope instability could result in mass movement of contaminants into the environment.
- Controlling surface water runoff.
- Resisting erosion.

- Controlling blowing debris.
- Minimizing noxious odors.
- Providing a more sightly appearance.

The planned usage of the site after landfill closure may generate additional design considerations. The final cover must provide structural support of the vegetative cover and sustain loadings imposed by site traffic such as maintenance vehicles. Covers over mining waste must minimize the rate of oxygen infiltration as well as water infiltration. The reactions of oxygen and water with the waste rock (tailings) result in acid drainage, a serious environmental problem for the mining industry.

The landfill cover must satisfy the needs of health and safety, aesthetics, and site usage after closure while simultaneously providing long-term performance and accommodating potential failure mechanisms. For example, the underlying landfill will subside because of consolidation of the landfill materials: Can the cover accommodate this deformation without cracking and thus compromising its integrity? Will seasonal variations in the weather cause wetting and drying of the cover, resulting in desiccation cracking or other degradation? Is the barrier layer of the cover system protected from cycles of freezing and thawing?

A typical cover is shown in Figure 13-6. The uppermost layer, a **vegetation support layer,** typically consists of an organic sandy loam (topsoil) material used to support vegetation. Vegetation provides several important functions in the performance of the landfill cover:

- Reduces erosion.
- Reduces precipitation infiltration.
- Enhances evapotranspiration, returning moisture that has been absorbed into the topsoil layer into the atmosphere to further reduce deeper infiltration.

In arid regions, vegetative covers are difficult to establish and maintain. As an alternative, a gravel/cobble layer can be used to armor the landfill. Unfortunately, this riprap armor permits easy infiltration of precipitation and no evapotranspiration capacity.

The vegetative support layer shown in Figure 13-6 also serves as a **protection layer.** Sometimes a separate and different material is used to provide protection

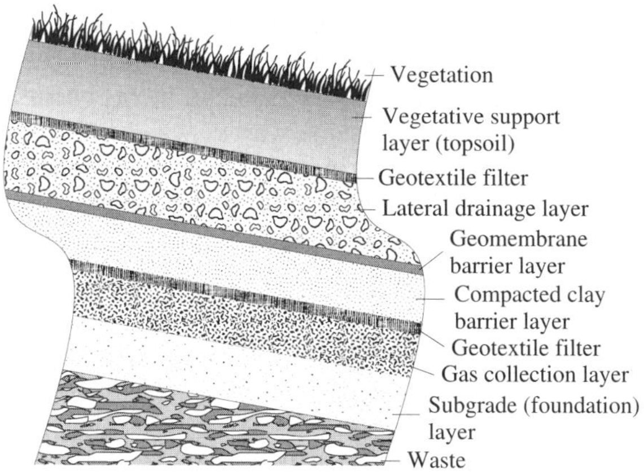

FIGURE 13-6
Schematic of hazardous waste landfill cover.

for underlying layers against the effects of freeze/thaw cycles, wet/dry/cycles, burrowing animals, and plant roots. The protection layer also stores water for later evapotranspiration.

The lateral drainage layer underlies the vegetative support layer and is a free-draining porous, gravelly material, geonet, or geocomposite. The purpose of this drainage layer is to enhance lateral drainage of any precipitation that infiltrates through the vegetative support zone. With proper grading of the lateral drainage layer, the infiltration can be collected. Collection of water from the lateral drainage layer also results in minimization of hydraulic head on the underlying barrier layer. As a result, the precipitation infiltration is reduced. The lateral drainage layer and the vegetative support layer function in combination to protect the underlying barrier layers from the environmental stresses of wetting and drying and freezing and thawing.

The lateral drainage layer may also include piping and water collection systems. A **geotextile filter** may be placed beneath the topsoil and above the underlying **lateral drainage layer.** The geotextile serves to maintain separation between the layers and to act as a filter to minimize migration of materials. Should fines from the topsoil migrate downward into the gravel, the topsoil layer will have a reduced capability for supporting vegetation and the gravel will have a reduced capability for lateral drainage. The presence of a geotextile between these layers thus reduces the risk of clogging of the drainage layer with topsoil fines.

Beneath the lateral drainage layer is one or more **barrier layers.** The barrier layers may be composed of any of the barriers previously discussed including geomembranes, geosynthetic clay liners, compacted clays, or admixed materials. The barrier layers represent the final impedance to precipitation infiltration. Considerable innovation is possible for landfills where regulations permit design flexibility, such as U.S. municipal waste landfills under Subtitle D or landfills managed by regulations structured differently from those in the United States (see Chapter 2). For example, a "green cap" was recently constructed over 45 acres of landfill employing a 36-mil reinforced polypropylene geomembrane.[27] This greenish geomembrane is the uppermost layer and is fully exposed. The green coloring was selected, in part, for aesthetics, as the landfill is the highest point in the county. This innovative design reduces infiltration and enhances runoff compared to a conventional clay cap system. Additional design considerations included faster and increased runoff, wind uplift, UV resistance, and thermal stability. Maintenance of the integrity of the barrier layer during and after settlement of the underlying landfill is a prime design consideration.

The rate of precipitation infiltration and leachate generation must not exceed the ability of the leachate collection system to remove the leachate. If this occurs, leachate levels build up within the landfill much the same way as water fills up a bathtub. Under this condition, the higher hydraulic head induces greater contaminant migration out of the base of the landfill. As a result, landfills lined with geomembrane liner systems are typically covered with geomembrane cover systems. The leachate collection system capacity is usually designed to handle worst case conditions—an open cell with no waste and high rainfall.

Underlying the barrier layer may be a **gas collection layer** used to collect generated gases migrating from the landfill for subsequent venting to the atmosphere. This layer is typically composed of porous sands and gravels and may include gas collection piping. A geotextile filter may also be needed between the gas

collection layer and the overlying compacted clay barrier layer to prevent clogging of the gas collection materials. A gas collection layer is virtually always included in the cover system for a landfill containing municipal solid waste or any other biodegradable materials. Gases that may be generated include carbon dioxide and methane from the anaerobic digestion of organic matter as well as volatile organics and hydrogen sulfide. In cases where active gas collection systems consisting of wells, laterals, headers, and pumps are employed, the overlying gas collection layer is not needed. Analytical solutions have been developed to determine gas flux in a landfill into which gas collection pipes have been installed.[28] Hazardous waste landfills contain significantly less organic matter than municipal solid waste landfills, thereby making it impractical to collect gases for resource recovery. Nevertheless, some gases can be expected and need to be managed.

The lowermost layer in a landfill cover system may be a **subgrade layer** to accommodate uneven and unstable landfill surfaces. This layer also aids construction of a cover with appropriate contours needed to enhance lateral drainage and minimize hydraulic head.

The cover system may include a geogrid layer to enhance the structural integrity of the cover system. Because some total and differential settlement is inevitable, a geogrid adds tensile capacity to the cover system to redistribute stresses and minimize differential settlements. As a result, the integrity of the cover layers is better protected.

The cover system may also include a biotic barrier layer below the vegetation support zone. This biotic layer could be installed in arid regions to prevent burrowing animals from destroying the integrity of the barrier layer. In humid regions, such a layer serves to prohibit penetration of roots into the barrier layers. The biotic layer is typically constructed of coarse-grained material such as cobbles and coarse gravel.

If the barrier layer is also to minimize the oxygen infiltration or gas exfiltration, a clayey soil material is normally utilized. It is essential to maintain a high degree of saturation in this barrier material, as the rate of gas migration is orders of magnitude slower in water than in air.

Cover System Performance

After the components of the cover system and their function are understood, calculations to predict their performance and to establish their design are required. Computer methods have been developed to assist in the evaluation of alternative cover designs such as the widely employed HELP model: Hydrologic Evaluation of Landfill Performance.[29] With this model, alternative designs can be investigated to study the impact of alternative cover designs on surface water runoff, lateral drainage above the barrier layer, and infiltration into the landfill.

As erosion is a common problem in cover systems, special attention should be given to cover slopes and selection of the soil to be used in the uppermost layer. Also, since clay shrinking and swelling is related to changes in moisture content, long-term performance monitoring may include such measures. On one cover system, moisture-sensitive radio-frequency cables were buried in the clayey barrier layer that can detect a 2 percent change in moisture content within a distance of 4 m.[30]

A sketch for the lateral drainage and barrier layers is shown in Figure 13-7. As steady-state seepage develops, a hydraulic head builds up on the barrier layer to provide the hydraulic energy for flow to the drain. Note that this schematic also represents the flow of leachate to the underdrains beneath the landfill. The analysis of flow to the drains considers the hydraulic conductivity of the materials, the spacing of the drains, the slope of the underlying barrier layer, and the head buildup between drains. Solutions to this problem, although complex, have been developed.[31]

One of the major factors affecting the long-term performance of the cover system is settlement, both total and differential. Movements and distortions of the constructed cover system have the potential to reduce its effectiveness. The extent and types of settlement depend largely on the nature of the material in the landfill and can vary significantly from landfill to landfill and from location to location within the same landfill. Mechanisms of landfill material settlement (compression) have been identified for municipal waste and apply (not necessarily to the same relative importance) to hazardous waste materials as well. The mechanisms are mechanical compression, degradation and ravelling.[32] **Mechanical compression** is due to the crushing, rearranging, bending, and distortion of the waste materials by the weight of the overlying materials. Mechanical compression may be enhanced by wetting of waste materials. **Degradation,** often the most important long-term mechanism, is due to the decomposition of the waste materials by biological or physicochemical processes such as oxidation. **Raveling** is the movement of finer materials from the material matrix into larger voids within the matrix and can occur anytimc throughout the long-term life of the landfill. To minimize damage to the cover system, sudden changes in the depth of the underlying landfill materials should be avoided and the cover system geometry should be such that settlement of the underlying waste puts the cover system in compression rather than tension.[33]

Surface Water Controls

The construction of a hazardous waste disposal facility alters the natural hydrology of the site. Prior to construction, surface water probably flowed onto the site. Surface water moves first as overland flow and ultimately through streams and larger surface water bodies. Hazardous waste disposal facilities must use surface water controls. The flowing surface water must be controlled to preclude erosion and

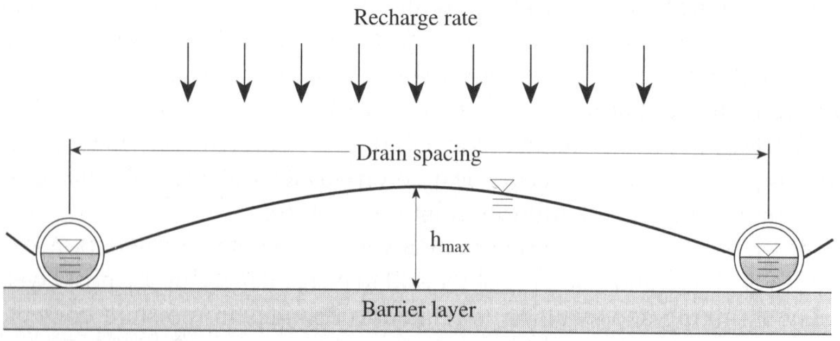

FIGURE 13-7
Schematic of drainage.

transport of sediment. A landfill section showing the deployment of surface water controls is shown in Figure 13-8.

To control surface water, routine technologies can be incorporated into the site development for hazardous waste disposal facilities.[34] The technologies include **surface water diversion and collection systems** to minimize any surface water entering active areas of the site. Surface water diversion technologies include dikes and berms, ditches and drainage ways, terraces and benches, and chutes and downpipes. Surface water can also be controlled by grading and revegetation. Areas need to be revegetated to enhance evapotranspiration and to minimize erosion and sediment transport. The surface water controls and their functions are summarized in Table 13-1. Successful application of surface water controls ensures that surface water originating in areas beyond the active portion of the site will not enter the site. If it were to enter the site, it could become contaminated and could contribute to the generation of leachate, and require treatment.

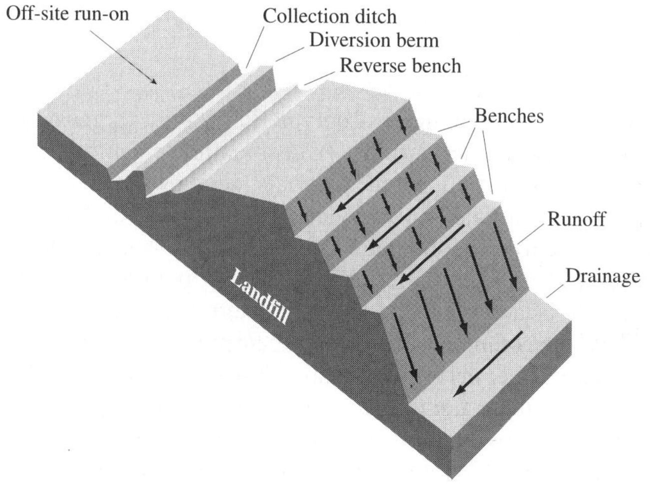

FIGURE 13-8
Landfill section showing surface water controls.

TABLE 13-1
Surface water controls for landfills

Grading	Revegetation	Surface water diversion and collection
Reshapes the surface to enhance runoff	Controls erosion Enhances evapotranspiration	Minimizes surface water entering landfill
Minimizes infiltration/leachate generation	Root penetration must be controlled	Channelizes runoff water to detention basins
May need periodic maintenance, e.g., annual	Maintenance required	May need periodic maintenance

Control of surface water continues when the final cover is constructed. Runoff from the cover system must be managed through the use of selected cover materials, collection channels, and other techniques as shown in Table 13-1 to prevent erosion. Sedimentation and detention basins must be included in the site drainage system to capture eroded sediment from the cover.

Many landfills have problems with sedimentation and erosion control. Typically, erosion is a result of settlement or poor maintenance of the landfill cover system. This performance record emphasizes the need for careful attention to what might be considered a mundane component of a landfill system.

13-6 MATERIALS

The cover, liner, and leachate collection system components can be constructed of a wide range of materials. The following sections of this chapter describe the material types and present the test methods available to evaluate the material performance.

Geomembranes

A geomembrane is an engineered polymeric material (a **geosynthetic**) that is fabricated to be virtually impermeable. A geomembrane can be made of any one of a number of plastic manufactured materials, such as butyl rubber, chlorinated polyethylene, chlorosulfonated polyethylene (Hypalon®), ethylene-propylene rubber (neoprene), polyethylene (both high-density and low-density polyethylene), and polyvinyl chloride (PVC). A number of different polymers with a wide range of chemical formulations are employed to manufacture geomembranes. Shown in Table 13-2 are the most common of these and their predominant uses. Compatibility studies and studies of liner performance have shown that high-density polyethylene (HDPE) is, at present, the material of choice for the mixed range of wastes typically encountered in land disposal facilities. HDPE can be made relatively thick, 100 mils or more; seams can be readily joined in the field (by extrusion or other welding techniques); and the material offers a high degree of chemical compatibility, resisting degradation to a wide range of chemicals at a wide range of concentrations. The design of geomembrane systems has emerged as a specialty in and of itself.[35]

To evaluate the performance of the geomembrane barrier layers, it is necessary to evaluate a wide range of properties, as shown in Table 13-3. Tensile properties include the tensile strength at break, the strain at failure, the stress level that causes 100 percent and 200 percent elongations, and modulus of elasticity. The modulus of elasticity is a measurement of the stiffness of the geomembrane and reflects the strains that will be induced by applied stresses. Hardness is a measure of the ability of the material to resist indentations. Tear resistance is a measure of how easily a rip is propagated once it begins. Puncture resistance is a measure of how easily the material is first ripped. It is also necessary to evaluate the strength of the seam as well as the strength of the material itself. The test methods for geosynthetics are presented in Table 13-4.[36]

Measures of the environmental performance include the rate of water vapor transmission, which is indicative of how quickly the gaseous phase will migrate

TABLE 13-2
Geomembrane types

Type	Use
High-density polyethylene (HDPE)	Excellent resistance to chemical degradation and widely used for municipal and hazardous waste landfills
Polyvinyl chloride (PVC)	OK if buried, subject to ozone and UV degradation, not good for oils
Chlorosulphonated polyethylene, Hypalon®	Fabric reinforced, good resistance to corrosive chemicals, not very good for oils and solvents
Chlorinated polyethylene	Chlorinating HDPE, good resistance as above
Low-density polyethylene (LDPE)	Easily punctured, hard to seam, used as a liner for 40 years, deficient for waste containment because of low chemical resistance
Neoprene	Excellent resistance to oils, weathering, ozone, ultraviolet, puncture, abrasion, and mechanical damage; expensive
Polyester elastomer	Good resistance to oil, fuel, and chemicals
Butyl rubber	Low gas and vapor permeability, thermally stable, ozone and weather resistant, chemical and moisture resistant, resistant to animal and vegetable oils; swollen by hydrocarbon solvents; more than 40 years in water impoundments
Elasticized polyolefin	Specialized use
Ethylene propylene rubber	Similar in chemistry and behavior to butyl rubber

TABLE 13-3
Geomembrane properties*

Physical properties	Thickness Density Water vapor transmission Solvent vapor transmission
Mechanical properties	Tensile behavior Seam behavior Tear resistance Impact resistance Puncture resistance Friction characteristics Stress cracking
Chemical properties	Swelling resistance Chemical compatibility Ozone resistance Ultraviolet light resistance
Biological properties	Resistance to animals Resistance to fungi Resistance to bacteria
Thermal properties	Warm temperature behavior Cold temperature behavior Thermal expansion

*Properties discussed in detail in Reference 35.

TABLE 13-4
Test methods for geosynthetics*

Type	Property (units)	Test procedure
Geotextile	Thickness (mils)	ASTM D-1777 (part 32)
	Mass per unit area (oz/sq yd)	ASTM D-3776
	Dimension (cm)	Direct measure machine and cross machine direction
	Grab tensile strength/elongation	ASTM D-4632
	Trapezoidal tear resistance	ASTM D-4533
	Hydraulic burst strength	ASTM D-3786
	Puncture resistance	ASTM D-3787
	Permittivity	ASTM D-4491
	Transmissivity (cm^2/s)	ASTM D-4716
Geonets	Mass per unit area	Direct measure (sample size $\geq$ 1 ft^2)
	Volatiles	Appendix G
	Extractables	Appendix E
	Thickness (mils)	ASTM D-1777 (Part 32)
	Rib dimensions	Direct
	Aperture size	Direct
	Dimensions of configuration (cm)	Direct measure machine and cross machine direction
	Specific gravity or density	ASTM D-792 Method A or ASTM D-1505 (dry sample before test)
	CBR puncture	GS-1
	Transmissivity (cm^2/s)	ASTM D-4716
	Compression behavior of geonets	GNI
	Strip tensile strength	Alternative method (see below)
Geogrid	Geogrid rib tensile strength	GG1
	Geogrid node junction strength	GG2
	Creep behavior and long-term design load of geogrids	GG3
	Rib dimension	Direct
	Aperture size	Direct
Pipe	Volatiles @ 105°C	Appendix G
	Extractables	Appendix E

**Source:* From Landreth 1990.[36]

through the specimen. Water absorption is indicative of potential property changes when the geosynthetic is in contact with the leachate. The greater the water absorption the more likely the properties of the liner will change. The most common test method employed for the measurement of compatibility of the membrane with the wastes is EPA Method 9090.[37] In this method, rectangular geomembrane samples are immersed in the expected leachate for a minimum of 120 days at room temperature of 22°C, and at an elevated temperature of 55°C. Physical properties, including hardness, tensile behavior, puncture resistance, specific gravity, volatile organics content, extractable organics content, hydraulic resistivity, dimensions, and weight, are measured after each 30-day period as well as before and after immersion. In addition, semicrystalline geosynthetics should be tested for modulus of elasticity, tear resistance, and environmental stress cracking, whereas fabric-reinforced geomem-

branes should be tested for ply adhesion. Additional discussion on the limitations of these compatibility tests is provided by White and Verschoor.[38]

Geotextiles

A **geotextile** is a geosynthetic which is fabricated to be permeable and has two categories of hydraulic properties:[39]

- *Filtration*—the removal of suspended solids from the flowing liquid.
- *Drainage*—the transportation of liquids across the plane of the fabric.

Geotextiles can be classified by the way they are manufactured as either woven or nonwoven. Woven geotextiles tend to be stronger in tension and have been used for reinforcement and for silt fences. Nonwoven geotextiles have been used for filtration, for separation, and as a protective layer over geomembranes.

Water flowing perpendicular to the plane of the geotextile is termed cross-plane flow. The ease by which cross-plane flow occurs is termed permittivity (cross-plane permeability). The permittivity Ψ is defined as:

$$\Psi = K_n/t \tag{13-1}$$

where Ψ = permittivity (s^{-1})
K_n = normal hydraulic conductivity (cm/s)
t = fabric (geotextile) thickness (cm)

The permittivity can be substituted into Darcy's law (Equation 4-3) to estimate flow rates through geotextiles under the imposed hydraulic head conditions. If the geotextile has inadequate permittivity, its function in providing a filtration medium is impaired.

Water can also flow within the plane of the geotextile. The ease in which in-plane flow occurs is termed the transmissivity θ (in-plane permeability). The transmissivity is defined as:

$$\theta = K_p t \tag{13-2}$$

where θ = transmissivity (cm^2/s)
K_p = in-plane hydraulic conductivity (cm/s)
t = fabric thickness (cm)

This can be substituted into Darcy's law to estimate flow rates within geotextiles to study their drainage characteristics and to design lateral drainage systems using geotextiles and geocomposites.[40] Note that ability of a geotextile to transmit water is stress dependent. Thus, laboratory tests must be conducted at stress levels expected in the field.

The durability of a geotextile for the long-term demands of a hazardous waste landfill is of critical importance. The properties for consideration of endurance include:[41]

- Ultraviolet light stability
- Abrasion resistance
- Chemical stability
- Thermal stability
- Creep resistance

Generally, the durability of geotextiles is not of concern except for chemical instability. They are protected from ultraviolet light and abrasion because they are buried. They are protected from thermal instability because they are stable over quite a wide range of temperatures. Although we can test the compatibility of geotextiles in accelerated lab tests designed to simulate the in-service environment, there is no substitute for the test of time. Unfortunately, a failure 30 years hence could have serious consequences unless adequate redundancies are built into the system.

The critical issue regarding long-term performance is the issue of clogging. How rapidly will the filter clog with sediment? What sort of biological activity will be present and how will this contribute to clogging? Will chemical precipitation contribute to clogging? These questions have no clear answers.

Compacted Clays

Compacted clay is widely used as a barrier layer in liner and cover systems. These soils typically consist of natural clays, silty clays, sandy clays, and clayey silts. An extensive body of literature has been developed regarding the compaction and resulting permeability in clay properties. [For example, see Murray (1998)[42] for a summary of properties and testing of clay liners that includes more than 50 additional references and an appendix citing both U.S. and European testing methods.] The summary of this information constitutes a book in itself.[43] This chapter can only hope to identify the critical issues and present the fundamental geotechnical engineering concepts governing the behavior of clay liners.

The hydraulic conductivity of a clay barrier layer depends on the type of clay as characterized by the clay mineralogy, grain size distribution, and plasticity limits. However, for a given clay, it has long been established that there is not a unique value of hydraulic conductivity. In one of the early studies of hydraulic conductivity, significant influence of compaction moisture content was established.[44] In fact, as shown on Figure 13-9, the hydraulic conductivity can vary, depending on compaction moisture content, by more than 2 orders of magnitude for the same clay compacted at the same compaction effort. The tremendous variation in hydraulic conductivity is attributed to differences in the microstructure of the compacted clay with varying compaction moisture content.[45] More recently, a design methodology has been developed to accommodate the variation in hydraulic conductivity resulting from variations in moisture and density.[46]

Let's assume for discussion purposes that we are able to obtain representative samples of a clay proposed for use in a clay barrier layer and test them in the laboratory. Is it reasonable to assume that the laboratory value of hydraulic conductivity is the same as the hydraulic conductivity of the field-constructed barrier layer? Probably not. In one study of liner systems with known high rates of leakage, the field hydraulic conductivity was found to be 10 to 1000 times greater than the laboratory value.[47] Although more difficult and costly to run, field hydraulic conductivity testing may be a more reliable method to determine the hydraulic conductivity of the barrier.[48] It has been shown that the macrostructure of the compacted soil (size of clods and pores between clods) can have a significant influence on the hydraulic conductivity.[49] If the lessons of more than thirty years of research into the hydraulic conductivity of compacted clays can be summarized in a few key points, they would be:

- Compact the soil wet of the optimum moisture content to achieve the desirable microstructure and macrostructure for low hydraulic conductivity.

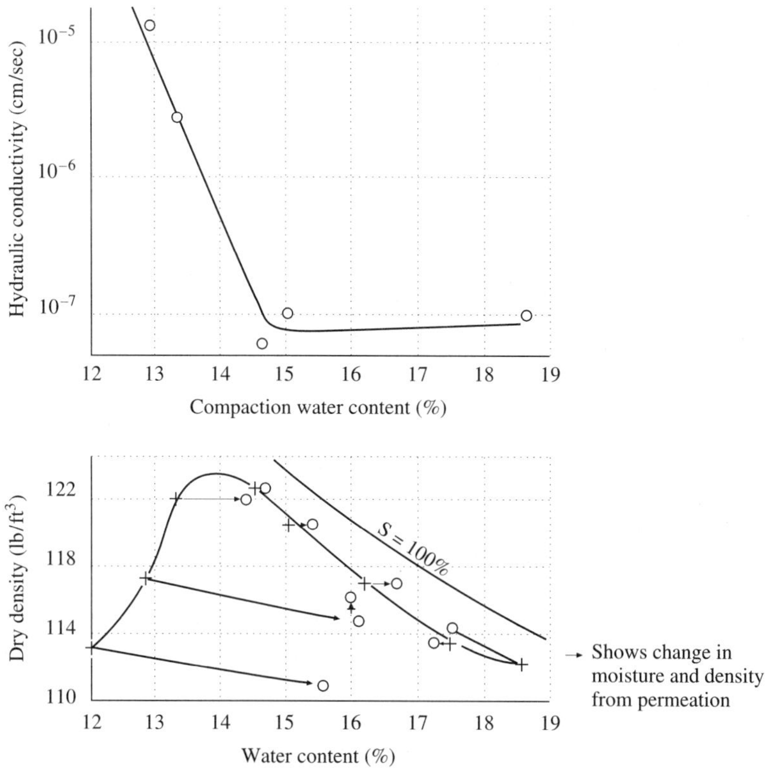

FIGURE 13-9
Influence of compaction moisture content on the hydraulic conductivity of compacted clay. (*After Lambe, 1958.*[45])

- Use adequate compaction effort to break down clods into a remolded mass relatively free of large pores between clods and to achieve the desirable microstructure.
- Construct the barrier layer in numerous thin lifts eventually forming a barrier layer that is relatively thick (a meter or more).

Examination of a database of 85 full-scale compacted clay liners that were intended to have a hydraulic conductivity of less than 1×10^{-7} cm/s found that only about 74 percent met this objective.[50] The data confirm the importance of ensuring compaction is primarily wet of optimum and of the benefits of liners that are thicker or have a greater number of lifts. This study also showed that a wide range of soils could be effectively used and that index properties and percent compaction are less important considerations.

How does the hydraulic conductivity of compacted clays change when they are permeated with landfill leachate? Permeation of compacted clay with concentrated organics, acids, and bases may cause dramatic increases in hydraulic conductivity. For example, shown in Figure 13-10 are the results of permeability tests where compacted clays were initially permeated with water and then with heptane, $CH_3(CH_2)_5CH_3$. As the curves show, the permeability increased 2 orders of magnitude or more because of heptane permeation.

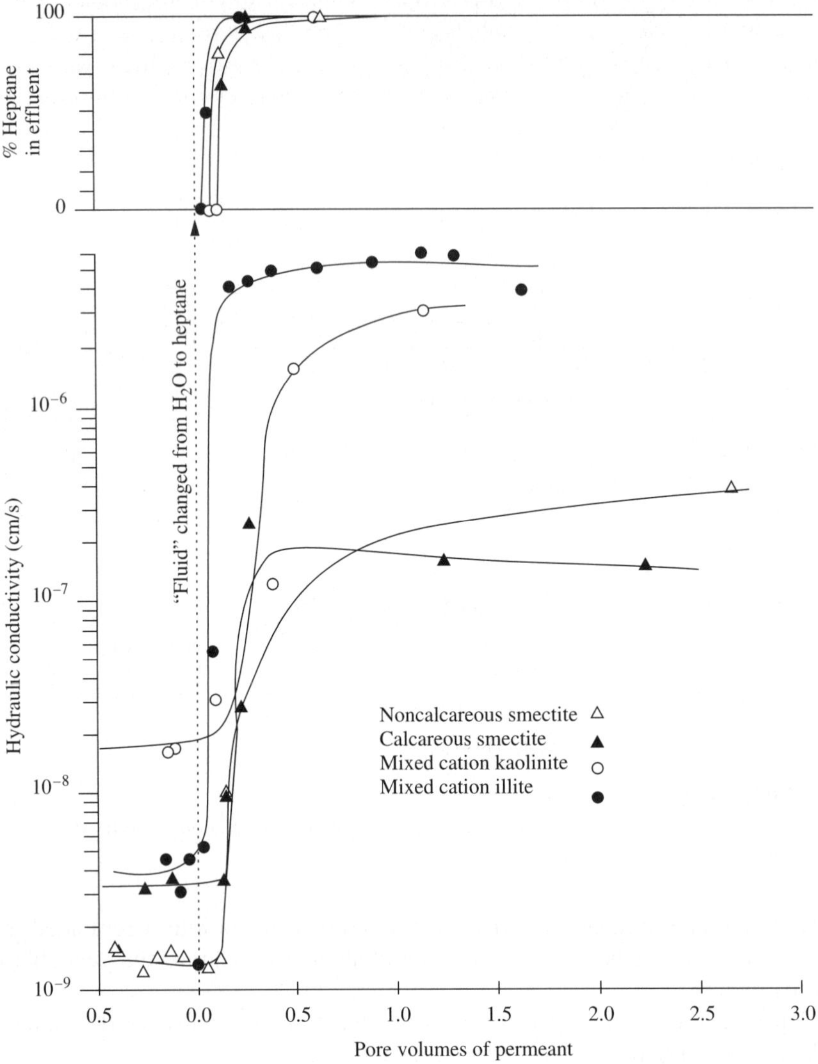

FIGURE 13-10
Influence of organic permeants on the hydraulic conductivity of compacted clay. (*After Brown and Anderson, 1983.*[51])

In contrast, permeation with more dilute solutions generally does not reveal such changes in permeability. An examination of the data from many researchers, coupled with an assessment of expected fundamental physicochemical changes based on clay mineralogy and the clay-water-electrolyte system, has led to the conclusions[52] that

- Dilute solutions of organics have essentially no effect on the hydraulic conductivity.
- However, dilute solutions of inorganic chemicals may affect the hydraulic conductivity.

A state-of-the-art compacted clay barrier does not stop the transport of contaminants. Containment systems simply delay transport; contaminants will still migrate through the compacted clay liner to the environment beneath the liner. As discussed earlier, even with absolute hydraulic containment, there is diffusive transport. The technical issues are: (1) At what rate will the contaminants migrate? (2) When does this migration begin? (3) Will the contaminant concentrations be detrimental to the environment?

Geosynthetic Clay Liners

Geosynthetic clay liners (GCLs) are a manufactured product combining geosynthetics with bentonite. Examples of GCLs are those with bentonite bound between two geotextiles (adhesively, stitch-bonded, or needle-punched), and bentonite adhesively bound onto a geomembrane. While quite thin (generally 4 to 5 mm), the hydraulic conductivity is quite low (generally less than 5×10^{-9} cm/s) because of the bentonite. Design concerns relate to slope stability and, for use in liners, chemical compatibility of the bentonite and leachate. While there are both advantages and disadvantages of GCLs compared to compacted clay, some conclude that GCLs are preferred over compacted clay for landfill final cover systems[53] assuming slope stability is adequate.

In laboratory tests, the hydraulic conductivity of 1×10^{-7} cm/s or less was not compromised when GCLs were subjected to tensile strains of up to 10 percent.[54] The same study demonstrated that the hydraulic conductivity was not compromised at the joints (overlapped panels) for slippage of 25 to 100 mm. Hence, for situations such as covers over landfills where differential settlement is likely, GCLs appear better suited to withstand the differential settlement than compacted clays. However, for situations where the GCL may be in the presence of contaminants that could alter the permeability of the bentonite, compacted clays may be a better choice.

In Situ Clay Formations

The principles of clay behavior previously described for compacted clays are, in many regards, similarly applicable to *in situ* clay formations. However, when assessing the integrity of underlying clay formations as barrier layers we do not have the luxury of inspecting each 6-in (150-mm) lift. Rather, we must evaluate the formation with the tools and methods described in Chapter 15. It is important to state here, however, that hydraulic pathways in clay deposits are not uncommon. For example, glacial till, a well-graded, dense clayey material deposited on the margins of glaciers, is often considered an excellent natural aquiclude. However, glacial till may be highly fractured. Hydraulic transport through these fractures is essentially unimpeded. Any dependence on *in situ* formations of clay requires careful exploration of the bulk properties of the formation.

Admixtures

Bentonite is added to natural soils to reduce the hydraulic conductivity and increase the adsorptive capacity of the native materials. Processed sodium montmorillonitic clay (bentonite) is added to the native soil, using 5 to 15 percent by dry weight. This added clay changes the natural soil to a clayey soil of high plasticity. The soil-bentonite admixture is compacted into a liner of relatively low hydraulic conductivity.

Bentonites are clays having montmorillonite (a smectite) as the major clay constituent (see Section 11-3 and Figure 11-7 for additional detail). Bentonite is further described by the principal exchangeable cation (typically 35 to 99 percent of the exchangeable cations) as calcium or sodium bentonite. The cation exchange capacities of calcium and sodium bentonites are approximately 80 and 150 milliequivalents per 100 grams of dry soil, respectively.[55] Calcium bentonite is commonly mined in Mississippi and Alabama, and sodium bentonite is found and mined in Wyoming and South Dakota.

EXAMPLE 13-5. BENTONITE ADMIXED LAYERS. A 200-mm (8-in) layer of soil-bentonite is planned for placement immediately beneath the secondary geomembrane liner. The layer is to be constructed in two 100 mm (4-in) lifts. The bentonite content is to be 6 percent by dry weight. The compacted dry unit weight of the soil, including bentonite, has been determined to be 15.7 kN/m^3 (100 lb/ft^3). How much dry bentonite needs to be spread for mixing on each lift?

Solution. Each lift is 100 mm (0.333 ft) thick yielding:

$$W_t = 100 \text{ mm} \times \frac{1 \text{ m}}{1000 \text{ mm}} \times 15.7 \text{ kN/m}^2 \text{ of lift}$$

or

$$W_t = 0.3333 \times 100 = 33.3 \text{ lb/ft}^2 \text{ of lift}$$

Of this, 6 percent is bentonite and the remainder is the base soil yielding:

$$W_{\text{bentonite}} = 0.06 \times 1.57 = 0.09 \text{ kN/m}^2 \text{ of lift}$$

Both sodium and calcium bentonite are hydrophylic, although sodium bentonite expands more than calcium bentonite in the presence of water. The monovalent sodium is more readily exchangeable than the divalent calcium. The exchangeable ions (calcium and sodium) in the clay readily exchange with heavy metals such as copper (Cu), nickel (Ni), and lead (Pb). Thus, as leachate migrates through a soil liner containing bentonite, the rate of advective transport is reduced as a result of the lower hydraulic conductivity, and the diffusive transport is reduced because of the exchange and adsorption of metal ions onto the clay surface.

Although not yet used in landfill liners, zeolites are naturally occurring or engineered aluminosilicates with a caged structure as shown on Figure 13-11. The

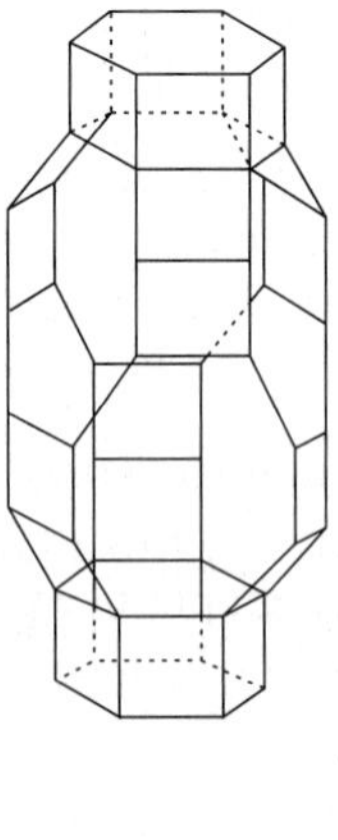

FIGURE 13-11
Zeolite structure.

cation exchange capacity of zeolites is about 250 milliequivalents per 100 grams with sodium and calcium being the major exchangeable cations.[56] The caged structure of the zeolite acts like a molecular sieve. Metal ions that pass into the sieve are trapped and held by ion exchange. The five principal zeolites found in nature are analcite, chabazite, clinoptilolite, erionite, and mordenite. The structure and behavior of zeolites have been studied extensively and are well established.[57] Zeolites already perform a wide range of waste management functions including purification of water contaminated with radioactive ions, removal of ammonium from the effluent of wastewater treatment plants, purification of water in fish tanks, odor control in kitty litter, and adsorption of metals from industrial wastewaters. Among the metal ions, zeolites preferentially adsorb arsenic, lead, and cadmium. The benefits of zeolite in enhancing landfill liner performance have been demonstrated on a laboratory scale but not in full-scale installations.

Organically modified clays have been shown to enhance liner performance in the laboratory but have not yet been incorporated into full-scale installations. The modification of naturally occurring clays into organically modified clays was described more than 40 years ago when Jordan published his original work on the subject.[58,59] Since that time, organically modified clays have been used in a number of environmental applications.[60] As described in Section 11-3, organically modified clays are naturally occurring clays in which a portion of the inorganic exchangeable cations are exchanged with suitable organic cations such as quaternary ammonium. Quaternary ammonium is an organic nitrogen compound in which a central nitrogen atom is joined to four organic groups along with an acid radical. The reactions that exchange the organic cations with the inorganic cations on the surface of the clay include adsorption, ion exchange, and intercalation. Different organically modified clays result from different base clays (e.g., montmorillonite and attapulgite) and different organic cations used in the exchange. The unmodified clay is hydrophilic whereas the modified clay is organophylic and hydrophobic.

Organically modified clays adsorb organic molecules as a result of adsorbate interactions and adsorbate-solvent interactions[61,62] The adsorption capacity of an organically modified clay is therefore dependent on both the properties of the organic cation used to make the organically modified clay as well as the properties of the fluids, such as temperature, pH, and organic waste type and concentration. A number of studies have clarified the behavior of organophylic clays.[63,64,65] For example, clays modified with strongly hydrophobic organic molecules tend to be excellent adsorbents of chlorophenols but are less effective for nonchlorinated phenols.

Lime has long been used to stabilize swelling clays and it is well established that the addition of small amounts of lime causes flocculation and results in an increase in the hydraulic conductivity. However, one study has shown that adding even larger amounts of lime can reverse this trend and ultimately lower the hydraulic conductivity.[66] At sufficient lime content (greater than about 10 percent for a mixture of montmorillonite and lime), a pozzolanic reaction occurs, resulting in a strong, low-permeability barrier material. In the referenced study, the lime-clay liner was also effective in delaying the breakthrough of zinc. Additionally, the precipitates further lowered the hydraulic conductivity. While such a liner of montmorillonite and lime may exist only in the laboratory, the opportunity continues to develop and evaluate alternatives to the conventional approaches to landfill liners.

Fly ash is the by-product of the combustion of coal burned for the production of electricity. The material is a silt-sized particulate, silicious in nature, containing various levels of unburnt carbon. A description of the fly ash properties best suited for landfill liner applications is found in Mott and Weber.[67] The addition of high-carbon fly ash within the barrier results in a substantial delay in breakthrough of organics as discussed later in this chapter. High-carbon fly ash shows an adsorptive preference for low molecular weight organics (similar to activated carbon) in contrast to organically modified clays which preferentially adsorb high molecular weight organics.

One study of landfill liner materials incorporating organophilic bentonite found the liner containing organically modified bentonite significantly retarded the transport of benzene relative to a conventional liner.[68] For the liner materials used in this study, modeling using laboratory results found that the maximum benzene concentration at a distance of 0.3 m into the conventional liner occurred after 4 years compared to 275 years for the liner including organically modified bentonite.

13-7 CONTAMINANT TRANSPORT THROUGH LANDFILL BARRIERS

Transport through Barriers

Contaminant transport through barriers includes contaminants carried with migrating fluid because of hydraulic gradients (advection) and molecular diffusion in response to chemical concentration gradients. The migration of contaminants by advection and molecular diffusion has been described in Section 4-2 and is well documented.[69] In an effort to better understand hydraulic transport, studies have included the hydraulic conductivity of clays in both the laboratory and field,[70,71,72] the transport of contaminants through geosynthetic liners,[73] and compatibility between membranes and the contained waste.[74,75] The absolute impermeability of geomembranes is well established as a myth.[76] Imperfections such as pinholes, improperly sealed seams and puncture tears as a result of construction are inevitable. The inevitability of hydraulic transport of contaminant through barrier layers is presently accounted for by the redundancy of multiple barrier layers. As we are able to build better and better liner systems with lower and lower effective hydraulic conductivities, molecular diffusion becomes the controlling mechanism for contaminant transport.

Amplification of the governing equations, dimensionless charts, and design methodology for contaminant transport through earthen barriers is presented in additional sources.[77,78] In this chapter, the diffusive transport is further examined, and transport through the liner materials is described.

The importance of contaminant transport through landfill liners by diffusion is illustrated by calculating the release rate for a contaminant-diffusing clay liner. Employing data from a landfill study[79] for a 1-m-thick liner, an assumed landfill area of 10^4 m^2, and a benzene concentration of 1 g/L in the landfill, a steady-state transport rate of 19 kg/year is computed. This transport rate has the potential to contaminate 3.8 billion liters of water at the EPA drinking water limit of 0.005 mg/L. Diffusion continues even with double liners and leachate collection systems, and the diffusion rate through a double liner system is similar to that of a single layer barrier of equal thickness.

Molecular diffusion occurs two ways: **liquid phase diffusion** through a saturated barrier layer or **gas phase diffusion** after **partitioning** from the liquid phase. For liners constructed of thin geomembranes, the contaminant migration rate due to molecular diffusion is surprisingly high. For example, the rate of diffusion of the low-octane gasoline through a 30-mil geosynthetic liner tested according to ASTM D1814-55 Inverted Cup Method was found equivalent to approximately 3000 gallons per acre in 24 hours.[80] Even though gasoline is not stored in geomembrane-lined landfills, this example illustrates the significant rate at which some organic contaminants of low molecular weight (volatile organics) can diffuse through a geosynthetic liner.

To examine liner systems to minimize diffusion, one can assume that the seepage velocity approaches zero and the molecular diffusion rates are characterized by Fick's second law describing one-dimensional diffusion of solutes in soil as:

$$\partial c/\partial t = (D^*/R)(\partial^2 c/\partial x^2) \tag{13-3}$$

where $c =$ concentration of a solute (mg/cm^3)
$t =$ time (s)
$D^* =$ effective diffusion coefficient (cm^2/s)
$R =$ dimensionless retardation factor
$x =$ direction of diffusion (cm)

According to Fick's law, the driving force (or gradient) for molecular diffusion is the chemical potential. The effective diffusion coefficient attempts to include real-world subsurface conditions including the tortuosity of the flow, temperature variations, and reactions occurring in the subsurface.[81] The apparent diffusion coefficient is defined as D^*/R and includes retardation effects such as those due to adsorption (see Section 4-3).

Contaminant partitioning from the water phase to the solid phase is described by the partition coefficient, and retardation is related to the partition coefficient. For fast reversible adsorption with a linear isotherm, the partition coefficient can be replaced by the **distribution coefficient** (see Section 4-3). **Retardation** R, expressed as a dimensionless retardation factor, is a function of the bulk density ρ, porosity η, and distribution coefficient K_d as:

$$R = 1 + (\rho/\eta)K_d \tag{13-4}$$

An increasing retardation factor results in a decreasing initial rate of diffusive transport. That is, more is adsorbed or otherwise retarded, and the **breakthrough** is delayed. In liners designed with diffusive transport in mind, this retardation is accomplished by using materials with higher distribution coefficients.

Transport through geomembranes

Geomembranes are manufactured as solid homogeneous nonporous materials. As such, they are not porous media materials, and cannot be permeated by liquids in the traditional sense. Geomembranes do, however, contain interstitial spaces between the large polymer molecules. Small molecules or gases can diffuse through the geomembrane within these interstitial spaces. Whereas flow through porous media can be described by Darcy's law (Equation 4-3), flow through geomembranes is

on a molecular level. Thus, even a perfectly constructed geomembrane, i.e., free of defects or holes, permits dissolved constituents of the contained fluids and gases to escape.

Such transport of contaminants through the geomembrane is on a molecular basis by the diffusion process. Liquids per se do not migrate through the geomembrane. The transport involves three steps[82]:

1. The solution or adsorption of the permeant into the upstream surface of the geomembrane (partitioning of molecule from leachate to gas).
2. Diffusion of the contaminant through the geomembrane.
3. Volatilization or desorption of the contaminant at the downstream surface of the geomembrane.

The steady-state transport is described by Fick's law (Equation 4-9). The permeability of a variety of geomembranes to carbon dioxide (CO_2), methane (CH_4), and nitrogen (N_2) gases is shown in Table 13-5.[83]

TABLE 13-5
Permeability of polymeric geomembranes to selected gases at 23°C, determined in accordance with ASTM D1434, Procedure V*

Geomembrane description				Gas transmission rate mL/(m^2 · day · atm) (STP[b])			Gas permeability coefficient $\bar{P}$, barrer[c]		
Base polymer	Density	Thickness,[a] mm	Compound type[d]	CO_2	CH_4	N_2	CO_2	CH_4	N_2
Butyl rubber	. . .	1.60	XL	512	120	19.7	12.5	2.92	0.480
Chlorosulfonated	. . .	0.82R	TP	122	21.6	26.2	1.52	0.27	0.33
polyethylene	. . .	0.86	TP	418	124	27.1	5.47	1.62	0.36
Elasticized polyolefin	. . .	0.58	CX	1450	280	125	12.8	2.47	1.10
Ethylene propylene rubber	. . .	0.89R	TP	2720[e]	. . .	. . .	36.8[e]	. . .	. . .
		0.90	XL	5260	1400	314	72.0	19.2	4.30
Neoprene	. . .	0.90	XL	716	80.9	31.1	9.81	1.11	0.43
Polybutylene	. . .	0.71	CX	818	248	62.3	8.84	2.68	0.67
Low-density polyethylene[f]	0.921	0.25	CX	6180[e]	1340[e]	. . .	23.5[e]	5.10[e]	. . .
Linear low-density polyethylene	0.923	0.46	CX	1370	322	. . .	9.59	2.25	. . .
High-density polyethylene	0.945	0.61	CX	729	138	. . .	6.77	1.28	. . .
	0.945	0.86	CX	467	104	. . .	6.11	1.28	. . .
Polyvinyl chloride	. . .	0.25	TP	7730[e]	1150[e]	. . .	29.4[e]	4.38[e]	. . .
(plasticized)	. . .	0.49	TP	3010	446	108	22.4	3.32	0.81
	. . .	0.81	TP	2840[e]	285[e]	. . .	35.0[e]	3.51	. . .
Polyester elastomer[g]	. . .	0.022	CX	357	. . .	. . .	0.119	. . .	. . .

[a] R = fabric-reinforced.
[b] STP = standard temperature and pressure.
[c] One barrer = 10^{-10} (mL · cm)/(cm^2 · s · cmHg)(STP).
[d] XL = cross-linked; TP = thermoplastic; CX = semicrystalline.
[e] Measured at 30°C.
[f] Natural resin (no carbon black).
[g] This sample is NTIS standard material 1470. The determination was made at 15.0 psi, under which condition the NTIS certified CO_2 transmission rate was calculated to be 338 mL/(m^2 · d · atm) (STP).
*After Haxo.[84]

The migration of "liquids" from one side of a geomembrane to another, by the multistep, complex process described above, can be quantified and expressed as a "permeability" value. This rate of migration can be measured and reported as the geomembrane's permeability (see Table 13-6).[84]

In an experiment employing various solvents, the transmission rates through selected liners were measured.[85] The results of this experiment are shown in Table 13-7.[86] These data can be used to assess contaminant transport through liner systems.

TABLE 13-6
Coefficient of permeability (hydraulic conductivity) to deaerated water (hydraulic conductivity) of several geomembranes at various pressures*

	Hydraulic pressure, kPa			
Base polymer	**100**	**400**	**700**	**1000**
Butyl rubber	22	8.0	4.1	2.8
Chlorosulfonated polyethylene	300	120	77	60
Ethylene propylene rubber	115	57	34	25
Polyvinyl chloride	86	37	20	10

*Data are reported in 10^{-15} m/s.
†After Haxo.[84]

TABLE 13-7
Permeability of polymeric geomembranes to various solvents, measured in accordance with ASTME96, Procedure B (modified to test solvents)*

Polymer[a]:	**ELPO**	**HDPE**	**HDPE**	**HDPE-A**	**LDPE**	**PB**
average thickness, mm:	**0.57**	**0.80**	**2.62**	**0.87**	**0.75**	**0.69**
Type of compound:	**CX**	**CX**	**CX**	**CX**	**CX**	**CX**
	(SVT)g/(m^2 · day)					
Methyl alcohol	2.10	0.16	...	0.50	0.74	0.35
Acetone	8.62	0.56	...	2.19	2.83	1.23
Cyclohexane	7.60	11.7	...	151	161	616
Xylene	359	21.6	6.86	212	116	178
Chloroform	3230	54.8	15.8	506	570	2120
	Solvent vapor permeability[b], 10^{-2} metric perms · cm					
Methyl alcohol	0.11	0.01	...	0.04	0.05	0.02
Acetone	0.23	0.02	...	0.09	0.10	0.04
Cyclohexane	0.49	1.05	...	14.7	13.6	47.8
Xylene	292	24.6	25.6	262	124	175
Chloroform	103	2.46	2.32	24.6	24.0	82.8

[a]ELPO = elasticized polyolefin; HPDE = high-density polyethylene; HDPE-A = high-density polyethylene alloy; LDPE = low-density polyethylene; PB = polybutylene.
[b]The median thickness value was used to calculate the permeability.
*After Haxo.[84]

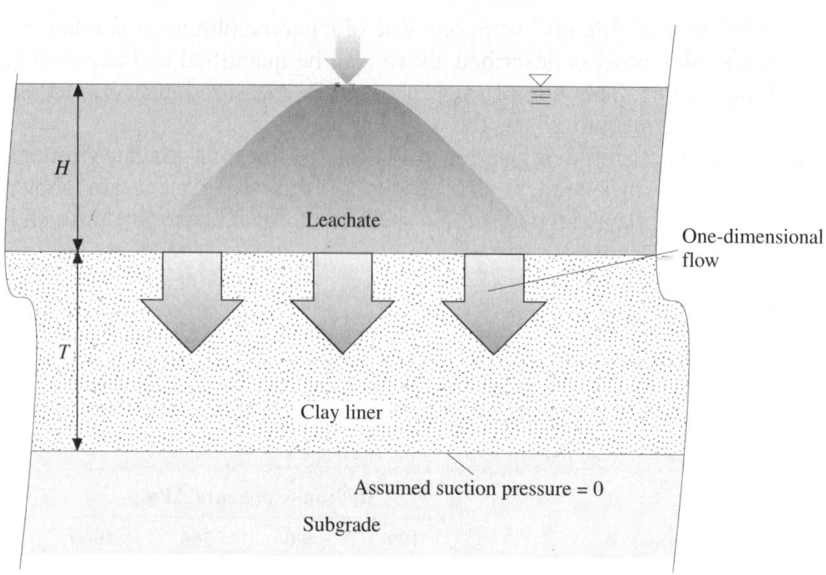

Hydraulic gradient $i = \frac{H + T}{T}$

Steady-state Darcy flux $V = K_L i$

FIGURE 13-12
Schematic of advective flow and transport.

Advective Transport through Compacted Clay

The rate of advective flow through clay liners depends on the hydraulic conductivity, gradient, and liner dimensions. A simplified, straightforward analysis (neglecting retardation) of the contaminant loading in the steady state can be made using Darcy's law (Equation 4-3) and the concentration of the contaminants in the leachate.

Identified in Figure 13-12 are the critical parameters for the analysis. The hydraulic head is equal to the height of free leachate above the liner plus the thickness of the clay liner, not the height of the leachate alone. The solution then is a direct application of Darcy's law to yield the flow rate. The advective transport rate can be determined as the product of the contaminant concentration and the flow rate.

EXAMPLE 13-6. CALCULATION OF CONTAMINANT MIGRATION THROUGH A COMPACTED CLAY LINER. An old landfill, 300 m by 500 m in plan, was constructed with a single clay liner 1 m thick having a hydraulic conductivity of 5×10^{-7} cm/s. The average leachate head is 0.5 m. An analysis of the leachate indicates a concentration C of carbon tetrachloride at 10 mg/L, among other contaminants. What is the annual steady-state flow and advective contaminant transport rate from this landfill? How much drinking water per year could be contaminated to the drinking water standard of 0.005 mg/L?

Solution. Calculate the hydraulic gradient:

From Figure 13-12, $i = (H + T)/T$

$$i = (0.5 + 1)/1 = 1.5 \text{ m/m}$$

Calculate the Darcy flux:

From Fig. 13-12, $v = ki$

$$v = 5 \times 10^{-7}\ \text{cm/s} \times (1.5\ \text{m/m})/(100\ \text{cm/m}) = 7.5 \times 10^{-9}\ \text{m/s}$$

Calculate the flow for the entire landfill on an annual basis:

$$Q = vAt = 7.5 \times 10^{-9}\ \text{m/s} \times 300\ \text{m} \times 500\ \text{m} \times 3.16 \times 10^{7}\text{sec/year}$$
$$= 3.6 \times 10^{7}\ \text{m}^3/\text{year}$$

Calculate the loading L to the subsurface environment:

$$L = Q \times C = 3.6 \times 10^{4}\ \text{m}^3/\text{year} \times 1.0\ \text{mg/L} \times 1000\ \text{L/m}^3$$
$$= 3.6 \times 10^{7}\ \text{mg/year} = 36\ \text{kg/year}$$

Calculate the quantity of contaminated groundwater, V_w:

$$V_w = (3.6 \times 10^{7}\ \text{mg/year})/(0.005\ \text{mg/L}) = 7.2 \times 10^{9}\ \text{L/year}$$

More than 2 billion liters per year from about 36 kg.

A comprehensive study employing both field and laboratory data examined the effect of sorption on diffusion rates for natural clay beneath a hazardous waste landfill.[87] The effective diffusion coefficient for the field and lab conditions was from 0.2 to 0.02 less than the theoretical diffusion coefficient for chloride, meaning movement is slower in field tests than in the laboratory. The decreased mobility of chloride through compacted clay results from sorption and molecular weight effects. However, in practice, much quicker rates have been noted. This is usually because transport has been through "defects" in the barrier systems, otherwise assumed to be homogeneous in the calculations. Diffusion of organics such as benzene is likely to be further retarded compared to chloride. To calculate the breakthrough time for high-toxicity organics such as benzene, small values of C/C_0 such as 0.001 are suggested. Thus, for the field data from a hazardous waste landfill, the time for benzene to break through a 1-m-thick clay layer (at $C/C_0 = 0.001$), without defects, was calculated to be 70 years. Note that reliably estimating the parameters from laboratory tests for predictions of contaminant transport through clay liners is difficult. To reliably estimate the seepage velocity, which depends on the effective porosity, tracer tests can be used.[88]

Transport through Minerally Enhanced Barriers

Bentonite has been found to reduce the rate of contaminant migration by lowering the hydraulic conductivity and increasing the retardation factor. A number of commercial bentonite products are available, and manufacturers provide guidance with respect to application rates. The ability of bentonite to retard contaminant migration is demonstrated by a study that used silica sand/bentonite mixtures and the reactive solute strontium-85. The relationships among the retardation factor R, the distribution coefficient K_d, and the bulk density/porosity ratio are shown in Figure 13-13.[89] Increasing the bentonite content in the admixed liner results in a corresponding increase in the sorptive capacity (and thus an increase in K_d). It would be expected that increasing K_d (by increasing the bentonite content) would result in an increasing retardation factor, but countering this expected relationship is the

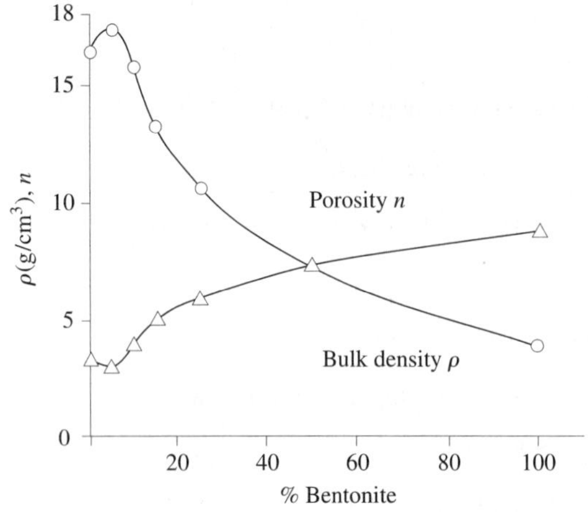

FIGURE 13-13
Strontium-85 coefficients, bulk density to porosity ratios and retardation factors versus bentonite content. (*From Gillham, et al.*[91])

fact that increasing the bentonite content also generally results in a decreasing density/porosity ratio. Thus, as shown in Figure 13-14,[90] the retardation factor reaches a maximum value at a bentonite content between 5 and 10 percent by weight. It is noted that a bentonite content of 5 to 10 percent typically results in the minimum hydraulic conductivity in admixed soil liners.

The effectiveness of a barrier consisting of soil-bentonite and high-carbon fly ash employed to reduce diffusive transport has also been investigated.[91] Because the diffusive transport rates are hindered by sorption to the media, the apparent diffusion coefficient (which incorporates both the effective diffusion coefficient and the retardation) was related to the sorption capacity of the fly ash. From experimental data, a barrier 1 m thick and having an area of 10^8 cm^2 was modelled by assuming a volatile organic solute at 1 mg/L on the interior and 0.0 mg/L on the outside. These results, shown on Figures 13-15[92] and 13-16,[93] are for the two limiting cases

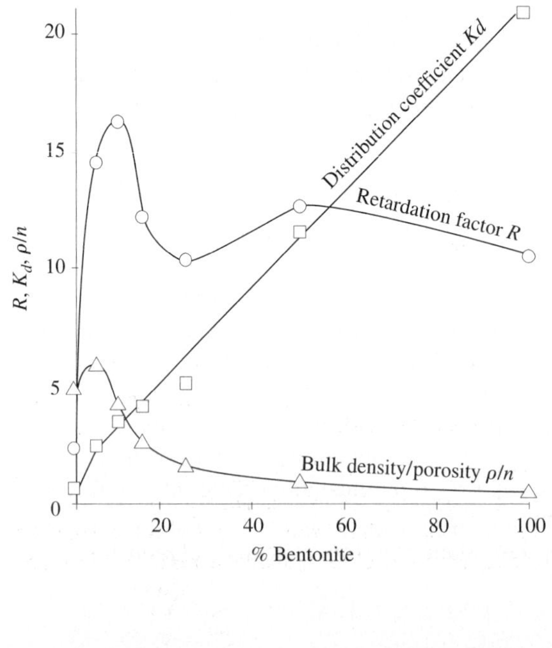

FIGURE 13-14
Strontium-85 diffusion coefficients: experimental and calculated versus bentonite content. (*From Gillham, et al.*[91])

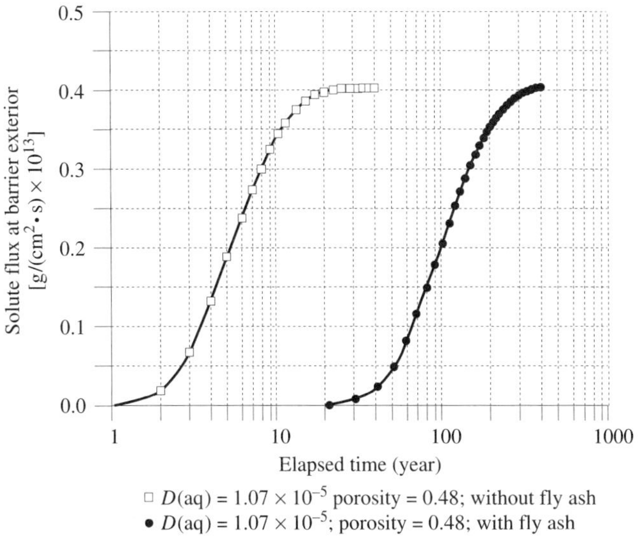

FIGURE 13-15
Model simulations of solute flux. (*From Mott and Weber.*[92])

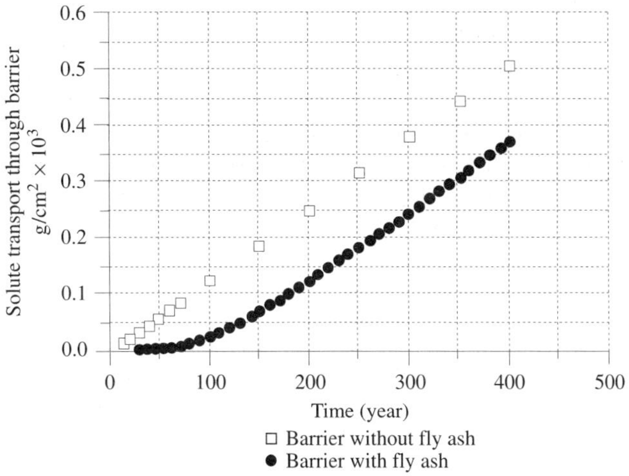

FIGURE 13-16
Model simulations of solute transport. (*From Mott and Weber.*[93])

with no sorptive capacity and equilibrium sorptive capacity. The total contaminant migration is calculated by multiplying the results in Figure 13-15 by the actual interior concentration and the area of the barrier wall.

EXAMPLE 13-7. CONTAMINANT TRANSPORT. Using the data from Figures 13-15 and 13-16 and assuming a liner beneath a 1-acre site: (1) What would be the total contaminant migration in the first 12 years with a soil-bentonite barrier layer? (2) What would be the total contaminant transport for a composite wall for the same period? (3) How long

would it take for the sorption capacity of the composite barrier to be exhausted and for the transport rate to be equal to that of the soil-bentonite barrier?

Solution

1. From Figure 13-16, the total solute flux through the barrier after 12 years is about $0.015\ \text{g/L} \times 10^3\ \text{cm}^2$.
   For one acre we have:

$$\text{Total} = \text{rate} \times \text{area}$$

$$= \frac{0.015\ \text{g}}{1 \times 10^3\ \text{cm}^2} \times 1\ \text{acre} \times 43{,}560\ \frac{\text{ft}^2}{\text{acre}} \times 1.44 \times 10^6\ \frac{\text{cm}^2}{\text{ft}^2} \times \frac{1\ \text{kg}}{1000}$$

$$\text{Total} = \underline{\underline{941\ \text{kg}}}$$

2. From Figure 13-16, the total flux is essentially zero.
3. From Figure 13-15, the flux rate for the barrier with fly ash approaches that without fly ash after about 400 years.

The process of sorbing organics onto organically modified clays from water is, in part, a partitioning process. The quaternary ammonium bound to the organically modified clay surface acts as a partitioning medium to remove organic molecules from water. When used in a composite liner, these organically modified clays retard the migration of organic contaminants that have diffused or otherwise passed through the geomembrane.

The adsorption characteristics of chabazite (a type of zeolite) as examined in triaxial permeability tests is shown in Figure 13-17.[94] In this figure, elapsed time is normalized with respect to hydraulic conductivity and presented as pore volume displacement. Pore volume displacement is the ratio between the volume of fluid permeated and the volume of the voids in the soil. For a solution of lead at 10 mg/L in distilled water, permeability tests on a silty clay and a silty clay chabazite composite were conducted. Figure 13-17 demonstrates the retardation of the lead in the composite liner as compared to the compacted silty clay. Comparison of the results for copper are also shown. Copper reaches breakthrough faster than the lead for both liners. This is due to the difference in retardation factors between copper and

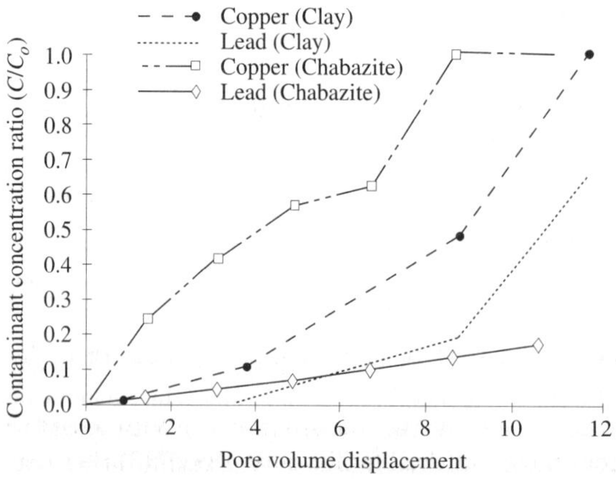

FIGURE 13-17
Permeation of chabazite and silty clay.

lead. Chabazite preferentially adsorbs lead as compared to the silty clay. Chabazite preferentially adsorbs lead over copper. These results emphasize the importance of designing composite liners to sorb specific contaminants of interest.

The foregoing examples of various liner systems are presented to encourage consideration of transport through the line systems during design. They are also presented to demonstrate the greater breakthrough time for contaminants that can be achieved. Laboratory studies frequently include column testing during which the contaminant concentration is measured as a function of time (resulting breakthrough curves shaped like those in Figure 13-15). These laboratory data are then used to determine critical parameters for design such as the retardation factor. An alternative approach for the analysis of laboratory column data, termed the cumulative mass approach, uses the measured data in terms of mass of the contaminant rather than the traditional measures of concentration.[95] This method allows for a reduced frequency of measurements without significant loss in the reliability of the determined parameters of Peclet number and retardation.

EXAMPLE 13-8. LONG-TERM PERMEABILITY TEST. A sample of clay barrier soil is permeated in the laboratory to study contaminant transport through the sample. The cylindrical sample is 70 mm in diameter, 70 mm high, and has a porosity of 0.4. What total volume of permeant has passed through the sample to achieve a pore volume displacement of 3? If the test is run at a gradient of 5 and the liner hydraulic conductivity is less than 1×10^{-7} cm/s, how long will the test take?

Solution. Total sample volume:

$$V_t = \pi \times r^2 \times h = 3.14 \times 3.5^2 \times 7.0 = 269 \text{ cm}^3$$

Void volume:

$$V_v = V_t \times n = 269 \times 0.4 = 107.6 \text{ cm}^3$$

Permeant volume:

$$V_p = 3 \times V_v = 3 \times 107.6 = 322.8 \text{ cm}^3$$

Test duration:

$$q = kiAt$$

$$t \geq q/kiA = 322.8/1 \times 10^{-7} \times 5 \times 3.14 \times 3.5^2$$

$$\geq 1.68 \times 10^7 \text{ s}$$

$$\geq 194 \text{ days}$$

Comment. These long-term permeability tests require considerable time to complete, necessitating special test precautions[96] and considerable advance planning.

13-8 LANDFILL STABILITY

A landfill can fail to meet performance requirements for a number of reasons. Many of them pertain to stability.

Failure mechanisms for cover systems include subsidence where total and differential settlements exceed the tolerance of the landfill cover system. In this case,

geotextiles or geogrids may be used to span expected zones of subsidence. Changes in the environment in the cover system may also alter its performance. For example, frequent wetting and drying may cause desiccation cracks in any clayey barrier layers. Freezing and thawing may likewise alter the performance of the drainage layer and cover layer. Frequently, therefore, cover systems are quite thick to provide adequate protection to the barrier layers from these environmental stresses. Failure of the cover system may also occur as a result of erosion. A cover is designed to enhance runoff, and the system must be able to withstand the erosional forces associated with this runoff. Of course, the cover materials must remain in place. Cover instability as a result of sliding between the dissimilar materials (such as geotextiles, granular drains, geomembranes, and clay barrier layers) is well documented[97] and the reader is advised to fully consider all potential failure surfaces and the strength along these surfaces.

EXAMPLE 13-9. LANDFILL FAILURE. An aboveground ash monofill was constructed over the natural sloping ground with little modification. Perimeter ditches were installed to control surface water. During operation and immediately after a rain, 40,000 m^3 of ash washed out of the landfill.[98] How could this happen?

Solution. The working area was too large, resulting in low runoff and high infiltration. As rainwater entered the ash and migrated downward and outward to drainage systems, the fill became unstable, causing the significant loss of ash from the landfill.

Landfills must be stable during construction, during operations, and for many years following closure. Some settlement is inevitable; however, the key measure is whether the waste can be considered **stable** is if it stays within the containment system or, more precisely, if there are no mass movements down slope. Stability, often termed **slope stability,** can be mathematically defined in a number of different ways such as by rotational force equilibrium, rotational moment equilibrium, or sliding wedge force equilibrium. Entire books have been written on the subject of slope stability,[99] and numerous papers have been written on the stability of waste fills.[100] The analysis of slope stability lies in the realm of the geotechnical engineer who is trained and experienced in the selection of both the method of analysis most appropriate and the selection of input parameters to that analysis. The following method of analysis is presented to give the student of hazardous waste management some insight into stability evaluations.

The analysis of stability necessitates the hypothesis of a failure surface, the determination of driving and resisting forces, and the definition of a factor of safety employing these force systems. Figure 13-18 depicts a cross section through a landfill and shows just one of many possible failure surfaces. The potential failure surface is often assumed circular, a common assumption that makes the mathematics easier and closely models observed failure surfaces. The driving forces are from the gravitational pull on the landfill materials, and the resisting forces are from the strength of the waste and the underlying soil. The strength of the material is expressed in terms of the sum of cohesion (stress independent) and friction (stress dependent). The strength of waste has received considerably less attention although it can be modeled for stability analysis in the same manner as soil. The computed factor of safety will vary with the assumed location of the failure surface. The factor of safety for the landfill is the lowest computed for the various assumed failure surface locations. The location that produces the lowest factor of safety is termed the **critical**

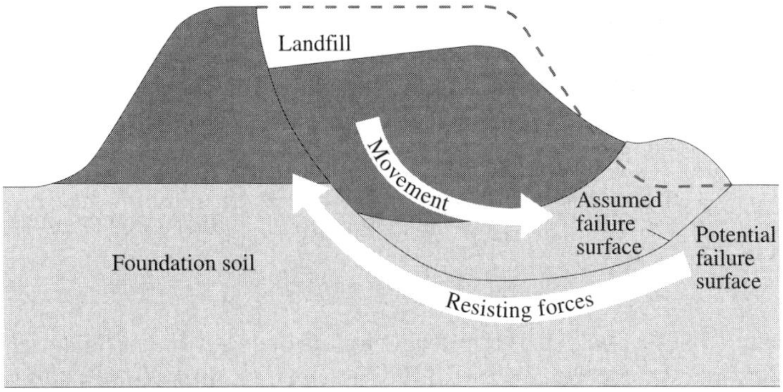

FIGURE 13-18
Schematic of landfill stability.

failure surface. For example, the sliding of the cover soils on the geomembrane barrier layer represents one such potential for failure surface.

In the next several paragraphs, an examination of a limit equilibrium analysis for evaluating slope stability is made. By focusing on one relatively simple case, the student can see one approach to slope stability analysis.

The maximum shearing stress that a soil can withstand without failure is defined by the Mohr-Coulomb failure criteria as:

$$T_{ff} = c + \sigma_{ff} \times \tan\ \phi \tag{13-5}$$

where T_{ff} = shear stress at failure on the failure plane (kPa)

c = cohesion (kPa)

σ_{ff} = applied normal stress at failure on the failure plane (kPa)

ϕ = angle of internal friction (degrees)

If the shear stress to failure is applied at a relatively rapid loading rate, it is considered an undrained loading. A landslide at the end of construction is represented by the undrained loading condition. The ability of the mass to remain stable for many years after construction is modeled as a drained loading condition. The analysis of landfill stability must demonstrate stable conditions at the end of construction (undrained shear strength parameters are used) and in the long term (drained shear strength parameters are used). The terms *drained* and *undrained* refer to the absence or presence of excess pore water pressures in the materials undergoing shearing strain. The landfill must be stable for potential failure surfaces that pass only through the waste and pass through both the waste and underlying foundation soil. The stability of the cover must also be considered with respect to the sliding of the vegetative and drainage layers over the barrier layer, particularly when the barrier layer is a geomembrane.

To illustrate just one of the methods of stability analysis, a special case is examined where the cohesion is zero, that is, for cohesionless soils (sands and gravels). This stability analysis is significantly simplified compared to that for cohesive soils. For this case, the critical failure surface is an infinitely small distance into the slope and the slope is assumed to be infinitely long. The resulting solution is termed the

infinite slope analysis. Force equilibrium equations can be written for the downslope movement of a soil mass infinitely thin. The factor of safety (FS) is defined as ratio of the resisting forces F_r to the driving forces F_d, written as:

$$FS = F_r/F_d \tag{13-6}$$

The resulting equation of equilibrium for a slope at angle β measured from the horizontal is:

$$FS = \tan\phi/\tan\beta \tag{13-7}$$

The infinite slope analysis is applicable to fills of cohesionless materials and can be used to examine the potential for sliding between geomembranes and soil zones in cover and liner systems. Example 13-10 illustrates its use.

EXAMPLE 13-10. LANDFILL STABILITY. An ash monofill is planned, employing fly ash from an incineration process. The angle of internal friction ϕ was measured to be 28° and the cohesion c was measured to be zero. A factor of safety of 2.0 is desired for stability, given the hazardous nature of the ash and the potential variability in ash properties. What is the maximum slope angle β for this ash monofill?

Solution. Rewriting Equation 13-7 yields:

$$\beta = \tan^{-1}\left(\frac{\tan\phi}{FS}\right)$$

Substituting gives

$$\beta = \tan^{-1}\left(\frac{\tan 28}{2.0}\right) = 14.9^\circ$$
$$= 3.76 \text{ horizontal to 1 vertical}$$

In developing or reviewing the design of a waste disposal facility, an often overlooked question is "Can it be built?" Instability during construction may render a design inadequate even though, if it could be built, it would have been stable.

EXAMPLE 13-11. LANDFILL CONSTRUCTABILITY. A landfill design such as shown on Figure 13-1 is planned. On behalf of Landfills-R-Us, a consulting firm employed a design using below-grade slopes of 1 horizontal to 1 vertical. The soil at the site is a medium dense, medium-to-fine sand. Is the landfill stable in the final configuration? Can it be built?

Solution. Yes and no. After the waste is placed in the landfill, the side slopes below grade are stable. However, during construction the evacuation slopes are not stable. Assume $\phi \approx 40^\circ$ for a medium dense sand. The factor of safety, by Equation 13-7, is:

$$FS = \frac{\tan\phi}{\tan\beta}$$

where $FS = \tan 40^\circ/\tan 45^\circ = 0.84 < 1$ Unstable!

$\beta =$ slope angle

$\phi =$ angle of internal friction

The infinite slope analysis illustrates just one of the many analyses and types of analyses that may be conducted to evaluate the stability of a landfill. The landfill

must be stable at the end of construction, during operation, and in the long term after closure.

One of the more complex aspects of the stability of landfills is determination of the appropriate shear strength parameters for use in the analysis and design. The shear strength parameters include "internal" shear strength parameters (such as the angle of internal friction in Example 14-10) as well as the "interface" shear strength between dissimilar materials. As shown in Figures 13-2 to 13-6, there may be many interfaces in the modern landfill liner and cover systems. Further, material choices are many and the interface friction between the selected materials must be assessed for the specific materials chosen. For example, textured HDPE geomembranes offer substantially increased interface friction over smooth HDPE geomembranes.[101] However, at high normal stresses (i.e., deep in the landfill) the texture may be damaged, resulting in a reduced friction on the surface. Similarly, the friction between a textured HDPE geomembrane and nonwoven geotextile may be reduced at high normal stresses (referred to as a nonlinear failure envelope). Thus, the determination of the shear strength parameters must be made over the range of stresses expected in the landfill.

As shown by the preceding sections of this chapter, there is much to consider in the design of waste disposal facilities. Let us imagine a perfect landfill design for a perfect landfill site. What are the chances that the constructed product is exactly as designed? Zero to none. However, with redundancies in design and adequate quality control, the design intent can be fulfilled. It must be first recognized that, although the parameters previously presented were described in a deterministic fashion, they are, in fact, stochastic parameters. To fully understand the results achievable in the field, it is likely that probabilistic methods need to be employed.[102]

13-9
SURFACE IMPOUNDMENTS AND DEEP WELL INJECTION

Hazardous waste placed in landfill facilities is considered managed. Hazardous waste placed in surface impoundments is placed there for treatment or storage on its way to treatment and/or ultimate waste management. Land-based storage facilities include surface impoundments such as pits, ponds, and lagoons.

In a study of the performance of surface impoundments, nine existing facilities were examined in an effort to improve design, construction, and operation of these types of facilities. Presented in Table 13-8[103] is a description of the general features and waste characteristics of the nine surface impoundments selected for detailed examination. This table illustrates to the student of hazardous waste management the range of wastes managed in surface impoundments. The ineffective performance of seven of the nine facilities studied in detail was attributed to inadequacies in design, subsurface investigations, and construction quality control. The reverse was true for the two facilities with successful performance records. It is intuitive that the level of environmental protection provided by surface impoundments depends on excellence in both design and construction.

Liquid hazardous waste has been and continues to be disposed of in deep wells. The Environmental Institute for Waste Management Studies concluded in a study of the performance of deep injection wells that most hazardous waste injection wells were properly sited, constructed, and operated.[105] It was found that many

TABLE 13-8
Case study facilities—general features and waste characteristics*

Case study no.	Type of facility	Type of SI (no. and function)	Year placed in service	SI size, acres	Waste type	Waste quantity
1	Electrolytic metal refining plant	Small disposal pond	1972	0.4	Acidic process liquor and sludge waste (pH < 2) high in heavy metal content	A total of 843,750 gal in 1982
		Disposal pond	1979	1.1	Acidic process liquor and sludge waste (pH < 2) high in heavy metal content	
2	Pesticide formulation and distribution plant	Pesticide washdown evaporation disposal pond	1979	<0.1	Pesticide rinsewater	Batch operation (400 gal/day maximum)
		Pesticide rinsewater evaporation disposal pond	1982	<0.1	Pesticide rinsewater	Batch operation (400 gal/day maximum)
3	Commercial hazardous waste disposal facility	Site A: 8 impoundments used for settling, storage, and sludge disposal	1951	15	Oily water and brines, alkaline and acid wastes, heavy metals, paint sludge, tank bottom sediments, cyanide, pesticides, and other chemical wastes	A total of 53 million gal in 1982
4	Agricultural fertilizer manufacturing plant	11 settling ponds used to remove gypsum	1965	14	Production water for ammonium phosphate/phosphoric acid plant with pH < 2 and high radionuclide content	20,000 gal/day
		1 evaporation pond (treatment)	1976	8	Wastewaters from plant boilers, water treaters, and nitric and sulfuric acid plants (pH < 2, high in radionuclides	130,000 gal/day
		1 cooling pond (treatment)	1976	38	Same as gypsum SI	10,000 gal/day
5	Mineral ore mining/manufacturing plant	5 low-head solar ponds (treatment and storage)	1972	90	Mineral liquor tailings with high arsenic and boron content	A total of 50 million gal/month to ponds A–E, 4, and 5
		High-head evaporation pond (treatment and storage)	1975	80	Mineral liquor tailings with high arsenic and boron content	
		High-head evaporation pond (treatment and storage)	1976	100	Mineral liquor tailings with high arsenic and boron content	

TABLE 13-8
Continued

Case study no.	Type of facility	Type of SI (no. and function)	Year placed in service	SI size, acres	Waste type	Waste quantity
		Evaporation pond (treatment and storage), high arsenic and boron content	1980	120	Acid plant wastewater with high arsenic and boron content	12 million gal/month
6	Commercial hazardous waste disposal facility	Evaporation pond (disposal)	1980	5	Geothermal muds and brines, wastewater treatment sludge, tank bottom sediments, cooling tower blowdown sludge and oil drilling muds	9.85 million gal in 1982
		Disposal pond (currently used for land treatment)	1980	5	Geothermal muds and brines, wastewater treatment sludge, tank bottom sediments, cooling tower blowdown sludge and oil drilling muds	
		Disposal pond (currently)	1981	5	Geothermal muds and brines, wastewater treatment sludge, tank bottom sediments, cooling tower blowdown sludge and oil drilling muds	
7	Agricultural fertilizer manufacturing plant	Cooling pond (treatment)	1974	100	Process water from phosphoric acid plant with pH < 2 and high flouride content	40,000 gal/min (maximum)
		Initial gypsum pond (disposal)	1974	150	Gypsum slurry with pH < 2 and flouride and phosphorus content	No data available
		Expansion gypsum pond (disposal)	1980	200	Gypsum slurry with pH < 2 and flouride and phosphorus content	No data available
8	Chemical production plant	2 equalization/ retention basins (treatment)	1976	35	Wastes high in organic nitrogen content and varying pH, resulting from synthetic fiber production	3000 gal/min
9	Uranium mining/ milling	Tailings pond (disposal)	1980	64	Acidic tailings slurry containing kerosene and radium 226	No data available

*After Ghassemi et al.[104]

of the problems with deep well disposal were operator related and could have been prevented with standardized operating procedures. A suitable site will generally have overlying and underlying confining strata sufficiently thick and laterally extensive to confine the waste to the injection interval.

The fate of contaminants after disposal in deep wells is not clear, although the combination of favorable chemical reactions and dilution afford protection to groundwater and surface water supplies. In fact, one facility owner, who operates more than 25 percent of the Class I injection wells, claims no known water supply contamination over an operational history of more than 35 years.[106] Nonetheless, monitoring of other nearby wells and fracture zones is typically conducted.

The injected wastes undergo chemical reactions both between waste constituents and between waste constituents and the formation materials. Reactions include carbonate dissolution in limestone and dolomite, sand dissolution, clay dissolution, hydrolysis, and coprecipitation. The fate of contaminants is predicted by modeling neutralization, hydrolysis, and precipitation reactions using reaction rates and equilibrium constants from the literature for solid-liquid equilibria for aqueous electrolyte solutions. Because these reactions occur deep in the bedrock formations, it is difficult to predict with certainty the extent of unreacted waste and the nature and distribution in the highly anisotropic and heterogeneous subsurface environment.

13-10
CLOSURE AND POSTCLOSURE CARE

The closure of a hazardous waste facility should "minimize or eliminate, to the extent necessary to protect human health and the environment, post-closure escape of hazardous waste, hazardous constituents, leachate, contaminated runoff, or hazardous waste decomposition products to the ground or surface waters or to the atmosphere. . . ."[107] The environmental protection should be accomplished with minimal need for further maintenance, although it must be recognized that landfills require some perpetual care, however minimal. The owners/operators of hazardous waste facilities must be financially responsible for properly closing and maintaining a closed facility as well as preclosure and postclosure sudden and nonsudden occurrences.

With the successful closing of a land disposal facility and construction of the final cover, monitoring of the system is required. Monitoring of the system includes measurement of the leachate head within the landfill to preclude the **bathtub effect.** Should the precipitation infiltration and leachate generation exceed the ability of the leachate collection system to remove the leachate, leachate levels build up within the landfill much the same as water fills up a bathtub. This higher hydraulic head causes greater contaminant migration out of the landfill.

Another major closure consideration is maintenance of the cover integrity. For example, burrowing animals dig holes through the cover and challenge its integrity. Erosion of slopes by heavy precipitation events also has been known to occur. Woody vegetation or vegetation with deep tap roots must be controlled so the root systems of vegetation do not challenge the integrity of the system. If the cover is subject to traffic, the effects need to be monitored.

Finally, monitoring of the groundwater and subsurface environment is a necessary component of the final closure and postclosure care. Monitoring wells in the

saturated zone and lysimeters in the unsaturated (vadose) zone are required. Discussion of monitoring well installations and their performance is included in Section 15-7. Sampling and analysis of groundwater from properly sited and installed monitoring wells provide a well-proved method to ensure that groundwater beneath a land disposal facility is not contaminated. However, should contaminants be found by this approach, the detection occurs *after* the groundwater is contaminated. The approach of using monitoring wells to detect contamination and employing cleanup techniques to remediate the site should contaminants be found is analogous to closing the barn door after the horse is out. To provide an improved level of environmental protection, methods to check the barn door are required.

Air monitoring may also be required. Air emissions can be monitored at the site boundaries to ensure limited risk to the public health and the environment.

The previous discussion of closure relates to landfill facilities closed in place. For facilities such as pits, ponds, and lagoons, closure may also be accomplished by total removal. In this case, liner systems, contaminated soils, and components of the temporary storage facility are excavated and removed for ultimate treatment and/or disposal.

It may be interesting to the student of hazardous waste management that closure and postclosure care has, in the past, not always proceeded along the technological lines previously described. In a 1986 study by the U.S. General Accounting Office (GAO), 74 hazardous waste facilities had filed for bankruptcy, hindering the efforts of the government to enforce a proper facility closure.[108] The GAO also found that about 75 percent of the firms employed the approved financial test required at the time to demonstrate they are financially strong and could pay closure and postclosure costs without set aside (a trust fund) money. However, because corporate financial strengths change rapidly, adequate funds were not available at the time of closure. Since this study, revisions in the financial tests and bonding and insurance requirements have been made to increase the likelihood that monies needed for closure and postclosure care are available.

DISCUSSION TOPICS AND PROBLEMS

13-1. Cite four reasons landfills remain a viable hazardous waste management alternative.

13-2. Redo Example 13-1 for three additional hazardous waste management techniques.

13-3. Explain the relationship between aquifers and the siting of landfills.

13-4. Considering the results of Example 13-3, discuss the advantages and disadvantages of employing 0.2 m of daily cover. Use calculations to support your answer.

13-5. Assume 17 percent of the infiltration is absorbed by the landfill and daily cover in Example 13-4. What will be the new rate of leachate generation?

13-6. Assume the landfill waste in Example 13-4 consolidates and releases liquids as leachate from the pore spaces equal to 1 percent of the landfill volume. No infiltration is absorbed by the landfill and daily cover. What will be the new rate of leachate generation?

13-7. Divide the class into two groups: environmental extremists and industrial party-liners. Discuss the following questions employing factual information.

(*a*) We are just making our 55-gallon drums bigger.

(*b*) All landfills leak, so they should be banned.

(*c*) When single liners were found inadequate, double liners were used beneath landfills. Now (2000) hazardous waste landfills have three liners, two of geosynthetic materials and one of natural materials. Are landfills now "safe" methods of hazardous waste disposal or will the future prove these systems inadequate?

13-8. Comment on the implications of an action leakage rate of 5 gal/acre/day versus the readily obtainable leakage rate of 22 gal/acre/day.

13-9. What are the factors which affect the rate of contaminant migration?

13-10. Explain the difference between the two types of molecular diffusion.

13-11. You are considering the selection of a geomembrane as secondary containment beneath a buried underground fuel (gasoline) storage tank. PVC, HDPE, and neoprene are being considered. Which would you recommend and why?

13-12. Identify and describe the three steps involved in the transport of a contaminant through a geomembrane.

13-13. Final cover over a hazardous waste landfill is typically installed to minimize the environmental risk by addressing potential contaminant pathways. Identify four of the contaminant pathways mitigated through the use of a cover system.

13-14. Explain the importance of the vegetative support layer in a landfill cover system.

13-15. Identify the three major components of a landfill cover system and describe their functions.

13-16. Sketch your conceptual design of a cover system for four regions of the United States: Minnesota, Arizona, Washington, and South Carolina.

13-17. Water treatment systems often use sand and gravel as filters, just as landfills use sand and gravel in leachate collection and filter zones. How are sand filters cleaned in water treatment plants? What are the implications for cleaning sand in leachate collection zones? What are implications for attempting to clean geotextiles?

13-18. A 12-inch layer of soil-bentonite is planned for placement immediately beneath the secondary geomembrane liner. The layer is to be constructed in two 6-inch lifts with a bentonite content of 9 percent by dry weight at a compacted unit weight of 95 lb/ft^3. By weight, how much bentonite is needed per square foot of liner per lift?

13-19. Using the data from Table 13-7:

(*a*) Calculate the total contaminant migration of xylene through a 10-acre site lined with a single 0.8 mm HDPE geomembrane in 1 year.

(*b*) Calculate the number of gallons of groundwater that could become contaminated with 1.0 μg/L for this contaminant in 1 year.

13-20. Using the data from Figures 13-15 and 13-16 and assuming a liner beneath a site of 10 acres, calculate the number of gallons of groundwater that could become contaminated per year to the drinking water limit for the contaminant during the first year, at 10 years, and at 100 years.

13-21. Describe the various ways by which liquids can migrate through a geomembrane liner.

REFERENCES

1. U.S. EPA: *Safer Disposal for Solid Waste, the Federal Regulations for Landfills,* EPA/530 SW-91 092, March 1993.
2. Ham, R. K.: "New Approaches to Technology—Landfills," *Fourth National Conference on Environmental Issues,* Academy of Natural Sciences, Philadelphia, January 1989.
3. Daniel, D. E., and R. M. Koerner: *Quality Assurance and Quality Control for Waste Containment Facilities,* EPA/600/R-93/182, 305 pp., September 1993.

4. Blum, J. A.: "Killing Two Birds with One Stone Quarry," *Geotechnical Fabrics Report,* vol. 17, no. 4, pp. 22–27, May 1999.
5. Koerner, R. M.: *Designing with Geosynthetics,* 4th ed., Prentice Hall, Upper Saddle River, N.J., 1998.
6. Reddy, K. R., and R. E. Saichek: "Characterization and Performance Assessment of Shredded Scrap Tires as Leachate Drainage Material in Landfills," *14th International Conference on Solid Waste Technology and Management,* Philadelphia, November 1998.
7. Kershaw, D. S., B. C. Kulik, and S. Pamukcu: "Ground Rubber: Sorption Media for Ground Water Containing Benzene and O-Xylene," *Journal of Geotechnical and Geoenvironmental Engineering,* vol. 123, no. 4, pp. 324–334, April 1997.
8. Lycoming Landfill, Lycoming Co., Pa., 1991.
9. Bonaparte, R., and B. A. Gross: "Field Behavior of Double Liner Systems," *Waste Containment Systems: Construction, Regulation, & Performance,* R. Bonaparte, ed., ASCE Geotechnical Special Publication No. 26, pp. 52–83, November 1990.
10. U.S. Enviromental Protection Agency: *Requirements for Hazardous Waste Landfill Design, Construction, and Closure,* EPA/625/4-89/022, 127 pp., August 1989.
11. 40 CFR 264.302.
12. U.S. EPA: *Action Leakage Rate for Leak Detection Systems,* EPA/530/R-92/004.
13. German Geotechnical Society, Geotechnics of Landfills and Contaminated Land: *Technical Recommendations, for International Society of Soil Mechanics and Foundation Engineering,* Ernst and Sohn, Berlin, 76 pp., 1990.
14. Evans, J. C., Y. Sambasivam, and S. J. Zarlinski: "Attenuating Materials in Composite Liners," *Waste Containment Systems: Construction, Regulation, & Performance,* R. Bonaparte, ed., ASCE Geotechnical Special Publication No. 26, pp. 246–263, November 1990.
15. Johnson, R. L., C. D. Palmer, and W. Fish: "Subsurface Chemical Processes," *Transport and Fate of Contaminants in the Subsurface,* EPA/625/4-89/019, September 1989.
16. Bagchi, A.: *Design, Construction & Monitoring of Sanitary Landfill,* John Wiley & Sons, New York, 1990.
17. Bagchi, A.: "Design of Attenuating Landfills," *Journal of Environmental Engineering Div.,* ASCE, vol. 110, no. EE-6, pp. 1211–1212, 1984.
18. Daniel, D. E., and C. D. Shackelford: Disposal Barriers that Release Contaminants Only by Molecular Diffusion, *Proceedings of the Oak Ridge Model Conference,* Oak Ridge National Laboratory, Oak Ridge, Tenn., vol. I, no. 3, 1987, pp. 315–332.
19. Evans, et al., "Attenuating Materials in Composite Liners."
20. Katz, L., D. N. Humphrey, P. T. Jankauskas, and F. A. DeMasicio, "Engineered Soils for Low-Level Radioactive Waste Disposal Facilities: Effects of Additives on the Adsorptive Behavior and Hydraulic Conductivity of Natural Soils," *Hazardous Waste and Hazardous Materials,* vol. 13, no. 2, pp. 283–306, 1996.
21. Smalstig, E.: "Comparison Study of Different Clay Liners Used as Barriers to Contaminant Transport," unpublished student independent laboratory report, Bucknell University, May 1990.
22. Mead, T.: "Composite Slurry Walls with Organo-clays" unpublished student independent laboratory report, Bucknell University, May 1990.
23. Paraseal, Paramount Technical Products, Inc., Bloomington, Minn., 55431.
24. Evans, J. C., and S. E. Pancoski: "Organic Waste Treatment with Organically Modified Clays," *Proceedings of the 3rd International Conference on New Frontiers for Hazardous Waste Management,* Pittsburgh, EPA/600/9-89/072, pp. 48–57, September 10–13, 1989.
25. Mitchell, J. K., and A. T. Yeung: "Electro-Kinetic Flow Barriers in Compacted Clay," *Transportation Research Record 1288,* Transportation Research Board, National Research Council, Washington, D.C., pp. 1–9, 1990.

26. Landreth, R. E., D. E. Daniel, R. M. Korenver, P. R. Schroeder, and G. H. Richardson: *Design and Construction of RCRA/CERCLA Final Covers,* EPA/625/4-91/025, May 1991.
27. Germain, A.: "Case History: Green Cap Reduces Leachate," *Solid Waste Technologies,* vol. 11, no. 5, pp. 37–39, September/October 1997.
28. Young, A.: "Mathematical Modeling of Landfill Gas Extraction," *Journal of Environmental Engineering,* vol. 115, no. 6, pp. 1073–1087, December 1989.
29. Schroeder, P. R., A. C. Gibson, and M. D. Smolen: *The Hydrologic Evaluation of Landfill Performance (HELP) Model,* EPA/830-SW-84-010, 1984.
30. Brandelik, K. A., and Huebner, C.: "Monitoring of Clay-Type Surface Barrier on Dump Sites," *Proceedings of the 6th International FZK/TNO Conference, ConSoil '98,* Thomas Telford, London, vol. 2, pp. 1055–1056, 1998.
31. McEnroe, B. M.: "Steady Drainage of Landfill Covers and Bottom Liners," *Journal of Environmental Engineering,* vol. 115, no. 6, pp. 1114–1122, December 1989.
32. Edil, T. B., V. J. Ranguette, and W. W. Wueller: "Settlement of Municipal Refuse," in *Geotechnics of Waste Fills—Theory and Practice,* ASTM STP 1070, A. Landva and G. D. Knowles, eds., American Society for Testing and Materials, Philadelphia, 1990.
33. Powrie, W., D. J. Richards, and R. P. Beaven: "Compression of Waste and Implications for Practice," in *Geotechnical Engineering of Landfills,* N. Dixon, E. J. Murray, and D. R. V. Jones, eds., Thomas Telford Publishing, London, 1998.
34. U.S. Environmental Protection Agency: *Remedial Action at Waste Disposal Sites,* EPA/625/6-85/006, October 1985.
35. Koerner, R.: *Designing with Geosynthetics,* 2nd ed., Prentice-Hall, Englewood Cliffs, N.J., 652 pp., 1990.
36. Landreth, Robert E.: "Chemical Resistance Evaluation of Geosynthetics Used in Waste Management Applications," *Geosynthetic Testing for Waste Containment Applications,* R. M. Koerner, ed., ASTM STP 1081, pp. 3–11, 1990.
37. EPA Method 9090; "Compatibility Tests for Wastes and Membrane Liners," EPA SW-846, *Test Methods for Evaluating Solid Waste,* U.S. Environmental Protection Agency, Washington, D.C.
38. White, D. F., and K. L. Verschoor: "Practical Aspects of Evaluating the Chemical Compatibility of Geomembranes for Waste Containment Applications," *Geosynthetic Testing for Waste Containment Applications,* R. M. Koerner, ed., ASTM STP 1081, pp. 25–36, 1990.
39. Carroll, R. G., Jr.: "Hydraulic Properties of Geotextiles," *Geotextile Testing and the Design Engineer, ASTM STP* 952, J. E. Fluet, Jr., ed., American Society of Testing and Materials, Philadelphia, pp. 7–20, 1987.
40. Koerner, R. M., and J. A. Bove: "Lateral Drainage Designs using Geotextiles and Geocomposites," *Geotextile Testing and the Design Engineer,* ASTM STP 952, J. E. Fluet, Jr., ed., American Society of Testing and Materials, Philadelphia, pp. 33–44, 1987.
41. Hodge, J.: "Durability Testing," *Geotextile Testing and the Design Engineer,* ASTM STP 952, J. E. Fluet, Jr., ed., American Society of Testing and Materials, Philadelphia, pp. 119–121, 1987.
42. Murray, E. J.: "Properties and Testing of Clay Liners," *Geotechnical Engineering of Landfills,* N. Dixon, E. J. Murray, and D. R. V. Jones, eds., Thomas Telford Publishing, London, pp. 37–60, 1998.
43. U.S. EPA: *Design, Construction, and Evaluation of Clay Liners for Waste Management Facilities,* EPA/530/SW-86/007F, November 1988.
44. Lambe, T. W.: "The Engineering Behavior of Compacted Clay," *J. Soil Mechanics and Foundation, Engineering Division, American Society of Civil Engineers,* vol. 84, SM2, May 1958.

45. Lambe, T. W.: "The Structure of Compacted Clay," *J. Soil Mechanics and Foundation Engineering Division, American Society of Civil Engineers,* vol. 84, SM2, May 1958.
46. Daniel, D. E., and C. H. Benson: "Water Content Density Criteria for Compacted Soil Liners," *Journal of the Geotechnical Engineering Division, ASCE,* 116(12), pp. 1811–30, 1990.
47. Daniel, D. E.: "Predicting Hydraulic Conductivity of Clay Liners," *Journal of the Geotechnical Engineering Division, ASCE,* vol. 110, no. 2, February 1984.
48. Daniel, D. E., and S. J. Trautwein: "Field Permeability Test for Earthen Liners," *Use of In Situ Tests in Geotechnical Engineering,* ASCE Geotechnical Special Publication No. 6, pp. 146–160, June 1986.
49. Benson, C. H., and D. E. Daniel: "Influence of Clods on Hydraulic Conductivity of Compacted Clay," *Journal of the Geotechnical Engineering Division, ASCE,* vol. 116, no. 8, August 1990.
50. Benson, C. H., D. E. Daniel, and G. P. Boutwell: "Field Performance of Compacted Clay Liners," *Journal of Geotechnical and Geoenvironmental Engineering,* vol. 125, no. 5, pp. 390–403, May 1999.
51. Brown, K. W., and D. C. Anderson. 1983. *Effects of Organic Solvents on the Permeability of Soils.* EPA 600/2-83-016. U.S. Environmental Protection Agency, Cincinnati, Ohio.
52. Mitchell, J. K., and F. T. Madsen: "Chemical Effects on Clay Hydraulic Conductivity," *Geotechnical Practice for Waste Disposal, ASCE,* Geotechnical Special Publication No. 13, pp. 87–116, June 1987.
53. Koener, R. M., and D. E. Daniel: *Final Covers for Solid Waste Landfills and Abandoned Dumps,* ASCE Press, Reston, Va., 226 pp., 1997.
54. LaGatta, M. D., T. B. Boardman, B. H. Cooley, and D. E. Daniel, "Geosynthetic Clay Liners Subjected to Differential Settlement," *Journal of Geotechnical and Geoenvironmental Engineering,* vol. 123, no. 5, pp. 402–410, May 1997.
55. Grim, R. E.: *Clay Mineralogy*. McGraw-Hill, New York, 1968.
56. Ibid.
57. Mumpton, F. A.: *Mineralogy and Geology of Zeolites,* Mineralogical Society of America, Short Course Notes, Southern Printing Co., Blacksburg, Va., vol. 4, 233 pp., 1977.
58. Jordan, J. W.: "Alteration of the Properties of Bentonite by Reaction with Amines," *Mineralogical Magazine and Journal of the Mineralogical Society,* vol. 28, no. 205, pp. 598–605, June 1949.
59. Jordan, J. W.: "Organophilic Bentonites: Swelling in Organic Liquids," *The Journal of Physical & Colloid Chemistry,* vol. 53, no. 2, pp. 294–306, February 1949.
60. Alther, G. R., J. C. Evans, and S. E. Pancoski: "Organically Modified Clays for Stabilization of Organic Hazardous Wastes," *Superfund '88, Proceedings of the 9th National Conference, Hazardous Materials Control Research Institute,* pp. 440–445, November 1988.
61. Chiou, C. T., P. E. Porter, and D. W. Schmedding: "Partition Equilibria of Nonionic Organic Compounds between Soil Organic Matter and Water," *Environmental Science and Technology,* vol. 17, pp. 227–231, 1983.
62. Chiou, C. T., L. J. Peters, and V. H. Freed: "A Physical Concept of Soil-Water Equilibria for Nonionic Organic Compounds," *Science,* vol. 213, pp. 684–685.
63. Boyd, S. A., J. F. Lee, and M. M. Mortland: "Attenuating Organic Contaminant Mobility by Soil Modification," *Nature,* vol. 33, pp. 345–347, May, 1988.
64. Mortland, M. M., S. Shaobai, and S. A. Boyd: "Clay-Organic Complexes as Adsorbents for Phenol and Chlorophenols," *Clays and Clay Minerals,* vol. 34, no. 5, pp. 581–585, 1986.
65. Wolf, T. A., T. Demirel, and R. E. Bauman: "Adsorption of Organic Pollutants on Montmorillonite Treate with Amines," *Journal of the Water Pollution Control Federation,* vol. 58, pp. 68–76, 1986.

66. Tsai, T. D., and P. A. Vesilind: "A New Landfill Liner to Reduce Ground-Water Contamination from Heavy Metals," *Journal of Environmental Engineering,* vol. 124, no. 11, pp. 1061–1065, November 1998.
67. Mott, H. V. and W. J. Weber: "Diffusive Transport and Attenuation of Organic Leachates in Cut-Off Wall Backfill Mixes," *Twelfth Annual Madison Waste Conference,* Department of Engineering Professional Development, University of Wisconsin-Madison, September 20–21, 1989.
68. Smith, J. A., and P. Jaffe: "Benzene Transport through Liners Containing Organophilic Bentonite," *Journal of Environmental Engineering,* vol. 120, no. 6, pp. 1559–1577, November/December, 1994.
69. Freeze, R. A., and J. A. Cherry: *Groundwater,* Prentice-Hall, Englewood Cliffs, N.J., 604 pp., 1979.
70. Daniel, D. E., S. J. Trautwein, S. S. Boynton, and D. E. Foreman: "Permeability Testing with Flexible Wall Permeameters," *Geolotechnical Testing Journal,* vol. 7, no. 3, pp. 113–122, 1984.
71. Dunn, R. J., and J. K Mitchell: "Fluid Conductivity Testing of Fine Grained Soils," *Journal of Geotechnical Engineering,* vol. 110, no. 11, pp. 1648–1665, 1984.
72. Evans, J. C. and H. Y. Fang: "Triaxial Permeability and Strength Testing of Contaminated Soils," *Advanced Triaxial Testing of Soil and Rock,* ASTM STP 977, American Society for Testing and Materials, Philadelphia, pp. 387–404, 1984.
73. Haxo, H. E. Jr., J. A. Miedema, and Nelson, N. A.: "Permeability of Polymeric Membrane Liner Systems," *Proceedings of the International Conference on Geomembranes,* Denver, pp. 151–156, 1984.
74. Verschoor, K., L. Britton, and R. Thomas: "Chemical Compatibility Testing of Geosynthetics to be Used as Containment Barriers in Hazardous Waste Management," *Hazardous Waste and Hazardous Materials,* pp. 205–209, 1988.
75. Haxo, H. E., R. S. Haxo, N. A. Nelson, P. D. Haxo, R. M. White, S. Dakessian, and M. A. Fong: *Liner Materials for Hazardous and Toxic Waste and Municipal Solid Waste Leachate,* Noyes Publications, 1985.
76. Giroud, J. P.: "Impermeabilty: The Myth and a Rational Approach," *Proceedings of the International Research Conference on Geomembranes,* Denver, pp. 157–162, 1984.
77. Shackelford, C. D.: "Transit-time Design of Earthen Barriers," *Engineering Geology,* vol. 29, pp. 79–94, 1990.
78. Rowe, R. K., R. M. Quigley, and J. R. Booker: *Clayey Barrier Systems for Waste Disposal Facilities,* E&FN SPON (an imprint of Chapman and Hall), London, 390 pp., 1995.
79. Johnson, R. L., J. A. Cherry, J. F. Pankow: "Diffusive Contaminant Transport in Natural Clay: A Field Example and Implications for Clay-Lined Waste Disposal Sites," *Environ. Sci. Technol.,* vol. 23, no. 3, pp. 340–349, 1989.
80. Seaman Corporation, XR5 Geomembrane literature, Wooster, Ohio.
81. Shackelford, C. D.: "Diffusion of Contaminants Through Waste Containment Barriers," Transportation Research Record 1219, *Geotechnical Engineering 1989,* Transportation Research Board, National Research Council, pp. 169–182.
82. Haxo, H. E.: "Determining the Transport through Geomembranes of Various Permeants in Different Applications," *Geosynthetic Testing for Waste Containment Applications,* R. M. Koerner, ed., ASTM STP 1081, pp. 75–94, 1990.
83. Haxo, "Determining the Transport through Geomembranes."
84. Ibid.
85. August, H., and R. Tatzky: "Permeabilities of Commercially Available Polymeric Liners for Hazardous Waste Landfill Leachate Organic Constituents," *Proceedings of the International Conference on Geomembranes,* Denver, vol. 1, Industrial Fabrics Association International, St. Paul, Minn., pp. 163–168, 1984.

86. Haxo, "Determining the Transport through Geomembranes."
87. Johnson, R. L., J. A. Cherry, and J. F. Pankow: "Diffusive Contaminant Transport In Natural Clay: A Field Example and Implications for Clay-Lined Waste Disposal Sites," *Environ. Sci. Technol.,* vol. 23, no. 3, pp. 340–349, 1989.
88. Kim, J. Y., T. B. Edil, and J. K. Park: "Effective Porosity and Seepage Velocity in Column Tests on Compacted Clay," *Journal of Geotechnical and Geoenvironmental Engineering,* vol. 123, no. 12, pp. 1135–1142, December 1997.
89. Gillham, R. W., M. J. L. Robin, D. J. Dytynyshyn, and H. M. Johnston: "Diffusion of Nonreactive and Reactive Solutes through Fine-Grained Barrier Materials," *Can. Geotech. J.,* vol. 21, pp. 541–550, 1984.
90. Ibid.
91. Mott, H. V., and W. J. Weber: "Solute Migration in Soil-Bentonite Containment Barriers," *Superfund '89, Proceedings of the 9th National Conference, Hazardous Materials Control Research Institute,* pp. 526–533, November, 1989.
92. Ibid.
93. Ibid.
94. Evans, J. C., Y. Sambasivam, and S. J. Zarlinski: "Attenuating Materials in Composite Liners," *Waste Containment Systems: Construction, Regulation, & Performance,* ASCE Geotechnical Special Publication No. 26, November, 1990, pp. 246–263.
95. Shackleford, C. D.: "Cumulative Mass Approach for Column Testing," *Journal of Geotechnical Engineering,* vol. 121, no. 10, October 1995.
96. Evans, J. C., and H. Y. Fang: "Triaxial Permeability and Strength Testing of Contaminated Soils," *Advanced Triaxial Testing of Soil and Rock,* American Society for Testing and Materials, Philadelphia, pp. 387–404, 1988.
97. Boschuck, J.: "Landfill Covers, An Engineering Perspective," *Geotechnical Fabric Report,* pp. 23–34, March 1991.
98. Morris, D. V. and C. E. Woods: "Settlement and Engineering Considerations in Landfill and Final Cover Design," *Geotechnics of Waste Fills—Theory and Practice,* ASTM STP 1070, A. Landva and G. D. Knowles, eds., American Society for Testing and Materials, Philadelphia, 1990.
99. Huang, Y. H.: *Stability Analysis of Earth Slopes,* Van Nostrand Reinhold, New York, 305 pp., 1983.
100. Landva, A., and G. D. Knowles, eds.: *Geotechnics of Waste Fills—Theory and Practice,* ASTM STP 1070, American Society for Testing and Materials, Philadelphia, 1990.
101. Stark, T. D., T. A. Williamson, and H. T. Eid: "HDPE Geomembrane/Geotextile Interface Shear Strength," ASCE, *Journal of Geotechnical Engineering,* vol. 122, no. 3, pp. 197–203, March 1996.
102. Benson, C.: "Reliability Analysis for Time of Travel in Compacted Soil Liners," *Geotechnical Engineering Congress 1991,* ASCE, Geotechnical Engineering Special Technical Publication No. 27, June 1991.
103. Ghassemi, M., M. Haro, and L. Fargo: *Assessment of Hazardous Waste Surface Impoundment Technology: Case Studies and Perspectives of Experts,* EPA-600/S2-84-173, January 1985.
104. Ibid.
105. Environmental Institute for Waste Management Studies: "Deep Well Injection Studies," *Newsletter,* vol. IX, no. 2, June 1989.
106. Scrivner, N. C., et al.: "Chemical Fate of Injected Wastes," *Ground Water Monitoring Review,* vol. 6, no. 3, pp. 53–58, Summer 1986.
107. U.S. Federal Regulations: "Closure Performance Standard," Sec. 265-111(b), September 23, 1988.
108. General Accounting Office: *Environmental Safeguards Jeopardized when Facilities Cease Operating,* GAO/RCED-86-77-Hazardous Waste, p. 3, February 1986.

ADDITIONAL READING

Bagchi, A.: *Design, Construction, & Monitoring of Sanitary Landfills,* John Wiley & Sons, New York, 284 pp., 1990.

Daniel, D. E., ed.: *Geotechnical Practice for Waste Disposal,* Chapman & Hall, New York, 632 pp., December 1992.

Fetter, C. W.: *Contaminant Hydrogeology,* 4th ed., MacMillan Publishers, 1998.

Koerner, R. M., and D. E. Daniel: *Final Covers for Solid Waste Landfills and Abandoned Dumps,* ASCE Press, Reston, Va., 256 pp., 1997.

Linsley, R. K., Kohler, M. A., and Paulhus, J. L.: *Hydrology for Engineers,* 2nd ed., McGraw-Hill, New York, 1975.

Oweiss, I. S., and Khera, R. P.: *Geotechnology of Waste Management,* 2nd ed., PWS Publishing, 472 pp., 1998.

Rowe, R. K., R. M. Quigley, and J. R. Booker: *Clayey Barrier Systems for Waste Disposal Facilities,* E&FN SPON (an imprint of Chapman and Hall), London, 390 pp., 1995.

Rumer, R. R., and M. E. Ryan: *Barrier Containment Technologies for Environmental Remediation Applications,* John Wiley & Sons, New York, 1995.

Vesilind, P. A., Peirce, T. T., and Weiner, R. F.: *Environmental Engineering,* 2nd ed., Butterworth, 1988, pp. 215–234.

ACKNOWLEDGMENTS

The authors wish to acknowledge the following individuals who assisted with the preparation of this chapter for the first edition: Craig Benson, *University of Wisconsin;* John Repasky and Robert Owens, *Bucknell University*. The principal reviewer for ERM was Alan F. Hassett, P.E.

The reviewer for the second edition was Joseph Fiacco, *ERM*.

CHAPTER 14

Quantitative Risk Assessment

Chaos umpire sits,
And by decision more embroils the fray
By which he reigns: next him high arbiter
Chance governs all.

Milton

The purpose of this chapter is to provide an understanding of risk, how to assess it at a specific site, and what the numerical estimates of risk represent. The National Research Council defines risk assessment as:

> . . . the characterization of the potential adverse health effects of human exposures to environmental hazards. . . . Risk assessment also includes characterization of the uncertainties inherent in the process of inferring risk.[1]

Risk assessment is a tool for understanding the health and environmental hazards associated with hazardous waste and can greatly improve the basis on which to make hazardous waste management decisions. Risk assessment can also assist the public in understanding the human health implications of the issues and available alternatives.

14-1 RISK

In the classical sense, **risk** is defined as the probability of suffering harm or loss. When the resulting harm is measurable (e.g., worker-days lost to accidents), risk may be calculated as the probability of an action occurring multiplied by the severity of the harm if the action does occur:

$$\text{Risk} = (\text{probability}) \times (\text{severity of consequence})$$

Using this calculation suggests that the following two activities have the same risk:

Activity no.	Severity of outcome, worker-days lost	Probability of outcome	Risk, worker-days lost
1	20	0.1	2
2	10	0.2	2

Often the undesirable consequence is not a quantifiable matter. An example of such a consequence is death; no one is ever slightly dead. Another example is the onset of cancer. For such consequences, risk is simply defined as the probability of the harm occurring. Because approximately 25 percent of the U.S. population will experience cancer in their lifetime, the lifetime risk of cancer is approximately 0.25.

An important distinction is whether the risk is background, incremental, or total risk. **Background risk** is what people are exposed to in the absence of the particular source of risk being studied, **incremental risk** is that caused by this source, and **total risk** is simply the sum of the two. The previous reference to 0.25 probability of cancer represents background risk. The U.S. Environmental Protection Agency (EPA) target for Superfund sites of 1×10^{-6} excess lifetime cancer risk represents an incremental risk—the probability of a person developing cancer from exposure to contaminants from a Superfund site is 1×10^{-6} (or "one in a million") in excess of that person's risk of cancer from all other sources combined. It also implies that for every million people potentially exposed to a site, one additional cancer case over their lifetimes would be expected. If the population exposed to a site is 10,000, this number can be interpreted as 0.01 additional occurrence of cancer.

Table 14-1 provides some estimates of some commonplace risks (death in the case of accidents and cancer for the other actions). These represent background risks on an annual basis.

The terms risk and hazard are often confused. **Hazard** is a descriptive term, referring to the intrinsic capability of the waste to cause harm; it is the source of the risk. The hazard posed by a waste is a function of such variables as its toxicity, mobility, and persistence as well as how it is contained. The release or threatened release of a hazardous waste thus represents a hazard—it is a source of risk; however, the waste does not represent a risk unless exposure has occurred or the possibility exists for future exposure.

TABLE 14-1
Common risks[2]

Action	Annual risk	Uncertainty
Motor vehicle accident		
Total	2.4×10^{-4}	10%
Pedestrian	4.2×10^{-5}	10%
Home accidents	1.1×10^{-4}	5%
Cigarette smoking (one pack per day)	3.6×10^{-3}	Factor of 3
Peanut butter (4 teaspoons per day)	8×10^{-6}	Factor of 3
Drinking water with EPA limit of TCE	2×10^{-9}	Factor of 10

Emergence as a Systematic Methodology

Undoubtedly, the concept of risk has a long history of influencing environmental policy making. However, the initial efforts relied more on intuition than on the scientific principles of toxicology, chemistry, and fate and transport modeling inherent in modern risk assessments. Only in the past few years has a *science-based* risk assessment played a major role in environmental decision making.

In this chapter we will use the term *quantitative* to describe the process of using scientific principles to calculate quantitative estimates of risk. The widely recognized approach is a four-stage process:

1. Hazard identification (which chemicals are important)
2. Exposure assessment (where the chemicals go, who might be exposed and how)
3. Toxicity assessment (determining numerical indices of toxicity for computing risk)
4. Risk characterization (estimating the magnitude of risk, and the uncertainty of the estimate)

Risk assessment, as performed under most hazardous waste regulatory programs, employs this concept of conducting the risk assessment process in these four stages. The U.S. National Academy of Sciences created the four-stage process and eventually the EPA codified it. The EPA first applied the concept in 1984 for carcinogens following a suit brought by the Natural Resources Defense Council regarding ambient water quality criteria. In response to this suit, the EPA issued slope factors (SFs) that could be utilized in a predictive capacity to estimate carcinogenic risk to potentially exposed populations. A similar approach was then applied to noncarcinogens by setting acceptable daily intake (ADI) levels, which are analogous to reference concentrations (RfCs). It should be noted that although SFs have also been named by various acronyms including CSF (carcinogenic slope factor) and CPF (carcinogenic potency factor), all terms refer to the expected lifetime cancer risk if someone is exposed daily for 70 years to 1 mg of substance per kilogram body weight.

Purposes of Risk Assessment

Using the estimates calculated by a risk assessment as a basis for making decisions is termed risk management. **Risk management** may be defined as:

> . . . evaluating alternative . . . actions and selecting among them . . . entails consideration of political, social, economic, and engineering information with risk-related information to develop, analyze, and compare regulatory options and to select the appropriate regulatory response to a potential chronic health hazard. The selection process necessarily requires the use of value judgments on such issues as the acceptability of risk and the reasonableness of the costs of control.[3]

In the hazardous waste field, risk assessment provides information to decision makers as to the consequences of possible actions. Important decisions that could use risk estimates include selecting waste treatment/disposal options, remediating contaminated sites, minimizing waste generation, siting new facilities, and developing new products. It should be emphasized that risk estimates are only one type

of information used, and hazardous waste decisions are often driven by political, social, economic, or other factors.

Risk assessment plays a major role in the decision making for the remediation of contaminated sites. It provides an important part of the basis for selecting from alternative remedies (see Section 17-6). Risks from exposure are calculated for each alternative under the assumption that its remedial actions have been implemented fully. This enables a balanced comparison of the effectiveness of each alternative at reducing risks.

Another application in site remediation is the establishment of cleanup standards. In many cases, definitive numerical standards for specific contaminants in soil or groundwater simply do not exist. Risk assessments are used to determine "How clean is clean?"[4] In this instance the calculation procedure is reversed. The process starts with a numerical definition of acceptable risk and works back to the level of contamination that will produce the acceptable risk level. An example would be a determination at a specific site that any soil containing greater than 588 mg/kg of toluene would pose an unacceptable risk and thus has to be remediated; soils containing smaller concentrations can be left in place.[5]

The EPA has proposed guidelines for the preparation of hazardous waste site risk assessments in a number of publications. The initial publications, *Handbook for Conducting Endangerment Assessments*[6] and the *Superfund Public Health Evaluation Manual,*[7] have been superseded by *Risk Assessment Guidance for Superfund* (RAGS),[8] the *Superfund Exposure Assessment Manual,*[9] and the *Exposure Factors Handbook.*[10]

RAGS, which comprises Volumes I and II, provides the site team—risk assessor, remedial project manager (RPM), and community involvement coordinator—with information to improve community involvement in the Superfund risk assessment process.

Volume I, Part A, provides suggestions for how Superfund staff and community members can work together during the early stages of Superfund cleanup. It identifies where, within the framework of the human health risk assessment methodology, community input can augment and improve the EPA's estimates of exposure and risk; recommends questions the site team should ask the community; and illustrates why community involvement is valuable during the human health risk assessment at Superfund sites.

Volume I, Part B, was prepared for risk assessors, remedial project managers, and others to assist them in developing preliminary remediation goals for National Priorities List (NPL) sites. Specifically, RAGS, Part B, provides guidance on using U.S. Environmental Protection Agency toxicity values and exposure information to derive risk-based preliminary remediation goals that are protective of human health.

Volume I, Part C, offers guidance on the human health risk evaluations of remedial alternatives conducted during the feasibility study, during selection and documentation of a remedy, and during and after remedy implementation. Part C assists users in preparing site-specific risk evaluations and maintaining flexibility during the analysis and decision-making process.

Volume I, Part D, provides guidance on standardized risk assessment planning, reporting, and review throughout the CERCLA remedial process, from scoping through remedy selection and completion and periodic review of the remedial action. Part D strives for effective and efficient implementation of Superfund risk

assessment practice described in Parts A, B, and C, and in supplemental Office of Solid Waste and Emergency Response (OSWER) directives.

These risk assessments are conducted to quantify potential human health and ecological risks stemming from toxic contaminants that may be transported to potential receptor populations in both present and future use scenarios. Human health risks for carcinogens are calculated as the probability of developing cancer in a lifetime and for noncarcinogens as hazard indices. In addition, an ecological assessment is attempted to quantify risk to on-site and off-site biota.

The following four sections describe each of the four stages of the risk assessment process as an overview of the state of the art in risk assessment. The methods in general follow those recommended by the EPA[11] because much of the risk assessment currently conducted adheres to these procedures. However, it must be noted that these procedures are constantly being revised, and it is our objective to explain the science of risk assessment and not to provide guidelines for regulatory compliance. To provide material for illustrating the process, the examples of exposure from a hazardous waste site are used.

14-2 HAZARD IDENTIFICATION

It is not uncommon to detect as many as 100 different chemicals at a contaminated site. The hazard identification stage examines the data for all contaminants detected at a site and consolidates the data to stress the chemicals of concern (i.e., those representing the vast majority of risk posed by the site).

Risk assessment requires a clear understanding of what chemicals are present at a site, their concentration and spatial distribution, and how they could move in the environment from the site to potential receptor points. The types of data typically sought at a hazardous waste site to address such needs are shown in Table 14-2, and the collection of such data is detailed in Chapter 15. A site investigation easily can produce a huge amount of data, necessitating that steps be taken in the hazard identification stage to facilitate subsequent analysis.

TABLE 14-2
Data needs

Site history
Land use
Contaminant levels in media:
Air, groundwater, surface water, soils and sediments
Environmental characteristics affecting chemical fate and transport:
Geologic
Hydrogeologic
Atmospheric
Topographic
Potentially affected population
Potentially affected biota

One approach is to examine data for all chemical contaminants detected in any media and select a subset of chemicals, consisting of the specific chemicals of concern and representative of all detected chemicals. These selected chemicals thus represent indicators or surrogates for all chemicals detected at the site. Their purpose is to limit the number of chemicals that must be modeled in fate and transport analyses and to focus efforts on the most significant hazards. The advent of modern computer methods that allow a large number of chemicals to be modeled simultaneously may obviate the need for a surrogate chemical selection step in many instances. Nevertheless, the EPA endorses the data consolidation approach for sites having a large number of chemicals detected in different media; however, professional judgment is emphasized. In general, most constituents at hazardous waste sites should be carried through a quantitative risk assessment.

The surrogate chemicals are selected on the basis of which compounds best represent the risk posed by the site:

- The most toxic, persistent, and mobile
- The most prevalent in terms of spatial distribution and concentration
- Those involved in the more significant exposures

The list of surrogate chemicals should encompass those chemicals that are estimated to account for 99 percent of the risk at the site. It should contain compounds that will support an adequate evaluation of both carcinogenic and noncarcinogenic risk.

Initial Screening

One approach to selecting this subset from all detected chemicals begins with an initial screening of data which may be summarized as follows:

1. Sort the contaminant data by medium (e.g., groundwater, soil, etc.) for both carcinogens and noncarcinogens.
2. Tabulate for each detected chemical the mean and range of concentration values observed at the site.
3. Identify the reference concentrations for noncarcinogens and slope factors for carcinogens (see Chapter 5) for each potential exposure route.
4. Determine the toxicity scores for each chemical in each medium. For noncarcinogens:

$$\mathrm{TS} = C_{\max}/\mathrm{RfC} \tag{14-1}$$

where TS = toxicity score

$C_{\max}$ = maximum concentration

RfC = chronic reference concentration (i.e., an estimate of acceptable daily intake) (see Section 5-4, and *www.epa.gov/iris/* for the latest qualitative and quantitative data)

For carcinogens:

$$\mathrm{TS} = \mathrm{SF} \times C_{\max} \tag{14-2}$$

where SF = slope factor (also called carcinogen potency factor) (see Section 5-5, and *www.epa.gov/iris/* for the latest qualitative and quantitative data)

5. For each exposure route, rank the compounds by toxicity scores.
6. For each exposure route, select those chemicals comprising 99 percent of the total score.

The process is illustrated by the following two examples. Data for surface soils appear in Table 14-3. Reference concentrations and slope factors may be found in the EPA's IRIS database, or in other sources of toxicological information.

EXAMPLE 14-1. SELECTION OF NONCARCINOGENIC CHEMICALS. Rank the chemicals for soil in Table 14-3.

Chemical	C_{max}, mg/kg	Oral RfC*	Toxicity score	Rank
Chlorobenzene	6.40	2.00×10^{-2}	3.20×10^{2}	3
Chloroform	4.10	1.00×10^{-2}	4.10×10^{2}	2
1,2-Dichloroethane	ND	NA	NA	NA
BEHP	2.30×10^{2}	2.00×10^{-2}	1.15×10^{4}	1

NA—not applicable.
**Source:* IRIS.

EXAMPLE 14-2. SELECTION OF CARCINOGENIC CHEMICALS. Rank the chemicals for soil in Table 14-3.

Chemical	C_{max}, mg/kg	Oral SF*	Toxicity score	Rank
Chlorobenzene	6.40	NA	NA	NA
Chloroform	4.10	6.10×10^{-3}	2.50×10^{-2}	2
1,2-Dichloroethane	ND	9.1×10^{2}	NA	NA
BEHP	2.30×10^{2}	1.40×10^{-2}	3.22	1

NA—not applicable.
**Source:* IRIS.

Note in Examples 14-1 and 14-2 that BEHP (*bis*-2-ethyl hexyl phthalate) and chloroform have both carcinogenic and noncarcinogenic toxicity so that they appear in both rankings.

TABLE 14-3
Chemical concentrations at the ABC landfill

	Air		Groundwater		Soil	
Chemical	Mean, mg/m³	Maximum, mg/m³	Mean, mg/L	Maximum, mg/L	Mean, mg/kg	Maximum, mg/kg
Chlorobenzene	4.09×10^{-8}	8.09×10^{-8}	2.50×10^{-4}	1.10×10^{-2}	1.39	6.40
Chloroform	1.12×10^{-12}	3.12×10^{-12}	4.30×10^{-4}	7.60×10^{-3}	1.12	4.10
1,2-Dichloroethane	1.40×10^{-8}	2.40×10^{-8}	2.10×10^{-4}	2.00×10^{-3}	ND	ND
BEHP	3.29×10^{-7}	8.29×10^{-7}	ND	ND	1.03×10^{2}	2.30×10^{2}

ND—not detected.

Further Screening of Chemicals

The ranking of toxicity scores as illustrated by the previous examples indicate which compounds pose the greatest hazard based solely on their maximum concentration and toxicity. The selection of chemicals of concern requires further evaluation to consider for each chemical its range of concentrations, its mobility in the environment, and other issues. These additional considerations include:

- Mean concentration
- Frequency of detection
- Mobility
- Persistence in the environment
- Chemicals associated with site operation
- Treatability

In general, chemicals with a low frequency of detection or extremely low mean concentration are not deemed sufficiently significant to receive direct evaluation. The evaluation of mobility in the environment uses information such as half-lives and physical-chemical properties (Henry's law constant, vapor pressure, octanol-water partition coefficient and solubility) (see Section 3-2). Chemicals whose persistence and mobility make them significant risks would remain in the list of chemicals of concern. Also, the risk analyst can selectively add compounds that are degradation products of others on the list (e.g., vinyl chloride from TCE).

Although the list of chemicals of concern should account for about 99 percent of the risk, it may account for less than 10 percent of the detected chemicals. It is not infrequent to reduce a total of 100 detected chemicals to 10 to 15 chemicals of concern.

14-3 EXPOSURE ASSESSMENT

The second stage of a quantitative risk assessment consists of estimating the exposure to the chemicals by the populations potentially at risk. To provide a comprehensive understanding of the sources of contamination, this stage begins with a delineation of the sources (e.g., sludge lagoons, contaminated soil) and the spatial distribution of contaminants at the site. The exposure assessment continues by analyzing how the contaminants might be released. Given the release of a chemical, it is necessary to estimate how it may migrate (e.g., via groundwater) to a potential receptor (e.g., users of groundwater). Having identified current and potential receptor points, considerable attention is given to (1) identification of general and sensitive populations of current and potential receptors and (2) estimation of both short- and long-term exposures in terms of doses by exposure route.

Environmental Pathways

Much of this stage consists of a fate and transport analysis as presented in depth in Chapter 4, and only a brief summary will be presented here. A chain of events must

occur to result in exposure. This chain in a collective sense is termed an environmental **pathway** (i.e., the environmental routes by which chemicals from the site can reach receptors). A pathway, consisting of the following elements, thus defines the framework of a fate and transport analysis:

- Source (e.g., a lagoon)
- Chemical release mechanism (e.g., leaching)
- Transport mechanisms (e.g., groundwater flow)
- Transfer mechanisms (e.g., sorption)
- Transformation mechanisms (e.g., biodegradation)
- Exposure point (e.g., residential well)
- Receptors (e.g., residential consumers of drinking water)
- Exposure route (e.g., ingestion)

An example of two simple but common pathways, one involving groundwater and the other atmospheric transport, is shown in Figure 14-1. Other examples of potential pathways, some of them quite complex, are discussed in Chapter 4 with various useful approaches for estimating fate and transport. It is necessary to establish potential pathways between sources and receptors for each chemical of concern. Figure 14-2 is a flowchart indicating the types of considerations necessary in analyzing fate and transport at hazardous waste sites.

Contaminant Release, Transport, Transfer, and Transformation

The release of contaminants from hazardous waste sites results from natural processes, such as leaching of soluble chemicals to groundwater; human activity, such as the construction of site drainage channels; and accidents such as chemical spills. Table 14-4 summarizes some potential mechanisms for release of chemicals.

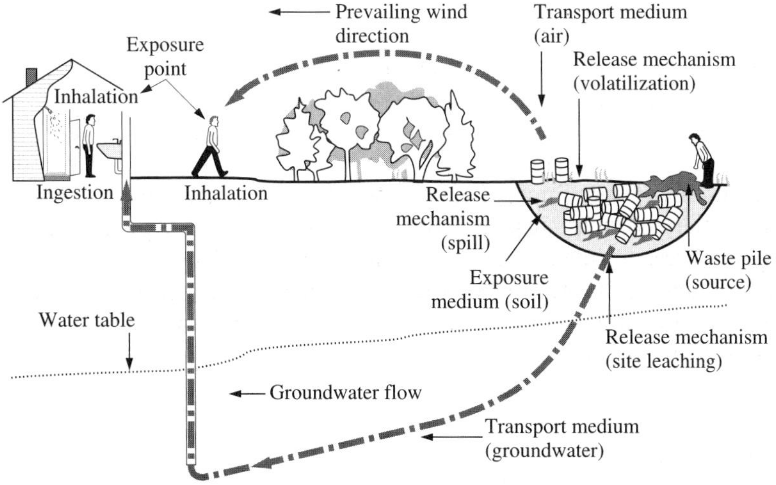

FIGURE 14-1
Examples of exposure pathways.[12]

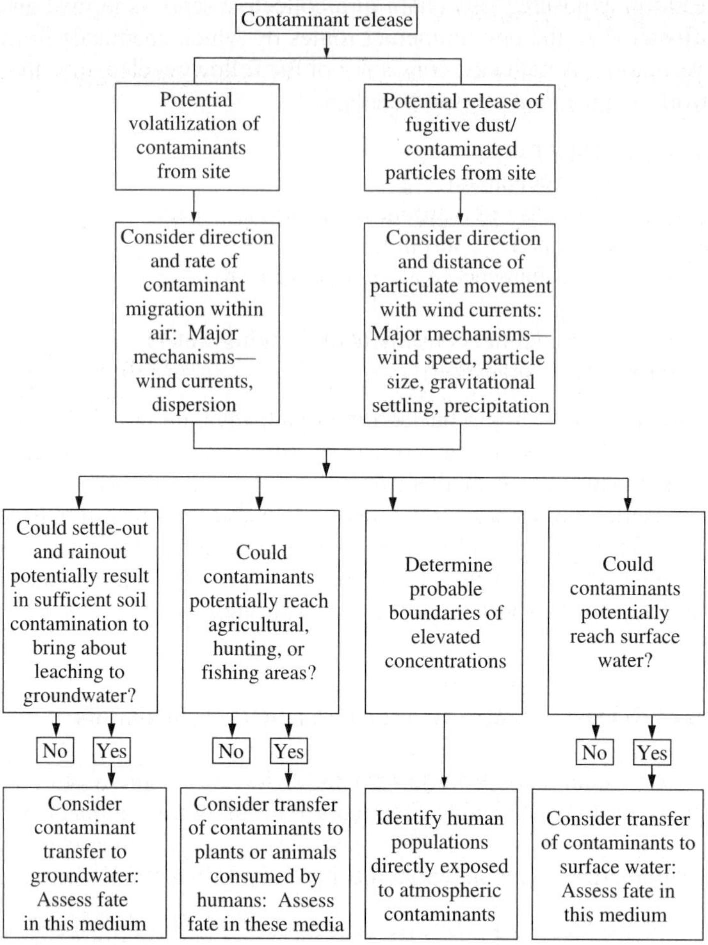

FIGURE 14-2A
Flowchart for fate and transport assessments of the atmosphere.[13]

In addition to determining what chemicals are released from a source, it is important to determine what happens to these chemicals: how they are transported, transferred, or transformed. Transport involves movement by advection and diffusion. A great number of mechanisms could act to transfer the contaminants to another medium or to storage sites (e.g., sorption onto soils). In some instances, organic contaminants transform under environmental mechanisms to CO_2 and water, while in other cases the degradation products may be more toxic than the original chemical (e.g., reductive dechlorination of tetrachloro- and trichloroethylene to vinyl chloride). Example mechanisms for the transfer and transformation of chemicals are presented in Chapter 4, and several examples are shown in Table 14-5.

Potentially Exposed Populations

The next step in the exposure assessment is to determine potentially exposed populations. These would include the following:

- Present population in vicinity of the site
- Future population in vicinity of the site
- Subpopulations of special concern (e.g., young children in the case of lead contamination)
- Potential on-site workers during any remediation

The present population may be initially identified as those living within specified distances from site boundaries (e.g., 1 and 3 miles). However, any distance could be specified, depending on the results of the fate and transport analysis. Determining

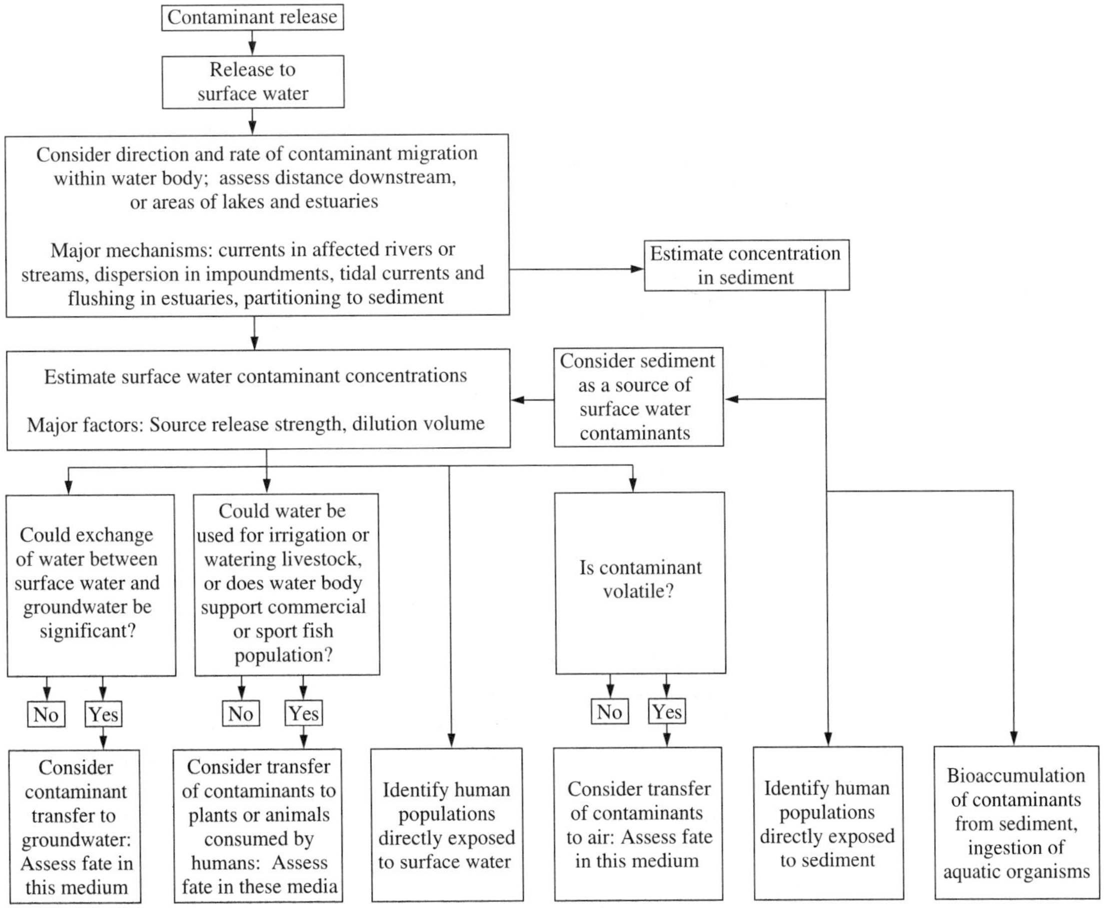

FIGURE 14-2B
Flowchart for fate and transport assessment in surface water and sediment.

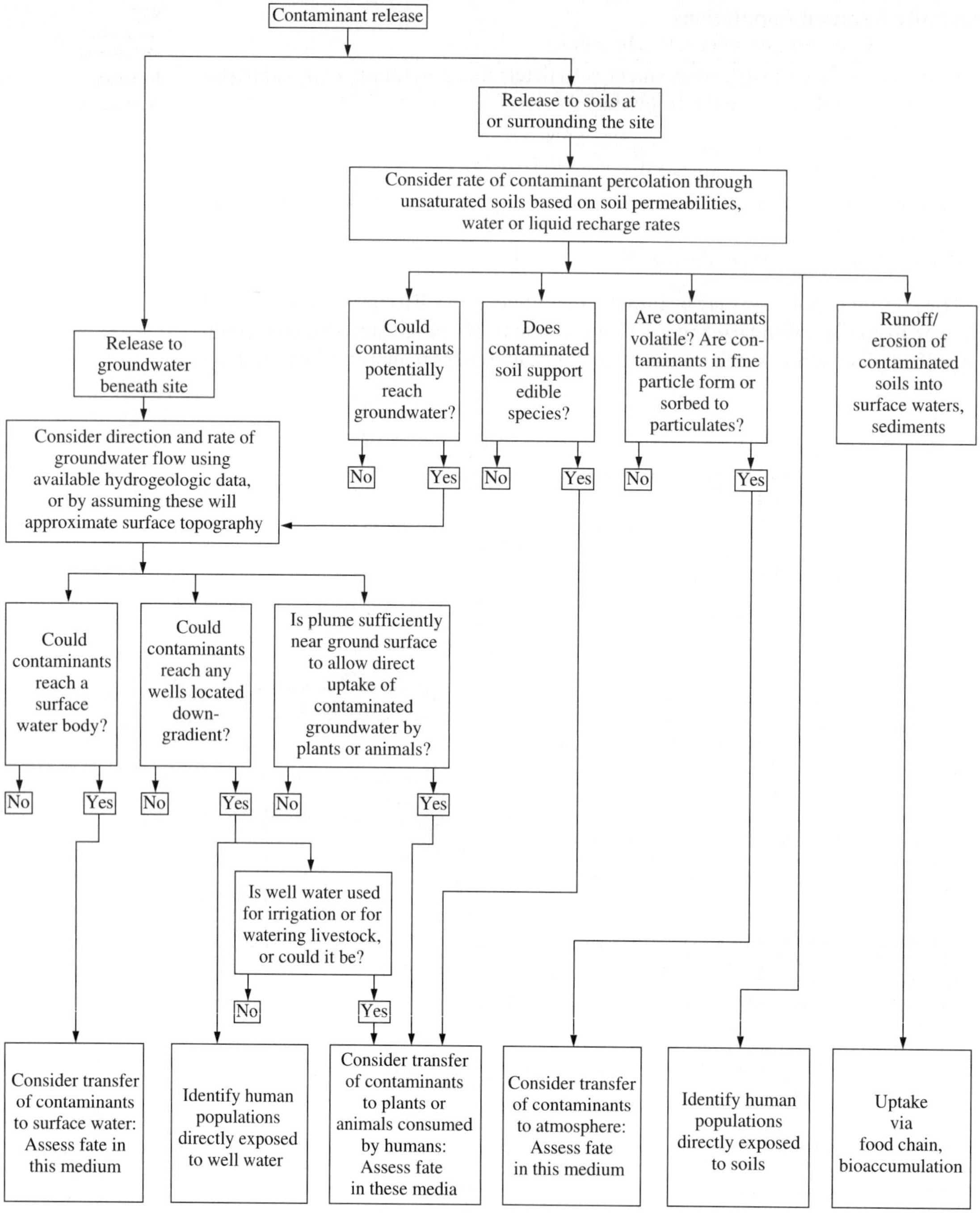

FIGURE 14-2C
Flowchart for fate and transport assessment for soils and groundwater.

TABLE 14-4
Contaminant release mechanisms

Media	Mechanism	Time frame
Air	Volatilization	C
	Fugitive dust generation	C, E
	Combustion	E
Soil	Erosion	C, E
	Leachate generation	C
	Spills	E
Surface and groundwater	Leachate generation	C
	Spills	E

C—Chronic E—Episodic

TABLE 14-5
Mechanisms affecting environmental transfer and transformation of chemicals*

	Fate mechanism	
Media	**Transfer**	**Transformation**
Water	Volatilization	Biodegradation
	Adsorption	Photochemical degradation
Soil	Uptake by plants	Biodegradation
	Dissolution in rainwater	
Atmosphere	Washout by rain	Oxidation by ozone
	Gravitational deposition	

*Adapted from R. A. Conway.[14]

potentially exposed populations is based on the surrounding land use and documented sources of demographic information such as summarized in Table 14-6.

After a preliminary overview of documents has been made, a thorough site visit to "ground truth" the current situation is critical. Patterns of human activity associated with various land and water uses should be determined, including answering questions such as:

- What sort of sensitive public facilities are located near the site (e.g., schools, hospitals, day care centers, nursing homes)?

TABLE 14-6
Sources of demographic information

Topographic maps	Tax maps
Census reports	County/regional demographic studies
Municipal zoning maps	Land use data
Projected future land use	Site visit
Human activity patterns	Aerial photos

- What outdoor activities occur (e.g., parks, playgrounds, recreational facilities nearby)?
- What fraction of time do various subpopulations spend in potentially contaminated area (e.g., children in a grade school)?
- What is likely to change (e.g., seasonally, immediate future, long-term future)?
- What secondary exposures are possible (e.g., contamination of crops at nearby farms)?
- What are the dietary habits (e.g., consumption of recreational fish containing bioaccumulated contaminants)?

Certain subpopulations may require special consideration because of their higher sensitivity to toxic substances—children, pregnant women, the elderly, and people with chronic illnesses. An initial step in identifying such subpopulations is to locate sensitive facilities such as schools and nursing homes. It is critical to evaluate future populations to address how changes could affect the estimates of risks. For example, if groundwater is not currently utilized as a drinking water source but could be used in the future, this may dramatically affect future risk calculations.

Development of Exposure Scenarios

After potential populations and exposure pathways have been defined, it is then necessary to characterize the conditions under which the populations may be potentially exposed. This involves an evaluation of both current and reasonable future uses of the site to establish a credible set of conditions under which the exposure could occur. After the scenarios have been established, then the specific parameters governing exposure (e.g., exposure frequency, duration, and intake rate) can be selected.

Some examples of commonly used exposure scenarios include a worker scenario, a trespasser scenario, a residential use scenario, a recreational use scenario, and a construction scenario. Each of these is discussed in the following listing, together with some pertinent questions for the risk assessor to consider in determining whether site conditions warrant evaluation of the given scenario.

- **Worker scenario:** Is the site currently used for industrial activities? Are workers exposed to site-related constituents under normal conditions? Could workers be exposed in the future, either because of a change in use of the site or because the workers would be involved in remedial activities? It should be noted that workers are strictly protected under Occupational Safety and Health Administration (OSHA) regulations. The estimation of potential risks for workers is usually done as a hypothetical evaluation to determine whether the potential exists for unacceptable risks, rather than as a mandated exercise designed to ensure that workers are not placed at risk during the normal course of their activities.
- **Trespasser scenario:** Is there evidence that trespassing may routinely occur at the site? Is there a fence that would limit access to the site? If so, is the fence in good condition? Have other measures been taken to limit access to the site? Is the site close to a school, shopping, or residential area where area children would have reason or inclination to play at the site? Is the site attractive to children?
- **Residential use scenario:** Is the site currently used for residential purposes? Will it or could it be used for residential purposes in the future? Are there any

zoning or deed restrictions that would limit its use for residential purposes? Are the residences single-family dwellings? Is there the potential for residential use of groundwater? A residential scenario is frequently evaluated as a hypothetical condition in an effort to estimate worst-case risks. Exposure under a residential scenario is generally the most restrictive condition for any age group, thus resulting in the greatest potential exposures and the highest potential risks.

- **Recreational use scenario:** This scenario is particularly applicable for evaluation of potential risks associated with surface water bodies where people may fish, swim, canoe, wade, etc. Observations made during site investigation activities can be useful in the development of a recreational exposure scenario. It may also be important to confirm the regulated use and classification of the water body with the appropriate state or federal agency.
- **Construction scenario:** Are construction activities planned or likely at the site? Will the construction result in potential exposures for both on-site receptors (e.g., direct contact with soils by construction workers) and off-site populations (e.g., exposure by off-site residents or workers to fugitive dust and volatiles released as a result of earth-moving activities)?

The development of scenarios involves making assumptions and must include subjective decisions. Ideally, this should produce credible scenarios, but in an effort to ensure conservative risk assessments, unrealistic scenarios have been used at some hazardous waste sites.

As an example, pica behavior is a real phenomenon and should be considered, but the assumption that children will eat 200 mg/day of soil for a 70-year lifetime is not realistic. Nondegradation policies, in many states, prohibit the discharge of contaminants into streams, independent of the present water quality in the stream. That policy has been used to suggest that a stream heavily contaminated with acid mine drainage be considered a primary drinking water source for the purpose of developing a credible scenario for a risk assessment. The point is that while nondegradation may be an appropriate public policy, drinking acid mine drainage is not realistic.

Exposure points define the locations of the receptors for the various scenarios. They are identified for each exposure scenario simply by overlaying the demographic information with the exposure pathway. An exposure point may be as close as the sources of waste at the site itself (e.g., the trespasser scenario) or at a considerable distance, particularly for pathways involving the food chain.

Exposure Point Concentrations

The risk analyst must next estimate the concentration of contaminants at the exposure points, including all pathways—air, ground and surface water, soils, sediments, and food (e.g., plants and fish). For present exposures, actual monitoring data at the exposure point should be used wherever possible. For example, contaminant concentrations should be obtained for drinking water wells in the vicinity of a site. For on-site exposure points, representative concentrations in soil or groundwater may be calculated as the arithmetic or geometric mean (depending on the statistical distribution of the site analytical concentration data).

Future conditions can differ starkly. A plume may not have yet migrated to a potential exposure point. Remediation will, of course, drastically reduce migration.

In a comprehensive risk assessment, the exposure concentrations for each remedial alternative would be estimated including the present situation (baseline conditions or no remedial action). Determining the concentration of a contaminant at the exposure point for future conditions often requires the use of fate and transport modeling methods of Chapter 4 and standard references.[15,16] The major effort with these models is calibration; once calibrated, successive runs can be made relatively easily to estimate concentrations for a range of conditions and assumptions.

For groundwater contaminants, hydrogeologic models can be used to estimate the future concentration at a downstream well. For volatile organic compounds released to the atmosphere, a gaussian diffusion model can be employed to estimate future downwind concentrations. In general, the level of effort employed in data collection and modeling will depend on the estimated severity of the risk. Nominal risks do not warrant the same level of analysis as the clearly significant risks.

All mathematical models require the making of assumptions. It is essential that the appropriateness of assumptions be reviewed carefully. The impact of assumptions on the exposure point concentrations and, ultimately, the risk values should be examined through a sensitivity analysis. A **sensitivity analysis** identifies which of the myriad of input variables have the most significant impact on the resulting risk value.

Receptor Doses

The final step in the exposure assessment stage is to estimate the doses of the different chemicals of concern to which receptors are potentially exposed at the exposure points. As with the previous stages, three exposure routes are considered—ingestion, inhalation, and dermal contact. Also, there are three types of doses—the administered dose (the amount ingested, inhaled, or in contact with the skin), the intake dose (the amount absorbed by the body), and the target dose (the amount reaching the target organ).

For purposes of calculating risks, the dose should be in the same form as that of the dose-response relationship reported for the specific chemical and the exposure route under study. This will almost always be either administered dose or absorbed dose. Given the concentration of the contaminant at the exposure point, the calculation of administered dose is straightforward. In contrast, the calculation of absorbed dose based on administered dose requires consideration of some complex factors (see Section 5-1). The key factors influencing the uptake of contaminants by the body are simplified as follows:

Ingestion	Contaminant concentration in the ingested media
	Amount of ingested material
	Bioavailability to the gastrointestinal system
Inhalation	Concentration in air and dust
	Particle size distribution
	Bioavailability to the pulmonary system
	Rate of respiration
Dermal contact	Concentration in soil and dust
	Rate of deposition of dust from air
	Direct contact with soil
	Bioavailability
	Amount of skin exposed

Other factors to be considered in determining the intake of contaminants include considerations of life style, frequency and duration of exposure (e.g., chronic, subchronic, or acute), and the body weight of the receptor. In the majority of hazardous waste sites, long-term (i.e., chronic) exposures are frequently of greatest concern.

The calculation of an administered dose is summarized in the following generic equation[17]:

$$I = \frac{C \times \mathrm{CR} \times \mathrm{EF} \times \mathrm{ED}}{\mathrm{BW} \times \mathrm{AT}} \tag{14-3}$$

where I = intake (mg/kg of body weight · day)

C = concentration at exposure point (e.g., mg/L in water or mg/m^3 in air)

CR = contact rate (e.g., L/day or m^3/day)

EF = exposure frequency (days/year)

ED = exposure duration (years)

BW = body weight (kg)

AT = averaging time (days)

Equation 14-3 is typically modified for specific exposure pathways. For example, the intake dose from the inhalation of fugitive dust may be calculated as:

$$I = \frac{C \times \mathrm{CR} \times \mathrm{EF} \times \mathrm{ED} \times \mathrm{RR} \times \mathrm{Abs}}{\mathrm{BW} \times \mathrm{AT}} \tag{14-4}$$

where RR = retention rate (decimal fraction)

Abs = absorption into bloodstream (decimal fraction)

For fugitive dust, the concentration in the air is determined by:

$$C = C_s \times P_c \tag{14-5}$$

where C_s = concentration of chemical in fugitive dust (mg/mg)

P_c = concentration of fugitive dust in air (mg/m^3)

Often, appropriate parameters may be found in the literature. Considerable research has been done in recent years to define many basic parameters, such as skin surface areas, soil ingestion rates, and inhalation rates. Other parameters, such as exposure frequency and duration, are often based on site-specific information (if available) or professional judgment.

Exposure duration is used to estimate the total time of exposure. National statistics are available on the upper-bound (90th percentile) and average (50th percentile) number of years spent by individuals at one residence, for example. The exposure duration selected must be appropriate for the exposure pathway being evaluated. Overestimating the exposure duration will overestimate the calculated risk level and underestimating the exposure duration will underestimate the calculated risk level. For example, in the evaluation of a trespasser scenario, observations of trespassers during site investigation activities may dictate the values used for exposure frequency (e.g., number of days per year or number of events per year) and exposure duration (e.g., number of years that the activity occurred). Common sense also plays an important role in the selection of exposure parameters. For example, if one were evaluating

TABLE 14-7
Standard parameters for calculation of dosage and intake determined for the ABC landfill

Parameter	Adults	Child age 6–12	Child age 2–6
Average body weight (kg)	70	29	16
Skin surface area (cm^2)	18,150	10,470	6,980
Water ingested (L/day)	2	2	1
Air breathed (m^3/h)	0.83	0.46	0.25
Retention rate (inhaled air)	100%	100%	100%
Absorption rate (inhaled air)	100%	100%	100%
Soil ingested (mg/day)	100	100	200
Bathing duration (minutes)	30	30	30
Exposure frequency (days/year)	365	365	365
Exposure duration (years)	30	6	4

residential use of two sites, one in North Dakota and one in Florida, it would not be reasonable to assume the exposure frequency (e.g., days per year that an activity occurred) would be the same in the two risk assessments for a child potentially exposed to surface soils as a result of their playing outdoors in the backyard.

Averaging time (AT) is another important parameter that must be defined in the intake equation. The averaging time selected will depend on the type of constituent being evaluated. For example, to assess long-term or chronic effects associated with exposure to noncarcinogens, the intake is averaged over the exposure duration (expressed in days). (The averaging time is equal to the exposure duration for noncarcinogens.) Exposure to carcinogens, however, is averaged over a lifetime (assumed to be 70 years or 25,550 days), to be consistent with the approach used to develop slope factors.

An example of the parameters used for this type of calculation are shown in Table 14-7. It should be noted that these values can vary greatly, depending on the assumed exposure conditions (i.e., the selected exposure scenario). As an example, the air breathing rate for adult males is 0.83 m^3/h in Table 14-7. However, this rate can vary by an order of magnitude from 0.6 m^3/h at rest to 7.1 m^3/h for vigorous physical exercise.[18] It should be noted that carcinogenic and noncarcinogenic human health risks are often computed by using both average and maximal chemical intakes. The purpose of using both values is to calculate most probable and worst-case risks, two end points that may have meaning to either regulators or potential receptor populations.

EXAMPLE 14-3A. CALCULATION OF CONTAMINANT INTAKE. Determine the chronic daily inhalation intake, by adults in a residential setting, of a noncarcinogenic chemical as a function of concentration in fugitive dust at the ABC Landfill (Table 14-3).

Solution. For an adult exposed to a noncarcinogenic constituent, the intake I_N may be calculated from Equation 14-3:

$$I_N = \frac{(C \times \mathrm{CR} \times \mathrm{EF} \times \mathrm{ED} \times \mathrm{RR} \times \mathrm{Abs})}{\mathrm{BW} \times \mathrm{AT}}$$

From Table 14-7, air breathing rate for adults = 0.83 m^3/h.

$$\mathrm{CR} = 0.83\ \mathrm{m^3/h} \times 24\ \mathrm{h/day} = 19.92\ \mathrm{m^3/day}$$

$$\mathrm{EF} = 365\ \mathrm{days/year}$$

For residential exposures, a default value of ED = 30 years is typically used. In the absence of better information, a conservative approach would assume the retention rate (RR) and the absorption into bloodstream would both equal 100 percent or 1.0.

$$\text{BW} = 70 \text{ kg (Table 14-7)}$$

$$\text{AT} = 365 \text{ days/year} \times 30 \text{ years} = 10{,}950 \text{ days}$$

$$I_N = \frac{C \times 19.92 \times 365 \times 30 \times 1.0 \times 1.0}{70 \times 10{,}950}$$

$$= \underline{\underline{0.285 \text{ m}^3/(\text{kg} \cdot \text{day}) \times C}}$$

where C = exposure point concentration (mg/m^3).

EXAMPLE 14-3B. CALCULATION OF CONTAMINANT INTAKE. Determine the chronic daily inhalation intake in children age 6–12 of a carcinogenic chemical (i.e., calculate I_C).

Solution

$$I_C = \frac{C \times \text{CR} \times \text{EF} \times \text{ED} \times \text{RR} \times \text{Abs}}{\text{BW} \times \text{AT}}$$

From Table 14-7:

$$\text{CR} = 0.46 \text{ m}^3/\text{h} \times 24 \text{ h/day} = 11.04 \text{ m}^3/\text{day}$$

$$\text{EF} = 365 \text{ days/year}$$

$$\text{ED} = 6 \text{ years}$$

$$\text{RR} = \text{Abs} = 1.0 \text{ as in Example 14-3a}$$

$$\text{BW} = 29 \text{ kg (Table 14-7)}$$

The averaging time (AT) for determining carcinogenic risk is 70 years or 25,550 days. As noted in the text, this is to be consistent with the approach used to develop carcinogenic slope factors.

$$I_C = \frac{C \times 11.04 \times 365 \times 6 \times 1.0 \times 1.0}{29 \times 25{,}550}$$

$$= \underline{\underline{3.26 \times 10^{-2} \text{ m}^3/(\text{kg} \cdot \text{day}) \times C}}$$

where C = exposure point concentration (mg/m^3).

EXAMPLE 14-4A. AVERAGE DAILY INTAKE FROM DERMAL CONTACT WITH SOIL. Determine the average daily intake of chlorobenzene over 1 year of exposure for on-site workers from dermal contact of the soils in Table 14-3.

Assume the following additional parameters:

A = skin exposed = 20% = $0.2 \times 18{,}150$ cm^2 = 3630 cm^2

DA = dust adherence = 0.51 mg/cm^2

Abs = skin absorption rate = 6%

SM = effect of soil matrix = 15% (i.e., because of soil matrix, only 15% of contamination is actually available for contact)

EF = two exposure events per day; 156 exposure days per year

ED = 1 year

BW = 70 kg

AT = 365 days or 1 year

Solution

$$I_N = \frac{C\left(\frac{\text{mg}}{\text{kg}}\right) \times A(\text{cm}^2) \times \frac{\text{DA}}{\text{exp. event}}\left(\frac{\text{mg}}{\text{cm}^2}\right) \times \text{Abs} \times \text{SM} \times \frac{2\ \text{exp. events}}{\text{day}} \times \frac{156\ \text{days}}{\text{year}} \times \text{ED} \times \frac{\text{kg soil}}{10^6\ \text{mg soil}}}{\text{BW} \times \text{AT}}$$

$$= \frac{C \times 3630 \times 0.51 \times 0.06 \times 0.15 \times 2 \times 156 \times 1 \times 10^{-6}}{70 \times 365}$$

$$= (2.03 \times 10^{-7} \times C)\ \text{mg/(kg} \cdot \text{day)}$$

From Table 14-3, the average concentration of chlorobenzene in soil is 1.39 mg/kg. The daily intake of chlorobenzene is

$$1.39\ \text{mg/kg} \times 2.03 \times 10^{-7}$$

Thus,

$$I_N = \underline{\underline{2.82 \times 10^{-7}\ \text{mg/(kg} \cdot \text{day)}}}$$

EXAMPLE 14-4B. AVERAGE DAILY INTAKE FROM DERMAL CONTACT WITH SOIL. Determine the average daily intake of chloroform for on-site workers from dermal contact of the soils in Table 14-3. Assume that the intake estimate will be used to assess the carcinogenic effects of chloroform (i.e., calculate I_C).

Solution

$$I_C = \frac{C \times A \times \frac{\text{DA}}{\text{exp. event}} \times \text{Abs} \times \text{SM} \times \frac{2\ \text{exp. events}}{\text{day}} \times \frac{156\ \text{days}}{\text{year}} \times \text{ED} \times \frac{\text{kg}}{10^6\ \text{mg}}}{\text{BW} \times \text{AT}}$$

$$= \frac{C \times 3630 \times 0.51 \times 0.06 \times 0.15 \times 2 \times 156 \times 10^{-6}}{70 \times 25{,}550}$$

$$= 2.91 \times 10^{-9} \times C$$

From Table 14-3, the daily intake of chloroform is 1.12 mg/kg $\times$ 2.91 $\times 10^{-9}$. Thus,

$$I_C = \underline{\underline{3.26 \times 10^{-9}\ \text{mg/(kg} \cdot \text{day)}}}$$

It should be noted that unlike the oral and inhalation routes, there are no generally accepted reference concentrations or slope factors for the dermal route. In lieu of these, oral RfCs and SF values may be utilized, but the advice of a qualified toxicologist should be sought.

Equations similar to those developed in Examples 14-3 and 14-4 are used in spreadsheets to calculate administered/absorbed doses for all chemicals of concern for all pathways. These are repeated for each exposure scenario.

14-4 TOXICITY ASSESSMENT

This stage of the risk assessment process defines the toxicity (i.e., the dose-response relationship) for each chemical of concern. The output takes the form of mathematical

constants for insertion into risk calculation equations. In addition to providing a set of mathematical constants for calculating risk, the toxicological assessment should also analyze the uncertainty inherent in these numbers, and describe how this uncertainty may affect the estimates of risk.

Chapter 5 explains in detail the scientific basis underlying the development and application of dose-response relationships. This section highlights some of the important concepts and quantitative methods involved in such undertakings and how they apply to the calculation of the risk from exposure to hazardous waste constituents.

Carcinogens versus Noncarcinogens

For the purposes of quantifying human health risks, chemicals are characterized as carcinogens (i.e., those with demonstrated propensity for cancer induction) and noncarcinogens. Some chemicals behave as both carcinogens and noncarcinogens and, hence, will appear in both types of calculation of potential human health risk.

Carcinogens tend to dominate public concerns about health risk; however, this is not the reason for distinguishing between chemicals that induce cancer and those that do not. The distinction is important because the two elicit toxic responses in different ways, giving rise to two quantitatively different models for induction of toxic response as a function of dose. Thus, the mathematical expressions of risk differ for the two categories of chemicals. Noncarcinogens have thresholds below which they fail to induce any discernible adverse health effect. In contrast, a linear nonthreshold model has been adopted for all carcinogens by the EPA (see Figure 5-24). According to this model, some risk is assumed for carcinogens at any dose, regardless of how small. There is increasing evidence, however, that some carcinogens do have thresholds, especially those that function as promoters.

Sources of Toxicity Data

The experimental research effort involved in developing a new dose-response relationship for a toxic substance takes considerable time, much longer than associated with starting and completing a study of most hazardous waste issues. Rather than conduct experimental research, the risk analyst defers to existing data found in standard sources of toxicological data, and selects from them the appropriate mathematical descriptors of toxicity.

Perhaps the most used source is IRIS (Integrated Risk Information System). The toxicologic indices in the IRIS database are updated continuously and available on-line (www.epa.gov/iris/). The IRIS database contains both qualitative and quantitative data regarding noncarcinogens as well as carcinogens.

It should be noted that regulatory agencies will frequently specify the particular mathematical constants to be used. Whether taken from regulatory guidance or other sources, the toxicity constants must apply to the range of doses predicted by the exposure assessment. Also, because toxic response for the same dose can vary, depending on the exposure route, a separate toxicity constant frequently must be selected for each exposure route predicted by the exposure assessment.

Slope Factors for Carcinogens

Dose-response relationships for carcinogens are conventionally reported as incidence of lifetime cancer (i.e., probability) versus dose (see Figure 5-23). The slope, known as the **slope factor,** represents the carcinogenic potency for the chemical. It is calculated as the 95 percent upper confidence limit of the cancer dose-response curve and is expressed as the inverse of dose, e.g., $[\text{mg/(kg} \cdot \text{day)}]^{-1}$ (see Section 5-5). Thus, the probability of cancer for a given exposure is calculated by multiplying the slope factor times the dose.

Substances are assigned carcinogenic status according to the weight of evidence from human epidemiologic studies along with data from laboratory experiments in test animals. Substances are classified by the EPA as **known, probable,** or **possible** human carcinogens on the basis of their apparent epidemiologic association with human cancer, induction of cancer in multiple species of test animals, or induction of cancer in one species, respectively. As the strength of a possible causal association declines from human epidemiology to carcinogenesis in a single species of test animal, so does the certainty of cancer risk assessment in exposed populations.

Of all carcinogens, only 14 have adequate epidemiologic evidence to be classified as known human carcinogens (Class A). The vast majority of the EPA's Priority Pollutants List (approximately 70 percent) is Class B2 (probable human carcinogens—insufficient human epidemiologic evidence but adequate evidence of carcinogenesis in at least two species of experimental test animals).

Reference Concentrations for Noncarcinogens

Unlike carcinogens, noncarcinogens exhibit a threshold effect. Namely, below a specific dose, they fail to induce any adverse health effect in exposed populations. This threshold is defined as the **reference concentration** (RfC), which is the estimated daily intake that is not believed to be associated with adverse health effects. The toxic end point varies from chemical to chemical as does the reference concentration.

Because a noncarcinogenic chemical can induce several different toxic end points, the RfC is set for the most sensitive toxic end point. For example, carbon disulfide, a commonly used industrial solvent, may affect the central nervous system, the cardiovascular system, and the auditory and visual systems. Effects on these target organs occur at high levels of exposure. A more sensitive end point, however, is fetal development. Long-term inhalation studies at lower levels of exposure to carbon disulfide demonstrated fetal toxicity and malformation. Therefore, this more sensitive end point provides the basis for a chronic oral RfC of 0.1 mg/(kg · day). Exposure below this level would protect against not only developmental toxicity, but also against the other target organs for carbon disulfide.

The hazard index is the standard unit for quantifying noncarcinogenic risk of a particular chemical. This simply is the ratio of the estimated daily intake of the toxic substance to its recommended RfC. An index greater than 1.0 would indicate the possibility for an adverse health effect from the exposure.

Chemicals of potential concern frequently do not have currently available toxicological indices, i.e., slope factors for carcinogens or RfCs for noncarcinogens. There are various ways for the risk assessor to proceed, however. If there are sur-

rogate compounds with similar toxic activities and published indices, these may be used (e.g., phenanthrene is often used as a surrogate for other noncarcinogenic PAHs). If no surrogates exist for a noncarcinogenic chemical, but toxicity data are available, there are methods for deriving an RfC. Normally, this means finding a rodent study in which a no observed adverse effect level (NOAEL) has been determined, expressing this concentration in mg/(kg · day), and then dividing by a 100-fold safety factor to obtain the RfC. Finally, even though some toxicological indices have been withdrawn by the EPA or state regulators, the last available RfC or CSF may be utilized with an accompanying explanation.

Protective Nature of EPA Approach

Although based on the best available information at present, many of the toxicologic indices utilized in calculations of carcinogenic and noncarcinogenic risks are highly uncertain. In a prudent approach to public health protection, the EPA has built several safety factors into its methods for establishing reference doses and carcinogenic slope factors. Virtually all errors of uncertainty are made in the direction of public health protection to ensure that risks are overestimated rather than underestimated. Four examples of this protective approach follow:

- For noncarcinogens, extrapolation of animal reference doses to humans utilizes at least two safety factors of 10: one for animal-to-human extrapolation, a second for variation for toxic sensitivities within the human population.
- For carcinogens, the linearized multistage model assumes the upper-bound 95 percent confidence level of extrapolated data (i.e., it is likely to be conservative 19 out of 20 times).
- For carcinogens, the linearized multistage model extrapolates data from the 10 to 90 percent carcinogenesis range observed in experimental animals to the regulatory target of 0.0001 percent carcinogenesis, a step that could overstate risk by several orders of magnitude.
- Although evidence indicates that, like noncarcinogens, nongenotoxic carcinogens have thresholds below which they fail to influence cellular differentiation or division, they are treated mathematically like genotoxic carcinogens according to the linearized multistage dose-response model. (Genotoxic carcinogens are discussed in Section 5-5.)

Risk assessments for exposure to complex mixtures of chemicals are less certain than for one or two chemicals at a time. Sometimes chemicals interact with target organs synergistically, and sometimes antagonistically. If these potential interactions are known, the risk assessor is free to depart from simple additive models.

14-5
RISK CHARACTERIZATION

The final stage of a four-stage human health risk assessment is to estimate risks. This consists in part of calculating quantitative estimates of both the carcinogenic and noncarcinogenic risks to receptors for all exposure scenarios considered. Estimates are typically calculated for all three exposure routes and for the maximally exposed

individual (MEI) as well as the most probable exposed population. Such calculations are straightforward, and yield quantitative estimates of risk. The challenge lies not in making the calculations but in interpreting the results such that they are applied properly in decision making. The overall effort is referred to as risk characterization.

Risk for Average and Maximum Exposures

In the exercises required for Superfund sites, both average and maximum exposure point concentrations are used to estimate risk. Performing a specific risk calculation using both values permits the estimation of a range of potential risks, which can frequently be useful in providing perspective regarding the potential hazards associated with a particular set of exposure conditions. The significance of either measurement depends, of course, on the amount of data and the associated confidence in that data. However, in general, calculation of potential risk by using an average concentration permits a better estimate of risk associated with chronic exposures, since the average value represents a more likely estimate of the exposure point concentration to which a receptor would be exposed over time. Use of a maximum value is best in the estimation of shorter-term, subchronic risks, although its use can provide a useful upper bound estimate of potential risk.

It should be noted that current risk assessment guidance[19] emphasizes the use of a single, upper bound estimate of exposure point concentration in the calculation of potential risks. This value is often taken to be the 95th percentile upper confidence limit of the arithmetic mean. Use of this number generally provides a worst-case estimate of risk, and can result in a significant overestimate of potential risk, especially when this value is used in combination with other worst-case assumptions to define a reasonable maximum exposure.

Carcinogenic Risk

Carcinogenic risk may be defined as the chronic daily intake dose (developed in the exposure assessment) multiplied by the carcinogenic slope factor (selected by the toxicity assessment). The product is a real term: the probability of excess lifetime cancer from exposure to this chemical. The computation is as follows:

$$\text{Risk} = I_c \times \text{SF} \tag{14-6}$$

where $I_c =$ chronic daily intake of carcinogen [mg/(kg · day)]
SF = carcinogen slope factor [(kg · day)/mg]

EXAMPLE 14-5. CALCULATION OF CARCINOGENIC RISK. Calculate the average carcinogenic risk from chloroform due to dermal contact of soil by on-site workers at the ABC Landfill.

Solution. From Example 14-4b: $I_C = 2.91 \times 10^{-9} \times C$, where $C =$ exposure point concentration. From Table 14-3: $C = 1.12$ mg/kg.

$$I_C = 3.26 \times 10^{-9} \text{ mg/(kg} \cdot \text{day)}$$

$$\text{Oral SF for chloroform} = 6.10 \times 10^{-3}\ \frac{\text{kg} \cdot \text{day}}{\text{mg}} \qquad \text{(IRIS)}$$

(Note that oral slope factors are typically used to assess dermal exposures; these slope factors may be adjusted for differences in absorption between the oral and dermal routes of exposure.)

$$\text{Risk} = (3.26 \times 10^{-9})(6.10 \times 10^{-3}) = \underline{\underline{1.99 \times 10^{-11}}} \qquad (14\text{-}7)$$

This would be termed the most probable risk since the average concentration value was used. If $C_{max} = 4.10$ mg/kg is used, then the maximum carcinogenic risk $= 7.27 \times 10^{-11}$.

Characterization of carcinogenic risk first involves performing calculations similar to this example for each chemical appropriate to the exposure route and pathway under study. The classical methodology for risk assessment assumes additivity of risks from individual toxicants. For carcinogens this means that the total carcinogenic risk equals the sum by exposure route of carcinogenic risks from all individual substances as shown in Table 14-8. This is repeated for each exposure scenario and exposed population. It must be emphasized that slope factors are specific to the exposure route (e.g., oral) and may only be used when the exposure data are for the same route.

Noncarcinogenic Risk

Noncarcinogenic risk is normally characterized in terms of a **hazard index.** This index is simply the ratio of the estimated intake dose from exposure to the reference concentration (RfC). Reference concentrations are dependent on the route of exposure and may only be used with exposure data for the same route. The hazard index is calculated as follows:

$$\text{HI} = I_N/\text{RfC} \qquad (14\text{-}8)$$

where HI = hazard index (dimensionless)

I_N = chronic daily intake of noncarcinogen [mg/(kg · day)]

RfC = reference concentration [mg/(kg · day)]

If the acceptable level of intake is deemed to equal the reference dose, then by definition, a hazard index less than 1.0 is acceptable. An exposure typically involves multiple chemicals, and an index must be calculated for each chemical of concern for all pathways and exposure routes. For exposure to multiple noncarcinogens, the hazard index scores for all noncarcinogens normally are summed to provide the final measure of the risk for noncarcinogenic toxic effects, as shown in Table 14-9. It should be noted that the acceptable target for the sum of hazard indices remains as less than 1.0.

TABLE 14-8
Carcinogenic risks—additive model

Carcinogen	Lifetime risk
Benzene	0.0005
Cadmium	0.0008
Vinyl chloride	0.0025
Total	$0.0038 = 3.8 \times 10^{-3}$

TABLE 14-9
Noncarcinogenic risks—additive model

Noncarcinogen	Hazard index
Acenaphthene	0.05
Cyanide	0.15
Mercury	0.03
Total	0.23 which is < 1.0

The standard additive model should not necessarily be applied to all contaminants as a single group because many toxic substances show a predilection for only one or two organ sites. If the mechanisms of toxic action of different noncarcinogenic chemicals are well known, it is preferable to sum hazard indices on an organ-specific basis [e.g., all compounds having a toxic effect on the liver (hepatotoxic compounds) would be added together but separate from those affecting the brain]. Again, this approach requires considerable understanding of the toxic mechanisms of the significant chemicals.

The importance of this approach is illustrated by the example of a hypothetical exposure yielding a hazard index of 0.7 for cyanide and 0.6 for cadmium. If summed in the conventional manner, this would yield an overall hazard index of 1.3, thereby indicating the potential for an adverse health effect. However, since cyanide affects the brain and cadmium is mainly toxic to kidneys, it is unlikely that the sum of their two hazard indices is, in fact, a viable measure of overall toxicity.

EXAMPLE 14-6. CALCULATION OF NONCARCINOGENIC RISK. Determine the hazard index for chloroform, based on the intake calculated in Example 14-4a.

Solution. From Example 14-4a: $I_N = 2.03 \times 10^{-7} \times C$, where C = exposure point concentration. From Table 14-3: $C = 1.12$ mg/kg.

$$I_N = 2.27 \times 10^{-7} \text{ mg/(kg} \cdot \text{day)}$$

Oral RfC $= 1.00 \times 10^{-2}$ mg/(kg · day). Most probable hazard index $= (2.27 \times 10^{-7})/(1.00 \times 10^{-2}) = \underline{\underline{2.27 \times 10^{-5}}}$.

Background Risk

In characterizing the risk from exposure involving hazardous waste, one consideration is whether to include or exclude background risk. This depends on the decision to be made. In comparing a set of remedial alternatives, the background risk of cancer is not relevant. It dwarfs the incremental risk and remains constant regardless of the remedy.

On the other hand, background risk is important in deciding what maximum allowable level of soil contamination is necessary to protect aquatic life in a nearby stream. In this latter case, the impact on aquatic life is determined by the total exposure, independent of the source. Total risk rather than the incremental risk drives the decision.

Iterative Nature of Risk Assessment

It is important to keep in mind that risk assessment is an iterative process. The four-stage process necessitates a number of assumptions. An analysis of the sensitivity of the magnitude of calculated risk to these assumptions will indicate whether the risk analyst should revisit these assumptions and refine them. This may even require additional data because the health risks are initially calculated from available soil, groundwater, and surface water monitoring data. Additional sampling may be required to fill data gaps or to present a more credible worst-case assessment.

Uncertainty Inherent in Calculated Risks

On calculating risks, it is essential to examine and explain the degree of uncertainty associated with these estimates. In many instances, this variability is not properly presented to decision makers who use the risk estimates. When risk numbers are reported in the public press, the uncertainty is rarely reported, much less explained.

The methods used in each of the four stages of risk assessment have deficiencies that can introduce a high degree of uncertainty and thus impair the validity of the results. The hazard identification stage is based on data for which detection, identification, and quantification limits could introduce errors. Selecting carcinogens as chemicals of concern requires much judgment and is made difficult by the facts that carcinogenic initiators are treated the same as promoters and experimental data are the basis for classifying all but 14 carcinogens.

Exposure assessments of future conditions depend heavily on fate and transport models, estimates of the performance of remedial alternatives, and projection of land use as well as assumptions about the frequency and duration of the exposure. Each is a potential source of uncertainty. For example, calibrating models to site-specific situations is a difficult task.

The toxicity assessment stage has a very high degree of uncertainty associated with the slope factors and reference doses. For many contaminants, there is a dearth of information regarding carcinogenicity for certain exposure routes. Even for those chemicals where data exist, extrapolation introduces a large measure of uncertainty (e.g., extrapolation from animal tests to human exposures and particularly extrapolation to the range of a 0.0001 percent carcinogenesis response).

Finally, the computation of risk is an exercise in applied probability of extremely rare events. It is not possible to enumerate every conceivable outcome, and credible worst-case exposure scenarios are used. This introduces an inherent conservatism that often results in assessing scenarios that will never be experienced.

In summary, the process by design is very protective of human health by ensuring that potential exposures and risks are not understated. Nevertheless, there are some sources of uncertainty that possibly are not fully taken into account from a conservative perspective. An example is that existing toxicity data typically derive from experiments with exposure of a single organism to a single chemical under a specified set of environmental conditions. Such data do not account for confounding factors such as synergism, unique species characteristics, and the multitude of factors that define a specific environment. Each factor serves to make the application of toxicity data to the complexity of real hazardous waste sites problematic. Simply stated,

the classic additive models may understate the risk of exposure to complex mixtures of toxic substances. Another factor that could prompt underestimating actual risk is the presence of sensitive subpopulations.

Mathematical techniques such as Monte Carlo simulation may be utilized to assess both uncertainty and the sensitivity of final answers to individual input parameters when estimating exposure. This is particularly important for situations with many pathways, such as found at hazardous waste sites. Monte Carlo methods are discussed in Section 14-8.

Quantitative assessment of risk is still in the developmental phase. As the field matures, some of these limitations, such as our knowledge of carcinogenicity, will become less severe through research. Other constraints, such as difficulty in communicating risk, may prove more challenging.

14-6 RISK COMMUNICATION

Having characterized the risk of exposure to constituents of hazardous waste and the degree of uncertainty associated with the risks, the next step is to use the information to improve the basis for making decisions. This often involves the public, or at least embraces public concerns and attitudes. It requires an examination of the question, "What is acceptable risk?" It enters the area of perception, and while not leaving the world of science, it is a much different world and one not relished by many scientists and engineers.

What Is Acceptable Risk?

At present the EPA has defined acceptable risks for carcinogens as within the range of 10^{-4} to 10^{-6} excess lifetime cancer risk and for noncarcinogens as a hazard index of less than 1.0. As discussed in Section 5-5, there have been precedents with other issues which have defined "acceptable risk" outside this range. The U.S. FDA deemed that less than 10^{-7} cancer risk for saccharin was "acceptable," and local citizens were willing to accept a risk much greater than 10^{-4} for a copper smelter. Clearly, "acceptability" is a personal concept and demands that the public, which ultimately must have jurisdiction over what level of risk is acceptable, be informed. At many sites it is ultimately the public that determines by its influence which levels of potential health risks are acceptable.

The EPA uses 10^{-6} excess lifetime cancer risk as a point of departure, meaning that a higher risk may be deemed acceptable only if there was special extenuating circumstances. For purposes of comparison, examples of actions that would increase the risk of death by a probability of 1×10^{-6} are illustrated in Table 14-10. Almost all of the actions shown in Table 14-10 seem highly acceptable if not commonplace; yet, the magnitude of their risk compares with that of the EPA's target risk for hazardous waste sites. It should be noted that risk of cancer is not the same as risk of death because not all cancer cases result in death. Still, cancer is the second greatest cause of death in the United States.

Another comparison is that of incremental risk and background risk. The 10^{-6} target represents an incremental risk of 0.0001 percent probability, an especially

TABLE 14-10
Actions increasing risk of death by one in a million[20]

Action	Nature of risk
Smoking 1.4 cigarettes	Cancer, heart disease
Drinking 0.5 liter of wine	Cirrhosis of the liver
Spending 1 hour in a coal mine	Black lung disease
Spending 3 hours in a coal mine	Accident
Living 2 days in New York or Boston	Air pollution/heart disease
Traveling 6 minutes by canoe	Accident
Traveling 10 miles by bicycle	Accident
Traveling 30 miles by car	Accident
Flying 1000 miles by jet	Accident
One chest X-ray taken in a good hospital	Cancer caused by radiation
Living 2 months with a cigarette smoker	Cancer, heart disease
Eating 40 tablespoons of improperly stored peanut butter	Liver cancer caused by Aflatoxin B1
Drinking heavily chlorinated water (e.g., Miami) for 1 year	Cancer caused by chloroform
Drinking thirty 12-oz cans of diet soda	Cancer caused by saccharin
Living 5 years at site boundary of a typical nuclear power plant in the open	Cancer caused by radiation
Living 150 years within 20 miles of a nuclear power plant	Cancer caused by radiation
Eating 100 charcoal-broiled steaks	Cancer from benzo(*a*)pyrene

small level in comparison to the 25 percent background risk of the very same disease for which this regulatory target is directed. The total risk to an individual exposed at the EPA's target would increase from 25 percent to 25.0001 percent. This increase is hardly meaningful from a scientific perspective, especially considering that the exposed population is not the whole nation but isolated pockets.

Risk Perception

While an increase in cancer risk by an increment of 10^{-6} may not be significant from a scientific viewpoint, it easily can alarm the community near a hazardous waste site. The reason is perception. Some of the factors that affect the perception of risk are indicated in Table 14-11 and are discussed in Section 1-4. The factors work in such a manner that, for example, an action *voluntarily* undertaken by an individual is perceived as posing a smaller risk than one *imposed* on that individual, all else being

TABLE 14-11
Risk perception

Less risky	More risky
Voluntary	Involuntary
Familiar	Unfamiliar
Controlled by self	Controlled by others
Chronic	Acute
Natural	Artificial
Fair	Unfair
Detectable	Undetectable
Not memorable	Memorable

equal. Almost all of the "more risky" items are associated with hazardous waste. Clearly, the four-stage risk assessment process described on the preceding pages can not incorporate such factors in a precise quantitative calculation.

A concept accompanying public perception is the notion of acceptable risk. Fischoff, et al.[21] identified five generic complexities that affect acceptable risk:

- Defining the decision problem
- Assessing the facts
- Assessing relative values
- Addressing the human element in decision-making processes
- Assessing the quality of the decisions

With each of these, there is a perception gap between the decision makers and the local citizens. Starting with the first item, a decision maker's responsibility in site remediation climaxes with the evaluation of several remedial alternatives and the selection of one. The public likely will view this evaluation as an unnecessary exercise. They will ask, "Why deliberate when it has been intuitively clear from the onset that the safest decision is to excavate the waste and haul it out of their community?"

The difficulties that surround assessing the facts are primarily those of dealing with the large magnitude of uncertainty inherent in any risk assessment. On the other hand, the public often expects that scientists and engineers should have absolute knowledge about risks before making decisions that affect them. The purposeful use of conservative safety factors at each stage to protect human health serves to inflate resulting risk numbers, and the public will often express more concern with the possible magnitude of the worst-case scenario, either not understanding or electing to ignore how unlikely if not incredible it may be. Merely learning that the assessment is filled with uncertainty may make the public uneasy about the estimates, believing that the true risk may be even greater than indicated.

It is important to note that almost all of the resources applied in risk assessment are devoted to the first two items of Fischoff's list. The local citizens are just as influenced, perhaps more so, by the other three. Concentrating resources on the first two items is thought by decision makers to portray their decisions as purely objective. On the contrary, assessing risk and selecting from remedial alternatives is fraught with value judgments. Acknowledging the existence of values is far simpler than identifying and explaining the set of values underlying the decisions.

It is critical that decision makers sufficiently address the human element. This probably represents the most important avenue for closing the perception gap between the decision maker and the public. To not do so is a failure to deal with risk in its broadest sense: "Thinking about risks may be more productive than calculating them."[22] Finally, rarely do decision makers revisit the issue long after the decisions have been made to determine if the decision was a good one and whether their decision-making process needs adjustment. Were the right questions asked? Did we solve the real problems?

Risk Communication

It is perhaps now understandable that, given the five complexities listed by Fischoff, "The public often sees proponents of risk assessment as trying to convince people

to accept risks that the proponents do not face rather than acting to remove them."[23] Not surprisingly, the remedy desired by the public is to remove the hazardous waste to "somewhere else" even though in scientific terms this may pose the largest risk.

While it is possible to provide reasonable quantitative estimates of risks, it is clear from the previous discussion that it is very difficult to explain these risks to the affected public. This derives only in part from the highly technical nature of the risk assessment process. Decision making at hazardous waste sites is a political as well as a technical process. At many sites, the selection of an alternative is driven by the demands of the citizens living in the vicinity of the site. It is therefore essential that consideration be given to communicating risks as well as calculating them.

Santos and Edwards[24] suggest that to achieve effective risk communication three questions must be answered:

- Is the communicator listening and acknowledging the concerns of the audience?
- How capable is the spokesperson?
- Can the objectives of the presentation be met while still meeting the information needs of the public?

14-7 ECOLOGICAL RISK ASSESSMENT*

The previous sections have addressed the assessment of risk to human health. The primary mission of hazardous waste management also encompasses the protection of the environment, and risk assessment has an equal role with this objective. The two applications of risk assessment share many concepts and approaches. However, there are also some stark differences.

The introduction of chemicals into an ecosystem can provide a direct hazard to individual organisms, or it may affect their ability to reproduce. These perturbations can also affect the overall structure and function of an ecosystem or its components. The framework of this response is illustrated in Figure 14-3.

This section introduces the current approaches used to estimate risks to the environment, broadly included under the category of **ecological risk assessment.** Ecological risk assessments are designed to evaluate the nature, magnitude, and transience or permanence of potential ecological effects. Somewhat similar to the four-stage process for human health risk assessment, the ecological risk assessment consists of four interrelated tasks:

- Characterization of the baseline ecology and identification of potential receptor populations
- Ecological toxicity assessment
- Evaluation of potential exposures
- Risk characterization

The following subsections provide a description of each task.

*Section 14-7 was prepared by Bruce Molhart and David Stout.

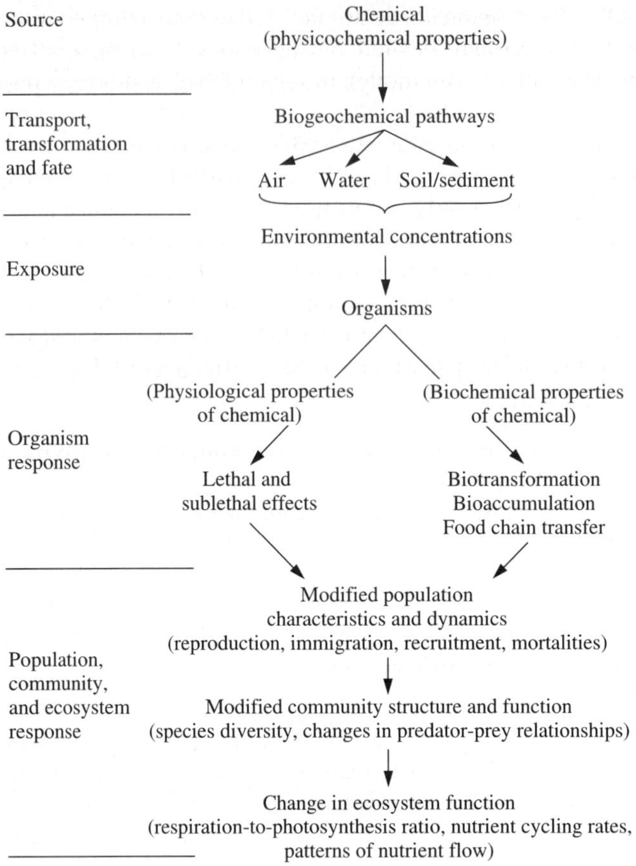

FIGURE 14-3
The ecotoxicological response to a chemical.[25]

Characterization of Baseline Ecology

This task begins with a review of previous site assessments and, if appropriate, a survey and evaluation of ecosystems and their components. This step serves as a synopsis of the prevailing ecology to provide an overview of the system diversity, as well as to identify potential sensitive subpopulations and any threatened or endangered species. As with human health risk assessment, it is often advantageous to select indicator species. Indicator species are selected for compelling reasons such as that they represent important commercial and/or recreational populations, sensitive species, or endangered species present near the affected area. The concept of indicator species may also be important for reducing uncertainty in the analysis; it may not be pertinent to evaluate a sensitive species when more tolerant species actually dominate the site.

Ecological Toxicity Assessment

An ecotoxicity assessment both qualitatively and quantitatively evaluates the potential for adverse effects from the possible exposure to the chemicals of concern. The

qualitative portion of the assessment provides a summary of ecotoxicity data for chemicals in the context of their presence in a sensitive ecosystem. This qualitative portion would draw on the information in the published literature. The quantitative portion entails identifying relevant indices of toxicity against which potential exposure concentrations can be compared in the risk characterization. Consideration is given to a number of end points (e.g., lethality, reproductive effects) and factors that may mitigate potential effects (e.g., temperature, water hardness).

Table 14-12 compares mammalian and aquatic/avian toxicology, indicating some of the concerns that must be addressed in ecotoxicology. It is particularly important to note that ecotoxicology is not nearly as well developed as human or mammalian toxicology, and is all the more difficult because of the diversity of species that must be protected. Ecotoxicology is discussed in Section 5-6.

Ecological Exposure Assessment

As with human health risk assessment, an exposure assessment is the process of measuring or estimating the intensity, frequency, and duration of exposure to an agent. More precisely, it evaluates the potential for exposure to each chemical of concern by the potential receptor populations (or indicator species) representing biotic components of any ecosystems in the site environs that may be affected. Two important components of an exposure assessment are an exposure pathway analysis and the estimation of exposure point concentrations.

Exposures are evaluated through the integration of baseline ecological data and information about the potential receptor populations. This information is then used

TABLE 14-12
Comparison of mammalian toxicology to ecotoxicology[26]

Mammalian toxicology	Ecotoxicology (aquatic/avian toxicology)
Objective: to protect humans	Objective: to protect populations of many diverse species
Must almost always rely on animal models since experimentation with humans is not feasible	Can experiment directly on species of concern
Species of interest (humans) is known; thus, degree of extrapolation is certain	Not able to identify and test all species of concern; thus, degree of extrapolation is uncertain
Test organisms are warm-blooded (body temperature is relatively uniform and nearly independent of environmental temperature); thus, toxicity is predictable	Test organisms (aquatic) live in a variable environment and are often cold-blooded (body temperature); thus, toxicity may not be sufficiently predictable
The dose of a test chemical usually can be measured directly and accurately and administered by a number of routes	The "dose" is known only in terms of the chemical's concentration in water (for aquatic organisms) and the length of exposure to it; the actual "absorbed dose" is sometimes determined experimentally by bioconcentration and metabolism studies
Extensive basic research has been conducted; emphasis has been on understanding mechanisms of toxic action	Little basic research has been conducted; emphasis has been on measuring toxic effects and generating numbers, with an eye toward regulatory needs
Test methods are well developed, their usefulness and limits are well understood	Test methods are either relatively new or generally not formalized (standardized); their usefulness in many cases is uncertain

to develop exposure scenarios. The exposure scenarios would consist of events that could occur in systems within or adjacent to the site. The routes of exposure considered for these scenarios would include ingestion and direct contact of contaminated surface water and sediments.

To evaluate these potential exposures, concentrations of the chemicals of concern in the exposure media (e.g., surface water and sediments) are required. These exposure point concentrations usually reflect collected environmental monitoring data. For example, a dissolved concentration in surface water may be used to represent the concentration of a chemical that is bioavailable to free-swimming organisms (i.e., nekton) inhabiting the water column.

Ecological Risk Characterization

The risk characterization task integrates information from all three other tasks to produce both a qualitative and quantitative estimate of risk. The risks to potential receptor populations that inhabit and/or use the affected ecosystems would be presented in a quantitative manner where quantitative indices are available, and a qualitative discussion would be presented regarding the possibility for adverse effects. Quantitative techniques involve the comparison for each compound of interest of the predicted and/or measured exposure point concentrations (from the exposure assessment) with the appropriate chemical-specific indices of toxicity (from the ecotoxicity assessment) to estimate the possibility of adverse effects. These chemical-specific comparisons represent ecological hazard indices and provide a means of assessing the potential significance of contamination.

Qualitative methods of characterization may not provide a definitive answer regarding potential risks, but they should provide a basis for professional judgment concerning the likelihood of adverse effects and yield important insight necessary for further evaluation. Finally, the limitations and uncertainties inherent to ecological risk assessment would be discussed to provide a proper perspective for use of the risk characterization results in decision making.

14-8 MONTE CARLO METHODS

Stochastic versus Deterministic Calculations

Up to this point, the calculations in this chapter have been deterministic (i.e., all of the variables are treated as known constants). Many of these constants are estimates of either average or conservative if not worst-case conditions. For example, the carcinogen slope factor (SF) typically used is the 95th percentile value, so that 19 times out of 20, the tabulated value is expected to exceed the actual potency.

Many of the values used in risk assessment are statistical point estimates (e.g., mean values, 95th percentile values) based on collected data. However, a point estimate provides only a portion of the information available in the data. This is illustrated in Figure 14-4, which shows two different probability distributions with the same point estimate for the 95th percentile. Clearly the underlying data sets are different.

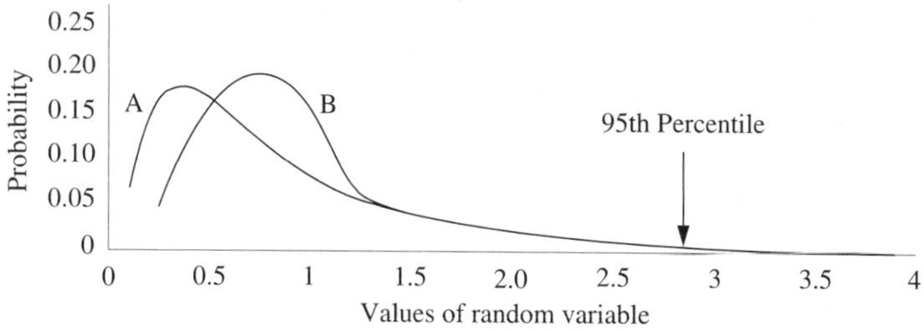

FIGURE 14-4
Example probability distributions.

Burmaster et al.[27] suggest three problems with the current practice in risk assessment to select point estimates so that the resulting risk is conservative:

- There is no way to know the degree of conservatism in the estimated risk (selecting 95th percentile values does not result in a 95th percentile of risk because the risk is calculated by numerous multiplications, divisions, summations, etc.).
- The resulting risk assessment will consider scenarios that will rarely if ever happen (e.g., every value at or near its respective maximum).
- Performing sensitivity analyses by varying input point estimates by ± 10 percent or ± 25 percent to determine the variability of the estimated risk is not meaningful because many of the inputs are at or near their maximum values.

These deficiencies can be addressed by using probability distributions in place of point estimates as input to the calculations. This is termed **stochastic modeling,** and it provides a mechanism to utilize the full range of information. Stochastic models treat some variables as random variables drawn from known (or estimated) probability distributions.

Method Description

The **Monte Carlo** method is a well-established simulation procedure that replaces point estimates with random variables drawn from probability density functions. A complete description of the method is beyond the scope of this text, and the reader is referred to standard texts on simulation.[28,29,30] The application of the method to risk assessment has been greatly simplified through the availability of commercial software that allows probability distributions to be input directly into spreadsheets.[31,32]

Once a probability distribution is incorporated into a spreadsheet cell, each time the spreadsheet is recalculated a new value of the random variable is selected from the distribution and used for calculations. The key to the method is to run the entire simulation hundreds or thousands of times. Each time new values of the random variables are selected and a new estimate of risk determined. The results of all of the calculations are summarized in a single histogram of risk values. This provides decision makers far more information than a single estimate of risk.

The process is summarized as follows:

1. Typical spreadsheet calculations are established for all chemicals and pathways to be modeled, following the methods used for deterministic calculations. For

each of the random variables, discrete or continuous probability density functions are placed in the appropriate cells.

2. Any correlations among the random variables must be identified. For example, body weight and skin surface area would be positively correlated whereas age and ingestion of surface soils would be negatively correlated. It is important to identify these correlations so that individual simulations avoid selecting values of two different random variables that simply do not make sense. Commercially available simulation software allows such dependencies to be easily input into the modeling process, but the risk analyst must identify the variables that are dependent. All other random variables are assumed to be independent.
3. A large number of simulations are run and the resulting risk values summarized. The summaries can include statistical tables and histograms of resulting risks and intermediate calculations.

A major difficulty in applying this method to quantitative risk assessment is estimating the probability distribution underlying many of the variables used. Table 14-13 provides some useful information.

Burmaster and von Stackelberg[33] suggest that SF data may vary by a factor of 10 above or below the mean. From this they infer that a lognormal distribution may be appropriate. By setting two standard deviations of the log SF at 10 and scaling the probability distribution to place the tabulated value of SF at the 95th percentile, they derive an estimate of the distribution as:

$$X1 \approx (\mathrm{SF}/6.645) \times \exp\,[1.1513 \times N(0, 1)] \tag{14-9}$$

where $X1$ = distribution of slope factor values
SF = tabulated point estimate
$N(0, 1)$ = normal distribution with zero mean and unit standard deviation

TABLE 14-13
Probability distributions for risk assessments[34]

Variable	Units	Point estimate	Distribution*	Point estimate location
Scenario-specific data:				
Average body weight	kg	47	Normal (47, 8.3)	Mean
Time soil stays on skin	h	8	Normal (6, 1)	95th percentile
Average body surface area	m^2	1.4	Normal (1.4, 0.17)	Mean
Fraction of skin area exposed	fraction	0.2	Lognormal (−2.15, 0.5)	85th percentile
Skin soil loading	mg/cm^2	1	Uniform (0.75, 1.25)	Mean
Soil ingestion rate	mg/day	50	Lognormal (3.44, 0.80)	75th percentile
Soil bulk density	kg/m^3	1600	Normal (1600, 80)	Mean
Cancer slope factors:				
SF, benzene	(kg · day)/mg	0.029	Lognormal (−4.33, 0.67)	88th percentile
SF, BaP	(kg · day)/mg	11.5	Lognormal (−0.79, 2.39)	91st percentile

*For a normal distribution, the mean and standard deviation are used to describe the distribution. For a lognormal distribution, the mean and standard deviation of the underlying normal are used to describe the distribution. For a uniform distribution, the low and high are used to describe the distribution.

Similarly, the reference concentrations (RfCs) for noncarcinogens are typically several orders of magnitude below levels that produce no observable adverse effects in animals (NOAEL). Burmaster and von Stackelberg use the assumption that 4 standard deviations below the logarithm of susceptibility equals a factor of 5. Two standard deviations above the mean of the logarithm of susceptibility is equivalent to the factor of 10 used in developing the tabulated values of RfC. From these assumptions the following probability distribution is derived:

$$X2 \approx (2.236 \times \text{RfC})/\exp\,[0.402 \times N(0, 1)] \qquad (14\text{-}10)$$

where $X2$ = distribution of reference concentration values

RfC = tabulated point estimate from the EPA's IRIS database (*www.epa.gov/iris/*)

The method is illustrated in the following section.

An Example

As a simple example of the method, the carcinogenic risk from the incidental ingestion of soil containing benzo(a)pyrene [B(a)P] is examined. In Section 14-5, carcinogenic risk was determined by using a constant called the carcinogenic slope factor (SF), with units (kg · day)/mg. As shown in Section 5-5, this constant was determined by selecting the 95th percentile value of the slope of a plot of tumor incidence (risk) versus dose (mg chemical per kg body weight per day).

Data for cancer potency tend to be distributed lognormally. Burmaster, et al.[35] suggest that the SF for B(a)P is distributed lognormally (−0.79, 2.39) where the mean (−0.79) and the standard deviation (2.39) are from the underlying normal distribution. A graph of this probability density function is illustrated in Figure 14-5.

The value of the point estimate of SF for benzo(a)pyrene is 11.5 (kg · day)/mg. This value occurs at the 91st percentile of the distribution shown in Figure 14-5. This simply indicates that there is a 9 percent chance that a value randomly selected from the density function of Figure 14-5 will exceed 11.5 (kg · day)/mg.

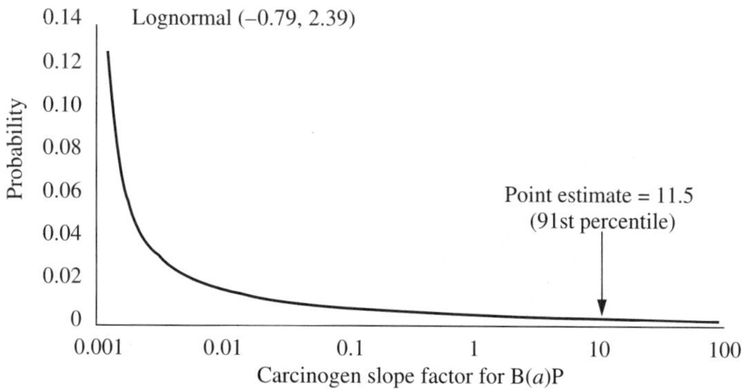

FIGURE 14-5
Probability density function of SF for benzo(a)pyrene.

Similarly, all of the variables in equations determining intake of contaminants, such as 14-3 and 14-4, are point estimates of random variables. In calculating the average daily intake, I (mg of chemical per kg body weight per day), conservative values of each of these variables were selected. In many instances the distribution of the data is available. For example, the concentration of a chemical in soil is estimated from many samples at a site. A histogram of these data provides the best estimate of the probability distribution underlying the data. Typical data for a site are shown in Figure 14-6.

The data in Figure 14-6 can be approximated by a distribution that is lognormal (0.165, 0.47). The 95th percentile of this distribution is approximately 2.5 mg/kg.

Suppose that the average daily intake at a specific site is related to the soil concentration by:

$$I = K \times C \qquad (14\text{-}11)$$

where I = average daily intake [mg/(kg · day)]

K = factor including variables such as body weight and frequency of exposure (i.e., all of the variables in Example 14-3) (day^{-1})

C = concentration in soil (mg/kg)

Additionally assume that the variables in Example 14-3 have been statistically analyzed and K has been determined to be distributed normally with a mean of 4.5×10^{-5} and a standard deviation of 1.2×10^{-5}. The 95th percentile value of this distribution is 6.5×10^{-5}. From Equation 14-6:

$$\text{Risk} = I \times \text{SF} = K \times C \times \text{SF} \qquad (14\text{-}12)$$

As illustrated in Table 14-14, if mean values are selected for the three variables in Equation 14-12 the resulting estimate of carcinogenic risk is 4.67×10^{-4} and if the point estimates are used the risk is 1.87×10^{-3}. However if random values are selected from the probability distributions, the resulting risk is a random variable. Five iterations are shown in Table 14-14.

The second column shows 5 values randomly selected for SF using the distribution in Figure 14-5, and the last column shows the resulting risks for that iteration.

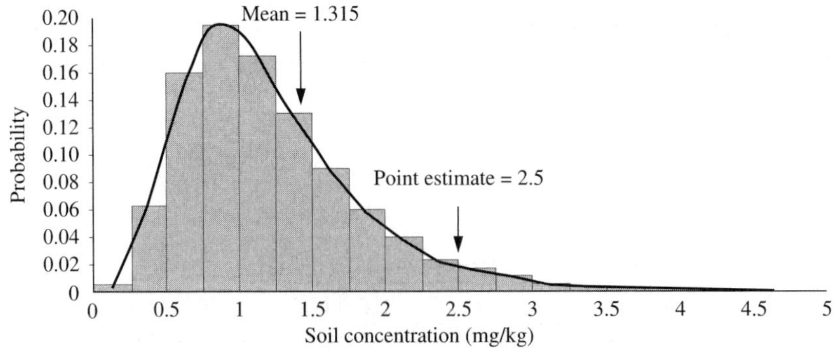

FIGURE 14-6
Histogram of benzo(a)pyrene concentrations in soil.

TABLE 14-14
Risk simulation

	SF, (kg · day)/mg	Soil concentration, mg/kg	K	Risk, day^{-1}
Mean value:	7.89	1.315	4.5×10^{-5}	4.67×10^{-4}
Point estimate:	11.5	2.500	6.5×10^{-5}	1.87×10^{-3}
Run				
1	3.771	1.822	2.71×10^{-5}	1.86×10^{-4}
2	0.243	1.336	4.47×10^{-5}	1.45×10^{-5}
3	0.016	1.467	3.96×10^{-5}	9.17×10^{-7}
4	0.098	0.763	4.98×10^{-5}	3.71×10^{-6}
5	2.945	0.642	3.58×10^{-5}	6.77×10^{-5}

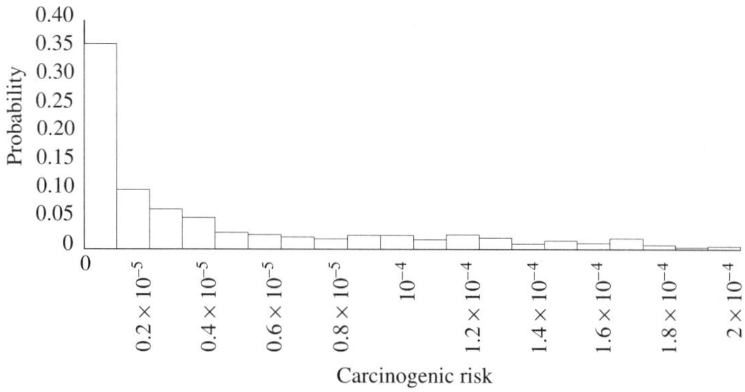

FIGURE 14-7
Histogram of carcinogenic risk simulation results.

In order to obtain a true picture of the distribution of the risks, hundreds of iterations are performed and a histogram of risks produced. This is illustrated in Figure 14-7.

Figure 14-7 is the result of 500 iterations of this calculation of risk from B(a)P in soil. The mean value of these 500 iterations was 4.675×10^{-4}, essentially the same as the mean value from Table 14-14. The median value of risk was 2.45×10^{-5} and for 90 percent of the iterations, the calculated risk was less than 5×10^{-4}. Clearly the histogram provides a clearer picture of estimated risks than a single point estimate.

14-9 CASE STUDY

At the Lackawanna Refuse Site in Old Forge, Pennsylvania, some waste transporters had dumped drums of hazardous waste illegally in a landfill licensed to receive only municipal waste. The EPA placed the site on the National Priorities List and completed the Remedial Investigation[36] /Feasibility Study[37] (RI/FS) necessary under its Superfund Program. Most of the drums had been crushed when placed in the

landfill; however, some remained intact. The existence of intact drums of unknown contents made the safety of normal drilling questionable and special procedures were required to investigate subsurface conditions.[38]

After the RI/FS, the EPA issued in March 1985 a Record of Decision (ROD), which indicated that

> During . . . hearings operators of the trucking firm testified that they brought drums of hazardous waste to the site. . . . Estimates ranged between 10,000 and 20,000 drums. . . . Other allegations [suggested] that bulk liquid wastes were disposed of.[39]

The ROD evaluated various remedial alternatives ranging from "No Action" to complete excavation and incineration. The EPA selected a remedy specifying the following:

> Removal of all drums and highly contaminated municipal refuse . . . for off-site disposal . . . Construction of a clay cap . . . [and] installation of surface water drainage.

Subsequent to the issuance of the ROD, a Quantitative Risk Assessment[40] was performed in 1987. The results of this risk assessment are summarized in the following sections.

Hazard Identification

Indicator chemicals were selected on the basis of an analysis of the data developed during the RI/FS on soils, groundwater, surface water, and leachate. All compounds detected were summed to one of the indicator compounds on the basis of having a similar effect. Of the more than 80 chemicals detected at the site, 13 final indicator compounds were selected as shown in Table 14-15.

Exposure Assessment

The exposure assessment modeled the migration of each of the indicator chemicals to the nearest receptors. This required developing a conceptual model of migration and exposure for the site using the geological cross section shown in Figure 14-8. In all cases highly conservative assumptions were made to simplify the modeling efforts. Some of the assumptions included:

TABLE 14-15
Indicator chemicals at the Lackawanna Refuse Site

Arsenic	Acetone
Barium	Ethylbenzene
Cresol	2-Hexanone
Trichloroethylene	4-Methyl-2-pentanone
1,2-Dichloroethane	Tetrachloroethylene
Methyl ethyl ketone	Xylene
Toluene	

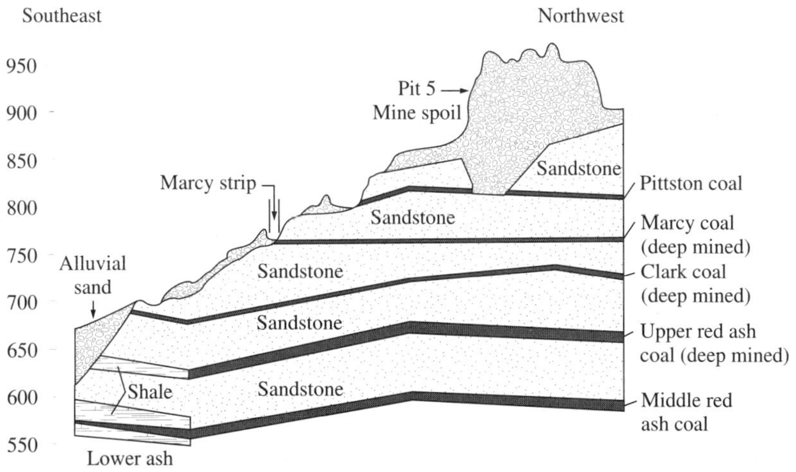

FIGURE 14-8
Geological cross section of the Lackawanna Refuse Site.[41]

- All contamination observed off-site is from the site being studied
- Neither attenuation nor degradation of contaminants occurs

To develop the worst-case exposure scenario, it was assumed that the mine pool under the site would be used as a future municipal water supply even though it had been heavily contaminated by acid mine drainage. The migration of contaminants to exposure points was estimated for each of the following remedial alternatives as developed in the ROD:

(*a*) No action (baseline on which to compare other actions)
(*b*) Capping and collecting and treating leachate
(*c*) Alternative B plus excavation and removal of drums
(*d*) Alternative C plus excavation and removal of all refuse that has been contaminated with hazardous waste.

The end product of the exposure assessment was the calculation of exposure level and dose incurred by the maximum exposed individual for each remedial alternative.

Toxicity Assessment

For each of the indicator chemicals, carcinogen slope factors (SFs) for carcinogens and reference concentrations (RfCs) for noncarcinogens were identified. Where standard references did not provide values for SFs or RfCs, estimates were made by toxicologists using the latest information available in the toxicology literature. In addition to the toxicity constants, appropriate environmental standards were identified to include the EPA's proposed maximum concentration limits (MCLs) and proposed maximum concentration limit guidelines (MCLGs) which had been established to protect drinking water. Other standards were the EPA Water Quality Criteria for Fish and Drinking Water, and Protection of Aquatic Life.

Risk Characterization

The final stage in the risk assessment brought together the results of the exposure assessment and toxicity assessment into a quantitative estimate of risk. For this site, the risk was determined for each remediation alternative. The results of the calculations for carcinogenic risk are presented as follows:

Alternative	Carcinogenic risk
A. No action	0.55×10^{-6}
B. Cap and leachate treatment	0.40×10^{-6}
C. B and drum removal	0.23×10^{-6}
D. C and removal of contaminated refuse	0.14×10^{-6}

As can be seen all of the alternatives, including the no action alternative, met the EPA's acceptable level of carcinogenic risk (10^{-4} to 10^{-7} at that time). This would indicate that perhaps this site was not of sufficient risk to warrant being placed on the National Priorities List. As noted in Section 2-4, seven other criteria in addition to cost and risk are used to determine appropriate cleanup strategies at Superfund sites.

DISCUSSION TOPICS AND PROBLEMS

14-1. Determine the average daily intake and maximum hypothetical intake for all compounds listed in Table 14-3 from dermal contact of soils by on-site workers. Note that some chemicals may require evaluation as both carcinogens and noncarcinogens.

14-2. Make a complete list of all of the pathways for which risk calculations would be required for a site identified by your instructor.

14-3. Determine the total carcinogenic risk from the results of Problem 14-1.

14-4. Determine the total noncarcinogenic risk from the results of Problem 14-1.

14-5. Determine the average daily intake and maximum hypothetical intake for all compounds listed in Table 14-3 from incidental ingestion of soils by on-site workers.

14-6. Determine the total carcinogenic risk from the results of Problem 14-5.

14-7. Determine the total noncarcinogenic risk from the results of Problem 14-5.

14-8. Prepare a two-page summary of the article by Brown.[42]

14-9. Explain the difference between risk assessment and risk management.

14-10. A risk assessment is to be performed for a proposed hazardous waste incinerator. Briefly describe the specific factors that should be considered in each of the four steps.

14-11. A risk assessment is to be performed for a proposed hazardous waste land disposal facility. Briefly describe the specific factors that should be considered in each of the four steps.

14-12. Explain, in terms an ordinary citizen could understand, what is meant by a one in a million risk.

14-13. Explain the difference between total risk and incremental risk.

14-14. Briefly describe some of the problems encountered in communicating risk from hazardous waste activities to the public.

14-15. Explain, in terms an ordinary citizen could understand, what is meant by acceptable risk.

14-16. Describe how risk assessment might be used at a hazardous waste site to assist in establishing a specific cleanup level for removal of trichloroethylene in groundwater. What other methods might be employed to determine this cleanup level?

14-17. Explain the difference between a hazard index and cancer risk.

14-18. Explain, in terms an ordinary citizen could understand, some of the weaknesses of the risk assessment process, and what steps may be taken to reduce the likelihood that unsafe decisions will be made on the basis of incorrect risk calculations.

14-19. Describe the type of data that would be required to perform a quantitative risk assessment for a Superfund site consisting of a former unlined municipal landfill where drummed and bulk liquid hazardous wastes were disposed. Assume the wastes were known to contain benzene, toluene, ethylbenzene, and xylene.

14-20. Describe the hazard identification stage of risk assessment as it might apply to a Superfund site.

14-21. What factors might be employed to reduce the number of chemicals for which fate and transport modeling will be required? For each factor explain the rationale for its use.

14-22. Review the remedial investigation report for a Superfund site, selected by your instructor, and identify what appear to be the major hazards; the most exposed populations; the most significant pathways.

14-23. Review the article "An Environmental Risk Assessment of a Pesticide" by G.M. Rand,[43] and provide a two-page summary of the requirements for ecological risk assessments under the Federal Insecticide, Fungicide, and Rodenticide Act (FIFRA) law.

14-24. Provide a preliminary estimate of the appropriate pathways for a site provided by your instructor using the flowchart in Figure 14-3.

14-25. Perform a qualitative risk assessment on a laboratory identified by your instructor. Specifically identify:

(*a*) Hazards (e.g., several chemicals of concern)
(*b*) Potential receptors
(*c*) Pathways
(*d*) Toxicity of the chemicals listed in (*a*).

14-26. Determine the potentially exposed population for a hypothetical hazardous waste site on the periphery of your campus. Specifically identify the most exposed individuals and subpopulations requiring special consideration.

14-27. Suppose that your campus water supply was contaminated with trichloroethylene at a concentration of 10 mg/L. Determine the total intake over a 4-year undergraduate program using Equation 14-3.

14-28. Repeat Problem 14-27, assuming the contamination was in beer rather than water.

14-29. Discuss some of the special problems in assessing risk to carcinogens as opposed to noncarcinogens.

14-30. Comment on the validity of the assumptions underlying the exposure assessment described in Section 14-9 (Lackawanna Refuse Site).

14-31. Review the article by Cox and Blayney[44] and briefly describe some of the difficulties in obtaining public input in risk decisions.

14-32. Review the article by S. Brett et al.[45] and discuss the special risks associated with excavation of wastes at a hazardous waste site.

14-33. Review the articles by Reinert[46] and Stephanatos.[47] Discuss how risk assessment may be used to help establish cleanup levels at hazardous waste sites.

14-34. Using simulation software provided by your instructor, create a histogram showing the distribution of carcinogenic risk from ingestion of benzene in groundwater. Assume the following site-specific information: Ingestion of water (L/day) is distributed

normally (1.34, 0.4) and benzene concentration in water (mg/L) is distributed lognormally (0.007, 0.76). Select distributions for body weight and slope factor from Table 14-13. Specifically calculate the probability of exceeding the 10^{-6} risk. Use EF = 365 days, ED = 170 year, AT = 25,550 days.

14-35. As part of a human health risk assessment evaluating exposure via the ingestion of the surficial soil exposure pathway, the risk assessor needs to use the current oral reference concentration (RfC_o) and oral cancer slope factor (CSF_o) for inorganic arsenic. What is the most efficient means of obtaining these toxicity criteria?

REFERENCES

1. *Risk Assessment in the Federal Government: Managing the Process,* National Research Council, Comm. on Institutional Means for Assessment of Risks to Public Health, National Academy Press, Washington, D.C., p. 18, 1983.
2. Wilson, W. R., and E. A. C. Crouch: "Risk Assessment and Comparisons: An Introduction," *Science,* vol. 236, pp. 267–270, April 1987.
3. *Risk Assessment in the Federal Government,* op. cit., pp. 18–19.
4. Brown, H. S.: "A Critical Review of Current Approaches to Determining 'How Clean is Clean' at Hazardous Waste Sites," *Hazardous Waste & Hazardous Materials,* vol. 3, no. 3, Mary Ann Liebert Publishers, pp. 233–260, 1986.
5. Huggins, A., and M. D. LaGrega: "Methods to Set Cleanup Goals to Protect Surface and Ground Water Quality at Hazardous Waste Sites," *Water Science & Technology,* vol. 23, pp. 405–412, 1991.
6. U.S. EPA: *Handbook for Conducting Endangerment Assessments,* U.S. EPA, Research Triangle Park, N.C., 1987.
7. U.S. EPA: *Superfund Public Health Evaluation Manual (SPHEM),* EPA 540/1-86/060, OSWER Directive 9285.4-1, U.S. EPA, October 1986.
8. U.S. EPA: *Risk Assessment Guidance for Superfund,* vol. I. Human Health Evaluation Manual, OERR, EPA/540/1-89/002, December 1989.
9. U.S. EPA: *Superfund Exposure Assessment Manual,* OERR, EPA/540/1-88/001, OSWER Directive 9285.5-1, 1988.
10. U.S. EPA: *Exposure Factors Handbook,*, vols. 1–3, Office of Health and Environmental Assessment, EPA/600/P-95/002Fa, b, and c, 1997.
11. U.S. Environmental Protection Agency: *Risk Assessment Guidance for Superfund,* 1989.
12. Ibid.
13. Ibid.
14. Conway, R. A., ed.: *Introduction to Environmental Risk Analysis,* Van Nostrand, p. 19, 1982.
15. U.S. EPA: *Superfund Exposure Assessment Manual,* EPA/540/1-88/001, OERR, OSWER Directive 9285.5-1, 1988.
16. U.S. EPA: *Exposure Assessment Methods Handbook,* Draft, Office of Health and Environmental Assessment, 1989.
17. U.S. EPA: *Risk Assessment Guidance for Superfund,* 1989.
18. U.S. EPA: *Superfund Exposure Assessment Manual,* 1988.
19. U.S. EPA: *Risk Assessment Guidance for Superfund,* 1989.
20. Wilson, R.: "Analyzing the Daily Risks of Life," *Technology Review* (M.I.T.), pp. 41–46, February 1979.
21. Fischoff, B., S. Lichtenstein, P. Slovic, S. L. Derby, and R. L. Keeney: *Acceptable Risk,* Cambridge University Press, 1981.
22. Hirschhorn, J. S., K. U. Oldenburg, and D. Dorau: "Using Risk Concepts in Superfund," *Superfund '87,* Hazardous Materials Control Research Inst., Washington, D.C., 1987.

23. Fischoff, B., S. Lichtenstein, P. Slovic, S. L. Derby, and R. L. Keeney: *Acceptable Risk.*
24. Santos, S. L., and S. Edwards: "Risk Communication: A Critical Part of the Public Participation Process," *Superfund '87,* Hazardous Materials Control Research Inst., Washington, D.C., pp. 254–259, 1987.
25. Rand, G. M.: "An Environmental Risk Assessment of a Pesticide," *The Risk Assessment of Environmental Hazards: A Textbook of Case Studies,* D. L. Paustenbach, ed., Wiley-Interscience, p. 903, 1989.
26. Rand, G. M.: "An Environmental Risk Assessment of a Pesticide," p. 902.
27. Burmaster, D. E., et al.: "Monte Carlo Techniques for Quantitative Uncertainty Analysis in Public Health Risk Assessments," *Superfund '90,* Hazardous Materials Control Research Inst., Washington, D.C., pp. 215–221, 1990.
28. Rubinstein, R. Y.: *Simulation and the Monte Carlo Method,* Wiley, 1981.
29. Bratley, P.: *A Guide to Simulation,* Springer-Verlag, 1983.
30. Kalos, M. H., and P. A. Whitlock: *Monte Carlo Methods,* vol. I: *Basics,* Wiley, 1986.
31. Salmento, J. S., E. S. Rubin, and A. M. Finkel: "A Review of @RISK®," *Risk Analysis,* vol. 9, no. 2, pp. 255–257, 1989.
32. Burmaster, D. E., and E. C. Udell: "A Software Review of Crystal Ball®," *Risk Analysis,* vol. 10, no. 2, pp. 343–345, 1990.
33. Burmaster, D. E., and K. E. von Stackelberg: "Quantitative Uncertainty Analysis in Exposure and Dose Response Assessments in Public Health Risk Assessments Using Monte Carlo Techniques," *Superfund '89,* Hazardous Materials Control Research Inst., Washington, D.C., pp. 82–85, 1989.
34. Burmaster, D. E., et al.: "Monte Carlo Techniques for Quantitative Uncertainty Analysis."
35. Ibid.
36. United States Environmental Protection Agency: *Remedial Investigation Report, Lackawanna Refuse Site,* U.S. EPA, August 1984.
37. United States Environmental Protection Agency: *Feasibility Study of Alternatives, Lackawanna Refuse Site,* U.S. EPA, February 1985.
38. O'Hara, P. F., K. J. Bird, and W. A. Baughman: "Exploratory Drilling into a Buried Uncontrolled Drum Disposal Pit," *Superfund '86—Proceedings of the Ninth National Conference,* Hazardous Materials Control Research Inst., Washington, D.C., pp. 126–131, 1986.
39. United States Environmental Protection Agency: "Record of Decision—Lackawanna Refuse Site, Old Forge, PA," March 22, 1985.
40. Environmental Resources Management, Inc.: "Risk Assessment for the Lackawanna Refuse Site," April 1987.
41. Blasko, M. J., et al.: "Design of Remedial Measures and Waste Removal Program—Lackawanna Refuse Superfund Site," *Superfund '87—Proceedings of the Ninth National Conference,* Hazardous Materials Control Research Inst., Washington D.C., pp. 367–370, 1987.
42. Brown, H. S.: "A Critical Review of Current Approaches to Determining 'How Clean is Clean,' " op. cit.
43. Rand, G. M.: "An Environmental Risk Assessment of a Pesticide," op. cit.
44. Cox, M., and E. Blayney: "Public Involvement in Acceptable Risk Decisions," *Mid-Atlantic Industrial Waste Conference,* G. D. Boardman, ed., pp. 454–461, Technomic Publishers, Lancaster, Pa., 1986.
45. Brett, S. M., et al.: "Assessment of the Public Health Risks Associated with the Proposed Excavation of a Hazardous Waste Site," *The Risk Assessment of Environmental Hazards: A Textbook of Case Studies,* D. L. Paustenbach, ed., Wiley-Interscience, pp. 427–460, 1989.
46. Reinert, K. H.: "Risk-Based Cleanup Levels for Soils," *Superfund '90,* Hazardous Materials Control Research Inst., Washington, D.C., pp. 185–188, 1990.

47. Stephanatos, B. N.: "How Clean is Clean? The Importance of Using Site-Specific Factors in Developing Cleanup Levels at Hazardous Waste Sites," *Superfund '90,* Hazardous Materials Control Research Inst., Washington, D.C., pp. 612–617, 1990.

ADDITIONAL READING

Andelman, J. B., and D. W. Underhill: *Health Effects from Hazardous Waste Sites,* Lewis Publishers, Chelsea, Mich., 1987.

Anon.: *Risk Assessment, Management, Communication: A Guide to Selected Sources,* NTIS No.: PB87-185500/GAR., NTIS, Springfield, Va., 1987.

Budd, W. W.: "A Comparison of Three Risk Assessment Techniques for Evaluating a Hazardous Waste Landfill," *Hazardous Waste and Hazardous Materials,* vol. 3, no. 3, pp. 309–320, 1986.

Clayson, D. B., et al.: *Toxicological Risk Assessment,* vols. 1 and 2, CRC Press, Boca Raton, Fla., 1985.

Cohrssen, J. J., and V. T. Covello: "Risk Analysis: A Guide to Principles and Methods for Analyzing Health and Environmental Risks," NTIS PB89-137772, Council on Environ. Qual., Washington, D.C., 1989.

Covello, V. T., et al.: "Risk Evaluation and Management," *Contemporary Issues in Risk Analysis,* vol. 1, Plenum Press, New York, 1986.

Federal Register: "Guidelines for Carcinogen Risk Assessment," vol. 51, no. 185, pp. 33992–34003, 1986.

Federal Register: "Guidelines for Development Toxicity Risk Assessment," vol. 51, no. 185, pp. 34028–34040, 1986.

Federal Register: "Guidelines for Exposure Assessment," vol. 51, no. 185, pp. 34041–34054, 1986.

Federal Register: "Guidelines for the Health Assessment of Chemical Mixtures," vol. 51, no. 185, pp. 34014–34025, 1986.

Fiksel, J.: "Quantitative Risk Analysis for Toxic Chemicals in the Environment," *Journal of Hazardous Materials,*, vol. 10, no. 2–3, pp. 227–240, 1985.

Glickman, T. S., and M. Gough, eds.: *Readings in Risk,* Resources for the Future, 1990.

Goldman, B. A.: "The Use of Risk Assessment During Selection of Off-Site Response Actions," *Hazardous Waste and Hazardous Materials,* vol. 3, no. 2, 1987.

Hallenbeck, W. H., and K. M. Cunningham: *Quantitative Risk Assessment for Environmental and Occupational Health,* Lewis Publishing, 1986.

Layton, D. W., B. J. Mallon, D. H. Rosenblatt, and M. J. Small: "Deriving Allowable Daily Intakes for Systemic Toxicants Lacking Chronic Toxicity Data," *Regulatory Toxicology and Pharmacology,* vol. 7, no. 1, p. 914–112, 1987.

Lowrance, W.: *Modern Science and Human Values,* Oxford University Press, London and New York, 1984.

Morgan, M. G., et al., "Communicating Risk to the Public," *Environ. Sci. & Tech.,* vol. 26, no. 11, November 1992, pp. 2048–2055.

Paustenbach, D. J., ed.: *The Risk Assessment of Environmental Hazards: A Textbook of Case Studies,* Wiley, 1989.

Scott, M. P.: "Applications of Risk Assessment Techniques to Hazardous Waste Management," *Waste Management Research,* vol. 5, no. 2, pp. 173–181, 1987.

Silbergeld, E. K.: "Five Types of Ambiguity: Scientific Uncertainty in Risk Assessment," *Hazardous Waste and Hazardous Materials,* vol. 4, no. 2, pp. 139–150, 1987.

Solomon, K. A., and P. Ricci, eds.: "Special Issue. Regulatory and Legal Aspects of Hazardous Materials Risks," *Journal of Hazardous Materials,* vol. 15, no. 1–2, 1987.

Tardiff, R. G., and J. V. Rodricks, eds.: *Toxic Substances and Human Risk—Principles of Data Interpretation,* Plenum Press, 1987.

Travis, C. C., et al.: "Cancer Risk Management: A Review of 132 Federal Regulatory Decisions," *Environ. Sci. Tech,* vol. 21, pp. 415–420, May 1987.

U.S. Environmental Protection Agency: *Health Effects Assessment Summary Tables (HEAST),* U.S. EPA, 1991 (revised annually).

Whipple, C.: *De Minimis Risk,* Macmillan, New York, 1988.

Wolf, K: "Hazardous Substances and Cancer Incidence: Introduction to the Special Issue on Risk Assessment and Risk Management," *Journal of Hazardous Materials,* vol. 10, no. 2-3, pp. 167–178, 1985.

ACKNOWLEDGMENTS

The authors wish to acknowledge the following individuals who assisted with the preparation of this chapter for the first edition: Section 14-7 was prepared by Bruce Molholt, Ph.D., *ERM,* and David Stout, *Blasland, Bouck and Lee, Inc*. Review assistance was provided by Andrew Huggins, Ph.D., Basilis Stephanatos, Ph.D., and Robin Streeter, *ERM*. The principal reviewer for ERM for the first edition was B. C. Robison, D.V. M., Ph.D., D.A.B.T.

The reviewers for the second edition were Diane Heim, Bruce Molholt, Ph.D., Randy Shuler, Ph.D., D.A.B.T., and Sandra Morales, *ERM*.

CHAPTER 15

Site and Subsurface Characterization

Bowed by the weight of century he leans
Upon his hoe and gazes on the ground
The emptiness of ages in his face
and on his back the burden of the world
The Man with the Hoe, Edwin Markham

15-1 INTRODUCTION

Whether designing a new facility or investigating potential contaminant migration at an existing site, proper characterization of the subsurface conditions and their interaction with surface features is critical to the process. For any hazardous waste management facility, it is absolutely essential to know what lies on and beneath the site. In the context of this book, **site and subsurface characterization** may be defined as the qualitative and quantitative description of the conditions on and beneath the site that are pertinent to the hazardous waste management application.

A simplified cross section developed by using site characterization tools and methodologies described in this chapter is shown in Figure 15-1. This cross section illustrates the type of information that is developed to define the site and subsurface conditions and answer the questions: What is there? Where is it going?

Site and subsurface characterization integrates a wide range of disciplines including geology, hydrology, hydrogeology, chemistry, biology and geotechnical engineering. Inadequate or, worse still, incorrect site and subsurface characterization can lead to ineffective remediation of contaminated sites and to inadequacies in design, engineering, and/or siting of new hazardous waste management facilities. For example, a new facility would be unwisely sited if a geologic fault were to go undetected. An undetected zone of groundwater flow beneath an existing site can result

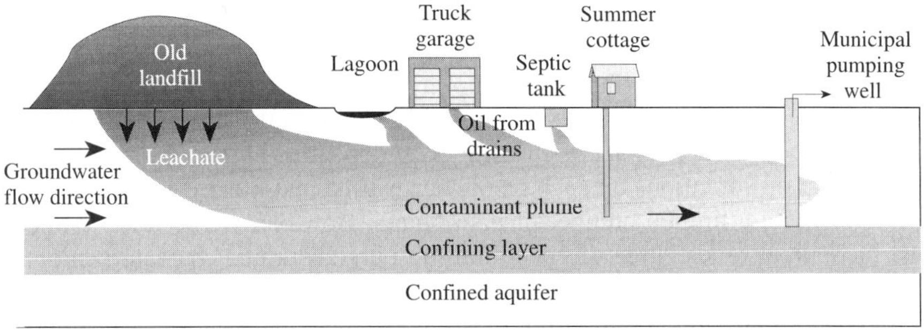

FIGURE 15-1
Simplified cross section developed from site and subsurface characterization.

in the remedial action functioning improperly. Subsurface investigations must generate a clear definition of the site's three-dimensional geologic and hydrogeologic characteristics, yet these investigations must assess more than the hydrogeology. The investigations must reveal the influence of site and subsurface conditions on the performance of the existing or planned hazardous waste management system. The investigations must therefore identify information on such things as prior site history, groundwater flow direction, water quality and quantity, potential contaminant migration, and constructability of remedial systems. In the case of site remediation, it is also necessary to estimate the total contaminant mass and distribution.

This chapter defines and describes an approach to site and subsurface characterization and identifies the tools available. The approach and techniques are suitable for siting and design of new facilities, and for site assessment for remediation of existing contaminated sites. It is the aim of this chapter to familiarize the student and hazardous waste manager with the available techniques and their applicability. Of the many hundreds of tools and techniques available for site and subsurface investigations (see Figure 15-2 for one of the earliest examples), the most useful (and

FIGURE 15-2
Early geophysicist (or modern diviner).

thus widely employed) are described in additional detail. This chapter places greater emphasis on subsurface characterization than on characterizing of site surface conditions. Air monitoring, surface soil sampling, water sampling of streams, springs and seeps, and aquatic surveys are examples of site surficial characterization studies that need to be conducted but which will not be described in this chapter.

Because of limitations on such resources as time and money, we must characterize sites with a limited number of data points. How can we do this? While we must recognize some degree of uncertainty in our subsurface characterization, we can at the same time recognize that there is a systematic nature to the distribution of geologic deposits. The more thoroughly we understand geologic processes and the site-specific geologic history, the more reliably we can interpolate between points of known information. It is this geologic system that adds a degree of order to the subsurface, and we should both acknowledge and use this to design and implement subsurface investigation programs.

Site and subsurface investigations are commonly implemented for CERCLA remedial investigations and RCRA corrective actions. The **remedial investigation** (RI) consists of the data collection and site characterization phase of a CERCLA remedial action and is begun in advance of the **feasibility study** (FS). In the FS, alternative remedial measures are developed and evaluated. The feasibility study phase of the CERCLA action, therefore, cannot precede the remedial investigation. However, the RI and FS must be interdependent, but generic FS alternatives can be considered with the RI, especially to ascertain that key data are obtained during the RI. Without an adequate site assessment, the selected remedial alternative resulting from the feasibility study may be inadequate. Further, as remedial alternatives are tentatively chosen, additional site characterization data may be needed to confirm the suitability of the alternatives. In addition to site characterization studies conducted to assess contaminant migration and remedial action, characterization is also conducted as part of the health and safety protection of site workers.

The techniques described in this chapter must be selected and employed with specific consideration given to the geologic and man-made setting of the site. For example, the necessity of avoiding buried underground utilities such as power and gas lines in an urban setting may dictate the use of different tools and techniques than those chosen in a rural setting, even with similar geologic conditions. The type and extent of the investigation will vary, depending on known or expected conditions. The chosen techniques must reflect the site-specific nature of site and subsurface, must be flexible to adapt as conditions are revealed, and must not evolve into a cookbook approach that is carried without additional thought from site to site.

15-2 METHODOLOGY

As with any comprehensive and complex environmental project, it is best to approach site and subsurface characterization in phases. It is recommended that these investigations be conceptualized in three phases as follows[1]:

1. Identify the problem, such as the range of possible remedial actions, and develop the scope of investigations (investigation planning).

2. Define the nature and extent of the site and subsurface conditions on a qualitative and quantitative basis.
3. Complete a detailed site and subsurface characterization.

During the initial phase, it is necessary to collect and evaluate existing site and subsurface information. As much as possible with the available data, this phase defines specific site problems that exist at present, or potentially exist as a result of the past, present, and/or proposed site activities. For example, site-specific problems influencing the investigation of site and subsurface conditions at an old chemical manufacturing facility could include former waste disposal lagoons (now filled, graded, and supporting a lawn), areas of buried underground storage tanks, or rail/truck loading and unloading areas. The initial phase also identifies site and subsurface features that may contribute to or reduce the severity of any problems or potential problems. For example, a limestone aquifer beneath the site may be overlain by fractured shale. In this case, the frequency and continuity of the shale fractures, as well as the hydraulic conductivity of the shale, govern the extent to which the shale can influence the rate of contaminant transport to the underlying limestone aquifer. These conditions draw attention to the need, in this example, to define the continuity and properties of the shale. Knowledge of the range of possible remedial alternatives will also improve the efficacy of the site investigations.

In the context of site characterization, how do we define a "problem"? As described in general terms above, the problem definition guides the nature and extent of subsurface characterizations. The problem can best be defined on a risk basis. The assessment of risk is preliminary in the early stages of site characterization and more detailed in the later stages of site characterization. For this, we need to understand contaminant fate, transport, and exposure. Detailed discussion of risk assessment is included in Chapter 14.

In the second phase, the collection of site-specific data through sampling and field studies begins. During this phase, initial site characterization is completed. Site characterization is the representation or depiction of the site and subsurface into discrete units having generalized physical and engineering properties and is expressed in both qualitative and quantitative terms (see Example 15-1). During this phase, exposure pathways are identified and data for the screening of remedial alternatives or for the site-specific design of new facilities are gathered.

EXAMPLE 15-1. INITIAL SITE CHARACTERIZATION. An example of the critical information developed for an initial site characterization of a former solid waste site in western Michigan is provided below.

Solution

1. The site was formally a sand borrow pit which excavated sands to the top of a clay layer.
2. At the site perimeter and underlying 6 to 12 cm of organic topsoil is a layer of fluvial "beach" sands averaging 10 m in thickness.
3. The sand has a relatively high hydraulic conductivity, perhaps on the order of 10^{-2} cm/s.
4. The sand is underlain by a 15-m-thick deposit of glacial lacustrine clays.
5. The hydraulic conductivity of the clay is estimated to be quite low, perhaps on the order of 10^{-7} cm/s.

6. The landfill accepted a wide range of municipal and industrial wastes from the surrounding communities and light industry.
7. The water table is encountered at a depth of 3 m below the ground surface.
8. The nearest residences are within 500 m and rely on groundwater for their potable water.

The above summary, developed from available information, can be used in the planning and design of the detailed site characterizations.

During the final phase, detailed site characterization, it is necessary to collect additional data required for the detailed analysis of the hazardous waste management alternatives. For example, feasibility studies could include tasks such as bench-scale treatability studies employing proposed treatment processes and site-specific wastes. The detailed site characterization could include analysis of the groundwater chemistry in the site vicinity as well as at the nearby residences.

Although it is human nature to desire more and more information, it is impossible to define the site and subsurface conditions without some degree of uncertainty. Regardless of the number of samples taken and the number of *in situ* (in place) tests run, extrapolation is always required. To acquire "total knowledge" requires, in theory, an infinite number of monitoring wells, soil samples, groundwater samples, and chemical analyses. It is necessary, and proper, to render informed scientific and engineering judgments regarding the site and subsurface conditions. It is cautioned that the inclination to define fully the subsurface conditions may be counterproductive, particularly in remedial investigations. In fact, additional data gathering and detailed site assessments have been employed to avoid or delay decisions regarding what remedy to implement and when to do it. The student is cautioned that a balance between additional study and moving forward with the appropriate action must be achieved. As of the time of writing this book, it is not unusual for a complete site and subsurface investigations in the United States to cost over a million dollars and require 2 or more years to complete. In the meantime, contaminants continue to migrate. While the continued contaminant migration may cause increased risk to the public health and environment, in some cases delays in active remediation have resulted in the natural attenuation of some plumes. Diligence is required in the design of investigations to reduce the time required, minimize the overall project costs, and yield the information that is **necessary and sufficient.**

15-3 PLANNING

It is safe to say that there is a wealth of information available regarding many sites under consideration. The information is available from a wide variety of sources, including those listed on Table 15-1. Much can be learned from these sources of information without ever leaving the comfort of your office. Although Table 15-1 identifies sources of geological, geotechnical, and operational information, it should be noted that many of these sources are accessible via electronic means such as the Internet and database services. In the following portion of this chapter, the planning process will be discussed in detail.

Although the site and subsurface investigations are conceptualized as phases, the investigation efforts are normally conducted along a time continuum. There must

TABLE 15-1
Sources of characterization information

Geological
Local universities
State agencies (e.g., Pennsylvania Geological Survey)
Federal agencies (e.g., U.S. Geological Survey, British Geological Society)
Agricultural agencies (e.g., Soil Conservation Service)
Professional organizations
Satellite imagery
Geotechnical
Prior site studies
Regional/local building permits
Consultants' reports
Construction records
Operational
Air photos
Plant records
Personnel interviews
Operational and disposal permits

be an ongoing evaluation of data as they become available and a reevaluation of the investigation plan in light of the increasing knowledge gained as the investigation progresses. A work plan developed for site and subsurface investigations must be flexible enough to permit necessary changes that will inevitably result from data evaluation during the early stages of the investigation program.

While defining site-specific data to gather during the field studies, it is essential to visit the site. The site inspection reveals information regarding topography, site access for investigation approaches, present drainage ways, seeps, existing landforms, and the potential for other interferences such as building foundations and site fill. The site visit should also include the area in the vicinity of the site, which may reveal potentially contaminating conditions or businesses. The field survey may also identify the potential hazardous waste sources, migration pathways, receptors, and any obvious impacts such as distressed vegetation. Further, during the initial phase of investigations of a prior hazardous waste site, the need for immediate remedial actions is identified. This initial remediation, which generally focuses on source control and containment, includes such action as immediate removal of leaking drums resulting in acute health and safety situations or planned removals where significant cost savings or minimizing risks of fire explosions could be accomplished. In addition, initial remedial measures often include fences and alternative water supplies to protect the public health and the environment.

Traditionally, the site assessment is thoroughly planned and includes the *in situ* testing, sampling, and analysis program as described in detail in the following sections. There is a sound basis for such detailed planning, as mistakes are minimized and the investigation can be optimized. However, such detailed planning must make assumptions about what is expected and may not be as flexible as desired to alter the plan on the basis of actual conditions encountered. An alternative approach, an expedited site characterization, emphasizes the use of noninvasive techniques[2] (such as geophysical methods discussed later in this chapter) and a dynamic work plan,[3] which focuses on the decision-making structure for sampling and analysis

rather than the sampling and analysis itself. One of the key features of a dynamic work plan is the ability to analyze for contaminants in the field rather than simply sampling and sending the samples to a remote laboratory for analysis. The EPA has recognized the need to accelerate the site assessment processes and issued guidance on "Abbreviated Preliminary Assessments"[4] and "combined" preliminary assessment and site inspection activities.[5] The abbreviated preliminary assessment makes use of the professional judgment and previous site assessment experience to make decisions earlier in the process than previously permitted. The combined preliminary assessment and site inspection streamlines the process by allowing the work to proceed as one continuous site investigation rather than in separate phases with overlapping and repetitive tasks. Progress at standardization of accelerated (expedited) site characterization has also been achieved.[6,7,8]

Investigation Plan

Why have a plan? Administratively, the objectives of an **investigation plan** are to:

1. Provide specific guidance (details and protocols) for all work
2. Provide mechanisms for planning and approving site activities
3. Provide a basis for schedule and cost estimates
4. Ensure that all the activities conducted are necessary and sufficient for planned activities

For remedial investigations conducted under CERCLA, documents and reports required are shown in Table 15-2. Although it may not be necessary to include all of these documents and reports for investigations not conducted under CERCLA, Table 15-2 does provide a comprehensive check list for the types of documentation and reports that should be considered. The U.S. Army Corps of Engineers has developed a manual for preparation of sampling and analysis plans that are intended to be functionally equivalent to the EPA requirements.[9]

The major elements of an investigation and sampling plan include:

1. Scope and objectives of the investigation
2. Summary of site background information
3. Summary assessment of existing data
4. Contaminants of interest
5. Sample types
6. Sampling locations and frequency
7. Sampling and testing procedures
8. Operation plan and schedule
9. Cost estimate
10. Other supporting documents such as:
    - (*a*) Quality assurance and quality control plans
    - (*b*) Health and safety plans
    - (*c*) Data management plans

It is not uncommon for the organization responsible for the investigations (e.g., consulting firm, government agency, or industry) to have some of the above items

TABLE 15-2
Work plan documents

Sampling plan
Quality assurance plan
Quality control plan
Health and safety plan
Operation plan
Operation schedule
Data management plan
Project management plan
Chain-of-custody procedures
Organizational chart
Field/laboratory documentation
Project/field log books
Sample tags
Sample data sheets and logs
Chain-of-custody records, seals
Receipt of sample forms
Laboratory log books
Laboratory data, calculations, graphs
RI task reports
Site description
Contamination assessment
Environmental assessment
Public health assessment
Endangerment assessment
Draft/final RI report
Technical progress and financial reports
Monthly technical progress report
Monthly financial progress report
Cumulative project cost report

as standard operating procedures. However, on the basis of the findings during the first phase of investigation, it is necessary to decide on the types and locations of samples required for the detailed site investigation.

The remainder of this section details the elements of the investigation and sampling plan. The remainder of the chapter then presents the methods and technologies appropriate for hazardous waste management site investigations.

Investigation and Information Objectives

What are the specific purposes of the proposed site investigations? Are they to define the distribution and migration of contaminants? Are they to assess the risk to the public health and the environment? Are the investigations being planned to develop remedial alternatives? Are they to design selected remediation techniques? Are the investigations to address all of the foregoing? What is the objective of the specific information that will result from the investigation? Will concentration of contaminants in a given groundwater sample be used in a risk assessment or in a treatment stream design? The success of hazardous waste site investigations begins with a well-thought-out definition of the objectives of the investigation and the uses of the information gathered in that investigation.

Site Background

Considerable information is typically available for both the operational history and geology of the site. It is absolutely essential that this background information be assimilated prior to initiation of detailed site and subsurface investigations. Sources of background information regarding the operational history include interviews of personnel and reviews of property records.

The regional geology information for most sites is readily available as listed in Table 15-1. Geological information can come from a variety of sources including the federal and state geological surveys where regional geology reports describe the predominant geologic materials and their characteristics. This regional geological information can be supplemented with local geologic information from such sources as the geology department at the nearby universities or the county soil conservation office.

Air photos are another particularly useful source of information. Air photos are available for many areas of the United States dating back to the 1930s. Interpretation of available air photos may show much about the site landforms, the engineering site history, past waste disposal operations, prior drainageways, potential for contaminant migration, topography, and vegetation. For example, shown in Figure 15-3 are the interpreted site conditions from an air photo for a former waste disposal site. Earlier photos revealed that the area began as predominantly grassy farmland with drainage proceeding from the northeast corner initially southwesterly and then southerly traversing the site. As time progressed, the early air photos reveal that, by 1958, the site was being developed as a landfill. By 1966, the date of the interpreted air photo presented in Figure 15-3, the site had been heavily developed as a landfill. This air photo showed that the generally north to south drainage had been intercepted along the northern boundary and diverted along the eastern edge of the site. Lagoons and storage areas are also revealed. By 1978, air photos show that the site had been essentially regraded with little evidence of prior landfill activity.

How can the results of air photo interpretation be used in this case? First they help to identify likely contaminant transport pathways, and this, in turn, helps de-

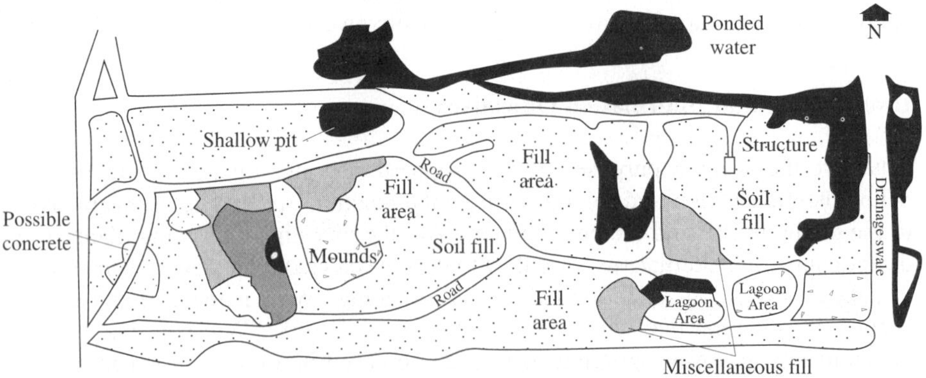

FIGURE 15-3
Interpretation of a historic air photo of a landfill.

sign the sampling plan. Specifically, the off-site drainage ways (old swales) would be more likely to contain contaminants transported from the landfill area than a randomly selected sampling location. Only from interpretation of air photos could this significant finding be developed. As a result, rather than **random** sampling, **focused** sampling can be undertaken to investigate contaminant levels and extent of migration in prior drainage ways. In addition, several lagoons are identified in the air photos. These lagoons constitute a likely source of contaminants within the landfill area. Interviews of plant personnel confirmed the lagoons identified on the air photos were used for waste disposal. As a result, the lagoons could be targeted as potential "hot spots" during the on-site investigations.

Aerial photography (photogrammetric topographic mapping) can also provide a cost-effective base map of the site. A base map of current topography is needed for engineering analysis of existing conditions, design of remedial or new facilities, and for construction and operation.

Assessment of Existing Data

Assessment of available data includes an analysis of the data validity, sufficiency, and sensitivity both temporally and areally. Data validity should first be considered. For example, records such as well logs prepared by drillers are frequently available for wells at the site. What is the validity of water quality data from one of these preexisting wells? A driller's log for a well installed many years ago may (or may not) include information about the depth of the casing, depth of the well, diameter of the borehole, depth and nature of different lithologies. Close examination of the driller's log commonly does not adequately characterize the geological conditions. (Why should they? That was often not their original purpose.) Thus a complete geologic or geotechnical description of all of the materials encountered would not normally be expected for a well installed in the distant past. The screened or monitored interval is often not fully defined, and as a result, water sampled from preexisting wells may come from more than one aquifer or from some unknown depth in the subsurface. Water levels measured during drilling are often obtained before the water level has stabilized. Consequently, the usefulness of such data is limited. Stated more generally, it is necessary to understand the limitations of the data. This brings to the forefront a basic tenet in hazardous waste management: *It is better to have no data than incorrect data.*

Since the subsurface environment is not static, existing data may not reflect the current situation, particularly as it relates to contaminant concentration and distribution. Similarly, existing data may not be areally distributed to the extent necessary. Another issue often associated with existing data is the question of water quality data validity. Much has been learned about sampling, transport, storage, and testing of water samples, and historic water quality samples may not measure up to the same standards as present-day data. However, in all cases, the trends and information provided by existing data are the only evidence pointing to what the subsurface contaminant conditions were in the past. Essentially, for better or worse, one must use historic data to reconstruct historic conditions.

Data sufficiency can be assessed in terms of anticipated analyses. For example, groundwater flow and contaminant transport analyses cannot be done without hy-

draulic conductivity and gradient information. The assessment of data helps shape the investigations.

Finally, it is necessary to look at data sensitivity. Will more data make any difference in the final outcome? Since generation of data takes time and costs money, evaluating the sensitivity of data is essential to a cost-effective investigation. Data sensitivity is illustrated by Example 15-2.

EXAMPLE 15-2. DATA SENSITIVITY. A review of existing data for the former solid waste site in western Michigan described in Example 15-1 included over 100 grain size distribution analyses on the formerly mined sand. In the samples the sand was found to be 30% ± 2% medium sand, 70% ± 2% fine sand, and less than 1% silt and clay fines. How many more grain size analyses are needed?

Solution. It would appear that the site is well characterized with respect to grain size distribution, yet the answer to how many more tests are needed is not zero. A few confirmatory tests (perhaps three to five) are recommended to ensure that the existing data are valid and free from systematic errors. However, once the previous laboratory results are confirmed, and provided field descriptions remain consistent, significant additional grain size testing would divert time, energy, and money from other perhaps more important concerns. A statistical analysis of the data might be employed to document the engineering judgment that only a few additional tests are required.

Contaminants of Interest

The work plan for site and subsurface investigations must identify contaminants of interest. Identifying the contaminants of interest in a work plan being developed to guide an investigation to determine the contaminants at the site may appear as a "Catch 22." In one sense it is, since the investigation must detect the unexpected. On the other hand, the analysis of existing data should reveal suspected contaminants and contaminant transport pathways. This preliminary contaminant identification provides a basis for fate and transport analysis and a preliminary risk assessment of the toxicological information and potential migration pathways. The toxicological and transport information is used to identify contaminants of particular interest for a site and subsurface investigation. With this information the types of samples, sampling locations and frequency, and sampling and analytical procedures may be better planned.

Sample Types

Samples of all media (air, water, soil, and waste) are typically obtained as part of the site and subsurface investigations. Water samples may include both surface and groundwater. Soil samples include those for chemical assays and for measurement of physical and engineering properties (strength, permeability, and compressibility). Sample types may be alternatively classified by whether or not they are samples of source (e.g., from an old waste disposal lagoon) or along transport pathways (e.g., groundwater at a down-gradient location).

Types of samples may also be indicated by the results of modeling studies that help identify those transport media that are most important for the particular site. As a result, samples could be of flora and fauna or of air quality as well as water quality.

Samples may be needed to look at the physical, chemical, and biological parameters in each medium. In some cases sampling of the human population may be included (not the whole human, just parts).

Sampling requires more than physical samples; it is meant to include other data as well. For example the sampling plan might include monitoring of meteorological conditions including temperature, pressure, precipitation, wind velocity and humidity.

Sampling Locations and Frequency

The work plan must define the location and frequency of samples. To establish sampling locations and frequency, the data objectives need to be considered. In other words, provide samples to fill data gaps, and provide better spatial coverage and chemical coverage. By planning, in advance, sample locations and frequency, the probability of getting the proper number and types of samples at the proper locations is increased. It is advisable to consider the statistical validity of the data at this point in the planning process. Regardless of the number of samples obtained, the importance of the miniscule look at the subsurface must be recognized (see Example 15-3).

> **EXAMPLE 15-3. LOOKING BENEATH THE SURFACE.** The work plan for site characterization of the relatively homogeneous clay beneath a former solid waste site in western Michigan calls for 20 undisturbed clay samples to be obtained for permeability testing. As described earlier, the sand is underlain by a 15-m-thick deposit of glacial lacustrine clays. Each sample is 74 mm in diameter and 144 mm long. The site is 900 m by 600 m. Calculate, on a volumetric basis, the percentage of soil actually sampled.
>
> **Solution.** Calculate the total volume of sample:
>
> $$V_s = 20 \times \left(\frac{3.14 \times 74^2}{4} \right) \times 144/1000^3 = 0.012 \text{ m}^3$$
>
> Calculate the total volume of clay layer:
>
> $$V_t = 900 \times 600 \times 15 = 8.1 \times 10^6 \text{ m}^3$$
>
> Calculate the ratio, in percent:
>
> $$R = 0.012/8.1 \times 10^6 \times 100 = \underline{0.00000015\%}$$
>
> Just think of the importance of the decisions based on viewing so little of the subsurface! In fact, some would say that 20 tests are too many for a uniform clay deposit as described for this problem.

Although we actually sample and test a very small portion of the subsurface, caution needs to be exercised to avoid obtaining and analyzing too many samples such that the information is unnecessary. For remedial investigations of sites with existing contamination, is it necessary to "define the plume"? It may not be necessary to know the location and concentration of all contaminants; a remedial system can often be based on well-developed knowledge of site and subsurface conditions without a full definition of the plume location. Balanced against the need to control the quantity

and quality of sampling information is the recognition that it is both expensive and time-consuming to return to the site to gather additional samples should these be needed. Thus, it is recommended that the sampling frequency exceeds the minimum necessary in order to provide flexibility at future stages of the investigations.

While it must be acknowledged that only a small portion of the subsurface is actually sampled and tested, the use of geostatistical methods may provide the investigator with insight into how much confidence can be placed in the data. For example, if the results of hydraulic conductivity tests conducted on the 20 samples in Example 15-3 indicate a small standard deviation and a narrow range of values, this could lead to the conclusion that the 20 samples are more than adequate to determine the hydraulic conductivity. Similarly, one might conclude that the degree of certainty and confidence in the data were high.

There is much debate over random sample locations versus sampling in accordance to a uniform grid. As mentioned earlier, sampling locations should be based on existing knowledge of the site conditions. Thus, where possible, sampling locations should be neither random nor based on a uniform grid; they should be purposely selected on the basis of site history and reconnaissance information, and most important, the geologic variability/uniformity.

Sampling and Testing Procedures

The work plan must include field sampling and testing procedures. The sampling procedures ensure that the proper type and quality sample is obtained for subsequent laboratory analysis. Field testing procedures ensure that *in situ* information is appropriately gathered. Since field explorations are costly, available opportunities for field testing must be properly anticipated and utilized. A list of procedures that have been standardized by ASTM and used in site investigations for field sampling and testing is presented in Table 15-3.

For samples collected during the field investigations, protocols for chemical analysis are presented in such sources as *Standard Methods*.[10] Procedures for physical and engineering tests are also readily available from ASTM.[11] Where defined

TABLE 15-3
Sampling and field testing procedures

ASTM number	Test method
D653	Terminology relating to soil, rock, and contained fluids
D1452	Practice for soil investigation and sampling by auger borings
D1586	Method for penetration test and split-barrel sampling of soils
D1587	Practice for thin-walled sampling of soils
D2113	Practice for diamond core drilling of soils for site investigations
D2487	Test method for classification of soils for engineering purposes
D2488	Practice for description and identification of soils (visual-manual procedure)
D2573	Test method for field vane shear test in cohesive soil
D2607	Classification of peats, mosses, humus, and related products
D2937	Test method for density of soil in-place by the drive-cylinder method
D3385	Test method for infiltration rate of soils in field using double-ring infiltrometers
D3441	Method for deep, quasi-static cone and friction-cone penetration tests of soils
D4220	Practices for preserving and transporting soil samples

procedures are not available for specific contaminants or sample types, it may be necessary to develop specific analytical procedures for the project. However, where possible, sampling and analytical procedures are specified in the work plan by reference to standardized procedures. A listing of commonly employed physical and engineering test methods is presented in Table 15-4.

It is important to note that the bulk of the sampling and testing methods for site characterization were originally developed and used as geotechnical engineering tests. As a consequence of the predominantly mechanical nature of the tests, specific tests to assess chemical or biological properties may also be needed. As these methods are employed in hazardous waste management studies, we must be conscious of the ever-present need to obtain samples without altering their chemical composition and without cross-contamination between locations. It is recommended that the analytical laboratory responsible for chemical testing be consulted during this planning stage to ensure all sample handling and storage procedures are suitable for the types of chemical analyses required.

Sampling and test procedures must also carefully describe acceptable methods for cleaning equipment between samples and between borehole locations. Since the investigation will undoubtedly encounter both "hot spots" and areas deemed relatively clean, it is essential that contaminants not be carried between locations with the drilling and sampling equipment. Such cross-contamination is inevitable if equipment is not suitably cleaned between samples. The sampling strategy needs to be designed to avoid a cross-contamination of samples, and thus ensure data integrity. Generally then, sampling proceeds from the least contaminated to most contaminated areas. As discussed later, field quality control and quality assurance measures are required.

Operation Plan and Schedule

This element of the work plan defines the tasks to be completed, the team members who will perform the work and their role, the equipment to be used, the sampling or-

TABLE 15-4
Geotechnical laboratory test procedures

ASTM number	Test method
D422	Method for particle size analysis of soils
D854	Test method for specific gravity of soils
D2166	Test method for unconfined compressive strength of cohesive soil
D2216	Method for laboratory determination of water (moisture) content of soil, rock and aggregate mixtures
D2434	Test method for permeability of granular soils
D2113	Test method for one-dimensional consolidation properties of soils
D2664	Test method for triaxial compressive strength of undrained rock core specimens without pore pressure measurements
D3080	Method for direct shear test of soils under consolidated drained conditions
D4318	Test method for liquid limit, plastic limit and plasticity index of soils
EPA 9090	Hydraulic Conductivity

der and the decontamination procedures. For example, who decides whether a boring should be terminated? The driller? The field inspector? The project manager? The owner's project manager? The regulatory agency? The operation plan will define the types of decisions expected and assigns authority for these decisions. The substitute for a well-developed operation plan is confusion and delay at best and anarchy at worst.

The schedule realistically defines the start time, the duration of activities, and the completion date. Contingencies, in terms of unexpected delays, must be accommodated.

Health and Safety

The protection of the personnel conducting site and subsurface investigations is too important to completely omit from this chapter, yet too extensive to cover in detail. The goal of the health and safety program is to avoid acute exposure and minimize the risk of long-term, chronic health problems. The potential health and safety problems for workers at hazardous waste sites include:

- Exposure to toxic chemicals
- Electrical, fire, explosion and other safety hazards
- Exposure to extremely high or extremely low temperatures
- Exposure to high noise levels

In many cases, the use of personal protective equipment (PPE) while providing protection against site contaminants, poses additional safety hazards. PPE is usually cumbersome and restricts peripheral vision, increasing the possibility of injury from slips, trips, and falls.

Addressing these and other hazards is the responsibility of a health and safety officer who will develop a site-specific health and safety plan. This plan should include[12]:

- A listing of key safety and project personnel, emergency phone numbers, and locations of nearest emergency facilities such as fire and police departments, ambulance, and hospitals.
- A description of the risks associated with proposed site operations, including the appropriate personal protective equipment (see Table 15-5) and mitigation measures to be taken.
- A description of required training and medical monitoring (see below).
- Monitoring requirements (personnel and environmental).
- Decontamination procedures for people and equipment.
- Site maps, including hospital route, site location, and site feature maps.
- Required safe work practices while performing assigned/scheduled tasks.

Any worker at a Superfund site must, by regulation (29 CFR 1910.120), undergo 40 hours of health and safety training. The training must explain the potential hazards of this type of work and provide the knowledge and skills required to use personal protective equipment. For example, the suits and face masks shown in Figure 15-4 are cumbersome and uncomfortable. Special training and practice is required to work

TABLE 15-5
Personal protective equipment at hazardous waste sites

Level	When required	Description
A	High degree of skin contact and respiratory hazard or chemical substances at levels requiring fully encapsulated suits or when contaminants are unknown.	Level B plus fully encapsulated suits. Level A is rarely used at hazardous waste sites.
B	Minimum level when hazards are generally characterized, such as during an initial site visit with limited background information. Required whenever toxic concentrations in air are a possibility, such as in enclosed spaces where volatile chemicals may be found.	Level C plus self-contained breathing apparatus (SCBA) such as typically worn by fire fighters, or other supplied-air system.
C	Where airborne hazards have been measured and it is known and within conditions such that canister air purifiers are safe and where there is little likelihood of skin contact.	Level D plus air purifying (canister respirator and Tyvek) suit with a hood. Rubber boots are worn over work shoes and taped to Tyvek suit to provide a seal. Respirators must have been individually fitted to assure an adequate face-to-mask seal.
D	Minimum level for any site.	Steel-toed work shoes. Hard hat with face mask if required. Work clothes with long-sleeve shirt.

FIGURE 15-4
Workers in Level C protective equipment sampling a tank truck.

safely in such equipment. It is possible to dress someone up in a way that reduces the risk of chemical exposure but significantly increases the risk of physical injury. Perhaps the most important part of the training is to explain the limitation of safety equipment.

Special training is required for the use of self-contained breathing apparatus (SCBA) where the worker breathes only bottled air that is carried on a back pack. This training would include the proper assembly and inspection of the air pack, adjustments to the equipment, fitting of the face mask, and practice in performing manual labor wearing an SCBA.

Another important aspect of a health and safety program is a medical surveillance program. This would include preemployment and annual medical examination, emergency and nonemergency treatment, and maintenance of records documenting the program. The annual physical should include a pulmonary function test to demonstrate that the employee can safely use an air-purifying respirator while performing assigned duties. It is a regulatory requirement to obtain medical approval for all personnel who are required to wear respiratory protection during the course of their job.

Health and safety programs are an integral part of any hazardous waste activity. Plans are required by government agencies involved in the project to ensure that health and safety risks are minimized. It is generally recommended that implementation of these plans be carried out by highly qualified health and safety professionals such as certified industrial hygienists and certified safety professionals.

Cost Estimate

On the basis of the foregoing elements, the cost estimate can be prepared for the project. This is necessary to establish the required level of funding and assure that a proper and sufficiently thorough investigation can be conducted. Every cost estimate should provide some flexibility to accommodate unexpected or troublesome site and subsurface conditions. To accommodate this flexibility, the cost estimate should always include a contingency, which will vary with the size of the site, amount of data available, and the nature of the investigations. Typical items included in an estimate are presented in Table 15-6.

Other Documents

In addition to the foregoing elements in the work plan, it is necessary to provide **quality assurance** (QA) and **quality control** (QC) procedures as well as a data management plan. The QA and QC procedures help ensure data validity in the analysis of the results. Specifically, a QC plan describes the standardized procedures by which the sampling and analysis will be conducted to meet the data quality objectives. The QA plan describes the procedure by which the QC implementation is audited to ensure that the work and documentation are being conducted in accordance with established QC procedures. QA attempts to validate the accuracy and precision of the QC program.

Data management plans, while including data quality objectives, provide details as to how the data will be utilized and what engineering analyses can and will be conducted for the site with the acquired data. For example, the data management plan will describe the procedure to be implemented so that all engineers are using the most recent spreadsheet summarizing the hydraulic conditions and not an outdated or incorrect version of the spreadsheet.

TABLE 15-6
Typical investigation cost elements

Preliminary data review
Field costs
Mobilization/demobilization
Drilling
Soil sampling—undisturbed
Soil sampling—disturbed
Air, water, and sediment sampling
Field supervision (labor and expenses)
Laboratory costs
Geotechnical tests
Chemical tests
Engineering and scientific staff
Geology
Hydrogeology
Chemistry
Biology
Civil engineering
Geotechnology
Hydrology
Site development
Computer modeling
Project management
Cost control
Client interface
Accounts
Meetings and presentations
Senior project review
Clerical/support staff graphics
Project data review
Report preparation

15-4 METHODS OF SITE CHARACTERIZATION

After the scope of the hazardous waste management problem has been defined and the preliminary investigations completed, the detailed quantification can proceed. In this phase, the detailed site and subsurface characteristics are developed and quantified. It is the detailed information developed during this phase of study that permits the development of the facility design or the evaluation of the remedial alternatives if a release has occurred. This phase of study has several tasks associated with it as described in the following sections of this chapter.

On completion of the work plans described in the problem identification and scoping phase, the collection of site-specific data can begin. Site-specific data are the result of sampling and field studies to characterize the site, to define exposure pathways, and to provide data for screening of remedial alternatives or the new facility design. Within these site-specific studies, both indirect (nonintrusive) and direct (intrusive) methods are used. In employing **indirect methods,** we measure some other parameter that can then be related to the parameter of interest. These methods could also be classed as nonintrusive because they do not generally require

significant intrusion into the subsurface. In contrast, there are **direct methods,** such as boring, sampling, and *in situ* testing, which generally more directly measure the parameters of interest. These methods could be classed as intrusive because they require substantial intrusion into the subsurface. Both direct and indirect methods are generally used for subsurface investigations.

The advantage of indirect nonintrusive methods is that they can be used for planning for intrusive methods, interpolation between points of hard data, and optimization of drilling and sampling costs. This is in contrast with direct methods, which, as noted, require boring, sampling, and testing and provide direct information. The direct methods are usually somewhat limited because of their intrusive nature and their associated time and cost. With coupled direct and indirect methods, interpolations between test borings and monitoring wells can be made with considerably more confidence, and the combination of both methods results in the most cost-effective and technically sound investigation program.

The methods discussed below are frequently presented in isolation from the geologic setting in which they are carried out. Recognize that a few pages on geophysics will not make one a geophysicist or a couple of paragraphs on well drill techniques will not qualify anyone as a well driller. However, the information will form a basis for chemical engineers to discuss alternatives with geophysicists, for biologists to consider optional methods with a driller, and the generalist to interact with the specialist. Hazardous waste management is multidisciplinary, and such dialogue between specialists is essential to the successful outcome of the project.

15-5 GEOPHYSICAL METHODS

Nonintrusive, indirect investigative methods are often called **geophysical** methods or methods of **remote sensing.** These techniques are useful for the reasons described above. Geophysical techniques provide information about the subsurface conditions through the measurement of certain properties in the media (e.g., arrival time of a seismic wave) and the deduction of parameters of interest (e.g., depth to bedrock). For example, terrain conductivity is often directly related to changes in groundwater chemistry. However, it is conductivity that is measured and changes in groundwater chemistry that are inferred. Geophysical methods do *not* normally directly measure the parameter of interest. Geophysical techniques generally rely on **contrasts** to be successful. For example, to differentiate two subsurface strata, the seismic velocity must be different in each. Similarly, groundwater containing inorganic contaminants will have a higher conductivity than nearby water with organic contamination. Interpretation of the meaning of the measure contrast is best left to those specifically trained for such judgments. Before employing any geophysical investigations, it is necessary to review the site history and geology, identify areas of cultural interference, and determine a potential target's response to a particular geophysical technique.

For most geophysical techniques, verification of interpretations is typically required by comparison with **ground truth** information. That is, confirmatory drilling,

sampling and/or testing are completed at selected locations to verify the interpretations. As noted earlier, geophysical techniques are often used to strategically locate borings and, as such, the resulting direct investigation serves to verify interpretations.

Finally, successful application of geophysical methods requires proper consideration of the technique's **resolution.** Resolution is the ability to differentiate to a certain precision. For example, a seismic refraction survey may be planned to determine the elevation of the top of the bedrock. The details of the survey (e.g., geophone spacing) will depend on the resolution needed (e.g., nearest 10 m, 1 m, 0.1 m, etc.). Increased resolution to some limit can often be obtained with increased time and effort and, therefore, cost.

The following sections describe in detail the most commonly employed surface geophysical methods utilized in site and subsurface investigations for hazardous waste management. Table 15-7 presents guidance on the applicability of each geophysical technique for specific purposes.[13]

Seismic

In seismic refraction surveys, a **seismic wave** is introduced into the subsurface as an energy impulse. The seismic wave velocity will vary, depending on material properties, mostly density. The seismic wave may be introduced by either a small explosive charge or by hitting a steel plate or ball into the ground with a sledge hammer. The detector of the seismic wave is termed a **geophone.** The seismic wave velocities generated by a sledge hammer are generally limited to survey depths of 15 m (50 ft) or less. The seismic velocity for various materials is shown in Table 15-8.

Figure 15-5a presents a schematic of a typical field set-up for the technique. As shown the seismic wave propagates radially in the subsurface from the source. By placing a series of geophones at varying distances from the source, the arrival time at each geophone can be measured.

For the stratigraphy shown in Figure 15-5a the arrival time at the geophones near the source depends only on the seismic velocity of the overburden. For the geophone near the source the arrival time is linearly related to the distance. However, at increasing distances from the source to the geophone the seismic wave will move faster along a path down through the overburden horizontally through the bedrock and back up through the overburden. This represents a change in the linear distance versus time relationship already established for seismic waves traveling directly through the overburden. Plotting the relationship between the distance of the geophones from the source and the arrival time for each geophone results in two straight lines of different slope as shown in Figure 15-5b. The inverse of the slope is velocity in length per unit time. The depth to bedrock (or any layer with a greater seismic velocity) is the thickness of the uppermost layer, t, and is calculated as:

$$t = (D/2)\sqrt{(v_2 - v_1)/(v_2 + v_1)} \tag{15-1}$$

where $D =$ distance to point of intersection

$v_1, v_2 =$ velocity of layer 1, layer 2

TABLE 15-7
Decision matrix of surficial geophysical methods for specific investigations[13]

	Lithology	Top of bedrock	Rippability	Detection of water surface	Fault detection	Suspected voids or cavity detection	*In situ* elastic moduli (velocities)	Material boundaries, dip, etc.	Linear subsurface water conduits	Landfill boundaries	Large ferrous bodies—Tanks	Conductive bodies, ores, plumes, etc.
Seismic refraction	S	W	W	S	S		W	S				
Seismic reflection	S	S	S		S	S		W				
Resistivity	S	S		S	S	S		S		W	S	S
Electromagnetics					S			S	W	S	S	W
Ground-penetrating radar		S		S	S	S		S	S	S	S	
Gravity					S	S		S				
Magnetics					S					W	W	

W: Works well in most materials and natural configurations.
S: Works under special circumstances of favorable materials or configurations.
Blank: Not recommended.

TABLE 15-8
Typical soil seismic velocities[14]

Material	Velocity m/s	Velocity ft/s
Weathered surface material	305–610	1,000–2,000
Gravel, rubble, or sand (dry)	468–915	1,500–3,000
Sand (wet)	610–1830	2,000–6,000
Clay	915–2750	3,000–9,000
Water (depending on temperature and salt content)	1430–1680	4,700–5,500
Sea water	1460-1530	4,800–5,000
Sandstone	1830–3970	6,000–31,000
Shale	2750–4270	9,000–41,000
Chalk	1830–3970	6,000–31,000
Limestone	2140–6100	7,000–20,000
Salt	4270–5190	14,000–17,000
Granite	4580–5800	15,000–19,000
Metamorphic rocks	3050–7020	10,000–23,000
Ice		12,050

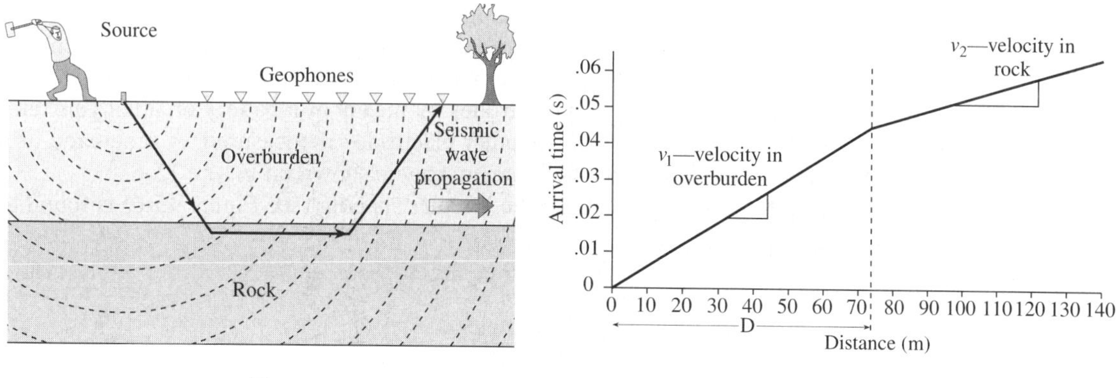

FIGURE 15-5
Schematic of seismic survey.

EXAMPLE 15-4. SEISMIC SURVEY. Using the graph in Figure 15-5b, calculate the depth to bedrock and render an opinion as to the overburden and bedrock condition/type.

Solution

$v_1 = 73 \text{ m}/0.045 \text{ s} = 1622 \text{ m/s}$ Medium stiff clay

$v_2 = (140 - 73)/(0.065 - 0.045) = 3350 \text{ m/s}$ Unweathered limestone

Substituting $D = 74$ m, $v_1 = 1622$ m/s, and $v_2 = 3350$ m/s into Equation 15-1 results in an overburden layer thickness of 35.7 m. Thus, the depth to bedrock is estimated to be 21.8 m.

Seismic reflection surveys work on the principle that a portion of the energy striking an interface between two materials of different density will be reflected to

the surface from the interface. The ratio between the reflected energy to the incident energy depends on the density and seismic velocity in the layers.

Resistivity

Resistivity is defined as the resistance to current flow as a result of an applied electrical potential. Resistance is measured in ohms and resistivity is resistance times length (typically ohm-m or ohm-ft). In geophysical surveys resistivity is determined as the electrical resistance per length l of a unit cross-section area (ohmsl/l^2 = ohms $\times$ l). Resistivity in the subsurface is a function of material type, porosity, water content, and concentration of dissolved constituents in the pore water. In general, dry soils and rock have very high resistance to electrical flow whereas inorganically contaminated saturated soils have a relatively low resistance. Soils contaminated with organic materials may have a relatively high resistance. Typical soil resistivity information for uncontaminated soils is shown in Table 15-9. Resistivity surveys are conducted to provide relatively detailed subsurface information to outline aquifer boundaries, depth to the water table and/or bedrock, and change in soil type or levels of contamination.

Resistivity surveys are conducted by applying an electrical current across two electrodes into the earth, and measuring the change in voltage (electrical potential) across two receiving electrodes. In order to conduct a resistivity survey, electrodes (metal stakes) are driven about 8 inches into the ground, typically along a straight line. The current is supplied to two electrodes by means of a battery or small generator and then the voltage between the other electrodes is measured. A schematic of a typical field setup known as the Wenner array is shown on Figure 15-6.

The soil resistivity, R, using the Wenner spacing (see Figure 15-6) is found to be:

$$R = 2\pi s V/I \tag{15-2}$$

where R = soil resistivity (ohm-meters)
s = electrode spacing (meters)
V = measured voltage (volts)
I = applied current (amperes)

There are a variety of spacings (arrays) that can be utilized, the most common being the Wenner and Schlumberger arrays. By selecting different spacings the electrical resistivity survey is able to target conditions at specific soil depths. Conceptually the contours of equal resistivity are analogous to those for groundwater flow potential. That is, the flow of electrical current is analogous to the flow of water through soil. For the flow of electricity, the voltage is the potential. For the flow of

TABLE 15-9
Typical values of soil resistivity

Soils	Resistivity range, ohm-m
Clays	1–150
Alluvium and sand	100–1500
Fractured bedrock	Low 1000s
Massive bedrock	High 1000s

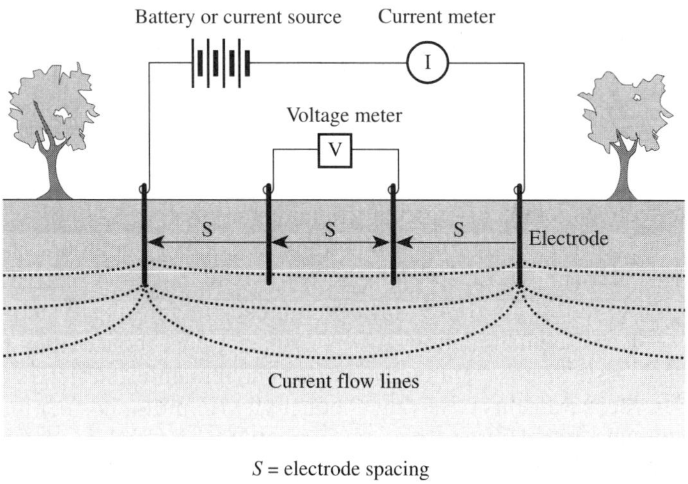

FIGURE 15-6
Schematic of resistivity survey.

water, the total hydraulic head is the potential. Interpretation of resistivity data is best left to those specifically educated for this task, geophysicists.

EXAMPLE 15-5. SOIL RESISTIVITY. Using Equation 15-2, calculate resistivity for three areas of a site under investigation and speculate as to soil type. In each case a current of 0.010 milliamperes was applied and the spacing was 5 meters. The measured voltage at each location is given below.

Location	Measured voltage, volts
1	0.467
2	0.034
3	0.492

Solution

$$R = 2\pi s V/I$$

Location 1: $R = 2 \times \pi \times 5 \times 0.467/0.01 = 1467$ Sand

Location 2: $R = 2 \times \pi \times 5 \times 0.034/0.01 = 107$ Clay

Location 3: $R = 2 \times \pi \times 5 \times 0.492/0.01 = 1546$ Sand

Electromagnetic Conductivity

Transmitting antennas are used to induce an electromagnetic current into the ground which is transmitted through the soil similar to the transmission of electrical current directly induced in the ground by the resistivity technique.[15] Electromagnetic methods commonly use a transmitter consisting of a closed loop of wire through which an AC current flows, thus generating a magnetic field. By indirectly inducing a current in the ground with the transmitter, a secondary magnetic field is generated and measured by a receiving antenna. The conductivity value is therefore an average conductivity over the effective depth of the survey. The effective survey depth is largely

a function of the distance between the sending and receiving coils. The technique is much faster than the ground-coupled resistivity technique. Terrain conductivity is influenced by material type, discontinuities (such as fractures), groundwater chemistry, intercoil spacings and coil orientation. It is also sensitive to electrical and magnetic interference (such as power lines and metal fences). Since groundwater chemistry has a significant influence on the measured conductivity, electromagnetics offer a rapid, cost-effective method of mapping contaminant plumes.

Shown in Figure 15-7 is a map of terrain conductivity contours. The survey was conducted at a site whose stratigraphy consists of saturated sand. The survey was conducted to aid in the design of a preliminary monitoring system to assess a potential source of contamination. It was suspected that the site may have been used for illegal waste disposal although there was no remaining surficial evidence of such disposal. Measurements of terrain conductivity (in millimhos per meter) were first obtained and plotted. From these data, contours of equal conductivity were plotted. What can we see in Figure 15-7? In this case, a conductivity anomaly (a high) is discovered. The center of the high can be interpreted to be the area of highest subsurface contamination. Next, observe the shape of the contours. In general, the contours are elongated toward the southeast. This elongation is interpreted to reflect the direction of groundwater flow and thus contaminant transport. The area of highest terrain conductivity is interpreted to be the area of greatest contamination. Finally, the area beyond the background conductivity of about 10 mmho/m is interpreted to be beyond the limits of contamination. Of course, drilling and sampling were required to verify the interpretations, but the direct investigation costs were minimized and the information maximized.

Ground-Penetrating Radar

In addition to seismic waves and electrical currents, it has been demonstrated that radio-frequency microwaves can penetrate the soil and be transmitted through soil layers. This technology may be utilized to identify buried materials at hazardous waste sites.[16] On entering the soil and progressing downward, the waves reflect off

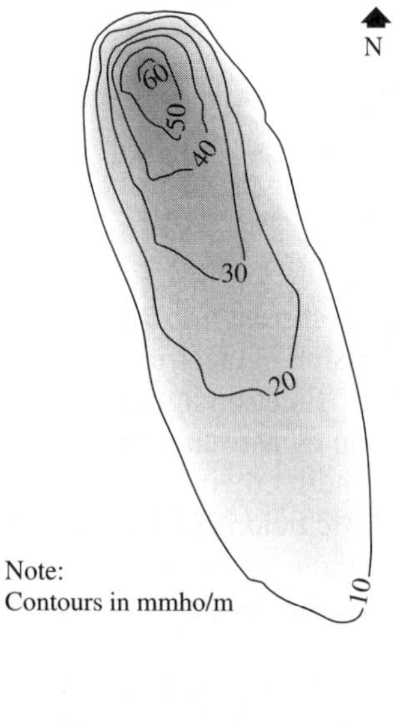

FIGURE 15-7
Terrain conductivity contours.

of objects or layers possessing contrasting density. Reflective waves or echoes are received and the resulting graph appears much like a sonar or sound profiling. An example of a ground-penetrating radar (GPR) survey is shown in Figure 15-8, where reflections from ground-penetrating radar indicate the presence of a buried gasoline tank.

The GPR techniques are therefore useful in detecting voids, underground trenches, buried drums of both plastic and metal, and floating hydrocarbon products in the subsurface. GPR is limited by rapid attenuation in conductive soils such as clays.

Gravimetric

Changes in density of subsurface materials result in differences in the earth's gravimetric field that can be measured with highly sensitive instrumentation. This technique detects differences in gravimetric response from subsurface deposits resulting from differences in density. It is typically used to detect features such as sink holes, fill zones, faults, or buried valleys in the subsurface. The technique has found only limited use for detailed definition of subsurface conditions for hazardous waste management and will not be discussed further in this text.

Magnetometer

A **magnetometer** measures magnetic fields in the subsurface. The most commonly used magnetometer, a proton precession magnetometer, uses spinning protons (the nuclei of hydrogen atoms) in a hydrocarbon fluid to generate a magnetic field. The spinning protons act as magnetic dipoles and, as such, can be oriented (aligned, po-

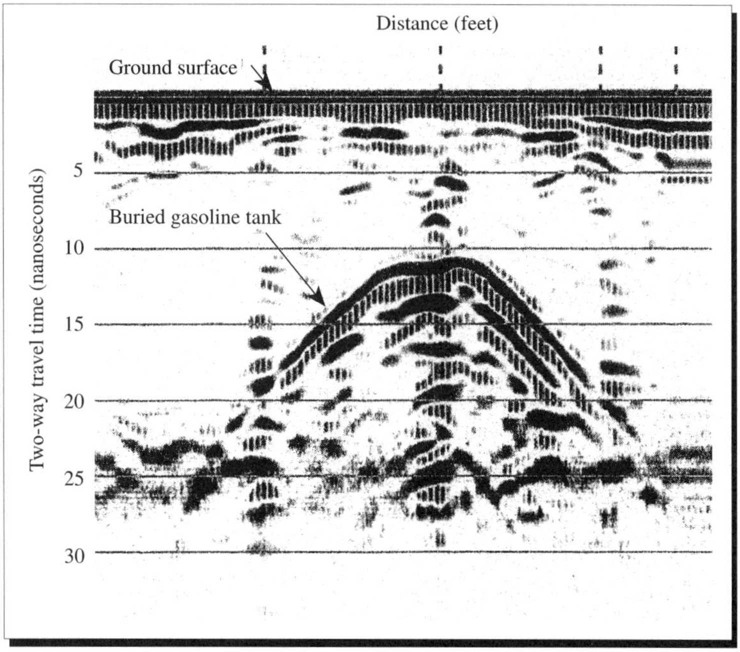

FIGURE 15-8
Ground-penetrating radar.[16]

larized) by application of a magnetic field. The magnetometer has a coil that is first used to align the spinning protons by applying a current. After aligning the spinning protons, the magnetic field in the subsurface causes the protons to precess, aligning themselves with respect to the subsurface magnetic field. This generates a current in the same coil used to originally align the protons. The frequency of the generated signal is proportional to the strength of the magnetic field (the total magnetic field intensity). The unit of measure is the gamma, with the earth's field equal to about 50,000 gammas in South Florida. The value depends on location. In the United States it ranges from 45,000 to 60,000 gammas. But in South America, the value is as low as 25,000 gammas. Local disturbances in the earth's magnetic field, such as by buried drums, give rise to magnetic anomalies. Thus this technique is especially useful for the identification of buried drums or other ferromagnetic (iron and steel) foreign materials. However, there are numerous sources of magnetic interference, including power lines, buried metal pipelines, and fences (and geologists' big belt buckles). Magnetometer surveys are particularly useful for determining the location of underground storage tanks.

The results of a magnetometer survey to investigate the location of buried wastes at a manufacturing facility are shown in Figure 15-9. The figure shows the results

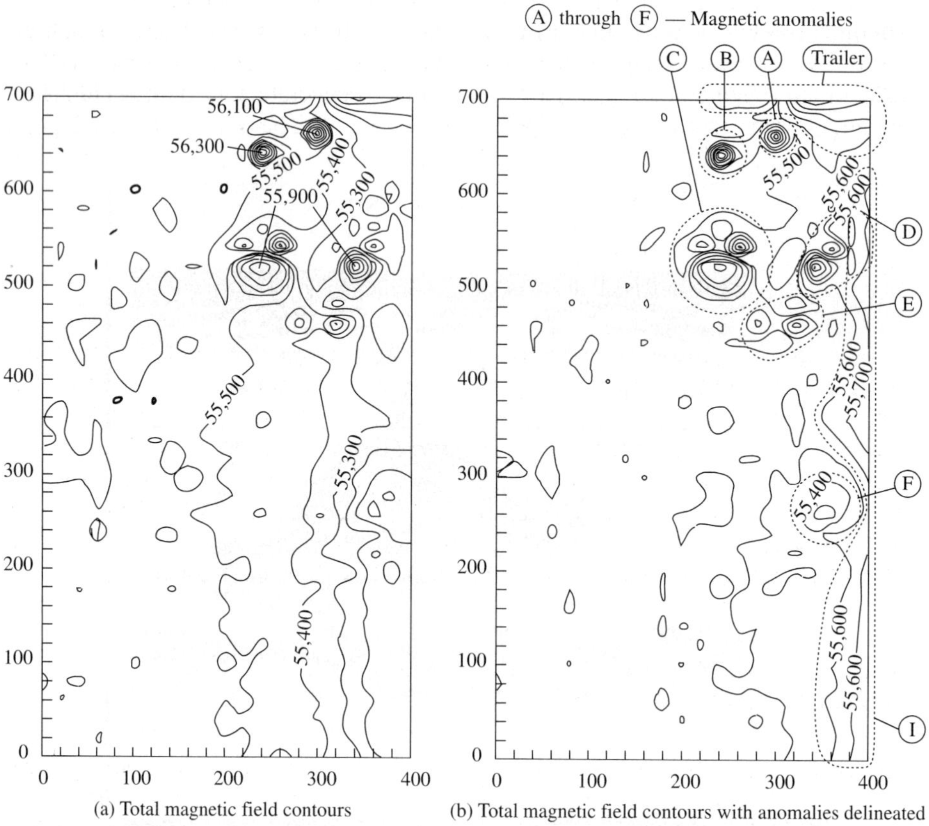

FIGURE 15-9
Magnetometer survey results.

TABLE 15-10
Borehole geophysics

Test technique	Measurement/application
Electrical techniques	
Spontaneous potential	Electrical potential/lithology
Resistance	Electrical resistivity/lithology
Single point resistance	Electrical resistivity/fracture
	Induction conductivity/lithology
Radiological techniques	
Gamma	Natural gamma radiation/lithology
Gamma gamma	Electron density/porosity, bulk density, moisture, lithology
Neutron	Hydrogen content/moisture, porosity
Miscellaneous techniques	
Acoustic velocity	Compression wave velocity/porosity, lithology, modulus
Caliper	Diameter/lithology, fractures
Temperature	Temperature/flow

as measured. However, as noted above, known interferences (known as cultural anomalies) must be accounted for. For this particular project the engineered surface features included a building (labeled as magnetic anomaly I on Figure 15-9), an on-site decontamination trailer, and a paved road. The data can be evaluated in light of these known interference, and represented as shown in Figure 15-9b to delineate anomalous areas. Test trenches verified the presence of waste pits at each of the anomalies (labeled as A through F on Figure 15-9) and the absence of waste pits in the areas without anomalies.[17]

Borehole Geophysics

Borehole geophysical methods are categorized as indirect methods in this text, as the techniques generally measure one parameter (e.g., natural radiation) to deduce another (e.g., stratigraphy). They do, however, require penetration into the subsurface via the borehole, making them more direct than the surface geophysical techniques described above. Borehole geophysical studies provide additional data to characterize the site and subsurface conditions.[18] The techniques available, the parameter actually measured, and the principal application of the measurement are shown in Table 15-10. Detailed discussion of these methods are beyond the scope of this text.

15-6 BORING AND SAMPLING

Intrusive field methods of subsurface exploration generally consist of boring, sampling, and *in situ* testing. One of the most commonly utilized techniques is the installation of groundwater monitoring wells. The diversity of subsurface investigation techniques available along with the specific information which they provide requires additional discussion. Intrusive subsurface investigation methods are necessary to provide *in situ* measurements of soil and rock properties, supply samples for

laboratory characterization of material properties, and serve as the ground truth for geophysical interpretations. The following sections of this chapter describe intrusive methods of field exploration methods and provide guidance for their use.

It is useful to classify intrusive exploration methods by their distinct individual components. First, there are a variety of drilling methods that are classified by the method by which the test boring is advanced into the subsurface. Next, there are distinct methods of soil and groundwater sampling. These sampling techniques may or may not overlap with the drilling methods. In addition, there are a number of *in situ* tests to measure specific soil properties. An *in situ* test may also result in the collection of a soil sample. Finally, any subsurface investigation must include collection and laboratory analysis of representative samples of soil and groundwater. Analysis can include geotechnical and chemical analysis. This chapter includes a discussion of the presentation of test results (i.e., data). The method of data presentation often reveals specific and informative aspects of the subsurface conditions.

Drilling Methods

A test boring may be advanced into the subsurface by any one of a number of techniques. The selection of the technique depends on the subsurface conditions, depth of investigation, types of samples required, type of *in situ* testing, and method of monitoring well construction, if any. Described in the following section is a brief overview of the most widely employed techniques used in hazardous waste management site and subsurface investigations. Conceptually, each drilling technique has three fundamental system requirements: (1) a method or tool to advance the borehole, (2) a method or tool to carry the cuttings to the surface, and (3) a method to maintain borehole stability. As each of the drilling techniques is discussed below, bear in mind the operational function of the tools and techniques with respect to these three fundamental system requirements.

The selected drilling method must also reflect the objectives of the test borings. In many cases, boreholes may be used for *in situ* testing and sampling as well as for the installation of monitoring wells. In some cases, particularly for monitoring wells having multiple monitoring levels, boreholes should be drilled for the sole purpose of monitoring well installation. In either case, the influence of the drilling technique on the borehole objectives must be considered.

Auger drilling methods

One of the most common methods of advancing test borings though unconsolidated materials (soil, to the geotechnical engineer) is through the use of **continuous flight auger** drillings as shown in Figure 15-10. Auger borings are drilled by rotating the auger and cutting bit with a drilling rig. The cutting bit loosens the soil and the auger flights carry the cuttings to the surface. Augers can vary in size from small (2-in diameter) augers that may be rotated by hand or with the help of a small engine (powered one or two person post hole digger) to large (up to 10-ft diameter or larger) requiring specially equipped cranes. Most commonly augers used in site investigations are in the range of 100 to 300 mm (4 to 12 in) in outside diameter, used with truck- or track-mounted drilling rigs. The choice of auger diameter is

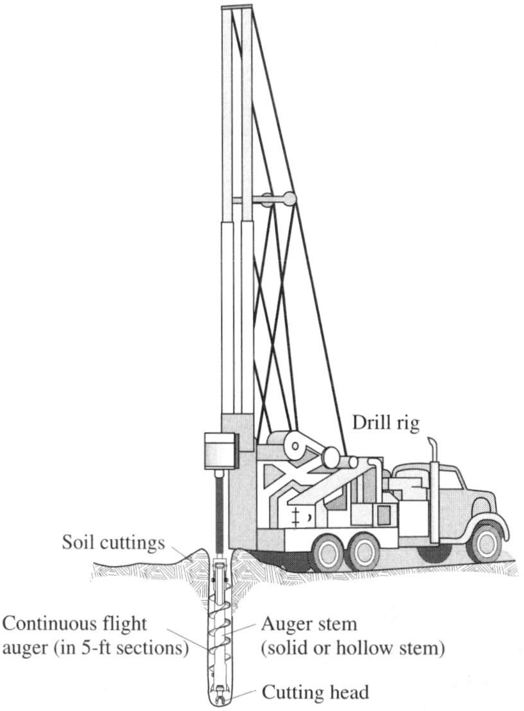

FIGURE 15-10
Continuous flight augers.

primarily influenced by the diameter of the monitoring well, if any, to be installed in the boring.

Two types of continuous flight augers are common: **solid-stem augers** and **hollow-stem augers.** The solid-stem auger is constructed by wrapping the auger flight on a solid pipe or steel rod as shown in Figure 15-11a. Samples can be collected from the cuttings carried to the surface by the auger. With auger samples, the precise sampling location in the subsurface cannot be determined. Small- and medium-diameter continuous flight solid-stem augers are typically used for shallow borings. Very large diameter holes are occasionally used to permit direct access by the engineer or geologist into the subsurface.

An improvement to solid-stem auger technology appeared in the 1950s when **hollow-stem augers,** as shown on Figure 15-11b, were developed.

In this case, the auger flight is connected to a pipe with an inside diameter varying from 50 to 300 mm (2 to 12 in). The auger penetrates the subsurface, advancing the test boring while permitting the engineer/driller to sample the subsurface in advance of the cutting bit through the center of the hollow stem auger. In this way the boring can be advanced to a desired and precise sampling or testing depth. The testing and/or sampling tool can be lowered into the hole through the center of the auger and the *in situ* test or sampling can be conducted. Hollow-stem augers also permit the installation of groundwater monitoring wells through the inside of the auger. As the auger is withdrawn, the monitoring well remains in place.

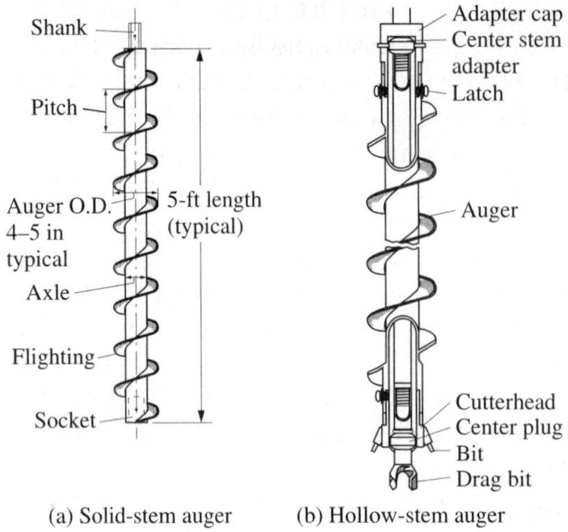

FIGURE 15-11
Solid- and hollow-stem augers. *(After Acker.*[19]*)*

Rotary drilling methods

Rotary drilling techniques, among the earliest developed, are widely used to advance test borings in both unconsolidated and consolidated deposits. In this technique a cutter head is rotated in the presence of drilling fluid. The cutting bit loosens the material and the drilling fluid (either water or drilling mud) carries the cuttings to the surface. Borehole stability during rotary drilling may be maintained through the use of a steel casing inserted in the bore hole or through the use of the drilling mud. Drilling mud functions by means of the formation of a thin impervious membrane along the walls of the borehole that, coupled with the unit weight of the fluid, maintains the borehole stability. A schematic of mud rotary drilling is shown in Figure 15-12.

The drilling of rock formations requires techniques and equipment different from that required for soil drilling. Rotary drilling is employed to advance a borehole through rock. Air rotary methods are relatively rapid and inexpensive. In this method

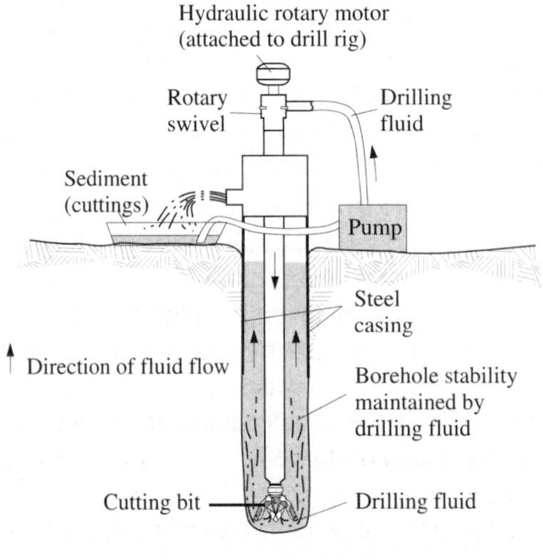

FIGURE 15-12
Mud rotary drilling.

air is the drilling fluid employed to transport the cuttings to the surface. Rotary wash techniques use water to carry the cuttings to the surface. Rotary core drilling techniques both advance the borehole and allow for the recovery of a sample (a rock core). In addition to these techniques, a variety of drilling techniques including reverse dual tube rotary, cable tool percussion, and reverse circulation become topics for detailed study.

The significance of knowledge of alternative drilling techniques lies in the ability to choose the most appropriate technique for the given site and geologic conditions. It is necessary to accommodate both the expertise of the driller with regard to specific drilling techniques and the expertise of the scientists and engineers with regard to the geology, site conditions, and hazardous waste management activities planned. The selection of drilling procedures should be mutually agreeable between the driller and the scientists and engineers.

As emphasized later in this chapter in the discussion of monitoring well installations, it is important that underlying formations not be contaminated during the drilling process. Auger and rotary drilling techniques all quite obviously begin at the ground surface. Drilling deep through shallow contaminated formations and into underlying formations may result in the transfer of contaminants from upper contaminated regions to deeper uncontaminated formations unless special precautions are taken to case off the overlying zones of contamination. Drilling through contaminated formations into underlying potentially uncontaminated formations requires the two-step process described later.

In Situ Tests

Drilling and sampling to obtain subsurface samples for laboratory testing and analysis provide one means of site characterization. No less important is testing ***in situ,*** that is, in place within the formation. Sampling and testing in the laboratory has the disadvantage that the original in-place properties of the material will be altered during sampling, transport, and preparing for laboratory tests. However, laboratory testing may permit more sophisticated tests to be conducted under controlled, reproducible laboratory conditions. *In situ* tests measure material properties as they exist in the subsurface; there are no sampling, transport, or handling costs. The cost lies in the time and difficulty in testing, the uncertainty of the actual conditions at the test interval and the often semiempirical methods of analysis associated with *in situ* test data. Nonetheless, *in situ* testing is usually an integral part of any thorough site characterization. The tests described below are among those most commonly and successfully employed.

Standard penetration test (SPT)

The most commonly employed *in situ* test, the **Standard Penetration Test** (SPT), is used to both secure a disturbed sample and to obtain a measure of the in-place density and/or strength of the material sampled. This test is conducted by using a standard split barrel sampler 51 mm (2 in) in outside diameter and 35 mm ($1\frac{3}{4}$ in) in inside diameter, driven into the ground with a 64-kg (140-lb) hammer falling 0.76 m (30 in), as shown in Figure 15-13, in accordance with ASTM requirements.[20] The sampler is a 2-in O.D. and $1\frac{3}{8}$-in I.D. The SPT value or N value is obtained by

FIGURE 15-13
Standard penetration test.

summing the total number of hammer blows for the second and third 150 mm (6-in) of sampler penetration in a 150-mm (18-in) drive. The *N* value is used for quantitative geotechnical engineering evaluations of subsurface conditions. The advantages of this test are that it incorporates information regarding the in-place density/strength of the materials and provides a sample for classification and chemical and/or physical property testing. The sample is considered a disturbed sample in the sense that the sample should not be used for engineering property tests of strength, permeability, or compressibility. The SPT is most widely used in hazardous waste site investigations to provide a sample for visual classification and laboratory testing.

Although the SPT is the most widely employed method of *in situ* testing and sampling, the potential variability of test results is not widely considered. Studies have shown that the SPT values depend on the energy imparted to the sampler, and a number of factors can influence the energy.[21,22] These factors include an inaccurate fall height, excessive rope friction caused by excessive turns of rope used to lift the hammer on the spinning cathead (rotating drum used by driller to help lift the hammer by wrapping the rope on the drums), and incomplete cleaning of the borehole prior to sampling and testing. Additional well-established factors[23] are shown in Table 15-11.

While the SPT data are widely obtained in the course of subsurface investigations conducted on hazardous waste sites, they are also widely underutilized. As a result of many years of standard penetration testing, a number of correlations to physical and engineering properties of soils have been developed. The SPT value may be correlated to relative density of sands,[24] angle of internal friction of sands,[25] undrained shear strength of clays, settlement of foundations in sands,[26] and liquefaction potential.[27] As a minimum, granular soils can be classified by their relative

TABLE 15-11
Factors affecting the results of the SPT

Test detail	Effect on *N* value
Inadequate cleaning of disturbed material in the borehole	Decreases
Failure to maintain sufficient hydrostatic head in the borehole	Decreases
Variations from the exact 30-in drop	Either
Extreme length of drill rods ($P > 175$ ft)	Increases
Any interference with free fall	Increases
Using deformed sample spoon	Increases
Excessive driving of sample spoon before the blow count	Decreases
Failure of driller to completely release the tension on the rope	Increases
Driving sample spoon above the casing bottom	Increases
Use of wire line rather than manila rope	Increases
Carelessness in recording blow count	Either
Insufficient lubrication of the sheave	Increases
Larger size of borehole	Decreases

density and cohesive soils by their consistency employing the standard penetration test *N* value. Presented in Table 15-12 are correlations between the *N* values and the soil relative density of a granular soil and the consistency of a cohesive soil.

Cone penetrometer test (CPT)

To supplement information obtained during drilling and SPT sampling and testing, the **cone penetrometer test,** as shown in Figure 15-14, is often utilized. The cone may be equipped with a piezometer to enable measurement of pore water

TABLE 15-12
Correlations to the N value from the SPT

Sands		
***N* value**	**Relative density (%)**	**Description**
0–4	0–15	Very loose
4–10	15–35	Loose
10–30	35–65	Medium
30–50	65–85	Dense
> 50	85–100	Very dense
Clays		
***N* value**	**Unconfined compressive strength (kN/m^2)**	**Description**
< 2	< 0.25	Very soft
2–4	0.25–0.50	Soft
4–8	0.5–1.0	Medium (firm)
8–15	1.0–2.0	Stiff
15–30	2.0–4.0	Very stiff
> 30	> 4.0	Hard

pressures and rate of generation and dissipation, which allows estimation of *in situ* permeability. The cone may also be modified to allow for pore water sampling. The cone may also be adapted to include environmental sensors such as conductivity or pH. The electronic cone penetrometer provides an additional advantage over the standard penetration test in that a continuous vertical profile is obtained as opposed to sampling at fixed intervals. The primary disadvantage is that no sample is obtained.

Shown in Figure 15-14a is a cross-sectional schematic of the cone penetrometer test equipment. The cone in this case includes a friction sleeve and a pore water pressure transducer along with the conical end (often termed a piezocone). Also shown is a cone modified to acquire pore water samples. As the cone is pushed with the cone rig, the tip resistance, friction sleeve resistance, and pore water pressure are measured. Shown in Figure 15-14b is a plot of the results. Since the cone produces a continuous profile, thin lenses of differing materials can be sensed. By calculating the ratio of the friction sleeve resistance to the cone tip resistance, a soil classification can be estimated. Finally, the pore water pressure profile can be determined from the pore pressure measurements. To summarize, the CPT is an excellent tool to deduce subsurface lithology.

Building on a long tradition of successful use in traditional applications, the CPT has been the focus of recent research and development to add environmental capabilities. For example, pH sensors and sensors for total petroleum hydrocarbons have been added.[28] Miniature video cameras have also been adapted to the cone to enable recording of a continuous stream of images and subsequent analysis of the data for soil grain size, and the presence of clay lenses, fissures, and sand seams as thin as 0.5 cm.[29]

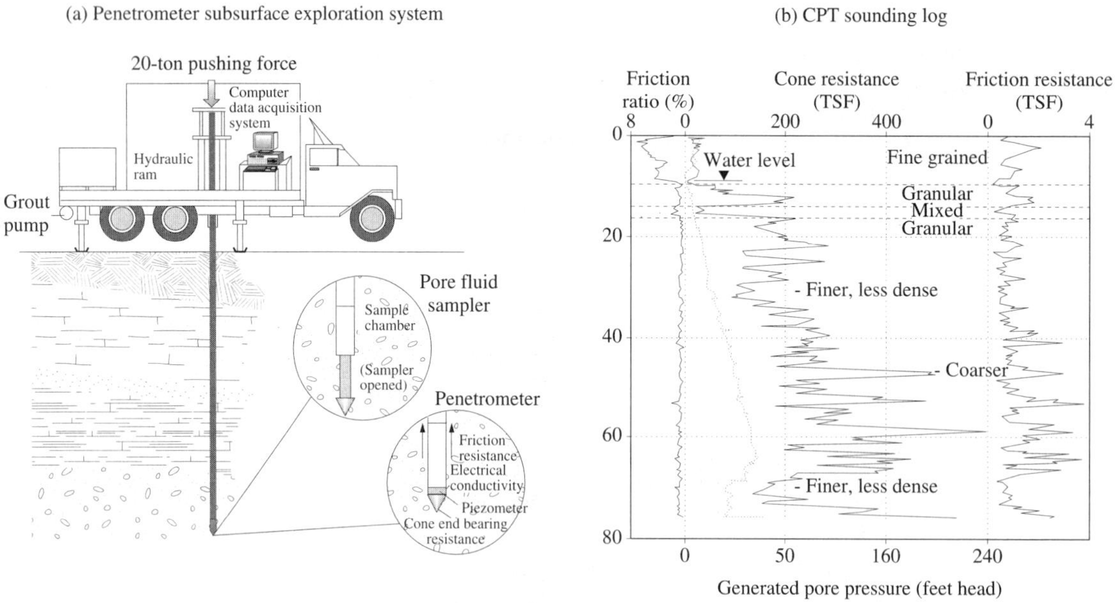

FIGURE 15-14
Cone penetration testing.[30]

Vane shear test

Facility design and, in particular, the analysis of slope stability require quantification of the soil's shear strength. For soft clays the shear strength can be measured with the **vane shear test,** as shown in Figure 15-15. In this test a vane is first pushed into the soil (always a clay-rich soil, never a sand or gravel). The vane is then rotated slowly, measuring the torque $T_{\max}$ required to shear the soil. The dimensions of the vane (height H by diameter D) and thus the surface area of failure are coupled with the applied torque to compute the shear strength of the clay. The vane shear test provides a reliable estimation of the *in situ* range shear strength of clays. The shearing resistance S_u is calculated as:

$$S_u = (3T_{\max})/(2\pi D^3) \qquad \text{for } H/D = 1 \tag{15-3}$$

or

$$S_u = (6T_{\max})/(7\pi D^3) \qquad \text{for } H/f D = 2 \tag{15-4}$$

Extensive studies have shown that the results from vane shear testing need to be corrected for variations in clay type. The clay type can best be characterized for this correction by the plasticity index. (The plasticity index is a measurement of the range of water contents over which a soil behaves as a plastic, deformable solid.) The correction to the computed value is shown in Figure 15-16. The field vane shear strength is multiplied by the correction factor to arrive at the actual field shear strength. For environmental applications, the principal use of the vane shear test is to characterize the shear strength of soft clay for landfill stability analysis and the like.

Dilatometer test (DMT)

When it is necessary to understand lateral stresses in the subsurface for the design of hazardous waste management facilities, it may be desirable to run a dilatometer test. The dilatometer, shown in Figure 15-17, is pushed into the ground with a conventional drilling rig to the desired test depth. The flexible membrane disk is

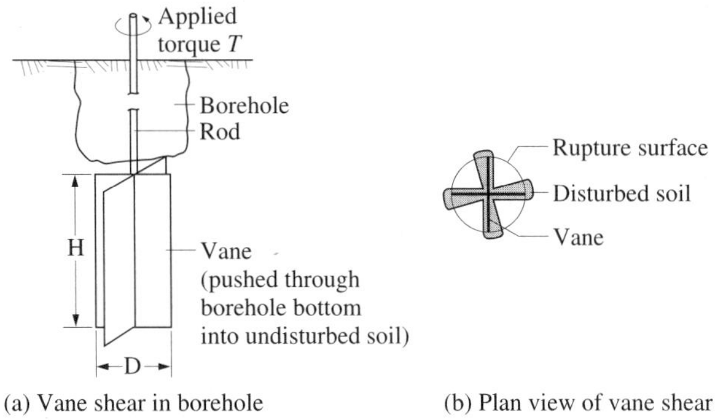

FIGURE 15-15
Vane shear testing.

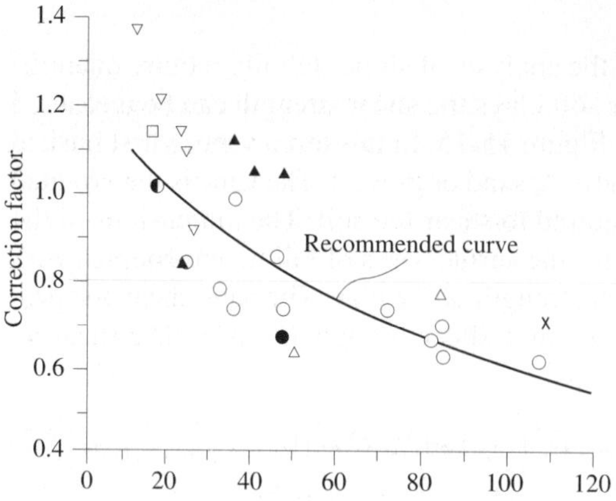

S_u (field) = S_u (vane) • correction factor

FIGURE 15-16
Correction to vane shear test results. (*After Holtz and Kovacs, 1981*).

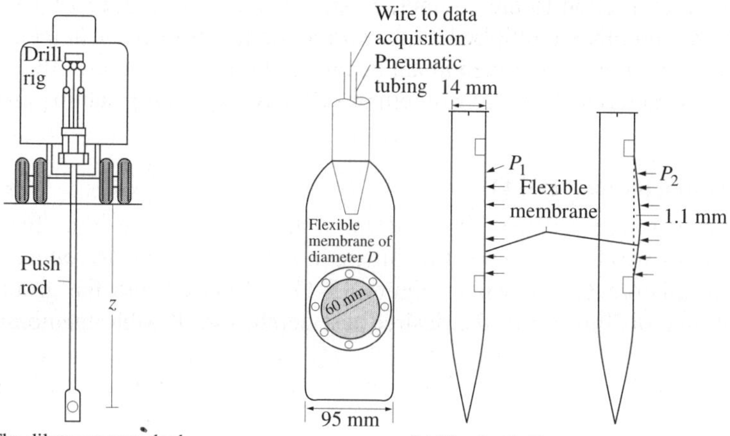

(a) The dilatometer pushed to depth z for test

(b) Marchetti dilatometer

FIGURE 15-17
Dilatometer testing.

then inflated with compressed air from the surface. As the pressure is increased to the horizontal earth pressure pushing against the disk, the disk lifts off its starting position. The pressure to cause this movement is considered the *in situ* horizontal earth pressure. The membrane is further expanded to cause movement through a fixed distance and the ratio of pressure to displacement is taken as a soil modulus (soil stiffness). Thus, the DMT allows for an estimation of lateral earth pressure and elastic modulus.[31]

In Situ Water Level Measurement

To understand contaminant transport at a site, it is necessary to understand groundwater flow characteristics. Wells to monitor and sample groundwater are discussed in Section 15-7. To know the hydraulic head at a given location, one can measure the depth to water in a monitoring well. A **water level indicator** such as that shown in Figure 15-18a is widely employed. This battery-operated device is essentially an electrical circuit continuity meter employing a weighted two-wire cable, calibrated in length, with an open circuit at the tip. The weighted tip is lowered into a monitoring well or piezometer (described in detail later in this chapter). When the tip encounters water, the circuit is completed and the current is registered on the ammeter. A measurement can be taken of the precise depth at which the circuit is completed and taken as the depth to water. The wire is spooled for ease of reeling out and in.

An economical alternative device (and, in the view of many, a technically superior method) is the **plopper.** The plopper, shown in Figure 15-18b, consists of a measuring tape (preferably fiberglass) to which a weighted acoustical generator is added (the plopper). The acoustical generator is constructed of a steel pipe cap and threaded eye bolt available from the hardware store. The plopper is lowered into the well and, on hitting the water surface, emits a highly audible "plop." Precise depth to water is then made by reading the measuring tape. The readily audible nature of the plop is seemingly amplified by the long well casing.

Also commonly used is the "wetted tape" method. In this method, a tape is inserted into the well to some distance below the water surface. By examining the tape, the location of the wet-dry interface can be determined.

In Situ Hydraulic Conductivity

In situ tests are commonly applied for determination of hydraulic properties of the formation as well as geotechnical properties previously described. These tests can include single-well permeability tests using water infiltration or exfiltration.

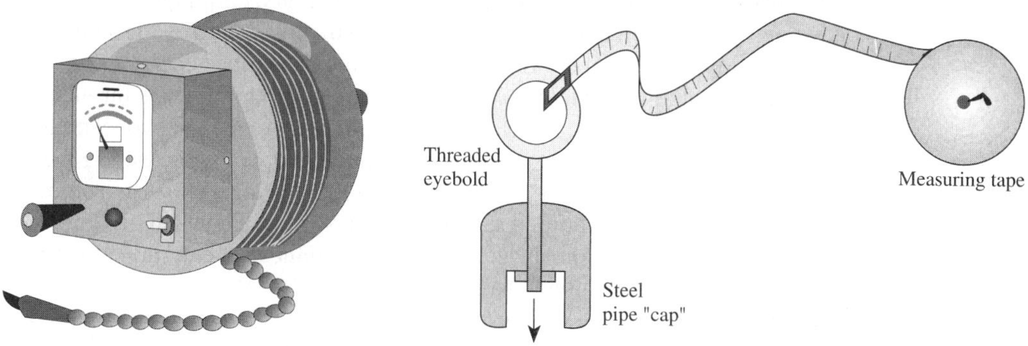

(a) Battery-powered water level indicator

(b) Acoustical generator for water level measurement (plopper)

FIGURE 15-18
Water level measurement.

Aquifer pumping tests can provide insight into the macro behavior of the formation. Additional information regarding hydraulic properties is presented in Chapter 4.

Aquifier pumping tests can be conducted to provide estimates of formation properties such as hydraulic conductivity, transmissivity, and storativity and to estimate potential capture zones and well efficiency.[32] Since aquifers are seldom homogeneous and isotropic, interpretation of results requires careful consideration of boundary conditions such as nearby recharge areas, secondary permeability from joints and/or fractures, artificial interferences such as underground utilities, and the integrity of the pumping and observation wells. There are a number of alternative configurations for pumping and observation wells but broadly speaking, water is pumped from the pumping well and measurements of water level are obtained over a period of time in the pumping well and in observation wells. The aquifer response to pumping will depend on the boundary conditions (for example, confined or unconfined flow) and the aquifer properties, primarily hydraulic conductivity. Depending on the data gathered and the response observed, there are a number of analytical methods available to analyze and interpret the data.[33]

As an alternative to aquifer pumping tests, the hydraulic conductivity and transmissivity of a formation may be estimated from the rate of fall (or rise) in water level within a single well in response to the insertion (or removal) of a "slug" that causes an instantaneous rise (or fall) of the equilibrium water level of the well. This single-well permeability test (or slug test) is less time-consuming and less expensive than aquifer pumping tests. On the other hand, the values of hydraulic conductivity derived from the test may be less reliable and represent formation behavior only in the immediate vicinity of the well. As a result, this test is frequently repeated many times in wells installed at locations throughout the site.

In Situ Water Quality

In situ tests to measure the concentrations of specific chemicals or analytes in water have been developed. For example, a remote-sensing groundwater monitor employing fiber optics is under development.[34] This device would measure *in situ* both the presence and concentration of contaminants such as mercury, chromium, and phenol. The development of a reliable *in situ* test device to measure analyte concentrations will result in more reliable and cost-effective site and subsurface characterizations.

One way to quantify volatile organic chemicals in groundwater is through the use of portable field analyzers. A field analyzer can measure the total concentration of organics in vapors in a sample "head space." For example, by measuring the air quality in the head space above the groundwater in a monitoring well, an indication of organic contaminant concentration groundwater is provided. Shown in Figure 15-19 is the use of an organic vapor analyzer (OVA) to determine concentration of organic vapors coming from an abandoned barrel.

Portable field devices are widely available in several types including photoionization detectors (PID), flame ionization detectors (FID), and portable gas chromatagraphs (GC). In order to detect volatile organic compound in a PID, a gas sample is drawn into an ionization chamber where it is exposed to ultraviolet light. The released ions from compounds ionized (those with ionization potentials less than or equal to those of the lamp) generate an electrical signal from a collection electrode. The sig-

FIGURE 15-19
An organic vapor analyzer.

nal strength is related to the concentration of ionized compounds. A flame ionization detector works as a carbon counting device that produces a current proportional to the number of ions formed in the flame gases and measured at the collector. Portable GCs employ an injection port, a separation column that contains a stationary phase, a carrier gas (a flowing mobile phase), and a detector. The organic compounds are separated because of differences in their partitioning behavior between the mobile gas phase and the stationary phase in the column. Sample components that partition (see Chapter 4) more strongly into the stationary phase move more slowly through the column and are separated from components that are less strongly partitioned and move more quickly through the column. Portable field instruments such as those described above are widely employed. As mentioned, they measure concentrations in the vapor phase and thus a head space analysis is conducted for a soil or water sample. Analyzing the head space above a soil or water sample involves the assumption that, as the concentration of a volatile organic contaminant increases in the soil or water, the concentration in the head space above the sample will also increase.

Soil and Rock Sampling Methods

Soil samples can be obtained from the cuttings of the drilling methods described above. However, cuttings provide a (1) disturbed sample and (2) difficulty in establishing exactly where in the formation the samples were obtained. Samples are more frequently obtained by using specifically designed samplers. The resulting samples are classified as either disturbed or undisturbed depending on the degree of alteration of the *in situ* stress state and soil structure. Disturbed samples are typically obtained through split barrel samplers during the standard penetration testing. Alternatively, undisturbed samplers have been developed to minimize disturbance.

The SPT and the accompanying split-barrel sampler was described above. It is relevant to note that the split-barrel sampler appears in a number of forms including those with one-piece liners, segmented liners, and differing dimensions. A larger split barrel, often known as a California sampler, may be driven with the 340-pound casing hammer. In all cases, split-barrel samplers yield disturbed samples for use in

physical or chemical property testing. It is likely that samples of subsurface soils for chemical analysis would be obtained by using the split barrel sampler.

When samples of cohesive materials for test of strength, permeability, or compressibility are required, it is necessary to use a thin-wall sampler to collect an "undisturbed" sample. For hazardous waste management applications, the most common use of these samplers is to secure clayey formation samples for hydraulic conductivity testing in the laboratory. Realize, however, that the sample is never totally undisturbed. The very act of sampling changes the state of stress on the sample, alters the sample temperature and may impart some physical disturbance to the soil. There are a number of thin-wall samplers in use including the Shelby tube and piston samplers (see Figure 15-20 for two examples). For "usual" cohesive soils, the thin-walled tube sample in Figure 15-20a would be chosen. For very soft cohesive soils, it may be necessary to use the piston sampler to help retain the sample in the tube during the withdrawal to the surface through the borehole.

The principle of these samplers is to:

1. Provide the least area of sampler as possible (hence the thin walls)
2. Provide a means to retain the sample in the tube (hence the check valves and pistons)
3. Minimize the sidewall friction as the sample enters the tube (hence the beveled cutting edge)

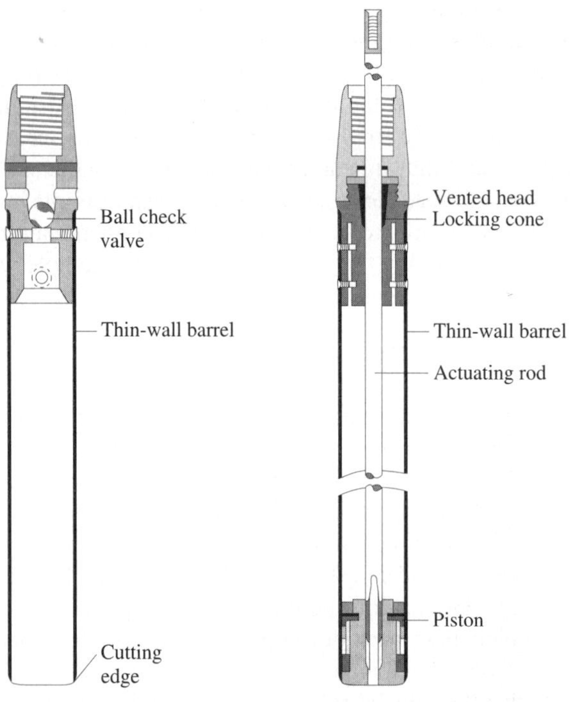

(a) Thin-wall tube sampler

(b) Stationary piston sampler

FIGURE 15-20
Thin-wall samplers.

Consolidated deposits (rock) in the subsurface is normally sampled by coring. Rotary rock coring employs a **core barrel,** made up of a tube and cutting bit. As the core barrel is rotated, a cylindrical hole is formed and the core of rock is recovered from within the tube. The cutting tip is usually made of industrial-grade diamonds. There are a number of core barrel designs, including both single- and double-tube. Schematics of typical bits and core barrels are shown in Figure 15-21. During coring, wash water is pumped down the hole to cool the cutting bit and to carry away the cuttings.

What good is a rock core for site characterization? First and foremost the core provides a sample for identification and laboratory testing. Much can be gleaned regarding the type of rock, the presence/absence of voids, bedding, fractures, and the overall rock quality. These characteristics are immensely important in understanding contaminant transport at a site. Quantitatively, the rock quality is measured by the **rock quality designation** (RQD). The RQD[35] expressed as a percent is calculated as:

$$\text{RQD} = 100 \times L/R \qquad (15\text{-}5)$$

where L = cumulative length of all pieces over 100 mm (4 in)
R = total recovery

The RQD provides data for an important assessment, that is, how representative the properties of a rock sample are of the properties of the rock mass.

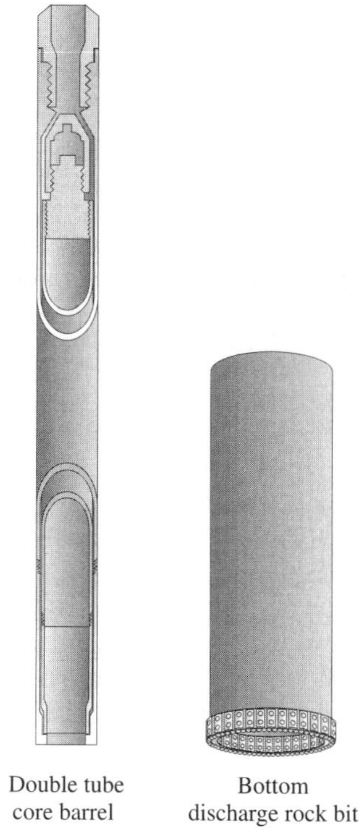

FIGURE 15-21
Rock coring equipment.

EXAMPLE 15-6. ROCK QUALITY DESIGNATOR. A 1.5-m core run was made from elevation 96 m to 94.5 m, and 1.4 m was recovered. The recovered core was found to be broken into 8 pieces which were measured as follows:

Piece no.	Length, mm
1	220
2	170
3	150
3	200
4	80
4	60
5	70
6	90
7	200
8	160

What is the RQD? What else can you say about rock quality?

Solution. The RQD is calculated from Equation 15-5:

$$\text{RQD} = 100 \times (220 + 170 + 150 + 200 + 200 + 160)/1400 = 78.6$$

The zone of poorer rock quality occurs approximately between elevation 95.26 m [96 − (220 + 170 + 150 + 200)] and 94.86 m (94.5 + 200 + 160). The precise depth range of the poorer quality rock cannot be known because of the loss (no recovery) of 0.1 m of rock. For the above estimate it is reasonable to assume that this 0.1-m core is missing from the zone of poorer quality.

15-7 MONITORING WELLS

It is probably fair to state that, for hazardous waste management applications, there can be no more important aspect to site and subsurface characterization than the understanding of the groundwater flow regime and the associated contaminant distribution and transport. The cornerstone of this assessment is the groundwater monitoring well. In this regard, the two principal uses of a groundwater monitoring well are to (1) permit determination of the hydraulic head and (2) permit water sampling for measurements of water quality.

In order to better understand how the water level measure in a well is interpreted, it is first useful to define what we mean by a confined aquifer and an unconfined aquifer. First a commonly accepted definition of an **aquifer** is a saturated geologic unit that is capable of transmitting significant quantities of water under ordinary hydraulic gradients.[36] An **unconfined aquifer** is one in which the upper boundary is defined by the **water table.** The water table is an underground surface (three-dimensional locus of points) where the pressure head is equal to atmospheric pressure. A **confined aquifer** is bounded both top and bottom by an **aquitard** (a less permeable layer in the geologic sequence). The term **confining layer** is often used synonymously with aquitard.

The measurement of pressure head in the subsurface is accomplished via **piezometers.** The standpipe piezometer is a tube or pipe open at the top to the atmo-

sphere and open at the bottom to limit a portion of the formation (usually less than a few feet). Piezometers can also take the form of devices employing pneumatic systems, pressure transducers, and other electronic components. A **monitoring well** is used for both measurement of pressure head (as a piezometer) and for sampling of groundwater from below the water table.

The piezometer (monitoring well) shown in Figure 15-22 in an unconfined aquifer allows the measurement of average pressure head along the screened interval (and nowhere else). That is because the water level will rise to an elevation in the standpipe to balance the hydraulic head outside the standpipe in the formation. Since the inside of the monitoring well is open to the atmosphere, the pressure head in the formation forces the water level in the casing to rise to an equilibrium level. That is, the height of the column of water in the well is equal to the pressure head in the formation at the screened interval. This is the essence of a piezometer. The water level in a piezometer completed in an unconfined aquifer defines one point on the water table. The water level in a piezometer in a confined aquifer defines one point on a potentiometric (or piezometric) surface.

With the establishment of a datum (often mean sea level), the elevation head at the screened interval can be determined and the total hydraulic head can be calculated. The total hydraulic head at a point (the location monitored by the well or piezometer) is found by first determining the height (with respect to the screened interval) of the water in the standpipe. This is the pressure head. The elevation head is the distance that the point of measurement (i.e., the screened interval) is above (or below) the datum. The total hydraulic head is then the sum of the pressure and elevation head.

In practical terms, on most sites a survey is conducted to determine the elevation of the top of the well. With a measuring device such as one of those discussed in the previous section, the depth to water is measured. By subtracting the water depth from the well elevation, the elevation of the water in the well is obtained. This is the total hydraulic head (based on whatever datum was chosen for the original survey).

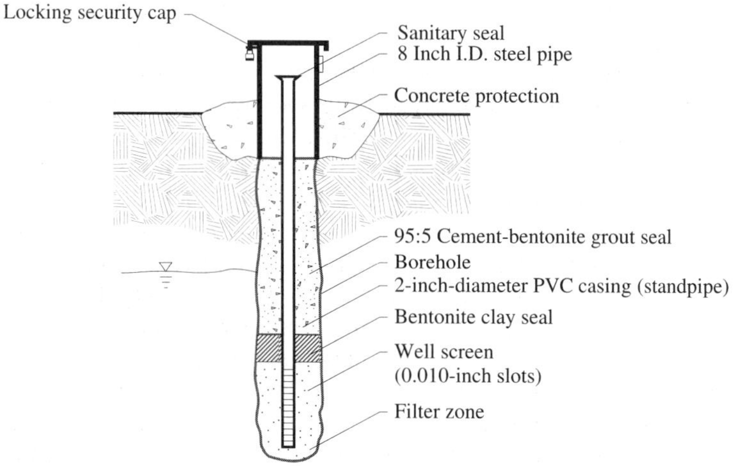

FIGURE 15-22
Typical monitoring well.

Monitoring wells often serve an additional and equally important purpose beyond measuring the piezometric pressure. A properly installed monitoring well can be used to collect representative samples of the groundwater for chemical analysis in the laboratory, or to measure ***in situ*** specific chemical properties of the groundwater such as pH and conductivity.

Lastly, monitoring wells can also be used in the measurement of hydraulic conductivity properties of the formation via single-well permeability (slug) tests. Monitoring wells can also be used in the field measurement of hydraulic conductivity properties by using a number of monitoring wells and pumping wells in a pumping test.

Monitoring Well Construction

Shown in Figure 15-22 is a typical **monitoring well** installed in an unconfined aquifer. Each of the components labeled on the monitoring well schematic serves a specific purpose in assuring that the monitoring well functions properly. The well is constructed of a standpipe **(riser pipe),** sealed along its length and connected to a short screen allowing formation water in and out of the riser pipe. The **screened interval** is that portion of the well that permits groundwater to flow into or out of the monitoring well. The screened interval is normally encased in a **filter zone** often known as a **sand pack** or **gravel pack.** The purpose of the filter zone is to permit the free flow of water from the formation into and out of the screened interval without the movement of fines from the formation. It is necessary to provide a filter between the screen openings and the formation to reduce the potential that the fine fraction of the formation (the silty and clayey portion of the formation) does not get into the well. If fines were to enter the well, physical plugging could occur. Since the plugging is caused by materials with a relatively fine grain size, interactions between this fine material and the formation water could alter the well water chemistry. Chemical and biological activity within the well could also alter the water chemistry. The **casing** (riser pipe) connected to the screened interval is used to isolate the monitoring system upward through the remainder of the formation to the surface.

To preclude vertical leakage along the annular space between the riser pipe and the sidewall of the boring, it is necessary to seal this space. Directly above the screened interval, this sealing is accomplished with a clay zone of low hydraulic conductivity. This clay zone is typically constructed of **bentonite,** a high-swelling clay with **sodium montmorillonite** as its primary clay mineral. The clay seal will typically have a thickness of about 0.5 m (1 to 2 ft) and is located immediately above the screened interval. This low-permeability, high-swelling clay affords a tight barrier to the transport of water or contaminants from other portions of the borehole into the screened interval. Above the clay seal is a **grout** seal typically constructed of a cement and bentonite mixture. This grout provides a further barrier preventing surface water or groundwater from elsewhere in the formation from migrating into the screened interval. The clay and grout seals are present to eliminate a preferential pathway for contaminant transport between zones of the subsurface. The hydraulic conductivity of the seal should be no greater than the hydraulic conductivity of the least pervious formation that the well penetrates.[37] Finally a vented **cap** on the riser pipe is normally provided to prevent rainwater or other surface water from entering into the well.

To provide security and to minimize the possibility of damage, a steel casing, embedded in a concrete pad, may also be employed. A locking cap may be used on the steel casing to limit access to the monitoring well.

Monitoring wells are constructed of a variety of materials including polyvinyl chloride (PVC), stainless steel, black steel, and Teflon®. The selection of the material type for monitoring wells should consider well depth, types of contaminants expected, design life of the well, and the geologic, biologic, and geochemical nature of the formation. For many wells polyvinyl chloride (PVC) is typically the material of choice. It is generally nonreactive, structurally sound, and relatively economical. In a subsurface environment contaminated with organic contaminants at high concentrations, the organic nature of the PVC may confound the analytical chemistry test results. In these cases Teflon or stainless steel is often the material of choice. Black steel is typically not utilized in hazardous waste management monitoring well installations because of its relatively low corrosion resistance. In terms of costs, stainless steel and Teflon impart considerably greater materials costs to the project than PVC. When drilling, long term sampling and monitoring, and analytical chemistry costs are included in the economic considerations, the cost differential between wells of Teflon, stainless steel and PVC may become inconsequential. Thus, the material selection is normally based on considerations of the geologic nature of the formation, the purposes of the monitoring well, and the nature of the expected groundwater chemistry.

Aquifer Monitoring

While most groundwater is encountered in confined or unconfined aquifers previously described, at times the water may be localized and discontinuous (related to geologic heterogeneity) as shown on Figure 15-23. This is termed a **perched water table.** It is important to ascertain whether a given water level measurement in a monitoring well represents the water table in an unconfined aquifer, a perched water table, or the hydraulic head in a confined aquifer. Measurements of hydraulic head in a confined aquifer can be contoured to develop a three-dimensional locus of points termed the **piezometric** or **potentiometric surface.** The "actual" water surface (not the one determined from water levels in monitoring wells) in a confined aquifer must be at the top of the aquifer, that is, the bottom of the upper confining layer.

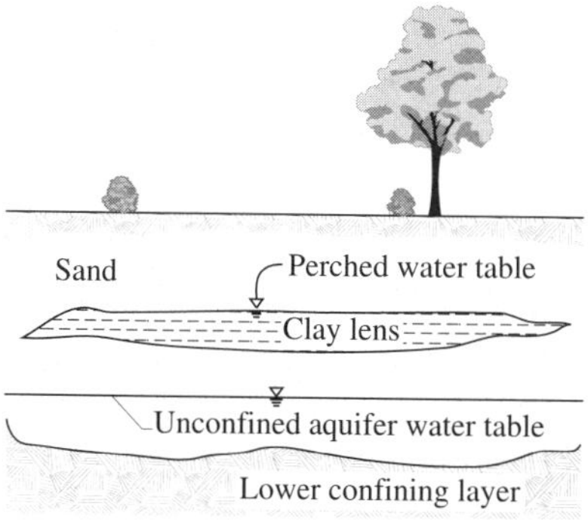

FIGURE 15-23
Perched water table.

In a confined aquifer, the water level in a piezometer open to the atmosphere will rise to a level that indicates the pressure in the formation at the screened interval, as previously discussed.

As described, the primary purpose of a piezometer is to measure hydraulic head in the formation. A piezometer cluster is a group of piezometers that act to measure the head at various depths at the same plan location. For example, the cluster of piezometers in Figure 15-24 is, for all practical purposes, located at the same location in plan view in an unconfined aquifer. In most cases, the wells in a cluster will be spaced within 2 to 3 meters of each other.

Note that two different water level elevations are observed in the piezometers in Figure 15-24. The screened interval for each piezometer is at a different depth (and in this figure, but not necessarily, in a different water-bearing geologic formation). These observations indicate that flow in this particular region is upward. There is a gradient or a change in head upward through the confining layer. In fact, the vertical gradient can be calculated as shown in Figure 15-24.

Shown in Figure 15-25 are two piezometers at different locations in a confined aquifer. Assuming that the flow in the confined aquifer is only horizontal, flow, as indicated by the varying water levels in the piezometers, is in a direction from left to right. As above, the hydraulic gradient can be calculated from the piezometer spacing and the water level elevations in the piezometers.

As in the example above, it is often necessary to measure the pressure head or to collect water samples at various depths within the formation at the same coordinate location. This is done with multiple wells in a cluster or nested wells as shown in Figure 15-26. This is accomplished in a cluster employing individual wells installed in separate but adjacent boreholes drilled to different depths in the formation. Alternatively, a multiple-well piezometer nest can be constructed in a single larger diameter borehole. Multilevel wells in a single borehole may offer advantages of lower cost. Because of the potential for cross-contamination as a result of an imperfect seal between screened intervals, the well nest with multiple boreholes is preferred over the single borehole installation, except for very deep wells.

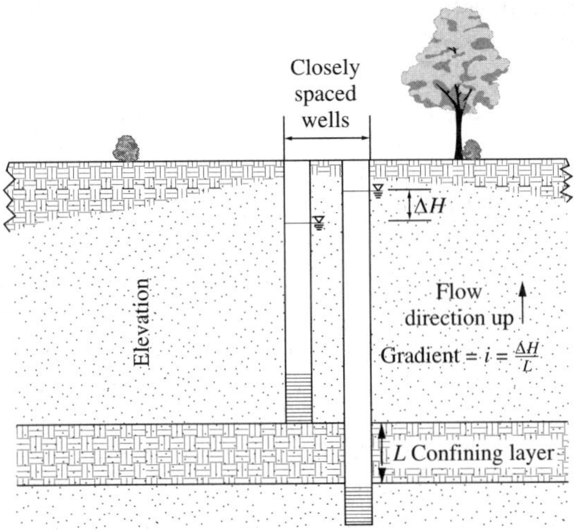

FIGURE 15-24
Piezometer cluster in unconfined/confined aquifer system.

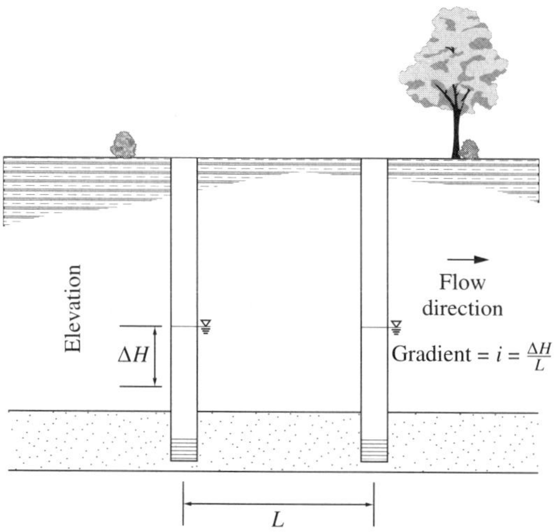

FIGURE 15-25
Piezometers in confined aquifer.

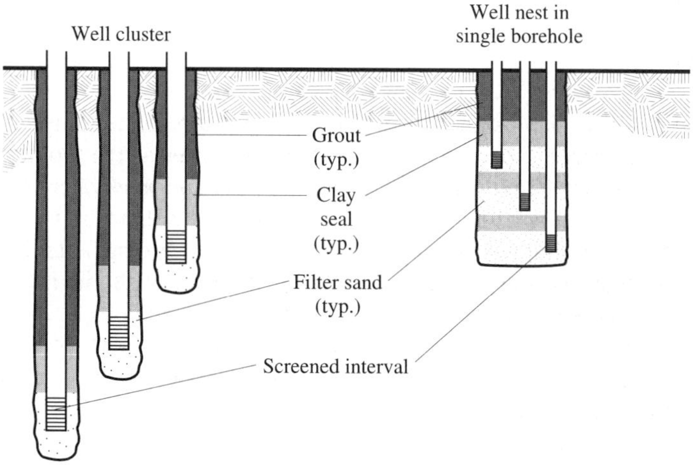

FIGURE 15-26
Multiple and nested wells.

As with unconsolidated formations described above, it is often necessary to monitor groundwater levels and obtain groundwater samples from rock formations as shown in Figure 15-27. In the rock wells where the rock formation has adequate strength to maintain borehole stability over the sampling interval, casing may not be needed. (Of course if regulations require casing independent of formation conditions, then one could say that casing is required.) Thus this rock formation may be monitored without the use of well screens. For stable rock formations, the borehole is advanced from the surface to the desired sampling interval (analogous to the screened interval in unconsolidated formations). At this point a steel casing is installed in the borehole, and the annulus between the borehole and the steel casing is sealed with grout to the surface. The casing serves as a riser pipe and lines the

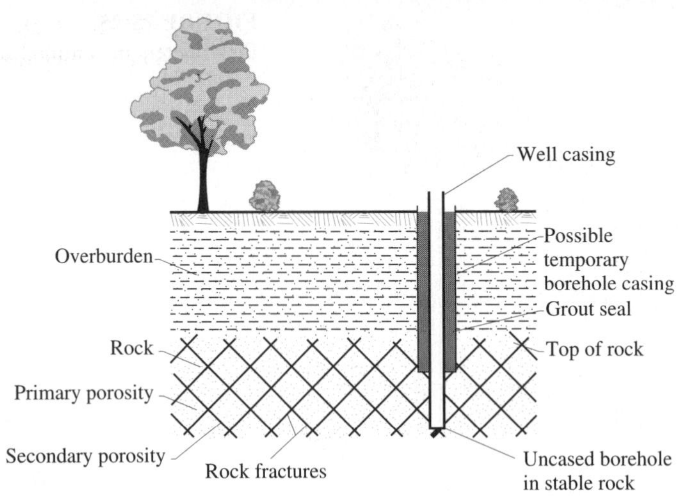

FIGURE 15-27
Monitoring rock formations.

borehole from above the ground surface down to the top of the sampling interval. With the annulus sealed, the completed installation serves as the monitoring well. The stable, uncased hole in the sampling interval allows groundwater to flow to the monitoring well. The installation allows groundwater levels to be monitored and groundwater to be withdrawn, sampled, and analyzed for groundwater quality. Uncased sampling intervals are commonly employed in rock, although the standard designs employing well screens are also used for rock formations.

Well Installation on Contaminated Sites

It is not unusual in studies of hazardous waste management sites to encounter a site that includes an unconfined aquifer separated from an underlying confined aquifer by a confining layer, as shown in Figure 15-28. In these cases it is often necessary to determine the groundwater flow and quality, in both the overlying unconfined aquifer and the underlying confined aquifer. To accomplish these objectives requires drilling the borehole through the potentially contaminated zone of the upper unconfined aquifer and into the confined aquifer. It is essential that precautions be taken to minimize the possibility of contaminating the underlying confined aquifer as a result of the investigation practices. Where possible, monitoring wells should be located beyond the limits of contamination. Many times this is not possible, however. Consider the following question. How deep has the contaminant plume migrated beneath the site? Can this question be answered by drilling only beyond the limits of the known contamination?

Installation of a monitoring well into the underlying confined aquifer without introducing contamination from above requires special steps, as shown in Figure 15-29. This techinique is termed the **telescoping method of well installation.** In drilling the borehole to install the monitoring well, the borehole should be advanced to the

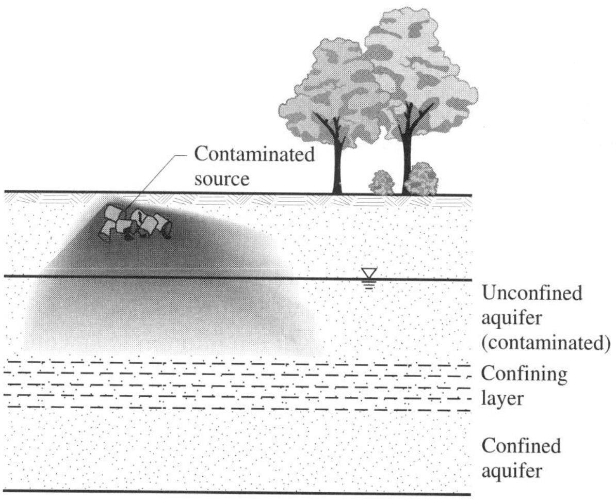

FIGURE 15-28
Multilayer aquifer system.

top of and slightly into the confining layer separating the unconfined and confined aquifers. A temporary steel casing to prevent fluids from the overlying aquifer from flowing down the borehole is then sealed in place by grouting. The inside of the casing is then flushed, cleaned, and washed to remove contamination within the temporary steel casing. After cleaning, drilling proceeds inside the temporary steel casing and into the underlying confined aquifer. This eliminates the potential for the drilling tools and fluids to carry contaminants from the overlying unconfined aquifer to the underlying confined aquifer. The monitoring well installation is then completed as previously described. Installing the well in this manner ensures that the hydraulic head and the groundwater quality in the installed well are representative of the underlying confined aquifer. This well installation process minimizes any potential for either (1) bringing contaminants downward as a result of drilling from

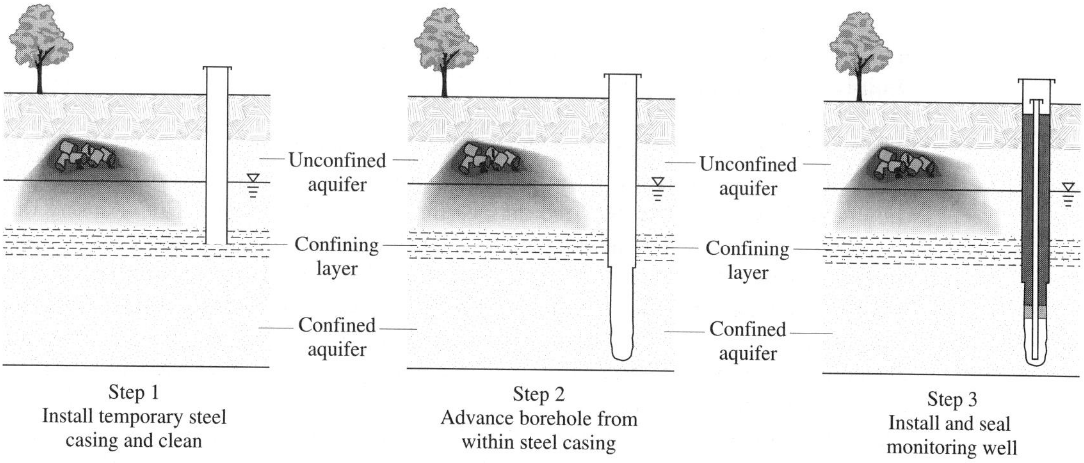

FIGURE 15-29
Drilling through contaminated aquifer.

the upper zone to the lower zone or (2) allowing contaminants to migrate downward after the monitoring well is in place.

In conducting investigations at hazardous waste management sites it is crucial that contaminants not be transported from one location to another as a result of the drilling operations. It has been shown above that special precautions must be taken when advancing a borehole through a contaminated zone into an underlying and potentially uncontaminated zone. Likewise, special precautions must be taken to decontaminate equipment as it is moved from one location to another on the site. Decontamination is accomplished by washing with detergent and steam cleaning the drilling tools at each point in the drilling operation where there is concern for transport of chemical contaminants by the drilling tools. At a minimum, this would be done between boring locations as the drill equipment moves from one location to another on the site. Cleaning of augers, wrenches, cutting bits, sampling tools, and drill rods is necessary to prevent contaminant transport by way of the drilling equipment. In addition, the use of lubricating oils by the drilling contractor must be carefully controlled (minimized or eliminated). Because groundwater quality parameters are analyzed to parts per billion concentrations, a few drops of lubricating oil entering the formation at the time of drilling may lead the investigators to erroneous conclusions about the concentration/presence of organic contaminants. For additional information, an excellent treatise on groundwater monitoring is provided by Driscoll.[38]

Test Pits

An excellent but sometimes overlooked technique for investigating the shallow subsurface condition is to use test pits. Test pits are typically advanced with a small backhoe, and they provide opportunity for continuous visual inspection and to obtain large bulk samples. They are, however, limited in depth by concern about the safety of the personnel entering the trench and limitations to the excavating depth of backhoe excavators. The implementation of robotics for test pit surveys can alleviate certain of these concerns.[39]

Soil Gas Surveys

The nature and extent of gases in the unsaturated (vadose) zone is often overlooked in efforts to obtain quality groundwater samples. However, on many sites the contaminants in the unsaturated zone may be the focus of the remediation. Among the newer developments for the investigation of the distribution and transport of contaminants in the subsurface are soil gas surveys.[40,41] Soil gas surveys are conducted to provide the quantitative information needed to assess the risks associated with the unsaturated zone. Soil gas surveys have also been used as a low-cost alternative to groundwater sampling and testing to assess the nature of the groundwater plume.

Soil gas surveys measure the vapor phase plume in the void space of the unsaturated soil. This information can then be interpreted in the context of source detection and plume delineation.[42] But how does it work? Soil gas surveys rely on the contaminant property of vapor pressure (see Section 3-2) for success. A contaminant with a high vapor pressure and a low water solubility will volatilize from the

groundwater into the vadose zone as the plume migrates, as shown in Figure 15-30. Detailed study has shown that the soil-gas sample results correlate reasonably well with groundwater sample results if the soil-gas sample is obtained within 5 feet of the water table for clayey soils and 10 feet of the water table for sandy soils.[43] The reliability of the technique is enhanced if the contaminant does not degrade rapidly into degradation products.

How are the contaminant concentrations measured in the soil voids? Although the details vary, soil gas surveys either directly withdraw the soil gas or withdraw a soil sample from which the gas concentration is measured. A sampler left in place[44] would seem appropriate where subsequent soil gas data are required to measure the time history of soil gas contaminant concentrations. After sampling, the gas is analyzed by an organic vapor analyzer, a photoionization detector, or gas chromatograph. The analytical end of the soil gas survey may be done with field instruments, in a field laboratory, or in a remote analytical laboratory. Shown in Figures 15-31 and 15-32 is a field laboratory using a GC/MS for the analysis of soil gas. Analyzing the samples in the field permits the site assessment to be adjusted as it progresses, on the basis of the data generated.

Unsaturated Zone Monitoring

Characterization of the unsaturated zone may require information regarding soil moisture potential, soil moisture content, soil salinity, temperature and soil pore water sampling. As described in Section 4-2, the pressure head is negative in the unsaturated zone and is dependent on the soil moisture content and pore size distribution. Thus the measurement device must be designed and installed so as not to alter these properties by the very act of measuring them. A thorough discussion of these topics constitute a separate textbook; the interested reader is referred to an excellent

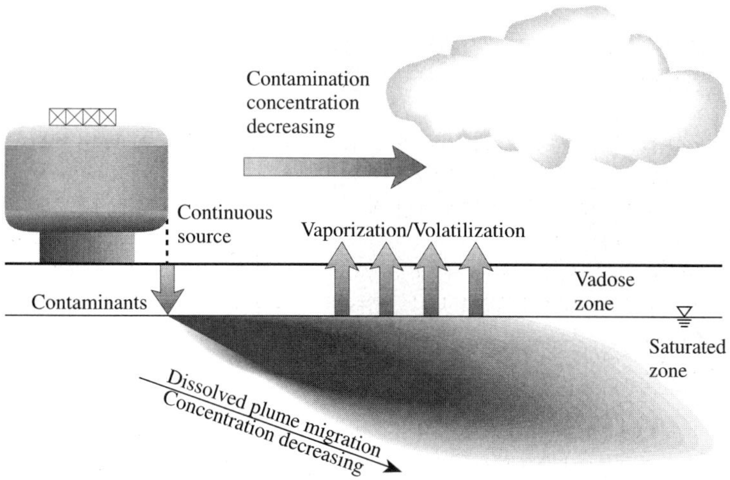

FIGURE 15-30
Principle of soil gas surveys.

FIGURE 15-31
Mobile laboratory used for soil analysis.

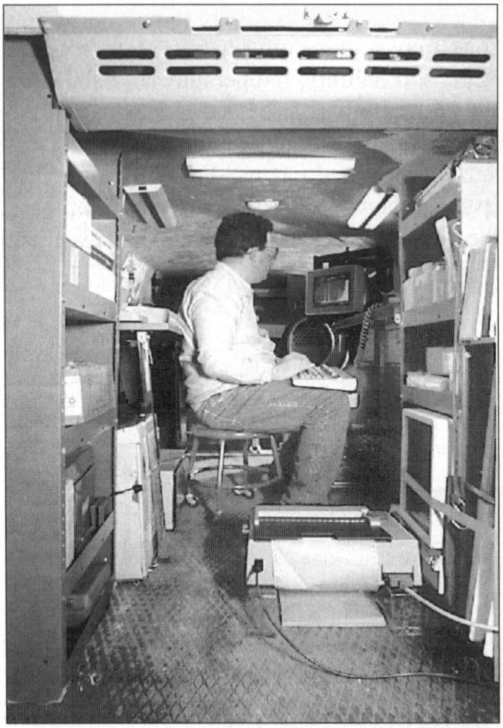

FIGURE 15-32
Interior of mobile laboratory.

reference by Morrison.[45] More recent developments include a cone penetrometer that was modified and a transient flow method developed to determine the hydraulic properties of the soil in the unsaturated zone.[46]

Groundwater Sampling

Obtaining a representative groundwater sample can be a difficult task at best. However, securing representative and reliable groundwater samples is perhaps the most critical element of a subsurface site characterization. Sampling and analysis of groundwater reveals chemical constituents present and their concentrations, and these define the lateral and vertical extent of contamination. Chemical characterizations obtained from groundwater samples permit the performance of a risk assessment to determine the magnitude of any adverse impact on human health and the environment and are needed to guide the evaluation and selection of appropriate remedial actions. In an ongoing monitoring program after the baseline is established, groundwater sampling and analysis can be used to evaluate temporal variability in water quality and assess whether existing subsurface contamination is migrating or whether it is being contained or abated. The entire RI/FS process depends on representative and reliable groundwater samples.

The decade of the 1980s saw explosive growth in the field of groundwater sampling and monitoring.[47] Extensive and detailed protocols were established[48] to ensure, as much as is possible, that proper sampling methods are employed. New and revised methodologies were developed to meet the increasing demand for measurement of contaminant concentrations in groundwater at lower and lower concentrations. This increasing sophistication is required to obtain reliable, repeatable, accurate, and precise results (as much as possible). For example, organic solvents in groundwater are of concern at concentrations measured in parts per billion (ppb) and are seemingly ubiquitous.

The most common method of groundwater sampling is from monitoring wells described earlier in this chapter. Numerous factors, both engineered and natural, will affect the reliability and representativeness of the groundwater sample. A number, but certainly not all, of the factors are the following:

1. Natural groundwater chemistry and geochemical interactions that occur on sample removal from *in situ* conditions such as temperature and stress
2. Well installation methods
3. Extent and method of purging of monitoring wells prior to sampling
4. Type of sampling device used
5. Ensuring decontaminated procedures (e.g., cleaning sampling devices between samples or using dedicated sampling devices for each well)
6. Methods of sample preservation
7. Methods of sample shipping and storage
8. Duration of sample storage prior to testing

Presentation and Interpretation of Data and Results

The findings and conclusions developed as a result of the site and subsurface investigations must be clearly, concisely, and accurately reported. Without such reporting, the value of investigations is considerably reduced. Those interested in the site and subsurface characterizations include owners, regulators, citizens, and scientists and engineers. The preparation of reports is an ongoing effort and should not be left in its entirety to the end of the project. Report reviews include several types, each with

different objectives. Editorial reviews are necessary to ensure the readability and grammatical accuracy of the product. Technical reviews are necessary to ensure the technical adequacy of the report. Details regarding methods of data preparation are incorporated into the case history at the end of this chapter.

15-8 GEOGRAPHIC INFORMATION SYSTEM*

GIS Overview

During the past several years extensive developments in the field of computer graphics have produced commercially available software systems that are applicable to the problems associated with characterizing hazardous waste sites. The use of digitized drawings for computer-aided drafting and design (CADD) has become common in the field of construction and civil engineering design. Likewise, automated mapping (AM) using digitized representations of map features has largely replaced conventional cartography, and facility/infrastructure management (FM) systems that associate facility drawings to text and numerical attributes have gained increasing application in both government and industrial organizations.

Each of these applications employs digitized drawings linked in varying degrees to data sets. The term **geographic information system** (GIS) is defined by Antenucci et al. as

> . . . a computer system that stores and links non-graphic attributes or geographically referenced data with graphic map features to allow a wide range of information processing and display operations, as well as map production, analysis, and modeling.[49]

A thorough discussion of GIS and its applications is provided in this reference. The linking of data to maps or drawings provides a powerful tool for increased visibility and insight into the areal and subsurface extent of contamination in soils and groundwater. GIS for site modeling is one particular application of the technology.

An industrial facility or any hazardous waste site where characterization and remediation are being conducted can be described by cartographic or map layers called planimetrics (e.g., roads, buildings, fences, property lines) and by data sets (e.g., sampling locations, drilling logs, groundwater elevations, topographic elevations, chemical concentrations). The cartographic layers in some systems may be composed of individual objects (e.g., buildings, sampling locations, rooms, waste sites), which have logical identity and spatial position. In other systems, it is more difficult or impossible to isolate the objects within the context of the layer or drawing phase in which they are stored.

Drawings or maps having objects linked to data are sometimes called *intelligent* because the objects are associated with information that is not necessarily graphical. The contamination level (mg/L) at a specific sampling location for a hazardous waste material, or for several chemicals is an example of information that can be geographically referenced and related to spatial objects. With GIS functions, the data sets can be accessed graphically to display the information associated with an object.

*Section 15-8 was prepared by William Douglas, Ph.D., ERM.

The technical data that are used to characterize the site (chemical, geological, geotechnical, geophysical, etc.) can be stored in a database that is accessible through conventional database management system functions. The linking of a database management system functionality with digitized cartography is another way of describing a GIS. Hence, a database of groundwater elevations, chemical concentrations in soil and water, and drilling logs can be related to the sampling or drilling locations (objects) in the digitized drawing for a hazardous waste site. The database management system can be used to extract the data from the database, and the automated mapping function can be used to display the data elements on a site drawing near the objects to which they relate. Data displays of this type (data posting) are commonly used in site investigations. GIS provides an efficient and flexible means for creating information products of this type. Computer-aided drafting functions provide a means for editing and modifying the display to make it more attractive or easier to understand. For example, one study used GIS to integrate groundwater modeling results with the location of municipal water supply wells and census information to estimate what portions of the city's population were exposed to TCE, the duration of exposure, and the concentration range of the exposure.[50]

Geographic data modeling is another element that has applicability to site investigation. Contamination plumes are simulated by modeling chemical concentrations in soil and groundwater on the basis of analytical laboratory results from actual samples. This approach is routinely used by scientists and engineers. A variety of contouring models are available that use different algorithms to estimate values at regular grid intervals from samples taken at irregular locations. These models develop lines of constant value (isopleths) from the gridded data through interpolation techniques. Linked to a GIS, they provide a means for defining a plume extent from database contents rapidly, and for displaying the isopleths as objects or as a layer to be used in conjunction with a site drawing. Topography and groundwater elevation contours can likewise be developed and stored for display along with other layers or objects. The set of coordinates and values (x, y, elevation, or x, y, contamination) are often called *surfaces*. The surfaces may be physical (topography) or logical (contamination). Data modeling can also be used to produce a cross section through a surface to display topography and subsurface layers or objects from the GIS. The integration of data modeling with the database management system and CADD functions in a GIS provides the site investigator with an effective means for analyzing, understanding, and conveying to others the extent and severity of site conditions, and for establishing a basis for defining remediation alternatives.

Applications

Two examples of applying GIS to site investigation are illustrated below. The first application relates to a site near the Washington D.C. National Airport contaminated with PCBs and heavy metals. A database management system named ORACLE®[51] was used to store the laboratory analytical results, which were transferred from EXCEL® spreadsheets into a technical database used for site investigations. A base map illustrating the property extent, fence line, nearby roads, buildings, north arrow, and sampling locations was digitized in AutoCAD.® The files were transferred to the Graphic Data System (GDS)®[52] which is an integrated GIS/CADD software

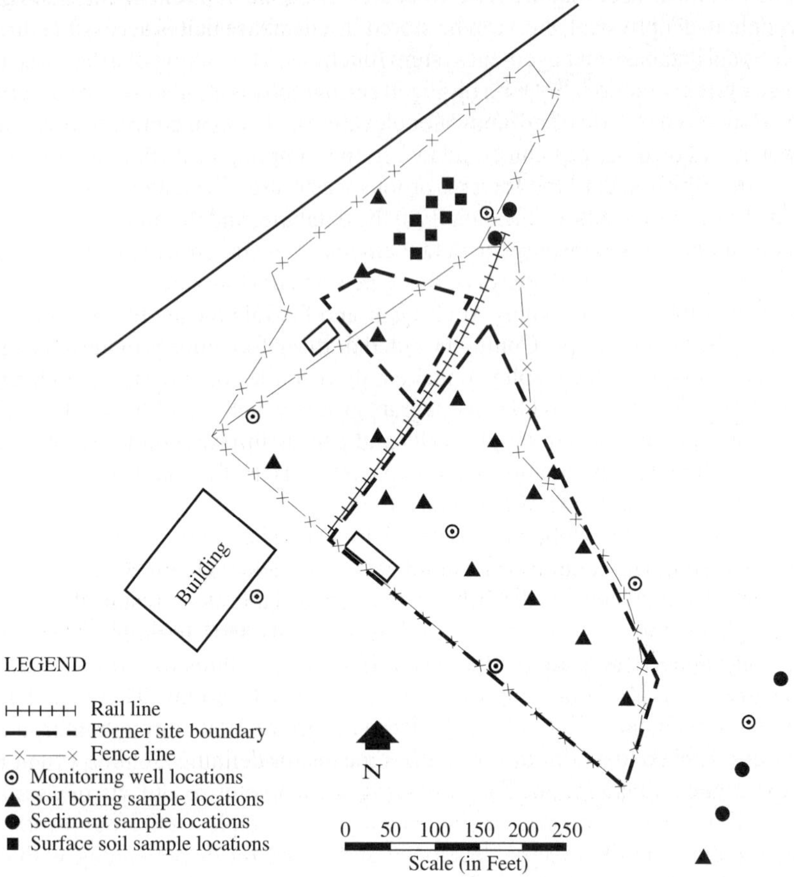

FIGURE 15-33
Site map.

system. The Site Modeler module of GDS® uses object-based intelligent drawings to describe and characterize hazardous waste sites.

The objective of the investigation was to illustrate the contamination plume using depth-dependent sample results and varying action levels for each chemical parameter, the intersection of the groundwater table with the plume, and the relationship of the plume to the property fence line. Figure 15-33 illustrates the base map for the property, and Figures 15-34 and 15-35 represent the contamination volume for two of eight scenarios that were investigated. The numerical values of the contaminated volume and the volume of excavation are also computed using GDS. Hard copy drawings of the results as well as slide photographs of the screen images are used to support the presentations to the decision-making group.

A second application of GIS relates to the Washington Metro rail line where construction design alternatives were affected by a municipal incinerator ash pile situated in the right-of-way. Aerial surveys were conducted in 1976, before ash dumping was initiated, and in 1990. These surveys provided the planimetric features

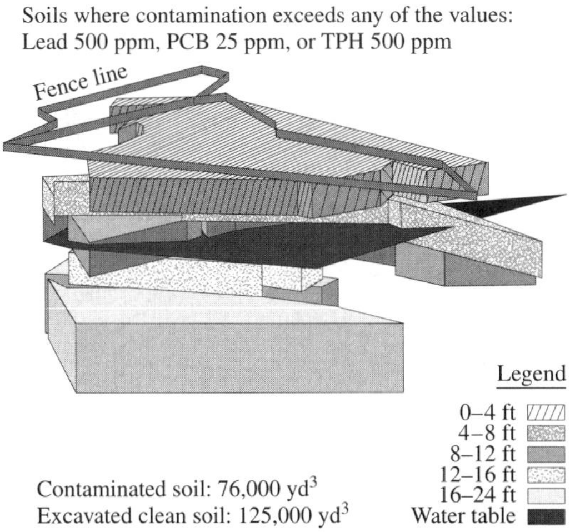

FIGURE 15-34
GIS representation of contaminated soil (Scenario II).

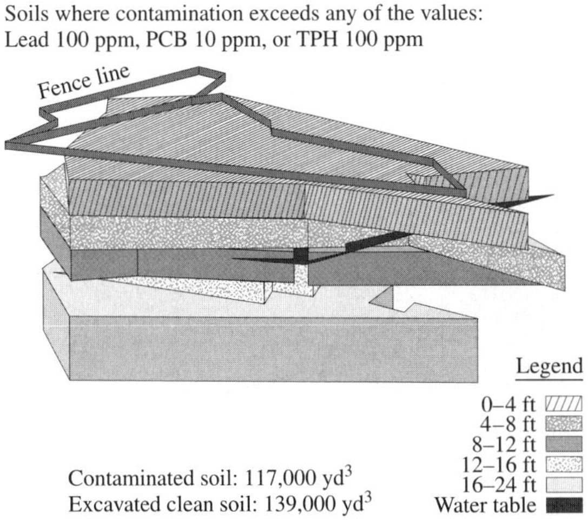

FIGURE 15-35
GIS representation of contaminated soil (Scenario V).

and topography around the site. The site investigation database included surface samples, soil borings, drilling logs, and soil and water sample results. A variety of information products were developed to support decisions relating to the hazardous classification of the ash pile, volume computations, cross sections, and evaluations of alternative alignments through and around the waste pile. Figure 15-36 illustrates a site map with three alignments and a set of transverse cross sections through the right-of-way showing the ash fill above the natural topography. Figure 15-37 is a three-dimensional view of the site topography, ash, and the three alignments. The darker shading in Figure 15-37 represents areas covered by ash, whereas the lighter shading represents the original topography. This information was used to assess the alternatives from a regulatory standpoint, and to evaluate tradeoffs between construction costs and remediation costs for the alternatives.

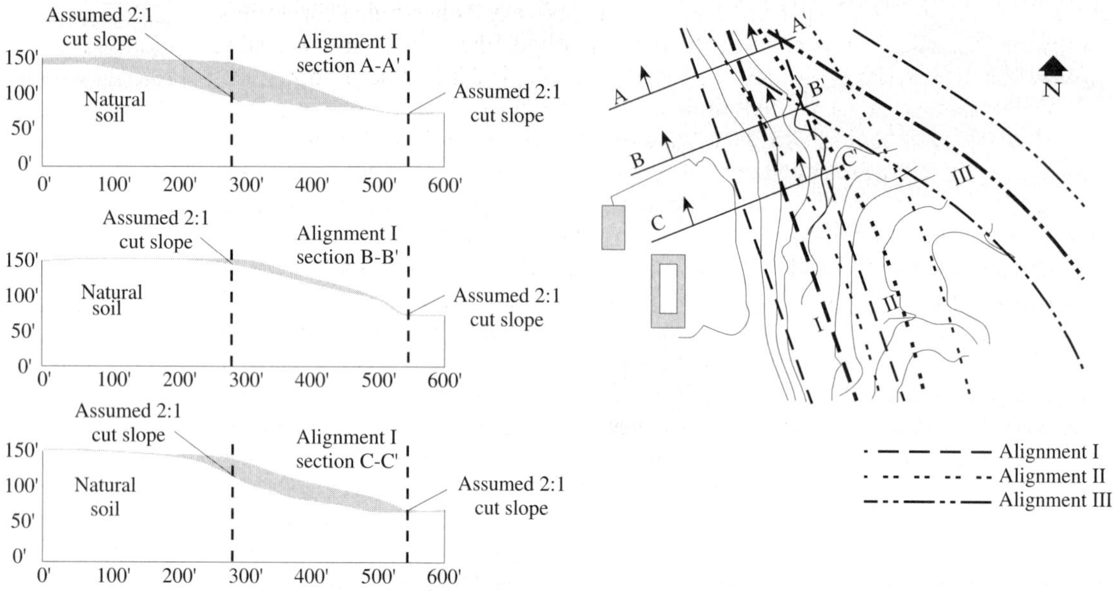

FIGURE 15-36
Contour map.

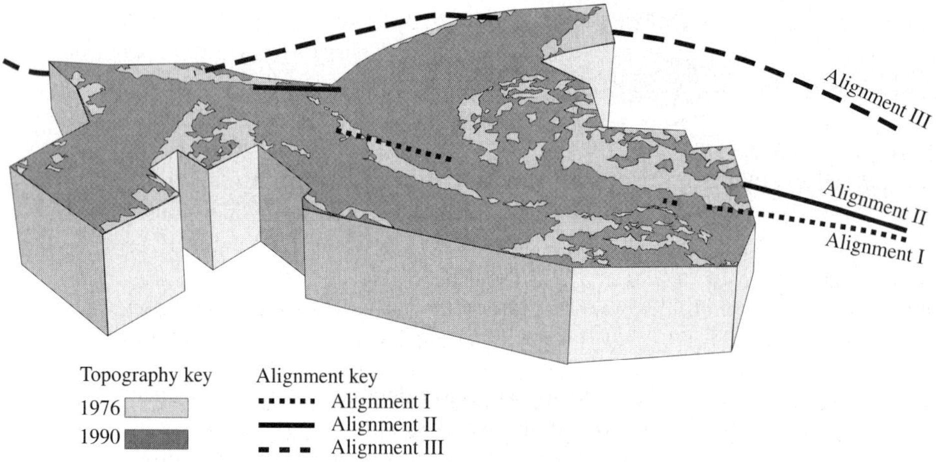

FIGURE 15-37
Three-dimensional representation of site topography.

15-9 CASE STUDY

An operating solid waste landfill site is located in the southeast of the United States. The owner wished to consider expansion of the site. Site and subsurface investigations were undertaken to provide the needed technical information to adequately

evaluate the suitability of expansion and to define design and operational constraints. As discussed earlier in this chapter, the effort first identified available preexisting information. For example, an investigation of the geological information showed the site to be located within the western limit of the coastal plain physiographic province. As a result the site could be expected to be underlain by Tertiary-Quaternary age sand and gravel deposits with an underlying silty clay deposit of Miocene age. The geologic studies further indicated that the silty clay is composed of quartz and montmorillonite. An examination of the site history showed that the site was formerly used to mine sand and gravel down to the Miocene clay. A lined landfill was then built on the northern portion and the owner was interested in expanding this landfill to the southern portion of the site. An industrial facility with the potential for groundwater contamination was located to the east of the site.

Using the available subsurface and geologic information, subsurface investigations were conducted that included 25 test borings at 18 locations with monitoring wells installed in 24 test borings at 17 locations. In order to minimize formation disturbance, rotary drilling techniques used Attapulgite clay drilling fluid to maintain borehole stability. Where the borings/monitoring wells penetrated the Miocene clay, special precautions were used as discussed earlier in the chapter so as to not carry any potential contaminants down from the overlying sand and gravel to the materials beneath the clay. *In situ* testing and sampling of all unconsolidated sediment formations were done by the standard penetration test. Thin-walled sampling of the Miocene clay was also carried out. Laboratory testing included physical property and classification tests of natural water content, liquid limit, plastic limit, unit weight, specific gravity, and grain size distribution by sieve and hydrometer (see Table 15-4 for the test methods). Engineering property tests included characterization of the undrained shear strength and the triaxial hydraulic conductivity of the Miocene clay.

How do we present all these data in a form that is understandable, usable, and, most important, revealing of the key site and subsurface features? For this project a number of presentation methods were used as presented in the following paragraphs. Normally, as for this project, a regional location plan and a boring and monitoring well site location plan are prepared. Each test boring is logged in the field and on the basis of the field log and the subsequent laboratory test data, a report form boring log is prepared as shown in Figure 15-38. The log of this particular test boring shows about 15 feet of fine-grained alluvium overlying about 23 feet of dense sand and gravel. The underlying Miocene clay was encountered at a depth of about 37 feet at this location. The boring log also presents the results of SPT testing and the natural water content data obtained in the laboratory. The log also notes that grain size data were obtained on two samples; the data are presented elsewhere in the complete report.

Although much is revealed in a boring log, the information relates to only one point in plan at the site. A subsurface geological profile can be constructed by combining the information from several borings along a line as shown in Figure 15-39. This section reveals the undisturbed natural section as well as the influence of mining operations on the subsurface conditions. Additional insight into the subsurface conditions results from constructing and studying a subsurface profile behind the study of individual boring logs.

As informative as subsurface profiles are in helping one visualize site and subsurface conditions, we can do more. Structural contour maps can be prepared to map

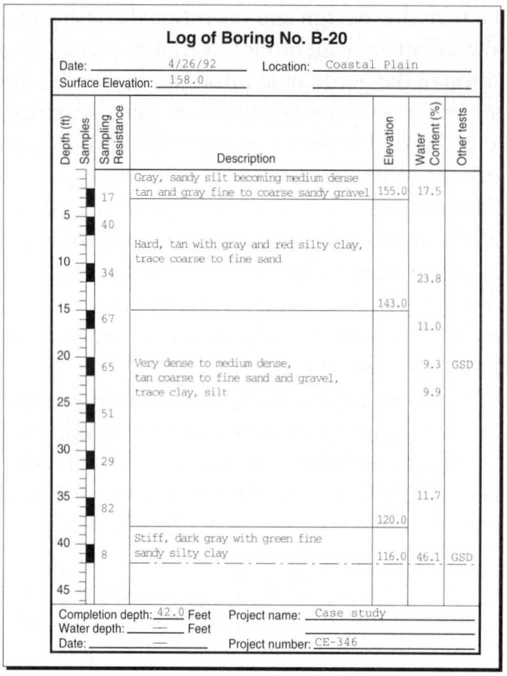

FIGURE 15-38
Typical boring log.

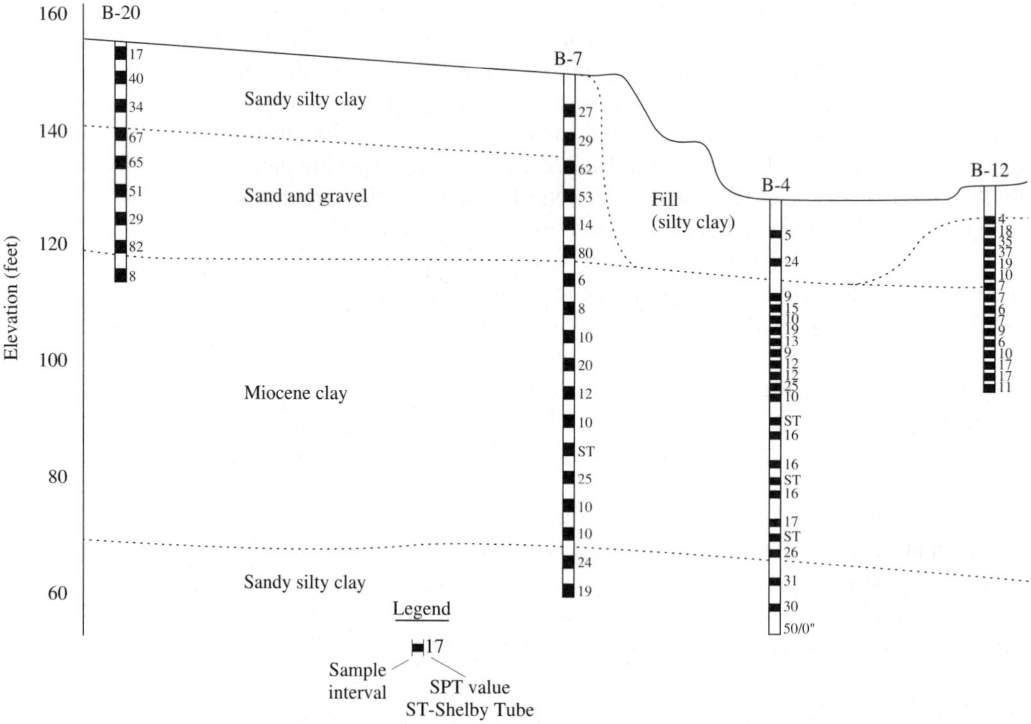

FIGURE 15-39
Geologic profile.

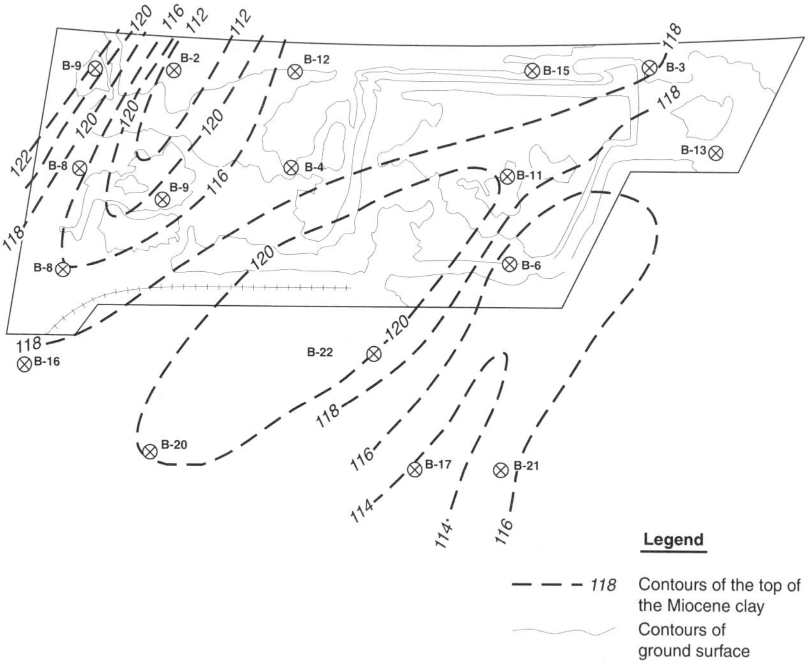

FIGURE 15-40
Structural contour.

individual geologic units. In this case study a structural contour of the top of the Miocene clay is particularly revealing. Shown in Figure 15-40 is a map prepared by taking the elevation of the interface between the sand and gravel and the Miocene from individual test borings and contouring these data. What can we see from this structural contour? Located at the southern end of the site is a buried valley in the Miocene clay, filled with sand and gravel and sloping toward the east northeast. We will look in a moment at how this information is used in the interpretation of groundwater contours. The structural contours also reveal a northeast trending ridge in the clay.

Employing data from test boring regarding the elevation of both the top and bottom of any geologic unit, **isopac** maps can be prepared. Isopac maps present lines representing equal thickness of a formation material. Shown in Figure 15-41 is an isopac map of the Miocene clay formation. This map is quite revealing with respect to the site characterization. We find that the clay is over 40 feet thick everywhere beneath the site and is as much as 50 feet in some locations. This thickness information coupled with the geologic information regarding the clay mineralogy (montmorillonite) and the hydraulic conductivity (varying from 3×10^{-7} to 2×10^{-8}, averaging 6×10^{-8} cm/s) indicate the suitability of this layer as a natural confining layer.

Our case study must concern itself with groundwater information as well as geologic information. To this end groundwater monitoring wells were installed and

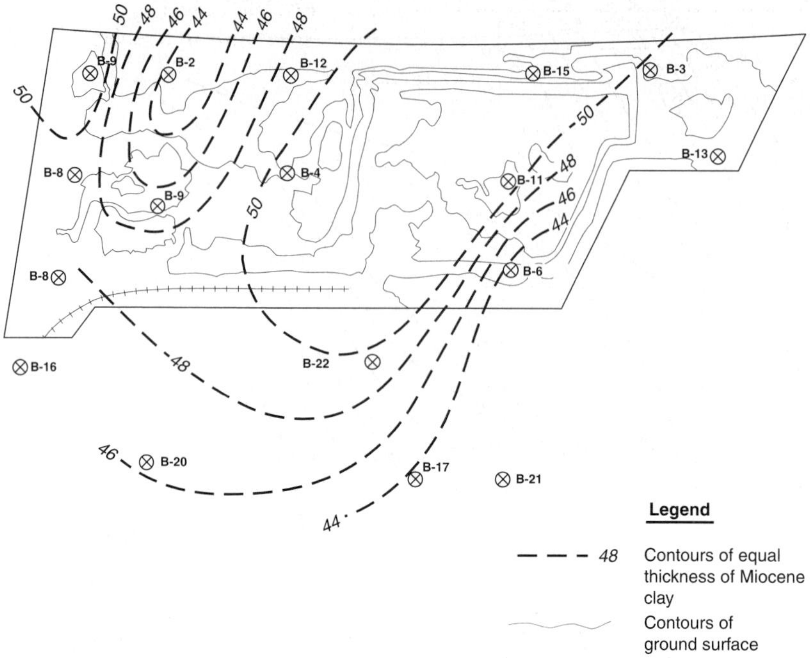

FIGURE 15-41
Isopac map.

the well installation details included on a monitoring well report (Figure 15-42). Of utmost importance is the clear definition of the **screened interval,** the zone from which water samples are obtained and piezometric levels are measured. A comparison of the elevations of the screened interval shown in Figure 15-42 with the information shown in Figure 15-38 reveals that the well is sampling the bottom 10 feet of the sand and gravel layer overlying the Miocene clay.

Groundwater contours for each aquifer can be constructed by taking water level readings from each of the wells using the methods described above. The water levels should be recorded "at the same time." Does this mean we need to be at all locations at once or use water level recorders to obtain this information? Sometimes yes and sometimes no. If water levels are rapidly changing, as during a pump test, we may need to go to such efforts. However, if water levels are steady state and do not change rapidly (over the course of hours) it is usually sufficient to record all the levels the same day. Shown in Figure 15-43 are the groundwater contours for the sand and gravel aquifer for our case study. These contours reveal that the buried valley exhibits control over the subsurface flow regime. Note the groundwater flow is toward the buried valley from all portions of the site. Next note that the flow is from the adjacent industrial site. Thus, if groundwater is contaminated beneath the adjacent site, the contaminated water would flow through our case study site.

Within the context of this case study, it is impossible to present all the data and interpretations available for this project. However, the case discussion above demonstrates the need to employ multiple means of data presentation to best understand the

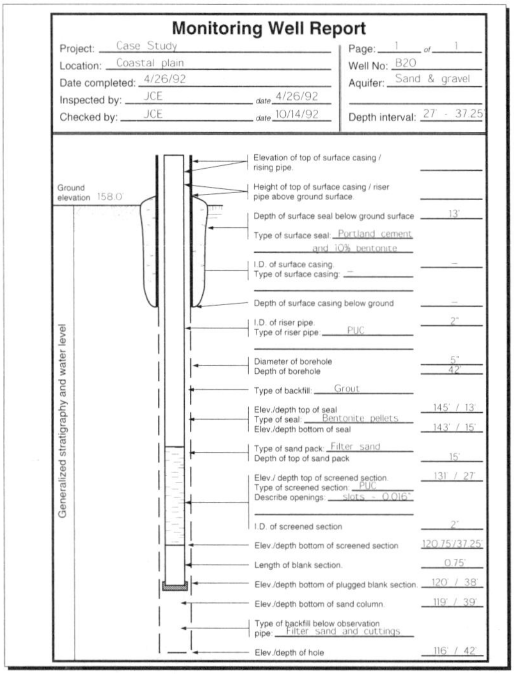

FIGURE 15-42
Monitoring well report.

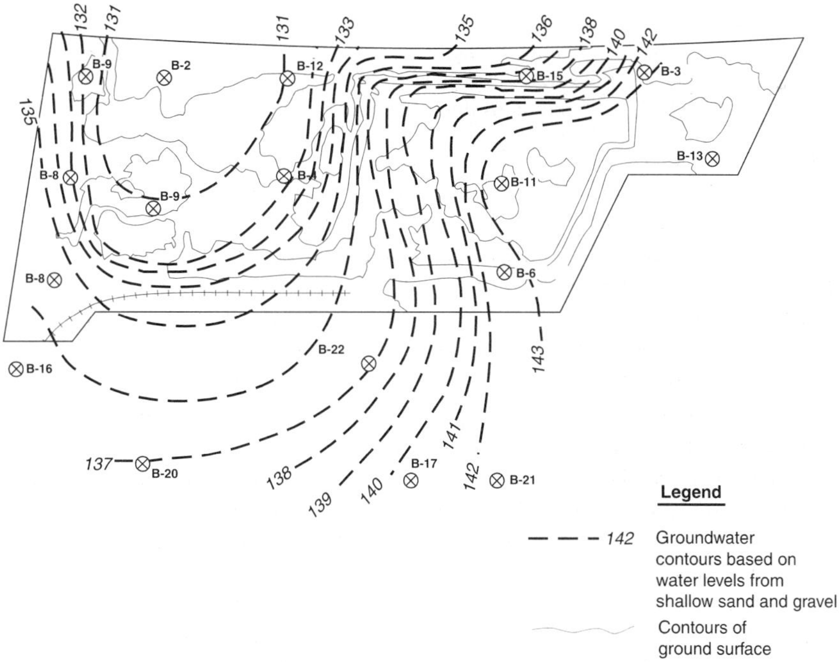

FIGURE 15-43
Groundwater contours.

site and subsurface conditions. The case study also demonstrates that the selection of the means of site characterization must be appropriate to the information need for the project. So as to not leave the reader wondering what happened, it is noted that the case study site was eventually developed as a solid waste land disposal facility with geomembrane liner systems. In addition, a soil-bentonite slurry trench (Chapter 16) was constructed (1) to control the influx of groundwater into the site to allow construction of a liner system (Chapter 13), (2) to eliminate/minimize the influx of potentially contaminated water from the adjacent industrial site, and (3) to provide a secondary containment for added confidence regarding the long-term security of the leachate containment system.

DISCUSSION TOPICS AND PROBLEMS

15-1. Describe what is meant by necessary and sufficient.

15-2. An investigation of subsurface conditions at a 200 m by 200 m site employs an average of one test boring for each 250 m^2 of site (but not on uniform 50-m centers). Split-barrel samples are taken each 1.5 m from the ground surface down to the depth of investigation of 15 m. By volume, what percent of the subsurface materials is available for examination and testing? Why not use a uniform grid of 50 m on center?

15-3. Visit the geology department at your school or your local soil conservation service office and study the air photos covering the nearest waste disposal facility. How do the existing drainage conditions influence the possible pathways for contaminant migration? Identify sites/areas you would target for purposive sampling and testing in the event of a spill at the facility.

15-4. In reference to groundwater, define the term plume.

15-5. Explain why direct and indirect methods should be used together and how this might be done for a specific site.

15-6. Seismic profiling is being considered for a site where dense material overlies soft material. Is this feasible? Why or why not?

15-7. Choose a geophysical technique and describe how the technique can be used to measure different subsurface properties in different situations.

15-8. For the following methods of subsurface characterization, state the measured parameter and the deduced parameter/information:

(*a*) Resistivity
(*b*) Seismic
(*c*) Magnetometry
(*d*) Electromagnetics

15-9. List the elements of a site and subsurface investigation plan.

15-10. Compare and contrast indirect and direct investigation methods.

15-11. Choose a direct investigation technique and describe how the technique can be used to measure different subsurface properties in different situations.

15-12. Describe three direct site investigation methods and the information each might provide pertaining to the site conditions.

15-13. An abandoned railroad car maintenance and refinishing facility is suspected of affecting the water supply of adjacent residential areas, but the specific impacts are uncertain. The facility operated from 1928 to 1983, and petroleum solvents and varied paints were both stored and directly used at the site. The location of those activities is poorly established. The site is situated on an unconsolidated sand unit

containing thin clay laminae; the unit is up to 50 feet thick, and overlies a sequence of interbedded sandstone and shale, these rocks being penetrated by fracture systems dipping southeast. The residential areas are northwest of the site. What steps would you recommend to:

(*a*) Locate buried metal (and other) containers on the site,

(*b*) Determine the location of the most heavily contaminated sands, if any, and the contaminants with them, and

(*c*) Establish the presence and (if within this unit) flow direction of the contaminants. Are they confined to the overlying unconsolidated sands, or flowing within the fracture system of the bedrock?

15-14. Why can't seismic profiling be conducted where hard or dense soil conditions overlie loose or soft soil conditions?

15-15. How does *in situ* testing differ from laboratory testing? What are the advantages and disadvantages of each approach?

15-16. An existing pesticide manufacturing plant is suspected of affecting the waters of a nearby stream. The site is set on a sequence of sandstone and arkose, with these units lying horizontally and being penetrated by a weakly developed system of perpendicular vertical fractures. What investigative methods should be undertaken, and what should the remedial investigation specifically aim at to determine the nature and extent of contaminants in the area?

15-17. You've unexpectedly been transported to a site (as if by helicopter) that has a long history of industrial activity, and is thought to harbor a variety of wastes in a greater variety of containers. In other words, a little is known, and quite a bit more is unknown, about the place. You're constrained by time and budgetary factors, and must limit yourself to two techniques to delineate the site conditions. What would you suggest be done? Why? Give pros and cons for each method you suggest. Based on whatever methods you have chosen, what would you not be able to tell about the site conditions? Which other sources of information, not field activities, would expand your knowledge of the site?

15-18. Identify and define the two Latin root words used to form isopac.

15-19. It has come to the attention of the dean that the environmental science laboratory previously dumped its wastes in the field next to a campus building. The range of wastes would be expected to include acids, bases, organic solvents, detergents, and biological wastes. It is rumored that drums also may have been buried. The practice was stopped in 1982.

You represent Country Club Consultants and have been asked to prepare a proposal to investigate this potential problem. The scope of work provided in the request for proposal is described below. Your assignment is to prepare the technical proposal for Country Club Consultants.

Scope of Work: Prepare a plan for remedial investigations for the aforementioned site. The plan should include your specific rationale and recommendations for the following:

- Remote sensing (geophysical methods)
- Borings: number, spacing, depth, type, locations, criteria, drilling method
- Monitoring wells
- Soil sampling
- Water sampling
- *In situ* tests
- Lab tests on soil and water samples
- Site characterization report

Basis of evaluation: The proposals will be evaluated on the basis of appropriateness of the technical scope of work, thoroughness, creativity, and proper selection of available techniques. Any available information that is developed as part of the proposal preparation (i.e., geology, etc.) should be included to strengthen the proposal. Listing of information for proposed work activities is suitable although such lists should only include relevant information. The proposal is limited to five pages plus tables and figures, if any.

REFERENCES

1. U.S. EPA: "Guidance for Conducting Remedial Investigations and Feasibility Studies Under CERCLA," EPA/540/G-89/004.
2. Kjartanson, B. H., J. D. Wonder, R. C. Benson, L. Yuhr, and A. J. Bevolo: "Role of Geophysics in Expedited Site Characterization: Case History," *Practice Periodical of Hazardous, Toxic and Radioactive Waste Management,* vol. 2, no. 1, January 1998, pp. 33–47.
3. Robbat, A., Jr., S. Smarason, and Y. Gankin: "Dynamic Work Plans and Field Analytics, the Keys to Cost-Effective Hazardous Waste Site Investigations," *Field Analytical Chemistry and Technology,* vol. 2, no. 5, pp. 253–265, 1998.
4. U.S. EPA: "Improving Site Assessment: Abbreviated Preliminary Assessments," EPA-540-F-98-037, Office of Solid Waste and Emergency Response, October 1999.
5. U.S. EPA: "Improving Site Assessment: Combined PA/SI Assessments," EPA-540-F-98-038, Office of Solid Waste and Emergency Response, October 1999.
6. ASTM: "Standard Practice for Expedited Site Characterization of Vadose Zone and Ground Water Contamination at Hazardous Waste Contaminated Sites," D6235-98, Conshohocken, Pa., 1998.
7. ASTM: "Standard Guide for Accelerated Site Characterization for Confirmed or Suspected Petroleum Releases," E1912-98, Conshohocken, Pa., 1998b.
8. ASTM: "Standard Provisional Guide for Expedited Site Characterization of Hazardous Waste Contaminated Sites," PS85-96, Conshohocken, Pa., 1996.
9. U.S. Army Corps of Engineers: "Requirements for the Preparation of Sampling and Analysis Plans," EM 200-1-3, September 1, 1994.
10. *Standard Methods for the Examination of Waterways,* 17th ed., American Public Health Association, 1989.
11. American Society for Testing and Materials: *Annual Book of Standards,* vol. 04.08, *Soil and Rock; Building Stones; Geotextiles,* Philadelphia.
12. *Occupational Safety and Health Guidance Manual for Hazardous Waste Site Activities,* National Institute of Occupational Safety and Health, Washington, D.C., October 1985.
13. U.S. Army Corps of Engineers: "Engineers and Design—Geophysical Exploration for Engineering and Environmental Investigations," EM 1110-1-1802, August 31, 1995.
14. Jakosky, J. J.: *Exploration Physics,* Trija Publishing, Newport Beach, Calif., p. 660.
15. McNeill, J. D.: Electromagnetic Terrain Conductivity Measurement at Low Induction Numbers, Technical Note TN-6, Geonics Limited, Ontario, Canada, 15 pp., 1980.
16. Merin, I. S.: "Identification of Previously Unrecognized Waste Pits Using Ground Penetrating Radar and Historical Aerial Photography," *Proceedings of the 11th National Superfund Conference,* HMCRI, Washington, D.C., November 1990, pp. 314–319.
17. Struttman, T., and T. Anderson: "A Comparison of Shallow Electromagnetic (EM-31) and Proton Precession Magnetometer Surface Geophysical Techniques to Effectively Delineate Pre-RCRA Buried Wastes at a Manufacturing Facility," *Proceedings of the 10th National Conference on the Management of Uncontrolled Hazardous Waste Sites,* Washington, D.C., pp. 27–34, November 1989.

18. Richardson, W. K., G. L. Kirkpatrick, and S. P. Cline: "Integration of Borehole Geophysics and Aquifer Testing to Define a Fractured Bedrock Hydrogeologic System," *Proceedings of the 10th National Conference on the Management of Uncontrolled Hazardous Waste Sites,* Washington, D.C., pp. 277–281, November 1989.
19. Acker, W. L. "Basic Procedures for Soil Sampling and Core Drilling," Acker Drill Co., Scranton, Pa., 1974, 246 pp.
20. ASTM: "Standard Test Method for Penetration Test and Split-Barrel Sampling of Soils," D 1586, American Society for Testing and Materials, Philadelphia, Pa.
21. Kovacs, W. D., A. H. Griffith, and J. C. Evans: "An Alternative to the Cathead and Rope for the Standard Penetration Test," *Geotechnical Testing Journal,* ASTM, vol. 1, no. 2, pp. 71–89, June 1978.
22. Kovacs, W. D., J. C. Evans, and A. H. Griffith: "Towards a More Standardized SPT," *Proceedings of the 9th International Conference on Soil Mechanics and Foundation Engineering,* Tokyo, vol. 2, pp. 269–276, 1977.
23. Fletcher, G. F. A.: "Standard Penetration Test: Its Uses and Abuses," *Journal of the Soil Mechanics and Foundations Division,* ASCE, vol. 91, no. SM4, pp. 67–75, July, 1965.
24. Gibbs, H. J. and W. H. Holtz: "Research on Determining the Density of the Sands by Spoon Penetration Testing," *4th International Conference on Soil Mechanics and Foundation Engineering,* London, vol. I, p. 35.
25. Peck, R. B., W. E. Hanson, and T. H. Thornburn: *Foundation Engineering,* 2nd ed., J. Wiley & Sons, New York, 464 pp., 1974.
26. Terzaghi, K., and R. B. Peck: *Soil Mechanics in Engineering Practice,*, 2nd ed., J. Wiley & Sons, New York, 1967.
27. Seed, H. B., and I. M. Idriss: "Simplified Procedure for Evaluating Soil Liquefaction Potential," *Journal of the Soil Mechanics and Foundations Division,* ASCE, vol. 97, no. SM9, pp. 1249–1273, Sept. 1971.
28. U.S. Department of Energy: "Cone Penetrometer, Innovative Technology Report," U.S. DOE, April 1996.
29. Hryciw, R. D., A. M. Ghalib, and S. A. Raschke: "In-Situ Soil Characterization by VisCPT." *Proceedings of the 1st International Conference on Site Characterization (ISC '98),* Balema Press, pp. 1081–1091, 1988.
30. Strutynsky, A. I., and T. J. Sainey: "Use of Piezometric Cone Penetration Testing and Penetrometer Groundwater Sampling for Volatile Organic Contaminant Plume Detection," *Proc. 1990 Conf. on Petroleum Hydrocarbons and Organic Petroleum Inst.,* Houston, Nov. 1990.
31. Marchetti, S.: "In situ Tests by Flat Dilatometer," *Journal of the Geotechnical Engineering Division,* ASCE, vol. 106, no. GT3, pp. 299–321, March, 1980.
32. Palmer, C. M.: *Principles of Contaminant Hydrogeology,* Lewis Publishers, Chelsea, MI, 1992.
33. Freeze, R. A., and J. A. Cherry: *Groundwater,* Prentice-Hall, Englewood Cliffs, N.J., 1979.
34. HMCRI: "Two Technologies under Development for Access to Subsurface Contaminants," *Focus,* October 1989.
35. Deere, D. V.: "Technical Description of Rock Cores for Engineering Purposes," *Rock Mechanical and Engineering Geology,* vol. 1, pp. 16–22, 1964.
36. Freeze, R. A., and J. A. Cherry: *Groundwater,* Prentice-Hall, Englewood Cliffs, N.J., 1979.
37. Pekarun, O. O., C. H. Benson, and T. B. Edil: "Significance of Defects in Annular Well Seals," *Practice Periodical of Hazardous, Toxic and Radioactive Waste Management,* vol. 2, no. 2, pp. 65–71, April 1998.
38. Driscoll: *Groundwater and Wells,* 2nd ed., 1986.
39. Pierdinock, M. J. et al.: "Remote Control Test Pit Excavation in a Shock Sensitive Drum Zone at a Superfund Site," *Haztech International 90,* sponsored by the Institute for International Research, Inc., Pittsburgh, October 2–4, 1990.

40. Kerfoot, H. B., and L. J. Barrows: "Soil-Gas Measurement for the Detection of Subsurface Organic Contamination," EPA/600/S2-87/027, June, 1987.
41. Burns, R. B.: "Field Screening of Soil Vapors at a Subsurface Gasoline Contaminated Site in West Liberty, Morgan County, Kentucky," *Great Lakes 90,* sponsored by the Hazardous Materials Control Research Institute, Cleveland, September 26–28, 1990.
42. Godoy, F. E., and Naleid, D. S.: "Optimizing the Use of Soil Gas Surveys," *Hazardous Materials Control,* vol. 3, no. 5, pp. 23–29, September–October, 1990.
43. Marks, B. J., and Singh, M.: "Comparison of Soil-Gas, Soil, and Groundwater Contaminant Levels of Benzene and Toluene," *Hazardous Materials Control,* vol. 3, no. 6, pp. 40–45, November–December, 1990.
44. Lucero, D. P.: "A Soil Gas Sampler Implant for Monitoring Dump Site Subsurface Hazardous Fluids," *Hazardous Materials Control,* vol. 3, no. 5, pp. 36–44, September–October, 1990.
45. Morrison, R. D.: "Ground Water Monitoring Technology: Procedures, Equipment and Applications," Timco Mfg., Inc., Prairie du Sac, Wisc., 1983, 111 pp.
46. Gribb, M. M., J. Simunek, and M. F. Leonar: "Development of Cone Penetrometer Method to Determine Soil Hydraulic Properties," *Journal of Geotechnical and Geoenvironmental Engineering,* vol. 124, no. 9, pp. 820–829, September 1998.
47. Nielsen, D. M., ed., *Practical Handbook of Ground-Water Monitoring,* Lewis Publishers, Inc., Chelsea, Mich., 1991.
48. U.S. EPA: "A Compendium of Superfund Field Operations Methods," EPA/540/P-87/001, 1987.
49. Autennucci, J. C., et al.: *Geographic Information Systems,* Van Nostrand Reinhold, New York, 1991.
50. Rodenbeck, S. E., and M. L. Maslia: "Groundwater Modeling and GIC to Determine Exposure to TCE at Tucson," *Practice Periodical of Hazardous, Toxic and Radioactive Waste Management,* vol. 2, no. 2, pp. 53–61, April 1998.
51. Dimmick, Shelly, Oracle RDBMS Database Administrators Guide, Oracle Corp., 1990.
52. GDS Operating Manual, Electronic Data Systems Corp., Maryland Heights, Mo., 1992.

ADDITIONAL READING

Benson, R. C.: "Geophysical Techniques for Subsurface Site Characterization," Chap. 14 in *Geotechnical Practice for Waste Disposal,* D. E. Daniel (ed.), Chapman and Hall, 1993.

Clark, Keith C.: *Analytical and Computer Cartography*, Prentice-Hall, Englewood Cliffs, N.J., 1990.

Falotico, R. J. et al.: "The Necessity of Hydrogeologic Analysis for Successful In Situ Bioremediation," *Haztech International 90,* sponsored by the Institute for International Research, Inc., Pittsburgh, October 2–4, 1990.

Findley, L. A.: "Supplied Air Versus SCBA Respiratory Protection at Hazardous Waste Sites," *Great Lakes 90,* sponsored by the Hazardous Materials Control Research Institute, Cleveland, September 26–28, 1990.

Jensen, John R.: *Introductory Digital Image Processing—A Remove Sensing Perspective,* Prentice-Hall, Englewood Cliffs, N.J., 1986.

Keith, L. H., ed.: *Principles of Environmental Sampling,* ACS Professional Reference Book, Washington, D.C., 1988.

Kerfoot, H. B.: "Soil-Gas Sampling for Detection and Delineation of Subsurface Organic Contamination," *Great Lakes 90,* sponsored by the Hazardous Materials Control Research Institute, Cleveland, September 26–28, 1990.

Medine, A. and Anderson, eds.: *Environmental Engineering—Proceedings of the ASCE Specialty Conference,* Boulder, Colo., ASCE, New York, 1983.

Pye, V. I., Patrick, R., and Quarles, J.: *Groundwater Contamination in the United States,* University of Pennsylvania Press, Philadelphia, 1983.

Wilburn, D. J. et al.: "Applications of Geotechnical Software to Remedial Investigations and Feasibility Studies," *Haztech International 90,* sponsored by the Institute for International Research, Inc., Pittsburgh, October 2–4, 1990.

ACKNOWLEDGMENTS

The authors wish to acknowledge the following individuals who assisted with the preparation of this chapter for the first edition: Steve Fulton, Scott McQuown, Mike Moskow, John Harsh, William Douglas, *ERM;* Gary Rabik, *Sun Refining and Marketing Co.;* John Link, Ben Myers, Jon Terry, *Bucknell University;* Don Fries, Jim Nash, *Villanova University;* and Ngayan Warren, *Pennsylvania State University—Harrisburg*. The principal reviewer for the first edition was Ronald Landon, *ERM*.

The reviewer for the second edition was Paul Gruber, *ERM*.

CHAPTER 16

Remedial Technologies

Πέμπτον ἐπέταξεν αὐτῷ ἆθλον τῶν Αὐγείου
βοσκημάτων ἐν ἡμέρᾳ μιᾷ μόνον ἐκφορῆσαι τὴν
ὄνθον ...
Ἡρακλῆς ... τῆς τε αὐγῆς τὸν θεμέλιον διεῖλε
καὶ τὸν Ἀλφειὸν καὶ τὸν Πηνειὸν σύνεγγυς
ῥέοντας παροχετεύσας ἐπήλγαγεν,
ἔκρουν δὶ ἄλλης ἐξόδου ποιήσας.

The fifth labor [Eurystheus] laid on [Hercules] was to carry out the dung
of the cattle of Augeas in a single day by himself. . . .
Hercules made a breach in the foundations of the cattle yard and then,
after diverting the courses of the Alpheus and Peneus rivers,
which flowed near one another, he turned them [into the yard],
having [first] made an outlet [for the water] through another opening.

The Library of Apollodorus, c. 100 B.C.

16-1 INTRODUCTION

Past industrial and waste management activities have contaminated soil and ground water at tens of thousands of sites in both the United States and abroad. As described in Section 2-4, a nationwide effort to clean up the "sins of the past" began in earnest as a result of the Comprehensive Environmental Response, Compensation and Liability Act (CERCLA),[1] commonly known as Superfund. In the United States, more than 60,000 sites have been identified that are potential threats to the public health and/or the environment. Many of these sites include some risk of contaminant migration in the subsurface. By 1997, over 1400 sites had been listed for cleanup by their addition

to the National Priorities List under Superfund. In the United Kingdom,[2] a registry of sites with contaminated land uses was expected to generate a list of up to 100,000 sites. Interestingly, it was never completed. It is clear that a significant portion of our hazardous waste management resources have been, are, and will continue to be devoted to the assessment and cleanup of existing hazardous waste sites.

The cleanup of an existing hazardous waste site requires thorough site characterization, as described in Chapter 15, and an assessment of risk as described in Chapter 14. The analysis of remedial alternatives is described in Chapter 17. But first, a knowledge of the remedial alternatives that are available is required. In some cases, removal of the source itself may be enough to eliminate contaminant transport into the environment. However, more frequently, it is necessary to minimize the rate of off-site contaminant migration by employing containment technologies to minimize risk to public health and the environment. Containment technologies may be associated with other technologies (such as *in situ* remedial alternatives) to implement a long-term cleanup strategy for the site.

Successful remediation of sites with contaminated groundwater is one of the toughest challenges environmental professionals face. One of the key steps in evaluating such sites is determining whether receptors have the potential to be adversely impacted by contaminated groundwater. These receptors range from users of drinking water from wells to aquatic life in surface water bodies. In some instances, contamination in the groundwater will attenuate before it reaches the receptor. In other cases, it is possible to design and implement a groundwater treatment system that will reduce the groundwater contaminant concentrations to acceptable levels. However, given the extent and magnitude of the contamination, the site geology and hydrogeology, the contaminant characteristics, and/or the proximity of the receptor, sometimes neither of the previous approaches are effective in protecting potential receptors. Under such scenarios an effective way to treat the groundwater contamination (in the short and possibly long terms) is to contain and/or limit its migration. This chapter focuses on the available site remediation technologies including containment. Prior to this focused discussion, an introduction to the establishment of remedial objectives is given in Section 16-2.

As discussed in this chapter, there are many different technologies that can be applied to achieve the necessary environmental protection. One of the more basic and proved ways is to impede or redirect the groundwater flow through the use of a passive physical barrier such as a soil-bentonite slurry wall. The implementation and design of passive-barrier systems for control of contaminant migration are discussed in Sections 16-3 through 16-6.

There are other, more active approaches to site remediation such as the use of groundwater recovery wells and treatment of the pumped groundwater. Although **pump-and-treat** systems may not always be effective in reducing contaminant concentrations to acceptable regulatory levels, they can be very effective in controlling the migration of groundwater plumes. This and other active technologies are discussed in Section 16-7.

There are sites where the contaminants are situated above the watertable. In these cases, technologies such as soil vapor extraction (Section 16-8) may be applicable.

Another recent approach to site remediation that combines containment with treatment is the use of **permeable reactive treatment walls** that integrate *in situ* treatment zones with conventional barrier wall systems. This technology has the

combined effect of controlling groundwater flow and achieving some measure of groundwater cleanup. This is covered further in Section 16-9.

As with many of the remediation technologies discussed elsewhere in this text, there are many conventional and innovative approaches to containment and site remediation that are currently being used and explored. These innovations are introduced and briefly discussed in Section 16-10.

While much of the emphasis of this chapter is on the control of contaminant migration, the application of containment is often used in conjunction with an active groundwater treatment system. For example, a containment wall may be installed to control the movement of a groundwater plume, while at the same time employing an *in situ* treatment technology (such as biological treatment or air sparging) on the contained groundwater. Therefore, the application of an optimum remedial approach to containment would likely involve a combination of different active remedial technologies.

The reader will notice that remediation of contaminated groundwater is the focus of the remedial technologies described in this chapter. Of course, subsurface contaminants can be associated with the vapor phase of the unsaturated zone and the solids of the soil. To further complicate the picture, there can be phase transfer from one to another either intentionally or unintentionally during the implementation of the selected remedial technology. Perhaps the emphasis on remedial technologies is associated with the U.S. experience of the writers where contaminated groundwater is often the route of exposure and transport. In contrast, much of the emphasis in the United Kingdom is on the remediation (treatment) of contaminated soils. Many of the technologies discussed in this chapter are amenable to the full-scale treatment of contaminated soil, as is the focus in the United Kingdom.[3]

In many cases, containment can be viewed as a final solution for addressing groundwater contamination, especially with the advent of risk-based approaches. The primary focus of risk-based remedial approaches is to mitigate all risks associated with exposure to groundwater contamination by elimination of exposure pathways (i.e., migration in groundwater) between contaminant sources and potential receptors. An effective method for achieving this is through the use of horizontal or vertical barrier walls, which represent institutional controls in the form of engineered barriers.

16-2
REMEDIAL OBJECTIVES

The specific remedial objectives vary from site to site, state to state, and country to country. In the broadest sense, the objective of any cleanup program is to render the site safe for its intended future use. To do this the remediation scheme must minimize or eliminate the hazard to human health and the environment. How then do we minimize or eliminate this hazard? Is it necessary to remove the contaminants from the soil, water, and waste to effect a cleanup? If removal is selected as the cleanup approach, does this mean that every molecule of contaminant must be removed? Approaches to site cleanup are many and varied. In some cases removal is the most cost-effective and environmentally protective solution. In others, on-site treatment may be selected. In still others, containment may suffice. In all cases, the question of

risk must be addressed and the selection of remedial alternatives will be influenced by their ability to reduce risk.

The cleanup or remediation program must be sufficient to accomplish the objective, that is, render the site and surrounding environment safe for the intended use. This objective will guide the selection of the appropriate remedial action. In turn, each component of the remedial program will have goals to meet. For some sites, complete removal of the contaminants is the only suitable remedial objective. For others, control of the contaminant migration pathways is adequate (and cost-effective).[4] In reality, no site will be completely cleaned of each molecule of contamination. Equally impractical is the digging up of all sites and hauling the materials to a landfill because this "dig and pump" merely transfers the contamination to another location. For many sites, mere containment is also inadequate. At present, the cleanup objectives depend on a myriad of social, technical, and political factors. This chapter focuses on the technical aspects of site remediation. Additional discussion of the establishment of objectives is included in Chapter 17.

One approach to the selection of the appropriate remedial action program is known as **risk-based corrective action** (RBCA). This approach merges the benefits of risk-based decision making (see Chapter 14) with the other elements of the corrective action process.[5] For petroleum release sites alone, it is believed that RBCA could save billions of unnecessary expenditures that do not contribute to risk reduction, while speeding the cleanup and closure process. In the United States alone, there are over 900,000 active petroleum storage tanks registered with the EPA and over 350,000 confirmed petroleum releases since 1980. Site remediation has been completed at less than 200,000 sites and currently (1999) there are about 30,000 new releases each year.[6] RCBA is a three-tiered process where each tier represents a different complexity and degree of conservativeness in assessing risk. The lowest tier represents the most conservative approach but requires the least tier of effort for the assessment. Advancement to a higher level requires increasing data and analysis, but a more economical corrective action may result as conservative assumptions of the lower level are replaced with site-specific information.

Europeans are also pursuing a risk-based approach to the assessment of contaminated land sites as a basis for determining the necessary remedial action. The Concerted Action on Risk Assessment for Contaminated Sites (CARACAS) program is led by regulatory authorities and is aimed at developing frameworks and procedures for assessing and managing the risks posed by contaminated sites.[7] A similar, industry-led initiative, Network of Industrially Contaminated Land in Europe (NICOLE), seeks to employ risk-based methods to protect human health and the environment while implementing such remediation as needed to make the site "fit for use."

How much data are needed from the remedial investigations to select, design, and implement a remedial action program? Suffice it to say that all the desired information is never available; the site is never fully characterized. In scientific terms, we must recognize a certain degree of **uncertainty.** How can uncertainty be dealt with? One approach, tried and proved in the geotechnical engineering area, is the **observational approach.**[8] Application of this approach to hazardous waste site remediation requires the recognition of uncertainty and as such suggests[9]: (1) remedial design based on the most probable site conditions, (2) identification of possible deviations from these conditions, (3) identification of parameters to monitor and confirm condi-

tions and performance of the remedial design, and (4) contingency plans for potential deviation. Recognizing uncertainty, and application of the observational approach, offers the best opportunity to achieve the desired environmental protection at the lowest cost with the least risk.

Containment is a potential component within remedial systems because this technology may be used for site remediation, either alone or with other remedial technologies to accomplish the desired objectives in a cost-effective manner. As the principles of each technology are described, their place in the universe of available methods is identified. To systematically discuss remediation in this text, the technologies are classified as either active or passive remedial system components.

- Active system components require considerable effort and ongoing energy input to operate (for example, pumping wells).
- Passive system components work without much attention, except maintenance (such as revegetating a cover soil).

This framework enables us to examine systems by function and differentiate between systems performing the same function.

16-3 PASSIVE CONTAMINANT CONTROL SYSTEMS

The function of passive contaminant control systems in site remediation is to eliminate exposure pathways (i.e., groundwater migration), minimizing contaminant transport rates. To examine this function, consider the potential pathways for contaminant migration at an uncontrolled hazardous waste site (see Figure 16-1). Pre-

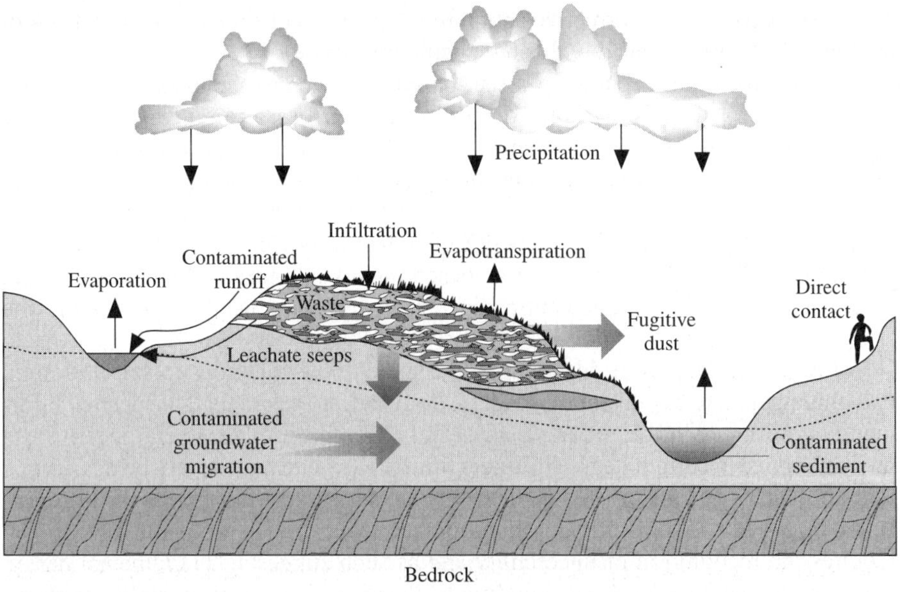

FIGURE 16-1
Pathways for contamination.

cipitation at the site either runs on or off, infiltrates, or returns into the atmosphere through evapotranspiration. Precipitation runoff that encounters waste or contaminated soil may transport contaminants or contaminated sediment into the surrounding environment. Precipitation may also infiltrate into and through the waste, generating leachate. Migration of leachate may introduce contaminants into the surface water and/or groundwater. Contaminants present in leachate migrating to the groundwater are then transported through the groundwater environment and may be discharged to surface water in rivers, lakes, ponds, etc.

Passive contaminant control technologies focus on controlling hydrologic pathways for contaminant migration. This approach is often termed **containment.** Because of uncertainties associated with design, construction, and long-term reliability of containment facilities, containment as a sole means of site remediation is not often used for sites with significant levels of contamination. For sites of high waste volume but low risk, containment may be the remedial alternative of choice. In fact, the EPA has established containment as the presumptive remedy for CERCLA municipal landfills.[10] Containment is also used at such sites in conjunction with source removal containment and institutional controls.

The technologies for the control of the contaminant transport pathways are discussed with respect to their function within the remedial program. For example, it is necessary to control the impact of precipitation to prevent it from becoming contaminated and to control the hydraulic transport of contaminants. Thus, the text first discusses surface water control technologies and then groundwater control technologies.

The selection, design, and construction of containment technologies must recognize each of the exposure and transport processes. Groundwater and contaminant transport processes, described in Chapter 4, include advection, dispersion (including both molecular diffusion and mechanical mixing), adsorption, retardation, and biological and chemical transformation. The selected remedial technology must consider these transport processes to assess its effectiveness in controlling contaminant migration to the surrounding environment.

16-4 SURFACE WATER CONTROL TECHNOLOGIES

Figure 16-1 depicts the need to prevent precipitation from transporting contaminants off site by surface water runoff or by leachate generation. Covers/caps, surface water diversions, and sedimentation and erosion control systems are employed for this purpose. Such systems that prevent/mitigate off-site transport of contamination are also incorporated into the design of new land disposal and storage facilities.

Although surface water control technologies are often just one component of the overall site remediation system, site conditions may occasionally allow surface water control technologies to be the sole means of site remediation, particularly in countries other than the United States where the remedial technology for a contaminated site is selected to be consistent with the intended site usage. For example, consider a site, underlain by deep chalk deposits, having concentrations of residual metals in the soils from prior site activities. Site redevelopment for continued industrial usage could involve covering the site to preclude direct contact, eliminate contaminated runoff,

and minimize/eliminate infiltration and leaching to reduce the risk and provide the appropriate level of protection for public health and the environment.

One type of cover system not discussed in Chapter 13 is the **capillary barrier.** Studies have indicated that capillary barriers may limit percolation in semiarid and arid regions.[11] A capillary barrier uses a finer-grained soil overlying a coarse-grained soil. As a result of the particle size difference, downward migration of moisture is limited. There are several benefits for arid and semiarid regions. Since a capillary barrier does not have the optimum moisture conditions of a compacted clay layer, its susceptibility to desiccation cracking is reduced. Further, since moisture control is not necessary, as it is for compacted clay barriers (see Section 13-5), construction costs are reduced. In a capillary barrier, moisture is stored in the upper layer of finer-grained material where it is later removed by evapotranspiration. Modeling of the capillary barrier performance requires the use of a model that considers the unsaturated flow conditions, precipitation, and evapotranspiration interactions. Studies have shown that field performance of a capillary barrier is reasonably well predicted by using the appropriate model and the appropriate laboratory and field data.[12]

How are surface water control technologies integrated into the overall site remediation scheme? The answer is that it depends on the other components of the remedial system. For example, the remedial program at Love Canal incorporates surface covers in conjunction with leachate/groundwater collection and treatment systems. The cover reduces infiltration and eliminates the problems of direct contact previously evident. Surface water control technologies have already been presented and discussed in the context of land disposal facilities (refer to Section 13-4 for additional detail regarding surface water control technologies). In a recent survey of remedial technologies employed in Europe and elsewhere, the use of cover systems is a commonly employed method to protect sites from contamination.[13]

16-5
GROUNDWATER CONTROL: VERTICAL BARRIERS

Subsurface vertical barriers are employed to contain contaminants and to redirect groundwater flow. These barriers to horizontal contaminant transport and groundwater flow used in hazardous waste management have developed from conventional engineering applications such as dewatering of excavations and control of groundwater flow through and beneath dams and levees. A number of vertical barrier techniques are available including slurry trench cutoff walls (made of soil-bentonite, cement-bentonite, or plastic concrete), grout curtains (made of cement-based or chemical grouts), and steel sheet piling.

It is useful to first establish the possible functions of a vertical barrier in the framework of the site remediation system. One function of a vertical barrier wall is to provide containment to prevent migration of contaminants in the groundwater. By employing a vertical barrier wall around the site, the rate of contaminant migration is slowed, particularly in anisotropic aquifers and formations having a horizontal hydraulic conductivity much higher than the vertical hydraulic conductivity. Passive containment systems have gained a great deal of acceptance in the United States in recent years as a method of hazardous waste site remediation. This trend is continuing as pump-and-treat performance data (see Section 16-7) indicates

significant limitations of pump-and-treat systems in site remediation. Passive barrier wall systems are used in conjunction with conventional pump-and-treat systems to minimize the required pumping rates as well as reduce the rates of chemical migration. For such combinations of pump-and-treat and containment, after contaminant levels have been reduced the vertical barrier wall left in place will in the long term serve as a passive barrier to reduce the rate at which any remaining contaminants may migrate into the subsurface environment.

A common function of a vertical barrier in a remediation system is to inhibit the flow of clean groundwater into the site. In this context, vertical barriers are commonly used in conjunction with groundwater extraction and treatment systems, as shown in Figure 16-2. Without the installation of vertical barrier, clean groundwater would flow beneath the site, become contaminated, and increase the quantity of water for the pump and treat component of the remedial system. With the installation of a vertical barrier, clean regional groundwater is prevented from entering the pump-and-treat system.

Vertical barriers also provide groundwater control during construction where excavation into the subsurface is required for direct waste treatment, waste removal, or construction of liner systems. The vertical barrier initially serves to expedite construction. Subsequently, the vertical barrier remains to impede off-site migration of any residual contamination.

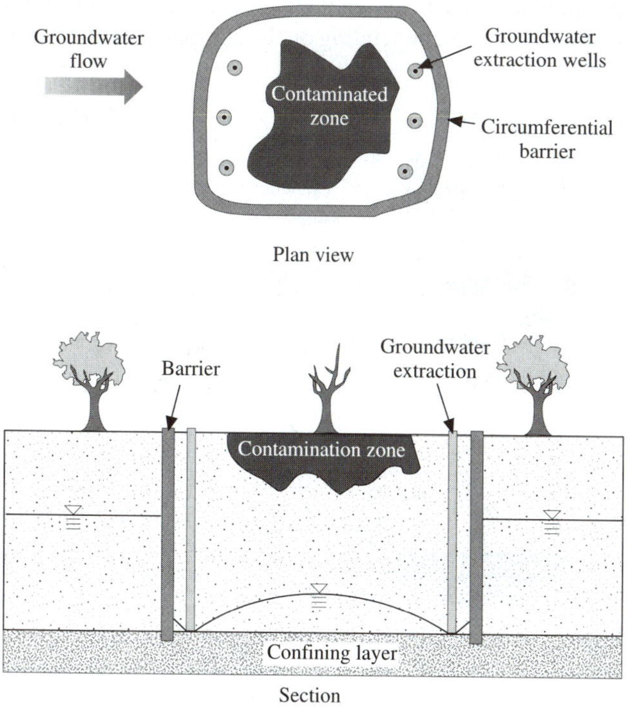

FIGURE 16-2
Circumferential cutoff wall.

Vertical barriers to control horizontal groundwater flow and contaminant migration can be configured in several ways. The most common configuration is circumferential, where the cutoff wall completely surrounds the waste or the waste site as shown in Figure 16-2. Vertical barrier walls may also be located upgradient or downgradient of the site, depending on the intended function of the barrier.

The principal purpose of an upgradient vertical barrier wall (Figure 16-3) is to provide a cutoff controlling the influx of clean groundwater flow from upgradient regions. When used in conjunction with a pumping system located on the down-gradient side (and no vertical barrier on the down-gradient side of the site), groundwater withdrawal may recover contaminants that have migrated down-gradient of the site. In this way, contaminants that have held site boundaries in total disregard may be captured and returned to the site contaminant treatment program.

The principal purpose of a down-gradient cutoff wall, as shown in Figure 16-4, is to utilize site groundwater flowing beneath the site to actively flush contaminants from beneath the site. A down-gradient wall is effective in aiding the pump-and-treat system to speed removal of contaminants from the subsurface. This method of vertical barrier wall deployment was used for remediation at the Rocky Mountain Arsenal Site.

Vertical barrier walls are typically embedded, or **keyed,** into a material of low permeability beneath the site (Figure 16-5), although this is not always necessary. In the case of product that is lighter than water (for example, from a leaking underground fuel oil storage tank) the vertical barrier does not need to fully penetrate the aquifer into the underlying aquitard (see Figure 16-5b). However, it is also important to consider that groundwater has the potential to mound either on the upgradient side of the barrier or within the barrier, depending on the comparative rates of infiltration and

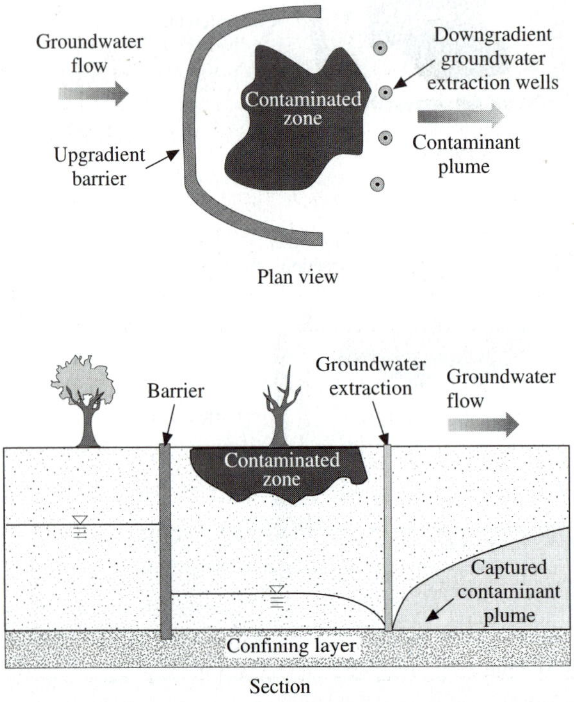

FIGURE 16-3
Upgradient vertical barrier.

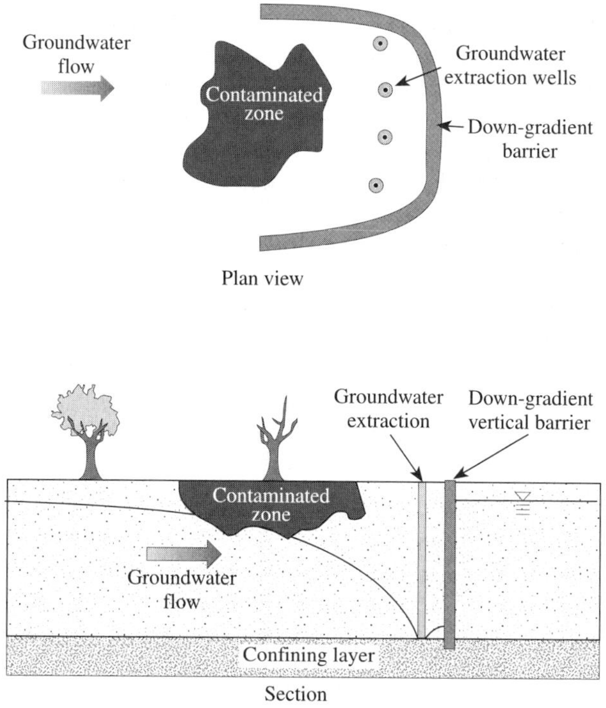

FIGURE 16-4
Down-gradient vertical barrier.

exfiltration. This phenomenon of groundwater mounding will increase the hydraulic head, resulting in enhanced downward migration of groundwater. The degree of mounding and subsequent vertical migration of groundwater, therefore, will need to be carefully considered in the barrier design (particularly if the barrier is not keyed into an underlying aquitard).

An evaluation of the permissible rate of contaminant transport through or across the barrier is essential to the considerations of barrier technology. At present, it is typical to simply specify a barrier as having a hydraulic conductivity of 1×10^{-7} centimeters per second (cm/s) or less. This conductivity is chosen because, with careful design and construction, it is the lowest that can be readily and economically achieved. Also, calculations for some representative chemical and gradient conditions have shown that contaminant transport is predominately diffusion controlled for low gradient conditions combined with hydraulic conductivity values lower than 1×10^{-7} cm/s. However, calculations of site-specific flow and contaminant transport rates through the barrier may demonstrate for many projects that a barrier having a hydraulic conductivity 1×10^{-6} or 1×10^{-5} cm/s may be suitable. This change in hydraulic conductivity requirement widens the choices available and potentially reduces the overall cost of the wall. It is usually necessary to employ groundwater modeling to determine the anticipated impact of the barrier. For example, in one study of a proposed slurry trench cutoff wall, groundwater modeling of the alternatives demonstrated that the presence of the vertical barrier was ineffective in reducing off-

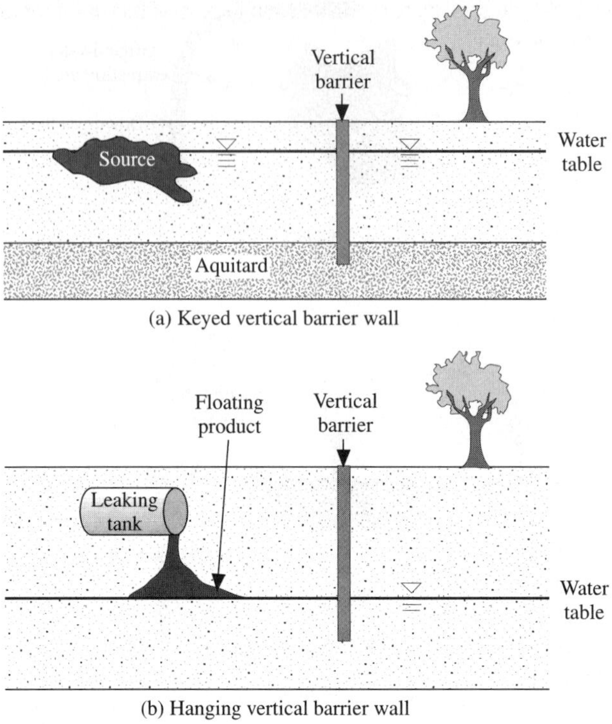

FIGURE 16-5
Embedment of vertical barrier walls.

site contaminant loadings because the contaminant transport was virtually entirely in the fractured rock underlying the overburden. This led the designers to conclude that alternative remedial technologies were more appropriate for the specific site and subsurface conditions.

Another example where site-specific modeling could affect the barrier wall design is the situation where molecular diffusion through the barrier wall represents the significant transport process (see Section 4-2). Recall molecular diffusion may be the dominant transport process for ionic or molecular constituents where high concentration gradients (and low hydraulic gradients and/or low hydraulic conductivity values) across the barrier are observed. This process is especially important for contaminants where the remedial objectives dictate very low acceptable concentrations.

EXAMPLE 16-1. VERTICAL BARRIER HYDRAULIC CONDUCTIVITY. A 0.75-m (30-in) thick circumferential barrier is planned to control the influx of clean groundwater to extraction wells. The cost-effective treatment plant capacity is 3.2×10^{-4} m^3/s (5 gal/min). The perimeter is 900 m (3000 ft) in length and the aquifer is 9 m (30 ft) thick. An inward gradient across the barrier wall will be maintained by lowering the water level within the containment by 1.5 m (5 ft). What must the hydraulic conductivity of the wall be to control the influx through the wall to 3.2×10^{-4} m^3/s (5 gal/min)?

Solution. Assume one-dimensional flow reasonably models the flow conditions and apply Darcy's law (Equation 4-3). (Attention nonbelievers—construct a flow net to

remove any doubt you may have in the reasonableness of this assumption.)

$$q = kiA$$

where $q =$ flow rate $= 3.2 \times 10^4\ m^3/s$

$k =$ hydraulic conductivity (cm/s)

$i = \Delta h/\Delta L = 1.5\ cm/0.75\ cm = 2.0$

$A = 9\ m \times 900\ m = 8100\ m^2$

Rearranging terms:

$$k = q/iA$$

Substituting:

$$k = (3.2 \times 10^{-4}\ m^3/s)/(2.0 \times 8100\ m^2)$$
$$= 2.0 \times 10^{-8}\ m/s = 2 \times 10^{-6}\ cm/s$$

At present, considerations of contaminant transport for the design of vertical barriers are usually limited to advective flow. As described in Section 4-2, a thorough analysis should include advective and diffusive flow and account for the soil-contaminant interactions utilizing diffusion and retardation coefficients. In one study the breakthrough time for carbon tetrachloride at an interior concentration of 1 mg/L was found to be about 2 years with a conventional soil-bentonite wall.[14,15] The study found that the breakthrough time increased to 30 years when the wall was designed with a high-carbon fly ash composite because of the adsorptive capacity of the carbon. These findings point to the importance of the consideration of contaminant transport in designing a cutoff wall.

The vertical barrier wall function and configuration are coupled with site and subsurface conditions to select the suitable types of vertical barrier. For example, grout curtains are appropriate for vertical barriers in fractured rock whereas slurry trench cutoff wall techniques typically require softer materials that lend themselves to soil excavation technologies. It is noted, however, that the use of soil-bentonite slurry trench cutoff walls in Superfund site remediation appears to be the most common.[16] Guidance as to the most appropriate type of vertical barrier for specific subsurface conditions will be provided later in the chapter as each of the following techniques is discussed in more detail:

- Soil-bentonite slurry trench cutoff walls
- Cement-bentonite slurry trench cutoff walls
- Plastic concrete cutoff walls
- Diaphragm (structural) cutoff walls
- Vibrating beam cutoff walls
- Deep soil-mixed walls
- Composite cutoff walls
- Grout curtains

Soil-Bentonite Slurry Trench Cutoff Walls

A soil-bentonite slurry trench cutoff wall, often known simply as a **slurry wall,** takes its name from the method of construction. To construct this vertical barrier

in the subsurface, a trench is first excavated below the existing grade utilizing a liquid slurry of bentonite and water to maintain the trench stability (Figure 16-6). The trench is then backfilled with a mixture of soil, bentonite, and slurry to form the completed cutoff.

The slurry is composed of approximately 95 percent water and 4 to 6 percent bentonite by weight. Bentonite is a sodium smectitic clay (the structure is shown in Figure 11-7). The resulting bentonite-water slurry is a viscous fluid with a unit weight of approximately 64 to 70 pounds per cubic foot (lb/ft^3) (a density of 1.03 to 1.12 g/cm^3).

Trench collapse is controlled by the resulting hydrostatic force system in which the slurry counteracts the active soil pressures. Because a positive fluid pressure is maintained within the trench, the slurry exfiltrates through the walls of the trench, a filter cake is formed, and the filtrate enters into the formation. The filter cake is a very thin layer of fully hydrated bentonite, forming an impermeable boundary. The fluid pressure of the slurry opposes the active earth pressure to maintain the trench stability. Figure 16-7 is a photograph of an excavation showing the construction of a slurry trench. The backhoe shown in this photograph is excavating to construct a vertical cutoff through an embankment surrounding a liquid impoundment. The excavated material is being case along the side of the trench for subsequent use in the preparation of the backfill. The excavation is made in the slurry-filled trench, and the excavation sidewalls are essentially vertical. The slurry maintains the trench stability and permits vertical excavations to depths of more than 30 m (100 ft).

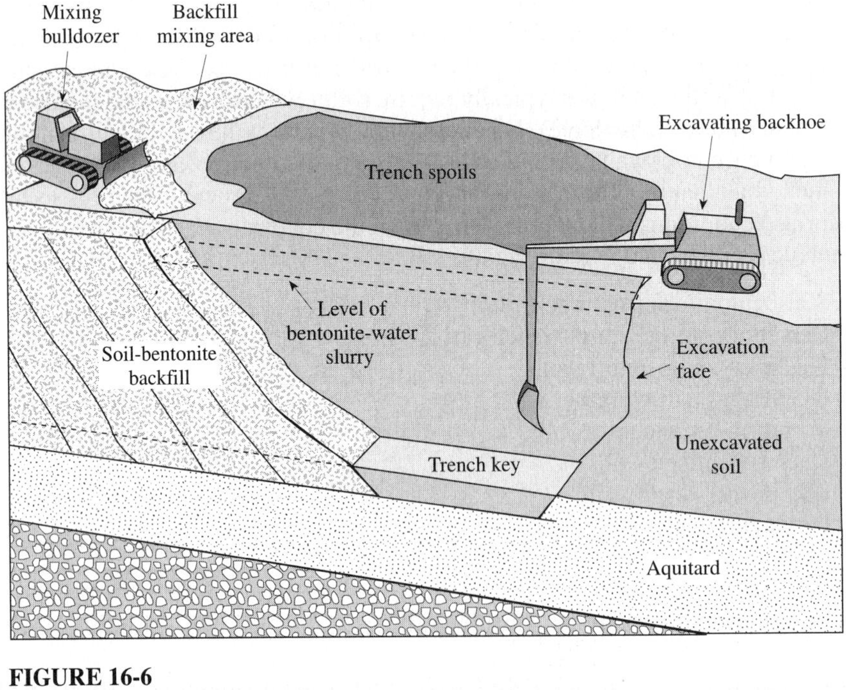

FIGURE 16-6
Soil-bentonite slurry trench cutoff wall construction schematic.

FIGURE 16-7
Slurry trench excavation.

EXAMPLE 16-2. SLURRY UNIT WEIGHT. Calculate the expected unit weight (lb/ft^3) of the bentonite-water slurry for a batch using one bag of bentonite (100 lb) per 200 gal water. The unit weight of water is 62.4 lb/ft^3 and the specific gravity G_s of the bentonite is 2.77. What is the bentonite content in percent as defined by $(W_{\text{bentonite}}/W_{\text{water}}) \times 100$? Why is the problem presented in such mixed units of pounds and gallons?

Solution. Bentonite content:

$$\% \text{ bentonite} = 100 \text{ lb}/[(200 \text{ gal}/7.48 \text{ gal/ft}^3) \times 62.4 \text{ lb/ft}^3] = 6.0\%$$

Determination of fluid unit weight:

$$G_s = W_s/V_s\gamma_w \qquad \text{(definition of specific gravity)}$$

where W_s = weight of solids
V_s = volume of solids
γ_w = specific weight of water

Rearranging and substituting gives

$$V_s = W_s/G_s\gamma_w = 100 \text{ lb}/(2.77 \times 62.4 \text{ lb/ft}^3) = 0.58 \text{ ft}^3$$

Volume of water

$$V_w = 200 \text{ gal}/7.48 \text{ gal/ft}^3 = 26.74 \text{ ft}^3$$

Weight of water

$$W_w = V_w/\gamma_w = 26.74 \text{ ft}^3 \times 62.4 \text{ lb/ft}^3$$
$$= 1668 \text{ lb}$$

Fluid unit weight

$$\gamma_f = W_T/V_T = (1668 + 100)/(26.74 + 0.58)$$
$$= \underline{\underline{64.7 \text{ lb/ft}^3}}$$

The units are representative of mix proportions used by contractors in the field.

Straightforward and well-documented methods of calculating trench stability are available.[17,18,19] Trench instability is modeled as an active soil wedge prone to slide into the trench. Resisting this instability is the strength of the soil adjacent to the excavation and the fluid pressure exerted by the bentonite-water slurry against the sidewalls of the trench. The filter cake is the vertical surface along which these forces must be balanced. Equations of equilibrium can be written for the driving forces (the active soil wedge) and the resisting forces (the soil strength and the bentonite fluid pressure).

Defining the factor of safety as the ratio of the resisting forces to the driving forces yields:

For clays:

$$F = 4S_u/H(\gamma - \gamma_f) \tag{16-1}$$

For sands:

$$F = 2\sqrt{(\gamma \times \gamma_f)} \times (\tan\phi)/(\gamma - \gamma_f) \tag{16-2}$$

where F = factor of safety
S_u = undrained shear strength (cohesion)
H = depth of the cutoff wall
γ = total unit weight of the soil
γ_f = unit weight of the bentonite-water slurry
ϕ = angle of internal friction

These equations can be used with either SI or U.S. customary units, as long as units are consistent.

EXAMPLE 16-3. FACTOR OF SAFETY. Calculate the factor of safety for a trench 9.1 m (30 ft) deep excavated in sand having an angle of internal friction of 30. The unit weight of the fluid is 10.2 kN/m^3 (64.7 lb/ft^3) as calculated above. The sand has a total unit weight of 20.4 kN/m^3 (130 lb/ft^3).

Solution

$$F = 2\sqrt{(\gamma \times \gamma_f)} \times (\tan\phi)/(\gamma - \gamma_f)$$

In U.S. customary units,

$$\begin{aligned}\gamma &= 130\\ \gamma_f &= 64.7\\ \phi &= 30\\ F &= 2\sqrt{(130 \times 64.7)} \times (\tan 30)/(130 - 64.7)\\ &= \underline{\underline{1.6}}\end{aligned}$$

In SI units,

$$\begin{aligned}\gamma &= 20.4\\ \gamma_f &= 10.2\\ \phi &= 30\\ F &= 2\sqrt{20.4 \times 10.2} \times (\tan 30)/(20.4 - 10.2)\\ &= \underline{\underline{1.6}}\end{aligned}$$

Note that, for sands, the factor of safety is independent of trench depth.

The trench excavation is commonly completed by a backhoe with a modified boom to depths of up to about 18 m (60 ft). Below this depth, it is common to use equipment such as drag lines and clam shells to excavate the trench. Table 16-1 summarizes the excavation equipment used for the construction of slurry cutoff walls.

The second part of the construction process for a soil-bentonite slurry trench cutoff wall involves placing soil-bentonite backfill (see Figure 16-6). After the trench excavation is completed to the entire depth, the bentonite-water slurry is displaced by the soil-bentonite backfill. The soil-bentonite backfill at a consistency of high slump concrete is placed from the surface and moves (oozes) forward (in the direction the excavation is proceeding). The soil-bentonite backfill is a low-permeability material composed of soil, water, and bentonite. The bentonite content of the backfill may be as low as 1 percent when there is an adequate presence of natural fines or may be as high as 4 to 5 percent.

Backfill excavated from the trench is often mixed along the side of the trench with bentonite-water slurry. For additional control over mixing, a remote batch process is utilized. The mixed backfill is transported by truck to the trench for placement. For waste containment applications, this is the recommended method of mixing because the backfill mixing can be better controlled in a remote mix area than along the trench.

On occasion, backfill has been mixed at a high enough water content to make the material pumpable. However, at the higher water content required to make the material pumpable,[16] the material is considerably more compressible. For hazardous waste containment applications, pumping of backfill is not recommended.

The hydraulic conductivity of soil-bentonite backfill of field mixed samples typically varies in laboratory tests from a high of 1×10^{-5} cm/s to a low of 1×10^{-9} cm/s. In general, however, the hydraulic conductivity of soil-bentonite backfill ranges from 1×10^{-7} cm/s to 1×10^{-8} cm/s as shown for an example project in Figure 16-8.

Hydraulic conductivity testing can be accomplished relatively rapidly in on-site laboratories using a rigid wall consolidometer/permeameter to achieve results comparable to triaxial tests[20] (see Figure 16-8). The determination of hydraulic conductivity must be at the same gradient and consolidation pressure as expected in the field because soil-bentonite backfill hydraulic conductivity is stress-dependent.[21]

TABLE 16-1
Slurry trench excavation techniques

Type	Trench width, m	Trench depth, m	Comments
Standard backhoe	0.5–1.5	10	Rapid, least costly
Backhoe with extended boom	0.5–1.5	18	Rapid and low cost (extended excavation boom on unmodified backhoe)
Modified backhoe with extended boom	0.5–1.5	30	Extended excavation boom; modified power and counterweights
Clamshell	0.3–1.5	75	Attached to a crane
Dragline	1.0–3.0	50	Slow; for wide, deep excavations
Rotary	0.5–1.5	75	Slow; for boulders and rock
Specialty	Varies	Varies	For example, Hydrofraize (from Soletanche)

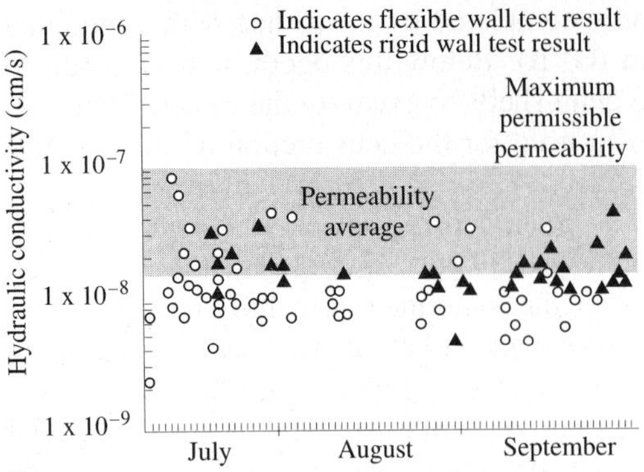

FIGURE 16-8
Soil-bentonite backfill permeability test results. (*From* G. K. N. Hayward-Baker.[20])

Note from Figure 16-9 that the hydraulic conductivity is not strongly related to the bentonite content. This is because soils of a well-graded nature (coarse to fine sands and gravels with silt and clay) tend to be quite impermeable when homogenized and mixed with bentonite-water slurry. In contrast, clean sands may require the addition of considerable dry bentonite to the mix to achieve a low hydraulic conductivity. For any particular soil, there is typically a relationship between bentonite content and hydraulic conductivity. The hydraulic conductivity is related to the presence and type of naturally occurring fine-grained materials as shown in Figure 16-9.[22] A base soil with 20 percent or more natural clay fines should be able to produce a soil-bentonite backfill with a hydraulic conductivity less than 1×10^{-7} cm/s.

In order of importance for environmental application, characteristics to be designed into the soil-bentonite backfill include:

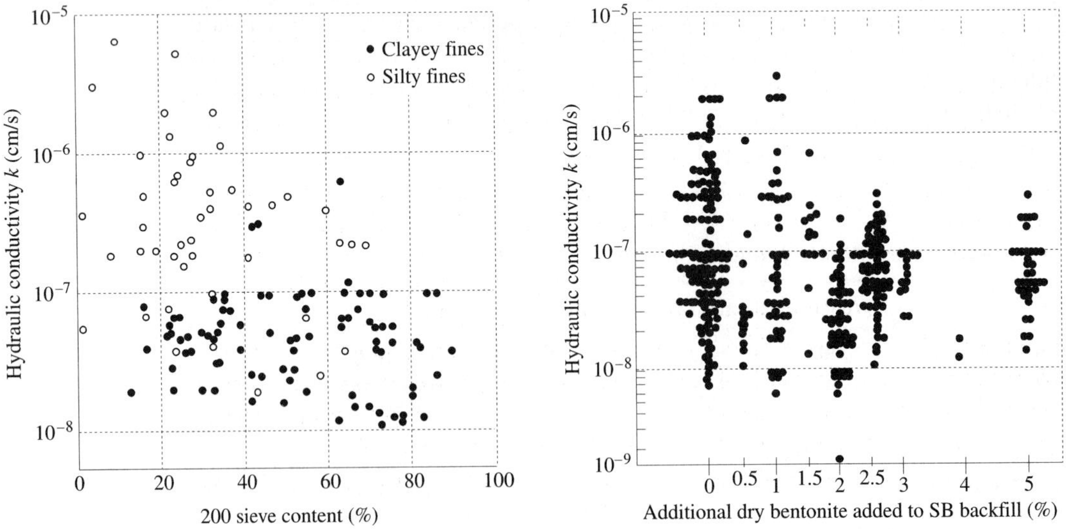

FIGURE 16-9
Soil-bentonite backfill hydraulic conductivity. (*From* Ryan, 1987.[22])

1. Chemical compatibility
2. Low permeability
3. Low compressibility
4. Moderate strength

In addition to these backfill characteristics, the backfill at the time of construction must flow freely into the trench and displace the bentonite-water slurry used to maintain trench stability. The most important soil parameters that affect these characteristics are grain size distribution and the water content of the backfill. Soil-bentonite backfills have been successfully fabricated from materials varying from clean sand to highly plastic clay. However, considering the containment of hazardous waste, backfill requirements are necessarily stricter than those used for conventional dewatering applications. The difference in requirements results from two fundamental differences in expected performance. First, conventional dewatering applications may not require the same degree of "perfection" because these systems are temporary. Small leaks may be inconsequential from a dewatering standpoint whereas from a hazardous waste containment standpoint they may be significant. Second, because these dewatering applications are short term, considerations of long-term changes in the hydraulic conductivity in the cutoff wall are not necessary. Specifically, for hazardous waste containment applications, compatibility of the subsurface contaminants with the cutoff wall material must be evaluated to determine the potential for alteration of the barrier permeability due to prolonged contact with the contaminants.

For environmental applications the backfill must be designed to minimize any potential increases in hydraulic conductivity. Generally, a lower hydraulic conductivity is required for hazardous waste containment than for conventional dewatering applications. These goals of long-term stability with low hydraulic conductivity can be best achieved by fabricating the backfill from a well-graded soil blended with soil-bentonite slurry. This well-graded soil should contain all particle size ranges including fine gravel; coarse, medium, and fine sand; silt; and clay. A recommended particle size range is shown in Figure 16-10. Because the coarser granular materials in a well-graded backfill material are in point-to-point contact, a relatively low compressibility results. Further, with the well-graded nature of the material, the pores (which become progressively finer) are filled with progressively finer material (silts and clays), resulting in a material with a relatively low hydraulic conductivity. Finally, by using a well-graded material, the potential for long-term changes in hydraulic conductivity due to soil-waste interactions is minimized. The components of a well-graded soil are relatively stable (sand, silt, and low-activity clay). Clay of low activity (such as kaolinite) is not subject to dramatic permeability increases[23] as those of high activity (such as bentonite) are.[24] The activity is defined as the plastic index normalized with respect to the clay fraction. For example, bentonite is considered to be a high-activity clay since it is subject to a high degree of shrinking and swelling. In contrast, kaolinite is a low-activity clay since it has a much smaller range of shrinking and swelling. This approach differs from simply adding more and more bentonite to a poorly graded soil to achieve a low-permeability backfill. The swelling of the bentonite is potentially reversible in the presence of contaminants (resulting in permeability increases over the long term).

As previously described, the use of a well-graded backfill coupled with the addition of bentonite, i.e., the bentonite-water slurry, to produce a backfill of low

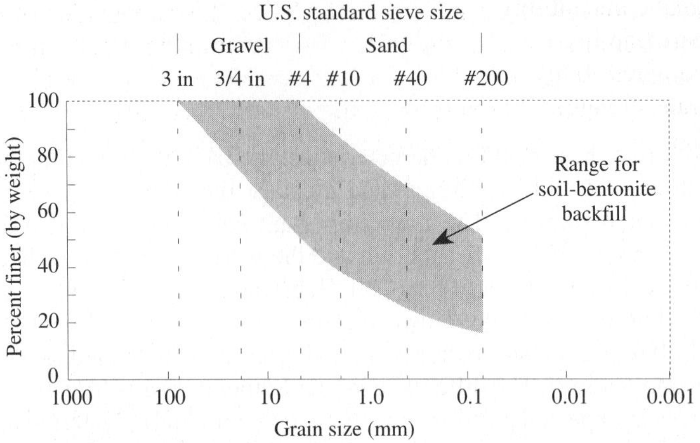

FIGURE 16-10
Grain size distribution for soil-bentonite backfill.

hydraulic conductivity is recommended. The reversible nature of the swelling of the bentonite may result in hydraulic conductivity increases in the presence of contaminants. As a worst case for hazardous waste containment applications, the hydraulic conductivity of the backfill may increase to the hydraulic conductivity of the soil without the bentonite. By selecting a well-graded material, a relatively low hydraulic conductivity is found in the base soil, similar to the relatively low hydraulic conductivity of naturally occurring well-graded glacial tills. The addition of bentonite-water slurry contributes bentonite and thus reduces the already low permeability of the material. Further, the slurry provides moisture and enhances the fluidity of the backfill to allow it to behave as the high-slump concrete necessary to displace the bentonite-water slurry used in the trench for trench stability.

Some designers historically considered filter cake as a significant factor in the low permeability aspect of the completed soil-bentonite slurry cutoff wall. However, for environmental applications, the added benefit of the low-permeability cake should be considered as an additional safety factor. This conservative approach results from several factors. First, the integrity of the filter cake cannot be ensured after backfill placement. As the backfill slides along, the backfill may scrape all or part of the filter cake off the trench wall. Second, the filter cake consists of high-swelling bentonite. It is well documented that bentonite swelling is reversible in the presence of many contaminants.[25] Thus, in the hazardous waste containment environment, the filter cake permeability may increase dramatically as a result of the presence of contaminants in the subsurface. Thus, it is recommended that any filter cake contribution to the low-permeability nature of the wall be considered an additional safety factor.

Cement-Bentonite Slurry Trench Cutoff Walls

Cement-bentonite slurry trench cutoff walls are excavated under a head of slurry to maintain trench stability. In the United States, the slurry is usually composed of water, cement, and bentonite. In the United Kingdom, both cement and slag are used in the mixture.[26] Unlike soil-bentonite slurry cutoff walls, however, the slurry

is left to harden in a trench. As the cement hydrates, the combination of the water, cement, bentonite, and sand in suspension from excavating results in a hardened material with a consistency of stiff clay or stronger. Thus, the cement-bentonite wall is a single-phase process. Because the excavated soils are not used in a backfill, the technique generates excess excavation materials for disposal. Although slurry trench cutoff walls of soil-bentonite can be consistently prepared with a hydraulic conductivity 1×10^{-7} cm/s or less, cement-bentonite cutoff walls in the United States generally have a hydraulic conductivity of approximately 1×10^{-5} to 1×10^{-6} cm/s (Figure 16-11). In the United Kingdom, where slag is used in the mixture, the hydrated hydraulic conductivity is typically less than 1×10^{-7} cm/s. As for the U.S. data shown in Figure 16-11, the cement/water ratio is typically between 0.15 and 0.25, resulting in a cutoff wall that has a very low solids content; the wall is mostly water. Thus, cement-bentonite may be more prone to increases in hydraulic conductivity in response to the contaminated fluids, although limited data in this regard are available. Generally speaking, for hazardous waste containment applications, cement-bentonite cutoff walls are rarely used in the United States.

In the literature, much has been written about the soil-bentonite slurry trench cutoff walls so widely used in the United States. In the United Kingdom, a specification for the construction of slurry trench cutoff walls as barriers to pollution migration has recently been published. The document summarizes the state of their practice in specification format.[27] A comparison of cement-bentonite slag walls widely used in the soil-bentonite barriers of U.S. practice reveals advantages and disadvantages of each approach.[28] Although the U.S. data on cement-bentonite cutoff walls constructed without slag indicate a permeability generally too high for hazardous waste applications (around 1×10^{-5} cm/s), the U.K. formulation with slag appears to overcome this deficiency. Studies of the performance of cementitious barriers in contaminated environments conducted by using immersion tests showed that some contaminants can degrade U.K.-style barrier walls.[29] This is comparable to findings with U.S.-style soil-bentonite barriers. It has been found that contamination resistance can be improved with the addition of fly ash to cement-bentonite mixtures.[30]

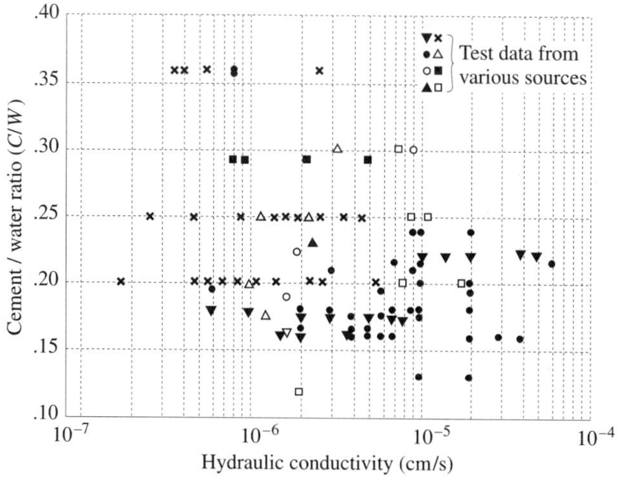

FIGURE 16-11
Cement-bentonite permeability test results.

Plastic Concrete Cutoff Walls

Another technique that may be utilized for containment systems is the plastic concrete slurry trench cutoff wall. How does a plastic concrete trench differ from a cement-bentonite trench? Plastic concrete is designed as a mixture of cement, bentonite, water and aggregate. Cement-bentonite does not include aggregate and has a much higher water content. A second difference is that the plastic concrete trench is usually excavated in panels (see the Diaphragm (Structural) Cutoff Wall section, below), and thus the plastic concrete is placed in the slurry-filled trench to replace the bentonite-water slurry used for excavation. For example, one project used a formulation of cement, pulverized fly ash, bentonite, and water to form the plastic concrete and employed the panel method of excavation to construct the cutoff to a depth of 50 m below grade.[31] Studies reveal that plastic concrete has relatively low hydraulic conductivity, less than 1×10^{-7} cm/s. Further, the shear strength is significantly greater than soil-bentonite or cement-bentonite. Finally, on the basis of the limited data available to date, plastic concrete may be more contaminant resistant than alternative techniques.[32] Although plastic concrete is stronger and less permeable than soil bentonite, it is more costly and has not found widespread use for hazardous waste applications to date in the United States.

Diaphragm (Structural) Cutoff Wall

The deepest cutoff wall built in the United States, 60 m, was constructed as a diaphragm wall.[33] In this technique the trench is excavated under a head of bentonite-water slurry to maintain trench stability. However, it is excavated in panels, each as much as 6 m long, using a clamshell or similar excavating tool. Steel reinforcement is then set into the bentonite-water slurry. The slurry is then displaced by high-slump concrete placed by the tremie method of concrete placement. A schematic of the method is shown in Figure 16-12.

The diaphragm technique is widely used in applications requiring significant wall strength, such as where a cutoff wall is to be part of a structural earth retention system.[34] The hydraulic conductivity is very low, 1×10^{-8} cm/s or lower. Because

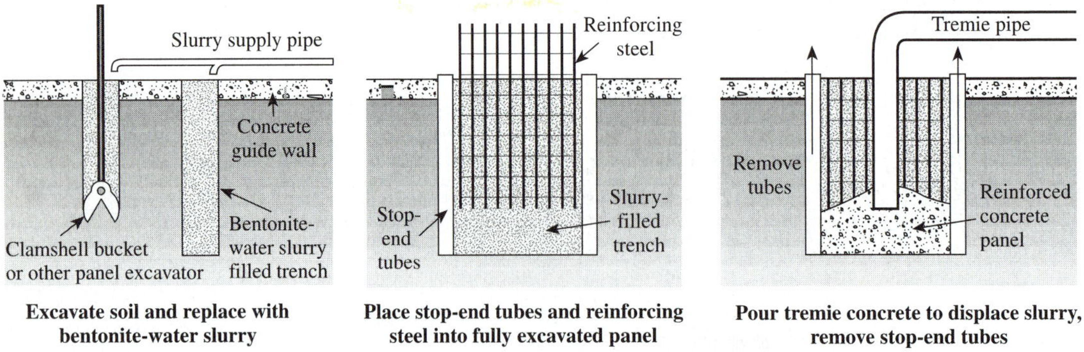

FIGURE 16-12
Schematic of diaphragm slurry trench cutoff wall construction.

of its high cost the diaphragm wall has not, as yet, been used in hazardous waste management applications in the United States.

Vibrated Beam Cutoff Walls

Vibrated beam cutoff walls have been employed to construct vertical barriers for the horizontal control of contaminant migration. The vibrated beam cutoff wall, constructed as shown in Figure 16-13, is essentially a grouting method appropriate for soils. A vibratory pile driver is used to advance a specially modified H beam into the subsurface. The pile is specially equipped with injection nozzles at the tip. During driving, grout is injected to lubricate the driven pile. As the beam is withdrawn, a void is left equal to the size of the beam. The void is filled with grout pumped in through the injection nozzles as the beam is withdrawn. Subsequent beam penetrations are overlapped, resulting in a continuous barrier with a typical thickness of 5 to 8 cm. The slurry may be cement-bentonite (the most commonly employed) or specialty grouts such as bituminous mixtures. The principal advantage of this system is the elimination of the need to excavate potentially contaminated materials. Excavation of contaminated soils as in the slurry trench methods described earlier results in associated health, safety, and disposal issues and costs.

The vibrated-beam method does not inspire confidence in the integrity of the completed cutoff wall. First, the cutoff wall is relatively thin, only a few centimeters. Second, the location of the beam tip is uncertain, particularly with deep penetration. Although it has been demonstrated to be successful in sands of shallow depths,[35] it is difficult to penetrate dense soils for any significant depth. Further, even sands that may be initially loose are densified by the vibratory action of the pile driver, resulting in difficulty in subsequent penetrations. This may lead to the need to preauger to achieve beam penetration.[36]

Deep Soil-Mixed Walls

Deep soil mixing has been employed to construct vertical barriers to horizontal groundwater flows.[22] This technique was originally developed in Japan and was subsequently employed in the United States. A special auger mixing shaft was de-

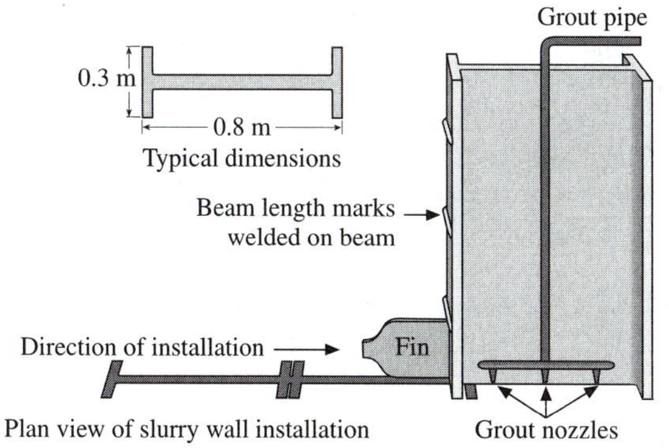

FIGURE 16-13
Vibrated-beam cutoff wall.

veloped that is rotated into the ground while it simultaneously enables the injection of a slurry of bentonite and water or cement, bentonite, and water (see Figure 16-14). The installation sequence, shown in Figure 16-15, results in a column of treated soil when multiple mixing shafts are employed. Reinforcement can be added to the treated soil columns if additional strength is required.[37] A continuous wall, typically 0.5 to 0.9 m wide, is obtained by overlapping penetrations. Because the bentonite is added as bentonite-water slurry during mixing, the quantity of bentonite may be limited to about 1 percent.

FIGURE 16-14
DSM augers used to construct a vertical barrier.

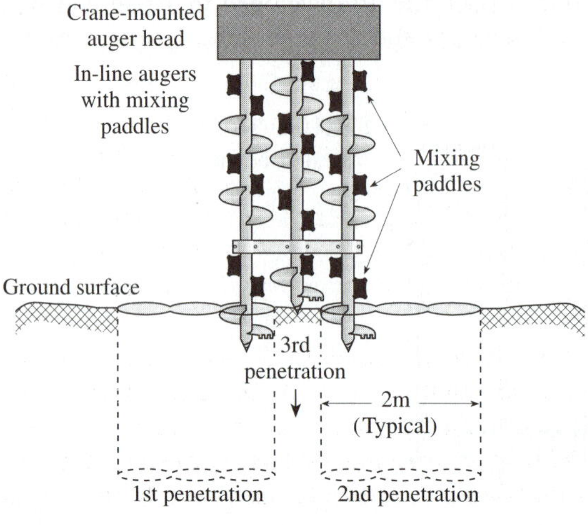

FIGURE 16-15
Deep soil-mixed cutoff wall.

The advantages of the deep soil-mixing technique are similar to those in the vibrated-beam method. Health and safety risks are minimized by elimination of the need to excavate materials as required for conventional slurry trench cutoff wall techniques. The deep soil-mixed wall can be constructed with added confidence because the wall is considerably wider than the vibrated-beam wall. The hydraulic conductivity approaches 10^{-7} cm/s or less.

EXAMPLE 16-4. DEEP SOIL-MIXED WALL. You are the field inspector for a deep soil-mixed wall constructed to a depth of 9 m and a width of 0.75 m. The final cement content is to be 5 percent by dry weight. The dry density of the soil *in situ* averages 1800 kg/m^3. The augers are configured to mix 2.3 m of wall per penetration. How much cement needs to be added in each penetration?

Solution. Calculate weight of soil to be mixed per penetration:

$$W_t = 9 \text{ m} \times 0.75 \text{ m} \times 2.3 \text{ m} \times 1800 \text{ kg/m}^3 = 27{,}000 \text{ kg}$$

Cement weight at 5 percent:

$$W_c = W_t \times 0.05 = 27{,}900 \text{ kg} \times 0.05 = \underline{\underline{1400 \text{ kg per penetration}}}$$

Composite and Active Vertical Barriers

Vertical barrier walls have been constructed in a combination of materials to result in a composite barrier. A geomembrane barrier within a slurry trench cutoff wall is one such composite. The composite cutoff wall properties include lower hydraulic conductivity and increased contaminant resistance.

Although construction techniques vary among contractors, two installation methods are:

1. The geomembrane sheet mounted on an installation frame is lowered into the trench, disconnected from the frame, and the frame is withdrawn.
2. The geomembrane sheet is pulled in by weights on the geomembrane bottom.

It is reported that placement of a geomembrane in a cement-bentonite slurry trench cutoff wall decreases the overall permeability by 2 orders of magnitude. If high-quality joining of geomembrane sheets can be achieved, this decrease in permeability can be 4 to 5 orders of magnitude.[38]

For example, to reduce the potential for the migration of methane gas from a landfill into adjacent land planned for residential construction, a slurry trench cutoff wall was built of soil-bentonite having a hydraulic conductivity of less than 1×10^{-6} cm/s and incorporating a 40-mil HDPE geomembrane liner added to the upper portion of the wall to further reduce gas migration.[39] The geomembrane was considered necessary above the water table where the soil-bentonite is subject to wetting and drying cycles as well as potential desiccation drying and cracking (see Figure 16-16).

While the composite vertical barriers described above focus on combinations of materials to reduce the hydraulic conductivity of the wall, other combinations of materials have been investigated to reduce the overall rate of contaminant transport through the walls. They are termed **low-permeability reactive barriers.** Low permeability vertical barriers discussed earlier in this chapter have long been used as

key components in site remediation. These cutoff walls have been installed widely as passive hydraulic barriers and their design is based predominantly on considerations of hydraulic conductivity. Recent work suggests that the enhancement of conventional soil-bentonite cutoff walls with mineral additives such as chabazite and attapulgite can improve their capacity as active treatment zones while maintaining their desirable characteristic of low hydraulic conductivity.[40,41] In another study, a soil-bentonite wall enhanced with zero-valent iron reduced the diffusive flux of TCE across the wall by over 10 orders of magnitude.[42] Low-permeability reactive barriers represent a major shift away from the present approach, where the wall is designed as a hydraulic barrier, to an improved approach, where the wall is designed as a treatment zone as well as hydraulic barrier. This approach is also markedly different from "funnel-and-gate" technology that includes high-permeability treatment zones.

Grout Curtains

Slurry trench cutoff wall technology is limited to soils that can be readily excavated. When it is necessary to install a vertical barrier to horizontal groundwater flow in material that cannot be readily excavated (e.g., rock), then alternative technologies must be examined. Grout curtains are the most appropriate and commonly used alternative for these conditions. A grout curtain is installed in rock through the pressure injection of a pumpable material into the pores and fractures within the rock as shown in Figure 16-17. The grout then gels or sets in place, resulting in a material of lower permeability filling the voids and fractures. Grout may be made from a variety of materials including cements, silicates, pozzolans, and various chemical formulations.

Grouting technology utilized in hazardous waste containment systems has been developed for conventional geotechnical engineering applications such as beneath dams and levees to reduce seepage.[43] Grouting has been used both successfully and unsuccessfully[44] in hazardous waste management applications. Detailed studies have been conducted to show that the resistance of grout to attack depends on the type of grout, the nature of the contaminant, and their compatibility.[45]

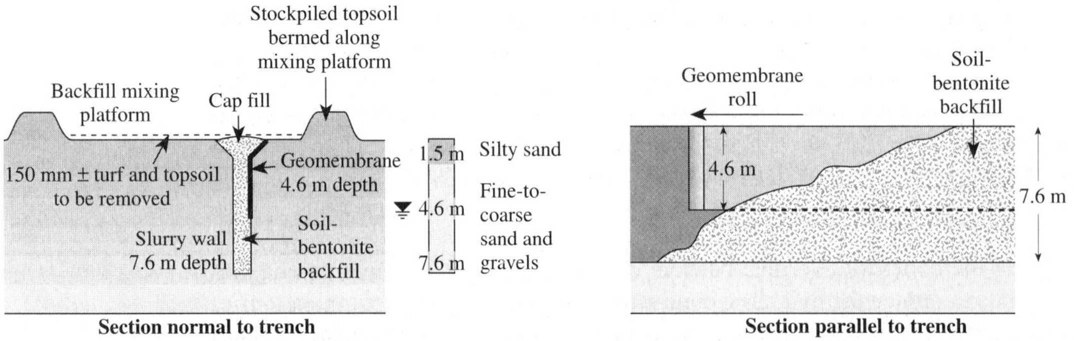

FIGURE 16-16
Composite cutoff wall.

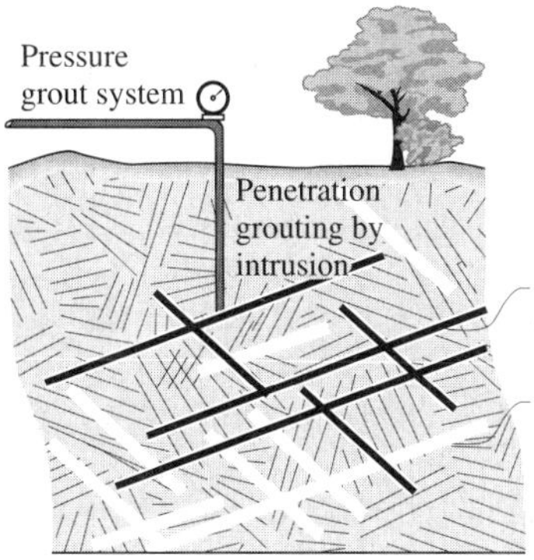

FIGURE 16-17
Grouting of rock fractures.

For example, in a laboratory study of sand grouted with colloidal silica, it was found that inclusion of up to 5 percent by volume of selected organic contaminants (tetrachloroethene, carbon tetrachloride, or aniline) had little effect on either the hydraulic conductivity or strength of the cured grouted sand samples.[46] These tests simulate the inevitable mixing of contaminants with the grout during the construction process. Immersion tests were also conducted, which indicated some degradation due to immersion in aniline. Immersion tests simulate the presence of NAPLs in contact with the barrier material. A site-specific testing program using site contaminants and proposed grouting materials should be conducted for projects where a grouted barrier is to be constructed.

Performance of Vertical Barrier Walls

The performance of vertical barrier walls is critical to the system performance, since the vertical barriers are a key component in assuring the design objectives of the containment system are being met. Evaluation of barrier wall performance is not only critical in the design and construction stages, but is also a key component of both short-term and long-term monitoring. The long-term performance of containment systems is an important component of the performance since the required design service lives of barrier walls typically range from as little as 10 years to more than 1000 years for radioactive waste storage structures.

The management of hazardous wastes requires an assessment of risk. Typically, the longer the service life of the containment system, the greater the probability for system failure (i.e., the greater likelihood for the migration of chemicals outside the barrier wall) and, therefore, the higher the overall risk. One must recognize the potential for construction defects and long-term property (i.e., wall hydraulic conductivity) changes. The resulting design and construction procedure must minimize these risks. The cause of failure of containment systems such as slurry trench cutoff walls can be classified as construction defects or postconstruction property changes.

Construction defects are those anomalies in the barrier wall that can result in localized greater rates of contaminant migration. Figure 16-18 depicts the potential construction defects for soil-bentonite slurry trench cutoff walls. Nonhomogeneous backfill results from improper mixing and includes lumps of unmixed soil and pockets of free slurry not fully blended with the soil. During backfill placement slurry

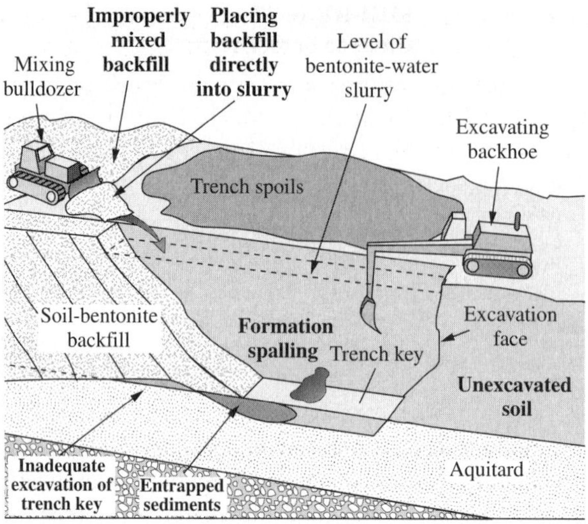

FIGURE 16-18
Potential defects in soil-bentonite slurry trench cutoff walls.

may become entrapped as a result of dropping of backfill into the trench directly through the slurry. Slurry can also become entrapped as the backfill moves forward,[47] and the potential for this slurry entrapment increases with increasing wall depth.[48] If the trench is allowed to remain open for too long, sediment may accumulate and become trapped as part of the backfill. The trench sediment is coarse-grained with higher permeability than the designed backfill. Trench sediment may accumulate during periods of no activity such as overnight or over weekends and holidays. In addition, there may be a cave-in of the trench side or spalling from the formation resulting in materials falling from the side walls that may not be visible from the surface. Any of these contributors to nonhomogeneous backfill (see Figure 16-18) results in a pocket or **window** within the wall that could act as a preferential pathway for contaminant migration across the wall.

Figure 16-18 also indicates the potential for inadequate excavation of the key. Regardless of the type of vertical barrier, an inadequate key can result in a preferential flow path beneath the wall. With regular soundings and inspection of the excavated soil, this defect can normally be avoided. Keying into rock is more difficult, and, unless measures are taken (such as preblasting along the alignment), reduced cutoff wall performance may result because of flow beneath the wall. The Sylvester site (also known as Gilson Road) is an example of this.[44]

In addition to construction defects, long-term property changes in the barrier material must be examined. Property changes can result from:

- Cycles of freezing and thawing
- Cycles of wetting and drying
- Desiccation of the cutoff wall material
- Chemical incompatibility

Cycles of freezing and thawing increase the hydraulic conductivity of the cutoff wall. This can be prevented with a properly designed and constructed final cover (see Section 13-4). Cycles of wetting and drying are inevitable with a fluctuating

water table and is of particular concern where groundwater fluctuations are evident (i.e., tidal areas). Drying can cause the bentonite to shrink, and increases in hydraulic conductivity will result. Since the swelling is reversible, it can be argued that the bentonite will again swell and the hydraulic conductivity drop on rewetting. Portions of the cutoff wall that are semipermanently above the water table may desiccate, resulting in increases in the hydraulic conductivity. Research has documented these permeability increases and results indicate that, even when the material is resaturated, the permeability remains higher than that of the materials that have remained saturated since placement.[49] For hazardous waste management applications, special attention must be given to chemical incompatibility. For example, significant increases in permeability have been seen in the laboratory when soil-bentonite backfill is permeated with concentrated organic permeants.[50] These tests simulate a barrier wall in the presence of dense nonaqueous-phase liquids (DNAPL). In these cases, designing the backfill to use well-graded materials as previously described is particularly important. In all cases, site-specific compatibility testing was performed using worst-case groundwater from the site with potential barrier wall mixtures.

Performance of vertical barrier walls is closely associated with the specific site remediation expectations and objectives. It is important to remember that each site remediation comprises a unique set of specific conditions and objectives that may impact the criteria by which the barrier wall will be constructed and evaluated. Site-specific criteria are also of key concern since barrier wall technologies are commonly used in conjunction with other remedial technologies at complex hazardous waste sites. Therefore, barrier wall performance may be influenced by other remedial system components. Key site-specific criteria that can affect both the construction and performance of barrier systems include:

- Site-specific land use
- Site topography
- Equipment accessibility
- Site-specific geologic characteristics
- Site-specific hydrogeologic characteristics
- Geotechnical considerations

Both the current and future land use of the site, as well as topographic relief, play directly into the feasibility of the design and construction of barrier wall systems. Certain land uses such as underground utilities, underground storage tanks, or high-traffic or congested areas will make the implementation of barrier wall systems difficult and limit their overall effectiveness. Topographic features such as steep slopes and surface water features can greatly impact the feasibility of implementing barrier wall systems. Site-specific considerations in terms of equipment accessibility are also a key concern. This would include the proximity of overhead electric lines, buildings, and underground utilities. Site-specific geologic characteristics include the heterogeneity of the subsurface geologic materials and the presence and characteristics of key formations such as the hydraulic conductivity of the key depth, likelihood of fractures, and the continuity of the key across the contaminated area. Hydrogeologic characteristics include the hydraulic conductivity of geologic units, depth to groundwater, fluctuations in groundwater levels (potential for dessication), and compatibility of the barrier wall materials with the general groundwater chemistry and site contaminants as previously discussed. Geotechnical considerations

would include the suitability of the *in situ* materials for use as backfill as well as the stability of the geologic material in the construction of the slurry wall trench.

Construction quality assurance (QA) and quality control (QC) are critical overall factors for the success of a barrier wall project. The purpose of QC is to generate sufficient data to enable the contractor to make judgments about the suitability of materials and workmanship and to allow for modifications in design/construction of the barrier wall based on site-specific information, if required. The responsibility of collecting and evaluating QC data typically rests with the contractor. QA provides a means by which the project stakeholders can verify that the contractor has satisfied the design criteria and minimize the overall risk involved with the performance of the barrier wall system. For example, QA/QC tests and observations for a soil-bentonite slurry trench cutoff wall can include the following:

- Tests of excavation stability (slurry viscosity, density and filtrate loss, soundings)
- Tests of backfill mix (slump, moisture, density, grain size, permeability)
- Observations of backfill mixing and placement (homogeneity, unsuitability)
- Observations of the excavation alignment and depth (surveys, soundings)
- Ground movement along the trench (elevation surveys, inclinometers)
- Connection with the key unit (soundings, observation of cuttings)

Of these parameters, it is especially critical that the barrier wall is keyed into the selected units, that the barrier material properties meet the specifications, and that there are no discontinuities in the barrier wall.

16-6
GROUNDWATER CONTROL: HORIZONTAL BARRIERS

Passive contaminant control systems include horizontal barriers as well as vertical barriers. Like vertical barriers, horizontal barriers in the subsurface are employed to contain contaminants and to redirect groundwater flow. The subsurface horizontal barriers to vertical contaminant transport and groundwater flow used in hazardous waste management can be categorized by their installation requirements. There are those that can and cannot be constructed *in situ*. "Constructed *in situ*" means that the barrier may be constructed beneath an existing waste site or contaminant plume. Most horizontal barriers must be constructed prior to the placement of overlying waste or contaminants. These horizontal barriers (or liner systems) were developed from conventional groundwater control applications such as canals, dams, and water storage impoundments. Liner systems for newly constructed waste management and facilities are discussed in Chapter 13. The number of horizontal barrier techniques available for *in situ* application at existing sites is quite limited.

It is necessary to establish the function of a horizontal barrier in the framework of the site remediation system. One function of a horizontal barrier wall is to provide containment for contaminants. By employing a horizontal barrier constructed *in situ* beneath the site, the rate of contaminant migration is reduced. This is particularly important in sites having dense, nonaqueous-phase liquids (DNAPL). Groundwater pumping systems may not adequately control the downward migration of DNAPL in response to gravitational forces. Passive containment of contaminants by horizontal

barriers (i.e., landfill liners) is widely employed as a method of hazardous waste management—site remediation can be achieved by excavating wastes and contaminated soil and disposing of them in a landfill. In cases where the quantities requiring disposal are large, the landfill can be newly designed and constructed on-site for the sole purpose of containing the site's contaminants.

Section 13-4, Linear and Leachate Collection Systems, provides guidance to the construction of horizontal barriers for the disposal of contaminants on site. The range of liner systems and the design and construction considerations are the same whether the liner is for a new facility to accept off-site waste or a new facility to accept only on-site materials from the remediation program.

By far, the bottom barrier materials most widely employed for site remediation are naturally occuring geologic formations of low hydraulic conductivity. As such, these barriers are not engineered. Rather, the emphasis is on the verification of the continuity of such strata and the evaluation of their *in situ* hydraulic properties. The ideal naturally occurring horizontal barrier is a continuous geologic formation of low hydraulic conductivity and great thickness. Where such stata can be found, they are the barriers of choice given the limitations of horizontal barriers construction *in situ* discussed below.

Grouted Liner Systems

The technology of grouting is well established for the control of water in conventional civil engineering applications such as seepage beneath dams and other water-retaining structures. The grouting technology for liners is the same as was previously discussed for grout curtains. The difference relates to the design of the grout holes. For a grout curtain, the grout is injected along a linear alignment. For a grouted liner, the grout is injected uniformly in plan view to the desired depth beneath the site. Grouting for horizontal barriers has been successfully used to control the influx of groundwater in excavations.[51] These applications make it possible to verify the effectiveness of the method. It is expected that grouted horizontal barriers will soon be used for waste containment applications. As noted above, where naturally occurring materials of low permeability, such as clayey soil or intact rock, underlie the site, the naturally low hydraulic conductivity is the preferred bottom containment barrier.

Block Displacement Technique

Another conceptual technique for the placement of a horizontal barrier is the block displacement technique. This method employs the principles of hydraulic fracturing commonly used to enhance oil recovery. Hydraulic fracturing of the subsurface materials occurs when the fluid pressure exceeds the total stress. The application of hydraulic fracturing to the installation of a horizontal barrier is shown in Figure 16-19. The block displacement technique is actually a special form of grouting—high pressures are used to fracture the soil, and grout is then pumped into the fracture to create the horizontal barrier. The technique has been demonstrated in pilot studies but has not been employed in actual site remediation.[52]

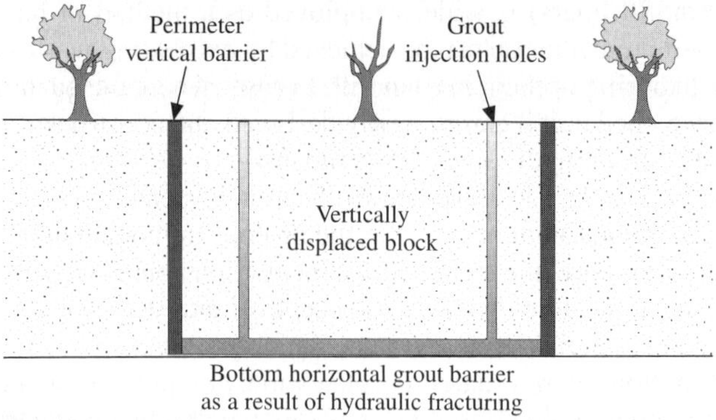

FIGURE 16-19
Block displacement technique.

Lagoon-Sealing Techniques

In some cases the leakage from the bottom of lagoons can be reduced by the installation of a horizontal barrier at the lagoon bottom constructed as the lagoon is still in service. The material used to construct this *in situ* horizontal barrier is bentonite (see Section 11-3). Bentonite, a high-swelling clay, can be produced in a granular form. Each granule is similar to a particle of sand in size but composed of thousands of colloidal-sized bentonite clay particles. When spread over the surface of a lagoon, the granules sink and form a blanket over the lagoon bottom. In time, and in the correct aqueous environment, the bentonite particles hydrate, resulting in significant swelling. The in-place swelling process yields a liner of low permeability.

The principal constraint on this system is the chemistry of the lagoon liquids. Liquids that are acidic, organic, or with high electrolyte concentrations inhibit the hydration of the bentonite and impede the formation of a low-permeability barrier. In any case the completed barrier is not protected from erosion should the lagoon experience significant turbulence. In addition the ability to inspect the completed liner and ensure its performance is limited. Consideration of these constraints usually limits the use of lagoon sealing techniques to nonhazardous applications.

16-7
ACTIVE REMEDIATION SYSTEMS

In addition to the various passive contaminant control systems (which are installed and then function without further ongoing energy input), active systems control and contain contaminant migration. These systems require ongoing energy input and include such systems as pump-and-treat, electrokinetics, *in situ* biotreatment, and soil washing. As noted earlier, active systems such as those described below are frequently used in conjunction with passive containment systems. The following sections of this chapter describe remedial systems that actively clean up the site.

Pump-and-Treat

Hazardous waste management sites have often employed **pump-and-treat** technologies to remediate contaminated groundwater. In this versatile approach, groundwater is first extracted by any one of a number of recovery methods including wells, well points, and drain tile collection systems.

The objectives of pump-and-treat systems include (1) hydraulic containment of contaminated groundwater and (2) removal of chemical mass from the groundwater. Hydraulic containment systems are designed to provide long-term containment of contaminated groundwaters at a low cost by optimizing well placement and minimizing pumping rates.

Once the groundwater is extracted from the subsurface, any number of water treatment methods, described in Chapters 9 and 10, such as air stripping, carbon adsorption, or biological treatment for organics and physical-chemical methods for inorganics may be employed. Thus, pump-and-treat schemes are developed on a site-specific basis for a range of site conditions and a range of contaminant types and concentrations.

Monitoring of active groundwater pump-and-treat systems, however, has revealed a number of limitations in this remedial method.[53] Chemical concentrations detected in groundwater extraction wells typically have shown an initial rapid decrease at early times of groundwater extraction followed by a decreased rate of chemical removal. This phenomenon is best illustrated in Figure 16-20, which shows the chemical concentrations in extraction wells at the Chem-Dyne site from 1987 through 1992.[54] Influent groundwater chemical concentrations show an initial rapid decline from approximately 10 to 2 mg/L over the first year of operation (1987 to 1988). Concentration from 1988 to 1991, however, indicated a minimal decline from 2 to 1 mg/L. These results indicate that the duration of operation of pump-

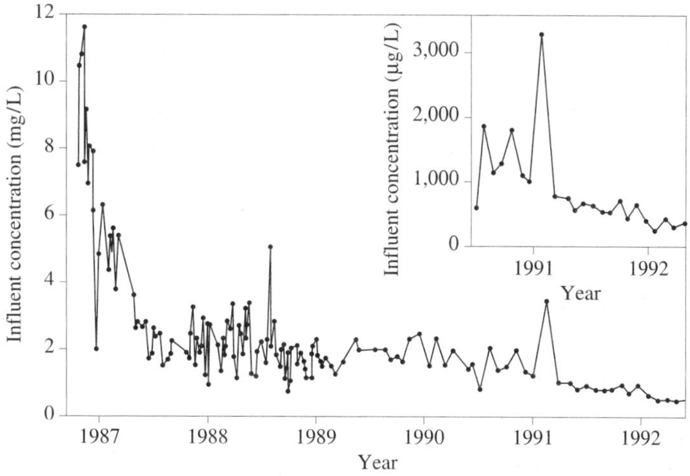

FIGURE 16-20
Pump-and-treat performance data for Chem-Dyne Site, 1987–1992 (not to scale).

and-treat systems could be extensive making the potential costs of these methods quite high. Another limitation of pump-and-treat systems is the rebound of chemical concentrations due to diffusion and desorption processes once the extraction wells are shut down. In many cases, this rebound effect is significant enough to warrant additional groundwater extraction. On the other hand, provided the annual cost is manageable, pump-and-treat techniques can protect human health and the environment adequately for the present and allows for development of improved remedial technologies for potential future remediation.

Although groundwater pump-and-treat systems have shown limited value as a mass removal remedial technology, they are often effective in maintaining hydraulic control and containment of contaminated groundwater. Hydraulic containment of groundwater plumes is often preferred when restoration of groundwater is technically impracticable, often because of the presence of subsurface NAPL, varied types of chemical contaminations, geologic heterogeneities, or other potential factors.[54]

The portion of the aquifer that is influenced by groundwater pump-and-treat systems is often referred to as the capture zone. The capture zone of an extraction well refers to that portion of the subsurface containing groundwater that will ultimately discharge to the well. The capture zone of the well is not coincident with the drawdown zone of the well's influence (i.e., zone of influence) since groundwater flow lines can be diverted by the influence by the pumping well yet not be captured by the well (Figure 16-21). The extent of the zone of influence depends largely on the transmissivity (hydraulic conductivity $\times$ saturated thickness) and pumping rate under steady-state conditions.

However, the extent of the capture zone depends on the regional hydraulic gradient as well as the pumping rate and transmissivity. Three critical parameters used to delineate a capture zone are: (1) the location of the extent of the capture zone down-gradient of the well (stagnation point), (2) the width of the capture zone opposite the well, and (3) the width of the capture zone infinitely far up-gradient from the well (Figure 16-22).

Capture zone delineation often involves the use of complex mathematical saturated groundwater flow models in order to define the capture zone geometry for

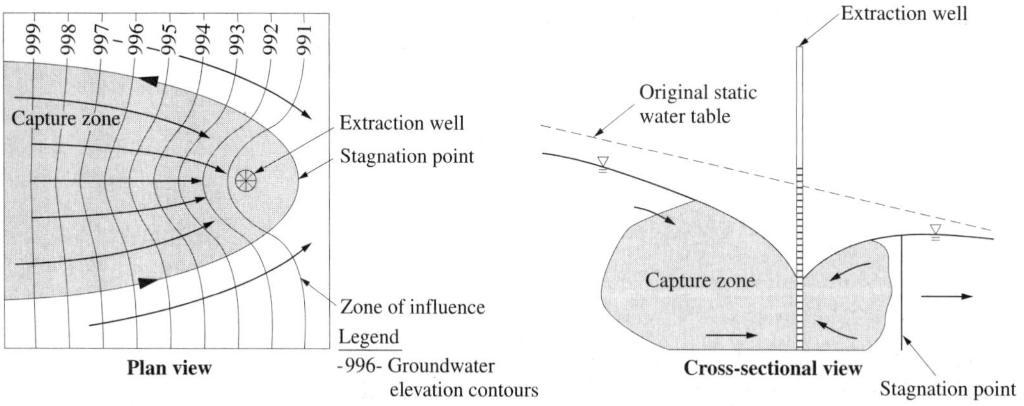

FIGURE 16-21
Single extraction well capture zone (not drawn to scale).

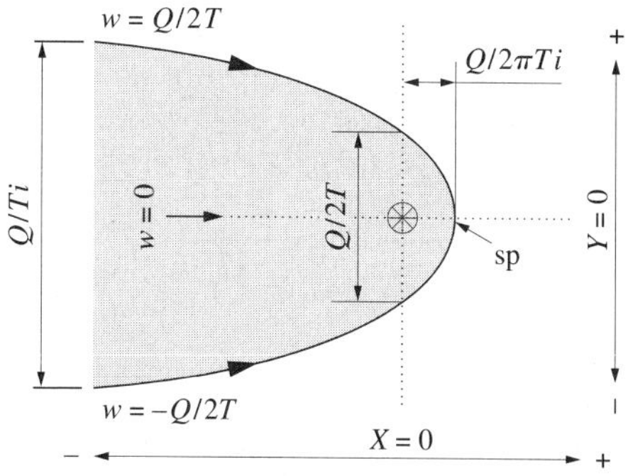

FIGURE 16-22
Single extraction well capture zone (not drawn to scale).

either single or multiple wells, particularly in groundwater aquifers that have a great deal of heterogeneity or that tend to be bounded by surface water bodies such as rivers or streams. These models provide insight into flow patterns generated by alternative pump-and-treat schemes and the selection of locations for performance monitoring wells.[55,56,57,58,59,60] These models are often combined with linear programming methods to optimize a pump-and-treat design[61,62,63,64] for different site-specific conditions by minimizing the total required pumping rate while maximizing the extraction well capture zone.

Flow model applications utilized in the design of pump-and-treat systems are often very complex because of the heterogeneities and uncertainties inherent in most hydrogeological environments. A simple calculation, however, can be used in relatively homogeneous uniform systems to delineate capture zones, as illustrated in Figure 16-22.

EXAMPLE 16-5. SINGLE WELL CAPTURE ZONE. Calculate the expected pumping rates for a single extraction well to contain a groundwater plume of trichloroethene (TCE) that is 76 m (250 ft) wide. The aquifer can be assumed to be homogeneous and isotropic, 12 m (40 ft) deep with a hydraulic conductivity of 4.2×10^{-4} m/s (120 ft/day). The regional hydraulic gradient is 0.005 m/m (ft/ft) according to regional water level data.

Assumptions: In addition to the assumption of a homogeneous isotropic aquifer, the calculation will also need to assume that the drawdown in the extraction well is insignificant compared to the saturated thickness of the aquifer (i.e., transmissivity is a constant). The capture zone will be assumed to be at steady state with negligible effects from recharge.

Solution. From the equations provided in Chapter 4 and in Figure 16-22, the following variables are defined:

Q = pumping rate (L^3/T)
k = hydraulic conductivity (L/T)
B = aquifer thickness (L)
W_{dg} = capture zone width down-gradient of well (L)
SP = stagnation point, down-gradient distance from the well (L)
i = regional hydraulic gradient (L/L)
W_w = capture zone at extraction well (L)

Calculate the transmissivity:

$$
\begin{aligned}
T &= kB \\
&= 4.2 \times 10^{-4}\ \text{m/s} \times 12\ \text{m} = 5.1 \times 10^{-3}\ \text{m}^2/\text{s} \\
&= (120\ \text{ft/day}) \times 40\ \text{ft} = 4800\ \text{ft}^2/\text{day}
\end{aligned}
$$

Calculate pumping rate to capture the down-gradient plume:

$$
\begin{aligned}
Q &= W_{\text{dg}} T i \\
&= 76\ \text{m} \times 5.1 \times 10^{-3}\ \text{m}^2/\text{s} \times 0.005\ \text{m/m} = 1.9 \times 10^{-3}\ \text{m}^3/\text{s} \approx 167\ \text{m}^3/\text{day} \\
&= 250\ \text{ft} \times 4800\ \text{ft}^2/\text{day} \times 0.005\ \text{ft/ft} = 6000\ \text{ft}^3/\text{day} \approx 31\ \text{gal/min}
\end{aligned}
$$

The capture zone width at the well and the down-gradient extent of the capture zone (stagnation point) can be calculated as follows:

$$
\begin{aligned}
W_w &= Q/2Ti \\
&= (6000\ \text{ft}^3/\text{day})/(2 \times 4800\ \text{ft}^2/\text{day} \times 0.005) \\
&= 125\ \text{ft} \\
\text{SP} &= Q/2\pi T i \\
&= (6000\ \text{ft}^3/\text{day})/(2 \times 4800\ \text{ft}^2/\text{day} \times 3.14 \times 0.005) \\
&\approx 40\ \text{ft to the down-gradient stagnation point}
\end{aligned}
$$

Example 16-5 assumes that the extraction well completely penetrates the aquifer. In cases where the extraction is partially penetrating (Figure 16-23), three-dimensional modeling is often required to account for the vertical hydraulic conductivity of the aquifer material.

Performance monitoring of pump-and-treat systems often employs a two-phase methodology to (1) monitor hydraulic control by showing inward groundwater gradients toward the extraction well (or wells) and (2) demonstrate that chemical concentrations in monitoring wells located down-gradient of the hydraulic containment system do not show chemical breakthrough. Demonstrating performance of hydraulic containment systems using pump-and-treat methods traditionally combines field monitoring with groundwater flow and capture zone analysis.

Successful performance is measured as the achievement of hydrodynamic control at the outer limits (horizontal and vertical extent) of the plume such that hydraulic

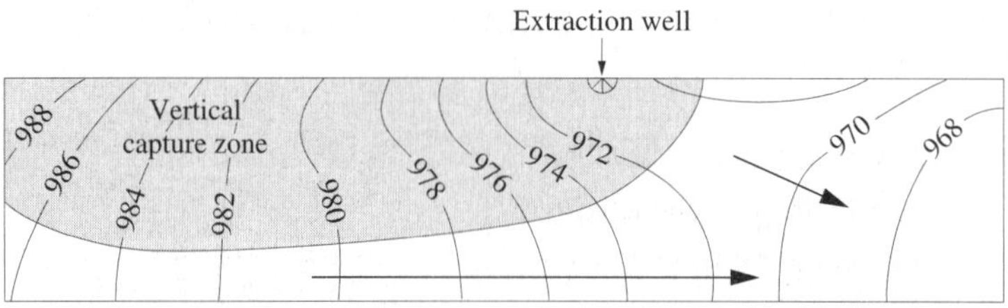

FIGURE 16-23
Partially penetrating single extraction well capture zone (not drawn to scale).

gradients are inward toward the well. Also, monitoring of chemical concentrations in selective performance monitoring wells outside the zone of influence is used to demonstrate capture of contaminants within the capture zone of the pump-and-treat system.

Air Sparging

For the remediation of groundwater contaminated with volatile organics, *in-situ* **air sparging**[65] may be considered as an alternative to pump-and-treat. The concept of air sparging involves the injection of air into the subsurface (in the zone of saturation) to encourage transfer of volatile organic contaminants from the aqueous phase to the vapor phase.[66] As shown in Figure 16-24, air sparging is normally conducted in conjunction with soil vapor extraction implemented in the unsaturated zone (as presented in Section 16-8). Early studies of air sparging revealed that the introduction of air into the subsurface not only resulted in volatilization but also stimulated aerobic biodegradation. When flow rates are designed to stimulate biotransformation, lower flow rates are used than for volatilization, and this practice is termed **biosparging.** The principal benefits of air sparging are the *in situ* volatilization of dissolved-phase contaminants, the desorption and volatilization of contaminants adsorbed to the soil matrix, and the enhanced aerobic biodegradation.

Principles that apply to air sparging are discussed earlier in this text. For example, the success of air sparging depends upon the contaminant volatilizing from the water phase to the vapor phase. Volatilization, in turn, depends upon Henry's constant for the contaminant (see Section 3-2). While many air sparging systems have been designed, constructed and operated, developments in modeling the contaminant mass removal are continuing.[67,68] One of the principal concerns regarding air sparging is the possibility of the spread of contaminants. The injected air can displace water in the soil pores and the volatilization of contaminants into the vapor phase results in contaminant movement. Hence the design of the injection and recovery system must be such that detrimental contaminant movement does not occur. Also of concern is

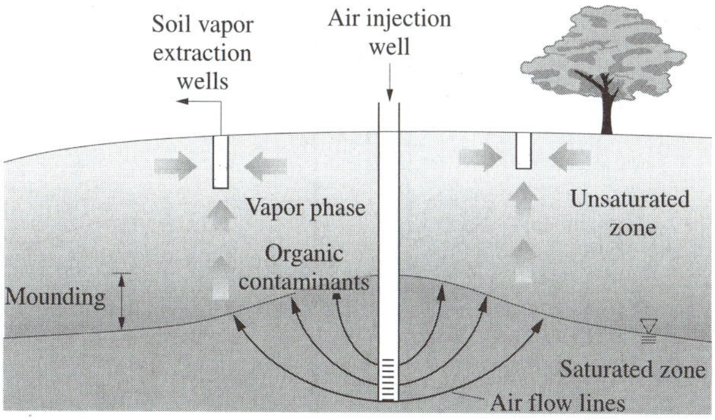

FIGURE 16-24
Schematic of air sparging.

the frequently observed "rebound" of contaminant concentrations on cessation of air sparging (similar to that observed on termination of pump-and-treat systems). This rebound can be attributed to the remaining presence of NAPL and the kinetic limitation of diffusion from discontinuous pores in the soil matrix.

For the remediation of groundwater contaminated with volatile organics, **in-well stripping** may be considered as an alternative to pump-and-treat. One such system, NoVOCs (U.S. patent nos. 5,180,503 and 5,389,267), uses a line placed in a well to deliver air under pressure. This results in a stream of air bubbles that rise through the water that enters the well through the screened interval near the bottom of the well. As the clean air passes upward though the contaminated groundwater, the volatile organic compounds undergo a phase change and are transferred from the water phase to the vapor phase (see Section 9-1 for discussion of the supporting theory). The air at the top of the well is then drawn off for treatment. The rising air bubbles also induce upward water flow in the well casing which can escape through an upper screened interval. A schematic of the process is shown on Figure 16-25.

Modeling and design tools have been developed and field studies of the effectiveness of the technique have been developed.[69] It has been shown that the system can develop a groundwater circulation pattern that covers an area having a radius over 30 m. The groundwater can be treated to levels of volatile organic contaminants in the low parts per billion range. Finally, the reduction in contaminant concentration occurs quite rapidly, with 90 percent reduction occurring in the first several months.

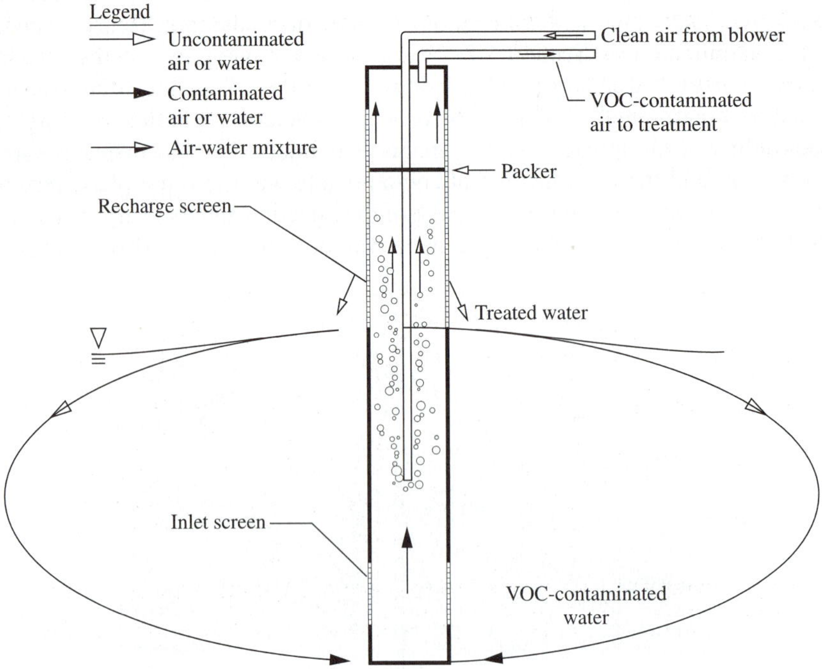

FIGURE 16-25
NoVOCs well construction.[69]

While widely employed, air stripping is not without its limitations. First, air stripping is simply the transfer of contaminants from one medium (water) to another (air). In some cases, the reduction in the contaminant concentrations in the groundwater provide the necessary protection to the public health while the increase of contaminants in the atmosphere may be inconsequential. In other cases, the air may require treatment by techniques such as activated carbon (see Section 9-2). Also, as discussed in the preceding portions of this section, air stripping is effective only for volatile and semivolatile compounds. Finally, stripping towers may be inclined to foul because of iron, algae, bacteria, fungi, and fine particulate matter in the water.

Drain Tile Collection Systems

A drain tile collection system is constructed by excavating a continuous trench in the subsurface and installing collection piping and filter media. By far Love Canal is the best known use of a drain tile collection system as a component in site remediation.[70] A cross section through the system is shown in Figure 16-26.

There are several advantages of a drain tile collection system over wells or well points. The continuous nature of the drain tile collection systems intercepts a variety of subsurface features that otherwise allow water to circumnavigate its way around a pumping well, including sand seams, root holes, buried conduits, and prior surface drainageways that have been filled in. Thus drain tile collection systems are more effective in the presence of subsurface heterogeneity and anisotropy. In the case of a floating contaminant, the drain tile collection system may require only a shallow interception of the groundwater level. Alternatively, if a very deep trench is required, the need for excavation bracing and concerns for worker safety may limit the use of this technique. (The biopolymer technique discussed below is better suited for deep excavations.)

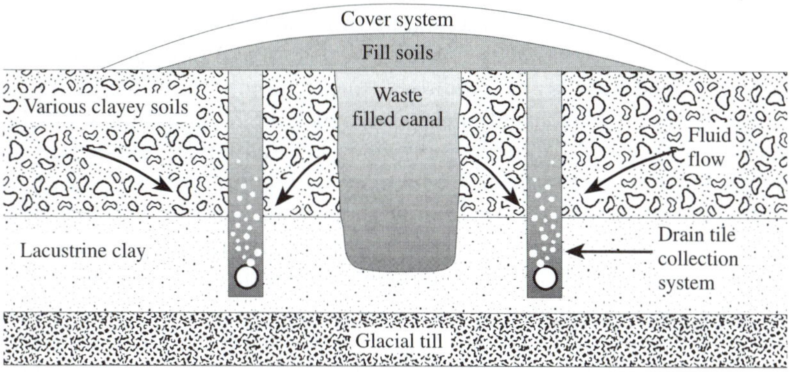

FIGURE 16-26
Section of containment system at Love Canal.

A drain tile collection system may be installed in the subsurface for several purposes:

- The system can be used as an interceptor trench. Contaminants flowing with the groundwater are intercepted by the trench, precluding further migration.
- The system can be used as a hydraulic barrier or containment system.
- The system can serve as a means of groundwater and contaminant extraction (pump-and-treat).
- The system can serve to capture a plume that may have already migrated beyond the collection system alignment by reversing the gradient and "recalling" contaminants that may have left the site.
- The system can function effectively in geologic units that have an insufficient saturated thickness for the use of extraction wells.

Biopolymer Extraction/Interception Trenches

The collection of groundwater for treatment using wells or well points is limited in two ways: (1) wells are costly and time-consuming to install, operate, and maintain. (2) groundwater may move past the barrier in the region between the wells, even with relatively close well spacings. Alternatively, drain tile collection systems provide a continuum from which water can be withdrawn, providing a more positive cutoff. However, drain tile collection systems are costly and dangerous because they must be installed with workers entering the excavation (unless they are installed by horizontal drilling techniques). Excavation requires temporary shoring, groundwater control, disposal of large volumes of soil, and risk to workers in the trench in a chemically and physically hazardous environment. The biopolymer slurry trench is a construction technique whereby a continuous subsurface drain can be installed relatively rapidly, safely, and cost-effectively.[71] The technique builds on long-known and employed slurry trench methods to install a subsurface drain.[72,73]

In the slurry trench method of excavation, a trench is excavated under a head of slurry to maintain trench stability. In the conventional slurry trenches previously described, the slurry is formed by a mixture of bentonite and water. In the biopolymer drain technique, the slurry is formed by a mixture of biodegradable materials (principally ground guar beans), additives (to extend the life of the biodegradable slurry), and water. In the conventional slurry trench method, the slurry-filled trench is backfilled with a soil-bentonite mixture of low hydraulic conductivity. In the biopolymer technique, the trench is backfilled with porous drainage media. Additives to the trench (to lower the pH to initiate enzyme action and to enhance rapid breakdown of the biopolymer slurry) result in the degradation of the slurry to water and natural carbohydrates. A sketch showing similarities and differences between the two techniques is in Figure 16-27.

The groundwater may be extracted from the trench through well casings lowered vertically in the trench. Gravel is tremied around the well to maintain its vertical position while the remainder of the trench gravel backfill is pushed into the trench. Horizontal drain pipes can be installed along the bottom of the trench by using flexible pipe and a pipe-laying machine that follows along after the excavation. A typical completed installation is shown in Figure 16-27.

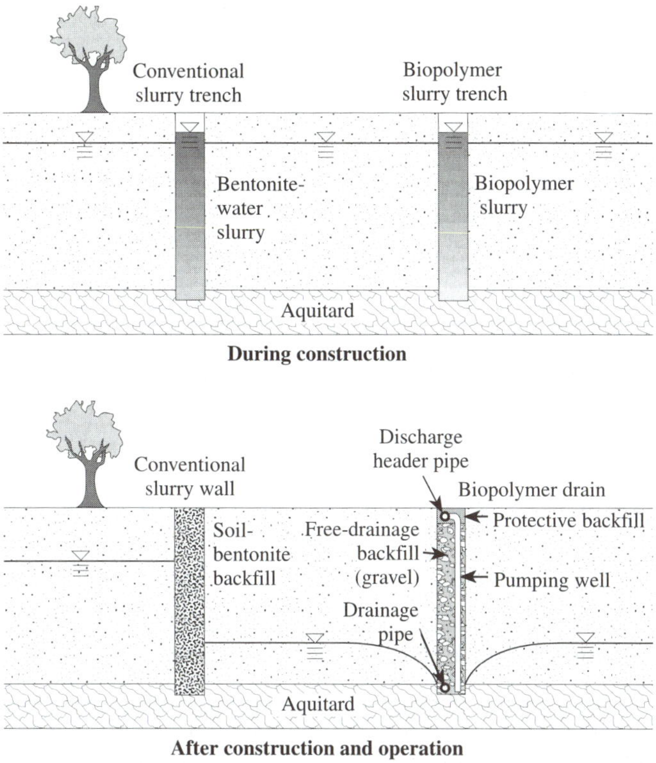

FIGURE 16-27
Biopolymer slurry trench.

16-8
SOIL VAPOR EXTRACTION*

Soil vapor extraction (SVE) is a remedial option utilized to remove volatile organic compounds (VOCs) from soils in the vadose zone (the unsaturated zone above the groundwater table) or from stockpiled excavated soils. The SVE process consists of passing an air stream through the soil, thereby transferring the contaminants from the soil (or soil/water) matrix to the air stream.

Modifications of the process are distinguished by the location of the treatment system (i.e., *in situ* or aboveground) and the method for developing airflow. SVE systems are implemented by installing vapor extraction wells or perforated piping in the zone of contamination and applying a vacuum to induce the movement of soil gases (refer to Figure 16-28). SVE systems typically include knockout drums to remove moisture from the soil gases, followed by vapor-phase treatment prior to discharge to the atmosphere. The effective radius for extraction wells is reported to vary from 6 m to over 45 m (20 to 150 ft), depending on soil conditions.[74] SVE systems have also been demonstrated to depths of 7 m (23 ft) in soils with an

*Section 16-8 was prepared by James D. Fitzgerald, P.E., and Kevin E. McHugh, ERM.

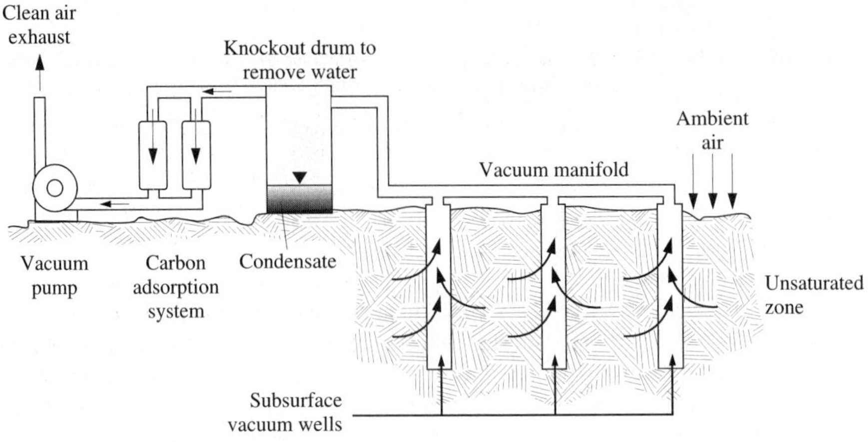

FIGURE 16-28
Schematic of SVE system.

average permeability (or hydraulic conductivity) of 10^{-4} cm/s.[75] SVE systems may be enhanced through the addition of alternative options/enhancements, including:

- Installation of groundwater extraction pumps into the vapor extraction wells to either lower the groundwater table (and therefore enlarge the vadose zone) and/or concurrently remove contaminated groundwater.
- Placing an impermeable barrier over the surface to minimize short circuiting of airflow from the surface and thereby increase the radius of influence. [Under these conditions the effective radius for the extraction wells has been reported to extend to 90 m (300 ft).[76]]
- Installing air recharge wells around the zone of contamination to enhance movement of soil gases through contaminated soils.
- Providing a compressor to force clean air into the recharge wells, enhancing soil gas movement. [Applied pressures are small and typically range from 3.5 to 85 kPa (0.5 to 12.5 psi).]
- Installing wells into the zone of contaminated groundwater and blowing air through the groundwater (air sparging). The induced airflow enhances the volatilization of the contaminants (especially compounds with low solubility in water), causing upward migration to the vadose zone, where it is captured by the SVE system.[77] (Care should be taken to also install hydraulic controls with the air sparging system to ensure that groundwater mounding does not result in migration of contaminants into uncontaminated zones.)

The above modifications represent a partial listing of the various alternatives that can be incorporated into an SVE system. Depending on site-specific conditions, the appropriate system might include various combinations of the above items.

Only SVE treatment of vadose zone contaminants will be discussed in detail in this section. However, the reader should be aware that the basic theory describing movement of soil gases and contaminants through the vadose zone applies to the various permutations of the SVE system presented above.

A VOC spill or release to the ground can be represented using a three-phase model of the vadose zone including: the soil matrix (solid phase), the soil moisture that is located in the voids between and on the surface of the soil matrix (liquid phase), and the soil gas that constitutes the remaining portion of the voids between the soil particles (gas phase). After a release has occurred, VOC contamination in the vadose zone will be distributed among all three phases. The mass distribution of contamination between the phases will depend on the soil conditions (e.g., water content, organic content) as well as the specific contaminants of concern.

Contaminant movement among the various phases in the vadose zone is ongoing, with the subsurface system moving toward equilibrium conditions. The removal of VOCs from the vadose zone may be modeled as the five-step process illustrated in Figure 16-29. Volatile compounds will desorb from the soil particle surface, transfer to the soil water (moisture in the vadose zone) and volatilize to the soil gas. The soil gas then migrates through the void spaces until reaching the ground surface and is finally released to the atmosphere. Assuming localized equilibrium (that is, the concentration of a volatile contaminant in the soil gas at any point in the system is in equilibrium with the chemical concentration in the stationary phase) may lead to an improper mass distribution.[78] However, equilibrium conditions are assumed on an area-wide scale in order to facilitate remedial design.

The movement of contaminants in the soil gas through the soil media can be described by two processes: advection and diffusion. Advection is the movement of a vapor-phase contaminant with the bulk airflow through the soil media and best describes the flow through permeable soils within the unsaturated zone. Diffusion is the movement of a contaminant through the soil medium via concentration gradients.[79] Diffusion tends to control in soils with lower permeability.

Spills or leaks of VOCs to the environment may result in surface or subsurface contamination. Provided the leak is of sufficient quantity, the VOC contaminants will tend to migrate downward through the unsaturated zone, leaving globules, films, and small droplets of the released material. Low-density contaminants will tend to collect in the capillary fringe or float on the groundwater surface. Dense contaminants (after transport through the vadose zone) will tend to pass through the groundwater until encountering an impermeable layer (see Figure 4-21).

A release of contaminants will result in residual contamination in the soil pores. This residual material in the unsaturated zone is the target contamination for cleanup

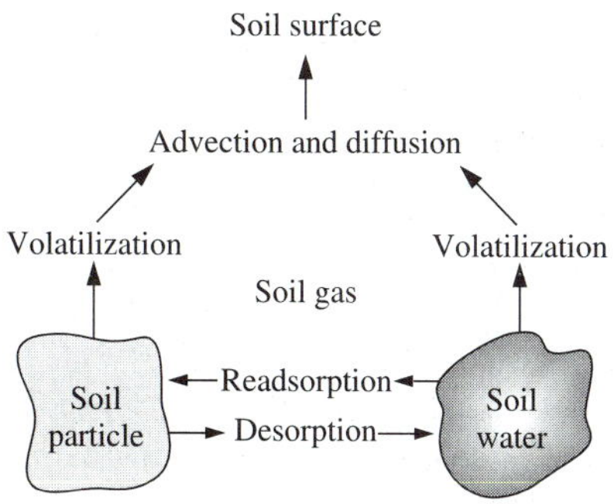

FIGURE 16-29
Schematic of soil/soil vapor model.

through SVE (with the addition of further processes to the SVE system, contaminants trapped in the capillary fringe and on the groundwater surface may also be remediated). The type (low or high density) of chemical released and therefore the location where the contaminants accumulate will affect the volatilization and movement of the contaminants and the design of a remediation system.

An SVE system removes volatile organic contamination by evacuating the soil gases, thereby enhancing the movements of contaminants from the solid and liquid phase to the gas phase as shown in Figure 16-27. The success of SVE remedial systems is therefore based on the effectiveness (both cost and performance) of moving the soil gases through the contaminated zone. Before the volatile compounds can be carried via advective flow to the SVE system for treatment, the contaminants must flow from lower permeability to higher permeability soils by diffusion. Therefore, diffusion may be the rate-limiting step for mass transfer.[80] It has been estimated that 40 to 60 percent of contaminant removal is controlled by advective flow, while the remaining amount is controlled by diffusion.[81]

Contaminant transport between different phases is affected by certain physical, environmental, and chemical properties as listed in Table 16-2. Physical properties that control SVE system performance include: moisture content of the soil in the vadose zone, organic content of the soil, grain size distribution of the soil, and porosity of the soil in the vadose zone. Environmental properties that affect SVE system performance include: area rainfall, air temperature, and ground cover. Chemical properties affecting SVE system performance include: the volatility of the contaminant (the contaminant's tendency to transfer from the liquid to the vapor phase), the relative polarity between the contaminant and the soil particles, and the density and viscosity of the liquid-phase contaminant. The latter properties dictate where the liquid phase contaminants will accumulate in the subsurface. For example, dense, nonaqueous-phase liquids (DNAPLs) will migrate downward through the groundwater. The physical-chemical properties that control contaminant movement in the environment are presented in Section 3-2, and contaminant transport is discussed in Section 4-2.

While theoretical models are available to attempt to predict the performance for each of the individual parameters identified in Table 16-2, actual soil conditions are too complex and nonhomogeneous for these models to be used in the design of SVE systems. Current practice is to utilize empirical models to select the most appropriate mechanical system and then to use field data to refine system design.

As previously discussed, movement of VOCs through the soil is controlled in part by diffusion (see Section 3-2). Fick's law best describes this type of vapor

TABLE 16-2
Properties affecting soil aeration systems

Soil properties	Contaminant properties	Environmental properties
Permeability	Henry's law constant	Temperature
Porosity	Solubility	Humidity
Grain size distribution	Adsorption coefficient	Wind speed
Moisture content	VOC concentration in soil	Solar radiation
pH	Polarity	Rainfall
Organic content	Vapor pressure	Terrain
Bulk density	Diffusion coefficient	Vegetation

transport (vapor flux) through the soil media:

$$J = -D_v \frac{dC}{dz} \tag{16-3}$$

where J = vapor flux [mol/(cm$^2 \cdot$ s)]
D_v = diffusion coefficient in the vadose zone (cm^2/s)
C = concentration of vapor in soil air (mol/cm^3)
z = distance in the direction of diffusion (cm)

Diffusion coefficients for pure compounds in air are tabulated in Appendix B. These values should be modified to account for the fact that air movement in a soil matrix is affected by soil properties. Grasso[82] suggests the Millington and Quirk[83] relationship:

$$D_v = \frac{a^{10/3} D}{\phi^2} \tag{16-4}$$

where a = portion of voids filled with air (dimensionless)
ϕ = soil porosity (volume voids/total volume)
= a + (portion filled with water) + (portion filled with immiscible fluid)
D = diffusion coefficient of chemical in free air (cm^2/s)

Dragun[84] suggests the following modification of the diffusion coefficient to account for soil properties:

$$D_v = T(\phi - \phi_B)D \tag{16-5}$$

where D_v = diffusion coefficient to be used in soil air
T = correction factor due to tortuous flow path
ϕ_B = volume of blocked pores (fraction)

Experimental values of T and ϕ_B are highly variable. Dragun quotes the following experimental values:

T	Range from 0.6 to 0.7; median 0.66
ϕ_B	Range from 0.0 to 0.25; median 0.1

The organic carbon partition coefficient (K_{oc}) (see Section 3-2) expresses a chemical's tendency to be adsorbed to soil particles in the soil (or soil/water) matrix. K_{oc} may be thought of as the ratio of the amount of chemical adsorbed per unit weight of organic carbon (oc) in the soil to the concentration of the chemical in the water phase at equilibrium.[85]

$$K_{oc} = \frac{\mu\text{g adsorbed/g organic carbon}}{\mu\text{g/mL of solution}} \tag{16-6}$$

Soils with relatively high organic content (e.g., peat) will tend to more readily adsorb organics than those soils with low organic content (e.g., silts). K_{oc} values of some volatile contaminants may be found in Appendix B.

The partition or distribution coefficient (K_d) for a particular compound refers to the preference of a contaminant for the soil or water phase. K_d was defined in Section 3-2 as:

$$K_d = \frac{X}{C} \tag{16-7}$$

where X = contaminant concentration on soil surface (μg contaminant/kg soil) at equilibrium

C = contaminant concentration remaining in water phase (μg contaminant/liter) at equilibrium

A higher K_d value indicates that a contaminant is more likely to remain on the soil and not be transmitted through soil moisture movement.

Gas and liquid phase equilibrium can be described for dilute aqueous solutions by Henry's law and for concentrated nonaqueous-phase liquid (NAPL) solutions by Raoult's law (see Section 3-2).

To determine the effect of Henry's law on the effectiveness of SVE, an expression relating the total concentration and the concentrations of the individual soil, water, and air phases is used along with the concept of partitioning[86]:

$$C_T = d_B C_s + \theta_w C_l + V_a C_v \tag{16-8}$$

where C_T = total concentration of contaminant (mass/volume)

d_B = soil dry bulk density (mass/volume)

C_s = mass of contaminant adsorbed per mass of soil

θ_w = volumetric fraction of water in the soil (volume/volume)

V_a = volumetric fraction of air in the soil (volume/volume)

C_l, C_v = concentration of contaminant associated with liquid and vapor phase respectively (mass/volume)

By substituting equilibrium equations into Equation 16-8, the vapor partition coefficients can be derived.

Two equilibrium equations describe the phase relationships between chemicals that exist in the soil matrix–water system and the relationship between total concentrations of a compound and the concentration of that compound in each phase. These relationships are as follows:

$$C_s = K_d C_l \qquad \text{(a linear adsorption model)} \tag{16-9}$$

where C_s = mass adsorbed per mass of soil

K_d = soil water partition coefficient

C_l = concentration in water (mass per volume)

$$C_v = H' C_l \qquad \text{(Henry's law)} \tag{16-10}$$

where C_v = concentration in air (mass per volume)

H' = Henry's constant (dimensionless, see Section 3-2)

By rearrangement of Equations 16-9 and 16-10 and substitution, Equation 16-8 becomes:

$$
\begin{aligned}
C_T &= d_B\,(K_d C_l) + \frac{\theta_w\,(C_v)}{H'} + V_a C_v = d_B \frac{(K_d C_v)}{H'} + \frac{(\theta_w C_v)}{H'} + V_a C_v \\
&= \left(d_B \frac{K_d}{H'} + \frac{\theta_w}{H'} + V_a\right) C_v \qquad (16\text{-}11) \\
&= R_v C_v
\end{aligned}
$$

where $R_v =$ vapor partition coefficient.

The liquid-phase equilibrium equations are the same as the vapor-phase equations above. By substitution, Equation 16-11 becomes:

$$
\begin{aligned}
C_T &= d_B K_d C_l + \theta_w C_l + V_a H' C_l \\
&= \left(d_B K_d + \theta_w + V_a H'\right) C_l \qquad (16\text{-}12) \\
&= R_l C_l
\end{aligned}
$$

where $R_l =$ the liquid partition coefficient.

EXAMPLE 16-6. SOIL VAPOR EXTRACTION. Three areas of a large site have been contaminated by similar concentrations of different compounds (area 1: ethylbenzene; area 2: dichlorobenzene; area 3: 1,1,1, trichloroethane). Assuming the physical and environmental conditions are the same at each area, which area would be best suited to vadose zone treatment by SVE? Explain.

Solution. From Appendix B:

Compound	Solubility, mg/L	Vapor pressure, mmHg	Henry's constant at 20°C, (atm · m³)/mol
Ethylbenzene	152	7.0	7.03×10^{-3}
Dichlorobenzene	79	2.28	2.86×10^{-3}
1,1,1, Trichloroethane	4400	100	1.34×10^{-2}

From Equations 16-10 through 16-12, compounds with higher values of H' will have a greater affinity for the vapor phase. On the basis of this information, Area 3 would appear to be the best candidate for SVE.

Thomas[87] summarizes a number of methods for estimating the volatilization of chemicals from soil with the caveat that no single method currently addresses all of the complexities of the process. The Hartley method provides an estimate based on the heat balance between the contaminant and air:

$$
J = \frac{A_{\text{sat}} \dfrac{(1-h)}{\delta}}{\dfrac{1}{D} + \dfrac{\lambda^2\,(A_{\text{sat}})\,M}{kRT^2}} \qquad (16\text{-}13)
$$

where $J =$ flux $[\text{g}/(\text{cm}^2 \cdot \text{s})]$

$A_{sat} =$ saturation concentration in air (at temperature of outer air) (g/cm^3)

$h =$ relative humidity $(0 < h < 1)$

$\delta =$ thickness of stagnant boundary layer (cm)

$D =$ diffusion coefficient in air (cm^2/s)

$\lambda =$ latent heat of vaporization (cal/g)

$M =$ molecular weight (g/mol)

$k =$ thermal conductivity of air $[61 \times 10^{-6}\ \text{cal}/(\text{s} \cdot \text{cm} \cdot \text{K})]$

$R =$ universal gas constant $[1.987\ \text{cal}/(\text{mol} \cdot \text{K})]$

$T =$ temperature (K)

The second term in the denominator indicates the resistance to volatilization due to thermal considerations and may be neglected for compounds significantly less volatile than water.

EXAMPLE 16-7. VOLATILIZATION OF TCE FROM SOIL. A tank containing 30,000 gallons of trichloroethylene (TCE) ruptures, contaminating 2000 ft^2 of soil. Use Hartley's equation to estimate the volatilization of TCE to the air and the time required for the entire 30,000 gallons to evaporate. Comment on the limitations of the method.

The following data are provided:

TCE saturation in air, $A_{sat} = 415\ \text{g/m}^3$ (Verschueren)[88]
Diffusion coefficient $D = 0.070\ \text{cm}^2/\text{s}$ (Lyman et al.)[89]
Humidity $h = 0.3$
Temperature $T = 20°\text{C} = 293$ K
Latent heat of vaporization, $\lambda = 7.52$ kcal/mol (Dean)[90]
Thickness of stagnant boundary layer, $\delta = 0.3$ cm
Specific gravity of TCE = 1.46 (Verschueren)[91]

Solution

1. Flux:

$$A_{sat} = 415\ \text{g/m}^3 \times 10^{-6}\ \text{m}^3/\text{cm}^3 = 4.15 \times 10^{-4}\ \text{g/cm}^3$$

$$M = 131.5\ \text{g/mol}$$

$$\lambda = \frac{7520\ \text{cal/ mol}}{131.5\ \text{g/mol}} = 57.2\ \text{cal/g}$$

$$k = 61 \times 10^{-6}\ \text{cal}/(\text{s} \cdot \text{cm} \cdot \text{K})$$

$$J = \frac{\dfrac{4.15 \times 10^{-4}\ (\text{g/cm}^3)\ (0.7)}{0.3\ \text{cm}}}{\dfrac{1}{0.07\ \text{cm}^2/\text{s}} + \dfrac{(57.2\ \text{cal/g})^2(4.15 \times 10^{-4}\ \text{g/cm}^3)(131.5\ \text{g/mol})}{[61 \times 10^{-6}\ \text{cal}/(\text{s} \cdot \text{cm} \cdot \text{K})][1.987\ \text{cal}/(\text{mol} \cdot \text{K})](293\ \text{K})^2}}$$

$$= \underline{\underline{3.08 \times 10^{-5}\ \text{g}/(\text{cm}^2 \cdot \text{s})}}$$

2. Area and mass of chemical:

$$\text{Area} = 2000 \text{ ft}^2 \times 929.034 \text{ (cm}^2/\text{ft}^2) = \underline{\underline{1.858 \times 10^6 \text{ cm}^2}}$$

$$\text{Volume} = 3 \times 10^4 \text{ gal} \times (0.003785 \text{ m}^3/\text{gal}) = 113.55 \text{ m}^3$$

$$\text{Mass} = 113.55(1.46)(1000 \text{ kg/m}^3)$$

$$= 165{,}783 \text{ kg} = \underline{\underline{1.658 \times 10^8 \text{ g}}}$$

3. Time required:

$$\text{Time} = \text{mass/(area} \times \text{flux)}$$

$$= \frac{1.658 \times 10^8 \text{ g}}{(1.858 \times 10^6 \text{ cm}^2)\left[3.08 \times 10^{-5} \text{ g}/(\text{cm}^2 \cdot \text{s})\right]}$$

$$= 2.897 \times 10^6 \text{ s} = 33.5 \text{ days; } \underline{\underline{\text{approximately 5 weeks}}}$$

4. Limitations of method: Hartley's method accounts for only a few of the parameters listed in Table 16-2. Specifically, this problem does not consider soil properties, depth of contamination, and many of the environmental properties. These limitations indicate that the results are probably accurate only to an order of magnitude.

Design Considerations

SVE's primary benefit is as an *in situ* remediation tool for the removal (and treatment) of VOCs from soils in the vadose zone. By utilizing *in situ* remediation, soil excavation and disposal costs, as well as site disruption that may be associated with large-scale excavation activities, are minimized. Also, the long-term liability that may result from off-site disposal of contaminated soils will be eliminated. SVE may be used to remediate soils beneath structures; does not require reagents; and employs conventional equipment, labor, and materials.

SVE is not an applicable remedial alternative at all locations. Its usefulness will be limited at sites with low-permeability soil, where contaminants with low vapor pressures (and low values of Henry's constant) are present, and at sites with high groundwater tables. Furthermore, because of the number of uncertainties that are inherent with *in situ* remediation, the time required to achieve the cleanup levels is unpredictable. However, through the collection of site-specific information, the anticipated effectiveness of SVE can be evaluated.

An SVE system relies on the effective movement of air and soil gas through the volume of contaminated soil. The flow rate of vapors per unit thickness of well screen through soil induced by SVE extraction wells can be estimated by using a steady-state radial flow solution for compressible flow[92]:

$$Q/H = \pi(K/\mu)P_w[1 - (P_{\text{atm}}/P_w)^2]/\ln(R_w/R_I)$$

TABLE 16-3
Soil permeability[90]

Soil type	Soil permeability, m/s	Soil permeability, darcy
Silt	10^{-8} to 10^{-7}	0.001 to 0.01
Silty sand	10^{-7} to 10^{-6}	0.01 to 0.1
Fine sand	10^{-6} to 10^{-5}	0.01 to 1.0
Medium sand	10^{-5} to 10^{-4}	1.0 to 10
Coarse sand	10^{-4} to $> 10^{-3}$	10 to > 100

where Q = airflow rate through vapor extraction well
H = length of extraction well screen
K = soil gas permeability
μ = viscosity of air
P_w = applied air pressure at extraction well (absolute)
P_{atm} = ambient pressure (absolute)
R_w = radius of extraction well
R_I = radius of influence of extraction well

The soil gas permeability K can be estimated from the soil type, as presented in Table 16-3. Then the only unknown parameter is R_I, the empirical radius of influence. The literature reports R_I values ranging from 30 to 100 ft.[93] Field data collected from SVE pilot studies may be used to determine site-specific values of R_I for a given extraction flow rate by plotting detected vacuum pressure (P_{atm}) versus distance from monitoring point to SVE extraction well. Examples of these methods using actual pilot test data are presented later in this section.

The number of SVE wells required for a specific full-scale SVE design may be estimated from the following equation:

$$N = A/\pi(R_I)^2$$

where N = number of SVE wells
A = area of contamination (ft^2)
R_I = radius of influence (ft)

Process Description

A typical SVE system consists of the following components:

Infrastructure:

- One or more vapor extraction wells.
- Manifold piping connecting extraction wells with the vacuum source.
- Vacuum monitoring wells.

- Vacuum gauges and flow control valves at each monitoring and extraction well (optional).
- Impermeable cover over the system site to minimize the downward infiltration of air and water (optional).
- One or more vent wells to enhance air movement through the vadose zone (optional).

Extraction Equipment:

- Vacuum/blower unit.
- Moisture knockout drum (optional).
- Off-gas treatment.

A schematic diagram of a typical SVE system was previously provided as Figure 16-28. A photo of an SVE system under construction is shown in Figure 16-30.

Three variables control the performance of an *in situ* SVE system and must be considered during system design: (1) the extraction well spacing, (2) the airflow rate induced in the subsurface remedial zone, and (3) the subsurface pressure. Additional variables relating to the specific contaminant that must be evaluated for the design include: (1) pressure gradients, (2) identity and concentration of VOCs in soil, (3) extracted air VOC concentration, (4) extracted air temperature, (5) extracted air moisture, and (6) power usage.[95]

Consideration must also be given to the size and type of blower/vacuum unit that produces the vacuum pressure. Typical blower/vacuum units generate a negative pressure of between 2 and 100 kPa (0.5 and 30 inHg). The blower/vacuum units may be regenerative centrifugal (the most common type, which provides a wide operating range), multistage centrifugal, rotary positive displacement, or liquid ring. *In situ* pilot studies have been conducted to evaluate the effect of site-specific variables on system performance.

FIGURE 16-30
An SVE system under construction.

Example Pilot Studies

Results of *in situ* pilot studies to evaluate the design variables on a site-specific basis have been summarized in the literature.[96] These studies were intended to determine the effects of soil type and varying source vacuum pressures on system performance (i.e., what is the effect of various source pressures on the vadose zone?). Two example pilot studies are presented below.

The results of one study are provided in Figure 16-31. The figures reveal that the change in negative pressure relative to distance from the extraction well is a decreasing exponential relationship. The pilot test consisted of the installation of one extraction well and five monitoring wells at the test site. The extraction well (VES-1) was installed in the area of highest-known soil VOC concentration, while the monitoring wells (VES-2 through VES-6) were installed at radial distances from VES-1. To evaluate the effects of different soil types (high and low permeability) on the SVE system, the monitoring wells were installed in two different hydrogeologic zones. Wells VES-1 through VES-4 were installed in sandy soils, while wells VES-5 and VES-6 were installed in an area of tight clay soils. All wells ranged in depth from 6 to 11 ft.

The results of this and similar studies indicate that SVE is highly dependent on specific site conditions. The results further reveal the importance of conducting an *in situ* pilot test prior to the design and implementation phases of a SVE project to ensure optimal well layout and blower selection.

A second study was conducted at the Twin Cities Army Ammunition Plant in New Brighton, Minnesota.[97] The study consisted of a 14-week field program to evaluate the feasibility of *in situ* air stripping as a remedial alternative. The target contaminant of the study was trichloroethylene (TCE). However, the information collected is applicable to VOCs with similar vapor pressures. It should be noted that the information gathered from the study is site-specific and should not be applied directly to other sites. Once again, pilot studies are typically performed at each site to evaluate the effectiveness of soil venting and to develop specific design information.

The Twin Cities pilot study consisted of the installation of three rows of vapor extraction wells around four air injection wells at two adjacent locations at the site. The extraction wells were manifolded to the vacuum intake of a vacuum/blower unit. The extracted air was treated by activated carbon and released to the atmosphere. The injection wells were manifolded to the exhaust of a separate vacuum/blower unit. Air was forced through the injection wells to induce soil stripping of VOCs in the subsurface.

The initial purpose of the pilot test was to determine the effectiveness of SVE on the removal of VOCs from the subsurface. The evaluation was conducted by collecting samples of the extracted air over the 14-week test period and performing a TCE analysis. A mass balance was then performed to determine the quantity of TCE removed from the soil.

Following the analysis of the system's removal effectiveness, an evaluation of the design parameters was performed. To evaluate the design parameters, the vacuum/blower unit was connected to a single extraction well, while the remaining wells were used as pressure monitoring points. A total of 26 pressure monitoring points (inclusive of the two remedial areas) were used while the airflow rate was

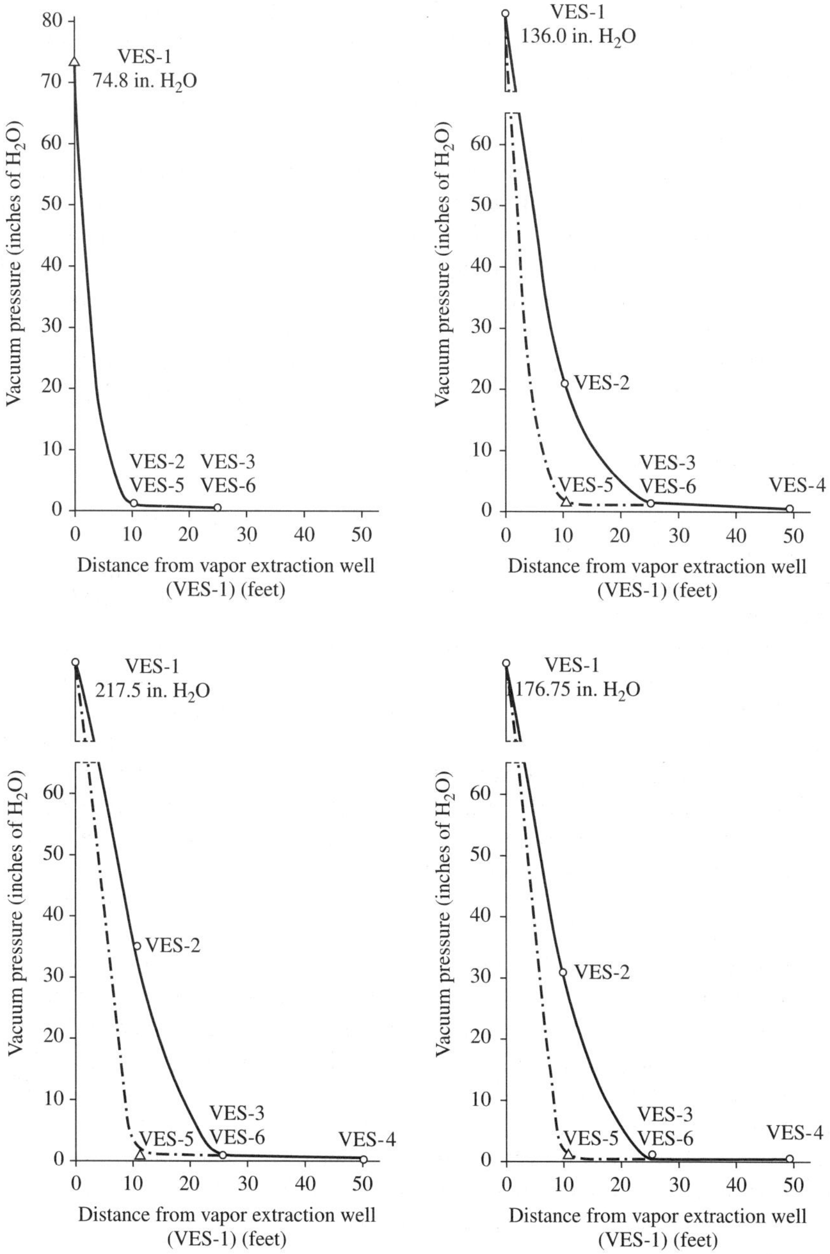

Note
VES 1 through VES 4: Sandy soils
VES 5 and 6 Tight clay soils

FIGURE 16-31
Effect of distance from vapor extraction well.[98]

adjusted between 50 and 250 ft^3/min. Figure 16-32 presents a relationship between the withdrawal rate at the extraction well and the vacuum head applied.

To determine the effects of distance on the vacuum developed, pressure readings were collected at varying flow rates. Figure 16-33 depicts vapor pressure drops at increasing distance from the extraction well for different vacuum/blower unit flow rates.

The relative head loss in the soil was determined by dividing the pressure readings at each monitoring well by the pressure developed in the extraction well. Again, these readings were collected at varying flows. The results revealed that although the increased flow rate expanded the radius of influence of the well, the change in the relative pressure drop due to varying the flow rate was insignificant. The relative pressure drop was found to be a function of the distance from the extraction wells, not flow rate.

The effect of distance on the resulting vacuum pressure is further depicted in Figure 16-34. This graph again reveals that vacuum pressure decreases with distance from the extraction well. This graph would be prepared from site-specific pilot test data and would then be used to design the full-scale remedial system.

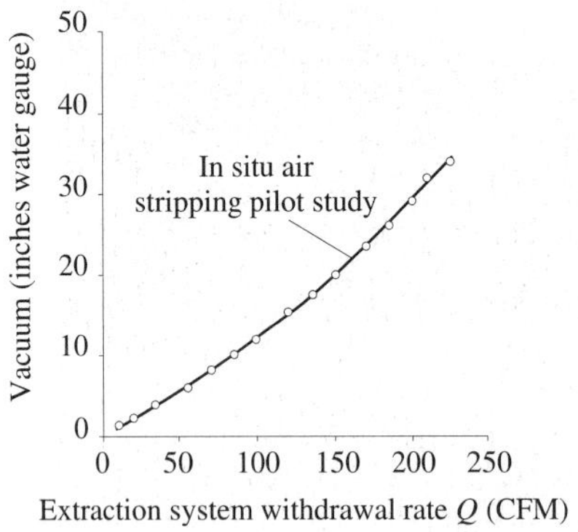

FIGURE 16-32
Twin Cities Army ammunition plant ratio of withdrawal rate versus vacuum head.[99]

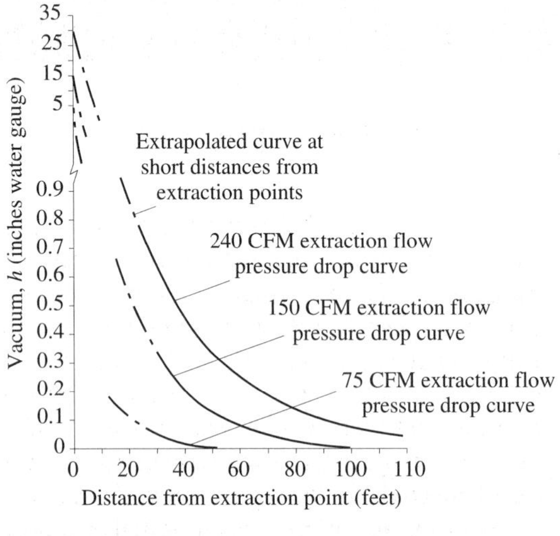

Note: Soil conditions consist of sandy materials with traces of silt or gravel and occasional gravelly layers.

FIGURE 16-33
Twin Cities army ammunition plant vapor pressure drop versus distance.[100]

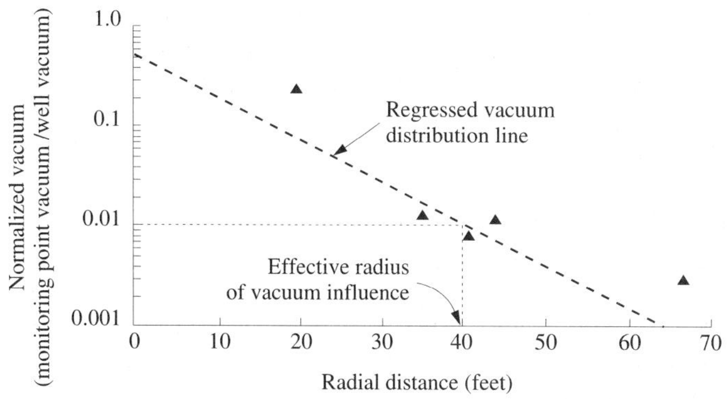

FIGURE 16-34
Effective radius of vacuum-influenced normalized pilot-test vacuum data plot.[102]

FIGURE 16-35
Trailer-mounted soil vapor extraction unit. (*Photo courtesy of ERM.*)

Buscheck and Peargin[101] recommend that SVE well spacing be determined as the radial distance corresponding to a percentage of the projected operating vacuum as interpolated from the fitted vacuum distribution line. The percentage used to determine well spacing will depend on how representative the pilot test data are of site conditions. That is, if subsurface site conditions are well known and consistent, then a higher confidence in the accuracy of the pilot test data will allow a higher percentage to be used in determining well spacing. Conversely, if site conditions are variable or unknown, then a lower percentage will be used in spacing extraction wells.

SVE studies have become less expensive through the use of trailer-mounted units containing vacuum pumps, controls, recorders, and air pollution control equipment (see Figure 16-35). A single vapor extraction well is drilled near the center of the contaminated area. Monitoring wells are placed at several distances along

perpendicular axes, and varying vacuum levels are applied. The resulting data are used to determine well spacing for site remediation.

EXAMPLE 16-8. PRELIMINARY DESIGN OF SVE SYSTEM. An underground storage tank (UST) at a gasoline station is leaking and the plume is about to migrate off the property. The plume currently extends south to a distance of 70 m and is 30 m wide at this distance. The depth of the plume is 3.65 m and the seasonal high groundwater table is at a depth of 6 m. Provide a preliminary layout of an SVE system to remediate the site.

Specifically determine:

1. The required number of extraction wells and their layout
2. The volumetric capacity in cubic feet per minute (ft^3/min) and operating system pressure of the blower
3. Equipment layout and any required ancillary facilities

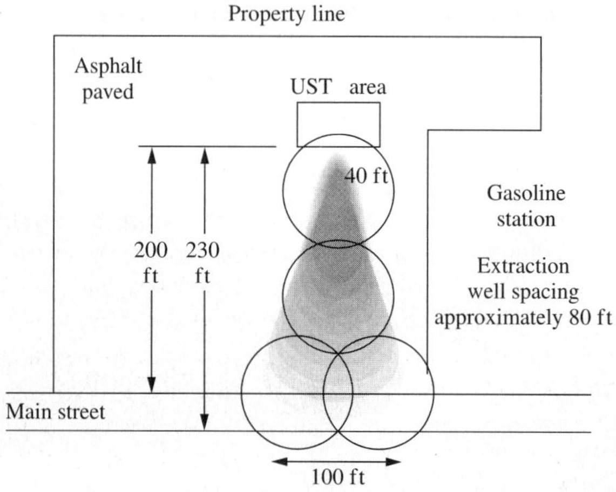

Assume the curves in Figures 16-32 and 16-33 are applicable to this site. Further assume that the minimum measurable vacuum pressure at the radius of influence of the extraction wells will be 0.5 inH_2O gauge (125 Pa).

Solution

1. From Figure 16-33, a vacuum of 0.5 inH_2O gauge can be obtained from either a 150-ft^3/min blower, with 25-ft radius of influence per extraction well, or a 240-ft^3/min blower with 40-ft radius of influence. Selecting a 240-ft^3/min blower yields an extraction well spacing of 80 ft. Four extraction wells should be adequate to cover the plume.
2. From Figure 16-32, a 240-ft^3/min blower per extraction well must be capable of generating a vacuum of 34 inH_2O gauge. This does not include any losses in the piping system, which would have to be considered after the piping system is designed. Because the blowers will be manifolded to all four extraction wells, the system must be capable of producing $4 \times 240 = 960$ ft^3/min at 34 inH_2O gauge (8450 Pa).
3. Ancillary equipment required includes manifold piping, a knockout drum to remove any water removed by the vacuum system, and a carbon adsorption system or other means of removing organics from the air before release to the atmosphere. A schematic of the equipment layout is shown in Figure 16-28. Many of the system options/enhancements discussed above may be considered in this preliminary design. Because the site is shown as paved, it would be inappropriate to suggest placing an impermeable barrier over the site, but air recharge wells may enhance the system.

16-9
PERMEABLE REACTIVE TREATMENT WALLS

Recent innovations in remedial technologies have resulted in the widespread application of permeable reactive treatment wall technologies (also known as permeable reactive barriers) for groundwater remediation. Treatment walls are permeable semipermanent or replaceable units that are installed across the flow path of a contaminant plume. Contaminants present in the groundwater are removed by physical, chemical, and/or biological processes that include fixation, degradation, oxidation/reduction, sorption, and precipitation. Each of these reactions depends on a number of parameters such as pH, oxidation/reduction potential, chemical concentration, and kinetics. Therefore, sufficient site characterization of contaminant properties, groundwater flow regimes, and subsurface geology are critical for the successful application of treatment wall systems.

Treatment walls are not designed to contain groundwater flow, but rather to contain the chemical mass present in groundwater while allowing uncontaminated (treated) groundwater flow to pass through the wall. A major advantage of treatment walls over pump-and-treat systems for groundwater remediation is that aboveground water treatment and the associated effluent requirements from pump-and-treat systems are avoided. Treatment walls are also passive remedial technologies and, therefore, do not require a continuous input of energy to run pumps or treatment systems. Operation and maintenance costs are thus significantly reduced; however, periodic replacement or rejuvenation of the reaction medium may be required after the capacity of the treatment is exhausted or clogged by precipitates and/or biofouling. The reactions that take place within the treatment wall are dependent on parameters such as pH, oxidation/reduction potential, contaminant concentration, and the reaction kinetics within the wall. Thus, treatment wall designs must consider characterization of the contaminants, groundwater flow rates through the site, and subsurface geology. Although the use of treatment walls has come more into the forefront of groundwater remedial technologies, these technologies have been in operation for only a few years and, therefore, the long-term performance of these systems has yet to be established.

Treatment wall designs can be classified as one of two types: *in situ* reaction curtains and funnel-and-gate systems. Figure 16-36 provides an illustration of plume management under these different configurations along with a comparison with pump-and-treat. Under normal pump-and-treat scenarios, groundwater withdrawn from extraction wells results in a capture zone that encompasses the contaminated groundwater by creating inward gradients toward a point source extraction well (see Figure 16-36a). *In situ* reaction curtains (see Figure 16-36b) involve the construction of a trench on the down-gradient portion of a groundwater plume, which extends the entire width of the plume. Reactive materials that compose the treatment wall are installed in the trench and contaminated groundwater is remediated as it passes through the treatment wall. The permeability of reaction curtains needs to be higher than or equal to the permeability of the surrounding geologic material to have groundwater flow converge on the curtain. In the event that hydraulic conductivity of the treatment wall is less than the surrounding aquifer material, the groundwater will tend to migrate around the treatment wall.

To successfully remediate a groundwater plume, the reaction curtain must be at least as wide as the groundwater plume. If the contaminant plume is extremely wide

or dispersed, the use of *in situ* reaction curtains could be impractical. To circumvent this problem, an alternative design to the curtain treatment wall has been developed that involves the installation of low hydraulic conductivity cutoff walls (see Section 16-5) in conjunction with highly permeable treatment zones. The combination of cutoff walls and reactive curtains is known as the funnel-and-gate system[103,104] (see Figure 16-36c). By combining the low-permeability cutoff walls (funnel) with the treatment wall (gate), the treatment wall is not required to extend the entire width of the contaminant plume, since the funnel acts to focus groundwater flow into the gate and prevent lateral movement of contaminants around the gate. The primary advantages of this method include control of groundwater flow and a reduced length of treatment wall. Variations on the funnel-and-gate design can also be employed under different site-specific conditions as illustrated in Figure 16-37.[37] Single-gate systems can be combined with additional barrier wall designs that can partially or completely enclose a contaminant plume (see Figure 16-37a) or multiple-gate systems for the control of relatively wide plumes (see Figure 16-37b).

For plumes with a mixture of different type of chemicals, multiple-reaction treatment walls can be effective if multiple gates are placed in series (see Figure 16-37b). This configuration will allow for the treatment of different chemicals that require different treatment technologies. For example, a groundwater plume originating from a manufacturing site might contain high levels of hexavalent chromium

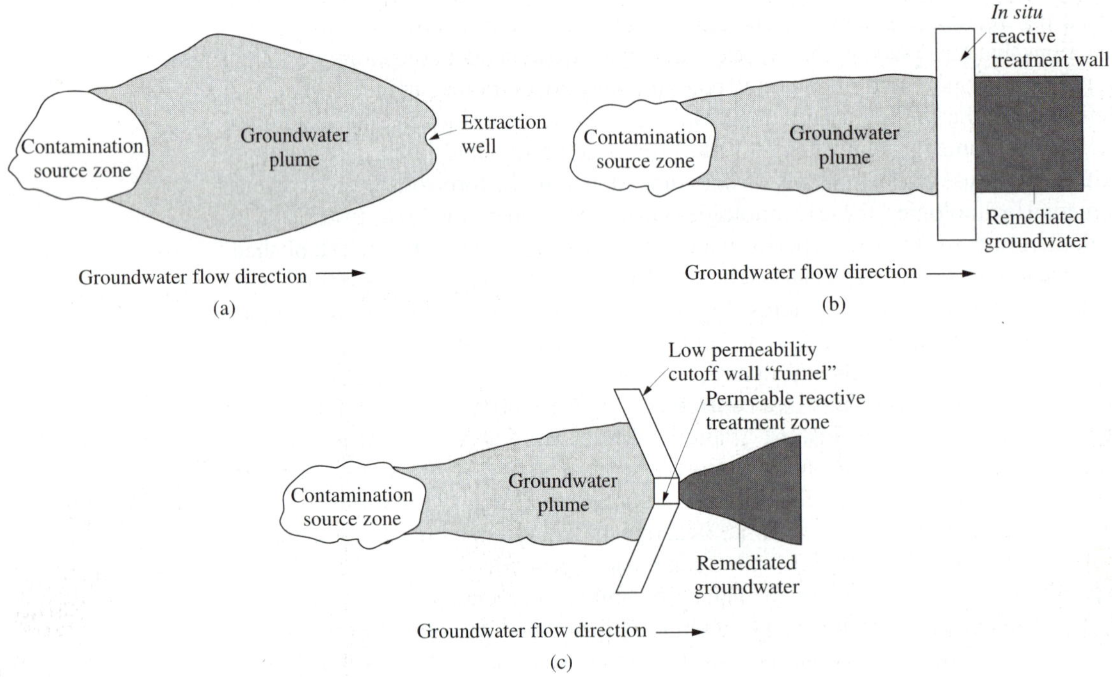

FIGURE 16-36
Plume management for single extraction well, treatment wall curtain, and funnel-and-gate systems. (*Modified from Star and Cherry, "In Situ Remediation of Contaminated Ground Water: the Funnel-and-Gate System,"* Ground Water, *vol. 32, no. 3, p. 465(12), May/June 1994.*)

and of benzene, toluene, ethylbenzene, and xylene (BTEX) in the groundwater. A multiple-gate system can be employed that allows for the reduction of the chromium from the more soluble hexavalent form to the less soluble trivalent form.[105,106,107] Because aromatics derived from gasoline readily biodegrade under aerobic conditions (i.e., the chromium will be transformed to the less soluble trivalent form under slightly reducing conditions), a second treatment wall can be constructed to create aerobic conditions that are conducive for the biological degradation of BTEX. In this way, different chemicals in the groundwater that require different redox conditions to degrade can be remediated within the same funnel-and-gate system.

Funnel-and-gate designs require a balance between a number of design parameters. Initially, the funnel-and-gate system must be designed to allow for all of the contaminated groundwater to flow through the gate. The flux of groundwater, termed the "relative discharge"[103] is influenced by factors such as the funnel width, gate width, the hydraulic conductivity of the gate, and the orientation of the funnel to regional groundwater gradients. The relative discharge of groundwater through the gate is proportional to the width of the funnel in two-dimensional systems.

The second criterion for designing funnel-and-gate systems deals with the residence time of contaminants within the gate. The required residence time within the gate depends on the influent chemical concentrations, the required remedial objective for the contaminants of concern, and the rate of the reaction within the gate (treatment zone). The residence time of contaminants is inversely proportional to the relative discharge through the gate. In other words, the larger the funnel (i.e., capture zone) the higher the relative discharge through the gate and, therefore, the lower the residence time of contaminants in the gate.

The design of a funnel-and-gate system should be optimized to maximize both the capture zone and the residence time of contaminants in the gate, while minimizing the amount of material needed for the construction of the gate (i.e., gate size). To design a funnel-and-gate system given the number of conflicting criteria, groundwa-

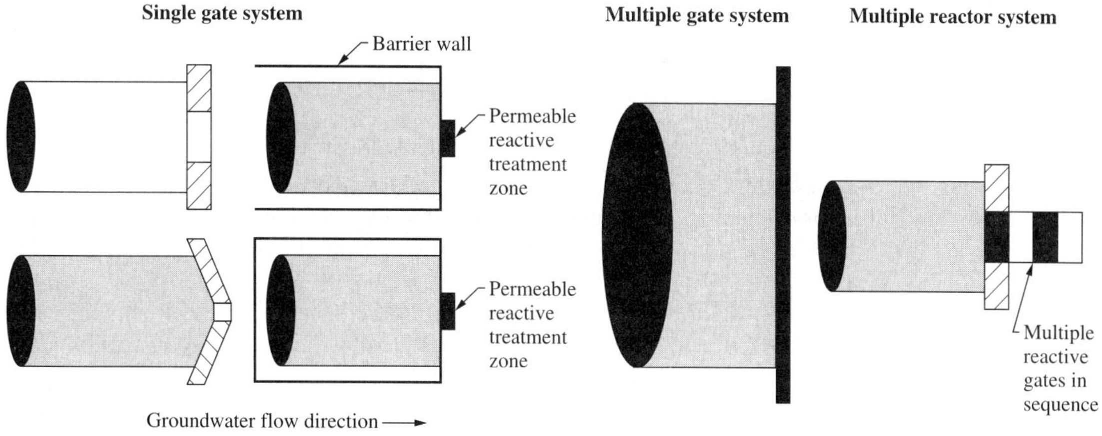

FIGURE 16-37
Funnel and gate system configurations. (*After Star and Cherry, "In Situ Remediation of Contaminated Ground Water: the Funnel-and-Gate System,"* Ground Water, *vol. 32, no. 3, p. 465(12), May/June 1994.*)

ter flows and transport modeling are usually needed to determine the optimal design configuration. This is especially true for hanging funnel-and-gate systems where vertical (three-dimensional) flow can significantly complicate the design.[108]

Treatment walls can be classified as degradation, precipitation, or sorption. Behavior and design characteristics of each type are described in the following sections.

Degradation Treatment Walls

Degradation barriers are designed to break down or degrade contaminants in groundwater into harmless products by employing either enhanced biological degradation or chemical destruction mechanisms. An excellent example of the degradation treatment wall involves the use of zero valence metallic iron filing granules (Fe^0)[109,110] to create reducing conditions that destroy chlorinated solvents. Biologic degradation walls are often permeable treatment walls in which nutrients are injected to enhance the biological activity within and down-gradient of the wall.[111,112] Typical biological enhancements would include air sparging[106] or bioventing techniques or compounds that release nutrients into the ground such as oxygen release compound (ORC) or hydrogen release compound (HRC).

One of the most widely utilized treatment walls in recent years has been the zero valence iron (Fe^0) filing wall for the degradation of chlorinated solvents such as TCE and perchloroethylene (PCE).[113] Under aerobic conditions, the preferred electron acceptor is typically dissolved oxygen. This acceptor can compete with chlorinated hydrocarbons that have similar oxidizing potentials to O_2. Aerobic groundwater enters the iron filings wall, and causes the oxidation of metallic iron (Fe^0) to ferrous iron (Fe^{2+}) with the subsequent release of two electrons (see Equation 16-14). Chlorinated solvents such as PCE also react as electron acceptors, resulting in dechlorination and release of a chloride ion (see Equation 16-15). Similar reaction mechanisms have also been found for inorganic constituents such as hexavalent chromium [Cr(VI)] with Fe^0.[105] This resulting reaction causes the subsequent precipitation of trivalent chromium [Cr(III)] and iron [Fe(III)] oxyhydroxides (see Equation 16-16).

$$Fe^0 \rightarrow Fe^{2+} + 2e^- \tag{16-14}$$

$$RCl + 2e^- + H^+ \rightarrow RH + Cl^- \tag{16-15}$$

$$CrO_4^{2-} + Fe^0 + 8H^+ \rightarrow Fe^{3+} + Cr^{3+} + 4H_2O \tag{16-16}$$

A number of chemicals that can be treated by zero valence iron are presented in Table 16-4, chemicals presently not treatable by zero valence iron are presented in Table 16-5, and chemicals with unknown treatability are presented in Table 16-6.[113]

Experimentally determined degradation rate constants for Fe^0 treatment walls are typically obtained by laboratory batch testing in which the rate constant is calculated from the slope of a semilog plot of the normalized concentration (C_0/C_t) versus time, where C_0 is the initial concentration and C_t is the concentration at time t. These estimated rate constants are typically decreased by 50 percent to adjust for subsurface temperatures in groundwater, which are lower than room temperature.[114,115]

The degradation rates of halogenated hydrocarbons with zero valence iron proceed via pseudo-first-order kinetics. Batch and column testing has indicated highly

TABLE 16-4
Contaminants Treatable by Fe^0 *

Organic compounds		Inorganic compounds	
Methanes	Tetrachloromethane	Trace metals	Chromium
	Trichloromethane		Nickel
	Dichloromethane		Lead
Ethanes	Hexachloroethane	Anion contaminants	Uranium
	1,1,1-trichloroethane		Technetium
	1,1,2-trichloroethane		Iron
	1,1-dichloroethane		Manganese
Ethenes	Tetrachloroethene		Selenium
	Trichloroethene		Copper
	cis-1,2-dichloroethene		Cobalt
	trans-1,2-dichloroethene		Cadmium
	1,1-dichloroethene		Zinc
	Vinyl chloride		Sulfate
Propanes	1,2,3-trichloropropane		Nitrate
	1,2-dichloropropane		Phosphate
Aromatics	Benzene		Arsenic
	Toluene		
	Ethylbenzene		
Other	Hexachlorobutadiene		
	1,2-dibromoethane		
	Freon 113		
	n-nitrosodimethylamine		

Source: Reference 113.

TABLE 16-5
Contaminants presently not treatable by Fe^0

Organic compounds	Inorganic compounds
1,2-dichloroethane	Chloride
Chloroethane	Perchlorate
Chloromethane	

Source: Reference 113.

TABLE 16-6
Contaminants with unknown treatability

Organic compounds	Inorganic compounds
Chlorobenzenes	Mercury
Chlorophenols	
Certain pesticides	
PCBs	

Source: Reference 113.

variable degradation rates due to operating conditions and experimental factors such as pH, metal surface area concentration, and mixing rates.

One of the most significant cost factors in the construction of zero valence treatment walls is the amount of reaction material required. A number of studies

TABLE 16-7
Surface-area-specific rate constant for selected organic chemicals with zero valence iron

Chemical	K_{sa}, $L/(m^2 \cdot h)$*
Tetrachloroethane (PCM)	$1.2 \pm 1.5 \times 10^{-4}$
Trichloromethane (TCM)	$9.2 \pm 7.3 \times 10^{-4}$
Tribromomethane (TBM)	1.7×10^{-2}
Hexachloroethane (HCA)	$3.1 \pm 3.3 \times 10^{-2}$
1,1,2,2-tetrachloroethane (1,1,2,2-TCA)	1.3×10^{-2}
1,1,1,2-tetrachloroethane (1,1,1,2-TCA)	1.4×10^{-2}
1,1,1-trichloroethane (1,1,1-TCA)	1.1×10^{-2}
1,2,3-trichloropropane (1,2,3-TCP)	$6.1 \pm 4.7 \times 10^{-6}$
Tetrachloroethene (PCE)	$2.1 \pm 2.7 \times 10^{-3}$
Trichloroethene (TCE)	$3.9 \pm 3.6 \times 10^{-4}$
1,1-dichloroethene (1,1-DCE)	$6.4 \pm 5.5 \times 10^{-5}$
trans-1,2,dichloroethene (T1,2-DCE)	$1.2 \pm 0.4 \times 10^{-4}$
cis-1,2-dichloroethene (C1,2-DCE)	$4.1 \pm 1.7 \times 10^{-6}$
Vinyl chloride (VC)	$5.0 \pm 1.5 \times 10^{-5}$

*L = barrier thickness.

have been conducted that tie together laboratory batch and column studies to actual rate constants observed in the field.[116,117] These studies provide rate data in terms of the iron surface area concentration that make it possible to calculate fairly reliable estimates of the iron requirement for a potential remedial design.

Surface-area-specific rate constants (K_{sa}) for selected organic chemicals are presented in Table 16-7.[116]

The mass of iron required to decrease chemical concentrations to acceptable levels can be calculated from the following relationship.

$$\frac{W}{A} = \frac{\mu n}{k_i} \ln\left(\frac{C_0}{C}\right) \tag{16-17}$$

where
W = weight of iron required
A = cross-sectional area of the plume
μ = groundwater flow velocity
n = aquifer porosity
k_i = chemical-specific degradation rate constant
C_0 = upgradient contaminant concentration
C = remedial objective down-gradient of the treatment wall

The first-order rate constant k_i can be calculated from the surface-area-specific rate parameter k_{sa} and the specific surface area for granular iron (typically 1.0 m^2/g). Equation 16-17 is often modified to include a safety factor F, which accounts for uncertainties in site conditions. A recommended safety factor of 3.5 has been proposed[118] and incorporated in Equation 16-18:

$$\frac{W}{A} = F\frac{\mu n}{k_i} \ln\left(\frac{C_0}{C}\right) \tag{16-18}$$

EXAMPLE 16-9. CALCULATION OF ZERO VALENCE IRON (Fe^0) FOR A TREATMENT WALL. Calculate the amount of Fe^0 needed to degrade a plume of PCE from 5000 mg/L

to 5 μg/L using a recommended safety factor of 3.5. Assume that the cross-sectional area of the plume is 92.9 m^2 (10,000 ft^2), the groundwater flow velocity is 0.15 m/day (0.5 ft/day), the aquifer porosity is 0.33, and the specific surface area of the granular iron is 1.0 m^2/g.

Solution. The k_{sa} for PCE is 2.1×10^{-3} L/($m^2 \cdot$ h). Assuming a correction of 50 percent reduction for temperature, and a specific surface area of 1.0 m^2/g, the PCE degradation rate constant k_{PCE} is

$$k_{PCE} = 2.1 \times 10^{-3} \frac{\text{L}}{\text{m}^2 \cdot \text{h}} \times 1.0 \frac{\text{m}^2}{\text{g}} \times 0.5 = 1.1 \times 10^{-3} \frac{\text{L}}{\text{g} \cdot \text{h}} = 1.0 \frac{\text{cm}^3}{\text{g} \cdot \text{h}}$$

$$= 0.024 \frac{\text{m}^3}{\text{kg} \cdot \text{day}}$$

Conversion to appropriate units of foot, pound, days gives

$$\left(1.1 \times 10^{-3} \frac{\text{L}}{\text{g} \cdot \text{h}}\right)\left(\frac{24\text{ h}}{1\text{ day}}\right)\left(\frac{1\text{ ft}^3}{28.3\text{ L}}\right)\left(\frac{454\text{ g}}{1\text{ lb}}\right) = 0.42 \frac{\text{ft}^3}{\text{lb} \cdot \text{day}}$$

Calculation of the weight of Fe^0 per cross-sectional area can be done using Equation 16-17 as follows:

$$\frac{W}{A} = F \frac{\mu n}{k_{PCE}} \ln\left(\frac{C_0}{C}\right) = 3.5 \frac{(0.5\text{ ft/day})(0.33)}{0.42 \dfrac{\text{ft}^3}{\text{lb} \cdot \text{day}}} \ln\left(\frac{5000}{5}\right) = 9.5 \frac{\text{lb}}{\text{ft}^2}$$

$$= \frac{3.5(0.15\text{ m/day})(.33)}{0.024\text{ m}^3} \ln\left(\frac{5000}{5}\right) = 50\text{ kg/m}^2$$

For a groundwater plume cross-sectional area of 10,000 ft^2, the total mass of Fe^0 required is:

$$(50\text{ kg/m}^2)(92.9\text{ m}^2) \cong 46.500\text{ kg} = 46.5\text{ metric tons}$$

or

$$\left(9.5 \frac{\text{lb}}{\text{ft}^2}\right)(10{,}000\text{ ft}^2) = 95{,}000\text{ lb or }47.5\text{ tons}$$

Table 16-8 presents a summary of the amount of iron used in selected full-scale zero valence treatment walls.[113]

TABLE 16-8
Summary of total Fe^0 used in selected treatment walls[113]

Site	Capture area, ft^2	W/A, lb/ft^2	Iron, tons
Coffeyville, Kansas	7,500	19	71
Belfast, N. Ireland	650	46	15
Denver, Colorado	11,700	99	580
Elizabeth City, North Carolina	2,900	210	300
Mountainview, California	660	270	90
Lowry AFB, Colorado	270	330	45
Upstate New York	270	380	45
Moffett AFB, California	475	400	96
Somersworth, New Hampshire	250	520	65
Sunnyvale, California	520	850	220

Sorption Treatment Walls

Sorption treatment walls contain reactive materials that remove contaminants from groundwater by physically absorbing the contaminants out of the groundwater onto surfaces of the barrier wall material.[113] Sorption treatment walls may consist of material such as activated carbon or cationic surfactants. These treatment walls are typically used for metals and hydrophobic organic chemicals that have an affinity to adsorb to material surfaces.

Precipitation Treatment Walls

Precipitation treatment walls contain materials that react with contaminants in groundwater as they migrate through the wall and cause these contaminants to precipitate out of solution as insoluble precipitates left within the treatment wall. Lead is a common example of a contaminant found at industrial sites that can be effectively removed by precipitation-type barrier wall systems using hydroxyapatite.[119]

A summary of installation details for active treatment wall systems already deployed is presented in Table 16-9.

16-10 OTHER REMEDIAL TECHNOLOGIES

All of the active and passive contaminant control technologies have a common effect—they control the rate of contaminant migration. In that sense they are not "cleanup" technologies, but rather containment technologies. Present innovations in treatment walls have resulted in the evolution of containment technologies into more active treatment technologies. As noted, they may be, and often are, used in conjunction with other technologies.

The remedial technologies described in the remainder of this chapter focus on additional remedial methods that are more directly aimed at cleanup of the site. They are designed to remove contaminants and render the site less contaminated. Additional treatment technologies are described in Chapters 9 through 12.

Excavation and Disposal

A simple, and often applied, cleanup technology is to excavate the site and dispose of the material at another location. In the United Kingdom, this approach is known as "dig-and-dump" and is a widely employed corrective action for existing contaminated land sites. Where land disposal costs are low (perhaps because of inadequate environmental protection measures taken at the land disposal site) this is often the most economical alternative. Other treatment technologies may be applied to the excavated materials including thermal (Chapter 12), biological (Chapter 10), and physicochemical (Chapter 9).

How is the excavation and disposal remedial alternative implemented? Excavation and disposal employs conventional earth-moving equipment to excavate and

TABLE 16-9

Summary of active treatment wall systems (permeable reactive barriers)

Site	Contaminants	Treatment wall design	Performance	References/website
Otis AFB, Mass.	TCE and PCE 5–150 ppb	Ni-Fe wall Plan 2 parallel 50-ft walls, 2-ft apart, perpendicular to flow path of 600-ft wide plume with 5 to 150 ppb TCE & PCE. Walls will begin 80-ft bgs to top of plume, extending to as deep as 150 ft. Induce fracture to fill with iron filings with slurry mixture moving down and outward, creating series of overlapping vertical planes thus becoming a "continuous" wall.	NA No information available	Appleton, E. L., *ES&T*, vol. 30, no. 12, p. 536A (1996) *www.magnet.state.ma.vs/dap/suro/mmr/files/fact966.htm* "Environmental Cleanup at the Massachusetts Military Reservation: An Overview"
Nickel Rim Mine Sudbury, Ontario, Canada	Acid mine drainage Mixture organics sulfate-reducing bacteria	Reactive wall 15 m long, perpendicular to groundwater flow, 3.6 m deep, 4 m thick Material: municipal compost, wood chips, pea gravel (permeability), sand buffer, clay cap	After 9 mo sulfate reduction and metal sulfide precipitation. Sulfate: 2400–4600 mg/L to 200–3600 mg/L; Fe: 250–1300 mg/L to 1–40 mg/L; pH: 5.8 to 7.0. Alkalinity rose 60–220 mg/L to 850–2700 mg/L as $CaCo_3$. Fe monosulfide precipitate. Cost approximately $30,000 (half materials/half installation) potential life 15 years.	Benner, S. G., et al., 1. *ACS Meeting,* San Francisco, April 13-17, p. 140 (1997) 2. *Internat'l Contain. Technol. Conf. & Exhib*, St. Petersburg, Fla., p. 835 (1997) 3. GWMR, Fall, pp. 99–107 (1997) *http://em-52.em.doe.gov/ifd/scfa/events/abstract.htm* "1997 International Technology Conference"
USCG, Elizabeth City, N.C.	Cr(VI), TCE	Fe^0 wall Models indicate reactive barrier most efficient/cost-effective. 46 m long × 0.6 m wide × 7.3 m deep wall installed <6 h using a continuous trencher. Wall of 100% Fe filings 46 m long, 0.6 m wide, 7.3 m deep installed in <6 h using a continuous trenching technique. Bench laboratory studies and flow and transport models used in wall design.	Sampling indicated 2.5 mg/L Cr(VI) declined to <MCL within barrier. Further sampling under way determining groundwater chemistry and organic concentrations. Site groundwater, reactive materials: Cr(VI) 11 mg/L to <0.01 mg/L and TCE 1700 μg/L to <1 μg/L. Decrease of Cr(VI) from influent 6 mg/L to <0.01 mg/L and TCE from 5600 μg/L to 5.3 μg/L within wall. TCE approaches or attains the MCL within the barrier.	Bennett, T. A., et al., *213th National ACS Meeting,* San Francisco, April 13–17, pp. 243–245 (1997) Blowes, D. W., et al., *Internat'l Contain. Tech. Conf. & Exhib.* St. Petersburg, Fla., Feb. 9–12, p. 851 (1997) *www.epa.gov/ahaazvuc/eliz/htm* *www.ecc.ksu.edm/HSRC/96proceed/pug/pll* "Remediation of Chromate-Contaminated Ground Water Using Zero Valent Iron: Field Test at USCG Support Center, Elizabeth City, NC"

TABLE 16-9
Continued

Site	Contaminants	Treatment wall design	Performance	References/website
Dept. of Energy site, Colo.	U	Fe^0 wall Pilot barrier using Fe to remove U from tailings effluent at UMTRA site scheduled 5/96 to 1999. Fe foam SA 0.1 and 5 m^2/g; Fe^0 SA 5.6 $\times 10^{-3}$ m^2/g. During 3 years U, Se, Mo, other elements monitored as well as costs and benefits.	Field coincides with experimental findings. In Fe foam batch U removed to <DL in 10 h. Metallic hydraulic conductivity maintained. Initially conduc. 6.4 $\times$ 10^{-3} cm/s; Fe foam was 0.53 cm/s. Capacity maintained after >700 PV oxygenated water.	Dwyer, B. P., and D. C. Marozas, *Internat'l Contain. Tech. Conf. & Exhib.*, St. Petersburg, Fla., February 9–12, pp. 844–850 (1997) *http://em-52.em.doe.gov/ifd/scfa/events/abstract.htm* "International Containment Technology Conference 1997"
Moffett AFB, Calif.	PCE TCE	Fe^0 wall Funnel and gate 50 ft long $\times$ 10 ft wide $\times$ 22 ft thick	TCE reduced from 1000 ppb within the first 2 ft of wall	Naval Facilities Eng. Service Center, Enviro. Restor. Div., www updated April 24, 1997 *http://enviro.nfesc.navy.mil/ps/projects/permeabl.htm* Technology Description
Industrial site, Calif.	VOCs	Fe^0 wall First commercial *in situ* iron wall 220 tons Fe^0, present in reaction canister	Water quality objectives are being met	Szerdy, F. S., et al., *ASCE National Convention,* November 12–14, Washington, D.C., pp. 245–256 (1996) *http://www.pubs.asce.org/www.display.cgi?9605142*
N.J. site pilot-scale test	PCE, TCE, cDCE, VC	Aboveground Fe^0 reactor flow rate 0.5 gal/min for 3 months	Assumptions: Time for PCE degradation sufficient for any TCE to degrade; 10% cDCE, 1% VC from PCE/TCE dehalogenation	Vogan, J., et al., *209th National ACS Meeting,* Anaheim, Calif., April 2–7, vol. 35, p. 800 (1995) *www.rtdf.org/TREATMENT WALLSUMMS/sglnj.htm*
Sunnyvale, Calif.	TCE, cDCE, Freon, VC	Funnel and gate 100% 4 ft thick, 40 ft wide, 20 ft deep retention time and 2 days for VC	TCE levels of 30–68 ppb entering wall reduced to <0.5 ppb; cDCE of 393–1916 ppb to <0.5 ppb	Yamane, C., et al., *209th National ACS Meeting,* Anaheim, Calif., April 2–7, vol. 35, p. 792 (1995) *www.environmental.com* *www.acs.org/meetings/boston/abook.htm*
DOE, Kansas	TCE	Fe^0 wall 1000-foot funnel and gate Reactive zone 30 ft wide, 17 ft deep, 3 ft thick 7900 tons of granule iron slurry wall		*www.beck.com/www/technologies/eti/fieldreports/kansas.htm*

TABLE 16-9
Continued

Site	Contaminants	Treatment wall design	Performance	References/website
New York	TCE, cDCE, VC	Fe^0 wall Funnel and gate 12 ft × 3.5 ft gate with 15 ft sheet piling on each side	VOC reduced to MCL within 1.5 feet of trench C_0 = 300 ppb TCE 500 ppb cDCE 80 ppb VC	*www.beck.com/www/technologies/efi/fieldreports/newyork.htm*
CFB Borden, Borden, Ontario, Canada	TCE, PCE	22% Fe^0, 78% concrete sand 5.5 m wide, 9.7 m deep	TCE reduced 95%, PCE 91% Cl—consistent with quantity degraded. Traces DCE, no VC	*sciburg.uwaterloo.cci/earth/borden*
CFB Borden, Borden, Ontario, Canada	TCE, PCE	22% Fe^0, 78% concrete sand 1.5 m wide	90% TCE and 88% PCE removed Cl—indicates dechlorination. cDCE product with peak at 2200 mg/L followed by cDCE and 1,1-DCE, VC	O'Hannesin, S. F., et al., *Emerg. Technol. in Haz. Waste Mangmt VII,* ACS, September 17–20, Atlanta, pp. 55–58 (1995) *sciburg.uwaterloo.cci/earth/borden*

TCE = trichloroethene
PCE = perchloroethylene
ppb = parts per billion
bgs = below ground surface
m = meters
h = hours
NA = Not applicable
MCL = maximum contaminant level
Fe = iron
Fe^0 = metallic iron
U = uranium
VOCs = Volatile organic compounds
cDCE = *cis*-dichloroethene
VC = vinyl chloride
DCE = dichloroethene
Cl^- = chloride
DL = detection limit
PV = pore volume
SA = surface area

transport the wastes and contaminated soils for treatment and/or disposal. Excavation equipment can include draglines, backhoes, clamshells, bulldozers, and loaders. Transportation is usually in containerized vehicles with adequate protection against spillage and leakage. In all cases, worker health and safety is a principal consideration. For contaminated sediment, hydraulic dredging equipment may be employed. Excavation of stream and river sediments typically requires handling a material that is about 90 percent water and 10 percent solids. Disposal and/or treatment may require dewatering of the sediment and treatment of the supernatant.

The clear advantage to excavation and disposal relates to future site usage. Where the contaminated site consists of primarily contaminated land (as opposed to contaminated groundwater), excavation and removal from the site eliminates any risk associated with the removed material. This may permit site redevelopment for residential and commercial uses that might be precluded by other remedial technologies. In addition, this approach may be the fastest method to remediate a site.

Soil Washing

What can be done about the thousands of tons of soil contaminated with a wide range of inorganic and organic contaminants? It does little good to clean groundwater as described earlier in this chapter without remediating the source, such as contaminated soil. Likewise, contaminated sediment, a primary nonpoint source of pollution, has been identified as one of the largest potential risks to water quality and the aquatic environment. In the Great Lakes region of the United States, contaminated sediments have been identified as a problem in 42 of 43 Areas of Concern.[120] Are there technologies available to cost-effectively handle such large volumes of contaminated material? One approach is the application of **soil-washing** techniques. Washing may be with water, aqueous extractive agents, solvents, or even air.[121] Washing may be done on the entire soil/sediment matrix or on selected portions containing the contaminants that are separated from the clean portion by fractionation.[122] In one soil-washing demonstration project, 82 percent of the input sediment was classified as clean sand output, with the contaminants concentrated in the residuals.[123]

What are the principles controlling the effectiveness of soil washing? What soil and chemical parameters are important? How does the process work? The objective of soil washing is to separate the contaminated soil into two output streams: one clean and one contaminated. Soil washing is therefore a concentrating technique that results in a volume reduction of the contaminated material. Because of the surface charges associated with clay particles (see Section 11-3 and Figure 11-7), inorganic contaminants are typically associated with the finer fraction of the soil matrix. Similarly, organic contaminants are typically associated with the humic and organic carbon fractions of the soil (see Section 4-3 and Figure 4-29). Thus, by separating the "fines" and/or humic fraction from the soil, the remaining "coarse" materials (sands and gravels) are relatively clean.

Further, for many soils, the output mass is predominantly coarse. This then leads to a much-reduced mass of soil requiring further treatment and/or disposal. Soil washing for the cleanup of contaminated soil and sediment developed from, and uses tools from, mineral processing. Thus, soil-washing treatment systems can effectively process large quantities of contaminated material.

The separation of soil into two fractions often relies on density and grain size differences within the soil matrix. For example, one process[125] first adds water to the soil and then directs the slurry through a series of separating devices including a trommel (screening) unit to separate materials coarser than 6 mm, cyclone separators to remove particles smaller than 45 μm, and a dense media separator to remove the less dense humic materials. This process also uses a scrubbing process to remove surficial contaminants from the separated coarse fraction, a partitioned dewatering screen to recover the washed coarse materials and humic substances from their slurries, and a flocculation clarifier to separate the contaminated fines from the fines slurry stream.

For the Bergman process just cited, the contaminated soil or sediment should be at least 60 percent coarse and the organic solids content should be less than 20 percent by volume. Thus, one can see that soil washing is a process, not an individual piece of equipment. Other standard separation methods from mineral processing sometimes employed in soil washing are spiral concentration and froth flotation. For example, at one site in Germany it was found that the use of froth flotation substantially improved the removal of zinc from the contaminated soil.[126] In froth flotation, air bubbles are introduced into the slurry (a liquid/solid suspension), and hydrophilic particles are captured by the air bubbles and float to the top of the slurry where they can be removed.

At another site, 2 million cubic yards of contaminated river sediment containing PCBs, pesticides, and heavy metals was dredged and treated in an integrated soil washing and thermal treatment process. Soil washing treatment consisted of wash-drums, sieves, hydrocyclones, and classifiers. The separated clean sediment fraction was reused as construction fill. The concentrated contaminant fraction of the sediment was blended with clay, treated in a rotary kiln, and reused as an aggregate feedstock.[127] Thus, the soil washing methods and equipment will vary with the site-specific contaminant, soil, and sediment conditions. For example, soil washing has historically been used to treat soils contaminated with metals and/or volatile organic contaminants. Studies have also shown its effectiveness for soils contaminated with mixed pollutants such as pesticides, metals, and volatile organics.[128]

Electrokinetic Soil Decontamination

Electroosmosis as a means to control groundwater movement is a phenomenon used and understood by civil engineers for more than 40 years.[124] In electroosmosis, pore fluid moves in response to the application of an electrical potential, that is, a voltage difference (a constant, low direct current). Chemical constituents also move in response to chemical potentials (chemical concentration gradient). Water and thus the dissolved chemical constituents also move in response to hydraulic gradients. Electrokinetics is the coupling of electrical, chemical, and hydraulic gradients to effect contaminant removal. This technique has been demonstrated in the laboratory as effective in the removal of contaminants.[129,130] In one study, 75 to 95 percent of the adsorbed lead was removed by the process.[131]

In another study, a simulated sludge and a specific site soil contaminated with a mixture of organic and metal contaminants was used and electrokinetic remediation was investigated.[132] The high buffering capacity of the soil resulted in high-pH

conditions that inhibited the migration of cationic metallic contaminants such as cadmium (Cd^{+}) while allowing the migration of metallic contaminants such as hexavalent chromium. In the sludge, low-pH conditions allowed simultaneous migration of both anionic and cationic contaminants in opposite directions. These studies add to our understanding of the impact of certain variables on the effectiveness of electrokinetic remediation. They also point to the need for careful laboratory bench-scale studies on the specific soils/sludges and contaminants under consideration for electrokinetic remediation. Once contaminants are concentrated by electrokinetic remediation, they may be recovered/recycled/reused.

Low-Temperature Soil Treatment

Soil contaminated with organics may be remediated by low-temperature thermal desorption. Soil is heated to 300 to 400°F, well below incineration temperatures. Air is blown through the soil as a carrier for the desorbed organics, then captured and treated, typically by condensation.

In one such low-temperature soil treatment unit, the system removed the volatile organic compounds to below the detection limit.[133] A removal efficiency of 88 percent was found for a less volatile organic compound.

Surfactant-Enhanced NAPL Remediation

lec pg 26

As noted earlier, the effectiveness of pump-and-treat (as well as other groundwater remediation alternatives) can be detrimentally affected by the presence of nonaqueous-phase liquids (NAPLs). NAPL is a long-term source of contaminants in groundwater as the contaminant partitions from the nonaqueous to the aqueous phase. One way to speed the remediation of NAPL-contaminated sites would be to speed the partitioning into the aqueous phase. Studies have found that the use of nonionic surfactants (such as polyethoxylate nonyl phenol) may significantly enhance the recovery of NAPL contaminants.[134] Studies of NAPL recovery are further complicated by what is essentially a change in the formation permeability when NAPL is present. Specifically, when discontinuous NAPL is present in a saturated soil, capillary forces may be stronger than opposing viscous and buoyancy forces, resulting in a permeability that is less than a water-saturated permeability.[135] Studies using gasoline (a light NAPL, or LNAPL) show that surfactants are effective in reducing these interfacial surface tensions and, hence, enhancing recovery.[136,137] The area of surfactant-enhanced remediation is still developing, but the work is underpinned by earlier work regarding multiphase fluid flow[138,139] and nonaqueous phase dissolutioning.[140,141] Surfactants have also been found to accelerate the bulk removal of NAPL in air sparging (surfactants did not facilitate removal of residual levels, however).[142]

DISCUSSION TOPICS AND PROBLEMS

16-1. The remedial investigation site characterization at an old industrial facility on Long Island revealed a sandy site with shallow groundwater contaminated with low levels

of VOCs. The contaminant plume was found to migrate a short distance beyond the property line but had not yet reached residences relying on groundwater for their drinking water supply. What might be the objective for cleanup of this site?

16-2. Cite three advantages and three disadvantages of containment as a means of hazardous waste site remediation.

16-3. Suppose that a section of the circumferential barrier wall described in Example 16-1 is constructed without adequately keying into an underlying layer of low hydraulic conductivity. A section of the original material 100 ft long and 3 ft deep remains with the original aquifer hydraulic conductivity of 1×10^{-2} cm/s (a shallow buried valley). What would the leakage through the wall be from this zone? Would this jeopardize the success of the associated pump-and-treat system?

16-4. A 1-m-thick up-gradient barrier is planned to control the influx of clean groundwater to extraction wells. The down-gradient groundwater extraction wells and treatment plant are designed to operate at 2 cm^3/s. The up-gradient barrier is planned to be 500 m long, and the aquifer is 10 m thick. It is estimated that the average drawdown across the wall will be 4 m. By calculating the flow through the wall at various values of hydraulic conductivity, recommend a design value for the barrier.

16-5. Calculate the volume of water to be mixed with a 100-lb bag of bentonite for a 5 percent bentonite-water slurry. What will be the unit weight of this slurry?

16-6. What is the factor of safety for a trench 10 m deep excavated in sand having an angle of internal friction of 30°? The unit weight of the fluid is as calculated in Problem 16-5. The sand has a total bulk density of 2.1 g/cm^3. How does the factor of safety change for a trench 20 m deep?

16-7. What is the factor of safety for a trench 10 m deep excavated in clay having a shear strength of 1 kg/cm^2? The unit weight of the fluid is as calculated in Problem 16-5. The clay has a total density of 2.2 g/cm^3. How does the factor of safety change for a trench 20 m deep?

16-8. Describe the steps to be taken prior to determining a containment alternative at a hazardous waste site.

16-9. Describe the hydrologic pathways for contaminant migration at a typical waste disposal site.

16-10. List and briefly describe the types of vertical barriers available as passive containment alternatives.

16-11. What are active containment systems? List some alternative techniques.

16-12. Construct the flow net for the conditions defined in Example 16-1.

16-13. Soil on a five-acre site has become contaminated with a mixture of volatile organic compounds. Prepare a brief outline indicating what type of data you would need to collect to begin analyzing remedial measures.

16-14. Use Hartley's method to estimate the rate of volatilization of trichloroethane. Use an estimated diffusion coefficient of 0.80 cm^2/s.

16-15. Repeat Problem 16-14 for benzene. Use an estimated diffusion coefficient of 0.087 cm^2/s.

16-16. Repeat Problem 16-14 for toluene. Use an estimated diffusion coefficient of 0.078 cm^2/s.

16-17. Briefly describe the alternative processes that could be considered for removing volatile organics from soils.

16-18. An underground storage tank (UST) containing trichloroethane at a chemical plant has ruptured. The plume currently extends west to a distance of 345 ft and is 120 ft wide at this distance. The depth of the plume is 12 ft and the seasonal high groundwater table is at a depth of 7 ft. Provide a preliminary design and layout of a soil vapor extraction system. Specifically determine:

(*a*) The required number of extraction wells and their layout.
(*b*) The volumetric capacity in cubic feet per minute (ft^3/min) and operating system pressure of the blower.
(*c*) Equipment layout and any required ancillary facilities.

State any assumptions made.

16-19. Identify additional parameters other than those presented in Table 16-2 that might affect SVE system performance. Further, describe how these parameters influence system performance. (Refer to Lyman, et al., Reference 85, for additional information.)

16-20. Given the following octanol water partition coefficients K_{ow}, indicate for which compounds vacuum extraction is a candidate process for removing the compound from the vadose zone of contaminated soil.

Compound	log K_{ow}
Trichloroethene	2.38
Pyrene	4.88
Polychlorinated biphenyl congeners	6.01
Vinyl chloride	1.38
Ethyl benzene	3.15
Phenol	1.46
Lindane	3.90
DDT	6.19
2-butanone (methyl ethyl ketone)	0.26

REFERENCES

1. The Comprehensive Environmental Responsibility, Compensation and Liability Act (Superfund), 42 USC 9601 et seq., 1980.
2. Denner, J.: "Contaminated Land: Policy Development in the UK," *Proceedings of the 3rd Annual Conference on Contaminated Land: Policy, Regulation, Technology,* IBC Tech. Services, Ltd., London, February 1992.
3. Martin, Ian, and Paul Bardos: "A Review of Full Scale Treatment Technologies for the Remediation of Contaminated Soil," *Report for the Royal Commission on Environmental Pollution,* EPP Publications, Surrey, U.K, 1996.
4. Loxham, M.: "The Control of Contaminant Migration Pathways as the Key to Cost Effective Remedial Measures for Contaminated Land," *Proceedings of the 2nd Conference on Contaminated Land: Policy, Regulation, and Technology,* IBC Technical Services, London, February 1991.
5. Rounds, Dennis: "The Rewards of Risk-Based Corrective Action," *ASTM Standardization News,* American Society for Testing and Materials, Conshohocken, Pa., pp. 22–25, February 1999.
6. Ibid.
7. Kasamas, H., V. Franzius, and C. Ferguson: "Concerted Action on Risk Assessment for Contaminated Sites in the European Union—CARACAS," *Proceedings of the 6th International FZK/TNO Conference (ConSoil '98),* Edinburgh, Thomas Telford Publishing, London, pp. 19–24, May 1998.
8. Peck, R. B.: "Advantages and Limitations of the Observational Method in Applied Soil Mechanics," *Geotechnique,* vol. 19, pp. 171–187, 1969.
9. Brown, S. M., D. R. Lincoln, and W. A. Wallace: "Application of the Observational Approach to Remediation of Hazardous Waste Sites," CH2M-Hill, Bellevue, Wash., April 1969.

10. U.S. EPA: "Presumptive Remedy for CERCLA Municipal Landfill Sites," Directive No. 9355, 0-49FS, EPA/540/93035, September 1993.
11. Gee, G., A. Ward, and M. Fayer: "Surface Barrier Research at the Hanford Site," *Proceedings of the International Containment Conference,* U.S. Dept. of Energy, Germantown, Md., pp. 305–311, 1997.
12. Khire, M. V., C. H. Benson, and P. J. Bosscher: "Field Data from a Capillary Barrier and Model Predictions with UNSAT-II," *Journal of Geotechnical and Geoenvironmental Engineering,* ASCE, vol. 125, no. 6, pp. 518–527, June 1999.
13. Visser, W., E. Elkenbracht, I. Martin, S. Johnson, and P. Bardos: "The Amsterdam Questionnaire," *Proceedings of the 6th International FZK/TNO Conference (ConSoil '98),* Edinburgh, Thomas Telford Publishing, London, pp. 729–730, May 1998.
14. Mott, H. V., and W. J. Weber: "Solute Migration in Soil-Bentonite Containment Barriers," *Proceedings of the 10th National Conference on the Management of Uncontrolled Hazardous Waste Sites,* HMCRI, Washington, D.C., pp. 526–533, November 1989.
15. Mott, H. V., and W. J. Weber: "Diffusion of Organic Contaminants Through Soil-Bentonite Cut-Off Barriers," *Research Journal WPCF,* vol. 63, no. 2, pp. 166–176, March/April 1991.
16. Ward, Linda M.: "Close-up on Cleanup at Lipari Waste Site," *Hazardous Materials & Waste Management Magazine*, pp. 40–42, May–June 1984.
17. Xanthakos, P. P.: *Slurry Walls,* McGraw-Hill, New York, 1974.
18. Spooner, Philip, et al.: *Slurry Trench Construction for Pollution Migration Control,* EPA-540/2-84-001, U.S. Environmental Protection Agency, Cincinnati, February 1984.
19. Nash, K. L. "Stability of Trenches Filled with Fluids," ASCE, *J. Const. Div.*, vol. 100, no 4, pp. 533–542, 1974.
20. GKN Hayward-Baker: *Case History—Mountain View Slurry Wall,* company literature, Odenton, Md., 1989.
21. McCandless, R. M., and A. Bodocsi: "Hydraulic Characteristics of Model Soil-Bentonite Slurry Cutoff Walls," *Proceedings of the 5th National Conference on Hazardous Wastes and Hazardous Materials,* Las Vegas, pp. 198–201, April 1988.
22. Ryan, C. R.: "Vertical Barriers in Soil for Pollution Containment," *Geotechnical Practice for Waste Disposal '87,* ASCE Geotechnical Special Publication No. 13, pp. 182–204, June 1987.
23. Acar, Y. B., A. Hamidon, S. Field, and L. Scott: "The Effects of Organic Fluids on Hydraulic Conductivity of Compacted Kaolinite," *Hydraulic Barriers in Soil and Rock,* ASTMSTP 874, A. I. Johnson, R. K. Frobel, and N. J. Cavalli, eds., American Society for Testing and Materials, Philadelphia, pp. 171–187, 1985.
24. Evans, J. C., H. Y. Fang, and I. J. Kugelman: "Containment of Hazardous Materials with Soil-Bentonite Slurry Walls," *Proceedings of the 6th National Conference on the Management of Uncontrolled Hazardous Waste Sites,* Washington, D.C., pp. 249–252, November 1985.
25. Alther, G. R., J. C. Evans, K. A. Witmer, and Hsai-Yang Fang: "Inorganic Permeant Effects Upon Bentonite," ASTM, *Hydraulic Barriers in Soil and Rock,* STP No. 874, Philadelphia, pp. 64–77, July 1985.
26. Jefferis, S. A.: "Bentonite-Cement Slurries for Hydraulic Cut Offs," *Proceedings of the 10th International Conference on Soil Mechanics and Foundation Engineering,* Stockholm, June 15–19, 1981, A. A. Balkema, Rotterdam, pp. 435–440.
27. "Specification for the Construction of Slurry Trench Cut-Off Walls as Barriers to Pollution Migration," Institute of Civil Engineers, Construction Industry and Information Association and Building/Research Establishment, Thomas Telford Publishing, London, 43 pp., 1999.
28. Evans, J. C., and A. R. Dawson: "Slurry Walls for the Control of Contaminant Migration: A Comparison of United Kingdom and United States Practices," *ASCE Specialty Con-*

ference on Geo-Engineering for Underground Facilities, ASCE Geotechnical Special Publication No. 90, June 1999.
29. Garvin, S. L., P. Tedd, and V. Paul: "Research on the Performance of Cement-Bentonite Containment Barriers in the United Kingdom," *Proceedings of the Second International Symposium on Environmental Contamination in Central and Eastern Europe,* Technical University of Budapest, Budapest, pp. 414–416, September 23–30, 1994.
30. Garvin, S. L., C. S. Hayles, and P. Tedd: "Cemetitious Cut-Off Walls with PFA Additions," *Proceedings of the 3rd International Symposium on Environmental Contamination in Central and Eastern Europe,* Warsaw, pp. 394–396, September 10–13, 1999.
31. Reina, P.: "Slurry Trench Cutoff Surrounds Plant Site," *Engineering News Record,* p. 44, October 1, 1987.
32. Evans, J. C., E. D. Stahl, and E. Droof: "Plastic Concrete Cutoff Walls," *Geotechnical Practice for Waste Disposal '87,* ASCE Geotechnical Special Publication No. 13, pp. 462–472, June 1987.
33. *Engineering News Record:* "Cutoff Wall Scores a Record," *ENR*, pp. 17–18, December 1990.
34. Boyes, R. G. H.: *Structural and Cut-Off Diaphragm Walls,* Applied Science Publishers, London, 1975.
35. Leonards, G. A., F. Schmednecht, J. L. Chameau, and S. Diamond: "Thin Slurry Cutoff Walls Installed by the Vibrated Beam Method," *Hydraulic Barriers in Soil and Rock,* ASTM STP 874, A. I. Johnson, R. K. Frobel, N. J. Cavalli, and C. B. Peterson, eds., American Society for Testing and Materials, Philadelphia, pp. 34–44, 1985.
36. McLay, D.: "Installation of a Cement-Bentonite Slurry Wall Using the Vibrating Beam Method—A Case History," *Proceedings of the Nineteenth Mid-Atlantic Industrial Waste Conference,* Bucknell University, Lewisburg, Pa., pp. 272–282, June 1987.
37. ASCE: "Deep Soil Mixing Builds Cut-off Wall," *Civil Engineering,* ASCE, p. 27, May 1990.
38. Jessberger, H. L. ed.: *Geotechnics of Landfills and Contaminated Land, Technical Recommendations,* European Technical Committtee, ETC 8, Ernst & Sohn, 1991.
39. GKN Hayward-Baker: *Case History—Oxnard Slurry Wall,* company literature, 1989.
40. Evans, J. C., and M. J. Prince: "Additive Effectiveness in Minerally Enhanced Slurry Walls," *ASCE Specialty Conference on In Situ Remediation of the Geoenvironment,* ASCE Geotechnical Special Publication No. 71, October 1997.
41. Evans, J. C., T. L. Adams, and M. J. Prince: "Metals Attenuation in Minerally Enhanced Slurry Walls," *Proceedings of the 1997 International Containment Technology Conference,* U. S. Department of Energy, February 1997.
42. Rabideau, A. J., P. Shen, and A. Khandelwal: "Feasibility of Amending Slurry Walls with Zero-Valent Iron," *Journal of Geotechnical and Geoenvironmental Engineering,* ASCE, vol. 125, no. 4, pp. 330–333, April 1999.
43. Houlsby, A. C.: "Cement Grouting for Dams," *Proceedings of the 10th National Conference of the Conference on Grouting in Geotechnical Engineering,* W. H. Baker, ed., ASCE, New Orleans, pp. 1–34, February 1982.
44. Trezek, G. J.: "Grout and Slurry Walls for Hazwaste Containment: The Down Side," *Hazardous Waste and Hazardous Materials,* vol. 3, no. 3, pp. 281–292, 1986.
45. Bodocsi, A., and M. T. Bowers: "Permeability of Acrylate, Urethane, and Silicate Grouted Sands with Chemicals," *Journal of the Geotechnical Engineering Division,* ASCE, vol. 117, no. GT8, pp. 1227–1244, August 1991.
46. Persoff, P., J. Apps, G. Moridis, and J. M. Whang: "Effect of Dilution and Contaminants on Sand Grouted with Colloidal Silica," *Journal of Geotechnical and Geoenvironmental Engineering,* ASCE, vol. 125, no. 6, pp. 461–469, June 1999.
47. Evans, J. C., G. P. Lennon, and K. A. Witmer: "Analysis of Soil-Bentonite Backfill Placement in Slurry Walls," *Proceedings of the 6th National Conference on the Management*

of Uncontrolled Hazardous Waste Sites, Washington, D.C., pp. 357–361, November 1985.
48. McCandless, R. M., and A. Bodocsi: "Investigation of Slurry Cutoff Wall Design and Construction Methods for Containing Hazardous Wastes," EPA 600-S2-87/063, U.S. Environmental Protection Agency, Cincinnati, November 1987.
49. Cooley, B.: "Behavior of Soil-Bentonite Slurry Trench Cutoff Walls," honors thesis, Bucknell University, December 1991.
50. Evans, J. C., Hsia-Yang Fang, and I. J. Kugelman: "Containment of Hazardous Materials with Soil-Bentonite Slurry Walls," *Proceedings of the 6th National Conference on the Management of Uncontrolled Hazardous Waste Sites,* Washington, D.C., pp. 249–252, November 1985.
51. Raabe, E. W., and S. Toth: "Producing Slurry Walls and Sealing Slabs by Applying Soilcrete Process," *Proceedings of the Deep Foundations Conference,* United Kingdom, pp. 111–118, 1989.
52. Brunsing, T. P., and W. E. Grube, Jr.: "A Block Displacement Technique to Isolate Uncontrolled Hazardous Waste Sites," *Proceedings of the National Conference on Management of Uncontrolled Hazardous Waste Sites,* HMCRI, pp. 249–253, November 1982.
53. Mackay, D. M., and J. A. Cherry: "Groundwater Contamination: Pump-and-Treat Remediation," *Env. Sci. Technology,* vol. 23, no. 6, pp. 630–636, 1989.
54. U.S. Environmental Protection Agency: "Methods for Monitoring Pump-and-Treat Performance," EPA/600/R-94/123, U.S. EPA, June 1994.
55. Javendel, I., C. Doughty, and C. F. Tsang: *Ground Water Transport: Handbook of Mathematical Models,* American Geophysical Union Water Resources Monograph No. 101, Washington, D.C., 1994.
56. Javendel, I., and C. F. Tsang: "Capture Zone-Type Curves: A Tool for Aquifer Cleanup," *Ground Water,* vol. 24, pp. 616–625, 1986.
57. Schaefer, J. M.: "Reversed Pathline Calculation of Time-Related Capture Zones in Nonuniform Flow," *Ground Water,* vol. 25, no. 3, pp. 283–289, 1987.
58. Schaefer, F. B.: "GWATH: Interactive Ground Water Flow Path Analysis," *Illinois State Water Survey Bulletin 69,* 1987.
59. Newson, J. P., and J. L. Wilson: "Flow of Ground Water to a Well Near a Stream—Effective of Ambient Ground Water Flow Direction," *Ground Water,* vol. 26, no. 6, pp. 703–711, 1988.
60. Fitts, C. R.: "Simple Analytic Functions for Modeling Three-Dimensional Flow in Layered Aquifers," *Water Resources Research,* vol. 25, no. 5, pp. 943–948, 1989.
61. Fitts, C. R.: "Well Discharge Optimization Using Analytical Elements," *Proceedings of the 1993 Ground Water Modeling Conference,* IGMWC, Golden, Colo., 1993.
62. Pollock, D. W.: *Documentation of Computer Programs to Compute and Display Pathlines Using Results from the USGS Modular 3-Dimensional Finite-Difference Ground Water Flow Model,* Open File Report 89-381, U.S. Geological Survey, 1989.
63. Hagemeyer, R. T., P. F. Anderson, R. M. Greenwald, and J. L. Clausen: "Evaluation of Alternative Plume Containment Designs at the Paducah Gaseous Diffusion Plan Using MODMAN, A Well Pumpage Optimization Module for MODFLOW," *Proceedings of the 1993 Ground Water Modeling Conference,* IGMWC, Golden, Colo., 1993.
64. Schweitzer, G.: "Assessing Soil Contamination at Love Canal," *Proceedings of the National Conference on Management of Uncontrolled Hazardous Waste Sites,* Hazardous Materials Control Research Institute, pp. 399–405, November 1982.
65. Brown, R. A.: "Air Sparging: A Primer for Application and Design," chap. 19 in *Subsurface Restoration,* C. H. Ward, J. A. Cherry, and M. R. Scalf, eds. Ann Arbor Press, Ann Arbor, Mich., 1997.

66. Johnson, R. L., P. C. Johnson, D. B. McWhorter, R. E. Hinchee, and I. Goodman: "An Overview of Air Sparging," *Ground Water Monitoring and Remediation,* vol. 13, no. 4, pp. 127–135.
67. Rabideau, A. J., and J. M. Blayden: "Analytical Model for Contaminant Mass Removal," *Ground Water Monitoring and Remediation,* vol. 18, no. 4, pp. 120–130.
68. Adams, J. A., and K. R. Reddy: "Laboratory Study of Air Sparging of TCE-Contaminated Saturated Soils and Ground Water," *Ground Water Monitoring Review,* pp. 182–190, Summer 1999.
69. McKeon, T., G. Dawson, and S. Peterson: "In-Situ Groundwater Remediation Using VOCs In-Well Stripping: Modeling, Design and Performance Data," *In Situ Remediation of the Geoenvironment, Proceedings of the Conference,* J. C. Evans, ed., ASCE Geotechnical Special Publication No. 71, pp. 98–112, February 1997.
70. Schweitzer, G.: "Assessing Soil Contamination at Love Canal," *Proceedings of the National Conference on Management of Uncontrolled Hazardous Waste Sites,* pp. 399–405, November 1982.
71. Ralston, M.: "Trenching with Vegetables," *Civil Engineering,* ASCE.
72. Day, S.: "Extraction/Interception Trenches by the Bio-Polymer Slurry Drainage Trench Technique," *Hazardous Materials Control,* vol. 4, no. 5, pp. 27–31, September/October 1991.
73. Day, S. R., and C. R. Ryan: "The State-of-the-Art in Biopolymer Drain Construction," *Slurry Walls: Design, Construction and Quality Control,* ASTM STP 1129, David B. Paul, Richard R. Davidson, and Nicholas J. Cavalli, eds., American Society for Testing and Materials, Philadelphia, 1992.
74. Hiller, D., and H. Gudeman: "Analysis of Vapor Extraction Data from Applications in Europe," *Proceedings of the 3rd International Conference on New Frontiers in Hazardous Waste Management,* U.S. EPA, pp. 434–441, September 1989.
75. Applegate, J., John K. Gentry, and James J. Malot: "Vacuum Extraction of Hydrocarbons from Subsurface Soils at a Gasoline Contamination Site," in *Superfund '87, Proceedings of the 8th National Conference,* The Hazardous Materials Control Research Institute, Greenbelt, Md., pp. 273–279, 1987.
76. Hiller, D., and H. Gudeman: "Analysis of Vapor Extraction Data from Applications in Europe."
77. Martin, L. M., Richard J. Sarnelli, and Matthew T. Walsh: "Pilot Scale Evaluation of Groundwater Air Sparging: Site-Specific Advantages and Limitations," *Proceedings of the National R&D Conference on the Control of Hazardous Materials,* Hazardous Materials Control Research Institute, pp. 318–327, February 4–6, 1992.
78. Anon.: "In Situ Vapor Stripping: Avoiding Problems Due to Simplifying Assumptions," *The Hazardous Waste Consultant,* pp. 1–5, September/October 1990.
79. Grasso, Dominic: "Soil Vapor Extraction (Module)," *Hazardous Waste Site Remediation,* The University of Connecticut under a grant from the U.S. Environmental Protection Agency, January 1993.
80. Multch, Robert D, Jr., Ann N. Clarke, James H. Clarke, and David J. Wilson: "In Situ Vapor Stripping: Preliminary Results of a Field-Scale U.S.EPA/Industry Funded Research Project," *Superfund '89, Proceedings of the 10th National Conference,* The Hazardous Materials Control Research Institute, Greenbelt, Md., November 1989.
81. Grasso, Dominic, "Soil Vapor Extraction (Module)."
82. Ibid., p. 9.
83. Millington, R. J., and J. M. Quirk: "Permeability of Porous Solids," *Trans. Faraday Soc.,* vol. 57, p. 1200, 1961.
84. Dragun, James: *The Soil Chemistry of Hazardous Materials,* Hazardous Materials Control Research Institute, Washington, D.C., p. 271, 1988.
85. Lyman, Warren J., Warren F. Reehl, and David H. Rosenblatt: *Handbook of Chemical Property Estimation Methods,* McGraw-Hill, New York, 1982.

86. SITE Program Demonstration Test, EPA 540 5–89-003a, U.S. Environmental Protection Agency, April 1989.
87. Thomas, R. G.: "Volatilization from Soil," in W. J. Lyman, W. F. Reehl, and D. H. Rosenblatt, *Handbook of Chemical Property Estimation Methods,* McGraw-Hill, New York, 1982.
88. Verschueren, Karel: *Handbook of Environmental Data on Organic Chemicals,* 2nd ed., Van Nostrand Reinhold, New York, p. 1132, 1983.
89. Lyman, Warren J., William F. Reehl, and David H. Rosenblatt: *Handbook of Chemical Property Estimation Methods,* Environmental Behavior of Organic Compounds, McGraw Hill, New York, pp. 16–35, 1982.
90. Dean, J. A. (ed.): *Lange's Handbook of Chemistry,* 12th ed. McGraw-Hill, 1979.
91. Verschueren, Karel: *Handbook of Environmental Data on Organic Chemicals,* 2nd ed., Van Nostrand Reinhold, New York, 1983.
92. Johnson, P. C., M. W. Kemblowski, J. D. Colthard, D. L. Byers, and C.C. Stanley: *A Practical Approach to the Design, Operation and Monitoring for In-Situ Soil Venting Systems,* Shell Development/Shell Oil Company, Westhollow Research Center, Houston, 1990.
93. Hutzler, N. J., B. E. Murphy, and J. S. Gierke: *State of Technology Review: Soil Vapor Extraction Systems,* U.S. EPA, CR/814319/01-1, 1988.
94. Mark-Brown, N. V.: "Aspects of Soil Venting Design," in *Hydrocarbon Bioremediation,* R. E. Hinchee, B. C. Alleman, R. E. Hoeppel, and R. N. Miller, eds. Lewis Publishers, Ann Arbor, Mich., 1994.
95. Fiss, Edward C., and Michael O. Smith: "The Application of Vapor Extraction for the Removal of Volatile Organic Compounds in Soils," presented at *Aquifer Reclamation and Source Control Conference,* New Jersey Institute of Technology, Newark, N.J., June 1991.
96. Ibid.
97. U.S. Army Toxic and Hazardous Materials Agency: "Installation Restoration General Environmental Technology Development Contract DAAK11-82-C-0017, Final Report," Task 11, In Situ Air Stripping of Soils Pilot Study, October 1985.
98. Fiss, E. C., and M. O. Smith, "The Application of Vapor Extraction for the Removal of Volatile Organic Compounds in Soil."
99. Anon.: "In Situ Air Stripping of Soils Pilot Study, Installation Restoration General Environmental Technology Development Contract DAAK11-82-C-0017, Final Report," prepared for U.S. Army Toxic and Hazardous Materials Agency, October 1985.
100. Ibid.
101. Buscheck, T. E., and R. G. Peargin: "Summary of a Nationwide Vapor Extraction System Performance Study," *Proceedings National Water Well Association,* Houston, pp. 205–219, 1991.
102. Chevron Corp: "General Procedures and Data Interpretation for Vapor Extraction System Pilot Tests," Chevron Corp., Houston, 1992.
103. McMurty, D. C., and R. O. Elton: "New Approach to In Situ Treatment of Contaminated Ground Waters," *Environmental Progress,* vol. 4, no. 3, pp. 168–170, 1985.
104. Blowes, D.W., and C. J. Ptacek: "Geochemical Remediation of Ground Water by Permeable Reaction Walls: Removal of Chromate by Reaction with Iron Bearing Solids," *Proceedings of the Subsurface Restoration Conference, Third International Conference on Ground Water Quality Research,* Dallas, pp. 21–24, June 1992.
105. Pankow, J. F., R. L. Johnson, and J. H. Cherry: "Air Sparging in Gate Walls in Cut Off Walls and Trenches for Control of Plumes of Volatile Organic Compounds (VOCs)," *Ground Water,* vol. 31, no. 4, pp. 654–663, July/August 1993.
106. Archer, W. L., and M. K. Harter: "Reactivity of Carbon Tetrachloride with a Series of Metals," *Corrosion,* vol. 34, no. 5, pp. 159–162, 1978.

107. Shikaze, S. G., C. D. Austins, D. J. A. Smyth, J. A. Cherry, J. F. Barker, and E. A. Sudicky: "Hydraulics of a Funnel-and-Gate System: Three-Dimensional Numbering Analysis," *Proceedings of the I.A.H. Conference,* Edmonton, Alberta, Canada, 1995.
108. Gillham, R. W., and S. F. O'Hannesin: "Enhanced Degradation of Halogenated Aliphatics by Zero-Valent Iron," *Ground Water,* vol. 32, no. 6, pp. 958–967, 1994.
109. Gillham, R. W., and D. R. Burris: "In Situ Treatment Walls—Chemical Dehalogenation, Denitrification, and Bioaugmentation," *Proceedings of the Subsurface Restoration Conference,* National Center for Ground Water Research, Dallas, June 21–24, 1992.
110. Devlin, F., and J. F. Barber: "A Semipassive Nutrient Injection Scheme for Enhanced In Situ Bioremediation," *Ground Water,* vol. 32, no. 3, pp. 374–380, 1994.
111. Devlin, F.: "Field Injection of Nutrient Pulse in an In Situ Bio-Stimulation Experiment," *Water Resources Research,* vol. 32, no. 9, pp. 2869–2877, 1996.
112. U.S. Environmental Protection Agency: "Permeable Reactive Barrier Technology for Contaminant Remediation," EPA/600/R-98/125, U.S. EPA, September 1998.
113. Sivavec, T. J., and D. P. Horney: "Reductive Dechlorination of Chlorinated Ethenes by Iron Metal," *Proceedings of the 209th ACS National Meeting,* Anaheim, Calif., April 2–7, 1995.
114. Sivavec, T. J., D. P. Horney, and S. S. Baghel: "Reduction Dechlorination of Chlorinated Ethenes by Iron Metal and Iron Sulfide," *Proceedings of Minerals, Emerging Technologies in Hazardous Waste Management VII,* Special Symposium of the American Chemical Society, Atlanta, September 17–20, 1995.
115. Johnson, T. L., M. M. Scherer, and P. G. Tratnyek: "Kinetics of Halogenated Organic Compound Degradation by Iron Metal," *Environmental Science and Technology,* vol. 30, no. 8, pp. 2634–2640, 1994.
116. Shoemaker, S. H., et al.: "Permeable Reactive Barriers," *Assessment of Barrier Containment Technologies: A Comprehensive Treatment for Environmental Remediation Applications,* chap. 11, R. R. Rumer and J. Mitchell, eds. U.S. Department of Energy, pp. 301–353, 1994.
117. Eykholt, G. R.: "Uncertainty-based Scaling of Iron Reactive Barriers," *In situ Remediation of the Geoenvironment,* American Society of Civil Engineers, New York, pp. 41–55, 1997.
118. Xu, Y., and F. W. Schwartz: "Immobilization of Lead in Groundwater with a Reactive Barrier System," *Presented at the 3rd International Conference on Groundwater Quality Research,* Dallas, June 21–24, 1992.
119. Raghavan, R., E. Coles, and D. Dietz: *Cleaning Excavated Soil Using Extraction Agents: A State-of-the-Art Review,* EPA/600/S2-89/034, U.S. Environmental Protection Agency, January 1990.
120. Zagula, S. J., and E.W. Beitinger, "Developing a Remediation Strategy for Contaminated Sediments: Selecting Removal, Treatment, Disposal, and Re-Use Alternatives," *Proceedings of the 48th Annual Purdue Industrial Waste Conference,* Purdue University, West Lafayette, Ind., 1993.
121. Raghavan, R., E. Coles, and D. Dietz: *Cleaning Excavated Soil Using Extraction Agents: A State-of-the-Art Review,* EPA/600/S2-89/034, January 1990.
122. Bardos, P., et. al.: "Contaminated Land Treatment Concepts under Development at Warren Spring Laboratory," *Euroenviron Contaminated Land Workshop,* September 1990.
123. U.S. Environmental Protection Agency, "Soil Sediment Washing System," EPA/540/MR-921075, U.S. EPA, October 1992.
124. Casagrande, L.: "The Application of Electro-osmosis to Practical Problems in Foundations and Earthwork," Technical Paper No. 30, Building Research Establishment, London.
125. U.S. EPA: "Bergman USA Soil Sediment Washing Technology: Applications Analysis Report," EPA/540/AR-92/075, September 1995.

CHAPTER 17

Evaluation and Selection of Remedial Actions and Corrective Measures

Who shall decide, when doctors disagree,
And soundest casuists doubt, like you and me?
Pope

17-1 OVERVIEW OF REMEDY SELECTION PROCESS

After investigating a site and defining the hazards that it poses, two basic questions remain:

1. To what level should the site be remediated to protect human health and the environment (i.e., "how clean is clean")?
2. How should this level of protection be achieved (i.e., what is the "best" remedy)?

The process of answering the question is commonly referred to as the remedy evaluation and selection process, which is the focus of this chapter.

Early in the history of the U.S. Environmental Protection Agency (EPA) Superfund and RCRA programs, this process involved a rigorous series of steps that required a detailed technical analysis to support the development and selection of a final remedial plan for a site. During the last decade (covering the period between the late 1980s and late 1990s), the process of selecting a site remedy evolved considerably. The degree to which the evaluation/selection of a remedy follows the more rigorous approach depends in part on the governing regulatory program. It is also driven by the need to comply with the requirements of the National Oil and Hazardous Substances Pollution Contingency Plan (referred to as the NCP and discussed in Chapter 2).

On one end of the spectrum, federal Superfund sites that commonly deal with highly complex technical and political issues strictly adhere to the NCP requirements.

As a result, remedy selection for Superfund sites typically involves numerous qualitative and semiquantitative evaluation procedures to select and justify a corrective measure. The application of a more rigorous methodology comes at a cost: increased time and expense.

Sites considered relatively simple, such as gasoline stations with leaking underground storage tanks (USTs), lie at the other end of the spectrum. In these cases, the extensive practical experience gained on thousands of sites over the last decade has allowed the evaluation/selection of a site remedy to become much more focused. Accordingly, the time and cost to develop and implement a remedy can be substantially reduced for many sites.

Public feedback on the successes and failures of environmental programs has indicated the desire to move rapidly toward focused evaluations and, where appropriate, the use of *interim* corrective measures. For candidate sites, interim measures take less time to implement and can begin to provide environmental benefits on a "fast track." Indeed, the literature is replete with examples of case studies of hazardous waste sites where *final* remedies have been perceived as failures because the lack of site-specific experience led to predicted remediation expectations and time frames that were technically unachievable.

Recent experience has demonstrated that more cost-effective approaches to site remediation often utilize a phased implementation. Thus, an interim corrective measure can provide the operational information and flexibility needed to support the design of an expanded system. This is more commonly applied at RCRA sites where the regulatory permitting process can ensure that the responsible party will follow through with the long-term operation and stewardship of the facility's remedial program.

To provide the reader with a fundamental understanding of the major aspects of evaluating and selecting remedial alternatives, this chapter will cover the subject with a broad view toward the more rigorous process. However, examples will also be presented to illustrate how the more focused approach can be applied where conditions warrant.

The remedy selection process can be broken down into four major steps:

- Establish remedial action objectives
- Develop alternatives
- Compare alternatives
- Select remedy

These major steps follow the general model for "problem solving" as illustrated in Figure 17-1.

The parties having a stake in the selection of a remedy come from diverse interests representing the local community, industry, environmentalists, the scientific and engineering professions, and government. In a perfect world, engineers and scientists would be objective and independent and, as a result, would not be stakeholders. Their concerns differ as do their values. There may be disagreement about what is the best remedy and even as to what level of protection is needed. Adversarial posturing can occur, making communication difficult, and a lack of trust can even enter the process, as discussed in Sections 1-4 and 8-5.

Such a climate invariably entangles and impedes the decision-making process. However, even if decision making were confined to the technical community, engi-

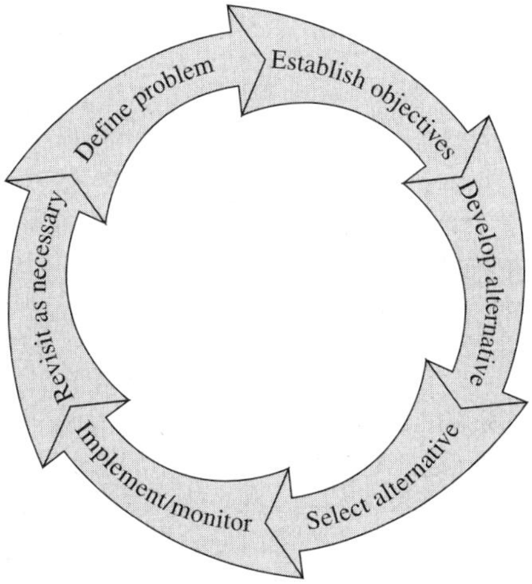

FIGURE 17-1
Problem solving approach to evaluation and selection of remedial actions.

neers and scientists with experience in the same field would probably offer differing opinions regarding the needed level of protection and the best remedy. There are several reasons for the technical difficulty in selecting a remedy, including:

1. Site conditions can be very complicated.
2. Remediation is not straightforward but takes place usually over a long period of time and involves a number of interdependent steps.
3. A great number of remedial alternatives can exist, each having its own technical tradeoffs.
4. Experience does not exist yet to show clearly how best to proceed.

Because of these complicating factors, the selection of a remedy involves a great deal of judgment. The process illustrated in Figure 17-2 was developed to reduce the potential errors that can be introduced by this inherent subjectivity. It is a stepped process and features increasingly intensive analysis of fewer, yet ever more detailed, alternatives. Regardless of whether one is evaluating a complex site or if the site is relatively simple, the selection process should follow the logical progression outlined in this chapter.

Worldwide, and within the United States, there are a number of separate and mostly independent remediation programs, with separate guidance for each. The fundamental components of the classic "problem solving" process are typically found in each program even though specific regulatory guidance may prescribe some variants. This chapter avoids using the specific guidance of any particular program. Regulatory guidance seems to change continuously. Yet, the *fundamental* process of selecting a remedy (i.e., the development and analysis of options) remains constant as it has in other planning endeavors. This chapter illustrates such a process. It probably does not comply with the exact requirements of any regulatory program, and a first step for practitioners in these endeavors is to obtain an understanding of the precise regulatory requirements and guidance applicable to the particular site.

The process covered in this chapter assumes that the site conditions are complicated—the site has complex geological characteristics, multiple sources of contamination, many different contaminants, or complicated pathways between sources and

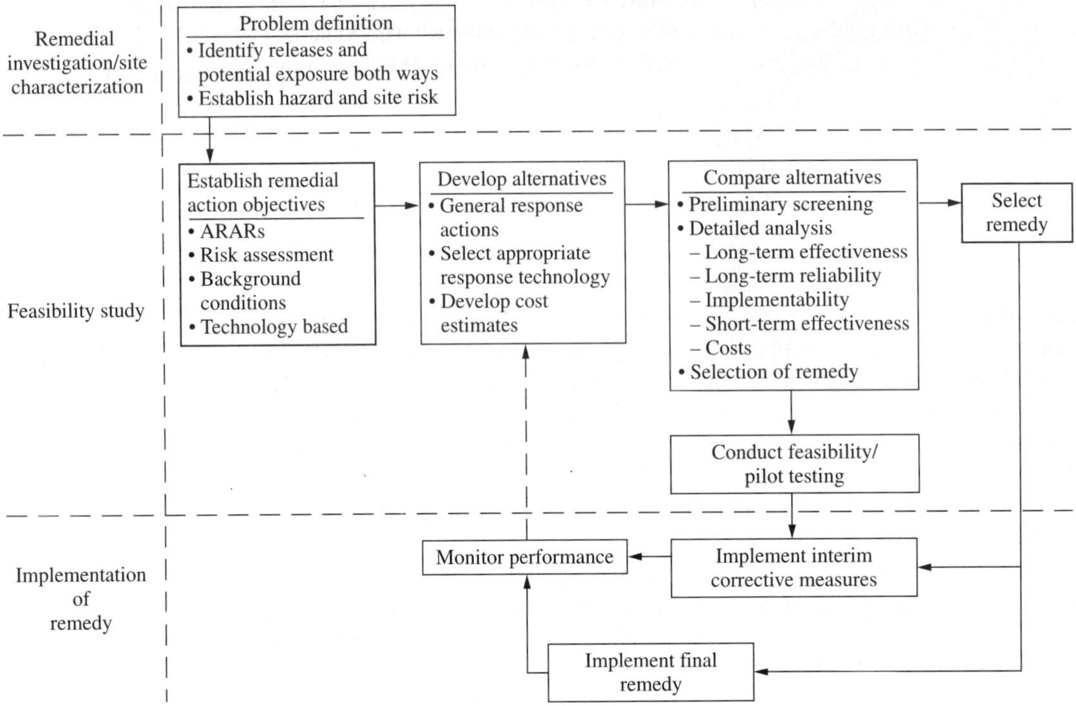

FIGURE 17-2
Remedy selection process overview.

receptors. The selection of a remedy for sites lacking such factors should use a streamlined approach (e.g., minimizing, if not eliminating, screening efforts and limiting the number of detailed alternatives). Nevertheless, the decision maker still should follow the same logical progression outlined in the remedy selection process of this chapter.

17-2 CASE STUDY 1—EVALUATION AND SELECTION OF A REMEDIAL ACTION FOR A "MODEL" SUPERFUND SITE

To illustrate the remedy selection process, examples taken from feasibility studies for several different Superfund sites have been compiled and used to create a generic or model site. The information presented in this case study is available through files and documentation records that are maintained in the public document repositories for the sites that were used to generate the model site. However, for purposes of discussion, some modifications to the details concerning the actual sites from which the model site was derived were made to accommodate presentation within this textbook.

The model site discussed in this case study[1] is a former industrial waste disposal facility that operated on a commercial basis under a state permit in the 1970s. It received a variety of industrial wastes including bulk liquids and drummed liquids and sludges. The wastes came from refineries, chemical plants, and other manufacturers.

It consisted mostly of organic contaminants; however, some wastes were primarily inorganic in nature. Containerized sludges were placed intact without treatment in a pit termed the Drum Pit. Management of bulk and containerized liquids featured three steps:

1. Partial evaporation in shallow ponds.
2. *In situ* mixing of soil with the concentrated waste remaining after evaporation.
3. Excavation and disposal of the soil-waste mixture in a pit, called the Sludge Pit, excavated to bedrock.

The model site was closed in the late 1970s because the operator was unable to comply with the more stringent regulations that were being promulgated for hazardous waste management. Closure of the site included scraping the visually stained soils from the mixing ponds and placing the soil in the Sludge Pit. The Sludge and Drum Pits were covered with a soil cap, and they were referred to collectively as the "main source area." The site also contained a stormwater pond in the channel flanking the eastern boundary of the site, and it had received runoff from waste management areas. This and other surface features, including a nearby creek, are shown in Figure 17-3.

The geology at the site consists primarily of fine-grained sandstones and mudstones plus some unconsolidated alluvial sediments along the creek. The sandstones and mudstones, to a depth of 220 feet, can be divided into five distinctive strata as shown in Figure 17-4. The upper three strata outcrop on the site. The uppermost

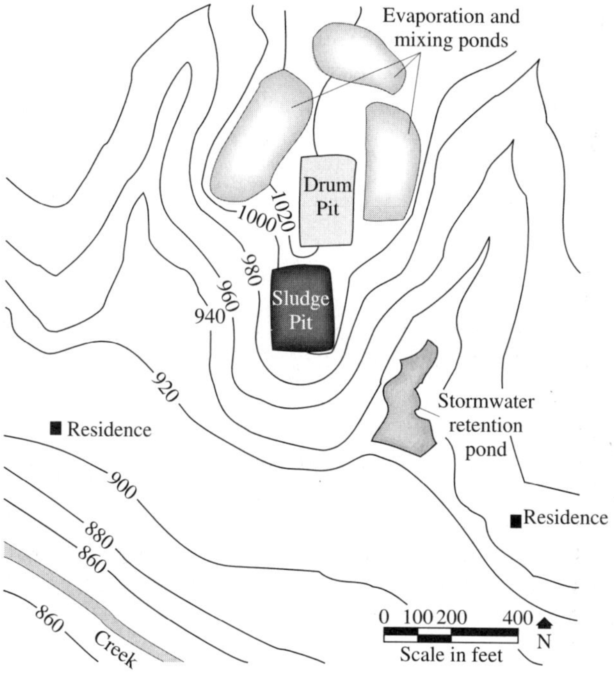

FIGURE 17-3
Plan view of model Superfund site for Case Study 1.

stratum (Stratum I) yields submarginal quantities of groundwater with the water table 15 to 30 feet below the surface. Stratum II is an aquitard. Stratum III resembles Stratum I, except that its yield is somewhat greater, serving as a marginal source of water supply to domestic wells in the region, although not near the site itself. Stratum IV yields only a few gallons per year of fresh to brackish water. Stratum V produces saline water only and has negligible hydraulic connection with overlying strata. In contrast with the bedrock strata, the alluvial deposits yield large quantities of groundwater and are the source of domestic water supply for some of the residents living in the region.

The historic migration of constituents from the main source area occurred primarily via stormwater runoff. Several areas of surficial soils and some sediments within site boundaries show significant levels of organic and inorganic constituents, including some pesticides and PCBs.

Leaching of contaminants from the site affected the shallow groundwater (the Stratum I aquifer) in two areas. A relatively widespread area exists east of the main source area, and affected groundwater moves to the southeast toward the stormwater pond. A smaller area exists southwest of the main source area, extending into the alluvium, and the affected groundwater moves to the southwest and south toward the creek. Constituent migration via groundwater pathways is limited to very slow movement in the shallow bedrock but moves rapidly through the alluvium, ultimately discharging to the creek. Organic constituents have not been detected in the creek, nor have inorganic constituents above background concentrations. The residences served by domestic wells in the alluvium near the site have been provided with an alternate water supply.

Remedial alternatives for the site were analyzed under the EPA Superfund program with two separate studies conducted—one for the sources (ponds, pits, soil,

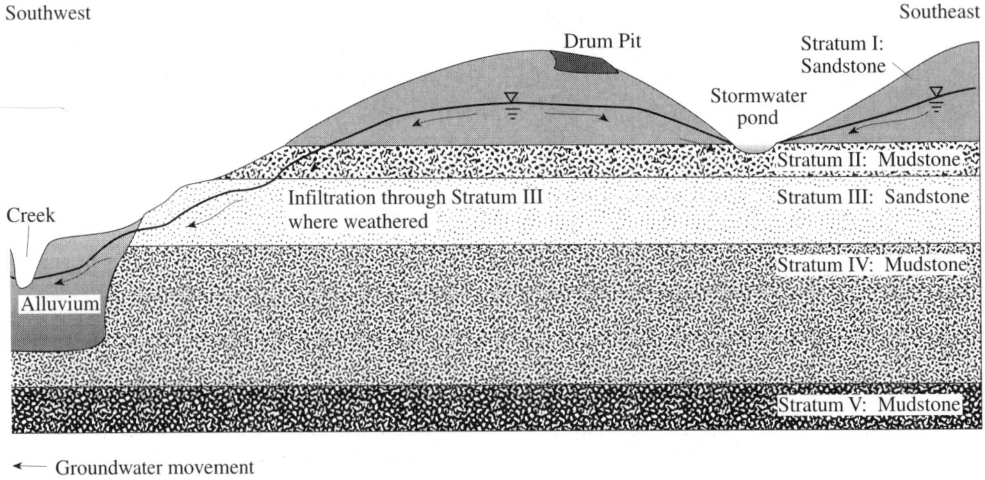

FIGURE 17-4
Simplified cross section illustrating geological strata and groundwater migration pathway.

and sediments) and the other for groundwater. The examples used in this chapter pertain to just the groundwater.

17-3
SITE CONCEPTUAL MODEL

As illustrated in Figure 17-3, the remedial investigation phase should culminate in the overall definition of the problem requiring remediation. In some cases, the first step of the feasibility study phase involves the development of a comprehensive site conceptual model. This approach is especially recommended for sites where remediation may be impracticable.[2,3] The objective of developing the conceptual model is to provide a concise summary of the components of a site into its most elemental parts. The site conceptual model should present the site in terms of the **contaminant sources, exposure pathways,** and **receptors** that exist at the site.

In its most basic form, the site conceptual model may be illustrated by a simplified hydrogeologic cross-section (e.g., Figure 17-4). For highly complex sites, more elaborate models may be needed. For example, at a site where intrinsic bioremediation may play an important role in limiting the migration of affected groundwater, the conceptual model might also include plan-view maps illustrating areas where anaerobic and aerobic degradation processes are occurring. Similarly, existing site features that serve to limit exposure to affected soils (say in the case of old disposal pits covered by a parking lot) may also be incorporated into the conceptual model.

By compiling the site data into a simplified model, the most critical conditions, where a complete exposure pathway from a source to a potential receptor exists, can be more readily identified. This is an important point because, for many sites, the shear volume of data, competing stakeholder interests, and conflicting regulatory requirements can make it very difficult to extract the critical technical issues from those that are perceived to be important.

In the case of the model site (Figure 17-4), potential migration of affected shallow groundwater to a local alluvial aquifer is a primary feature of the site conceptual model. Other elements of the conceptual model for the model site are listed in Table 17-1.

Although no residents currently use the affected portion of the alluvium for drinking water, the alluvium does discharge into the surface water of the creek itself. Likewise, constituents in the affected groundwater would discharge to the creek if not attenuated. Cattle may drink from the creek, and the human ingestion of milk and beef from such cattle does pose a risk, however remote. Contaminants may be taken up by crops irrigated with surface water or shallow groundwater and then ingested by humans directly or through animal products as described previously. The possibility also exists for dermal contact or incidental ingestion of the water in the creek.

Thus, the shallow groundwater pathway requires some form of remedial action. In contrast, vertical migration from the source areas (the Drum Pit) into deeper groundwater is unlikely because the intervening low permeability mudstones limit the downward movement of contamination. Other aspects of the conceptual model concerning groundwater at the model site are described in Table 17-1.

TABLE 17-1
Elements of the site conceptual model for the model site

Release mechanism	Leaching from source area to groundwater
Environmental transport	Movement of constituents in groundwater
Potential receptor points	Surface waters of the creek and pond; potential future drinking water wells into alluvium
Routes of human exposure	Incidental dermal absorption and incidental ingestion of surface water of the creek; ingestion of beef and milk from cattle drinking from the creek; ingestion of drinking water from potential future wells into alluvium
Potentially exposed human populations	Landowners of property on both sides of the creek downstream from the site and population served by dairy
Route of environmental exposure and exposed biota	Aquatic biota of the creek

17-4 REMEDIATION OBJECTIVES

The fundamental goal of any remedial alternative is to provide adequate protection of human health and the environment. Beyond this broad goal, it is necessary to develop criteria that are measurable and that address specific aspects of the site. Following the flowchart in Figure 17-2: After completing the site characterization and problem definition activities, the first step in conducting a feasibility study is to establish the remedial action objectives (RAOs) for the site. RAOs should specify:

- The constituents of concern
- The exposure pathways and receptors
- Acceptable concentrations for each exposure pathway

This will allow for performance monitoring of the selected alternative during the implementation stage of the remedial project.[4]

There are at least four methods for establishing remedial action objectives:

Method	Example
Regulatory requirements and guidelines	Reduce trichloroethylene in all soil to not greater than the hypothetical regulatory limit of 10 mg/kg
Site-specific risk assessment	Reduce carcinogenic risk posed by all contaminants to not greater than 1×10^{-6}
Background conditions	Reduce all chemical constituents in soil and groundwater to naturally occurring levels
Technology-based solutions	Excavate and incinerate all soil contaminated with greater than 10 mg/kg trichloroethylene

Early in the history of dealing with hazardous waste sites, RAOs were commonly aimed at attaining full site restoration, thereby providing the highest achievable level

of protection. In many cases, regulatory agencies took the approach of imposing specific standards (such as the EPA's maximum contaminant levels, or MCLs) for groundwater or no detectable organics in soils as cleanup goals.

In the early 1990s it became recognized that such "black and white" standards are:

- Not always achievable (for example, because of the presence of NAPLs in the soil and groundwater; see discussion in Section 4-2)
- Sometimes not necessary (for example, this is often the case where current and future industrial use of the affected property indicates that there is no potential for exposure to the general public and that future spills will occur in the same areas)

Thus, the concept of adopting RAOs based on principles of risk assessment began to drive the feasibility study process.[5,6] The important point is that the site characterization must include some understanding of current and future land use activities and risk so that the RAOs can appropriately address the issues specific to each site.

For example, consider a Superfund site located in the middle of a neighborhood where future land use may involve the redevelopment of the property for homes, day-care centers, and other residential purposes. In that situation, it may be necessary to adopt conservatively low cleanup standards for soil and groundwater (i.e., cleanup to MCLs). In contrast, for a chemical spill located within a multifacility industrial complex, where other contamination sources have impacted the environment, attempting to completely restore soil and groundwater to background levels (i.e., pristine, uncontaminated conditions) would be inappropriate. In this case, the cost to restore the affected media, and create an "island of clean in a sea of contamination," would far exceed the potential benefits of the remedial action.

During the mid-1990s, regulatory agencies began to develop and implement regulations specifying remediation criteria that incorporate risk management policies and that would be uniformly applied across all environmental programs (Superfund, RCRA, UST, spill sites, etc.). These programs commonly establish different tiers for developing cleanup criteria based on different exposure assumptions (see, for example, the ASTM Standard E 1739-95: *Guide for Risk-Based Corrective Action Applied at Petroleum Release Sites*[7]).

Texas was one of the earlier states to promulgate such risk-based cleanup standards. Termed the Texas Risk Reduction Rules,[8] this law allows the responsible party to select a remediation approach following one of three risk reduction standards:

Standard 1—requires full site restoration (i.e., cleanup to background conditions). This standard is the most protective and expensive remedial approach, but requires no postclosure care.

Standard 2—requires remediation to a set of conservative default criteria for soil and groundwater that are not site-specific, yet do consider residential and nonresidential exposure settings. This standard requires deed recordation, may require some degree of postclosure care (e.g., monitoring), and would apply to any site regardless of location or other unique conditions.

Standard 3—allows remediation to risk-based cleanup levels calculated by using site-specific exposure assumptions. This approach would potentially yield the

least conservative cleanup levels and thus yield lower remediation costs, and would also require deed recordation and postclosure care.

Objectives Based on Risk Assessment

The development of a remediation objective should take into account the hazards posed by the site, the potential for exposure, and the consequences if exposure were to occur. All of these are the elements embodied in the systematic framework of a quantitative risk assessment (Chapter 14). The concept requires establishing a target level of risk that becomes the remediation objective. An acceptable level of lifetime carcinogenic risk is generally considered to be in the range of 1×10^{-4} to 1×10^{-6}, with 1×10^{-6} recognized as the point of departure for the purposes of conducting a comparative analysis. Similar guidelines exist for noncarcinogenic risk. The risks posed by a site for all source-pathway-receptor combinations under current and future exposures are calculated by receptor and compared with the objective—the target level of risk. If the risks posed by the site exceed the objective, remediation is required to reduce the risk to the target level.

The target risk level thus drives the selection of remedy. A remedy can reduce risk by reducing the hazard (e.g., removing highly contaminated soil) or reducing the potential for exposure (e.g., providing alternate drinking water supplies if the groundwater were contaminated). In fact, by defining an acceptable level of risk, the risk assessment process can be used in reverse to calculate the acceptable concentration or mass of chemical contaminants that could remain on site.

To use risk assessment requires making some far-reaching decisions about future exposure scenarios. Although most sites have not resulted in significant exposure of off-site receptors, the potential is there. The potential exposure depends in part on the future uses made of the site and surrounding areas. To calculate risk, hypothetical scenarios must be developed about future uses of the site, the type of development that could occur on nearby properties, and the future use made of natural resources, particularly groundwater. Starkly different amounts of residual soil or groundwater contamination could be allowed to remain, depending on assumptions such as whether groundwater will be consumed or whether houses will be built on or adjacent to the site. Risk assessment is an excellent tool for examining such scenarios.

The groundwater migration pathway at the model site provides an example of using risk assessment as a basis for setting remediation objectives. As stated above, the calculation of risks requires developing exposure scenarios. The key features of exposure scenarios assumed for the model site are as follows:

- **Exposure Scenario I—Current site conditions.** This scenario appears to present no unacceptable risks because of the fencing of the site, the absence of detectable target contaminants in the creek, and the provision of an alternate water supply to prevent the domestic use of affected alluvial groundwater.
- **Exposure Scenario II—Most probable land and water use over the next 70 years.** This scenario assumes more intensive use patterns and, hence, more potential exposure to hazardous substances than current conditions. Some contaminants are predicted to migrate, and exposure could occur from recreational use of the creek and consumption of beef and dairy products from cattle drinking from the creek.

- **Exposure Scenario III—Hypothetical worst-case scenario.** Future domestic use of affected alluvial ground water is assumed prior to it discharging to the creek. This scenario is characterized as unrealistic in that no credible future circumstances could be identified wherein this particular groundwater would be used as a drinking water supply. For example, it would involve the adjacent residences knowingly consuming water from their affected wells rather than using the alternate water supply.

On the basis of these scenarios, the carcinogenic risks posed by exposure to groundwater from the model site were calculated as shown in Table 17-2.

It should be noted that Exposure Scenarios II and III do not represent expected risks but are based on conservative assumptions and safety factors and are useful only in selecting a remedy. Even so, Exposure Scenario II does not pose unacceptable risk. Only when assuming future domestic use of the alluvial groundwater in Scenario III do the estimated risks enter the debatable range as to what level of risk is acceptable.

Establishment of a risk-derived remedial objective involves deciding what level of risk is acceptable and what is the appropriate exposure scenario. This is a policy decision that should consider "likely future land and resource use, the proximity of residents, the use of surrounding properties, the level of contamination of surrounding properties, and specific environmental issues."[9] To do this is to hypothesize, and it adds a degree of uncertainty that has hindered public acceptance of risk-derived objectives. Nevertheless, risk assessment remains a primary tool for setting objectives. Without it, hypotheses must still be made but on an implicit, intuitive basis labeled as "professional judgment."

Objectives Based on Regulatory Requirements

In some cases there may exist under federal or state environmental law a ". . . standard, requirement, criterion or limitation" that is ". . . legally applicable or relevant and appropriate . . ." under the circumstances of the release or threatened release of hazardous substances at a site.[10] In recognition of this as discussed in Chapter 2,

TABLE 17-2
Summary of carcinogenic risks posed by groundwater pathway at the model site

	Assuming upper bound predicted 70-year average concentrations	Assuming best estimate 70-year average concentrations
Exposure Scenario I	Risk $< 10^{-12}$	Risk$< 10^{-12}$
Exposure Scenario II		
Child 0–6 years	8×10^{-12}	3×10^{-12}
Child 6–12 years	1×10^{-10}	4×10^{-11}
Adult	7×10^{-11}	2×10^{-11}
Exposure scenario III		
Child 0–6 years	1×10^{-5}	7×10^{-6}
Child 6–12 years	8×10^{-6}	3×10^{-6}
Adult	1×10^{-4}	8×10^{-5}

these requirements represent remediation objectives for the actions taken to remedy the site. That is, the selected remedy must attain a level or standard of control equal to them.

In the context of the Superfund program, one task in developing RAOs is to conduct an ARARs analysis. This involves identifying the various **A**pplicable or **R**elevant and **A**ppropriate **R**equirements (see Section 2-4) that will influence the development of the remedial objectives. ARARs can be divided into two categories based on how specifically they were promulgated for hazardous substances encountered in site remediation. Requirements specifically addressing hazardous waste remediation are termed "applicable" requirements. They would be legally enforceable for virtually any remediation program. An example of an applicable standard would be one expressing the maximum allowable concentrations of a specific contaminant in soil. Few such explicit standards exist. These standards would apply to all sites regardless of location. The difficulty in developing national or even statewide standards is the enormous variation in geology, climate, population, environmental setting, and other site factors that determine risk. A standard may not provide adequate protection for one set of conditions and may be overly protective in another case.

The Texas Risk Reduction Standards described above are regulatory requirements and are considered to be "applicable" standards that a selected remedy would be expected to meet. In addition to such applicable standards, there are potentially many "relevant" requirements. Environmental law is replete with requirements in the form of standards, criteria, and guidelines that have been developed as part of programs directed at issues other than hazardous waste (e.g., air quality). A requirement may address a constituent (e.g., lead), an action (e.g., incineration), or a location (e.g., wetlands). If a requirement generally pertains to the related conditions (i.e., constituent, action, or location) of the site or its remedy, the requirement is considered relevant. Many relevant requirements, however, simply are not appropriate for application to Superfund sites. Most simply "were developed to regulate very different types of problems than those involved in the Superfund program."[11] Some requirements, such as zero contaminant concentration goals, as discussed previously, are technically infeasible. Many are arbitrary and bear little resemblance to site conditions. Their application to Superfund sites requires extensive judgment and this often creates disagreement as to whether the interpretation is correct.

EXAMPLE 17-1. RELEVANT AND APPROPRIATE REQUIREMENTS. Name the factors to be considered in determining whether a relevant requirement is appropriate.

Solution. The factors include[12]:

1. The purpose of the requirement
2. The media regulated or affected
3. The substances regulated
4. The entities or interests affected or protected
5. The actions or activities regulated
6. Any variances, waivers, or exemptions to the requirements
7. The type of place regulated
8. The type and size of structure or facility regulated
9. The use or potential use of affected resources

At the model site, a number of ARARs were identified. The key requirements are water quality standards for the surface water bodies that receive discharges of affected groundwater, air quality regulations for fugitive dust control, and ambient air quality standards.

Other Methods for Setting Remediation Objectives

One simple method for establishing objectives is to have a single objective that the site be restored to the **background** conditions that existed before the site was contaminated. For the model site, this would require elimination of all contamination other than that from cattle ranching. Restoration to background conditions is a flawed, if not infeasible, approach at many sites. An example is a site surrounded by heavy industrialization where risk to receptors may not diminish measurably unless the surrounding properties also restore their sites to background conditions. Further, the surrounding industrial use may conflict with the possible recreational and residential uses implied by returning an industrial site to background conditions.

Using **technology-based standards** is another method for developing objectives. Actually, the method would use a subset of the regulatory requirements discussed early in this section, and it has been applied in the United States in the past (e.g., the regulatory program for controlling water pollution). Under this concept, contamination would be reduced to a level consistent with what could be achieved by a specified available technology. For example, if the specified technology were incineration, the concentration of organic contaminants would have to be reduced by 99.99 percent. The decision would remain open whether this was achieved by incineration or an equivalent alternative. This required level of reduction would apply even if it were overly protective considering specific site conditions. If it were deemed underprotective, additional measures would be necessary. A variant of this concept is to develop a set of minimal remedial requirements based on general site conditions and potentially exposed populations.

Example of Remediation Objectives

A number of remedial action objectives for the groundwater pathway at the model site could be considered:

1. Prevent spread of affected groundwater into unaffected areas.
2. Protect drinking water supply at affected residences.
3. Control affected groundwater migration to the alluvium and surface water bodies.
4. Reduce groundwater constituent levels sufficiently to satisfy appropriate regulatory requirements.

Style and content are both important when writing objectives because all subsequent steps focus on meeting those objectives. It is important to note that a remediation objective should express implicitly, if not explicitly: exposure route, receptors, contaminants of concern, and a measurable standard of performance or control. Ideally, the standard of performance could be expressed as an acceptable contaminant

level in quantitative terms (e.g., concentration or amount). Objective 4 above is an example of such an approach in that regulatory standards as they exist provide the quantitative terms. It is not always possible or desirable to use or imply quantitative terms. In such a case, the standard of performance still should be definitive.

EXAMPLE 17-2. IMPLICATION OF REMEDIATION OBJECTIVE. What are the implications of the following three remediation objectives for a hypothetical site where leachate from contaminated soil is migrating via groundwater to potential future locations for drinking water wells?

1. Reduce contamination of soil sufficiently such that levels of contaminants at wells represent a carcinogenic risk no greater than 1×10^{-6}.
2. Prevent spread of contamination to wells.
3. Provide safe drinking water to users of wells.

Solution. Objective 1 is possibly too specific because it necessitates source control, either treatment or removal of contaminated soil. Objective 2 is fully protective of human health but may be accomplished via various measures including containment and groundwater treatment, as well as source control. By not using quantifiable terms, Objective 2 is absolute and would imply to some parties that "zero" contamination is required. Objective 3 could be accomplished via institutional controls and an alternate water supply, without treating, removing, or containing any contamination.

A remediation objective should not specify a technology, or even a general class of technologies. As illustrated in Figure 17-2, that comes later. Specifying a technology may preempt a more effective one, or it could even direct the study toward an infeasible one. For example, at the model site, a sole objective stating "pump and treat bedrock groundwater" is not technically feasible because even aggressive pumping would yield only negligible quantities of groundwater because of hydrogeologic constraints and physicochemical limitations posed by the bedrock formation. This illustrates the need to keep an eye toward what is achievable to prevent establishing impossible objectives.

In some instances, ARARs clearly mandate that a technology be applied. As described earlier, however, this may create problems because even if a specific technology is deemed "appropriate" by some regulation, it may not be feasible at a given site.

Although it is not practical to pump and treat the shallow groundwater at the model site, it is appropriate to limit the amount of affected groundwater discharging to surface water. Because possible future domestic use of the alluvial groundwater poses what could be considered an unacceptable risk, it is also appropriate to prevent this from happening. Thus, the following remediation objectives were established for the groundwater pathway for the model site:

1. Monitor/control affected groundwater discharge to the creek to ensure that surface water quality standards continue to be met.
2. Prevent domestic and agricultural use of contaminated groundwater.
3. Control the spread of the plume of contaminated groundwater.

Remediation objectives represent steadfast requirements that should be considered mandatory for any analysis of alternatives—all alternatives developed for evaluation should, as a minimum, meet the objectives. If an alternative is found unable to meet the objectives, it should be excluded from further consideration.

17-5 DEVELOPMENT OF ALTERNATIVES

Having established remediation objectives, there are six steps in developing alternatives (Figure 17-5):

1. Identify general response actions for each objective.
2. Define volume and area of media requiring remediation.
3. Identify potential technologies.
4. Screen the potential technologies.
5. Identify and acquire data needed to complete screening.
6. Assemble the screened technologies into alternatives.

General Response Actions

The task of developing technical alternatives begins by identifying **general response actions** for each remediation objective. A general response action is a simplified

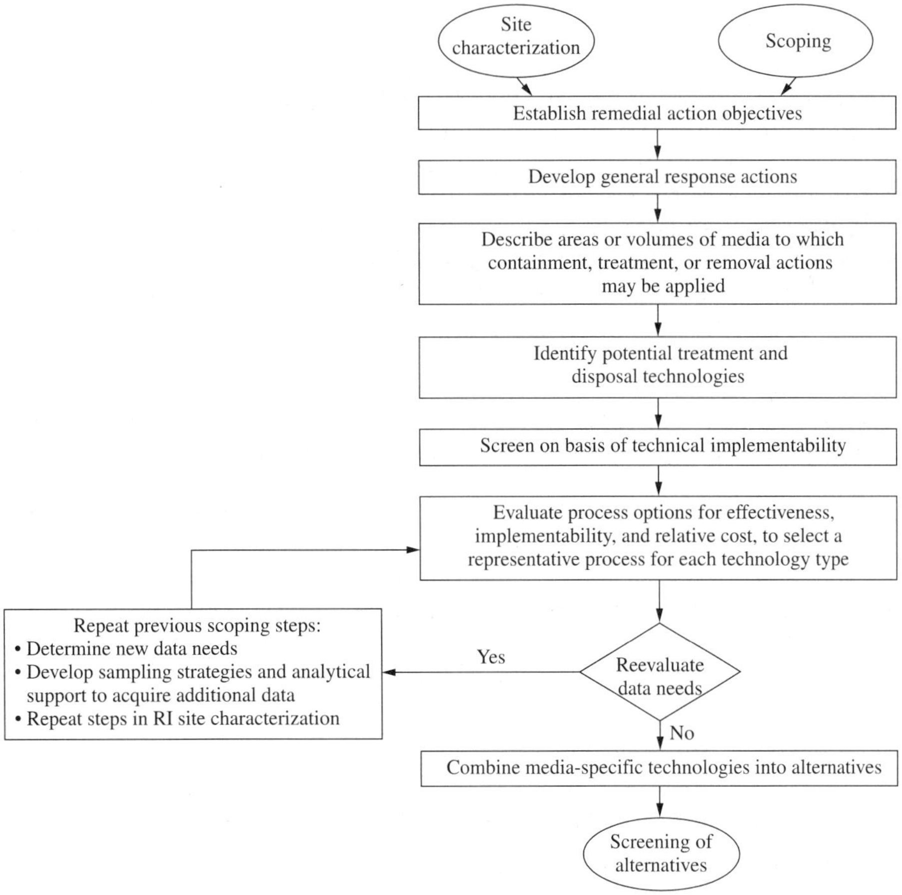

FIGURE 17-5
Development of alternatives. (*After U.S. EPA, Reference 13.*)

description of the remedial actions that will satisfy the objective. It could be thought of as a broad category of technologies. For a site where the objective is to control leaching of constituents from contaminated soils, an example of a general response action could be "excavation and treatment of contaminated soil." This stops short of defining how the soil would be excavated or what treatment technology would be used. Other examples for the same objective are "control infiltration" and "contain migration of constituents." General response actions provide direction and focus to the subsequent tasks.

Define Extent of Media to be Remediated

The general response actions, since they originate from a source-pathway-receptor analysis, are already grouped by medium (e.g., groundwater, soil). The step following the identification of general response actions is to quantify the extent to which the various media must undergo remediation. This need be done only in approximate terms, consistent with the general response action and sufficiently accurate to allow the general sizing of technologies. Examples of this activity for contaminated soil include the following:

1. If the general response action was excavation and treatment of contaminated soil, it is necessary to estimate the volume of soil with contaminants exceeding the levels identified in a remediation objective.
2. If the general response action was controlling infiltration, it is necessary to estimate the areal content of contaminated soil.
3. If the general response action was containment, it is necessary to estimate the lateral and areal extent of contaminated soil.

EXAMPLE 17-3. EXCAVATION OF CONTAMINATED SOIL. Soil samples were collected at a depth of 0.6 m, 1.2 m, and 1.8 m from the property of a former battery-recycling plant. The property measures 300 m by 100 m and is suspected to be contaminated with lead. The sample locations are in a uniform grid as shown in Figure 17-6. Table 17-3 lists the soil depth and the lead concentrations for each soil sample.

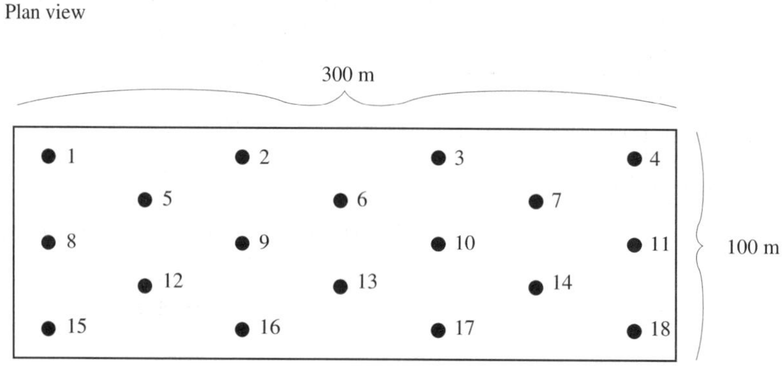

FIGURE 17-6
Contaminated soil sample location.

TABLE 17-3
Soil depth and lead concentrations

Soil sample location	Sample depth, m	Lead concentration, mg/kg	Soil sample location	Sample depth, m	Lead concentration, mg/kg
1	0.6	20	10	0.6	2000
2	0.6	150	10	1.2	1200
3	0.6	25	10	1.8	1000
4	0.6	20	11	0.6	200
5	0.6	300	12	0.6	2000
6	0.6	1200	12	1.2	250
6	1.8	800	13	0.8	800
7	1.2	1100	14	0.8	1400
7	1.8	800	14	1.2	1100
8	0.6	200	14	1.8	900
9	0.6	1800	15	0.6	100
9	1.2	1250	16	0.6	50
9	1.8	850	17	0.6	50
			18	0.6	100

The nature of such contamination is that the concentration of lead decreases with depth. One response action is to excavate for off-site disposal all soil containing greater than 1000 mg/kg of lead. What is the approximate volume of affected soil requiring excavation under this alternative?

Solution. At each sample location, the depth to a concentration of 1000 mg/kg is approximated by interpolation. As an example, such depths for sample locations 6 and 14 are 0.8 m and 1.5 m, respectively. Next, around each sample location exceeding the 1000-mg/kg target, a polygon is delineated extending midway between any two sample stations. Because in this case the samples were located in a uniform grid, each represents an area of approximately 1700 m^2. With this information, the volume of soil to be excavated can be estimated as follows:

Soil sample location	Approximate depth of soil exceeding target level, m	Approximate volume of soil exceeding target level, m^3
6	1.2	2,000
7	1.5	2,500
9	1.5	2,500
10	1.8	3,000
12	1.0	1,700
13	0.3	500
14	1.5	2,500
		Total volume = 14,700 m^3

For ease of calculation in this example a uniform grid was used. As discussed in Chapter 15, a uniform sampling grid is not normally recommended.

TABLE 17-4
Potential groundwater technologies at the generic site

General response action	Types of potential technologies
In situ treatment	Solvent flushing
	Bioreclamation
	Permeable treatment beds
Containment system	Cutoff walls or barriers
	Recovery wells
	Recovery trenches
	Horizontal barriers
	Vacuum extraction wells

Identification and Screening of Technologies

After characterizing the media, potentially applicable response technologies are identified. Many technologies are covered in Part 3 of this book as well as in Chapter 16. Examples of two general response actions applicable to groundwater at the model site, and some types of potential technologies that conform to each general response action, are shown in Table 17-4.

The next step is to evaluate the identified technologies against a set of screening criteria. Important criteria are **suitability** (general technical ability to address contaminants at the site), **implementability** (primarily the ability to construct and operate the technology), **performance** (effectiveness and reliability), **environmental and health concerns, institutional concerns** (permitting), and **cost** (in relative, qualitative terms).

Table 17-5 shows examples of some of the technologies identified for groundwater containment and *in situ* treatment and the results of the screening evaluation at the model site. At this point in the feasibility study (FS) process, the evaluation is very qualitative and must be based largely on judgment and experience. Thus, each technology is simply rated as being favorable (+), neutral (0), or unfavorable (−) with regard to the criteria. The usual reasons for eliminating technologies include unfavorable performance, difficulty in implementation, or excessive cost. For the model site, more than 80 technologies were evaluated with about half retained for incorporation into alternatives. Table 17-5 provides summary remarks explaining the reason for eliminating a technology.

Combining Technologies Into Alternatives

Technologies not eliminated are retained for further consideration. The retained technologies are assembled and configured into one or more sets of alternatives covering the full range of actions that could be undertaken to remediate the site. The goal is to develop a sufficiently wide range of alternatives, each of which accomplishes the remediation objectives, to support a comparative analysis of tradeoffs.

The remediation of a site typically takes place over a long period of time and features a number of technologies that are implemented in an interdependent manner.

TABLE 17-5
Selected results of technology screening for the model site

Technologies	Suitability	Implement-ability	Performance	Environmental and health concerns	Cost	Retained	Remarks
Groundwater containment							
Slurry wall	+	+	+	+	0	Yes	Can be applied to impede down-gradient groundwater movement or to reduce the recovery and treatment of clean groundwater.
Cement diaphragm wall	+	+	+	+	–	Yes	Only where a slurry wall is inappropriate. Higher construction cost than a standard slurry wall.
Vibrating beam	0	+	–	+	0	No	Difficult to achieve a uniform low-permeability barrier. Installation equipment can easily be deflected by buried debris.
Grout curtain	0	0	–	+	–	No	Difficult to achieve a uniform low-permeability barrier. Prohibitive installation cost.
Sheet piling cutoff wall	0	0	0	+	0	No	Limited to cohesive soil applications. Difficult to achieve an effective low-permeability barrier.
Recovery well system	+	+	0	0	+	Yes	Primarily used for control and recovery of affected groundwater and light nonaqueous phase liquids. Can be used as an ancillary component to *in situ* treatment systems.
Recovery trench system	+	+	+	0	0	Yes	Implemented to control and recover affected shallow groundwater and light nonaqueous phase liquids.
Recharge well system	0	0	0	0	0	Yes	Adjunct system to dispense *in situ* treatment fluids.
Well vacuum system	+	+	0	0	0	Yes	Enhances fluid recovery efficiency of a standard recovery well system. Application limited to fine-grained soil or rock formations and most confined aquifers.
***In situ* treatment**							
Permeable treatment beds	0	0	–	+	–	No	Unproved technology, and implementation costs are prohibitive.
In situ bioreclamation	0	0	0	–	0	Yes	Suitable for treating low concentrations of dissolved or organic compounds. Probably only applicable in alluvium because of low permeabilities of bedrock strata.
Soil ventilation	–	–	0	0	0	No	Not suitable to site conditions. Low-permeability formations at site would inhibit successful implementation.
Soil flushing	–	–	0	0	–	No	Not suitable to site conditions. Low-permeability formations at site would inhibit successful implementation.

An alternative is thus composed of various component technologies that represent building blocks in the alternative; some are **key components** while others are **ancillary.** A key component is one that establishes, by itself, a general approach for remediating one or more pathways. Ancillary components simply contribute to the overall performance of a key component or its approach. An alternative directed at groundwater control for a hypothetical site may feature a cutoff wall (as the key component). A cutoff wall would require recovery of groundwater that could be accomplished by either recovery wells or a recovery trench as ancillary components.

To simplify the effort, ancillary components are tied together with key components to form larger building blocks. Some technologies are a necessary component of all alternatives, because no remedy would be protective without them. Typical examples include monitoring, removal of highly contaminated sources (i.e., "hot spots"), and control of surface water. They are usually grouped and referred to as "primary controls" for inclusion in all alternatives.

Alternatives may feature different technologies. In one alternative, treatment of affected soils may be the primary component, while in another alternative, containment by capping may be proposed as the primary component. In other situations, a single technology may be the primary component of different alternatives. This might occur in a situation where cutoff walls are the primary component for two alternatives: The first alternative could have multiple cutoff walls designed to contain separate source areas across a site, while the second alternative might propose a single, very long cutoff wall that would surround the whole site.

Two or more types of general treatment concepts may be retained as separate alternatives (e.g., incineration and soil vapor extraction). However, the range of alternatives should not be so narrow as to consist entirely of comparing one unit operation with another, both representing essentially the same concept (e.g., two types of incineration). This comes later as part of the design of the selected remedy.

Early approaches to site remediation were sometimes piecemeal rather than integrated. A typical case is that each of the contaminated media probably have been considered separately, as was done at the model site. In assembling alternatives it is important to examine comprehensively the entire site and its remediation. For example, the approach to remediating a contaminated groundwater plume needs to be integrated with the steps taken to address the source of that contamination because the time frame for remediating groundwater, if not the minimum concentration levels attainable, depends in large measure on such source control actions.

The alternatives developed for groundwater at the model site were assembled from the following components:

Primary controls

Common to all alternatives:

- Groundwater monitoring
- Surface water monitoring
- Security fence

Common to all alternatives except "no action":

- Excavation of affected soils outside the main source area
- Surface cover over main source area

- Surface water controls
- Alternate water supply to residences

Optional key components

- Alluvial groundwater recovery (by wells or trench) and treatment
- Alluvial *in situ* bioremediation
- Bedrock groundwater recovery (by wells or trench) and treatment in the following areas:
  - Southeast area
  - Southwest area
- Cutoff wall (with recovery wells or trench) in the following areas:
  - Around main source area (Drum Pit and Sludge Pit)
  - Along southwest boundary
  - Along southeast boundary
- Deep recovery trench around main source area (Drum Pit and Sludge Pit)

These components were combined and assembled to yield 21 separate alternatives for the generic site, representing varied combinations of technical approaches, spatial coverage, and volumes of media remediated. In summary, the process of developing alternatives started with problem definition and the establishment of remediation objectives, and it has proceeded through the screening of technologies to assembling them into comprehensive alternatives. With examples from the model site, this process is illustrated in Figure 17-7.

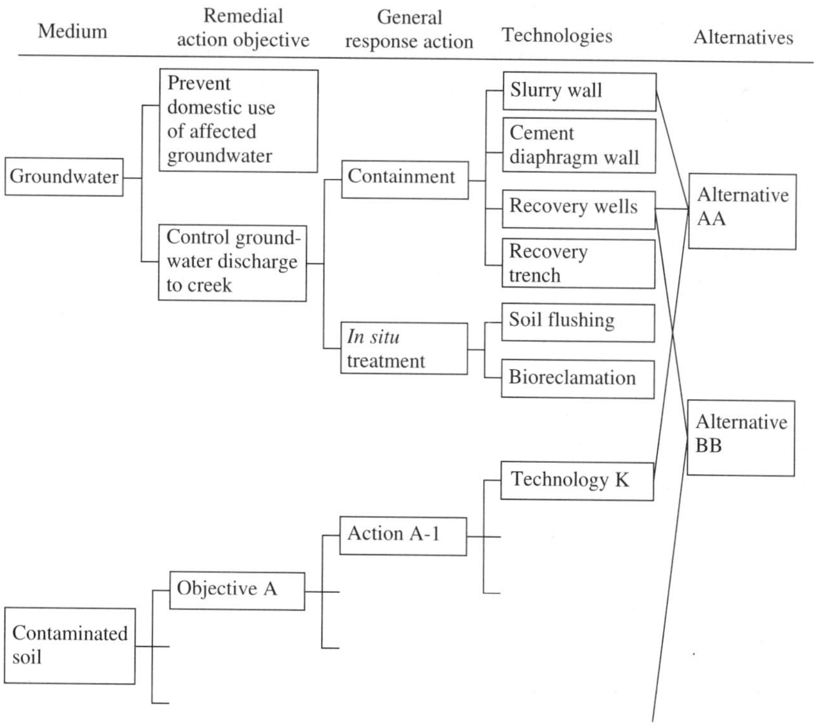

FIGURE 17-7
Development of alternatives for model site.

17-6 ANALYSIS OF ALTERNATIVES

Preliminary Screening

As a result of combining this assemblage of the technologies into alternatives, a large number of different remedial options can emerge. This large number, such as the 21 alternatives in the case of the model site, needs to be screened against basic criteria to limit the total to a manageable level of the more promising alternatives for the final, detailed analysis. The screening criteria include effectiveness, implementability, and cost.

Information needs to be developed on important factors (e.g., the need for off-site disposal, required permits, time frame for realizing beneficial results) necessary to apply the screening criteria. For example, each alternative should be defined as to size and configuration:

- *Treatment*—flow rates and process sequence
- *Containment*—location and general dimensions

At this point, cost estimates can be of a preliminary nature but must include both capital cost and the cost of operation and maintenance. The cost estimates may derive from prior experience, cost curves, and vendor information. Nevertheless, it is still important to ensure that all estimates share the same degree of accuracy.

The preliminary screening should yield the final set of discrete alternatives covering a wide range of remedial actions that could be implemented at the site. The number of final alternatives depends primarily on the degree of complexity presented by the site; whether compelling information shows that certain alternatives are likely to be chosen, and the possible need to evaluate certain alternatives because of regulatory requirements; preferences of regulatory agencies, persuasive public opinion, or institutional constraints.

Usually, four to six alternatives are selected, and certainly not more than 10. One of the alternatives should be a "no action" alternative as a baseline for measuring the effectiveness of the others. After screening of the 21 alternatives developed for the model site, six were selected for further analysis as shown in Table 17-6. It should be

TABLE 17-6
Summary of alternatives (model site)

		Groundwater recovery and treatment			
Alternative	Primary controls	Alluvium	Southwest area	Southeast area	Cutoff wall/trench
A	Partial				None
B	X				None
C	X	X			None
E	X		X	X	None
E1	X		X	X	Cutoff wall along southwest boundary.
L	X	X			Main source area surrounded by either cutoff wall or deep recovery trench.

noted that the six alternatives do not address source control. Source control, because of regulatory procedures, was the subject of a separate study. Instead, each of the six address only groundwater, and they had to assume a certain level of source control.

Technical Data Needed for Final Alternatives

Each alternative contained in the final set of alternatives should be reexamined and refined so that a thorough comparative analysis can be made prior to final selection. The level of detail should be consistent with that of a concept design but short of what would be required for a detailed design to implement an alternative. The particular technical details needed for each alternative are essentially as follows:

- Preparation of a basic component diagram and conceptual layout
- Major equipment needs and utility requirements
- Special engineering considerations for implementation of the alternative
- Potential environmental impacts and some methods for mitigating adverse impacts
- Operation, maintenance, and long-term monitoring requirements
- Implementation requirements including safety considerations, regulatory and permit requirements, temporary storage, off-site disposal needs, and transportation
- Phasing and segmenting options
- Preliminary implementation schedule

EXAMPLE 17-4. CONCEPT DESIGN VERSUS ENGINEERING DESIGN. The types of information developed as part of an engineering design are unnecessarily detailed for the comparative analysis required to select an alternative. Instead, the level of detail should conform with that of a concept design. Illustrate some differences between the level of detail needed for concept design, preliminary design, and final design (use various categories of information as examples).

Solution

Category of information	Concept design	Preliminary design	Final plans and specs
Layout	Component diagram	Piping and instrumentation diagrams	Piping drawings
Cost estimate	±50%	±30%	±10%
Required equipment	List of major equipment	Preliminary specifications for major equipment	Purchase or technical specifications for all equipment and material

Schedule requirements for completing the study can constrain the level of information obtained for each alternative. As an example, treatability tests may be necessary to confirm the applicability of a unit process or to define the parameters that have a major influence on costs. However, time frames may not allow such studies, suggesting that promising technologies should be identified and tested as early as possible during the field investigation. If not, the study may have to proceed with literature data, and that may prove inconclusive and thus inhibit the selection of that particular remedy.

Criteria

The selection of the preferred alternative is based on a comparative analysis using a set of criteria. The specific criteria will vary, depending on the regulatory program and the latest guidance from the responsible agency. These criteria tend to fall in the following broad categories:

1. Long-term effectiveness
2. Long-term reliability
3. Implementability
4. Short-term effectiveness
5. Cost

EXAMPLE 17-5. COMPARISON OF CRITERIA. What are the differences between the five criteria listed above and the nine required under the EPA's National Contingency Plan required under SARA[14] (see Section 2-4 for a list of the nine criteria)?

Solution. The EPA criteria are divided into three groups. The five EPA criteria referred to as the "Primary Balancing Criteria" group are essentially the same as the five previously listed. In fact, three read the same. A fourth is close; instead of "long-term effectiveness and permanence," it deletes the term "permanence." The fifth criterion in the EPA's list is "reduction of toxicity, mobility, or volume through treatment" and is replaced herein by "long-term reliability." The reason for this replacement is that treatment is simply a means for achieving long-term reliability, with permanence the ultimate achievement. As expressed by Clean Sites, "It is not the use of treatment, but the result of using treatment, that is important."[15]

The EPA's criteria include two other groups that are not addressed by the five criteria presented in this textbook. The EPA's first group consists of two criteria referred to as "threshold criteria" (i.e., they must be satisfied for an alternative remedy to be eligible for selection). These two criteria are redundant because the alternatives analysis process presented in this text develops only alternatives that do indeed meet these requirements. The last two EPA criteria, "modifying criteria," are not true criteria; instead, they ensure that the selected remedy reflects the concerns of the state and local community, something that a proper public participation process would achieve.

Long-term effectiveness

Long-term effectiveness can be analyzed in two ways. The simple way is a largely qualitative assessment of how well an alternative meets the remedial action objective over the long term. On the other hand, a complete analysis would calculate the residual risk; this is the quantitative risk represented by untreated contaminants or treated residuals remaining at the site at the conclusion of the remedy's construction and operation period.

Many regulatory agencies require calculation of residual risk. This may seem unnecessary because a remedy that achieves the remedial objectives for a site is, by definition, resulting in acceptable residual risks. Is it meaningful, then, to calculate how acceptable is the residual risk? These calculations have value when tradeoffs among alternatives are examined, provided the differences warrant the expense of such calculations. An example of such calculations was provided in Section 14-9 for the Lackawanna Refuse Site.

"Long-term" corresponds to a period of 70 years (i.e., the life expectancy for calculating lifetime exposure to adults). Because residual risk depends on the toxicity, mobility, and mass of contamination, an obviously effective solution is one that treats

the contaminants such as to reduce any or all of these partial components of risk. However, if a containment-type alternative provides barriers and controls sufficiently suitable in regard to the specific site conditions to meet the remedial action objective, the alternative is considered to be effective, even though it may not reduce the toxicity, mobility, or mass of contamination. The question remains whether a containment alternative provides long-term reliability, and that depends on the adequacy of its controls and is the subject of the next criterion.

Long-term reliability

Long-term reliability is an issue only with the alternatives that leave untreated contaminants or treatment residuals at the site after the remedy is completed. It is obvious that the alternatives providing the greatest degree of reliability are those providing "permanence." Stated in other words, if long-term reliability were measured on a continuum, permanence would represent the absolute maximum. "Permanence" has received much debate among practitioners and is even listed as a statutory preference in the Superfund law, which unfortunately does not define the term. A definition proposed by one study seems to capture the spirit of the term "permanence":

> Permanent solutions are those that will endure indefinitely without posing the threat of any future release that would increase the risk above levels established for the site.[16]

Although such a definition in many cases would not consider alternatives that feature containment as permanent, it would allow some level of residual risk to remain at the site implementation.

EXAMPLE 17-6. DEFINITION OF LONG-TERM. For what period should "long-term reliability" be evaluated?

Solution. As with "long-term effectiveness," it is at least 70 years, the life expectancy of humans. Whether to analyze for a longer period depends on the expected life of the technological components of the remedy. For example, a slurry wall built to contain contaminated groundwater without remediation of the contamination should be expected to provide effective containment for at least 70 years; however, it inevitably will allow contaminants through at some time beyond that period because of natural factors. Therefore, the period of analysis for such remedies should extend beyond their expected life to allow an examination of replacement provisions or other contingencies.

When the selection of alternatives is done on a comparative basis, it encounters tradeoffs. **One tradeoff that requires careful consideration at most sites is whether to treat or to contain.** Treatment typically costs significantly more than containment but can diminish the hazard posed by the waste. Containment literally means that waste, whether untreated or treatment residuals, will be left on site. Containment requires technical controls, and sometimes controls of an institutional nature. An assessment of the long-term adequacy and reliability of these controls is crucial, and must address the following questions:

Containment/reduction of contamination

- How consistent over time is the alternative in effectively controlling migration of contaminants?
- How long will residual groundwater constituent concentration levels persist?
- What is the amount of reduction with time?

Operation and maintenance

- What operation and maintenance functions must be performed?
- What difficulties and uncertainties may be associated with long-term operation and maintenance?

Management and monitoring

- What type and degree of long-term management is required?
- What are the requirements for long-term monitoring?
- What risk of exposure exists should monitoring be insufficient to detect failure?

Safety factors

- Do controls adequately handle potential problems?
- What is the relative safety factor (or amount of redundancy) incorporated into the alternative?
- What is the possibility that future remedial actions, not otherwise included in the remedy, may be required?
- How difficult would it be to implement the additional remedial actions, if required?
- What is the magnitude of the risk should the remedial action fail without being replaced?

Some alternatives feature treatment and leaving the residuals of this treatment in place without containment. Many of the questions previously listed also apply to such alternatives in regard to how the residuals are managed. A critical question that must be addressed for treatment alternatives is "To what degree will the treatment be irreversible?"

EXAMPLE 17-7. LONG-TERM RELIABILITY. How should a remedy involving stabilization of organics be evaluated for "long-term reliability"?

Solution. Stabilization of organics can be reversed if the stabilized material deteriorates—for example, from repeated freezing and thawing. Specific durability testing can estimate the impact of wet/dry or freeze/thaw cycles (see Section 11-4). Treatment, as shown by this example, does not necessarily result in "permanence."

Implementability

Implementability is a function of three factors: (1) the history of the demonstrated performance of a technology, (2) the ability to construct and operate it given the existing conditions at the particular site, and (3) the ability to obtain the necessary permits from regulatory agencies. The types of questions raised by this issue include:

Demonstrated performance

- Are methods and equipment used in implementing the alternative common to commercial operations?
- Are technologies under consideration generally available and sufficiently demonstrated for the specific application?

- Will technologies require further development before they can be applied full-scale to the type of waste at the site?
- When should the technology be available for full-scale use?
- Are the qualified contractors limited in number?

Constructability

- How difficult is it to construct the alternative at this site?
- What uncertainties are related to construction?
- Are specialty contractors required?
- What is the likelihood that problems will develop and lead to schedule delays?

Operability

- What operating problems are anticipated and what are their impacts?
- Are the necessary equipment supplies and specialists available?
- For alternatives proposing off-site treatment and disposal, are such services available in adequate capacity?
- What difficulties and uncertainties may be associated with maintenance?

Regulatory feasibility

- What steps are required to coordinate with other agencies during the construction and operational period?
- What steps are required to set up long-term coordination among agencies?
- Can permits for on-site and off-site activities be obtained if and when required?

EXAMPLE 17-8. IMPLICATION OF STRESSING IMPLEMENTABILITY. What implication does stressing ease of implementation have for the type of remedy selected?

Solution. It implies a preference for proved technologies, thus inhibiting the selection of innovative technologies. The EPA's Superfund program recognizes this as a potential obstacle to the development of new, improved technologies and requires the evaluation of innovative as well as proved technologies. However, overemphasizing innovation can turn cleanup efforts into research projects, thus delaying the original goal to remedy the site.

As stated earlier, a general policy in remedy selection is to examine the risk posed in the future under a hypothetical exposure scenario. This raises the issue of what investment should be made in the present to prevent future, hypothetical problems. The tradeoff boils down to whether to implement preventive measures now to eliminate the possibility of a future, hypothetical problem, or implement them at a future date if and when monitoring shows that without those measures the hypothetical problem would become real. This tradeoff could be examined in a stochastically derived present-value basis, which typically would show that it is far less costly to implement them on a contingency basis—the problem may never materialize.

This decision, however, is not a strictly economic issue. The saga of Superfund shows many examples of why it is far better to prevent a problem now than to try to correct it in the future. Therefore, the first question to be asked is: Will monitoring provide ample warning in the future to implement preventive measures? Also, will far more expensive remedial measures be necessary? Another question is: Will funds

be available in the future for the contingency responses? These questions and a number of others (listed under long-term effectiveness, long-term reliability, and implementability) require careful examination to decide this issue.

Short-term effectiveness

Short-term effectiveness deals primarily with the effects on human health and the environment of the remediation itself during its implementation phase. This would include the construction period and its operation period, up until the remedial action objectives are achieved. Even though the expression "short-term" is used, there is no limit to the time frame for evaluating short-term effectiveness. As an example for a remedy involving excavation of soil, the risk to nearby residents from volatilization, dust, and other releases would continue for however long the excavation takes place, even if it occurs for 15 years.

Short-term effectiveness encompasses two issues other than health and environmental effects. It also examines worker safety and the time frame required to achieve the remedial action objectives. The questions for these issues include:

Health and environmental risks

- What emissions and discharges are released by the construction and operation of the alternative?
- What measures must be used during implementation and operation to mitigate health and environmental risk?
- How effective are the mitigation measures?
- Do migrational or exposure pathways exist that cannot be monitored adequately?

Worker safety

- Will the alternative threaten the safety of workers during implementation?
- Can these hazards be monitored adequately?

Implementation time

- What is the likelihood that treatment technologies will meet required process efficiencies and performance specifications?
- How long will it take to complete the construction after site work is initiated?
- How long until the remedial action objectives are achieved?

Cost

The weight given to cost in evaluating alternatives depends on the particular guidance of the agency involved in regulating the remediation effort. Some practitioners have contended that the way SARA skirts the issue of cost means that cost does not play a major role in the selection of alternatives under the Superfund program. Others contend that cost is indeed an important factor and should be considered directly in the selection of alternatives.

Engineers recognize that practically any risk level can be achieved if cost is not a factor. Thus, cost must be a factor and should be considered as a last step in remedy selection. A comparison of costs inevitably ends up balancing noncost criteria, such as long-term reliability and effectiveness, with cost. One way to illustrate this

is to compare costs with the degree to which each alternative reduces risk posed by contamination at a site. Such a comparison is illustrated in Figure 17-8 (Lackawanna Site) as a curve showing the risk reduction achieved by various alternatives versus their cost. The risk calculations for this particular example were discussed in Section 14-9. Risk calculations almost always assume that the alternative meets its objective. A highly sophisticated risk assessment would factor into its risk calculations the probability of failure represented by the long-term reliability of an alternative.

Costs consist of capital costs (the cost to construct the remedy) and annual operating and maintenance (O&M) costs (postconstruction expenditures). Table 17-7 lists the types of costs included in these two categories.

Capital costs consist of direct and indirect costs. Direct costs consist primarily of expenditures for the materials, labor, and equipment necessary to construct the remedies. Direct costs also include the cost to transport off-site and dispose of any waste (e.g., commercial incineration of excavated soil). Indirect costs include

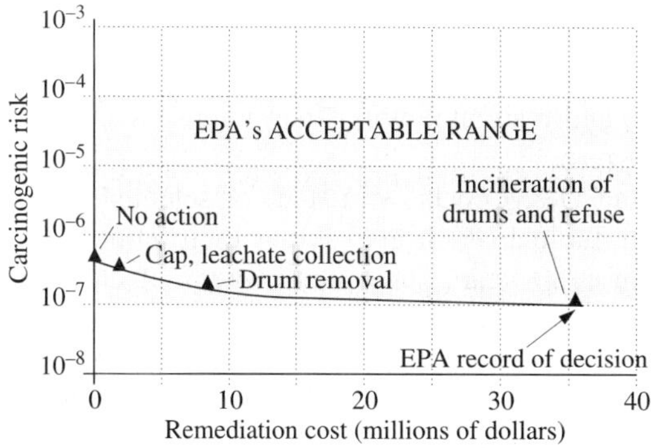

FIGURE 17-8
Carcinogenic risk versus cost for various remedial options.

TABLE 17-7
Types of costs

Capital costs	Annual operating and maintenance (O&M) costs
Construction costs	Operating labor costs
Equipment costs	Maintenance materials and labor costs
Land and site development costs	Auxiliary materials and energy
Building and services costs	Treatment residues disposal
Contractors' insurance and bonds	Environmental sampling and analysis
Relocation expenses	Purchased services
Surveying costs	Administrative costs
Disposal costs	Insurance, taxes, and licensing costs
Environmental sampling and analysis	Maintenance reserve and contingency funds
Engineering expenses	Rehabilitation costs
Construction material testing	Cost of periodic site reviews
License or permit costs	
Construction administration and reporting	
Construction interest	
Start-up and shakedown costs	
Agency oversight	
Contingency allowances	

engineering design services and legal/financial services related to the installation of the remedy. Experience has shown the former amounts to be about 15 percent to 20 percent of the total direct costs while the latter can be as much as 10 percent. A contingency of at least 25 percent should be added to the total of direct and indirect costs.

The comparison of the costs of different alternatives should be done on a present-worth analysis. For alternatives involving long-term operating and monitoring costs, the selection of a period of performance is necessary to develop comparative costs. In such cases, a period of performance is usually considered to be 30 years. Most remedies featuring treatment, management, and monitoring can be accomplished in 10 to 15 years, usually far less. However, the measurement of the effectiveness of some remedies can require management and monitoring of a site for a substantial period of time following implementation of the remedy.

A present-worth analysis, simply stated, defines the amount of money needed at the beginning (base year) to (1) fund the construction of the remedy and (2) provide an annual payment for operation and maintenance over the project life. The analysis involves discounting future costs to a common base year, usually the present. This requires the selection of a discount rate. The standard approach is to discount future costs at a rate equal to the interest paid on long-term bonds. Another approach is to assume that long-term inflation, taxes, and insurance will offset interest such that the effective discount rate would equal zero.

Table 17-8 tabulates the capital and O&M costs estimated for selected alternatives for the generic site as listed in Table 17-6. It also shows their values discounted at 0 percent and 10 percent for 30 years. The present-worth factor for 10 percent and 30 years is 9.427 (see Section 7-2). In the case of alternative C, the annual O&M cost is \$1,027,000. At an effective discount rate of zero, the present worth $= 1.027 \times 30 =$ \$30.8 million. At a rate of 10 percent, the present worth $= 1.027 \times 9.427 =$ \$9.68 million.

TABLE 17-8
Summary of cost of selected alternatives for the model site

Alternative	Description	Capital cost*	Annual O&M cost*	O&M cost $I = 0\%$	O&M cost $I = 10\%$	Total cost* $I = 0\%$	Total cost* $I = 10\%$
A	No action	0.0	.067	2.0	0.6	2.0	0.6
B	Primary controls	3.9	.103	3.1	1.0	7.0	4.9
C	Primary controls and alluvial recovery	5.2	1.03	30.8	9.7	36.0	14.9
E	B plus southwest and southeast area recovery	6.1	1.019	32.6	10.2	38.7	16.4
E1	E plus southwest boundary cutoff wall	11.3	1.07	32.3	10.1	43.6	21.4
L	Primary controls, alluvial recovery, plus main source area cutoff wall	26.6	.31	9.3	2.9	35.9	29.5

*All costs in million dollars.

EXAMPLE 17-9. DETERMINING PRESENT VALUES. The following cost estimate is provided for an alternative remedy:

Cost element	Cost
Initial capital cost	\$3,000,000
Replacement capital cost, year 20	3,000,000
Annual O&M cost	600,000

The following are given:

- The useful life for the capital items is 20 years.
- Capital items have a salvage value at the end of their useful life of 10 percent of their initial cost.
- The depreciated values of the capital items are a linear function of time, but the entire depreciation is taken on the first day of the year.
- Engineering costs are 15 percent of the direct capital costs.
- Contingencies are 25 percent of the direct cost.
- The effective discount rate is 10 percent.
- Inflation is neglected.

Determine the present-value cost of this alternative using a 30-year planning period.

Solution. The solution begins by calculating the present worth of individual elements.

The factor for determining the present worth of a single expenditure in 20 years may be determined for Equation 7-3:

$$\text{Cost factor} = \frac{1}{(1+i)^n} = \frac{1}{(1+0.1)^{20}} = 0.1486$$

The present-worth factor (PWF) to determine the present worth of 10 equal annual payments may be determined from Equation 7-4:

$$\text{PWF} = \frac{(1+i)^n - 1}{i(1+i)^n} = \frac{(1+0.1)^{10} - 1}{(0.1)(1+0.1)^{10}} = 6.1446$$

Dealing with depreciation is a bit more complicated. In this example, the capital equipment is replaced at the end of 20 years, but the planning period is 30 years. At the end of year 30 (the beginning of year 31), the capital equipment would still have approximately half of its useful life. Assume the value at this point is \$1.35 million or 9/20 of the original value. To calculate the present worth we again use Equation 7-3:

$$\text{PW} = 1.35\left[\frac{1}{(1+0.1)^{30}}\right] = 1.35(0.0573) = \$0.077 \text{ million}$$

The remaining calculations are shown in the following table.

Item	Cost	Cost factor	Present value
Capital costs:			
Initial capital cost	$3,000,000	1.0000	$3,000,000
Replacement cost	3,000,000	0.1486	445,800
Salvage value	(300,000)	0.1486	(44,600)
Undepreciated value	(1,350,000)	0.0573	(77,000)
Subtotal			3,324,200
Engineering (15%)			498,600
Contingencies (25%)			831,100
Subtotal, capital costs			4,653,900
O&M costs:			
Annual O&M (year 1–10)	600,000	6.1446	3,686,800
Annual O&M (year 11–20)	600,000	6.1446	
	3,686,800	0.3855	1,421,300*
Annual O&M (year 21–30)	600,000	6.1446	
	3,686,800	0.1486	547,700†
Subtotal			5,655,800
Contingencies (25%)			1,414,000
Subtotal, O&M			7,069,800
Total estimated present-value cost			11,700,000

*Present value at year 10 discounted to year 0.

†Present value at year 20 discounted to year 0.

Selection of Preferred Alternative

The selection of the preferred alternative involves a thorough evaluation of all alternatives on the basis of the criteria discussed earlier. The evaluation process is not structured. One reason is that the criteria actually are considerations rather than true criteria. A consideration is a specific characteristic or issue deemed important in evaluating alternatives. A wide range of quantitative data and qualitative information can be developed for each consideration. A criterion is a standard for making a decision with regard to a particular consideration; it would define a threshold between acceptable and unacceptable.

This distinction illustrates the nature of the process of evaluating the final alternatives. Thus, there should be no thresholds in the selection criteria; standards are reflected in the objectives and each alternative meets the objectives and is supposedly protective of human health and the environment, with the possible exception of the "no action" or any intentionally limited alternatives.

The process differs from a structured approach in other key ways. Alternatives are not ranked; hence, rigid methods such as a weighted summation model are avoided because it would be difficult to establish the validity of weights and ratings. Finally, there should be no prescriptive approach on how to apply criteria. A prescriptive approach does not allow the full use of common sense.

Instead of a structured approach, the evaluation of alternatives should be done on a flexible, comparative basis featuring the following steps:

- Interactions (e.g., conflicts and synergies) among criteria are considered.
- Advantages and disadvantages of alternatives relative to one another become clear.

- Tradeoffs among alternatives are examined and balanced.
- This information converges to compelling reasons on which a consensus can build for selecting one remedy.

If the analysis is made by knowledgeable persons under the scrutiny of interested parties, this approach should yield the most appropriate and defensible decisions.

17-7 REGULATORY GUIDANCE

A governmental agency usually provides guidance specifying the process and criteria to be used in conducting an analysis of alternatives.[17,18] The guidance varies from agency to agency, and the guidance from the same agency usually changes over time as policies change and more experience is obtained. For example, the component of remedial action that seeks to characterize the nature, extent, and rate of contaminant release is termed the remedial investigation (RI) under CERCLA, and the RCRA Facility Investigation (RFI) under RCRA. To clarify some of the confusion that may have arisen from differences in terminology in the regulatory guidance, Table 17-9 identifies the objective of each major step in the remedial action process and provides the CERCLA and RCRA terms applied. Further, the guidance can have a significant impact on the results of the analysis, ranging from making the study complicated and tedious to influencing the range of alternatives considered, possibly even changing the ultimate outcome.

An example is the EPA Superfund process as specified in its guidance document.[19] A critical study, drawing on the experience of more than 100 professionals familiar with the issues of site remediation, reported that the EPA process tended to work in a reverse order from that of the classic planning model.[20] The study found that instead of explicitly defining the required level of cleanup at the start of the process, the EPA process frequently identified this " . . . very late in the process and then only implicitly by what the selected remedy would achieve." In effect, alternatives were evaluated before clear objectives had been established.

TABLE 17-9
Comparison of CERCLA and RCRA terminology

Objective	RCRA	CERCLA
Identify possible further releases	RCRA Facility Assessment (RFA)	Preliminary Assessment and Site Investigation (PA/SI)
Characterize nature, extent, and rate of release	RCRA Facility Investigation (RFI)	Remedial investigation (RI)
Develop and evaluate alternatives	Corrective Measures Study (CMS)	Feasibility study (FS)
Propose remedy	Draft Permit Modification	Proposed remedial plan
Public participation in the process	Public comment	Public comment
Authorize remedy	RCRA permit	Record of Decision (ROD)
Design, construct, and operate selected remedy	Corrective Measures Implementation (CMI)	Remedial Design and Remedial Action (RD/RA)

The EPA approach also promotes developing and retaining a wide range of alternatives. The range extends from those failing to meet remediation objectives to some greatly exceeding them. In that way the final detailed analysis compares diverse courses of actions:

- "No action"
- Alternatives that fail, meet, or exceed ARARs
- An alternative that involves off-site treatment or disposal
- An alternative that features treatment
- An alternative that eliminates, to the extent feasible, the need for any long-term management, including monitoring (i.e., the "permanence" alternative that requires virtually complete elimination or destruction of hazardous substances at the site)
- An alternative that features containment
- An alternative that features innovative technology

Retaining this wide range serves to emphasize stark differences in the degree of effectiveness and the cost of the alternative remedies. Such is useful for fomenting participation by interested parties. However, it also requires expending effort in refining and evaluating specific alternatives that will not be selected, either because they are not protective or they are overly protective at an unrealistically high cost.

The EPA approach does provide a special capability. There is a great deal of uncertainty at Superfund sites:

- The existing site conditions may not be fully defined.
- It is not evident what future uses of the site and its resources need to be protected.
- The effectiveness and performance of many remedial technologies are not clear.

Under such constraints and incomplete knowledge, the selection of a remedy must be subjective in large part. It must be dependent on a large measure of common sense and have the flexibility to evaluate technologies on what they truly can achieve. Therefore, the EPA concept of delaying the setting of objectives, particularly those implicit in the hypothetical assumptions of future conditions, avoids preempting at too early of a stage some potentially viable alternatives.

17-8
CASE STUDY 2—FOCUSED EVALUATION AND SELECTION OF AN INTERIM CORRECTIVE MEASURE AT AN OPERATING CHEMICAL PLANT*

To augment the example feasibility study for the model Superfund site presented throughout this chapter (Case Study 1), this section presents the results of a focused evaluation and remedy selection for a group of solid waste management units (SWMUs) at an operating chemical plant. The subject site for this case study is an operating facility, governed by RCRA regulations. The operator will retain long-term responsibility for the site. In developing an interim corrective measure for the

---

*Section 17-8 was prepared by Reif Hedgcoxe, ERM.

SWMUs, several of the steps illustrated in Figure 17-2 were eliminated or significantly streamlined. This approach saved substantial time and reduced the total cost of the project.

Problem Definition

The SWMUs in this case study are a group of three brick-lined, in-ground pits that were used in a neutralization process for acid wastewater. The pits, which were located near the plant's boundary, were operated beginning in the early 1960s and "closed" during construction of a new chemical production unit in the late 1970s.

As a requirement of the facility's hazardous waste permit, the neutralization pits were assessed with about 20 other SWMUs as part of the RCRA Facility Investigation (RFI) process. The findings from the RFI revealed that the pits were closed by simply backfilling them with rocks and covering with fill dirt. Chlorinated solvents, principally 1,2-dichloroethane, were present as free phase fluids (DNAPLs) within the pits. Undocumented, historical releases from the pits had impacted the soil and groundwater near the pits.

As a follow-up to the initial RFI, a preliminary risk evaluation was conducted to provide a basis for ranking the more than 20 SWMUs. Because of the nature of the releases and the potential for affected groundwater to migrate off-site, the neutralization pits were ranked as high-priority SWMUs and slated for an expedited Phase II RFI and implementation of an interim corrective measure. The results of the Phase II RFI showed that groundwater affected with 1,2-dichloroethane and its associated degradation products extended beyond the plant fence line, under a highway and under a vacant lot owned by the plant. The total plume migration distance was estimated to be approximately 1100 feet from the area of the pits.

Identification of Potential Exposure Pathways

The RFI data were used to generate a site conceptual model. The model is presented in the plan view map and cross-sectional diagram in Figures 17-9 and 17-10, respectively. The nature of the constituents identified in the pits suggested the following exposure pathways were of potential concern:

- On-site exposure via skin adsorption or ingestion due to direct contact with affected soil and/or liquids in the pits
- On-site and near off-site exposure via inhalation of vapors or dust from exposed affected soil, liquids, and/or affected groundwater
- Off-site exposure via ingestion due to consumption of affected groundwater

Access by the general public to the area near the pits is restricted by the plant's security system. Therefore, exposures to the general public via direct contact or inhalation were determined to be incomplete pathways. The plant's supply for process water is from a surface impoundment, and additional bottled water is provided for drinking purposes. Therefore, there is no on-site use of groundwater for industrial or drinking water supplies, and ingestion of affected groundwater by plant workers was also determined to be an incomplete pathway.

During the Phase II RFI work, a survey of public water wells in the vicinity of the plant was conducted. This survey indicated that a private well completed about 150 feet below the deepest affected groundwater zone was located adjacent to the plant's off-site property and within about 200 feet of the known extent of the groundwater plume.

From this site conceptual model, it was determined that the following exposure pathways could be considered complete under certain conditions:

1. Potential exposure to affected soils or groundwater by plant construction or maintenance workers. This could occur if plant activities required excavation or other subsurface work in the immediate area of the pits that could result in direct contact with affected soils and DNAPLs in or near the pits.
2. Potential exposure via inhalation of vapors from exposed affected soil, liquids and/or affected groundwater by plant construction or maintenance workers. This may occur if plant activities required excavation or other subsurface work in the immediate area of the pits that could result in direct atmospheric contact with exposed affected soil, liquids, and/or affected groundwater.
3. Potential exposure to affected groundwater by the off-site general public. This could occur if individuals owning property adjacent to the plant's off-site vacant lot were to install and use a shallow water well for domestic purposes.

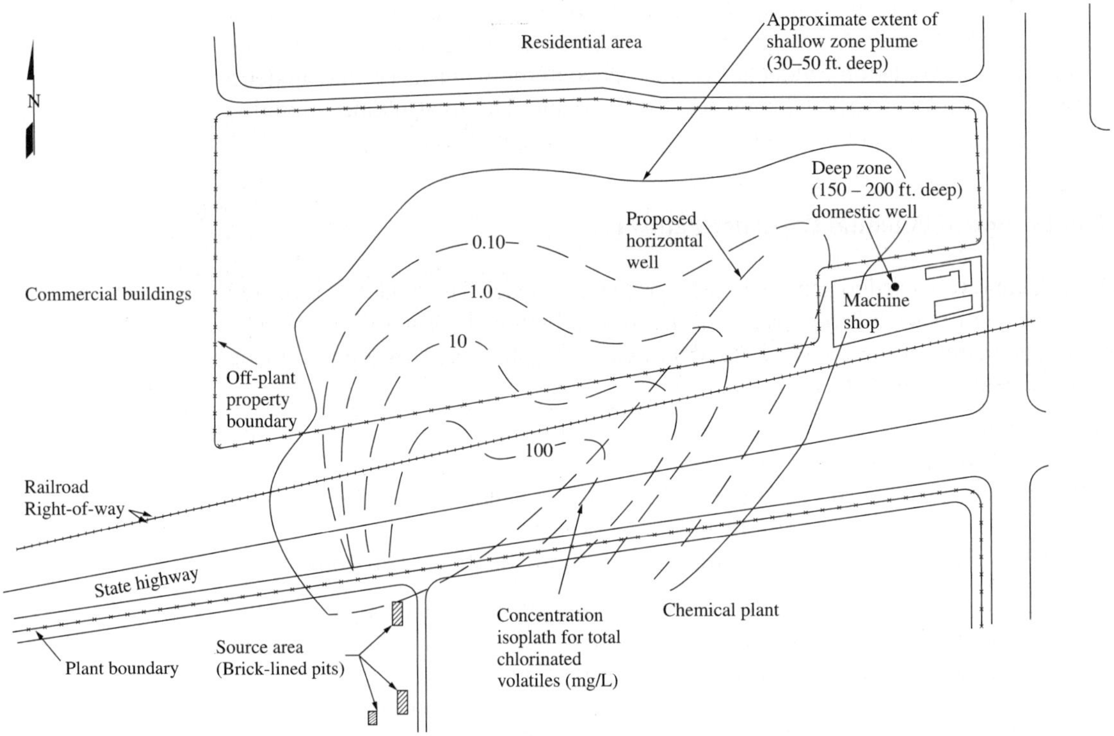

FIGURE 17-9
Plan view map.

Summary of Focused Evaluation of Alternatives

Given the three identified exposure pathways, a focused evaluation of alternatives was conducted. In this case, establishing the remedial action objectives and screening the general response actions for the SWMUs was a relatively straightforward process with a goal of containment of the affected media. Full restoration of affected media was not pursued since the facility is an operating plant and the chemical company will retain long-term responsibility for the site. In addition, with the exception of that portion of the plume beneath the highway, none of the release had impacted property that was not owned by the plant.

As summarized in Tables 17-10 and 17-11, the evaluation process addressed the identified exposure pathways by defining the following general response actions:

1. **Source control** to substantially reduce or eliminate the potential for worker inhalation of vapors and exposure to affected soils and DNAPLs at or near the pits.
2. **Plume control** to limit the potential for future migration of affected groundwater into areas beyond the off-plant property.

On selection of the general response actions, potential remediation technologies were qualitatively assessed on the basis of cost, potential for eliminating exposure pathways, and overall plant objectives. Since the short-term goal of a focused

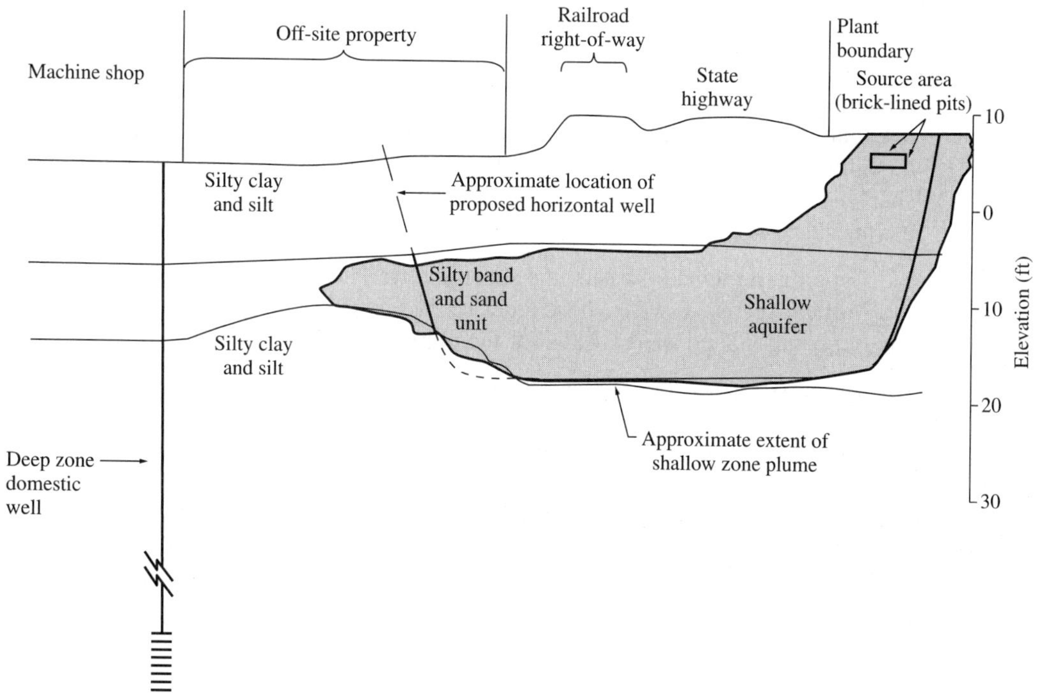

FIGURE 17-10
Cross-section diagram.

TABLE 17-10
Source area response actions

Media: soil and DNAPL source
Remedial action driver: worker exposure to soil and DNAPL
General response action: eliminate potential for worker exposure

Viable technologies	Screening rank*	Remarks
Capping	4	Does not remove source and thus would not address long-term leaching to groundwater
Extract pumpable fluids	2	Easy to implement, but may require controls to ensure volatilization; does not impact air quality
Excavate soil and waste	2	Decreases amount of material leachable to groundwater Potential exposure during removal due to fluids could increase costs and cause safety concerns Reduces future risk to construction workers
Cutoff wall	3	Reduces potential for continued leaching and lateral migration from source area Potentially hinders future plant construction activities Difficult to install because of subsurface obstructions and utilities
Sequenced removal: fluid extraction, then excavation and backfill with clean soil	1	Allows bulk of materials to be removed while limiting short-term exposure risks

*Ranking included a qualitative assessment of costs, implementability and operability, and overall effectiveness.

screening of remedial alternatives was to implement an interim corrective measure, quantitative assessment of all available alternatives was not required. Accordingly, a qualitative screening was conducted to identify a short list of alternatives. The short list of viable potential technologies, summarized in Tables 17-10 and 17-11, includes the following:

1. Capping, removal of waste, and sheet pile walls for the source area
2. Vertical recovery wells, horizontal recovery wells, treatment walls, intrinsic remediation, and sparge walls for the plume

The next step in the focused screening of remedial alternatives was to rank the viable potential technologies according to cost of implementation/operation and effectiveness in addressing goals and concerns. As shown in Tables 17-10 and 17-11, the potential technologies were ranked for each medium of concern. Of the potential technologies considered viable for the shallow groundwater, the highest-ranked options were horizontal recovery wells and intrinsic remediation. The main drivers that influenced the high ranking of horizontal wells for the control of shallow groundwater were:

1. The ability of one or two wells to capture the entire plume
2. The ability to incorporate the site's geology and plume characteristics as an integral part of the remedial plan
3. The need to transport the recovered groundwater to the plant's wastewater treatment system

The location of the plume beneath a major highway and railroad right-of-way made the use of off-site vertical recovery wells cost-prohibitive because of the added requirement to pipe water back to the plant. The horizontal well provides the unique ability to recover directly from the "hot" portion of the plume within and beyond the plant boundary and manage the water at the plant without the need for surface piping. Intrinsic remediation was also ranked high for its ability to address plume remediation at a relatively low cost.

Alternatives for soil and DNAPL remediation in the source areas (pits) were also ranked according to costs and benefits. Waste removal received the higher ranking because of the high cost-to-benefit ratio as indicated in Tables 17-10 and 17-11. The other alternatives were either prohibitively expensive or did not control leaching of the material to groundwater.

TABLE 17-11
Groundwater plume response actions

Medium: Shallow groundwater plume
Remedial action driver: public exposure to groundwater
General response action: contain groundwater.

Potential technologies*	Screening ranking†	Remarks
Vertical recovery wells	2	Requires multiple wells both on and off property to control plume migration Installation and O&M of multiple pumps and piping has prohibitive cost Additional cost to secure multiple wells and transfer affected groundwater from off-site to the plant wastewater treatment system is prohibitive
Horizontal recovery wells	1	One or two wells can achieve capture of majority of plume Well can be designed to capitalize on site geology and plume characteristics (shape and orientation of hot zones) Piping can be installed with wells to transport recovered off-site groundwater into plant wastewater treatment system
Treatment walls (funnel-and-gate)	4	Costly to install because of underground obstructions, utilities, and depth constraints Treatment walls installed on plant would address only source area and treatment walls off site would address only distal portion of plume
Intrinsic remediation (natural attenuation)	1	Lowest-cost alternative, no equipment O&M, no treatment costs Works well assuming successful source removal/control Requires long-term commitment to own the off-site area and monitor slow, natural process
Sparge wall	3	Costly to install in areas with many underground utilities Because of specific layout of the subject property, only the outer edges of the plume and the source area could be treated, or if placed off-site, would also require O&M of off-site facilities

*Other soil technologies such as soil flushing and SVE were qualitatively eliminated on the basis of cost of implementation.
†Ranking included a qualitative assessment of costs, implementability and operability, and overall effectiveness.

In addressing remedial alternatives for sites that do not pose a significant risk to public health and the environment (such as the site described in the above case study), a qualitative remedial alternative analysis may be appropriate. Especially if the goal of the analysis is selection of an interim measure, then testing and observation can be used to confirm the effectiveness of the interim measure or modify it to improve attainment of remediation goals.

DISCUSSION TOPICS AND PROBLEMS

17-1. What general response actions could be implemented to protect human health from contaminated groundwater, other than treating the groundwater?

17-2. What is the difference between "long-term effectiveness" and "long-term reliability"?

17-3. Which would require a greater cost contingency: excavation of soil determined in the field to have contamination above a certain concentration or incineration of the sludge from a lagoon?

17-4. When should treatability tests be conducted? Why?

17-5. What are five technologies for achieving the general response action "treat soil contaminated with chlorinated solvents"?

17-6. On the basis of the limited knowledge provided in this chapter, rank each of the final six alternatives selected for the generic site (see Table 17-6) with regard to long-term effectiveness and for long-term reliability.

17-7. For Problem 17-6, how would *in situ* biotreatment rank if the contaminants were degradable organics?

17-8. Which of the alternatives in Table 17-6 would you consider to be the preferred alternative on the basis of information available in this chapter?

17-9. What additional information would you feel to be necessary for selecting a remedy from the six alternatives in Table 17-6?

17-10. What would this additional information need to show to convince you to select an alternative other than the one selected in Problem 17-8?

17-11. Assuming that the alluvial groundwater at the generic site could be restored such that residences would not require an alternative water supply, up to what cost would you consider this the preferred alternative?

17-12. Provide the following data for a hypothetical or actual contaminated site: simplified plan view and one page of text and tables listing contaminant sources and concentrations and quantitative risks of exposure.

The remaining problems pertain to the site that you have described in Problem 17-12.

17-13. What are some possible remediation objectives for this site?

17-14. What is an example of a remediation objective that would be inappropriate?

17-15. What are some possible general response actions that could be used to achieve these objectives?

17-16. For one of the general response actions, which technologies are possibly applicable?

17-17. Which technologies should be retained for further analysis on the basis of your preliminary screening of their effectiveness and reliability?

17-18. How could some of the technologies be combined to create different alternatives?

REFERENCES

1. "Feasibility Study Report, Management of Migration Second Operable Unit, Royal Hardage Site, Criner, Oklahoma," report prepared for the Hardage Steering Commit-

tee and the U.S. Environmental Protection Agency, by ERM-Southwest, Inc., Houston, 30 May 1989.
2. U.S. EPA, Office of Solid Waste and Emergency Response (OSWER): "Guidance for Evaluating the Technical Impracticability of Ground Water Restoration," Directive 9234.2-25, September 1993.
3. U.S. EPA OSWER: "Ground Water Issue—Light Nonaqueous Phase Liquids," EPA/540/S-95/500, July 1995.
4. U.S. EPA: "Guidance for Conducting Remedial Investigations and Feasibility Studies under CERCLA," Interim Final, OSWER Directive 9355.3-01, August 1988.
5. U.S. EPA, Science Advisory Board: "Reducing Risk: Setting Priorities and Strategies for Environmental Protection," SAB-EC-90-021, September 1990.
6. U.S. EPA OSWER Directive: "Role of the Baseline Risk Assessment in Superfund Remedy Selection Decisions," OSWER Directive 9355.0-30, April 1991.
7. ASTM, Standard E 1739-95: *Guide for Risk-Based Corrective Action Applied at Petroleum Release Sites,* American Society for Testing and Materials, September 1995.
8. U.S. EPA: RI/FS Guidance, August, 1988, interim final.
9. Clean Sites, Inc.: "Improving Remedy Selection: An Explicit and Interactive Process for the Superfund Program," Alexandria, Va., October 1990.
10. *Federal Register:* "National Oil and Hazardous Substances Pollution Contingency Plan; Final Rule," 40 CFR Part 300.430, vol. 55, no. 46, March 1990.
11. Clean Sites, Inc., "Improving Remedy Selection."
12. *Federal Register:* "National Oil and Hazardous Substances Pollution Contingency Plan."
13. U.S. EPA: "Guidance for Conducting Remedial Investigations and Feasibility Studies Under CERCLA," OSWER Directive 9355.3-01, October 1988.
14. *Federal Register:* "National Oil and Hazardous Substances Pollution Contingency Plan."
15. Clean Sites, Inc., "Improving Remedy Selection," op. cit.
16. Ibid.
17. U.S. Environmental Protection Agency (EPA): "Guidance for Conducting Remedial Investigations and Feasibility Studies under CERCLA," EPA 540/G-89/004, Washington, D.C., October 1988.
18. U.S. Air Force Occupational & Environmental Health Laboratory: "Handbook to Support the Installation Restoration Program (IRP) Statements of Work for Remedial Investigation/Feasibility Studies," Brooks Air Base, Texas, April 1988.
19. U.S. Environmental Protection Agency (EPA): "Guidance for Conducting Remedial Investigations and Feasibility Studies under CERCLA."
20. Clean Sites, Inc., "Improving Remedy Selection," op. cit.

ADDITIONAL READING

Bennett, G. F., F. S. Feates, and I. Wilder: *Hazardous Materials Spills Handbook,* McGraw-Hill, New York, 1982.

Blackman, W. C.: *Basic Hazardous Waste Management,* Lewis Publishers, Boca Raton, Fla., 1993.

Boutwell, S. H.: *Selection of Models for Remedial Action Assessment,* U.S. EPA, Washington, D.C., 1984.

Cohen, R. M., and W. J. Miller III: "Use of Analytical Models for Evaluating Corrective Actions at Hazardous Waste Sites," *Proceedings of the 3rd National Symposium of Aquifer Restoration and Groundwater Monitoring,* National Water Well Association, Dublin, Ohio, May 1983.

Cole, C. R., R. W. Bond, S. M. Brown, and G. W. Dawson: *Demonstration/Application of Groundwater Technology for Evaluation of Remedial Action Alternatives,* U.S. EPA, Washington, D.C., 1983.

Dawson, G. W., and B. W. Mercer: *Hazardous Waste Management,* John Wiley & Sons, New York, 1986.

Ford, P. J., P. J. Turina, and D. E. Seely, *Characterization of Hazardous Waste Sites—A Methods Manual,* vol. II, *Available Sampling Methods,* EPA 600/4-83-040, NTIS no. PB-83-126920, U.S. EPA, Springfield, Va., 1983.

Hallstedt, P. A., M. A. Puskar, and S. P. Levine: "Application of the Hazard Ranking System to the Prioritization of Organic Compounds Identified at Hazardous Waste Remediation Action Sites," *Hazardous Waste and Hazardous Materials,* vol. 3, no. 2, 1986.

O'Brien & Gere Engineers, Inc.: *Hazardous Waste Site Remediation,* Van Nostrand-Reinhold, 1988.

U.S. Congress: *Superfund Strategy,* Office of Technology Assessment, Washington, D.C., April 1985.

U.S. Environmental Protection Agency: "Guidance for Conducting Remedial Investigation and Feasibility Studies under CERCLA, Interim Final," EPA/540/6-89/004, OSWER Directive 9355.3-01, October 1988.

U.S. Environmental Protection Agency: "Report on the Usage of Computer Models on Hazardous Waste/ Superfund Programs: Phase II, Final Report," U.S. EPA, OSWER, December 1990.

U.S. Environmental Protection Agency: "The Cost Digest: Cost Summaries for Selected Environmental Control Technologies," Office of Env. Engineering and Technology, EPA-600-8-84-010, October 1984.

U.S. Environmental Protection Agency: *Case Studies 1–23: Remedial Response at Hazardous Waste Sites,* U.S. EPA, Washington, D.C., March 1984.

U.S. Environmental Protection Agency: *EPA Guide for Identifying Cleanup Alternatives at Hazardous Waste Sites and Spills: Biological Treatment,* EPA-600/3-83-063, U.S. EPA, Springfield, Va., 1983.

U.S. Environmental Protection Agency: *Handbook for Evaluating Remedial Action Technology Plans,* EPA 600/2-83-076, U.S. EPA, Cincinnati, 1983.

U.S. Environmental Protection Agency: *Handbook: Remedial Action at Waste Disposal Sites (Revised),* EPA/625/6-85/006, U.S. EPA, Cincinnati, October 1985.

U.S. Environmental Protection Agency: *Modeling Remedial Actions at Uncontrolled Hazardous Waste Sites,* EPA 540/2-85-001, U.S. EPA, Springfield, Va., 1985.

U.S. Environmental Protection Agency: *Remedial Response at Hazardous Waste Sites, Summary Report,* EPA 540/2-82-002a, *Case Studies,* EPA 540/2-84-002b, U.S. EPA, Cincinnati, 1984.

Wagner, K., et al.: *Remedial Action Technology for Waste Disposal Sites,* Noyes, Park Ridge, N.J., 1986.

ACKNOWLEDGMENTS

The authors wish to acknowledge the following individuals who assisted with the preparation of this chapter for the first edition: Doyan Main, Gary Donna, Alan F. Hassett, Mark Schmittle, Howie Wiseman, *ERM;* Chris Edgar, *Bucknell University.* The principal reviewer for ERM was John Iannone.

The reviewer for ERM for the second edition was Reif Hedgcoxe.

APPENDIX A

Basel Convention

A-1
ANNEX VIII: LIST A

Note: Wastes contained in this Annex are characterized as hazardous under Article 1, paragraph 1 (a), of this Convention, and their designation on this Annex does not preclude the use of Annex III to demonstrate that a waste is not hazardous.

A1 Metal and metal-bearing wastes

A1010 Metal wastes and waste consisting of alloys of any of the following:

Antimony
Arsenic
Beryllium
Cadmium
Lead
Mercury
Selenium
Tellurium
Thallium

but excluding such wastes specifically listed on list B.

A1020 Waste having as constituents or contaminants, excluding metal waste in massive form, any of the following:

- Antimony; antimony compounds
- Beryllium; beryllium compounds
- Cadmium; cadmium compounds
- Lead; lead compounds
- Selenium; selenium compounds
- Tellurium; tellurium compounds

A1030 Wastes having as constituents or contaminants any of the following:

- Arsenic; arsenic compounds
- Mercury; mercury compounds
- Thallium; thallium compounds

A1040 Wastes having as constituents any of the following:

- Metal carbonyls
- Hexavalent chromium compounds

A1050 Galvanic sludges
A1060 Waste liquors from the pickling of metals
A1070 Leaching residues from zinc processing, dust and sludges such as jarosite, hematite, etc.
A1080 Waste zinc residues not included on list B, containing lead and cadmium in concentrations sufficient to exhibit Annex III characteristics
A1090 Ashes from the incineration of insulated copper wire
A1100 Dusts and residues from gas cleaning systems of copper smelters
A1110 Spent electrolytic solutions from copper electrorefining and electrowinning operations
A1120 Waste sludges, excluding anode slimes, from electrolyte purification systems in copper electrorefining and electrowinning operations
A1130 Spent etching solutions containing dissolved copper
A1140 Waste cupric chloride and copper cyanide catalysts
A1150 Precious metal ash from incineration of printed circuit boards not included on list B
A1160 Waste lead-acid batteries, whole or crushed
A1170 Unsorted waste batteries excluding mixtures of only list B batteries. Waste batteries not specified on list B containing Annex I constituents to an extent to render them hazardous.
A1180 Waste electrical and electronic assemblies or scrap containing components such as accumulators and other batteries included on list A, mercury-switches, glass from cathode-ray tubes and other activated glass and PCB-capacitors, or contaminated with Annex I constituents (e.g., cadmium, mercury, lead, polychlorinated biphenyl) to an extent that they possess any of the characteristics contained in Annex III (note the related entry on list B, B1110)

A2 Wastes containing principally inorganic constituents, which may contain metals and organic materials

A2010 Glass waste from cathode-ray tubes and other activated glasses
A2020 Waste inorganic fluorine compounds in the form of liquids or sludges but excluding such wastes specified on list B
A2030 Waste catalysts but excluding such wastes specified on list B
A2040 Waste gypsum arising from chemical industry processes, when containing Annex I constituents to the extent that it exhibits an Annex III hazardous characteristic (note the related entry on list B B2080)
A2050 Waste asbestos (dusts and fibres)
A2060 Coal-fired power plant fly-ash containing Annex I substances in concentrations sufficient to exhibit Annex III characteristics (note the related entry on list B B2050)

A3 Wastes containing principally organic constituents, which may contain metals and inorganic materials

A3010 Waste from the production or processing of petroleum coke and bitumen

A3020 Waste mineral oils unfit for their originally intended use

A3030 Wastes that contain, consist of or are contaminated with leaded anti-knock compound sludges

A3040 Waste thermal (heat transfer) fluids

A3050 Wastes from production, formulation and use of resins, latex, plasticizers, glues/adhesives excluding such wastes specified on list B (note the related entry on list B B4020)

A3060 Waste nitrocellulose

A3070 Waste phenols, phenol compounds including chlorophenol in the form of liquids or sludges

A3080 Waste ethers not including those specified on list B

A3090 Waste leather dust, ash, sludge and flours when containing hexavalent chromium compounds or biocides (note the related entry on list B B3100)

A3100 Waste paring and other waste of leather or of composition leather not suitable for the manufacture of leather articles containing hexavalent chromium compounds or biocides (note the related entry on list B B3090)

A3110 Fellmongery wastes containing hexavalent chromium compounds or biocides or infectious substances (note the related entry on list B B3110)

A3120 Fluff—light fraction from shredding

A3130 Waste organic phosphorous compounds

A3140 Waste non-halogenated organic solvents but excluding such wastes specified on list B

A3150 Waste halogenated organic solvents

A3160 Waste halogenated or unhalogenated non-aqueous distillation residues arising from organic solvent recovery operations

A3170 Wastes arising from the production of aliphatic halogenated hydrocarbons (such as chloromethane, dichloro-ethane, vinyl chloride, vinylidene chloride, allyl chloride and epichlorhydrin)

A3180 Wastes, substances and articles containing, consisting of or contaminated with polychlorinated biphenyl (PCB), polychlorinated terphenyl (PCT), polychlorinated naphthalene (PCN) or polybrominated biphenyl (PBB), or any other polybrominated analogues of these compounds, at a concentration level of 50 mg/kg or more

A3190 Waste tarry residues (excluding asphalt cements) arising from refining, distillation and any pyrolitic treatment of organic materials

A4 Wastes which may contain either inorganic or organic constituents

A4010 Wastes from the production, preparation and use of pharmaceutical products but excluding such wastes specified on list B

A4020 Clinical and related wastes; that is wastes arising from medical, nursing, dental, veterinary, or similar practices, and wastes generated in hospitals or other facilities during the investigation or treatment of patients, or research projects

A4030 Wastes from the production, formulation and use of biocides and phytopharmaceuticals, including waste pesticides and herbicides which are off-specification, outdated, or unfit for their originally intended use

A4040 Wastes from the manufacture, formulation and use of wood-preserving chemicals

A4050 Wastes that contain, consist of or are contaminated with any of the following:

- Inorganic cyanides, excepting precious-metal-bearing residues in solid form containing traces of inorganic cyanides
- Organic cyanides

A4060 Waste oils/water, hydrocarbons/water mixtures, emulsions

A4070 Wastes from the production, formulation and use of inks, dyes, pigments, paints, lacquers, varnish excluding any such waste specified on list B (note the related entry on list B B4010)

A4080 Wastes of an explosive nature (but excluding such wastes specified on list B)

A4090 Waste acidic or basic solutions, other than those specified in the corresponding entry on list B (note the related entry on list B B2120)

A4100 Wastes from industrial pollution control devices for cleaning of industrial off-gases but excluding such wastes specified on list B

A4110 Wastes that contain, consist of or are contaminated with any of the following:

- Any congenor of polychlorinated dibenzo-furan
- Any congenor of polychlorinated dibenzo-dioxin

A4120 Wastes that contain, consist of or are contaminated with peroxides

A4130 Waste packages and containers containing Annex I substances in concentrations sufficient to exhibit Annex III hazard characteristics

A4140 Waste consisting of or containing off specification or outdated chemicals corresponding to Annex I categories and exhibiting Annex III hazard characteristics

A4150 Waste chemical substances arising from research and development or teaching activities which are not identified and/or are new and whose effects on human health and/or the environment are not known

A4160 Spent activated carbon not included on list B (note the related entry on list B B2060)

A-2
ANNEX IX: LIST B

Wastes contained in the Annex will not be wastes covered by Article 1, paragraph 1 (a), of this Convention unless they contain Annex I material to an extent causing them to exhibit an Annex III characteristic.

B1 Metal and metal-bearing wastes

B1010 Metal and metal-alloy wastes in metallic, non-dispersible form:

- Precious metals (gold, silver, the platinum group, but not mercury)
- Iron and steel scrap
- Copper scrap
- Nickel scrap
- Aluminium scrap
- Zinc scrap
- Tin scrap
- Tungsten scrap

- Molybdenum scrap
- Tantalum scrap
- Magnesium scrap
- Cobalt scrap
- Bismuth scrap
- Titanium scrap
- Zirconium scrap
- Manganese scrap
- Germanium scrap
- Vanadium scrap
- Scrap of hafnium, indium, niobium, rhenium and gallium
- Thorium scrap
- Rare earths scrap

B1020 Clean, uncontaminated metal scrap, including alloys, in bulk finished form (sheet, plate, beams, rods, etc), of:

- Antimony scrap
- Beryllium scrap
- Cadmium scrap
- Lead scrap (but excluding lead-acid batteries)
- Selenium scrap
- Tellurium scrap

B1030 Refractory metals containing residues

B1040 Scrap assemblies from electrical power generation not contaminated with lubricating oil, PCB or PCT to an extent to render them hazardous

B1050 Mixed non-ferrous metal, heavy fraction scrap, not containing Annex I materials in concentrations sufficient to exhibit Annex III characteristics

B1060 Waste selenium and tellurium in metallic elemental form including powder

B1070 Waste of copper and copper alloys in dispersible form, unless they contain Annex I constituents to an extent that they exhibit Annex III characteristics

B1080 Zinc ash and residues including zinc alloys residues in dispersible form unless containing Annex I constituents in concentration such as to exhibit Annex III characteristics or exhibiting hazard characteristic H4.3

B1090 Waste batteries conforming to a specification, excluding those made with lead, cadmium or mercury

B1100 Metal-bearing wastes arising from melting, smelting and refining of metals:

- Hard zinc spelter
- Zinc-containing drosses:
  -Galvanizing slab zinc top dross (>90% Zn)
  -Galvanizing slab zinc bottom dross (>92% Zn)
  -Zinc die casting dross (>85% Zn)
  -Hot dip galvanizers slab zinc dross (batch)(>92% Zn)
  -Zinc skimmings
- Aluminium skimmings (or skims) excluding salt slag
- Slags from copper processing for further processing or refining not containing arsenic, lead or cadmium to an extend that they exhibit Annex III hazard characteristics

- Wastes of refractory linings, including crucibles, originating from copper smelting
- Slags from precious metals processing for further refining
- Tantalum-bearing tin slags with less than 0.5% tin

B1110 Electrical and electronic assemblies:

- Electronic assemblies consisting only of metals or alloys
- Waste electrical and electronic assemblies or scrap (including printed circuit boards) not containing components such as accumulators and other batteries included on list A, mercury-switches, glass from cathode-ray tubes and other activated glass and PCB-capacitors, or not contaminated with Annex I constituents (e.g., cadmium, mercury, lead, polychlorinated biphenyl) or from which these have been removed, to an extent that they do not possess any of the characteristics contained in Annex III (note the related entry on list A A1180)
- Electrical and electronic assemblies (including printed circuit boards, electronic components and wires) destined for direct reuse, and not for recycling or final disposal

B1120 Spent catalysts excluding liquids used as catalysts, containing any of:

- Transition metals, excluding waste catalysts (spent catalysts, liquid used catalysts or other catalysts) on list A: Scandium, Vanadium, Manganese, Cobalt, Copper, Yttrium, Niobium, Hafnium, Tungsten, Titanium, Chromium, Iron, Nickel, Zinc, Zirconium, Molybdenum, Tantalum, Rhenium,
- Lanthanides (rare earth metals): Lanthanum, Praseodymium, Samarium, Gadolinium, Dysprosium, Erbium, Ytterbium, Cerium, Neody, Europium, Terbium, Holmium, Thulium, Lutetium

B1130 Cleaned spent precious-metal-bearing catalysts

B1140 Precious-metal-bearing residues in solid form which contain traces of inorganic cyanides

B1150 Precious metals and alloy wastes (gold, silver, the platinum group, but not mercury) in a dispersible, non-liquid form with appropriate packaging and labelling

B1160 Precious-metal ash from the incineration of printed circuit boards (note the related entry on list A A1150)

B1170 Precious-metal ash from the incineration of photographic film

B1180 Waste photographic film containing silver halides and metallic silver

B1190 Waste photographic paper containing silver halides and metallic silver

B1200 Granulated slag arising from the manufacture of iron and steel

B1210 Slag arising from the manufacture of iron and steel including slags as a source of TiO_2 and vanadium

B1220 Slag from zinc production, chemically stabilized, having a high iron content (above 20%) and processed according to industrial specifications (e.g., DIN 4301) mainly for construction

B1230 Mill scaling arising from the manufacture of iron and steel

B1240 Copper oxide mill-scale

B2 Wastes containing principally inorganic constituents, which may contain metals and organic materials

B2010 Wastes from mining operations in non-dispersible form:

- Natural graphite waste
- Slate waste, whether or not roughly trimmed or merely cut, by sawing or otherwise
- Mica waste
- Leucite, nepheline and nepheline
- Syenite waste
- Feldspar waste
- Fluorspar waste
- Silica wastes in solid form excluding those used in foundry operations

B2020 Glass waste in non-dispersible form:

- Cullet and other waste and scrap of glass except for glass from cathode-ray tubes and other activated glasses

B2030 Ceramic wastes in non-dispersible form:

- Cermet wastes and scrap (metal ceramic composites)
- Ceramic based fibres not elsewhere specified or included

B2040 Other wastes containing principally inorganic constituents:

- Partially refined calcium sulphate produced from flue-gas desulphurization (FGD)
- Waste gypsum wallboard or plasterboard arising from the demolition of buildings
- Slag from copper production, chemically stabilized, having a high iron content (above 20%) and processed according to industrial specifications (e.g., DIN 4301 and DIN 8201) mainly for construction and abrasive applications
- Sulphur in solid form
- Limestone from the production of calcium cyanamide (having a pH less than 9)
- Sodium, potassium, calcium chlorides
- Carborundum (silicon carbide)
- Broken concrete
- Lithium-tantalum and lithium-niobium containing glass scraps

B2050 Coal-fired power plant fly-ash, not included on list A (note the related entry on list A A2060)

B2060 Spent activated carbon resulting from the treatment of potable water and processes of the food industry and vitamin production (note the related entry on list A A4160)

B2070 Calcium fluoride sludge

B2080 Waste gypsum arising from chemical industry processes not included on list A (note the related entry on list A A2040)

B2090 Waste anode butts from steel or aluminium production made of petroleum coke or bitumen and cleaned to normal industry specifications (excluding anode butts from chlor alkali electrolyses and from metallurgical industry)

B2100 Waste hydrates of aluminium and waste alumina and residues from alumina production excluding such materials used for gas cleaning, flocculation or filtration processes

B2110 Bauxite residue ("red mud") (pH moderated to less than 11.5)

B2120 Waste acidic or basic solutions with a pH greater than 2 and less than 11.5, which are not corrosive or otherwise hazardous (note the related entry on list A A4090)

B3 Wastes containing principally organic constituents, which may contain metals and inorganic materials

B3010 Solid plastic waste: The following plastic or mixed plastic materials, provided they are not mixed with other wastes and are prepared to a specification:

- Scrap plastic of non-halogenated polymers and co-polymers, including but not limited to the following:
  -ethylene
  -styrene
  -polypropylene
  -polyethylene terephthalate
  -acrylonitrile
  -butadiene
  -polyacetals
  -polyamides
  -polybutylene terephthalate
  -polycarbonates
  -polyethers
  -polyphenylene sulphides
  -acrylic polymers
  -alkanes C10-C13 (plasticiser)
  -polyurethane (not containing CFCs)
  -polysiloxanes
  -polymethyl methacrylate
  -polyvinyl alcohol
  -polyvinyl butyral
  -polyvinyl acetate
- Cured waste resins or condensation products including the following:
  -urea formaldehyde resins
  -phenol formaldehyde resins
  -melamine formaldehyde resins
  -epoxy resins
  -alkyd resins
  -polyamides
- The following fluorinated polymer wastes
  -perfluoroethylene/propylene (FEP)
  -perfluoroalkoxy alkane (PFA)
  -perfluoroalkoxy alkane (MFA)
  -polyvinylfluoride (PVF)
  -polyvinylidenefluoride (PVDF)

B3020 Paper, paperboard and paper product wastes The following materials, provided they are not mixed with hazardous wastes:

- Waste and scrap of paper or paperboard of: unbleached paper or paperboard or of corrugated paper or paperboard
- Other paper or paperboard, made mainly of bleached chemical pulp, not coloured in the mass
- Paper or paperboard made mainly of mechanical pulp (for example, newspapers, journals and similar printed matter)

- Other, including but not limited to 1) laminated paperboard 2) unsorted scrap.

B3030 Textile wastes The following materials, provided they are not mixed with other wastes and are prepared to a specification:

- Silk waste (including cocoons unsuitable for reeling, yarn waste and garnetted stock), not carded or combed, other
- Waste of wool or of fine or coarse animal hair, including yarn waste but excluding garnetted stock, noils of wool or of fine animal hair, other waste of wool or of fine animal hair, waste of coarse animal hair
- Cotton waste (including yarn waste and garnetted stock), yarn waste (including thread waste), garnetted stock, other
- Flax tow and waste
- Tow and waste (including yarn waste and garnetted stock) of true hemp (Cannabis sativa L.)
- Tow and waste (including yarn waste and garnetted stock) of jute and other textile bast fibres (excluding flax, true hemp and ramie)
- Tow and waste (including yarn waste and garnetted stock) of sisal and other textile fibres of the genus Agave
- Tow, noils and waste (including yarn waste and garnetted stock) of coconut
- Tow, noils and waste (including yarn waste and garnetted stock) of abaca (Manila hemp or Musa textilis Nee)
- Tow, noils and waste (including yarn waste and garnetted stock) of ramie and other vegetable textile fibres, not elsewhere specified or included
- Waste (including noils, yarn waste and garnetted stock) of man-made fibres, of synthetic fibres, of artificial fibres
- Worn clothing and other worn textile articles
- Used rags, scrap twine, cordage, rope and cables and worn out articles of twine, cordage, rope or cables of textile materials, sorted, other

B3040 Rubber wastes The following materials, provided they are not mixed with other wastes: Waste and scrap of hard rubber (e.g., ebonite); other rubber wastes (excluding such wastes specified elsewhere)

B3050 Untreated cork and wood waste: Wood waste and scrap, whether or not agglomerated in logs, briquettes, pellets or similar forms; cork waste: crushed, granulated or ground cork

B3060 Wastes arising from agro-food industries provided it is not infectious:

- Wine lees
- Dried and sterilized vegetable waste, residues and by-products, whether or not in the form of pellets, of a kind used in animal feeding, not elsewhere specified or included
- Degras: residues resulting from the treatment of fatty substances or animal or vegetable waxes
- Waste of bones and horn-cores, unworked, defatted, simply prepared (but not cut to shape), treated with acid or degelatinised
- Fish waste
- Cocoa shells, husks, skins and other cocoa waste
- Other wastes from the agro-food industry excluding by-products which meet national and international requirements and standards for human or animal consumption

B3070 The following wastes:

- Waste of human hair
- Waste straw
- Deactivated fungus mycelium from penicillin production to be used as animal feed

B3080 Waste parings and scrap of rubber

B3090 Paring and other wastes of leather or of composition leather not suitable for the manufacture of leather articles, excluding leather sludges, not containing hexavalent chromium compounds and biocides (note the related entry on list A A3100)

B3100 Leather dust, ash, sludges or flours not containing hexavalent chromium compounds or biocides (note the related entry on list A A3090)

B3110 Fellmongery wastes not containing hexavalent chromium compounds or biocides or infectious substances (note the related entry on list A A3110)

B3120 Wastes consisting of food dyes

B3130 Waste polymer ethers and waste non-hazardous monomer ethers incapable of forming peroxides

B3140 Waste pneumatic tyres, excluding those destined for Annex IVA operations

B4 Wastes which may contain either inorganic or organic constituents

B4010 Wastes consisting mainly of water-based/latex paints, inks and hardened varnishes not containing organic solvents, heavy metals or biocides to an extent to render them hazardous (note the related entry on list A A4070)

B4020 Wastes from production, formulation and use of resins, latex, plasticizers, glues/adhesives, not listed on list A, free of solvents and other contaminants to an extent that they do not exhibit Annex III characteristics, e.g., water-based, or glues based on casein starch, dextrin, cellulose ethers, polyvinyl alcohols (note the related entry on list A A3050)

B4030 Used single-use cameras, with batteries not included on list A

APPENDIX B

Contaminant Properties

It is a capital mistake to theorize before one has data.
[Sherlock Holmes] Sir Arthur Conan Doyle

The following tables were developed by starting with Table C-1 of the **Superfund Public Health Evaluation Manual** (EPA 540/1-86/060), October, 1986. This material was obtained from sources A through N (see sources of data below). Additional data were added from the references listed, and the source is denoted by the column labeled S.

Compounds are identified by their commonly used chemical names, chemical formulas, and the Chemical Abstract Service Registry Numbers (CASRN). Many chemicals are listed as toxic substances for specific regulatory purposes. The four columns labeled P, C, N, and H designate:

- P Priority Pollutants (designated under the Clean Water Act)
- C Hazardous Substance List (designated under CERCLA)
- N Appendix IX Chemicals (designated under RCRA)
- H Hazardous Air Pollutant (designated under the 1991 Clear Air Act Amendment)

These designations are continuously updated, and the latest *Federal Register* listing should be consulted.

Definitions and examples of the uses of the properties tabulated in this appendix may be found in Chapter 3. Use of the Freundlich parameters is discussed in Section 9-2.

In general the data included in this appendix are measured values. Where calculated values are utilized, the source is designated by the symbol "◇ ". Where solubility in water is reported as "infinitely soluble" the designation "Total" is shown in the table. The organic carbon partition coefficient K_{oc} is estimated from water

solubility by the following equation:

$$K_{oc} = (-0.55 \log S) + 3.64 \qquad \text{(B-1)}$$
$$S = \text{solubility (mg/L)}$$

For parameters such as water solubility and vapor pressure, an additional column has been added to indicate the temperature at which the parameter was measured. The sources of data for Henry's law constants are Z and AA. Both provide H as a function of temperature, but the form of the equation is different. For data source Z:

$$\ln H = A/T + B \qquad T \text{ in kelvins} \qquad \text{(B-2)}$$

For data source AA:

$$H = \exp\,(A - B/T) \qquad \text{(B-3)}$$

where H = Henry's law constant [(atm · m^3)/mol]
T = temperature (kelvins)
A, B = tabulated constants

When two estimates were given in source Z, the one with the higher coefficient of determination (r^2) was selected.

While these tables were prepared with care, no warranty is intended or implied, and the user is advised to independently verify information prior to use. Most of these parameters were measured in pure water and may not be appropriate for environmental conditions different from those used in the measurement. The reader is strongly urged to consult the original reference before applying this information.

SOURCES OF DATA

A. U.S. Environmental Protection Agency: Health Effects Assessment for [Specific Chemical]. [Note: 58 individual documents available for specific chemicals or chemical groups], U.S. EPA, ECAO, 1985.

B. Jaber, et al.: *Data Acquisition for Environmental Transport and Fate Screening,* EPA 600/6-84-009, Office of Health and Environmental Assessment, U.S. EPA, Washington, D.C., 1984.

C. Mabey, et al.: *Aquatic Fate Process Data for Organic Priority Pollutants,* prepared by SRI International, EPA Contract Nos. 68-01-3867 and 68-03-2981, prepared for Monitoring and Data Support Division, Office of Water Regulations and Standards, Washington, D.C., 1982.

D. Callahan, et al.: *Water-Related Environmental Fate of 129 Priority Pollutants,* vols. I and II, Office of Water Planning and Standards, Office of Water and Waste Management, U.S. EPA, EPA Contract Nos. 68-01-3852 and 68-01-3867, 1979.

E. U.S. Environmental Protection Agency: *Treatability Manual,* vol. I, EPA 600/2-82-001a, ORD, 1981.

F. Dawson, et al.: *Physical/Chemical Properties of Hazardous Waste Constituents,* prepared by Southeast Environmental Research Laboratory for U.S. EPA, 1980.

G. Lyman, et al.: *Handbook of Chemical Property Estimation Methods,* McGraw-Hill, New York, 1982.

H. U.S. Environmental Protection Agency: Ambient Water Quality Criteria Documents for [Specific Chemical], OWRS, 1980.

I. Weast, et al.: *CRC Handbook of Chemistry and Physics,* 1979.

J. Verschueren, Karel: *Handbook of Environmental Data for Organic Chemicals,* 2nd ed., Van Nostrand Reinhold, New York, 1983.

L5. Perry, R. H., and Chilton: *Chemical Engineers' Handbook*, 5th ed., McGraw-Hill, New York, 1973.

L6. Perry, R. H., and D. Green: *Chemical Engineers' Handbook,* 6th ed., McGraw-Hill, New York, 1984.

M. U.S. Environmental Protection Agency: *Exposure Profiles for RCRA Risk-Cost Analysis Model,* prepared by Environ Corporation, 1984.

N. U.S. Environmental Protection Agency: *Characterization of Constituents from Selected Waste Streams Listed in 40 CFR Section 261,* prepared by Environ Corporation, 1984.

O. Dobbs, Richard A., and Jesse M. Cohen: *Carbon Adsorption Isotherms for Toxic Organics,* EPA-600/8-80-023, Wastewater Research Division, Cincinnati, OH.

P. Sax, N. Irving, and Richard J. Lewis, Sr.: *Dangerous Properties of Industrial Materials,* 7th ed., Van Nostrand Reinhold, New York, 1989.

R. The Merck Index: Merck & Co., Inc., Rahway, New Jersey, NJ, 1989.

T. Yaws, Carl L., et al.: "Organic Chemicals: Water Solubility Data." *Chemical Engineering,* July 1990.

U. Schultz, H. Lee, et al.: *Superfund Exposure Assess Manual,* U.S. EPA, EPA/540/1-88/001, OSWER Directive 9285.5-1, April 1988.

V. U.S. Coast Guard: *Chemical Hazards Response Information System (CHRIS Manual),* Hazardous Chemical Data, M.16465.12A, November 1984.

W. Haarhoff, J., and J. L. Cleasby: "Evaluation of Air Stripping for the Removal of Organic Drinking Water Contaminants," *Water South Africa,* January 1990.

X. Yaws, Carl L., et al.: "Hydrocarbons: Water Solubility Data," *Chemical Engineering,* April 1990.

Y. Howard, P. H.: *Handbook of Environmental Fate and Exposure Data for Organic Chemicals,* Lewis Publishers, 1989, 1990.

Z. Howe, G. B., M. E. Mullins, and T. N. Rogers: "Evaluation and Production of Henry's Law Constants and Aqueous Solubilities for Solvents and Hydrocarbon Fuel Components—Volume I: Technical Discussion," Final Report, Engineering and Services Lab, Tyndall Air Force Base, Report No. ESL-86-66, 1987.

AA. Gossett, J. M.: "Measurement of Henry's Law Constants for C1 and C2 Chlorinated Hydrocarbons," *Environmental Science and Technology,* vol. 21, pp. 202–208, 1987.

AB. Patterson, J. W.: *Industrial Wastewater Treatment Technology,* Butterworth Publishers, Stoneham, Mass., pp. 331–334, 1985.

AC. Shen, T.: "Estimating Hazardous Air Emissions from Disposal Sites," *Pollution Engineering,* vol. 13, no. 8, pp. 31–34, 1981.

AD. Lighton, David T., and Joseph M. Calo: "Distribution Coefficients of Chlorinated Hydrocarbons in Dilute Air-Water Systems for Ground Water Contamination Applications," *Journal of Chemical Engineering Data,* 1981.

ADDITIONAL REFERENCES

Aquatic Fate Process Data for Organic Priority Pollutants, prepared by SRI International, EPA 440/4-81-09014 [NTIS: PB 87-169 090], December 1982.

American Mutual Insurance Alliance, *Handbook of Organic Industrial Solvents,* AMIA, Chicago, IL, 1988.

Kenaga, E., "Correlation of Bioconcentration Factors of Chemicals in Aquatic and Terrestrial Organisms with their Physical and Chemical Properties," *Ecotoxicol. Environ. Saf.,* vol. 4, pp. 26–38, 1980.

Patniak, Pradyot, *A Comprehensive Guide to the Hazardous Properties of Chemical Substances,* Van Nostrand-Rheinhold, New York, 1992.

Chemical name	Chemical formula	P	C	N	H	CASRN	Molecular weight, g/mol	Water solubility		
								mg/L	°C	S
Acenaphthene	$C_{12}H_{10}$	P	C	N		83-32-9	154.21	3.42	25	C
Acenaphthylene	$C_{12}H_8$	P	C	N		208-96-8	152.2	3.93	20	J
Acetone	C_3H_6O		C	N		67-64-1	58.09	Total	20	T
Acetonitrile	C_2H_3N	P		N	H	75-05-8	41.06	Total	20	T
2-(Acetylamino) fluorene	$C_{15}H_{13}NO$			N	H	53-96-3	223.29	6.50×10^3		B
Acrylic acid	$C_3H_4O_2$				H	79-10-7	72.07	Total	20	T
Acrylonitrile	C_3H_3N				H	107-13-1	53.07	7.35×10^4	20	T
Aldicarb	$C_7H_{14}N_2O_2S$					116-06-3	190.29			
Aldrin	$C_{12}H_8Cl_6$	P	C	N		309-00-2	364.9	1.80×10^{-1}	25	C
Allyl alcohol	C_3H_6O			N		107-18-6	58.09	Total	20	T
4-Aminobiphenyl	$C_{12}H_{11}N$			N	H	92-67-1	169.24	8.42×10^2		B
Amitrole	$C_2H_4N_4$					61-82-5	84.1	2.80×10^2	25	J
Ammonia	H_3N					7664-41-7	17.04	5.30×10^2	20	J
Anthracene	$C_{14}H_{10}$	P	C	N		120-12-7	178.24	4.50×10^{-2}	25	C
Antimony	Sb				H	7440-36-0	121.75			
Arsenic	As				H	7440-38-2	74.92			
Auramine	$C_{17}H_{21}N_3{\cdot}ClH{\cdot}H_2O$					2465-27-2	321.89	2.10		B
Azaserine	$C_5H_7N_3O_4$					115-02-6	173.15	1.36×10^5		B
Aziridine	C_2H_5N					151-56-4	43.08	Total	20	T
Barium	Ba					7440-39-3	137.36	10		P
Barium cyanide	C_2BaN_2					542-62-1	189.38	6.86×10^5	14	V
Benefin	$C_{13}H_{16}F_3N_3O_4$					1861-40-1	335	7.00	25	J
Benz(*c*)acridine	$C_{17}H_{11}N$					225-51-4	229.29	1.40		B
Benz(*a*)anthracene	$C_{18}H_{12}$	P	C	N		56-55-3	228.3	5.70×10^{-3}	20	C
Benzene	C_6H_6	P	C	N	H	71-43-2	78.12	1.78×10^2	25	C
Benzidine	$C_{12}H_{12}N_2$	P		N	H	92-87-5	184.26	4.00×10^2	12	J
Benzo(*b*)fluoranthene	$C_{20}H_{12}$					205-99-2	252.32	1.40×10^{-2}	20	C
Benzo(*k*)fluoranthene	$C_{20}H_{12}$					207-08-9	252.32	4.30×10^{-3}	25	C
Benzo(*ghi*)perylene	$C_{22}H_{12}$					191-24-2	276.34	2.60×10^{-4}	25	J
Benzo(*a*)pyrene	$C_{20}H_{12}$					50-32-8	252.32	3.80×10^{-3}	25	C
Benzotrichloride	$C_7H_5Cl_3$				H	98-07-7	195.47			
Benzyl chloride	$C_6H_5CH_2Cl$				H	100-44-7	126.59	2.57×10	25	V
Beryllium	Be				H	7440-41-7	9.01			
1,1′-Biphenyl	$C_{12}H_{10}$				H	92-52-4	154.22	7.50	25	J
Bis(2-chloroethyl)ether	$C_4H_8Cl_2O$	P	C	N		111-44-4	143.02	1.20×10^4	25	C
Bis(2-chloroisopropyl)ether	$C_6H_{12}Cl_2O$	P	C	N		108-60-1	171.08	1.70×10^3	25	C
Bis(chloromethyl)ether	$C_2H_4Cl_2O$				H	542-88-1	114.96	2.20×10^4	25	C
Bis(2-ethylhexyl)phthalate	$C_{24}H_{38}O_4$	P	C	N	H	117-81-7	390.62	4.00×10	25	C
Bromomethane	CH_3Br					74-83-9	94.95	9.00×10^2	20	J
1,3-Butadiene	C_4H_6				H	106-99-0	54.1	7.35×10^2	23	J
n-Butanol	$C_4H_{10}O$					71-36-3	74.14	6.69×10^4	20	K
Butyl phthalyl butyl glycolate	$C_{18}H_{24}O_6$					85-70-1	336.42	5.92×10^5	20	K
Cacodylic acid	$C_2H_7AsO_2$					75-60-5	138.01	5.88×10^5	20	V

S = source of data; P = priority poll; C = hazardous substance list; N = Appendix IX; H = hazardous air pollutant

Vapor pressure mmHg	°C	S	Diffusion coefficient in air cm^2/s	S	Henry's constant, H* ($atm \cdot m^3$)/mol; T(K) A	B	S	K_{oc} mL/g	S	log K_{ow}	S	pH	Freundlich parameters K, mg/g	$1/n$	S	Fish BCF L/kg	S
1.55×10^{-3}	25	C						4.50×10^{3}	C	4.00	C	5.3	190	0.36	O	242	H
2.90×10^{-2}	20	C						2.50×10^{3}	C	3.70	C	5.3	115	0.37	O		
2.70×10^{2}	30	J	0.1093	U				2.20	◇	-2.40×10^{-1}	J						
7.40×10	20	J						2.20	◇	-3.40×10^{-1}	F						
								1.60×10^{3}	◇	3.28		7.1	318	0.12	O		
3.20	20	J						1.61×10^{-1}	Y							0	F
1.00×10^{2}	23	C						8.50×10^{-1}	C	2.50×10^{-1}	C	5.3	1	0.51	AB	48	G
1.00×10^{4}	25	J															
6.00×10^{6}	25	C						9.60×10^{4}	C	5.30	C	5.3	651	0.92	O	28	H
2.00×10	20	J						1.70×10^{-1}	J	-2.20×10^{-1}	◇						
1.00×10	157	I						1.07×10^{2}	◇	2.78	B	7.2	200	0.26	O		
								4.40	◇	-2.08							
7.60×10^{3}	26	J						3.10	◇	0.00	F					0	F
1.70×10^{-5}	25	C						1.40×10^{4}	C	4.45	A	5.3	376	0.70	O		
1.00	886	P														1	H
1.00	372	P														44	H
								2.90×10^{3}	◇	4.16	B						
								6.60	◇	-1.08	B						
4.45×10^{2}	20	J						1.30	◇	-1.01							
1.00×10	1049	P															
4.00×10^{-7}	25	J															
								1.00×10^{3}	◇	4.56	B						
2.20×10^{-8}	20	C						1.38×10^{6}	G	5.60	C						
9.52×10	25	C	0.09234	U	-3.19×10^{3}	5.53	Z	8.30×10	G	2.12	A	7	1	2.90	O	5.2	H
1.00×10	176	I						1.05×10	C	1.30	C	7-9	220	0.35	O	87.5	H
5.00×10^{-7}	20	C						5.50×10^{5}	C	6.06	A						
5.10×10^{-7}	20	C						5.50×10^{5}	C	6.06	G	7.1	181	0.57	O		
1.03×10^{-10}	25	C						1.60×10^{6}	C	6.51	A	7	11	0.37	O		
5.60×10^{-9}	25	C						5.50×10^{6}	C	6.06	C	7.1	34	0.44	O		
2.30×10	20	Y								2.92	Y						
1.00	22	J						5.00×10	◇	2.63	F						
1.00×10	1860	I															
1.00×10	117	I															
7.10×10^{-1}	20	J						1.39×10	C	1.50	C	5.3	0	1.84	AB	6.9	H
8.50×10^{-1}	20	J						6.10×10	C	2.10	C	5.4	24	0.57	O	0	D
3.00×10	22	Y						1.20	C	3.80×10^{-1}	C					0.63	H
												5.3	11,300	1.50	O		
2.33×10^{3}	37	J	0.1083	U													
2.50	20	J						1.20×10^{2}	◇	1.99	F						
4.76	21	V								8.80×10^{-1}	Y						
								2.40	◇	0.00	F						

◇ = Calculated value.

*When the source (S) is Z, then $\ln H = A/T + B$, T in kelvins; when the source (S) is AA, then $H = \exp(A - B/T)$, T in kelvins.

Chemical name	Chemical formula	P	C	N	H	CASRN	Molecular weight, g/mol	Water solubility		
								mg/L	°C	S
Cadmium	Cd				H	7740-43-9	112.4			
Captan	$C_9H_8Cl_3NO_2S$				H	133-06-2	300.59	5.00×10^{-1}	20	J
Carbaryl	$C_{12}H_{11}NO_2$				H	63-25-2	201.24	4.00×10	30	J
Carbon disulfide	CS_2		C	N	H	75-15-0	76.13	2.30×10^3	22	J
Carbon tetrachloride	CCl_4	P	C	N	H	56-23-5	153.81	8.00×10^2	20	J
Chlordane	$C_{10}H_6Cl_8$	P	C	N	H	12789-03-6	409.76	5.60×10^{-1}	25	C
Chlorobenzene	C_6H_5Cl	P	C	N	H	108-90-7	112.56	5.00×10^2	20	J
Chlorobenzylate	$C_{16}H_{14}Cl_2O_3$			N	H	510-15-6	325.2	2.19×10		B
Chlorodibromomethane	$CHBr_2Cl$	P	C	N		124-48-1	208.29	4.75×10^3	20	W
Chloroform	$CHCl_3$	P	C	N	H	67-66-3	119.37	8.00×10^3	20	J
Chloromethane	CH_3Cl					74-87-3	50.49			
Chloromethyl methyl ether	C_2H_5ClO				H	107-30-2	80.52			
Chromium	Cr				H	7440-47-3	52			
Chrysene	$C_{18}H_{12}$	P	C	N		218-01-9	228.3	1.80×10^{-3}	25	C
Copper	Cu					7440-50-8	63.55			
Cresol	C_7H_8O				H	1319-77-3	108.15	2.15×10^4	20	K
Crotonaldehyde	C_4H_6O					4170-30-3	70.09	1.33×10^5	20	V
Cyanogen	C_2N_2					460-19-5	52.04	9.49×10^{-3}	20	T
Cyanogen chloride	CClN					506-77-4	61.47	1.00×10^{-3}	20	J
Cyclohexane	C_6H_{12}					110-82-7	84-18			
Cyclophosphamide	$C_7H_{15}Cl_2N_2O_2P$					50-18-0	261.1	1.31×10^9		B
Dalapon	$C_3H_4Cl_2O_2$					75-99-0	142.97			
4,4′-DDD	$C_{14}H_{10}Cl_4$	P	C	N		72-54-8	320.04	0.02–0.09	25	C
4,4′-DDE	$C_{14}H_8Cl_4$	P	C	N	H	72-55-9	318.02	4.00×10^{-2}	20	J
4,4′-DDT	$C_{14}H_9Cl_5$	P	C	N		50-29-3	354.48	5.50×10^{-3}	25	C
Decalin	$C_{10}H_{18}$					91-17-8	138.28			
Diallate	$C_{10}H_{17}Cl_2NOS$					2303-16-4	270.24	1.40×10		B
2,4-Diaminotoluene	$C_7H_{10}N_2$					95-80-7	122.19	4.77×10^4		B
1,2,7,8-Dibenzopyrene	$C_{24}H_{14}$					189-55-9	302.38	1.10×10^{-1}		B
Dibenz(*a*, *h*)anthracene	$C_{22}H_{14}$	P	C	N		53-70-3	278.36	5.00×10^{-4}	25	C
Dibromochloromethane	$CHBr_2Cl$					124-48-1	208.29			
1,2-Dibromo-3-Chloropropane	$C_3H_5Br_2Cl$			N	H	96-12-8	236.36	1.00×10^3	20	W
Di-*n*-butyl phthalate	$C_{16}H_{22}O_4$				H	84-74-2	278.38	1.30×10	25	C
1,2-Dichlorobenzene	$C_6H_4Cl_2$	P	C	N		95-50-1	147	1.00×10^2	20	J
1,3-Dichlorobenzene	$C_6H_4Cl_2$	P	C	N		541-73-1	147	1.23×10^2	25	C
1,4-Dichlorobenzene	$C_6H_4Cl_2$	P	C	N	H	106-46-7	147	7.90×10	25	C
3,3′-Dichlorobenzidene	$C_{12}H_{10}Cl_2N_2$	P	C	N	H	91-94-1	253.14	3.99	22	J
Dichlorodifluoromethane	CCl_2F_2			N		75-71-8	120.91	2.80×10^2	25	C
1,1-Dichloroethane	$C_2H_4Cl_2$	P	C	N		75-34-3	98.96	5.50×10^3	20	J
1,2-Dichloroethane	$C_2H_4Cl_2$	P	C	N		107-06-2	98.96	8.69×10^3	20	J
1,1-Dichloroethylene	$C_2H_2Cl_2$	P	C	N		75-35-4	96.94	2.25×10^3	20	W
trans-1,2-Dichloroethene	$C_2H_2Cl_2$					156-60-5	96.94	6.00×10^2	20	J

S = source of data; P = priority poll; C = hazardous substance list; N = Appendix IX; H = hazardous air pollutant.

Vapor pressure mmHg	°C	S	Diffusion coefficient in air cm²/s	S	Henry's constant, H* (atm · m³)/mol; T(K) A	B	S	K_{oc} mL/g	S	log K_{ow}	S	Freundlich parameters pH	K, mg/g	1/n	S	Fish BCF L/kg	S
1.00	394	P														81	H
6.00×10^{-5}		E						6.40×10^{3}	◇	2.35	F						
5.00×10^{-3}	26	J								2.36	F						
2.60×10^{2}	20	J						5.40×10	◇	2.00	F					0	
9.00×10	20	J	0.08451	U	3.95×10^{3}	9.74	Z	1.10×10^{2}	◇	2.64	A	7	40	0.84	O	19	H
1.00×10^{-5}	25	C						1.40×10^{5}	C	3.32	A	5.3	245	0.38	O	14,000	H
1.17×10	20	W	0.07627	U	-2.69×10^{3}	3.47	Z	3.30×10^{2}	C	2.84	A	7.4	91	0.99	O	10	H
1.20×10^{-6}		B						8.00×10^{2}	◇	4.51							
1.50×10	20	W			-6.37×10^{3}	1.46×10	Z			2.09	D	7	63	0.93	O		
1.60×10^{2}	20	J	0.09404	U	9.84	4.61×10^{3}	AA	3.10×10	◇	1.97	G	7	11	0.84	O	3.75	H
					9.36	4.22×10^{3}	AA										
1.92×10^{2}	21	V								0.00	F						
1.00×10	1840	I															
																16	
6.30×10^{-9}	25	C						2.00×10^{5}	C	5.61	A		716	0.46	26		
1.00	1628	P															
																200	
1.00	53	J						5.00×10^{2}	G	1.97	F					0	F
1.90×10	20	J															
3.92×10^{3}	21	V															
1.00×10^{3}	20	J								6.40×10^{-1}	J						
					-3.24×10^{3}	9.14	Z										
								4.20×10^{-2}	◇	−3.22	B						
5.07	71	V															
1.89×10^{-6}	30	C						7.70×10^{5}	C	6.20	C						
6.50×10^{-6}	20	C						4.40×10^{6}	C	7.00	C	5.3	232	0.37	O	51,000	G
1.90×10^{-7}	25	C						2.43×10^{5}	G	6.19	J	5.3	322	0.50	O	54,000	H
					-4.13×10^{3}	1.18×10	Z										
6.40×10^{-3}		B						1.00×10^{3}	◇	7.30×10^{-1}	B						
3.80×10^{-5}		B						1.20×10	◇	3.50×10^{-1}	B						
								1.20×10^{3}	◇	6.62	B						
1.00×10	20	C						3.30×10^{6}	C	6.80	C	7.1	69	0.75	O		
					1.46×10	6.37×10^{3}	Z										
1.00	20	W						9.80×10	◇	2.29	B						
1.00×10^{-5}	25	C						1.70×10^{5}	C	5.60	C	3	220	0.45	AB		
1.00	20	C						1.70×10^{3}	C	3.60	C	5.5	129	0.43	O	56	H
2.28	25	C			-2.56×10^{3}	2.88	Z	1.70×10^{3}	C	3.60	C	5.1	118	0.45	O	56	H
1.18	25	C			-2.72×10^{3}	3.37	Z	1.70×10^{3}	C	3.60	C	5.1	121	0.47	O	56	H
1.00×10^{-5}	22	C						1.55×10^{3}	C	3.50	C	7.2-9.1	300	0.20	O	312	H
4.87	25	C						5.80×10	C	2.16	D						
1.80×10^{2}	20	J	0.09643	U	8.64	4.13×10^{3}	AA	3.00×10	C	1.79	A	5.3	2	0.53	O		
6.10×10	20	J			-4.02×10^{3}	6.94	Z	1.40×10	C	1.48	A	5.3	4	0.83	O	1.2	H
6.00×10^{2}	20	W	0.08386	U	8.85	3.73×10^{3}	AA	6.50×10	C	1.84	A	5.3	5	0.54	O	5.6	H
3.26×10^{2}	20	C			9.34	4.18×10^{3}	AA	5.90×10	C	4.80×10^{-1}		6.7	3	0.51	O	1.6	

◇ = Calculated value.

*When the source (S) is Z, then $\ln H = A/T + B$, T in kelvins; when the source (S) is AA, then $H = \exp(A - B/T)$, T in kelvins.

Chemical name	Chemical formula	P	C	N	H	CASRN	Molecular weight, g/mol	Water solubility		
								mg/L	°C	S
cis-1,2-Dichloroethene	$C_2H_2Cl_2$					540-59-0	96.94	8.00×10^2	20	J
Dichloromethane	CH_2Cl_2	P	C	N		75-09-2	84.93	2.00×10^4	20	J
2,4-Dichlorophenol	$C_6H_4Cl_2O$	P	C	N		120-83-2	163	4.50×10^3	25	J
2,4-Dichlorophenoxyacetic acid	$C_8H_6Cl_2O_3$			N		94-75-7	221.04	8.90×10^2	25	J
Dichlorophenylarsine	$C_6H_5AsCl_2$					696-28-6	221.93			
1,2-Dichloropropane	$C_3H_6Cl_2$	P	C	N		78-87-5	112.99	2.70×10^3	20	J
1,3-Dichloropropene	$C_3H_4Cl_2$				H	542-75-6	110.97	2.80×10^3	25	C
Dieldrin	$C_{12}H_8Cl_6O$	P	C	N		60-57-1	380.9	1.95×10^{-1}	25	C
1,2-Diethylhydrazine	$C_4H_{12}N_2$					1615-80-1	88.18	2.88×10^7		B
Diethylnitrosamine	$C_4H_{10}N_2O$					55-18-5	102.16			
Diethyl-*o*-phthalate	$C_{12}H_{14}O_4$	P	C	N		84-66-2	222.26	8.96×10^2	25	C
Diethylstilbcstcrol	$C_{18}H_{20}O_2$					56-53-1	268.38	9.60×10^{-3}		B
Dihydrosafrole	$C_{10}H_{12}O_2$					94-58-6	164.22	1.50×10^3		B
Dimethoate	$C_5H_{12}NO_3PS_2$					60-51-5	229.27	2.50×10^4	23	J
Dimethylamine	C_2H_7N					124-40-3	45.1	6.20×10^5	20	T
Dimethyl phthalate	$C_{10}H_{10}O_4$				H	131-11-3	194.2			
Dimethyl sulfate	$C_2H_6O_4S$				H	77-78-1	126.13	3.24×10^5		B
Dimethyl terephthalate	$C_{10}H_{10}O_4$					120-61-6	194.2			
4-Dimethylaminoazobenzene	$C_{14}H_{15}N_3$					60-11-7	225.32	1.36×10		B
7,12-Dimethylbenz(*a*) anthracene	$C_{20}H_{16}$					57-97-6	256.36	4.40×10^{-3}		B
Dimethylcarbamoyl chloride	C_3H_6ClNO				H	79-44-7	107.55	1.44×10^7		B
1,1-Dimethylhydrazine	$C_2H_8N_2$				H	57-14-7	60.12	1.24×10^8		B
1,2-Dimethylhydrazine	$C_2H_8N_2$					540-73-8	60.12			
Dimethylnitrosamine	$C_2H_6N_2O$					62-75-9	74.1	1.00×10^6		◇
1,3-Dinitrobenzene	$C_6H_4N_2O_4$					99-65-0	168.12	8.50	20	V
4,6-Dinitro-*o*-cresol	$C_7H_6N_2O_5$	P	C	N	H	534-52-1	198.15	2.90×10^2	25	C
2,4-Dinitrophenol	$C_6H_4N_2O_5$	P	C	N	H	51-28-5	184.12	5.60×10^3	18	C
2,3-Dinitrotoluene	$C_7H_6N_2O_4$					602-01-7	182.15	3.10×10^3		B
2,4-Dinitrotoluene	$C_7H_6N_2O_4$	P	C	N	H	121-14-2	182.15	2.70×10^2	22	C
2,5-Dinitrotoluene	$C_7H_6N_2O_4$					619-15-8	182.15	1.80×10^2	20	C
2,6-Dinitrotoluene	$C_7H_6N_2O_4$	P	C	N		606-20-2	182.15	1.80×10^2	20	C
3,4-Dinitrotoluene	$C_7H_6N_2O_4$					610-39-9	182.15	1.08×10^3		B
Dinoseb	$C_{10}H_{12}N_2O_5$					88-85-7	240.24	5.00×10	25	J
Dioxane-1,4	$C_4H_8O_2$			N	H	123-91-1	88.11	4.31×10^5		B
N,N-diphenylamine	$C_{12}H_{11}N$			N		122-39-4	169.24	3.00×10^2	25	J
1,2-Diphenylhydrazine	$C_{12}H_{12}N_2$	P		N	H	122-66-7	184.26	1.84×10^3	20	C
Dipropylnitrosoamine	$C_6H_{14}N_2O$	P	C	N		621-64-7	130.22	9.90×10^3	25	C
Disulfoton	$C_8H_{19}O_2PS_3$			N		298-04-4	274.42	1.72×10	20	V
Endosulfan	$C_9H_6Cl_6O_3S$					115-29-7	406.91	5.30×10^{-1}	25	C

S = source of data; P = priority poll; C = hazardous substance list; N = Appendix IX; H = hazardous air pollutant.

Vapor pressure mmHg	°C	S	Diffusion coefficient in air cm^2/s	S	Henry's constant, H* $(atm \cdot m^3)/mol$; T(K) A	B	S	K_{oc} mL/g	S	log K_{ow}	S	pH	Freundlich parameters K, mg/g	$1/n$	S	Fish BCF L/kg	S
2.08×10^2	20	W			8.48	4.19×10^3	AA	4.90×10	C	7.00×10^{-1}						1.6	
3.49×10^2	20	J			6.65	3.82×10^3	AA	8.80	C	1.30	C					5	H
5.90×10^{-2}	20	C						3.80×10^2	C	2.90	C	5.3	157	0.15	O	41	H
4.00×10^{-1}		F						2.00×10	G	2.81	F						
2.10×10^{-2}	20	P															
4.20×10	20	J			-4.71×10^3	9.84	Z	5.10×10	C	2.00	C	5.3	6	0.60	O		
3.80×10	25	J						4.80×10	C	2.00	C	5.3	8	0.46	O	1.9	H
1.78×10^{-7}	20	C						1.70×10^3	C	3.50	C	5.3	606	0.51	O	4760	H
								3.00×10^{-1}	◇	−1.68	B						
5.00		F								4.80×10^{-1}	F						
3.50×10^{-3}	25	C						1.42×10^2	C	2.50	C	5.4	110	0.27	O	117	G
								2.80×10	◇	5.46	B						
								7.80×10	◇	2.56	B						
2.50×10^{-2}	25	J								2.71	J						
1.52×10^3	10	F	0.12577	U				2.20	◇	-3.80×10^{-1}	F					0	F
6.80×10^{-1}	20	B						4.10	◇	−1.24	B						
1.60×10	100	J															
3.30×10^{-7}		B						1.00×10^3	◇	3.72	B						
								4.76×10^5	G	6.94	B						
1.95		B						5.00×10^{-1}	◇	−1.32	B						
1.57×10^2	25	J						2.00×10^{-1}	◇	−2.42	B						
1.57×10^2	25	J															
8.10	25	C						1.00×10^{-1}	C	-6.80×10^{-1}	C					0	D
								1.50×10^2	◇	1.62	F						
5.00×10^{-2}	20	C						2.40×10^2	C	2.70	C	3.5	237		O	0	E
1.49×10^{-5}	18	C						1.66×10	C	1.50	C	7	33	0.61	O	0	D
								5.30×10	◇	2.29	B					3.8	H
5.10×10^{-3}	20	C						4.50×10	C	2.00	C	5.4	146	0.31	O	3.8	H
2.00×10^{-5}	25	Y						8.40×10	◇	2.28	B					3.8	H
1.80×10^{-2}	20	C						9.20×10	C	2.00	C	5.4	145	0.32	O	3.8	H
								9.40×10	◇	2.29	B					3.8	H
1.00	151																
3.70×10	25	B						3.50	◇	0.01	B						
1.00	108	P						4.70×10^2	◇	3.60	B					30	G
2.60×10^{-5}	25	C						4.18×10^2	C	2.90	C	5.3	16,000	2.00	O	25	H
1.00×10^{-1}	25	C						1.50×10	C	1.50	C						
1.00×10^{-5}	25	C										5.3	194	0.50	O		

◇ = Calculated value.

*When the source (S) is Z, then $\ln H = A/T + B$, T in kelvins; when the source (S) is AA, then $H = \exp(A - B/T)$, T in kelvins.

Chemical name	Chemical formula	P	C	N	H	CASRN	Molecular weight, g/mol	Water solubility		
								mg/L	°C	S
1-Chloro-2,3-epoxypropane	C_3H_5ClO				H	106-89-8	92.53	6.00×10^4	20	J
Ethanol	C_2H_6O					64-17-5	46.08	6.52×10^4	20	T
2-Ethoxyethanol	$C_4H_{10}O_2$					110-80-5	90.14	1.00×10^6		F
Ethyl acetate	$C_4H_8O_2$					141-78-6	88.12	7.40×10^4	35	J
Ethyl benzene	C_8H_{10}	P	C	N	H	100-41-4	106.18	1.52×10^2	20	J
Ethyl carbitol	$C_6H_{14}O_3$					111-90-0	134.2			
Ethylene dibromide	$C_2H_4Br_2$				H	106-93-4	187.88	4.30×10^3	20	J
Ethylene glycol monobutyl ether	$C_6H_{14}O_2$					111-76-2	118.2	1.00×10^6		F
Ethylene oxide	C_2H_4O				H	75-21-8	44.06	3.80×10^8	20	B
Ethylene thiourea	$C_3H_6N_2S$				H	96-45-7	102.17	2.00×10^3		F
Ethyl methanesulfonate	$C_3H_8O_3S$					62-50-0	124.17	3.69×10^5		B
1-Ethyl-1-nitrosourea	$C_3H_7N_3O_2$					759-73-9	117.13	3.31×10^8		B
Fluoranthene	$C_{16}H_{10}$	P	C	N		206-44-0	202.26	2.06×10^{-1}	25	J
Fluorene	$C_{13}H_{10}$					86-73-7	166.22	1.69	25	C
Formaldehyde	CH_2O					50-00-0	30.03	4.00×10^5		
Formic acid	CH_2O_2					64-18-6	46.03	1.00×10^6		
Furan	C_4H_4O					110-00-9	68.08	9.90×10^3	20	T
Glycidaldehyde	$C_3H_4O_2$					765-34-4	72.1	1.70×10^8		B
Heptachlor	$C_{10}H_5Cl_7$	P	C	N	H	76-44-8	373.3	1.80×10^{-1}	25	C
Heptachlor epoxide	$C_{10}H_5Cl_7O$					1024-57-3	389.3	3.50×10^{-1}	25	C
Hexachlorobenzene	C_6Cl_6	P	C	N	H	118-74-1	284.76	6.00×10^{-3}	25	C
Hexachlorbutadiene	C_4Cl_6	P	C	N	H	87-68-3	260.74	2.00	20	C
Hexachlorocyclopentadiene	C_5Cl_6	P	C	N	H	77-47-4	272.75	1.80	25	C
α-Hexachlorocyclohexane	$C_6H_6Cl_6$					319-84-6	290.82	1.63	25	C
β-Hexachlorocyclohexane	$C_6H_6Cl_6$					319-85-7	290.82	2.40×10^{-1}	25	C
γ-Hexachlorocyclohexane	$C_6H_6Cl_6$					58-89-9	290.82	7.80	25	C
δ-Hexachlorocyclohexane	$C_6H_6Cl_6$					319-86-8	290.82	3.14×10	25	C
Hexachloroethane	C_2Cl_6	P	C	N	H	67-72-1	236.72	4.20×10^4	20	C
Hexachlorophene	$C_{13}H_6Cl_6O_2$			N		70-30-4	406.89	4.00×10^{-3}		F
Hexane	C_6H_{14}				H	110-54-3	86.17			
Hydrazine	H_4N_2				H	302-01-2	32.06	3.41×10^8		B
Hydrogen cyanide	CHN					74-90-8	27.03	1.00×10^6		◇
Hydrogen sulfide	H_2S				H	7783-06-4	34.08	4.13×10^3		
Indeno(1,2,3-cd)pyrene	$C_{22}H_{12}$	P	C	N		193-39-5	276.34	5.30×10^{-4}	25	C
Iodomethane	CH_3I			N		74-88-4	141.94	1.40×10^4	20	J
Iron(II)	Fe					15438-31-0	55.85			
Isobutanol	$C_4H_{10}O$					78-83-1	74.14	7.29×10^4	20	K
Isophorone	$C_9H_{14}O$	P	C		H	78-59-1	138.23	1.04×10^4	20	V
Isoprene	C_5H_8					78-79-5	68.11			
Isosafrole	$C_{10}H_{10}O_2$			N		120-58-1	162.2	1.09×10^3		B
Kepone	$C_{10}Cl_{10}O$			N		143-50-0	490.6	7.60	24	J
Lasiocarpine	$C_{21}H_{33}NO_7$					303-34-4	411.55	1.60×10^3		B
Lead	Pb				H	7439-92-1	207.19			
Malathion	$C_{10}H_{19}O_6PS_2$					121-75-5	330.38	1.45×10^2	20	J

S = source of data; P = priority poll; C = hazardous substance list; N = Appendix IX; H = hazardous air pollutant.

Vapor pressure mmHg	°C	S	Diffusion coefficient in air cm^2/s	S	Henry's constant, H* $(atm \cdot m^3)/mol$; T(K) *A*	*B*	S	K_{oc} mL/g	S	log K_{ow}	S	Freundlich parameters pH	*K*, mg/g	1/*n*	S	Fish BCF L/kg	S
1.20×10	20	J						1.00×10	◇	0.15	B						
7.50×10	30	J	0.1273	U				2.20	◇	−0.32	J	5.3					
3.80	20	J								0.00	F						
1.15×10^2	30	J	0.09005	U						0.73	Y						
7.00	20	J	0.0707	U	-4.99×10^3	1.19×10	Z	1.10	C	3.15	A	7.3	53	0.79	O	37.5	H
5.20×10^{-2}	16	V															
1.17×10	20	J			-3.88×10^3	5.70	Z	4.40×10	G	1.76	B						
6.00×10^{-1}	20	J								0.00	F						
1.10×10^3	20	J						2.20	◇	−0.22	B						
								6.70×10	◇	−0.66	J					0	F
2.06×10^{-1}		B						3.80	◇	0.21	B						
								1.00×10^{-1}	◇								
5.00×10^{-6}	25	C						3.80×10^4	C	4.90	A	5.3	664	0.61	O	1150	H
7.10×10^{-4}	20	C						7.30×10^3			C	5.3	330	0.28	O	1300	G
3.88×10^3	25	Y						3.60	◇	0.35	Y					0	F
3.50×10	20	J								−0.54	F					0	F
6.02×10^2	25	Y								1.34	Y						
1.97×10		B						1.00×10^{-1}	◇	−1.55	B						
3.00×10^{-4}	25	C						1.20×10^4	C	4.40	C	5.3	1220	0.95	O	15,700	H
3.00×10^{-4}	25	C						2.20×10^2	G	2.70	C	5.3	1038	0.70	O	14,400	G
1.90×10^{-5}	20	C						3.90×10^3	C	5.23	A	5.3	450	0.60	O	8690	H
1.50×10^{-1}	20	C						2.90×10^4	C	4.78	A	7	360	0.63	O	2.8	H
8.00×10^{-2}	25	J						4.80×10^3	C	5.04	A	5.3	370	0.17	O	4.3	H
2.50×10^{-5}	20	C						3.80×10^3	C	3.90	C					130	H
2.80×10^{-7}	20	C						3.80×10^3	C	3.90	C					130	H
1.70×10	20	C						1.08×10^3	G	3.90	C					130	H
1.70×10^{-5}	20	C						6.60×10^3	C	4.10	C					130	H
4.00×10^{-1}	20	J			-2.55×10^3	3.74	Z	2.00×10^4	C	4.60	C	5.3	97	0.38	O	87	H
								9.10×10^4	◇	7.54	F						
					-7.53×10^3	2.53×10	Z										
1.60×10	20	J						1.00×10^{-1}	◇	−3.08	B						
6.20×10^2	20	J								-2.50×10^{-1}	F					0	F
1.52×10^4	26	P															
1.00×10^{-1}	20	C						1.60×10^6	C	6.50	C						
4.00×10^2	25	J						2.30×10	◇	1.69	J						
1.00×10	2040	I															
1.04×10	25	Y								0.76	Y						
3.80×10^{-1}	20	C								2.22	Y	5.5	32	0.39	O		
4.93×10^2	20	J															
3.36	66	V						9.30×10	◇	2.66	B						
								5.50×10^4	◇	2.00	B					8400	G
								7.60×10	◇	0.99	B						
1.00×10	1160	I														49	H
4.00×10^{-5}	20	J								2.89	J					0	F

◇ = Calculated value.

*When the source (S) is Z, then $\ln H = A/T + B$, T in kelvins; when the source (S) is AA, then $H = \exp(A - B/T)$, T in kelvins.

Chemical name	Chemical formula	P	C	N	H	CASRN	Molecular weight, g/mol	Water solubility		
								mg/L	°C	S
Manganese	Mn				H	7439-96-5	54.94			
Mercury	Hg					7439-97-6	200.59			
Methanol	CH_4O				H	67-56-1	32.05	Total	20	T
2-Methoxyethanol	$C_3H_8O_2$					109-86-4	76.11	1.00×10^6		
2-Methyl aziridine	C_3H_7N				H	75-55-8	57.11	9.44×10^5		B
4,4′-Methylene(*bis*) -chloroaniline	$C_{13}H_{12}Cl_2N_2$			N	H	101-14-4	267.17			
Methyl ethyl ketone	C_4H_8O		C	N	H	78-93-3	72.12	3.50×10^5	10	J
Methyl isobutyl ketone	$C_6H_{12}O$				H	108-10-1	100.18	1.72×10^4	20	V
Methyl methacrylate	$C_5H_8O_2$			N	H	80-62-6	100.13	1.28×10^4	20	V
Methylnitrosourea	$C_2H_5N_3O_2$					684-93-5	103.1	6.89×10^8		B
Methyl parathion	$C_8H_{10}NO_5PS$			N		298-00-0	263.23	1.72×10	20	F
Methylvinylnitrosamine	$C_3H_6N_2O$					4549-40-0	86.11	7.60×10^5		B
Mustard gas	$C_4H_8Cl_2S$					505-60-2	159.08	8.00×10^2		B
1-Naphthylamine	$C_{10}H_9N$			N		134-32-7	143.2	1.46×10^3	20	K
2-Naphthylamine	$C_{10}H_9N$			N		91-59-8	143.2	5.86×10^2		B
Nickel	Ni					7440-02-0	58.71			
Nickel cyanide, (solid)	C_2N_2Ni					557-19-7	110.75	5.15×10	18	V
Nitric oxide	NO					10102-43-9	30.01			
Nitrobenzene	$C_6H_5NO_2$	P	C	N	H	98-95-3	123.12	1.90×10^3	20	J
Nitrogen dioxide	NO_2					10102-44-0	46.01			
N-nitrosopiperidine	$C_5H_{10}N_2O$			N		100-75-4	114.17	1.90×10^6		B
N-nitrosopyrrolidine	$C_4H_8N_{20}$			N		930-55-2	100.14	7.00×10^6		B
Osmium tetroxide	O_4Os					20816-12-0	254.2			
Pentachlorobenzene	C_6HCl_5			N		608-93-5	250.32	2.40×10^{-1}	22	J
Pentachloronitro benzene	$C_6Cl_5NO_2$			N	H	82-68-8	295.32	7.11×10^{-2}		B
Pentachlorophenol	C_6HCl_5O	P	C	N	H	87-86-5	266.32	1.40×10	20	J
Phenanthrene	$C_{14}H_{10}$	P	C	N		85-01-8	178.24	1.00	15	J
Phenobarbital	$C_{12}H_{12}N_2O_3$					50-06-6	232.26	1.00×10^3	25	C
Phenol	C_6H_5OH				H	108-95-2	94.12	8.20×10^4	20	T
m-Phenylenediamine	$C_6H_8N_2$					108-45-2	108.16	3.51×10^5	25	J
Phenyl mercuric acetate	$C_8H_8HgO_2$					62-38-4	336.75	1.67×10^3		
Polychlorinated biphenyls	NA				H	1336-36-3	NA	3.10×10^{-2}	25	C
Propyl benzene	C_9H_{12}					103-65-1	120.21			
Pyrene	$C_{16}H_{10}$	P	C	N		129-00-0	202.26	1.30×10^{-1}	25	C
Pyridine	C_5H_5N			N		110-86-1	79.11	1.00×10^6		F
Safrole	$C_{10}H_{10}O_2$			N		94-59-7	162.2	1.50×10^3		B
Selenious acid	H_2O_3Se					7783-00-8	128.98			
Selenium	Se					7782-49-2	78.96			
Silver	Ag					7440-22-4	107.87			
Strychnine	$C_{12}H_{22}N_2O_2$					57-24-9	334.45	1.43×10^2		E
Styrene	C_8H_8		C	N	H	100-42-5	104.16	3.00×10^2	20	J

S = source of data; P = priority poll; C = hazardous substance list; N = Appendix IX; H = hazardous air pollutant.

Vapor pressure mmHg	°C	S	Diffusion coefficient in air cm^2/s	S	Henry's constant, H* (atm · m^3)/mol; T(K) A	B	S	K_{oc} mL/g	S	log K_{ow}	S	Freundlich parameters pH	K, mg/g	$1/n$	S	Fish BCF L/kg	S
1.00	1292	P															
2.00×10^{-3}	25	P														5500	H
1.00×10^{2}	21	P	0.16686	U						−0.77	Y						
6.20	20	J															
1.18×10^{2}	20	V						2.30	◇	−0.48	B						
												7.5	190	0.64	O		
7.75×10	20	K	0.09485	U	-2.31×10^{3}	-4.84×10	Z	4.50	◇	0.26	A					0	F
1.56×10	21	V								1.19	Y						
2.80×10	20	J						8.40	◇	0.79	F						
								1.00×10^{-1}	◇	−3.81	B						
9.70×10^{-6}		E						4.60×10^{2}	◇	1.91	F					45	F
1.23×10		B						2.50	◇	−0.23	B						
1.70×10^{-1}		B						1.10	◇	1.37	B						
1.00	104	P						6.10×10		2.07	B						
1.00	108	P						1.30×10^{2}		2.07	B						
1.00	1810	P														47	H
1.00×10	−178	I															
1.50×10^{-1}	20	J						3.60×10	C	1.85	D				O		
4.00×10^{2}	80	P															
1.40×10^{-1}		B						1.50	◇	−0.49	B						
1.10×10^{-1}		B						8.00×10^{-1}	◇	−1.06	B						
1.00×10	26	P															
								1.30×10^{4}	◇	5.19	F					2125	H
1.30×10^{-2}	25	P						1.90×10^{4}	◇	5.45	B						
1.10×10^{-4}	20	J						5.30×10^{4}	C	5.00	C	7	150	0.42	O	770	G
9.60×10^{-4}	25	C						1.40×10^{4}	C	4.46	A	5.3	215	0.44	O	2630	G
								9.80×10	◇	−0.19	B						
2.00×10^{-1}	20	J	0.08924	U	-4.66×10^{3}	1.13×10	Z	1.42×10	C	1.46	A	3.0-9	21	0.54	O	1.4	H
1.00	100	P															
												*	270	0.44	O		
7.70×10^{-5}	25	C	0.05571	U				5.30×10^{5}	C	6.04	C					1×10^{5}	G
					-3.68×10^{3}	7.84	Z										
2.50×10^{-6}	25	C						3.80×10^{4}	C	4.88	A		389	0.39	26		
2.00×10	25	J								0.66	F						
1.00	64	P						7.80×10	◇	2.53	B						
2.00	15	P															
1.00	356	P														16	H
1.00×10	1540	I														3080	D
5.01	270	V															
5.12	20	V	0.0746	U						2.95	Y		120	0.56	O		

◇ = Calculated value.

*When the source (S) is Z, then $\ln H = A/T + B$, T in kelvins; when the source (S) is AA, then $H = \exp(A - B/T)$, T in kelvins.

Chemical name	Chemical formula	P	C	N	H	CASRN	Molecular weight, g/mol	Water solubility		
								mg/L	°C	S
1,2,4,5-Tetrachlorobenzene	$C_6H_2Cl_4$			N		95-94-3	215.88	3.00×10^{-1}	22	F
2,3,7,8-Tetrachlorodi-benzo-*p*-dioxin	$C_{12}H_4Cl_4O_2$	P		N		1746-01-6	321.96	2.00×10^{-4}	25	C
1,1,1,2-Tetrachloroethane	$C_2H_2Cl_4$			N		630-20-6	167.84	2.90×10^{3}	20	J
Tetrachloroethene	C_2Cl_4					127-18-4	165.82	1.50×10^{2}	25	J
2,4,4,6-Tetrachlorophenol	$C_6H_2Cl_4O$					58-90-2	231.88	1.00×10^{3}		F
Tetraethyl lead	$C_8H_{20}Pb$					78-00-2	323.47	8.00×10^{-1}	20	J
Thallium	Tl					7440-28-0	204.37			
Thallium chloride	TlCl					7791-12-0	239.82	2.90×10^{3}		E
Thallium sulfate	$Tl_2{\cdot}O_4S$					7446-18-6	504.8	4.88×10^{4}	21	V
Thioacetamide	C_2H_5NS					62-55-5	75.14			
Thiourea	CH_4N_2S					62-56-6	76.13	9.18×10^{5}	13	J
Tolidine	$C_{14}H_{16}N_2$					119-93-7	212.32	7.35×10		B
Toluene	C_7H_8	P	C	N	H	108-88-3	92.15	5.15×10^{2}	20	J
o-Toluidine hydrochloride	$C_7H_9N{\cdot}ClH$					636-21-5	143.63	1.50×10^{4}		J
Toxaphene	$C_{10}H_{10}Cl_8$ (approx)	P	C	N	H	8001-35-2	413.8	5.00×10^{-1}	25	C
Tribromomethane	$CHBr_3$	P	C	N		75-25-2	252.75	3.19×10^{3}	30	J
1,2,4-Trichlorobenzene	$C_6H_3Cl_3$	P	C	N	H	120-82-1	181.44	3.00×10	25	C
1,1,1-Trichloroethane	$C_2H_3Cl_3$	P	C	N		71-55-6	133.4	4.40×10^{3}	20	J
1,1,2-Trichloroethane	$C_2H_3Cl_3$	P	C	N	H	79-00-5	133.4	4.50×10^{3}	20	J
Trichloroethylene	C_2HCl_3					79-01-6	131.38	1.10×10^{3}	25	J
Trichlorfon	$C_4H_8Cl_3O_4P$					52-68-6	257.44	1.54×10^{5}	20	V
Trichlorofluoromethane	CCl_3F			N		75-69-4	137.36	1.10×10^{3}	20	T
2,4,5-Trichlorophenol	$C_6H_3Cl_3O$		C	N	H	95-95-4	197.44	1.19×10^{3}	25	J
2,4,6-Trichlorophenol	$C_6H_3Cl_3O$	P	C	N	H	88-06-2	197.44	8.00×10^{2}	25	J
2,4,5-Trichlorophenoxy-acetic acid	$C_8H_5Cl_3O_3$					93-76-5	255.48	2.06×10^{2}	20	V
1,2,3-Trichloropropane	$C_3H_5Cl_3$			N		96-18-4	147.43	1.90×10^{3}	20	T
1,1,2-Trichloro-1,2,2,-trifluoroethane	$C_2Cl_3F_3$					76-13-1	187.37	1.70×10	20	T
1,3,5-Trimethylbenzene	C_9H_{12}					108-67-8	120.21			
syn-Trinitrotoluene (TNT)	$C_7H_5N_3O_6$					118-96-7	227.15	2.00×10^{2}	15	J
tris(2,3-dibromopropyl) phosphate	$C_9H_{15}Br_6O_4P$			N		126-72-7	697.67	8.00	24	J
Uracil mustard	$C_8H_{11}Cl_2N_3O_2$					66-75-1	252.12	6.41×10^{2}		B
Urethane	$C_3H_7NO_2$					51-79-6	89.11			
Vinyl chloride	C_2H_3Cl	P	C	N	H	75-01-4	62.5	4.27×10^{3}	20	W
m-Xylene	C_8H_{10}				H	108-38-3	106.18	1.75×10^{2}	20	J
o-Xylene	C_8H_{10}				H	95-47-6	106.18	1.30×10^{2}	20	W
p-Xylene	C_8H_{10}				H	106-42-3	106.18	1.98×10^{2}	20	W
Xylene	C_8H_{10}				H	1330-20-7	106.18	1.98×10^{2}		F
Zinc	Zn					7440-66-6	65.37			

S = source of data; P = priority poll; C = hazardous substance list; N = Appendix IX; H = hazardous air pollution.

Vapor pressure mmHg	°C	S	Diffusion coefficient in air cm^2/s	S	Henry's constant, H* $(atm \cdot m^3)/mol$; T(K) A	B	S	K_{oc} mL/g	S	log K_{ow}	S	Freundlich parameters pH	K, mg/g	$1/n$	S	Fish BCF L/kg	S
1.00×10^{-1}	25	P						1.60×10^3		4.67	F					1125	H
1.00×10^{-6}	25	C						3.30×10^6		6.72	A					5000	H
5.00	20	J	0.07729	U				5.40×10		3.03	Y						
6.00×10	190	J	0.07729	U	1.25×10	4.92×10^3	AA	3.64×10^2		2.60	A	5.3	51	0.56	O	31	H
								9.80×10		4.10	F					240	H
1.50×10^{-1}	20	J						4.90×10^3	◇								
1.00	825	P															
1.00×10	517	P															
0.00		E															
										−0.46	J						
								1.60	◇	−1.14	G						
								4.10×10^2	◇	2.88	B						
2.20×10	20	J	0.08301	U	-3.02×10^3	5.13	Z	3.00×10^2	C	2.73	A	5.6	26	0.44	O	10.7	H
1.00×10^{-1}		J						2.20×10	◇	1.29	J						
4.00×10^{-1}	25	C						9.64×10^2	C	3.30	C					13,100	H
5.60	25	J						1.16×10^2	C	2.40	C						
2.90×10^{-1}	25	C			-4.03×10^3	7.36	Z	9.20×10^3	C	4.30	C	5.3	157	0.31	O	2800	G
1.00×10^2	20	J			9.78	4.13×10^3	AA	1.52×10^2	C	2.50	C	5.3	2	0.34	O	5.6	H
3.20×10	30	J						5.60×10	C	2.47	A	5.3	6	0.60	O	5	H
6.00×10	20	J			1.14×10	4.78×10^3	AA	1.26×10^2	C	2.38	A					10.6	H
7.80×10^{-6}	20	J						6.10	◇	2.29	A						
6.67×10^2	25	C	0.08329	U	-3.51×10^3	9.48	Z	1.59×10^2	C	2.53	D	5.3	5.6	0.24	AB		
4.00×10^2	25	J						8.90×10	◇	3.72	A					110	H
1.20×10^{-2}	25	C						2.00×10^3	C	3.87	A	3	219	0.29	O	150	H
2.00	20	J															
2.70×10^2	20	J			-3.24×10^3	9.65	Z			2.00	F						
					-3.63×10^3	7.24	Z										
								3.10×10^2	◇	4.12	B					2.7	G
								1.20×10^2	◇	−1.09	B	*	11	0.63	O		
1.00×10	78	P															
2.66×10^3	20	W	0.11375	U	7.39	3.29×10^3	AA	5.70×10	◇	1.38	A					1.17	H
1.00×10	20	W			-3.34×10^3	6.28	Z			2.95	F						
1.00×10	20	W			-3.22×10^3	5.54	Z			3.26	F						
1.00×10	20	W			-3.52×10^3	6.93	Z			3.15	F	7.3	85	0.19	O		
1.00×10	20	W	0.07597	U				2.40×10^2	◇	3.26	F						
0.00		D														47	H

◇ = Calculated value.

*When the source (S) is Z, then $\ln H = A/T + B$, T in kelvins; when the source (S) is AA, then $H = \exp(A - B/T)$, T in kelvins.

APPENDIX C

Thermodynamic Properties

The laws of thermodynamics
1st law: You can't win, you can only break even.
2nd law: You can only break even at absolute zero.
3rd law: You can't reach absolute zero.

TABLES IN APPENDIX C

FIGURES IN APPENDIX C

This appendix presents some of the thermodynamic properties needed to work the problems in Chapters 3 and 12. The source for the data is given at the end of each table. For further thermodynamic data the following sources are also suggested:

Dean, J. A., ed.: *Lange's Handbook of Chemistry,* 13th ed., McGraw-Hill, New York, 1992.
Felder, R. M., and R. W. Rousseau: *Elementary Principles of Chemical Processes,* 2nd ed., John Wiley and Sons, New York, 1986.

TABLE C-2
Continued

Abs. press. kPa (sat. temp. °C)		Sat. water	Sat. steam	Temperature, °C (K) 100 (373.15)	125 (398.15)	150 (423.15)	175 (448.15)	200 (473.15)	225 (498.15)	250 (523.15)	300 (573.15)	350 (623.15)	400 (673.15)	450 (723.15)	500 (773.15)	550 (823.15)	600 (873.15)	650 (923.15)
	V'	1.061	885.44	—	897.47	959.54	1020.4	1080.4	1139.8	1198.9	1316.2	1432.8	1549.2	1665.3	1781.2	1897.1	2012.9	2128.6
200	U'	504.489	2529.2	—	2536.9	2576.6	2615.7	2654.4	2692.9	2731.4	2808.8	2887.2	2966.9	3048.0	3130.8	3215.3	3301.4	3389.2
(120.23)	H'	504.701	2706.3	—	2716.4	2768.5	2819.8	2870.5	2920.9	2971.2	3072.1	3173.8	3276.7	3381.1	3487.0	3594.7	3704.0	3815.0
	S'	1.5301	7.1268	—	7.1523	7.2794	7.3971	7.5072	7.6110	7.7096	7.8937	8.0638	8.2226	8.3722	8.5139	8.6487	8.7776	8.9012
	V'	1.064	792.97	—	795.25	850.97	905.44	959.06	1012.1	1064.7	1169.2	1273.1	1376.6	1479.9	1583.0	1686.0	1789.0	1891.9
225	U'	520.465	2533.2	—	2534.8	2575.1	2614.5	2653.5	2692.2	2730.8	2808.4	2886.9	2966.6	3047.8	3130.6	3215.1	3301.2	3389.1
(123.99)	H'	520.705	2711.6	—	2713.8	2766.5	2818.2	2869.3	2919.9	2970.4	3071.5	3173.3	3276.3	3380.8	3486.8	3594.4	3703.8	3814.8
	S'	1.5705	7.0873	—	7.0928	7.2213	7.3400	7.4508	7.5551	7.6540	7.8385	8.0088	8.1679	8.3175	8.4593	8.5942	8.7231	8.8467
	V'	1.068	718.44	—	—	764.09	813.47	861.98	909.91	957.41	1051.6	1145.2	1238.5	1331.5	1424.4	1517.2	1609.9	1702.5
250	U'	535.077	2536.8	—	—	2573.5	2613.3	2652.5	2691.4	2730.2	2808.0	2886.5	2966.3	3047.6	3130.4	3214.9	3301.1	3389.0
(127.43)	H'	535.343	2716.4	—	—	2764.5	2816.7	2868.0	2918.9	2969.6	3070.9	3172.8	3275.9	3380.4	3486.5	3594.2	3703.6	3814.6
	S'	1.6071	7.0520	—	—	7.1689	7.2886	7.4001	7.5050	7.6042	7.7891	7.9597	8.1188	8.2686	8.4104	8.5453	8.6743	8.7980
	V'	1.071	657.04	—	—	693.00	738.21	782.55	826.29	869.61	955.45	1040.7	1125.5	1210.2	1294.7	1379.0	1463.3	1547.6
275	U'	548.564	2540.0	—	—	2571.9	2612.1	2651.6	2690.7	2729.6	2807.5	2886.2	2966.0	3047.3	3130.2	3214.7	3300.9	3388.8
(130.60)	H'	548.858	2720.7	—	—	2762.5	2815.1	2866.8	2917.9	2968.7	3070.3	3172.4	3275.5	3380.1	3486.2	3594.0	3703.4	3814.4
	S'	1.6407	7.0201	—	—	7.1211	7.2419	7.3541	7.4594	7.5590	7.7444	7.9151	8.0744	8.2243	8.3661	8.5011	8.6301	8.7538
	V'	1.073	605.56	—	—	633.74	675.49	716.35	756.60	796.44	875.29	953.52	1031.4	1109.0	1186.5	1263.9	1341.2	1418.5
300	U'	561.107	2543.0	—	—	2570.3	2610.8	2650.6	2689.9	2729.0	2807.1	2885.8	2965.8	3047.1	3130.0	3214.5	3300.8	3388.7
(133.54)	H'	561.429	2724.7	—	—	2760.4	2813.5	2865.5	2916.9	2967.9	3069.7	3171.9	3275.2	3379.8	3486.0	3593.7	3703.2	3814.2
	S'	1.6716	6.9909	—		7.0771	7.1990	7.3119	7.4177	7.5176	7.7034	7.8744	8.0338	8.1838	8.3257	8.4608	8.5898	8.7135

TABLE C-2
Continued

Abs. press. kPa (sat. temp. °C)		Sat. water	Sat. steam	Temperature, °C (K) 150 (423.15)	175 (448.15)	200 (473.15)	220 (493.15)	240 (513.15)	260 (533.15)	280 (553.15)	300 (573.15)	325 (598.15)	350 (623.15)	400 (673.15)	450 (723.15)	500 (773.15)	550 (823.15)	600 (873.15)	650 (923.15)
325 (136.29)	V'	1.076	561.75	583.58	622.41	660.33	690.22	719.81	749.18	778.39	807.47	843.68	879.78	951.73	1023.5	1095.0	1166.5	1237.9	1309.2
	U'	572.847	2545.7	2568.7	2609.6	2649.6	2861.2	2712.7	2744.0	2775.3	2806.6	2845.9	2885.5	2965.5	3046.9	3129.8	3214.4	3300.6	3388.6
	H'	573.197	2728.3	2758.4	2811.9	2864.2	2905.6	2946.6	2987.5	3028.2	3069.0	3120.1	3171.4	3274.8	3379.5	3485.7	3593.5	3702.9	3814.1
	S'	1.7004	6.9640	7.0363	7.1592	7.2729	7.3585	7.4400	7.5181	7.5933	7.6657	7.7530	7.8369	7.9965	8.1465	8.2885	8.4236	8.5527	8.6764
350 (138.87)	V'	1.079	524.00	540.58	576.90	612.31	640.18	667.75	695.09	722.27	749.33	783.01	816.57	883.45	950.11	1016.6	1083.0	1149.3	1215.6
	U'	583.892	2548.2	2567.1	2608.3	2648.6	2680.4	2712.0	2743.4	2774.8	2806.2	2845.6	2885.1	2965.2	3046.6	3129.6	3214.2	3300.5	3388.4
	H'	584.270	2731.6	2756.3	2810.3	2863.0	2904.5	2945.7	2986.7	3027.6	3068.4	3119.6	3170.9	3274.4	3379.2	3485.4	3593.3	3702.7	3813.9
	S'	1.7273	6.9392	6.9982	7.1222	7.2366	7.3226	7.4045	7.4828	7.5581	7.6307	7.7181	7.8022	7.9619	8.1120	8.2540	8.3892	8.5183	8.6421
375 (141.31)	V'	1.081	491.13	503.29	537.46	570.69	596.81	622.62	648.22	673.64	698.94	730.42	761.79	824.28	886.54	948.66	1010.7	1072.6	1134.5
	U'	594.332	2550.6	2565.4	2607.1	2647.7	2679.6	2711.3	2742.8	2774.3	2805.7	2845.2	2884.8	2964.9	3046.4	3129.4	3214.0	3300.3	3388.3
	H'	594.737	2734.7	2754.1	2808.6	2861.7	2903.4	2944.8	2985.9	3026.9	3067.8	3119.1	3170.5	3274.0	3378.8	3485.1	3593.0	3702.6	3813.7
	S'	1.7526	6.9160	6.9624	7.0875	7.2027	7.2891	7.3713	7.4499	7.5254	7.5981	7.6856	7.7698	7.9296	8.0798	8.2219	8.3571	8.4863	8.6101
400 (143.62)	V'	1.084	462.22	470.66	502.93	534.26	558.85	583.14	607.20	631.09	654.85	684.41	713.85	772.50	830.92	889.19	947.35	1005.4	1063.4
	U'	604.237	2552.7	2563.7	2605.8	2646.7	2678.8	2710.6	2742.2	2773.7	2805.3	2844.8	2884.5	2964.6	3046.2	3129.2	3213.8	3300.2	3388.2
	H'	604.670	2737.6	2752.0	2807.0	2860.4	2902.3	2943.9	2985.1	3026.2	3067.2	3118.5	3170.0	3273.6	3378.5	3484.9	3592.8	3702.3	3813.5
	S'	1.7764	6.8943	6.9285	7.0548	7.1708	7.2576	7.3402	7.4190	7.4947	7.5675	7.6552	7.7395	7.8994	8.0497	8.1919	8.3271	8.4563	8.5802
425 (145.82)	V'	1.086	436.61	441.85	472.47	502.12	525.36	548.30	571.01	593.54	615.95	643.81	671.56	726.81	781.84	836.72	891.49	946.17	1000.8
	U'	613.667	2554.8	2562.0	2604.5	2645.7	2678.0	2709.9	2741.6	2773.2	2804.8	2844.4	2884.1	2964.4	3045.9	3129.0	3213.7	3300.0	3388.0
	H'	614.128	2740.3	2749.8	2805.3	2859.1	2901.2	2942.9	2984.3	3025.5	3066.6	3118.0	3169.5	3273.3	3378.2	3484.6	3592.5	3702.1	3813.4
	S'	1.7990	6.8739	6.8965	7.0239	7.1407	7.2280	7.3108	7.3899	7.4657	7.5388	7.6265	7.7109	7.8710	8.0214	8.1636	8.2989	8.4282	8.5520

TABLE C-2
Continued

Abs. press. kPa (sat. temp. °C)		Sat. water	Sat. steam	Temperature, °C (K) 150 (423.15)	175 (448.15)	200 (473.15)	220 (493.15)	240 (513.15)	260 (533.15)	280 (553.15)	300 (573.15)	325 (598.15)	350 (623.15)	400 (673.15)	450 (723.15)	500 (773.15)	550 (823.15)	600 (873.15)	650 (923.15)
450 (147.92)	V'	1.088	413.75	416.24	445.38	473.55	495.59	517.33	538.83	560.17	581.37	607.73	633.97	686.20	738.21	790.07	841.83	893.50	945.10
	U'	622.672	2556.7	2560.3	2603.2	2644.7	2677.1	2709.2	2741.0	2772.7	2804.4	2844.0	2883.8	2964.1	3045.7	3128.8	3213.5	3299.8	3387.9
	H'	623.162	2742.9	2747.7	2803.7	2857.8	2900.2	2942.0	2983.5	3024.8	3066.0	3117.5	3169.1	3272.9	3377.9	3484.3	3592.3	3701.9	3813.2
	S'	1.8204	6.8547	6.8660	6.9946	7.1121	7.1999	7.2831	7.3624	7.4384	7.5116	7.5995	7.6840	7.8442	7.9947	8.1370	8.2723	8.4016	8.5255
475 (149.92)	V'	1.091	393.22	393.31	421.14	447.97	468.95	489.62	510.05	530.30	550.43	575.44	600.33	649.87	699.18	748.34	797.40	846.37	895.27
	U'	631.294	2558.5	2558.6	2601.9	2643.7	2676.3	2708.5	2740.4	2772.2	2803.9	2843.6	2883.4	2963.8	3045.4	3128.6	3213.3	3299.7	3387.7
	H'	631.812	2745.3	2745.5	2802.0	2856.5	2899.1	2941.1	2982.7	3024.1	3065.4	3116.9	3168.6	3272.5	3377.6	3484.0	3592.1	3701.7	3813.0
	S'	1.8408	6.8365	6.8369	6.9667	7.0850	7.1732	7.2567	7.3363	7.4125	7.4858	7.5739	7.6585	7.8189	7.9694	8.1118	8.2472	8.3765	8.5004
500 (151.84)	V'	1.093	374.68	—	399.31	424.96	444.97	464.67	484.14	503.43	522.58	546.38	570.05	617.16	664.05	710.78	757.41	803.95	850.42
	U'	639.569	2560.2	—	2600.6	2642.7	2675.5	2707.8	2739.8	2771.7	2803.5	2843.2	2883.1	2963.5	3045.2	3128.4	3213.1	3299.5	3387.6
	H'	640.116	2747.5	—	2800.3	2855.1	2898.0	2940.1	2981.9	3023.4	3064.8	3116.4	3168.1	3272.1	3377.2	3483.8	3591.8	3701.5	3812.8
	S'	1.8604	6.8192	—	6.9400	7.0592	7.1478	7.2317	7.3115	7.3879	7.4614	7.5496	7.6343	7.7948	7.9454	8.0879	8.2233	8.3526	8.4766

Abs. press. kPa (sat. temp. °C)		Sat. water	Sat. steam	Temperature, °C (K) 225 (498.15)	250 (523.15)	275 (548.15)	300 (573.15)	325 (598.15)	350 (623.15)	375 (648.15)	400 (673.15)	425 (698.15)	450 (723.15)	475 (748.15)	500 (773.15)	525 (798.15)	550 (823.15)	600 (873.15)	650 (923.15)
2400 (221.78)	V'	1.193	83.199	84.149	91.075	97.411	103.36	109.05	114.55	119.93	125.22	130.44	135.61	140.73	145.82	150.88	155.91	165.92	175.86
	U'	949.066	2600.7	2608.6	2665.6	2717.3	2765.4	2811.1	2855.4	2898.8	2941.7	2984.5	3027.1	3069.9	3112.9	3156.1	3199.6	3287.7	3377.2
	H'	951.929	2800.4	2810.6	2884.2	2951.1	3013.4	3072.8	3130.4	3186.7	3242.3	3297.5	3352.6	3407.7	3462.9	3518.2	3573.8	3685.9	3799.3
	S'	2.5343	6.2690	6.2894	6.4338	6.5586	6.6699	6.7714	6.8656	6.9542	7.0384	7.1189	7.1964	7.2713	7.3439	7.4144	7.4830	7.6152	7.7414
2500 (223.94)	V'	1.197	79.905	80.210	86.985	93.154	98.925	104.43	109.75	114.94	120.04	125.07	103.04	134.97	139.87	144.74	149.58	159.21	168.76
	U'	958.969	2601.2	2603.8	2662.0	2714.5	2763.1	2809.3	2853.9	2897.5	2940.6	2983.4	3026.2	3069.0	3112.1	3155.4	3198.9	3287.1	3376.7
	H'	961.962	2800.9	2804.3	2879.5	2947.4	3010.4	3070.4	3128.2	3184.8	3240.7	3296.1	3351.3	3406.5	3461.7	3517.2	3572.9	3685.1	3798.6
	S'	2.5543	6.2536	6.2604	6.4077	6.5345	6.6470	6.7494	6.8442	6.9333	7.0178	7.0986	7.1763	7.2513	7.3240	7.3946	7.4633	7.5956	7.7220

TABLE C-2

Continued

Abs. press. kPa (sat. temp. °C)		Sat. water	Sat. steam	Temperature, °C (K) 225 (498.15)	250 (523.15)	275 (548.15)	300 (573.15)	325 (598.15)	350 (623.15)	375 (648.15)	400 (673.15)	425 (698.15)	450 (723.15)	475 (748.15)	500 (773.15)	525 (798.15)	550 (823.15)	600 (873.15)	650 (923.15)
	V'	1.201	76.856	—	83.205	89.220	94.830	100.17	105.32	110.33	115.26	120.11	124.91	129.66	134.38	139.07	143.74	153.01	162.21
2600	U'	968.597	2601.5	—	2658.4	2711.7	2760.9	2807.4	2852.3	2896.1	2939.4	2982.3	3025.2	3068.1	3111.2	3154.6	3198.2	3286.5	3376.1
(226.04)	H'	971.720	2801.4	—	2874.7	2943.6	3007.4	3067.9	3126.1	3183.0	3239.0	3294.6	3349.9	3405.3	3460.6	3516.2	3571.9	3684.3	3797.9
	S'	2.5736	6.2387	—	6.3823	6.5110	6.6249	6.7281	6.8236	6.9131	6.9979	7.0789	7.1568	7.2320	7.3048	7.3755	7.4443	7.5768	7.7033
	V'	1.205	74.025	—	79.698	85.575	91.036	96.218	101.21	106.07	110.83	115.52	120.15	124.74	129.30	133.82	138.33	147.27	156.14
2700	U'	977.968	2601.8	—	2654.7	2708.8	2758.6	2805.6	2850.7	2894.8	2938.2	2981.2	3024.2	3067.2	3110.4	3153.8	3197.5	3285.8	3375.6
(228.07)	H'	981.222	2801.7	—	2869.9	2939.8	3004.4	3065.4	3124.0	3181.2	3237.4	3293.1	3348.6	3404.0	3459.5	3515.2	3571.0	3683.5	3797.1
	S'	2.5924	6.2244	—	6.3575	6.4882	6.6034	6.7075	6.8036	6.8935	6.9787	7.0600	7.1381	7.2134	7.2863	7.3571	7.4260	7.5587	7.6853
	V'	1.209	71.389	—	76.437	82.187	87.510	92.550	97.395	102.10	106.71	111.25	115.74	120.17	124.58	128.95	133.30	141.94	150.50
2800	U'	987.100	2602.1	—	2650.9	2705.9	2756.3	2803.7	2849.2	2893.4	2937.0	2980.2	3023.2	3066.3	3109.6	3153.1	3196.8	3285.2	3375.0
(230.05)	H'	990.485	2802.0	—	2864.9	2936.0	3001.3	3062.8	3121.9	3179.3	3235.8	3291.7	3347.3	3402.8	3458.4	3514.1	3570.0	3682.6	3796.4
	S'	2.6106	6.2104	—	6.3331	6.4659	6.5824	6.6875	6.7842	6.8746	6.9601	7.0416	7.1199	7.1954	7.2685	7.3394	7.4084	7.5412	7.6679
	V'	1.213	68.928	—	73.395	79.029	84.226	89.133	93.843	98.414	102.88	107.28	111.62	115.92	120.18	124.42	128.62	136.97	145.26
2900	U'	996.008	2602.3	—	2647.1	2702.9	2754.0	2801.8	2847.6	2892.0	2935.8	2979.1	3022.3	3065.5	3108.8	3152.3	3196.1	3284.6	3374.5
(231.97)	H'	999.524	2802.2	—	2859.9	2932.1	2998.2	3060.3	3119.7	3177.4	3234.1	3290.2	3346.0	3401.6	3457.3	3513.1	3569.1	3681.8	3795.7
	S'	2.6283	6.1969	—	6.3092	6.4441	6.5621	6.6681	6.7654	6.8563	6.9421	7.0239	7.1024	7.1780	7.2512	7.3222	7.3913	7.5243	7.6511
	V'	1.216	66.626	—	70.551	76.078	81.159	85.943	90.526	94.969	99.310	103.58	107.79	111.95	116.08	120.18	124.26	132.34	140.36
3000	U'	1004.7	2602.4	—	2643.2	2700.0	2751.6	2799.9	2846.0	2890.7	2934.6	2978.0	3021.3	3064.6	3107.9	3151.5	3195.4	3284.0	3373.9
(233.84)	H'	1008.4	2802.3	—	2854.8	2928.2	2995.1	3057.7	3117.5	3175.6	3232.5	3288.7	3344.6	3400.4	3456.2	3512.1	3568.1	3681.0	3795.0
	S'	2.6455	6.1837	—	6.2857	6.4228	6.5422	6.6491	6.7471	6.8385	6.9246	7.0067	7.0854	7.1612	7.2345	7.3056	7.3748	7.5079	7.6349

Smith, J.M., and H.C. Van Ness: *Introduction to Chemical Engineering Thermodynamics,* 4th ed., McGraw-Hill, New York, 1987.

TABLE C-3

Temperature-dependent heat capacities (constant-pressure values)

Form C_p [J/(mol ·°C)] or [J/(mol · K)] $= a + bT + cT^2 + dT^3$

Compound	Formula	State*	Temp. unit	a	$b \times 10^2$	$c \times 10^5$	$d \times 10^9$	Range (units of T)
Acetone	CH_3COCH_3	l	°C	123.0	18.6			−30–60
Air		g	°C	28.94	0.4147	0.3191	−1.965	0–1500
Benzene	C_6H_6	l	K	62.55	23.4			279–350
		g	°C	74.06	32.95	−25.20	77.57	0–1200
Carbon tetrachloride	CCl_4	l	K	93.39	12.98			273–343
Cumene	C_9H_{12}	g	°C	139.2	53.76	−39.79	120.5	0–1200
Formaldehyde	CH_2O	g	°C	34.28	4.268	0.0000	−8.694	0–1200
n-Hexane	C_6H_{14}	l	°C	216.3				20–100
Hydrogen sulfide	H_2S	g	°C	33.51	1.547	0.3012	−3.292	0–1500
Methane	CH_4	g	°C	34.31	5.469	0.3661	−11.00	0–1200
		g	K	19.87	5.021	1.268	−11.00	273–1500
Methyl alcohol	CH_3OH	l	°C	75.86				0
				82.59				40
Methyl cyclopentane	C_6H_{12}	g	°C	98.83	45.857	−30.44	83.81	0–1200
Toluene	C_7H_8	l	°C	148.8				0
		l	°C	181.2				100
Water	H_2O	l	°C	75.4				0–100

Note: Formulas for gases are strictly applicable at pressures low enough for the ideal gas laws to apply.

*l = liquid; g = gas.

Source: Adapted from Felder, R. M., and R. W. Rousseau: *Elementary Principles of Chemical Processes,* 2nd ed., John Wiley & Sons, New York, 1986.

TABLE C-4

Latent heats of fusion at 1 atm and vaporization at stated temperatures

Compound	Formula	Melting point, °C	Heat of fusion, cal/g	Temperature, °C	Heat of vaporization, cal/g
Acetone	C_3H_6O	−95.5	23.42	56	124.41
Benzene	C_6H_6	5.533	30.100	25	103.57
				80.10	94.14
Benzoic acid	$C_7H_8O_2$	122.45	33.90	—	—
Carbon tetrachloride	CCl_4	—	—	0	52.06
				76.75	46.42
				200	32.73
Chlorobenzene	C_6H_5Cl	—	—	130.6	77.59
Cresol (p-)	C_7H_8O	34.6	26.28	202	100.58
Cyclohexane	C_6H_{12}	6.67	7.569	25	93.81
				80.74	85.6
Dibromobenzene (o-)	$C_6H_4Br_2$	1.8	12.78	—	—
Diphenyl amine	$C_{12}H_{11}N$	52.98	25.23	—	—
Ethyl amine	C_2H_7N	—	—	15	145.97
Ethylbenzene	C_8H_{10}	−94.950	20.629	25	95.11
				136.19	81.0
Methane	CH_4	−182.48	14.03	−161.6	121.87
Methanol	CH_4O	−97.8	23.7	0	284.29
				64.7	262.79
				100	241.29
Naphthalene	$C_{10}H_8$	80.0	36.0	218	75.49
Toluene	C_7H_8	−94.991	17.171	25	98.55
				110.62	86.8
Urethane	$C_3H_7NO_2$	48.7	40.85	—	—

Source: Perry, R. H., and D. W. Green, eds.: *Perry's Chemical Engineer's Handbook,* 7th ed., McGraw-Hill, New York, pp. 2-154 to 2-160, 1997.

1 cal/g × 0.004184 = KJ/g.

To convert to a molar basis multiply by mol. wt., e.g. for acetone, mol. wt. = 58.08 g/g-mol.

Heat of vaporization for acetone @ 56° C = 124.41 cal/g = 124.41 cal/g × 58.08 g/g-mol × 0.00419 1 × J/g = 30.2 1 × J/g-mol.

TABLE C-5
Heat of combustion for common waste elements at standard state*

Hazardous constituent	Heat of combustion, kcal/g	Hazardous constituent	Heat of combustion, kcal/g
Trichloromonofluoromethane	0.11	Octachlorobiphenyl	2.72
Tribromomethane	0.13	Acetyl chloride	2.77
Dichlorodifluoromethane	0.22	Trichloropropane, N.O.S.	2.81
Tetrachloromethane	0.24	1,2,3-Trichloropropane	2.81
Tetranitromethane	0.41	Dichloropropanol, N.O.S.	2.84
Hexachloroethane	0.46	Dimethyl sulfate	2.86
Dibromomethane	0.50	2,4,5-T	2.87
Pentachloroethane	0.53	2,4,5-Trichlorophenol	2.88
Hexachloropropene	0.70	2,4,6-Trichlorophenol	2.88
Chloroform	0.75	N-Nitroso-N-methylurea	2.89
Chloral (trichloroacetaldehyde)	0.80	Heptachlorobiphenyl	2.98
Cyanogen bromide	0.81	1,1-Dichloroethane	3.00
Trichloromethanetiol	0.84	1,2-Dichloroethane	3.00
Hexachlorocyclohexane	1.12	trans-1,2-Dichloroethane	3.00
Tetrachloroethene (Tetrachloroethylene)	1.19	Phenyl dichloroarsine	3.12
		N-Nitrosoarcosine	3.19
Cyanogen chloride	1.29	Azaserine	3.21
Formic acid	1.32	2-Fluoroacetamide	3.24
Iodomethane	1.34	Chloromethane	3.25
Tetrachloroethane, N.O.S.	1.39	Hexachlorobiphenyl	3.28
1,1,1,2-Tetrachloroethane	1.39	*bis*(2-chloroethyl) ether	3.38
1,1,2,2-Tetrachloroethane	1.39	1,2,3,4,10,10-Hexachloro-1,4,4a,5,7,8a-hexahydro-1,4:5,8-endo, endo-dimethanonaphthalene	3.38
1,2-Dibromomethane	1.43		
1,2-Dibromo-3-chloropropane	1.48		
Pentachloronitrobenzene	1.62		
Bromomethane	1.70	Benzenearsonic acid	3.40
Dichloromethane	1.70	Maleic anhydride	3.40
Trichloroethene (Trichloroethylene)	1.74	1,2,4-Trichlorobenzene	3.40
		TCDD	3.43
Hexachlorobenzene	1.79	Dichloropropene, N.O.S.	3.44
bis (chloromethyl) ether	1.97	1,3-Dichloropropene	3.44
1,1,1-Trichloroethane	1.99	Endrin	3.46
1,1,2-Trichloroethane	1.99	Chloromethyl methyl ether	3.48
Pentachlorobenzene	2.05	2,4-Dinitrophenol	3.52
Pentachlorophenol	2.09	Nitrogen mustard N-oxide and hydrochloride salt	3.56
Hexachlorocyclopentadiene	2.10		
Hexachlorobutadiene	2.12	Parathion	3.61
Kepone	2.15	2,4-D	3.62
2,3,4,6-Tetrachlorophenol	2.23	Pentachlorobiphenyl	3.66
Dichlorophenylaraine	2.31	1,3-Propane sultone	3.67
Decachlorobiphenyl	2.31	Methyl methanosulfonate	3.74
Endosulfan	2.33	Aldrin	3.75
Nonachlorobiphenyl	2.50	Nitroglycerine	3.79
Toxaphene	2.50	2,4-Dichlorophenol	3.81
1,2,4,5-Tetrachlorobenzene	2.61	2,6-Dichlorophenol	3.81
Bromoacetone	2.66	Hexachlorophene	3.82
Dichloroethylene, N.O.S.	2.70	Trypan blue	3.84
1,1-Dichloroethylene	2.70	Benzotrichloride	3.90
Chlordane	2.71	Cycasin	3.92
Heptachlor epoxide	2.71	*N*-Nitroso-*N*-ethylurea	3.92
Phenylmercury acetate	2.71	Cyclophosphamide	3.97

TABLE C-5
Continued

Hazardous constituent	Heat of combustion, kcal/g
Dichloropropane, N.O.S.	3.99
1,2-Dichloropropane	3.99
Methylparathion	4.00
Uracil mustard	4.00
Amitrole	4.01
Dimethoate	4.02
Tetraethyl lead	4.04
4,6-Dinitro-*o*-cresol and salts	4.06
N-Methyl-*N*-nitro-*N*-nitrosoguanidine	4.06
Mustard gas	4.06
Maleic hydrazide	4.10
Dinitrobenzene, N.O.S.	4.15
N-Nitroso-N-methylurethane	4.18
1,4-Dichloro-2-butene	4.27
Nitrogen mustard and hydrochloride salt	4.28
Tetrachlorobiphenyl	4.29
Hydrazine	4.44
Vinyl chloride	4.45
Formaldehyde	4.47
Saccharin	4.49
3-Chloropropionitrile	4.50
DDT	4.51
Thiourea	4.51
1-Acetyl-2-thiourea	4.55
Thiosemicarbazide	4.55
Dichlorobenzene, N.O.S.	4.57
Ethyl cyanide	4.57
bis (2-chloroethoxy) methane	4.60
2,4-Dinitrotoluene	4.68
Isocyanic acid, methyl ester	4.69
7-Oxabicyclo (2.2.1)-heptane-2,3-dicarboxylic acid	4.70
Ethyl carbamate	4.73
5-(Aminomethyl)-3-isoxazolol	4.78
Methylthiouracil	4.79
4,4′-Methylene-*bis*-(2-chloroaniline)	4.84
bis (2-chloroisopropyl) ether	4.93
4-Nitrophenol	4.95
DDE	5.05
Dimethylcarbamoyl chloride	5.08
p-Chloro-*m*-cresol	5.08
Dichloromethylbenzene	5.09
Trichlorobiphenyl	5.10
DDD	5.14
Dimethylnitrosoamine	5.14
N-Nitrosodimethylamine	5.14
Diethylarsine	5.25
Phthalic anhydride	5.29
1-(*o*-chlorophenyl) thiourea	5.30
2-Methyl-2-(methylthio) propionaldehyde-*o*-(methylcarbonyl) oxime	5.34
2-*sec*-Butyl-4,6 dinitrophenol (DNBP)	5.46
p-Nitroaniline	5.50
Chlorobenzilate	5.50
Dieldrin	5.56
2,4,5-TP	5.58
Methoxychlor	5.59
4-Nitroquinoline-1-oxide	5.59
Diallate	5.62
Daunomycin	5.70
Ethylenebisdithiocarbamate	5.70
3,3′-Dichlorobenzidine	5.72
Pronamide	5.72
Aflatoxins	5.73
Disulfoton	5.73
Diepoxybutane	5.74
Dimethyl phthalate	5.74
Glycidylaldehyde	5.74
Acrylamide	5.75
3,3-Dimethyl-1-(methylthio)-2-butanone-0-(methylamino) carbonyl oxime	5.82
4-Bromophenyl phenyl ether	5.84
Thiuram	5.85
Methanethiol	5.91
Tolulene diisocyanate	5.92
Chlorambucil	5.93
Thioacetamide	5.95
Ethylenethiourea	5.98
Malononitrile	5.98
5-Nitro-o-toluidine	5.98
Nitrobenzene	6.01
3,4-Dihydroxy-alpha-(methylamino) methyl benzyl alcohol	6.05
Benzoquinone	6.07
N-Nitrosomethylethylamine	6.13
p-Chloroaniline	6.14
Benzyl chloride	6.18
Resorcinol	6.19
Propylthiouracil	6.28
Paraldehyde	6.30
Dichlorobiphenyl	6.36
Diethyl phthalate	6.39
Dioxane	6.41
2-Methylacetonitrile	6.43
N-Nitrosopyrrolidone	6.43
Methyl methacrylate	6.52
Chlorobenzene	6.60
o-Toluidine hydrochloride	6.63

TABLE C-5
Continued

Hazardous constituent	Heat of combustion, kcal/g	Hazardous constituent	Heat of combustion, kcal/g
N,*N*-*bis*(2-chloroethyl)-		Methyl ethyl ketone (MEK)	8.07
2-naphthylamine	6.64	Cresylic acid	8.09
2,6-Dinitrotoluene		Cresol	8.18
di-*n*-octyl phthalate	6.67	Toluene diamine	8.24
Reserpine	6.70	Acetophenone	8.26
Methyl hydrazine	6.78	Butyl benzyl phthalate	8.29
Cyanogen	6.79	Ethyl cyanide	8.32
Ethylene oxide	6.86	*bis*(2-ethylhexyl) phthalate	8.42
N-Nitrosodiethylamine	6.86	Benzenethiol	8.43
2-Chlorophenol	6.89	N-Nitrosodi-N-butylamine	8.46
N-Phenylthiourea	6.93	2,4-Dimethylphenol	8.51
Acrolein	6.96	Indenol (1,2,3-c,d) pyrene	8.52
2-Butanone peroxide	6.96	Diethylstilbestrol	8.54
p-Dimethylaminoazobenzene	6.97	1-Naphthylamine	8.54
1,4-Naphthoquinone	6.97	2-Naphthylamine	8.54
3-(alpha-Acetonylbenzyl)-		Methacrylonitrile	8.55
4-hydroxy-coumarin		Isobutyl alcohol	8.62
and salts (Warfarin)	7.00	1,2-Diethylhydrazine	8.68
N-Nitrosodiethanolamine	7.02	2-Picoline	8.72
N-Nitrosopiperidine	7.04	Aniline	8.73
N-Nitrosonornicotine	7.07	1,2-Diphenylhydrazine	8.73
Phenacetin	7.17	3,3′-Dimethoxybenzidine	8.81
Ethyl methacrylate	7.27	7N-Dibenzo (c,g) carbazole	8.90
Di-*n*-butyl phthalate	7.34	Benz(*c*)acridine	8.92
3,3′-Dimethoxybenzidine	7.36	Nicotine and salts	8.92
Acetonitrile	7.37	4-Amino biphenyl	9.00
4-Aminopyridine	7.37	Diphenylamine	9.09
2-Chloronaphthalene	7.37	2-Methylaziridine	9.09
2 Propyn-1-ol	7.43	Benzidine	9.18
1-Naphthyl-2-thiourea	7.50	Benzo(*b*)fluoranthene	9.25
Isosafrole	7.62	Benzo(*j*)fluoranthene	9.25
Dihydrosafrole	7.66	Benzo(*a*)pyrene	9.25
Safrole	7.68	Dibenzo(*a*, *e*)pyrene	9.33
Auramine	7.69	Dibenzo(*a*, *h*)pyrene	9.33
Crotonaldehyde	7.73	Dibenzo(*a*, *i*)pyrene	9.33
Allyl alcohol	7.75	Fluoranthene	9.35
Monochlorobiphenyl	7.75	Benz(*a*)anthracene	9.39
Phenol	7.78	Dibenz(*a*, *h*)anthracene	
Phenylenediamine	7.81	(Dibenzo(*a*, *h*)anthracene)	9.40
Di-*n*-propylnitrosoamine	7.83	Dibenz(*a*, *h*)acridine	9.53
Pyridine	7.83	Dibenz(*a*, *j*)acridine	9.53
Ethyleneimine	7.86	alpha, alpha-Dimethyl-	
1,1-Dimethylhydrazine	7.87	phenethylamine	9.54
1,2-Dimethylhydrazine	7.87	3-Methylcholanthrene	9.57
N-Nitrosomethylvinylamine	7.91	*n*-Propylamine	9.58
2-Acetylaminofluorine	7.82	7,12-Dimethylbenz(*a*)anthracene	9.61
Acrylonitrile	7.93	Naphthalene	9.62
Methapyrilene	7.93	Benzene	10.03
Strychnine and salts	8.03	Toluene	10.14

Source: U.S. EPA: *Hazardous Waste Incinerator Inspection Manual,* 1989.

*Note: Standard state = 298.15 K and 101.325 kPa.

TABLE C-6

Heats of formation and free energies of formation

Compound	State	Heat of formation at 25°C, kcal/mol	Free energy of formation at 25°C, kcal/mol
Cyclohexane	gas	−29.43	7.59
	liquid	−37.34	6.39
Benzene	gas	19.820	30.989
Ethylene	gas	12.496	16.282
Acetone	gas	−51.79	−36.45
Isobutanol	gas	−69.05	−38.25
SO	gas	19.02	12.75
SO_2	gas	−70.94	−71.68
SO_3	gas	−94.39	−88.59
	liquid	−103.03	−88.28
CO	gas	−26.416	−32.808
CO_2	gas	−94.052	−94.260
NO	gas	21.600	20.719
N_2O	gas	19.55	24.82

Source: Perry, R. H., and D. W. Green, eds.: *Perry's Chemical Engineer's Handbook,* 7th ed., McGraw-Hill, New York, Tables 2-220 and 2-221, 1997.

TABLE C-7
Saturated steam: temperature table*

Temp., °F. t	Abs. pressure, lb./sq. in. p	Volume, cu. ft./lb. Liquid v_f	Volume, cu. ft./lb. Vapor v_g	Enthalpy, B.t.u./lb. Liquid h_f	Enthalpy, B.t.u./lb. Vapor h_g	Entropy, B.t.u./(lb.)(°R.) Liquid s_f	Entropy, B.t.u./(lb.)(°R.) Vapor s_g	Temp., °F. t	Abs. pressure, lb./sq. in. p	Volume, cu. ft./lb. Liquid v_f	Volume, cu. ft./lb. Vapor v_g	Enthalpy, B.t.u./lb. Liquid h_f	Enthalpy, B.t.u./lb. Vapor h_g	Entropy, B.t.u./(lb.)(°R.) Liquid s_f	Entropy, B.t.u./(lb.)(°R.) Vapor s_g
32	0.08854	0.01602	3306	0.00	1075.8	0.0000	2.1877	250	29.825	.01700	13.821	216.48	1164.0	.3675	1.6998
35	.09995	.01602	2947	3.02	1077.1	.0061	2.1770	260	35.429	.01709	11.763	228.64	1167.3	.3817	1.6860
40	.12170	.01602	2444	8.05	1079.3	.0162	2.1597	270	41.858	.01717	10.061	238.84	1170.6	.3958	1.6727
45	.14752	.01602	2036.4	13.06	1081.5	.0262	2.1429	280	49.203	.01726	8.645	249.06	1173.8	.4096	1.6597
50	.17811	.01603	1703.2	18.07	1083.7	.0361	2.1264	290	57.556	.01735	7.461	259.31	1176.8	.4234	1.6472
60	.2563	.01604	1206.7	28.06	1088.0	.0555	2.0948	300	67.013	.01745	6.466	269.59	1179.7	.4369	1.6350
70	.3631	.01606	867.9	38.04	1092.3	.0745	2.0647	310	77.68	.01755	5.626	279.92	1182.5	.4504	1.6231
80	.5069	.01608	633.1	48.02	1096.6	.0932	2.0360	320	89.66	.01765	4.914	290.28	1185.2	.4637	1.6115
90	.6982	.01610	468.0	57.99	1100.9	.1115	2.0087	330	103.06	.01776	4.307	300.68	1187.7	.4769	1.6002
100	.9492	.01613	350.4	67.97	1105.2	.1295	1.9826	340	118.01	.01787	3.788	311.13	1190.1	.4900	1.5891
110	1.2748	.01617	265.4	77.94	1109.5	.1471	1.9577	350	134.63	.01799	3.342	321.63	1192.3	.5029	1.5783
120	1.6924	.01620	203.27	87.92	1113.7	.1645	1.9339	360	153.04	.01811	2.957	332.18	1194.4	.5158	1.5677
130	2.2225	.01625	157.34	97.90	1117.9	.1816	1.9112	370	173.37	.01823	2.625	342.79	1196.3	.5286	1.5573
140	2.8886	.01629	123.01	107.89	1122.0	.1984	1.8894	380	195.77	.01836	2.335	353.45	1198.1	.5413	1.5471
150	3.718	.01634	97.07	117.89	1126.1	.2149	1.8685	390	220.37	.01850	2.0836	364.17	1199.6	.5539	1.5371
160	4.741	.01639	77.29	127.89	1130.2	.2311	1.8485	400	247.31	.01864	1.8633	374.97	1201.0	.5664	1.5272
170	5.992	.01645	62.06	137.90	1134.2	.2472	1.8293	410	276.75	.01878	1.6700	385.83	1202.1	.5788	1.5174
180	7.510	.01651	50.23	147.92	1138.1	.2630	1.8109	420	308.83	.01894	1.5000	396.77	1203.1	.5912	1.5078
190	9.339	.01657	40.96	157.95	1142.0	.2785	1.7932	430	343.72	.01910	1.3499	407.79	1203.8	.6035	1.4982
200	11.526	.01663	33.64	167.99	1145.9	.2938	1.7762	440	381.59	.01926	1.2171	418.90	1204.3	.6158	1.4887
210	14.123	.01670	27.82	178.05	1149.7	.3090	1.7598	450	422.6	.0194	1.0993	430.1	1204.6	.6280	1.4793
212	14.696	.01672	26.80	180.07	1150.4	.3120	1.7566	460	466.9	.0196	0.9944	441.4	1204.6	.6402	1.4700
220	17.186	.01677	23.15	188.13	1153.4	.3239	1.7440	470	514.7	.0198	.9009	452.8	1204.3	.6523	1.4606
230	20.780	.01684	19.382	198.23	1157.0	.3387	1.7288	480	566.1	.0200	.8172	464.4	1203.7	.6645	1.4513
240	24.969	.01692	16.323	208.34	1160.5	.3531	1.7140	490	621.4	.0202	.7423	476.0	1202.8	.6766	1.4419

Temp., °F. t	Abs. pressure, lb./sq. in. p	Volume, cu. ft./lb. Liquid v_f	Volume, cu. ft./lb. Vapor v_g	Enthalpy, B.t.u./lb. Liquid h_f	Enthalpy, B.t.u./lb. Vapor h_g	Entropy, B.t.u./(lb.)(°R.) Liquid s_f	Entropy, B.t.u./(lb.)(°R.) Vapor s_g	Temp., °F. t	Abs. pressure, lb./sq. in. p	Volume, cu. ft./lb. Liquid v_f	Volume, cu. ft./lb. Vapor v_g	Enthalpy, B.t.u./lb. Liquid h_f	Enthalpy, B.t.u./lb. Vapor h_g	Entropy, B.t.u./(lb.)(°R.) Liquid s_f	Entropy, B.t.u./(lb.)(°R.) Vapor s_g
500	680.8	.0204	.6749	487.8	1201.7	.6887	1.4325	620	1786.6	.0247	.2201	646.7	1150.3	.8398	1.3062
520	812.4	.0209	.5594	511.9	1198.2	.7130	1.4136	640	2059.7	.0260	.1798	678.6	1130.5	.8679	1.2789
540	962.5	.0215	.4649	536.6	1193.2	.7374	1.3942	660	2365.4	.0278	.1442	714.2	1104.4	.8987	1.2472
560	1133.1	.0221	.3868	562.2	1186.4	.7621	1.3742	680	2708.1	.0305	.1115	757.3	1067.2	.9351	1.2071
580	1325.8	.0228	.3217	588.9	1177.3	.7872	1.3532								
								700	3093.7	.0369	.0761	823.3	995.4	.9905	1.1389
600	1542.9	.0236	.2668	617.0	1165.5	.8131	1.3307	705.4	3206.2	.0503	.0503	902.7	902.7	1.0580	1.0580

*Abridged from Keenan and Keyes, *Thermodynamic Properties of Steam,* Wiley, New York, 1936. Copyright, 1937, by Joseph H. Keenan and Frederick G. Keyes.

TABLE C-8

Saturated water: temperature table

Temp., T °C	Sat. press., P_{sat} kPa	Specific volume, m³/kg: Sat. liquid, v_f	Sat. vapor, v_g	Internal energy, kJ/kg: Sat. liquid, u_f	Evap., u_{fg}	Sat. vapor, u_g	Enthalpy, kJ/kg: Sat. liquid, h_f	Evap., h_{fg}	Sat. vapor, h_g	Entropy, kJ/(kg · K): Sat. liquid, s_f	Evap., s_{fg}	Sat. vapor, s_g
0.01	0.6113	0.001000	206.14	0.0	2375.3	2375.3	0.01	2501.3	2501.4	0.000	9.1562	9.1562
5	0.8721	0.001000	147.12	20.97	2361.3	2382.3	20.98	2489.6	2510.6	0.0761	8.9496	9.0257
10	1.2276	0.001000	106.38	42.00	2347.2	2389.2	42.01	2477.7	2519.8	0.1510	8.7498	8.9008
15	1.7051	0.001001	77.93	62.99	2333.1	2396.1	62.99	2465.9	2528.9	0.2245	8.5569	8.7814
20	2.339	0.001002	57.79	83.95	2319.0	2402.9	83.96	2454.1	2538.1	0.2966	8.3706	8.6672
25	3.169	0.001003	43.36	104.88	2304.9	2409.8	104.89	2442.3	2547.2	0.3674	8.1905	8.5580
30	4.246	0.001004	32.89	125.78	2290.8	2416.6	125.79	2430.5	2556.3	0.4369	8.0164	8.4533
35	5.628	0.001006	25.22	146.67	2276.7	2423.4	146.68	2418.6	2565.3	0.5053	7.8478	8.3531
40	7.384	0.001008	19.52	167.56	2262.6	2430.1	167.57	2406.7	2574.3	0.5725	7.6845	8.2570
45	9.593	0.001010	15.26	188.44	2248.4	2436.8	188.45	2394.8	2583.2	0.6387	7.5261	8.1648
50	12.349	0.001012	12.03	209.32	2234.2	2443.5	209.33	2382.7	2592.1	0.7038	7.3725	8.0763
55	15.758	0.001015	9.568	230.21	2219.9	2450.1	230.23	2370.7	2600.9	0.7679	7.2234	7.9913
60	19.940	0.001017	7.671	251.11	2205.5	2456.6	251.13	2358.5	2609.6	0.8312	7.0784	7.9096
65	25.03	0.001020	6.197	272.02	2191.1	2463.1	272.06	2346.2	2618.3	0.8935	6.9375	7.8310
70	31.19	0.001023	5.042	292.95	2176.6	2469.6	292.98	2333.8	2626.8	0.9549	6.8004	7.7553
75	38.58	0.001026	4.131	313.90	2162.0	2475.9	313.93	2321.4	2635.3	1.0155	6.6669	7.6824
80	47.39	0.001029	3.407	334.86	2147.4	2482.2	334.91	2308.8	2643.7	1.0753	6.5369	7.6122
85	57.83	0.001033	2.828	355.84	2132.6	2488.4	355.90	2296.0	2651.9	1.1343	6.4102	7.5445
90	70.14	0.001036	2.361	376.85	2117.7	2494.5	376.92	2283.2	2660.1	1.1925	6.2866	7.4791
95	84.55	0.001040	1.982	297.88	2102.7	2500.6	397.96	2270.2	2668.1	1.2500	6.1659	7.4159
	Sat. press., MPa											
100	0.10135	0.001044	1.6729	418.94	2087.6	2506.5	419.04	2257.0	2676.1	1.3069	6.0480	7.3549
105	0.12082	0.001048	1.4194	440.02	2072.3	2512.4	440.15	2243.7	2683.8	1.3630	5.9328	7.2958
110	0.14327	0.001052	1.2102	461.14	2057.0	2518.1	461.30	2230.2	2691.5	1.4185	5.8202	7.2387
115	0.16906	0.001056	1.0366	482.30	2041.4	2523.7	482.48	2216.5	2699.0	1.4734	5.7100	7.1833
120	0.19853	0.001060	0.8919	503.50	2025.8	2529.3	503.71	2202.6	2706.3	1.5276	5.6020	7.1296
125	0.2321	0.001065	0.7706	524.74	2009.9	2534.6	524.99	2188.5	2713.5	1.5813	5.4962	7.0775
130	0.2701	0.001070	0.6685	546.02	1993.9	2539.9	546.31	2174.2	2720.5	1.6344	5.3925	7.0269
135	0.3130	0.001075	0.5822	567.35	1977.7	2545.0	567.69	2159.6	2727.3	1.6870	5.2907	6.9777
140	0.3613	0.001080	0.5089	588.74	1961.3	2550.0	589.13	2144.7	2733.9	1.7391	5.1908	6.9299
145	0.4154	0.001085	0.4463	610.18	1944.7	2554.9	610.63	2129.6	2740.3	1.7907	5.0926	6.8833
150	0.4758	0.001091	0.3928	631.68	1927.9	2559.5	632.20	2114.3	2746.5	1.8418	4.9960	6.8379
155	0.5431	0.001096	0.3468	653.24	1910.8	2564.1	653.84	2098.6	2752.4	1.8925	4.9010	6.7935
160	0.6178	0.001102	0.3071	674.87	1893.5	2568.4	675.55	2082.6	2758.1	1.9427	4.8075	6.7502
165	0.7005	0.001108	0.2727	696.56	1876.0	2572.5	697.34	2066.2	2763.5	1.9925	4.7153	6.7078
170	0.7917	0.00114	0.2428	718.33	1858.1	2576.5	719.21	2049.5	2768.7	2.0419	4.6244	6.6663
175	0.8920	0.001121	0.2168	740.17	1840.0	2580.2	741.17	2032.4	2773.6	2.0909	4.5347	6.6256
180	1.0021	0.001127	0.19405	762.09	1821.6	2583.7	763.22	2015.0	2778.2	2.1396	4.4461	6.5857
185	1.1227	0.001134	0.17409	784.10	1802.9	2587.0	785.37	1997.1	2782.4	2.1879	4.3586	6.5465
190	1.2544	0.001141	0.15654	806.19	1783.8	2590.0	807.62	1978.8	2786.4	2.2359	4.2720	6.5079
195	1.3978	0.001149	0.14105	828.37	1764.4	2592.8	829.98	1960.0	2790.0	2.2835	4.1863	6.4698

Source: Cengel, Y. A., and M. A. Boles: *Thermodynamics: An Engineering Approach,* McGraw-Hill, Table A-4, pp. 904–905, 1998.

TABLE C-8
Continued

Temp., T°C	Sat. press., P_{sat} MPa	Specific volume, m³/kg: Sat. liquid, v_f	Specific volume, m³/kg: Sat. vapor, v_g	Internal energy, kJ/kg: Sat. liquid, u_f	Internal energy, kJ/kg: Evap., u_{fg}	Internal energy, kJ/kg: Sat. vapor, u_g	Enthalpy, kJ/kg: Sat. liquid, h_f	Enthalpy, kJ/kg: Evap., h_{fg}	Enthalpy, kJ/kg: Sat. vapor, h_g	Entropy, kJ/(kg · K): Sat. liquid, s_f	Entropy, kJ/(kg · K): Evap., s_{fg}	Entropy, kJ/(kg · K): Sat. vapor, s_g
200	1.5538	0.001157	0.13736	850.65	1744.7	2595.3	852.45	1940.7	2793.2	2.3309	4.1014	6.4323
205	1.7230	0.001164	0.11521	873.04	1724.5	2597.5	875.04	1921.0	2796.0	2.3780	4.0172	6.3952
210	1.9062	0.001173	0.10441	895.53	1703.9	2599.5	897.76	1900.7	2798.5	2.4248	3.9337	6.3585
215	2.104	0.001181	0.09479	918.14	1682.9	2601.1	920.62	1879.9	2800.5	2.4714	3.8507	6.3221
220	2.318	0.001190	0.08619	940.87	1661.5	2602.4	943.62	1858.5	2802.1	2.5178	3.7683	6.2861
225	2.548	0.001199	0.07849	963.73	1639.6	2603.3	966.78	1836.5	2803.3	2.5639	3.6863	6.2503
230	2.795	0.001209	0.07158	986.74	1617.2	2603.9	990.12	1813.8	2804.0	2.6099	3.6047	6.2146
235	3.060	0.001219	0.06537	1009.89	1594.2	2604.1	1013.62	1790.5	2804.2	2.6558	3.5233	6.1791
240	3.344	0.001229	0.05976	1033.21	1570.8	2604.0	1037.32	1766.5	2803.8	2.7015	3.4422	6.1437
245	3.648	0.001240	0.05471	1056.71	1546.7	2603.4	1061.23	1741.7	2803.0	2.7472	3.3612	6.1083
250	3.973	0.001251	0.05013	1080.39	1522.0	2602.4	1085.36	1716.2	2801.5	2.7927	3.2802	6.0730
255	4.319	0.001263	0.04598	1104.28	1596.7	2600.9	1109.73	1689.8	2799.5	2.8383	3.1992	6.0375
260	4.688	0.001276	0.04221	1128.39	1470.6	2599.0	1134.37	1662.5	2796.9	2.8838	3.1181	6.0019
265	5.081	0.001289	0.03877	1152.74	1443.9	2596.6	1159.28	1634.4	2793.6	2.9294	3.0368	5.9662
270	5.499	0.001302	0.03564	1177.36	1416.3	2593.7	1184.51	1605.2	2789.7	2.9751	2.9551	5.9301
275	5.942	0.001317	0.03279	1202.25	1387.9	2590.2	1210.07	1574.9	2785.0	3.0208	2.8730	5.8938
280	6.412	0.001332	0.03017	1227.46	1358.7	2586.1	1235.99	1543.6	2779.6	3.0668	2.7903	5.8571
285	6.909	0.001348	0.02777	1253.00	1328.4	2581.4	1262.31	1511.0	2773.3	3.1130	2.7070	5.8199
290	7.436	0.001366	0.02557	1278.92	1297.1	2576.0	1289.07	1477.1	2766.2	3.1594	2.6227	5.7821
295	7.993	0.001384	0.02354	1305.2	1264.7	2569.9	1316.3	1441.8	2758.1	3.2062	2.5375	5.7437
300	8.581	0.001404	0.02167	1332.0	1231.0	2563.0	1344.0	1404.9	2749.0	3.2534	2.4511	5.7045
305	9.202	0.001425	0.019948	1359.3	1195.9	2555.2	1372.4	1366.4	2738.7	3.3010	2.3633	5.6643
310	9.856	0.001447	0.018350	1387.1	1159.4	2546.4	1401.3	1326.0	2727.3	3.3493	2.2737	5.6230
315	10.547	0.001472	0.016867	1415.5	1121.1	2536.6	1431.0	1283.5	2714.5	3.3982	2.1821	5.5804
320	11.274	0.001499	0.015488	1444.6	1080.9	2525.5	1461.5	1238.6	2700.1	3.4480	2.0882	5.5362
330	12.845	0.001561	0.012996	1505.3	993.7	2498.9	1525.3	1140.6	2665.9	3.5507	1.8909	5.4417
340	14.586	0.001638	0.010797	1570.3	894.3	2464.6	1594.2	1027.9	2622.0	3.6594	1.6763	5.3357
350	16.513	0.001740	0.008813	1641.9	776.6	2418.4	1670.6	893.4	2563.9	3.7777	1.4335	5.2112
360	18.651	0.001893	0.006945	1725.2	626.3	2351.5	1760.5	720.3	2481.0	3.9147	1.1379	5.0526
370	21.03	0.002213	0.004925	1844.0	384.5	2228.5	1890.5	441.6	2332.1	4.1106	0.6865	4.7971
374.14	22.09	0.003155	0.003155	2029.6	0	2029.6	2099.3	0	2099.3	4.4298	0	4.4298

TABLE C-9

Properties of water in the range 0 to 100°C

t, °C	Density, g/cm^3	Specific heat capacity, J/(g · °C)	Vapor pressure, kPa	Viscosity, μPa · s
0	0.99987	4.2177	0.6113	1781
10	0.99973	4.1922	1.2281	1307
20	0.99823	4.1819	2.3388	1002
30	0.99567	4.1785	4.2455	797.7
40	0.99224	4.1786	7.3814	653.2
50	0.98807	4.1807	12.344	547.0
60	0.98324	4.1844	19.932	466.5
70	0.97781	4.1896	31.176	404.0
80	0.97183	4.1964	47.373	354.4
90	0.96534	4.2051	70.117	314.5
100	0.95838	4.2160	101.325	281.8

Source: Lide, D. R. ed.: *CRC Handbook of Chemistry and Physics,* 78th ed., CRC Press, Boca Raton, Fla., 1997, p. 6-3.

TABLE C-10

Properties of air

Saturated air				
T, K	P^*_{boil}, bar	$P^\dagger_{con}$, bar	$\rho^\ddagger_{liquid}$, g/cm^3	$\rho^\ddagger_{gas}$, g/L
65	0.1468	0.0861	0.939	0.464
70	0.3234	0.2052	1.917	1.033
75	0.6366	0.4321	0.894	2.048
80	1.146	0.8245	0.871	3.709
85	1.921	1.453	0.845	6.258
90	3.036	2.397	0.819	9.980
95	4.574	3.748	0.792	15.21
100	6.621	5.599	0.763	22.39
110	12.59	11.22	0.699	45.15
120	21.61	20.14	0.622	87.34
130	34.16	33.32	0.487	184.33
132.55	37.69	37.69	0.313	312.89

Gaseous air		
Pressure, bar	Temperature, °K	ρ_{gas},‡ g/L
1	100	3.556
1	200	1.746
1	300	1.161

Source: Lide, D. R. ed.: *CRC Handbook of Chemistry and Physics,* 87th ed., CRC Press, Boca Raton, Fla., 1997, p. 6-1.

*Bubble point temperature (pressure at which boiling begins).

†Dew point temperature (pressure at which condensation begins).

‡ ρ = density.

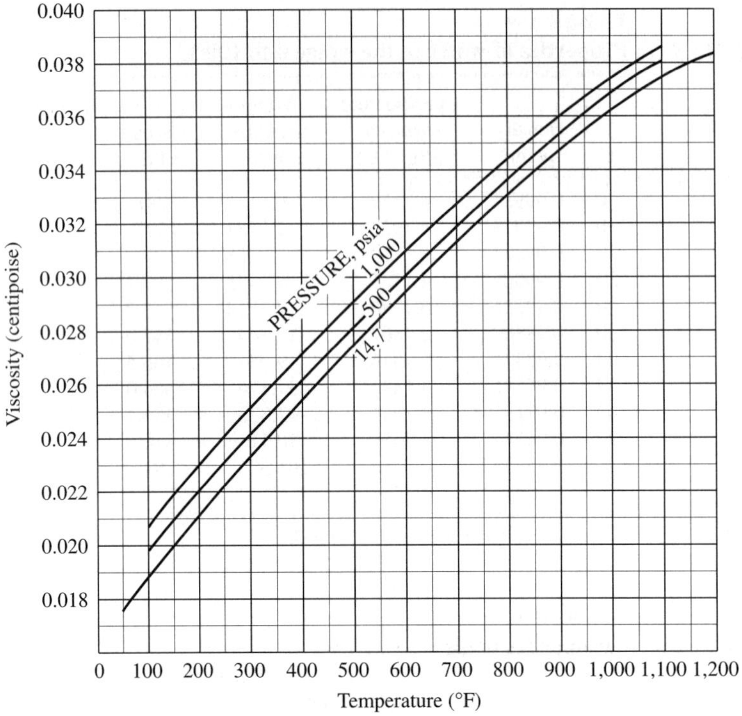

FIGURE C-1
Viscosity of air. (*Engineering Data Book, 10th ed., Gas Processors and Suppliers Association, Tulsa, Okla., p. 23.41, 1994.*)

APPENDIX D

Conversion Factors

Bring out number, weight, and measure in a year of dearth.
Blake

TABLES IN APPENDIX D

TABLE D-1
Metric conversion factors (U.S. customary units to SI units)

Multiply U.S. customary unit		By	To obtain the SI unit	
Name	**Symbol**		**Symbol**	**Name**
Acceleration				
feet per second squared	ft/s^2	0.3048[a]	m/s^2	meters per second squared
inches per second squared	in/s^2	0.0254[a]	m/s^2	meters per second squared
Area				
acre	acre	0.4047	ha	hectare
acre	acre	4.0469×10^{-3}	km^2	square kilometer
square foot	ft^2	9.2903×10^{-2}	m^2	square meter
square inch	in^2	6.4516[a]	cm^2	square centimeter
square mile	mi^2	2.5900	km^2	square kilometer
square yard	yd^2	0.8361	m^2	square meter
Concentration/density				
slugs per cubic foot	$slugs/ft^3$	0.51541	g/cm^3	grams per cubic centimeter
pounds per cubic foot	lb/ft^3	16.0185	kg/m^3	kilograms per cubic meter
pounds per acre-foot	lb/acre-ft	3.6773×10^{-1}	g/m^3	grams per cubic meter
grains per cubic foot	$grain/ft^3$	2.2883	g/m^3	grams per cubic meter
pounds per million gallons	lb/Mgal	0.119826	mg/L	milligrams per liter
Diffusion coefficient				
square inches per second	in^2/s	6.4516	cm^2/s	square centimeter per second
square feet per second	ft^2/s	5.57418	m^2/min	square meter per minute
Energy				
British thermal unit	Btu	1.0551	kJ	kilojoule
foot-pound (force)	$ft \cdot lb_f$	1.3558	J	joule (newton-meter)
horsepower-hour	$hp \cdot h$	2.6845	MJ	megajoule
kilowatt-hour	$kW \cdot h$	3,600[a]	kJ	kilojoule
kilowatt-hour	$kW \cdot h$	3.600×10^{6}[a]	J	joule
watt-hour	$W \cdot h$	3.600[a]	kJ	kilojoule
watt-second	$W \cdot s$	1.000[a]	J	joule
Force				
pound force	lb_f	4.4482	N	newton ($kg \cdot m/s^2$)
Flow rate				
cubic feet per second	ft^3/s	2.8317×10^{-2}	m^3/s	cubic meters per second
gallons per day	gal/d	4.3813×10^{-5}	L/s	liters per second
gallons per day	gal/d	3.7854×10^{-3}	m^3/d	cubic meters per day
gallons per minute	gal/min	6.3090×10^{-5}	m^3/s	cubic meters per second
gallons per minute	gal/min	6.3090×10^{-2}	L/s	liters per second
million gallons per day	Mgal/d	43.8126	L/s	liters per second
million gallons per day	Mgal/d	3.7854×10^{3}	m^3/d	cubic meters per day
million gallons per day	Mgal/d	4.3813×10^{-2}	m^3/s	cubic meters per second
Length				
foot	ft	0.3048[a]	m	meter
inch	in	2.54[a]	cm	centimeter
inch	in	0.0254[a]	m	meter
inch	in	25.4[a]	mm	millimeter
mile	mi	1.6093	km	kilometer
yard	yd	0.9144[a]	m	meter

Tables D-1 through D-3 adapted from Metcalf and Eddy, *Wastewater Engineering,* 3d. ed., McGraw-Hill, New York, 1991, by permission.

[a] Indicates exact conversion.

TABLE D-1
Continued

Multiply U.S. customary unit		By	To obtain the SI unit	
Name	**Symbol**		**Symbol**	**Name**
Loading rate				
pounds per square feet per second	$lb_f/ft^2 \cdot s$	4.88	$kg/m^2 \cdot s$	kilograms per square meter per second
Mass				
ounce	oz	28.3495	g	gram
pound	lb	4.5359×10^2	g	gram
pound	lb	0.4536	kg	kilogram
ton (short: 2,000 lb)	ton	0.9072	Mg	megagram (10^3 kilogram) (metric ton)
ton (long: 2,240 lb)	ton	1.0160	Mg	megagram (10^3 kilogram) (metric ton)
Power				
British thermal units per second	Btu/s	1.0551	kW	kilowatt
foot-pounds (force) per second	$ft \cdot lb_f/s$	1.3558	W	watt
horsepower	hp	0.7457	kW	kilowatt
Pressure (force/area)				
atmosphere (standard)	atm	1.0133×10^2	kPa (kN/m^2)	kilopascal (kilonewtons per square meter)
inches of mercury (60°F)	inHg (60°F)	3.3768×10^3	Pa (N/m^2)	pascal (newtons per square meter)
inches of water (60°F)	inH_2O (60°F)	2.4884×10^2	Pa (N/m^2)	pascal (newtons per square meter)
pounds (force) per square foot	lb_f/ft^2	47.8803	Pa (N/m^2)	pascal (newtons per square meter)
pounds (force) per square inch	lb_f/in^2	6.8948×10^3	Pa (N/m^2)	pascal (newtons per square meter)
pounds (force) per square inch	lb_f/in^2	6.8948	kPa (kN/m^2)	kilopascal (kilonewtons per square meter)
Temperature				
degrees Fahrenheit	°F	0.555 (°F - 32)	°C	degrees Celsius (centigrade)
degrees Fahrenheit	°F	0.555 (°F + 459.67)	K	kelvin
Velocity				
feet per second	ft/s	0.3048[a]	m/s	meters per second
miles per hour	mi/h	4.4704×10^{-1}[a]	m/s	meters per second
Absolute viscosity				
pounds second per square foot	$lb \cdot s/ft^2$	47.88	$kg/m \cdot s$	kilograms per meter per second
pounds (mass) per inches per second	$lb_m/in \cdot s$	17.86	$kg/m \cdot s$	kilograms per meter per second
pounds per feet per second	$lb/ft \cdot s$	14.88	poise ($g/cm \cdot s$)	poise
pounds second per square inch	$lb_f \cdot s/in^2$	6895	$kg/m \cdot s$	kilograms per meter per second

[a] Indicates exact conversion.

TABLE D-1
Continued

Multiply U.S. customary unit		By	To obtain the SI unit	
Name	**Symbol**		**Symbol**	**Name**
Kinematic viscosity				
square feet per second	ft^2/s	929	stoke (cm^2/s)	stoke
square feet per second	ft^2/s	0.0929	m^2/s	square meters per second
square inches per second	in^2/s	6.4516	cm^2/s	square centimeters per second
Volume				
acre-foot	acre-ft	1.2335×10^3	m^3	cubic meter
cubic foot	ft^3	28.3168	L	liter
cubic foot	ft^3	2.8317×10^{-2}	m^3	cubic meter
cubic inch	in^3	16.3871	cm^3	cubic centimeter
cubic yard	yd^3	0.7646	m^3	cubic meter
gallon	gal	3.7854×10^{-3}	m^3	cubic meter
gallon	gal	3.7854	L	liter
ounce (U.S. fluid)	oz (U.S. fluid)	2.9573×10^{-2}	L	liter

[a] Indicates exact conversion.

TABLE D-2
Metric conversion factors (SI units to U.S. customary units)

Multiply the SI unit		By	To obtain the U.S. customary unit	
Name	**Symbol**		**Symbol**	**Name**
Acceleration				
meters per second squared	m/s^2	3.2808	ft/s^2	feet per second squared
meters per second squared	m/s^2	39.3701	in/s^2	inches per second squared
Area				
hectare (10,000 m^2)	ha	2.4711	acre	acre
square centimeter	cm^2	0.1550	in^2	square inch
square kilometer	km^2	0.3861	mi^2	square mile
square kilometer	km^2	247.1054	acre	acre
square meter	m^2	10.7639	ft^2	square foot
square meter	m^2	1.1960	yd^2	square yard
Concentration/density				
grams per cubic centimeter	g/cm^3	1.9422	$slugs/ft^3$	slugs per cubic foot
kilograms per cubic meter	kg/m^3	0.06243	lb/ft^3	pounds per cubic foot
grams per cubic meter	g/m^3	2.7192	lb/acre-ft	pounds per acre-foot
grams per cubic meter	g/m^3	0.4370	$grain/ft^3$	grains per cubic foot
grams per liter	g/L	8.3454×10^{-3}	lb/gal	pounds per gallon
Diffusion coefficient				
square centimeters per second	cm^2/s	0.155	in^2/s	square inches per second
square meters per minute	m^2/min	0.1794	ft^2/s	square feet per second
Energy				
kilojoule	kJ	0.9478	Btu	British thermal unit
joule (newton-meter)	J	2.7778×10^{-7}	$kW \cdot h$	kilowatt-hour
joule	J	0.7376	$ft \cdot lb_f$	foot-pound (force)
joule	J	1.0000	$W \cdot s$	watt-second
joule	J	0.2388	cal	calorie
kilojoule	kJ	2.7778×10^{-4}	$kW \cdot h$	kilowatt-hour
kilojoule	kJ	0.2778	$W \cdot h$	watt-hour
megajoule	MJ	0.3725	$hp \cdot h$	horsepower-hour
Force				
newton	N	0.2248	lb_f	pound force
Flow rate				
cubic meters per day	m^3/d	264.1720	gal/d	gallons per day
cubic meters per day	m^3/d	2.6417×10^{-4}	Mgal/d	million gallons per day
cubic meters per second	m^3/s	35.3147	ft^3/s	cubic feet per second
cubic meters per second	m^3/s	22.8245	Mgal/d	million gallons per day
cubic meters per second	m^3/s	15,850.3	gal/min	gallons per minute
liters per second	L/s	22,824.5	gal/d	gallons per day
liters per second	L/s	0.0228	Mgal/d	million gallons per day
liters per second	L/s	15.8508	gal/min	gallons per minute
Length				
centimeter	cm	0.3937	in	inch
kilometer	km	0.6214	mi	mile
meter	m	39.3701	in	inch
meter	m	3.2808	ft	foot
meter	m	1.0936	yd	yard
millimeter	mm	0.03937	in	inch

TABLE D-2
Continued

Multiply the SI unit		By	To obtain the U.S. customary unit	
Name	**Symbol**		**Symbol**	**Name**
Loading rate				
kilograms per square meter per second	$kg/m^2 \cdot s$	0.2048	$lb_f/ft^2 \cdot s$	pounds per square feet per second
Mass				
gram	g	0.0353	oz	ounce
gram	g	0.0022	lb	pound
kilogram	kg	2.2046	lb	pound
kilogram	kg	0.068521	slug	slug
megagram (10^3 kilogram)	Mg	1.1023	ton	ton (short: 2,000 lb)
megagram (10^3 kilogram)	Mg	0.9842	ton	ton (long: 2,240 lb)
Power				
kilowatt	kW	0.9478	Btu/s	British thermal units per second
kilowatt	kW	1.3410	hp	horsepower
watt	W	0.7376	$ft \cdot lb_f/s$	foot-pounds (force) per second
Pressure (force/area)				
pascal (newtons per square meter)	Pa (N/m^2)	1.4504×10^{-4}	lb_f/in^2	pounds (force) per square inch
pascal (newtons per square meter)	Pa (N/m^2)	2.0885×10^{-2}	lb_f/ft^2	pounds (force) per square foot
pascal (newtons per square meter)	Pa (N/m^2)	2.9613×10^{-4}	inHg	inches of mercury (60°F)
pascal (newtons per square meter)	Pa (N/m^2)	4.0187×10^{-3}	inH_2O	inches of water (60°F)
kilopascal (kilonewtons per square meter)	kPa (kN/m^2)	0.1450	lb_f/in^2	pounds (force) per square inch
kilopascal (kilonewtons per square meter)	kPa (kN/m^2)	0.0099	atm	atmosphere (standard)
Temperature				
degrees Celsius (centigrade)	°C	1.8 (°C) + 32	°F	degrees Fahrenheit
kelvin	K	1.8 (K) − 459.67	°F	degrees Fahrenheit
Velocity				
meters per second	m/s	2.2369	mi/h	miles per hour
meters per second	m/s	3.2808	ft/s	feet per second
Absolute viscosity				
kilograms per meter per second	$kg/m \cdot s$	0.02089	$lb \cdot s/ft^2$	pounds second per square feet
kilograms per meter per second	$kg/m \cdot s$	0.05599	$lb_m/in \cdot s$	pounds (mass) per inches per second
poise	poise ($g/cm \cdot s$)	0.0672	$lb/ft \cdot s$	pounds per feet per second
kilograms per meter per second	$kg/m \cdot s$	1.4503×10^{-4}	$lb \cdot s/in^2$	pounds second per square inch
Kinematic viscosity				
stoke	stoke (cm^2/s)	1.076×10^{-3}	ft^2/s	square feet per second
square meters per second	m^2/s	10.764	ft^2/s	square feet per second
square centimeter per second	cm^2/s	0.155	in^2/s	square inches per second

TABLE D-2
Continued

Multiply the SI unit		By	To obtain the U.S. customary unit	
Name	**Symbol**		**Symbol**	**Name**
Volume				
cubic centimeter	cm^3	0.0610	in^3	cubic inch
cubic meter	m^3	35.3147	ft^3	cubic foot
cubic meter	m^3	1.3079	yd^3	cubic yard
cubic meter	m^3	264.1720	gal	gallon
cubic meter	m^3	8.1071×10^{-4}	acre-ft	acre-foot
liter	L	0.2642	gal	gallon
liter	L	0.0353	ft^3	cubic foot
liter	L	33.8150	oz	ounce (U.S. fluid)

TABLE D-3
Common environmental unit conversion
To convert, multiply in direction shown by arrows

U.S. units	→	←	SI units
acre/(Mgal/d)	0.1069	9.3536	ha/(10^3m^3/d)
Btu	1.0551	0.9478	kJ
Btu/lb	2.3241	0.4303	kJ/kg
Btu/ft^2 · °F · h	5.6735	0.1763	W/m^2 · °C
bu/acre · yr	2.4711	0.4047	bu/ha · yr
ft/h	0.3048	3.2808	m/h
ft/min	18.2880	0.0547	m/h
ft^2/capita	0.0929	10.7639	m^2/capita
ft^3/capita	0.0283	35.3147	m^3/capita
ft^3/gal	7.4805	0.1337	m^3/m^3
ft^3/ft · min	0.0929	10.7639	m^3/m · min
ft^3/lb	0.0624	16.0185	m^3/kg
ft^3/Mgal	7.04805×10^{-3}	133.6805	m^3/10^3m^3
ft^3/Mgal · d	407.4611	0.0025	m^2/10^3m^3 · d
ft^3/ft^2 · h	0.3048	3.2808	m^3/m^2 · h
ft^3/10^3gal · min	7.04805×10^{-3}	133.6805	m^3/m^3 · min
ft^3/min	1.6990	0.5886	m^3/h
ft^3/s	2.8317×10^{-2}	35.3145	m^3/s
ft^3/10^3ft^3 · min	0.001	1,000.0	m^3/m^3 · min
gal	3.7854	0.2642	L
gal/acre · d	0.0094	106.9064	m^3/ha · d
gal/ft · d	0.0124	80.5196	m^3/m · d
gal/ft^2 · d	0.0407	24.5424	m^3/m^2 · d
gal/ft^2 · d	0.0017	589.0173	m^3/m^2 · h
gal/ft^2 · d	0.0283	35.3420	L/m^2 · min
gal/ft^2 · d	40.7458	2.4542×10^{-2}	L/m^2 · d
gal/ft^2 · min	2.4448	0.4090	m/h
gal/ft^2 · min	40.7458	0.0245	L/m^2 · min
gal/ft^2 · min	58.6740	0.0170	m^3/m^2 · d
gal/min · ft	12.4193	8.052×10^{-2}	L/min · m
hp/10^3gal	0.1970	5.0763	kW/m^3
hp/10^3ft^3	26.3342	0.0380	kW/10^3m^3
in	25.4	3.9370×10^{-2}	mm
inHg (60°F)	3.3768	0.2961	kPa Hg (60°F)
lb	0.4536	2.2046	kg
lb/acre	1.1209	0.8922	kg/ha
lb/10^3gal	0.1198	8.3452	kg/m^3
lb/hp · h	0.6083	1.6440	kg/kW · h
lb/Mgal	0.1198	8.3454	g/m^3
lb/Mgal	1.1983×10^{-4}	8345.4	kg/m^3
lb/ft^2	4.8824	0.2048	kg/m^2
lb$_f$/in^2 (gage)	6.8948	0.1450	kPa (gage)
lb/ft^3 · h	16.0185	0.0624	kg/m^3 · h
lb/10^3ft^3 · d	0.0160	62.4280	kg/m^3 · d
lb/ton	0.5000	2.0000	kg/tonne
Mgal/acre · d	0.9354	1.0691	m^3/m^2 · d
Mgal/d	3.7854×10^3	0.264×10^{-3}	m^3/d
Mgal/d	4.3813×10^{-2}	22.8245	m^3/s
min/in	3.9370	0.2540	min/10^2mm
tons/acre	2.2417	0.4461	Mg/ha
yd^3	0.7646	1.3079	m^3

TABLE D-4
Equivalent values

Quantity	Equivalent values
Mass	1 kg = 1,000 g = 0.001 metric ton = 2.20462 lb_m = 35.27392 oz 1 lb_m = 16 oz = 5 × 10^{-4} ton = 453.593 g = 0.453593 kg 1 ton = 2,000 lb
Length	1 m = 100 cm = 1,000 mm = 10^6 microns (μ) = 10^{10} angstroms (A) = 39.37 in = 3.2808 ft = 1.0936 yards = 0.0006214 mile 1 ft = 12 in = 1/3 yd = 0.3048 m = 30.48 cm 1 mile = 5,280 ft
Volume	1 m^3 = 1,000 liters = 10^6 cm^3 = 10^6 mL = 35.3145 ft^3 = 220.83 imperial gallons = 264.17 gallons = 1056.68 quarts 1 ft^3 = 1,728 in^3 = 7.4805 gallons = 0.028317 m^3 = 28.317 liters = 28,317 cm^3 1 acre-ft = 43,560 ft^3 = 1613.3 yd^3 = 0.325851 Mgal = 1.876 × 10^{-2} mi^2 · in
Force	1 N = 1 kg · m/s^2 = 10^5 dynes = 10^5 g · cm/s^2 = 0.22481 lb_f 1 lb_f = 32.174 lb_m · ft/s^2 = 4.4482 N = 4.4482 × 10^5 dynes
Pressure	1 atm = 1.01325 × 10^5 N/m^2 (Pa) = 1.01325 bars = 1.01325 × 10^6 $dynes/cm^2$ = 760 mmHg @ 0°C (torr) = 10.333 m H_2O @ 4°C = 14.696 lb_f/in^2 (psi) = 33.9 ft H_2O @ 4°C = 29.921 inHg @ 0°C
Energy	1 J = 1 N · m = 10^7 ergs = 10^7 dyne · cm = 2.778 × 10^{-7} kW · h = 0.23901 cal = 0.7376 ft-lb_f = 9.486 × 10^{-4} Btu
Power	1 W = 1 J/s = 0.23901 cal/s = 0.7376 ft · lb_f/s = 9.486 × 10^{-4} Btu/s = 1.341 × 10^{-3} hp
Concentrations	1 mg/L = $10^3 \mu$ g/L = 10^{-3} g/L = 10^{-6} kg/L = 1 g/m^3 ppm = parts per million = mg/L* ppb = parts per billion = 10^{-3} ppm = μg/L* lb_f/ft^3 = 7,000 grains/ft^3
Thermal conductivity	cm^2/s = 10^{-4} m^2/s
Emission rate	lb/1,000 ft^3 · day = 43.56 lb/acre-ft · day = 133.7 lb/Mgal · day lb_f/ft^3 = 7,000 grains/ft^3

*Converting dimensionless ppm or ppb to mass/volume assumes water is the solvent.

Table D-4 adapted from *Elementary Principles of Chemical Processes,* John Wiley & Sons, 1978, by permission.

Index

Notes:

1. Page numbers in *italic type* refer to figures or tables.
2. Names that have commonly used acronyms are usually found under the acronym instead of the spelled-out name. *Example:* TCE (trichloroethylene).